Calculus

Calculus

Main Author : **Sang-Gu Lee**

Co-Authors : Eung-Ki Kim · Yoonmee Ham · Ajit Kumar · Robert Beezer
Quoc-Phong Vu · Suk-Geun Hwang · Lois Simon

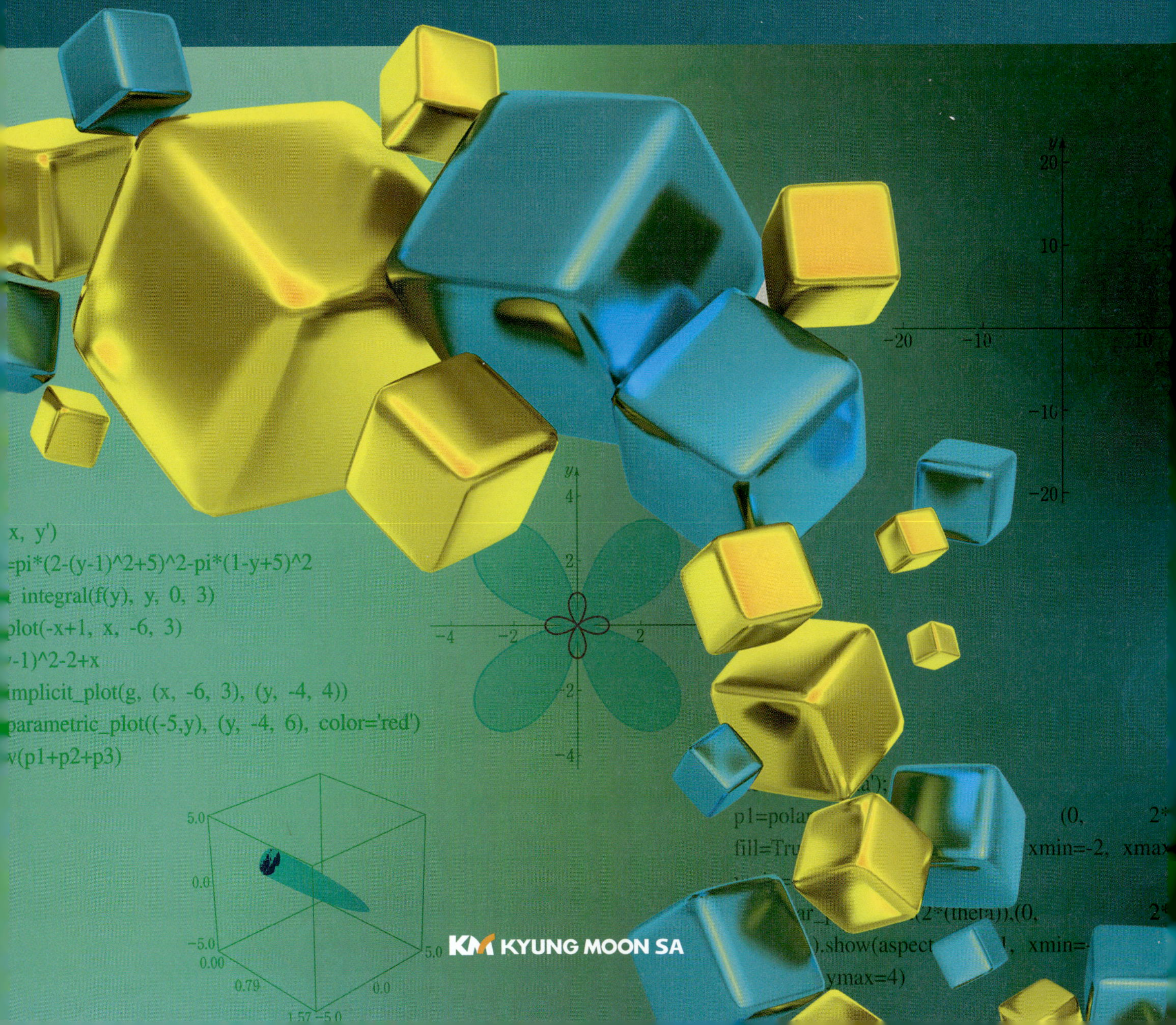

KYUNG MOON SA

Preface

Calculus is the mathematical foundation for much of the university curriculum in mathematics, science, and engineering. For students of mathematics, it is often their first exposure to rigorous mathematics. For the future engineer, it is an introduction to the modeling and approximation techniques used throughout the engineering curriculum. It is an introduction to the mathematical language that expresses many fundamental important concepts in science.

We begin our study of differential and integral calculus with functions of a single variable and topics such as L'Hospital's theorem, concavity, convexity, inflection points, optimization, simple ordinary differential equations, parametric equations, polar coordinates, infinite sequences and infinite series. We then turn to multivariate and vector calculus, covering vector-valued functions, coordinate systems, partial derivatives and multiple integrals. This book has many problems presenting calculus as the foundation of modern mathematics, science and engineering, using concepts, definitions, terminology, and interpretation of the situation presented in the problem to discover solutions.

Many of today's calculus textbooks use computer algebra systems (CAS) and a variety of visualization tools. But in most cases, use of these aids by students is limited. Therefore, we have adopted the usage of a wonderful free and open-source mathematics visualization program, Sage. With the new learning environment at many universities, students can take full advantage this 21st-century, state-of-the-art technology to learn calculus easily in preparation for their future careers in any applicable area. Sage can be accessed easily on any web browser, without the need for additional software installation, and on smaller devices like tablets and smartphones. More content and related materials will be continually added to the website for the book. So when you see CAS(CAS) or a web address in the book, you will be able to find relevant information from http://matrix.skku.ac.kr/Cal-Book/ and http://math1.skku.ac.kr/ . This will allow you (the student) to easily and quickly edit and manipulate existing examples to help with creating and solving new examples.

Finally, we would like to emphasize that the creation of a technology-rich calculus textbook, which also reflects both the established traditions of calculus instruction and recent reforms, requires a team of dedicated people. Our effort is no exception. We greatly appreciate the valuable contributions of many of our colleagues to this project and the efforts of hundreds of contributors to the Sage software project as well.

We would also like to express thanks to Hyunsoo Kim, Jaedong Sim, Insung Hwang, Mohit Kumbhat, L. Shapiro, R. Sakthivel, K. Das, Victoria Lang and all mathematics professors at Sungkyunkwan University who carefully reviewed the preprint of this book for many years.

2014. 1. 1.

1. Introduction and Use of Sage [1)]

Mathematical tools have long held an important place in the classroom. With the innovation of information and communication technologies(ICT), many tools have appeared and been adapted for educational purposes. Sage is popular mathematical software which was released in 2005. This software has efficient features which utilize the internet and can handle most mathematical problems, including linear algebra, abstract algebra, combinatorics, number theory, symbolic computation, numerical mathematics and calculus. In this book, we will introduce this powerful software and discuss how it can be used in classes.

Sage is a mathematical Computer Algebra System (CAS) that can be easily used online. The Sage notebook was released in April 2008 at University of Washington, USA. It is free and has powerful capabilities that can be compared with expensive commercial software such as Mathematica, Maple or Matlab. However, it does not require separate installation of the program. It is more like Web-Mathematica, but has some better features. When you connect with any web browser, you can solve almost all calculus problems in the book by using resources that we are offering. You can easily find pre-existing commands to modify for your own problems. The Sage notebook is a platform that allows you to interface with many open-source and commercial programs (if you have a license).

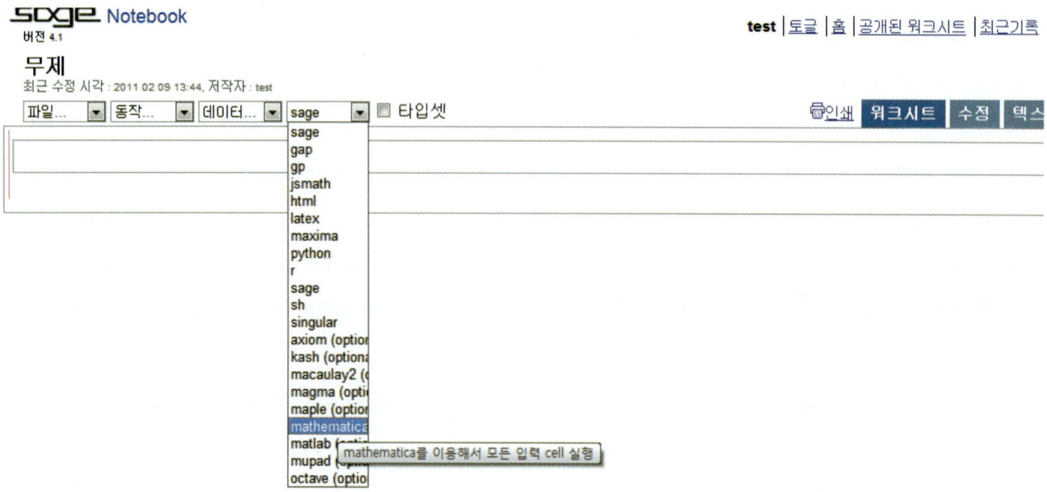

Figure 1 Sage notebook can interface to many programs

1) Mobile Sage for Linear Algebra and its Application, Duk-Sun Kim, Sang-Gu Lee*, Greg Markowsky, Electronic Journal of Mathematics & Technology (eJMT) https://php.radford.edu/~ejmt/ContentIndex.php V.4, No. 3, p. 1-13, Oct 2010. ISSN 1933-2823.
Korean Sage model was developed and relevant experiments were done by BK 21 Mathematical Modelling HRD division at Sungkyunkwan University. We have built Korean servers that you can use as you see below.

Online Sage Servers:

- http://sagenb.skku.edu/ (Sign in) or http://www.sagenb.org (Sign in)
- http://math1.skku.ac.kr (ID: math2013, Password: math32304)

    http://sagenb.skku.edu (Make your own accounts)

    http://math3.skku.ac.kr (Mobile Server, No need for login)

    http://sage.skku.edu (SKKU Sage Cell Server)

    https://salv.us (University of Washington Sage Server)

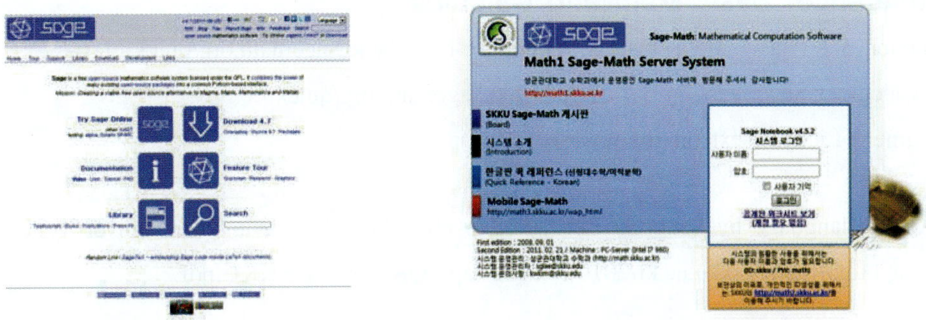

Figure 2 Sage Community (Left) http://www.sagemath.org,
Sage Korean Version (Right) http://math1.skku.ac.kr

Use Firefox or Chrome to connect to one of the above servers. Then register your ID and password to get started.

Steps to connect to http://math1.skku.ac.kr server

1. ID : math2013, Password : math32304.
2. Click "New Worksheet" button in the upper left corner.
3. Change the name of a worksheet.
4. It is ready to use.

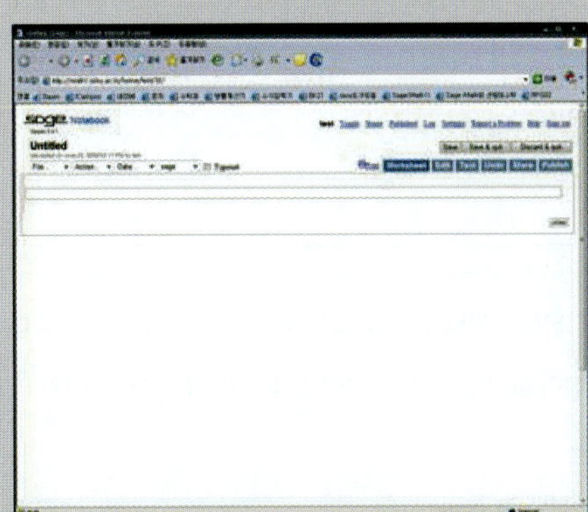

Figure 3 Sage Worksheet

2. Development of the Korean Version of Sage

There are Quick Reference (simple manual) for Sage in English and Korean. Those Quick References can be downloaded from the web site address below.

http://wiki.sagemath.org/quickref

http://matrix.skku.ac.kr/2010-Album/Sage-QReference-SKKU.pdf

Sage Quick Reference: Basic Math and Calculus

Peter Jipsen, version 1.1 (Basic Math) Latest Version at http://wiki.sagemath.org/quickref

William Stein (Calculus) Sage Version 3.4 http://wiki.sagemath.org/quickref

GNU Free Document License, extend for your own use

Translated to Korean language by Sang-Gu Lee and Jae Hwa Lee (Sungkyunkwan University)

Korean Version at http://matrix.skku.ac.kr/2010-Album/Sage-QReference-SKKU.pdf

Notebook (and Command line)

Evaluate cell: <shift-enter>

com <tab> tries to complete command

command?<tab> shows documentation

command??<tab> shows source

a.<tab> shows all methods for object a (more: dir(a))

search_doc('string or regexp') shows links to docs

search_src('string or regexp') shows links to source

lprint() toggle LaTeX output mode

version() print version of Sage

Insert cell: click on blue line between cells

Delete cell: delete content then backspace

Numerical types

Sage has many different number systems. ZZ represents set of integers, QQ represents set of rational numbers, RDF represents floating-point real numbers, CDF represents floating-point complex numbers.

Example for Rationals:

sage: a=1/2; b=QQ(4/2); print a+b; b.parent()

output: 5/2

Rational Field

Builtin constants and functions:

Constants: π=pi e=e i=I=i ∞=oo

∞=oo=infinity NaN=NaN log(2)=log2

ϕ=golden_ratio γ=euler_gamma

...

Builtin functions:

sin cos tan sec csc cot sinh cosh tanh sech csch coth log ln exp

ab=a*b $\frac{a}{b}$=a/b a^b=a^b $\sqrt{x}$=sqrt(x) $\sqrt[n]{x}$=x^(1/n) $|x|$ =abs(x) $\log_b(x)$=log(x,b)

Symbolic variables: e.g. t, u, v, y, z = var('t u v y z')

Define functions: e.g. $f(x)=x^2$ f(x)=x^2 or f=lambda x: x^2 or def f(x): return x^2

Operations and equations:

Relations: $f=g$: f == g, $f\neq g$: f != g, $f\leq g$: f <= g, $f\geq g$: f >= g, $f<g$: f < g, $f>g$: f > g

Solve $f(x)=g(x)$: solve(f(x)==g(x),x)

Solve $f(x,y)=g(x,y)$: solve([f(x,y)==0, g(x,y)==0], x,y)

Exact roots: (x^3+2*x+1).roots(x)

Real roots: (x^3+2*x+1).roots(x, ring=RR)

Complex roots: (x^3+2*x+1).roots(x, ring=CC)

$\sum_{i=1}^{10} f(i)$: var('i'); f(i)=i; sum(f(i), i, 1, 10)

$\prod_{i=1}^{10} f(i)$: var('i'); f(i)=i; prod(f(i), i, 1, 10)

Defining symbolic expressions

Create symbolic variables: var("t u theta") var('t, u, theta') ...

Symbolic functions

Symbolic function (can integrate, differentiate, etc.): f(a, b, theta) = a + b*theta^2

Also, a "formal" function of theta: f = function('f', theta)

Piecewise symbolic functions: Piecewise([[(0, pi/2), sin(1/x)], [(pi/2, pi), x^2+1]])

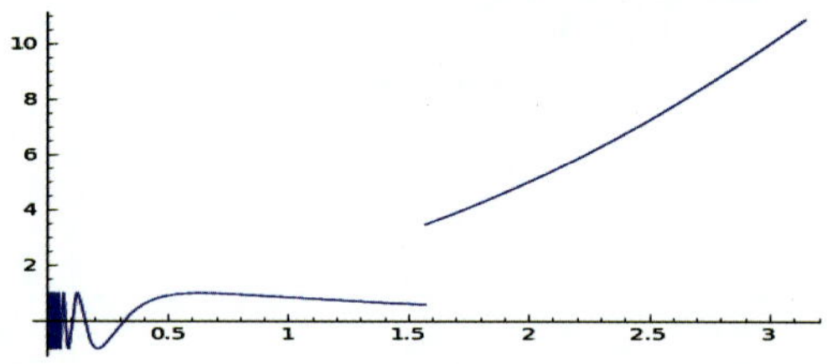

Python functions

Defining:

```
def f(a, b, theta=1):
    c = a + b*theta^2
    return c
```

Inline functions:

```
    f = lambda a, b, theta = 1: a + b*theta^2
  ...
```

Factorization

Factored form: (x^3-y^3).factor()

List of (factor, exponent) pairs: (x^3-y^3).factor_list()

Limits

$\lim_{x\to a} f(x)$: limit(f(x), x=a)

...

Derivatives

$\frac{d}{dx}(f(x))$: diff(f(x),x) or f.diff(x)

$\frac{\partial}{\partial x}(f(x,y))$: diff(f(x,y),x)

e.g. diff(x*y + sin(x^2) + e^(-x), x)

Integrals

$\int f(x)dx$: integral(f(x),x)

e.g. integral(x*cos(x^2), x)

$\int_a^b f(x)dx$: integral(f(x),x,a,b)

e.g. integral(x*cos(x^2), x, 0, sqrt(pi))

$\int_a^b f(x)dx$: numerical_integral(f(x),a,b)[0]

e.g. numerical_integral(x*cos(x^2),0,1)[0]

assume(...): use if integration asks a question

e.g. assume(x>0)

Multivariable calculus

Gradient: f.gradient() or f.gradient(vars)

e.g. (x^2+y^2).gradient([x, y])

Hessian: f.hessian()

e.g. (x^2+y^2).hessian()

Jacobian matrix: jacobian(f, vars)

e.g. jacobian(x^2 - 2*x*y, (x,y))

2D graphics

Polynomial: plot(x^2, (x,-2,5))

Functions: plot(f(x), (x, $x_{\min}$, $x_{\max}$), options)

Line: line([(x_1,y_1), (x_2, y_2)], options)

Polygon: polygon([(x_1,y_1),...,(x_n, y_n)], options)

Circle: circle((x, y), r, options)

Parametric functions: parametric_plot((f(t),g(t)), (t, $t_{\min}$, $t_{\max}$), options)

Polar functions: polar_plot(f(t), (t, $t_{\min}$, $t_{\max}$), options)

Animate: animate(list of graphics objects, options).show(delay=20)

3D graphics

Line: line3d([(x_1, y_1, z_1), (x_2, y_2, z_2)], options)

Sphere: sphere((x, y, z), r, options)

Tetrahedron: tetrahedron((x, y, z), size, options)

Functions: plot3d(f(x, y), (x, $x_{\min}$, $x_{\max}$), (y, $y_{\min}$, $y_{\max}$), options)

add option plot_points=[m, n]

Parametric functions: parametric_plot3d((f(t), g(t), h(t)), (t, $t_{\min}$, $t_{\max}$), options) or

parametric_plot3d((f(u, v), g(u, v), h(u, v), (u, $u_{\min}$, $u_{\max}$), (v, $v_{\min}$, $v_{\max}$), options)

Linear algebra

$\begin{pmatrix} 1 \\ 2 \end{pmatrix}$: vector([1, 2])

$A = \begin{pmatrix} 1 & 2 \\ 3 & 4 \end{pmatrix}$: A=matrix([[1, 2], [3, 4]]); A

$\begin{vmatrix} 1 & 2 \\ 3 & 4 \end{vmatrix}$: det(matrix([[1, 2], [3, 4]]))

Av: A*v

A^{-1}: A^-1 or A.inverse()

A^T: A.transpose()

Other methods: A.<tab>

CAS Examples Sage Problem: Exercise 4 in Section 8.2 (Area of a Surface of Revolution)

★ Find the area of the surface when k is a positive even integer.

$y = \sin(kx)$. $0 \le x \le \pi$

Sol Published at http://math1.skku.ac.kr/home/pub/77

```
k,x=var('k,x');
@interact
def _(y = input_box(sin(k*x), label="y="), kappa=slider(0,100,1,default=1,label=' k')):
    html('$$AREA=%s$$'%(
  lim(integral(2*pi*sin(k*x)*sqrt(1+diff(sin(k*x),x)),x,0,pi),k=kappa) ))
    plot(sin(kappa*x), x,0,pi, color='purple', fill=true).show(aspect_ratio=1, xmin=0, xmax=pi, ymin=-1, ymax=1)
```

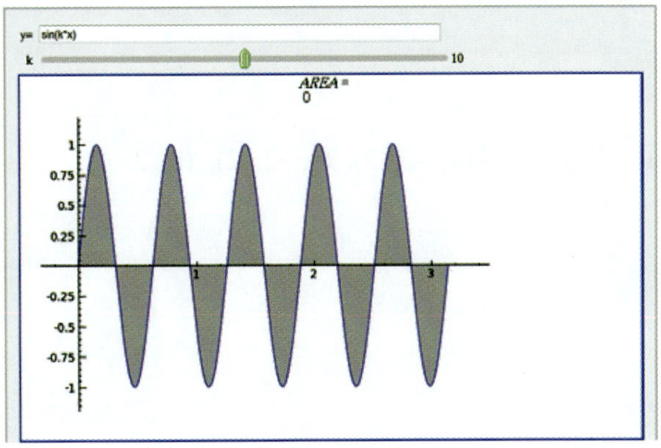

Answer : 0.

CAS Examples Sage Problem: Exercise 22 in Section 9.3 (Polar Coordinates)

★ Sketch the curve with the given polar equation.

$r = \sin(4\theta)$

Sol Published at http://math1.skku.ac.kr/home/pub/78

```
theta=var('theta');
polar_plot(sin(4*theta), (0, 2*pi)).show(aspect_ratio=1, xmin=-1, xmax=1, ymin=-1, ymax=1)
```

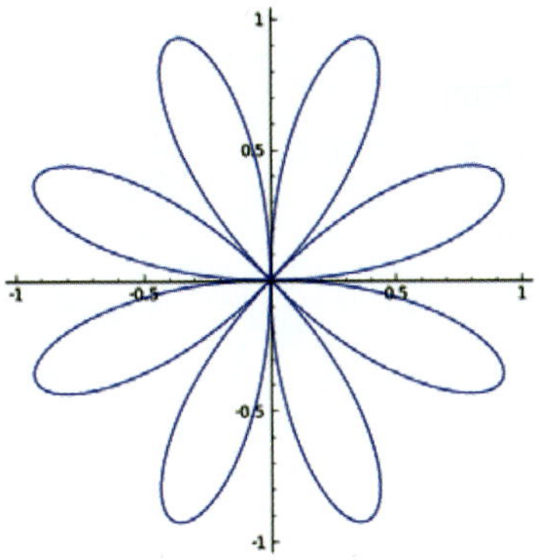

CAS Examples Sage Problem: Exercise 16 in Section 11.6 (Cylinders and Quadric Surfaces)

★ Sketch the region bounded by the surfaces $z - \sqrt{x^2 + y^2}$ and $x^2 + y^2 = 4$ for $2 \le z \le 4$.

Sol Published at http://matrix.skku.ac.kr/cal-lab/cal-11-6-16.html

http://math1.skku.ac.kr (ID/PW: math2013/math32304)

```
var('x, y, z')
z=(x^2+y^2)^(1/2)
plot3d(z, (x, -2, 2), (y, -2, 2), (z, 2, 4))+implicit_plot3d(x^2+y^2==4,(x, -2, 2), (y, -2, 2), (z, 2, 4),
opacity=0.5)
```

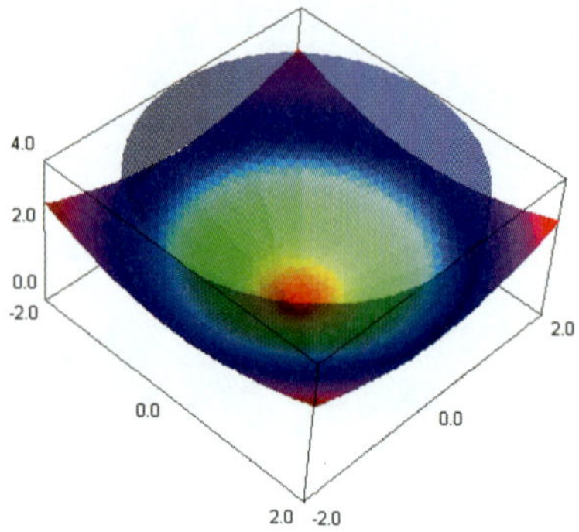

All of the above Sage commands and many more in the published section of each Sage server (e.g. http://math1.skku.ac.kr/pub/) can be copied and pasted, so you can modify them. Sungkyunkwan University has made more than 3,000 Sage commands. We hope everybody can take full advantage of Sage in learning and teaching Calculus.

* http://matrix.skku.ac.kr/cal-lab/Math-CAS.htm
Sage-Calculus--Grapher : http://matrix.skku.ac.kr/cal-lab/sage-grapher.html
Sage-Parametric Equation Grapher : http://matrix.skku.ac.kr/cal-lab/sage-grapher-para.html
Sage-Polar Equation Grapher : http://matrix.skku.ac.kr/cal-lab/sage-grapher-polar.html
Sage-Implicit Function Grapher : http://matrix.skku.ac.kr/cal-lab/sage-grapher-imp.html

Calculus Map

http://matrix.skku.ac.kr/Cal-Book/

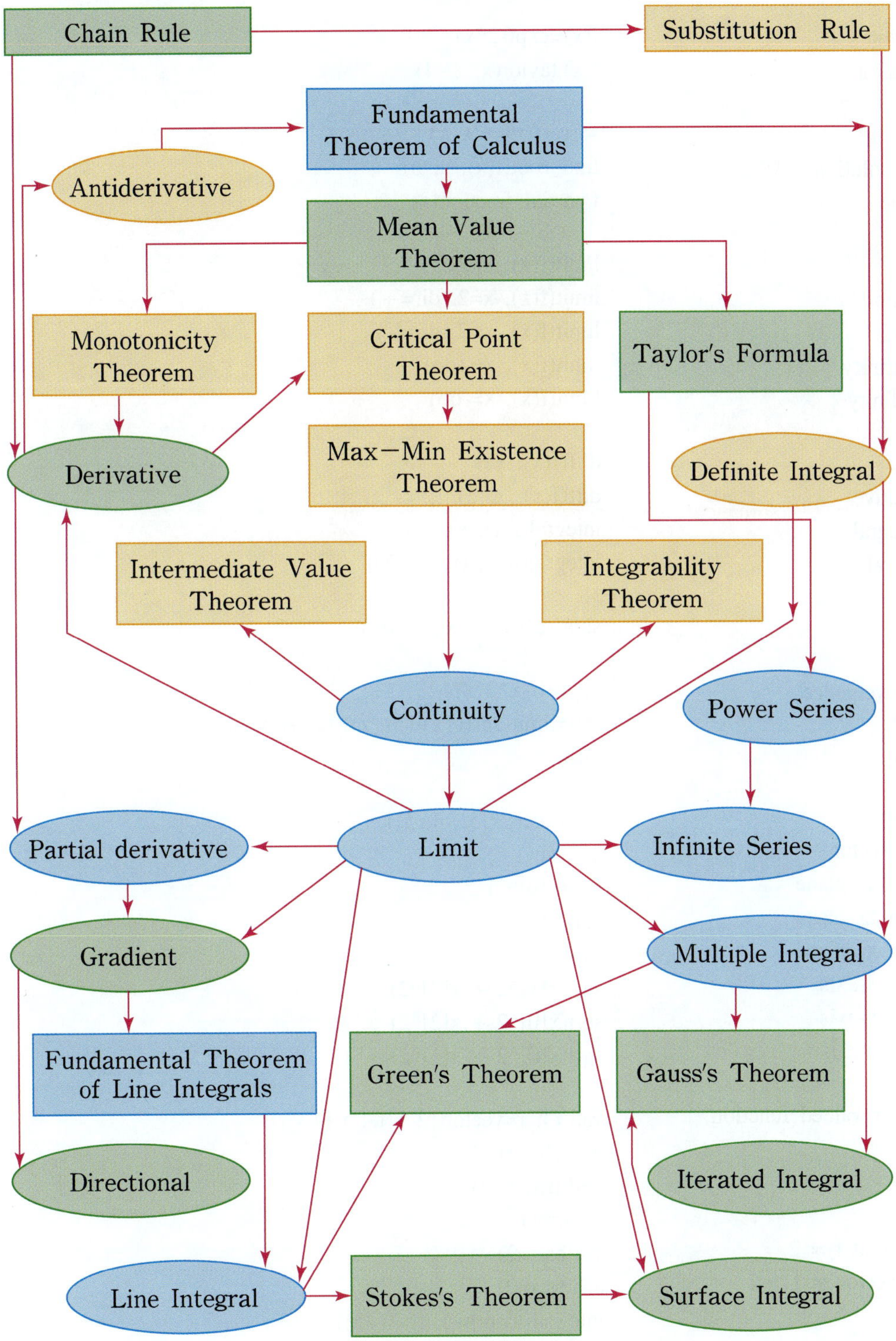

Sage Code

http://matrix.skku.ac.kr/Cal-Book/ http://sage.skku.edu/

http://matrix.skku.ac.kr/Mobile-Sage-G/sage-grapher.html

Define Function	f(x)=exp(-2*x)
Taylor expansion	f(x).taylor(x, 2, 4)
Equation solving	solve(f(x)==0, x)
Approximate solution	find_root(f(x), a, b)
Partial fraction	f.partial_fraction(x)
Limit	limit(f(x), x=2)
Right hand limit	limit(f(x), x=2, dir='+')
Left hand limit	limit(f(x), x=2, dir='-')
Limit at + infinity	limit(f(x), x=+oo)
Limit at - infinity	limit(f(x), x=-oo)
Derivative	diff(f(x), x)
Second derivative	diff(f(x), x, 2)
Indefinite integral	integral(f(x), x)
Definite integral	integral(f(x), (x, -2, 3))
Series	sum((e/10)^x, x, 1, +oo)
Radius of convergence	u(x)=1/(x*2^x) rho=limit(abs(u(x+1)/u(x)), x=+oo) 1/rho
Define a vector	x = vector([3, -4, 5])
Norm of the vector	x.norm()
Distance from yz-plane	sqrt(x[0]^2)
Distance from zx-plane	sqrt(x[1]^2)
Distance from xz-plane	sqrt(x[2]^2)
Distance from x-axis	sqrt(x[1]^2 + x[2]^2)
Distance from y-axis	sqrt(x[0]^2 + x[2]^2)
Distance from z-axis	sqrt(x[0]^2 + x[1]^2)
Define a vector valued function	var('t'); r=vector([2-5*t,4*t,1+3*t])
Velocity	v=diff(r, t)
Acceleration	a=diff(r, t, 2)
Speed	v.norm()
Velocity vector at t = 2	v.subs(t=2)
Acceleration vector at t = 2	a.subs(t=2)
Arc length	integral(v.norm(), (t, 0, 5))

Dot product Cross product	v.dot_product(a) v.cross_product(a)
Find gradient	var('x, y, z') f=x*y+y*z^2+x*z^3 f.gradient()
Double integral Triple integral	integral(integral(f, (x, 0, y)), (y, 0, 1)) integral(integral(integral(f, (x, 0, y)), (y, 0, 1)), (z, 0, 1))
Double integral (in Polar coordinate)	var('r, t') f=arctan(tan(t)) integral(integral(f*r, (r, 1, 2)), (t, 0, pi/4))
Cylindrical to Rectangular	T = Cylindrical('height', ['radius', 'azimuth']) T.transform(radius=1, azimuth= pi, height=2)
Spherical to Rectangular	T = Spherical('radius', ['azimuth', 'inclination']) T.transform(radius=3, azimuth=pi/6, inclination=pi/6)
Graph in plane Parametric function in plane Implicit function in plane Line segment in plane	plot(sin(x), (x, -4, 4)) parametric_plot((sin(x), cos(x)), (x, 0, 2*pi)) var('x, y'); implicit_plot(sin(x)-y==1, (x, -2, 2), (y, -2, 2)) line([(1, 1), (2, 2)], color='red')
Graph in $\mathbb{R}^3$ Parametric function in $\mathbb{R}^3$ Implicit function in $\mathbb{R}^3$	var('x, y'); plot(x^2+y^2, (x, -2, 2), (y, -2, 2)) parametric_plot3d((sin(x), cos(x), x), (x, 0, 2*pi)) var('x, y, z'); implicit_plot3d(x^2+y^2==5, (x, -3, 3), (y, -3, 3), (z, -3, 3))
Level curve	var('x, y'); contour_plot(x^2+y^2, (x, -1, 1), (y, -1, 1), cmap='hsv', labels=True)
Vector field	var('x, y'); plot_vector_field((x+y, x), (x, -3, 3), (y, -3, 3))
Vector field in $\mathbb{R}^3$	var('x, y, z'); plot_vector_field3d((0, 0, 1), (x, -3, 3), (y, -3, 3), (z, -3, 3))
Line integral	var('t'); r=vector([t-t^3, t^2, 0]) integral(r.dot_product(diff(r, t)), (t, 0, pi))

curl (A)
div (A)
	A(x,y,z) = P*i+Q*j+R*k # conservative if curl(F)=0.
	curlA=(diff(R,y)-diff(Q,z))*i+(diff(P,z)-diff(R,x))*j+(diff(Q,x)-diff(P,y))*k
	divA = diff(P,x)+diff(Q,y)+diff(R,z) # divergence of A

Green's theorem Stokes' theorem Divergence theorem	integral(integral((diff(N, x)-diff(M, y))*r, (r, 0, 3)), (t, 0, 2*pi)) integral(integral(curl(F).dot_product(-n), (r, 0, 1)), (t, 0, 2*pi)) integral(integral(integral(Div, 0, 3), (y, 0, 2)), (z, 0, 1))

CONTENTS

Part II Multivariate Calculus

Chapter 10. Parametric Equations and Polar Coordinates

Chapter 11. Vectors and the Geometry of Space

Chapter 12. Vector Valued Functions

Chapter 13. Partial Derivatives

About this book : **http://matrix.skku.ac.kr/Cal-Book/**

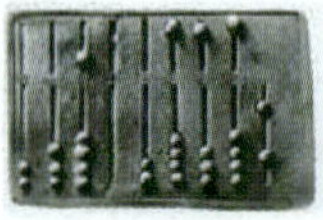

Abacus

Calculus with Sage

('x, y')
)=pi*(2-(y-1)^2+5)^2-pi*(1-y+5)^2
nt integral(f(y), y, 0, 3)
=plot(-x+1, x, -6, 3)
y-1)^2-2+x
=implicit_plot(g, (x, -6, 3), (y, -4, 4))
=parametric_plot((-5,y), (y, -4, 6), color='red')
w(p1+p2+p3)
p1=polar
fill=Tru
xmin=-2, xma
).show(aspect
xmin=-
ymax=4)

Chapter 1

Functions

Pythagoras of Samos

BC580-BC500

Pythagorean Theorem

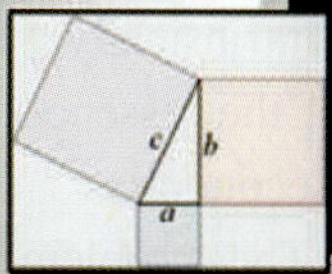

[Pythagorean Theorem]

Pythagoras was a Greek philosopher and founder of the religious movement called Pythagoreanism. He is often revered as a great mathematician, mystic and scientist, and he is best known for the Pythagorean theorem which bears his name. His followers pursued the religious rites and practices developed by Pythagoras.

YouTube http://youtu.be/cl8GqIWIRD0

In this chapter, we will work with various functions and algebraic operations involving functions that may be viewed geometrically. We will work with polynomial, exponential, logarithmic, and trigonometric functions. We will consider some of the symmetry properties of these functions with respect to the origin, the x-axis and the y-axis. We will also consider these functions with respect to translations, rotations, dilations and compressions.

1.1 Functions and Graph

http://matrix.skku.ac.kr/Cal-Book/part1/CS-Sec-1-1.htm

The growth of mathematics parallels the development of human civilization. Numbers were used for counting, while fractions were used for measuring and trading. The relationship or dependence between one object or quantity on another is represented by functions.

Calculus is one of the branches of the mathematics focused on functions, series, limits, derivatives, integrals, and their relations. Intrinsically, the mathematical study of change, motion, growth or decay is the essence of calculus. Calculus has broad applications in sciences, economics, and engineering. Functions are fundamental objects in calculus. The problem of computing the area enclosed by curve was the main thema in the integral calculus, whose origin goes back to 2,500 years ago to the ancient Greeks. The differential calculus, which had not been invented until more than 2,000 years after integral calculus, deals with problems with derivatives. The main ideas behind differential calculus are attributed to Pierre Fermat (1601-1665, French mathematician) and were developed by John Wallis (1616-1703, English mathematician), Isaac Barrow (1630-1677, English mathematician), Isaac Newton (1642-1727, English mathematician) and Gottfried Leibniz (1646-1716, German mathematician).

Before going into details of calculus, we give a brief introduction on sets, functions and numbers which are essential for this book.

Sets

In this section, we denote some common terminology about sets that will be occasionally used throughout the textbook. Note that we only restrict ourselves with the main terminology, we refer students who are interested in a deeper study of set theory, to the books in the reference.

A set A is a collection of elements. If an element x belongs to A, then we write $x \in A$. If x is not an element of A, then we write $x \not\in A$.

An empty set, denoted by $\varnothing$, is the set that does not contain any elements.

A set B is called a subset of a set A, if each element of B is also an element of A, we write $B \subset A$. Two sets A and B are equal, if they have the same elements, that is $A = B$ if and only if $A \subset B$ and $B \subset A$.

There are important operations over sets; namely the union, intersection and complement.

The union of two sets A and B is a set denoted by $A \cup B$, which consists of elements x such that x belongs to either A or B, that is $A \cup B = \{x : x \in A \text{ or } x \in B\}$.

The intersection of two sets A and B is a set denoted by $A \cap B$, which consists of elements x such that x belongs to both A and B, that is

$A \cap B = \{x : x \in A \text{ and } x \in B\}$.

If B is a subset of A, then the complement of B in A, denoted by $A \setminus B$, is the set of elements which belong to A but do not belong to B, that is $A \setminus B = \{x : x \in A, \ x \notin B\}$.

Numbers

A number is a mathematical object which is used for counting and measuring.

The set of natural numbers is denoted by $\mathbb{N}$. The integers are the collection of natural numbers, negative and zero. The set of integers is denoted by $\mathbb{Z}$. The rational numbers are numbers that can be expressed as fractions with an integer numerator and a non-zero natural number denominator. The set of rational numbers is denoted by $\mathbb{Q}$. The set $\mathbb{R}$ of all real numbers is frequently encountered in calculus. Intuitively, the set of real numbers is the set of all points on the axis without exception. The complex numbers, $\mathbb{C}$ are an extension of the real numbers, $\mathbb{R}$.

Functions

Real world phenomena can be represented by mathematical models using functions. The dependence of one quantity on another can be represented by a function. Let us consider a few examples.

(1) The temperature T during the day at a given place depends on the time t. For a given time there is a unique temperature. So, the temperature T is a function of time t.

It is denoted by

$$T = T(t) \text{ or } T = f(t).$$

Here, the notation $T(t)$ or $f(t)$ on the right-hand side means "function of t or dependence on t." However, two **different** temperatures **cannot** be associated with the **same** time. Whereas the same temperature can occur at two or more different times. So we call t an independent variable, which varies independently, while we call T a dependent variable because T depends on time t.

(2) The area A of a rectangle depends on its length L and width W, $A = A(L, W)$ or $A = f(L, W)$. Note that here the area A depends on L and W. That is a function of two variables.

(3) Production cost of bread C depends on the cost of sugar S, the cost of wheat W and the cost of labor L. Thus $C = C(S, W, L) = f(S, W, L)$.

In fact, in real world problems, a quantity may depend on several variables. Try to think some examples in everyday life.

Let C and D be two non empty sets and for each element $x \in D$ a function f assigns exactly one element, in a set C known as $f(x)$, the value of f at x

(read as "f of x"). Here D and C denote the domain and co-domain of the function f. A function may be described in many ways, algebraically by an equation or visually by a graph, or numerically in tabular form as a set of ordered n-tuples or verbally through description in words.

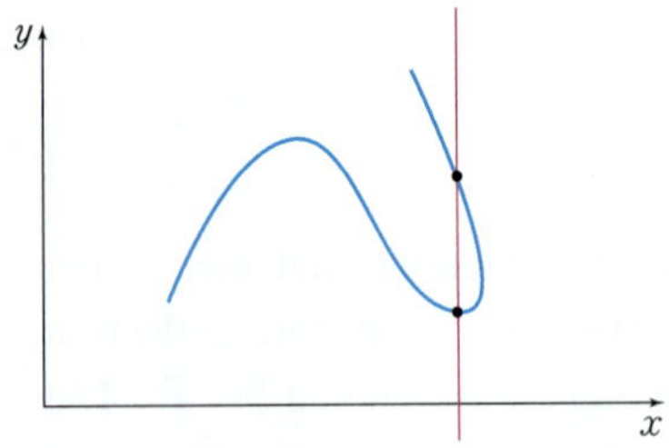

Not a Function
(a vertical line crosses 2 values)

Figure 1
http://www.mathsisfun.com/sets/images/vertical-line-test.gif

DEFINITION 1

A function f is a rule that associates to each value x in a set D one and only one value y in a set C. In this case, we write $y=f(x)$. The set D is called domain of the function f, the set $C=\{y|y=f(x) \text{ for some } x \in D\}$ is called the range of f.

DEFINITION 2

Two functions f and g, with domains $D(f)$ and $D(g)$, respectively, are said to be equal, if $D(f)=D(g)$ and $f(x)=g(x)$ for all x in $D(f)=D(g)$.

Thus a curve in the xy-plane is the graph of a function of x if and only if no vertical line intersects the curve more than once. For example, if each vertical line $x=a$ intersects a curve only once at (a,b), then exactly one functional value is defined by $f(a)=b$. But if a line $x=a$ intersects the curve twice at (a,b) and (a,c), then the curve cannot be represented by a function because a function cannot assign two different values to a as shown in the Figure 1.

A function defined by different formulas in different parts of its domain is known as a **piecewise defined function**.

EXAMPLE 1

An example of a piecewise defined function is shown in the Figure 2. The algebraic representation is given below.

$$f(x)=\begin{cases} -x, & \text{when } x<0 \\ x^3, & \text{when } 0 \le x \le 2 \\ 3, & \text{when } x>2 \end{cases}$$

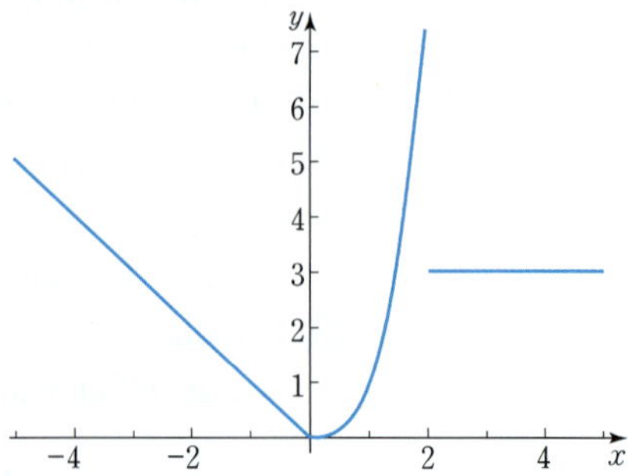

Figure 2

■

Now we illustrate the use of "Sage" which is the computer algebra system (CAS) used throughout this text book. "Sage" is a free open source software. For clarity, Sage commands are shown in bold face green. A brief introduction of how to use Sage is available at

http://matrix.skku.ac.kr/Cal-Book/
http://matrix.skku.ac.kr/2012-LAwithSage/interact/

For example, the above piecewise defined function can be plotted using the following command: http://matrix.skku.ac.kr/cal-lab/cal-1-1-Fig-1.html.

```
f1(x) = -x
f2(x) = x^3
f3(x) = 3
f = Piecewise([[(-5, 0), f1], [(0, 2), f2], [(2, 5), f3]])
f.plot()
```

A function is called a **step function** if it jumps from one value to the next at certain values of x. For example, the function whose graph shown in Figure 3 is an example of a step function.

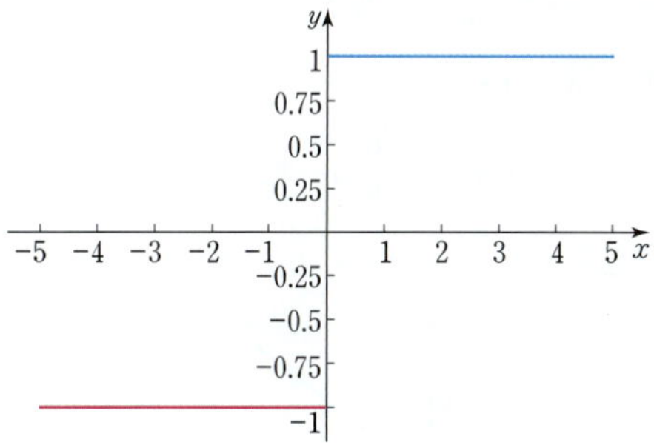

Figure 3 Plot of function g

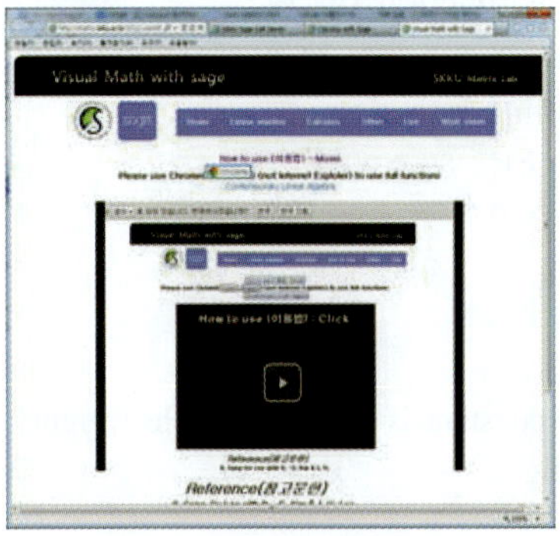


```
g1(x) = -1
g2(x) = 1
g = Piecewise([[(-5, 0), g1], [(0, 5), g2]])
g.plot()
```

■

1.1 EXERCISES (Functions and Graph)

http://matrix.skku.ac.kr/Cal-Book/part1/CS-Sec-1-1.htm

YouTube http://youtu.be/rQ2CB8EvkoE

1. (Celsius and Fahrenheit). If the temperature is C degrees Celsius, then the temperature is also F degrees Fahrenheit, where $F(C)=\frac{9}{5}C+32$

(a) Find $F(-24)$, $F(1)$, $F(15)$ and $F(38)$.

(b) Suppose the outside temperature is 20 degrees Celsius. What is the temperature in degrees Fahrenheit?

(c) What temperature is the same in both degrees Fahrenheit and in degrees Celsius?

http://matrix.skku.ac.kr/cal-lab/cal-1-1-1.html

Solution (a) $F(-24)=\frac{9}{5}(-24)+32=-\frac{56}{5}=-11.2$

$F(1)=\frac{9}{5}(1)+32=\frac{169}{5}=33.8$

$F(15)=\frac{9}{5}(15)+32=59$

$F(38)=\frac{9}{5}(38)+32=\frac{502}{5}=100.4$

(b) $F(20)=\frac{9}{5}(20)+32=68$

(c) Let $F(C)=C=A$. Then $A=\frac{9}{5}A+32$.

That is $\frac{4}{5}A+32=0$. Hence $A=-40$.

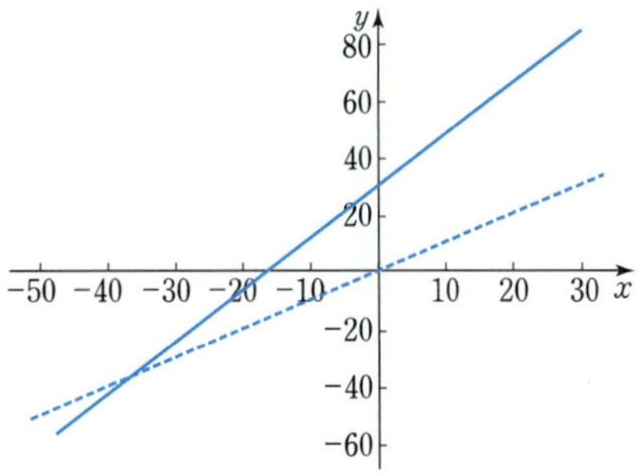

Therefore the temperature that the same in both degrees Fahrenheit and in degrees Celsius is -40.

```
F(x)=9/5*x+32
G(x)=x
show(plot(F(x), (x, -50, 30))+plot(G(x), (x, -50, 30), linestyle='--'))
```

2. (Brain Weight Problem) The weight B of a human's brain is directly proportional to his/her body weight W.

(a) It is known that a person whose weight is 200 lb has a brain that weighs 5 lb. Find an equation of variation expressing B as a function of W.

(b) Express the variation constant as a percent and interpret the resulting equation.

(c) What is the weight of the brain of a person who weighs 120 lb?

Solution (a) $B=kW$.

Since $5=200k$, $k=0.025$.

Hence the required equation is $B=0.025W$.

(b) $B=0.025W\times 100(\%)$.

That is, $B=2.5W(\%)$.

Hence the variation constant is 2.5 and brain weight is 2.5% of body weight.

(c) Since $B=0.025W$,

$B=0.025\times 120\text{lb}=3\text{lb}$

3. (Muscle Weight) The weight M of the muscles in a human is directly proportional to his/her body weight W.

(a) It is known that a person who weighs 200 lb has 80 lb of muscles. Find an equation of variation expressing M as a function of W.

(b) Express the variation constant in percentage and interpret the resulting equation.

Solution (a) Since $M=kW$, $80=200k$.

So, we know that $k=0.4$.

Hence $M=0.4W$.

(b) $B=0.4W\times 100(\%)$.

That is, $B=40W(\%)$.

Hence the variation constant is 40 and muscle weight is 40% of body weight.

4. (Estimating Heights) An anthropologist can use certain linear functions to estimate the height of a male or female, given the length of certain bones. The humerus is the bone from the elbow to the shoulder. Let x be the length of the humerus, in centimeters. Then the height, in centimeters, of a male with a humerus of length x is given by

$$M(x)=2.89x+70.64.$$

The height in centimeters of a female with a

humerus of length x is given by

$$F(x) = 2.75x + 71.483$$

A 26cm humerus was uncovered in some ruins.

(a) If we assume it was from a male, how tall was he?

(b) If we assume it was from a female, how tall was she?

Solution (a) $M(26) = 2.89 \times 26 + 70.64 = 145.78$

(b) $F(26) = 2.75 \times 26 + 71.48 = 142.98$

5. (Urban Population) The population of a town is P. After a growth of 2%, its new population is N.

(a) Assuming that N is directly proportional to P, find an equation of variation.

(b) Find N when $P = 200{,}000$.

(c) Find P when $N = 367{,}200$.

Solution (a) $N = P + P \times 2\% = P + \frac{2P}{100} = 1.02P$

(b) $N = 1.02 \times 200{,}000 = 204{,}000$

(c) $367{,}200 = 1.02P$.

That is $P = \frac{367{,}200}{1.02} = 360{,}000$.

6. (Average Age of Women at First Marriage) In general, our society is marrying at a later age. The average age of women at first marriage can be approximated by the linear function

$$A(t) = 0.08t + 19.7,$$

where $A(t)$ is the average age of women at 1st marriage t years after 1950. Thus, $A(0)$ is the average age of women at 1st marriage in the year 1950, $A(50)$ is the mean age in 2000, and so on.

(a) Find $A(0)$, $A(1)$, $A(10)$, $A(30)$ and $A(50)$.

(b) What will be the mean age of women at first marriage in 2020?

(c) Graph $A(t)$.

Solution (a) $A(0) = 0.08 \times 0 + 19.7 = 19.7$

$A(1) = 0.08 \times 1 + 19.7 = 19.78$

$A(10) = 0.08 \times 10 + 19.7 = 20.5$

$A(30) = 0.08 \times 30 + 19.7 = 22.1$

$A(50) = 0.08 \times 50 + 19.7 = 23.7$

(b) $A(70) = 0.08 \times 70 + 19.7 = 25.3$

(c)

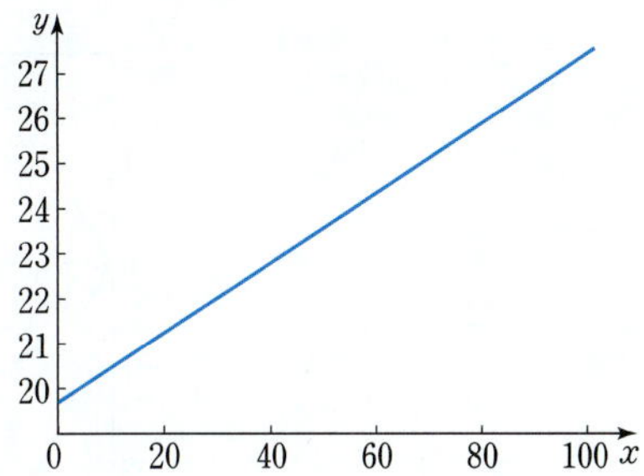

* Your tools to draw graphs (in Chrome or Firefox browser) :

Calculus-Grapher :
http://matrix.skku.ac.kr/cal-lab/sage-grapher.html

Parametric Equation Grapher :
http://matrix.skku.ac.kr/cal-lab/sage-grapher-para.html

Polar Equation Grapher :
http://matrix.skku.ac.kr/cal-lab/sage-grapher-polar.html

Implicit Function Grapher :
http://matrix.skku.ac.kr/cal-lab/sage-grapher-imp.html

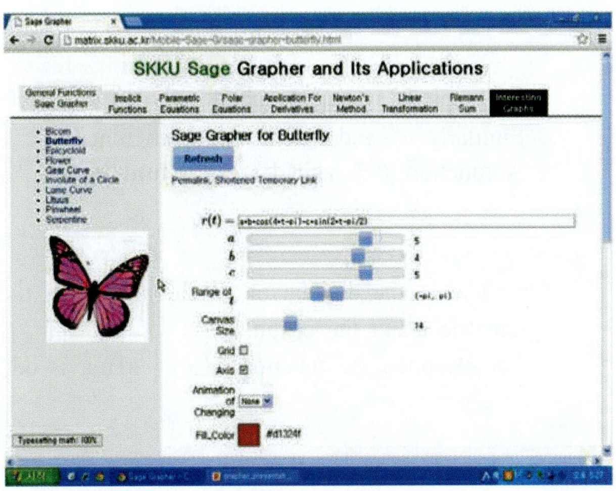

<Sage Grapher>

1.2 Symmetry

http://matrix.skku.ac.kr/Cal-Book/part1/CS-Sec-1-2.htm

Symmetry

The points $(x,\ y)$, $(-x,\ y)$, $(x,\ -y)$ and $(-x,\ -y)$ form the corners of a rectangle. The points $(x,\ y)$ and $(x,\ -y)$ are said to be **symmetric about the x-axis,** the $(x,\ y)$ and $(-x,\ y)$ are said to be **symmetric about the y-axis,** and the $(x,\ y)$ and $(-x,\ -y)$ are said to be **symmetric about the origin,**

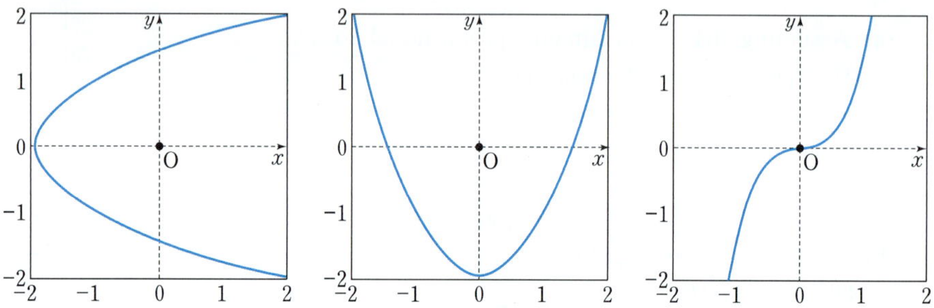

Symmetric about x-axis　　Symmetric about y-axis　　Symmetric about the origin

Figure 1

Even and Odd function

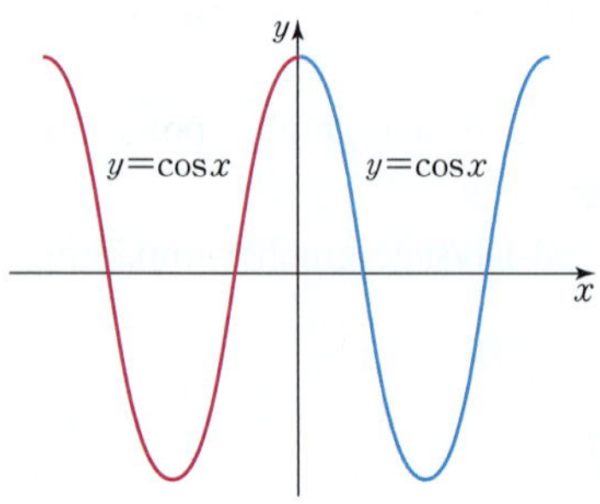

Figure 2 Symmetry (y–axis) Even function

A function f is said to be an **even function** if

$$f(-x)=f(x)$$

for all x in the domain of f. Geometrically, the graph of an even function is symmetric about the y-axis.

For example, the function $f(x)=\cos x$ is even because

$$f(-x)=\cos(-x)=\cos x=f(x).$$

Its graph is symmetric about the y-axis (see Figure 2).

Similarly x^2 and $x\sin x$ are even functions.

A function f is called an **odd function** if

$$f(-x)=-f(x)$$

for all x in the domain of f. Geometrically, the graph of an odd function is symmetric about the origin.

For example, the function $f(x)=\sin x$ is odd because

$$f(-x)=\sin(-x)=-\sin x=f(x)$$

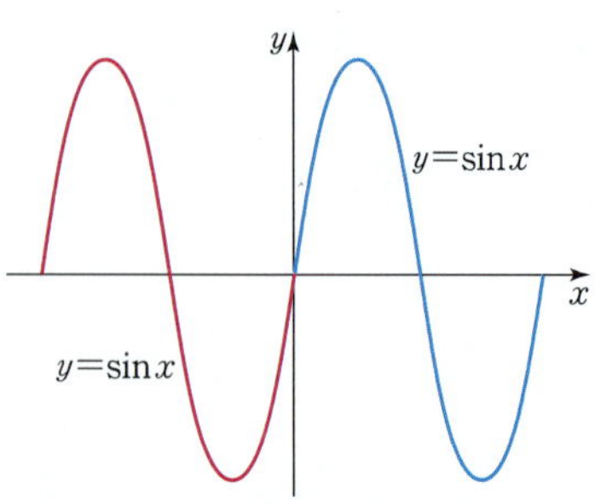

Figure 3 Symmetry (origin) Odd function

Its graph is symmetric about the origin (see Figure 3).

Similarly x^5 and $x^3\cos x$ are odd functions.

However, a function may be neither even nor odd. The graph of such a function is neither symmetric about the y-axis, nor about the origin. For example, $f(x)=\sin x+\cos x$ is neither even nor odd (see Figure 4).

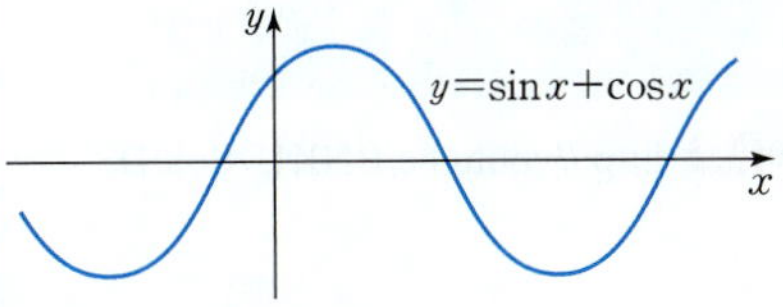

Figure 4 $f(x)=\sin x+\cos x$

Increasing and Decreasing Function

Let f be defined on an interval I.

(a) f is increasing on I if, whenever x_1, x_2 in I, and

$$x_1 < x_2 \text{ , we have } f(x_1) \le f(x_2).$$

(b) f is decreasing on I if, whenever x_1, x_2 in I, and

$$x_1 < x_2 \text{ , we have } f(x_1) \ge f(x_2).$$

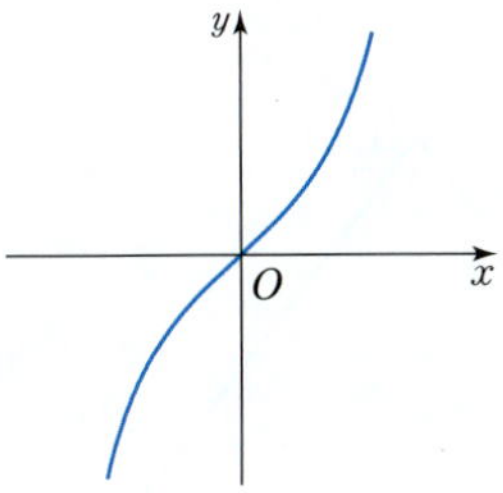

Figure 5 $f(x)= x^3$

For example, the function in Example 1 in section 1.1 is decreasing in the interval $(-\infty, 0)$ and increasing in the interval $(0, 2)$ and is neither increasing nor decreasing in the interval $(2, \infty)$.

If the inequality in the above definition is strict, we call the function (a) strictly increasing and (b) strictly decreasing, respectively. The function $f(x)= x^3$ is an example of strictly increasing function in Figure 5.

Try to find a strictly decreasing function on an interval.

1.2 EXERCISES (Symmetry)

http://matrix.skku.ac.kr/Cal-Book/part1/CS-Sec-1-2-Sol.htm

YouTube http://youtu.be/BNKUzSohiD8

1-4. Graph the following functions.

1. $f(x)=\begin{cases} x, & 0 \le x \le 1 \\ 2-x, & 1 < x \le 2 \end{cases}$

Solution

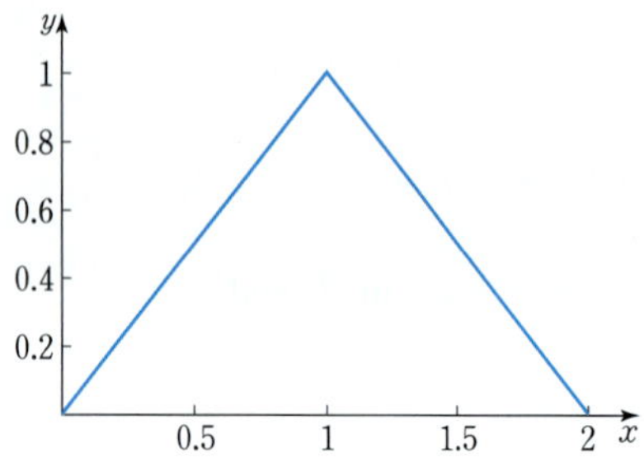

2. $g(x)=\begin{cases} 1-x, & 0 \le x \le 1 \\ 2-x, & 1 < x \le 2 \end{cases}$

Solution

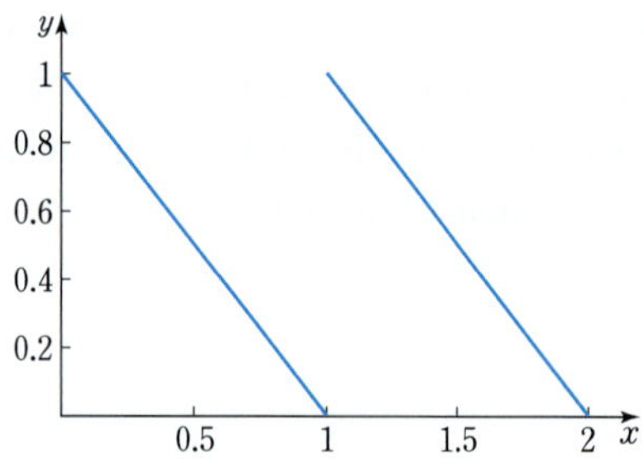

3. $F(x)=\begin{cases} 4-x^2, & x \le 1 \\ x^2+2x, & x > 1 \end{cases}$

Solution

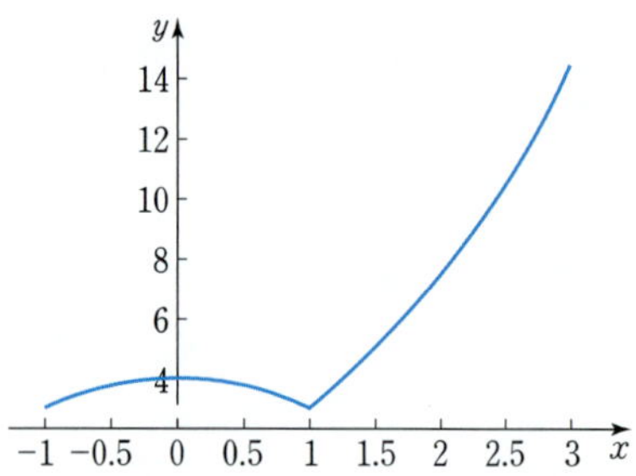

4. $G(x)=\begin{cases} 1/x, & x < 0 \\ x, & 0 \le x \end{cases}$

Solution

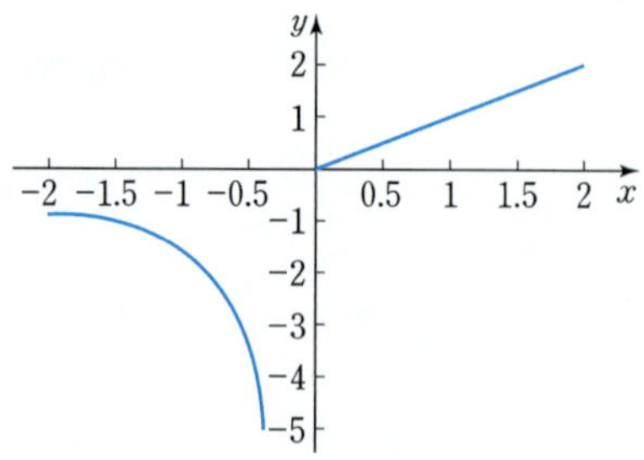

5. Find the formula for graph of funstion.

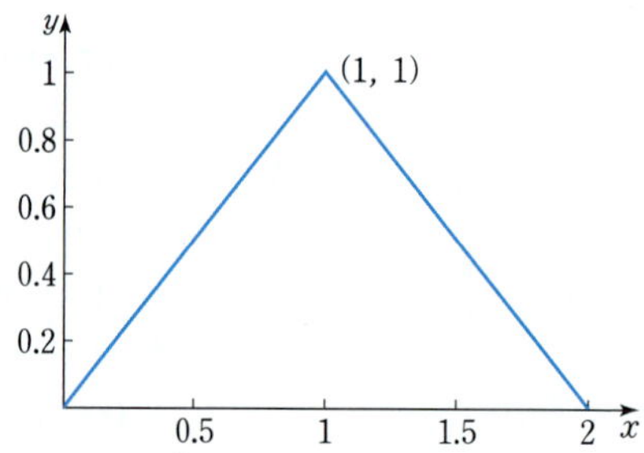

Solution $f(x)=\begin{cases} x, & 0 \le x \le 1 \\ 2-x, & 1 < x \le 2 \end{cases}$

6-10. Graph the functions in Exercises 6-10. What symmetry, if any, do the graphs have? Specify the intervals where the function is increasing and where it is decreasing.

6. $y=-x^3$

Solution

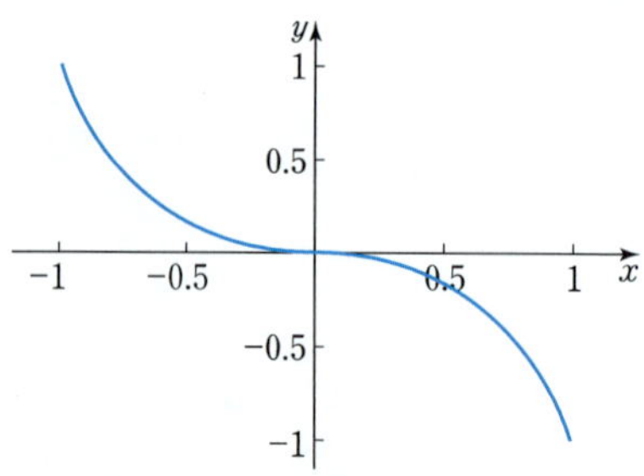

(i) symmetric about the origin, and

(ii) decreasing on $\mathbb{R}$

7. $y=-\dfrac{1}{x}$

http://matrix.skku.ac.kr/cal-lab/cal-1-2-7.html

Solution

```
plot(-1/x, -2, 2, exclude=[0], ymax=5, ymin=-5)
```

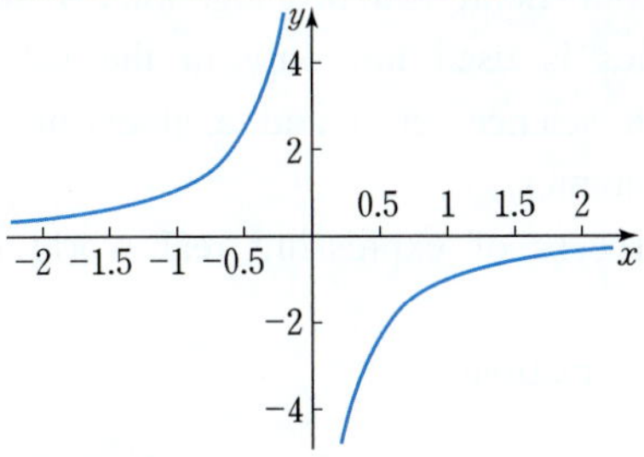

(i) symmetric about the origin,

(ii) increasing on $(0, \infty)$,

(iii) increasing on $(-\infty, 0)$

8. $y=\sqrt{|x|}$

Solution

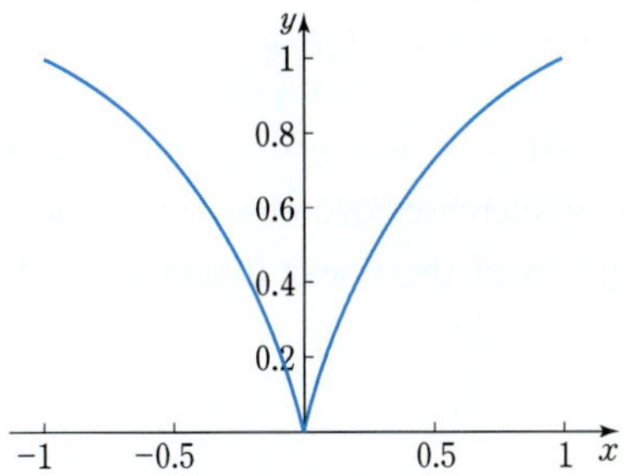

(i) symmetric with respect to the y-axis

(ii) decreasing on $(-\infty, 0)$,

(iii) increasing on $(0, \infty)$

9. $y=\dfrac{x^3}{8}$ (similar to 6)

Solution

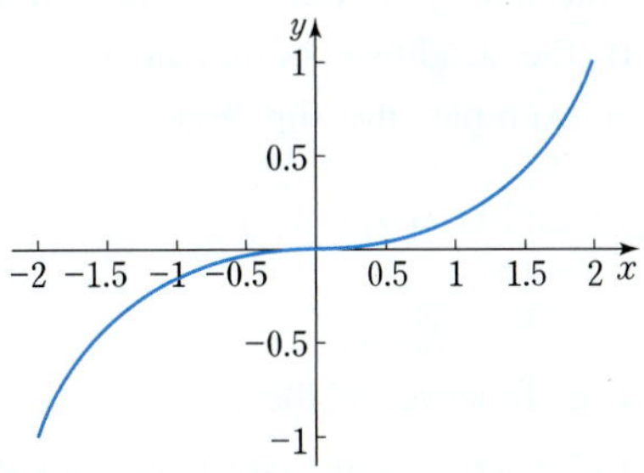

(i) symmetric about the origin

(ii) increasing on $\mathbb{R}$

10. $y=-x^{3/2}$

Solution

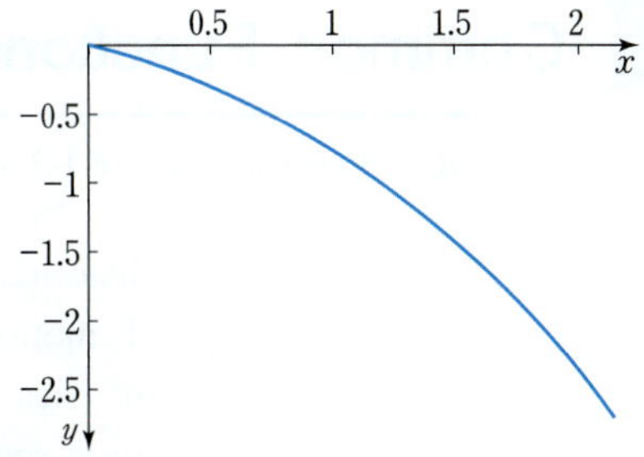

(i) it has no symmetries

(ii) decreasing on $(0, \infty)$

11-16. Determine whether the given function is even, odd, or neither. Give reasons for your answer.

11. $f(x)=3$

Solution Since $f(-x)=3=f(x)$ for all x, f is even function.

12. $f(x)=x^2+1$

Solution Since $f(-x)=(-x)^2+1=(x)^2+1=f(x)$ for all x, f is an even function.

13. $g(x)=x^3+x$

Solution Since

$g(-x)=(-x)^3-x=-(x)^3-x=-((x)^3+x)=-g(x)$

for all x, g is an odd function.

14. $g(x)=\dfrac{1}{x^2-1}$

Solution Since $g(-x)=\dfrac{1}{(-x)^2-1}=\dfrac{1}{(x)^2-1}=g(x)$

for all x, g is an even function.

15. $h(t)=\dfrac{1}{t-1}$

Solution $h(-t)=\dfrac{1}{-t-1}=-\dfrac{1}{t+1}$.

This means $h(-t)\neq h(t)$ and also $h(-t)\neq -h(t)$. Hence h is neither an odd nor even function.

16. $h(t)=2t+1$

http://matrix.skku.ac.kr/cal-lab/cal-1-2-Rational-Fts.html

Solution $h(-t)=-2t+1$.

h is neither an odd nor even function.

```
# [Note] Sage code for Graph of rational functions:
plot((1+x^2)/((x-1)*(x+1)*(x-2)), -3, 4, exclude=[-1, 1, 2], ymax=20, ymin=-20, figsize=6)
```

1.3 Common Functions

http://matrix.skku.ac.kr/Cal-Book/part1/CS-Sec-1-3.htm

Historically, calculus developed from theoretical investigations of physical laws of motion. Nowadays, mathematics is used not only in the natural sciences (physics, chemistry, biology, earth science, etc.) and engineering, but also in some social sciences such as economics.

Mathematical modeling is a process of expressing real world problems or phenomena in terms of functions.

Now we introduce some useful functions.

Linear Function

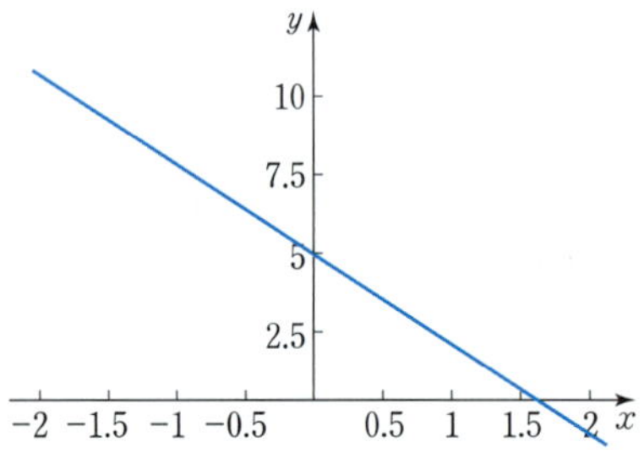

Figure 1 $f(x)=-3x+5$

If y is a **linear function** of x, then the graph of y is a straight line. In slope-intercept form the equation of a straight line can be written as

$$y=f(x)=mx+b,$$

where m is the slope of the line and b is the y-intercept. One characteristic feature of a linear function is that it increases/decreases at a constant rate.

For example, Figure 1 shows a graph of the linear function $f(x)=-3x+5$.

Polynomials

A **polynomial** $P(x)$ over $\mathbb{R}$ is a function defined as

$$P(x)=a_nx^n+a_{n-1}x^{n-1}+\cdots+a_2x^2+a_1x+a_0,$$

where n is a non-negative integer and the real numbers a_0, a_1, a_2, ..., a_n are constants called the **coefficients** of the polynomial. The domain of any polynomial is $\mathbb{R}=(-\infty, \infty)$. If the leading coefficient $a_n \neq 0$, then the **degree** of the polynomial is n. For example, the function

$$P(x)=3x^5-7x^4+12x^2-32$$

is a polynomial of degree 5.

A polynomial of degree 1 is a **linear function** of the form $P(x)=mx+b$. A polynomial of the form $P(x)=ax^2+bx+c$, $a\neq 0$, is of degree 2 and is called a **quadratic function**. The graph of a quadratic function can be obtained by shifting the parabola $y=ax^2$, because we can write $ax^2+bx+c=a'(x-b')^2+c'$ for some a', b', c'. The parabola opens upward if $a>0$(e.g. $2x^2+4x-5$) and downward if $a<0$(e.g. $-2x^2+7x+1$).

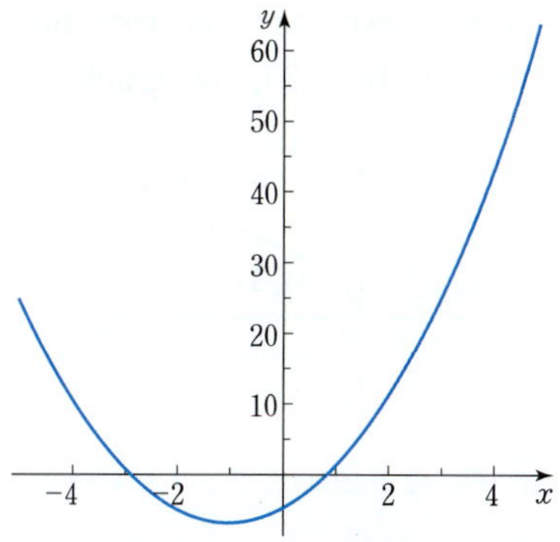

Figure 2 $P(x)=2x^2+4x-5$

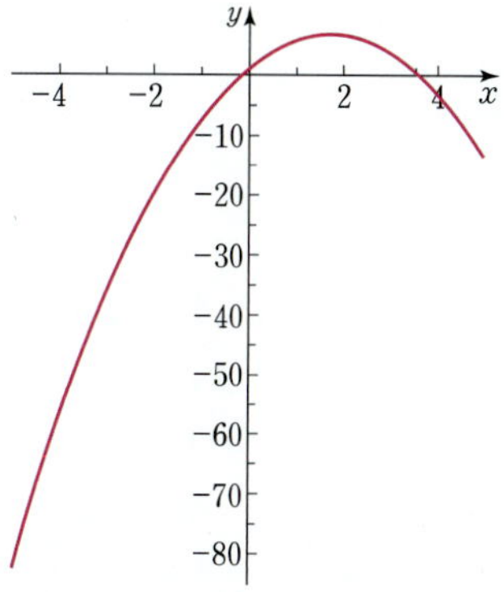

Figure 3 $P(x)=-2x^2+7x+1$

A **cubic function** is a polynomial of degree 3

$$P(x)=ax^3+bx^2+cx+d,\ a\neq 0.$$

Power Functions

A **power function** is a function of the form $f(x)=x^a$, defined on an appropriate domain where a is a constant. We consider the following cases where $a=n$, $a=\dfrac{1}{n}$, $n\in\mathbb{Z}$, $a=-1$.

Case I : $a=n$, where n is a positive integer.
The graphs of $f(x)=x^n$ for $n=1$, 2, 3, 4, and 5, are shown in Figure 4, which are polynomials with only one term.

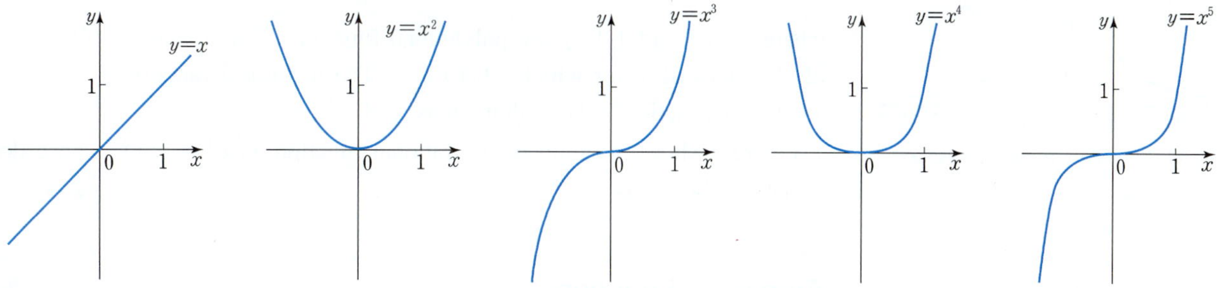

Figure 4 Graphs of $f(x)=x^n$ for $n=1, 2, 3, 4, 5$

The general shape of the graph of $f(x)=x^n$ depends on whether n is even or odd. If n is even, then $f(x)=x^n$ is an even function and its graph is similar to the graph of $y=x^2$. If n is odd, then $f(x)=x^n$ is an odd function and its graph is similar to graph of $y=x^3$.

Case II : $a=1/n$, where n is a positive integer.
The function $f(x)=x^{1/n}=\sqrt[n]{x}$ $(x\geq 0)$ is a **root function**.

For $a=1/2$ it is the square root function $f(x)=\sqrt{x}$ that has domain $[0,\infty)$ and whose graph is the upper half of the parabola $x=y^2$. For other even values of n, the graph of $y=\sqrt[n]{x}$ is similar to the graph of $y=\sqrt{x}$.

For $n=3$ we have the cube root function $f(x)=\sqrt[3]{x}$ whose domain is $\mathbb{R}$. When n is odd $(n>3)$, the graph of $y=\sqrt[n]{x}$ is similar to that of $y=\sqrt[3]{x}$.

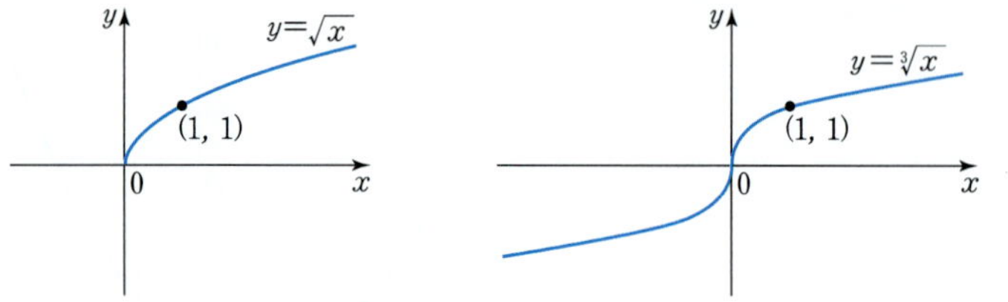

Figure 5 Graphs of root functions

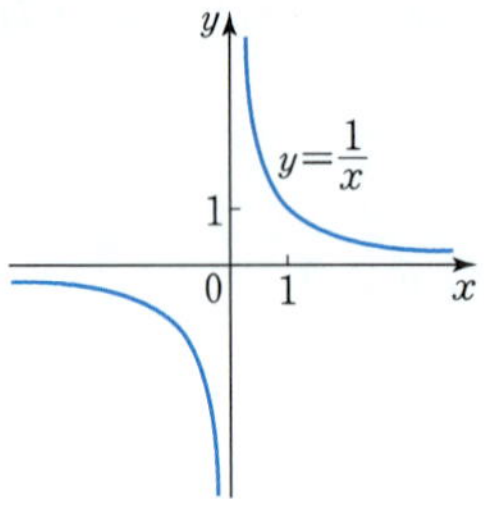

Figure 6
The reciprocal function

Case Ⅲ: $a=-1$, then the function $f(x)=x^{-1}=1/x$ is known as a **reciprocal function**, whose graph is shown in Figure 6.

For example, Boyle's Law is represented by the reciprocal function below. It states that, when the temperature is constant, the volume V of gas is inversely proportional to the pressure P:

$$V=\frac{C}{P},$$

where C is a constant.

Rational Functions

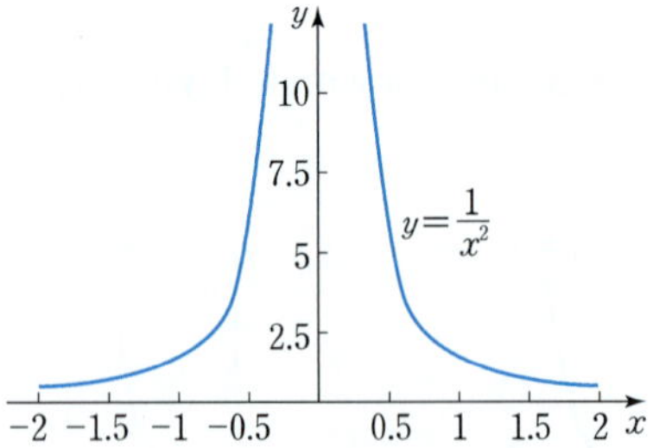

Figure 7 Graph of a rational function

A **rational function** $y=f(x)$ is a function of the form

$$f(x)=\frac{P(x)}{Q(x)}$$

where $P(x)$ and $Q(x)$ are polynomial functions. The domain of $f(x)$ consists of all values of x for which $Q(x)\neq 0$. The reciprocal function $f(x)=1/x$, is a rational function, whose domain is $\{x\in\mathbb{R}\mid x\neq 0\}$.

Figure 7 shows a graph of the rational function $f(x)=1/x^2$, whose domain is $\{x\in\mathbb{R}\mid x\neq 0\}$.

Algebraic Functions

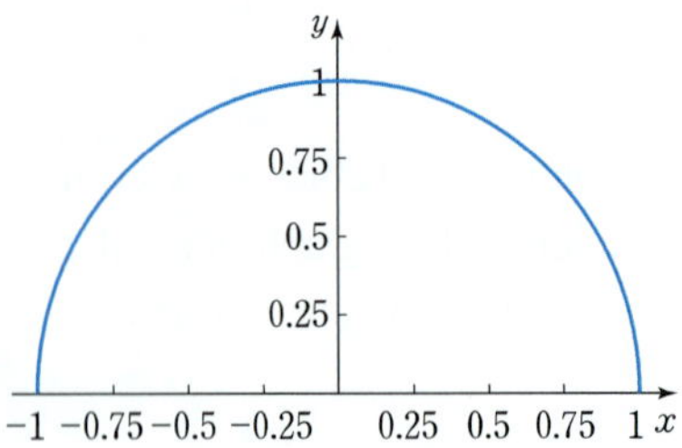

Figure 8 $y=\sqrt{1-x^2}$

An **algebraic function** is a function constructed by using the algebraic operations (such as addition, subtraction, multiplication, division and taking roots) on polynomials.

For example, $y=\sqrt{1-x^2}$ is an algebraic function. See Figure 8 for the graph of $y=\sqrt{1-x^2}$.

Trigonometric and Periodic Functions

Trigonometric functions (sometimes called circular functions) are functions of an angle, which is usually measured in radians. The most familiar trigonometric functions are the sine, cosine and tangent functions. A periodic function is a function that repeats its values in regular intervals or periods. The most important examples are the trigonometric functions, which are periodic over intervals of length 2π radians. They are important in the study of triangles and many periodic motions that occur in nature. The graphs of the sine and cosine functions are shown Figure 9 and 10.

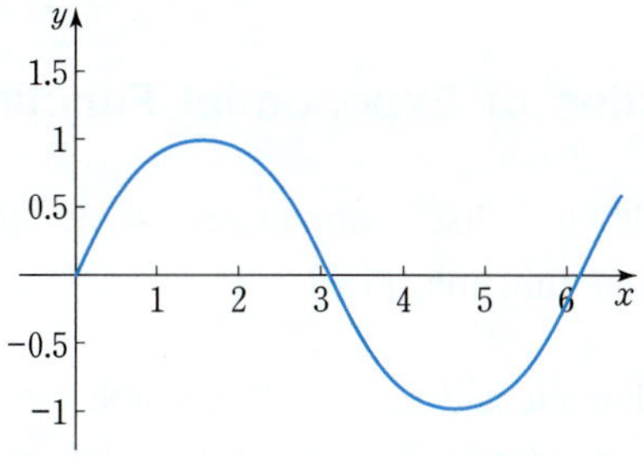

Figure 9 Graph of $y = \sin x$

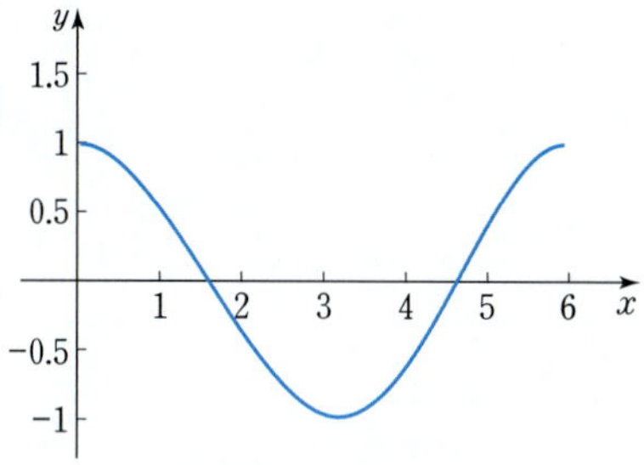

Figure 10 Graph of $y = \cos x$

We have $-1 \leq \sin x \leq 1$, $-1 \leq \cos x \leq 1$ (for all x). The domain of $\sin x$ and $\cos x$ is $(-\infty, \infty)$. The range of $\sin x$and $\cos x$ is $[-1, 1]$.
Since

$$\sin(x+2\pi) = \sin x \text{ and } \cos(x+2\pi) = \cos x$$

for all values of x, both sine and cosine functions are periodic functions with period 2π.

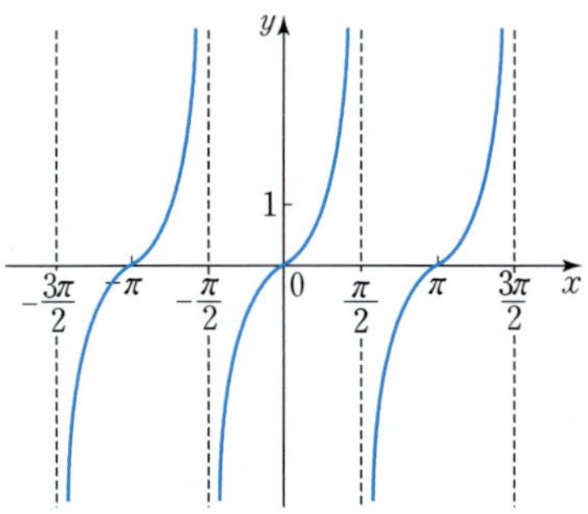

Figure 11 Graph of $y = \tan x$

The tangent function is defined as $\tan x = \dfrac{\sin x}{\cos x}$, except for when $\cos x = 0$, and its graph is shown in Figure 11. When $x = \pm\pi/2, \pm 3\pi/2, \ldots$, it is undefined. Its range is $(-\infty, \infty)$. Since $\tan(x+\pi) = \tan x$ for all x it is periodic with period π.

Cosecant, secant, and cotangent are the reciprocals of the sine, cosine, and tangent functions, respectively and their graph are shown in Figurs 12-14.

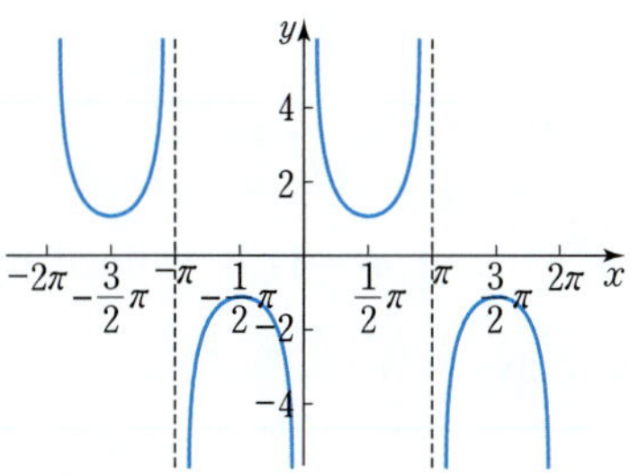

Figure 12 Graph of $y = \csc x$

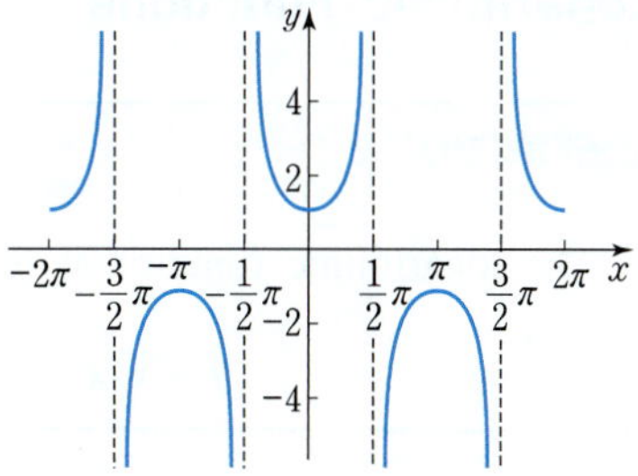

Figure 13 Graph of $y = \sec x$

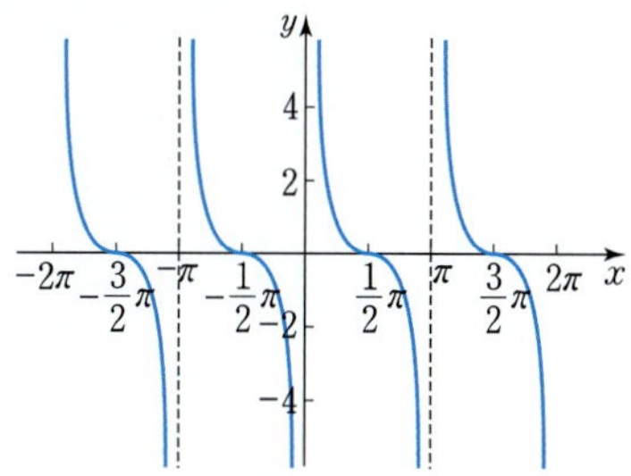

Figure 14 Graph of $y = \cot x$

Exponential Functions

> **DEFINITION 1**
>
> If $a > 0$ and $a \neq 1$, then the function
>
> $$f(x) = a^x$$
>
> is called **exponential function.**

The number a is called the base and x is called the exponent.

Properties of Exponential Functions

The following list summarizes some of the important properties of the exponential function $f(x) = a^x$.

- The domain is $(-\infty, \infty)$ and the range is $(0, \infty)$.
- The y-intercept of f is $(0, 1)$. The graph of f has no x-intercept.
- The function f is increasing on $(-\infty, \infty)$ for $a > 1$. (see Figure 15.)
- The function f is decreasing on $(-\infty, \infty)$ for $0 < a < 1$ (see Figure 16.)
- The x-axis is a horizontal asymptote for the graph of f.
- The function f is one-to-one.

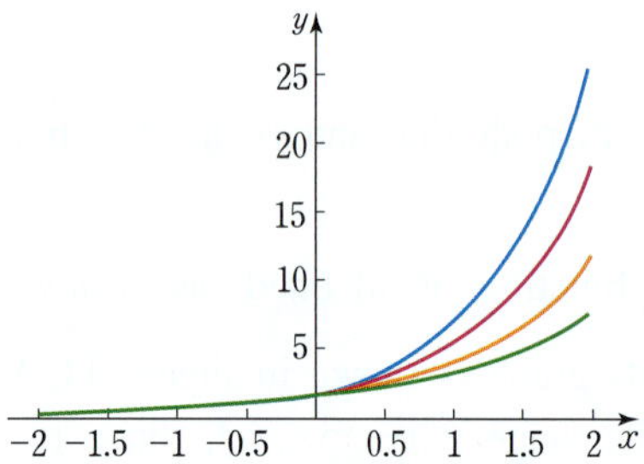

Figure 15 Exponential functions of $a > 1$

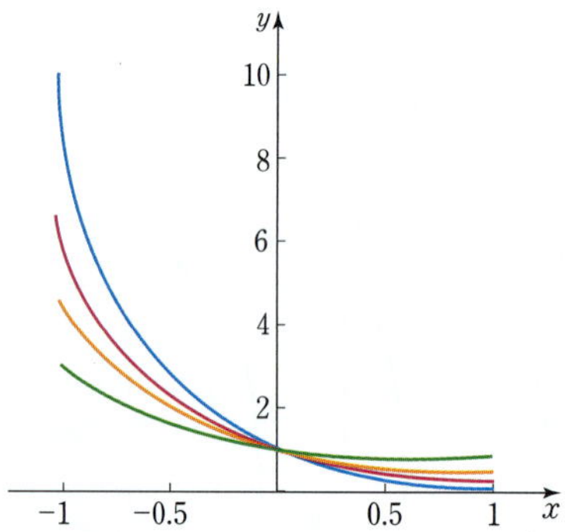

Figure 16 Exponential functions of $0 < a < 1$

Logarithmic Functions

> **DEFINITION 2**
>
> The **logarithmic function** with base $a > 0$ and $a \neq 1$, is defined by
>
> $$y = \log_a x \quad \text{if and only if} \quad x = a^y$$

The logarithmic equation $y = \log_a x$ and the exponential equation $x = a^y$ are equivalent.

Properties of Logarithmic Functions

The following list summarizes some of the important properties of logarithmic function $y = \log_a x$.

- The domain is $(0, \infty)$ and the range is $(-\infty, \infty)$.
- The x-intercept of f is 1. The function of f has no y-intercept.
- The function f is increasing on $(0, \infty)$ for $a > 1$ (see Figure 17.)
- The function f is decreasing on $(0, \infty)$ for $0 < a < 1$. (Sketch)
- The y-axis is a vertical asymptote for the graph of f.
- The function f is one-to-one.

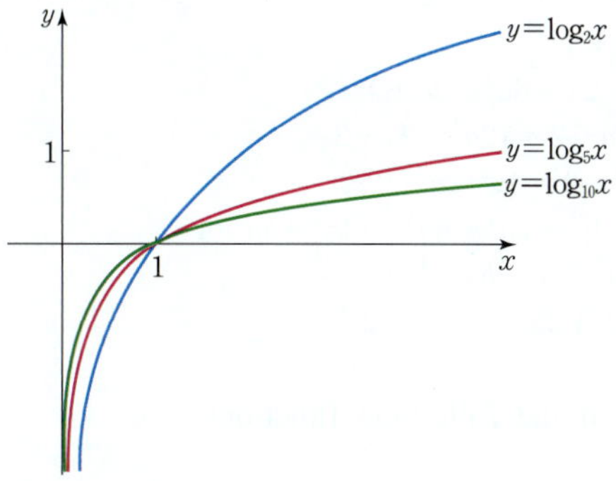

Figure 17 Logarithmic functions, $a > 1$

Transcendental Functions

Functions that are not algebraic are called **transcendental functions**. For example, trigonometric, exponential and logarithmic functions are transcendental functions.

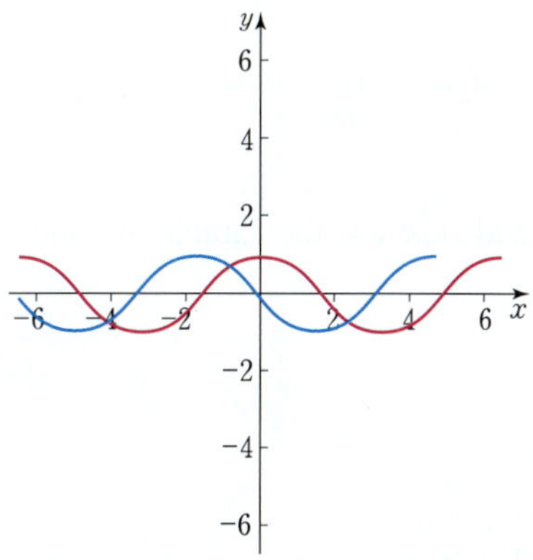

Figure 18 Cosine and Sine functions

http://matrix.skku.ac.kr/cal-lab/sage-grapher.html

1.3 EXERCISES (Common Functions)

http://matrix.skku.ac.kr/Cal-Book/part1/CS-Sec-1-3-Sol.htm   YouTube http://youtu.be/x0E0ZMxZ3Og

1. If $f(x)=3x^2-x+2$, find $f(2)$, $f(-2)$, $f(a)$, $f(-a)$, $f(a+1)$, $2f(a)$, $f(2a)$, $f(a^2)$, $[f(a)]^2$ and $f(a+h)$.

Solution $f(2)=3\cdot 2^2-2+2=12$

$f(-2)=3(-2)^2+2+2=16$

$f(a)=3a^2-a+2$

$f(-a)=3(-a)^2+a+2$

$f(a+1)=3(a+1)^2-(a+1)+2$

$=3a^2+5a+4$

$2f(a)=2(3a^2-a+2)=6a^2-2a+4$

$f(2a)=3(2a)^2-2a+2=12a^2-2a+2$

$f(a^2)=3(a^2)^2-a^2+2=3a^4-a^2+2$

$[f(a)]^2=[3a^2-a+2]^2=9a^4-6a^3+13a^2-4a+4$

$f(a+h)=3(a+h)^2-(a+h)+2$

$=3a^2+6ah+3h^2-a-h+2$

2. Find the domain of the following functions.

(a) $f(x)=\dfrac{3}{3x-1}$

(b) $f(x)=\dfrac{5x+4}{x^2+3x+2}$

(c) $f(x)=\sqrt{t}+\sqrt[3]{t}$

(d) $g(u)=\sqrt{u}+\sqrt{4-u}$

(e) $h(x)=\dfrac{1}{\sqrt[4]{x^2-5x}}$

Solution (a) $D=(-\infty,\ 1/3)\cup(1/3,\ \infty)$

(b) $D=(-\infty,\ -2)\cup(-2,\ -1)\cup(-1,\ \infty)$

(c) $D=[0,\ \infty)$

(d) $D=[0,\ \infty)\cap(-\infty,\ 4]=[0,\ 4]$

(e) $D=(-\infty,\ 0)\cup(5,\ \infty)$

3. Find the domain and sketch the graph of the following functions.

(a) $f(x)=\dfrac{1}{\sqrt{5x-1}}$

(b) $f(x)=\dfrac{x^2+4x+3}{5x+4}$

(c) $f(x)=\begin{cases} x, & \text{if } x<0 \\ x+1, & \text{if } x>0 \end{cases}$

Solution (a) $5x-1>0$ so $x>\dfrac{1}{5}$. So it's

Domain: $\left(\dfrac{1}{5},\ \infty\right)$

```
var('x,y')
p1=plot((1/(5*x-1)^(1/2)),    x,    -10,    10,
color='blue');
show(p1, aspect_ratio=1, ymax=5)
```

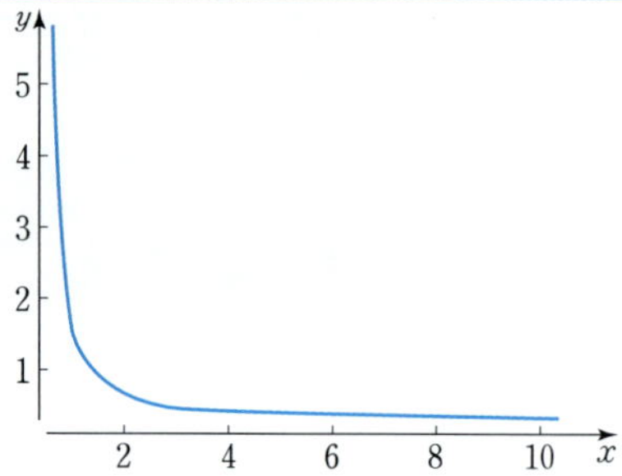

(b) $5x+4\neq 0$ so $x\neq -\dfrac{4}{5}$. So it's domain:

$\left(-\infty,\ -\dfrac{4}{5}\right)\cup\left(-\dfrac{4}{5},\ \infty\right)$

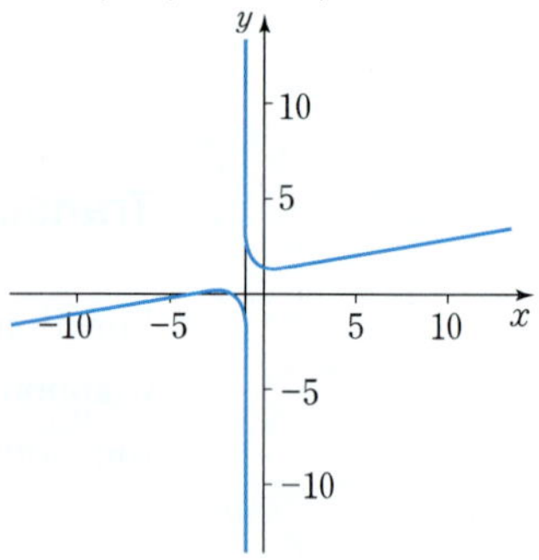


```
var('x, y')
p1=plot((x^2+4*x+3)/(5*x+3),    x,    -10,    10,
color='blue');
show(p1,   aspect_ratio=1,   ymax=10,   ymin=-10)
```

(c) Domain : $(-\infty,\ 0)\cup(0,\ \infty)$

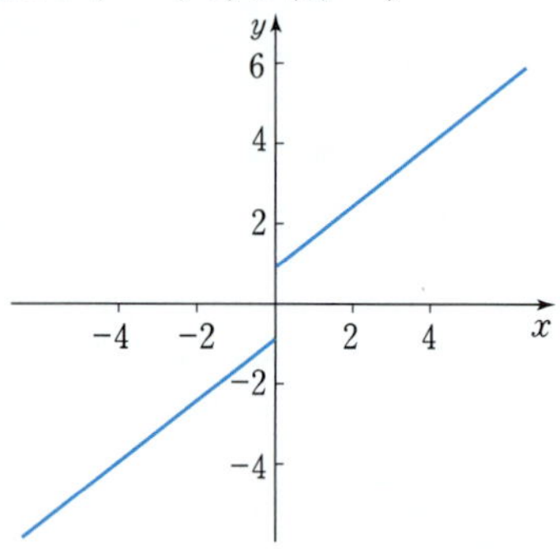

4. Determine whether f is even, odd or neither. If f is even or odd, use symmetry to sketch its graph.

(a) $f(x) = x^{-2}$

(b) $f(x) = x^{-3}$

(c) $f(x) = x^2 + x$

(d) $f(x) = x^4 - 4x^2$

(e) $f(x) = x^3 - x$

(f) $f(x) = 3x^3 + 2x^2 + 1$

Solution (a) even function

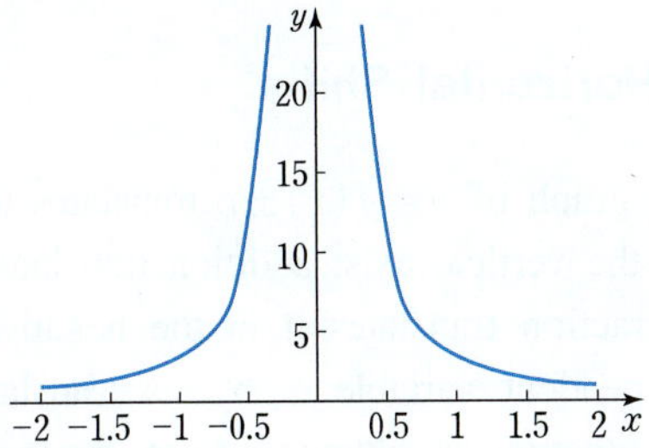

(b) odd function

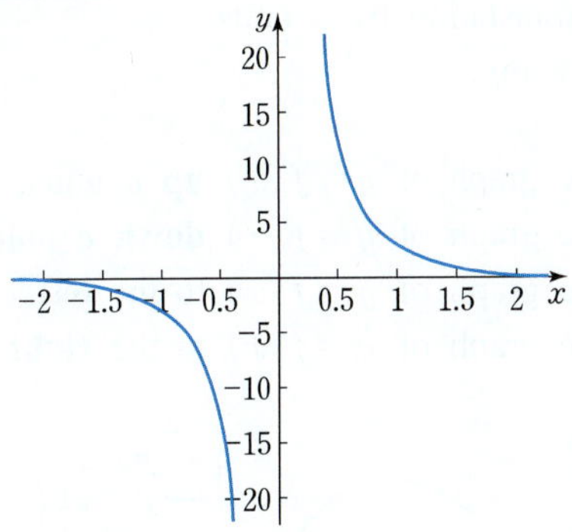

(c) neither

(d) even function

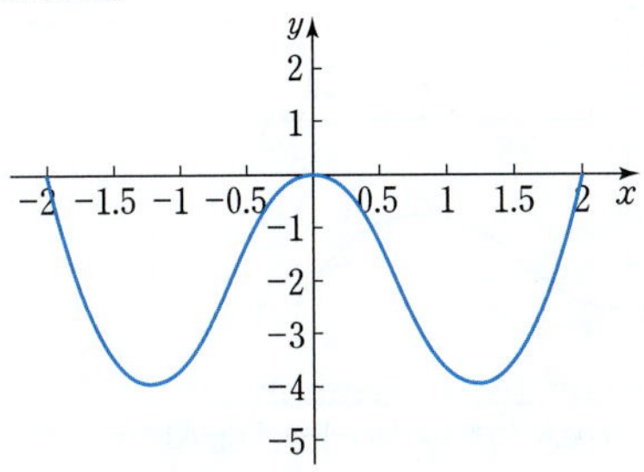

(e) odd function

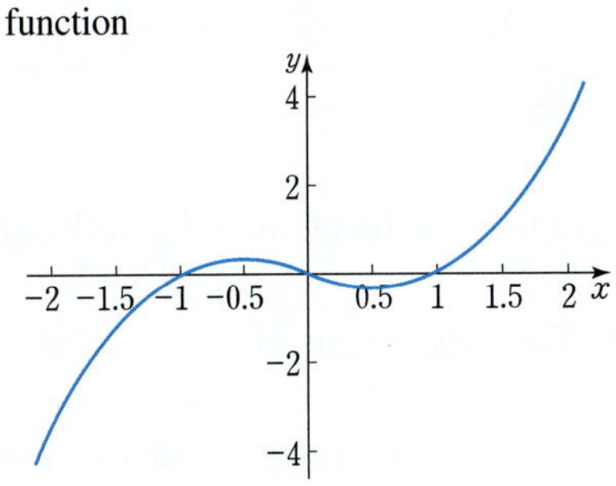

(f) neither

5-6. Draw the graph of given functions.

CAS **5.** $f(x) = e^{-x^2/2}$

http://matrix.skku.ac.kr/cal-lab/cal-1-3-5.html

```
p1=plot(exp(-(x^2)/2), x, -3, 3, color='blue')
show(p1, aspect_ratio=1, ymax=2)
```

Solution

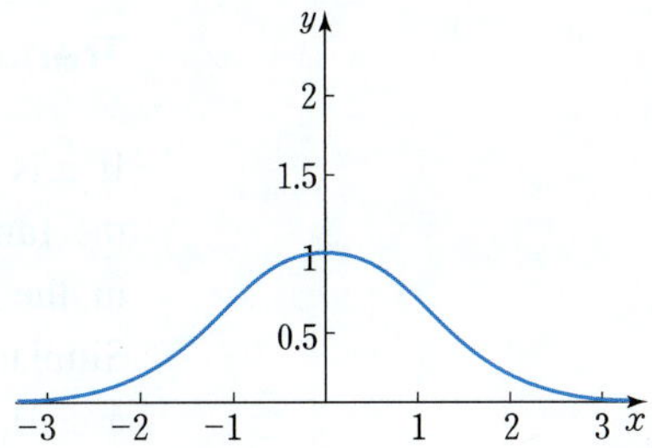

6. $f(x) = e^{-1/x}, \ x > 0$

Solution

```
plot(exp(-1/x), x, 0, 10, color='blue')
```

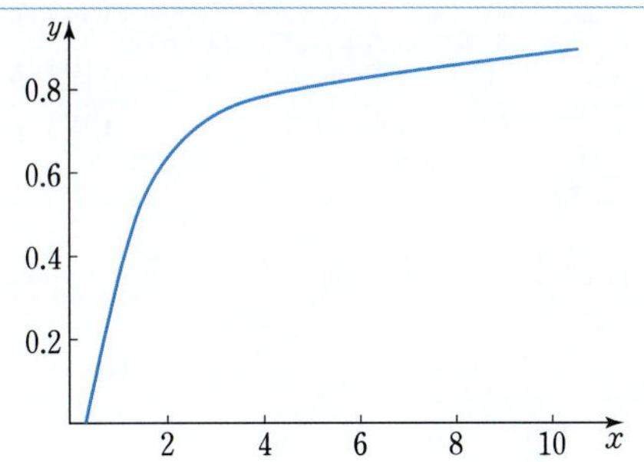

*Smartphone App. of World Mathematicians

https://play.google.com/store/apps/details?id=korean.mathematicians.en

1.4 Translation, Stretching and Rotation of Functions

http://matrix.skku.ac.kr/Cal-Book/part1/CS-Sec-1-4.htm

If we apply certain translations to the graph of a given function, the graphs of certain related functions can be obtained. This will enable us to write equations for the graphs and to sketch the graphs of many functions quickly.

Translation: Vertical and Horizontal Shifts:

If c is a positive number, then the graph of $y=f(x)\pm c$ translates the graph of the function $y=f(x)$ parallel to the vertical axis; addition translates the graph in the positive direction and subtraction translates it in the negative direction. Similarly, if we replace the independent variable x by its translation (x by $x+c$). The graphs of the new function is obtained from the graph of the original function by a horizontal translation by c units.
Thus for $c>0$, we have the following:

(i) $y=f(x)+c$, translates the graph of $y=f(x)$ **up** c units.
(ii) $y=f(x)-c$, translates the graph of $y=f(x)$ **down** c units.
(iii) $y=f(x+c)$, translates the graph of $y=f(x)$ to the **left** c units.
(iv) $y=f(x-c)$, translates the graph of $y=f(x)$ to the **right** c units.

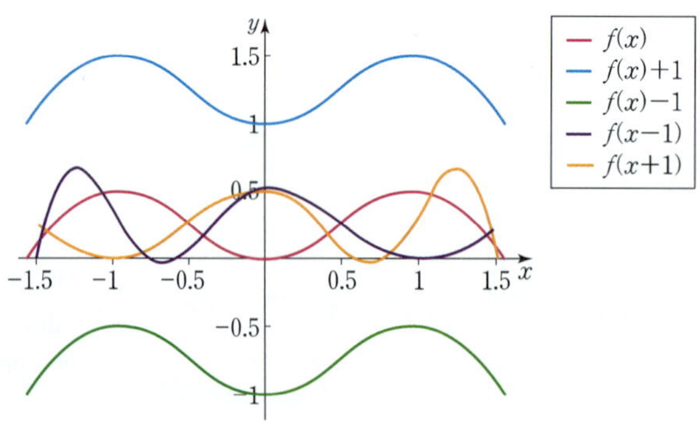

Figure 1 $f(x)=\cos x\ \sin x^2$ and its translations
http://matrix.skku.ac.kr/cal-lab/cal-1-4-Fig-A.html

Reflection:

Another way of transforming the graph of a function is by reflecting about a coordinate axis.

Suppose $y=f(x)$ is a function. Then the graph of

(i) $y=-f(x)$ is the graph of $y=f(x)$ reflected about the x-**axis**.
(ii) $y=f(-x)$ is the graph of $y=f(x)$ reflected about the y-**axis**.

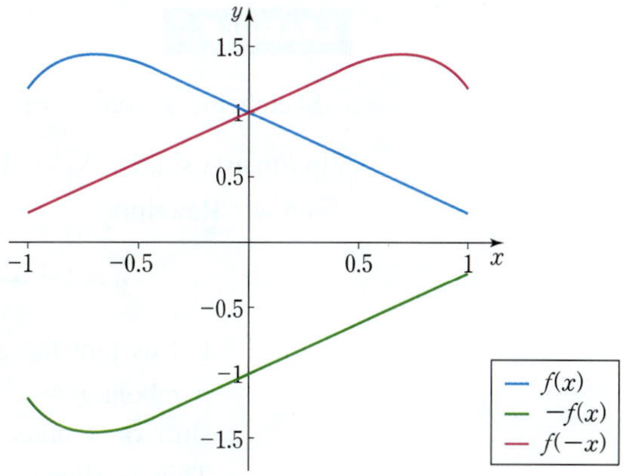

Figure 2 Reflections of $f(x) = \cos x^2 * \exp(-x)$
http://matrix.skku.ac.kr/cal-lab/cal-1-4-Fig-B.html

Vertical and Horizontal Stretching:

If a function $f(x)$ is multiplied by a constant $c > 0$, the shape of the graph is changed but retains roughly its original shape. The graph of $y = cf(x)$ is the graph of $y = f(x)$ distorted vertically; the graph $f(x)$ is either stretched or compressed vertically depending on the value of c. In other words, a vertical stretch is a stretch of the graph of $y = f(x)$ toward or away from the x-axis. The graph of the function $y = f(cx)$ is distorted horizontally, either by stretching the graph $y = f(x)$ away from the y-axis or by squeezing the graph of $y = f(x)$ toward the y-axis.

Suppose $y = f(x)$ is a function and c is a positive constant. Then

(i) $y = cf(x)$ is the graph of $y = f(x)$ **vertically stretched** by a factor c.
(ii) $y = f(x)/c$ is the graph of $y = f(x)$ **vertically compressed** by a factor c.
(iii) $y = f(cx)$ is the graph of $y = f(x)$ **horizontally compressed** by a factor c.
(iv) $y = f(x/c)$ is the graph of $y = f(x)$ **horizontally stretched** by a factor c.

Thus for $c > 1$, the graph is scaled.

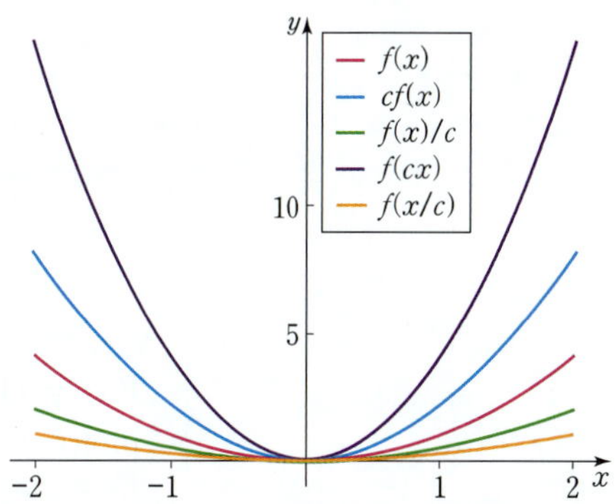

Figure 3 Vertical and Horizontal Stretching of $f(x) = x^2$
http://matrix.skku.ac.kr/cal-lab/cal-1-4-Fig-C.html

EXAMPLE 1

Sketch the graph of the function $y = \left| \frac{x^2 + 8x}{2} \right|$.

http://matrix.skku.ac.kr/cal-lab/cal-1-4-Exm-1.html

Solution Rewriting

$$y = (x^2 + 8x)/2 = (x^2 + 8x + 16)/2 - 8 = (x+4)^2/2 - 8$$

Let us plot the graph of $y = (x^2 + 8x)/2$. Start with the graph of the parabola $y = x^2$, compress it vertically by a factor 2, followed by a shift of 4 units to the left, and then finally shift 8 units downward. This is shown in Figures 5-8. If $-8 < x < 0$, then $y < 0$. So the graph lies below the x-axis. To draw the graph of the function $y = |(x^2 + 8x)/2|$, we reflect the negative part of the graph about the x-axis. This is shown in Figure 4.

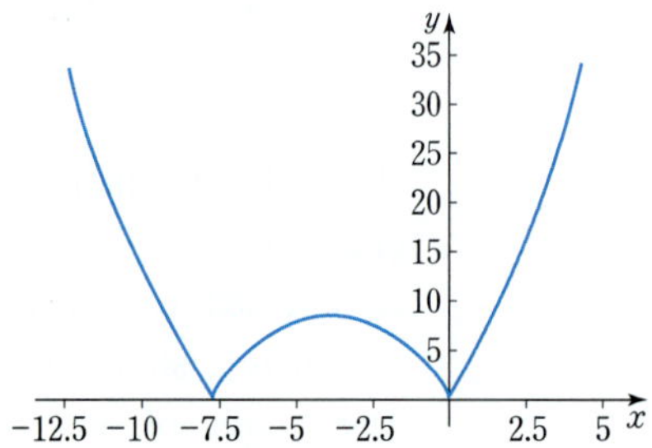

Figure 4 $y = \left| \frac{x^2 + 8x}{2} \right|$

```
f(x) = x^2
plot(f, (x, -13, 5))
```

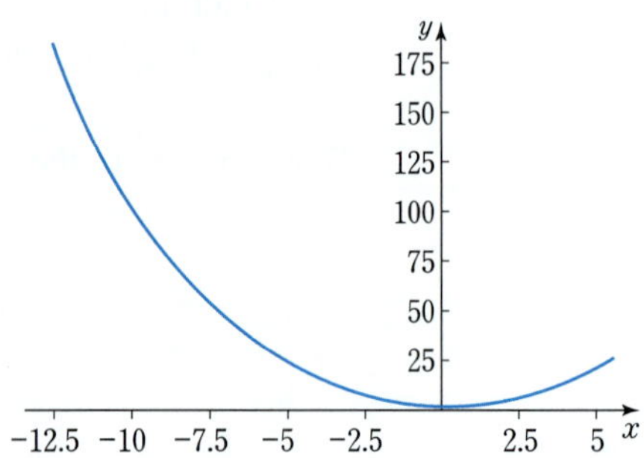

Figure 5 $y = x^2$

```
f(x) = (1/2)*x^2
plot(f, (x, -13, 13))
```

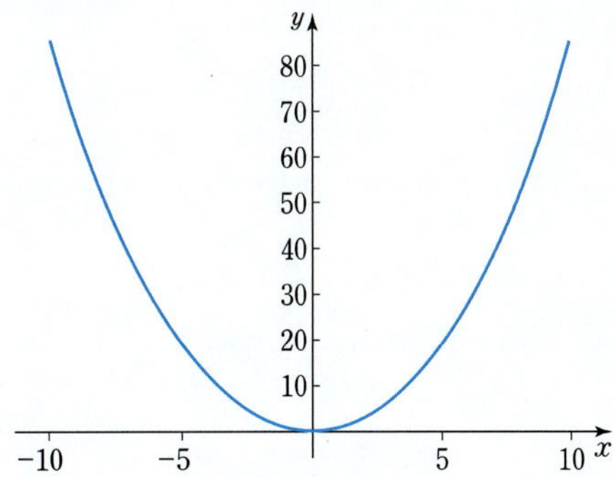

Figure 6 $y=\dfrac{x^2}{2}$

```
f(x) = (1/2)*(x+4)^2
plot(f, (x, -13, 5))
```

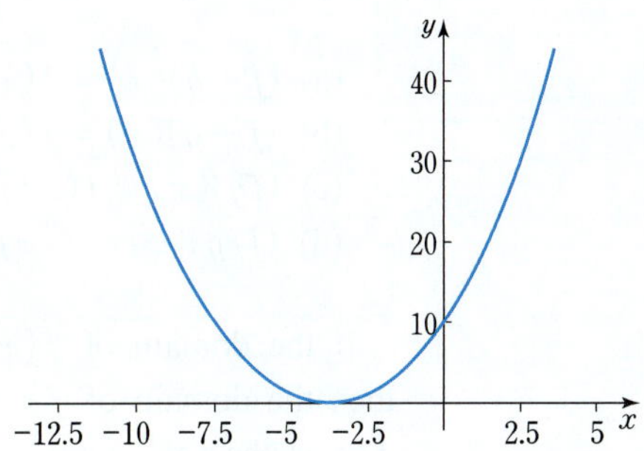

Figure 7 $y=\dfrac{(x+4)^2}{2}$

```
f(x) = (1/2)*(x+4)^2-8
plot(f, (x, -13, 5))
```

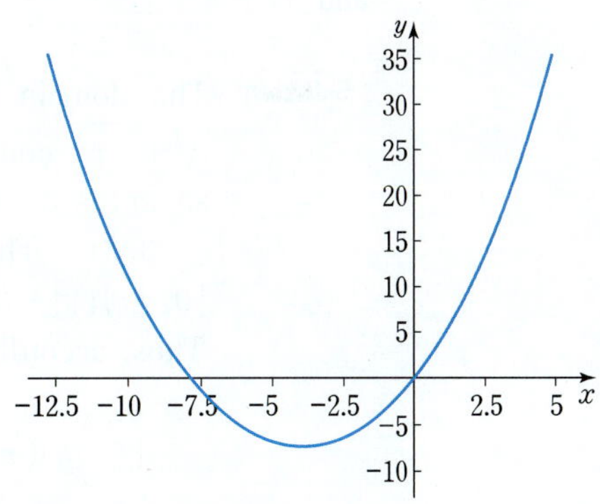

Figure 8 $y=\dfrac{(x+4)^2}{2}-8$

```
f(x) = abs((1/2)*(x+4)^2-8)
plot(f, (x, -13, 5))
```

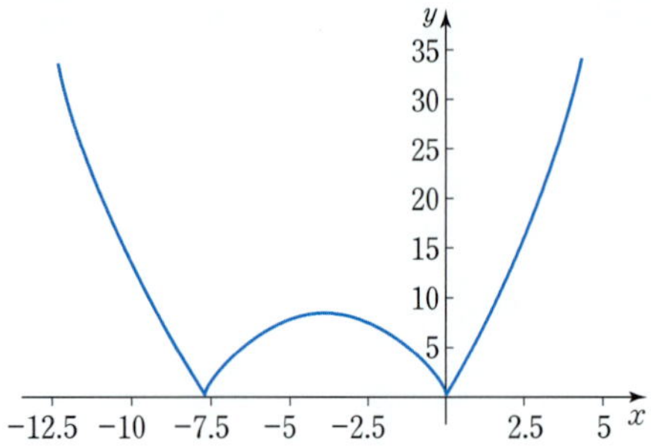

Figure 9 $y=\left|\frac{(x+4)^2}{2}-8\right|$

■

Arithmetic Combinations:

If $f(x)$ and $g(x)$ are two functions, then the sum $f+g$, the difference $f-g$, the product fg, and the quotient f/g are defined as follows:

(a) $(f+g)(x)=f(x)+g(x)$
(b) $(f-g)(x)=f(x)-g(x)$
(c) $(fg)(x)=f(x)g(x)$
(d) $(f/g)(x)=f(x)/g(x),\ g(x)\neq 0$

If the domain of $f(x)$ is the set A and the domain of $g(x)$ is the set B, then the domain of $f+g$, $f-g$, and fg is the set $A\cap B$, and the domain of f/g is the set

$$\{x\,|\,x\in A\cap B,\ g(x)\neq 0\}.$$

EXAMPLE 2

If $f(x)=\sqrt{x}$ and $g(x)=\sqrt{9-x^2}$, find the functions $f+g$, $f-g$, fg and f/g.

Solution The domain of $f(x)=\sqrt{x}$ is $[0,\infty)$. The domain of $g(x)=\sqrt{9-x^2}$ consists of all numbers x such that $9-x^2\geq 0$, or $x^2\leq 9$ so $|x|\leq 3$, or $-3\leq x\leq 3$, Thus the domain of g is the interval $[-3,3]$. The intersection of the domains of f and g is $[0,\infty)\cap[-3,3]=[0,3]$.
Thus, according to the above definitions, we have the following

$$(f+g)(x)=\sqrt{x}+\sqrt{9-x^2} \qquad (0\leq x\leq 3)$$
$$(f-g)(x)=\sqrt{x}-\sqrt{9-x^2} \qquad (0\leq x\leq 3)$$
$$(fg)(x)=\sqrt{x}\,\sqrt{9-x^2}=\sqrt{9x-x^3} \qquad (0\leq x\leq 3)$$
$$\left(\frac{f}{g}\right)(x)=\frac{\sqrt{x}}{\sqrt{9-x^2}}=\sqrt{\frac{x}{9-x^2}} \qquad (0\leq x<3)$$

Note that the interval $[0, 3)$ is the domain of f/g and should not include $x = 3$ because $g(3) = 0$. ■

Composite Functions

Given any two functions f and g, if we find the image $g(x)$ of a number x in the domain of g. If this value $g(x)$ is in the domain of f, then we find the value of $f(g(x))$. The new function $h(x)$, is known as the composite function $h(x) = (f \circ g)(x) = f(g(x))$. It is obtained by substituting g into f, and is known as the composition (or composite) of f and g. It is denoted by $f \circ g$ ("f circle g").

In general, $f \circ g \neq g \circ f$. For example, let $f(x) = \sin x$ and $g(x) = 2x$. Then $(g \circ f)(x) = 2\sin x$ and $(f \circ g)(x) = \sin(2x)$. Clearly $f \circ g \neq g \circ f$.

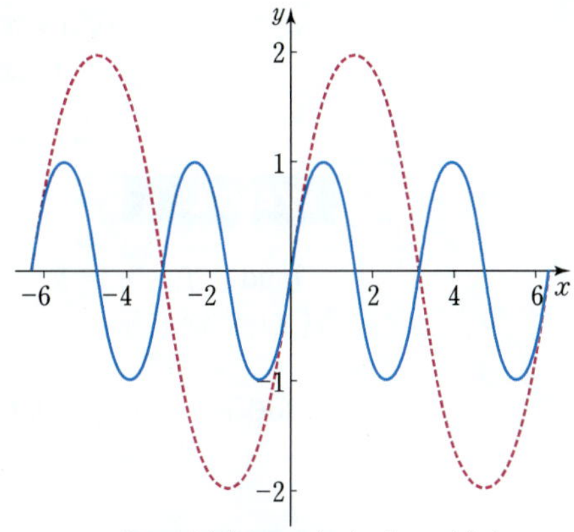

Figure 10 $(f \circ g)(x) = \sin(2x)$, $(g \circ f)(x) = 2\sin x$

The domain of $f \circ g$ is the set of all x in the domain of g such that $g(x)$ is in the domain of f. In other words, $(f \circ g)(x)$ is defined whenever both $g(x)$ and $f(g(x))$ are defined.

The composition of three or more functions say, the composite function $f \circ g \circ h$ is found by first applying, h then g, and then f:$(f \circ g \circ h)(x) = f(g(h(x)))$.

http://matrix.skku.ac.kr/cal-lab/sage-grapher.html

EXAMPLE 3

Given $f(x) = x^4$, $g(x) = x - 5$. Find the composite functions $f \circ g$ and $g \circ f$.

Solution We have $(f \circ g)(x) = (f(g(x)) = f(x-5) = (x-5)^4$
$(g \circ f)(x) = (g(f(x)) = x^4 - 5$. ■

EXAMPLE 4

Given $f(x) = \sqrt{x}$ and $g(x) = \sqrt{4-x}$, Find for (a)-(d) each function and its domain.

(a) $f \circ g$ (b) $g \circ f$ (c) $f \circ f$ (d) $g \circ g$

Solution (a) $(f \circ g)(x) = f(g(x)) = f(\sqrt{4-x}) = \sqrt{\sqrt{4-x}} = \sqrt[4]{4-x}$
The domain of $f \circ g$ is $\{x \mid 4-x \geq 0\} = (-\infty, 4]$.

(b) $(g \circ f)(x) = g(f(x)) = g(\sqrt{x}) = \sqrt{4-\sqrt{x}}$
For $\sqrt{x}$ to be defined, we must have $x \geq 0$. For $\sqrt{4-\sqrt{x}}$ to be defined, we must have $4-\sqrt{x} \geq 0$, that is, $\sqrt{x} \leq 4$, or $x \leq 16$. Thus, we have $0 \leq x \leq 16$, so the domain of $g \circ f$ is the closed interval $[0, 16]$.

(c) $(f \circ f)(x) = f(f(x)) = f(\sqrt{x}) = \sqrt{\sqrt{x}} = \sqrt[4]{x}$
The domain of is $f \circ f$ is $[0, \infty)$.

(d) $(g \circ g)(x) = g(g(x)) = g(\sqrt{4-x}) = \sqrt{4-\sqrt{4-x}}$
This expression is defined when $4-x \geq 0$, that is, $x \leq 4$ and $4-\sqrt{4-x} \geq 0$. This latter inequality is equivalent to $\sqrt{4-x} \leq 4$, or $4-x \leq 16$, that is, $x \geq -12$. Thus, $-12 \leq x \leq 4$, so the domain of $g \circ g$ is the closed interval $[-12, 4]$. ■

EXAMPLE 5

Find $f \circ g \circ h$, where $f(x) = \sin x/(\tan x - 2)$, $g(x) = x^{20}$, and $h(x) = 3x - 5$.

Solution $(f \circ g \circ h)(x) = f(g(h(x))) = f(g(3x-5)) = f((3x-5)^{20})$
$= \sin(3x-5)^{20}/(\tan(3x-5)^{20} - 2)$ ■

EXAMPLE 6

Given $F(x) = \sin^3(x+8)$, find functions f, g and h such that $F = f \circ g \circ h$.

Solution Since $F(x) = \sin^3(x+8)$ the formula for F says: First add 8, then take the sine of the result, and finally take cube. So we let $h(x) = x+8$, $g(x) = \sin x$, $f(x) = x^3$.
Then

$$\begin{aligned}(f \circ g \circ h)(x) &= f(g(h(x))) \\ &= f(g(x+8)) = f(\sin(x+8)) \\ &= \sin^3(x+8) = F(x).\end{aligned}$$ ■

1.4 EXERCISES (Translation, Stretching and Rotation of Functions)

http://matrix.skku.ac.kr/Cal-Book/part1/CS-Sec-1-4-Sol.htm

YouTube http://youtu.be/vx7GCWY68Zw

1-6. Exercises 1-6 show how many units and in what directions the graphs of the given equations are to be shifted. Find an equation for the shifted graph.

1. $y=2x-7$. Up 7.

Solution $y=2x-7+7=2x$

2. $y=x^3$. Left 1, Down 1.

Solution $y=(x+1)^3-1$

3. $y=\sqrt{x}$. Left 0.81.

Solution $y=\sqrt{x+0.81}$

4. $y=1/x$. Up 1, Right 1.

Solution $y=\dfrac{1}{(x-1)}+1$

5. $y=\dfrac{1}{2}(x+1)+5$. Down 5, Right 1.

Solution $y=\dfrac{1}{2}(x+1-1)+5-5=\dfrac{1}{2}x$

6. $x^2+y^2=49$. Down 3, Left 2.

Solution $(x+2)^2+(y+3)^2=49$

CAS **7-16.** Graph the functions in Exercises 7-16.

7. $y=\sqrt{x+4}$

http://matrix.skku.ac.kr/cal-lab/cal-1-4-7.html

Solution

```
var('x')
plot(sqrt(x+4), x, -4, 2, color='blue')
```

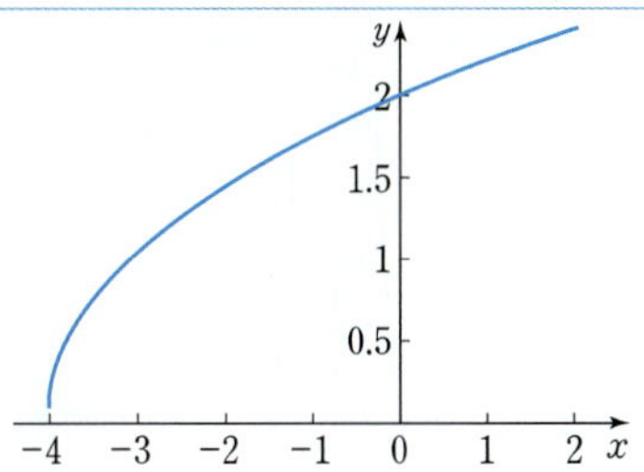

8. $y=|x-2|$

http://matrix.skku.ac.kr/cal-lab/cal-1-4-8.html

Solution

```
var('x')
plot(abs(x-2), x, 0, 4, color='blue')
```

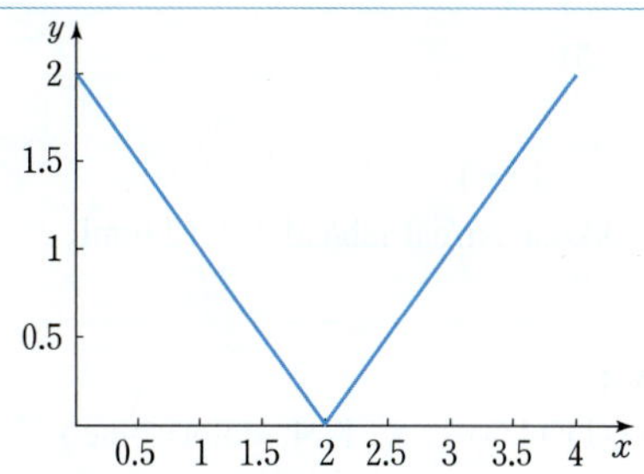

9. $y=1+\sqrt{x-1}$

http://matrix.skku.ac.kr/cal-lab/cal-1-4-9.html

Solution

```
var('x')
plot(1+sqrt(x-1), x, 1, 4, color='blue')
```

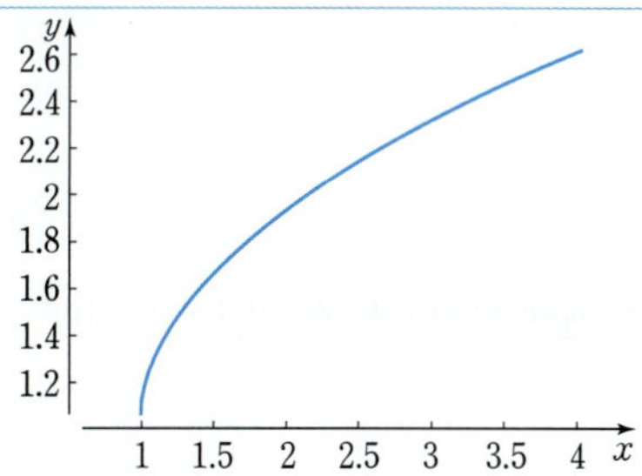

10. $y=(x+1)^{2/3}$

http://matrix.skku.ac.kr/cal-lab/cal-1-4-10.html

Solution

```
var('x')
plot((x+1)^(2/3), x, -1, 4, color='blue')
```

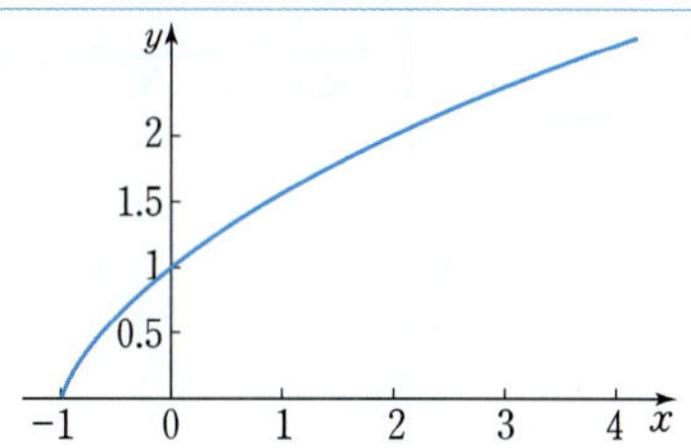

11. $y=1-x^{2/3}$

http://matrix.skku.ac.kr/cal-lab/cal-1-4-11.html

Solution

```
var('x')
plot(1-x^(2/3), x, 0, 4, color='blue')
```

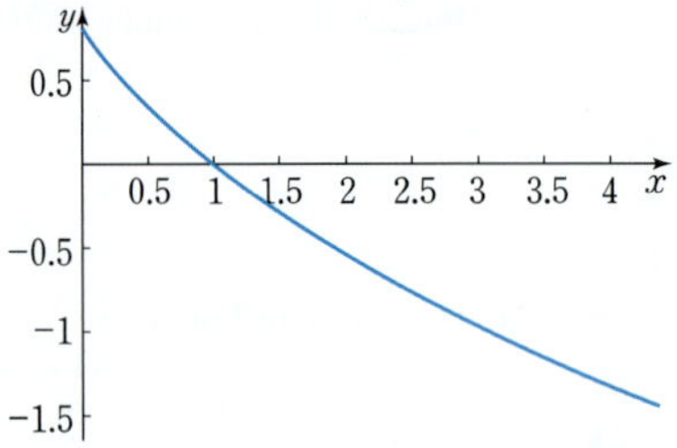

12. $y=\sqrt[x]{x-1}-1$

http://matrix.skku.ac.kr/cal-lab/cal-1-4-12.html

Solution

```
var('x');
plot((x-1)^(1/3)-1, x, 1, 4, color='blue')
```

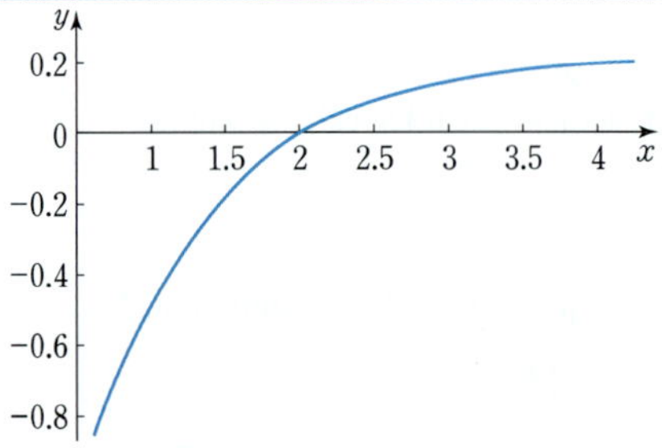

13. $y=\dfrac{1}{x-2}$

http://matrix.skku.ac.kr/cal-lab/cal-1-4-13.html

Solution

```
var('x')
show(plot(1/(x-2)  -1,  x,  1,  4,  color='blue'),
ymax=10, ymin=-10)
```

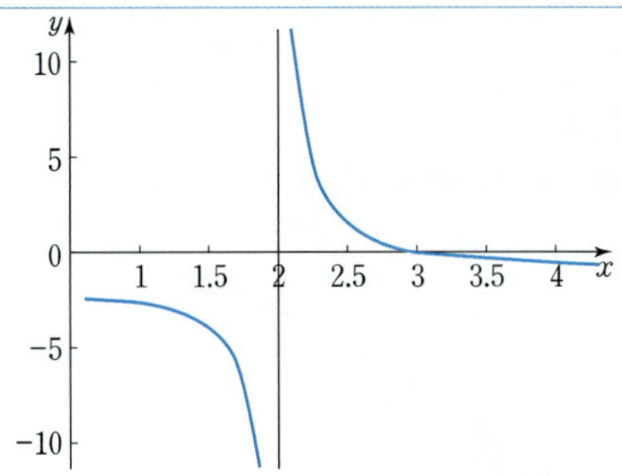

14. $y=\dfrac{1}{x}+2$

http://matrix.skku.ac.kr/cal-lab/cal-1-4-14.html

Solution

```
var('x')
show(plot(1/x  +2,  x,  -2,  2,  color='blue'),
ymax=10, ymin=-10)
```

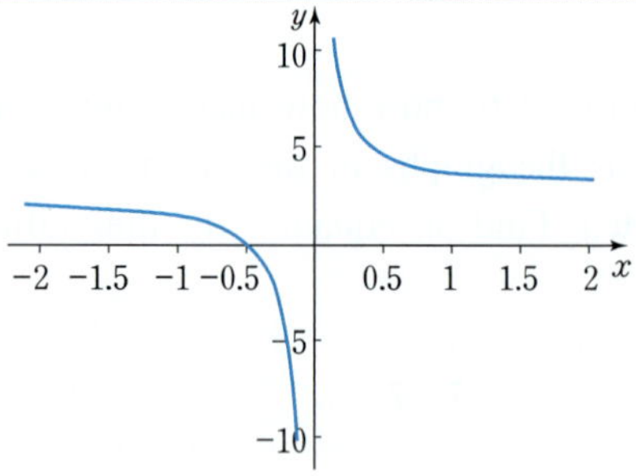

15. $y=\dfrac{1}{(x-1)^2}$

http://matrix.skku.ac.kr/cal-lab/cal-1-4-15.html

Solution

```
var('x')
show(plot(1/(x-1)^2,  x,  -2,  2,  color='blue'),
ymax=10, ymin=-10)
```

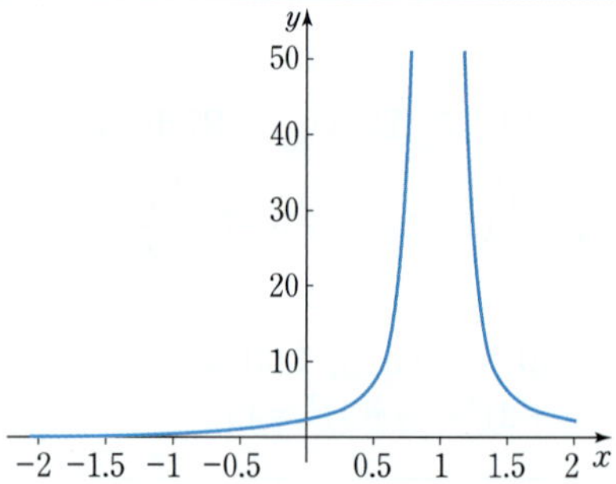

16. $y=\dfrac{1}{x^2}+1$

http://matrix.skku.ac.kr/cal-lab/cal-1-4-16.html

Solution

```
var('x')
show(plot(1/(x-1)^2, x, -2, 2, color='blue'),
ymax=10, ymin=-10)
```

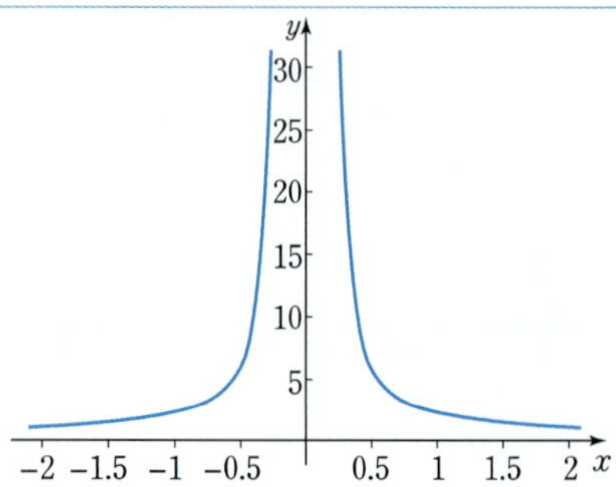

Seoul Central Railway Station and Graph

Chapter 2

Limits and Continuity

Johann Carl Friedrich Gauss

1777–1855

Math : Queen of Sciences

"Painted by Christian Albrecht Jensen"

Gauss contributed significantly to many fields, including number theory, statistics, analysis, differential geometry, geodesy, geophysics, electrostatics, astronomy and optics. Gauss had a remarkable influence in many fields of science and is ranked as one of history's most influential mathematicians. He referred to mathematics as "the queen of sciences."

YouTube http://youtu.be/VBCeAllP1M0

In this chapter, we will work with functions and their limits. We will use limits to introduce the concept of continuity. We will consider the effect on $y=f(x)$ as x approaches a constant value (positive or negative) or an infinite value (negative or positive.) The concept of a limit, and being able to evaluate limits, is useful in both differential and integral calculus. The concept is used to define both the derivative and the integral.

2.1 Limits of Functions

http://matrix.skku.ac.kr/Cal-Book/part1/CS-Sec-2-1-Sol.html

In this section, we introduce the notion of the "limit." Intuitively, this notion is easy to understand. But mathematically, we should prepare for situations that are not intuitive at all. As the chapter progresses we will introduce a more rigorous definition of the limit. Thus, for completeness, we will define the "limit" by 'ε-δ' arguments at the end of this section. We also will study many limit laws and most of them can be proved rigorously with 'ε-δ' arguments.

First, consider one example.

EXAMPLE 1

Investigate the behavior of the function $f(x) = \dfrac{x^2-4}{x-2}$, $x \neq 2$, near the point $x = 2$.

Solution Note that $f(2)$ is not defined because the denominator is zero when $x = 2$. Compute some values of $f(x)$ when x is close to 2, and smaller than 2. (We say "from the left" of 2.)

x	1.9	1.99	1.999	1.9999	1.99999	1.999999
$f(x)$	3.9	3.99	3.999	3.9999	3.99999	3.999999

From the above table, we observe that as x is getting closer and closer to 2 from the left ($x < 2$), the function value is getting closer and closer to 4.

On the other hand, compute some other values of $f(x)$ when x is close to 2 and larger than 2 from the right.

x	2.1	2.01	2.001	2.0001	2.00001	2.000001
$f(x)$	4.1	4.01	4.001	4.0001	4.00001	4.000001

From the above table, we observe that as x is getting closer and closer to 2 from the right ($x > 2$), the function value is also getting closer and closer to 4 but not equal to 4. So if x approaches 2, then we have only two cases, that is, (i) $x > 2$ and x approaches 2, or (ii) $x < 2$ and x approaches 2. From above arguments, we find that the function value approaches 4 in both cases. ■

The above example suggests the following definition of the limit.

DEFINITION 1

The expression "$\lim_{x \to a} f(x) = L$" is read as "the limit of $f(x)$, as x approaches a, equals L." This means that we can make the values of $f(x)$ to be arbitrarily close to L by taking x to be sufficiently close to a but not equal to a. In other words, $f(x)$ gets closer and closer to L as x gets closer and closer to a (but not equal to a).
Alternatively, we write '$f(x) \to L$ as $x \to a$.'

In finding the limit of $f(x)$ as x approaches a, the phrase "but $x \neq a$" means that $x = a$ is never considered. Thus we do not care about the value of $f(a)$ for finding the limit of $f(x)$. It does not matter even if $f(x)$ is not defined at $x = a$. Our interest is in how f is defined near a. Figure 1 shows the graphs of three functions. Note that in part (b), $f(a) \neq L$, in part (c), $f(a)$ is not defined. But in each case, regardless of what happens to $f(x)$ at $x = a$, it is true that $\lim_{x \to a} f(x) = L$.

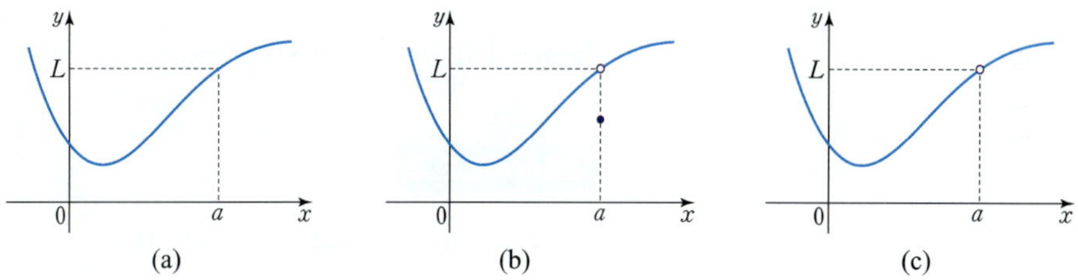

Figure 1 $\lim_{x \to a} f(x) = L$ three cases

DEFINITION 2

The expression "$\lim_{x \to a^-} f(x) = L$" is read as "the left-hand limit of $f(x)$, as x approaches a, equals L". This means that we can make the values of $f(x)$ arbitrarily close to L by taking x to be sufficiently close to a and x is less than a. In other words, $f(x)$ gets closer and closer to L as x gets closer and closer to a with restriction $x < a$.
Similarly, if we restrict that x to be greater than a, then we have "the right-hand limit of $f(x)$ as x approaches a, equals L." For notation, we use

$$\lim_{x \to a^+} f(x) = L.$$

Thus, the symbol "$x \to a^+$" means that we consider only $x > a$ (only values of x that are greater than a, that is, to the right of a are considered). Similarly "$x \to a^-$" means that we consider only $x < a$ (only values of x that are less than a, that is, to the left of a are considered).

In the Figure 1, the limit of $f(x)$ exists if the limits of $f(x)$ 'from the right and left' coincide. We will call the limit from the left and from the right, 'left-hand limit' and 'right-hand limit', respectively.

We have the following theorem.

THEOREM 3

The limit of $f(x)$ as x approaches a exists if and only if both left and right limits as x approaches a exist and coincide. That is,

$$\lim_{x \to a} f(x) = L \text{ if and only if } \lim_{x \to a^-} f(x) = L = \lim_{x \to a^+} f(x).$$

In some cases, both the left-hand and the right-hand limits exist, but they are not equal, so the limit does not exist. Consider the Heaviside function $H(x)$: (The Heaviside function is very important in ordinary differential equations and electromagnetism because this function represents turning on a switch at time 0).

$$H(x) = \begin{cases} 0, & x < 0, \\ 1, & x \geq 0. \end{cases}$$

For the Heaviside function, we have $\lim_{x \to 0^-} H(x) = 0$ and $\lim_{x \to 0^+} H(x) = 1$. Hence $\lim_{x \to 0} H(x)$ does not exist.

EXAMPLE 2

Determine whether the limit $\lim_{x \to 0} \frac{x}{|x|}$ exists or not.

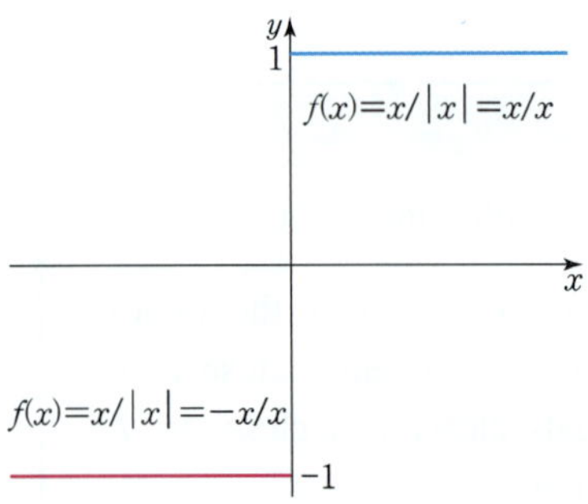

Figure 2 $f(x) = \frac{x}{|x|}$

Solution Here we use the Theorem 3 above. That is, we consider the left -hand and the right-hand limits. If we consider the left-hand limit, then we have $|x| = -x$, because $x < 0$. Thus we have the left-hand limit $\lim_{x \to 0^-} \frac{x}{|x|} = \lim_{x \to 0^-} \frac{x}{-x} = -1$. On the other hand, if we consider the right-hand limit, then we have $|x| = x$, because $x > 0$. Hence we have the right-hand limit $\lim_{x \to 0^+} \frac{x}{|x|} = \lim_{x \to 0^+} \frac{x}{x} = 1$. Because these two limits do not coincide, $\lim_{x \to 0} \frac{x}{|x|}$ does not exist by Theorem 3. ■

Here we describe what it means for the values of a function $f(x)$ to increase (or decrease) without bound or to approach infinity and what it means for f to have a limit as x approaches infinity (∞). The symbol ∞ does not represent any real number. Also we cannot use ∞ in arithmetic in the usual way.

Infinite Limits

The following definition of infinite limits is informal, but it is adequate for most functions encountered in this book.

DEFINITION 4

Suppose f is defined for all x near a. If $f(x)$ grows arbitrarily large for all x sufficiently close(but not equal) to a, we write

$$\lim_{x\to a} f(x) = \infty.$$

We say the limit of $f(x)$ as x approaches a is infinity.

If $f(x)$ is negative and grows arbitrarily large in magnitude for all x sufficiently close(but not equal) to a, we write

$$\lim_{x\to a} f(x) = -\infty.$$

In this case, we say the limit of $f(x)$ as x approaches a is negative infinity. In both cases, the limit does not exist.

For example consider the function $f(x) = 1/x$, which is defined for all real numbers except at $x = 0$.

Let us consider the limit of $f(x)$, and observe that:

(i) $1/x$ approaches 0 as x tends to ∞.
(ii) $1/x$ tends to ∞ as x approaches 0 from the right.
(iii) $1/x$ tends to $-\infty$ as x approaches 0 from the left.
(iv) $1/x$ approaches 0 as x tends to $-\infty$.

Let a function f be defined on both sides of a, except possibly at a itself. Then $\lim_{x\to a} f(x) = \infty$ means that the values of $f(x)$ can be made arbitrarily large (as large as we want) (or "increase without bound") by taking x sufficiently close to a, but of course not equal to a. In this case we say that the limit does not exist.

Similarly, $\lim_{x\to a} f(x) = -\infty$ means that the values of $f(x)$ can be made arbitrarily large and negative by taking x sufficiently close to a, but not equal to a.

Similar definitions can be given for the following one-sided limits

$$\lim_{x\to a^-} f(x) = \infty,\ \lim_{x\to a^+} f(x) = \infty,\ \lim_{x\to a^-} f(x) = -\infty,\ \lim_{x\to a^+} f(x) = -\infty.$$

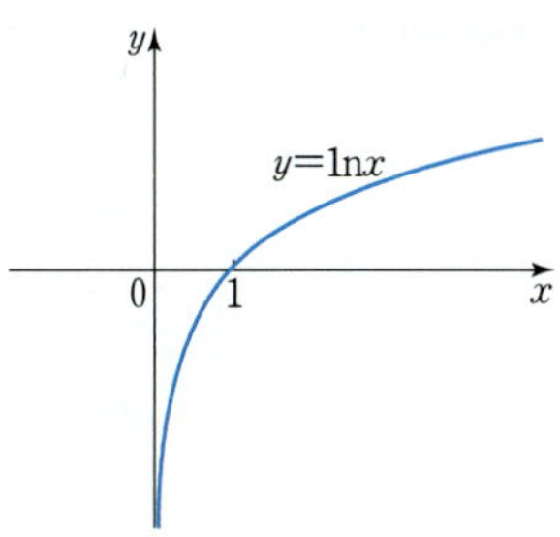

Figure 3 $y = \ln x$

Consider the natural logarithm function $y = \ln x$. (See Figure 3.) As $x \to 0^+$, $\ln x$ tends to $-\infty$ and the distance between curve $(x, \ln x)$ and y-axis approaches 0.

We will call the straight line $x = 0$ the vertical asymptote of $y = \ln x$.

THEOREM 5

A straight line L is said to be an asymptote to a curve C if the distance between the curve C and the line L decreases to zero as we traverse along the curve C. Asymptotes may be horizontal, or vertical.

If $f(x) \to L$ (a finite number) as $x \to \infty$ or $x \to -\infty$, then $y = L$ is a horizontal asymptote.

Similarly if $f(x) \to \infty$ (or $f(x) \to -\infty$) as $x \to a$ (finite number), then $x = a$ is a vertical asymptote. The calculation of slant asymptotes is more involved. In general, a knowledge of asymptotes is very useful when sketching graphs.

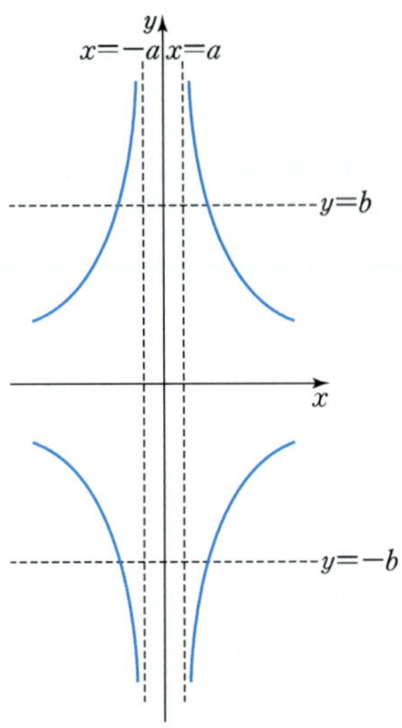

Figure 4
$(y^2 - b^2) = \dfrac{a^2b^2}{(x^2 - a^2)}$

EXAMPLE 3

Find vertical and horizontal asymptotes of

$$y^2 - b^2 = \frac{a^2b^2}{x^2 - a^2}.$$

Solution Now

$$\lim_{x \to \pm a} (y^2 - b^2) = \lim_{x \to \pm a} \frac{a^2b^2}{(x^2 - a^2)} = \pm\infty.$$

So $x = \pm a$ are two vertical asymptotes.

Similarly $y = \pm b$ are two horizontal asymptotes. ■

For calculations of limits, the following limit laws are very useful. Proofs of the limit laws can be constructed using ε-δ arguments.

Properties of Limit

Suppose that the limits $\lim_{x \to a} f(x)$ and $\lim_{x \to a} g(x)$ exist and c is a constant. Then

1. $\lim_{x \to a} [f(x) + g(x)] = \lim_{x \to a} f(x) + \lim_{x \to a} g(x)$
2. $\lim_{x \to a} [f(x) - g(x)] = \lim_{x \to a} f(x) - \lim_{x \to a} g(x)$
3. $\lim_{x \to a} [cf(x)] = c \lim_{x \to a} f(x)$
4. $\lim_{x \to a} [f(x)g(x)] = \lim_{x \to a} f(x) \lim_{x \to a} g(x)$
5. $\lim_{x \to a} \dfrac{f(x)}{g(x)} = \dfrac{\lim_{x \to a} f(x)}{\lim_{x \to a} g(x)}$ if $\lim_{x \to a} g(x) \neq 0$
6. $\lim_{x \to a} [f(x)]^n = [\lim_{x \to a} f(x)]^n$, where n is a positive integer.

7. $\lim_{x \to a} \sqrt[n]{f(x)} = \sqrt[n]{\lim_{x \to a} f(x)}$, where n is a positive integer.
(If n is even, we require $\lim_{x \to a} f(x) \geq 0$.)

http://www.milefoot.com/math/calculus/limits/GenericLimitLawProofs04.htm

Note The Limit Laws also hold for one-sided limits.

Next, we state the Squeeze Theorem (or Sandwich Theorem), which is useful in finding many important limits.

THEOREM 6 The Squeeze Theorem (or Sandwich Theorem)

If $f(x) \leq g(x) \leq h(x)$, when x is near a (except possibly at a) and

$$\lim_{x \to a} f(x) = \lim_{x \to a} h(x) = L, \quad \text{then } \lim_{x \to a} g(x) = L.$$

This says that if $g(x)$ is squeezed between $f(x)$ and $h(x)$ near a, and if f and h have the same limit L at a, then g is forced to have the same limit L at a. (See Figure 5 and also imagine a sandwich.)

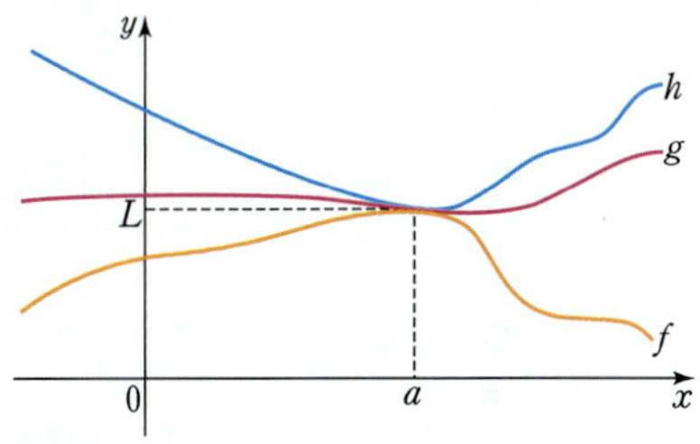

Figure 5 Squeeze Theorem

EXAMPLE 4

Find $\lim_{x \to 0} x^2 \cos\left(\frac{1}{x}\right)$.

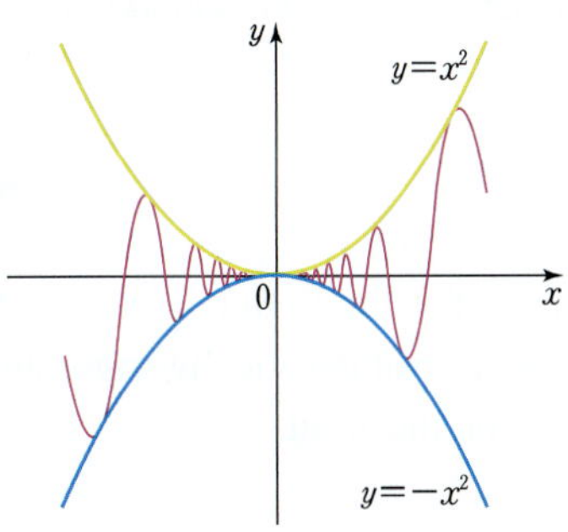

Figure 6 $y = x^2 \cos\left(\frac{1}{x}\right)$

Solution Since $\lim_{x \to 0} \cos\left(\frac{1}{x}\right)$ does not exist (the functions oscillates between -1 and 1), the fourth property of limit cannot be applied here. However, since $-1 \leq \cos\left(\frac{1}{x}\right) \leq 1$, we have

$$-x^2 \leq x^2 \cos\left(\frac{1}{x}\right) \leq x^2. \text{ (See Figure 6.)}$$

But we know that $\lim_{x \to 0} x^2 = \lim_{x \to 0} (-x^2) = 0$.

Take $f(x)=-x^2$, $g(x)=x^2\cos\left(\frac{1}{x}\right)$, $h(x)=x^2$, then by the Squeeze Theorem, we have $\lim_{x\to 0} x^2\cos\left(\frac{1}{x}\right)=0$. ■

CAS EXAMPLE 5

Find $\lim_{x\to 2} f(x)=\lim_{x\to 2}\frac{x^3-8}{x-2}$ numerically by filling out the following form using the given Sage code.

http://matrix.skku.ac.kr/cal-lab/cal-2-1-Exm5.html

Solution

```
var('x')
f(x)=(x^3-8)/(x-2)
a=[1.9, 1.99, 1.999, 1.9999, 1.99999, 1.999999, 1.9999999]
b=[2.1, 2.01, 2.001, 2.0001, 2.00001, 2.000001, 2.0000001]
print [f(k) for k in a]
print
print [f(k) for k in b]
```

x	1.9	1.99	1.999	1.9999	1.99999	1.999999	1.9999999
$f(x)$							

x	2.1	2.01	2.001	2.0001	2.00001	2.000001	2.0000001
$f(x)$							

[11.4100000000000, 11.9401000000000, 11.9940010000001, 11.9994000099957, 11.9999400000839, 11.9999940003548, 11.9999993960387]

[12.6100000000000, 12.0601000000000, 12.0060009999991, 12.0006000099972, 12.0000600001826, 12.0000060005334, 12.0000006039613]

Answer : It converges to 12. ■

Until now, we have used an intuitive definition of the limit. But if we want to prove the previous theorems and limit laws with mathematically rigorous deduction, we need the following "ε-δ" definition for the limit.

1)http://mathworld.wolfram.com/Epsilon-DeltaDefinition.html

DEFINITION 7 Epsilon-Delta Definition[1]

An epsilon-delta definition is a mathematical definition in which a statement on a real function of one variable f having, for example, the form "for all neighborhoods U of y_0 there is a neighborhood V of x_0 such that, whenever $x \in V$, then $f(x) \in U$ " is rephrased as "for all $\varepsilon > 0$ there is $\delta > 0$ such that, whenever $0 < |x - x_0| < \delta$, then $|f(x) - y_0| < \varepsilon$." These two statements are equivalent formulations of the definition of the limit $\lim_{x \to x_0} f(x) = y_0$. In the second one, the neighborhood U is replaced by the open interval $(y_0 - \varepsilon, y_0 + \varepsilon)$, and the neighborhood V by the open interval $(x_0 - \delta, x_0 + \delta)$.

2)http://www.youtube.com/watch?v=-ejyelI0i5c

For a function of several variables, the absolute value would be replaced by the norm $\|\cdot\|$ of $\mathbb{R}^n$.[2] These statements express the fact that for all x which lie close enough to x_0, $f(x)$ lies as close to y_0 as desired. This facilitates the task of proving limits since the fundamental formula are actually shown by constructing a δ with the required property for any given ε.

Let us prove the sum law using the above definition.

Limit Law (Sum Law)

If $\lim_{x \to a} f(x) = L$ and $\lim_{x \to a} g(x) = M$ both exist, then

$$\lim_{x \to a} [f(x) + g(x)] = L + M.$$

Proof The Sum Law Let $\varepsilon > 0$ be given. We must find $\delta > 0$ such that

$$|f(x) + g(x) - (L + M)| < \varepsilon \text{ whenever } 0 < |x - a| < \delta.$$

Using the Triangle Inequality we have

$$\begin{aligned} |f(x) + g(x) - (L + M)| &= |(f(x) - L) + (g(x) - M)| \\ &\le |f(x) - L| + |g(x) - M|. \end{aligned}$$

We make $|f(x) + g(x) - (L + M)|$ less than ε by making each of the terms $|f(x) - L|$ and $|g(x) - M|$ less than $\varepsilon/2$, respectively.

Since $\varepsilon/2 > 0$ and $\lim_{x \to a} f(x) = L$ there exists a number $\delta_1 > 0$ such that

$$|f(x) - L| < \frac{\varepsilon}{2} \text{ whenever } 0 < |x - a| < \delta_1.$$

Similarly, since $\lim_{x \to a} g(x) = M$ there exists a number $\delta_2 > 0$ such that

$$|g(x) - M| < \frac{\varepsilon}{2} \text{ whenever } 0 < |x - a| < \delta_2.$$

Let $\delta = \min\{\delta_1, \delta_2\}$. Notice that if $0 < |x-a| < \delta$ then $0 < |x-a| < \delta_1$ and $0 < |x-a| < \delta_2$, and so $|f(x) - L| < \dfrac{\varepsilon}{2}$ and $|g(x) - M| < \dfrac{\varepsilon}{2}$. Therefore,

$$\begin{aligned} |f(x)+g(x)-(L+M)| &\le |f(x)-L| + |g(x)-M| \\ &< \frac{\varepsilon}{2} + \frac{\varepsilon}{2} = \varepsilon. \end{aligned}$$

To summarize, $|f(x)+g(x)-(L+M)| < \varepsilon$ whenever $0 < |x-a| < \delta$. Thus, by the definition of a limit, $\lim_{x\to a}[f(x)+g(x)] = L+M$. ■

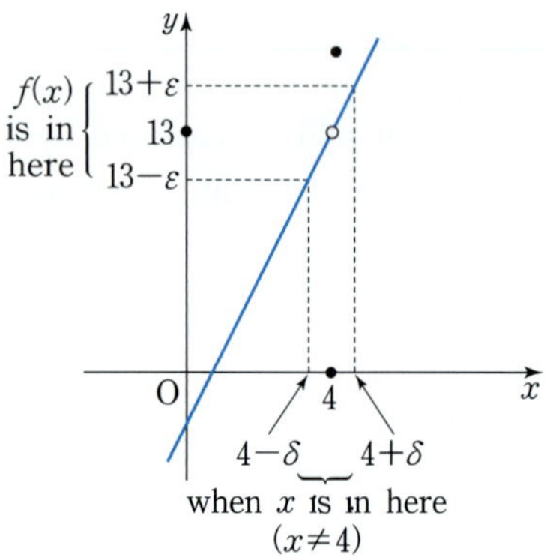

Figure 7 "ε-δ" definition

Consider the function $f(x) = 3x+1$ if $x \neq 4$ and $f(4) = 13$ (see Figure 7). When x is close to 4 but $x \neq 4$, it is clear from the definition that $f(x)$ is close to 13, and so $\lim_{x\to 4} f(x) = 13$.

If $f(x)$ should differ from 13 by less than 0.1, determine how much x should differ from 4. Since the distance from x to 4 is $|x-4|$ and the distance from $f(x)$ from 13 is $|f(x)-13|$, the problem is to find a number δ such that if $|x-4| < \delta$ but $x \neq 4$, then $|f(x) - 13| < 0.1$.

By the side calculation, $|f(x)-13| = |(3x+1)-13| = 3|x-4|$ gives as a possible candidate of δ.

Notice that if $0 < |x-4| < \dfrac{1}{30}$ then $|f(x)-13| = |(3x+1)-13| = 3|x-4| < 0.1$. Thus, $\delta = \dfrac{1}{30}$. In other words, if x is within a distance of $\dfrac{1}{30}$ from 4, then $f(x)$ will be within a distance of 0.1 from 13. Similarly for $\varepsilon = 0.01$, we get $\delta = \dfrac{0.01}{3} = \dfrac{1}{300}$, for $\varepsilon = 0.001$, we get $\delta = \dfrac{0.001}{3} = \dfrac{1}{3000}$ and so on. The numbers $\varepsilon = 0.1$, 0.01, and 0.001 are known as the error tolerances. Thus for any arbitrary positive number ε, there exists a positive number δ, namely $\delta = \dfrac{\varepsilon}{3}$, such that

[1] $$|f(x) - 13| < \varepsilon, \quad \text{if } 0 < |x-4| < \delta = \frac{\varepsilon}{3}.$$

Note that [1] can be rewritten as $13 - \varepsilon < f(x) < 13 + \varepsilon$ whenever $4 - \delta < x < 4 + \delta$, $(x \neq 4)$ (see Figure 7). By taking the values of x to lie in the interval $(4-\delta, 4+\delta)$, we can make the values of $f(x)$ lie in the interval $(13-\varepsilon, 13+\varepsilon)$.

A precise definition of a limit can be given using [1] as a model. Let f be a function defined on some open interval that contains the number a, except possibly at a itself. Then we say that the limit of $f(x)$ is L as x approaches a, and we write

$$\lim_{x\to a} f(x) = L.$$

If, for every number $\varepsilon > 0$, there is a number $\delta > 0$ such that

$$|f(x) - L| < \varepsilon \text{ whenever } 0 < |x - a| < \delta$$

or

$$\text{if } 0 < |x - a| < \delta, \text{ then } |f(x) - L| < \varepsilon.$$

In other words, $\lim_{x \to a} f(x) = L$ means that the distance between $f(x)$ and L can be made arbitrarily small by taking the distance from x to a sufficiently small (but not 0).

Alternatively, $\lim_{x \to a} f(x) = L$ means that the values of $f(x)$ can be made as close as to L by taking x close enough to a (but not equal to a).

Therefore, in terms of intervals, it can be stated as follows:

$\lim_{x \to a} f(x) = L$ means that for every $\varepsilon > 0$ (no matter how small ε is) we can find $\delta > 0$ such that if x lies in the open interval $(a - \delta,\ a + \delta)$ and $x \neq a$ then $f(x)$ lies in the open interval $(L - \varepsilon,\ L + \varepsilon)$.

Let the function f map a subset of $\mathbb{R}$ onto another subset of $\mathbb{R}$ (see the arrow diagram in Figure 8).

Figure 8

The definition of limit says that if any small interval $(L - \varepsilon,\ L + \varepsilon)$ is given around L, then we can find an interval $(a - \delta,\ a + \delta)$ around a such that f maps all the points in $(a - \delta,\ a + \delta)$ (except possibly a) into the interval $(L - \varepsilon,\ L + \varepsilon)$. (See Figure 9.)

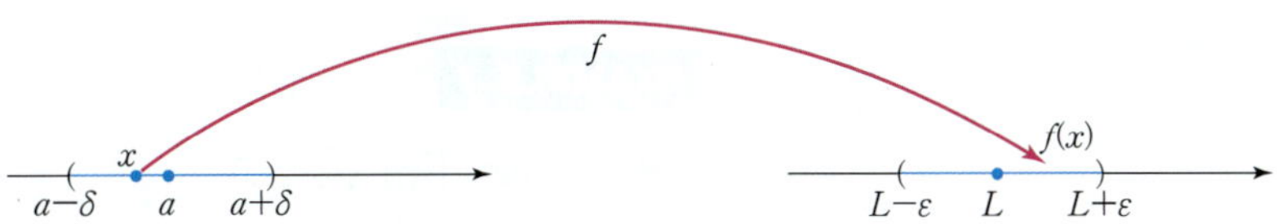

Figure 9

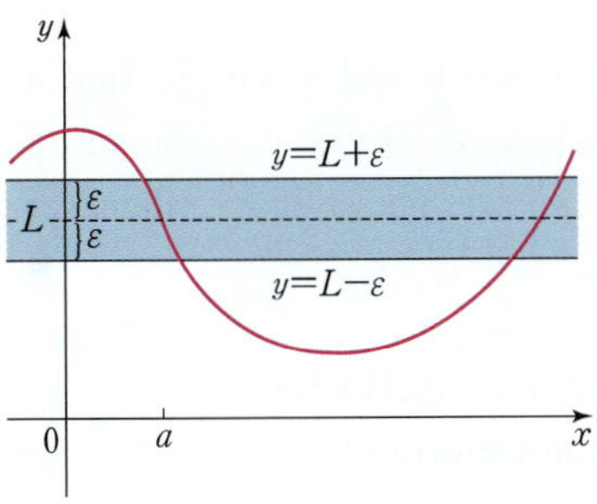

Figure 10

Another geometric interpretation of limits can be given in terms of the graph of a function. If $\varepsilon > 0$ is given, then we draw the horizontal lines $y = L + \varepsilon$ and $y = L - \varepsilon$ and the graph of f (see Figure 10). If $\lim_{x \to a} f(x) = L$, then we can find a number $\delta > 0$ such that if we restrict to lie in the interval $(a - \delta,\ a + \delta)$ but not equal to a, then the curve $y = f(x)$ lies between the lines $y = L + \varepsilon$ and $y = L - \varepsilon$. (See Figure 11.)

If a smaller ε is chosen, then a smaller δ may be required. (See Figure 12.)

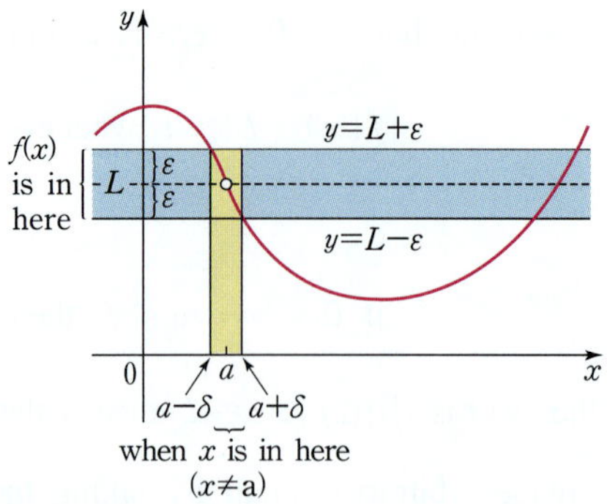

Figure 11

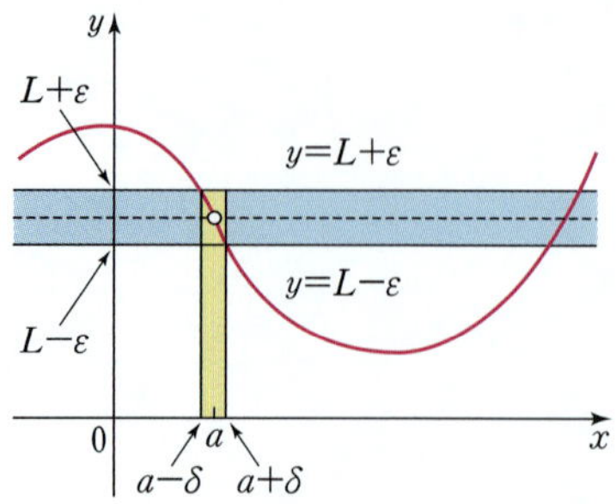

Figure 12

EXAMPLE 6

Use the above definition to prove that $\lim_{x \to 2}(3x-2)=4$.

Solution Let ε be a given positive number. Find a number δ such that $|(3x-2)-4|<\varepsilon$ whenever $0<|x-2|<\delta$. Since the choice of δ depends on ε, we can establish a connection between the absolute values $|(3x-2)-4|$ and $|x-2|$.

Consider $|(3x-2)-4|=3|x-2|$. So for a given $\varepsilon>0$, choose $\delta=\frac{\varepsilon}{3}$. For this choice, if $0<|x-2|<\delta=\frac{\varepsilon}{3}$, then

$$|(3x-2)-4|=3|x-2|<3\left(\frac{\varepsilon}{3}\right)=\varepsilon.$$

Therefore, by the definition of a limit we have

$$\lim_{x \to 2}(3x-2)=4.$$

■

For a one-sided limit, we have a similar ε-δ definition.

EXAMPLE 7

Prove that $\lim_{x \to 0^+}\sqrt{x}=0$.

Solution Let ε be a given positive number. Here $a=0$ and $L=0$. To find a number δ such that $|\sqrt{x}-0|<\varepsilon$ whenever $0<x<\delta$. Squaring both sides of the inequality $\sqrt{x}<\varepsilon$, we get $x<\varepsilon^2$ whenever $0<x<\delta$. This suggests us to choose $\delta=\varepsilon^2$. ■

(Movie on 'ε-δ' : http://www.youtube.com/watch?v=-ejyeII0i5c)
http://matrix.skku.ac.kr/cal-lab/SKKU-Cell-Epsilon-Delta.html

Precise Definition of Infinite Limits

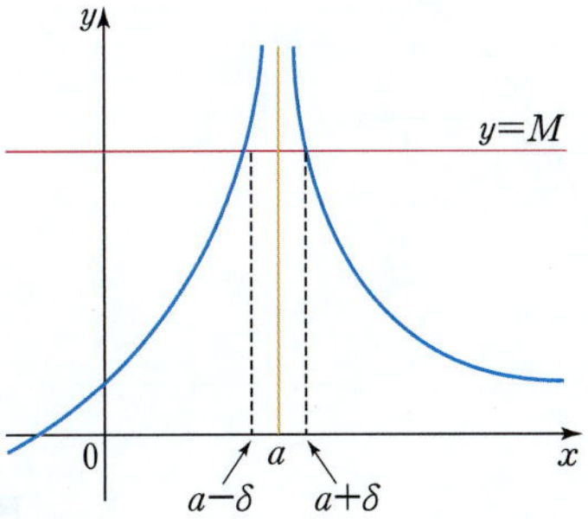

Figure 13

Let f be a function defined on some open interval that contains the number a, except possibly at a itself. Then $\lim_{x \to a} f(x) = \infty$ means that for every positive number M there is a positive number δ such that $f(x) > M$ whenever $0 < |x-a| < \delta$.

So the values of $f(x)$ can be made arbitrarily large (larger than any given number M) by taking x close enough to a (within a distance where δ depends on M, but with $x \neq a$). (See Figure 13.)

Given any horizontal line $y = M$, we can find a number $\delta > 0$ such that when x is restricted to lie in the interval $(a-\delta, a+\delta)$ but $x \neq a$, then the curve $y = f(x)$ lies above the line $y = M$. If another M, lager value, is chosen, a smaller δ may be required.

EXAMPLE 8

Show that $\lim_{x \to 0} \frac{1}{x^2} = \infty$.

Solution Given any (large) number $M > 0$, find $\delta > 0$ such that $\frac{1}{x^2} > M$ whenever $0 < |x-0| < \delta$.

Since both M and x^2 are positive, $x^2 < \frac{1}{M}$ whenever $0 < |x| < \delta$.

Taking the square root of both sides and recalling that $\sqrt{x^2} = |x|$, we get $|x| < \frac{1}{\sqrt{M}}$ whenever $0 < |x| < \delta$.

So for any $M > 0$, choose $\delta = \frac{1}{\sqrt{M}}$.

Now if $0 < |x-0| < \delta$, then $x^2 < \delta^2$, that is, $\frac{1}{x^2} > \frac{1}{\delta^2} = M$.

Thus $\frac{1}{x^2} > M$ whenever $0 < |x-0| < \delta$.

Therefore, $\lim_{x \to 0} \frac{1}{x^2} = \infty$. ■

CAS **EXAMPLE 9**

Find the following limit or explain why the limit does not exist.

$$\lim_{x \to 0} \cos \frac{1}{|x|}$$

Solution

```
var('x')
plot(cos(1/abs(x)), x, -1,1)
limit(cos(1/abs(x)), x=0)
```

Answer: Diverge, its limit does not exist.

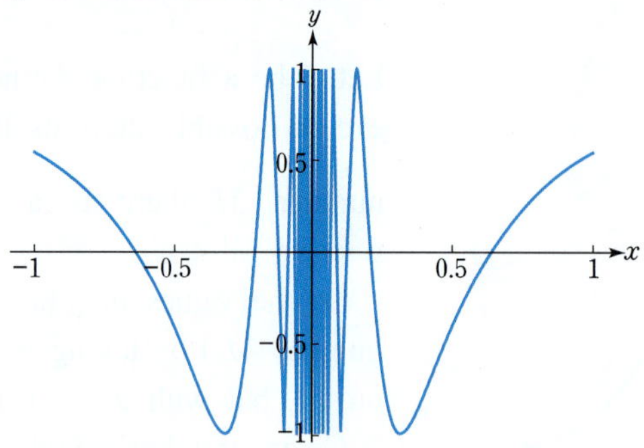

Figure 14 $\cos\dfrac{1}{|x|}$

■

CAS **EXAMPLE 10**

Plot the graph of the function $f(x) = \dfrac{\sin x}{2\,|\sin x|}$ near 0 and check if the left-hand limit, right-hand limit and limit at $x = 0$ exists.

Solution

```
f(x) = sin(x)/(2*abs(sin(x)))
       f.plot(exclude=[0])
       f.limit(x=0, dir='+')
       f.limit(x=0, dir='-')
       f.limit(x=0)                #    x |--> und
```

Answer : Diverge, its limit does not exist.

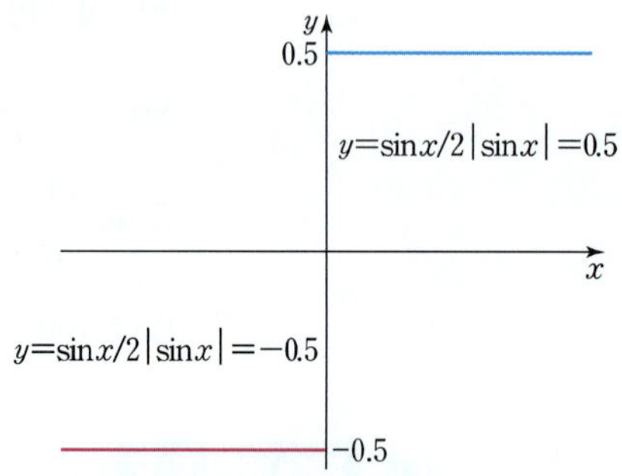

Figure 15 $f(x) = \dfrac{\sin x}{2\,|\sin x|}$

■

2.1 EXERCISES (Limits of Functions)

http://matrix.skku.ac.kr/Cal-Book/part1/CS-Sec-2-1-Sol.html

YouTube http://youtu.be/LZSmRPAAXME

1-4. Find the following limits or explain why the limit does not exist.

1. $\lim\limits_{x\to 2^-}\dfrac{x^2-4}{|x-2|}$

Solution $\lim\limits_{x\to 2^-}\dfrac{x^2-4}{|x-2|}=\lim\limits_{x\to 2^-}\dfrac{(x-2)(x+2)}{|x-2|}$

$=\lim\limits_{x\to 2^-}\{-(x+2)\}=-4$

2. $\lim\limits_{x\to 0}\dfrac{\sin x}{|\sin x|}$

Solution

Since $\lim\limits_{x\to 0^+}f(x)=\lim\limits_{x\to 0^+}\dfrac{\sin x}{|\sin x|}=1$, and

$\lim\limits_{x\to 0^-}f(x)=\lim\limits_{x\to 0^-}\dfrac{\sin x}{|\sin x|}=-1$, so

$\lim\limits_{x\to 0^+}f(x)\neq\lim\limits_{x\to 0^-}f(x)$. Then $\lim\limits_{x\to 0}\dfrac{\sin x}{|\sin x|}$ does not exist.

CAS **3.** $\lim\limits_{x\to \frac{\pi}{2}^+}\dfrac{\sin x}{|\cos x|}$

http://matrix.skku.ac.kr/cal-lab/cal-2-1-3.html

Solution We draw $f(x)=\dfrac{\sin x}{|\cos x|}$ with the following Sage command.

We see it diverge $(+\infty)$ as $x\to\dfrac{\pi}{2}^+$.

```
var('x')
p1=plot(sin(x)/abs(cos(x)),     x,     -pi,pi/2,
color='blue');
p2=plot(sin(x)/abs(cos(x)), x, pi/2,pi, color='red');
show(p1+p2, ymax=50, ymin=-10)
```

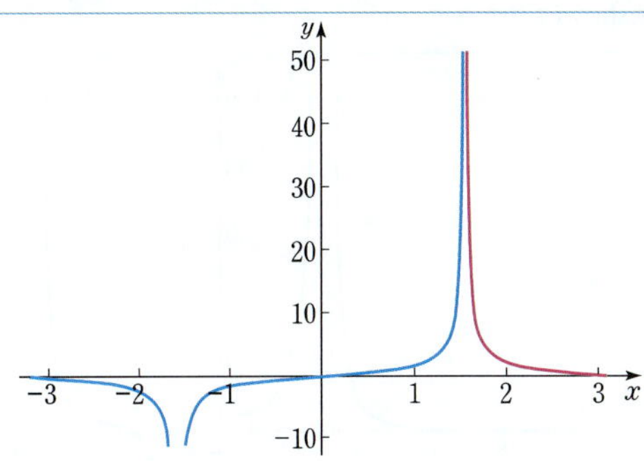

We can find $\lim\limits_{x\to \frac{\pi}{2}^+}\dfrac{\sin x}{|\cos x|}$ with the following Sage command.

```
limit(sin(x)/abs(cos(x)), x=pi/2, dir='plus')
```

Answer : +Infinity

4. $\lim\limits_{x\to 0^+}\left(\dfrac{1}{|x|}-\dfrac{1}{x}\right)$

Solution Since $|x|=x$ when $x\to 0^+$,

$\lim\limits_{x\to 0^+}\left(\dfrac{1}{|x|}-\dfrac{1}{x}\right)=\dfrac{1}{x}-\dfrac{1}{x}=0.$

---

5. $\lim\limits_{x\to 0^-}\left(\dfrac{1}{|x|}-\dfrac{1}{x}\right)$

http://matrix.skku.ac.kr/cal-lab/cal-2-1-5.html

Solution $\lim\limits_{x\to 0^-}\left(\dfrac{1}{|x|}-\dfrac{1}{x}\right)=\lim\limits_{x\to 0^-}-\dfrac{1}{x}-\dfrac{1}{x}=\lim\limits_{x\to 0^-}-\dfrac{2}{x}=\infty$

CAS **6.** $\lim\limits_{x\to 0}\left(\dfrac{1}{|x|}-\dfrac{1}{x}\right)$

http://matrix.skku.ac.kr/cal-lab/cal-2-1-6.html

Solution

```
P1=plot(1/abs(x)-1/x,   x,   -5,-0.1,   fontsize=20,
color='blue')
P2=plot(1/abs(x)-1/x,   x,   0.1,2.5,   fontsize=20,
color='red')
```

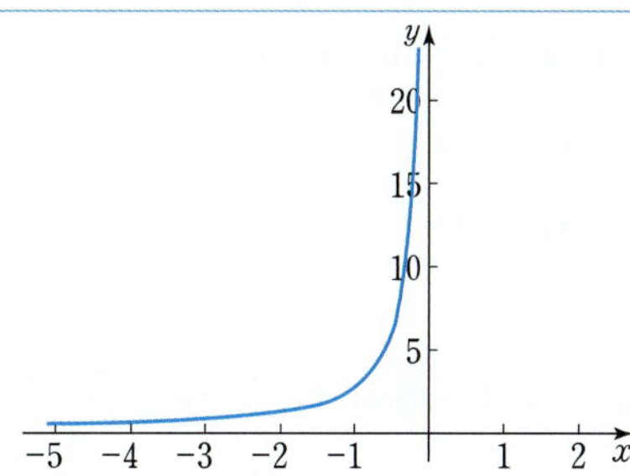

CAS **7.** $\lim\limits_{x\to 0}x\sin\dfrac{3}{x}$

http://matrix.skku.ac.kr/cal-lab/cal-2-1-7.html

Solution We draw $f(x)=x\sin\dfrac{3}{x}$ with the following Sage command.

We see the function converge to 0 as $x \to 0$.

```
p1=plot(x*sin(3/x), x, -0.5, 0, color='blue')
p2=plot(x*sin(3/x), x, 0, 0.5, color='red');
show(p1+p2, ymax=0.3, ymin=-0.3, aspect_ratio=1)
```

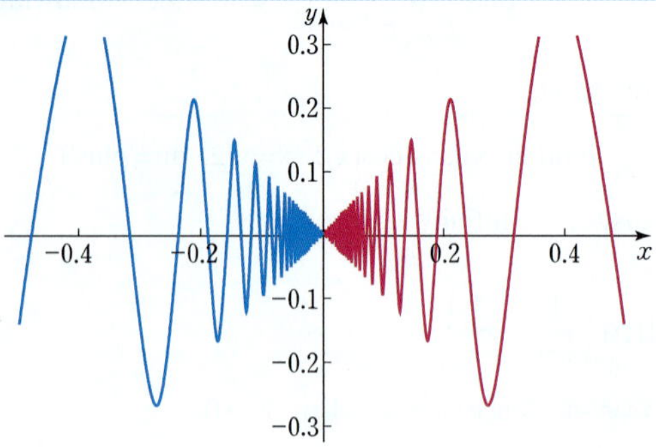

We can find $\lim_{x \to 0} x \sin \frac{3}{x}$ with the following Sage command.

```
limit(x*sin(3/x), x=0)                    # 0
```

8. The sign function, denoted by sgn, is defined by the following piecewise function.

$$\operatorname{sgn}(x) = \begin{cases} 1, & x > 0 \\ 0, & x = 0 \\ -1, & x < 0 \end{cases}$$

Find the following limits or explain why the limit does not exist.

Solution (a) $\lim_{x \to 0^+} \operatorname{sgn}(x) = 1$

(b) $\lim_{x \to 0} \operatorname{sgn}(x^3)$ does not exist.

$(\lim_{x \to 0^-} \operatorname{sgn}(x^3) = -1, \lim_{x \to 0^+} \operatorname{sgn}(x^3) = 1,$

$\lim_{x \to 0^-} \operatorname{sgn}(x^3) \neq \lim_{x \to 0^+} \operatorname{sgn}(x^3))$

(c) $\lim_{x \to 0} (\operatorname{sgn} x)^2 = 1$

(d) $\lim_{x \to 0} x \operatorname{sgn}(x) = 0$

9. Consider the function $f(x) = \dfrac{x^2 + 2x}{x^2 + x - 2}$.

(a) Find $\lim_{x \to -2} f(x)$ and $\lim_{x \to 1^+} f(x)$.

(b) Find the asymptotes of f; vertical and horizontal.

(c) Sketch the graph.

Solution (a) $\lim_{x \to -2} f(x) = \lim_{x \to -2} \dfrac{x(x+2)}{(x-1)(x+2)} = -\dfrac{8}{3}$

(b) $x = 1$: vertical asymptote

(c) $y = 1$: horizontal asymptote

```
var('x, y')
p1=plot((x^2+2*x)/(x^2+x-2), x, -10, 10, color='blue');
show(p1, aspect_ratio=1, ymax=10, ymin=-10)
```

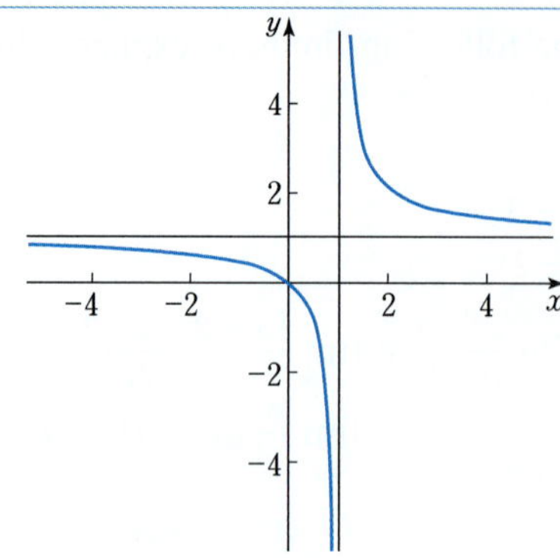

10. Draw the graph of a function $f(x)$ with all of the following properties:

(a) its domain is $(-\infty, -3) \cup (-3, 3) \cup (3, 5]$

(b) there is a vertical asymptote at $x = -3$

(c) $\lim_{x \to -3} f(x) = \infty$

(d) $\lim_{x \to 3} f(x) = 3$

(e) $\lim_{x \to 0} f(x)$ does not exist.

(f) $\lim_{x \to 5^-} f(x)$ does not exist.

(g) $\lim_{x \to -\infty} f(x) = 1$

```
var('x')
f(x)=(-x^2 + 21)/(x + 3)^2 + 2
p=plot(f(x), (x, -10, 0), ymax=100)
pl=plot(f(x), (x, -10, 0), ymax=100, color="blue")
p2=plot(x^2 -6 , (x, 0, 3), ymax=100, color="blue")
p3=plot((-x-3)/(x-5), (x, 3, 5), ymax=100, color="blue")
g=p+pl+p2+p3
g.show()
```

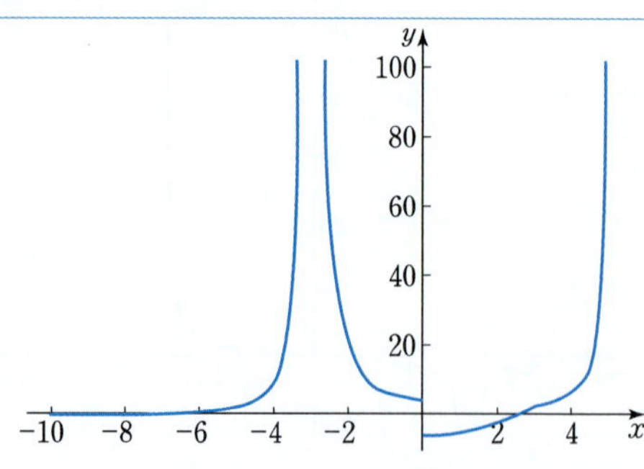

11. Let $g(x) = \dfrac{3\cos x}{x^4} - 1$.

(a) Find $\lim\limits_{x \to 0} g(x)$ or explain why the limit does not exist.

(b) Find $f(x)$ and $h(x)$ such that $f(x) \le g(x) \le h(x)$ for all $x \neq 0$.

(c) Use the Squeeze Theorem to find $\lim\limits_{x \to -\infty} g(x)$.

Solution (a) $\lim\limits_{x \to 0} \cos x = 1$, $\lim\limits_{x \to 0} x^4 = 0$

$\Rightarrow \lim\limits_{x \to 0} \dfrac{3\cos x}{x^4} - 1 = \infty - 1 = \infty$.

(b) Since $-1 \le \cos x \le 1$, we have

$$-\frac{3}{x^4} - 1 \le \frac{3\cos x}{x^4} - 1 \le \frac{3}{x^4} - 1.$$

Take, $f(x) = -\dfrac{3}{x^4} - 1$ and $h(x) = \dfrac{3}{x^4} - 1$.

(c) We know that $\lim\limits_{x \to -\infty} -\dfrac{3}{x^4} - 1 = -1$ and $\lim\limits_{x \to -\infty} \dfrac{3}{x^4} - 1 = -1$, so by the Squeeze Theorem, $\lim\limits_{x \to -\infty} \left(-\dfrac{3}{x^4} - 1\right) \le \lim\limits_{x \to -\infty} g(x) = \lim\limits_{x \to -\infty} \dfrac{3\cos x}{x^4} - 1 \le \lim\limits_{x \to -\infty} \left(\dfrac{3}{x^4} - 1\right) = -1$.

CAS **12.** Use the Squeeze Theorem to find

$$\lim_{x \to 0^+} \sqrt{5x}\left(4 - \cos\left(\frac{3}{\sqrt{x}}\right)\right).$$

http://matrix.skku.ac.kr/cal-lab/cal-2-1-12.html

Solution

```
P1=plot(sqrt(5*x)*(4-cos(3/sqrt(x))), 0.01, 2)
P2=plot(sqrt(5*x)*(5), 0.01, 2, color='red')
P3=plot(sqrt(5*x)*(3), 0.01, 2, color='red')
show(P1+P2+P3)
```

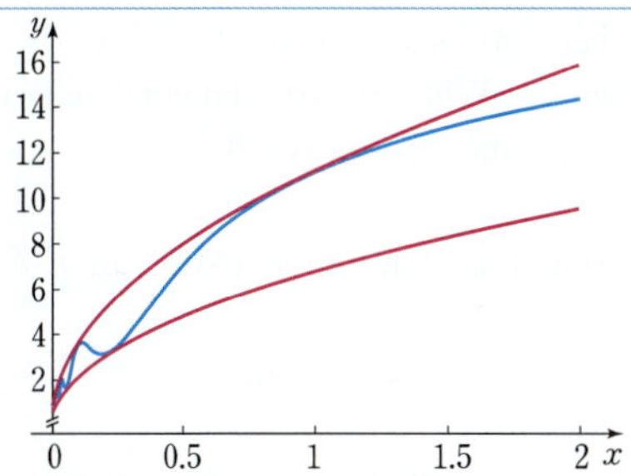

CAS **13.** Use the Squeeze Theorem to find $\lim\limits_{x \to 0} \dfrac{\sin(x)}{x}$.

http://matrix.skku.ac.kr/cal-lab/cal-2-1-13.html

Solution

```
P1=plot(sin(x)/x, -4, 4, color='red')
P2=plot(cos(x), -4, 4, color='green')
P3=plot(1, -4, 4, color='blue')
show(P1+P2+P3)
```

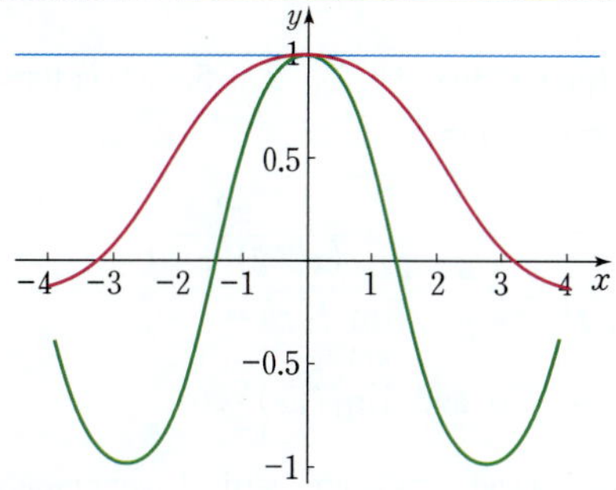

(※ $\cos(x)$ is not always smaller than $\dfrac{\sin(x)}{x}$, but it is when near 0. So the Squeeze Theorem can be used.)

14. Let $f(x) = \dfrac{5x^6 - 3x^3 + 11}{3x^n - 7x + 14}$. Find all positive integer n such that $\lim\limits_{x \to -\infty} f(x) = \infty$.

Solution $\lim\limits_{x \to -\infty} f(x) = \lim\limits_{x \to -\infty} \dfrac{5 - \dfrac{3}{x^3} + \dfrac{11}{x^6}}{3x^{n-6} - \dfrac{7}{x^5} + \dfrac{14}{x^6}} = L$

(i) $n = 6$; $L = \dfrac{5}{3}$

(ii) $n < 6$, $n = 2k - 1$ (k is positive integer) ; $L = -\infty$

(iii) $n < 6$, $n = 2k$ (k is positive integer) ; $L = \infty$

(iv) $n > 6$; $L = 0$

Therefore $n = 2, 4$

```
var('x, y')
p1=plot((5*x^6-3*x^3+11)/(3*x^n-7*x+14), x, -5,
5, color='blue');
show(p1, aspect_ratio=1, ymax=10, ymin=-1
```

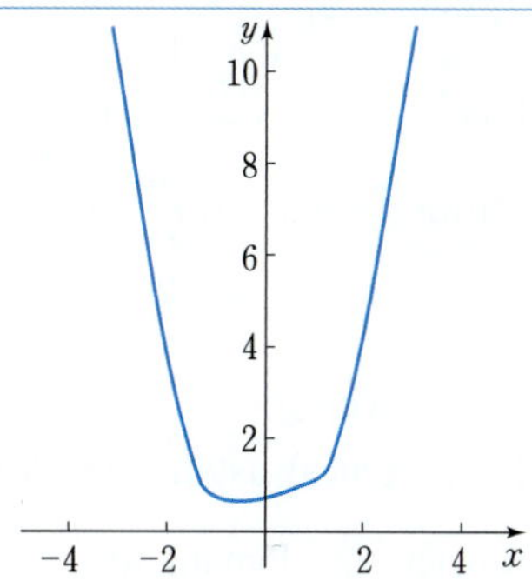

When $n = 4$.

15. Find all the asymptotes (vertical and horizontal) of the function $f(x)=\dfrac{x^3}{x^2-x-2}$.

http://matrix.skku.ac.kr/cal-lab/cal-2-1-15.html

Solution

```
var('x');
show(plot(x^3/(x^2-x-2), x, -5, 5, color='blue'), ymax=5, ymin=-5)
```

$y=f(x)=\dfrac{x^3}{x^2-x-2}=\dfrac{x^3}{(x-2)(x+1)}$,

so $\lim\limits_{x\to -1^-} f(x)=\infty$, $\lim\limits_{x\to -1^+} f(x)=-\infty$,

$\lim\limits_{x\to 2^-} f(x)=-\infty$ and $\lim\limits_{x\to 2^+} f(x)=\infty$.

Thus, $x=-1$ and $x=2$ are vertical asymptotes.

$$y=f(x)=\frac{x^3}{x^2-x-2}=x+1+\frac{3x+2}{x^2-x-2}.$$

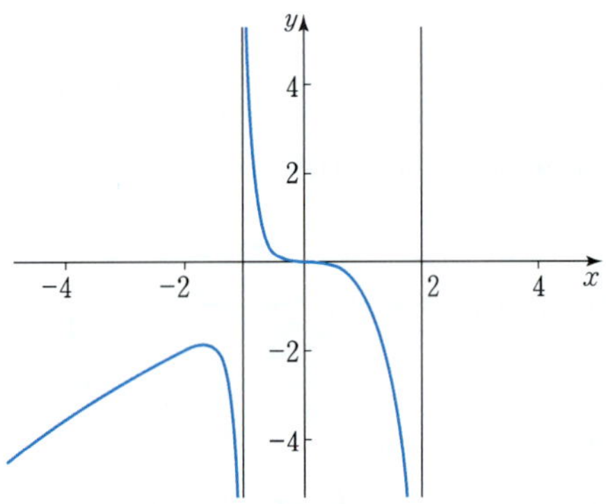

16. Find the limit $\lim\limits_{x\to 1}\dfrac{x^6-1}{\sqrt[3]{x}-1}$.

Solution $\lim\limits_{x\to 1}\dfrac{x^6-1}{\sqrt[3]{x}-1}=\lim\limits_{x\to 1}\dfrac{(x^3+1)(x^3-1)}{\sqrt[3]{x}-1}$

$=\lim\limits_{x\to 1}\dfrac{(x^3+1)(x^2+x+1)(x-1)}{\sqrt[3]{x}-1}$

$=\lim\limits_{x\to 1}\dfrac{(x^3+1)(x^2+x+1)(\sqrt[3]{x}-1)(\sqrt[3]{x^2}+\sqrt[3]{x}+1)}{\sqrt[3]{x}-1}$

$=\lim\limits_{x\to 1}(x^3+1)(x^2+x+1)(\sqrt[3]{x^2}+\sqrt[3]{x}+1)$

$=\lim\limits_{x\to 1}(x^3+1)\lim\limits_{x\to 1}(x^2+x+1)\lim\limits_{x\to 1}(\sqrt[3]{x^2}+\sqrt[3]{x}+1)$

$=(1+1)(1+1+1)(1+1+1)=2\times 3\times 3=18$

17. Consider $f(x)=\tan x$.

(a) Find all the vertical asymptotes for $f(x)$.

(b) If we restrict the domain to $\left(-\dfrac{\pi}{2},\dfrac{\pi}{2}\right)$, then show that there exists an inverse function defined on $(-\infty,\infty)$.

(c) If the above inverse function is $g(x)$, then find all the vertical asymptotes.

Solution (a) $f(x)=\tan x$,

so $\lim\limits_{x\to \frac{\pi}{2}^-}\tan x=\infty$, $\lim\limits_{x\to \frac{\pi}{2}^+}\tan x=-\infty$.

$\lim\limits_{x\to \frac{(2n+1)\pi}{2}^-}\tan x=\infty$, $\lim\limits_{x\to \frac{(2n+1)\pi}{2}^+}\tan x=-\infty$

Thus, $x=\dfrac{(2n+1)\pi}{2}$, $(n\in Z)$ are vertical asymptotes.

18. Find $\delta>0$ such that $|\sqrt{4x+1}-3|<0.5$ whenever $|x-2|<\delta$.

Solution $-\delta<x-2<\delta \Leftrightarrow -\delta+2<x<\delta+2$

$\Leftrightarrow -4\delta+9<4x+1<4\delta+9$

$\Leftrightarrow \sqrt{-4\delta+9}<\sqrt{4x+1}<\sqrt{4\delta+9}$

$\Leftrightarrow \sqrt{-4\delta+9}-3<\sqrt{4x+1}-3<\sqrt{4\delta+9}-3$

$=\dfrac{4\delta}{\sqrt{4\delta+9}+3}$

$\Leftrightarrow |\sqrt{4\delta+9}-3|<4\delta=0.5 \Leftrightarrow \delta=\dfrac{1}{8}$

19. Use an ε-δ argument to prove that $\lim\limits_{x\to 2}\dfrac{1}{x}=\dfrac{1}{2}$.

Solution Let ε be a given positive number. Here $a=2$ and $L=\dfrac{1}{2}$. We must find a number δ such that $\left|\dfrac{1}{x}-\dfrac{1}{2}\right|<\varepsilon$ whenever $0<|x-2|<\delta$. With easy computation, we may choose $\delta=\varepsilon$ to get the desired result.

20. Use the ε-δ argument to prove that

$$\lim_{x\to a}\sqrt{x}=\sqrt{a} \text{ if } a>0.$$

Solution Let ε be a given positive number. We must find a number δ such that $|\sqrt{x}-\sqrt{a}|<\varepsilon$ whenever $0<|x-a|<\delta$. With an easy computation, we may choose $\delta=\varepsilon$ to get the desired result.

21-26. Prove the statements using an ε-δ argument.

21. $\lim\limits_{x\to -1}(x^2+2x-8)=-9$

Solution $\forall\varepsilon>0$ Find δ. Let $\delta=\sqrt{\varepsilon}$.

If $0<|x+1|<\delta$, then $|f(x)-(-9)|=|x^2+2x+1|$

$=|(x+1)^2|=|(x+1)|^2<\delta^2=\varepsilon$.

$f(x)-f(-1)=x^2+2x-8+9=x^2+2x+1=(x+1)^2$.

22. $\lim_{x\to 3}\sqrt{7-x}=2$

Solution Let ε be a given positive number. Here $a=3$ and $L=2$. We must find a number δ such that $|\sqrt{7-x}-2|<\varepsilon$ whenever $0<|x-3|<\delta$. With an easy computation, we may choose $\delta=4\varepsilon$ to get the desired result.

23. $\lim_{x\to\frac{\pi}{2}}\sin x=1$

Solution $\forall\varepsilon>0$ Find δ. Let $\delta=\varepsilon$. If $\left|x-\frac{\pi}{2}\right|<\delta$, then

$|f(x)-1|=\left|\sin x-\sin\frac{\pi}{2}\right|\le\left|x-\frac{\pi}{2}\right|<\delta=\varepsilon.$

$$\left|\sin x-\sin\frac{\pi}{2}\right|=\left|2\cos\frac{x+\frac{\pi}{2}}{2}\sin\frac{x-\frac{\pi}{2}}{2}\right|$$

$$\le 2\left|\sin\frac{x-\frac{\pi}{2}}{2}\right|\le 2\left|\frac{x-\frac{\pi}{2}}{2}\right|=\left|x-\frac{\pi}{2}\right|<\delta.$$

24. $\lim_{x\to 3}(3x+2)=11$

Solution $\forall\varepsilon>0$ Find δ. Let $\delta=\frac{\varepsilon}{3}$.

If $0<|x-3|<\delta$, then

$|f(x)-11|=|3x+2-11|=3|x-3|<3\delta=3\frac{\varepsilon}{3}=\varepsilon.$

25. $\lim_{x\to 0}x^2+4x+1=1$

Solution $\forall\varepsilon>0$ Find δ. Let $\delta=\min\{\frac{\varepsilon}{5},1\}$.

If $0<|x-0|<\delta\le 1$, then

$|f(x)-1|=|x^2+4x|\le|x|\,|x+4|<\delta\,(|x|+|4|).$

$f(x)-f(0)=x^2+4x+1-1=x^2+4x.$

$|x+4|\le|x|+|4|\le 1+4=5.$

26. $\lim_{x\to 3}x^3=27$

Solution $\forall\varepsilon>0$ Find δ. Let

$\delta=\min\left\{\delta_1,\ \frac{1}{37}\varepsilon\right\}=\min\left\{1,\ \frac{1}{37}\varepsilon\right\}.$

If $0<|x-3|<\delta$, then

$|f(x)-27|=|x^3-27|=|(x-3)(x^2+3x+9)|$

$\le|x-3||(x-3)^2+9(x-3)+27|$

$<|x-3|\,|1^2+9+27|=37|x-3|<37\delta=\varepsilon.$

$f(x)-f(3)=x^3-27=(x-3)(x^2+3x+9)$

Take $\delta_1=1$.

If $0<|x-3|<\delta_1=1$, then $(x-3)^2<1$.

$|(x^2+3x+9)|=|(x-3)^2+9(x-3)+27|$

$<|1^2+9+27|.$

27. $\lim_{x\to 0+}\sqrt{3x^3}\left(4-\cos\left(\frac{3}{\sqrt{x^3}}\right)\right)=0$

http://matrix.skku.ac.kr/cal-lab/cal-2-1-27.html

Solution $\forall\varepsilon>0$ Find δ. Let $\delta=\frac{1}{\sqrt{108}}\sqrt[3]{\varepsilon^2}$

$\Rightarrow 108^{1/2}|\delta|^{3/2}=\varepsilon$. If $0<|x-0|=|x|<\delta$, then

$|f(x)-0|=\left|\sqrt{3x^3}\left(4-\cos\left(\frac{3}{\sqrt{x^3}}\right)\right)\right|$

$=|\sqrt{3x^3}|\left|4-\cos\left(\frac{3}{\sqrt{x^3}}\right)\right|<6\sqrt{3x^3}$

$=(108x^3)^{(1/2)}=108^{1/2}|x|^{3/2}<108^{1/2}|\delta|^{3/2}=\varepsilon.$

$3\le\left|4-\cos\frac{3}{\sqrt{x^3}}\right|\le 5\Rightarrow\left|4-\cos\frac{3}{\sqrt{x^3}}\right|<6$ and $6\sqrt{3x^3}$

$=(108x^3)^{(1/2)}$. $\delta=\frac{1}{\sqrt{108}}\sqrt[3]{\varepsilon^2}\Rightarrow 108^{1/2}|\delta|^{3/2}=\varepsilon.$

28. $\lim_{x\to 5}\frac{1}{(x-5)^2}=\infty$

http://matrix.skku.ac.kr/cal-lab/cal-2-1-28.html

Solution $\forall M>0$, Find δ. Let $\delta=\frac{1}{\sqrt{M}}$ $(\Rightarrow\frac{1}{\delta^2}=M)$.

If $0<|x-5|<\delta$, then $|f(x)|=\frac{1}{(x-5)^2}>\frac{1}{\delta^2}=M.$

Since $M>0$, $(x-5)^2>0$, $\sqrt{(x-5)^2}=|x-5|>0$

$\Rightarrow|x-5|<\delta.$

$\frac{1}{(x-5)^2}>\frac{1}{\delta^2}=M\Rightarrow\delta=\frac{1}{\sqrt{M}}.$

29. $\lim_{x\to 2}\frac{1}{(x-2)^2}=\infty$

Solution Given any (large) number $M>0$ we must find $\delta>0$ such that $\frac{1}{(x-2)^2}>M$ whenever $0<|x-2|<\delta$.

Since both M and $(x-2)^2$ are positive, $(x-2)^2<\frac{1}{M}$ whenever $0<|x-2|<\delta$. Taking the square root of both sides and recalling that $\sqrt{(x-2)^2}=|x-2|$, we get $|x-2|<\frac{1}{\sqrt{M}}$ whenever $0<|x-2|<\delta$.

So for any $M>0$, choose $\delta=\frac{1}{\sqrt{M}}$.

Now if $0<|x-2|<\delta$, then $(x-2)^2<\delta^2$, that is,

$\frac{1}{(x-2)^2}>\frac{1}{\delta^2}=M.$

Thus $\frac{1}{(x-2)^2}>M$ whenever $0<|x-2|<\delta$.

Therefore, $\lim_{x\to 0}\frac{1}{(x-2)^2}=\infty$.

30. $\lim_{x \to 0} e^x = 1$

Solution Let ε be a given positive number. Here $a = 0$ and $L = 1$. We must find a number δ such that $|e^x - 1| < \varepsilon$ whenever $0 < |x| < \delta$. With an easy computation, we may choose $\delta = \ln\varepsilon$ to get the desired result.

31. Use an ε-δ argument to prove that

$$\lim_{x \to 0^+} \ln x = -\infty.$$

Solution Given any (large) number $M > 0$ we must find $\delta > 0$
such that $\ln x < -M$ whenever $0 < x < \delta$.
Since both M and x are positive, $x < e^{-M}$ whenever $0 < x < \delta$.
So for any $M > 0$, choose $\delta = e^{-M}$.
Now if $0 < x < \delta$, then $\ln x < \ln e^{-M} = -M$.
Thus, $\lim_{x \to 0^+} \ln x = -\infty$.

32. If $\lim_{x \to a} f(x) = \infty$ and $\lim_{x \to a} g(x) = M$, where M is a real number. Show that

(a) $\lim_{x \to a} [f(x) + g(x)] = \infty$

(b) $\lim_{x \to a} (f(x) \cdot g(x)) = \infty$ if $M > 0$

Solution (a) $\lim_{x \to a} f(x) + \lim_{x \to a} g(x) = \infty + M = \infty$

(b) $\lim_{x \to a} f(x) \cdot \lim_{x \to a} g(x) = M \cdot \lim_{x \to a} f(x) = \infty$

2.2 Continuity

http://youtu.be/zGxx3PUCTnM

A function f is continuous at $x=a$ if $\lim_{x\to a} f(x)=f(a)$.

So if a function f is continuous at $x=a$, then f will satisfy the following three conditions :

1. $f(a)$ is defined (that is, $x=a$ is in the domain of f)
2. $\lim_{x\to a} f(x)$ exists
3. $\lim_{x\to a} f(x)=f(a)$

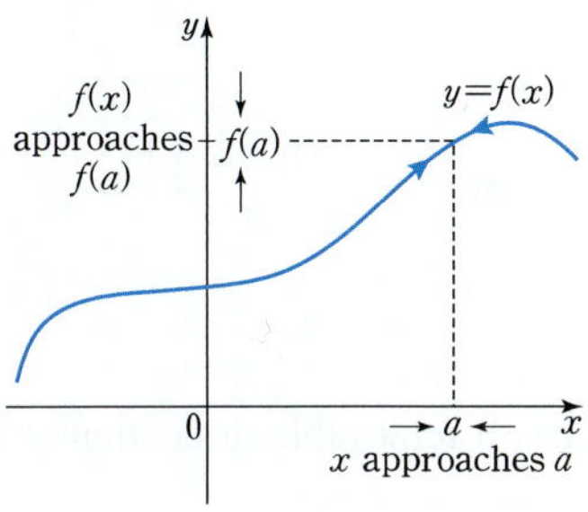

Figure 1

In other words, f is continuous at $x=a$ if $f(x)$ approaches $f(a)$ as x approaches a (see Figure 1). Thus, a continuous function f has the property that a small change in x produces a small change in the corresponding value $f(x)$. Intuitively the graph of a continuous function f has no break. So the graph can be drawn without removing the pen from the paper.

A function f is said to be discontinuous at $x=a$, or f has a discontinuity at $x=a$, if f is not continuous $x=a$. This can happen in three possible ways shown in Figure 2.

Thus, a function $f(x)$ has a discontinuity at $x=a$ if one of the following three possible scenarios occurs:

i) $f(x)$ is not defined at a;
ii) $f(x)$ is defined at a but has no limit as $x \to a$;
iii) $f(x)$ is defined at a and $\lim_{x\to a} f(x)$ exists, but is not equal $f(a)$.

$$\left(\lim_{x\to a} f(x) \neq f(a)\right)$$

Discontinuities fall into two categories: removable and non-removable. A discontinuity at $x=a$ is called removable if f can be made continuous by defining (as in Figure 2(a)) or redefining f at $x=a$ (as in Figure 2(c)). The discontinuity is in Figure 2(b) and Figure 2(d) non-removable.

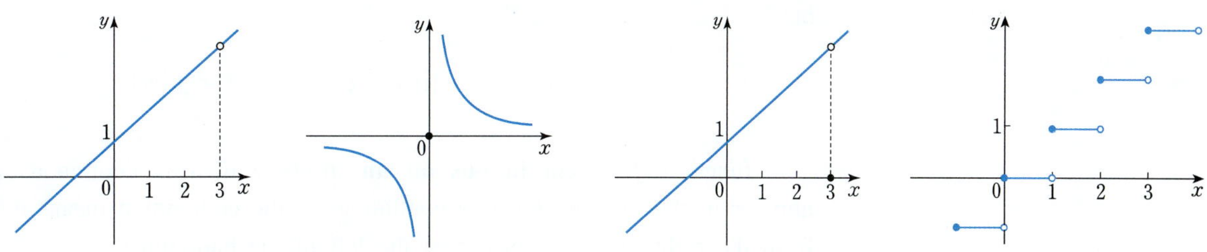

(a) $f=\dfrac{x^2-2x-3}{x-3}$

(b) $f=\begin{cases} \dfrac{1}{x} & \text{if } x\neq 0 \\ 0 & \text{if } x=0 \end{cases}$

(c) $f=\begin{cases} \dfrac{x^2-2x-3}{x-3} & \text{if } x\neq 3 \\ 0 & \text{if } x=3 \end{cases}$

(d) $f=[\![x]\!]$

Figure 2 Graphs of the functions in example 1

EXAMPLE 1

Classify the points of discontinuity of the following functions (see Figure 2).

(a) $f(x)=\dfrac{x^2-2x-3}{x-3}$ (b) $f(x)=\begin{cases}\dfrac{1}{x} & \text{if } x\neq 0\\ 0 & \text{if } x=0\end{cases}$

(c) $f(x)=\begin{cases}\dfrac{x^2-2x-3}{x-3} & \text{if } x\neq 3\\ 0 & \text{if } x=3\end{cases}$ (d) $f(x)=[\![x]\!]$

Solution (a) It is easy to check that $\lim_{x\to 3} f(x)=4$. Since $f(3)$ is not defined so f has a removable discontinuity at $x=3$.

(b) At $x=0$ f has an discontinuity because $\lim_{x\to 0} f(x)=\lim_{x\to 0}\dfrac{1}{x}$ does not exist, although $f(0)=0$ is defined.

(c) Here $f(3)=0$ is defined and

$$\lim_{x\to 3} f(x)=\lim_{x\to 3}\frac{x^2-2x-3}{x-3}=\lim_{x\to 3}\frac{(x-3)(x+1)}{x-3}$$
$$=\lim_{x\to 3}(x+1)=4$$

exists. But $\lim_{x\to 3} f(x)\neq f(3)$ so f has a removable discontinuity at 3.

(d) The greatest integer function $f(x)=[\![x]\!]$ has jump discontinuities ("jumps" from one value to another) at all of the integers because $\lim_{x\to n}[\![x]\!]$ does not exist if n is an integer. ■

A function f is **continuous from the right at** $x=a$ if $\lim_{x\to a^+} f(x)=f(a)$ and f is **continuous from the left at** $x=a$ if $\lim_{x\to a^-} f(x)=f(a)$.

For example, the function $f(x)=[\![x]\!]$ is continuous from the right but discontinuous from the left at each integer n because

$$\lim_{x\to n^+} f(x)=\lim_{x\to n^+}[\![x]\!]=n=f(n)$$

but

$$\lim_{x\to n^-} f(x)=\lim_{x\to n^-}[\![x]\!]=n-1\neq f(n).$$

A function f is **continuous on an interval** if it is continuous at every number in the interval. (If f is continuous at the endpoint it means continuous from the right or continuous from the left of the endpoint.)

THEOREM 1

If f and g are continuous at $x=a$ and c is a constant, then the following functions are also continuous at a:

1. $f+g$
2. $f-g$
3. cf
4. fg
5. $\dfrac{f}{g}$ if $g(a)\neq 0$

Proof Each of the five parts of this theorem follows from the corresponding Limit Law. For instance, we give the proof of part 1. Since f and g are continuous at $x=a$, we have

$$\lim_{x\to a} f(x)=f(a) \quad \text{and} \quad \lim_{x\to a} g(x)=g(a).$$

Therefore,

$$\begin{aligned}\lim_{x\to a}(f+g)(x) &= \lim_{x\to a}[f(x)+g(x)]\\ &= \lim_{x\to a} f(x)+\lim_{x\to a} g(x)\\ &= f(a)+g(a)\\ &= (f+g)(a).\end{aligned}$$

■

This shows that $f+g$ is continuous at a.

THEOREM 2

(a) Any polynomial is continuous on $\mathbb{R}=(-\infty,\ \infty)$.

(b) Any rational function is continuous on its domain.

Proof (a) Let the polynomial be a function of the form

$$P(x)=c_n x^n + c_{n-1}x^{n-1}+\cdots+c_1 x + c_0$$

where $c_0,\ c_1,\ \ldots,\ c_n$ are constants. We know that $\lim_{x\to a} c_0 = c_0$ and $\lim_{x\to a} x^m = a^m \ (m=1,\ 2,\ \ldots,\ n)$.

Thus the function $f(x)=x^m$ is a continuous function, and so is the function $g(x)=cx^m$. Therefore, P is continuous because P is a sum of continuous functions.

(b) A rational function is a function of the form

$$f(x)=\frac{P(x)}{Q(x)}$$

where P and Q are polynomials. The domain of f is $D=\{x\in\mathbb{R}$

$\mid Q(x) \neq 0\}$. It follows from (a) that P and Q are continuous everywhere. It follows from Limit Law 5, that f is continuous at every number in D. ■

Note The inverse function of any continuous function is also continuous. (The graph of f^{-1} is obtained by reflecting the graph of f about the line $y = x$. So if the graph of f has no break in it, neither does the graph of f^{-1}).

EXAMPLE 2

Evaluate $\lim_{x \to -1} \dfrac{x^4 + 3x^2 - 5x + 8}{2 - 5x}$.

Solution The function $f(x) = \dfrac{x^4 + 3x^2 - 5x + 8}{2 - 5x}$, is rational, so it is continuous on its domain, which is $\{x \mid x \neq 2/5\}$. Therefore,

$$\lim_{x \to -1} f(x) = f(-1) = 17/7.$$ ■

THEOREM 3

The polynomials, rational functions, root functions, trigonometric functions, inverse trigonometric functions, exponential functions and logarithmic functions, are continuous on their domains.

EXAMPLE 3

Find where the function $f(x) = \dfrac{\ln x + \sin x}{x^2 - 1}$ is continuous.

Solution Since the function $y = \ln x$ is continuous for $x > 0$ and $y = \sin x$ is continuous on $\mathbb{R}$, so $y = \ln x + \sin x$ is continuous on $(0, \infty)$. The denominator, $y = x^2 - 1$, is a polynomial, so it is continuous on $\mathbb{R}$. Therefore, f is continuous at all positive numbers x except where $x^2 - 1 = 0$. So f is continuous on the intervals $(0, 1)$ or $(1, \infty)$. ■

THEOREM 4 Limit of Continuous Functions

If f is continuous at b and $\lim_{x \to a} g(x) = b$, then $\lim_{x \to a} f(g(x)) = f(b)$.

In other words, $\lim_{x \to a} f(g(x)) = f\left(\lim_{x \to a} g(x)\right)$.

Thus if the function is continuous and the limit of its argument exists then a limit symbol can be "moved through the function symbol".

Consider $f(x) = \sqrt[3]{x}$.

Then $f(g(x)) = \sqrt[3]{g(x)}$ and $f\left(\lim_{x\to a} g(x)\right) = \sqrt[3]{\lim_{x\to a} g(x)}$.

Thus (Limit Law 7),

$$\lim_{x\to a} \sqrt[3]{g(x)} = \sqrt[3]{\lim_{x\to a} g(x)}.$$

EXAMPLE 4

Find $\lim_{x\to 2} \ln\left(\dfrac{\sqrt{1+x}}{x}\right)$.

Solution Since $\ln x$ is a continuous function, using Theorem 3 we have

$$\begin{aligned}\lim_{x\to 2} \ln\left(\frac{\sqrt{1+x}}{x}\right) &= \ln\left(\lim_{x\to 2} \frac{\sqrt{1+x}}{x}\right)\\ &= \ln\left(\frac{\sqrt{1+2}}{2}\right) = \ln\frac{\sqrt{3}}{2}.\end{aligned}$$

■

THEOREM 5 Composite of Continuous Functions

If g is continuous at a and f is continuous at $g(a)$, then the composite function $f \circ g$ given by $f \circ g(x) = f(g(x))$ is continuous at a.
In other words "**the composition of continuous functions is also a continuous function**."

Proof Since g is continuous at a, we have

$$\lim_{x\to a} g(x) = g(a).$$

Since f is continuous at $b = g(a)$, we to obtain

$$\lim_{x\to a} f(g(x)) = f(g(a)).$$

This means that the function $h(x) = f(g(x))$ is continuous at a.
Thus, $f \circ g$ is continuous at a. ■

EXAMPLE 5

Determine where the functions

(i) $F(x) = \cos(\exp(x^3))$
(ii) $F(x) = \ln(1 + \cos x)$

are continuous.

Solution (i) With $h(x) = x^3$, $g(x) = e^x$ and $f(x) = \cos x$ we have

$$F(x) = f(g(h(x))).$$

Since h is a polynomial it is continuous on $\mathbb{R}$, the exponential function g is continuous on $\mathbb{R}$, and f is also continuous on $\mathbb{R}$. Thus $F(x) = f(g(h(x)))$ is continuous on $\mathbb{R}$.

(ii) We know that $f(x) = \ln x$ is continuous and $g(x) = 1 + \cos x$ is continuous (because both $y = 1$ and $y = \cos x$ are continuous). Therefore, $F(x) = f(g(x))$ is continuous wherever it is defined. Now $\ln(1 + \cos x)$ is defined when $1 + \cos x > 0$. It is undefined when $\cos x = -1$, and this happens when $x = \pm\pi, \pm 3\pi, \ldots$. Thus, F has discontinuities when x is an odd multiple of π and is continuous on the intervals between these values. (See Figure 3.) ■

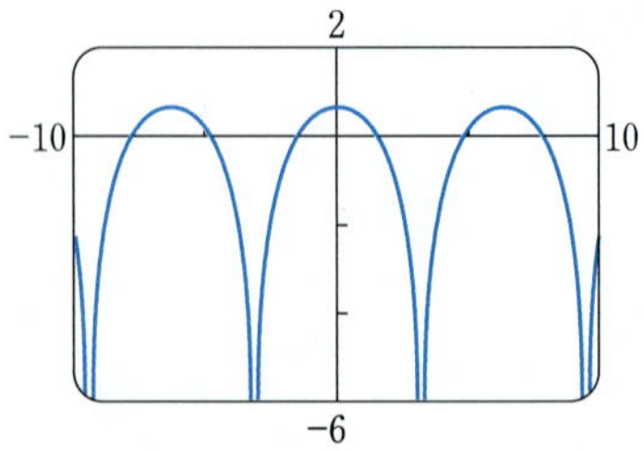

Figure 3 $y = \ln(1 + \cos x)$

THEOREM 6 The Intermediate Value Theorem

Suppose that f is continuous on the closed interval $[a, b]$ and let N be any number between $f(a)$ and $f(b)$, where $f(a) \neq f(b)$. Then there exists a number c in (a, b) such that $f(c) = N$.

Thus a continuous function on $[a, b]$ must take on every intermediate value between the function values $f(a)$ and $f(b)$ at least once. (See Figure 4.)

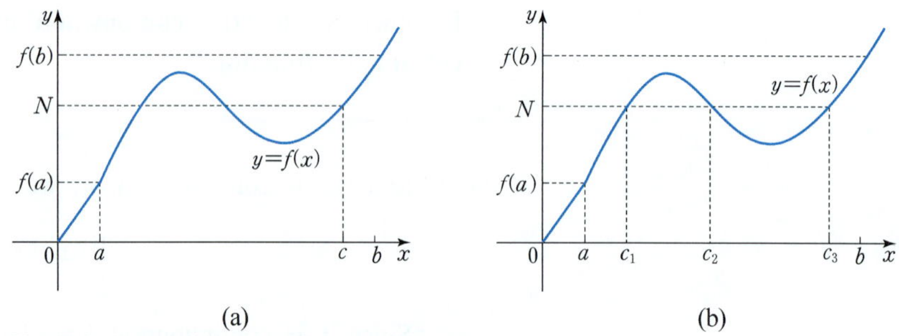

Figure 4 Intermediate Value Theorem illustrated

Geometrically, the graph of a continuous function f cannot skip over any horizontal line $y = N$ given between $y = f(a)$ and $y = f(b)$. It must intersect $y = N$.

Observations:

(i) The value N can be taken on once [as in Figure 4 part (a)] or more than once [as in part (b)].

(ii) The Intermediate Value Theorem is not true in general for discontinuous functions.

(iii) The Intermediate Value Theorem is useful in locating roots of equations (see corollary below).

The following theorem is an immediate and useful application of the Intermediate Value Theorem.

THEOREM 7 **The Intermediate Value Theorem**

If f is continuous on $[a, b]$ and $f(a)$ and $f(b)$ have opposite signs [*i.e.*, $f(a)f(b) < 0$], then there exists at least one number $c \in (a, b)$ such that $f(c) = 0$. (Recall that c is known as a zero or root of $f(x)$.) (See Figure 5.)

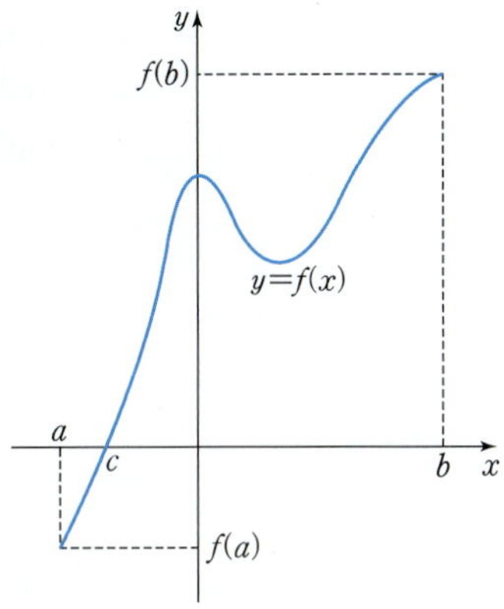

Figure 5

Observe that this corollary is simply the special case of the Intermediate Value Theorem where $N = 0$.

The Intermediate Value Theorem and the above corollary are classical examples of *existence theorems* which tell the existence of a number c, but do not specify what c is.

EXAMPLE 6

Locate the zeros of $f(x) = x^5 + 4x^2 - 9x + 3$.

Solution Let us find the values of $f(x)$ for several values of x as follows.

x	$-\infty$	−10	−5	−3	−2	−1	0	1	2	3	5	10	$+\infty$
$f(x)$	$-\infty$	−99507	−2977	−177	5	15	3	−1	33	255	3183	100313	$+\infty$

From the above data and using the above corollary we conclude that there is at least one root of $f(x)$ in each of the intervals $(-3, -2)$, $(0, 1)$ and $(1, 2)$ since $f(x)$ changes sign in these intervals [i.e., for example $f(-3)f(-2) < 0$]. To locate a more accurate value of the root, we have to apply either the method of bisection, secant method or Newton's method from numerical analysis. ■

CAS **EXAMPLE 7**

Prove that there is a root of the given equation in the specified interval by using the Intermediate Value Theorem.

$$2x^3 - 5x + 10\cos(2x) = 0, \ (1, 3)$$

http://matrix.skku.ac.kr/cal-lab/cal-2-2-Exm7.html

Solution

```
var('x')
f(x)= 2*x^3 - 5*x +10*cos(2*x)
plot(f, 0, 3)
bool ( f(1) < 0 )    #   True
bool ( f(3) > 0 )    #   True
```

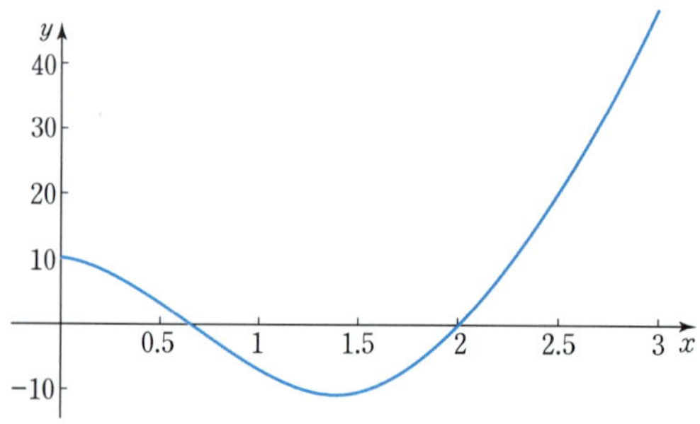

Figure 6

Thus by the Intermediate Value Theorem, there exists at least one number $c \in (1, 3)$ such that $f(c) = 0$. ■

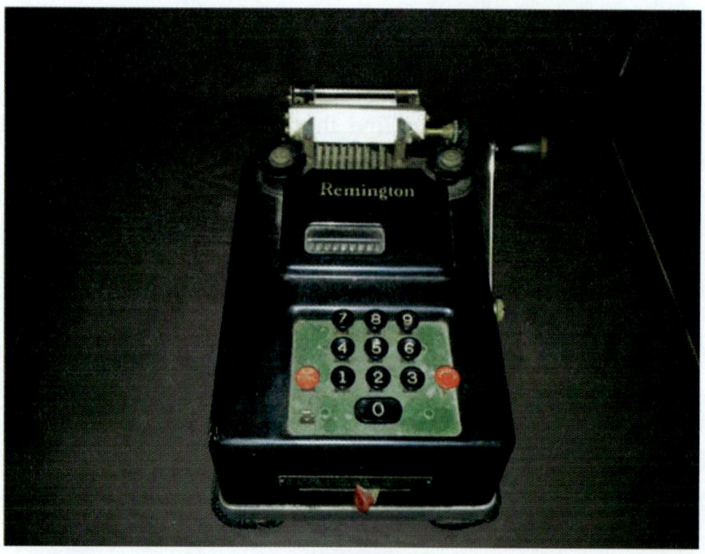

<Calculator 1>

2.2 EXERCISES (Continuity)

http://matrix.skku.ac.kr/Cal-Book/part1/CS-Sec-2-2-Sol.html

YouTube http://youtu.be/azrkT1RP4-c, http://youtu.be/hj8d-j_DGf4

CAS **1.** Given continuous functions f and g with $f(4)=3$ and $\lim_{x\to 4}[(f(x))^2-4f(x)g(x)]=5$, find $g(4)$.

http://matrix.skku.ac.kr/cal-lab/cal-2-2-1.html

Solution Since f and g are continuous functions.

$$\begin{aligned}5 &= \lim_{x\to 4}[(f(x))^2-4f(x)g(x)] \\ &= \left[\lim_{x\to 4}f(x)\right]^2-4\left[\lim_{x\to 4}f(x)\right]\left[\lim_{x\to 4}g(x)\right] \\ &= (f(4))^2-4f(4)g(4) \\ &= 3^2-4\cdot 3\cdot g(4)=9-12g(4)\end{aligned}$$

$$\Rightarrow 12g(4)=4 \Rightarrow g(4)=\frac{1}{3}$$

```
var('g')
solve(3^2-4*3*g==5, g )
```

Answer : g(4)=1/3

2. Given continuous functions f and g with $f(4)=3$ and $\lim_{x\to 4}[g((f(x))^2)-(f(x))^2]=12$, find $\lim_{x\to 9}g(x)$.

Solution $\lim_{x\to 4}[g((f(x))^2)-(f(x))^2]=g(3^2)-3^2=12$

Since g is continuous, $\lim_{x\to 9}g(x)=21$.

CAS **3.** Show that the function $f(x) = |\ln(x-2)^2|$ is discontinuous at $x=2$.

http://matrix.skku.ac.kr/cal-lab/cal-2-2-3.html

Solution We may draw $f(x)=|\ln(x-2)^2|$ with the following Sage command. It shows the function diverges (+infinity) at $x=2$. This shows $f(x)$ is discontinuous at $x=2$.

```
p1=plot(abs(ln((x-2)^2)), x, -3, 2, color='blue');
p2=plot(abs(ln((x-2)^2)), x, 2, 3, color='red');
show(p1+p2, ymax=5, ymin=-1)
```

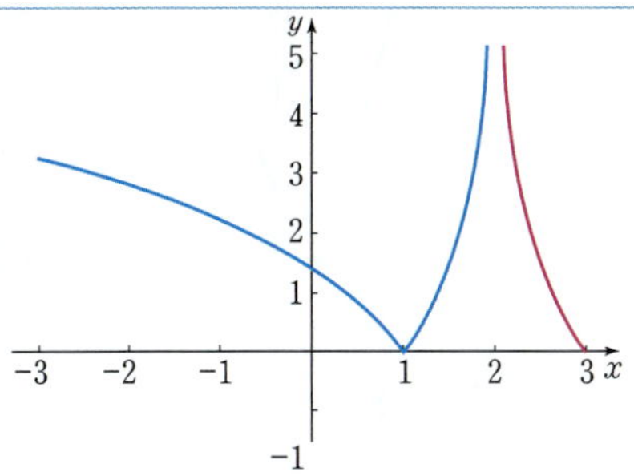

If we use the following Sage command, we may get limit(abs(ln((x-2)^2))) at $x=2$ directly.
It shows $f(x)$ is discontinuous at $x=2$.

```
limit(abs(ln((x-2)^2)), x=2)
```

Answer : +Infinity

4-7. Determine the points of discontinuity of f. At which of these numbers is f continuous from the right, from the left or neither? Sketch the graph of f.

CAS **4.** $f(x)=\begin{cases}\sin x & \text{if } x\le 0 \\ \cos x & \text{if } 0<x\le\pi \\ -1 & \text{if } x>\pi\end{cases}$

http://matrix.skku.ac.kr/cal-lab/cal-2-2-4.html

Solution We may draw $f(x)$ with the following Sage command. It shows the function is discontinuous at $x=0$.

```
plot(Piecewise([[(-2*pi, 0), sin(x)], [(0, pi), cos(x)], [(pi, 2*pi), -1]]))
```

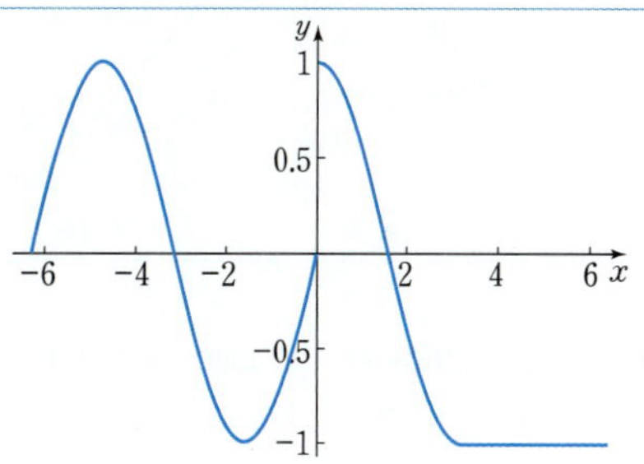

5. $f(x)=\begin{cases}1+x^2 & \text{if } x\le 0 \\ 2-x & \text{if } 0<x\le 2 \\ (x-2)^2 & \text{if } x>2\end{cases}$

Solution We see that $\lim_{x\to a}f(x)$ exists for all a except $a=0$. Notice that the right and left limits are different $a=0$.

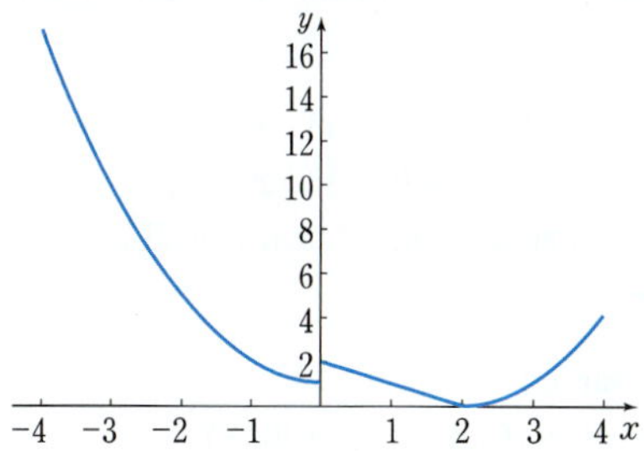

6. $f(x)=\begin{cases} x+1 & \text{if } x\le 1 \\ 1/x & \text{if } 1<x<3 \\ \sqrt{x-3} & \text{if } x\ge 3 \end{cases}$

Solution We see that $\lim_{x\to a} f(x)$ exists for all a except $a=1$ and $a=3$. Notice that the right and left limits are different at $a=1$ and $a=3$.

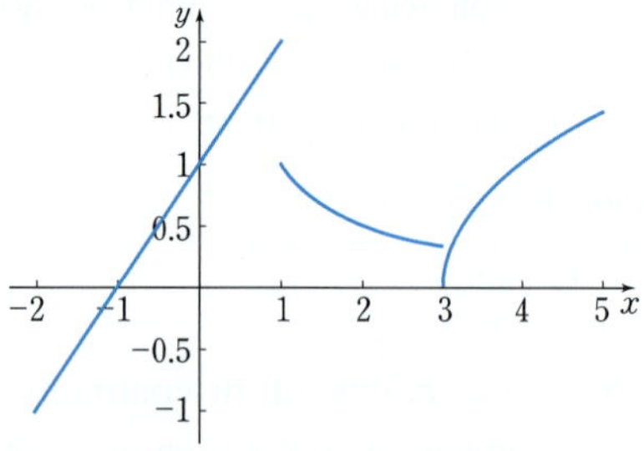

7. $f(x)=\begin{cases} x+2 & \text{if } x<0 \\ e^x & \text{if } 0\le x\le 1 \\ 2-x & \text{if } x>1 \end{cases}$

Solution We see that $\lim_{x\to a} f(x)$ exists for all a except $a=0$ and $a=1$. Notice that the right and left limits are different at $a=0$ and $a=1$.

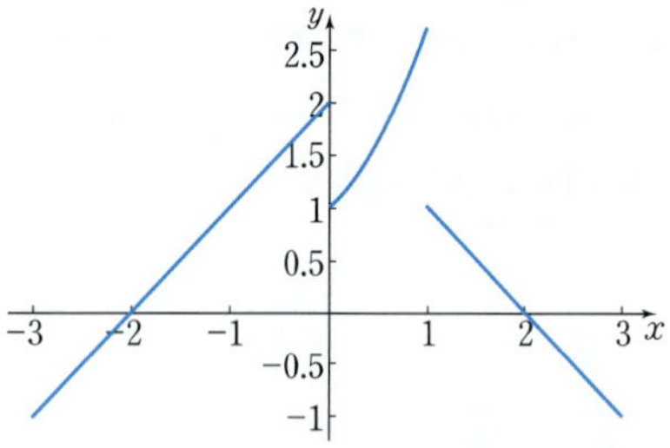

8-10. For what values of the constant c is the function f continuous on $(-\infty,\infty)$?

8. $f(x)=\begin{cases} cx+1 & \text{if } x\le 3 \\ cx^2-1 & \text{if } x>3 \end{cases}$

Solution $\lim_{x\to 3^-} f(x)=3c+1=9c-1=\lim_{x\to 3^+} f(x)$

$\Rightarrow 6c=2 \Rightarrow c=\dfrac{1}{3}$

Thus, for f to be continuous on $(-\infty,\infty)$, $c=\dfrac{1}{3}$.

CAS **9.** $f(x)=\begin{cases} x^2-c^2 & \text{if } x<4 \\ cx+20 & \text{if } x\ge 4 \end{cases}$

http://matrix.skku.ac.kr/cal-lab/cal-2-2-9.html

Solution

```
var('c')
solve(16-c^2==4*c+20, c)
```

Answer : c=-2

10. $f(x)=\begin{cases} \dfrac{5x}{c-x} & \text{if } x<2 \\ cx+4 & \text{if } x\ge 2 \end{cases}$

http://matrix.skku.ac.kr/cal-lab/cal-2-2-10.html

Solution Suppose that function f is continuous on $(-\infty,\infty)$. Find C such that $\lim_{x\to 2^-}\left(\dfrac{5x}{c-x}\right)=\lim_{x\to 2^+}(cx+4)$

$\Leftrightarrow$ Find C such that $\dfrac{10}{c-2}=2c+4 \Rightarrow c=\pm 3$. $f(x)$ is continuous at $c=3$ (But $f(x)=\dfrac{5x}{-3-x}$ is not continuos at $x=-3$). The solution is $c=3$.

```
var('c')
solve(10/(c-2)==2*c+4, c)
```

Answer : c=-3, c=3

11-13. Show that the following functions f has removable discontinuity at a. Also find a function g that agrees with f for $x\ne a$ and is continuous on $\mathbb{R}$.

11. $f(x)=\dfrac{x^3-2x-4}{x-2}$, $a=2$

http://matrix.skku.ac.kr/cal-lab/cal-2-2-11.html

Solution

```
var('x,y')
p1=plot((x^3-2*x-4)/(x-2), x, -10, 10, color='blue');
show(p1, aspect_ratio=1, ymax=10, ymin=-10)
```

$\lim_{x\to 2^+} f(x)=\lim_{x\to 2^-} f(x)=10$, $f(a)$ does not exist. Hence f has removable discontinuity at a. Then we can change the expression $\dfrac{x^3-2x-4}{x-2}\Rightarrow x^2+2x+2$ to remove the discontinuity. Let $g=x^2+2x+2$. It agrees with f for $x\ne a$ and is continuous on $\mathbb{R}$.

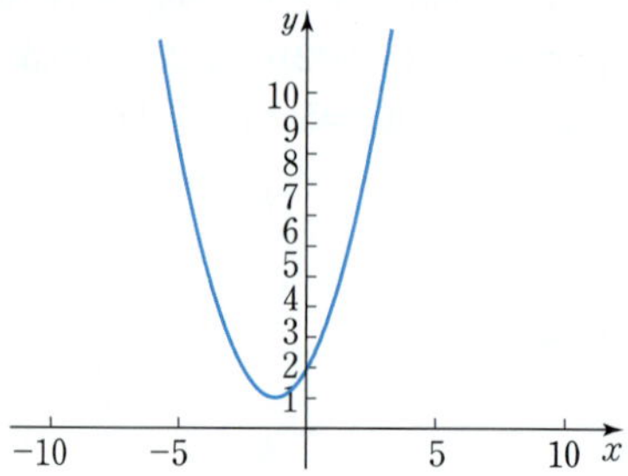

12. $f(x)=\dfrac{x^3+64}{x+4}$, $a=-4$

Solution $f(x)=\dfrac{x^3+64}{x+4}=\dfrac{(x+4)(x^2-4x+16)}{x+4}$ $=x^2-4x+16$ for $a \neq -4$. The discontinuity is removable and $f(x)=x^2-4x+16$ agrees with f for $a \neq -4$ and is continuous on $\mathbb{R}$.

13. $f(x)=\dfrac{3-\sqrt[3]{x}}{27-x}$, $a=27$

Solution $f(x)=\dfrac{3-\sqrt[3]{x}}{27-x}=\dfrac{3-\sqrt[3]{x}}{(3-\sqrt[3]{x})(9+3\sqrt[3]{x}+\sqrt[3]{x^2})}$ $=\dfrac{1}{9+3\sqrt[3]{x}+\sqrt[3]{x^2}}$ for $a \neq 27$. The discontinuity is removable and $f(x)=\dfrac{1}{9+3\sqrt[3]{x}+\sqrt[3]{x^2}}$ agrees with f for $a \neq 27$ and is continuous on $\mathbb{R}$.

---

14. Let $f(x)=\dfrac{x-7}{|x-7|}$.

Is $x=7$ removable discontinuity?

Solution Since $\lim\limits_{x\to 7^-} f(x)=\lim\limits_{x\to 7^-}\dfrac{x-7}{-(x-7)}=-1$, $\lim\limits_{x\to 7^+} f(x)=\lim\limits_{x\to 7^+}\dfrac{x-7}{x-7}=1$, $x=7$ is not a removable discontinuity.

15. If $f(x)=x^3-x^2+x$, show that there is a number c such that $f(c)=10$.

Solution $f(x)=x^3-x^2+x$ is continuous on the interval $[2, 3]$, $f(2)=6$ and $f(3)=21$. Since $6<10<21$, there is a number c in $(2, 3)$ such that $f(c)=10$ by the Intermediate Value Theorem.

16. Prove using Intermediate Value Theorem that there is a positive number c such that $c^2=2$.

Solution Let $f(x)=x^2$. We know that $f(x)=x^2$ is continuous on the interval $[1, 2]$, $f(1)=1$ and $f(2)=4$. Since $1<2<4$, there is a number c in $(1, 2)$ such that $f(c)=2$ by the Intermediate Value Theorem.

17-22. Prove that there is a root of the given equation in the specified interval by using the Intermediate Value Theorem.

CAS **17.** $x^4+x-3=0$, $(1, 2)$

http://matrix.skku.ac.kr/cal-lab/cal-2-2-17.html

Solution Let $f(x)=x^4+x-3=0$.

Then $f(1)=-1$, $f(2)=15$. Since f is a real-valued continuous function on the interval $[1, 2]$, there is a root $c\in[1, 2]$ such that $f(c)=0$ by the intermediate value theorem.

```
var('x'); f(x)=x^4+x-3
plot(f, 1, 2)
```

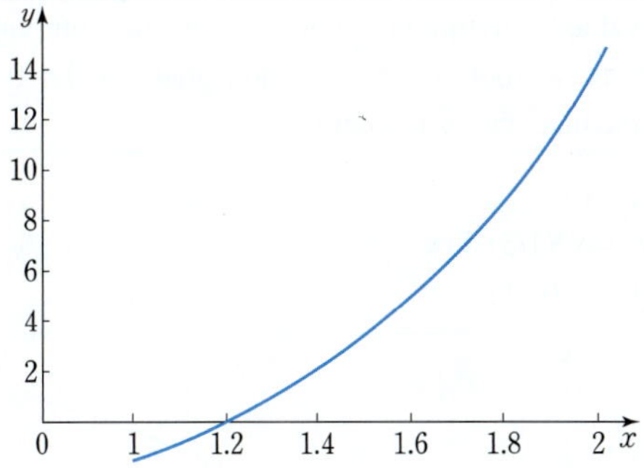

```
f(1)
```

Answer : -1

```
f(2)
```

Answer : 15

18. $\sqrt{6x-x^2}=1$, $(0, 2)$

http://matrix.skku.ac.kr/cal-lab/cal-2-2-18.html

Solution Let $f(x)=\sqrt{6x-x^2}-1$.

Then $f(0)=-1<0$, $f(2)=\sqrt{8}-1>0$. Since f is a real-valued continuous function on the interval $[0, 2]$, there is a root $c\in[0, 2]$ such that $f(c)=0$ by the intermediate value theorem.

```
var('x')
f(x)=sqrt(6*x-x^2)-1
P=plot(f, 0, 6)
show(P, aspect_ratio=1)
```

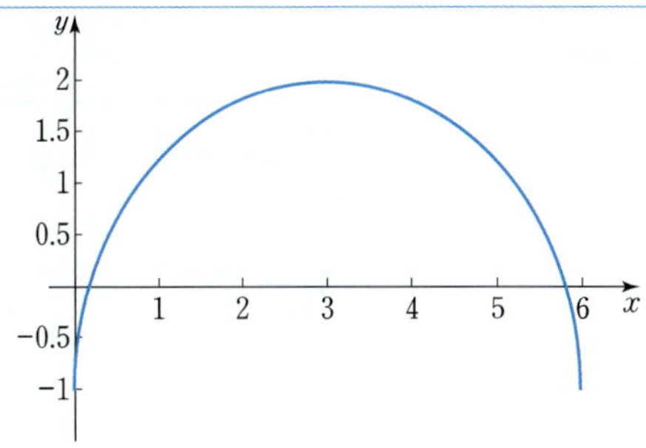

```
bool(f(0)<0)
```

Answer : True

```
bool(f(2)<0)
```

Answer : True

CAS **19.** $\sqrt[3]{x} = 1 - x$, $(0,\ 1)$

http://matrix.skku.ac.kr/cal-lab/cal-2-2-19.html

Solution Let $f(x) = \sqrt[3]{x} + x - 1$.

Then $f(0) = -1 < 0$, $f(2) = 1 > 0$. Since f is a real-valued continuous function on the interval $[0,\ 1]$, there is a root $c \in [0,\ 1]$ such that $f(c) = 0$ by the intermediate value theorem.

```
var('x')
f(x)=x^(1/3)-1+x
plot(f, 0, 1)
```

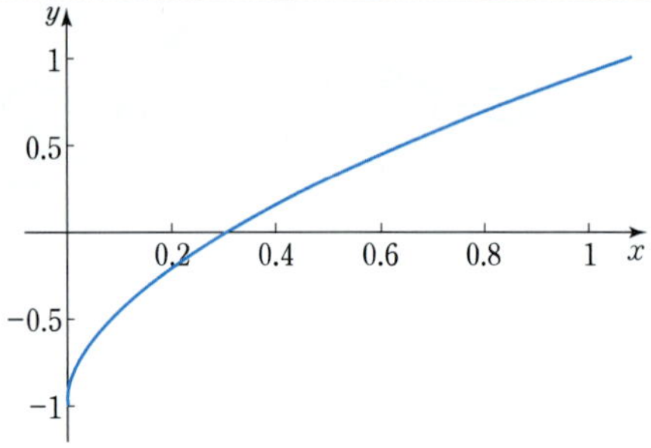


```
f(0)
```

Answer : -1

```
f(1)
```

Answer : 1

CAS **20.** $e^{x^3} = x^6$, $(-1,\ 1)$

http://matrix.skku.ac.kr/cal-lab/cal-2-2-20.html

Solution

```
var('x')
f(x)=exp(x^3)-x^6
plot(f, -1, 1)
```

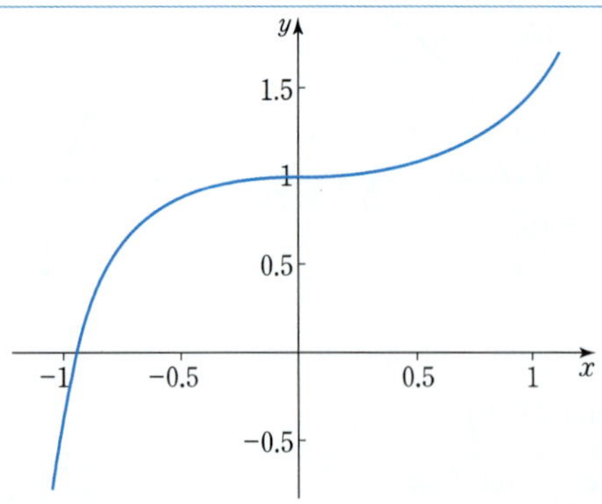


```
bool(f(-1)<0)
```

Answer : True

```
bool(f(1)>0)
```

Answer : True

21. $\cos x = x$, $(0,\ 1)$

Solution $f(x) = \cos x - x$ is continuous on the interval $[0,\ 1]$, $f(0) = 1$ and $f(1) = \cos 1 - 1 \approx -0.00015$.
Since $-0.00015 < 0 < 1$, there is a number c in $(0,\ 1)$ such that $f(c) = 0$ by the Intermediate Value Theorem. Thus, there is root of the equation $\cos x = x$ in the interval $(0,\ 1)$.

22. $\ln x = e^x$, $(1,\ 2)$

Solution $f(x) = \ln x - e^x$ is continuous on the interval $[1,\ 2]$, $f(1) = e$ and $f(2) = \ln 2 - e^2 \approx -6.696$. Since $-6.696 < 0 < e$, there is a number c in $(1,\ 2)$ such that $f(c) = 0$ by the Intermediate Value Theorem. Thus, there is root of the equation $\ln x = e^x$ in the interval $(1,\ 2)$.

23-26. Show that each of the following equations has at least one real root.

23. $e^x = 4\sin x$

http://matrix.skku.ac.kr/cal-lab/cal-2-2-23.html

http://math2.skku.ac.kr/home/pub/37

Solution We may draw both $f(x) = e^x$ and $g(x) = 4\sin x$ in one graph to find intersections. It shows the function has two real roots in $0 < x < 1$ and $1 < x < 2$.

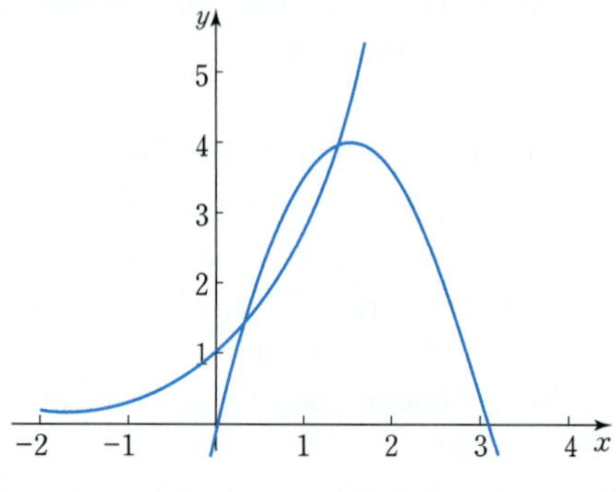


```
p1=plot(exp(x), x, -2, 4, color='blue')
p2=plot(4*sin(x), x, -2, 4, color='red')
show(p1+p2, ymax=5, ymin=0)
```

The following Sage commands give the value of the intersection in each interval, $0 < x < 1$ or $1 < x < 2$.

```
find_root(exp(x)==4*sin(x), 0, 1)
```

Answer : 0.37055809596982464

```
find_root(exp(x)==4*sin(x), 1, 2)
```

Answer : 1.3649584337330951

24. $x^5 - x^2 + 2x + 3 = 0$

Solution Let $f(x) = x^5 - x^2 + 2x + 3$.
Then $f(-1) = -1 < 0$ and $f(0) = 3 > 0$. So by the Intermediate Value Theorem, there is a number c in $(-1, 0)$ such that $f(c) = 0$. This implies that $c^5 - c^2 + 2c + 3 = 0$.

25. $x^5 - x^2 - 4 = 0$

Solution Let $f(x) = x^5 - x^2 - 4$.
Then $f(1) = -4 < 0$ and $f(2) = 24 > 0$. So by the Intermediate Value Theorem, there is a number c in (1, 2) such that $f(c) = 0$. This implies that $c^5 - c^2 - 4 = 0$.

26. $\sqrt{x-5} = \dfrac{1}{x+3}$

Solution Let $f(x) = \sqrt{x-5} - \dfrac{1}{x+3}$.

Then $f(5) = -\dfrac{1}{8} < 0$ and $f(6) = \dfrac{8}{9} > 0$, and f is continuous in $[5, \infty)$ So by the Intermediate Value Theorem, there is a number c in $(5, 6)$ such that $f(c) = 0$. This implies that $\sqrt{c-5} = \dfrac{1}{c+3}$.

27-28. Find the values of x for which f is continuous.

27. $f(x) = \begin{cases} 0 & \text{if } x \text{ is rational} \\ 1 & \text{if } x \text{ is irrational} \end{cases}$

Solution Suppose $f(x)$ is continuous at a in $\mathbb{R}$, then $\lim\limits_{x \to a^-} f(x) = f(a) = \lim\limits_{x \to a^+} f(x)$.
But if a is rational, then $f(a) = 0$ but $\lim\limits_{x \to a^-} f(x)$ or $\lim\limits_{x \to a^+} f(x)$ can be 1. if a is irrational, then $f(a) = 1$ but $\lim\limits_{x \to a^-} f(x)$ or $\lim\limits_{x \to a^+} f(x)$ can be 0, this is a contradiction. so no such value a in $\mathbb{R}$ exists.

28. $f(x) = \begin{cases} 0 & \text{if } x \text{ is rational} \\ x & \text{if } x \text{ is irrational} \end{cases}$

Solution f is continuous only at $x = 0$ since $\lim\limits_{x \to 0^-} f(x) = f(0) = \lim\limits_{x \to 0^+} f(x)$. But there is no nonzero a in $\mathbb{R}$, such that $\lim\limits_{x \to a^-} f(x) = f(a) = \lim\limits_{x \to a^+} f(x)$.

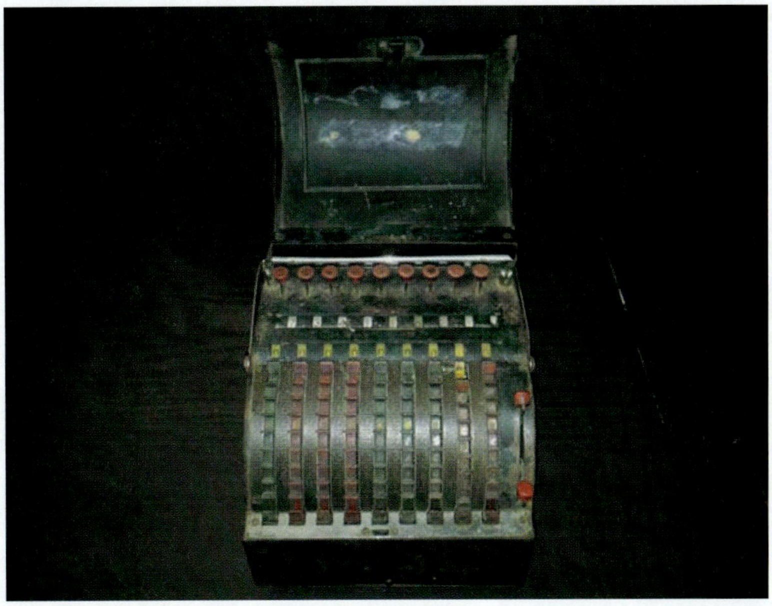

<Calculator 2>

Chapter 3

Derivatives

3.1 Definition of Derivatives, Differentiation

3.2 Derivatives of Functions, The Product and Quotient Rule

3.3 The Chain Rule and Inverse Functions

*3.4 Approximation and Related Rates

Gottfried Leibniz

1646-1716

Infinitesimal Calculus

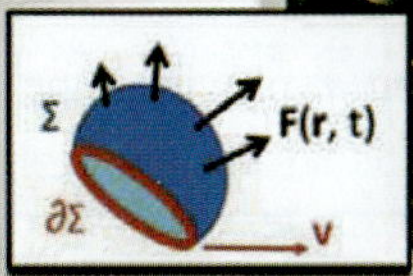

[Leibniz Integration]

A true Renaissance man, Leibniz had many interests including law, philosophy, natural sciences, linguistics and mathematics. He and Newton are both credited with inventing infinitesimal calculus. He also created the binary system, which is the foundation of virtually all modern computer architectures.

YouTube http://youtu.be/A-vDsF9uITs

In this chapter, we will introduce the definition of the derivative of a function. We will also introduce and evaluate the derivatives of some elementary functions, such as polynomials and trigonometric functions. Some differentiation techniques (rules) such as the Product Rule, Quotient Rule, and Chain Rule will also be addressed in this chapter, along with linear approximations and related rates.

3.1 Definition of Derivatives, Differentiation

DEFINITION 1

Let $f:(a,b)\to\mathbb{R}$ be a function defined on an open interval (a,b). The function f' defined by the formula

$$f'(x)=\lim_{h\to 0}\frac{f(x+h)-f(x)}{h}$$

is called the **derivative with respect to** x of the function f provided the limit exists.

Note f' assigns the number $f'(x)$ to each x, so that we may regard $f'(x)$ as a new function.
Also we note that the domain of f' is the set $\{x\,|\,f'(x) \text{ exists}\}$.

EXAMPLE 1

Find $g'(x)$ if $g(x)=\dfrac{2x}{x+2}$.

Solution From the definition, we have

$$\begin{aligned} g'(x)&=\lim_{h\to 0}\frac{g(x+h)-g(x)}{h}\\ &=\lim_{h\to 0}\frac{1}{h}\left\{\frac{2(x+h)}{(x+h)+2}-\frac{2x}{x+2}\right\}=\frac{4}{(x+2)^2}. \end{aligned}$$ ■

Notation Some common notation for the derivative are

$$f'(x)=y'=\frac{dy}{dx}=\frac{df}{dx}=\frac{d}{dx}f(x)=D_x y=D_x f(x)$$

The operator $Df=f'$, which transforms f to its derivative, is called the differentiation operator.

The value of a derivative $\dfrac{dy}{dx}$ at a specific number a, is denoted by $\left.\dfrac{dy}{dx}\right|_{x=a}$ or $f'(a)$.

DEFINITION 2

A function f is differentiable at a if $f'(a)$ exists.

A function is differentiable on (a, b) if it is differentiable at every number in (a, b). One of the typical examples of a non-differentiable function is $f(x)=|x|$ at $x=0$.

There is a relationship between continuity and differentiability. The relationship is indicated in the following theorem.

THEOREM 3 Differentiability Implies Continuity

If f is differentiable at a, then f is continuous at a.

Proof Assume f is differentiable at a point a, which implies that

$$f'(a)=\lim_{x\to a}\frac{f(x)-f(a)}{x-a}$$

exists. To show that f is continuous at a, we must show that $\lim_{x\to a}f(x)=f(a)$. The key is the identity

$$f(x)=\frac{f(x)-f(a)}{x-a}(x-a)+f(a), \qquad x\neq a.$$

Taking the limit as x approaches a on both sides of this formula and simplifying, we have

$$\begin{aligned}\lim_{x\to a}f(x)&=\lim_{x\to a}\left[\frac{f(x)-f(a)}{x-a}(x-a)+f(a)\right]\\&=\lim_{x\to a}\left(\frac{f(x)-f(a)}{x-a}\right)\lim_{x\to a}(x-a)+\lim_{x\to a}f(a)\\&=f'(a)\cdot 0+f(a)\\&=f(a).\end{aligned}$$

Therefore, $\lim_{x\to a}f(x)=f(a)$, which means that f is continuous at a. ■

Remark The converse of Theorem 3 is not true. For example, $f(x)=|x|$ is continuous at $x=0$ but it is not differentiable at $x=0$.

3.1 EXERCISES (Definition of Derivatives, Differentiation)

http://matrix.skku.ac.kr/Cal-Book/part1/CS-Sec-3-1-Sol.html

YouTube http://youtu.be/7wTBWuk2CzU

1-7. Differentiate each of the following functions using Definition 3.1.1, if it exists.

1. $f(x) = 15$

Solution $f'(x) = \lim_{h\to 0}\frac{f(x+h)-f(x)}{h} = \lim_{h\to 0}\frac{15-15}{h} = 0$

2. $f(x) = 4x - 3$

Solution
$$f'(x) = \lim_{h\to 0}\frac{f(x+h)-f(x)}{h} = \lim_{h\to 0}\frac{(4(x+h)-3)-(4x-3)}{h} = \lim_{h\to 0}\frac{4h}{h} = 4$$

3. $f(x) = \pi x$

Solution
$$f'(x) = \lim_{h\to 0}\frac{f(x+h)-f(x)}{h} = \lim_{h\to 0}\frac{\pi(x+h)-\pi x}{h} = \lim_{h\to 0}\frac{\pi h}{h} = \pi$$

4. $f(x) = x^4$

Solution
$$f'(x) = \lim_{h\to 0}\frac{f(x+h)-f(x)}{h}$$
$$= \lim_{h\to 0}\frac{(x+h)^4 - x^4}{h}$$
$$= \lim_{h\to a}\frac{1}{h}\left(\left({}_4C_4x^4 + {}_4C_3x^3h + {}_4C_2x^2h^2 + {}_4C_1xh^3 + {}_4C_0h^4\right) - x^4\right)$$
$$= 4x^3$$

5. $g(t) = (3t-4)^8$

Solution
$$g'(t) = \lim_{h\to 0}\frac{g(t+h)-g(t)}{h}$$
$$= \lim_{h\to 0}\frac{(3t-4+3h)^8 - (3t-4)^8}{h}$$
$$= \lim_{h\to 0}\frac{1}{h}[({}_8C_8(3t-4)^8 + {}_8C_7(3t-4)^7 3h + {}_8C_6(3t-4)^6(3h)^2 + \cdots + {}_8C_1(3t-4)(3h)^7 + {}_8C_0(3h)^8) - (3t-4)^8]$$
$$= 24(3t-4)^7$$

6. $g(t) = t^8 + 3t^4 + 2t + 1$

Solution
$$g'(t) = \lim_{h\to 0}\frac{g(t+h)-g(t)}{h}$$
$$= \lim_{h\to 0}\frac{[(t+h)^8 + 3(t+h)^4 + 2(t+h) + 1] - (t^8 + 3t^4 + 2t + 1)}{h}$$
$$= \lim_{h\to 0}\frac{1}{h}[({}_8C_7t^7h + \cdots + {}_8C_1th^7 + {}_8C_0h^8) + 3({}_4C_3t^3h + \cdots + {}_4C_0h^4) + 2h]\}$$
$$= 8t^7 + 12t^3 + 2$$

7. $g(t) = \dfrac{t}{t+1}$

Solution
$$g'(t) = \lim_{h\to 0}\frac{g(t+h)-g(t)}{h}$$
$$= \lim_{h\to 0}\frac{\frac{t+h}{t+h+1} - \frac{t}{t+1}}{h}$$
$$= \lim_{h\to 0}\frac{1}{h}\left\{\frac{(t+h)(t+1) - t(t+h+1)}{(t+1)(t+h+1)}\right\}$$
$$= \lim_{h\to 0}\frac{1}{h}\left\{\frac{h}{(t+1)(t+h+1)}\right\}$$
$$= \lim_{h\to 0}\frac{1}{(t+1)(t+h+1)}$$
$$= \frac{1}{(t+1)^2}$$

---

8. Is the function

$$f(x) = \begin{cases} x|x|\sin\frac{1}{x^2}, & x \neq 0 \\ 0, & x = 0 \end{cases}$$

differentiable at $x = 0$?

Proof
$$\lim_{x\to 0}\frac{f(x)-f(0)}{x} = \lim_{x\to 0}\frac{x\,|\,x\,|\,\sin\frac{1}{x^2}}{x}$$
$$= \lim_{x\to 0}|\,x\,|\,\sin\frac{1}{x^2} = 0 \quad \left(-1 \le \lim_{x\to 0}\sin\frac{1}{x^2} \le 1\right)$$

Hence $f(x)$ is differentiable at $x = 0$ and $f'(0) = 0$.

9. Discuss a geometric and physical interpretation of $f'(x)$.

Solution 1. Geometric interpretation: The slope of the tangent to the curve at a point is equal to the derivative of the function at that point.

2. Physical interpretation: The average velocity is the ratio between distance travelled (Δd) and the time elapsed (Δt).

3.2 Derivatives of Functions, The Product and Quotient Rule

YouTube http://youtu.be/XXMnCESesfQ

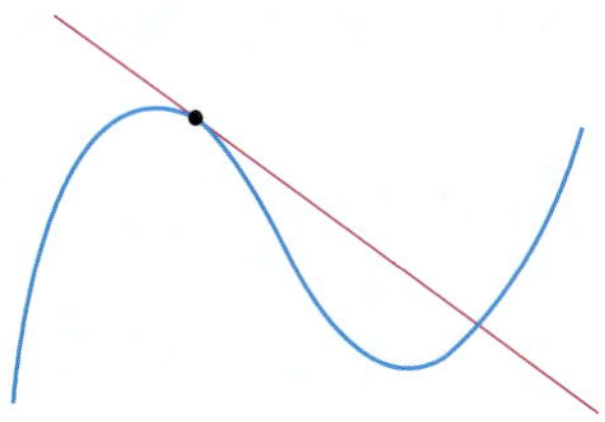

Figure 1

In this section, we shall consider derivatives of some of the basic functions such as polynomials, exponential, trigonometric and inverse trigonometric functions. The product rule and quotient rule are helpful in derivations of derivatives of functions which are expressed as the product and/or quotient of other functions. We also consider applications of the derivative to other sciences.

For the constant function $f(x)=c$, the graph is the horizontal line $y=c$, which has slope 0. So we expect $f'(x)=0$. From the definition of a derivative,

$$\lim_{h\to 0}\frac{f(x+h)-f(x)}{h}=\lim_{h\to 0}\frac{c-c}{h}=\lim_{h\to 0}0=0.$$

Thus, $f'(x)$ exists for every x and its value is zero, that is, we have the following fact and theorem.

$$\frac{d}{dx}(c)=0$$

THEOREM 1 **The Power Rule**

If n is a positive integer, then

$$\frac{d}{dx}(x^n)=nx^{n-1}.$$

Proof Let $f(x)=x^n$. Using the formula

$$x^n-a^n=(x-a)(x^{n-1}+x^{n-2}a+\cdots+xa^{n-2}+a^{n-1})$$

we have

$$\begin{aligned}f'(a)&=\lim_{x\to a}\frac{f(x)-f(a)}{x-a}=\lim_{x\to a}\frac{x^n-a^n}{x-a}\\&=\lim_{x\to a}(x^{n-1}+x^{n-2}a+\cdots+xa^{n-2}+a^{n-1})\\&=a^{n-1}+a^{n-2}a+\cdots+aa^{n-2}+a^{n-1}\\&=na^{n-1}.\end{aligned}$$

Hence $f'(x)=nx^{n-1}$. ■

EXAMPLE 1

Find $f'(x)$ where

$$f(x)=x^{\frac{1}{n}},\quad n\neq 0.$$

Solution Let us define $t=x^{1/n}$ and $s=a^{1/n}$. Then $x=t^n$ and $a=s^n$. Note that $t^n-s^n=(t-s)(t^{n-1}+t^{n-2}s+\cdots+s^{n-1})$. This implies

$$\frac{t-s}{t^n-s^n}=\frac{1}{t^{n-1}+t^{n-2}s+\cdots+s^{n-1}}$$

and

$$\begin{aligned}\frac{x^{1/n}-a^{1/n}}{x-a}&=\frac{1}{t^{n-1}+t^{n-2}s+\cdots+s^{n-1}}\\&=\frac{1}{x^{\frac{n-1}{n}}+x^{\frac{n-2}{n}}a^{\frac{1}{n}}+\cdots+a^{\frac{n-1}{n}}}.\end{aligned}$$

Hence $f(x)=x^{1/n}$ is differentiable and

$$f'(x)=\left(\frac{1}{n}x^{(\frac{1}{n}-1)}=\frac{1}{n}x^{\frac{1-n}{n}}\right)$$

since $\lim_{x\to a}\dfrac{x^{1/n}-a^{1/n}}{x-a}=\dfrac{1}{nx^{(n-1)/n}}=\dfrac{1}{n}x^{(1-n)/n}$.

Note that $f(x)$ might not be differentiable at $x=0$. ■

1)http://faculty.mdc.edu/mmontane/chain-rule.pdf

THEOREM 2 The General Power Rule[1]

If a is any nonzero real number, then

$$\frac{d}{dx}(x^a)=ax^{a-1}.$$

Theorem 1, $\dfrac{d}{dx}(x^n)=nx^{n-1}$, is a special case of Theorem 2.

EXAMPLE 2

For the curve $y=x^{\frac{1}{3}}$ find an equation of the tangent line at the point $(1,\ 1)$.

Solution Since $y'=\dfrac{1}{3\sqrt[3]{x^2}}$, the slope of the tangent line at (1, 1) is $\dfrac{1}{3}$.

Thus the equation of the tangent line is

$$y-1=\frac{1}{3}(x-1)\quad\text{or}\quad y=\frac{1}{3}x+\frac{2}{3}.$$

■

THEOREM 3 The Constant Multiple Rule

If c is a constant and f is a differentiable function at x, then cf is differentiable at x and

$$\frac{d}{dx}[cf(x)] = c\frac{d}{dx}f(x).$$

That is, the derivative of a constant times a function is the constant times the derivative of the function.

Proof Let $g(x) = cf(x)$ for some constant c. By the limit definition of the derivative: $f'(x) = \lim_{h\to 0}\frac{f(x+h)-f(x)}{h}$ and

$$cf'(x) = c\lim_{h\to 0}\frac{f(x+h)-f(x)}{h} = \lim_{h\to 0}\frac{cf(x+h)-cf(x)}{h}$$

To prove the proposition, it suffices to show that $g'(x) = cf'(x)$.

$$g'(x) = \lim_{h\to 0}\frac{g(x+h)-g(x)}{h} = \lim_{h\to 0}\frac{cf(x+h)-cf(x)}{h} = cf'(x).$$ ■

THEOREM 4 The Sum Rule

If f and g are both differentiable at x, then $f+g$ is differentiable at x and

$$\frac{d}{dx}[f(x)+g(x)] = \frac{d}{dx}f(x) + \frac{d}{dx}g(x).$$

That is, the derivative of a sum of functions is the sum of the derivatives.

Proof Left as an exercise. ■

The sum rule can be extended to the sum of finitely many number of functions, for example

$$(f+g+h)' = [(f+g)+h]' = (f+g)' + h' = f' + g' + h'.$$

In general $(f_1 + \cdots + f_n)' = f_1{}' + \cdots + f_n{}'$.

THEOREM 5 The Difference Rule

If f and g are both differentiable at x, then $f-g$ is differentiable at x and

$$\frac{d}{dx}[f(x)-g(x)] = \frac{d}{dx}f(x) - \frac{d}{dx}g(x).$$

This simple proof is omitted.[2)]

2)http://jwilson.coe.uga.edu/EMAT6680/Horst/derivativesum/derivativesum.html

EXAMPLE 3

$$\frac{d}{dx}(x^{20}+9x^{10}+8x-4)=\frac{d}{dx}(x^{20})+9\frac{d}{dx}(x^{10})+8\frac{d}{dx}(x)-\frac{d}{dx}(4)$$
$$=20x^{19}+90x^{9}+8.$$

■

EXAMPLE 4

Find the points on the curve $y=x^4-3x^2+4$ where the tangent line is horizontal.

Solution The tangent line is horizontal where the derivative is zero. We have

$$\frac{dy}{dx}=2x(2x^2-3).$$

Thus, $\frac{dy}{dx}=0$ if $x=0$ or $x=\pm\frac{\sqrt{6}}{2}$. The corresponding points are $(0,4)$ and $\left(\pm\frac{\sqrt{6}}{2},\frac{7}{4}\right)$. ■

Exponential Function

Consider the exponential function $f(x)=a^x$ where $a>0$, $a\neq 1$. From the definition of the derivative:

$$f'(x)=\lim_{h\to 0}\frac{f(x+h)-f(x)}{h}=\lim_{h\to 0}\frac{a^{x+h}-a^x}{h}$$
$$=\lim_{h\to 0}\frac{a^x a^h-a^x}{h}=\lim_{h\to 0}\frac{a^x(a^h-1)}{h}.$$

Thus,

$$f'(x)=a^x\lim_{h\to 0}\frac{a^h-1}{h}=a^x\lim_{h\to 0}\frac{f(h)-f(0)}{h}=a^x f'(0).$$

That is, the rate of change of any exponential function is proportional to the function itself.

Definition of the Number $e\approx 2.71828$

The number e is an important mathematical constant, it is an irrational number which approximately equals to 2.71828. This number arises in the study of compound interest, and can also be calculated as the sum of the infinite series

$$e = 1 + \frac{1}{1} + \frac{1}{1 \cdot 2} + \frac{1}{1 \cdot 2 \cdot 3} + \frac{1}{1 \cdot 2 \cdot 3 \cdot 4} + \cdots.$$

The constant may be defined in many ways; for example, e is the unique real number such that the value of the derivative (slope of the tangent line) of the function $f(x) = e^x$ at the point $x = 0$ is equal to 1. The number $e = 2.71828182845905\cdots$ is defined so that $f(x) = a^x \to e$ as $x \to 1$ when $a = e$.

Derivative of the Natural Exponential Function $\frac{d}{dx}(e^x) = e^x$

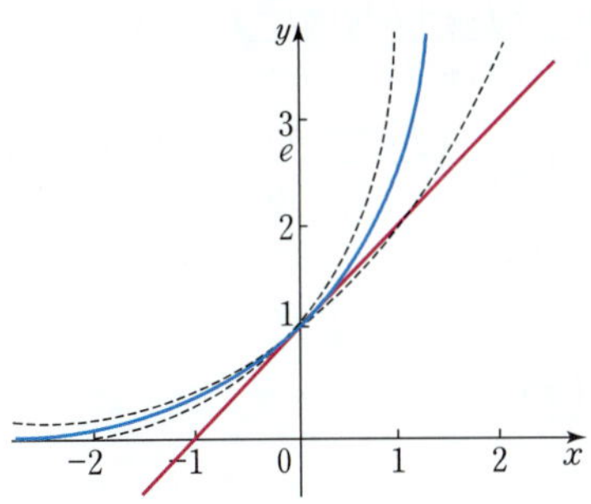

Figure 2 e^x

There is a very important exponential function that arises naturally in many places. This function is called the natural exponential function. However, for most people this is simply the exponential function. Then for $f(x) = e^x$, $\lim_{h \to 0} \frac{e^h - 1}{h} = 1$ implies $f'(0) = 1$, and from the general derivative above we have $f'(x) = e^x$.

Thus the slope of a tangent line to the curve $y = e^x$ is equal to the y -coordinate of the point.

EXAMPLE 5

Find the point on the curve $y = e^x - 2x$ where the tangent line is parallel to the line $y = 2x - 15$.

Solution We have $y' = e^x - 2$. If the tangent line is parallel to $y = 2x - 15$, then the slope of the tangent line is 2. If $e^x - 2 = 2$, then $x = 2\ln 4$. The point is $(2\ln 2,\ 4 - 4\ln 2)$. ∎

The Product Rule

In calculus, the product rule is a formula used to find the derivatives of products of two or more differentiable functions. It may be stated as:

THEOREM 6 The Product Rule

If f and g are both differentiable at x, then fg is differentiable at x and

$$\frac{d}{dx}(f(x)g(x)) = f(x)\frac{d}{dx}(g(x)) + g(x)\frac{d}{dx}(f(x))$$

or

$$(fg)' = f'g + fg'.$$

That is, the derivative of a product of two functions is the first function times the derivative of the second function plus the second function times the derivative of the first function.

Proof Let $F(x)=f(x)g(x)$. We use the continuity of $f(x)$ in the proof. Then

$$\begin{aligned}
F'(x) &= \lim_{h\to 0}\frac{F(x+h)-F(x)}{h} \\
&= \lim_{h\to 0}\frac{f(x+h)g(x+h)-f(x)g(x)}{h} \\
&= \lim_{h\to 0}\frac{f(x+h)g(x+h)-f(x+h)g(x)+f(x+h)g(x)-f(x)g(x)}{h} \\
&= \lim_{h\to 0}\left[f(x+h)\frac{g(x+h)-g(x)}{h}+g(x)\frac{f(x+h)-f(x)}{h}\right] \\
&= \lim_{h\to 0}f(x+h)\cdot\lim_{h\to 0}\frac{g(x+h)-g(x)}{h} \\
&\quad +\lim_{h\to 0}g(x)\cdot\lim_{h\to 0}\frac{f(x+h)-f(x)}{h} \\
&= f'(x)g(x)+f(x)g'(x).
\end{aligned}$$

■

EXAMPLE 6

Differentiate the function $f(x)=x^{20}e^{x}$.

Solution Using the Product Rule, we have

$$f'(x)= x^{20}e^{x}+20\,x^{19}e^{x}.$$

■

The Quotient Rule

THEOREM 7 The Quotient Rule

If f and g are differentiable at x and $g\neq 0$, then $\frac{f}{g}$ is differentiable at x and

$$\frac{d}{dx}\left(\frac{f(x)}{g(x)}\right)=\left\{f(x)\frac{d}{dx}(g(x))-g(x)\frac{d}{dx}(f(x))\right\}/\{g(x)\}^2$$

or

$$\left(\frac{f}{g}\right)'=\frac{f'g-fg'}{g^2}.$$

That is, derivative of a quotient is the denominator times the derivative of the numerator minus the numerator times the derivative of the denominator, all divided by the square of the denominator.

Proof Let $F(x)=f(x)/g(x)$. Then

$$
\begin{aligned}
F'(x) &= \lim_{h\to 0}\frac{\dfrac{f(x+h)}{g(x+h)}-\dfrac{f(x)}{g(x)}}{h} \\
&= \lim_{h\to 0}\frac{f(x+h)g(x)-f(x)g(x+h)}{hg(x+h)g(x)} \\
&= \lim_{h\to 0}\frac{f(x+h)g(x)-f(x)g(x)+f(x)g(x)-f(x)g(x+h)}{hg(x+h)g(x)} \\
&= \lim_{h\to 0}\frac{g(x)\cdot\dfrac{f(x+h)-f(x)}{h}-f(x)\cdot\dfrac{g(x+h)-g(x)}{h}}{g(x+h)g(x)} \\
&= \frac{f'(x)g(x)-f(x)g'(x)}{\{g(x)\}^2}.
\end{aligned}
$$

■

EXAMPLE 7

For the curve $y=\dfrac{e^x+1}{x^3+1}$, find the equation of the tangent line at the point $(0,\ 2)$.

Solution By the Quotient Rule, we have

$$\frac{dy}{dx}=\frac{e^x(x^3+1)-3x^2(e^x+1)}{(x^3+1)^2}.$$

Thus the slope at the given point is 1. The equation of the tangent line is $y=x+2$. ■

Table of Differentiation Formulas

$\dfrac{d}{dx}(c)=0$	$\dfrac{d}{dx}(x^a)=ax^{a-1}$	$\dfrac{d}{dx}(e^x)=e^x$
$(cf)'=cf'$	$(f+g)'=f'+g'$	$(f-g)'=f'-g'$
$(fg)'=f'g+fg'$	$\left(\dfrac{f}{g}\right)'=\dfrac{f'g-fg'}{g^2}\quad(g\neq 0)$	

Derivatives of Trigonometric Functions

In this section, we make use of the following limits.

THEOREM 8 Trigonometric Limits

$$\lim_{x\to 0}\frac{\sin x}{x}=1, \quad \lim_{x\to 0}\frac{\cos x-1}{x}=0$$

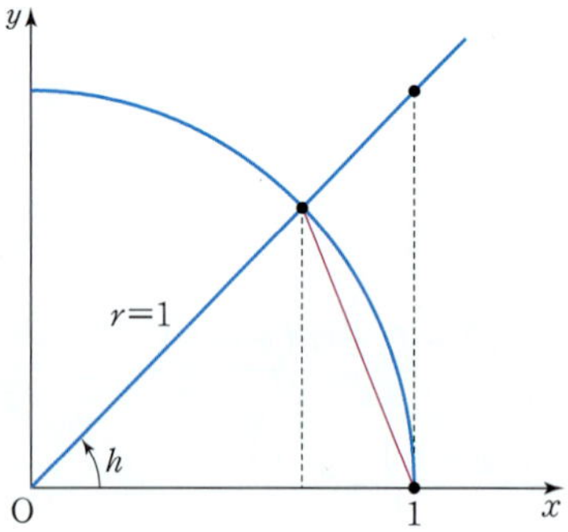

Figure 3

Using the Squeeze Theorem, we have

$\lim_{h\to 0}\frac{\sin h}{h}=1$. (Note that $\cos h<\frac{\sin h}{h}<1$ if $0<h<\frac{\pi}{2}$ in Figure 3)

$$\begin{aligned}\text{Now, } \lim_{h\to 0}\frac{\cos h-1}{h}&=\lim_{h\to 0}\frac{\cos h-1}{h}\cdot\frac{\cos h+1}{\cos h+1}\\ &=\lim_{h\to 0}\frac{-\sin^2 h}{h(\cos h+1)}\\ &=-\lim_{h\to 0}\frac{\sin h}{h}\ \lim_{h\to 0}\frac{\sin h}{\cos h+1}\\ &=0.\end{aligned}$$

Vibrations, waves, elastic motion and other quantities that vary in a periodic manner can be described using trigonometric functions.

Consider $f(x)=\sin x$. From the definition of a derivative, we have

$$\begin{aligned}f'(x)&=\lim_{h\to 0}\frac{\sin(x+h)-\sin x}{h}\\ &=\lim_{h\to 0}\frac{\sin x\cos h+\cos x\sin h-\sin x}{h}\\ &=\lim_{h\to 0}\left(\frac{\sin x\cos h-\sin x}{h}+\frac{\cos x\sin h}{h}\right)\\ &=\sin x\cdot\lim_{h\to 0}\frac{\cos h-1}{h}+\cos x\cdot\lim_{h\to 0}\frac{\sin h}{h}\\ &=\sin x\cdot 0+\cos x\cdot 1=\cos x.\end{aligned}$$

Thus we have $f'(x)=\frac{d}{dx}(\sin x)=\cos x$.

We note that $\cos x=\sin\left(\frac{\pi}{2}-x\right)$. Similarly, we can show

$$\frac{d}{dx}(\cos x)=-\sin x.$$

To find the derivative of $\tan x$, we will use the quotient rule.

We have

$$\begin{aligned}\frac{d}{dx}(\tan x)&=\frac{d}{dx}\left(\frac{\sin x}{\cos x}\right)\\ &=\frac{\cos x\cos x-\sin x(-\sin x)}{\cos^2 x}=\frac{1}{\cos^2 x}=\sec^2 x.\end{aligned}$$

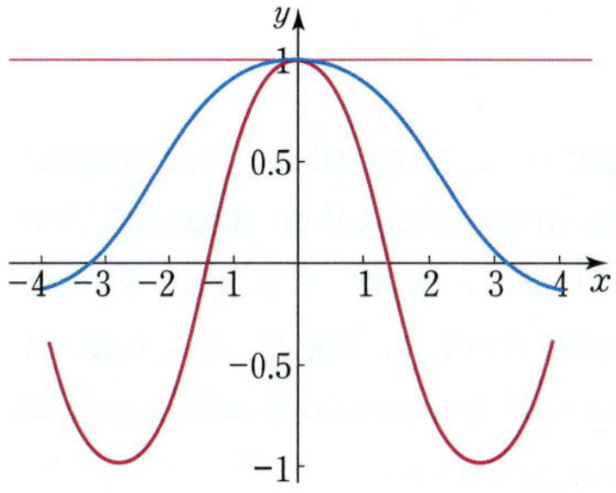

Figure 4

We may deduce derivatives of the remaining three trigonometric functions in a similar manner.

So in summary, we have the following.

Derivatives of Trigonometric Functions

$$\frac{d}{dx}(\sin x) = \cos x \qquad \frac{d}{dx}(\csc x) = -\csc x \cot x$$

$$\frac{d}{dx}(\cos x) = -\sin x \qquad \frac{d}{dx}(\sec x) = \sec x \tan x$$

$$\frac{d}{dx}(\tan x) = \sec^2 x \qquad \frac{d}{dx}(\cot x) = -\csc^2 x$$

For the "**cofunctions**" namely cosine, cosecant, and cotangent, notice the appearance of the minus signs before their derivatives.

EXAMPLE 8

If $y = e^x \sin x$ find $\dfrac{dy}{dx}$.

Solution Differentiation gives us that $\dfrac{dy}{dx} = e^x(\sin x + \cos x)$. ■

EXAMPLE 9

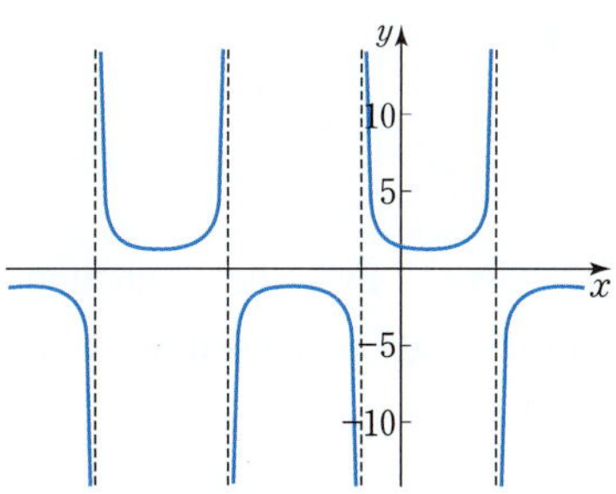

Figure 5 $f(x) = \dfrac{\csc x}{1 + \cot x}$

Find the values of x where the graph of $f(x) = \dfrac{\csc x}{1 + \cot x}$ has a horizontal tangent.

http://matrix.skku.ac.kr/cal-lab/cal-3-2-Exm10.html

Solution Using the Quotient Rule, we have

$$f'(x) = \frac{-(1+\cot x)\csc x \cot x + \csc^3 x}{(1+\cot x)^2}$$

$$= \frac{+\csc x(\csc^2 x - \cot x - \cot^2 x)}{(1+\cot x)^2} = \frac{+\csc x(1-\cot x)}{(1+\cot x)^2}.$$

When $\cot x = 1$, $f'(x) = 0$ (and $\csc x$ is not zero for all x). Thus, for $x = n\pi + \dfrac{\pi}{4}$, the curve has horizontal tangents.

Use http://sage.skku.edu/ or http://www.sagemath.org/eval.html

```
f(x) = csc(x)/(1 + cot(x))
plot(f(x),-3*pi, pi, ymax=10, ymin= -10,ticks=[[n*pi+pi/4 for n in [-3,3]],[-10,-5,0,5,10]])
```

■

Applications

In this section, we consider applications of the rate of change of y with respect to x, dy/dx, in physics, chemistry, biology, economics, and other sciences. We have mentioned the following, rates of change that are important in physics are velocity, density, current, power (the rate at which work is done), the rate of heat flow, temperature gradient (the rate of change of temperature with respect to position) and the rate of decay of a radioactive substance.

If $y=f(x)$ and x changes from x_1 to x_2, then the change in x is written as $\Delta x = x_2 - x_1$ and the corresponding change in y is $\Delta y = f(x_2) - f(x_1)$.

So, the **average rate of change of y with respect to x** over the interval $[x_1, x_2]$

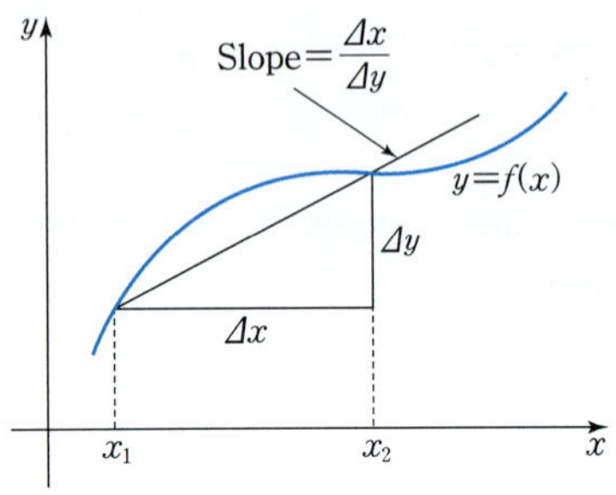

Figure 6

$$\frac{\Delta y}{\Delta x} = \frac{f(x_2) - f(x_1)}{x_2 - x_1}$$

which is the slope of the secant line through the points $(x_1, f(x_1))$ and $(x_2, f(x_2))$. Its limit as $\Delta x \to 0$ is the derivative $f'(x_1)$ which can therefore be interpreted as the instantaneous rate of change of y with respect to x, or the slope of the tangent line at $(x_1, f(x_1))$.

Physics: Let the position function of a particle that is moving in a straight line be denoted $s=f(t)$. Then, the average velocity over a time period is the average rate of change and the instantaneous **velocity** is $v = ds/dt$ (the rate of change of displacement with respect to time).

EXAMPLE 10

The position function of a particle is given by the function $s(t) = 3t^3 - 9t^2 + 6t$, where time t is measured in seconds and position s in meters.

(a) Find the velocity v at time t.

(b) When is the particle moving in the positive direction?

(c) Find the total distance traveled by the particle during the time $[1, 3]$.

(d) Find the total displacement traveled by the particle during the time $[1, 3]$.

Solution (a) The velocity function is the derivative of the position function:

$$v(t) = 9t^2 - 18t + 6 = 3(3t^2 - 6t + 2).$$

(b) When $v(t) > 0$, the particle is moving in the positive direction. Thus,

$$t < 1 - \frac{\sqrt{3}}{3} \text{ or } t > 1 + \frac{\sqrt{3}}{3}.$$

(c) The total distance is

$$\left[s(1)-s\left(1+\frac{\sqrt{3}}{3}\right)\right]+\left[s(3)-s\left(1+\frac{\sqrt{3}}{3}\right)\right]=18.$$

(d) The total displacement is

$$\left[s\left(1+\frac{\sqrt{3}}{3}\right)-s(1)\right]+\left[s(3)-s\left(1+\frac{\sqrt{3}}{3}\right)\right]=18.$$ ■

Biology: Let $n=f(t)$ be the number of individuals in an animal or plant population at time t. The change in the population size between the times t_2 and t_1 is $f(t_2)-f(t_1)$. So, the average rate of growth during the time period is $\dfrac{f(t_2)-f(t_1)}{t_2-t_1}$.

The Instantaneous Rate of Growth is $\dfrac{dn}{dt}$

The instantaneous rate of change is important in many sciences. Consider a population of bacteria in a homogeneous nutrient medium. In the ideal situation, the bacteria's population has an exponential rate of growth. For example, if the bacteria's population is $n(1)=10000$, $n(2)=20000$, then we have $n(3)=40000$.

We note that the derivatives of the demand, revenue, and profit functions are known as marginal demand, marginal revenue, and marginal profit.
In psychology, if $p(t)$ represents the performance of someone learning a skill as a function of the training time t, then the rate at which performance improves as time passes is, dp/dt. In sociology, if $p(t)$ denotes the proportion of a population that knows a rumor (innovations, fads or fashions) by time t, then the derivative dp/dt represents the rate of spread of the rumor.

In other sciences, a meteorologist is concerned with the rate of change of atmospheric pressure with respect to height, an urban geographer is interested in the rate of change of the population density in a city as the distance from the city center increases, a geologist may be interested in knowing the rate at which an intruded body of molten rock cools by conducting heat to surrounding rocks, and an engineer wants to determine the rate at which water flows into or out of a reservoir.

CAS **EXAMPLE 11**

Consider the function

$$f(t)=t(1-t)e^{-\frac{t}{2}}.$$

Find $f'(t)$ and plot the graph of $f(t)$ and $f'(t)$ together. Also plot the tangent at $t=-1$

http://matrix.skku.ac.kr/cal-lab/cal-3-2-Exm12.html

Solution

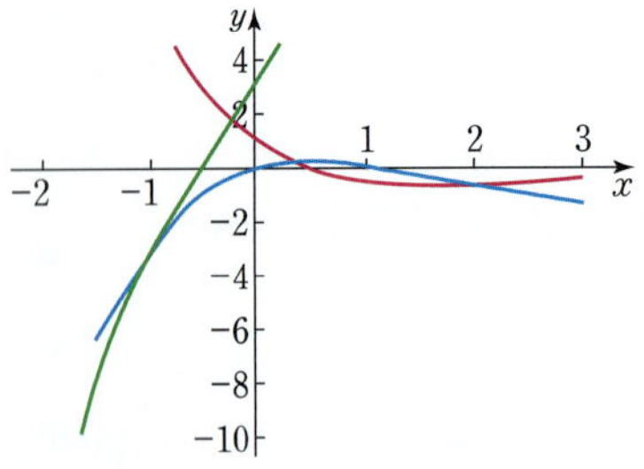

Figure 4

```
var('t')
f(t)=t*(1-t)*exp(-t/2)
f1(t)=f.diff()  # derivative
p1=plot(t*(1-t)*exp(-t/2),-2,3)
p2=plot(f1(t),-2,3,ymin=-10,ymax=5,color='red')
T(t)=f(-1)+f1(-1)*(t+1)  # tangent line at
pt=plot(T(t),-1.5,0.2,color='green')
p1+p2+pt
```

Derivative: $1/2(t-1)te^{-t/2}-(t-1)e^{-t/2}-te^{-t/2}$ ■

<Curve>

3.2 EXERCISES (Derivatives of Functions, The Product and Quotient Rule)

http://matrix.skku.ac.kr/Cal-Book/part1/CS-Sec-3-2-Sol.html

YouTube http://youtu.be/Ei5KGW9vZhE

CAS **1.** Find $f'(x)$ where $f(x) = x^{\frac{15}{14}} + 5e^x$.
http://matrix.skku.ac.kr/cal-lab/cal-3-1-6.html

Solution $f'(x) = \frac{15}{14} x^{\frac{1}{14}} + 5e^x$

```
var('x');
diff(x^(15/14)+5*e^x,x)
```

Answer : 15/14*x^(1/14) + 5*e^x

2-5. [You may have to use a Chain Rule see Section 3.3] Find the derivative $\frac{dy}{dx}$ where y is

2. $e^{\sqrt{x^3+1}}$

Solution $\frac{d}{dx}\left(e^{\sqrt{x^3+1}}\right) = e^{\sqrt{x^3+1}} \frac{d}{dx}\left(\sqrt{x^3+1}\right)$

$= \frac{3x^2}{2\sqrt{x^3+1}} e^{\sqrt{x^3+1}}$

3. $\sqrt{\frac{1+x^2}{1-x^2}}$

Solution $\frac{d}{dx}\left(\sqrt{\frac{1+x^2}{1-x^2}}\right) = \frac{\frac{d}{dx}\left(\frac{x^2+1}{1-x^2}\right)}{2\sqrt{\frac{x^2+1}{1-x^2}}}$

$= \frac{(2x)(1-x^2)-(x^2+1)(-2x)}{2(1-x^2)^2\sqrt{\frac{x^2+1}{1-x^2}}}$

$= \frac{2x}{(1-x^2)^2\sqrt{\frac{x^2+1}{1-x^2}}}$

4. $\frac{1+\cos x}{1-\cos x}$

Solution $\frac{d}{dx}\left(\frac{1+\cos x}{1-\cos x}\right)$

$= \frac{-\sin x(1-\cos x)-(1+\cos x)\sin x}{(1-\cos x)^2}$

$= \frac{-2\sin x}{(1-\cos x)^2}$

5. $(1+\tan e^x)^{2020}$

Solution $\frac{d}{dx}(1+\tan e^x)^{2020}$

$= 2020\,(1+\tan e^x)^{2019} \frac{d}{dx}(1+\tan(e^x))$

$= 2020\,(1+\tan e^x)^{2019}\, e^x \sec^2(e^x)$

$= 2020\, e^x (1+\tan e^x)^{2019} \sec^2 e^x$

CAS **6.** Find the equation of the tangent line to the curve $y = x^2\sqrt{x}$ at $(4, 3)$.
http://matrix.skku.ac.kr/cal-lab/cal-3-1-7.html

Solution $y' = \frac{5}{2}\sqrt{x^3} \Rightarrow y'(4) = 20$.

So the slope of the tangent line is 20.
$y - 32 = 20(x-4) \Rightarrow y = 20x - 48$.
(Since y passes through $(4, 32)$)

```
f(x)=x^2*sqrt(x)
df(x)=diff(f(x),x)
y(x)=df(4)*(x-4)+32
y(x)
```

Answer : 20*x-48

```
p1=plot(f(x),x,0,10, color='blue')
p2=plot(y(x),x,0,10, color='red')
show(p1+p2,ymax=50,ymin=-10)
```

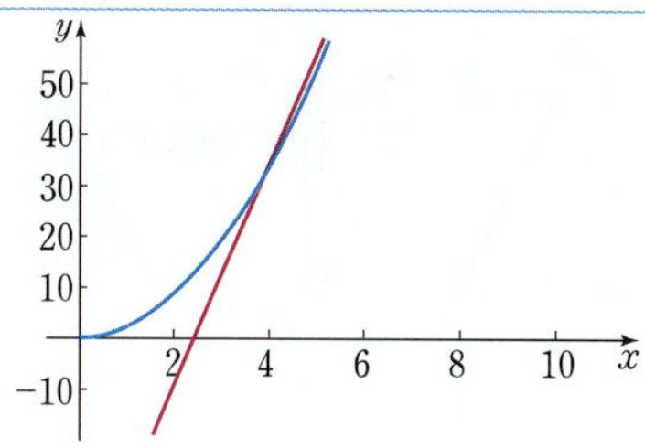

CAS **7.** The normal line to a curve C at a point P is the line that passes through P and is perpendicular to the tangent line to C at P. Find an equation of the normal line to the curve $y = 1 + e^x$ at the point $(0, 2)$.
http://matrix.skku.ac.kr/cal-lab/cal-3-1-8.html

Solution $y = 1 + e^x \Rightarrow y' = e^x$

So the slope of the tangent line is 1. (Since $y'(0) = 1$)
Then the slope of the normal line is -1.
Thus, $y - 2 = -x \Rightarrow y = -x + 2$.
Therefore, the normal line is $y = -x + 2$.

```
f(x)=1+e^x
df(x)=diff(f(x),x)
```

```
y(x)=-1/df(0)*x+2
y(x)
```

Answer : -x+2

```
p1=plot(f(x),x,-5,5, color='blue');
p2=plot(y(x),x,-5,5, color='red');
show(p1+p2,ymax=5,ymin=-5)
```

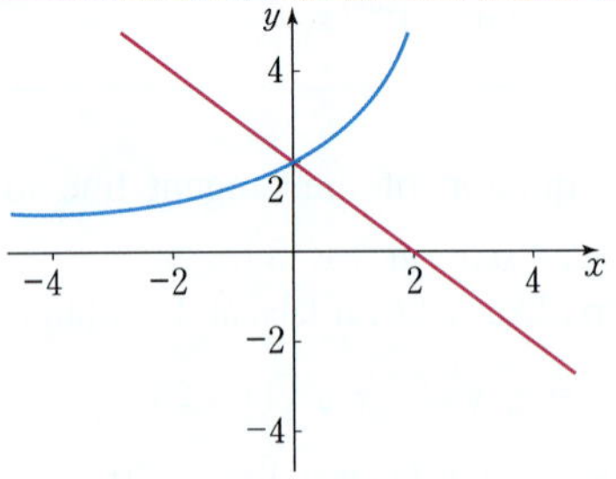

8. Where is the function $y=|x-3|+|x+4|$ differentiable? Give a formula for y'.

Solution $y=\begin{cases} -2x-1 & \text{if} \quad x \le -4 \\ 7 & \text{if} \ -4 < x \le 3 \\ 2x+1 & \text{if} \quad 3 < x \end{cases}$,

then $y(-4)=y(3)=7$.

So y is continuous on $(-\infty, +\infty)$.

$y'=\begin{cases} -2 & \text{if} \quad x \le -4 \\ 0 & \text{if} \ -4 < x \le 3 \\ 2 & \text{if} \quad 3 < x \end{cases}$,

then y is not differentiable at $x=-4,\ 3$.

Therefore, y is differentiable on $(-\infty, -4) \cup (-4, 3) \cup (3, +\infty)$.

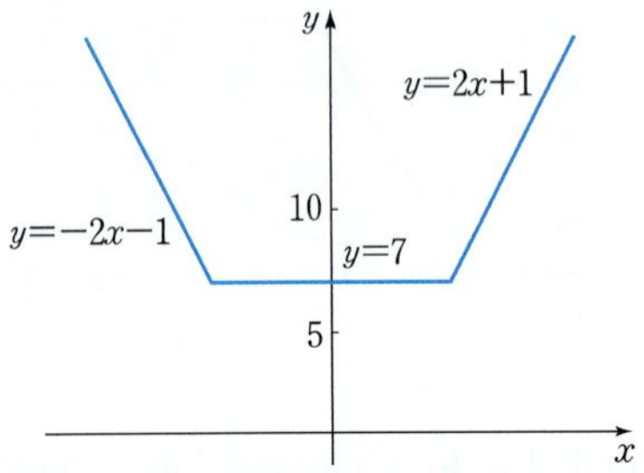

9. Let $f(t)=\begin{cases} 2x^2, & x \le 2 \\ mx+b, & x > 2 \end{cases}$. Find the values of m and b that make f differentiable everywhere.

Solution Note that f is differentiable everywhere except $x=2$. We must check at $x=2$. For f to be differentiable at 2,

$f'(x)=\begin{cases} 4x, & x \le 2 \\ m, & x > 2 \end{cases}$, so $f'(2)=8=m$. And also f have to be continuous at 2. $f(2)=\begin{cases} 8 & x \le 2 \\ 2m+b & x > 2 \end{cases}$,

since $m=8,\ 2m+b=8 \Rightarrow b=-8$.

Therefore, $m=8,\ b=-8$.

CAS **10.** Evaluate $\lim_{x \to 1} \dfrac{x^{2020}-1}{x-1}$.

http://matrix.skku.ac.kr/cal-lab/cal-3-1-11.html

Solution Method (1) Let $f(x)=x^{2020}$, and $f'(x)=2020x^{2019}$.

Then by the definition of derivative,

$$f'(1)=\lim_{x \to 1}\frac{f(x)-f(1)}{x-1}=\lim_{x \to 1}\frac{x^{2020}-1}{x-1}=2020.$$

Method (2) Note that

$(x^{2020}-1)=(x-1)\ (x^{2019}+x^{2018}+\cdots+x^2+x+1)$.

So $\lim_{x \to 1}\dfrac{x^{2020}-1}{x-1}$

$$=\lim_{x \to 1}\frac{(x-1)(x^{2019}+x^{2018}+\cdots+x^2+x+1)}{x-1}$$

$$=\lim_{x \to 1}(x^{2019}+x^{2018}+\cdots+x^2+x+1)$$

$$=\underbrace{1+1+1+\cdots+1+1+1}_{2020 \text{ one's}}=2020$$

```
lim((x^2020-1)/(x-1),x=1)
```

Answer : 2020

11-12. Differentiate the following functions.

11. $\dfrac{1+e^x}{1-e^x}$

Solution $\dfrac{d}{dx}\left(\dfrac{1+e^x}{1-e^x}\right)$

$$=\frac{\left[\dfrac{d}{dx}(1+e^x)\right](1-e^x)-(1+e^x)\left[\dfrac{d}{dx}(1-e^x)\right]}{(1-e^x)^2}$$

$$=\frac{e^x(1-e^x)-(1+e^x)(-e^x)}{(1-e^x)^2}$$

$$=\frac{e^x-e^{2x}+e^x+e^{2x}}{(1-e^x)^2}=\frac{2e^x}{(1-e^x)^2}$$

12. $(t+e^{2t})(1-\sqrt{t})$

Solution $\dfrac{d}{dt}(t+e^{2t})(1-\sqrt{t})$

$$=(1-\sqrt{t})\left(\frac{d}{dt}(t+e^{2t})\right)+(t+e^{2t})\left(\frac{d}{dt}(1-\sqrt{t})\right)$$

$$=(1-\sqrt{t})(1+2e^{2t})+(t+e^{2t})\left(-\frac{1}{2\sqrt{t}}\right)$$

$$=1+2e^{2t}-\sqrt{t}-2\sqrt{t}\,e^{2t}-\frac{\sqrt{t}}{2}-\frac{e^{2t}}{2\sqrt{t}}$$

$$=e^{2t}\left(2-2\sqrt{t}-\frac{1}{2\sqrt{t}}\right)-\frac{3\sqrt{t}}{2}+1$$

13. Show that if $f(x)$ is differentiable and satisfies the identity $f(x+y)=f(x)f(y)$ for all x and y and $f'(0)=1$, then $f(x)$ satisfies $f'(x)=f(x)$ for all x.

Solution $f(x+y)=f(x)f(y) \Rightarrow f(0)=f(0)f(0)$

Thus, either $f(0)=0$ or $f(0)=1$.

Since $f(x+0)=f(x)f(0)$ for all x and $f'(0)=1$, $f(0)=1$. Now,

$\Rightarrow f(x+y)-f(x)=f(x)(f(y)-1)$

$\Rightarrow \dfrac{f(x+y)-f(x)}{y}=\dfrac{f(x)(f(y)-1)}{y}$

$\Rightarrow \lim\limits_{y\to0}\dfrac{f(x+y)-f(x)}{y}=\lim\limits_{y\to0}\dfrac{f(x)(f(y)-1)}{y}$

$\Rightarrow f'(x)=\lim\limits_{y\to0}\dfrac{f(x)(f(y)-1)}{y}$

Since $f(0)=1$, then

$\lim\limits_{y\to0}\dfrac{f(y)-1}{y}=\lim\limits_{y\to0}\dfrac{f(y)-f(0)}{y}=f'(0)=1.$

Therefore, $f'(x)=f(x)$.

14-16. Find derivatives of the following functions.

14. $f(x)=\csc^2 x\cot x$

Solution $f'(x)=\left(\dfrac{d}{dx}\csc^2 x\right)\cot x+\csc^2 x\left(\dfrac{d}{dx}\cot x\right)$

$=(2\csc x(-\cot x\csc x)\cot x+\csc^2 x(-\csc^2 x)$

$=-2\csc^2 x\cot^2 x-\csc^4 x$

```
f(x)= (csc(x))^2  * cot(x)
df(x)=diff(f(x),x)
print df(x)
```

Answer : -2*csc(x)^2*cot(x)^2-csc(x)^4

15. $f(x)=\dfrac{\cos x\cot x}{x^4}$

Solution

$f'(x)=\dfrac{-(\sin x\cot x+\cos x(-\csc^2 x))x^4-4x^3(\cos x\cot x)}{x^8}$

$=\dfrac{-\cos x(1+\csc^2 x)}{x^4}-\dfrac{4\cos x\cot x}{x^5}$

16. $\phi(x)=\dfrac{e^x\cos x}{x^{10}+1}$

Solution $\phi'(x)=\dfrac{(e^x\cos x-e^x\sin x)(x^{10}+1)-(e^x\cos x)(10x^9)}{(x^{10}+1)^2}$

$=\dfrac{e^x((x^{10}-10x^9+1)\cos x-(x^{10}+1)\sin x)}{(x^{10}+1)^2}$

---

17. Show that the curve $y=4x^3+3x+2\sin x$ has no tangent line with slope 0.

Solution $y=4x^3+3x+2\sin x \Rightarrow y'=12x^2+3+2\cos x$. Since $12x^2+3+2\cos x$ is always positive, there is no x such that $y'(x)=0$.

So $y=4x^3+3x+2\sin x$ has no tangent line with slope 0.

18. A stone is thrown into a pond, creating a wave whose radius increases at the rate of 0.5 meter per second. In square meters per second, how fast is the area of the circular ripple increasing 4 seconds after the stone hits the water?

Solution Let radius $=r$, time $=t$. $r(t)=0.5t$. $\dfrac{dr}{dt}=0.5$ when $t=4$, $r=2$. Area $=A(t)=\pi r^2$, $A(t)=\pi(0.5t)^2$ $=0.25\pi t^2$, $A'(t)=0.5\pi t$. Therefore, $A'(4)=2\pi$.

19. A particle moves along a straight line with equation of motion $s(t)=4t^5-16t^2+7$.

(a) When is the particle moving forward?

(b) When is (the acceleration) $a(t)=\dfrac{d}{dt}v(t)$ zero?

(c) When is the particle speeding up? Slowing down?

Solution (a) $v(t)=\dfrac{d}{dt}s(t)=\dfrac{d}{dt}(4t^5-16t^2+7)$

$=20t^4-32t$, $v(t)\le 0$, when $0\le t\le \dfrac{2}{\sqrt[3]{5}}$.

When $t>\dfrac{2}{\sqrt[3]{5}}$, the particle moves forward.

(b) $a(t)=\dfrac{d}{dt}v(t)=\dfrac{d}{dt}(20t^4-32t)=80t^3-32$.

Therefore the acceleration $a(t)$ is 0, when $t=\sqrt[3]{\dfrac{2}{5}}$.

(c) When the particle is speeding up, the acceleration is positive and when it is slowing down, the acceleration is negative. $a(t)>0$ when $t>\sqrt[3]{\dfrac{2}{5}}$ and $a(t)<0$ when $t<\sqrt[3]{\dfrac{2}{5}}$. Therefore the particle speeds up when $t>\sqrt[3]{\dfrac{2}{5}}$, and the particle slows down when $t<\sqrt[3]{\dfrac{2}{5}}$.

20. A particle moves in a straight line with equation of motion $s(t)=e^t(10-t)-10$, where t is measured in seconds and s in meters.

(a) What is the position of the particle at $t=0$, $t=2$, and $t=9$?

(b) Find the velocity of the particle at time t.

(c) When is the particle moving forward?

(d) Find the total distance traveled by particle on the time interval $[0, 10]$.

(e) Find the acceleration of the particle at time t.

Solution (a) $s(0)=0,\ s(2)=8e^2-10,\ s(9)=e^9-10$

(b) $\frac{d}{dt}s(t)=v(t)=e^t(9-t)$

(c) When $v(t)>0,\ 0<v(t)=e^t(9-t)$.
Therefore, $t\in[0,\ 9)$.

(d) $|s(9)-s(0)|+|s(10)-s(9)|$
$=e^9-10+|-10-(e^9-10)|=2e^9-10$

(e) $a(t)=\frac{d}{dt}v(t)=e^t(8-t)$

21. The population of a bacteria colony after t hours is $50+10t+4t^2$. Find the growth rate when $t=2$.

Solution $y=50+10t+4t^2$.
$y'=10+8t.\ y'(2)=26.$

22. A cost function is given by

$$C(x)=10{,}000\ +50x+2x^{3/2}.$$

(a) Find the marginal cost function.

(b) Find $C'(10)$.

Solution (a) $\frac{d}{dx}C(x)=50+3x^{1/2}$

(b) $C'(x)=50+3\sqrt{x}\ \Rightarrow C'(10)=50+3\sqrt{10}$

23. If a stone is thrown vertically upward with a velocity $10\,\mathrm{m/s}$, then its height after t seconds is $s(t)=10t-4.9t^2$.

(a) What is the maximum height reached by the stone?

(b) What is the velocity of the stone when it is $5\,m$ above the ground on its way up? On its way down?

Solution (a) $\frac{d}{dt}s(t)=v(t)=10-9.8t$

$0=v(t_0)=10-9.8t_0\ \Rightarrow\ t_0=\frac{10}{9.8}$

$\Rightarrow\ s(t_0)=10t_0-4.9t_0^2$ and

$s\left(\frac{10}{9.8}\right)=\frac{100}{19.6}(\mathrm{m})=\frac{250}{49}(\mathrm{m})\fallingdotseq 5.1020\mathrm{m}$

(b) $5=s(t_1)=10t_1-4.9t_1^2$

$\Rightarrow 0=4.9t_1^2-10t_1+5$

$\Rightarrow t_1=\frac{5\pm\sqrt{(-5)^2-(4.9)(5)}}{4.9}$

$\Rightarrow t_1=\frac{5\pm\sqrt{0.5}}{4.9}$

The time is $t_1=-5\pm\sqrt{0.5}$ when the height of the stone is $5\mathrm{m}$.

$v(t_1)=10-9.8t_1=v\left(\frac{5\pm\sqrt{0.5}}{4.9}\right)=\pm\sqrt{2}\,(\mathrm{m/s})$

The velocity of the stone is $\sqrt{2}\,\mathrm{m/s}$ on its way up. The velocity of the stone is $-\sqrt{2}\,\mathrm{m/s}$ on its way down.

24. $P(x)$ is the total value of the production when there are x workers in a plant, then the average productivity is

$$A(x)=\frac{P(x)}{x}.$$

Find A'. Explain why the company wants to hire more workers if $A'(x)>0$?

Solution $A'(x)=\frac{P'(x)x-P(x)}{x^2}$

Since $x^2>0,\ A'(x)>0$
implies $P'(x)x-P(x)>0$.
This means $P'(x)>\frac{P(x)}{x}=A(x)$.
This means the rate of productivity $P'(x)$ is larger than the average productivity $A(x)=\frac{P(x)}{x}$.
Therefore, if the company hires more workers, then they can expect to have a better productivity.

25. Let $p(t)=\frac{5000(e^{0.3t}+1)}{e^{0.3t}+10}$ be the population of a bacteria colony at time t hours. Find the growth rate of the bacteria after 10 hours.

Solution

$\frac{d}{dt}p(t)=\frac{1500e^{0.3t}(10+e^{0.3t})-5000(e^{0.3t}+1)(0.3e^{0.3t})}{(10+e^{0.3t})^2}$

$=\frac{13500e^{0.3t}}{(10+e^{0.3t})^2}$

26. The angular displacement θ of a simple pendulum is given by $\theta(t)=\theta_0\sin(\omega t+\phi)$ with an angular amplitude θ_0, an angular frequency ω and a phase constant ϕ. Find $\frac{d\theta}{dt}$.

Solution $\frac{d\theta}{dt}=[\theta_0\cos(\omega t+\phi)](\omega t+\phi)'=\theta_0\omega\cos(\omega t+\phi)$

27. Show that $\lim_{n\to\infty}\left(1+\frac{x}{n}\right)^n=e^x$.

Solution We note $\lim_{m\to\infty}\left(1+\frac{1}{m}\right)^m=e$. Let $m=\frac{n}{x}$.
Then $n=xm$, and as $n\to\infty,\ m\to\infty$.

Therefore, $\lim_{n\to\infty}\left(1+\frac{x}{n}\right)^n=\lim_{m\to\infty}\left(1+\frac{1}{m}\right)^{mx}$
$=\left[\lim_{m\to\infty}\left(1+\frac{1}{m}\right)^m\right]^x=e^x.$

3.3 The Chain Rule and Inverse Functions

YouTube http://youtu.be/HfScHEsPfKI

In this section, we consider the derivative of the composition of two functions, derivatives of implicitly defined functions, higher order derivatives, the inverse trigonometric functions, logarithmic functions and the hyperbolic functions.

THEOREM 1 The Chain Rule

If f and g are both differentiable and $F = f \circ g$ is the composite function defined by $F(x) = f(g(x))$, then $F(x)$ is differentiable and

$$F'(x) = f'(g(x))g'(x).$$

Differentiate the outer function f with respect to the inner function $g(x)$ and then multiply by the derivative of the inner function.

If $y = f(u)$ and $u = g(x)$ are both differentiable functions, then the rate of change of y with respect to x is $\dfrac{dy}{dx} = \left(\dfrac{dy}{du}\right)\left(\dfrac{du}{dx}\right)$ $=$ *(the rate of change of y with respect to u) (rate of change of u with respect to x).*

EXAMPLE 1

Find $F'(x)$ if $F(x) = \sqrt{x^2 + e^x}$.

Solution $F(x) = (x^2 + e^x)^{\frac{1}{2}}$

$$F'(x) = \frac{1}{2}(x^2 + e^x)^{-\frac{1}{2}}\frac{d}{dx}(x^2 + e^x) = \frac{2x + e^x}{2\sqrt{x^2 + e^x}}.$$

Suppose the outer function f is a power function. If $y = [g(x)]^n$, then we can write $y = f(u) = u^n$, where $u = g(x)$. By using the Chain Rule and the Power Rule, we get

$$\frac{dy}{dx} = n[g(x)]^{n-1} \cdot g'(x).$$

■

THEOREM 2 The Power Rule Combined with the Chain Rule

Let n be any real number and $g(x)$ be differentiable. If $y(x) = [g(x)]^n$, then $y'(x) = n[g(x)]^{n-1} \cdot g'(x)$.

EXAMPLE 2

Find y' if $y(x) = (x^4 - 1)^9$.

Solution $y'(x) = 9(x^4 - 1)^8\, 4x^3 = 36x^3(x^4 - 1)^8$. ■

EXAMPLE 3

Find the derivative of the function $g(t) = \left(\dfrac{5t-3}{t+2}\right)^7$.

Solution Combining the Power Rule, Chain Rule, and Quotient Rule, we get

$$g'(t) = 7\left(\frac{5t-3}{t+2}\right)^6 \frac{d}{dt}\left(\frac{5t-3}{t+2}\right) = \frac{91(5t-3)^6}{(t+2)^8}.$$ ■

EXAMPLE 4

Find the derivative of the function $e^{\csc x}$ $\left(\text{using } \dfrac{d}{dx}(e^u) = e^u \dfrac{du}{dx}\right)$.

Solution $(e^{\csc x})' = e^{\csc x}\dfrac{d}{dx}\csc x = -\csc x \cot x\, e^{\csc x}$. ■

Suppose that $y = f(u)$, $u = g(x)$, and $x = h(t)$, where f, g, and h are differentiable functions. Then, to compute the derivative of y with respect to t, we use the Chain Rule twice:

$$\frac{dy}{dt} = \frac{dy}{du}\frac{du}{dx}\frac{dx}{dt}$$

EXAMPLE 5

Find $f'(x)$ if $f(x) = \csc(\cos(\tan x))$.

Solution Apply the chain rule twice,

$$f'(x) = \csc(\cos(\tan x))\cot(\cos(\tan x))\sin(\tan x)\sec^2 x.$$ ■

EXAMPLE 6

Differentiate $y = e^{\sec^2\theta}$.

Solution $\dfrac{dy}{d\theta} = e^{\sec^2\theta}\dfrac{d}{d\theta}\sec^2\theta = 2\sec^2\theta\, \tan\theta\, e^{\sec^2\theta}$. ■

Tangents to Parametric Curves

Now, consider the particle which is moving along the curve in the plane. Then, the trajectory of the particle can be described by the parametric equations

$$x = f(t), \quad y = g(t).$$

The chain rule allows as to find $\frac{dy}{dx}$, in particular, to find equations of the tangent lines to the parametric curves.

Since $\frac{dy}{dt} = \frac{dy}{dx} \cdot \frac{dx}{dt}$, we have

$$\frac{dy}{dx} = \frac{\frac{dy}{dt}}{\frac{dx}{dt}} \quad \text{when} \quad \frac{dx}{dt} \neq 0.$$

Remark When $\frac{dx}{dt} = 0$ and $\frac{dy}{dt} \neq 0$, we have a vertical tangent line. When $\frac{dx}{dt} \neq 0$ and $\frac{dy}{dt} \neq 0$, we have a horizontal tangent line.

EXAMPLE 7

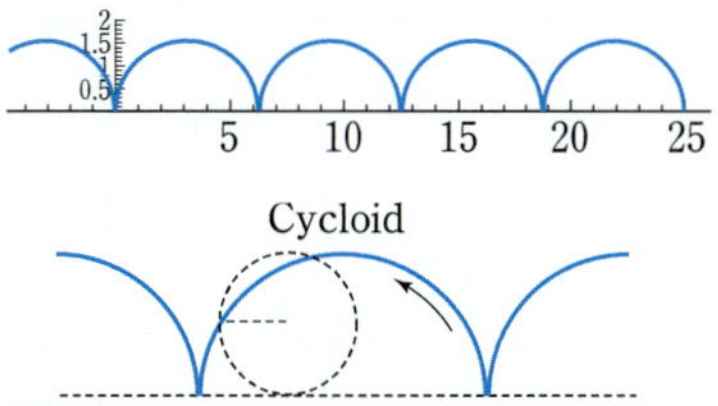

Figure 1

Consider the cycloid

$$x = t - \sin t, \ y = 1 - \cos t.$$

Find the equation of the tangent line which is tangent at $\left(\frac{\pi}{4} - \frac{\sqrt{2}}{2},\ 1 - \frac{\sqrt{2}}{2}\right)$.

Solution At $t = \frac{\pi}{4}$, the point is $\left(\frac{\pi}{4} - \frac{\sqrt{2}}{2},\ 1 - \frac{\sqrt{2}}{2}\right)$.

From the above formula, we have the slope of the tangent line

$$\frac{dy}{dx} = \frac{\sin t}{1 - \cos t}.$$

The slope at $t = \frac{\pi}{4}$ is $\sqrt{2} + 1$. The equation of the tangent is

$$y - 1 + \frac{\sqrt{2}}{2} = (\sqrt{2} + 1)\left(x - \frac{\pi}{4} + \frac{\sqrt{2}}{2}\right)$$

or

$$y = (\sqrt{2} + 1)x + 2 - \frac{\sqrt{2} + 1}{4}\pi.$$

■

CAS **EXAMPLE 8**

Find an equation of the tangent line to the curve at the given point.

$$\begin{cases} x = \sin t \\ y = \cos 2t \end{cases}, \quad t = \frac{\pi}{4}$$

http://matrix.skku.ac.kr/cal-lab/cal-3-3-Exm-8.html

Solution

```
var('t, x, y, u')
x=sin(t)
y=cos(2*t)
dxdt=diff(x,t)
dydt=diff(y,t)
dydx=dydt/dxdt
yu=y(pi/4) + dydx(pi/4)*(u-x(pi/4))
p1=parametric_plot((x,y),(t,0,pi/2), color='blue');
p2=plot(yu, u, 0, 1.5, color='red');
show(p1+p2)
```

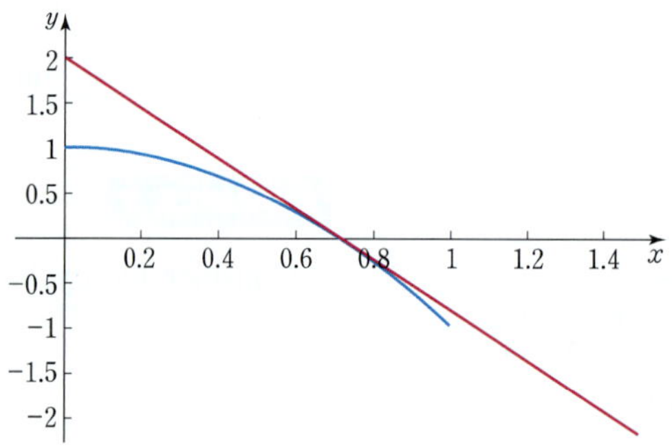

Figure 2

■

Higher Order Derivatives

The **second order derivative** of f, denoted by f'', $\dfrac{d^2y}{dx^2}$ or $D^2f(x)$ is the derivative of the derivative of f. Thus, $f'' = \dfrac{d^2f}{dx^2} = \dfrac{d}{dx}\left(\dfrac{df}{dx}\right)$.

If $s = s(t)$ is the position function of an object that moves in a straight line, then its first derivative represents the velocity $v(t)$ of the object as a function of time:

$$v(t) = s'(t) = \frac{ds}{dt}.$$

The instantaneous rate of change of velocity with respect to time is called the **acceleration** $a(t)$ of the object. Thus, the acceleration function is the derivative of the velocity function and is therefore the second derivative of the position function:

$$a(t) = v'(t) = s''(t)$$

or

$$a = \frac{dv}{dt} = \frac{d^2s}{dt^2}.$$

In general, **a second derivative is the rate of change of a rate of change.**

The **third order derivative** is the derivative of the second order derivative: So f''' can be interpreted as the slope of the curve $y = f''(x)$ or as the rate of change of $f''(x)$.

Similarly the fourth order derivative is usually denoted by $f^{(4)}$. In general, the nth order derivative of f is denoted by $f^{(n)}$ and is obtained from f by differentiating n times. If $y = f(x)$, we write

$$y^{(n)} = f^{(n)}(x) = \frac{d^n y}{dx^n} = D^n f(x).$$

EXAMPLE 9

If $f(x) = \sin x$, find $f^{(n)}(x)$.

> Lagrange's notation: $f''(x)$
> Euler's notation : $D^2 f(x)$

Solution We have $f(x) = \sin x$, $f'(x) = \cos x = \sin\left(x + \frac{\pi}{2}\right)$,

$f''(x) = -\sin x = \sin\left(x + \frac{2}{2}\pi\right)$,

$f'''(x) = -\cos x = \sin\left(x + \frac{3}{2}\pi\right)$, $\cdots$ repeatedly.

Thus using induction on n, we obtain one formula,

$$f^{(n)}(x) = \sin\left(x + \frac{n\pi}{2}\right).$$ ■

EXAMPLE 10

The position of a particle is given by the equation $s(t) = e^{-\alpha t}\sin(\omega t + \beta)$ $(\alpha > 0)$, where t is measured in seconds and s in meters.

(a) Find the velocity and acceleration at time t.

(b) Draw the graph of $s(t)$.

Solution (a) The velocity function is the derivative of the position function,

$$v(t) = e^{-\alpha t}(-\alpha\sin(\omega t + \beta) + \omega\cos(\omega t + \beta)).$$

The acceleration is the derivative of the velocity function.

$$a(t) = e^{-\alpha t}((\alpha^2 - \omega^2)\sin(\omega t + \beta) - 2\alpha\omega\cos(\omega t + \beta)).$$

(b)

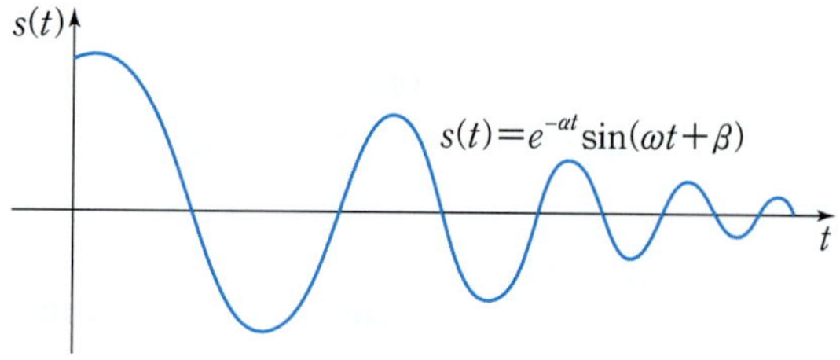

Figure 3 Velocity

■

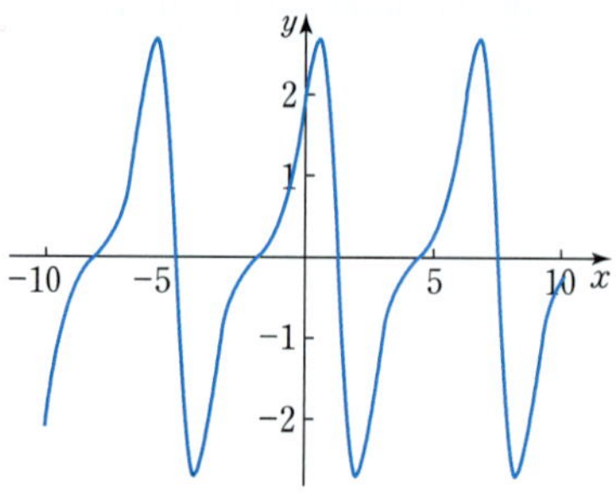

Figure 4 Acceleration

Leibniz's Rule

When Gottfried Wilhelm von Leibniz worked out his version of the calculus he derived the rules governing the process of differentiation (and introduced the notation that we use today). Among those rules we find a theorem concerning multiple differentiations of a product of two functions.

If $f(x)$ and $g(x)$ are (smooth, differentiable) functions of x on I, we have

$$\begin{aligned}(f(x)g(x))^{(n)} &= \sum_{k=0}^{n} {}_nC_k f^{(n-k)} g^{(k)}(x) \\ &= {}_nC_0 f^{(n)}(x)g(x) + {}_nC_1 f^{(n-1)}(x)g'(x) + \cdots + {}_nC_n f(x)g^{(n)}(x)\end{aligned}$$

where $f^{(0)}(x) = f(x)$, $g^{(0)}(x) = g(x)$, ${}_nC_k = \dfrac{n!}{k!(n-k)!}$.

EXAMPLE 11

Find the nth derivative of x^2e^x.

Solution
$$\begin{aligned}(x^2e^x)^{(n)} &= \sum_{k=0}^{n} {}_nC_k (x^2)^{(k)}(e^x)^{(n-k)} \\ &= {}_nC_0\, x^2(e^x)^{(n)} + {}_nC_1 (x^2)'(e^x)^{(n-1)} \\ &\quad + {}_nC_2 (x^2)''(e^x)^{(n-2)} + \cdots \\ &= x^2e^x + 2nxe^x + \frac{n(n-1)}{2} \cdot 2e^x \\ &= \{x^2 + 2nx + n(n-1)\}e^x\end{aligned}$$

■

CAS **EXAMPLE 12**

Find $f^{(n)}(x)$ of the function $f(x) = xe^x$.

Solution
$$\begin{aligned}f'(x) &= x'e^x + x(e^x)' = e^x + xe^x = (x+1)e^x \\ f''(x) &= (x+1)'e^x + (x+1)(e^x)' \\ &= e^x + (x+1)e^x = (x+2)e^x \\ f'''(x) &= (x+2)'e^x + (x+2)(e^x)' \\ &= e^x + (x+2)e^x = (x+3)e^x\end{aligned}$$

Similarly,

$$f^{(n)}(x) = (x+n)e^x .$$ ■

Implicit Differentiation

So far we are dealing with explicit functions, that is, with functions of the form $y = f(x)$. We have noted that this function has the property that each value of x is associated with one and only one value y. But in practice, sometimes the relation between two variables x and y is expressed in a more general form case, we can regard y as an implicit function of x, that is, regard x as independent variable and y as the dependent variable. Of course, we can also regard y as the independent variable and x as a function of y. Whether one determines x or y as independent variable depends either on the problem or on the concrete application.

Note that, while explicit functions are a particular case of implicit functions, since equation $y = f(x)$ can be written as $y - f(x) = 0$ (So it has form $g(x, y) = 0$, with $g(x, y) = y - f(x)$), not every implicit function can be written as explicit function, that is, it is not always possible to solve the equation $g(x, y) = 0$ with respect to y in order to write $y = f(x)$. Also, note that, for implicit functions $g(x, y) = 0$, one value of x can be associated with more than one value of y.

EXAMPLE 13

The equation $x^5 + y^5 = 1$ shows y is as an implicit function of x, but it is not an explicit function, since it is impossible to find an expression for y in terms of x (that is, it is impossible to solve this equation for y). ■

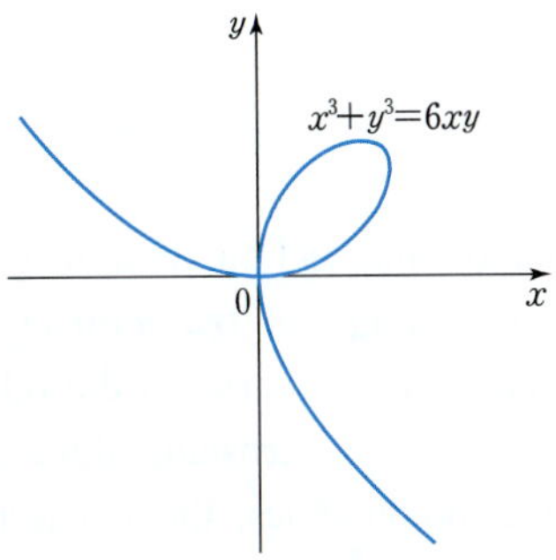

Figure 5 The folium of Descartes

EXAMPLE 14

The equation $x^3 + y^3 = 6xy$ is another example of an implicit function. This curve is called the **folium of Descartes** shown in Figure 5. ■

In order to find the derivative of y it is not necessary to solve an equation for y in terms of x. Instead we may use the method of implicit differentiation which consists of differentiating both sides of the equation with respect to x and then solving the resulting equation for y'. Assume that the given equation determines y implicitly as a differentiable function of x so that the method of implicit differentiation may be applied.

EXAMPLE 15

If $2x^3 + 3x^2y - y^3 = 1$, find $\dfrac{dy}{dx}$.

Solution Differentiate both sides with respect to x:

$$6x^2+6xy+3x^2\frac{dy}{dx}-3y^2\frac{dy}{dx}=0$$

Solving for $\frac{dy}{dx}$, we obtain $\frac{dy}{dx}=\frac{2x^2+2xy}{y^2-x^2}$. ■

EXAMPLE 16

Given $x^3+y^3=6xy$, find the tangent line at the point $(3,\ 3)$. Also find $\frac{d^2y}{dx^2}$ at $(3,\ 3)$.

Solution Differentiate both sides with respect to x:

$$3x^2+3y^2\frac{dy}{dx}=6x\frac{dy}{dx}+6y$$

Solving for $\frac{dy}{dx}$, we obtain $\frac{dy}{dx}=\frac{2y-x^2}{y^2-2x}$.
At $(3,\ 3)$, the slope of the tangent line is -1.
Thus the equation of the tangent line is $y=-x+6$.

$$\frac{d^2y}{d^2x}=\frac{d}{dx}\left(\frac{2y-x^2}{y^2-2x}\right)$$
$$=\frac{\left(2\frac{dy}{dx}-2x\right)(y^2-2x)-(2y-x^2)\left(2y\frac{dy}{dx}-2\right)}{(y^2-2x)^2}$$

Thus, $\left.\frac{d^2y}{dx}\right|_{(3,3)}=\frac{(-8)3-(-3)(-8)}{3^2}=-\frac{16}{3}$. ■

Orthogonal Trajectories

Two curves are called **orthogonal** if at each point of intersection their tangent lines are perpendicular. Two families of curves are **orthogonal trajectories** of each other if every curve in one family is orthogonal to every curve in the other family. For example in physics, the lines of force in an electrostatic field are orthogonal to the lines of constant potential. In thermodynamics, the isotherms (curves of equal temperature) are orthogonal to the flow lines of heat. In aerodynamics, the streamlines (curves of direction of airflow) are orthogonal trajectories of the velocity-equipotential curves.

EXAMPLE 17

Show that the family of curves $x=ky^2$, (k is an arbitrary constant) and $x^2+\frac{y^2}{2}=C$ (C is an arbitrary constant) are orthogonal **trajectories**.

Solution Differentiating $x = ky^2$ gives $1 = 2ky\dfrac{dy}{dx}$

$\Rightarrow y = 2ky^2\dfrac{dy}{dx} \Rightarrow y = 2x\dfrac{dy}{dx} \Rightarrow \dfrac{dy}{dx} = \dfrac{y}{2x}$.

Differentiating $x^2 + \dfrac{y^2}{2} = C$ gives $2x + y\dfrac{dy}{dx} = 0 \Rightarrow \dfrac{dy}{dx} = -\dfrac{2x}{y}$.

So at any of the intersection points, the slopes of the tangents are negative reciprocals of each other. Therefore, the two families are orthogonal families.

```
p=circle((0,0), 0.1)
for i in srange(0.2, 2, 0.2):
  p=p+circle((0, 0), i)
q=plot(x, 0, 2, color='red')
r=plot(-x, -2, 0)
for j in srange(-4, 4, 0.4):
  q=q+plot(j*x, 0, 2, color='red')
  r=r+plot(-j*x, -2, 0, color='red')
(p+q+r).show(ymin=-2, ymax=2)
```

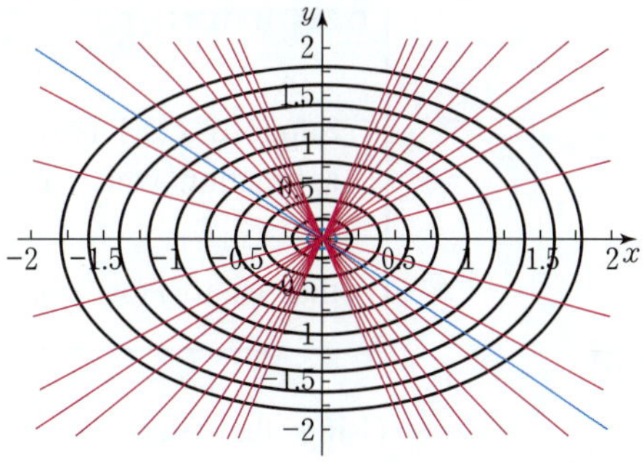

Figure 6

■

Note http://math1.skku.ac.kr/home/pub/1001

```
var('x,y,k')
p=plot(x^2, -2, 2)
for i in srange(-4, 4, 0.4):
p=p+plot(i*x^2, -2, 2)
q=implicit_plot(x^2+2*y^2==1, (x,-2,2), (y,-2,2), color='red')
for j in srange(-4, 4, 0.4):
q=q+implicit_plot(x^2+2*y^2==j,(x, -2, 2), (y, -2, 2),
color='red')
p+q
```

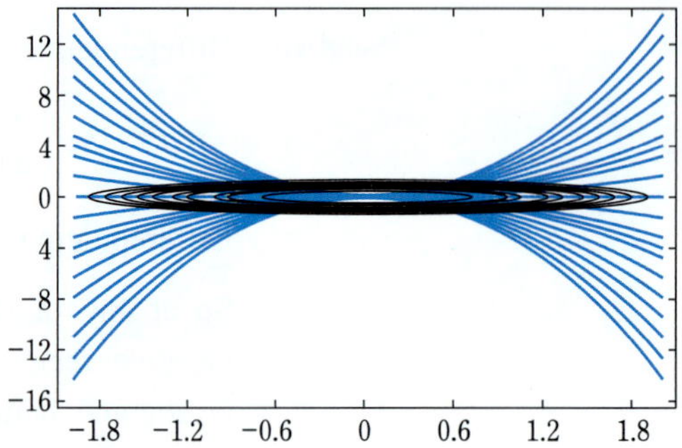

Figure 7

Derivatives of Inverse Functions

DEFINITION 3

A function $f(x)$ with domain D is said to be one-to-one, if $x \neq y$ $(x, y \in D)$ then $f(x) \neq f(y)$.

DEFINITION 4

Suppose $f(x)$ is one-to-one function on D. Then its inverse, $f^{-1}(x)$, is a function which is defined on the range R of $f(x)$ by

$$f^{-1}(f(x)) = x,$$

that is, if $f(x) = y$, then $f^{-1}(y) = x$.

Using the Chain Rule, we can find the derivative of the inverse function. Namely, from

$$f^{-1}(f(x)) = x$$

it follows

$$f^{-1}(f(x)) \cdot [f^{-1}(x)]' = 1$$

so that

$$[f^{-1}(x)]' = \frac{1}{f^{-1}(f(x))}.$$

Derivatives of Inverse Trigonometric Functions[3]

3)http://www.youtube.com/watch?v=osUEjDZ2fg0&feature=relmfu

The derivatives of the inverse trigonometric functions are found using implicit differentiation. If f is any one-to-one differentiable function, it can be proved that its inverse function f^{-1} is also differentiable, except where its tangents are vertical.

The definition of the inverse sine (arcsine) function:

$$\text{For } x \in [-1, 1],\ \sin^{-1}x = y \text{ if } \sin y = x \text{ and } -\frac{\pi}{2} \leq y \leq \frac{\pi}{2}.$$

(Implicit Differentiation) Differentiating $\sin y = x$ implicitly with respect to x, we obtain

$$\cos y \frac{dy}{dx} = 1 \Rightarrow \frac{dy}{dx} = \frac{1}{\cos y}.$$

Now $\cos y > 0$, since $-\frac{\pi}{2} < y < \frac{\pi}{2}$, so

$$\cos y = \sqrt{1 - \sin^2 y} = \sqrt{1 - x^2}.$$

Therefore $\frac{dy}{dx} = \frac{1}{\cos y} = \frac{1}{\sqrt{1 - x^2}}$.

$$\frac{d}{dx}(\sin^{-1}x) = \frac{1}{\sqrt{1 - x^2}}, \qquad -1 < x < 1$$

The formula for the derivative of the inverse tangent function (arctangent) is derived in a similar way. If $y = \tan^{-1}x$, then $\tan y = x$. Differentiating the latter equation implicitly with respect to x, we have

$$\sec^2 y \frac{dy}{dx} = 1$$

$$\frac{dy}{dx} = \frac{1}{\sec^2 y} = \frac{1}{1 + \tan^2 y} = \frac{1}{1 + x^2}.$$

$$\frac{d}{dx}(\tan^{-1}x) = \frac{1}{1 + x^2}$$

Similarly, the derivative of any inverse function may be found using the above procedure.

EXAMPLE 18

Differentiate $f(x) = \sin^{-1}(1 - x)$.

Solution $f'(x) = \{\sin^{-1}(1-x)\}' = \frac{(1-x)'}{\sqrt{1-(1-x)^2}} = \frac{-1}{\sqrt{2x - x^2}} = -\frac{1}{\sqrt{2x - x^2}}$ ■

4)http://www.math.ucdavis.edu/~kouba/CalcOneDIRECTORY/invtrigderivdirectory/InvTrigDeriv.html

Derivatives of Inverse Trigonometric Functions[4]

$$\frac{d}{dx}(\sin^{-1}x) = \frac{1}{\sqrt{1-x^2}} \qquad \frac{d}{dx}(\csc^{-1}x) = -\frac{1}{x\sqrt{x^2-1}}$$

$$\frac{d}{dx}(\cos^{-1}x) = -\frac{1}{\sqrt{1-x^2}} \qquad \frac{d}{dx}(\sec^{-1}x) = \frac{1}{x\sqrt{x^2-1}}$$

$$\frac{d}{dx}(\tan^{-1}x) = \frac{1}{1+x^2} \qquad \frac{d}{dx}(\cot^{-1}x) = -\frac{1}{1+x^2}$$

Derivatives of Logarithmic Functions

The derivatives of the logarithmic functions $y = \log_a x$ and the natural logarithmic function $y = \ln x$ are obtained using implicit differentiation.

1

$$\frac{d}{dx}(\log_a x) = \frac{1}{x \ln a}$$

Proof Let $y = \log_a x$. Then, $a^y = x$.
Differentiating this equation implicitly with respect to x, we get

$$a^y(\ln a)\frac{dy}{dx} = 1$$

and so $\frac{dy}{dx} = \frac{1}{a^y \ln a} = \frac{1}{x \ln a}$.

■

If we put $a = e$ in Formula 1, then the factor $\ln a$ on the right side becomes $\ln e = 1$ and we get the formula for the derivative of the natural logarithmic function $\log_e x = \ln x$:

2

$$\frac{d}{dx}(\ln x) = \frac{1}{x}$$

In general, we have

3

$$\frac{d}{dx}(\ln u) = \frac{1}{u}\frac{du}{dx} \quad \text{or} \quad \frac{d}{dx}[\ln g(x)] = \frac{g'(x)}{g(x)}$$

EXAMPLE 19

Find $f'(x)$ if $f(x) = \ln|x|$.

Solution Since

$$f(x)=\begin{cases}\ln x & \text{if } x>0\\ \ln(-x) & \text{if } x<0,\end{cases}$$

it follows that

$$f'(x)=\begin{cases}\dfrac{1}{x} & \text{if } x>0\\ \dfrac{1}{-x}(-1)=\dfrac{1}{x} & \text{if } x<0.\end{cases}$$

Thus, $f'(x)=1/x$ for all $x\neq 0$.
Hence,

4
$$\frac{d}{dx}\ln|x|=\frac{1}{x}$$
■

EXAMPLE 20

Differentiate $y=\ln\left|\tan\dfrac{x}{2}\right|$.

Solution
$$\frac{dy}{dx}=\frac{\sec^2\frac{x}{2}}{2\tan\frac{x}{2}}=\frac{1}{2}\,\frac{1/\cos^2\frac{x}{2}}{\sin\frac{x}{2}/\cos\frac{x}{2}}=\frac{1}{2\sin\frac{x}{2}\cos\frac{x}{2}}$$
$$=\frac{1}{\sin x}=\csc x.$$
■

EXAMPLE 21

Differentiate $y=(x+1)^x$.

Solution Applying the logarithm to both sides, we get $\ln y=x\ln(x+1)$
Differentiating with respect to x, we obtain

$$\frac{1}{y}\frac{dy}{dx}=\left(\ln(x+1)+\frac{x}{x+1}\right)$$

$$\Rightarrow\frac{dy}{dx}=y\left(\ln(x+1)+\frac{x}{x+1}\right)=(x+1)^x\ln(x+1)+x(x+1)^{x-1}.$$
■

EXAMPLE 22

Let $y=f(x)$ be implicitly defined by $x^{\cos y}=y^{\sin x}$. Compute y' in terms of x and y.

Solution Taking logarithms gives $\cos y\ln x=\sin x\ln y$.
Differentiating implicitly, we have

$$(\cos y)\frac{1}{x} - \sin y\frac{dy}{dx}\ln x = (\sin x)\frac{1}{y}\frac{dy}{dx} + \cos x\ln y,$$

and solving for $\dfrac{dy}{dx}$ yields $\dfrac{dy}{dx} = \dfrac{y\cos y - xy(\cos x)(\ln y)}{x\sin x + xy(\sin y)(\ln x)}$. ■

Remark Apply the definition to the derivative to the logarithmic function, $f(x) = \ln x$, we have

$$\begin{aligned} f'(1) &= \lim_{h\to 0}\frac{f(1+h)-f(1)}{h} = \lim_{x\to 0}\frac{f(1+x)-f(1)}{x} \\ &= \lim_{x\to 0}\frac{\ln(1+x)-\ln(1)}{x} = \lim_{x\to 0}\frac{1}{x}\ln(1+x) \\ &= \lim_{x\to 0}\ln\left[(1+x)^{1/x}\right]. \end{aligned}$$

So, $\lim\limits_{x\to 0}\ln(1+x)^{1/x} = 1$. Thus,

$$e = e^1 = e^{\lim\limits_{x\to 0}\ln(1+x)^{1/x}} = \lim_{x\to 0} e^{\ln(1+x)^{1/x}} = \lim_{x\to 0}(1+x)^{1/x}.$$

5

$$e = \lim_{x\to 0}(1+x)^{1/x}$$

Formula 5 can be rewritten in the following way, by substituting x by $\dfrac{1}{n}$,

6

$$e = \lim_{n\to\infty}\left(1+\frac{1}{n}\right)^n.$$

Hyperbolic Functions

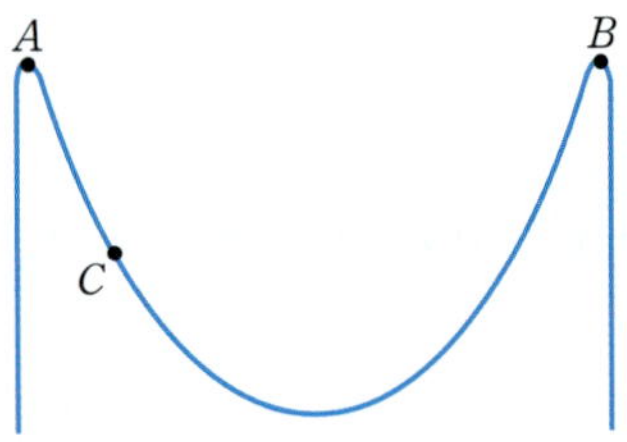

Figure 7 Catenary curve

Hyperbolic functions which arise frequently in mathematics and its applications, are certain combinations of the exponential functions e^x and e^{-x}. In many ways they are analogous to the trigonometric functions, and they have the same relationship to the hyperbola that the trigonometric functions have to the circle. For this reason they are called **hyperbolic Functions**.

5) http://en.wikipedia.org/wiki/Hyperbolic_function

Definition of the Hyperbolic Functions[5)]

$$\sinh x = \frac{e^x - e^{-x}}{2} \qquad \cosh x = \frac{e^x + e^{-x}}{2}$$

$$\tanh x = \frac{\sinh x}{\cosh x} = \frac{e^x - e^{-x}}{e^x + e^{-x}} \qquad \coth x = \frac{\cosh x}{\sinh x} = \frac{e^x + e^{-x}}{e^x - e^{-x}}$$

$$\operatorname{sech} x = \frac{1}{\cosh x} = \frac{2}{e^x + e^{-x}} \qquad \operatorname{csch} x = \frac{1}{\sinh x} = \frac{2}{e^x - e^{-x}}$$

The graphs of the hyperbolic sine and cosine can be sketched using graphical addition as in Figures 8-13.

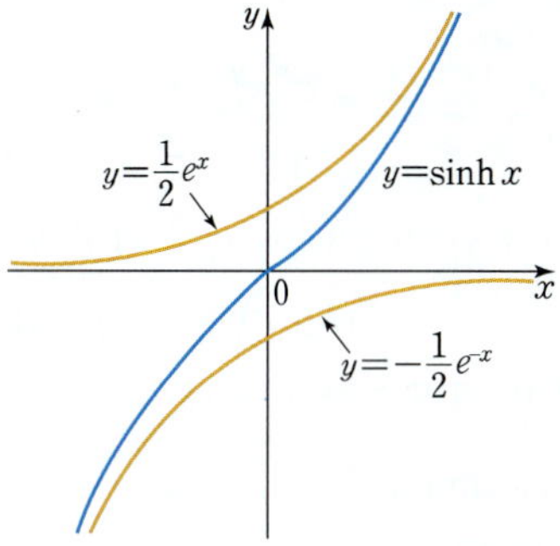

Figure 8 $y = \sinh x = \frac{1}{2}e^x - \frac{1}{2}e^{-x}$

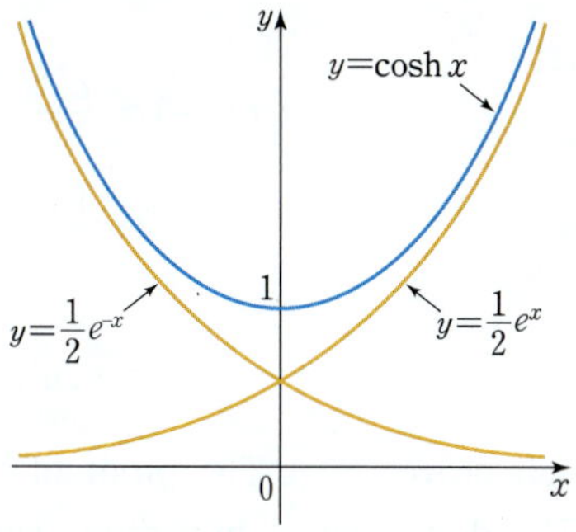

Figure 9 $y = \cosh x = \frac{1}{2}e^x + \frac{1}{2}e^{-x}$

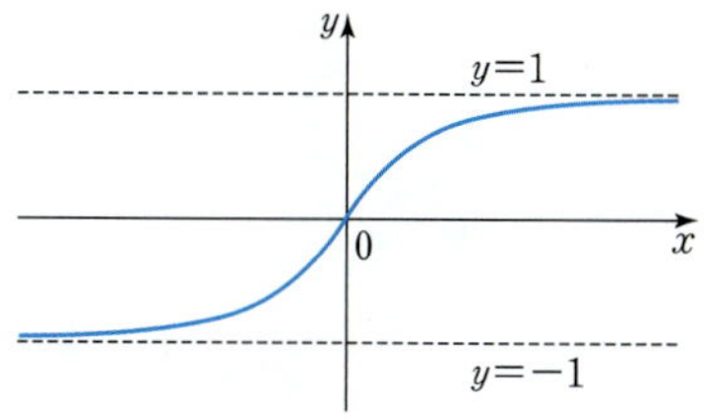

Figure 10 $y = \tanh x$

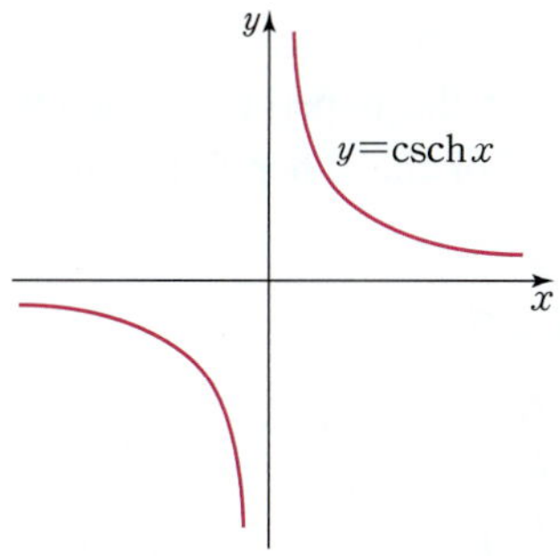

Figure 11 $y = \operatorname{csch} x$

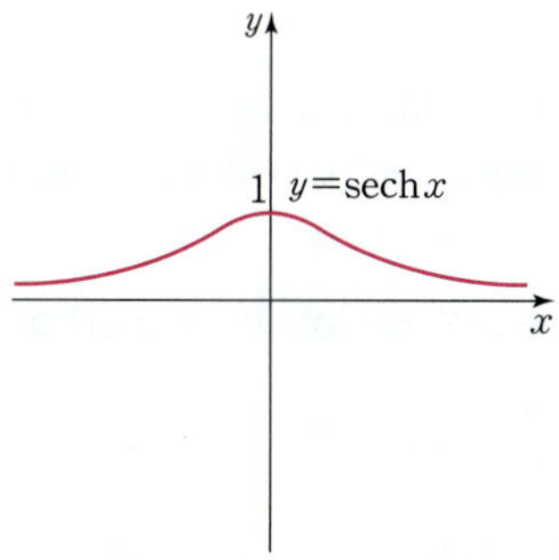

Figure 12 $y = \operatorname{sech} x$

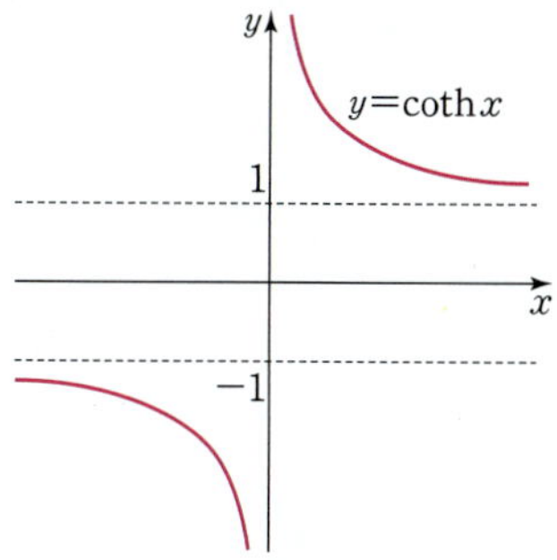

Figure 13 $y = \coth x$

Hyperbolic Identities

The hyperbolic functions satisfy various identities similar to the identities for the trigonometric functions. The most fundamental of these is

$$\boxed{7} \qquad \cosh^2 x - \sinh^2 x = 1$$

If we divide $\boxed{7}$ by $\cosh^2 x$, we obtain

$$\boxed{8} \qquad 1 - \tanh^2 x = \operatorname{sech}^2 x$$

and if we divide $\boxed{7}$ by $\sinh^2 x$, we obtain

$$\coth^2 x - 1 = \operatorname{csch}^2 x$$

9

The addition formulas for $\sinh$ and $\cosh$ are

10a $$\sinh(x+y) = \sinh x \cosh y + \cosh x \sinh y$$
10b $$\cosh(x+y) = \cosh x \cosh y + \sinh x \sinh y$$

For example, to prove 10a we use the relations

11a $$\cosh x + \sinh x = e^x$$
11b $$\cosh x - \sinh x = e^{-x}$$

From 11a and 11b it follows that

$$\begin{aligned}\sinh(x+y) &= \frac{e^{(x+y)} - e^{-(x+y)}}{2} = \frac{e^x e^y - e^{-x} e^{-y}}{2} \\ &= \frac{1}{2}[(\cosh x + \sinh x)(\cosh y + \sinh y) \\ &\qquad - (\cosh x - \sinh x)(\cosh y - \sinh y)] \\ &= \sinh x \cosh y + \cosh x \sinh y\end{aligned}$$

which proves 10a. The proof of 10b is similar.

The derivative of hyperbolic sine function,

$$\frac{d}{dx}\sinh x = \frac{d}{dx}\frac{e^x - e^{-x}}{2} = \frac{1}{2}[e^x - (-e^{-x})] = \frac{e^x + e^{-x}}{2} = \cosh x$$

follows immediately from the definition of the hyperbolic functions. By similar techniques we can deduce the derivatives of the other five hyperbolic function.

Derivatives of the Hyperbolic Functions

$$\frac{d}{dx}(\sinh x) = \cosh x \qquad \frac{d}{dx}(\coth x) = -\operatorname{csch}^2 x$$
$$\frac{d}{dx}(\cosh x) = \sinh x \qquad \frac{d}{dx}(\operatorname{sech} x) = -\operatorname{sech} x \tanh x$$
$$\frac{d}{dx}(\tanh x) = \operatorname{sech}^2 x \qquad \frac{d}{dx}(\operatorname{csch} x) = -\operatorname{csch} x \coth x$$

If we use 10a and 10b, we obtain the following analogs of the trigonometric double-angle formulas

12a $$\sinh 2x = 2\sinh x \cosh x$$
12b $$\cosh 2x = \cosh^2 x + \sinh^2 x$$

To obtain subtraction formulas for the hyperbolic function we shall use the addition formulas and the relations

13a $$\sinh(-x) = -\sinh x$$
13b $$\cosh(-x) = \cosh x$$

13c $$\tanh(-x) = -\tanh x$$

If we replace y by $-y$ in 10a and 10b and then apply 13a and 13b, we obtain

$$\sinh(x-y) = \sinh x \cosh y - \cosh x \sinh y$$
$$\cosh(x-y) = \cosh x \cosh y - \sinh x \sinh y$$

EXAMPLE 23

Differentiate $\sinh(x^2)$.

Solution $\frac{d}{dx}\sinh(x^2) = \cosh(x^2)\cdot\frac{d}{dx}x^2 = 2x\cosh(x^2)$. ■

EXAMPLE 24

Differentiate $\ln(\coth x)$.

Solution $\frac{d}{dx}\ln(\coth x) = \frac{1}{\coth x}\cdot\frac{d}{dx}\coth x = -\frac{\operatorname{csch}^2 x}{\tanh x}$. ■

EXAMPLE 25

Evaluate the derivative of $y = \frac{3x}{4+\sinh x^2}$ at $x=0$.

Solution Using the Quotient Rule

$$\frac{dy}{dx} = \frac{(4+\sinh x^2)\cdot 3 - 3x(2x\cosh x^2)}{(4+\sinh x^2)^2} = \frac{12+3\sinh x^2 - 6x^2\cosh x^2}{(4+\sinh x^2)^2}.$$

Because $\sinh(0)=0$ and $\cosh(0)=1$, we obtain

$$\left.\frac{dy}{dx}\right|_{x=0} = \frac{12}{16} = \frac{3}{4}.$$ ■

CAS **EXAMPLE 26**

Find $\frac{dy}{dx}$.

$$y = (1+\cos x)^{\frac{1}{x}}$$

http://matrix.skku.ac.kr/cal-lab/cal-3-3-Exm24.html

Solution

```
var('x')
diff((1+ cos(x))^(1/x),x)
```

Answer : -(sin(x)/((cos(x)+1)*x)+log(cos(x)+1)/x^2)*(cos(x)+1)^(1/x)

■

Inverse Hyperbolic Functions

Because the hyperbolic functions are expressed in terms of combinations of e^x, it should not be surprising that the inverse hyperbolic function are expressed in terms of the natural logarithms. For example, $y=\tanh^{-1}x$ $(|x|<1)$ is equivalent to $x=\tanh y$, so that

$$x=\tanh y=\frac{e^y-e^{-y}}{e^y+e^{-y}} \Rightarrow e^y(1-x)=(1+x)e^{-y} \Rightarrow e^{2y}=\frac{1+x}{1-x}$$

$$2y=\ln\left(\frac{1+x}{1-x}\right)$$

Hence, $y=\tanh^{-1}x=\dfrac{1}{2}\ln\dfrac{1+x}{1-x}$.

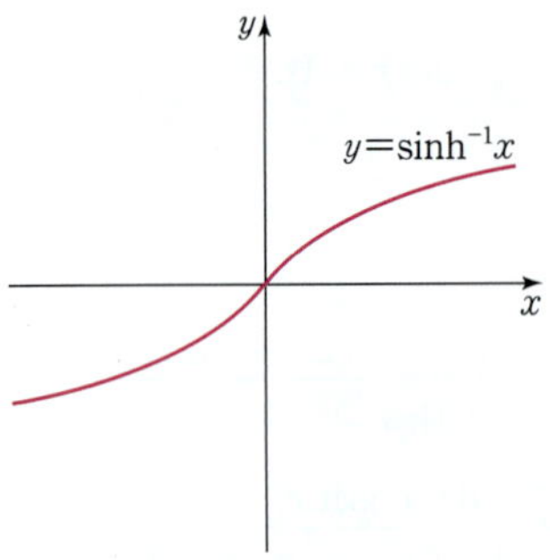

Figure 14 $y=\sinh^{-1}x$

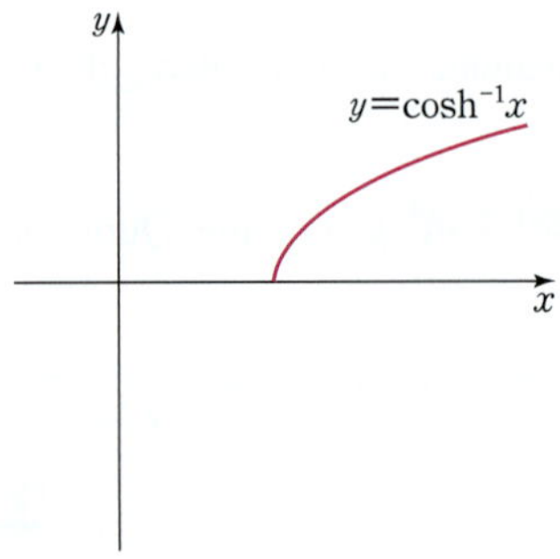

Figure 15 $y=\cosh^{-1}x$

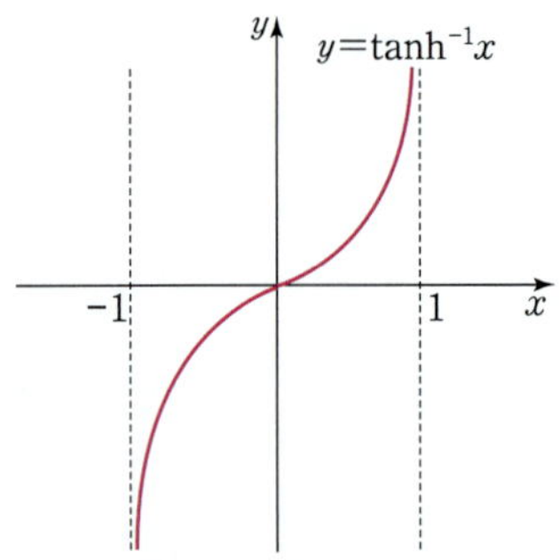

Figure 16 $y=\tanh^{-1}x$

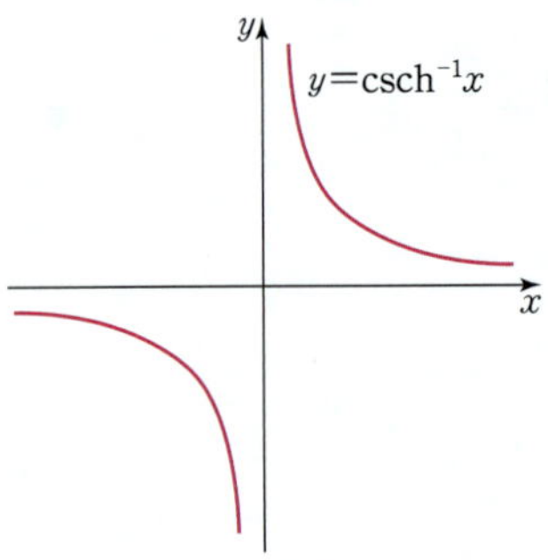

Figure 17 $y=\operatorname{csch}^{-1}x$

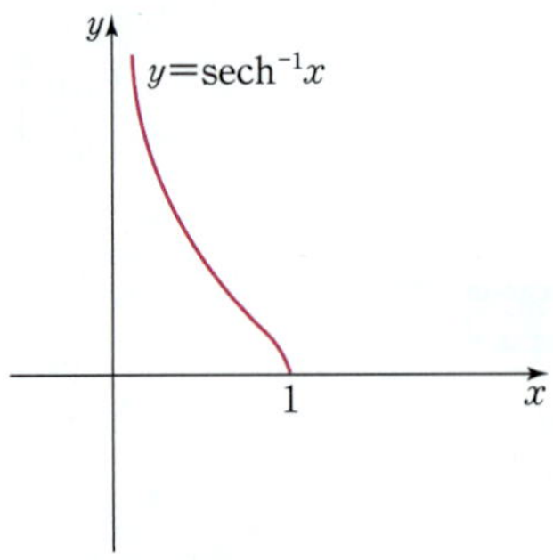

Figure 18 $y=\operatorname{sech}^{-1}x$

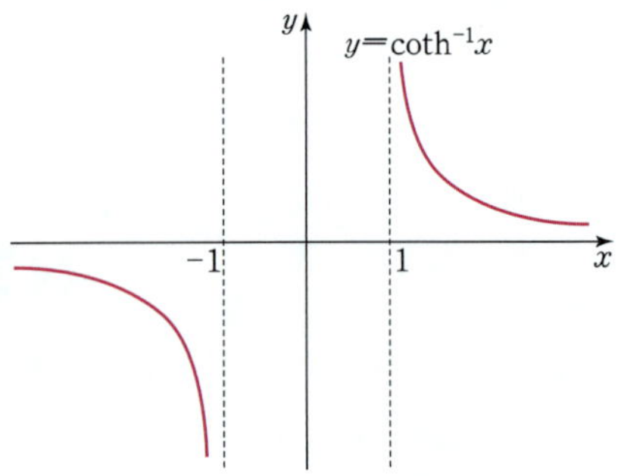

Figure 19 $y=\coth^{-1}x$

The Inverse Hyperbolic Functions

$$\sinh^{-1}x = \ln(x+\sqrt{x^2+1}) \qquad (x\in\mathbb{R})$$
$$\cosh^{-1}x = \ln(x+\sqrt{x^2-1}) \qquad (x \geq 1)$$
$$\tanh^{-1}x = \frac{1}{2}\ln\frac{1+x}{1-x} \qquad (|x|<1)$$
$$\coth^{-1}x = \frac{1}{2}\ln\frac{x+1}{x-1} \qquad (|x|>1)$$
$$\operatorname{sech}^{-1}x = \ln\left(\frac{1+\sqrt{1-x^2}}{x}\right) \qquad (0<x<1)$$
$$\operatorname{csch}^{-1}x = \ln\left(\frac{1}{x}+\frac{\sqrt{1+x^2}}{|x|}\right) \qquad (x\neq 0)$$

The proofs of some important identities, relative to derivatives and the inverse hyperbolic functions will be presented in the exercises.[6]

6)http://www.efunda.com/math/derivatives/HyperbolicD.cfm

Derivatives of the Inverse Hyperbolic Functions

The derivatives of inverse hyperbolic functions may be obtained by differentiating the logarithmic expression for the inverse hyperbolic functions. For example,

$$\frac{d}{dx}(\sinh^{-1}x) = \frac{d}{dx}\left[\ln(x+\sqrt{x^2+1})\right] = \frac{1}{x+\sqrt{x^2+1}}\left(1+\frac{x}{\sqrt{x^2+1}}\right)$$
$$= \frac{\sqrt{x^2+1}+x}{(x+\sqrt{x^2+1})(\sqrt{x^2+1})} = \frac{1}{\sqrt{x^2+1}}.$$

Derivatives of the Inverse Hyperbolic Functions

$$\frac{d}{dx}(\sinh^{-1}x) = \frac{1}{\sqrt{1+x^2}} \qquad (x\in\mathbb{R})$$
$$\frac{d}{dx}(\cosh^{-1}x) = \frac{1}{\sqrt{x^2-1}} \qquad (x>1)$$
$$\frac{d}{dx}(\tanh^{-1}x) = \frac{1}{1-x^2} \qquad (|x|<1)$$
$$\frac{d}{dx}(\coth^{-1}x) = \frac{1}{1-x^2} \qquad (|x|>1)$$
$$\frac{d}{dx}(\operatorname{sech}^{-1}x) = -\frac{1}{x\sqrt{1-x^2}} \qquad (0<x<1)$$
$$\frac{d}{dx}(\operatorname{csch}^{-1}x) = -\frac{1}{|x|\sqrt{1+x^2}} \qquad (x\neq 0)$$

EXAMPLE 27

Find derivative of the function $y=\sinh^{-1}(x^3+5)$.

Solution $\dfrac{dy}{dx}=\dfrac{1}{\sqrt{(x^3+5)^2+1}}\cdot\dfrac{d}{dx}(x^3+5)=\dfrac{3x^2}{\sqrt{x^6+10x^3+26}}.$ ∎

EXAMPLE 28

Find derivative of the function $y=\operatorname{sech}^{-1}x^2$.

Solution $\dfrac{dy}{dx}=-\dfrac{1}{\sqrt{x^4(1-(x^2)^2)}}\cdot\dfrac{d}{dx}x^2=-\dfrac{2}{x\sqrt{1-x^4}}.$ ∎

EXAMPLE 29

Find derivative of the function $y=\ln x\operatorname{sech}^{-1}x$.
http://matrix.skku.ac.kr/cal-lab/cal-3-3-Exm-27.html

Solution By the Product Rule, we obtain

$$\begin{aligned}\frac{dy}{dx}&=\ln x\cdot\left(\frac{-1}{x\sqrt{1-x^2}}\right)+\frac{1}{x}\operatorname{sech}^{-1}x\\&=-\frac{\ln x}{x\sqrt{1-x^2}}+\frac{\operatorname{sech}^{-1}x}{x}\\&=\frac{\ln x\tanh x}{\operatorname{sech}x}+\frac{1}{x\operatorname{sech}x}.\end{aligned}$$ ∎

3.3 EXERCISES (The Chain Rule and Inverse Functions)

http://matrix.skku.ac.kr/Cal-Book/part1/CS-Sec-3-3-Sol.html

YouTube http://youtu.be/aSKm12922FE

1-3. Find the derivative dy/dx.

1. $y = \sec^2 x + \sin x$

Solution $\dfrac{dy}{dx} = 2(\sec x)(\sec x \tan x) + \cos x$

$= 2\sec^2 x \tan x + \cos x = 2\sec^3 x \sin x + \cos x$

2. $y = \tan^{-1}(e^x)$

Solution $\dfrac{dy}{dx} = \dfrac{1}{(e^x)^2 + 1}(e^x)' = \dfrac{e^x}{e^{2x} + 1}$

3. $y = \cosh(\cos x)$

Solution $\dfrac{dy}{dx} = -\sinh(\cos x)\sin x$

4-5. Find $f'(x)$ of these functions.

4. $f(x) = \tan^{-1}\left(\dfrac{x}{a}\right) - \tan^{-1}\left(\dfrac{x}{b}\right)$

Solution $y = \tan^{-1}\left(\dfrac{x}{a}\right)$

$\tan y = \dfrac{x}{a}$

$\sec^2 y \dfrac{dy}{dx} = \dfrac{1}{a}$

$\dfrac{dy}{dx} = \dfrac{1}{a\left(\dfrac{x^2}{a^2} + 1\right)} = \dfrac{a}{x^2 + a^2}$, so

$\dfrac{d}{dx}\left(\tan^{-1}\dfrac{x}{b}\right) = \dfrac{b}{x^2 + b^2}$

Therefore, $f'(x) = \dfrac{a}{x^2 + a^2} - \dfrac{b}{x^2 + b^2}$.

5. $f(x) = \sqrt{\tan^{-1}(3x)}$

Solution $f'(x) = \dfrac{1}{2}\dfrac{1}{\sqrt{\tan^{-1}(3x)}}\dfrac{3}{1 + (3x)^2}$

$= \dfrac{3}{2(9x^2 + 1)\sqrt{\tan^{-1}(3x)}}$

---

6. Find derivative of the function $\tan(\sin(\cos x^2))$.

Solution $\dfrac{d}{dx}(\tan(\sin(\cos x^2)))$

$= -\sec^2(\sin(\cos x^2))(\cos(\cos x^2))(2x)\sin x^2$

$= -2x\sin x^2(\cos(\cos x^2))\sec^2(\sin(\cos x^2))$

7. Find $f^{(n)}(x)$ where $f(x) = \ln(1 + x)$.

Solution $f'(x) = \dfrac{1}{x+1}$

$f''(x) = \dfrac{-1}{(x+1)^2}$

$f'''(x) = \dfrac{2}{(x+1)^3}$

$f^{(4)}(x) = \dfrac{-6}{(x+1)^4} = \dfrac{(-1)^3 \cdot 3!}{(x+1)^4}$

Using induction on n, we get

$$f^{(n)}(x) = \frac{(-1)^{n-1}(n-1)!}{(x+1)^n}.$$

8. Find $y^{(k)}$ for an integer $n \geq 2$ if $y = x^{n-1}$.

Solution $y = x^{n-1}$

$y' = (n-1)x^{n-2}$

$y'' = (n-1)(n-2)x^{n-3}$

Therefore, using induction

$$y^{(k)} = (n-1)(n-2)\cdots(n-k)x^{(n-1-k)}.$$

9. Show that the curves $x^2 + y^2 = 2x$ and $x^2 + y^2 = 3y$ are orthogonal.

Solution (i) $2x + 2yy' = 2 \Rightarrow y' = \dfrac{1-x}{y}$

(ii) $2x + 2yy' = 3y' \Rightarrow y' = \dfrac{2x}{3-2y}$

At a point of intersection,

$2x = 3y \Rightarrow$ (i) $\left(\dfrac{9}{4}\right)y^2 + y^2 = 3y$

$\Rightarrow y = 0,\ 12/13$

$\Rightarrow x = 0,\ 18/13$.

Thus, the intersections of the two curves are $(0, 0)$ and $\left(\dfrac{18}{13}, \dfrac{12}{13}\right)$.

At $\left(\dfrac{18}{13}, \dfrac{12}{13}\right)$, (i) $y'\left(\dfrac{18}{13}, \dfrac{12}{13}\right) = -\dfrac{5}{12}$

(ii) $y'\left(\dfrac{18}{13}, \dfrac{12}{13}\right) = \dfrac{12}{5}$

$-\dfrac{12}{5} \cdot \dfrac{5}{12} = -1$

Therefore, the curves are orthogonal.

At $(0, 0)$ the derivative $\dfrac{dy}{dx}$ for the equation $x^2 + y^2 = 2x$ does not exist (or, in other words, it is equal to infinity). Hence the tangent line to the curve $x^2 + y^2 = 2x$ at $(0, 0)$ is vertical. At the same point, the derivative $\dfrac{dy}{dx}$ for the equation $x^2+y^2=3y$ is equal 0, hence the tangent line for the curve $x^2+y^2=3y$ at $(0,0)$ is horizontal. Thus, the two curves are orthogonal at $(0,0)$.

10-13. Find $f^{(n)}(x)$.

10. $f(x)=\dfrac{1}{1-x^2}$; Note $f(x)=\dfrac{1}{2}\left(\dfrac{1}{1+x}+\dfrac{1}{1-x}\right)$

Solution $f'(x)=\dfrac{1}{2}\left(\dfrac{-1}{(1+x)^2}+\dfrac{1}{(1-x)^2}\right)$,

$f''(x)=\dfrac{1}{2}\left(\dfrac{2}{(1+x)^3}+\dfrac{2}{(1-x)^3}\right)$,

$f'''(x)=\dfrac{1}{2}\left(\dfrac{-6}{(1+x)^4}+\dfrac{6}{(1-x)^4}\right)$, ...,

$\Rightarrow f^{(n)}(x)=\dfrac{1}{2}\left(\dfrac{(-1)^n n!}{(1+x)^{n+1}}+\dfrac{n!}{(1-x)^{n+1}}\right)$

11. $f(x)=\sin^3 x$

Solution $f'(x)=3\sin^2 x\cos x=\dfrac{3}{4}(\cos(x)-\cos(3x))$

$f''(x)=\dfrac{3}{4}(-\sin(x)+3\sin(3x))$

$f'''(x)=\dfrac{3}{4}(-\cos(x)+9\cos(3x))$

$f^{(4)}(x)=\dfrac{3}{4}(\sin(x)-27\sin(3x))$

$\Rightarrow f^{(n)}(x)$

$$=\begin{cases}\dfrac{3}{4}(\sin(x)-3^{n-1}\sin(3x)) & \text{for } n=4l\\ \dfrac{3}{4}(\cos(x)-3^{n-1}\cos(3x)) & \text{for } n=4l+1\\ \dfrac{3}{4}(-\sin(x)+3^{n-1}\sin(3x)) & \text{for } n=4l+2\\ \dfrac{3}{4}(-\cos(x)+3^{n-1}\cos(3x)) & \text{for } n=4l+3\end{cases},\quad l\in N$$

12. $f(x)=\dfrac{x^3}{1-x}$

Solution $f'(x)=\dfrac{(3-2x)x^2}{(x-1)^2}$

$f''(x)=-\dfrac{2x(x^2-3x+3)}{(x-1)^3}$

$f'''(x)=\dfrac{6}{(x-1)^4}$

$f^{(4)}(x)=-\dfrac{24}{(x-1)^5}$

$$\Rightarrow f^{(n)}(x)=\begin{cases}\dfrac{(3-2x)x^2}{(x-1)^2} & \text{for } n=1\\ \dfrac{-2x(x^2-3x+3)}{(x-1)^3} & \text{for } n=2\\ \dfrac{(-1)^{n+1}n!}{(x-1)^{n+1}} & \text{for } n\ge 3\end{cases}$$

13. $f(x)=\dfrac{x}{x+1}$

Solution $f'(x)=\dfrac{(x+1)-x}{(x+1)^2}=\dfrac{1}{(x+1)^2}$

$f''(x)=\dfrac{-2(x+1)}{(x+1)^4}=\dfrac{-2}{(x+1)^3}$

$f'''(x)=\dfrac{-(-2)3(x+1)^2}{(x+1)^6}=\dfrac{6}{(x+1)^4}$

$f^{(4)}(x)=\dfrac{-6(4(x+1)^3)}{(x+1)^8}=\dfrac{-24}{(x+1)^5}$

We observe that the denominator is $(x+1)^{n+1}$, and the numerator is $(-1)^{(n-1)}n!$.

Therefore $f^{(n)}(x)=\dfrac{(-1)^{n-1}n!}{(x+1)^{n+1}}$.

---

14. Find the nth derivative of $f(x)=\cos kx$.

Solution $f'(x)=-k\sin(kx)$

$f''(x)=-k^2\cos(kx)$

$f'''(x)=k^3\sin(kx)$

$f^{(4)}(x)=k^4\cos(kx)$

$f^{(5)}(x)=-k^5\sin(kx)$

$$\Rightarrow f^{(n)}(x)=\begin{cases}k^n\cos(kx) & \text{for } n=4l\\ -k^n\sin(kx) & \text{for } n=4l+1\\ -k^n\cos(kx) & \text{for } n=4l+2\\ k^n\sin(kx) & \text{for } n=4l+3\end{cases},\quad l\in N$$

15. Differentiate $y=\sec^{-1}x$.

Solution By differentiating $\sec y=x$ implicitly, we have $\sec y\tan y\dfrac{dy}{dx}=1$.

Thus, $\dfrac{dy}{dx}=\dfrac{1}{\sec y\tan y}=\dfrac{1}{\sec y\sqrt{\sec^2 y-1}}=\dfrac{1}{x\sqrt{x^2-1}}$.

16. Use differentiation to show that

$$\tan^{-1}x+\tan^{-1}\left(\frac{1}{x}\right)=\begin{cases}\dfrac{\pi}{2} & \text{if } x>0\\ -\dfrac{\pi}{2} & \text{if } x<0\end{cases}$$

Solution $\dfrac{d}{dx}\tan^{-1}(x)=\dfrac{1}{x^2+1}$

$\dfrac{d}{dx}\tan^{-1}\left(\dfrac{1}{x}\right)=\dfrac{1}{\left(\dfrac{1}{x^2}+1\right)}\dfrac{d}{dx}\left(\dfrac{1}{x}\right)$

$=\dfrac{1}{\left(\dfrac{1}{x^2}+1\right)}\dfrac{-1}{x^2}=\dfrac{-1}{x^2+1}$

$\dfrac{d}{dx}\tan^{-1}(x)+\dfrac{d}{dx}\tan^{-1}\left(\dfrac{1}{x}\right)$

$=\dfrac{1}{x^2+1}-\dfrac{1}{x^2+1}=0$

$$\Rightarrow \tan^{-1}x+\tan^{-1}\left(\frac{1}{x}\right)=c$$

for some constant c

$$\lim_{x\to 0+} \tan^{-1}x+\tan^{-1}\left(\frac{1}{x}\right)=\frac{\pi}{2}$$

$$\lim_{x\to 0-} \tan^{-1}x+\tan^{-1}\left(\frac{1}{x}\right)=-\frac{\pi}{2}$$

$$\lim_{x\to +\infty} \tan^{-1}x+\tan^{-1}\left(\frac{1}{x}\right)=\frac{\pi}{2}$$

$$\lim_{x\to -\infty} \tan^{-1}x+\tan^{-1}\left(\frac{1}{x}\right)=-\frac{\pi}{2}$$

Therefore

$$\tan^{-1}x+\tan^{-1}\left(\frac{1}{x}\right)=\begin{cases}\frac{\pi}{2} & \text{if } x>0\\ -\frac{\pi}{2} & \text{if } x<0\end{cases}$$

17-20. In problems below, find $\frac{dy}{dx}$.

17. $y=\ln\frac{\sqrt{3x-5}}{x^2+4x}$

Solution $\frac{d}{dx}\left(\ln\left(\frac{\sqrt{3x-5}}{x^2+4x}\right)\right)$

$$=\frac{d}{dx}\left(\frac{1}{2}\ln(3x-5)-\ln(x^2+4x)\right)$$

$$=\frac{1}{2}\frac{3}{3x-5}-\frac{2x+4}{x^2+4x}=\frac{-9x^2+8x+40}{2x(x+4)(3x-5)}$$

18. $y=(\ln(\csc x))^4$

Solution $\frac{d}{dx}(\ln(\csc x))^4=4((\ln(\csc x))^3)(-\cot x)$

$$=-4\cot x(\ln(\csc x))^3$$

CAS **19.** $y=(\sin x)^{\sqrt{x}}$

http://matrix.skku.ac.kr/cal-lab/cal-3-2-12.html

Solution

```
var('x')
diff((sin(x))^(sqrt(x)))
```

Answer : 1/2*(2*sqrt(x)*cos(x)/sin(x)+ log(sin(x))/sqrt(x))*sin(x)^sqrt

CAS **20.** $y=(\cos^{-1}x)^{\tan^{-1}x}$

http://matrix.skku.ac.kr/cal-lab/cal-3-2-13.html

Solution

```
var('x')
diff((arccos(x))^arctan(x))
```

Answer : (log(arccos(x))/(x^2+1)-arctan(x)/(sqrt(-x^2+1)*arccos(x)))*arccos(x)^arctan(x)

CAS **21.** Find y'' if $y=x^{\tan^{-1}x}$.

http://matrix.skku.ac.kr/cal-lab/cal-3-2-14.html

Solution $\ln y=(\tan^{-1}x)\ln x \Rightarrow \frac{1}{y}\frac{dy}{dx}$

$$=\frac{1}{1+x^2}\ln x+\frac{1}{x}\tan^{-1}x$$

$$\Rightarrow \frac{dy}{dx}=\left(\frac{1}{1+x^2}\ln x+\frac{1}{x}\tan^{-1}x\right)x^{\tan^{-1}x}$$

```
var('x')
df(x)=diff(x^(arctan(x)),x);
df(x)
```

Answer : (log(x)/(x^2 + 1) + arctan(x)/x)*x^arctan(x)

```
ddf(x)=diff(df(x),x)
ddf(x)
```

Answer : (log(x)/(x^2 + 1) + arctan(x)/x)^2*x^arctan(x) - (2*x*log(x)/(x^2 + 1)^2 + arctan(x)/x^2- 2/((x^2 + 1)*x))*x^arctan(x)

22-23. Find $\frac{dy}{dx}$ and $\frac{d^2y}{dx^2}$ of the following parametric equations (a and b are constants).

22. $x=a\cos t,\ y=b\sin t,\quad (a>0,\ b>0)$.

Solution $\frac{dy}{dt}=b\cos t,\quad \frac{dx}{dt}=-a\sin t$

$$\frac{dy}{dx}=\frac{dy}{dt}\frac{dt}{dx}=\frac{b\cos t}{-a\sin t}=\left(-\frac{b}{a}\right)\cot t$$

$$\frac{d^2y}{dt^2}=-b\sin t,\quad \frac{d^2x}{dt^2}=-a\cos t$$

$$\frac{d^2y}{dx^2}=\frac{d^2y}{dt^2}\left(\frac{dt}{dx}\right)^2+\frac{dy}{dt}\frac{d^2t}{dx^2}$$

$$=\frac{-b\sin t}{a^2\sin^2 t}+\frac{b\cos t}{-a\cos t}$$

$$=\left(-\frac{b}{a}\right)\left(\frac{1}{\sin t}+1\right).$$

23. $x=\frac{3at}{1+t^3},\ y=\frac{3at^2}{1+t^3}$.

Solution $\frac{dy}{dt}=\frac{-3at(t^3-2)}{(t^3+1)^2}$,

$$\frac{dx}{dt}=\frac{-3a(2t^3-1)}{(t^3+1)^2},\quad \frac{dy}{dx}=\frac{dy}{dt}\frac{dt}{dx}=\frac{t(t^3-2)}{2t^3-1}$$

$$\frac{d^2y}{dt^2}=\frac{6a(t^6-7t^3+1)}{(t^3+1)^3},\quad \frac{d^2x}{dt^2}=\frac{18at^2(t^3-2)}{(t^3+1)^3},$$

$$\frac{d^2y}{dx^2}=\frac{d^2y}{dt^2}\left(\frac{dt}{dx}\right)^2+\frac{dy}{dt}\frac{d^2t}{dx^2}$$

$$= \frac{6a(t^6-7t^3+1)}{(t^3+1)^3} \cdot \frac{(t^3+1)^4}{6a^2(2t^3-1)^2} + \frac{-3at(t^3-2)}{(t^3+1)^2} \cdot \frac{(t^3+1)^3}{18at^2(t^3-2)}$$

$$= \frac{(t^6-7t^3+1)(t^3+1)}{a(2t^3-1)^2} - \frac{t^3+1}{6t}$$

---

24. Given $f(x)=\sin^{-1}x$, show that it satisfies the following identity.

$$(1-x^2)f^{(n+2)}(x)-(2n+1)xf^{(n+1)}(x)-nf^{(n)}(x)=0$$

Using this identity, find $f^{(n)}(0)$.

Solution $f'(x)=\dfrac{1}{\sqrt{1-x^2}}=(1-x^2)^{(-1/2)}$

$$f''(x) = \left(-\frac{1}{2}\right)(1-x^2)^{(-3/2)}(-2x) = x(1-x^2)^{(-3/2)}$$

$$f'''(x)=(1-x^2)^{(-3/2)}+\left(-\frac{3}{2}\right)(1-x^2)^{(-5/2)}(-2x) = (1-x^2)^{(-3/2)}+(3x)(1-x^2)^{(-5/2)}$$

$$f^{(4)}(x)=(3x)(1-x^2)^{(-5/2)}+3(1-x^2)^{(-5/2)}+(5x)(1-x^2)^{(-7/2)}$$

$$\begin{aligned} f^{(5)}(x) &= 3(1-x^2)^{(-5/2)}+(3x)(5x)(1-x^2)^{(-7/2)} +3(5x)(1-x^2)^{(-7/2)}+5(1-x^2)^{(-7/2)} +(7x)(1-x^2)^{(-9/2)} \\ &= 3(1-x^2)^{(-5/2)}+(15x^2+15x+5) +(1-x^2)^{(-7/2)}+(7x)(1-x^2)^{(-9/2)} \\ &= 3f'''(x)/(1-x^2)+(7x)f^{(4)}(x)/(1-x^2) \end{aligned}$$

$$f^{(n)}(x)=(n-2)f^{(n-2)}(x)/(1-x^2) +(2n-3)(x)f^{(n-1)}(x)/(1-x^2)$$

Therefore,

$(1-x^2)f^{(n+2)}(x)-(2n+1)xf^{(n+1)}(x)-nf^{(n)}(x)=0.$

$f^{(n+2)}(0)-nf^{(n)}(0)=0,$

$$f^{(n)}(0)=\begin{Bmatrix}(n-2)(n-4)\cdots 2f''(0) & \text{for } n=2l \\ (n-2)(n-4)\cdots f'(0) & \text{for } n=2l-1\end{Bmatrix},\ l\in N.$$

Therefore, $f^{(n)}(0)=0$ for n is even,

$f^{(n)}(0)=(n-2)(n-4)\cdots 2\cdot 1$ for n is odd.

25. Given $f(x)=\begin{cases} e^{-\frac{1}{x^2}}, & x\neq 0 \\ 0, & x=0\end{cases}$, show that $f''(0)=0$.

Solution Since $f(0)=0$ so $f(x)$ is continuous on $(-\infty,+\infty)$.

And $f'(x)=\begin{cases}\dfrac{2}{x^3}e^{\frac{-1}{x^2}} & x\neq 0 \\ 0 & x=0\end{cases} \Rightarrow f'(0)=0,$

$f''(x)=\begin{cases}\dfrac{(4-6x^2)}{x^6}e^{\left(\frac{-1}{x^2}\right)} & x\neq 0 \\ 0 & x=0\end{cases} \Rightarrow f''(0)=0.$

Therefore, $f''(0)=0.$

26-28. Find $\dfrac{dy}{dx}$ for the following implicit functions.

26. $x\sin y+y\cos x=0$

Solution $\sin y+x\cos y\dfrac{dy}{dx}+\cos x\dfrac{dy}{dx}-y\sin x=0$

$\Rightarrow (x\cos y+\cos x)\dfrac{dy}{dx}=y\sin x-\sin y$

$\dfrac{dy}{dx}=\dfrac{y\sin x-\sin y}{x\cos y+\cos x}.$

27. $x^3-xy^2-x+1=0$

Solution $x^3-xy^2-x+1=0.$

$3x^2-\left(y^2+2xy\dfrac{dy}{dx}\right)-1=0.$

$3x^2-y^2-1=2xy\dfrac{dy}{dx}.$

$\dfrac{dy}{dx}=\dfrac{3x^2-y^2-1}{2xy}.$

28. $x\sin y+\cos 2y+xe^{x^4}=4$

Solution $\sin y+x\cos y\dfrac{dy}{dx}-2\sin 2y\dfrac{dy}{dx}+e^{x^4}+4x^4e^{x^4}=0$

$\dfrac{dy}{dx}=\dfrac{(4x^4+1)e^{x^4}+\sin y}{2\sin 2y-x\cos y}$

---

29. If $y=(f\circ g)(x)=f(g(x))$, where f and g are three times differentiable, find expressions for y'' and y'''.

Solution $y'=f'(g(x))g'(x)$

$y''=f''(g(x))(g'(x))^2+f'(g(x))g''(x)$

$y'''=f'''(g(x))(g'(x))^3+3f''(g(x))g'(x)g''(x)+f'(g(x))g'''(x)$

30. Given $x^2y^4+x^3y=2$, find y'' at the point $(1,1)$.

Solution $2xy^4+4x^2y^3y'+3x^2y+x^3y'=0$

$y'=-\dfrac{x^3+4x^2y^3}{3x^2y+2xy^4}$

$y'(1,1)=-1$

$2y^4+8xy^3y'+8xy^3y'+12x^2y^2(y')^2+4x^2y^3y''+6xy+3x^2y'+3x^2y'+x^3y''=0$

$2-8-8+12+4y''(1,1)+6-3-3+y''(1,1)=0$

Therefore, $y''(1,1)=\dfrac{2}{5}.$

31. Determine the points on the curve $x^3+2y^3=1$ at which the tangent is either vertical or horizontal.

32. Find an equation of the tangent line to the curve at $(1, f(1))$ for an arbitrary value $a \in (0, 1) \cup (1, \infty)$.

Solution $f(1) = \log_a 1 = 0$, so we have to find the $f(x) = \log_a x$ tangent line at $(1, 0)$

$$f'(x) = \frac{1}{x\ln a} \Rightarrow y = \frac{1}{\ln a}(x-1)$$

$$\left(\text{Because } f'(1) = \frac{1}{\ln a}\right)$$

33. Establish the following derivative rules.

(a) $\frac{d}{dx}(\cosh x) = \sinh x$

(b) $\frac{d}{dx}(\tanh x) = \operatorname{sech}^2 x$

Solution (a) $\frac{d}{dx}(\cosh x) = \frac{d}{dx}\left(\frac{e^x + e^{-x}}{2}\right)$

$$= \frac{e^x - e^x}{2} = \sinh x$$

(b) $\frac{d}{dx}(\tanh x) = \frac{d}{dx}\left(\frac{\sinh x}{\cosh x}\right)$

$$= \frac{\cosh x \cosh x - \sinh x \sinh x}{(\cosh x)^2}$$

$$= \frac{\cosh^2 x - \sinh^2 x}{\cosh^2 x} = \frac{1}{\cosh^2 x} = \operatorname{sech}^2 x$$

34-46. Prove the following identities.

34. $\cosh^2 x - \sinh^2 x = 1$

Solution $\cosh^2 x - \sinh^2 x = \left(\frac{e^x + e^{-x}}{2}\right)^2 - \left(\frac{e^x - e^{-x}}{2}\right)^2$

$$= \frac{e^{2x} + 2 + e^{-2x}}{4} - \frac{e^{2x} - 2 + e^{-2x}}{4} = \frac{4}{4} = 1$$

35. $1 - \tanh^2 x = \operatorname{sech}^2 x$

Solution $1 - \left(\frac{e^x - e^{-x}}{e^x + e^{-x}}\right)^2 = 1 - \left(\frac{e^{2x} - 1}{e^{2x} + 1}\right)^2$

$$= 1 - \frac{e^{4x} - 2e^{2x} + 1}{(e^{2x}+1)^2}$$

$$= \frac{e^{4x} + 2e^{2x} + 1 - e^{4x} + 2e^{2x} - 1}{(e^{2x}+1)^2}$$

$$= \frac{(2e^x)^2}{(e^{2x}+1)^2} = \frac{4}{(e^x + e^{-x})^2} = \frac{1}{\cosh^2 x} = \operatorname{sech}^2 x$$

or $1 - \tanh^2 x = 1 - \left(\frac{\sinh x}{\cosh x}\right)^2$

$$= \frac{\cosh^2 x - \sinh^2 x}{\cosh^2 x}$$

$$= \frac{1}{\cosh^2 x} = \operatorname{sech}^2 x$$

36. $\coth^2 x - 1 = \operatorname{csch}^2 x$

Solution $\coth^2 x - 1 = \left(\frac{e^x + e^{-x}}{e^x - e^{-x}}\right)^2 - 1 = \left(\frac{e^{2x}+1}{e^{2x}-1}\right)^2 - 1$

$$= \frac{e^{4x} + 2e^{2x} + 1}{(e^{2x}-1)^2} - 1 = \frac{e^{4x} + 2e^{2x} + 1 - e^{4x} + 2e^{2x} - 1}{(e^{2x}-1)^2}$$

$$= \frac{(2e^x)^2}{(e^{2x}-1)^2} = \frac{4}{(e^x - e^{-x})^2} = \frac{1}{\sinh^2 x} = \operatorname{csch}^2 x$$

or

$$\coth^2 x - 1 = \left(\frac{\cosh x}{\sinh x}\right)^2 - 1 = \frac{\cosh^2 x - \sinh^2 x}{\sinh^2 x}$$

$$= \frac{1}{\sinh^2 x} = \operatorname{csch}^2 x$$

37. $\sinh(2x) = 2\sinh x \cosh x$

Solution $\frac{e^{2x} - e^{-2x}}{2} = \frac{(e^x - e^{-x})(e^x + e^{-x})}{2}$

$$= 2\frac{(e^x - e^{-x})}{2}\frac{(e^x + e^{-x})}{2} = 2\sinh x \cosh x$$

38. $\cosh(2x) = \cosh^2 x + \sinh^2 x$

Solution $\frac{e^{2x} + e^{-2x}}{2} = \frac{(e^x + e^{-x})^2 + (e^x - e^{-x})^2}{4}$

$$= \frac{(e^x + e^{-x})^2}{4} + \frac{(e^x - e^{-x})^2}{4} = \cosh^2 x + \sinh^2 x$$

39. $(\cosh x + \sinh x)^a = \cosh ax + \sinh ax,\ a \in \mathbb{R}$

Solution Use Mathematical Induction. If $n = 1$, trivial.

Let $(\cosh x + \sinh x)^k$

$= \cosh kx + \sinh kx\,(\cosh x + \sinh x)^{k+1}$

$= (\cosh x + \sinh x)^k (\cosh x + \sinh x)$

$= (\cosh kx + \sinh kx)(\cosh x + \sinh x)$

$= \cosh kx \cosh x + \cosh kx \sinh x$

$\quad + \sinh kx \cosh x + \sinh kx \sinh x$

$= \sinh(kx + x) + \cosh(kx + x)$

$= \sinh(k+1)x + \cosh(k+1)x$

So, $(\cosh x + \sinh x)^a = \cosh ax + \sinh ax,\ a \in \mathbb{R}$

40. $\frac{d}{dx}(\operatorname{sech} x) = -\operatorname{sech} x \tanh x$

Solution $\frac{d}{dx}(\operatorname{sech} x) = \frac{d}{dx}\left(\frac{2}{e^x + e^{-x}}\right) = \frac{-2(e^x - e^{-x})}{(e^x + e^{-x})^2}$

$$= -\frac{2}{(e^x + e^{-x})}\frac{e^x - e^{-x}}{(e^x + e^{-x})} = -\operatorname{sech} x \tanh x$$

41. $\sinh^{-1} x = \ln(x + \sqrt{x^2+1}),\ x \in \mathbb{R}$

Solution Let $y = \sinh^{-1} x$. Then $x = \sinh y = \frac{e^y - e^{-y}}{2}$,

so $e^y - 2x - e^{-y} = 0 \Rightarrow e^{2y} - 2xe^y - 1 = 0$.

$\Rightarrow (e^y)^2 - 2x(e^y) - 1 = 0$.

$$\Rightarrow e^y = \frac{2x \pm \sqrt{4x^2+4}}{2} = x \pm \sqrt{x^2+1}$$

Note that $e^y > 0$, but $x - \sqrt{x^2+1} < 0$ because $x < \sqrt{x^2+1}$.

Thus, $e^y = x + \sqrt{x^2+1}$.

So, $y = \ln(e^y) = \ln\left(x + \sqrt{x^2+1}\right)$.

42. $\cosh^{-1}x = \ln(x + \sqrt{x^2-1}),\ x \ge 1$

Solution Let $y = \cosh^{-1}x$. Then $x = \cosh y = \dfrac{e^y + e^{-y}}{2}$,

so $e^y - 2x + e^{-y} = 0 \Rightarrow e^{2y} - 2xe^y + 1 = 0$.

$$\Rightarrow (e^y)^2 - 2x(e^y) + 1 = 0$$

$$\Rightarrow e^y = \frac{2x \pm \sqrt{4x^2-4}}{2} = x \pm \sqrt{x^2-1}$$

Note that $e^y > 0$, but $x - \sqrt{x^2-1} < 0$ because $x < \sqrt{x^2-1}$ for $x \ge 1$.

Thus, $e^y = x + \sqrt{x^2-1}$ for $x \ge 1$

So $y = \ln(e^y) = \ln\left(x + \sqrt{x^2-1}\right)$.

43. $\dfrac{d}{dx}(\tanh^{-1}x) = \dfrac{1}{1-x^2}$

Solution Let $y = \tanh^{-1}x$. Then

$$\tanh y = x \Rightarrow (\operatorname{sech}^2 y)y' = 1 \Rightarrow y' = \frac{1}{\operatorname{sech}^2 y}$$

$$= \frac{1}{1-\tanh^2 y} = \frac{1}{1-x^2}$$

44. Find $f'(x)$, where $f(x) = \sin^{-1}(1-x)$.

Solution $f'(x) = \{\sin^{-1}(1-x)\}' = \dfrac{(1-x)'}{\sqrt{1-(1-x)^2}}$

$$= \frac{-1}{\sqrt{2x-x^2}} = -\frac{1}{\sqrt{2x-x^2}}$$

45. Find $f'(x)$, where $f(x) = \cot^{-1}\left(\dfrac{1+x}{1-x}\right)$.

Solution $f'(x) = \left(\cot^{-1}\left(\dfrac{1+x}{1-x}\right)\right)'$

$$= -\frac{1}{1+\left(\frac{1+x}{1-x}\right)^2} \cdot \left(\frac{1+x}{1-x}\right)'$$

$$= -\frac{(1-x)^2}{(1-x)^2+(1+x)^2} \cdot \frac{(1+x)'(1-x)-(1+x)(1-x)'}{(1-x)^2}$$

$$= -\frac{(1-x)^2}{(1-x)^2+(1+x)^2} \cdot \frac{(1-x)-(1+x)(-1)}{(1-x)^2}$$

$$= -\frac{1}{1+x^2}$$

46. Find $f'(x)$, where $f(x) = \sec^{-1}(e^x)$.

Solution $f'(x) = (\sec^{-1}(e^x))'$

$$= \frac{(e^x)'}{e^x\sqrt{(e^x)^2-1}} = \frac{1}{\sqrt{e^{2x}-1}}$$

---

47. Find $\dfrac{dy}{dx}$, where $y^2 \sin x + y = \tan^{-1}x$.

Solution $2yy'\sin x + y^2\cos x + y' = \dfrac{1}{1+x^2}$

$$\Rightarrow y'(2y\sin x + 1) = \frac{1}{1+x^2} - y^2\cos x$$

$$\Rightarrow y' = \frac{1-(1+x^2)y^2\cos x}{(1+x^2)(2y\sin x+1)}$$

3.4 Approximation and Related Rates

YouTube http://youtu.be/ViRwEJ0Wfkw

In this section, we deal with approximating a given function by means of a class of functions which are simple and easy to deal with. One such class of functions is polynomials.

Linear Approximation

Given a function $f(x)$, sometimes it is desirable to choose another function $g(x)$ from a class A of functions which are convenient to work with as an approximation of $f(x)$. For the approximation $g(x)$ to be a good approximation of the function $f(x)$, the difference $|f(x)-g(x)|$ which is called the error of the approximation must be small in some definite sense.
In this section, we consider linear approximations, that is, the class A of functions from which we choose an approximation consists of linear functions. If we require $g(x)$ to be a linear function, such that its value of the its first derivative at a given point $x=a$ coincides with the value of f and f' at a, respectively, that is, $g(a)=f(a)$, $g'(a)=f'(a)$, then we have

$$g(x)=f(a)+f'(a)(x-a).$$

This function is the linear approximation of the function $f(x)$ at the point $x=a$, and is denoted by $L_a(x)$ (often we simply write $L(x)$ instead of $L_a(x)$, if it is clear at which point we consider the linear approximation). Thus,

$$L_a(x)=f(a)+f'(a)(x-a).$$

Note that from the Mean Value Theorem, we have

$$f(x)-L_a(x)=f(x)-f(a)-f'(a)(x-a)=[f'(c)-f'(a)](x-a),$$

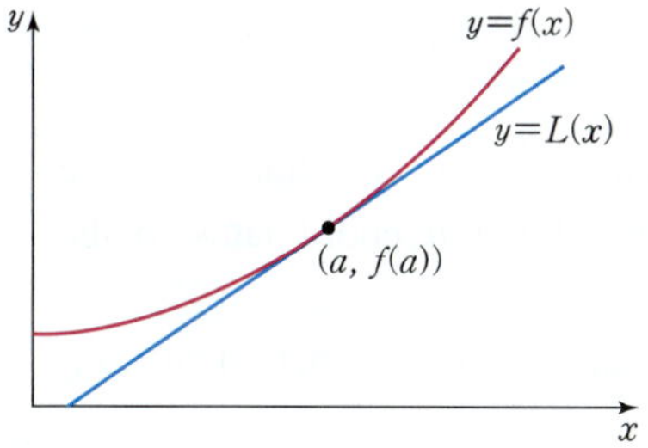

Figure 1

where c is some number between x and a.

Under the assumption that $f'(x)$ is continuous (on an interval contain x and a), then $|f'(c)-f'(a)|$ is small for all values x which are in some small interval containing a, hence the error of the linear approximation is small for all values of x which are close to a. This fact is illustrated in Figure 1. For values x which are far from a, since $|f'(c)-f'(a)|$ can be large, the error of the linear approximation can be large. Thus, linear approximation does not give a good approximation for x which are far from a.

EXAMPLE 1

Determine the linear approximation for $\sin\theta$ at $\theta=0$.

Solution All we need to do is find the tangent line to $\sin\theta$ at $\theta = 0$.

$$f(\theta) = \sin\theta \qquad f'(\theta) = \cos\theta$$
$$f(0) = 0 \qquad f'(0) = 1$$

The linear approximation is,

$$L(\theta) = f(0) + f'(0)(\theta - 0) = 0 + (1)(\theta - 0) = \theta$$

So, as long as θ remains small we can say that $\sin\theta \approx \theta$. ■

Differentials

In this section, we introduce some notation that we will use frequently the next chapter.

Given a function $y = f(x)$, if differential dy (or df) is defined by

$$df = dy = f'(x)\,dx$$

Let Δx denote an actual change in the variable x, and Δy denote the corresponding actual change in the variable y, that is,

$$\Delta y = f(x + \Delta x) - f(x)$$

The idea of using differential is to give an approximation of the value Δy, so that we can use it to approximate $f(x + \Delta x)$ (since $f(x + \Delta x) = f(x) + \Delta y$).

If the function $f(x)$ is continuous, as in our case, then a small change in x produces a small change in y.

Geometrically, the differential may be explained as follows:

Let $P(x, y)$ and $Q(x + \Delta x,\ y + \Delta y)$ be two points on the curve $y = f(x)$. Let $dx = \Delta x$. Since the derivative at P is the same as the slope of the tangent line PT, we have

$$dy = f'(x)dx = \frac{|RT|}{|PR|} \cdot |PR| = |RT|.$$

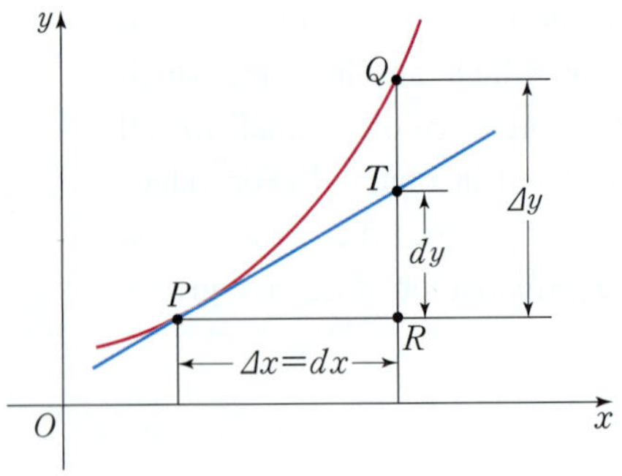

Figure 2

In other words, $dy(= |RT|)$ is the differential of the y coordinates to the tangent line while $\Delta y(= |RQ|)$ is the increment of the y coordinates to the curve.

Therefore, the differential dy is an approximation for the increment Δy. More formally

$$\lim_{\Delta x \to 0} \frac{\Delta y - dy}{\Delta x} = \lim_{\Delta x \to 0} \frac{\Delta y - f'(x)\Delta x}{\Delta x}$$
$$= \lim_{\Delta x \to 0} \frac{\Delta y}{\Delta x} - f'(x) = f'(x) - f'(x) = 0.$$

In other words, df can be used for the approximation of the increment

$\Delta f(x) = f(a+\Delta x) - f(a)$ near $x = a$. We now write $f(a+\Delta x) \approx f(a) + f'(a)\Delta x$. From the above formula, we can find the approximating value of $f(a+\Delta x)$ if we know $f(a)$ and $f'(a)\Delta x = df|_{x=a}$.

EXAMPLE 2

Use differentials to approximate $\sqrt{98}$.

Solution Let $f(x) = \sqrt{x}$. Set $x = 100$ and $\Delta x = -2$.

Since $df \approx \Delta x$, $df = \dfrac{dx}{2\sqrt{x}}$, we have $df|_{x=100} = -\dfrac{1}{10}$.

Hence approximately,

$$\begin{aligned}\sqrt{98} &= f(100+\Delta x) \approx f(100) + df|_{x=100} \\ &= \sqrt{100} - \frac{1}{10} = \frac{99}{10} = 9.9.\end{aligned}$$

■

EXAMPLE 3

Let $y = x^4 - 2x^3 + 9x + 7$. Use differentials to approximate the value of y at $x = 1.997$.

Solution $1.997 = x + \Delta x = 2 - 0.003$.

Thus, we set $x = 2$ and $\Delta x = -0.003$. Since we have

$\dfrac{dy}{dx} = 4x^3 - 6x^2 + 9$, the differential is $dy = (4x^3 - 6x^2 + 9)dx$.

As an approximation at $x = 2$, $dy = (32 - 24 + 9)(-0.003) = -0.051$. When $x = 2$ (and $y = 25$),

$$y + dy = 25 - 0.051 = 24.949.$$

■

(In this case, we can compute $y(1.997) = 24.9491078380810$ and the true error, which is $\Delta y - dy = 0.000107838081$. Hence, the approximating value is accurate within an error of 0.0002, $\Delta y \approx dy$.)

Sometimes, a better description of the error can be given by the relative error, which is computed as the error dy divided by the total quantity y, i.e., the **relative error** is dy/y. Also, sometimes **percentage error** is more suitable. The **percentage error** is the (relative error)$\times 100$ (%).

EXAMPLE 4

The radius of the sphere was measured and found to be 5 cm with a possible error in measurement of at most 0.02 cm. What is the maximum error and the relative error in using this value of radius in computing the volume of the sphere?

Solution If the radius of the sphere is r, then its volume is $V(r) = \dfrac{4\pi}{3}r^3$. If the error in r is denoted by $dr = \Delta r$, then the corresponding error in

V will be denoted by dV. Since $r=s$, $dr=\pm 0.02$, we have

$$dV=4\pi r^2 dr=4\pi(25)(\pm 0.02)=\pm 2\pi\ (\text{cm}^3),\quad V=\frac{500\pi}{3}.$$

Hence, the maximum error is 2π and the relative error is $\dfrac{dV}{V}=\pm\dfrac{2\pi}{500\pi/3}=\pm 0.012$. ■

Quadratic Approximation

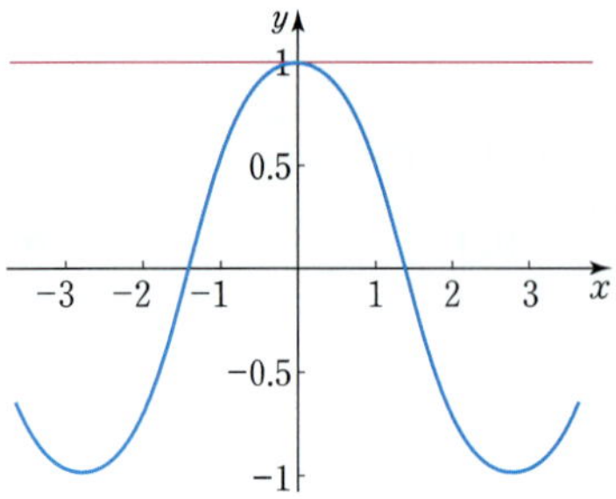

Figure 3

In the above, we considered the linear (tangent line) approximation for a curve. The graph of the cosine function is a very nice looking curve. But it is just that, a curve. The best linear approximation to the cosine function near 0 is quite unexciting; you can check that for $f(x)=\cos(x)$, the best linear approximation near 0 is given by $L_0(x)=1$. One reason that linear approximations are popular is because linear functions are easy to work with. But as we see from the cosine function, linear approximations have their limitations. After linear functions, the next simplest functions are the quadratics. For a better approximation, we approximate a curve by using a parabola instead of a straight line. We will look for a second-degree polynomial in the following form:

$$P(x)=A+B(x-a)+C(x-a)^2,$$

this is a better approximation of $y=f(x)$ near $x=a$ than the linear approximation. Similarly, as with the linear approximation, we require $P(a)=f(a)$, $P'(a)=f'(a)$ and $P''(a)=f''(a)$. Then, $A=f(a)$, $B=f'(a)$ and $C=\dfrac{f''(a)}{2}$.

Hence if $f(x)$ is at least twice differentiable, then

$$P(x)=f(a)+f'(a)(x-a)+\frac{f''(a)}{2}(x-a)^2$$

is the "quadratic approximation" for $y=f(x)$ near $x=a$.

We note that linear approximations at given point a, while quadratic approximation at a, at first and second derivative at a.

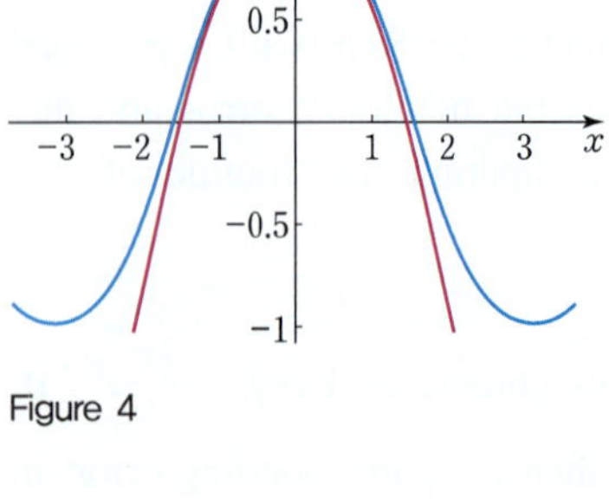

Figure 4

EXAMPLE 5

Consider the cosine function $f(x)=\cos(x)$ and let $a=0$. It is not hard to verify that if we let $P(x)=f(0)+f'(0)x+\dfrac{f''(0)}{2}x=1-\dfrac{x^2}{2}$, then $f(x)=\cos(x)$ and $P(x)=1-\dfrac{x^2}{2}$ have the same first two derivatives at 0. In addition, the graph of $P(x)=1-\dfrac{x^2}{2}$ is very close to the graph of $f(x)=\cos(x)$ (very nicely) near 0.

EXAMPLE 6

Find the quadratic approximation for $y = \sqrt{x+1}$ near $x = 0$. Use it to approximate $\sqrt{0.99}$.

Solution $f'(x) = \dfrac{1}{2\sqrt{x+1}}$ and $f''(x) = -\dfrac{1}{4}(x+1)^{-3/2}$.

The quadratic approximation near $x = 0$ is $P(x) = 1 + \dfrac{1}{2}x - \dfrac{1}{8}x^2$.

$$\sqrt{0.99} \approx 1 + \frac{1}{2}(-0.01) - \frac{1}{8}(0.01)^2 = \frac{79599}{80000} = 0.9949875.$$

A more accurate approximation is $\sqrt{0.99} \approx 0.9949875$ so we have found agreement in 6 decimal places. ■

Related Rates

In the real world, many quantities are related with each other. In turn, their derivatives (rates of change) will be related with each other. This is also a useful approach to the solution of real-world problems. Let us consider some simple examples.

EXAMPLE 7

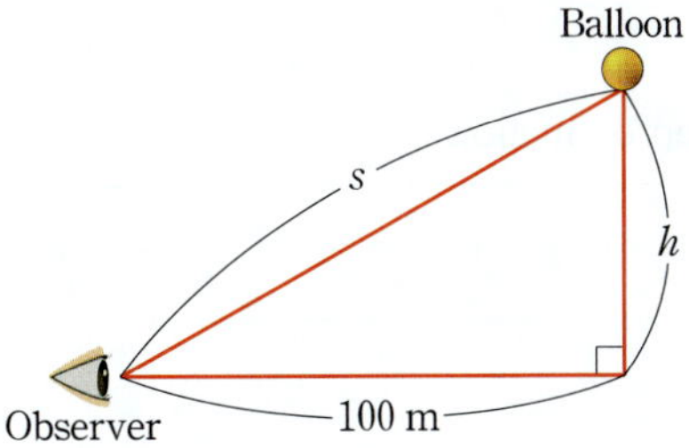

Figure 5

A balloon is released at a point 100m away from an observer, who is on level ground. (See Figure 5.) If the balloon goes straight up at a rate of 5 meters per second, how fast is the distance from the observer to the balloon increasing when the balloon is 50 meters high?

Solution We have a relationship between s and h, i.e., $s^2 = h^2 + 100^2$. Implicit differentiation gives us that $s\dfrac{ds}{dt} = h\dfrac{dh}{dt}$ Note that $h = 50$ and $\dfrac{dh}{dt} = 5$. Since $s = \sqrt{50^2 + 100^2} = 50\sqrt{5}$, we have $\dfrac{ds}{dt} = \sqrt{5}$. At $h = 50$, the distance between the balloon and the observer is increasing at a rate of $\sqrt{5}\ \mathrm{m/s}$. ■

EXAMPLE 8

A rotating camera is focusing on a car driving on the straight road as depicted below. The camera lies 100 meters away from the road. Right now, the car is 200 meters away from the camera, and the angle θ is increasing at a rate of 1/2 radian per second. How fast is the car moving?

Solution We have $\tan\theta = \dfrac{s}{100}$. Differentiating, with respect to t we have

$$\sec^2\theta \frac{d\theta}{dt} = \frac{1}{100}\frac{ds}{dt}.$$

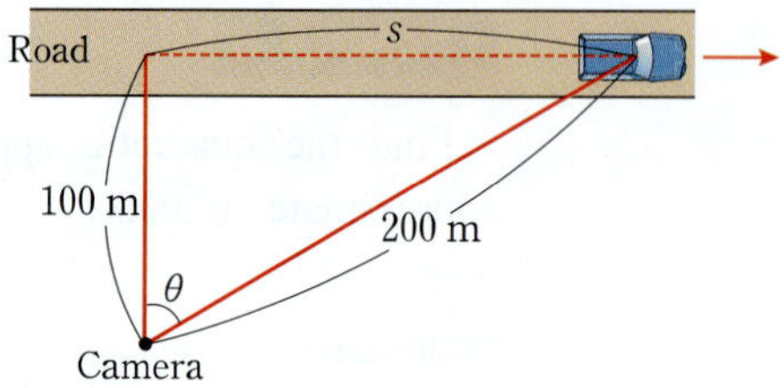

Figure 6

If $s = 100\sqrt{3}$, then $\theta = \dfrac{\pi}{3}$.

Thus, we have $\dfrac{ds}{dt} = 200\,\mathrm{m/s}$. So, the car is moving at the speed of $200\ \mathrm{km/hr}$. ■

CAS **EXAMPLE 9**

Use differential to approximate $\sin(30^\circ 30')$
http://matrix.skku.ac.kr/cal-lab/cal-3-4-Exm-9.html

Solution $30^\circ 30'$ is $30^\circ + 30' = \left(\dfrac{30}{180} + \dfrac{30}{1800}\right)\pi = \dfrac{\pi}{6} + \dfrac{\pi}{60}$ and

$\sin 30^\circ 30' = \sin\left(\dfrac{\pi}{6} + \dfrac{\pi}{60}\right) \approx \sin\dfrac{\pi}{6} + \left(\cos\dfrac{\pi}{6}\right)\dfrac{\pi}{60}$.

```
var('x,dx');
f(x)=sin(x)
f(pi/6 + pi/60).n()  # Numerical approximation
```

Answer : 0.544639035015027

```
dy(x)=diff(f(x),x)*dx
f(pi/6)+dy(x=pi/6,dx=pi/60).n()
# Approximation by derivative
```

Answer : 0.545344984105855 ■

3.4 EXERCISES (Approximations and Related Rates)

http://matrix.skku.ac.kr/Cal-Book/part1/CS-Sec-3-4-Sol.html

YouTube http://youtu.be/JmBOv6_D6qA

CAS **1-3.** Use differential to approximate the following quantities.

1. $\sqrt{27}$

http://matrix.skku.ac.kr/cal-lab/cal-3-3-2.html

Solution

```
var('a,b,x,dx');
f(x)=(x+25)^(1/2);
dy(x)=diff(f(x),x)*dx;
dy(x)
```

Answer : 1/2*dx/sqrt(x+25)

```
(f(0)+dy(x=0, dx=2)).n()
```

Answer : 5.20000000000000

2. $\sqrt[4]{36.01}$ (and $\sqrt[3]{61}$)

http://matrix.skku.ac.kr/cal-lab/cal-3-3-3.html

(http://matrix.skku.ac.kr/cal-lab/cal-3-4-exs-2.html)

Solution In order to find an approximate of $\sqrt[4]{36.01}$, we define a function $f(x)=\sqrt[4]{x+36}$.

Then $f(0.01)=\sqrt[4]{36.01}$

Now we use differentials to approximate. An approximate of $\sqrt[4]{36.01}$ can be found by using $f(0)+dy$.

$$dy=f'(0)\,dx=\frac{1}{4}\left(\frac{1}{0+36}\right)^{\frac{3}{4}}dx.$$

Set $x=0$ and $dx=0.01$ at $dy=f'(x)dx$

$$dy=\frac{1}{4}\left(\frac{1}{36}\right)^{\frac{3}{4}}0.01,\ f(0)=\sqrt[4]{36}.$$

Therefore $\sqrt[4]{36.01}=f(0)+dy\approx 2.44965982852130$.

```
var('a,b,x,dx')
f(x)=(x+60)^(1/3);
f(x);
f(1.05)
```

Answer : 1/3*dx/(x + 60)^(2/3)

```
dy(x)=diff(f(x),x)*dx
(f(1)+dy(x=1, dx=0)).n()
```

Answer : 3.93649718310217

3. $\sqrt[5]{32.05}$

http://matrix.skku.ac.kr/cal-lab/cal-3-3-1.html

Solution In order to find an approximation of $\sqrt[5]{32.05}$, we define a function $f(x)=\sqrt[5]{x+32}$.

Then $f(0.05)=\sqrt[5]{32.05}$.

```
var('a,b,x,dx');
f(x)=(x+32)^(1/5);
f(x);
```

Answer : (x+32)^(1/5)

Now we use differentials to approximate. Set $x=1$ and $dx=0.05$ at $dy=f'(x)dx$. Find an approximation of $\sqrt[5]{32.05}$ by using $f(1)+dy$

```
dy(x)=diff(f(x),x)*dx;
dy(x)
```

Answer : 1/5*dx/(x+32)^(4/5)

```
f(1)+dy(x=1,dx=0.05)
```

Answer : 2.00062500000000

There are Sage built in command .n() for finding such approximations.

```
(32.05)^(1/5).n()
```

Answer : 2.00062460974081

4. The height of a circular cone is the same as the radius of its circular bottom. The height and radius were measured and found to be 4 cm with a possible error in measurement of at most 0.04 cm. What is the relative error in using this value to compute the volume?

Solution $V=\frac{1}{3}\pi r^2 h=\frac{1}{3}\pi r^3$ (Because $r=h$)

$\Rightarrow dV=\pi r^2 dr$

$$\Rightarrow \frac{dV}{V}=\frac{\pi r^2 dr}{\frac{1}{3}\pi r^3}=\frac{3}{r}dr=\frac{3}{4}(0.04)=0.03$$

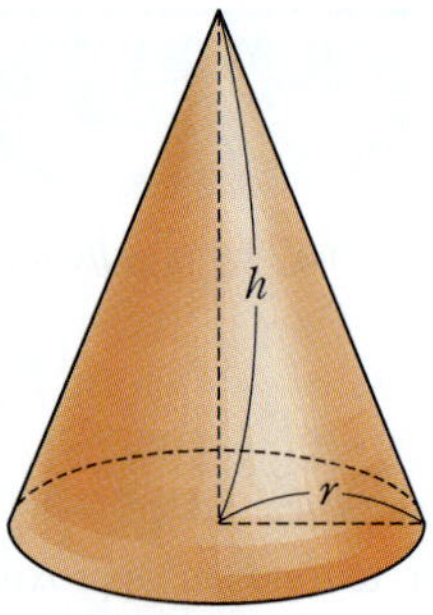

5. Find an approximation of the difference between the surface areas of two spheres whose radii are 5 cm and 5.05 cm, respectively.

Solution $5.05 = x + \Delta x = 5 + 0.05$. Thus $\Delta x = 0.05$.

Since we have $S = 4\pi r^2 \Rightarrow S'(r) = \dfrac{dS}{dr} = 8\pi r$.

$\Rightarrow dS = S'(5)dr = 8\pi(5)(0.05) = 2\pi$

6. The period of a pendulum is given by the formula $T = 2\pi\sqrt{\dfrac{L}{g}}$, where L is the length of the pendulum measured in meters and $g = 9.8\,\mathrm{m/s^2}$ is the gravitational constant. If the length of the pendulum is measured to be 3 m with a possible error in measurement 1 cm. What is the approximate percentage error in calculating the period T?

Solution $T = 2\pi\sqrt{\dfrac{L}{g}}$, $g = 9.8\mathrm{m/s^2}$,

$T = \dfrac{2\pi}{\sqrt{g}}\sqrt{L} \Rightarrow dT = \dfrac{2\pi}{\sqrt{g}}\dfrac{1}{2}\dfrac{1}{\sqrt{L}}dL$

$$\Rightarrow \frac{dT}{T} = \frac{\frac{2\pi}{\sqrt{g}}\frac{1}{2}\frac{1}{\sqrt{L}}}{\frac{2\pi}{\sqrt{g}}\sqrt{L}}dL = \frac{1}{2L}dL = \frac{0.01}{6}$$

The approximate percentage error = (relative error)×100%
= 0.16 %.

7. A ladder 10 meters long is leaning against a wall. If the foot of the ladder is being pulled away from the wall at 4 m/s, how fast is the top of the ladder sliding down the wall when the foot of the ladder is 8 meter away from the wall?

Solution Let r be the distance between the foot of the ladder to the wall and h be the distance between the top of the ladder to the bottom.

Then from the Pythagorean Theorem,

$r^2 + h^2 = 10^2 = 100$, $h = \sqrt{100 - r^2}$, $dh = \dfrac{1}{2\sqrt{100-r^2}}(-2r)dr$.

Since $r = 8$, $dr = 4$, so $dh = \dfrac{1}{2}\dfrac{1}{6}(-16)4 = -\dfrac{16}{3}$.

Therefore, the speed of the ladder sliding down from the top is $\dfrac{16}{3}\mathrm{m/s}$.

8. A ladder 10 meters long is leaning against a wall. If the top of the ladder is sliding down the wall at 3 m/s, how fast is the foot of the ladder being pulled away from the wall when the foot of the ladder is 6 meter away from the wall?

Solution $r^2 = 100 - h^2 \Rightarrow dh = \dfrac{1}{2}\dfrac{1}{\sqrt{100-r^2}}(-2r)dr$

Since $dh = -3$, $r = 6$, $-3 = \dfrac{1}{2}\dfrac{1}{8}(-12)dr \Rightarrow dr = 4$.

9. A ladder 10m long is leaning against a wall. If the top of the ladder is sliding down the wall at 1m/s, how fast is the angle between the top of the ladder and the wall changing when the foot of the ladder is 5m away from the wall?

Solution $h = 10\cos\theta \Rightarrow \dfrac{dh}{dt} = -10\sin\theta\dfrac{d\theta}{dt}$

$\Rightarrow \sin\theta = \dfrac{5}{10} = \dfrac{1}{2}$ when $h = 5$.

$\dfrac{dh}{dt} = -1$, $-1 = -10 \times \dfrac{1}{2}\dfrac{d\theta}{dt} \Rightarrow \dfrac{d\theta}{dt} = \dfrac{1}{5}$.

Therefore, when the foot of ladder is 5m away, angle at the top of the ladder changes is $\dfrac{1}{5}$.

10. Two cars start moving from the same point. One travels south at A km/hr and the other travels west at B km/hr. How fast is the distance changing between the two cars?

Solution Distance of travel south is a, distance of travel west is b, and the distance between the two travelers is l at time t

$l^2 = a^2 + b^2$

$\Rightarrow l = \sqrt{a^2 + b^2} = \sqrt{(At)^2 + (Bt)^2}$

$\Rightarrow \dfrac{dl}{dt} = \sqrt{A^2 + B^2}$

11. Water is being pumped at a rate of 20 liters per minute into a tank shaped like a frustrum of a right circular cone. The tank has an altitude of 8 meters and lower and upper radii of 2 and 4 meters, respectively. How fast is the water level rising when the depth of the water is 3 meters?

Solution $8+x:4=x:2 \Rightarrow 4x=16+2x \Rightarrow x=8$.

Let $h'=h+8$.

$$V=\frac{1}{3}\pi r^2 h' = \frac{\pi}{48}(h')^3 \Rightarrow \frac{dV}{dt}=\frac{\pi}{16}(h')^2\frac{dh'}{dt}$$

If $\frac{dV}{dt}=20 \Rightarrow \frac{\pi}{16}(h')^2\frac{dh'}{dt} = \frac{\pi}{16}11^2\times\frac{dh'}{dt}=20$

$\Rightarrow \frac{dh'}{dt}=\frac{320}{121\pi}$

Therefore the water level rises $\frac{320}{121\pi}\mathrm{m/s}$ when the depth of the water is 3 meters.

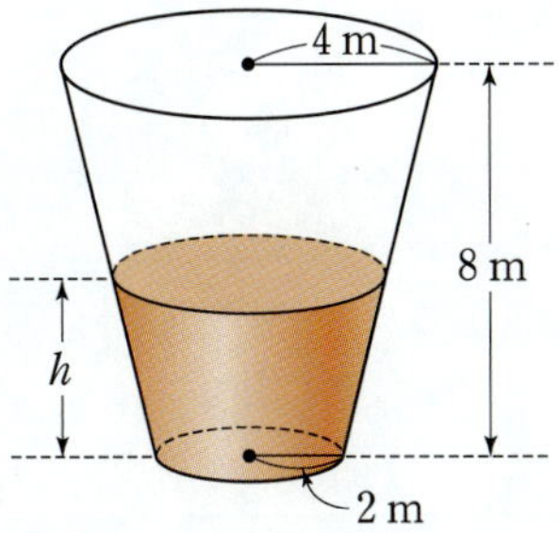

12. Water is being pumped at a rate of 20 liters per minute into a tank shaped like a hemisphere. The tank has a radius of 8 meters. How fast is the water level rising when the depth of the water is 3 meters?

Solution $V=\frac{2}{3}\pi r^3 \Rightarrow dV=2\pi r^2 dr \Rightarrow 20=2\pi(3)^2 dr$.

$\Rightarrow dr=\frac{10}{9\pi}$

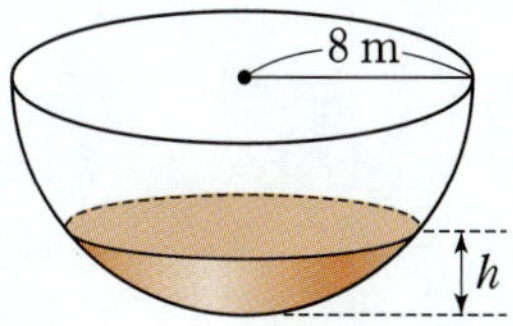

13. A snowball melts at a rate proportional to its surface area. Does the radius shrink at a constant rate? If it melts to 1/4 its original volume in one hour, how long does it take to melt completely?

Solution $S=4\pi r^2$, $V=\frac{4}{3}\pi r^3$. The rate of melting is described by $\frac{dV}{dt}=kS$. Since the snowball melts at a rate proportional to its surface area. Notice that

$$k4\pi r^2 = kS=\frac{dV}{dt}=4\pi r^2\frac{dr}{dt}$$

so $\frac{dr}{dt}=k$ is constant. Let the volume of first time be $V_1=\frac{4}{3}\pi r_1^3$, after one hour,

$$V_2=\frac{4}{3}\pi r_2^3=\frac{1}{4}V_1 \Rightarrow r_2^3=\frac{1}{4}r_1^3 \Rightarrow r_2=\frac{1}{\sqrt[3]{4}}r_1.$$

$$(r_1-r_2):1=r_1:t \Rightarrow \left(1-\frac{1}{\sqrt[3]{4}}\right)r_1:1=r_1:t \Rightarrow t$$

$$=\frac{\sqrt[3]{4}}{\sqrt[3]{4}-1}\approx 2.7024.$$

The snowball melts completely at time ($t\approx 2.7024$).

Chapter 4

Applications of Derivatives

John von Neumann

1903-1957

Inventor of Modern Computer

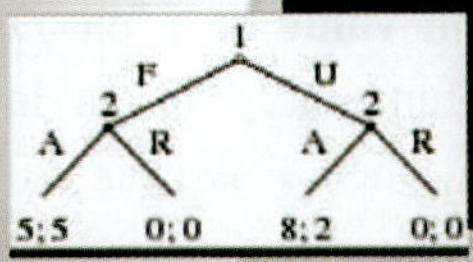

[An Extensive Form of Game Theory]

Regarded as one of the foremost mathematicians of the twentieth century, von Neumann made major contributions to a vast range of mathematical fields. Along with Edward Teller and Stanislaw Ulam, von Neumann worked out key steps in the nuclear physics involved in thermonuclear reactions and the hydrogen bomb.

YouTube http://youtu.be/mXVU8OqIHJY

In this chapter, we will look at applications of derivatives. We will apply derivatives to the problem of finding the maximum and minimum values of a function, whether they are local (relative) or global (absolute.) Derivatives may also be applied to problems about maximizing profit, and/or revenue, or minimizing costs, and/or the amount of material used in construction. Derivatives will also be utilized to determine and/or discover the shape of a function's graph.

4.1 Extreme values of a function

Basic concept of extreme values

Let f be a function defined on a domain D. A function f has an **absolute maximum** (or **global maximum**) at c if $f(c) \geq f(x)$ for all x in D. The number $f(c)$ is called the **absolute maximum value** of f on D. Similarly, f has an **absolute minimum (**or **global minimum)** at c if $f(c) \leq f(x)$ for all x in D; the number $f(c)$ is called the **absolute minimum value** of f on D.

Let us now consider only values of x near c. A function f has a **local maximum** (or **relative maximum**) at c if $f(c) \geq f(x)$ when x is near c, that is, there exists some open interval $I = (a, b)$ containing c such that $f(c) \geq f(x)$ for all x in $D \cap I$. The number $f(c)$ is called the **local maximum value** of f on D. Similarly, f has a **local minimum** (or **relative minimum**) at c if $f(c) \leq f(x)$ when x is near c, and the number $f(c)$ is called the **local minimum value** of f on D.

By the definition, the global maximum and global minimum values are also local maximum and local minimum values of f, respectively. However, the converse is not true in general. In general, the local maximum and local minimum values are called the **extreme values of a function** $\boldsymbol{f}$.

Figure 1 shows the graph of a function f defined on $[a, b]$ with **local maxima** (plural in Latin) at c_2, c_4 and **local minima** at c_1, c_3, c_5. Note that the function f has the absolute maximum value $f(c_2)$ on $[a, b]$ and the absolute minimum value $f(c_3)$ on $[a, b]$.

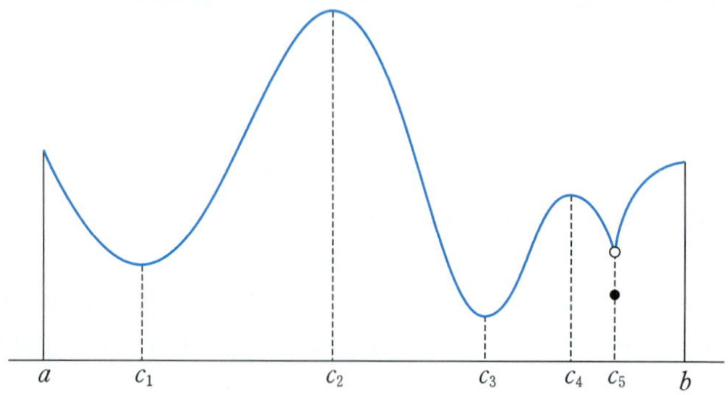

Figure 1

EXAMPLE 1

Let $f(x) = \cos x$. Since $-1 \leq \cos x \leq 1$ for all x and $\cos 2n\pi = 1$ for any integer n, f has maximum value of 1 which is achieved at infinitely many points ($x = 2n\pi$, $n = 0, \pm 1, \ldots$) infinitely many times. Similarly, since $\cos(2n+1)\pi = -1$ for any integer n, f has minimum value of -1 infinitely many times, achieved at infinitely many points ($x = (2n+1)\pi$, $n = 0, \pm 1, \ldots$). ■

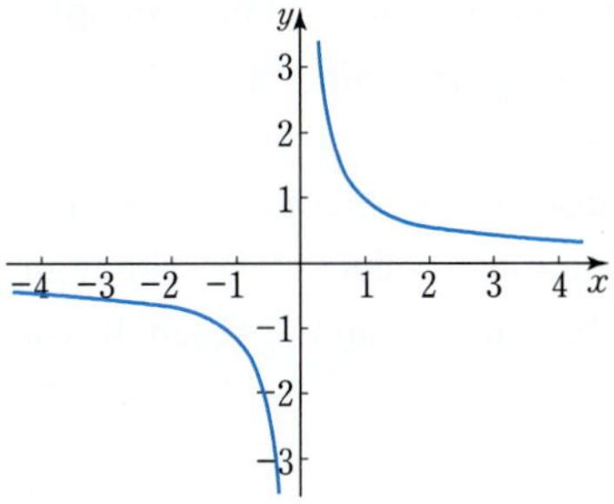

Figure 2 $f(x)=1/x(x \neq 0)$

EXAMPLE 2

Let $f(x)=\dfrac{1}{x}(x \neq 0)$. Then f has no extreme values as we see in Figure 2. ■

As shown in the examples above, some functions have extreme values whereas others do not. In the following theorem we obtain some conditions under which a function has extreme values.

> **THEOREM 1 The Extreme Value Theorem**
>
> If f is continuous on a closed interval $[a, b]$, then f attains an absolute maximum value $f(c)$ and an absolute minimum value $f(d)$ at some numbers c and d in $[a, b]$.

If either hypothesis (continuity or closed interval) of the Extreme Value Theorem is not satisfied, then the function need not possess an absolute maximum value nor an absolute minimum value. For instance, the function f in Figure 3 has minimum value $f(2)=0$ but no maximum value. The function g in Figure 4 has no maximum and minimum values.

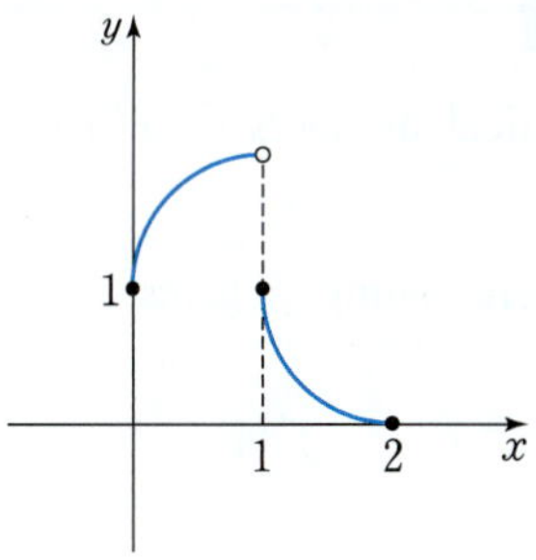

Figure 3 Graph of f

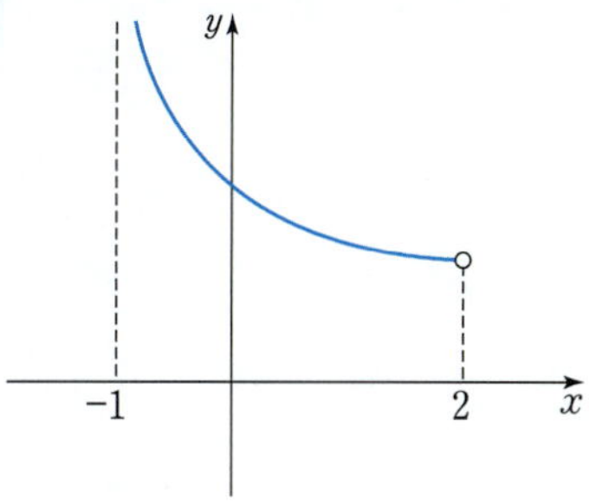

Figure 4 Graph of g

Let us now consider extreme values for a differentiable function. As illustrated in Figure 1, the tangent lines of f at the points $x=c_1, c_2, c_3, c_4$ are horizontal. It suggests that if f has a local maximum or minimum at $x=c$ and $f'(c)$ exists then $f'(c)=0$. In general, we have the following theorem.

> **THEOREM 2 Fermat's Theorem**
>
> If f is differentiable at $x=c$ and has an extreme value at c, then $f'(c)=0$.

This theorem can be easily proven by applying the definition of extreme values and of a derivative. (See Exercise 28.)

The converse of Fermat's theorem is not true in general, that is, although

$f'(c)=0$, f does not necessarily have extreme values at c. For example, $f(x)=x^3$ has neither local maximum nor local minimum value at 0 although $f'(0)=0$.

We should note that f may have an extreme value at c where $f'(c)$ does not exist. For instance, $f(x)=|x|$ has its local minimum value at 0, although $f'(0)$ does not exist in Figure 5. Also see the function f at c_5 which has a removable discontinuity in Figure 1.

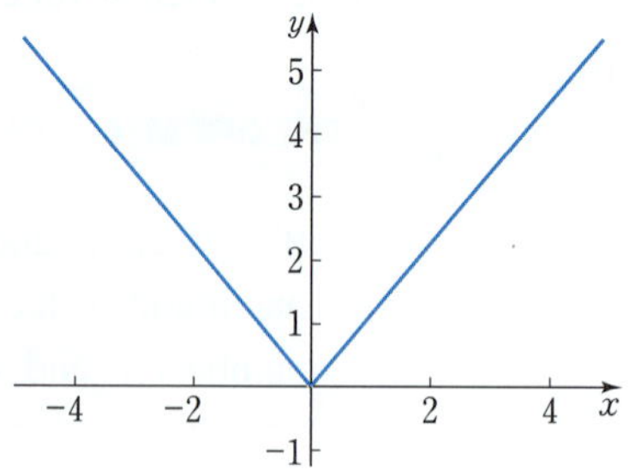

Figure 5 $f(x)=|x|$

A **critical number (point)** of a function f is a number c in the domain of f such that either $f'(c)=0$ or $f'(c)$ does not exist.[1)]

To find extreme values of f, we should find critical numbers of f.

1)http://tutorial.math.lamar.edu/Classes/CalcI/CriticalPoints.aspx

EXAMPLE 3

Find the critical numbers of $f(x)=x^2+\dfrac{16}{x}$.

Solution Differentiating f gives

$$f'(x)=2x-\frac{16}{x^2}=\frac{2(x-2)(x^2+2x+4)}{x^2}$$

Therefore, $f'(x)=0$ if $x=2$ and $f'(x)$ does not exist when $x=0$. Thus, the critical numbers are 2 and 0. ■

To find the absolute maximum and minimum values of a differentiable function f on a closed interval $[a, b]$, we suggest the following three steps:

Step 1. Find the values of f at the critical numbers of f in (a, b).

Step 2. Find the values of f at the endpoints of the interval.

Step 3. The largest of the values from Steps 1 and 2 is the absolute maximum value; the smallest of these values is the absolute minimum value. This method is called "The Closed Interval Method."

EXAMPLE 4

Let $f(x)=x^2-2|x|+2$, $\left[-\dfrac{1}{2}, \dfrac{3}{2}\right]$. Find the absolute maximum and absolute minimum values of f on the given interval.

Solution (Step 1) Differentiating f gives

$$f'(x) = \begin{cases} 2x+2 & \text{if } -\dfrac{1}{2} < x < 0 \\ 2x-2 & \text{if } 0 < x < \dfrac{3}{2} \end{cases}$$

Therefore, $f'(x)=0$ if $x=1$ and $f'(x)$ does not exist when $x=0$. Thus, the critical numbers on the given interval are 1 and 0. At these critical numbers, $f(1)=1$, $f(0)=2$.

(Step 2) At the endpoints of the interval, $f\left(-\dfrac{1}{2}\right)=\dfrac{5}{4}$, $f\left(\dfrac{3}{2}\right)=\dfrac{9}{4}-3+2=\dfrac{9}{4}-1=\dfrac{5}{4}$.

(Step 3) Comparing these four function values, the absolute maximum value is $f(0)=2$ and the absolute minimum value is $f(1)=1$.

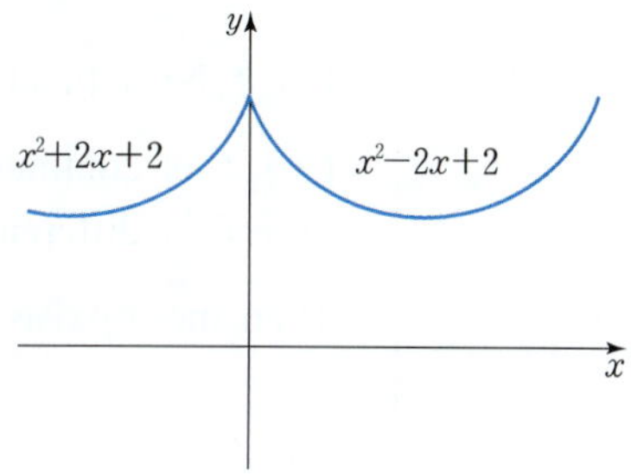

Figure 6 $f(x)=x^2-2|x|+2$

■

The Mean Value Theorem

We will now study how the signs of the first and second derivatives f' and f'' influence the extreme values of f. The Mean Value Theorem will be useful not only presently but also for explaining why some of the basic properties of calculus are true. We begin with Rolle's theorem. (For proof refer to Exercise 29.)

THEOREM 3 Rolle's Theorem

Let f be a function satisfying the three conditions:

(i) f is continuous on the closed interval $[a, b]$;
(ii) f is differentiable on the open interval (a, b);
(iii) $f(a)=f(b)$.

Then there exists a number c in (a, b) such that $f'(c)=0$.

We can see that this theorem is reasonable by interpreting it geometrically. Geometrically, Rolle's Theorem tells us that at some point between a and b

the tangent line of the graph of f is horizontal whenever f satisfies Rolle's hypotheses. See Figure 7(a).

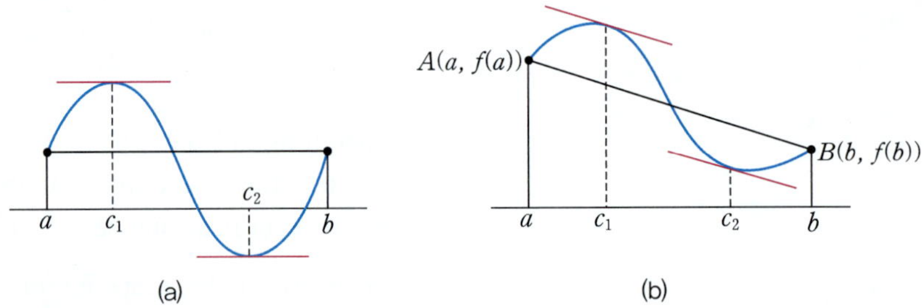

Figure 7

Rolle's theorem can be used to prove another well known theorem in calculus the **Mean Value Theorem.**

THEOREM 4 Mean Value Theorem

Let f be a function satisfying the following two conditions:

(i) f is continuous on the closed interval $[a, b]$;
(ii) f is differentiable on the open interval (a, b).

Then there exists a number c in (a, b) such that

1 $$f'(c) = \frac{f(b)-f(a)}{b-a}$$

or, equivalently,

2 $$f(b)-f(a) = f'(c)(b-a).$$

Proof Let us define a new function h as

3 $$h(x) = f(x) - f(a) - \frac{f(b)-f(a)}{b-a}(x-a).$$

By the assumptions (i) and (ii) for f, we see that h is continuous on $[a, b]$ and is differentiable on (a, b). In addition, it can be easily seen that $h(a) = h(b)$. Thus h satisfies Rolle's hypotheses. By Rolle's Theorem, there exists a number c in (a, b) such that $h'(c) = 0$, that is,

$$0 = h'(c) = f'(c) - \frac{f(b)-f(a)}{b-a}$$

and so

$$f'(c) = \frac{f(b)-f(a)}{b-a}.$$ ■

We note that the hypotheses of the Mean Value Theorem do not require f to be differentiable at either a or b. It is enough to be continuous at a and b. Geometrically, the Mean Value Theorem states that there is at least one point $P(c, f(c))$ on the graph of $f(x)$ where the slope of the tangent line is parallel to the secant line AB joining the points $A(a, f(a))$ and $B(b, f(b))$. (See Figure 7(b).)

In general, the Mean Value Theorem says that there is a number at which the instantaneous rate of change is equal to the average rate of change over an interval. This enables us to obtain information about a function from information about its derivative. For instance, if an object moves in a straight line with position function $s = f(t)$, then the velocity at $t = c$ is $f'(c)$ and the average velocity between $t = a$ and $t = b$ is $\dfrac{f(b) - f(a)}{b - a}$. Thus by the Mean Value Theorem we see that the instantaneous velocity $f'(c)$ is equal to that average velocity at some time $t = c$ between a and b.

Extreme Value Tests at a Critical Number

Recall that by Fermat's Theorem if f has a local maximum or minimum at c then c must be a critical number of f. But not every critical number gives rise to an extreme value. Thus we need a test that will tell us whether or not f has a local maximum or minimum at a critical number. First, it is important to use the derivative to determine intervals on which a function f is increasing or decreasing.

Definition of Increasing and Decreasing Test

(i) If $f'(x) > 0$ on an interval then f is increasing on that interval.
(ii) If $f'(x) < 0$ on an interval then f is decreasing on that interval.

This can be easily proven by the Mean Value Theorem (see Exercise 30).

The following two tests are useful in determining whether a critical number gives rise to a local maximum or minimum.

The First Derivative Test

Let c be a critical number of a function f that is differentiable on some open interval containing c, except possibly at c.

(i) If f' changes from positive to negative at c, then f has a local maximum at c.
(ii) If f' changes from negative to positive at c, then f has a local minimum at c.
(iii) If f' does not change sign at c, then f has neither local maximum nor local minimum at c.

Proof (i) Let us choose a, b sufficiently near to c such that $a < c < b$. By the Mean Value Theorem, there exist c_1, c_2 with $a < c_1 < c$ and $c < c_2 < b$ such that

$$\frac{f(c)-f(a)}{c-a} = f'(c_1) > 0 \text{ and } \frac{f(b)-f(c)}{b-c} = f'(c_2) < 0.$$

Since $f(a) < f(c)$ for $a < c$ and $f(b) < f(c)$ for $b < c$, f is increasing on (a, c) and decreasing on (c, b). Hence f has a local maximum at c. This proves the first case of the theorem. The second and third cases can be proven in a similar way. (See Exercise 31.) ■

EXAMPLE 5

Find the local minimum and maximum values of $f(x) = |x^2 - 1|$.

Solution Differentiating f gives

$$f'(x) = \begin{cases} 2x & \text{if } x < -1 \\ -2x & \text{if } -1 < x < 1 \\ 2x & \text{if } 1 < x \end{cases}$$

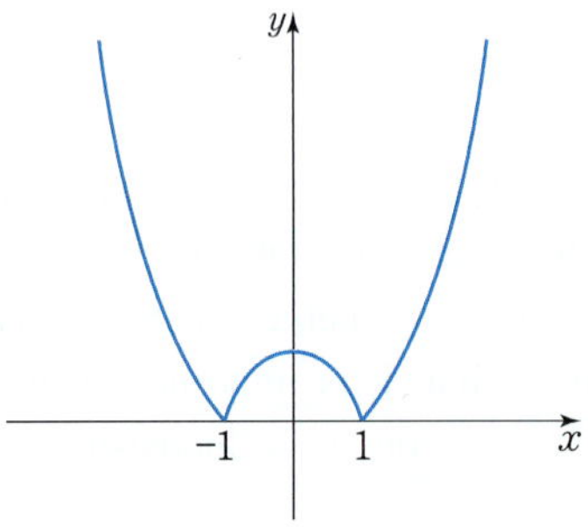

Figure 8 $f(x) = |x^2 - 1|$

Therefore, $f'(x) = 0$ if $x = 0$ and $f'(x)$ does not exist when $x = -1$, and 1. Thus, the critical numbers are -1, 0 and 1. f' changes from negative to positive at -1, so $f(-1) = 0$ is a local minimum value by the First Derivative Test. Similarly, f'changes from positive to negative at 0, so $f(0) = 1$ is a local maximum value, and, f'changes from negative to positive at 1, so $f(1) = 0$ is also a local minimum. (See Figure 8.) ■

> **The Second Derivative Test**
>
> Let c be a critical number such that $f'(c) = 0$ and f'' exists in an open interval containing c.
>
> (i) If $f''(c) > 0$, then f has a local minimum at c.
> (ii) If $f''(c) < 0$, then f has a local maximum at c.

Proof (i) Since $f''(c) > 0$, by the Increasing and Decreasing Test, $f'(x)$ is increasing on some open interval (a, b) containing c. Hence $f'(x) < f'(c) = 0$ for $a < x < c$ and $f'(x) > f'(c) = 0$ for $c < x < b$. Thus by the First Derivative Test (ii), f has a local minimum at c. This proves the first case of the theorem. The second case can be proved in a similar way. (See Exercise 32). ■

CAS **EXAMPLE 6**

Find all critical numbers for the following function.

$$f(x) = 9x^4 + 26x^3 - 27x^2 - 60x + 4.$$

Solution

```
var('x')
f(x)=9*x^4 + 26*x^3 - 27*x^2 - 60*x +4
df(x)=diff(f) # 6*(x - 1)*(2*x + 5)*(3*x + 2) : factored
solve(df==0,x)
```

Answer: [x == (-5/2), x == (-2/3), x == 1] ■

EXAMPLE 7

Find the local minimum and maximum values of $f(x) = x^3 - x$.

Solution Differentiating f gives $f'(x) = 3x^2 - 1$. Thus $x = -1/\sqrt{3},\ 1/\sqrt{3}$ are critical numbers such that $f'(x) = 0$. Since $f''(x) = 6x$, we have $f''(-1/\sqrt{3}) < 0$ and $f''(1/\sqrt{3}) > 0$. Thus by the Second Derivative Test, $f(-1/\sqrt{3}) = 2/(3\sqrt{3})$ is a local maximum value and $f(1/\sqrt{3}) = -2/(3\sqrt{3})$ is a local minimum value.

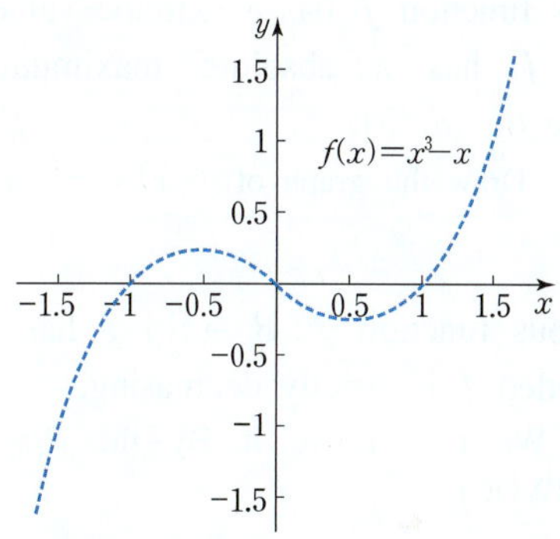

Figure 9 $f(x) = x^3 - x$

■

4.1 EXERCISES (Extreme Values of a Function)

http://matrix.skku.ac.kr/Cal-Book/part1/CS-Sec-4-1-Sol.html

YouTube http://youtu.be/_V4MryNEzWY

1-4. Determine if the following statement is true or false. Explain your answer.

1. If $f:[a,\ b]\to\mathbb{R}$ is a continuous function, then f has a maximum at only one point in $[a,\ b]$.

Solution False. Consider $f(x)=0$ that has more than one maximum.

2. One can apply the Mean Value Theorem to $f(x)=|x|$ on $[-1,\ 1]$.

Solution False, since f is not differentiable on the open interval $(-1,\ 1)$.

3. If a continuous function f has a extreme value on $(a,\ b)$, then f has a absolute maximum or minimum value on $(a,\ b)$.

Solution False. Draw the graph of $f(x)=x^3-x^2$ on $(-1,\ 2)$.

4. For a continuous function $f:\mathbb{R}\to\mathbb{R}$, f has only one zero provided f is strictly decreasing.

Solution True. We may prove it By the Way Of Contradiction (BWOC).

5-8. Find all critical numbers of the given function.

CAS **5.** $f(x)=9x^4-62x^3+159x^2-180x$

http://matrix.skku.ac.kr/cal-lab/cal-4-1-5.html

Solution

```
var('x')
f=9*x^4-62*x^3+159*x^2-180*x
df=diff(f)
solve(df==0,x)
```

Answer : [x == 2, x == (3/2), x == (5/3)]

6. $f(x)=\sqrt{3}\,\sin x-\cos x$

Solution $f'(x)=\sin x+\sqrt{3}\cos x=\sqrt{3+1}\,\sin l\left(x+\dfrac{\pi}{3}\right)$

$=2\sin(x+\dfrac{\pi}{3})\,f'(x)=0\Rightarrow\ x=-\dfrac{\pi}{3}\pm n\pi(n=0,\,1,\,2,\,\cdots)$

Therefore critical numbers of $f(x)$ are

$x=-\dfrac{\pi}{3}\pm n\pi(n=0,\ 1,\ 2,\ \cdots)$

CAS **7.** $f(x)=|x|+|x^2-1|$

http://matrix.skku.ac.kr/cal-lab/cal-4-1-7.html

Solution

```
var('x')
f=abs(x)+abs(x^2 -1)
df=diff(f, x)
solve(df==0,x)
```

Answer : [x == 0, x^2 == -1/2*(abs(x^2 - 1) - 2*abs(x))/abs(x)]

CAS **8.** $f(x)=x^2e^x$

http://matrix.skku.ac.kr/cal-lab/cal-4-1-8.html

Solution

```
var('x')
f=x^2*exp(x)
print f
plot(f, xmin=-3, xmax=3, ymax=5)
```

To find critical numbers, we differentiate $f(x)$.

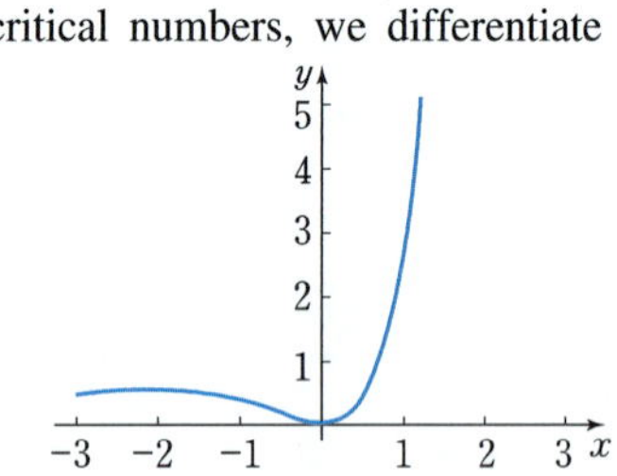


```
df=diff(f)
print df
```

Answer : x^2*e^x + 2*x*e^x # $f'(x)=2xe^x+x^2e^x$

Therefore, $f'(x)=0\Rightarrow e^x(x^2+2x)=0$ and $x^2+2x=0$ (since $e^x>0$)

```
solve((df/e^x)==0, x)
```

Answer : [x == -2, x == 0]

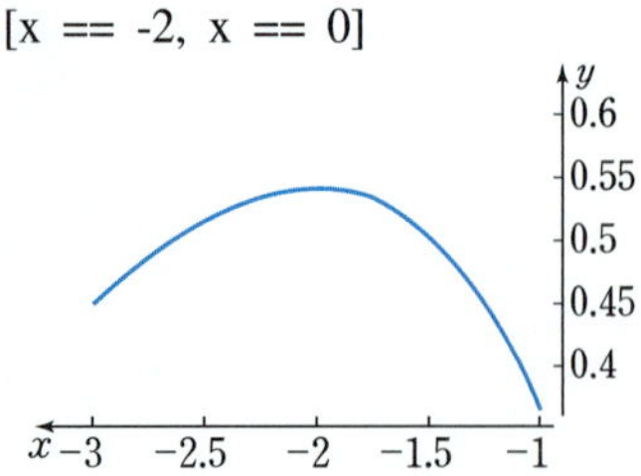

```
print f(-2)
plot(f, xmin=-3, xmax=-1, ymax=0.6)
```

```
print f(0)
plot(f, xmin=-1, xmax=1, ymax=0.6)
```

Critical numbers of $f(x)$ are $x=0$ or $x=-2$.

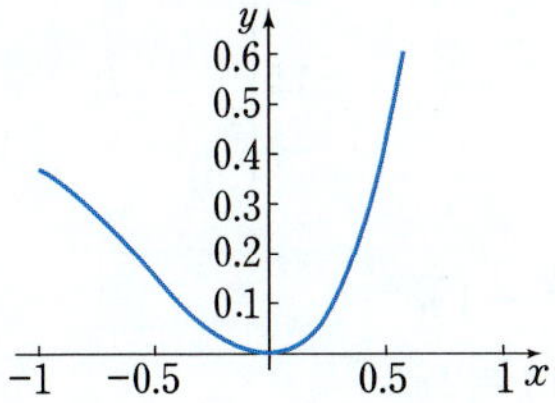

9-13. Find all local maxima and minima if any exist.

CAS **9.** $f(x)=\sqrt{5-x}$, $x<5$

http://matrix.skku.ac.kr/cal-lab/cal-4-1-9.html

Solution No local minimum and maximum when

```
var('x');
f(x)=sqrt(5-x)
P = plot(f,x, xmax=5, xmin=0, linestyle="--", color='red')
Q = text("$f(x)=sqrt(5-x) $", (1,1), fontsize=20)
show(P+Q)
```

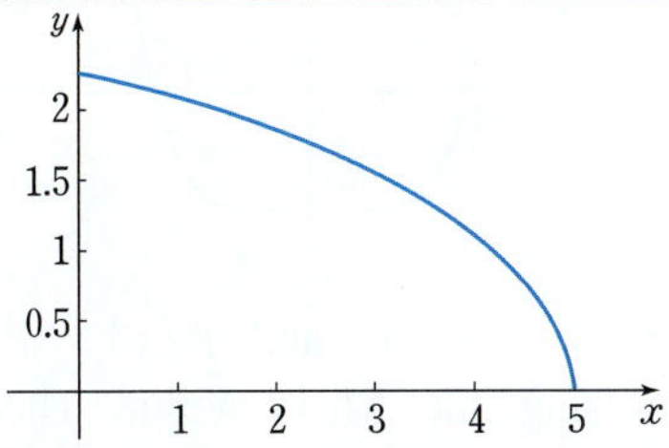

CAS **10.** $f(x)=|x|$

http://matrix.skku.ac.kr/cal-lab/cal-4-1-10.html

Solution Local minimum $f(0)=0$, but no local maximum.

```
var('x')
f=abs(x)
plot(f,-1,1)
```

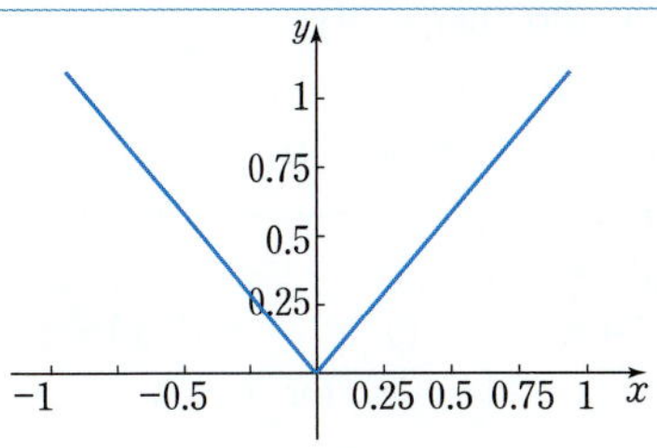

CAS **11.** $f(x)=|\sin x|$, $0\le x\le \pi$

http://matrix.skku.ac.kr/cal-lab/cal-4-1-11.html

Solution

```
var('x');
f=abs(sin(x))
plot(f,0,pi)
```

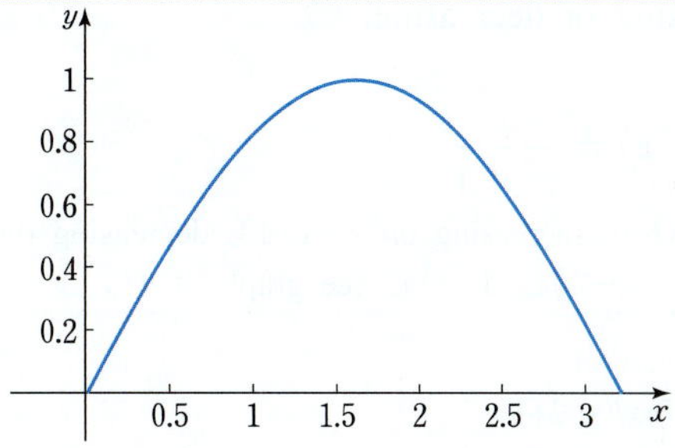

```
df=diff(f)
df
```

Answer : sin(x)*cos(x)/abs(sin(x))

```
plot(df,0,pi)
```

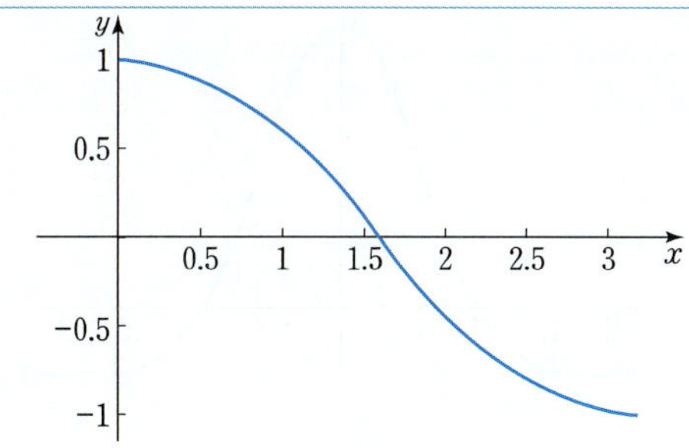

```
plot(df,0,pi)
```

Answer : [x == 0, x == 1/2*pi, x == pi]

12. $f(x)=x^4-8x^3-2x^2+120x-112$

Solution $f'(x)=4x^3-24x^2-4x+120$

$f'(x)=0 \Rightarrow x=-2,\ 3,\ 5$

$f''(x)=12x^2-48x-4$

if $c=-2$, $f''(c)=140>0$, then f has a local minimum at $c=-2$.

if $c=5$, $f''(c)=56>0$, then f has a local minimum at $c=5$.

if $c=3$, $f''(c)=-40<0$, then f has a local maximum at $c=3$.

13. $f(t)=\sin t+\cos t,\ 0\le t\le \dfrac{\pi}{2}$

Solution $f'(t)=\cos t-\sin t$

$f'(t)=0 \Rightarrow t=\dfrac{1}{4}\pi$

$f''\left(\dfrac{1}{4}\pi\right)<0$ then f has a local maximum at $\dfrac{1}{4}\pi$

14-17. Find the intervals where the function is increasing or decreasing.

14. $f(x)=\dfrac{x}{x^2+1}$

Solution Increasing on $(-1,\ 1)$, decreasing on $(-\infty,\ -1)\cup(1,\ \infty)$. See graph of $f(x)$

```
var('x')
f=x/(x^2+1)
plot(f,-2,2)
df=diff(f);
show(df)
plot(df,-2,2)
```

$f'(x)=-2x^2/(x^2+1)^2+1/(x^2+1)$

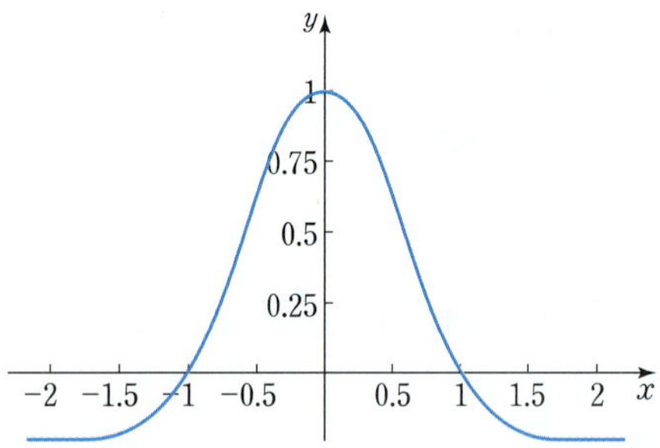

CAS **15.** $f(x)=2e^{-(x-1)^2}$

http://matrix.skku.ac.kr/cal-lab/cal-4-1-15.html

Solution Increasing on $(-\infty,\ 1)$, decreasing on $(1,\ \infty)$

```
var('x')
f=2*exp(-(x-1)^2)
F=plot(f,-5,5)
df=diff(f);
show(F+plot(df,-5,5, linestyle="--", color='red'))
print solve(df==0,x)
```

$f'(x)=-4(x-1)e^{-(x-1)^2}$

Answer : [x == 1]

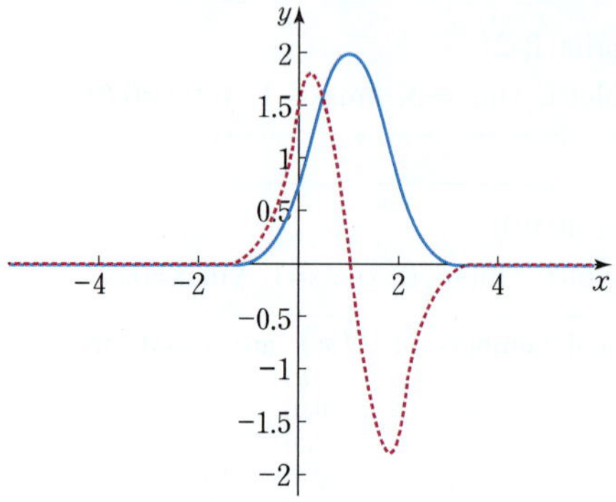

16. $f(x)=(x^2-1)e^{x^2}$

Solution $f'(x)=2x^3e^{x^2}$

```
var('x')
f=(x^2-1)*exp(x^2)
g=plot(f, -2, 2, ymax=3, ymin=-1)
df=diff(f)
show(g)
show(df)
```

Answer : Increasing on $(-\infty,0)$, decreasing on $(0,\infty)$

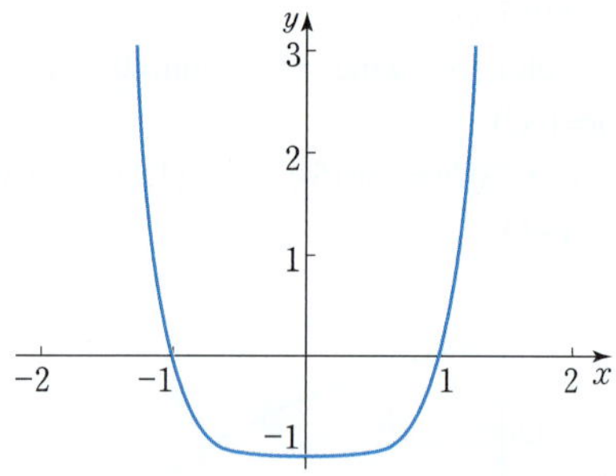

17. Suppose $f(a)=g(a)$ and $f'(x)>g'(x)$ for all $x>a$. Using the Mean Value Theorem show $f(x)>g(x)$ for all $x>a$.

Solution By the Mean Value Theorem, $c\in[a,\ x]$

$\dfrac{f(x)-f(a)}{x-a}=f'(c)$ and $\dfrac{g(x)-g(a)}{x-a}=g'(c)$

Since $f'(c)>g'(c)$, we have

$\dfrac{f(x)-f(a)}{x-a}>\dfrac{g(x)-g(a)}{x-a} \Rightarrow f(x)>g(x).$

18-20. Prove the inequality.

18. $xe^x>e^x-1>x,\ x>0$

Solution $f(x)=x,\ g(x)=e^x-1,\ h(x)=xe^x$

$f(0)=g(0)=h(0)=0$

$h'(x)=e^x+xe^x>g'(x)=e^x>f'(x)=1$

So, $h(x)>g(x)>f(x)$ for $x>0$.

```
var('x')
p1=plot(x*exp(x))
p2=plot(exp(x) -1)
p3=plot(x)
show(p1+p2+p3)
```

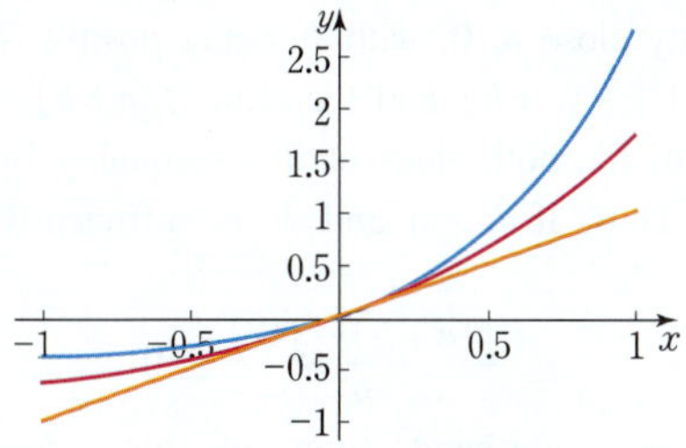

19. $x > \tan^{-1}x > \dfrac{x}{x^2+1},\ x>0$

Solution Let $f(x)=x,\ g(x)=\tan^{-1}x,\ h(x)=\dfrac{x}{x^2+1}$.

Then $f'(x)=1,\ g'(x)=\dfrac{1}{x^2+1}$,

$h'(x)=\dfrac{1}{x^2+1}-\dfrac{2x^2}{(x^2+1)^2}$.

$$f'(x)=1>g'(x)=\frac{1}{1+x^2}>h'(x)$$
$$=\frac{1}{x^2+1}-\frac{2x^2}{(x^2+1)^2}$$

From the inequality $f'(x)>g'(x)>h'(x)$ and the Mean Value Theorem,

$$f(x)>g(x)>h(x),\ x>0.$$

Therefore $x>\tan^{-1}x>\dfrac{x}{x^2+1}$ for all $x>0$.

20. $\dfrac{1}{1+x}<\log(1+x)+1,\ x>0$

Solution $f(x)=\dfrac{1}{1+x},\ g(x)=\log(1+x)+1$

$\Rightarrow f(0)=1=g(0)$ and

$f'(x)=\dfrac{-1}{(1+x)^2},\ g'(x)=\dfrac{1}{1+x}\Rightarrow g'(x)>f'(x)$

```
var('x, y')
p1=plot(-1/(1-x)^2)
p2=plot(1/(1+x))
show(p1+p2, ymax=2, ymin=-2)
```

So, $g(x)>f(x)$ for all $x>0$.

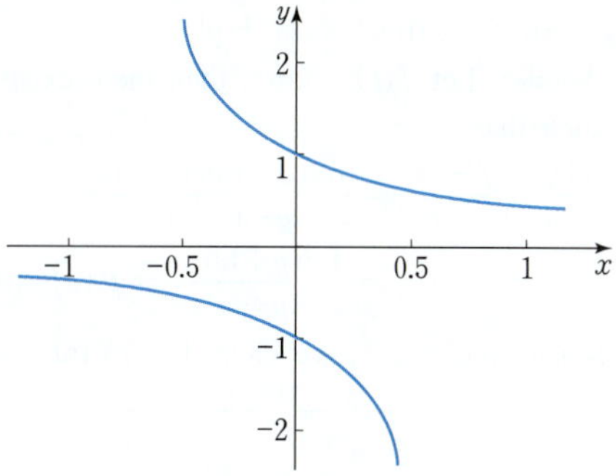

CAS **21.** Let $f(x)=(x-1)+10\sin x$. Find the number of zeros and give a proper interval containing all zeros.

http://matrix.skku.ac.kr/cal-lab/cal-4-1-20.html

Solution 7 zeros in $(-7,\ 12)$

```
var('x,y')
p1=plot.
((x-1+ 10*sin(x)),x,-8,13,color='blue');
show(p1, aspect_ratio=1, ymax=20)
```

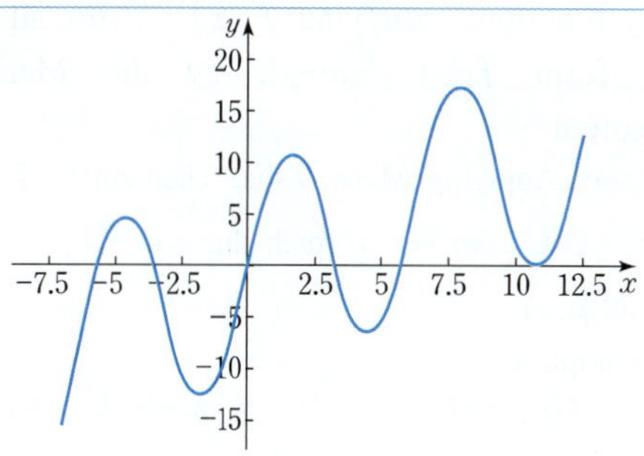

22. Show that $x>\sin x$ for $x\in\left(0,\ \dfrac{\pi}{2}\right)$.

Solution $f(x)=x,\ g(x)=\sin x$

$f(0)=0=g(0)$

$f'(x)=1,\ g'(x)=\cos x$

$f'(x)>g'(x)$

Then, $f(x)>g(x)$.

23-24. Prove the inequality using the Mean Value Theorem.

23. $|\cos 6-\cos 5|\le 1$

Solution Let $f(x)=\cos x$, then by the Mean Value Theorem there exists c in $[5,6]$ such that

$$\frac{f(6)-f(5)}{6-5}=\cos 6-\cos 5=f'(c)=-\sin c.$$

Since $-1\le\sin x\le 1$ for x, we have

$|\cos 6-\cos 5|\le 1$.

24. $|\tan x+\tan y| \geq |x+y|$

Solution Let $f(t)=\tan t$, then there exists c in $(-x, y)$ such that

$$\frac{f(y)-f(-x)}{y-(-x)}=\frac{\tan y-\tan(-x)}{y-(-x)}$$
$$=\frac{\tan y+\tan x}{y+x}=f'(c)=\sec^2 c.$$

Since, $\sec^2 x \geq 1$, we have $|\tan y+\tan x| \geq |x+y|$.

---

25. Prove that the inequality $|\cos x-\cos y| \leq |x-y|$ for all x and y using the Mean Value Theorem.

Solution Let $f(t)=\cos t$, $f'(t)=-\sin t$, then there exists c in (x, y) such that

$$\frac{f(x)-f(y)}{x-y}=\frac{\cos x-\cos y}{x-y}=f'(c)$$
$$\Rightarrow \cos x-\cos y=f'(c)(x-y)$$
$$\Rightarrow |\cos x-\cos y|=|f'(c)(x-y)|$$
$$\Rightarrow |\cos x-\cos y|=|-\sin c(x-y)| \leq |x-y|$$

Thus, $|\cos x-co\, y| \leq |x-y|$.

26. Let f be a continuous function. Prove that the only functions satisfying $f'(x)=c$ for all x are of the form $f(x)=cx+d$, by the Mean Value Theorem.

Solution Applying Mean Value Theorem to f on $[a, b]$, there exists c in (a, b) such that $c=f'(x)=\frac{f(a)-f(b)}{a-b}$ for all a, b.

This implies

$$f(a)=c(a-b)-f(b)=ca+(-bc-f(b)).$$

Now set $a=x$, $d=-bc-f(b)$,

then $f(x)=cx+d$ for all x.

27. Suppose $f(x)$ is continuous on $[a, b]$ and differentiable on (a, b). Prove that there exists $c\in(a, b)$ such that

$$f(b)-f(a)=c\left(\log\frac{b}{a}\right)f'(c).$$

Solution Let $g(x)=f(x)(\log b-\log a)\ -\log x(f(b)-f(a))$,

then $g'(x)=f'(x)(\log b-\log a)-\frac{1}{x}(f(b)-f(a))$.

By Rolle's theorem,

there exist $c\in(a, b)$ such that $g'(c)=0$, so

$$0=g'(c)=f'(c)(\log b-\log a)-\frac{1}{c}(f(b)-f(a))$$
$$f(b)-f(a)=cf'(c)(\log b-\log a)$$
$$f(b)-f(a)=c\log\frac{b}{a}f'(c).$$

28. Prove Fermat's Theorem: If f is differentiable at $x=c$ and has an extreme value at c, then $f'(c)=0$.

http://en.wikipedia.org/wiki/Fermat's_theorem_ (stationary_points)

Solution Suppose, for the sake of definiteness, that f has a local maximum at c. Then, $f(c) \geq f(x)$ if x is sufficiently close to c. This implies that if h is sufficiently close to 0, with h being positive or negative, then $f(c) \geq f(c+h)$ and therefore $f(c+h)-f(c) \leq 0$. We can divide both sides of this inequality by a positive number. Thus, if $h>0$ and h is sufficiently small, we have

$$\frac{f(c+h)-f(c)}{h} \leq 0.$$

Taking the right-hand limit of both sides of this inequality, we get

$$\lim_{h\to0^+}\frac{f(c+h)-f(c)}{h} \leq \lim_{h\to0^+}0=0$$

Since $f'(c)$ exists, we have

$$f'(c)=\lim_{h\to0}\frac{f(c+h)-f(c)}{h}$$
$$=\lim_{h\to0^+}\frac{f(c+h)-f(c)}{h}$$

and so we have shown that $f'(c) \leq 0$.

If $h<0$, then the direction of the inequality (1) is reversed. We have

$$\frac{f(c+h)-f(c)}{h} \geq 0,\ h<0.$$

So, taking the left-hand limit, we have

$$f'(c)=\lim_{h\to0}\frac{f(c+h)-f(c)}{h}$$
$$=\lim_{h\to0^-}\frac{f(c+h)-f(c)}{h} \geq 0$$

We have shown that $f'(c) \geq 0$ and also that $f'(c) \leq 0$. Since both of these inequalities must be true, the only possibility is that $f'(c)=0$.

29. Prove Rolle's Theorem.

http://en.wikipedia.org/wiki/Rolle's_theorem

Solution Since $f(a)=f(b)$, we can use the above result due to Fermat's Theorem to conclude this.

Suppose that the maximum is obtained at an interior point c of (a, b). We shall examine the above right- and left-hand limits separately. For a real h such that $c+h$ is in $[a, b]$, the value $f(c+h)$ is smaller or equal to $f(c)$ because f attains its maximum at c. Therefore, for every $h>0$, $\frac{f(c+h)-f(c)}{h} \leq 0$, hence

$$f'(c)=\lim_{h\to0^+}\frac{f(c+h)-f(c)}{h} \leq 0,$$

where the limit exists by assumption (it may be negative infinity).

Similarly, for every $h<0$, the inequality is reversed because the denominator is now negative and we get

$$\frac{f(c+h)-f(c)}{h} \geq 0,$$

hence

$$f'(c)=\lim_{h\to 0^-}\frac{f(c+h)-f(c)}{h} \geq 0,$$

where the limit might be positive infinity.
Finally, because the above right- and left-hand limits agree, the derivative of f at c must be zero.

30. Prove the Increasing and Decreasing Test.

Solution (a) Let x_1 and x_2 be any two numbers in the interval with $x_1<x_2$. According to the definition of an increasing function we have to show that $f(x_1)<f(x_2)$. Applying the Mean Value theorem to f on the interval $[x_1, x_2]$, there exists $c\in[x_1, x_2]$ such that

$$f(x_2)-f(x_1)=f'(c)(x_2-x_1).$$

By assumption, $f'(c)>0$ and $x_2-x_1>0$, we get $f(x_2)-f(x_1)>0$. In particular, $f(x_2)>f(x_1)$.
Since x_1 and x_2 were arbitrary, we conclude that $f(x_2)>f(x_2)$ whenever $x_2>x_1$. (b) is left to the reader.

31. Prove the cases (ii), (iii) of the First Derivative Test.

http://en.wikipedia.org/wiki/First_derivative_test

Solution (ii) Let us choose a, b sufficiently near to c such that $a<c<b$. By the Mean Value Theorem, there exist c_1, c_2 with $a<c_1<c$ and $c<c_2<b$ such that

$$\frac{f(c)-f(a)}{c-a}=f'(c_1)<0 \text{ and } \frac{f(b)-f(c)}{b-c}=f'(c_2)>0.$$

This implies $f(c)<f(a)$ for $a<c$ and $f(c)<f(b)$ for $b<c$. In particular f is decreasing on (a, c) and increasing on (c, b). Hence f has a local minimum at c.
(iii) Let us choose a,b sufficiently near c such that $a<c<b$. By the Mean Value Theorem, there exist c_1, c_2 with $a<c_1<c$ and $c<c_2<b$ such that

$$\frac{f(c)-f(a)}{c-a}=f'(c_1)=0 \text{ and } \frac{f(b)-f(c)}{b-c}=f'(c_2)=0.$$

Since $f(c)=f(a)$ for $a<c$ and $f(c)=f(b)$ for $b<c$. Hence f has neither local maximum nor local minimum at c.

32. Prove the case (b) of the Second Derivative Test.

Solution (b) Suppose we have $f''(x)<0$. Then

$$f''(x)=\lim_{h\to 0}\frac{f'(x+h)-f'(x)}{h}=\lim_{h\to 0}\frac{f'(x+h)-0}{h}$$
$$=\lim_{h\to 0}\frac{f'(x+h)}{h}<0$$

Thus, for h sufficiently small we get

$$\frac{f'(x+h)}{h}<0$$

which means that $f'(x+h)>0$ if $h<0$ so that f is decreasing to the left of x, and that $f'(x+h)<0$ if $h>0$ so that f is increasing to the right of x. Now, by the first derivative test we know that f has a local maximum at x.

4.2 Shapes of Curves

YouTube http://youtu.be/cZrAF_77On4

In section 4.1, we discussed how the signs of the first and second derivatives f', f'' can be used to determine the extreme values of a function f. Moreover, we have studied the following basic concepts of calculus that will be useful in plotting graphs of functions:

- ◦ Domain and range
- ◦ x-intercepts and y-intercepts
- ◦ Critical numbers
- ◦ Infinite limits
- ◦ Limits as $x \to \pm\infty$
- ◦ Local maximum and minimum
- ◦ Symmetry
- ◦ Continuity
- ◦ Vertical and horizontal asymptotes
- ◦ Absolute maximum and minimum
- ◦ Increasing and decreasing

To determine the shape of a graph, here we will revisit those concepts above and we will study more basic concepts of calculus such as

- ◦ Concavity
- ◦ Oblique asymptotes
- ◦ Inflection points

Concavity

First, we will see how locating the intervals in which f' increases or decreases can be used to determine where the graph of f is concave upward or concave downward.

Let f be differentiable on an open interval I. The graph of f is called **concave upward** on I if f' is increasing on the interval, and is called **concave downward** on I if f' is decreasing on the interval.

Geometrically, it means that if the graph of f lies above all of its tangents on an interval I, then it is concave upward on I (see Figure 1(a)), and if the graph of f lies below all of its tangents on I, it is concave downward on I (see Figure 1(b)).

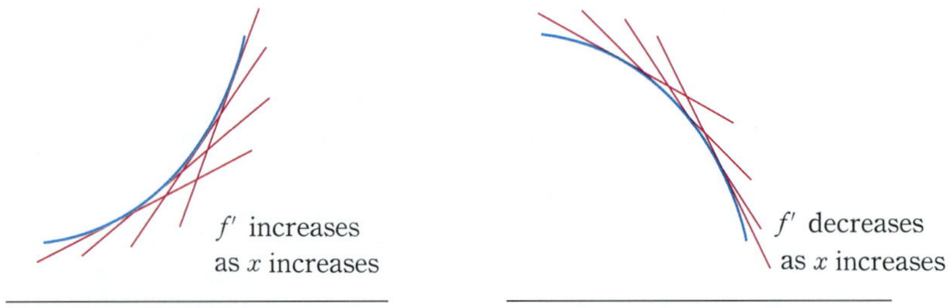

(a) Concave Upward (b) Concave downward

Figure 1

> **Concavity Test**
>
> Let f be a function whose second derivative exists on an open interval I.
>
> (i) If $f''(x) > 0$ for all x in I, then the graph of f is concave upward on I.
>
> (ii) If $f''(x) < 0$ for all x in I, then the graph of f is concave downward on I.

A proof of this test follows directly from the Increasing and Decreasing Test and the definition of concavity. (See Exercise 24.)

EXAMPLE 1

Determine the open intervals on which the graph of $f(x) = x^{1/3}$ is concave upward and concave downward.

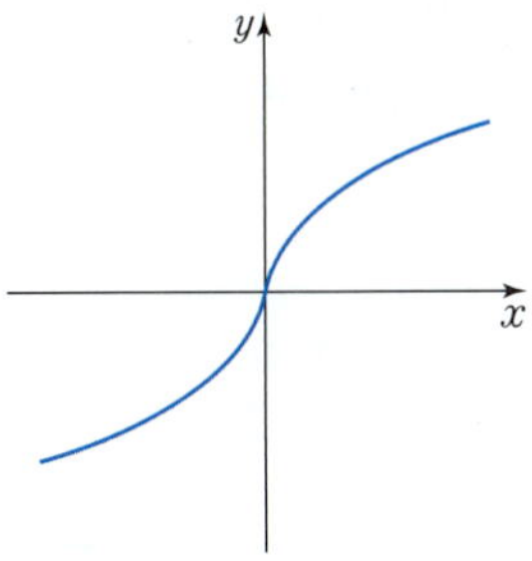

Figure 2 The graph of $x^{\frac{1}{3}}$

Solution Differentiating f twice gives

$$f''(x) = \frac{d}{dx}\left(\frac{1}{3}x^{-2/3}\right) = -\frac{2}{9}x^{-5/3}$$

Clearly, $f''(x) < 0$ for $x < 0$ and $f''(x) > 0$ for $x > 0$. Hence $f''(0) = 0$. There is no point at which $f''(x) = 0$, but by the concavity test the graph of f is concave upward on $(-\infty, 0)$ and is concave downward on $(0, \infty)$. (See Figure 2.) ■

Inflection Point

As shown in Figure 2, we see that the curve changes from concave upward to concave downward at $x = 0$. A point where the graph of a function has a tangent line and where the concavity changes is called an inflection point. There are four types of **inflection points** illustrated in Figure 3.

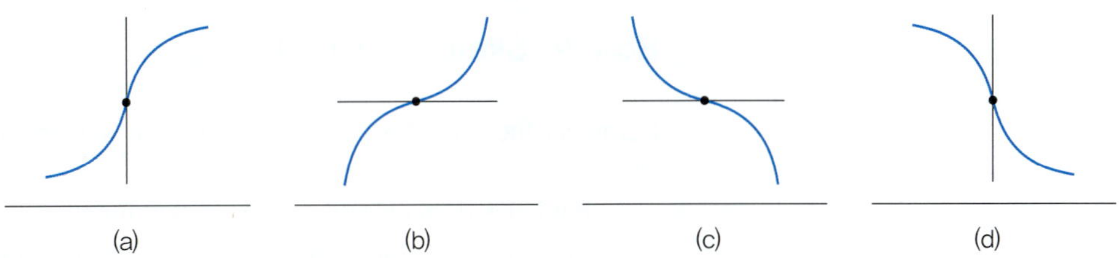

Figure 3

Oblique Asymptotes

In some cases, a function has a slant asymptote. Assume that $f(x)$ can be written as

$$f(x) = ax + b + h(x).$$

If $\lim_{x \to \infty} h(x) = 0$ or $\lim_{x \to -\infty} h(x) = 0$, then $y = ax + b$ is called an **equation of the oblique asymptote (or slant asymptote)** of the graph of f. We note that if $f(x) = \dfrac{g(x)}{h(x)}$ is a rational function such that

$$\deg(g) = \deg(h) + 1$$

then the graph of f has an oblique asymptote.

EXAMPLE 2

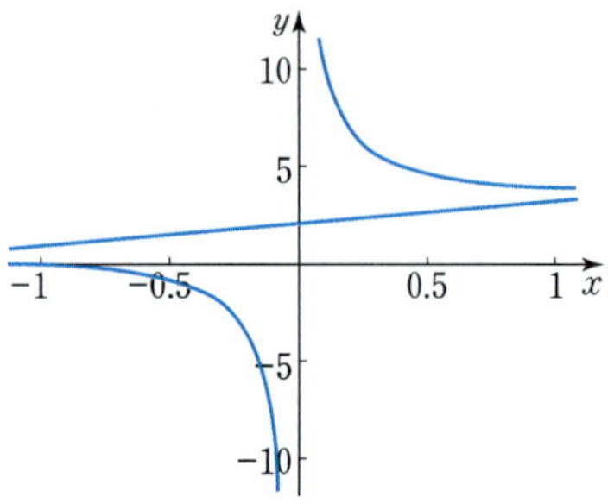

Figure 4

(a) If $y = \dfrac{(x+1)^2}{x}$ then $y = x + 2 + \dfrac{1}{x}$ and $\lim_{x \to \pm\infty} \dfrac{1}{x} = 0$.
Hence $y = x + 2$ is the oblique asymptote in Figure 4.

```
var('x, y')
p1=plot(x + 2 + 1/x)
p2=plot(x + 2)
show(p1+p2, ymax=10, ymin=-10)
```

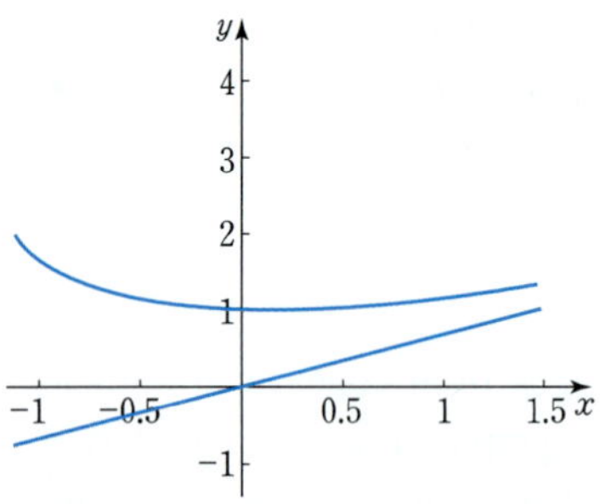

Figure 5

(b) If $e^x(y - x) = 1$ then $y = x + e^{-x}$ and $\lim_{x \to \infty} e^{-x} = 0$.

Hence $y = x$ is the oblique asymptote in Figure 5.

```
var('x, y')
p1=plot(x + e^(-x))
p2=plot(x)
show(p1+p2, ymax=4, ymin=-1, xmax=1.5)
```

■

How to Sketch a Graph

Below is the list of steps we will use to sketch the graph of a function.

1. Find the domain and range of the function.
2. Find the x, y-intercepts, symmetry of the graph.
3. Find vertical, horizontal, oblique asymptotes of the graph.
4. Compute f' and f'' to find critical numbers, inflection points, extreme values, intervals for increasing and decreasing, and intervals for concave upward and concave downward.
5. Sketch the shape of a graph, consistent with all of the information obtained.

EXAMPLE 3

Sketch the graph of the function $f(x) = \dfrac{x^2-1}{x^3}$.

Solution We will follow the above guidelines:

1. The domain is $-\infty < x < \infty,\ x \neq 0$.
2. The graph has two intercepts $(-1,\ 0),\ (1,\ 0)$. Since $f(-x) = -f(x)$, the graph is symmetric about the origin.
3. It has no oblique asymptotes, but it has one vertical asymptote $y=0$ and one horizontal asymptote $x=0$.
4. Computing the derivatives gives

$$f'(x) = \frac{3-x^2}{x^4} \quad \text{and} \quad f''(x) = \frac{2(x^2-6)}{x^5}.$$

The function f has three critical numbers $x = \pm\sqrt{3},\ x=0$, and two possible inflection points $x = \pm\sqrt{6}$. The other information for the graph is shown in the table.

	f	f'	f''	Information
$x=0$	undefined	undefined	undefined	vertical asymptote
$0 < x < \sqrt{3}$		$+$	$-$	increasing, concave downward
$x = \sqrt{3}$	$\dfrac{2}{3\sqrt{3}}$	0	$-$	local maximum
$\sqrt{3} < x < \sqrt{6}$		$-$	$-$	decreasing, concave downward
$x = \sqrt{6}$	$\dfrac{5}{6\sqrt{6}}$	$-$	0	inflection point
$\sqrt{6} < x < \infty$		$-$	$-$	decreasing, concave upward

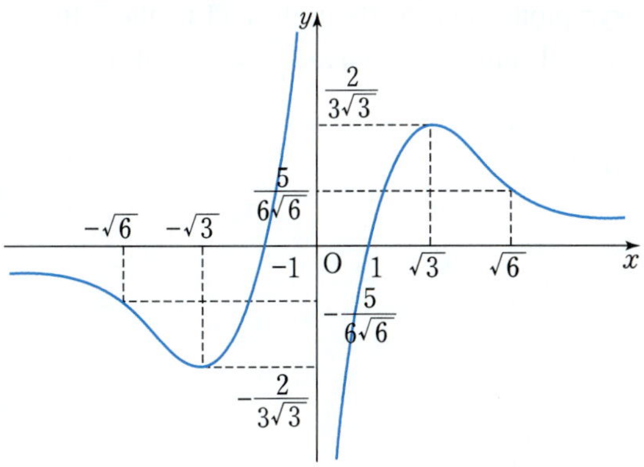

Figure 6 $f(x) = \dfrac{x^2-1}{x^3}$

5. With symmetry about the origin, the graph is shown in Figure 6. ∎

CAS **EXAMPLE 4**

Let $f(x) = \operatorname{sech} x$,

(a) Find the vertical and horizontal asymptotes.

(b) Sketch the graph using f' and f''.

Solution We now show the graph of $f(x) = \operatorname{sech} x$ as a blue curve.

```
C=plot(f, -5, 5, rgbcolor='blue')
show(C)
```

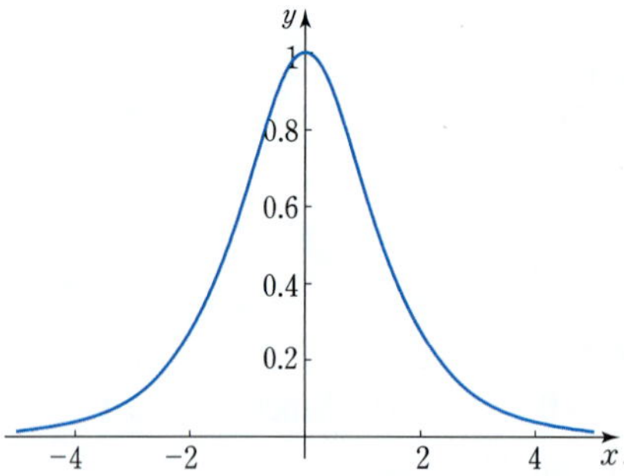

Figure 7 $y = \text{sech}\ x$

Then we find f' and f'':

```
x=var('x')
f=sech(x)
df=diff(f,x)
ddf=diff(df,x)
df, ddf
solve(f==0,x)
```

df= -tanh(x)*sech(x), ddf=tanh(x)^2*sech(x)+(tanh(x)^2 - 1)*sech(x))
[ ]   # No solution

We see that there is no x such that $f(x) = 0$.
Sage input/output

```
limit(f, x = infinity)
```

0

```
limit(f, x = -infinity)
```

0
So $y = 0$ is a horizontal asymptote, both to the right and to the left. We now plot f' and f'' as red and green curves (respectively).

```
A=plot(df,-5,5,rgbcolor='red')
show(A)
```

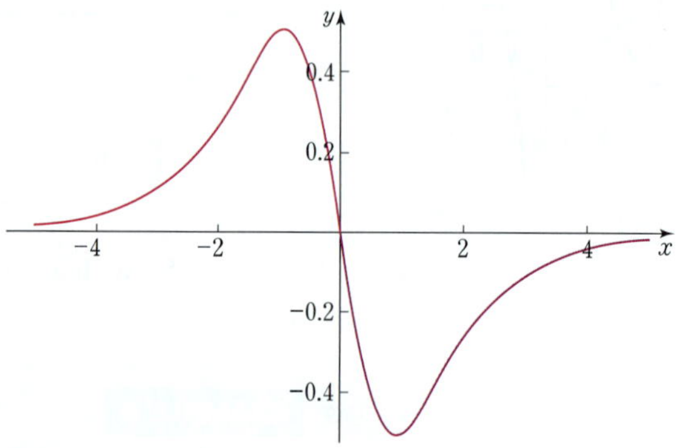

Figure 8

```
B=plot(ddf,-5,5,rgbcolor='green')
show(B)
```

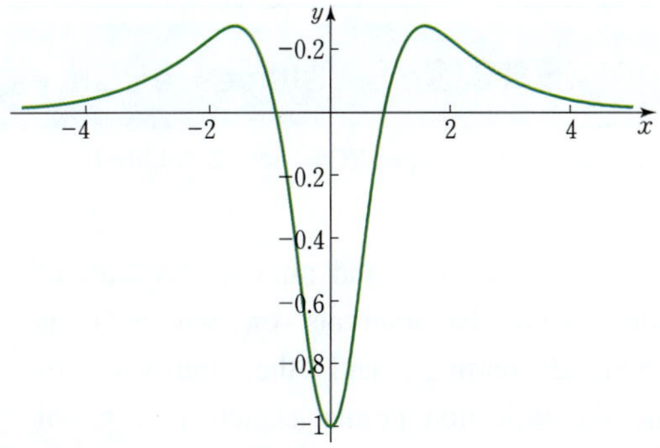

Figure 9

Notice how each curve's values are the slopes of tangent lines to the subsequent curve. Also interpret the sign of f' as increasing or decreasing for f and the sign of f'' as concavity of f.
We have:

```
show(A+B+C)
```

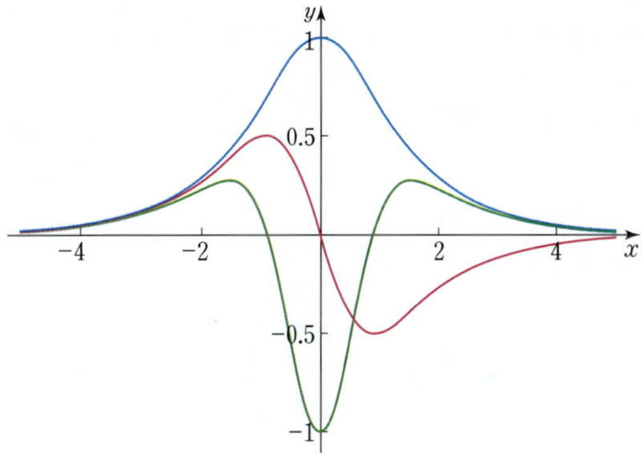

Figure 10

Answer : (a) $y=0$
(b)

x	$<$	$-\frac{\ln(3+2\sqrt{2})}{2}$	$<$	0	$<$	$\frac{\ln(3+2\sqrt{2})}{2}$	$<$
f				1			
f'	+	+	+	0	−	−	−
f''	+	0	−	−	−	0	+

■

4.2 EXERCISES (Shapes of Cuves)

http://matrix.skku.ac.kr/Cal-Book/part1/CS-Sec-4-2-Sol.html

YouTube http://youtu.be/SVOWADHlzV8

1-5. Find the local maximum and minimum values of f. In addition, find the intervals on which f is increasing and decreasing, and the intervals of concavity and the inflection points, sketch a graph of f.

1. $f(x) = 2x^3 + 3x^2 - 36x + 7$

http://matrix.skku.ac.kr/cal-lab/cal-4-2-1.html

Solution $f'(x) = 6x^2 + 6x - 36$

$f'(x) = 0 \Rightarrow x = 2, -3$

$f''(x) = 12x + 6$

(a) local maximum: $f(-3) = 88$

local minimum: $f(2) = -37$

(b) increasing on $(-\infty, -3) \cup (2, \infty)$

decreasing on $(-3, 2)$

(c) inflection point at $x = -1/2$

concave down on $(-\infty, -1/2)$

concave up on $(-1/2, \infty)$

```
var('x')
f=2*x^3 +3*x^2 -36*x +7
plot(f, (x-6, 6))
```

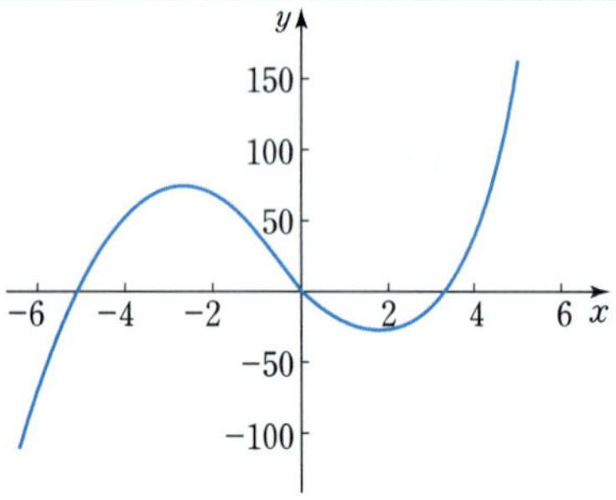

CAS **2.** $f(x) = x^4 - 34x^2 + 225$

http://matrix.skku.ac.kr/cal-lab/cal-4-2-2.html

Solution (a) local minimum: $f(\pm\sqrt{17}) = -64$

(b) increasing on $(-\sqrt{17}, 0) \cup (\sqrt{17}, \infty)$

decreasing on $(-\infty, -\sqrt{17}) \cup (0, \sqrt{17})$

(c) inflection point at $x = \pm\dfrac{\sqrt{51}}{3}$

concave down on $\left(-\dfrac{\sqrt{51}}{3}, \dfrac{\sqrt{51}}{3}\right)$

concave up on $\left(-\infty, -\dfrac{\sqrt{51}}{3}\right) \cup \left(\dfrac{\sqrt{51}}{3}, \infty\right)$

```
var('x')
f=x^4-34*x^2+225
plot(f,-6,6)
```

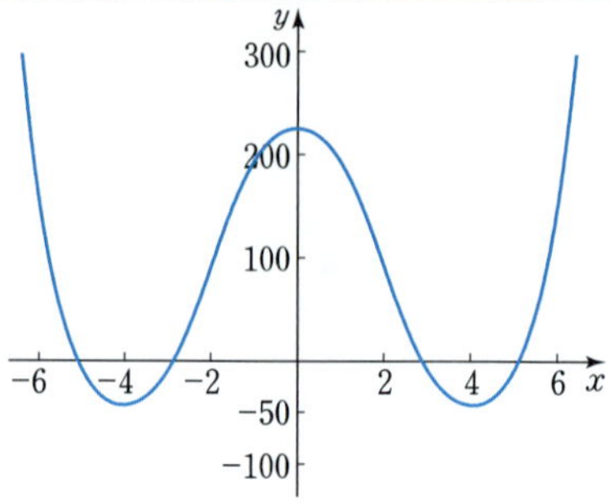


```
df=diff(f,x)
df
```

Answer : 4*x^3-68*x

```
solve(df==0,x)
```

Answer : [x == -sqrt(17), x == sqrt(17), x == 0]

```
df=diff(df,x)
df
```

Answer : 12*x^2-68

```
solve(dff==0,x)
```

Answer : [x == -1/3*sqrt(3)*sqrt(17), x == 1/3*sqrt(3)* sqrt(17)]

3. $f(x) = \sin x + \cos x,\ -\pi \le x \le \pi$

Solution $f'(x) = \cos x - \sin x$

$f'(x) = 0 \Rightarrow x = \dfrac{1}{4}\pi, -\dfrac{3}{4}\pi$

$f''(x) = -\sin x - \cos x$

(a) local maximum: $f\left(\dfrac{1}{4}\pi\right) = \sqrt{2}$

local minimum: $f\left(-\dfrac{3}{4}\pi\right) = -\sqrt{2}$

(b) increasing on $\left(-\dfrac{3}{4}\pi, \dfrac{1}{4}\pi\right)$

decreasing on $\left(-\pi, -\frac{3}{4}\pi\right) \cup \left(\frac{1}{4}\pi, \pi\right)$

(c) inflection point at $x = -\frac{3}{4}\pi$

concave down on $\left(-\frac{3}{4}\pi, \infty\right)$

concave up on $\left(-\infty, -\frac{3}{4}\pi\right)$

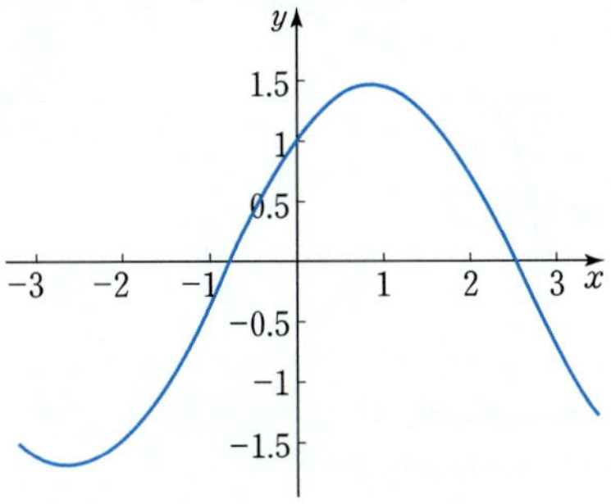

4. $f(x) = \sec^2 x, \ -\frac{\pi}{2} < x < \frac{\pi}{2}$

Solution (a) local maximum : No

local minimum : $f(0) = 1$

(b) increasing on $\left(0, \frac{\pi}{2}\right)$

decreasing on $\left(-\frac{\pi}{2}, 0\right)$

(c) inflection point : No

concave up on $\left(-\frac{\pi}{2}, \frac{\pi}{2}\right)$

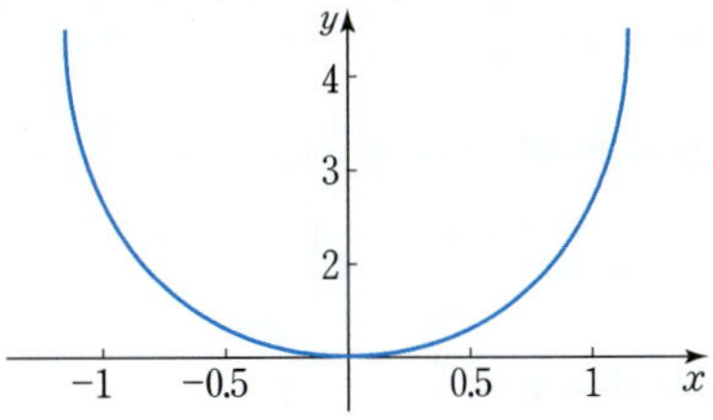

CAS **5.** $f(x) = \tanh(x^2), \ -\frac{3\pi}{2} \le x \le \frac{3\pi}{2}$

http://matrix.skku.ac.kr/cal-lab/cal-4-2-5.html

Solution (a) Maximum : 1, Minimum : 0

(b) interval of increase : $0 \le x \le \frac{3\pi}{2}$

interval of decrease : $-\frac{3\pi}{2} \le x \le 0$

(c) inflection point at $x = 0$

concave down on $(-\infty, 0) \cup (0, \infty)$

```
var('x')
f=tanh(x^2)
plot(f,-3*pi/2,3*pi/2)
```

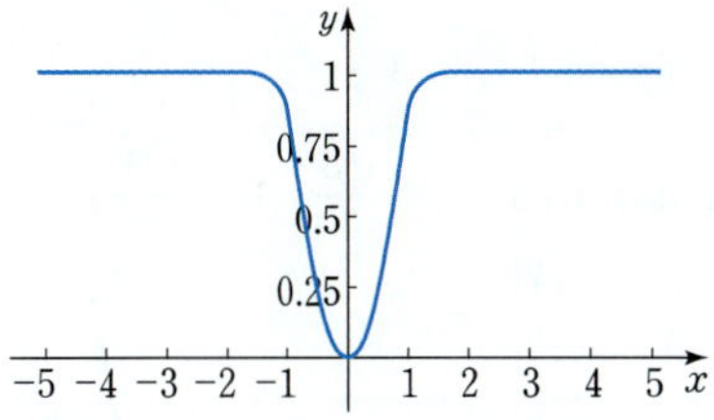

```
df=diff(f)
df
```

Answer : -2*(tanh(x^2)^2 - 1)*x

```
plot(df,-3*pi/2,3*pi/2)
```

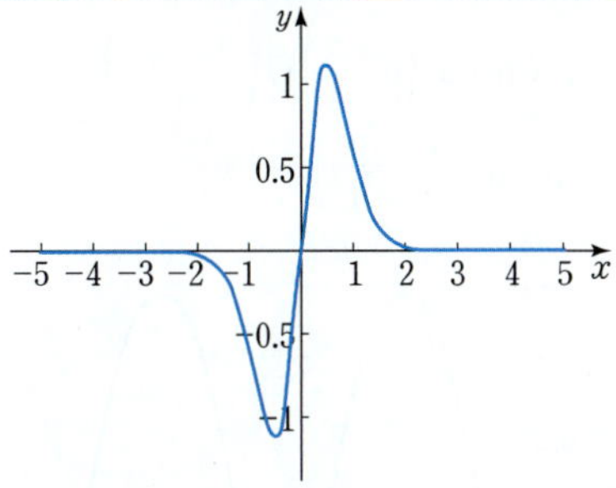

6-11. Find the inflection points of f. In addition, find intervals in which the graph of f is concave upward or concave downward.

6. $f(x) = 18x^3 - 54x^2 + 8x - 2$

Solution (a) inflection point at $x = 1$

(b) concave down on $(-\infty, 1)$

concave up on $(1, \infty)$

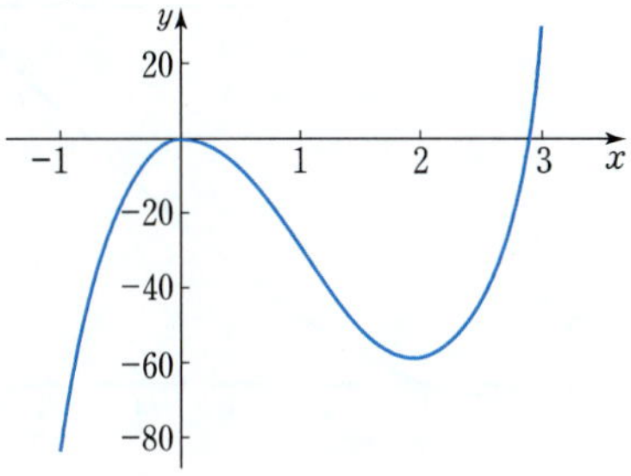

7. $f(x) = x^4 - 4x^3 + 4x^2$

Solution $f'(x) = 4x^3 - 12x^2 + 8x$

$f'(x) = 0 \Rightarrow x = 0, 1, 2$

$f''(x) = 12x^2 - 24x + 8$

$f''(x) = 0 \Rightarrow x = 1 \pm \frac{\sqrt{3}}{3}$

(a) inflection point at $x = 1 \pm \frac{\sqrt{3}}{3}$

(b) concave up on $\left(-\infty,\ 1-\dfrac{\sqrt{3}}{3}\right)\cup\left(1+\dfrac{\sqrt{3}}{3},\ \infty\right)$

concave down on $\left(1-\dfrac{\sqrt{3}}{3},\ 1+\dfrac{\sqrt{3}}{3}\right)$

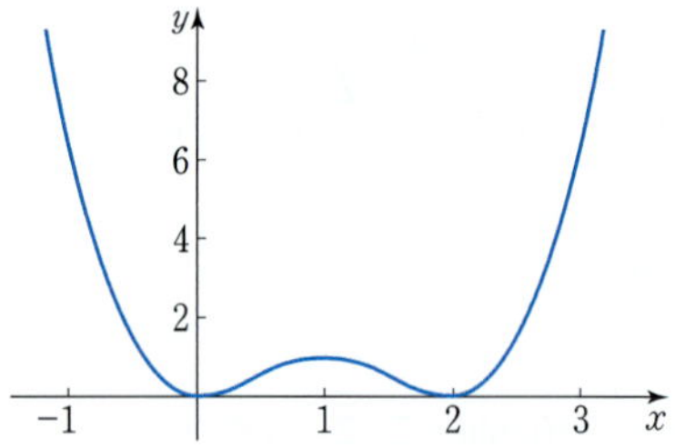

8. $f(x)=\sin^2 x,\ 0\le x\le 2\pi$

Solution (a) inflection point at $x=\dfrac{\pi}{4},\ \dfrac{3\pi}{4}$

(b) concave down on $\left(0,\ \dfrac{\pi}{4}\right)\cup\left(\dfrac{3\pi}{4},\ 2\pi\right)$

concave up on $\left(\dfrac{\pi}{4},\ \dfrac{3\pi}{4}\right)$

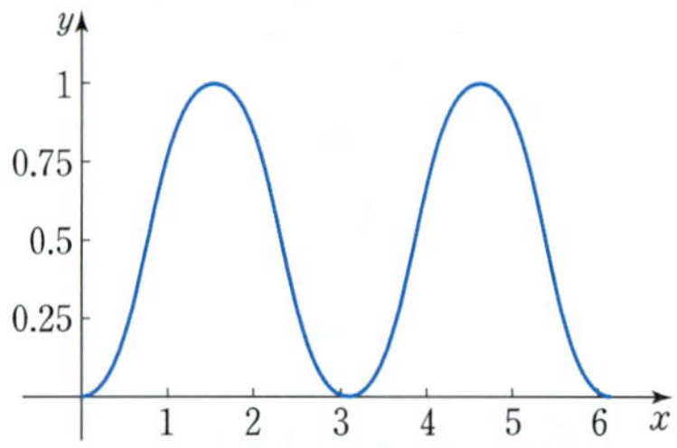

CAS **9.** $f(x)=\sqrt{x}\,(x-1)$

http://matrix.skku.ac.kr/cal-lab/cal-4-2-9.html

Solution

```
var('x')
f(x)=sqrt(x)*(x-1);
plot(f(x), -5, 5)
```

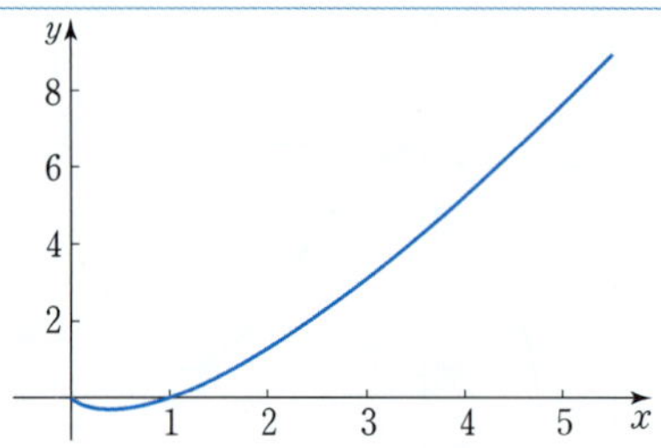

```
var('x')
f(x)=sqrt(x)*(x-1)
plot(f(x), -5, 5)
```

```
df=diff(f(x));
dff=diff(df, x)
show(df)
plot(df, -5, 5)
```

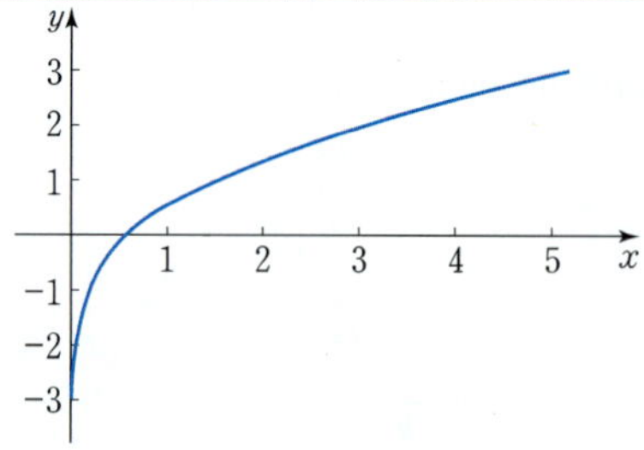

```
print solve(df==0, x)
print solve(dff==0, x)
```

Answer : [x == (1/3)], [x == (-1/3)]

(a) critical point at $x=\dfrac{1}{3}$, inflection point at $x=-\dfrac{1}{3}$

(b) concave down on $\left(\dfrac{1}{3},\ \infty\right)$ concave up on $\left(0,\ \dfrac{1}{3}\right)$

10. $f(x)=e^x\sin x,\ -\pi\le x\le\pi$

Solution $f'(x)=e^x(\sin x+\cos x)$

$f'(x)=0\Rightarrow x=-\dfrac{1}{4}\pi$

$f''(x)=2e^x\cos x$

$f''(x)=0\Rightarrow x=\dfrac{1}{2}\pi$

(a) inflection point at $x=\dfrac{1}{2}\pi$

(b) concave up on $\left(\dfrac{1}{2}\pi,\ \infty\right)$

concave down on $\left(-\infty,\ \dfrac{1}{2}\pi\right)$

```
var('x')
f=plot(e^(x)*sin(x),    x,    ymax=8,    ymin=-1,
xmax=3)
g=plot(e^(x)*(sin(x)+cos(x)),        x,linestyle="--",
color='red', ymax=8, ymin=-1, xmax=3)
h=plot(2*e^(x)*cos(x),              x,linestyle="--",
color='green', ymax=8, ymin=-1, xmax=3)
show(f+g+h)
```

11. $f(x)=e^{-x^2/2}$

Solution (a) inflection points : $x=-1$ and $x=1$

(b) concave down on $(-1,\ 1)$

concave up on $(-\infty,\ -1)\cup(1,\ \infty)$

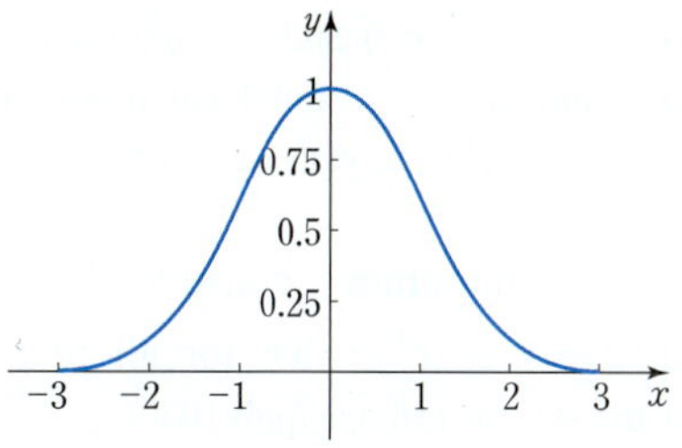

12-15. Find the vertical and horizontal asymptotes of f.

12. $f(x)=\dfrac{x}{x-3}$

Solution $\lim\limits_{x\to a^{\pm}} f(x)=\pm\infty$ vertical asymptote $a=3$

$\lim\limits_{x\to\infty} f(x)=1$ horizontal asymptote $y=1$

13. $f(x)=\left(\dfrac{x}{x^2-3}\right)^2$

http://matrix.skku.ac.kr/cal-lab/cal-4-2-13.html

Solution vertical asymptote: $x=\pm\sqrt{3}$

horizontal asymptote: $y=0$

```
var('x')
plot((x/(x^2 - 3))^2, -5, 5, thickness=2,
color='goldenrod', figsize=4, ymax=20)
```

14. $f(x)=\tanh x$

Solution $\lim\limits_{x\to\infty} f(x)=1$ horizontal asymptote $y=1$

15. $f(x)=\ln(\sec x)$

http://matrix.skku.ac.kr/cal-lab/cal-4-2-15.html

Solution vertical asymptote: $x=-\dfrac{\pi}{2},\ \dfrac{\pi}{2}$

horizontal asymptote: No

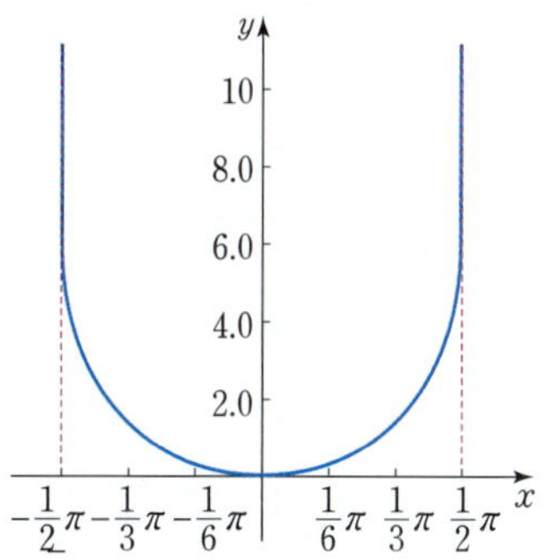


```
var('x')
p1=plot(ln(sec(x)),-pi/2,pi/2,ymax=6,aspect_ratio=1
,tick_formatter=pi)
l1=line([(-pi/2,0),(-pi/2,6)],linestyle="--",color='red'
)
l2=line([(pi/2,0),(pi/2,6)],linestyle="--",color='red')
p1+l1+l2
```

16-18. Sketch the graph of f using the following information.

(a) Find the local maximum and minimum values of f.

(b) Find the intervals of increase or decrease.

(c) Find the inflection points of f and intervals of concavity.

(d) Find the vertical and horizontal asymptotes.

16. $f(x)=\dfrac{x^3}{x^3-3}$

Solution (a) local maximum: No

local minimum: No

(b) decreasing on $(-\infty,\ \sqrt[3]{3})\cup(\sqrt[3]{3},\ \infty)$

(c) inflection point at $x=0,\ -\sqrt[3]{\dfrac{3}{2}}$

concave up on $\left(-\infty,\ -\sqrt[3]{\dfrac{3}{2}}\right)\cup(0,\ \sqrt[3]{3})$

concave down on $\left(-\sqrt[3]{\dfrac{3}{2}},\ 0\right)\cup(\sqrt[3]{3},\ \infty)$

(d) vertical asymptote: $x=\sqrt[3]{3}$

horizontal asymptote: $y=1$

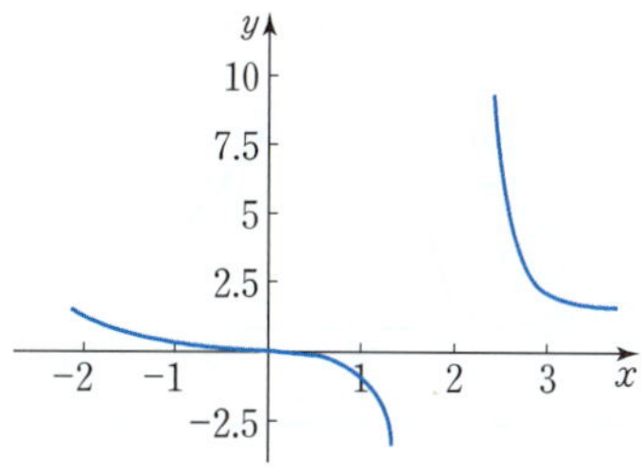

17. $f(x)=\dfrac{1}{1-\sin x}+\dfrac{1}{1+\sin x},\ -\pi/2\le x\le \pi/2$

Solution $f'(x)=\dfrac{\cos x}{(1-\sin x)^2}-\dfrac{\cos x}{(1+\sin x)^2}$

$f'(x)=0,\ x=0$

(a) local minimum at $x=0$

(b) increasing on $(0,\ \pi/2)$

decreasing on $(-\pi/2,\ 0)$

(c) inflection point: No

concave up on $\left(-\frac{1}{2}\pi, \frac{1}{2}\pi\right)$

(d) vertical asymptote: $x = \pm\frac{\pi}{2}$

horizontal asymptote: No

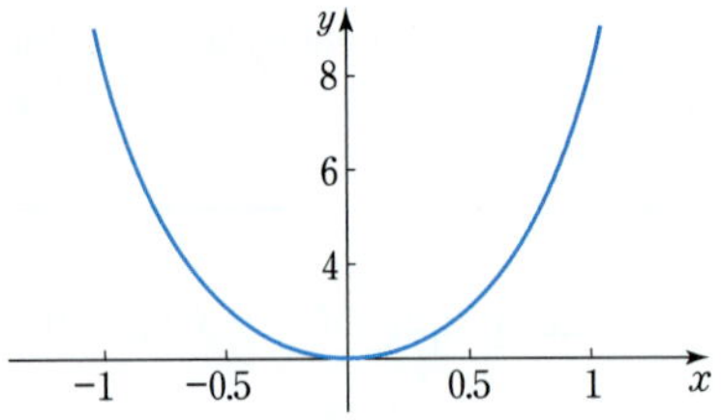

18. $f(x) = \frac{x^3}{e^x}$, $x \geq 0$

Solution $f'(x) = \frac{3x^2}{e^x} - \frac{x^3}{e^x}$ and

$f''(x) = x^3e^{-x} - 6x^2e^{-x} + 6xe^{-x}$

$f'(x) = 0 \Rightarrow x = 0$ and $x = 3$;

$f''(x) = 0 \Rightarrow x = 3 - \sqrt{3}$ and $x = 3 + \sqrt{3}$

(a) local minimum at $x = 0$

local maximum : $f(3) = \frac{27}{e^3}$

(b) increasing on $(0, 3)$

decreasing on $(3, \infty)$

(c) inflection point at $x = 0, 3 + \sqrt{3}, 3 - \sqrt{3}$

concave up on $(0, 3 - \sqrt{3}) \cup (3 + \sqrt{3}, \infty)$

concave down on $(3 - \sqrt{3}, 3 + \sqrt{3})$

(d) vertical asymptote: No

horizontal asymptote: $y = 0$

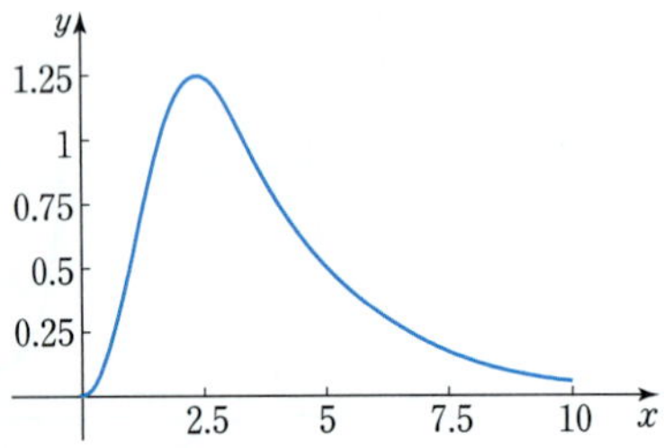

19. Find all the values of a such that $x^3 - 6x = a$ has 3 different roots.

Solution Let $f(x) = x^3 - 6x - a$. Then will have one local maximum and one local minimum. There will be 3 roots if and only if the maximum is positive and the minimum is negative.

$f'(x) = 3x^2 - 6$ and $f'(x) = 6x$.

$f'(x) = 0,\ x = \pm\sqrt{2}$

$f(\sqrt{2}) = -4\sqrt{2} - a$

$f(-\sqrt{2}) = 4\sqrt{2} - a$

Note that $f''(-\sqrt{2}) < 0$ and $f''(\sqrt{2}) > 0$. Hence f has local maximum at $-\sqrt{2}$ and local minimum at $\sqrt{2}$. Therefore, if $-4\sqrt{2} < a < 4\sqrt{2}$ there will be 3 roots.

20. Find the minimum constant c for which $x^4 - 17x^2 + c \geq x^3 - 33x$ for all real x.

http://math1.skku.ac.kr/home/pub/1040/

Solution $x^4 - x^3 - 17x^2 + 33x \geq -c$

$f(x) = x^4 - x^3 - 17x^2 + 33x$

$f'(x) = 4x^3 - 3x^2 - 34x + 33$

$f'(x) = 0,\ x = 1, -3, \frac{11}{4}$

$f''(x) = 12x^2 - 6x - 34$

local minimum at -3

$f(-3) = -144$

Therefore, $c = 144$.

21. Find a, b, c so that $f(x) = x^4 + ax^3 + bx^2 + c$ has two inflection points at $(1, 0)$ and $(2, 9)$.

Solution $f'(x) = 4x^3 + 3ax^2 + 2bx$

$f''(x) = 12x^2 + 6ax + 2b$

$0 = f''(1) = 12 + 6a + 2b$

$0 = f''(2) = 48 + 12a + 2b$ and so $a = -6,\ b = 12$

$0 = f(1) = 14 + (-6)(1)^3 + 12(1)^2 + c = 7 + c$ and so $c = -7$

22. Let $f(x) = e^{-1/(x+1)}$.

(a) Find f' and f''.

(b) Find the vertical and horizontal asymptotes of f.

(c) Sketch the graph of f using (a) and (b).

Solution (a) $f'(x) = e^{(-1/(x+1))}/(x+1)^2$

$f''(x) = -2e^{(-1/(x+1))}/(x+1)^3 \quad + e^{(-1/(x+1))}/(x+1)^4$

(b) $\lim_{x \to a} f(x) = \pm\infty$, $\lim_{x \to \infty} f(x) = 1$

vertical asymptote $x = -1$

horizontal asymptote $y = 1$

(c)

		-1	
f		∞	
f'	$+$	∞	$+$
f''	$+$	∞	$-$

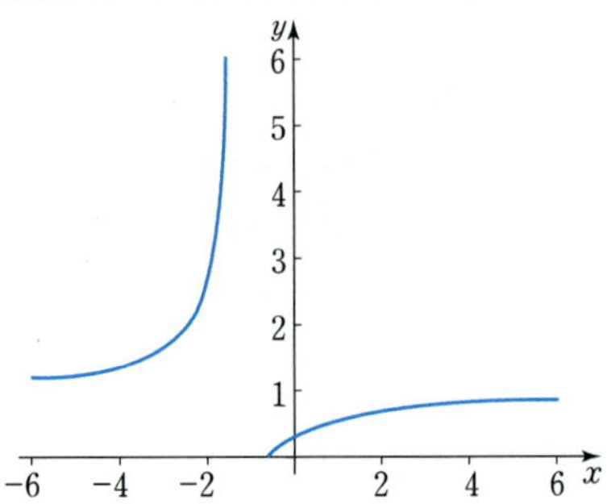

23. Let f and g be increasing functions. Prove that $f+g$ is increasing function.

Solution For $x<y$, $f(x)<f(y)$, $g(x)<g(y)$

$(f+g)(x)=f(x)+g(x)<f(y)+g(y)<(f+g)(y)$

Therefore, $f+g$ is increasing.

24. Prove the Concavity Test.

Solution (a) By Increasing Test

If $f''(x)>0$ on an interval I then f' is increasing on I.

So f is concave upward on I.

Part (b) is proved similarly.

CAS **25.** Use CAS to find f' and f'' when

$$f(x)=6x^{1/3}+3x^{4/3}.$$

http://matrix.skku.ac.kr/cal-lab/cal-4-2-25.html

Solution

```
var('x')
f=6*x^(1/3)+3*x^(4/3)
df=diff(f,x)
dff=diff(df,x)
```

Answer : [df == 4*x^(1/3) + 2/x^(2/3), dff == 4/3/x^(2/3) - 4/3/x^(5/3)]

CAS **26.** Let $f(x)=\dfrac{x+2}{(x+5)^2}$. Find the local maximum, minimum values and inflection points of f. Sketch the graph of f.

http://matrix.skku.ac.kr/cal-lab/cal-4-2-26.html

Solution local maximum: $f(1)=\dfrac{1}{12}$

local minimum: No

inflection point: $x=4$

```
var('x')
f=(x+2)/((x+5)^(2))
df=diff(f, x);
dff=diff(f, x, 2);
print "The first derivative is:"
show(df(x))
print "The second derivative is:"
show(dff(x))
s=solve(df(x)==0, x, solution_dict=true)
```

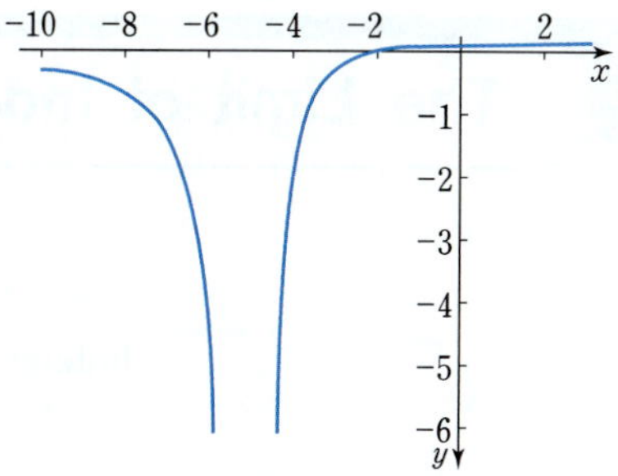


```
plot(f(x), -10, 2, thickness=1.2, color='blue', figsize=5, ymin=-10, dpi=120)
```

```
print "f has only one critical point at x =", s[0][x]
print "f has local maximum at x=", s[0][x]
print "The maximum value of f is ", f(s[0][x])
s1=solve(dff(x)==0, x, solution_dict=true); s1
print "The inflection point of f is at ", s1
```

f has only one critical point at x = 1

f has local maximum at x= 1

The maximum value of f at x=1 is 1/12

The inflection point of f is at x = 4

CAS **27.** Let $f(x)=\dfrac{\ln x}{\sqrt[4]{x}}$. Find the local maximum, minimum values and inflection points of f.

http://matrix.skku.ac.kr/cal-lab/cal-4-2-27.html

Solution local maximum: $f(e^4)=\dfrac{\ln e^4}{e}$

local minimum: No

inflection point: $x=e^{24/5}$

```
var('x')
f(x)=(ln(x))/(x^(1/4))
df=diff(f,x)
dff=diff(df,x)
plot(f,x,0,900)
```

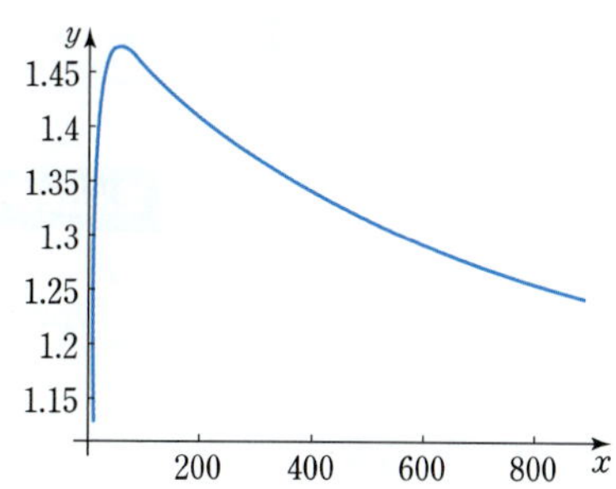

4.3 The Limit of Indeterminate Forms and L'Hospital's Rule

YouTube http://youtu.be/vp-gck5-gKE

Some functions have certain limits which appear to be one of the following **indeterminate forms**.

$$\frac{0}{0},\ \frac{\infty}{\infty},\ 0\cdot\infty,\ \infty-\infty,\ 0^0,\ 1^\infty,\ \infty^0$$

In some cases, we can use **L'Hospital's Rule to find these limits**.

L' Hospital' s Rule

Let f and g be differentiable functions on an open interval containing a and $g'(x)\neq 0$ near a (except possibly at a).

If $\lim_{x\to a}\frac{f(x)}{g(x)}$ has an indeterminate form of type $\frac{0}{0}$ or $\frac{\infty}{\infty}$, then

$$\lim_{x\to a}\frac{f(x)}{g(x)}=\lim_{x\to a}\frac{f'(x)}{g'(x)}$$

provided the limit on the right side exists (or is ∞ or $-\infty$).

The result follows immediately from

$$\frac{f(x)}{g(x)}=\frac{f(x)-f(a)}{g(x)-g(a)}=\frac{\dfrac{f(x)-f(a)}{x-a}}{\dfrac{g(x)-g(a)}{x-a}}\left(\approx\frac{f'(x)}{g'(x)}\right)^{1)}.$$

1)http://www.mathdb.org/resource_sharing/analysis/se_hospital_rule.pdf

We note that indeterminate forms of type $0\cdot\infty,\ \infty-\infty,\ 0^0,\ 1^\infty,\ \infty^0$ can be converted into a form to $\frac{0}{0}$ or $\frac{\infty}{\infty}$ by suitable algebraic transformations.

EXAMPLE 1

Type 1. $\frac{0}{0}$

$$\lim_{x\to 0}\frac{\sinh^{-1}x}{x}=\lim_{x\to 0}\frac{\dfrac{1}{\sqrt{1+x^2}}}{1}=1.$$ ■

EXAMPLE 2

Type 2. $\frac{\infty}{\infty}$

$$\lim_{x\to 0^+}x\ln x=\lim_{x\to 0^+}\frac{\ln x}{\dfrac{1}{x}}=\lim_{x\to 0^+}\frac{\dfrac{1}{x}}{-\dfrac{1}{x^2}}=\lim_{x\to 0^+}(-x)=0.$$ ■

EXAMPLE 3

Type 3. $0 \cdot \infty$

$$\lim_{x \to 2}\left[\ln\left(2-\frac{x}{2}\right)\right]\cot\frac{\pi}{2}x = \lim_{x \to 2}\frac{\ln\left(2-\dfrac{x}{2}\right)}{\tan\dfrac{\pi}{2}x}$$

$$= \lim_{x \to 2}\frac{\dfrac{-1/2}{2-x/2}}{\dfrac{\pi}{2}\sec^2\dfrac{\pi}{2}x} = -\frac{1}{\pi}. \quad \blacksquare$$

EXAMPLE 4

Type 4. $\infty - \infty$

$$\lim_{x \to 1}\left(\frac{x}{x-1}-\frac{1}{\ln x}\right) = \lim_{x \to 1}\frac{x\ln x - x + 1}{(x-1)\ln x} = \lim_{x \to 1}\frac{\ln x + 1 - 1}{\ln x + \dfrac{x-1}{x}}$$

$$= \lim_{x \to 1}\frac{x\ln x}{x\ln x + x - 1} = \lim_{x \to 1}\frac{\ln x + 1}{\ln x + 1 + 1} = \frac{1}{2}. \quad \blacksquare$$

EXAMPLE 5

Type 5. 0^0

Find $\lim_{x \to 0^+}(\sin x)^x$. $\blacksquare$

Solution Let $y = \lim_{x \to 0^+}(\sin x)^x$. Then

$$\ln y = \ln\left[\lim_{x \to 0^+}(\sin x)^x\right] = \lim_{x \to 0^+}\ln(\sin x)^x = \lim_{x \to 0^+}x\ln(\sin x)$$

$$= \lim_{x \to 0^+}\frac{\ln(\sin x)}{1/x} = \lim_{x \to 0^+}\frac{\cot x}{-1/x^2} = \lim_{x \to 0^+}\frac{-x^2}{\tan x}$$

$$= \lim_{x \to 0^+}\frac{-2x}{\sec^2 x} = 0.$$

Since $\ln y = 0$, $y = e^0 = 1$. Hence $\lim_{x \to 0^+}(\sin x)^x = 1$. $\blacksquare$

EXAMPLE 6

Type 6. 1^∞

Find $\lim_{x \to 0^+}(1-\sin x)^{1/x}$.

Solution Let $y = \lim_{x\to 0^+} (1-\sin x)^{1/x}$. Then

$$\begin{aligned}\ln y &= \ln\left[\lim_{x\to 0^+} (1-\sin x)^{1/x}\right]\\ &= \lim_{x\to 0^+} \ln(1-\sin x)^{1/x}\\ &= \lim_{x\to 0^+} \frac{\ln(1-\sin x)}{x}\\ &= \lim_{x\to 0^+} \frac{-\cos x/(1-\sin x)}{1}\\ &= -1.\end{aligned}$$

Hence $\lim_{x\to 0^+} (1-\sin x)^{1/x} = e^{-1} = \dfrac{1}{e}$. ■

EXAMPLE 7

Type 7. ∞^0

Find $\lim_{x\to\infty} x^{1/x}$.

Solution Let $y = \lim_{x\to\infty} x^{1/x}$. Then

$$\begin{aligned}\ln y &= \ln\left[\lim_{x\to\infty} x^{1/x}\right]\\ &= \lim_{x\to\infty} \ln x^{1/x}\\ &= \lim_{x\to\infty} \frac{\ln x}{x}\\ &= \lim_{x\to\infty} \frac{1/x}{1}\\ &= 0.\end{aligned}$$

Hence $\lim_{x\to\infty} x^{1/x} = e^0 = 1$. ■

CAS **EXAMPLE 8**

Evaluate the limit of these indeterminate forms.

(a) $\lim_{x\to 0} \dfrac{\sin(\sin x)}{x}$ (b) $\lim_{x\to\infty} \left(1 - \dfrac{1}{x}\right)^{\sqrt{x}}$

Solution (a)

```
var('x')
f=sin(sin(x))/x
plot(f, (x,-1,1))
```

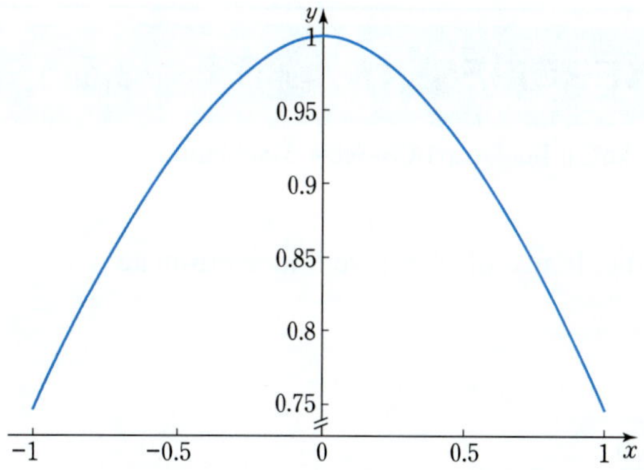

limit(f, x=0)

Answer : 1

(b)

```
var('x')
f=(1-1/x)^(sqrt(x))
show(plot(f, 0,250))
print limit(f, x=infinity)
```

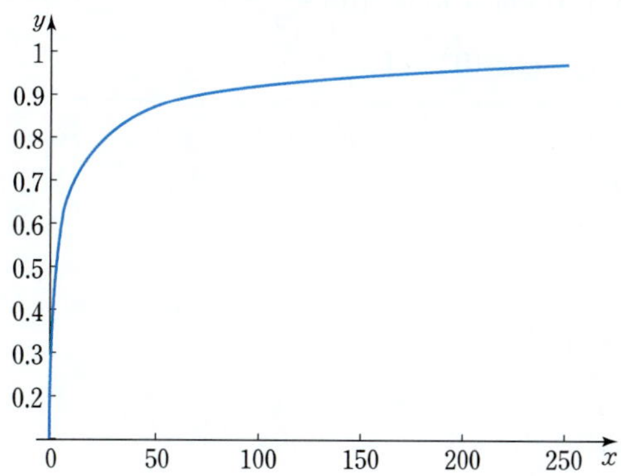

Answer : 1 ■

4.3 EXERCISES (The Limit of Indeterminate Forms and L'Hospital's Rule)

http://matrix.skku.ac.kr/Cal-Book/part1/CS-Sec-4-3-Sol.html

YouTube http://youtu.be/gR2luDDPsMY

1-11. Evaluate the limits of the given indeterminate forms.

1. $\lim_{x\to\infty}\frac{x^2-4}{x^2+x-6}$

Solution 1 Apply L'Hospital Rule twice.

2. $\lim_{x\to\infty}\frac{4x^2-x^3}{5x^3+2x^2}$

Solution $-\frac{1}{5}$

3. $\lim_{x\to 0}\frac{\sin x}{x}$

Solution By L'Hospital Rule, $\lim_{x\to 0}\frac{\sin x}{x}$ $=\lim_{x\to 0}\cos x=\cos(0)=1.$

4. $\lim_{x\to\infty}(1+x^2)^{(\ln x)^{-1}}$

Solution e^2

5. $\lim_{x\to 0}(e^x+2x)^{1/x}$

Solution Let $y=(e^x+2x)^{\frac{1}{x}}\Rightarrow \ln y=\frac{\ln(e^x+2x)}{x}$.

Note $\lim_{x\to 0}x=0$ and $\lim_{x\to 0}\ln(e^x+2x)=0$

$\Rightarrow \lim_{x\to 0}\ln y=\lim_{x\to 0}\frac{\ln(e^x+2x)}{x}=\frac{0}{0}$.

By L'Hospital's Rule,

$\lim_{x\to 0}\ln y=\lim_{x\to 0}\frac{e^x+2}{e^x+2x}=3$

$(\lim_{x\to 0}\ln y=\lim_{x\to 0}\frac{e^x+2}{1}=3)$

$\Rightarrow \lim_{x\to 0}y=\lim_{x\to 0}e^{\frac{e^x+2}{e^x+2x}}=e^{\lim_{x\to 0}\frac{e^x+2}{e^x+2x}}=e^3$

$\Rightarrow \lim_{x\to 0}(e^x+2x)^{\frac{1}{x}}=e^3$

CAS **6.** $\lim_{x\to 1}\frac{\ln x}{x-\sqrt{x}}$

http://matrix.skku.ac.kr/cal-lab/cal-4-3-6.html

Solution

```
var('x')
f=ln(x)/(x-sqrt(x))
plot(f,x,0.5,1.5,ymin=0)
```

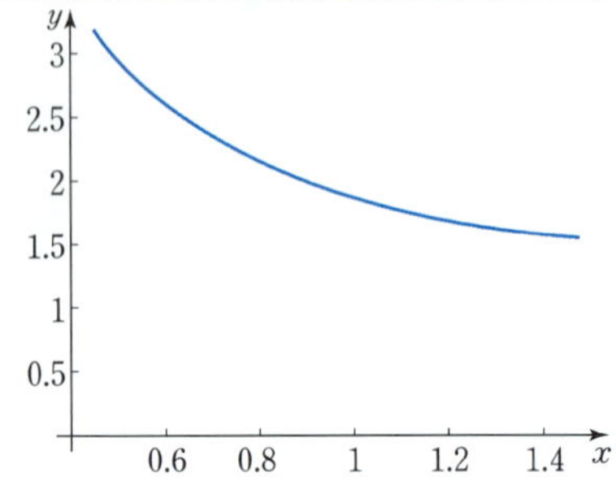


```
limit(f,x=1)
```

Answer : 2

7. $\lim_{x\to\infty}x\sin\frac{1}{x}$

Solution 1

8. $\lim_{x\to 0}\frac{x-\tan x}{x-\sin x}$

Solution By L'Hospital's Rule

$\lim_{x\to 0}\frac{1-\sec^2 x}{1-\cos x}=\lim_{x\to 0}\left(-\frac{\cos x+1}{\cos^2 x}\right)=-2.$

9. $\lim_{x\to 0^+}\frac{\sin x}{x^2}$

Solution ∞

CAS **10.** $\lim_{x\to 0^+}\frac{\sin x}{\sqrt{x}}$

http://matrix.skku.ac.kr/cal-lab/cal-4-3-10.html

Solution $\lim_{x\to 0^+}\frac{\sin x}{\sqrt{x}}=\lim_{x\to 0^+}\frac{\cos x}{\frac{1}{2\sqrt{x}}}=\lim_{x\to 0^+}2\sqrt{x}\cos x=0$

```
var('x')
f(x)=sin(x)/sqrt(x)
limit(f(x),x=0)
plot(f(x),x,0,5)
```

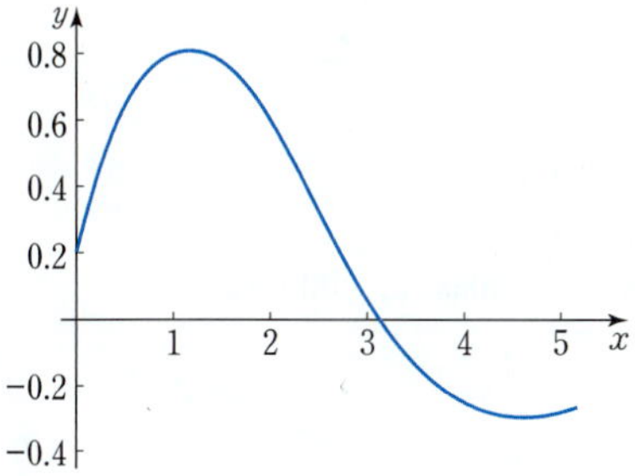

11. $\lim_{x\to 0^+} \tan x \ln x$

Solution $\lim_{x\to 0^+}\dfrac{\ln x}{1/\tan x}=\lim_{x\to 0^+}\left(-\dfrac{1/x}{\sec^2 x/\tan^2 x}\right)$

$=\lim_{x\to 0^+}\left(-\dfrac{\sin^2 x}{x}\right)=0$

12. Let $n\geq 0$. Use L'Hospital's Theorem to show that $\lim_{x\to -\infty} x^n e^{-x}=\infty$.

Solution $\lim_{x\to -\infty}\dfrac{x^n}{e^x}=\lim_{x\to -\infty}\dfrac{nx^{n-1}}{e^x}=\lim_{x\to -\infty}\dfrac{n(n-1)x^{n-2}}{e^x}$

$=\cdots=\lim_{x\to -\infty}\dfrac{n!}{e^x}=\infty$, since $n!$ is constant and the denominator grows without bound.

13. For what values of a and b is the following equation correct?

$$\lim_{x\to 0}\frac{a+\cos bx}{x^2}=-2$$

Solution $\lim_{x\to 0}\dfrac{a+\cos bx}{x^2}=-2$, $\lim_{x\to 0}\dfrac{-b\sin bx}{2x}=-\dfrac{b^2}{2}$

By L'Hospital's Rule for a form of type $\dfrac{0}{0}$

$$\lim_{x\to 0}\frac{-b^2\cos bx}{2}=-2$$

$$\frac{-b^2}{2}=-2 \Rightarrow b=2,\ -2.$$

and we may take a to be any value.

14. $\lim_{x\to 1}\dfrac{x^3-2x^2+2x-1}{x^3-x^2}$

http://matrix.skku.ac.kr/cal-lab/cal-4-3-14.html

Solution By L'Hospital's, Rule for a form of type $\dfrac{0}{0}$

```
var('x')
f(x)=(x^3-2*x^2+2*x-1)/(x^3-x^2)
show(plot(f,(x,-0.5,1.5),
ymax=2)+plot(1,(x,-0.5,1.5),linestyle='--'))
limit(f(x), x=1)
```

15. $\lim_{x\to 0}\left(\dfrac{1}{\tan x}-\dfrac{1}{x}\right)$

Solution $\lim_{x\to 0}\left(\dfrac{1}{\tan x}-\dfrac{1}{x}\right)=\lim_{x\to 0}\dfrac{x-\tan x}{x\tan x}$

Note $\lim_{x\to 0} x\tan x=\lim_{x\to 0} x-\tan x=0$.

Therefore, by L'Hospital's Rule,

$\lim_{x\to 0}\dfrac{1}{\tan x}-\dfrac{1}{x}=\lim_{x\to 0}\dfrac{1-\sec^2 x}{\tan x+x\sec^2 x}$

$=\lim_{x\to 0}\dfrac{-\tan^2 x}{\tan x+x\sec^2 x}$

$=\lim_{x\to 0}\dfrac{-\tan^2 x}{\tan x\left(1+x\dfrac{\sec^2 x}{\tan x}\right)}$

$=\lim_{x\to 0}\dfrac{-\tan x}{1+\dfrac{x}{\sin x\cos x}}=0$

16-17. Find the following limit.

16. $\lim_{\theta\to 0}\dfrac{\tan\theta}{\theta+\sin\theta}$

Solution $\lim_{\theta\to 0}\dfrac{1}{\cos^2\theta\,(\cos\theta+1)}=\dfrac{1}{1(1+1)}=\dfrac{1}{2}$

(by L'Hospital's Rule) or

$$\lim_{\theta\to 0}\frac{\dfrac{\tan\theta}{\theta}}{1+\dfrac{\sin\theta}{\theta}}=\frac{1}{1+1}=\frac{1}{2}$$

17. $\lim_{\theta\to 0}\dfrac{\tan(2020\,\theta)}{\sin\theta+2020\theta}$

Solution $\lim_{\theta\to 0}\dfrac{\tan(2020\,\theta)}{\sin\theta+2020\,\theta}=\lim_{\theta\to 0}\dfrac{2020\,(\sec^2(2020\,\theta))}{\cos\theta+2020}$

$=\dfrac{2020}{1+2020}=\dfrac{2020}{2021}$

(by L'Hospital's Theorem) or

$\lim_{\theta\to 0}\dfrac{\dfrac{\tan(2020\,\theta)}{\theta}}{\dfrac{\sin\theta}{\theta}+2020}=\dfrac{2020}{1+2020}=\dfrac{2020}{2021}$

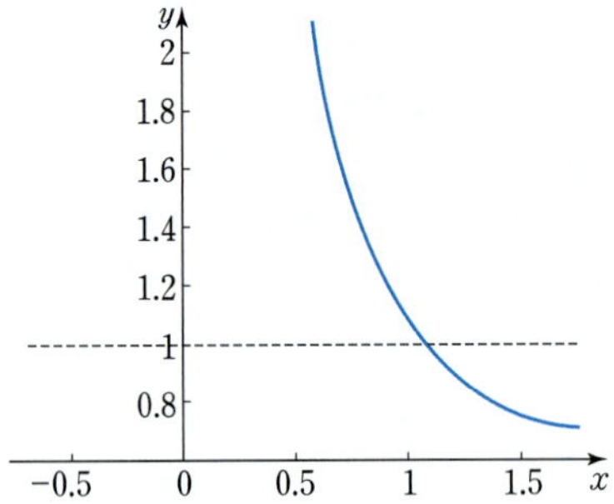

18-27. Find the limit by using L'Hospital's rule. If you cannot apply L'Hospital's rule, explain why and then find the limit by another method.

18. $\lim\limits_{x\to 0^+}\dfrac{\ln x}{x}$

Solution $(-\infty)\cdot\infty$, limit does not exist.

19. $\lim\limits_{x\to \infty}\dfrac{\ln x}{x}$

http://matrix.skku.ac.kr/cal-lab/cal-4-3-19.html

Solution $\lim\limits_{x\to\infty}\dfrac{\ln x}{x}=\lim\limits_{x\to\infty}\dfrac{\frac{1}{x}}{1}=\lim\limits_{x\to\infty}\dfrac{1}{x}$

(by L'Hospital's Rule) $\Rightarrow \lim\limits_{x\to\infty}\dfrac{\ln x}{x}=0$

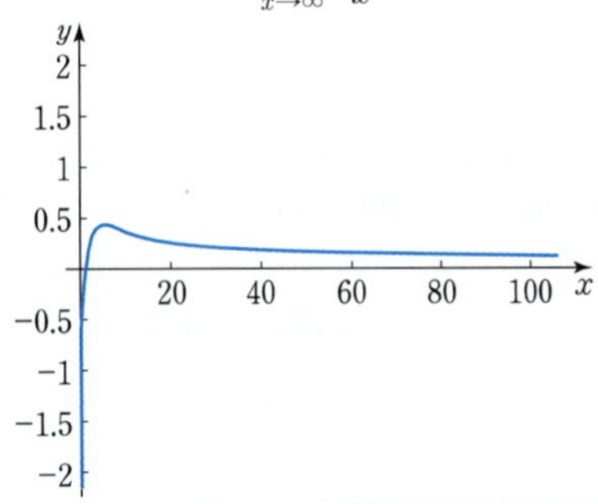


```
var('x');
f(x)=ln(x)/x;
show(plot(f(x),x,0,100, ymax=2, ymin=-2))
limit(f, x=Infinity)
```

20. $\lim\limits_{x\to\infty}\dfrac{e^x}{x}$

Solution Apply L'Hospital's rule, $\lim\limits_{x\to\infty}e^x=\infty$

CAS **21.** $\lim\limits_{x\to 0}\dfrac{x^3}{e^x}$

http://matrix.skku.ac.kr/cal-lab/cal-4-3-21.html

Solution 0

```
var('x')
f(x)=x^3/exp(x);
limit(f(x),x=0)
plot(f(x),x,-5,5, ymax=2, ymin=-2)
```

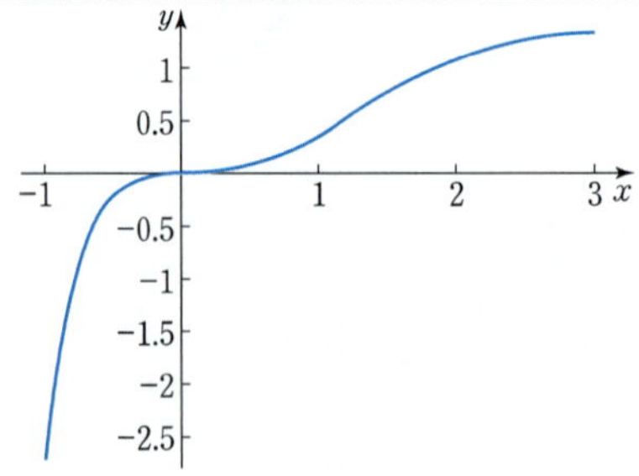

22. $\lim\limits_{x\to 0^+}\dfrac{e^x}{x}$

Solution $\dfrac{1}{0}=\infty$

23. $\lim\limits_{x\to\infty}e^{-x}\ln x$

Solution 0

24. $\lim\limits_{x\to 0^+}e^{-x}\ln x$

Solution $\lim\limits_{x\to 0^+}\dfrac{\ln x}{e^x}=\dfrac{-\infty}{1}$ does not exist.

25. $\lim\limits_{x\to 0}\dfrac{x^2\sin\frac{1}{x}}{\sin x}$

Solution 0

CAS **26.** $\lim\limits_{n\to\infty}\log\left(1+\dfrac{2}{n}\right)^n$

http://matrix.skku.ac.kr/cal-lab/cal-4-3-26.html

Solution

```
var('n, a');
a=2
f=log((1 + a/n)^n)
plot(f, (n,0,40)).show(ymax=3)
limit(f,n=oo)
```

Answer: 2

27. $\lim\limits_{x\to\infty}\dfrac{\log(a+x)}{x}$

Solution 0

28. The function $f,\ g:[0,\ 1]\to R$ is defined by

$$g(x)=\sin x \quad \text{and} \quad f(x)=\begin{cases} x^2\sin\dfrac{1}{x} & x\neq 0 \\ 0 & x=0 \end{cases}.$$

Answer the following questions.

(a) Find $\lim_{x\to 0^+}\dfrac{f(x)}{g(x)}$.

(b) Does $\lim_{x\to 0^+}\dfrac{f'(x)}{g'(x)}$ exist?

Solution (a) $\lim_{x\to 0^+}\dfrac{x^2\sin(1/x)}{\sin x}=\lim_{x\to 0^+}\dfrac{x\sin(1/x)}{\dfrac{\sin x}{x}}=\dfrac{0}{1}=0$

(b) $\lim_{x\to 0^+}\dfrac{f'(x)}{g'(x)}=\lim_{x\to 0^+}\dfrac{2x\sin(1/x)-\cos(1/x)}{\cos x}$

does not exist.

29. Let f be a continuous function with $f(3)=0$ and $f'(3)=3$. Find $\lim_{x\to 0}\dfrac{f(3+2x)-f(3+x)}{x}$.

Solution $\lim_{x\to 0}[f(3+2x)-f(3+x)]=0$ and $\lim_{x\to 0}x=0$.

Hence, by L'Hospital's Theorem

$$\lim_{x\to 0}\frac{f(3+2x)-f(3+x)}{x}=\lim_{x\to 0}2f'(3+2x)-f'(3+x)$$

$$=2f'(3)-f'(3)=f'(3)=3.$$

30. Let θ be the angle of a sector of a circle. Find $\lim_{\theta\to 0^+}\dfrac{P(\theta)}{Q(\theta)}$ where $P(\theta)$ and $Q(\theta)$ are the area of the segment shaded region between the chord AB and arcAB, the area of the triangle OAC respectively.

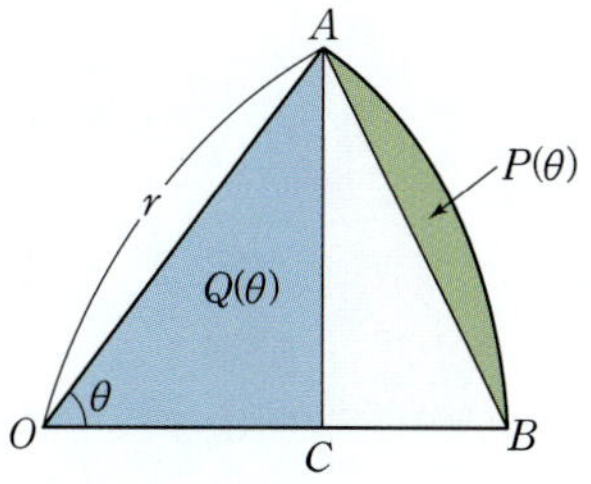

Solution $P(\theta)=\dfrac{1}{2}r^2(\theta-\sin\theta)$,

$$Q(\theta)=\frac{1}{2}|OC|\cdot|AC|=\frac{1}{2}(r\cos\theta)(r\sin\theta)$$

$$=\frac{1}{2}r^2\cos\theta\sin\theta$$

$$\lim_{\theta\to 0^+}\frac{P(\theta)}{Q(\theta)}=\lim_{\theta\to 0^+}\frac{(1/2)r^2\theta-(1/2)r^2\sin\theta}{(1/2)r^2\cos\theta\sin\theta}$$

$$=\lim_{\theta\to 0^+}\frac{\theta-\sin\theta}{\cos\theta\sin\theta}=\lim_{\theta\to 0^+}\frac{1-\cos\theta}{\cos^2\theta-\sin^2\theta}$$

$$=\lim_{\theta\to 0^+}\left[\frac{1}{\cos\theta\sin\theta}-\frac{1}{\cos\theta}\right]=\frac{0}{1}=0$$

4.4 Mathematical Optimization Problems

YouTube http://youtu.be/k0NtkmZFnh8

In this section, we apply calculus to various optimization problems which occur in the real life. There are many applications to optimizing: maximizing areas, volumes, and minimizing distances, times. An example of such a problem is maximum profit and minimum cost in business.

Guidelines for Solving Optimization Problems

Step 1. Understand the Problem: First understand the given problem, read carefully. What is unknown? What are the given quantities and the given conditions?

Step 2. Draw a Picture with Variables: Drawing a picture about the given and required quantities such as A for areas, V for volume, h for height and t for time is very useful.

Step 3. Express the Quantity as a Function: Express the quantity that is to be maximized or minimized as a function with one variable. If the function is of more than one variable then use mathematical ideas or given conditions to obtain a function of one variable. Do not forget to include the domain of this function.

Step 4. Determine the absolute maximum or minimum value of the function.

Step 5. Finally, interpret your solution obtained from step 4 in the context of the original problem.

EXAMPLE 1

A waterway is to be made so that its cross section is a trapezoid with equally sloping sides. The width and two sides of the waterway are of the same length and equal to 5 m. How should the sloping angle θ be chosen so that the waterway carries the maximum amount of water?

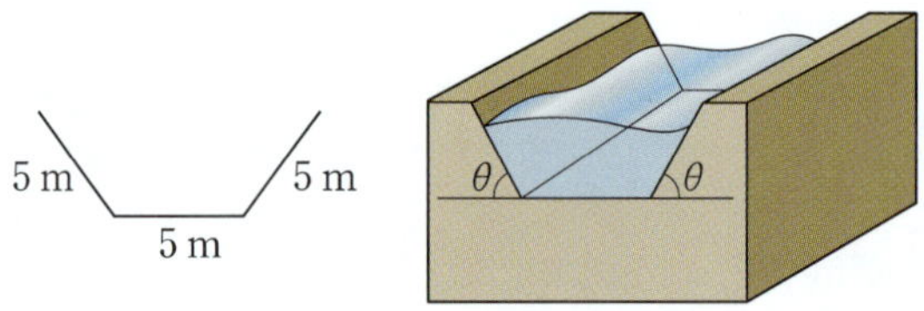

Figure 1

http://matrix.skku.ac.kr/cal-lab/cal-4-4-example-1.html

Solution 1. Unknown is the cross-sectional area of the waterway, and the shape of cross-section is given as the trapezoid with sides sloping by the angle θ, with the length of each side equal to 5 m.

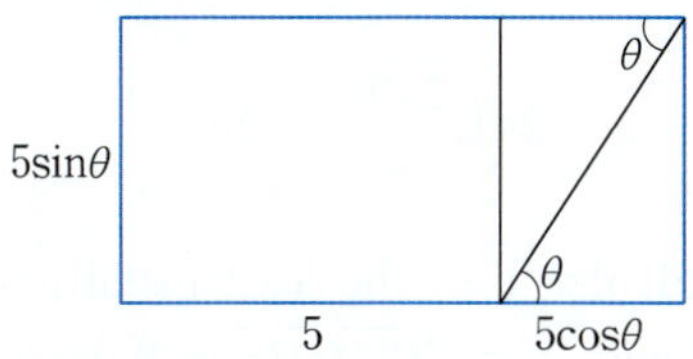

Figure 2

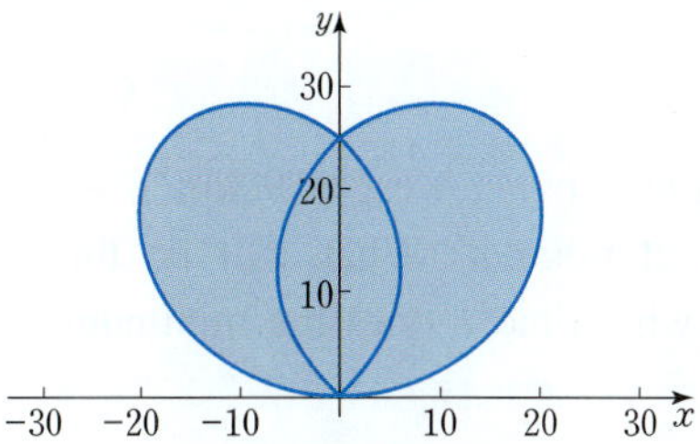

Figure 3 $A(\theta)=25\sin\theta(1+\cos\theta)$

2. The cross-sectional area is same as the area of the rectangle.

3. In order to maximize the cross-sectional area, we express the area as,

$$A(\theta)=5\sin\theta\,(5+5\cos\theta)=25\sin\theta\,(1+\cos\theta).$$

4. Our goal is to find a value of θ that maximizes $A(\theta)$. Since $A'(\theta)=25(2\cos\theta-1)(\cos\theta-1)$ and $0<\theta\le\dfrac{\pi}{2}$, we obtain $\cos\theta=\dfrac{1}{2}$, that is, $\theta=\dfrac{\pi}{3}$.
   ($\theta=0$ is a critical number that does not make physical sense in the context of the problem.)

■

EXAMPLE 2

A closed cylindrical can is to be made to hold 5 L (Liter) of oil. What dimensions will use the least amount of metal to make the can?
http://matrix.skku.ac.kr/cal-lab/cal-4-4-example-2.html

Solution We will follow the strategy above.

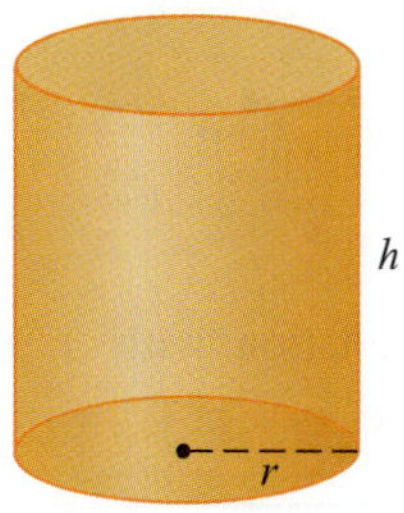

Figure 4

1. Unknowns are the dimension of a cylindrical can; knowns are the shape of the can is given as a cylinder together with a volume of 5L.
2. Let r be the radius and h the height (both in centimeters) of the can. Draw the picture labeled r and h as shown in Figure 4.
3. In order to minimize the amount of the metal, we express the total surface area A of the cylinder consisting of top, bottom and sides as a function with r and h. So the surface area is

$$A=2\pi r^2+2\pi rh.$$

   To eliminate a variable we use the fact that the volume is 5 L which is equal to $5000\,\text{cm}^3$. Thus $\pi r^2h=5000$ which gives $h=5000/\pi r^2$. Therefore,

$$A(r)=2\pi r^2+\frac{10000}{r},\ r>0$$

4. Our goal is to find a value of $r>0$ that minimizes the value of $A(r)$. Since

$$A'(r)=4\pi r-\frac{10000}{r^2}=\frac{4(\pi r^3-2500)}{r^2}$$

   the critical number of intersect as we search for an extreme value is $r=\sqrt[3]{2500/\pi}$. Applying the First Derivative Test, we see that $A(r)$ has an absolute minimum at $r=\sqrt[3]{2500/\pi}$. The

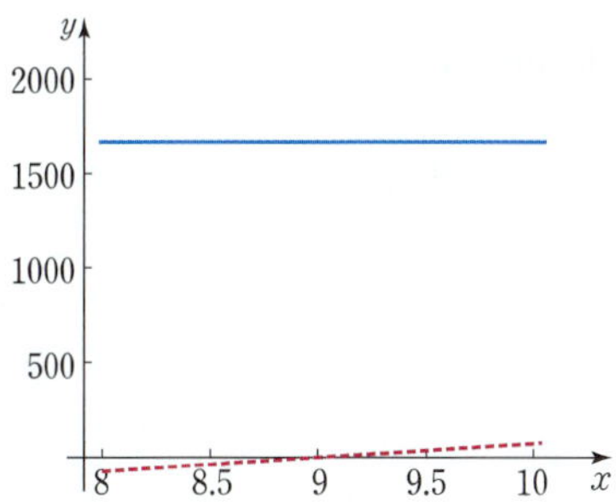

Figure 5 $A'(r)=4\pi r-\dfrac{10000}{r^2}$

corresponding value of h is

$$h = \frac{5000}{\pi r^2} = \frac{5000}{\pi(2500/\pi)^{2/3}} = 2\sqrt[3]{\frac{2500}{\pi}} = 2r.$$

5. Therefore, the cylindrical can of 5L that uses the least metal has height equal to twice the radius with $r = \sqrt[3]{2500/\pi} \approx 9.2668$ cm and $h = 2\sqrt[3]{2500/\pi}$ cm. ■

EXAMPLE 3

Car B is 30km directly east of car A and begins moving west at 90km/h. At the same moment car A begins moving north at 60km/h. What will be the minimum distance between the two cars, and at what time t does the minimum distance occur?

Solution

1. Unknown is the distance between the two cars A and B. The speeds of the cars are given as 60 km/h and 90 km/h, respectively.
2. Let x, y be the distance car A and B travel in t hours respectively, and let L be the distance between two cars A and B after t hours as shown in Figure 6.

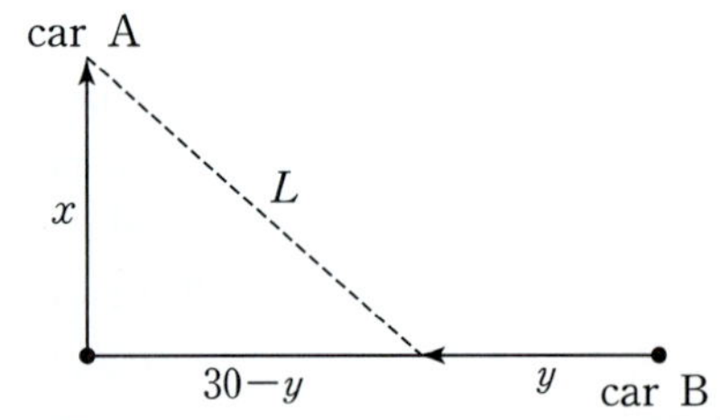

Figure 6

3. We express the distance L as a function with x and y by the Pythagorean theorem. Thus we have,

$$L(x,\ y) = \sqrt{x^2 + (30 - y)^2}.$$

To express L as a function of t, we recall that if travel is at a constant rate then

(distance traveled) = (speed) (time).

So substituting $x = 60t$ and $y = 90t$ into $L(x, y)$ gives

$$L(t) = \sqrt{3600t^2 + (30 - 90t)^2}.$$

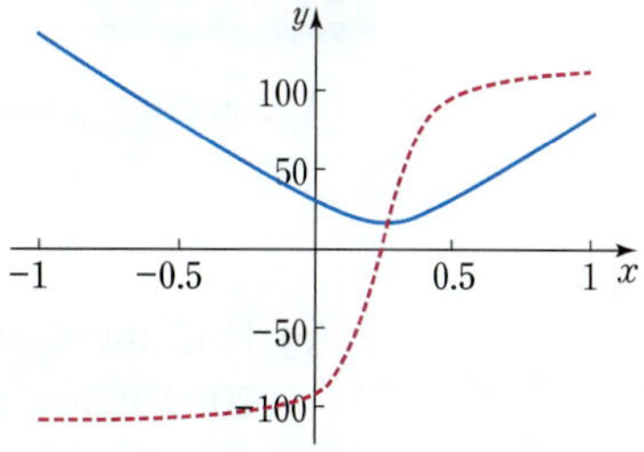

Figure 7 $L(t)=\sqrt{3600t^2+(30-90t)^2}$

4. Our goal is to find $t>0$ that minimizes $L(t)$. Since,

$$L'(t)=\frac{30(13t-3)}{\sqrt{13t^2-6t+1}},$$

we have $t=3/13\,\text{hours}\approx 0.23\,\text{hours}\,(\approx 13.8\,\text{min})$.
If $t=0.23$ hours then $x\approx 13.85$ km. and $y\approx 20.77$km.

5. Therefore the shortest distance between the two cars is $L\approx 24.96$ km. ■

*Application to Economics

We consider an optimization problem arising in economic theory. This problem is about maximizing profit and minimizing average cost.

Let $C(x)$ be the **cost** of producing x units of a certain product. Then the **marginal cost** is the rate of change of $C(x)$ with respect to x, that is, the marginal cost function is $C'(x)$. The **average cost function** is given by

$$\overline{C}(x)=\frac{C(x)}{x}$$

which represents the cost per unit when x units are produced. To minimize the average cost, differentiating $\overline{C}(x)$ gives

$$\overline{C}'(x)=\frac{xC'(x)-C(x)}{x^2}.$$

Thus $\overline{C}'(x)=0$ when $xC'(x)-C(x)=0$, that is,

$$C'(x)=\frac{C(x)}{x}=\overline{C}(x).$$

Therefore by the First Derivative Test we see that if the marginal cost is equal to the average cost then the average cost is a minimum. This principle is plausible because if the marginal cost is smaller than the average cost, then we should produce more, thereby lowering average cost. Similarly, if the marginal cost is larger than the average cost, then we should produce less in order to raise the average cost.

EXAMPLE 4

A company estimates that the cost in dollars of producing x items is

$$C(x) = 2600 + 2x + 0.001x^2.$$

(a) Find the cost, average cost, and marginal cost of producing 1000 items, 2000 items, and 3000 items.

(b) At what production level will the average cost be lowest, and what is this minimum average cost?

Solution (a) The average cost function is

$$\overline{C}(x) = \frac{C(x)}{x} = \frac{2600 + 2x + 0.001x^2}{x}.$$

The marginal cost function is $C'(x) = 2 + 0.002x$.

The following table gives the cost, average cost, and marginal cost:

x	$C(x)$	$\overline{C}(x)$	$C'(x)$
1000	5,600.00	5.60	4.00
2000	10,600.00	5.30	6.00
3000	17,600.00	5.87	8.00

(b) To minimize the average cost we must have

$$\text{marginal cost} = \text{average cost}$$
$$C'(x) = \overline{C}(x).$$

Hence $2 + 0.002x = \dfrac{2600}{x} + 2 + 0.001x.$

This equation simplifies to $0.001x = 2600/x$, that is,

$$x^2 = \frac{2600}{0.001} = 2600000 \quad \text{and} \quad x = \sqrt{26000000} \approx 1612.$$

Since $C''(x) = 5200/x^3 > 0\,(x > 0)$, C is concave upward on its entire domain. Thus the minimum average cost at $x = 1612$ is

$\overline{C}(1612) = \dfrac{2600}{1612} + 2 + 0.001(1612) = \$5.22/1612\,\text{items}.$ ■

4.4 EXERCISES (Mathematical Optimization Problems)

http://matrix.skku.ac.kr/Cal-Book/part1/CS-Sec-4-4-Sol.html

YouTube http://youtu.be/AELEV2ElaeQ

1. Find the point on the curve $y=\sin x$ that is closest to the point $\left(\frac{\pi}{2},\ 0\right)$.

Solution The distance d from $\left(\frac{\pi}{2},\ 0\right)$ to an arbitrary point $P(x,\ y)$ on the curve $y=\sin x$ is

$$d=\sqrt{(x-\pi/2)^2+(y-0)^2}$$

and the square of the distance is

$$S=d^2=(x-\pi/2)^2+(\sin x)^2,$$
$$S'=2(x-\pi/2)+2(\sin x)\cos x.$$

Graphing S' on $(-\pi,\ \pi)$ gives us a zero at $x=\frac{\pi}{2}$, and so $\sin\frac{\pi}{2}=1$. The point on $y=\sin x$ that is closest to $\left(\frac{\pi}{2},\ 0\right)$ is $\left(\frac{\pi}{2},\ 1\right)$.

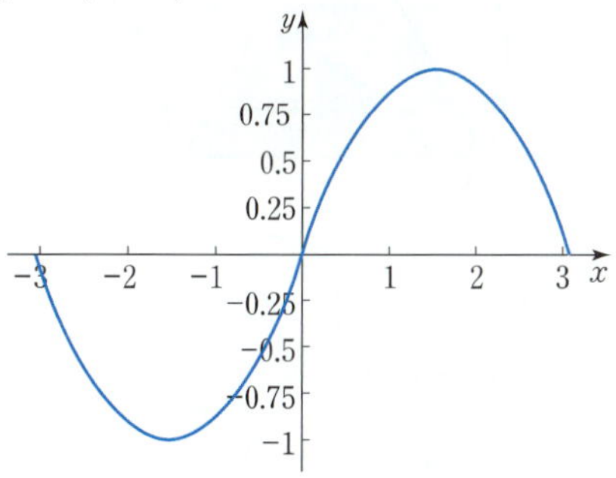

2. A shop sells 200 MP3 players per week while each costs \$350. According to the market research, sales will increase by 20 MP3 players per week for each \$10 discount. How much should they discount to maximize profits?

Solution If \$10 discount is made for each MP3 player sale,

$$\text{profit}=(350-10x)(200+20x)$$
$$=100(-2x^2+50x+700).$$

The profit is maximized when $x\approx 12.5$. They should discount by \$125.

```
var('x')
f=100*(-2*x^2 + 50*x + 700)
df=diff(f, x);
show (df)
solve (df==0, x)
```

3. The height of a safe is y meters and its bottom is in the shape of a square whose side is x meters. It costs 20000 Korean won per $1\ \text{m}^2$ to make the top and the bottom, and 10000 Korean won per $1\ \text{m}^2$ to make the side. Find the maximum volume of the safe which can be made by using 480000won.

Solution $\cos t = 40000x^2+40000xy$

$=40000(x^2+xy)\le 480000$

$x^2+xy\le 12$

$xy=12-x^2$

$V=x^2y=x(12-x^2)=-x^3+12x$

$V'=-3x^2+12$ and so $x=\pm 2$

$x=2,\ y=4$

$V=16\ \text{m}^3$

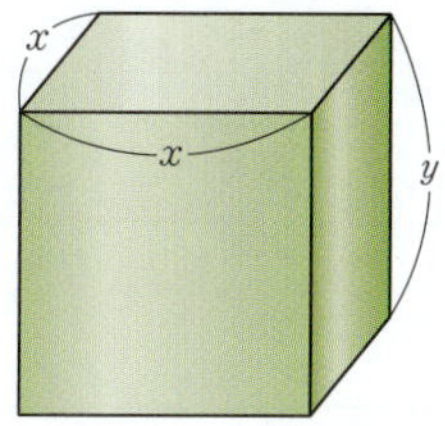

4. Two particles have locations at time t on the xy-plane given by $A=(t,-t^2)$ and $B=(-2-t,\ t)$. Find the minimum distance between A and B.

Solution $\overline{AB}^2=(2+2t)^2+(t+t^2)^2$

$=t^4+2t^3+5t^2+8t+4$

$(\overline{AB}^2)'=4t^3+6t^2+10t+8=2(t+1)(2t^2+t+4)=0$

Therefore, $t=-1$.

Therefore, the minimum distance between the two particle is 0.

5. A closed cylindrical can is to hold $500\ \text{cm}^3$ of liquid. Find the height and radius that minimize the amount of material needed to manufacture the can.

http://matrix.skku.ac.kr/cal-lab/cal-4-4-exs-5.html

Solution $\pi r^2h=500$

$S=2\pi r^2+2\pi rh=2\pi r^2+2\frac{500}{r}$

$S'=4\pi r-\frac{1000}{r^2}$

$$S'=0\Rightarrow r=\sqrt[3]{\frac{250}{\pi}}\ \text{cm}$$

$$h=\frac{500}{\pi(250/\pi)^{2/3}}=10\left(\frac{2}{\pi}\right)^{1/3}\text{ cm}$$

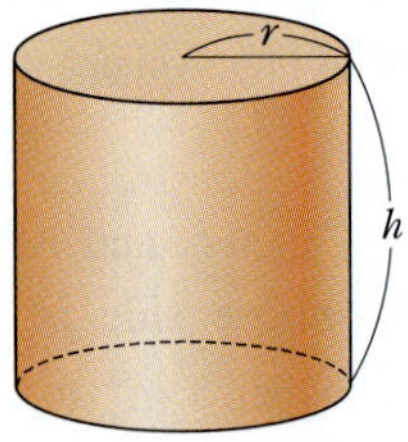


```
var('x,y');
f=2*pi*x^2 + 1000/x
P= plot(f,x, xmin=-10, xmax=10, ymin=-60, ymax=60)
df=diff(f, x)
Q=plot(df, x, xmin=-10, xmax=10, ymin=-60, ymax=60,linestyle="--", color='red')
solve(df==0, x)    # (250/pi)^(1/3)
numerical_approx( (250/pi)^(1/3), 100) # 4.3
show(P+Q)
```

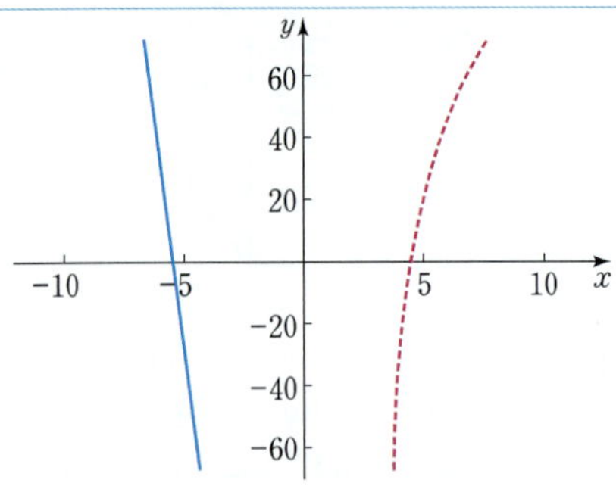

6. Determine where the point P between A and B should be located to maximize the angle θ.

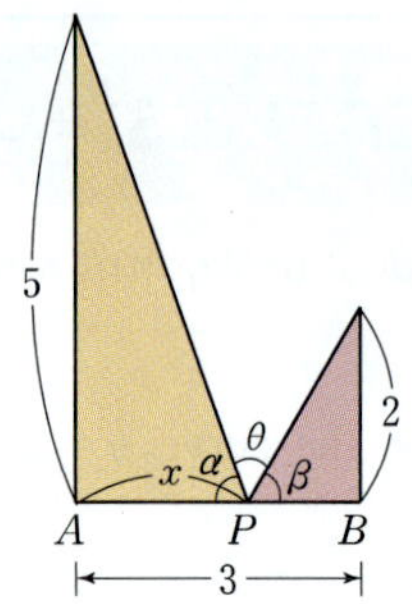

Solution Let $\overline{AP}=x,\ \alpha=\tan^{-1}\frac{5}{x},\ \beta=\tan^{-1}\frac{2}{3-x}$.

Then $\tan\theta=\tan(\pi-(\alpha+\beta))=-\tan(\alpha+\beta)$

$$=-\frac{\tan\alpha+\tan\beta}{1-\tan\alpha\tan\beta}=-\frac{\frac{5}{x}+\frac{2}{3-x}}{1-\left(\frac{5}{x}\cdot\frac{2}{3-x}\right)}$$

$$=\frac{3x-15}{-x^2+3x-10},$$

$$\frac{d}{dx}(\tan\theta)=\frac{d}{dx}\left(\frac{3x-15}{-x^2+3x-10}\right)$$

$$=\frac{3x^2-30x+15}{(-x^2+3x-10)^2}.$$

Solving $\frac{d}{dx}(\tan\theta)=0$, we get $x=5\pm2\sqrt{5}$.

Since $0\le x\le 3$, we have $x=5-2\sqrt{5}$.

7. Find the largest area of a triangle which is inscribed in a circle of radius r.

Solution Let r and t be as shown in the figure. The area of the triangle is

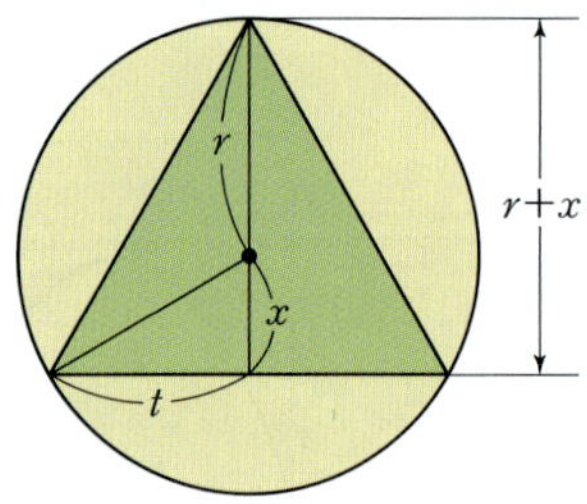

$$A(x)=\frac{1}{2}(2t)(r+x)=t(r+x)$$

$$=\sqrt{r^2-x^2}\,(r+x)$$

Then,

$$0=A'(x)=r\frac{-2x}{2\sqrt{r^2-x^2}}+x\frac{-2x}{2\sqrt{r^2-x^2}}+\sqrt{r^2-x^2}$$

$$=-\frac{x^2+rx}{\sqrt{r^2-x^2}}+\sqrt{r^2-x^2}$$

$$=\frac{-x^2-rx+r^2-x^2}{\sqrt{r^2-x^2}}=\frac{-2x^2-rx+r^2}{\sqrt{r^2-x^2}}$$

$$\Rightarrow 0=2x^2+rx-r^2=(2x-r)(x+r)$$

$$\Rightarrow x=\frac{1}{2}r \text{ or } x=-r$$

Now $A(r)=0=A(-r)$.

The maximum area occurs when $x=\frac{1}{2}r$ with the height $r+\frac{1}{2}r=\frac{3}{2}r$ and the base $2\sqrt{r^2-\left(\frac{r}{2}\right)^2}=2\sqrt{\frac{3}{4}r^2}=\sqrt{3}r$.

The largest area of a triangle is

$$A(x)=\frac{\sqrt{3}r}{2}\cdot\frac{3}{2}r=\frac{\sqrt{3}}{4}r^2.$$

8. Let V_1 be the volume of the right circular cone and V_2 be the volume of the right circular cylinder that can be inscribed in the cone. Find the ratio $\dfrac{V_2}{V_1}$ when the cylinder has the greatest volume.

Solution Let r and h be the radius of the base and height of the right circular cone let. Let x and y as show in the figure. $V_1 = \frac{1}{3}\pi r^2 h$.

By similar triangles, $y/x = h/r$, so $y = hx/r$.

The volume of the cylinder is

$$\pi x^2(h-y) = \pi h x^2 - (\pi h/r)x^3 = V_2.$$

Now $V_2{}'(x) = 2\pi h x - (3\pi h/r)x^2 = \pi h x(2 - 3x/r)$.

So $V_2{}'(x) = 0 \Rightarrow x = 0$ or $\frac{2}{3}r$.

The maximum clearly occurs when $x = \frac{2}{3}r$ and then the volume is

$$\begin{aligned}\pi h x^2 - (\pi h/r)x^3 &= \pi h x^2(1 - x/r)\\ &= \pi\left(\frac{2}{3}r\right)^2 h\left(1-\frac{2}{3}\right)\\ &= \frac{4}{27}\pi r^2 h = V_2,\end{aligned}$$

$$\frac{V_2}{V_1} = \frac{\frac{4}{27}\pi r^2 h}{\frac{1}{3}\pi r^2 h} = \frac{4}{9}.$$

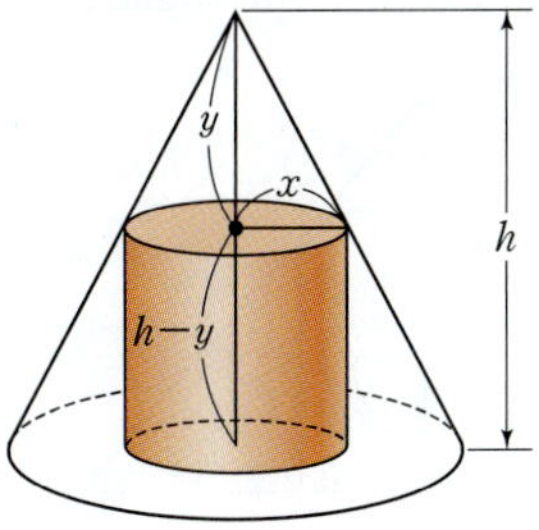

9. Consider an ellipse $\dfrac{x^2}{a^2} + \dfrac{y^2}{b^2} = 1$. Find the area of the rectangle of greatest area that can be inscribed in the ellipse.

Solution Without loss of generality, choose

$$\left(x,\ \sqrt{b^2 - \frac{b^2x^2}{a^2}}\right)$$

on the ellipse in first quadrant.

Let A= area of the rectangle and S = area of the ellipse.

Since $A = \left(\frac{S}{4}\right)^2$, $A = \left(\frac{S}{4}\right)^2 = b^2x^2 - \dfrac{b^2x^4}{a^2}$

$$A' = 2b^2x - 4\frac{b^2}{a^2}x^3 = 2b^2x\left(1 - \frac{2}{a^2}x^2\right)$$

$$A' = 0 \Rightarrow 1 - \frac{2}{a^2}x^2 = 0$$

$$x^2 = \frac{a^2}{2} \Rightarrow x = \frac{a}{\sqrt{2}}$$

$$A = \frac{1}{4}a^2b^2 \Rightarrow 4a^2b^2 = 16A = S^2$$

Thus, $S = 2ab$.

10. A closed cylindrical can is to be made to hold $100\ \text{cm}^3$ of oil. What dimensions will use the least metal?

Solution Let r be the radius and h the height (both in centimeters) of the can. The total surface area A of the cylinder consisting of top, bottom and sides as a function with r and h is $A = 2\pi r^2 + 2\pi r h$. We eliminate the variable h by using the given condition $\pi r^2 h = 100$. Therefore

$$A(r) = 2\pi r^2 + \frac{200}{r},\ r > 0$$

and $A'(r) = 4\pi r - \dfrac{200}{r^2} = \dfrac{4\pi r^3 - 200}{r^2}$.

The only critical number is $r = \sqrt[3]{50/\pi}$. Applying the First Derivative Test, we see that $A(r)$ has an absolute minimum at $r = \sqrt[3]{50/\pi}$ cm. The corresponding value of h is

$$h = \frac{100}{\pi r^2} = \frac{100}{\pi(50/\pi)^{2/3}}\ \text{cm}.$$

11. A cone-shaped paper cup is to be made to hold 24cm^3 of water. Find the height and radius of the cup that minimizes the amount of paper needed to make the cup.

Solution The volume and surface area of a cone with radius r and height h are given by $V = \frac{1}{3}r^2 h\pi$ and $S = \pi r\sqrt{r^2 + h^2}$. We will minimize $A = S^2$ subject to $V = 24$.

$$V = 24 \Rightarrow \frac{1}{3}\pi r^2 h = 24 \Rightarrow r^2 = \frac{72}{\pi h}$$

$$\begin{aligned}A &= \pi^2 r^2(r^2 + h^2) = \pi^2\left(\frac{72}{\pi h}\right)\left(\frac{72}{\pi h} + h^2\right)\\ &= \frac{72^2}{h^2} + 72\pi h\end{aligned}$$

so $A' = 0$

$$\Rightarrow \frac{-2\cdot 72^2}{h^3} + 72\pi = 0$$

$$\Rightarrow 72\pi = \frac{2\cdot 72^2}{h^3} \Rightarrow h^3 = \frac{144}{\pi}$$

$$\Rightarrow h = \left(\frac{144}{\pi}\right)^{1/3} \approx 3.579\text{cm}$$

$$r^2 = \frac{72}{\pi h} = \frac{72}{\pi(144/\pi)^{1/3}}$$

$$\Rightarrow r = \left(\frac{72}{\pi (144/\pi)^{1/3}} \right)^{1/2} \approx 2.531\text{cm}$$

$A'' = \dfrac{6 \cdot 72^2}{h^4} > 0$, so A and hence S has an absolute minimum at these values of r and h.

12. A pipe is being carried horizontally around a corner from a hallway 3 m wide into a hallway 2 m wide. What is the longest length the pipe may have?

Solution Let L be the length of the line ACB going from wall to wall touching the inner corner C. As $\theta \to 0$ or $\theta \to \frac{\pi}{2}$, we have $L \to \infty$ and there will be an angle that makes L a minimum. A pipe of this length will just fit around the corner. From the diagram,

$L = L_1 + L_2$ where $L_1 = AC$ and $L_2 = BC$.

$= 2\csc\theta + 3\sec\theta$

$\Rightarrow dL/d\theta = -2\csc\theta\cot\theta + 3\sec\theta\tan\theta = 0$

when $3\sec\theta\tan\theta = 2\csc\theta\cot\theta \Leftrightarrow \tan^3\theta = \dfrac{2}{3}$

$\Leftrightarrow \tan\theta = \left(\dfrac{2}{3}\right)^{1/3}$.

Then $\sec^2\theta = 1 + \left(\dfrac{2}{3}\right)^{2/3}$ and $\csc^2\theta = 1 + \left(\dfrac{2}{3}\right)^{-2/3}$, so the longest pipe has length

$$L = 2\left[1 + \left(\frac{2}{3}\right)^{-2/3}\right]^{1/2} + 3\left[1 + \left(\frac{2}{3}\right)^{2/3}\right]^{1/2}$$

$$\approx 7.02 \text{ m}.$$

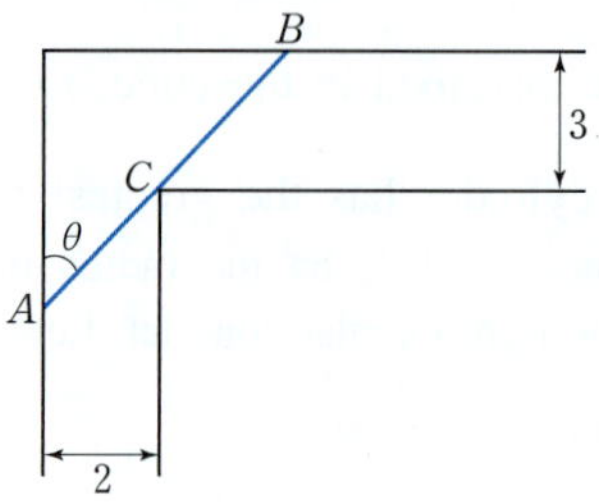

13. Find the length of the shortest ladder that reaches over an 8ft high fence to a wall that is 3ft behind the fence.

Solution $L = 8\csc\theta + 3\sec\theta,\ 0 < \theta < \dfrac{\pi}{2}$,

$\dfrac{dL}{d\theta} = -8\csc\theta\cot\theta + 3\sec\theta\tan\theta = 0$ when

$3\sec\theta\tan\theta = 8\csc\theta\cot\theta \Leftrightarrow \tan^3\theta = \dfrac{8}{3}$

$\Leftrightarrow \tan\theta = (8/3)^{1/3} \Leftrightarrow \theta = \tan^{-1}(8/3)^{1/3}$

$dL/d\theta < 0$ when $0 < \theta < \tan^{-1}(8/3)^{1/3}$,

$dL/d\theta > 0$ when $\tan^{-1}(8/3)^{1/3} < \theta < \dfrac{\pi}{2}$,

so L has an absolute minimum when $\theta = \tan^{-1}(8/3)^{1/3}$, and the shortest ladder has length

$$L = 8\csc\theta + 3\sec\theta \approx 46.31,\ \theta \approx 0.186.$$

(*Label theta and L in the figure.)

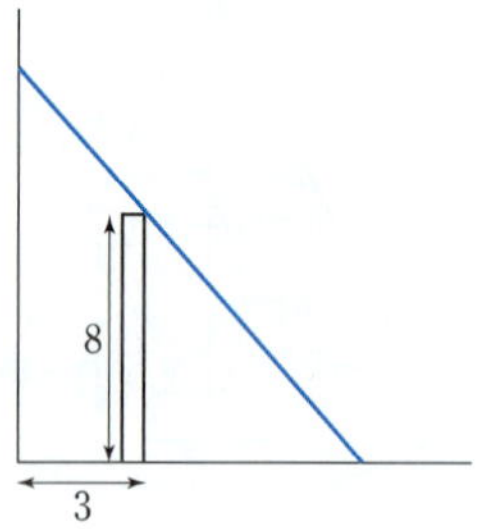

4.5 Newton's Method

YouTube http://youtu.be/VxCfl2JzMYU

The simple formulas for solving linear equations $ax+b=0$ and quadratic equations $ax^2+bx+c=0$ are well-known. Also, there are somewhat more complicated formulas for equations of degree 3 and 4. But there is no such formula for equations of degree 5 or greater. Likewise, there is no formula that will enable us to find the exact solutions of a transcendental equation such as $\sin x+\cos x=x$. That means if there is no exact formula for solving $f(x)=0$, so we turn to numerical techniques. One of these techniques is called **Newton's method.** This method is based on the idea of a linear approximation with tangent lines to the graph of $y=f(x)$.

Procedure for Newton's Method

Let us consider the graph of $y=f(x)$ and we are trying to solve $f(x)=0$. We start with the initial approximation x_1, which may be obtained by guessing, or examining the graph of f. Then we use the tangent line L to the curve $y=f(x)$ at the point $(x_1, f(x_1))$ to approximate the curve and look at the x-intercept of L, labeled x_2. The equation of the tangent line L is

$$y-f(x_1)=f'(x_1)(x-x_1).$$

Thus, we can obtain x-intercept x_2, by replacing $y=0$;

$$0-f(x_1)=f'(x_1)(x_2-x_1).$$

If $f'(x_1)\neq 0$, we can solve this equation for x_2:

$$x_2=x_1-\frac{f(x_1)}{f'(x_1)}$$

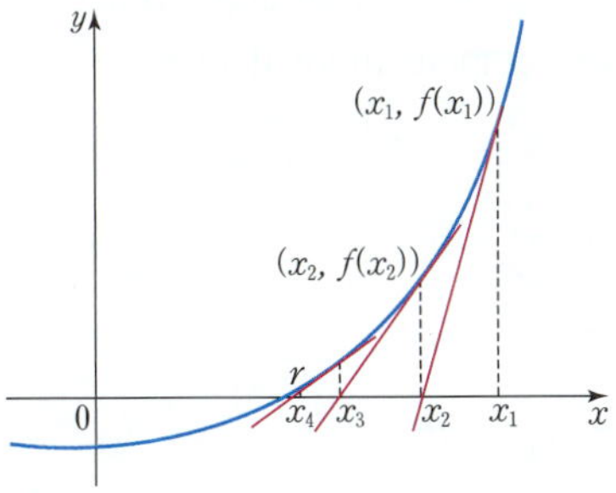

Figure 1

Under certain conditions, x_2 is usually a better approximation to the solution than x_1. Then we repeat this procedure with x_1 replaced by x_2, using the tangent line at $(x_2, f(x_2))$. This gives a third approximation:

$$x_3=x_2-\frac{f(x_2)}{f'(x_2)}$$

Continuing this process, we obtain a sequence of approximations $x_1, x_2, x_3, \cdots$ as shown in Figure 1. In general, if $f'(x_n)\neq 0$ then we have

1

$$x_{n+1}=x_n-\frac{f(x_n)}{f'(x_n)}$$

The number x_n becomes closer and closer to the solution if the sequence $\{x_n\}$ converges as $n \to \infty$. We note that if $f'(x_1) \to 0$ then the sequence may not converge. In this case, we have to choose a different initial x_1.

EXAMPLE 1

Compute the third approximation x_3 of the root of the equation $x^4 - 7x + 1 = 0$, starting with $x_1 = 2$.
http://math1.skku.ac.kr/home/pub/1028

Solution Apply Newton's method with

$$f(x) = x^4 - 7x + 1 \text{ and } f'(x) = 4x^3 - 7$$

Choose, $x_1 = 2$.
Note that

$$x_{n+1} = x_n - \frac{x_n^4 - 7x_n + 1}{4x_n^3 - 7}.$$

With $n = 1$ we have

$$\begin{aligned} x_2 &= x_1 - \frac{x_1^4 - 7x_1 + 1}{4x_1^3 - 7} \\ &= 2 - \frac{2^4 - 7(2) + 1}{4(2)^3 - 7} = \frac{47}{25} = 1.88. \end{aligned}$$

Then with $n = 2$ we obtain

$$\begin{aligned} x_3 &= x_2 - \frac{x_2^4 - 7x_2 + 1}{4x_2^3 - 7} \\ &= 1.88 - \frac{(1.88)^4 - 7(1.88) + 1}{4(1.88)^3 - 7} \approx 1.8630 \end{aligned}$$

It turns out this third approximation $x_3 \approx 1.8630$ is an approximated root. We may not know this unless we compute more iterates.

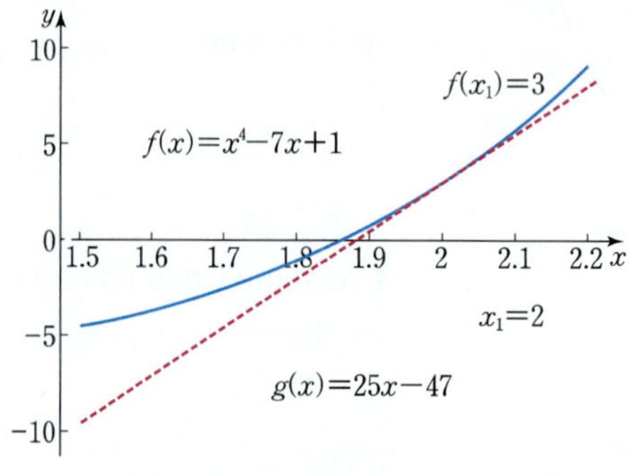

Figure 2 Newton's method

■

EXAMPLE 2

Find $\sqrt[5]{40}$ correct to four decimal places using Newton's method.

Solution Finding $\sqrt[5]{40}$ is equivalent to finding the positive root of the equation $x^5 - 40 = 0$, so take $f(x) = x^5 - 40$.
Then $f'(x) = 5x^4$. Newton's formula becomes

$$x_{n+1} = x_n - \frac{x_n^5 - 40}{5x_n^4}.$$

Using $x_1 = 2$ as the initial approximation, we obtain

$$\begin{aligned} x_2 &= 2.1 \\ x_3 &\approx 2.0914 \\ x_4 &\approx 2.0913 \\ x_5 &\approx 2.0913 \end{aligned}$$

Since x_4 and x_5 agree to four decimal places, we conclude that $\sqrt[5]{40} \approx 2.0913$ to four decimal places. ■

Newton's Method by Sage:

In Sage define

```
newton(x)=x-(x^5-40)/(5x^4)
```

```
x1=2
x2=newton(x1);N(x2)
```

2.1000

```
x3=newton(x2);N(x3)
```

2.09135

```
x4=newton(x3);N(x4)
```

2.0913

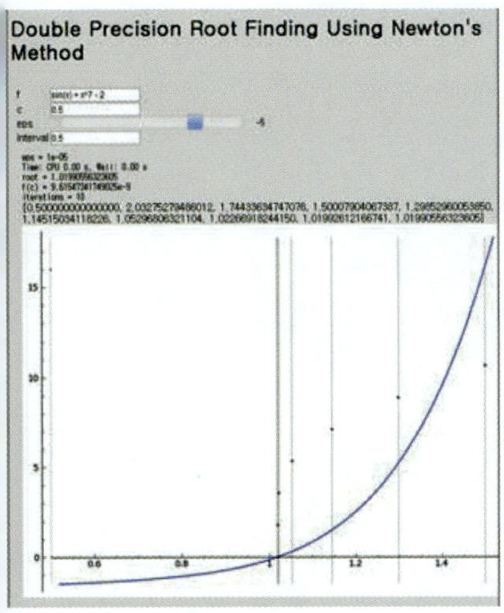

Figure 3 Newton's method

Practice in http://matrix.skku.ac.kr/cal-lab/cal-Newton-method.html
Root Finding Using Bisection:
http://matrix.skku.ac.kr/2012-LAwithSage/interact/4/1.html

4.5 EXERCISES (Newton's Method)

http://matrix.skku.ac.kr/Cal-Book/part1/CS-Sec-4-5-Sol.html

YouTube http://youtu.be/fdBHQ46g9RE

1-3. Compute x_3, the third approximation to the root of the given equation using Newton's method with the specified initial approximation x_1.

1. $x^3+x^2-2x+1=0,\ x_1=-2$

Solution $f(x)=x^3+x^2-2x+1$

$f'(x)=3x^2+2x-2$

$x_1=-2$

$x_2\approx-2.166666667$

$x_3\approx-2.148148148$

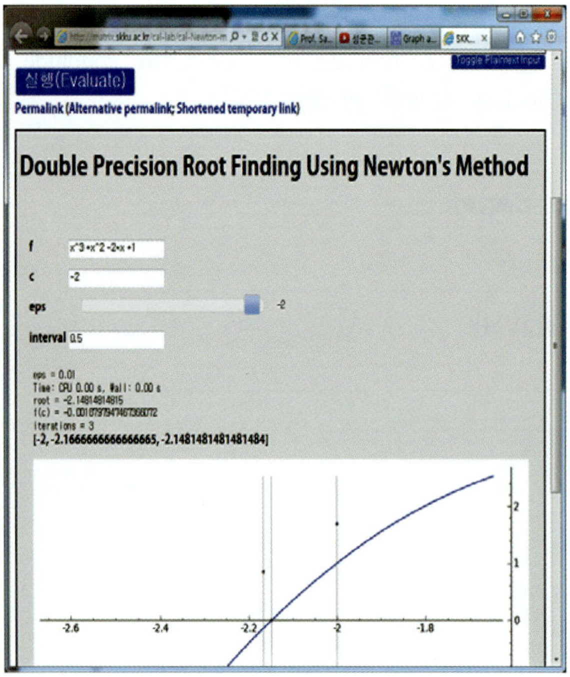

2. $x-\cos^2 x=0,\ x_1=1$

Solution $f(x)=x-\cos^2 x$

$f'(x)=1+2\cos x\sin x$

$x_2\approx0.6291445177$

$x_3\approx0.6417386152$

3. $x^7-1000=0,\ x_1=2$

Solution $f(x)=x^7-1000$

$f'(x)=7x^6$

$x_2\approx3.946428571$

$x_3\approx3.420469138$

---

4. Calculate the second approximation x_2 to the root of the equation $x^7+x+3=0$ by using Newton's method with initial approximation $x_1=-1$.

http://matrix.skku.ac.kr/cal-lab/cal-Newton-method.html

Solution $f(x)=x^7+x+3$

$f'(x)=7x^6+1$

$$x_{n+1}=x_n-\frac{f(x_n)}{f'(x_n)}$$

$$x_2=(-1)-\frac{1}{8}=-1.125$$

5-6. Approximate the given number using Newton's method.

5. $\sqrt[4]{30}$.

http://matrix.skku.ac.kr/2012-LAwithSage/interact/4/3.html

Solution Let $x=\sqrt[4]{30}\Rightarrow x^4=30$. Let $f(x)=x^4-30$ and find x by Newton's method with $x_1=2$.

$f'(x)=4x^3$

$$x_{n+1}=x_n-\frac{f(x_n)}{f'(x_n)}$$

$x_1=2$

$x_2=2.4375$

$x_3\approx2.346002914$

$x_4\approx2.340367738$

$x_5\approx2.340347320$

$x_6\approx x_7\approx2.340347319$

6. $\sqrt[9]{1000}$

Solution $f(x)=x^9-1000$

$f'(x)=9x^8$

$$x_{n+1}=x_n-\frac{f(x_n)}{f'(x_n)}$$

$x_1=2$

$x_2\approx2.211805556$

$x_3\approx2.160040540$

$x_4\approx2.154492533$

$x_5\approx2.154434696$

$x_6\approx2.154434690$

---

7. $\sin x+\cos x=0$

Solution $f(x)=\sin x+\cos x$

$f'(x)=\cos x-\sin x$

$x_1 = -0.5$
$x_2 \approx -0.7934079931$
$x_3 \approx -0.7853979921$
$x_4 \approx x_5 \approx -0.7853981634$

8. $e^x = x^2 - x$

Solution $f(x) = e^x - x^2 + x$

$f'(x) = e^x - 2x + 1$

$x_1 = -0.5$
$x_2 \approx -0.4449577392$
$x_3 \approx -0.4441304127$
$x_4 \approx x_5 \approx -0.4441302288$

9. $x^2\sqrt{x^2 + 2x - 2} = 4$

Solution $f(x) = x^2\sqrt{x^2+2x-2} - 4$

$f'(x) = 2x\sqrt{x^2+2x-2} + \dfrac{x^2(2x+2)}{2\sqrt{x^2+2x-2}}$

(i) $x_1 = -3$
$x_2 \approx -2.791666667$
$x_3 \approx -2.804627598$
$x_4 \approx -2.805108088$
$x_5 \approx x_6 \approx -2.805108613$

(ii) $x_1 = 1$
$x_2 = 1.75$
$x_3 \approx 1.527428725$
$x_4 \approx 1.494091762$
$x_5 \approx 1.493377700$
$x_6 \approx x_7 \approx 1.493377377$

10. $\ln(16 - x^2) = \sqrt{9 - x^2}$

Solution $f(x) = \ln(16 - x^2) - \sqrt{9 - x^2}$

$f'(x) = -\dfrac{2x}{16 - x^2} + \dfrac{x}{\sqrt{9 - x^2}}$

(i) $x_1 = 1.5$
$x_2 \approx x_3 \approx 1.436067288$

(ii) $x_1 = -1.5$
$x_2 \approx -1.436067288$
$x_3 \approx -1.434166254$
$x_4 \approx x_5 \approx -1.434164615$

11. Apply Newton's method to the equation

$$\frac{1}{x^3} - k = 0$$

to derive the following reciprocal algorithm:

$$x_{n+1} = \frac{4}{3}x_n - \frac{k}{3}x_n^4.$$

Hence compute $(1.1246)^{-3}$.

Solution $f(x) = \dfrac{1}{x^3} - k$

$f'(x) = -\dfrac{3}{x^4}$

$x_{n+1} = x_n - \dfrac{1/x_n^3 - k}{-3/x_n^4} = \dfrac{4}{3}x_n - \dfrac{k}{3}x_n^4$

(i) $x_1 = 1$
$x_2 \approx 0.9584666663$
$x_3 \approx 0.9615931426$
$x_4 \approx 0.9616136951$
$x_5 \approx x_8 \approx \cdots \approx 0.9616136956$
$x_6 \approx x_9 \approx \cdots \approx 0.9616136959$
$x_7 \approx x_{10} \approx \cdots \approx 0.9616136966$

12. Use Newton's method to find the absolute minimum value of the function $f(x) = x^2 - \cos x$.

Solution $f'(x) = 2x + \sin x$

$f''(x) = 2 + \cos x$

$x_1 = 0.5$
$f(x_1) = -0.6275825619$
$x_2 \approx -0.0141209703$
$f(x_2) \approx -0.9997008990$
$x_3 \approx 3.1286 \cdot 10^7$
$f(x_3) \approx -1$
$x_4 \approx x_5 \approx 0$
$f(x_4) \approx f(x_5) \approx -1$

Newton-Raphson Root Finding:
http://matrix.skku.ac.kr/2012-LAwithSage/interact/4/3.html

Chapter 5

Integrals

Isaac Newton

1642-1727

Scientist and Mathematician

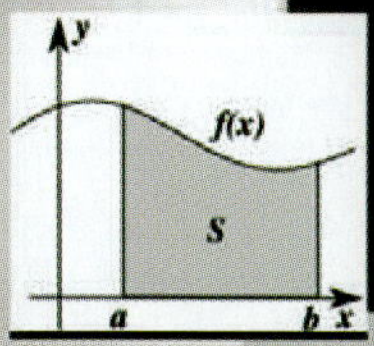

[Calculus] "This photograph is from Wikimedia Commons"

Considered one of the most influential men in history, Newton is considered the father of calculus. In 1665, he discovered the generalized binomial theorem. A year later, in 1666, he wrote his first work on calculus. His 1687 publication, Principia, is among the most noted books in the field of science history and laid the groundwork for classical mechanics.

YouTube http://youtu.be/mT_oxlD6RSA

In this chapter, we introduce the definite integral as a Riemann sum and show the connection between the integral and derivative of a function via the Fundamental Theorem of Calculus. As simple applications to the definite integral, we deal with area bounded by curves, distance traveled by particles, population prediction etc. We also introduce the natural logarithm and natural exponential functions in applications to definite integral.

5.1 Areas and Distances

Area Problem: The area of a region is computed as the limit of the sum of areas of rectangles, as the number of rectangles become large.

Let S be the area of the region that lies under the curve $y = f(x)$ of a continuous function f [where $f(x) \geq 0$] and bounded by the vertical lines $x = a$ and $x = b$, and the x-axis as in Figure 1.

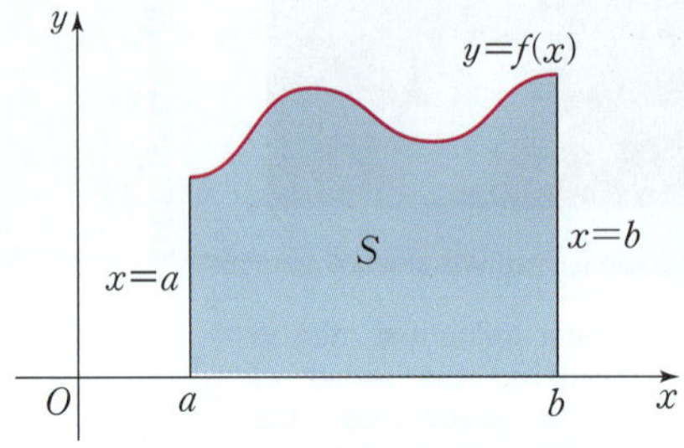

Figure 1 $S = \{(x, y) | a \leq x \leq b,\ 0 \leq y \leq f(x)\}$

For regions with straight sides such as a rectangle, the area is defined as the product of the length and the width. The area of a triangle is half of the base times the height. The area of a polygon is found by dividing it into triangles (as in Figure 2) and summing the areas of the triangles.

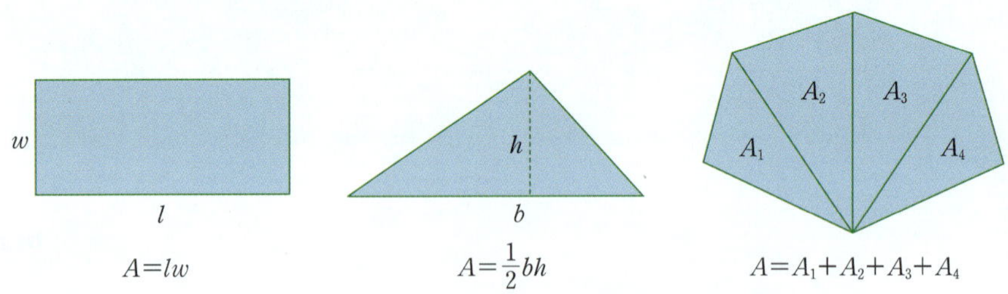

Figure 2

In the case of area of a region with curved sides, we first approximate the region S by rectangles and then take the limit of the sum of areas of these rectangles as the number of rectangles increase.

Consider a general region S. Subdivide S into n strips S_1, S_2, ..., S_n of equal width $\Delta x = \dfrac{b-a}{n}$ as in Figure 3 with n subintervals

$$[x_0,\ x_1],\ [x_1,\ x_2],\ [x_2,\ x_3],\ ...,\ [x_{n-1},\ x_n],$$

where $x_0 = a$ and $x_n = b$, and with the right endpoints of the subintervals given by

$$x_1 = a + \Delta x,\ x_2 = a + 2\Delta x,\ x_3 = a + 3\Delta x,\ ...$$

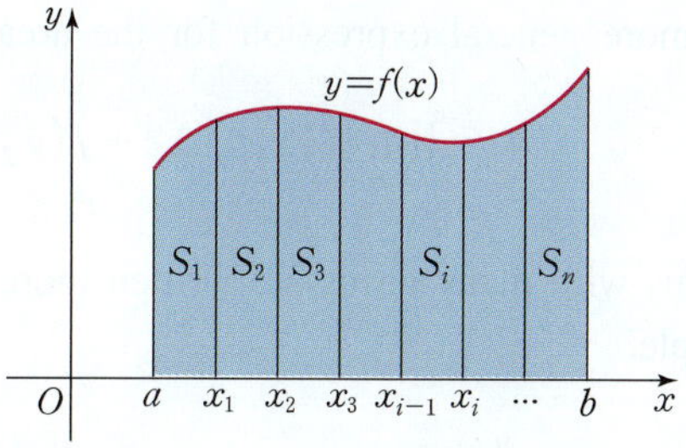

Figure 3

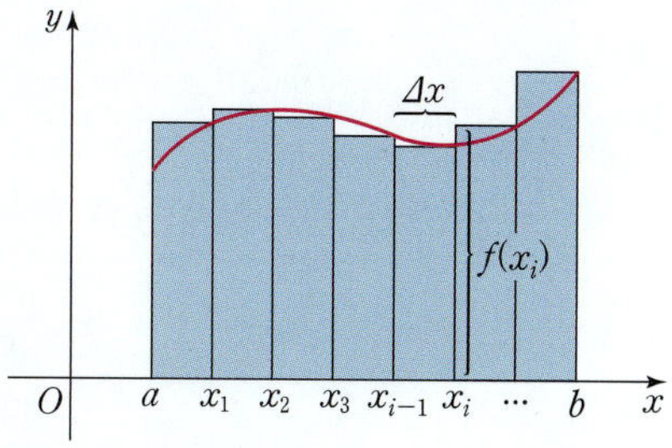

Figure 4

The ith strip S_i is approximated by $f(x_i)\Delta x$, which is the area of the ith rectangle with width Δx and height $f(x_i)$. (See Figure 4.)

We approximate the area of S by the sum of the areas of these rectangles, which is

$$R_n = f(x_1)\Delta x + f(x_2)\Delta x + \cdots + f(x_n)\Delta x.$$

As $n \to \infty$, this approximation appears to become better and better. Therefore, the area A of the region S is defined as follows: The **area** A of the region S that lies under the graph of the continuous function f is the limit of the sum of the areas of approximating rectangles:

1 $$A = \lim_{n\to\infty} R_n = \lim_{n\to\infty} [f(x_1)\Delta x + f(x_2)\Delta x + \cdots + f(x_n)\Delta x].$$

It can also be shown that we get the same value if we use left endpoints provided the right hand limit exists:

2 $$A = \lim_{n\to\infty} L_n = \lim_{n\to\infty} [f(x_0)\Delta x + f(x_1)\Delta x + \cdots + f(x_{n-1})\Delta x].$$

Instead of using left endpoints or right endpoints, the height of the ith rectangle can be taken as the value of f at any **sample point** $x_i{}^*$ in the ith subinterval $[x_{i-1}, x_i]$. Figure 5 shows approximating rectangles when the sample points are not chosen to be endpoints.

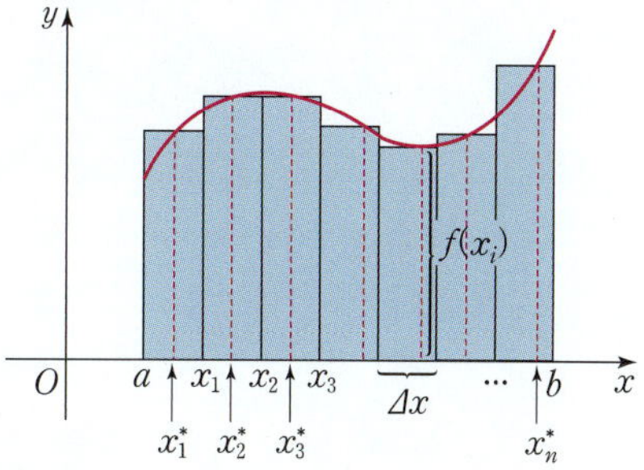

Figure 5

http://matrix.skku.ac.kr/cal-lab/Area-Sum.html

A more general expression for the area of S is

3 $$A = \lim_{n \to \infty} [f(x_1^*)\Delta x + f(x_2^*)\Delta x + \cdots + f(x_n^*)\Delta x]$$

Sums with many terms are written more compactly using **sigma notation**. For example,

$$\sum_{i=1}^{n} f(x_i)\Delta x = f(x_1)\Delta x + f(x_2)\Delta x + \cdots + f(x_n)\Delta x$$

Then, 1, 2 and 3 can be rewritten as

$$A = \lim_{n \to \infty} \sum_{i=1}^{n} f(x_i)\Delta x, \text{ (on right endpoints)}$$

$$A = \lim_{n \to \infty} \sum_{i=1}^{n} f(x_{i-1})\Delta x, \text{ (on left endpoints)}$$

$$A = \lim_{n \to \infty} \sum_{i=1}^{n} f(x_i^*)\Delta x. \text{ (some specific point in each subinteravls)}$$

Later, these areas, represent the Riemann sum in section 5.2 of right endpoints, left endpoints and midpoints in each subinterval, respectively.

5.1 EXERCISES (Areas and Distances)

http://matrix.skku.ac.kr/Cal-Book/part1/CS-Sec-5-1-Sol.html YouTube http://youtu.be/Y_nCn76RPmY

1. Find the area under the curve $y=x^5$ from 0 to 2.
http://matrix.skku.ac.kr/cal-lab/cal-5-1-exs-1.html

Solution We divide the interval $[0,\ 2]$ into n equal parts. Thus the length of each sub-interval is $\dfrac{2}{n}$ and the ith sub-interval is given by $\left[\dfrac{2(i-1)}{n}, \dfrac{2i}{n}\right]$. Now we apply the right end formula to find required area.

$$A=\lim_{n\to\infty}\sum_{i=1}^{n}\frac{2}{n}f(x_i)=\lim_{n\to\infty}\sum_{i=1}^{n}\frac{2}{n}f\left(\frac{2i}{n}\right)=\lim_{n\to\infty}\sum_{i=1}^{n}\frac{2^6}{n^6}i^5$$

$$=\lim_{n\to\infty}\frac{2^6}{n^6}\sum_{i=1}^{n}i^5=\lim_{n\to\infty}\frac{2^6}{n^6}\left[\frac{1}{6}n^6+\frac{1}{2}n^5+\frac{5}{12}n^4-\frac{1}{12}n^2\right]$$

$$=\lim_{n\to\infty}2^6\left[\frac{1}{6}+\frac{1}{2n}+\frac{5}{12n^2}-\frac{1}{12n^4}\right]=\frac{2^6}{6}=\frac{32}{3}.$$

```
plot(x^5, 0, 2, ymax=33, fill="axis")+ plot(x^5, 0, 2.2, ymax=33, color='red')
```

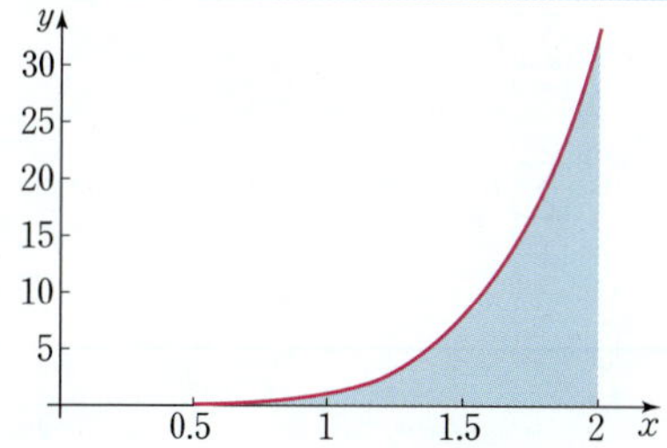

CAS 2. Find the area of the region under the graph of $y=e^{-x}$ from 0 to 2.
http://matrix.skku.ac.kr/cal-lab/cal-5-1-exs-2.html

Solution $-e^{-2}+1$

```
var('i,n')
A=sum(2/n*exp(-2*i/n), i, 1, n)
A.simplify()
l=limit(A.simplify(), n=oo)
l.expand()
```

Answer: -e^(-2) + 1

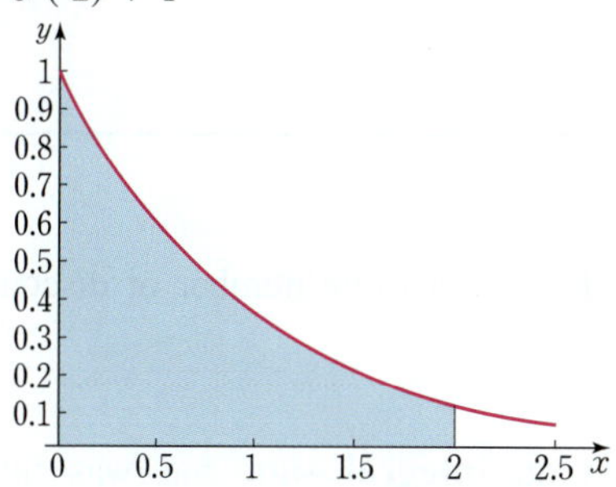

3. Find the area under the curve $y=\cos x$ from $x=0$ to $x=b$, where $0\le b\le \pi/2$.

Solution $\sin b$ (Since $b\le \pi/2$, the value what we evaluate is equal to the area.)

4. (a) Let A_n be the area of a polygon with n equal sides inscribed in a circle with radius r. By dividing the polygon into n congruent triangles with central angle $2\pi/n$, show that

$$A_n=\frac{1}{2}n\,r^2\sin(2\pi/n).$$

(b) Show that $\lim_{n\to\infty}A_n=\pi r^2$. [Hint: Use Equation 3.4.2]

Solution

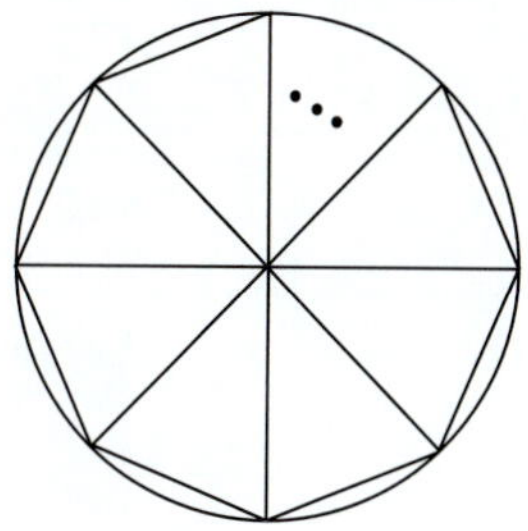

(a) The area of one piece of

$$A_n=\frac{1}{2}r^2\sin(2\pi/n)$$

A_n consists of n such pieces so the area of

$$A_n=n\times\frac{1}{2}r^2\sin(2\pi/n)$$

Remark To estimate area under the graph of $y=f(x)$, one can take the sample point x^* as the point x such that $f(x^*)=\max f(x)$, $x\in[x_{i-1},x_i]$. Similarly we can choose the sample point x^* as the point x^* such that $f(x^*)=\min f(x)$, $x\in[x_{i-1},x_i]$.

(b) Using $\lim_{x\to 0}\dfrac{\sin x}{x}=1$,

$$\lim_{n\to\infty}A_n=\lim_{n\to\infty}\frac{nr^2\sin(2\pi/n)}{2}=\lim_{n\to\infty}r^2\frac{\pi\sin(2\pi/n)}{2\pi/n}$$

$$=\pi r^2\lim_{x\to 0}\frac{\sin x}{x}=\pi r^2.$$

(Let $\dfrac{2\pi}{n}=x$. Then $n\to\infty$ implies $x\to 0$.)

Think of the lower and upper sums of the graph.

5.2 The Definite Integral

YouTube http://youtu.be/GIm3Oz58Ti8

Certain summation formulas will be helpful when using Riemann sums to evaluate definite integrals:

1 $$\sum_{i=1}^{n} i = \frac{n(n+1)}{2}$$

2 $$\sum_{i=1}^{n} i^2 = \frac{n(n+1)(2n+1)}{6}$$

3 $$\sum_{i=1}^{n} i^3 = \left[\frac{n(n+1)}{2}\right]^2$$

4 $$\sum_{i=1}^{n} c = nc$$

5 $$\sum_{i=1}^{n} ca_i = c\sum_{i=1}^{n} a_i, \text{ where } c \text{ is a constant.}$$

6 $$\sum_{i=1}^{n} (a_i + b_i) = \sum_{i=1}^{n} a_i + \sum_{i=1}^{n} b_i$$

7 $$\sum_{i=1}^{n} (a_i - b_i) = \sum_{i=1}^{n} a_i - \sum_{i=1}^{n} b_i$$

DEFINITION 1

If f is a function defined for $a \le x \le b$, divide the interval $[a, b]$ into n subintervals of equal width $\Delta x = (b-a)/n$. Let $x_0 = a$, x_1, x_2, $\cdots$, $x_n = b$ be the endpoints of these sub-intervals and we let $x_1{}^*$, $x_2{}^*$, $\cdots$, $x_n{}^*$ be any **sample points** in these subintervals, so $x_i{}^*$ lies in the ith sub-interval, $[x_{i-1}, x_i]$. Then the **definite integral of f from a to b** is

$$\int_a^b f(x)\,dx = \lim_{n\to\infty}\sum_{i=1}^{n} f(x_i{}^*)\Delta x$$

provided the right-hand limit exists. The function f is called integrable on $[a, b]$ if $\int_a^b f(x)dx$ exists.

Note 1 Definition 1 is valid even if f has a finite number of discontinuities the limit still exists.

Note 2 The symbol $\int$ is called an **integral sign** and was introduced by

Leibniz. Here $f(x)$ is called the **integrand** and a and b are called the **limits of integration**; a is the **lower limit** and b is the **upper limit**. The x in dx is the **variable of integration;** $\int_a^b f(x)dx$ is all one symbol. **Integration** is the process of evaluating an integral.

Note 3 The definite integral $\int_a^b f(x)dx$ is a number; it is independent of the **dummy** variable x. In fact, any letter can be used in place of x without changing the value of the integral:

$$\int_a^b f(x)\,dx = \int_a^b f(t)\,dt = \int_a^b f(r)\,dr$$

B. Riemann
(1826~1866)

Note 4 The sum $\sum_{i=1}^{n} f(x_i^*)\Delta x$ is called a **Riemann sum** (after the German mathematician Bernhard Riemann (1826–1866)). If f is positive, then the Riemann sum can be interpreted as a sum of areas of approximating rectangles. (See Figure 1.) The definite integral $\int_a^b f(x)dx$ can be interpreted as the area under the curve $y = f(x)$ from a to b. (See Figure 2.)

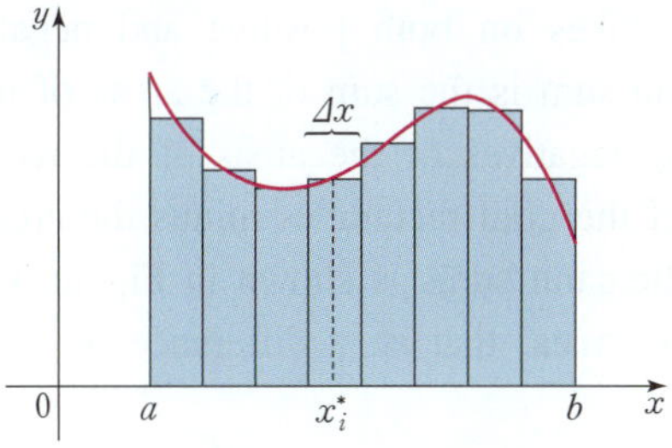

Figure 1 If $f(x) \geq 0$, the Riemann sum $\Sigma f(x_i^*)\Delta x$ is the sum of areas of rectangles.

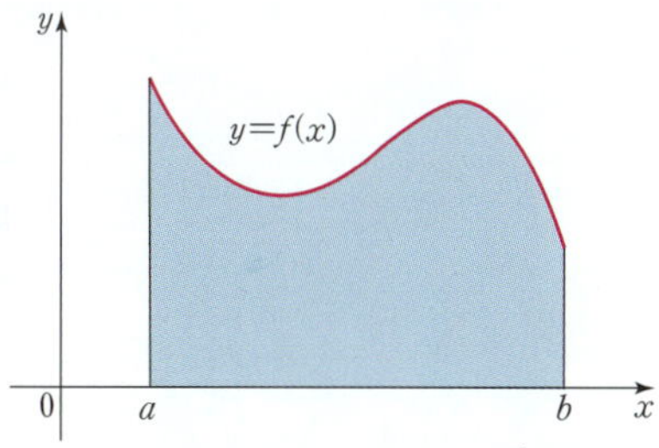

Figure 2 If $f(x) \geq 0$, the integral $\int_a^b f(x)dx$ is the area under the curve $y = f(x)$ from a to b.

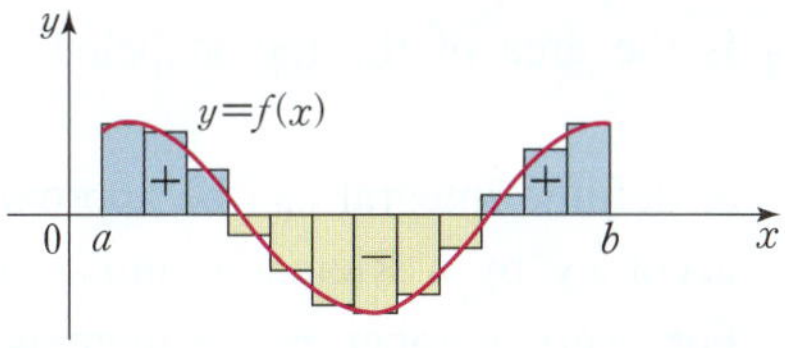

Figure 3 $\Sigma f(x_i^*)\Delta x$ is an approximation to the net area.

http://matrix.skku.ac.kr/cal-lab/Area-Sum.html

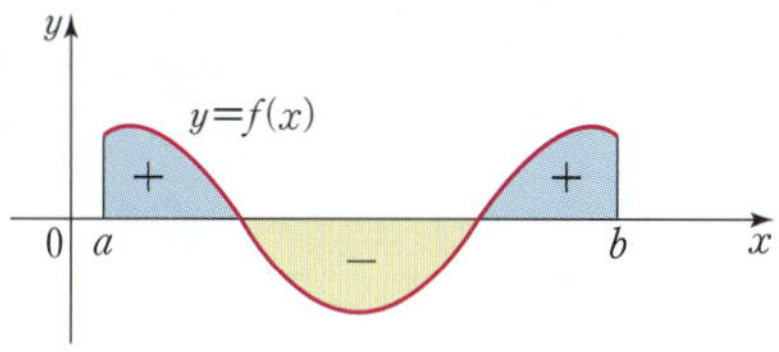

Figure 4 $\int_a^b f(x)dx$ is the net area.

EXAMPLE 1

(a) Evaluate the Riemann sum for $f(x) = 5x$ taking the sample points to be right endpoints and $a = 0$, $b = 3$, and $n = 6$.

(b) Evaluate $\int_0^3 5x\,dx$.

Solution (a) With $n=6$ the interval width is

$$\Delta x=\frac{b-a}{n}=\frac{3-0}{6}=\frac{1}{2}$$

and the right endpoints are $x_1=0.5$, $x_2=1.0$, $x_3=1.5$, $x_4=2.0$, $x_5=2.5$, and $x_6=3.0$. So the Riemann sum is

$$\begin{aligned}R_6&=\sum_{i=1}^{6} f(x_i)\Delta x\\&=f(0.5)\Delta x+f(1.0)\Delta x+f(1.5)\Delta x+f(2.0)\Delta x\\&\qquad+f(2.5)\Delta x+f(3.0)\Delta x\\&=\frac{1}{2}(5/2+5+15/2+10+25/2+15)\\&=26.25.\end{aligned}$$

$$\begin{aligned}\int_0^3(5x)dx&=\lim_{n\to\infty}\sum_{i=1}^{n} f(x_i)\Delta x\\&=\lim_{n\to\infty}\sum_{i=1}^{n} f\left(\frac{3i}{n}\right)\frac{3}{n}=\lim_{n\to\infty}\frac{45(n+1)}{2n}=22.5.\end{aligned}$$ ■

If f takes on both positive and negative values, as in Figure 3, then the Riemann sum is the sum of the areas of the rectangles that lie above the x-axis and the negatives of the areas of the rectangles that lie below the x-axis (the areas of the gold rectangles minus the areas of the blue rectangles). The limit of such Riemann sums is shown in Figure 4. A definite integral can be interpreted as a **net area**, that is, a difference of areas:

$$\int_a^b f(x)dx=A_1-A_2$$

where A_1 is the area of the region above the x-axis and below the graph of f, and A_2 is the area of the region below the x-axis and above the graph of f.

Note 5 A definite integral can be approximated to within any desired degree of accuracy by a Riemann sum as follows:
For every number $\varepsilon>0$ there is an integer N such that

$$\left|\int_a^b f(x)dx-\sum_{i=1}^{n} f(x_i^*)\Delta x\right|<\varepsilon$$

for every integer $n>N$ and for every choice of x_i^* in $[x_{i-1}, x_i]$.

Note 6 We can use unequal subintervals Δx_1, Δx_2, ..., Δx_n of $[a, b]$. We construct the limit so the width of the widest rectangle, $\max_i \Delta x_i$, approaches zero. So in this case,

$$\boxed{8} \quad \int_a^b f(x)dx = \lim_{\max \Delta x_i \to 0} \sum_{i=1}^{n} f(x_i^*)\Delta x_i.$$

The expression on the left side of this equation is called the definite integral of f from a to b.

In general, we write

$$\lim_{n \to \infty} \sum_{i=1}^{n} f(x_i^*)\Delta x = \int_a^b f(x)\,dx.$$

EXAMPLE 2

(a) Evaluate the Riemann sum for $f(x) = 4 - x^2$ taking the sample points to be right endpoints and $a = 0$, $b = 3$, and $n = 6$.

(b) Evaluate $\int_0^3 (4 - x^2)dx$.

http://matrix.skku.ac.kr/cal-lab/cal-5-2-3.html

Solution (a) With $n = 6$ the interval width is

$$\Delta x = \frac{b-a}{n} = \frac{3-0}{6} = \frac{1}{2}$$

and the right endpoints are $x_1 = 0.5$, $x_2 = 1.0$, $x_3 = 1.5$, $x_4 = 2.0$, $x_5 = 2.5$, and $x_6 = 3.0$. So the Riemann sum is

$$\begin{aligned} R_6 &= \sum_{i=1}^{6} f(x_i)\Delta x \\ &= f(0.5)\Delta x + f(1.0)\Delta x + f(1.5)\Delta x + f(2.0)\Delta x \\ &\qquad + f(2.5)\Delta x + f(3.0)\Delta x \\ &= \frac{1}{2}(3.75 + 3 + 1.75 + 0 - 2.25 - 5) \\ &= 0.625. \end{aligned}$$

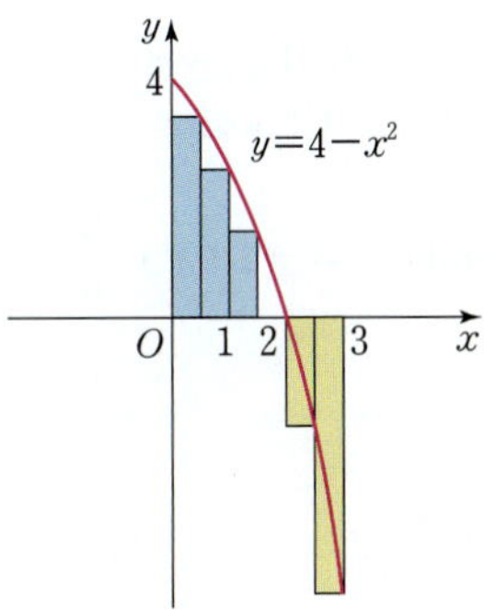

Figure 5 $y = 4 - x^2$

Notice that f is not a positive function and so the Riemann sum does not represent a sum of areas of rectangles. But it does represent the sum of the areas of the gold rectangles (above the x-axis) minus the sum of the areas of the blue rectangles (below the x-axis) in Figure 5.

(b) With n sub-intervals we have

$$\Delta x = \frac{b-a}{n} = \frac{3}{n}$$

Thus $x_0 = 0$, $x_1 = 3/n$, $x_2 = 6/n$, $x_3 = 9/n$, and, in general, $x_i = 3i/n$. Since we are using right endpoints, we can use $\boxed{2}$:

$$\int_0^3 (4-x^2)dx$$
$$= \lim_{n\to\infty}\sum_{i=1}^{n} f(x_i)\Delta x = \lim_{n\to\infty}\sum_{i=1}^{n} f\left(\frac{3i}{n}\right)\frac{3}{n}$$
$$= \lim_{n\to\infty}\frac{3}{n}\sum_{i=1}^{n}\left[4-\left(\frac{3i}{n}\right)^2\right] \quad \text{(by Equation 5 with } c=3/n)$$
$$= \lim_{n\to\infty}\frac{3}{n}\sum_{i=1}^{n}\left(4-\frac{9}{n^2}i^2\right)$$
$$= \lim_{n\to\infty}\left(\frac{12}{n}\sum_{i=1}^{n}1-\frac{27}{n^3}\sum_{i=1}^{n}i^2\right) \quad \text{(by Equations 5 and 7)}$$
$$= \lim_{n\to\infty}\left[12-\frac{27}{6}\left(1+\frac{1}{n}\right)\left(2+\frac{1}{n}\right)\right]$$
$$= 12-9=3.$$

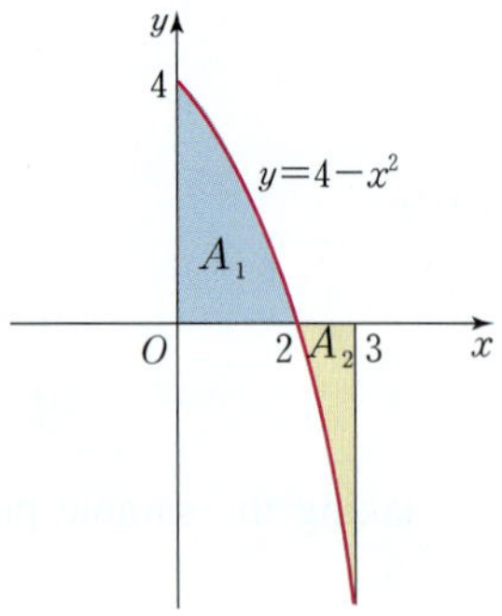

Figure 6 $\int_0^3 (4-x^2)dx = A_1 - A_2 = 0.625$

Since f takes on both positive and negative values this integral cannot be interpreted as an area. But it can be interpreted as the difference in areas $A_1 - A_2$, where A_1 and A_2 are shown in Figure 6.

```
p1=plot(4-x^2, x, 0, 2, fill="axis", fillcolor='goldenrod')
p2=plot(4-x^2, x, 2, 3, fill="axis", fillcolor='blue')
p1+p2
```

■

EXAMPLE 3

Evaluate $\int_{-2}^{1} x^3\,dx$.

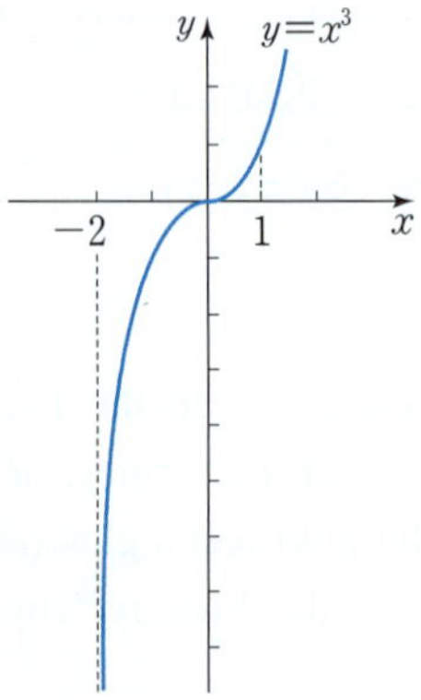

Figure 7 $y=x^3$

Solution Since $f(x)=x^3$ is continuous on $[-2, 1]$, we know from Definition 1 that the definite integral exists. Choosing

$$\Delta x = \frac{1-(-2)}{n} = \frac{3}{n} \qquad \text{and } x_i = -2 + i\cdot\frac{3}{n}$$

we have

$$\int_{-2}^{1} x^3\,dx = \lim_{n\to\infty}\sum_{i=1}^{n} f(x_i)\Delta x = \lim_{n\to\infty}\sum_{i=1}^{n} f\left(-2+\frac{3i}{n}\right)\frac{3}{n}$$
$$= \lim_{n\to\infty}\frac{3}{n}\sum_{i=1}^{n}\left[-8+36\left(\frac{i}{n}\right)-54\left(\frac{i}{n}\right)^2+27\left(\frac{i}{n}\right)^3\right]$$
$$= \lim_{n\to\infty}\frac{3}{n}\left[-8n+\frac{36}{n}\frac{n(n+1)}{2} - \frac{54}{n^2}\frac{n(n+1)(2n+1)}{6}+\frac{27}{n^3}\frac{n^2(n+1)^2}{4}\right]$$
$$= \lim_{n\to\infty}\left[-24+54\left(1+\frac{1}{n}\right)-27\left(1+\frac{1}{n}\right)\left(2+\frac{1}{n}\right)\right.$$

$$+\frac{81}{4}\left(1+\frac{1}{n}\right)\left(1+\frac{1}{n}\right)\Bigg]$$
$$=-24+54-27\cdot 2+\frac{81}{4}=-\frac{15}{4}.$$ ■

EXAMPLE 4

Evaluate the following integrals by interpreting each in terms of areas.

(a) $\int_0^2 \sqrt{4-x^2}\,dx$ (b) $\int_0^2 (x+1)dx$

Solution (a) Since $f(x)=\sqrt{4-x^2}\geq 0$, this integral can be interpreted as the area under the curve $y=\sqrt{4-x^2}$ from 0 to 2. But, since $y^2=4-x^2$, we get $x^2+y^2=4$, which shows that the graph of f is the quarter-circle with radius 2 in Figure 8. Therefore

$$\int_0^2 \sqrt{4-x^2}\,dx=\frac{1}{4}\pi(2)^2=\pi.$$

(b) The graph of $y=x+1$ is the line with slope 1 shown in Figure 9. The integral is computed as the area of trapezoid.

$$\int_0^2 (x+1)dx=A=2\cdot 1+\frac{1}{2}\cdot 2\cdot 2=4.$$

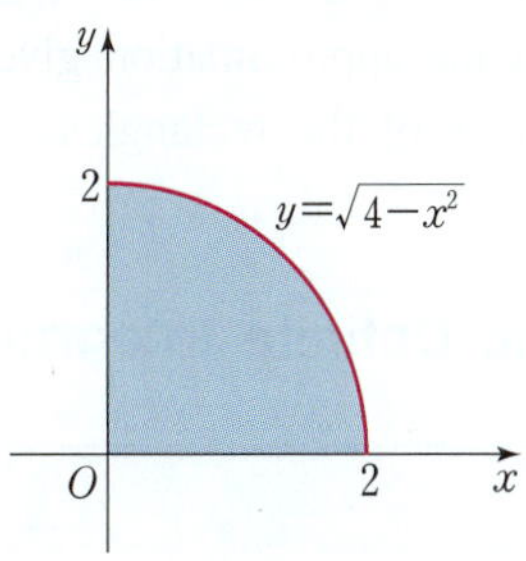

Figure 8 $y=\sqrt{4-x^2}$

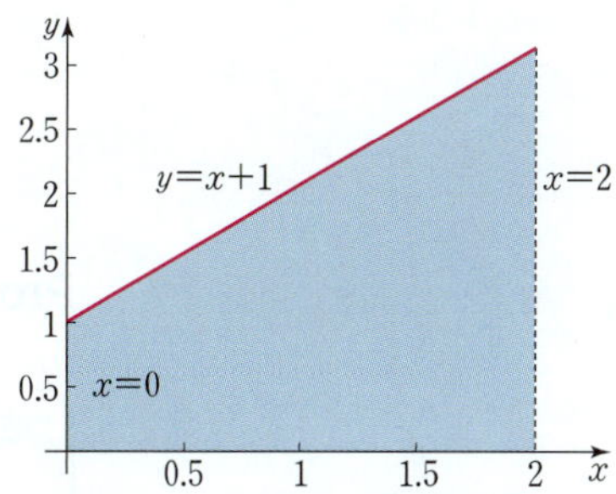

Figure 9 $y=x+1$

■

The Midpoint Rule

In approximating a definite integral by Riemann sums, if we denote the sample points x_i^* to be the middle points of the subinterval $[x_{i-1}, x_i]$, then the corresponding method of approximation is called the **Midpoint Rule**.

Midpoint Rule

$$\int_a^b f(x)dx \approx \sum_{i=1}^{n} f(\overline{x_i})\Delta x = \Delta x[f(\overline{x_1})+\cdots+f(\overline{x_n})]$$

where $\Delta x = \dfrac{b-a}{n}$ and $\overline{x}_i = \dfrac{1}{2}(x_{i-1}+x_i)=$ midpoint of $[x_{i-1}, x_i]$.

http://matrix.skku.ac.kr/cal-lab/cal-RiemannSum.html

EXAMPLE 5

Use the Midpoint Rule with $n=5$ to approximate $\displaystyle\int_1^2 \frac{1}{x^2}dx$.

Solution The endpoints of the five subintervals are 1, 1.2, 1.4, 1.6, 1.8, and 2.0, so the midpoints are 1.1, 1.3, 1.5, 1.7, and 1.9. The width of the subintervals is $\Delta x = (2-1)/5 = 1/5$, so the Midpoint Rule gives

$$\begin{aligned}\int_1^2 \frac{1}{x^2}dx &\approx \Delta x[f(1.1)+f(1.3)+f(1.5)+f(1.7)+f(1.9)]\\ &= \frac{1}{5}\left(\frac{1}{(1.1)^2}+\frac{1}{(1.3)^2}+\frac{1}{(1.5)^2}+\frac{1}{(1.7)^2}+\frac{1}{(1.9)^2}\right)\\ &\approx 0.497127.\end{aligned}$$

Since $f(x)=1/x^2>0$ for $1 \le x \le 2$, the integral represents an area, and the approximation given by the Midpoint Rule is the sum of the areas of the rectangles. ■

Properties of the Definite Integral

DEFINITION 2

If f is integrable on $[a, b]$, then

$$\int_b^a f(x)dx = -\int_a^b f(x)dx.$$

If a is in the domain of f, then

$$\int_a^a f(x)dx = 0.$$

Properties of the Definite Integral

If f and g are integrable on $[a, b]$ and if c is a constant, then cf, $f+g$, and $f-g$ are integrable on $[a, b]$ and

1. $\displaystyle\int_a^b c\,dx = c(b-a)$

2. $\displaystyle\int_a^b cf(x)dx = c\int_a^b f(x)dx$

3. $\displaystyle\int_a^b [f(x)+g(x)]dx = \int_a^b f(x)dx + \int_a^b g(x)dx$

4. $\displaystyle\int_a^b [f(x)-g(x)]dx = \int_a^b f(x)dx - \int_a^b g(x)dx$

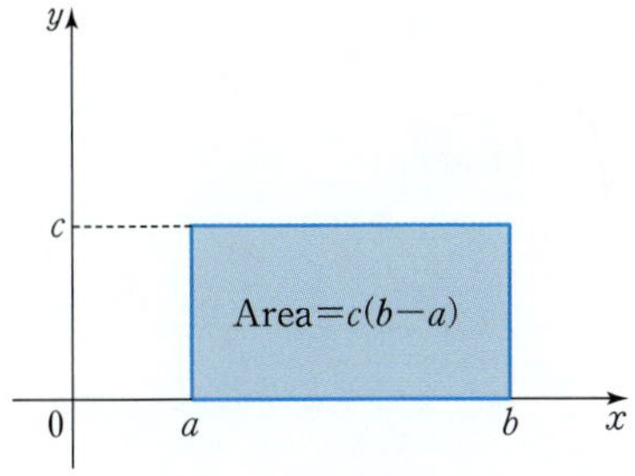

Figure 10

From Property 1 we have that the integral of a constant function $f(x)=c$ is the constant times the length of the interval. If $c>0$ and $a<b$, then $c(b-a)$ is the area of the shaded rectangle in Figure 10.

From Property 3 we have that the integral of a sum is the sum of the integrals. Since the limit of a sum is the sum of the limits, the area under $f+g$ is the area under f plus the area under g, for positive functions.

$$\begin{aligned}\int_a^b [f(x)+g(x)]dx &= \lim_{n\to\infty}\sum_{i=1}^n [f(x_i)+g(x_i)]\Delta x \\ &= \lim_{n\to\infty}\left[\sum_{i=1}^n f(x_i)\Delta x + \sum_{i=1}^n g(x_i)\Delta x\right] \\ &= \lim_{n\to\infty}\sum_{i=1}^n f(x_i)\Delta x + \lim_{n\to\infty}\sum_{i=1}^n g(x_i)\Delta x \\ &= \int_a^b f(x)dx + \int_a^b g(x)dx.\end{aligned}$$

EXAMPLE 6

Evaluate $\displaystyle\int_0^1 (3-2x)dx$.

Solution Using Properties 2 and 3 of integrals, we have

$$\int_0^1 (3-2x)dx = \int_0^1 3dx - \int_0^1 2xdx = \int_0^1 3dx - 2\int_0^1 xdx.$$

We know from Property 1 that

$$\int_0^1 3dx = 3(1-0) = 3.$$

Since $\displaystyle\int_0^1 xdx = \frac{1}{2}$,

$$\int_0^1 (3-2x)dx = \int_0^1 3dx - 2\int_0^1 x dx = 3 - 2 \cdot \frac{1}{2} = 2.$$ ■

The next property tells us how to combine integrals of the same function over adjacent intervals:

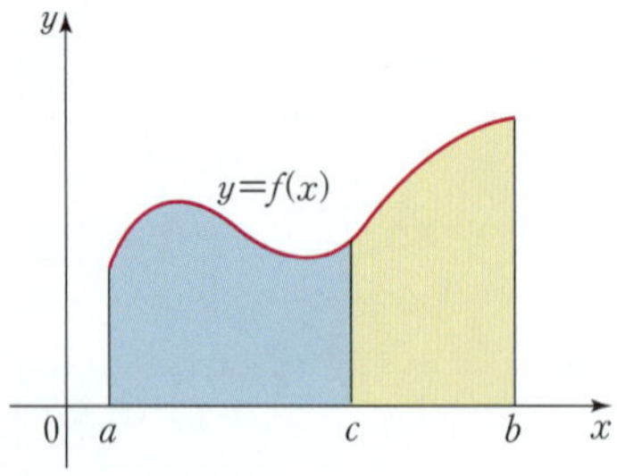

Figure 11

Additive Interval Property

If f is an integrable function on a closed interval containing the numbers a, b and c, then

5. $\int_a^c f(x)dx + \int_c^b f(x)dx = \int_a^b f(x)dx$. (See Figure 11.)

EXAMPLE 7

If $\int_2^8 f(x)dx = 2$ and $\int_5^8 f(x)dx = 3$, find $\int_2^5 f(x)dx$.

Solution By Property 5, we have

$$\int_2^5 f(x)dx + \int_5^8 f(x)dx = \int_2^8 f(x)dx.$$

So, $\int_2^5 f(x)dx = \int_2^8 f(x)dx - \int_5^8 f(x)dx$

$= 2 - 3 = -1.$ ■

Comparison Properties of the Integral

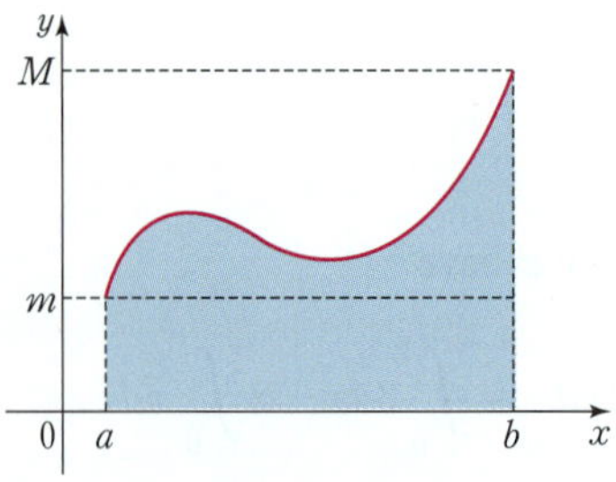

Figure 12

6. If $f(x) \geq 0$ for $a \leq x \leq b$, then $\int_a^b f(x)dx \geq 0$.

7. If $f(x) \geq g(x)$ for $a \leq x \leq b$, then $\int_a^b f(x)dx \geq \int_a^b g(x)dx$.

8. If $m \leq f(x) \leq M$ for $a \leq x \leq b$, then

$$m(b-a) \leq \int_a^b f(x)dx \leq M(b-a).$$

(See Figure 12.)

Proof of Property 8. Since $m \leq f(x) \leq M$, Property 7 gives

$$\int_a^b m\,dx \leq \int_a^b f(x)dx \leq \int_a^b M\,dx.$$

Using Property 1 to evaluate the integrals on the left and right sides, we obtain

$$m(b-a) \leq \int_a^b f(x)dx \leq M(b-a).$$

■

EXAMPLE 8

Find a lower bound and an upper bound for the integral

$$\int_0^1 \sqrt{1+x^4}\,dx.$$

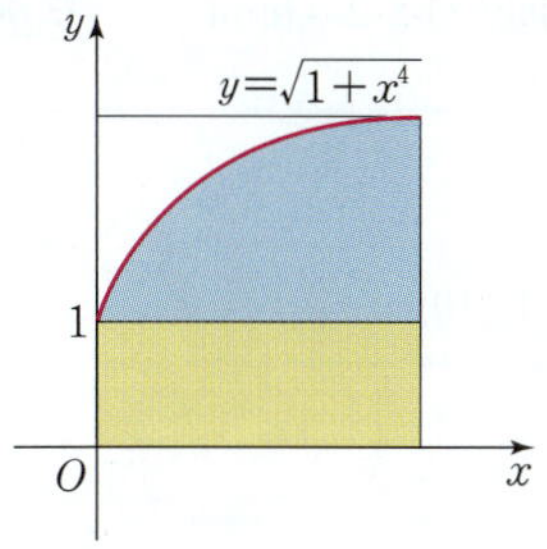

Figure 13 $f(x)=\sqrt{1+x^4}$ on $[0,\ 1]$

Solution Since $f(x)=\sqrt{1+x^4}$ is an increasing function on $[0,\ 1]$, its absolute maximum value is $M=f(1)=\sqrt{2}$ and its absolute minimum value is $m=f(0)=1$. Thus, by Property 8,

$$1\,(1-0) \leq \int_0^1 \sqrt{1+x^4}\,dx \leq \sqrt{2}\,(1-0),$$

$$1 \leq \int_0^1 \sqrt{1+x^4}\,dx \leq \sqrt{2}.$$

The integral is greater than the area of the lower small rectangle (square) and less than the area of the large rectangle. (See Figure 13.)

■

CAS **EXAMPLE 9**

Evaluate $\displaystyle\int_{\frac{\pi}{4}}^{\frac{\pi}{3}} \frac{1}{\sin^2 x}dx.$

http://matrix.skku.ac.kr/cal-lab/cal-5-2-Exm9.html

Solution

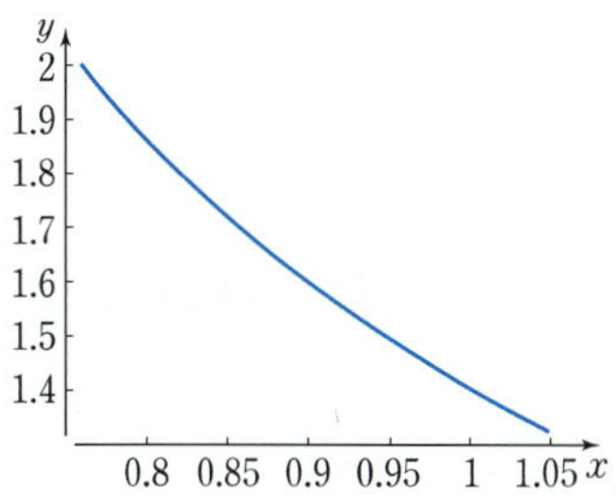

Figure 14

```
var('x')
f=1/(sin(x)^2)
show(plot(f,  (x, pi/4, pi/3)))
print integral(f, (x, pi/4, pi/3))
print integral(f, (x, pi/4, pi/3)).n()
```

Answer : -1/3*sqrt(3) + 1
0.422649730810374

■

5.2 EXERCISES (The Definite Integral)

http://matrix.skku.ac.kr/Cal-Book/part1/CS-Sec-5-2-Sol.html

YouTube http://youtu.be/iUsf1h_hTAE

1-4. Find the Riemann sum by using the Midpoint Rule with the given value of n to approximate the integral.

1. $\displaystyle\int_1^9 \left(\frac{3}{2}\sqrt{x}+1\right)dx,\ n=4$

http://matrix.skku.ac.kr/cal-lab/cal-RiemannSum.html

Solution

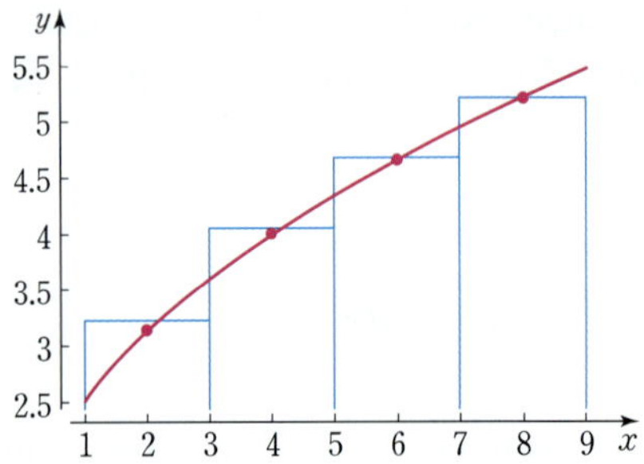

Let $f(x)=\dfrac{3}{2}\sqrt{x}+1$. With $n=4$ the interval width is $\Delta x=\dfrac{9-1}{4}=2$ and midpoints are $\overline{x_i}=2i$ for $i=1,\ 2,\ 3,\ 4$. So the Riemann sum is

$$R_4=\sum_{i=1}^{4} f(\overline{x_i})\Delta x=\sum_{i=1}^{4} f(2i)2$$
$$=(3\sqrt{2}+2)+(3\sqrt{4}+2)+(3\sqrt{6}+2)+(3\sqrt{8}+2)$$
$$=9\sqrt{2}+3\sqrt{6}+14\approx 34.076.$$

```
a=9*sqrt(2)+3*sqrt(6)+14
a.n()
```

2. $\displaystyle\int_0^{\pi} \sin^2 x\, dx,\ n=6$

Solution Let $f(x)=\sin^2 x$. With $n=6$ the interval width is $\Delta x=\dfrac{\pi-0}{6}=\dfrac{\pi}{6}$ and midpoints are $x_i=\dfrac{2i-1}{12}\pi$ for $i=$ 1, 2, 3, 4, 5, 6. So the Riemann sum is

$$R_6=\sum_{i=1}^{6} f(x_i)\Delta x=\sum_{i=1}^{6} f\left(\frac{2i-1}{6}\pi\right)\frac{\pi}{6}$$
$$=\sum_{i=1}^{6}\sin^2\left(\frac{2i-1}{6}\pi\right)\frac{\pi}{6}=\sum_{i=1}^{6}\frac{1-\cos\left(\dfrac{2i-1}{3}\pi\right)}{2}\frac{\pi}{6}$$
$$=\frac{\pi}{2}.$$

3. $\displaystyle\int_1^5 e^{2x}dx,\ n=4$

Solution $R_4=\displaystyle\sum_{i=1}^{4} e^{2i+1}=e^3+e^5+e^7+e^9\approx 0.671$

CAS **4.** $\displaystyle\int_0^2 \frac{1}{\sqrt{1+4x^2}}dx,\ n=10$

http://matrix.skku.ac.kr/cal-lab/cal-5-2-4.html

Solution

```
var('i');
f(x)=1/sqrt(1+4*x^2);
sum(f(0.2*i-0.1)*0.2, i, 1, 10)
```

Answer : 1.04754618408

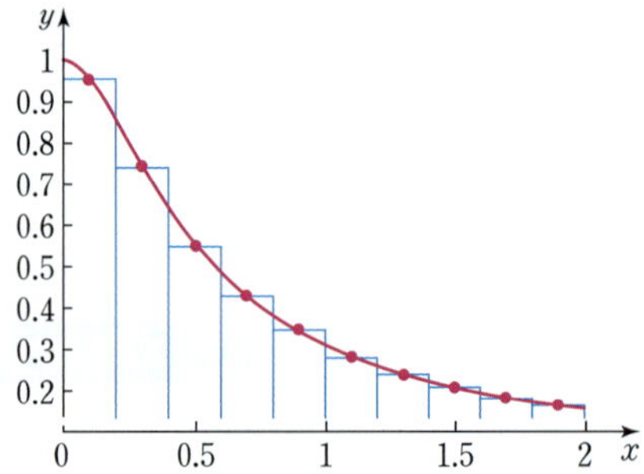

5-8. Express the limit as a definite integral on the given interval.

5. $\displaystyle\lim_{n\to\infty}\sum_{i=1}^{n} x_i\cos x_i\,\frac{\pi}{n},\ [0,\ \pi]$

Solution $\displaystyle\lim_{n\to\infty}\sum_{i=1}^{n} x_i\cos x_i\,\Delta x=\int_0^{\pi} x\cos x\,dx.$

6. $\displaystyle\lim_{n\to\infty}\sum_{i=1}^{n}\frac{e^{x_i}}{x_i-1}\frac{3}{n},\ [2,\ 5]$

Solution $\displaystyle\lim_{n\to\infty}\sum_{i=1}^{n}\frac{e^{x_i}}{x_i-1}\Delta x=\int_2^5\frac{e^x}{x-1}dx.$

7. $\displaystyle\lim_{n\to\infty}\sum_{i=1}^{n}\sqrt{(x_i)^2-2x_i}\,\frac{8}{n},\ [2,\ 10]$

Solution $\displaystyle\lim_{n\to\infty}\sum_{i=1}^{n}\sqrt{(x_i)^2-2x_i}\,\Delta x=\int_2^{10}\sqrt{x^2-2x}\,dx.$

8. $\lim_{n\to\infty}\sum_{i=1}^{n}[4-3(x_i)^2+6(x_i)^5]\frac{2}{n},\ [0,\ 2]$

Solution

$\lim_{n\to\infty}\sum_{i=1}^{n}[4-3(x_i)^2+6(x_i)^5]\Delta x=\int_0^2(4-3x^2+6x^5)dx.$

9-18. Determine whether the statement is true or false. If it is true, explain why. If it is false, give a counterexample.

9. If f and g are continuous on $[a,\ b]$, then

$$\int_a^b[f(x)-g(x)]dx=\int_a^b f(x)dx-\int_a^b g(x)dx.$$

Solution True by Definition

10. If f and g are continuous on $[a,\ b]$, then

$$\int_a^b[f(x)g(x)]\,dx=\left(\int_a^b f(x)dx\right)\left(\int_a^b g(x)dx\right).$$

Solution False: A counterexample is $f(x)=g(x)=x$ and $a=0,\quad b=1$.

11. If f and g are continuous on $[a,\ b]$ and $g(c)\neq 0$ for all $c\in[a,\ b]$, then $\int_a^b\frac{f(x)}{g(x)}dx=\frac{\int_a^b f(x)dx}{\int_a^b g(x)dx}$.

Solution False: counterexample. $f(x)=x^2-2$, $g(x)=x+1$ and $a=0,\quad b=1$.

12. If f is continuous on $[a,\ b]$, then

$$\int_a^b 3f(x)dx=3\int_a^b f(x)dx.$$

Solution True: Property 2

13. If f is continuous on $[a,\ b]$, then

$$\int_a^b xf(x)dx=x\int_a^b f(x)dx.$$

Solution False: Counterexample $f(x)=1$

14. If f is continuous on $[a,\ b]$ and $f(x)\geq 0$, then

$$\int_a^b\ln f(x)dx=\ln\left(\int_a^b f(x)dx\right).$$

Solution False

Let $f(x)=e^x$.

Then $\int_a^b\ln e^x dx=\int_a^b xdx=\frac{b^2-a^2}{2}$

and $\ln\left(\int_a^b e^x dx\right)=\ln(e^b-e^a)$.

Hence $\int_a^b\ln f(x)dx\neq\ln\left(\int_a^b f(x)dx\right)$.

15. If $\int_a^b f(x)dx\geq 0$ then $f(c)\geq 0$ for all $c\in[a,b]$.

Solution False. $\int_0^3(4-x^2)dx$

16. If f and g are continuous and $f(x)\leq g(x)$ and $a\leq x\leq b$ then $\int_a^b f(x)dx\leq\int_a^b g(x)dx$.

Solution True: Let $\Delta x=\frac{b-a}{n},\ x_i=a+\frac{b-a}{n}i,$

and $x_i^*\in[x_{i-1},\ x_i]$. $\quad(x\in[a,\ b])$.

Then $\int_a^b f(x)dx=\lim_{n\to\infty}\sum_{i=1}^{n}f(x_i^*)\Delta x,$

$\int_a^b g(x)dx=\lim_{n\to\infty}\sum_{i=1}^{n}g(x_i^*)\Delta x.$

Because $f(x)\leq g(x)$, $f(x_0^*)\leq g(x_0^*)$, $f(x_1^*)\leq g(x_1^*)$, ..., $f(x_n^*)\leq g(x_n^*)$.

So $\sum_{i=1}^{n}f(x_i^*)\Delta x\leq\sum_{i=1}^{n}g(x_i^*)\Delta x.$

Therefore $\int_a^b f(x)dx\leq\int_a^b g(x)dx.$

17. If f and g are differentiable and $f(x)\leq g(x)$ for $a<x<b$, then $f'(x)\leq g'(x)$ for $a<x<b$.

Solution False: $f(x)=x\quad g(x)=1\quad 0<x<1$

$f'(x)=1\quad g'(x)=0$

18. All continuous functions are integrable.

Solution True

19-22. Evaluate the integral. (You should mention which method you are using.)

19. $\int_{-2}^{4}(2+5x)dx$

http://matrix.skku.ac.kr/cal-lab/cal-5-2-19.html

Solution $\int_{-2}^{4}(2+5x)dx=\left[2x+\frac{5x^2}{2}\right]_{-2}^{4}=48-6=42.$

```
var('x');
f(x)=2+5*x
p1=plot(f(x), (x, -3, 5))
p2=parametric_plot((-2, x), (x, -14, 30), linestyle='--')
p3=parametric_plot((4, x), (x, -14, 30), linestyle='--')
show(p1+p2+p3)
```

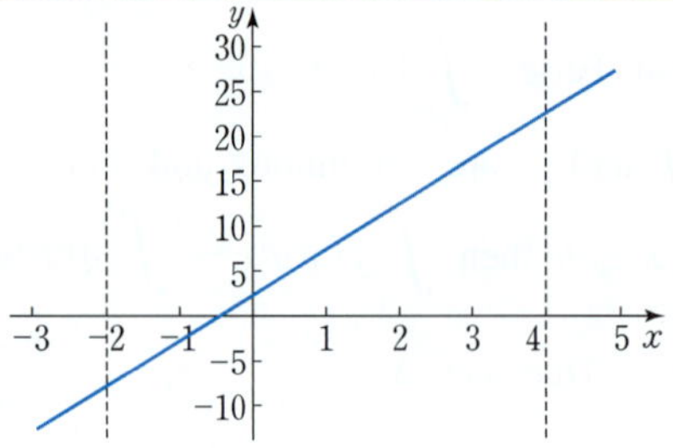

20. $\int_1^3 (3x^2 - 2x + 3)dx$

http://matrix.skku.ac.kr/cal-lab/cal-5-2-20.html

Solution $\int_1^3 (3x^2 - 2x + 3)dx = [x^3 - x^2 + 3x]_1^3$

$= (27-1) - (9-1) + 3(3-1) = 24.$

```
var('x, t');
f(x)=3*x^2-2*x+3
p1=plot(f(x), (x, -1, 1))+plot(f(x), (x, 1, 3), fill="axis")+plot(f(x), (x, 3, 4))
p2=parametric_plot((1, t), (t, 0, 45), linestyle='--')
p3=parametric_plot((3, t), (t, 0, 45), linestyle='--')
show(p1+p2+p3)
print integral(f(x),x,1,3)
```

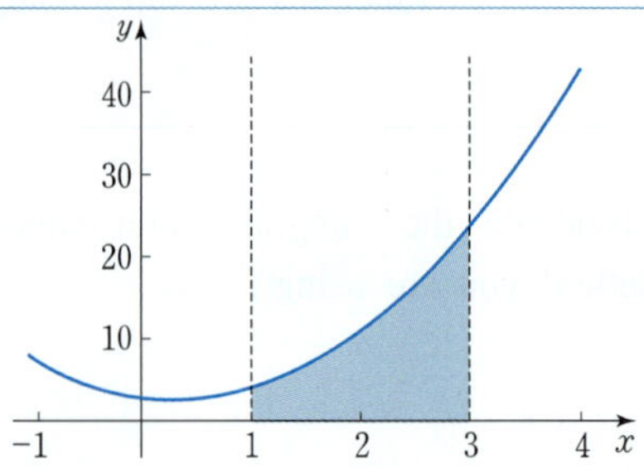

21. $\int_0^3 (4 - x^2)dx$

Solution $\int_0^3 (4-x^2)dx = \left[4x - \frac{1}{3}x^3\right]_0^3$

$= 4(3-0) - \frac{1}{3}(27-0) = 3.$

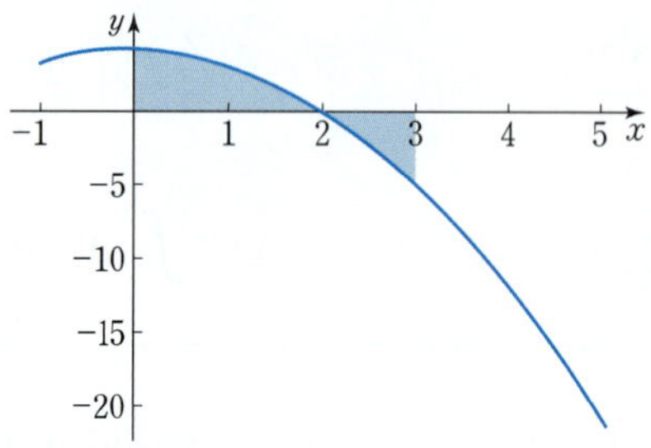

```
var('x');
f(x)=4-x^2
print integral(f(x), x, 0, 3)
plot(f(x), (x, -1, 0))+plot(f(x), (x, 0, 3), fill=true)+plot(f(x), (x, 3, 5))
```

22. $\int_1^2 (x^3 - 1)dx$

Solution $\int_1^2 (x^3-1)dx = \left[\frac{1}{4}x^4 - x\right]_1^2$

$= \frac{1}{4}(16-1) - (2-1)$

$= \frac{11}{4}$

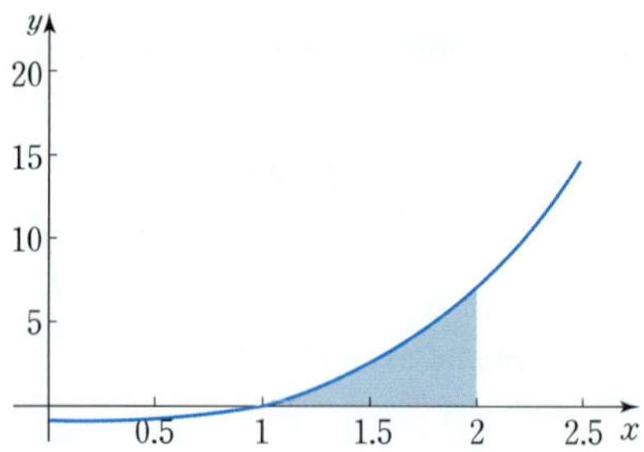

```
f(x)=x^(3)-1
print integral(f(x), x, 1, 2)
plot(f(x), 0, 1, ymax=2)+plot(f(x), 1, 2, ymax=10, fill="axis")+plot(f(x), 2, 5/2, ymax=20)
```

23-26. Evaluate the integral by interpreting it as a sum of the areas.

23. $\int_0^6 (x-2)dx$

Solution $S = A - B = \frac{1}{2} \times 4 \times 4 - \frac{1}{2} \times 2 \times 2 = 6$

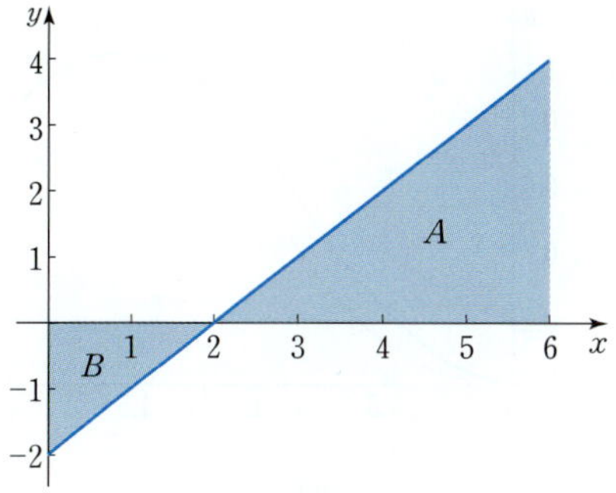

24. $\displaystyle\int_0^8 |x-2|dx$

Solution $S=A+B=\dfrac{1}{2}\times 2\times 2+\dfrac{1}{2}\times 6\times 6=20$

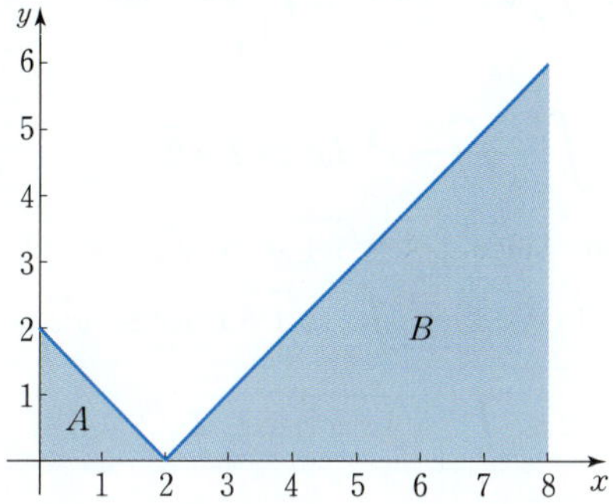

25. $\displaystyle\int_{-2}^{2} \sqrt{4-x^2}\,dx$

Solution Let $y=\sqrt{4-x^2}$. Then $x^2+y^2=4$.
Since $-2\le x\le 2$ and $y\ge 0$, we have
$S=\dfrac{1}{2}\pi * 2^2=2\pi$.

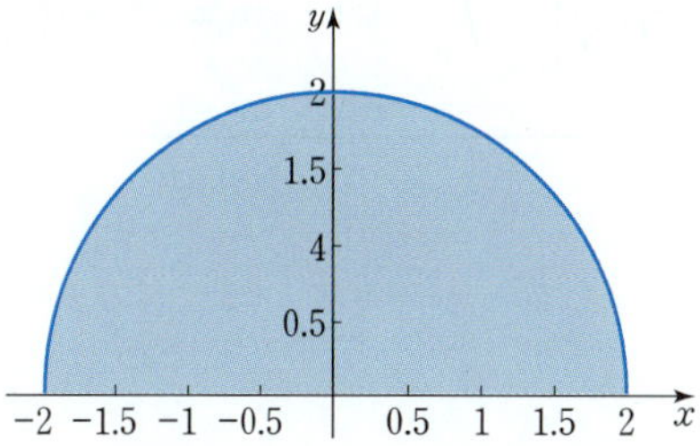

26. $\displaystyle\int_{-1}^{1} \sqrt{3-2x-x^2}\,dx$

Solution Let $y=\sqrt{3-2x-x^2}$.
Then $(x+1)^2+y^2=4$.
Since $-1\le x\le 1$ and $y\ge 0$, we have
$S=\dfrac{1}{4}\pi 2^2=\pi$.

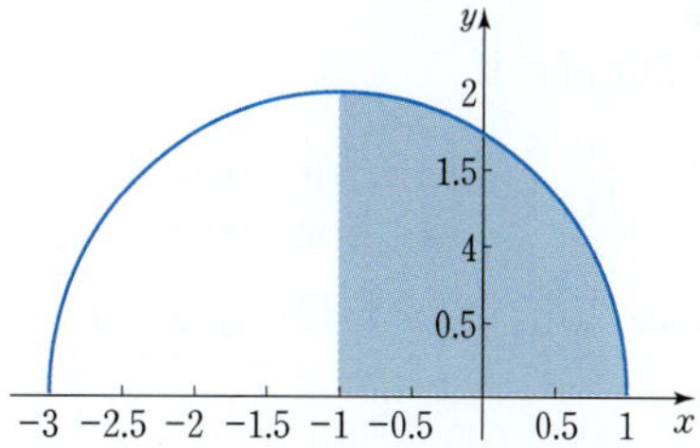

27. Prove that $\displaystyle\int_a^b x\,dx=\frac{b^2-a^2}{2}$.

http://matrix.skku.ac.kr/cal-lab/cal-RiemannSum.html

Solution By using the endpoint rule,

$$\int_0^a x\,dx=\lim_{n\to\infty}\sum_{i=1}^{n}\left(\frac{a}{n}i\right)\frac{a}{n}=\lim_{n\to\infty}\frac{a^2}{n^2}\sum_{i=1}^{n}i$$
$$=\lim_{n\to\infty}\frac{a^2}{n^2}\frac{n(n+1)}{2}=\frac{a^2}{2}.$$

Hence, $\displaystyle\int_a^b x\,dx=\int_0^b xdx-\int_0^a xdx=\frac{b^2-a^2}{2}$.

28. Prove that $\displaystyle\int_a^b x^2\,dx=\frac{b^3-a^3}{3}$.

Solution By using the endpoint rule,

$$\int_0^a x^2\,dx=\lim_{n\to\infty}\sum_{i=1}^{n}\left(\frac{a}{n}i\right)^2\frac{a}{n}=\lim_{n\to\infty}\left(\frac{a}{n}\right)^3\sum_{i=1}^{n}i^2$$
$$=\lim_{n\to\infty}\left(\frac{a}{n}\right)^3\frac{n(n+1)(2n+1)}{6}$$
$$=\lim_{n\to\infty}\frac{a^3(2n^3+3n^2+n)}{6n^3}=\frac{a^3}{3}.$$

Hence, $\displaystyle\int_a^b x^2\,dx=\int_0^b x^2dx-\int_0^a x^2dx=\frac{b^3-a^3}{3}$.

29. If $\displaystyle\int_0^6 f(x)dx=10$ and $\displaystyle\int_0^4 f(x)dx=7$, find

$$\int_4^6 f(x)dx.$$

Solution $\displaystyle\int_4^6 f(x)dx=\int_0^6 f(x)dx-\int_0^4 f(x)dx$
$=10-7=3$

30. If $\displaystyle\int_0^7 f(x)dx=19$ and $\displaystyle\int_0^7 g(x)dx=4$, find

$$\int_0^7 [2f(x)+3g(x)]dx.$$

Solution $\displaystyle\int_0^7 [2f(x)+3g(x)]dx$

$\displaystyle=2\int_0^7 f(x)dx+3\int_0^7 g(x)dx=38+12=50$

31. Find $\int_2^6 f(x)dx$ if

$$f(x)=\begin{cases} x & \text{for} \quad x<4 \\ 4 & \text{for} \quad x \geq 4. \end{cases}$$

Solution Since $\lim_{x\to 4^-} f(x)=4=\lim_{x\to 4^+} f(x)$, $f(x)$ is continuous.

$$\begin{aligned}\int_2^6 f(x)dx &= \int_2^4 f(x)dx+\int_4^6 f(x)dx \\ &= \int_2^4 x dx+\int_4^6 4dx \\ &= \left[\frac{1}{2}x^2\right]_2^4+[4x]_4^6 \\ &= 8-2+24-16=14.\end{aligned}$$

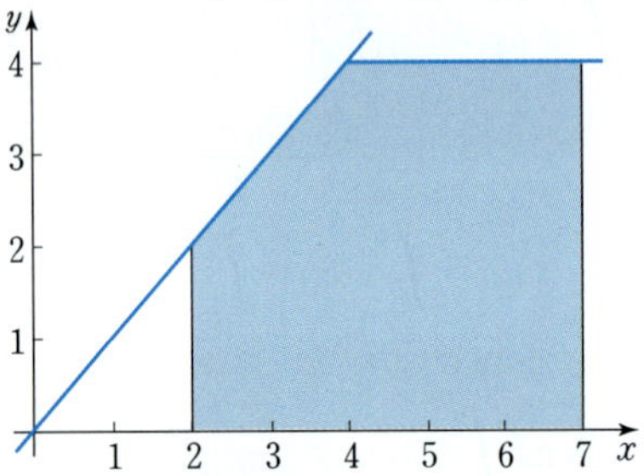

32-35. Verify the inequality without evaluating the integrals.

32. $\int_0^{\pi/3} \cos^3 x\,dx \leq \int_0^{\pi/3} \cos^2 x\,dx$

http://matrix.skku.ac.kr/cal-lab/cal-5-2-32.html

Solution Since $0 \leq \cos x \leq \frac{1}{2}$ for $0 \leq x \leq \frac{\pi}{3}$,

we have $\cos^3 x \leq \cos^2 x$.

Hence, $\int_0^{\pi/3} \cos^3 x\,dx \leq \int_0^{\pi/3} \cos^2 x\,dx.$

```
var('x, y')
p1=plot(sin(x)^3, (x, 0, 2), rgbcolor=(1, 0, 0));
p2=plot(sin(x)^2, (x, 0, 2), rgbcolor=(0, 0, 1));
show(p1+p2)
```

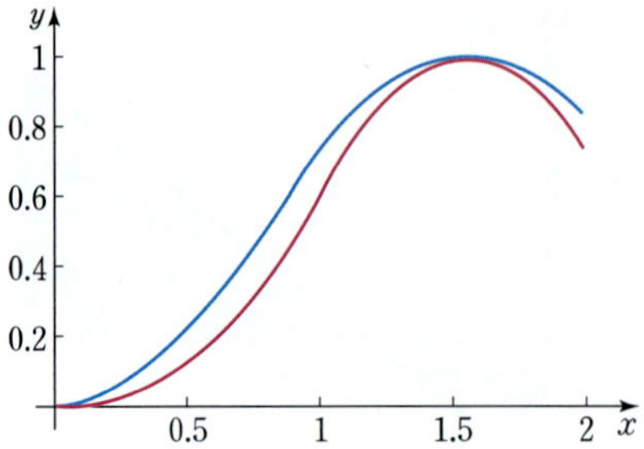

33. $\int_1^4 \sqrt{9-x}\,dx \geq \int_1^4 \sqrt{x+1}\,dx$

Solution Since $9-x \geq x+1 > 0$ for $1 \leq x \leq 4$, we obtain $\sqrt{9-x} \geq \sqrt{x+1}$.

Hence $\int_1^4 \sqrt{9-x}\,dx \geq \int_1^4 \sqrt{x+1}\,dx.$

34. $2 \leq \int_{-1}^1 \sqrt{1+x^2}\,dx \leq 2\sqrt{2}$

Solution Since $1 \leq \sqrt{1+x^2} \leq \sqrt{2}$ for $-1 \leq x \leq 1$, we obtain $1-(-1) \leq \int_{-1}^1 \sqrt{1+x^2}\,dx \leq \sqrt{2}(1-(-1))$.

Hence, $2 \leq \int_{-1}^1 \sqrt{1+x^2}\,dx \leq 2\sqrt{2}$.

35. $\frac{\pi}{6} \leq \int_0^{\pi/3} \sqrt{1-\sin^2 x}\,dx \leq \frac{\pi}{3}$

Solution Since $1-\sin^2 x=\cos^2 x$ and $\frac{1}{2} \leq \cos x \leq 1$ for $0 \leq x \leq \frac{\pi}{3}$, we obtain

$$\frac{1}{2}\left(\frac{\pi}{3}\right) \leq \int_0^{\pi/3} \sqrt{1-\sin^2 x}\,dx \leq 1\left(\frac{\pi}{3}\right).$$

5.3 The Fundamental Theorem of Calculus

You Tube http://youtu.be/Zf1HT2H2fbA

The Fundamental Theorem of Calculus (FTC) is the most important theorem in calculus and, is ranked as one of the greatest accomplishments of the human mind. The FTC gives the precise inverse relationship between the derivative and the integral, establishing the connection between differential calculus (which arose from the tangent problem) and integral calculus (which arose from the area problem). The two parts of the theorem say that differentiation and integration are inverse processes. Each undoes what the other does. This fact enables us to compute areas and integrals very easily without having to compute them as limits of sums. The second part of the FTC provides a very powerful and much simpler method for evaluating the definite integral of a function.

Antiderivatives

Let $f(x)$ be a function defined on an interval $[a, b]$. Suppose we can find a function $F(x)$ such that $F'(x) = f(x)$. If such a function $F(x)$ exists, it is called an **antiderivative** of $f(x)$. In each case, we are required **to find a function, given knowledge about its derivatives**.

DEFINITION 1

A function F is called an antiderivative of f on an interval I if $F'(x) = f(x)$ for all x in I.

For example, an antiderivative of $f(x) = 3x^2$ is $F(x) = x^3$, since $F'(x) = 3x^2$.

Recall that if two functions have identical derivatives on an interval, then they must differ by a constant. To see this, suppose F and G are any two antiderivatives of f, then $F'(x) = f(x) = G'(x)$. This implies $(F - G)'(x) = 0$ on the interval I. Hence $(F - G)(x) = C$ is a constant function. Thus $F(x) = G(x) + C$.

THEOREM 1 Antiderivaitves Differ by a Constant

If F is an antiderivative of f on an interval I, then the most general antiderivative of f on I is

$$F(x) + C,$$

where C is an arbitrary constant. By assigning specific values to the constant C, we obtain a family of functions whose graphs are vertical translates of one another.

The Fundamental Theorem of Calculus

We now recall the following facts:

1. If f is continuous on $[a, b]$, then $\int_a^b f(x)dx$ exists.

2. If f is continuous on $[a, b]$ and $c \in [a, b]$, then

$$\int_a^c f(x)\,dx + \int_c^b f(x)\,dx = \int_a^b f(x)\,dx.$$

3. If $m \leq f \leq M$ on $[a, b]$, then

1 $$(b-a)m \leq \int_a^b f(x)\,dx \leq (b-a)M.$$

THEOREM 2 Mean Value Theorem for Integration

If $f(x)$ is continuous on $[a, b]$, then there exists a $\xi \in [a, b]$ such that

$$\int_a^b f(x)\,dx = f(\xi)(b-a).$$

Proof Since $f(x)$ is continuous on $[a, b]$, there exist a maximum M and minimum m (Extreme Value Theorem)[1]. So $m \leq f(x) \leq M$ implies

$$\int_a^b m\,dx \leq \int_a^b f(x)\,dx \leq \int_a^b M\,dx$$

$$m(b-a) \leq \int_a^b f(x)\,dx \leq M(b-a)$$

$$m \leq \frac{\int_a^b f(x)dx}{b-a} \leq M.$$

By the Mean Value Theorem, there exist ξ in $[a, b]$ such that

$$f(\xi) = \frac{1}{b-a}\int_a^b f(x)\,dx.$$ ■

1)http://en.wikipedia.org/wiki/Extreme_value_theorem.html

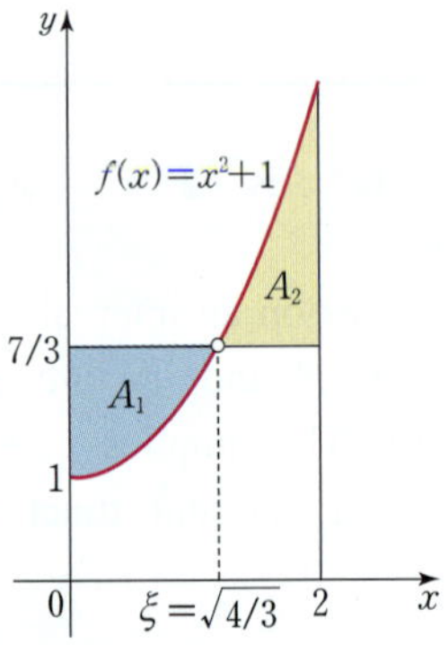

Figure 1 $f(x) = x^2+1$ and MVT $A_1 = A_2$.

Let f be a continuous function on $[a, b]$. For $a \leq x \leq b$ define a new function $F(x)$ by

2 $$F(x) = \int_a^x f(t)\,dt.$$

If x is a fixed number, then the integral $\int_a^x f(t)\,dt$ is a definite number. If f happens to be a positive function, then $g(x) := F(x)$ can be interpreted as the area under the graph of f from a to x, where x can vary from a to b.

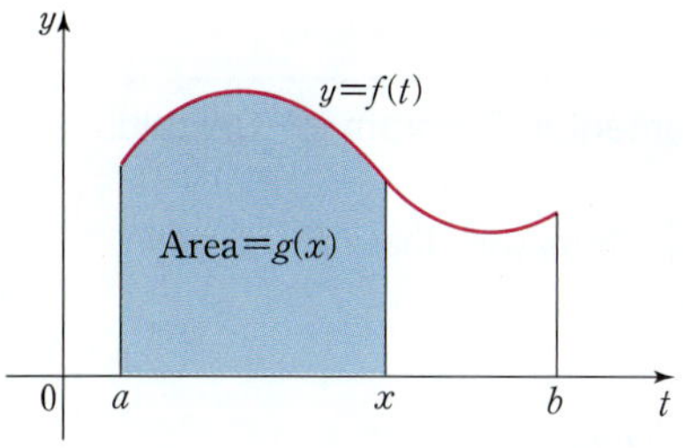

Figure 2

Thus g is the "area so far" function; (See Figure 2.)

The Fundamental Theorem of Calculus has two parts:

THEOREM 3 The First Part of the Fundamental Theorem of Calculus

If f is a continuous function on $[a, b]$, then

$$F(x)=\int_a^x f(t)\,dt$$

is continuous on $[a, b]$, differentiable on (a, b), and satisfies

$$F'(x)=f(x).$$

Proof Let x be a point in (a, b).

$$\frac{F(x+\Delta x)-F(x)}{\Delta x}=\frac{1}{\Delta x}\left(\int_a^{x+\Delta x} f(t)dt-\int_a^x f(t)dt\right)$$
$$=\frac{1}{\Delta x}\int_x^{x+\Delta x} f(t)dt.$$

By Mean Value Theorem for integration, there exist ξ in $[x, x+\Delta x]$ such that

$$\frac{1}{\Delta x}\int_x^{x+\Delta x} f(t)dt=f(\xi).$$

Since $\xi\to x$ as $\Delta x\to 0$ and $f(x)$ is continuous

$$F'(x)=\lim_{\Delta x\to 0}\frac{F(x+\Delta x)-F(x)}{\Delta x}=\lim_{\xi\to x} f(\xi)=f(x).$$ ■

An alternate notation for the derivative portion is,

$$\frac{d}{dx}\int_a^x f(t)\,dt=f(x). \qquad \boxed{3}$$

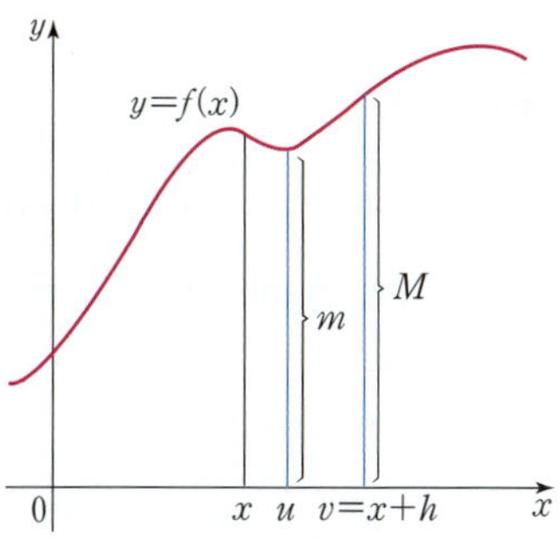

Figure 3

At the end points, F has a one-sided derivative, and the same formula holds. That is, the right-handed derivative of F at a is $f(a)$, and the left-handed derivative of F at b is $f(b)$.

That is if we first integrate f and then differentiate the result, we get back to the original function f. This first part of the Fundamental Theorem of Calculus says that our earlier observation is correct. Namely if $f(x)$ is a continuous function and $F(x)=\int_a^x f(t)dt$, then $F'(x)=f(x)$.

The first part of the Fundamental Theorem of Calculus establishes a remarkable relationship between the operation of differentiation and integration, the two processes initially appear to be unrelated. The second part of the Fundamental Theorem of Calculus is given below.

THEOREM 4 The Second Part of the Fundamental Theorem of Calculus

Let f be a continuous function on $[a, b]$. Suppose that F is any antiderivative of f that satisfy $F' = f$. Then

$$\int_a^b f(x)\,dx = F(b) - F(a).$$

Proof Consider the function $g(x) = \int_a^x f(t)dt$. Since g is differentiable on $[a, b]$, it is continuous there (including at the end points, where the one-sided derivatives exist). We also know that g and F are differentiable on (a, b), and by the first Fundamental Theorem of Calculus their derivatives are equal. Recall that we had (as a consequence of the mean value theorem for derivatives) that F and g differ by a constant. That is, there is a number C such that $F(x) = g(x) + C$ for all $x \in [a, b]$. Then

$$\begin{aligned} F(b) - F(a) &= [g(b) + C] - [g(a) + C] \\ &= g(b) - g(a) = \int_a^b f(x)dx - \int_a^a f(x)dx = \int_a^b f(x)dx. \end{aligned}$$ ■

Visual Proof : http://archives.math.utk.edu/visual.calculus/4/ftc.9/

Because $F'(x) = f(x)$, Part 2 can be rewritten as

$$\int_a^b F'(x)dx = F(b) - F(a).$$

Recall that $F(b) - F(a)$ is a number, not a function.

The derivative of the function $g(x) = \int_0^x \sqrt{1+t}\,dt$ is $g'(x) = \sqrt{1+x}$ by part 1 of the FTC with the non-negative continuous function $f(t) = \sqrt{1+t}$. So $g(x) = \int_a^x f(t)dt$ can be interpreted as the area under the graph of f from a to x, as in Figure 4.

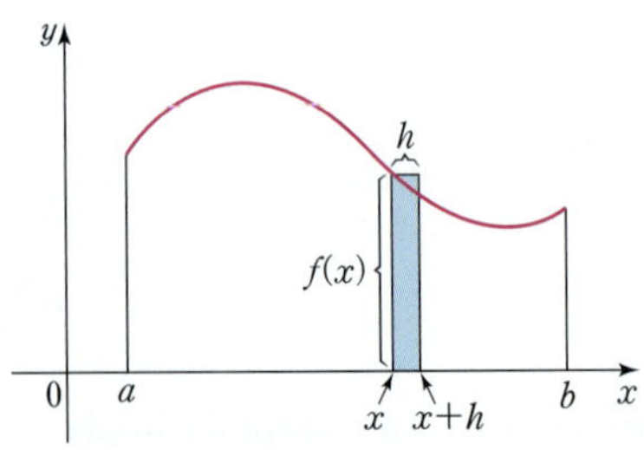

Figure 4

EXAMPLE 1

Find the derivative of the function $F(x) = \int_0^x (t^2 + 1)dt$.

Solution Since $f(t) = t^2 + 1$ is continuous, the First Part of the Fundamental Theorem of Calculus gives $F'(x) = x^2 + 1$. ■

EXAMPLE 2

Find $\dfrac{d}{dx}\displaystyle\int_1^{x^2} t\sin t^3\,dt.$

Solution Let $u = x^2$. Then

$$\begin{aligned}\frac{d}{dx}\int_1^{x^2} t\sin t^3\,dt &= \frac{d}{dx}\int_1^{u} t\sin t^3\,dt \\ &= \frac{d}{du}\left(\int_1^{u} t\sin t^3\,dt\right)\frac{du}{dx} \qquad \text{(by the Chain Rule)} \\ &= u\sin u^3\frac{du}{dx} = x^2\sin(x^6)\cdot 2x \\ &= 2x^3\sin(x^6).\end{aligned}$$

■

EXAMPLE 3

Compute $\displaystyle\int_1^3 x^2 dx.$

Solution The function $f(x) = x^2$ is continuous everywhere. An antiderivative is $F(x) = \dfrac{1}{3}x^3$. So from part 2 of the FTC we have

$$\int_1^3 x^2 dx = F(3) - F(1) = \frac{1}{3}\cdot 3^3 - \frac{1}{3}\cdot 1^3 = \frac{26}{3}.$$

■

EXAMPLE 4

Determine the area under the parabola $y = x^3$ from 0 to 2.

Solution An antiderivative of $f(x) = x^3$ is $F(x) = \dfrac{1}{4}x^4$. The required area A is found using Part 2 of the FTC.

$$A = \int_0^2 x^3 dx = \left[\frac{x^4}{4}\right]_0^2 = \frac{2^4}{4} - \frac{0^4}{4} = 4.$$

■

EXAMPLE 5

Find $\displaystyle\int_3^6 \frac{dx}{x}.$

Solution An antiderivative of $f(x) = 1/x$ is $F(x) = \ln|x|$ since $3 \le x \le 6$, we can write $F(x) = \ln x$. So

$$\int_3^6 \frac{1}{x}dx = [\ln x]_3^6 = \ln 6 - \ln 3 = \ln\frac{6}{3} = \ln 2.$$

■

EXAMPLE 6

Compute the area under the sine curve from 0 to b, where $0 \le b \le \pi/2$.

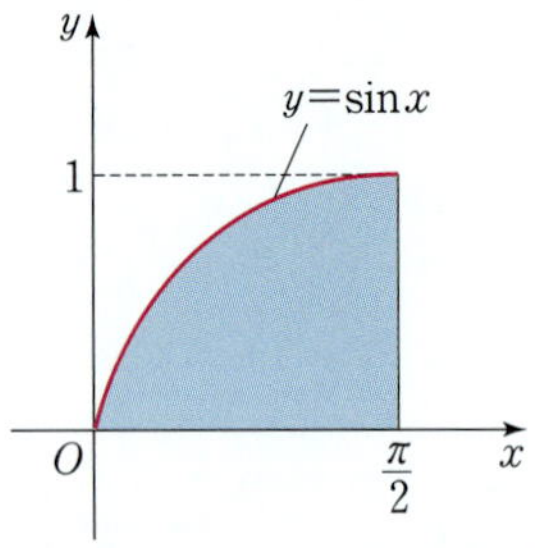

Figure 5

Solution Since $F(x) = -\cos x$ is an antiderivative of $f(x) = \sin x$, so

$$A = \int_0^b \sin x\, dx = [-\cos x]_0^b = -\cos b + \cos 0 = 1 - \cos b.$$

With $b = \pi/2$, the area under the sine curve from 0 to $\pi/2$ is $1 - \cos\frac{\pi}{2} = 1$. (See Figure 5.) ■

CAS **EXAMPLE 7**

Estimate the value of the integral.

$$\int_1^2 (\ln x)^2 dx$$

http://matrix.skku.ac.kr/cal-lab/cal-5-3-exm-6.html

Solution

```
var('x')
f=ln(x)^2
show(plot(f,  x, 1, 2))
print integral(ln(x)^2,  x,  1,  2)
print numerical_integral(f,  1,  2)
```

The integrand : [ln(x)^2 - 2ln(x) + 2] x
Answer : 2log(2)^{2}-4log(2)+2=0.18831730
(2.0907420860874118e-15, error bound)

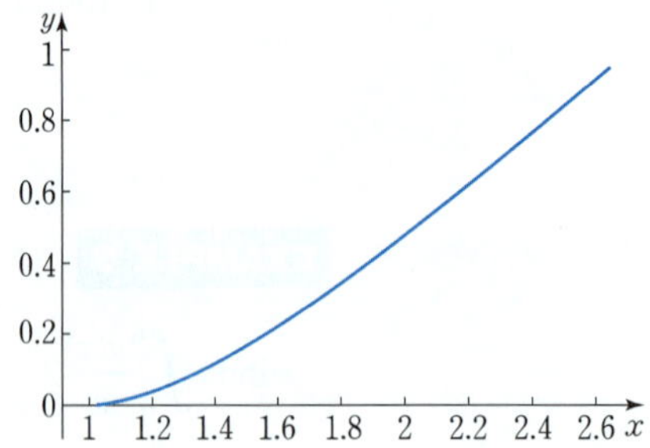

Figure 6

■

5.3 EXERCISES (The Fundamental Theorem of Calculus)

http://matrix.skku.ac.kr/Cal-Book/part1/CS-Sec-5-3-Sol.html

YouTube http://youtu.be/Pa4Z38KkDVY

1. Let $g(x)=\int_{-3}^{x} f(t)dt$, where f is the function whose graph is shown below.

(a) Evaluate $g(-3)$ and $g(3)$.

(b) Estimate $g(-2)$, $g(-1)$ and $g(0)$.

(c) On what interval is g increasing?

(d) Where does g have a maximum value?

(e) Sketch a rough graph of g.

(f) Use the given graph to sketch the graph of $g'(x)$. Compare with the graph of f.

Solution (a) $g(-3)=0,\ g(3)=0$

(b) $g(-2)\approx 1,\ g(-1)\approx \frac{7}{2},\ g(0)\approx 5$

(c) $x\in[-3,\ 0]$

(d) $x=0$

(e) (f is blue, g is red)

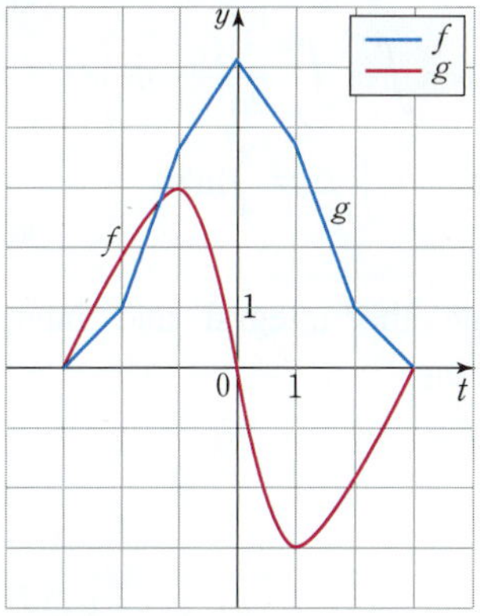

(f) $g'(x)=f(x)$

2-3. Draw the area represented by $g(x)$. Then find $g'(x)$ in two ways:

(a) by using Part 1 of the FTC and

(b) by evaluating the integral using Part 2 and then differentiating.

2. $g(x)=\int_{1}^{x} t^2 dt$

Solution (a) $g'(x)=x^2$

(b) $g'(x)=\frac{d}{dx}\left(\frac{1}{3}x^3-\frac{1}{3}\right)=x^2$

CAS **3.** $g(x)=\int_{1}^{x}(1+\sqrt{t})dt$

Solution (a) $g'(x)=1+\sqrt{x}$

(b) $g'(x)=\frac{d}{dx}\left(\frac{2}{3}x^{\frac{3}{2}}+x-\frac{5}{3}\right)=1+\sqrt{x}$

4-7. Find the derivative of the function using part 1 of the FTC.

4. $g(u)=\int_{3}^{u}\frac{1}{x+x^2}dx$

Solution $g'(u)=\frac{1}{u+u^2}$

5. $F(x)=\int_{x}^{2}\cos(t^2)dt$

[Hint: $\int_{x}^{2}\cos(t^2)=-\int_{2}^{x}\cos(t^2)dt$]

Solution $F'(x)=-\cos(x^2)$

CAS **6.** $h(x)=\int_{2}^{1/x} 2t^2\arctan t\,dt$

Solution $h'(x)=2\left(\frac{1}{x}\right)^2\arctan\left(\frac{1}{x}\right)\cdot\left(\frac{-1}{x^2}\right)$

7. $y=\int_{3}^{\sqrt{x}}\frac{\cos t}{t}dt$

Solution $y'=\frac{\cos\sqrt{x}}{\sqrt{x}}\cdot\frac{1}{2\sqrt{x}}=\frac{\cos\sqrt{x}}{2x}$

8-10. Evaluate the integral using Part 2 of the FTC.

8. $\int_{0}^{4}(3x^2+4x+2)dx$

Solution $\int_{0}^{4}(3x^2+4x+2)dx=\left[x^3+2x^2+2x\right]_0^4$

$=(64+32+8)-(0+0+0)$

$=104-0=104$

CAS **9.** $\displaystyle\int_{1/2}^{\sqrt{3}/2} \frac{6}{1-t^2}dt$

Solution

$$-3\log\left(-\frac{\sqrt{3}}{2}+1\right)+3\log\left(\frac{\sqrt{3}}{2}+1\right)-3\log\frac{3}{2}-3\log2$$
$$=3\log\left(\frac{2}{2-\sqrt{3}}\times\frac{2+\sqrt{3}}{2}\times\frac{2}{3}\times\frac{1}{2}\right)$$
$$=3\log\left(\frac{7+4\sqrt{3}}{3}\right).$$

10. $\displaystyle\int_0^1\left(x^3-\frac{3}{2}x^2+\frac{3}{4}x-\frac{1}{8}\right)dx$

Solution $\displaystyle\int_0^1\left(x^3-\frac{3}{2}x^2+\frac{3}{4}x-\frac{1}{8}\right)dx$

$$=\int_0^1\left(x-\frac{1}{2}\right)^3dx=\left[\frac{1}{4}\left(x-\frac{1}{2}\right)^4\right]_0^1=0$$

---

11. Let $g(x)=\displaystyle\int_1^x te^{2t}dt$. Use Part 1 of the FTC to find $g'(1)$.

Solution $g'(x)=xe^{2x}$ so $g'(1)=e^2$.

```
var('t,x ');
g(x)=integral(t*e^(2*t), t, 1, x)
f(x)=diff(g(x), x)
print g(x)
print f(x)
print f(1)
```

12. Give a non-polynomial function $f(x)$ $(\neq 0)$ such that $f(2)=0$ and $f'(0)=0$.

Solution For any function $g(t)$ set $f(x)=\displaystyle\int_2^x tg(t)dt$.

Then clearly $f(2)=\displaystyle\int_2^2 tg(t)dt=0$ and $f'(x)=xg(x)$ so $f'(0)=0$.

For example $f(x)=\displaystyle\int_2^x t\sin t\,dt$.

13. Let $g(x)=\displaystyle\int_0^{x^3} f(t)dt$ and $g'(2)=24$. Find $f(8)$.

Solution $g'(x)=f(x^3)d(x^3)=f(x^3)\times 3x^2$ and

$24=g'(2)=f(2^3)(3\cdot 2^2)$

Hence $f(8)\times 12=24$, so $f(8)=2$.

14. Let $g(x)=\cos x=\displaystyle\int_0^{f(x)} f^{-1}(t)dt$ defined on $[0,\ \pi]$. Find $f\left(\dfrac{\pi}{2}\right)-f'\left(\dfrac{\pi}{2}\right)$.

Solution $g'(x)=-\sin x=xf'(x)$.

So $-1=-\sin\dfrac{\pi}{2}=\dfrac{\pi}{2}f'\left(\dfrac{\pi}{2}\right)$.

Note $\cos\dfrac{\pi}{2}=0=\displaystyle\int_0^{f\left(\frac{\pi}{2}\right)} f^{-1}(t)dt$.

and $\cos x\neq 0$ for $x\neq\dfrac{\pi}{2}$.

Hence $f\left(\dfrac{\pi}{2}\right)$ should be 0.

Therefore $f\left(\dfrac{\pi}{2}\right)-f'\left(\dfrac{\pi}{2}\right)=0-\left(-\dfrac{2}{\pi}\right)=\dfrac{2}{\pi}$.

15. Let $f(x)=\sqrt{2x+1}=\displaystyle\int_1^x g'(f(t))dt$.

Find $\displaystyle\int_0^1 g(f(x))dx$.

Solution Differentiate both sides to get $f'(x)=g'(f(x))$.

$\Leftrightarrow (f'(x))^2=g'(f(x))f'(x)=\dfrac{d}{dx}(g(f(x)))$.

Hence,

$$\int_0^1 g(f(x))dx=\int_0^1\left(\int_0^x\frac{1}{2x+1}\,dx\right)dx$$
$$=\int_0^1\left(\frac{1}{2}\ln(2x+1)\right)\,dx=\frac{3}{4}\ln3-\frac{1}{2}.$$

16-17. Evaluate the integral and interpret it as a difference of areas.

16. $\displaystyle\int_{-1}^2 x^3dx$

Solution $\displaystyle\int_{-1}^2 x^3dx\ =0+\int_1^2 x^3dx=\left[\frac{x^4}{4}\right]_1^2=\frac{15}{4}$

since $\displaystyle\int_{-1}^1 x^3dx\ =0$.

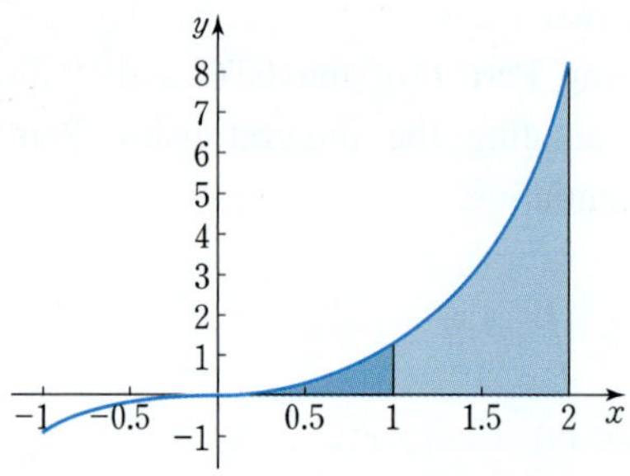

CAS **17.** $\displaystyle\int_{\pi/4}^{5\pi/2}\sin x\,dx$

Solution $\int_0^{5\pi/2} \sin x\, dx - \int_0^{\pi/4} \sin x\, dx$

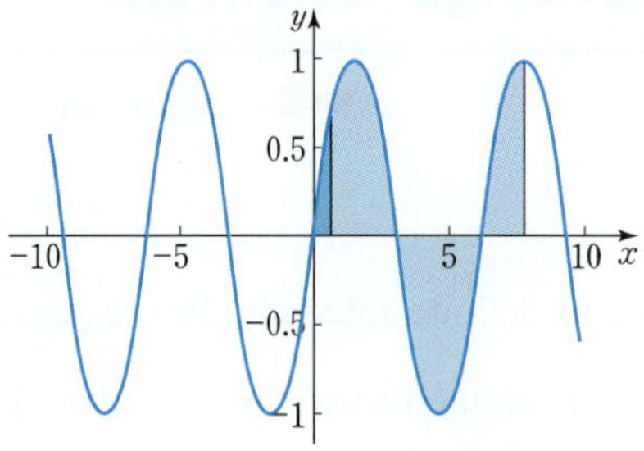

18. If $F(x) = \int_1^x f(t)dt$,

where $f(t) = \int_1^{t^2} \frac{\sqrt{1+u^4}}{u} du$, find $F''(2)$.

Solution $F''(2) = f'(2)$

$$f'(t) = \frac{\sqrt{1+t^8}}{t^2} \times 2t = \frac{2\sqrt{1+t^8}}{t}$$

Therefore, $F''(2) = f'(2) = \sqrt{257}$.

CAS **19.** Find the value of $f(3)$ if $f(1) = 10$, f' is continuous, and $\int_1^3 f'(x)dx = 5$.

Solution $5 = \int_1^3 f'(x)dx = f(3) - f(1) \Rightarrow f(3) = 5 + f(1) = 15$

20. If f is continuous and g and h are differentiable functions, find a formula for $\frac{d}{dx}\int_{g(x)}^{h(x)} f(t)dt$.

Solution $\frac{d}{dx}\int_{g(x)}^{h(x)} f(t)dt$

$$= \frac{d}{dx}\left(\int_0^{h(x)} f(t)dt - \int_0^{g(x)} f(t)dt\right)$$

$$= f(h(x))h'(x) - f(g(x))g'(x).$$

<String Art>

5.4 Indefinite Integrals and the Net Change Theorem

YouTube http://youtu.be/E6I3EDzAVuU

Indefinite Integrals

The FTC connects antiderivatives and definite integrals. Part 1 says that if f is continuous, then $\int_a^x f(t)dt$ is an antiderivative of f. Part 2 says that $\int_a^b f(x)dx$ can be found by evaluating $F(b)-F(a)$, where F is an antiderivative of f.

If $F(x)$ is an anti-derivative of $f(x)$, then the **indefinite integral is**

$$\int f(x)dx = F(x)+C$$

where C is an arbitrary constant.

$$\int f(x)dx = F(x)+C \text{ means } F'(x)=f(x)$$

Thus an indefinite integral represents an entire family of functions $F(x)+C$. They are related by the FTC

$$\int_a^b f(x)dx = \left[\left(\int f(x)dx\right)\right]_a^b = F(b)-F(a).$$

The "richness" of FTC depends on the "availability" of antiderivatives of function.

Table of Indefinite Integrals

$$\int cf(x)dx = c\int f(x)dx$$

$$\int [f(x)+g(x)]dx = \int f(x)dx + \int g(x)dx$$

$\int k\,dx = kx + C$, where k is a constant.

$\int x^n dx = \dfrac{x^{n+1}}{n+1} + C \quad (n \neq -1)$ $\qquad \int \dfrac{1}{x}dx = \ln|x| + C$ for nonzero x.

$\int e^x dx = e^x + C$ $\qquad \int a^x dx = \dfrac{a^x}{\ln a} + C$, $a > 0$.

$\int \sin x\,dx = -\cos x + C$ $\qquad \int \cos x\,dx = \sin x + C$

$\int \sec^2 x\,dx = \tan x + C$ $\qquad \int \csc^2 x\,dx = -\cot x + C$

$$\int \sec x \tan x \, dx = \sec x + C \qquad \int \csc x \cot x \, dx = -\csc x + C$$

$$\int \frac{1}{x^2+1} dx = \tan^{-1} x + C \qquad \int \frac{1}{\sqrt{1-x^2}} dx = \sin^{-1} x + C$$

Remember that the most general antiderivative on a given interval is obtained by adding a constant to a particular antiderivative. A **formula for a general indefinite integral is assumed to be valid only on an interval (finite or infinite)**.

The following example shows that the general antiderivative of the function $f(x) = \frac{1}{x}$, $x \neq 0$, is

$$F(x) = \begin{cases} \ln(-x) + C_1 & \text{if } x < 0 \\ \ln x + C_2 & \text{if } x > 0 \end{cases}, \quad C_1 \text{ and } C_2 \text{ are constants.}$$

EXAMPLE 1

Compute $\int_0^2 (4 - x^2) dx$.

Solution We have

$$\begin{aligned} \int_0^2 (4 - x^2) dx &= \left[4x - \frac{x^3}{3} \right]_0^2 \\ &= \left(4 \cdot 2 - \frac{2^3}{3} \right) - \left(4 \cdot 0 - \frac{0^3}{3} \right) \\ &= 8 - \frac{8}{3} = \frac{16}{3}. \end{aligned}$$

■

EXAMPLE 2

Obtain the general indefinite integral $\int (8x^3 - 2\csc^2 x) dx$.

Solution We have

$$\begin{aligned} \int (8x^3 - 2\csc^2 x) dx &= 8 \int x^3 dx - 2 \int \csc^2 x \, dx \\ &= 8 \frac{x^4}{4} + 2\cot x + C \\ &= 2x^4 + 2\cot x + C. \end{aligned}$$

■

EXAMPLE 3

Evaluate $\int_1^4 \left(3t^2 - 3\sqrt{t} + \frac{1}{t} \right) dt$.

Solution $\int_1^4 \left(3t^2 - 3\sqrt{t} + \frac{1}{t}\right)dt = \left[t^3 - 2t^{\frac{3}{2}} + \ln t\right]_1^4$

$$= (4^3 - 2 \cdot 4^{\frac{3}{2}} + \ln 4) - (1^3 - 2 \cdot 1^{\frac{3}{2}} + \ln 1)$$
$$= 49 + 2\ln 2.$$ ■

EXAMPLE 4

Find $\int \frac{\sin\theta}{\cos^2\theta} d\theta$.

Solution Rewrite the function and integrate:

$$\int \frac{\sin\theta}{\cos^2\theta} d\theta = \int \left(\frac{1}{\cos\theta}\right)\left(\frac{\sin\theta}{\cos\theta}\right) d\theta$$
$$= \int \sec\theta \tan\theta \, d\theta = \sec\theta + C.$$ ■

Applications of The Net Change Theorem

$$\int_a^b F'(x)dx = F(b) - F(a)$$

Here $F'(x)$ represents the rate of change of $y = F(x)$ with respect to x and $F(b) - F(a)$ is the change in y when x changes from a to b. Although y might change in both directions, increase or decrease, $F(b) - F(a)$ represents the *net* change in y. So the FTC 2 can restated in words as follows, the integral of a rate of change is the net change.

1. If $\frac{dQ}{dt} = f(t)$ is the rate of change of quantity $Q = F(t)$, then

$\int_{t_1}^{t_2} f(t)\,dt$ = Net Change over $[t_1, t_2]$ $= F(t_2) - F(t_1)$.

2. If $V'(t)$ gives the velocity of a moving object whose position function is $s = V(t)$, then

$$\int_{t_1}^{t_2} V'(t)dt = \text{Displacement over } [t_1, t_2] = V(t_2) - V(t_1)$$

is the net change of V between time t_1 and t_2.

3. (a) If an object moves along a straight line with position function $s(t)$, then its velocity is $v(t) = s'(t)$, so

1 $$\int_{t_1}^{t_2} v(t)dt = s(t_2) - s(t_1)$$

is the net change of position, or displacement, of the particle during the time period from t_1 to t_2.

(b) To calculate the distance traveled during the time interval, consider the intervals when $v(t) \geq 0$ (the particle moves to the right) and also the intervals when $v(t) \leq 0$ (the particle moves to the left). In both cases the distance is computed by integrating $|v(t)|$ the speed. Therefore

2 $$\int_{t_1}^{t_2} |v(t)|dt = \text{total distance traveled}$$

Both displacement and distance traveled can be interpreted in terms of areas under a velocity curve. (See Figure 1.)

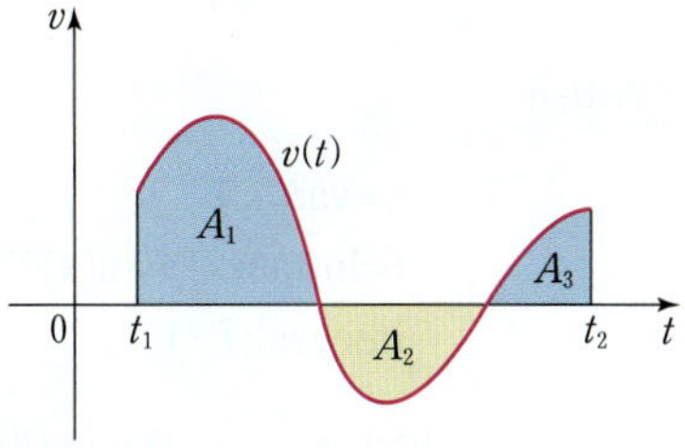

$\text{displacement} = \int_{t_1}^{t_2} v(t)dt = A_1 - A_2 + A_3$

$\text{distance} = \int_{t_1}^{t_2} |v(t)|dt = A_1 + A_2 + A_3$

Figure 1

EXAMPLE 5

A particle moves along a line so that its velocity at time t is $v(t) = t^2 - t - 6$ (measured in meters per second).

(a) Find the displacement of the particle during the time period $1 \leq t \leq 4$.

(b) Find the distance traveled during this time period.

Solution (a) The displacement is

$$s(4) - s(1) = \int_1^4 v(t)dt = \int_1^4 (t^2 - t - 6)dt$$
$$= \left[\frac{t^3}{3} - \frac{t^2}{2} - 6t\right]_1^4 = -\frac{9}{2}.$$

This means that the particle moved 4.5 m toward the left.

(b) Note that $v(t) = t^2 - t - 6 = (t-3)(t+2)$ and so $v(t) \leq 0$ on the interval $[1, 3]$ and $v(t) \geq 0$ on $[3, 4]$. Thus, the distance traveled is

$$\int_1^4 |v(t)|dt = \int_1^3 [-v(t)]dt + \int_3^4 v(t)dt$$
$$= \int_1^3 (-t^2 + t + 6)dt + \int_3^4 (t^2 - t - 6)dt$$
$$= \left[-\frac{t^3}{3} + \frac{t^2}{2} + 6t\right]_1^3 + \left[\frac{t^3}{3} - \frac{t^2}{2} - 6t\right]_3^4$$
$$= \frac{61}{6} \approx 10.17\,\text{m}.$$

■

Note The unit of measurement for $\int_a^b f(x)dx$ is the product of the units for $f(x)$ and the units for x.

CAS **EXAMPLE 6**

Find the general indefinite integral

$$\int \frac{\ln \tan x}{\sin^2 x} dx.$$

http://matrix.skku.ac.kr/cal-lab/cal-5-4-exm-7.html

Solution

```
x=var('x')
f=ln(tan(x))/sin(x)^2
integral(f,x)
```

Answer : -log(tan(x))/tan(x) - 1/tan(x) ■

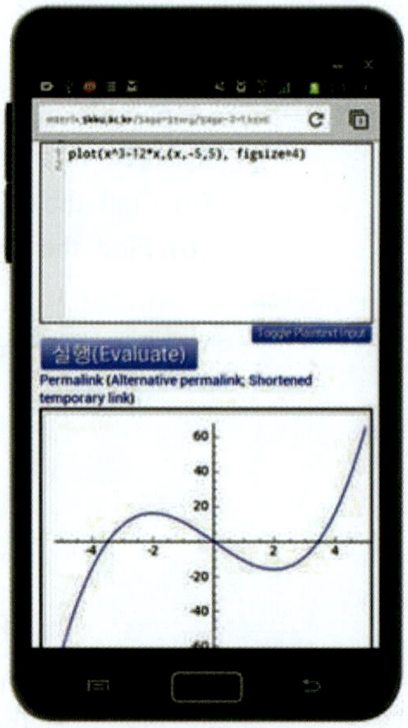

Figure 2 Mobile mathematics

5.4 EXERCISES (Indefinite Integrals and the Net Change Theorem)

http://matrix.skku.ac.kr/Cal-Book/part1/CS-Sec-5-4-Sol.html

YouTube http://youtu.be/Pa4Z38KkDVY

1-4. Verify by differentiation that the formula is correct.

1. $\int \frac{x}{\sqrt{x^2+1}}dx = \sqrt{x^2+1}+C$

Solution We take derivative of the right hand side and show that it is integrand.

2. $\int x\cos x\, dx = x\sin x + \cos x + C$

Solution We take derivative of the right hand side and show that it is integrand.

CAS **3.** $\int \frac{1}{(\sqrt{x^2+a^2})^3}dx = \frac{x}{a^2\sqrt{x^2+a^2}}+C$

Solution Let $x = a\tan\theta$. Then $dx = a\sec^2\theta d\theta$ and

$$\int \frac{1}{(\sqrt{x^2+a^2})^3}dx = \int \frac{1}{a^3\sqrt{(\tan^2\theta+1)^3}}a\sec^2\theta d\theta$$

$$= \int \frac{1}{a^2\sec\theta}d\theta \quad (\text{because } x = a\tan\theta \text{ and } \frac{a}{x} = \cot\theta)$$

$$= \int \frac{1}{a^2}\cos\theta d\theta$$

$$= \frac{\sin\theta}{a^2}+C$$

But $1+\cot^2\theta = \csc^2\theta \Rightarrow \sin\theta = \frac{1}{\sqrt{1+\cot^2\theta}}$

$$= \frac{1}{\sqrt{1+\frac{a^2}{x^2}}} = \frac{x}{\sqrt{x^2+a^2}}$$

Therefore $\int \frac{1}{(\sqrt{x^2+a^2})^3}dx = \frac{\sin\theta}{a^2}+C$

$$= \frac{x}{a^2\sqrt{x^2+a^2}}+C.$$

4. $\int \frac{1}{x^2\sqrt{x^2+a^2}} = \frac{\sqrt{x^2+a^2}}{a^2x}+C$

Solution Let $x = a\tan\theta$. Then $dx = a\sec^2\theta d\theta$ and

$$\int \frac{1}{x^2\sqrt{(x^2+a^2)}}dx = \int \frac{1}{a^3\tan^2\theta\sqrt{\tan^2\theta+1}}a\sec^2\theta d\theta$$

$$= \int \frac{\sec\theta}{a^2\tan^2\theta}d\theta = \int \frac{\cos\theta}{a^2\sin^2\theta}d\theta.$$

Substitute $\sin\theta = t$, then $\cos\theta d\theta = dt$.

Therefore, $\int \frac{1}{a^2t^2}dt = -\frac{1}{a^2t}+C = -\frac{1}{a^2\sin\theta}+C$

$$= -\frac{\sqrt{x^2+a^2}}{a^2x}+C.$$

5-13. Find the general indefinite integral.

5. $\int x^{-3/4}dx$

Solution $\int x^{-3/4}dx = 4x^{1/4}+C$

6. $\int \sqrt[3]{x}\,dx$

Solution $\int x^{1/3}dx = \frac{3}{4}x^{4/3}+C$

CAS **7.** $\int (x^3+6x+1)dx$

Solution $\int (x^3+6x+1)\,dx = \frac{1}{4}x^4+3x^2+x+C$

```
var('x');
f(x)=x^3+6*x+1
F(x)=integral(f(x), x)
print F(x)
```

8. $\int (1-t)(2+t^2)dt$

Solution $\int (-t^3+t^2-2t+2)\,dt = -\frac{1}{4}t^4+\frac{1}{3}t^3-t^2+2t+C$

CAS **9.** $\int \left(x^2+1+\frac{1}{x^2+1}\right)dx$

Solution $\int \left(x^2+1+\frac{1}{x^2+1}\right)dx = \frac{1}{3}x^3+x+\arctan x+C$

10. $\int (2-\sqrt{x})^2dx$

Solution $\int (x-4\sqrt{x}+4)\,dx = \int (x-4x^{1/2}+4)\,dx$

$$= \frac{1}{2}x^2-4\frac{2}{3}x^{3/2}+4x+C$$

$$= \frac{1}{2}x^2-\frac{8}{3}x^{3/2}+4x+C$$

11. $\int (\cos x - 2\sin x)du$

Solution $\int (\cos x-2\sin x)du = \sin x+2\cos x+C$

12. $\int (3e^u + \sec^2 u)\,du$

Solution $\int (3e^u + \sec^2 u)\,du = 3e^u + \tan u + C$

CAS **13.** $\int \frac{\sin x}{1-\sin^2 x}\,dx$

Solution $\int \frac{\sin x}{1-\sin^2 x}\,dx = \int \frac{\sin x}{\cos^2 x}\,dx$

$= \int (\tan x \sec x)\,dx = \sec x + C$

14-22. Evaluate the integral.

14. $\int_0^2 (3x^2 - 4x + 5)\,dx$

Solution $\int_0^2 (3x^2 - 4x + 5)\,dx = [x^3 - 2x^2 + 5x]_0^2$

$= 8 - 8 + 10 = 10$

CAS **15.** $\int_{-1}^{1} (3u+1)^2\,du$

Solution $\int_{-1}^{1} (3u+1)^2\,du = \int_{-1}^{1} (9u^2 + 6u + 1)\,du$

$= [3u^3 + 3u^2 + u]_{-1}^{1} = 2[3u^3 + u]_0^1$

$= 2(3+1) = 8$

16. $\int_{-2}^{-1} \left(2y^3 + \frac{4}{y^3}\right) dy$

Solution $\int_1^2 \left(2y^3 + \frac{4}{y^3}\right) dy = \left[\frac{1}{2}y^4 - 2y^{-2}\right]_1^2$

$= \left(8 - \frac{1}{2}\right) - \left(\frac{1}{2} - 2\right) = 9$

17. $\int_1^4 \sqrt{\frac{3}{x}}\,dx$

Solution $\int_1^4 \sqrt{\frac{3}{x}}\,dx = \int_1^4 \sqrt{3}\,x^{-1/2}\,dx$

$= [2\sqrt{3}\,x^{1/2}]_1^4 = 4\sqrt{3} - 2\sqrt{3} = 2\sqrt{3}$

CAS **18.** $\int_0^{\pi/4} \frac{1 + 4\cos^2\theta}{\cos^2\theta}\,d\theta$

Solution $\int_0^{\pi/4} \frac{1 + 4\cos^2\theta}{\cos^2\theta}\,d\theta = \int_0^{\pi/4} (\sec^2\theta + 4)\,d\theta$

$= [\tan\theta + 4\theta]_0^{\pi/4} = 1 + \pi$

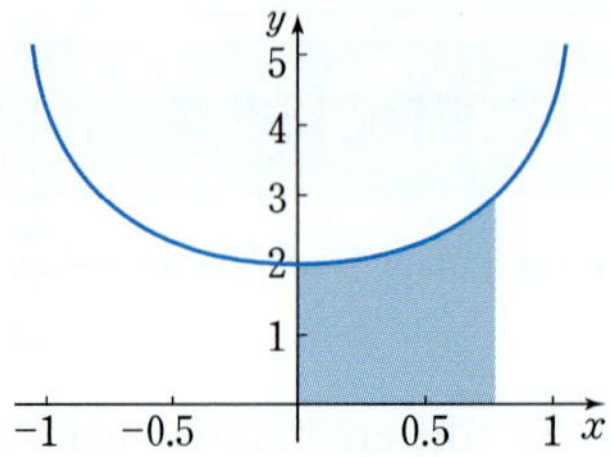

19. $\int_1^{e^2} \frac{4x^2 + x + 1}{x}\,dx$

Solution $\int_1^{e^2} \frac{4x^2 + 2x + 1}{x}\,dx = \int_1^{e^2} \left(4x + 2 + \frac{1}{x}\right) dx$

$= \int_1^{e^2} 4x\,dx + \int_1^{e^2} 2\,dx + \int_1^{e^2} \frac{1}{x}\,dx$

$= [2x^2]_1^{e^2} + [2x]_1^{e^2} + [\ln x]_1^{e^2}$

$= 2e^4 - 2 + 2e^2 - 2 + \ln e^2 - \ln 1 = 2e^4 - 2 + 2e^2 - 2 + 2 - 0$

$= 2e^4 + 2e^2 - 2$

```
var('x');
f(x)=(4*x^(2)+2*x+1)/x
print integral(f(x), x)
print integral(f(x), x, 1, e^2)
plot(f(x), (x, 0, 1), ymax=50)+plot(f(x), (x,
1, e^2), fill=true)+plot(f(x), (x, e^2, 10))
```

CAS **20.** $\int_0^2 \sqrt{\frac{2}{x} + 2}\,dx$

http://matrix.skku.ac.kr/cal-lab/cal-5-4-20.html

Solution

```
integral(sqrt(2/x+2),x,0,2)
```

Answer : -1/2*sqrt(2)*log(-(2*sqrt(2))*sqrt(3)+5)+2*sqrt(3)

21. $\int_1^3 \left(\sqrt{x} + \frac{1}{\sqrt{x}}\right)^2 dx$

Solution $\int_1^3 \left(\sqrt{x} + \frac{1}{\sqrt{x}}\right)^2 dx = \int_1^3 \left(x + 2 + \frac{1}{x}\right) dx$

$= \left[\frac{1}{2}x^2 + 2x + \ln x\right]_1^3 = \left(\frac{9}{2} + 6 + \ln 3\right) - \left(\frac{1}{2} + 2\right)$

$= 8 + \ln 3$

22. $\int_{\pi/4}^{\pi/6} (\sec x \tan x)\,dx$

Solution $\int_{\pi/4}^{\pi/6} \tan x \sec x\,dx = [\sec x]_{\pi/4}^{\pi/6} = -\sqrt{2} + \frac{2}{3}\sqrt{3}$

CAS **23.** Estimate the area of the region that lies under the curve $y = x^3 + 2x^7 - 2x^9$ and above the x-axis.

http://matrix.skku.ac.kr/cal-lab/cal-5-4-23.html

Solution It is easy to see that the graph has a real root 1.1914.

```
f(x)=x^3+2*x^7 -2*x^9
c=find_root(f(x), 1, 2);
print integral(f(x), x, 0, c)
plot(f(x), (x, -0.5, 0), ymax=2, ymin=-2)+plot(f(x), (x, 0, c), fill="axis")+plot(f(x), (x, c, 2))
```

Answer : 0.393951914046

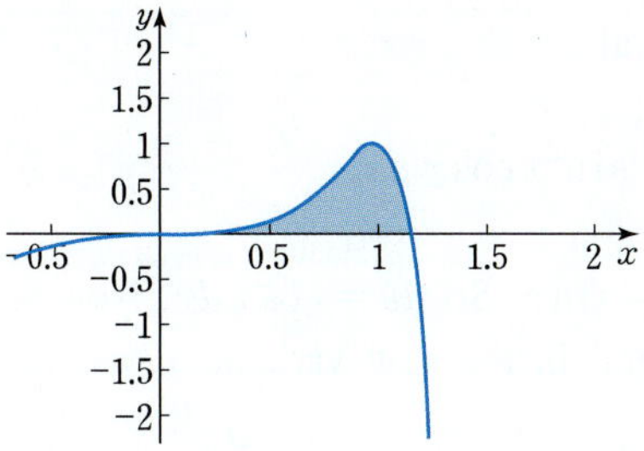

24. Water is being added to a tank at a rate of $r(t)=100-4t-t^2$ per minute. How much water is added to the tank from $t=1$ to $t=6$?

http://matrix.skku.ac.kr/cal-lab/cal-5-4-24.html

Solution The amount of water added for $1 \le t \le 6$ is

$$\int_1^6 r(t)dt = R(6)-R(1) = 456-\left(97+\frac{1}{3}\right)=\frac{1075}{3}$$

where $r(t)=R'(t)$ and $R(t)=100t-2t^2-\frac{1}{3}t^3$.

```
x=var('x')
f(x)=100-4*x-x^2
g= integral(f(x), x)
R=integral(f(x), (x, 1, 6))
show (g)
show (R)
p1=plot(f(x), 0, 7)
p2=parametric_plot((1, x), (x, -14, 105), linestyle='--')
p3=parametric_plot((6, x), (x, -14, 100), linestyle='--')
p4=plot(f(x), 1, 6, ymax=100, fill="axis")
show(p1+p2+p3+p4)
plot(f(x),1,6,ymax=110,fill="axis")+plot(f(x),1,6,ymax=110,color='blue')
```

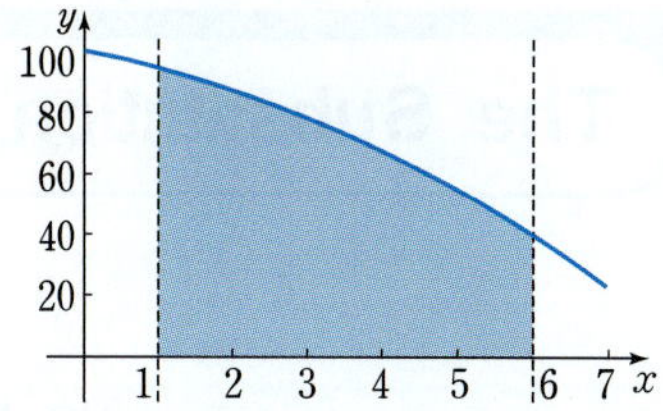

25. Find the area of the region that lies to the right of the y-axis and to the left of the parabola $x=-4+5y-y^2$.

Solution Since the parabola meets the y-axis at $(0, 1)$ and $(0, 4)$, the area is given by

$$\int_1^4 (-4+5y-y^2)dy = \frac{8}{3}-\left(-\frac{11}{6}\right)=\frac{9}{2}.$$

26. A particle is moving along a line with the velocity function $v(t)=t^3-6t^2+9t-4$. Find the displacement and the distance traveled by the particle during the time $2 \le t \le 6$.

Solution The displacement$=\int_2^6 v(t)dt = 32$,

The distance$=\int_2^6 |v(t)|dt = \int_2^4 [-v(t)]dt + \int_4^6 v(t)dt$
$=44$.

27. A honeybee population starts with 30 bees and increases at a rate of $n(t)=t^3-2t$ bees per week. How many honeybees are there after 10 weeks?

Solution Since the net change in population during 10 weeks is $\int_0^{10} n(t)dt = 2400$, the total number of honeybees after 10 weeks is $30+2400=2430$.

28. The acceleration function (in $\mathrm{m/s^2}$) of a particle is given by $a(t)=2t+1$ and the initial velocity is $v(0)=3$. Find the velocity of the particle at time t and determine the total distance traveled for $0 \le t \le 3$.

Solution Since $a(t)=v'(t)$, the velocity at time t is $v(t)=\int (2t+1)dt = t^2+t+C$.

And since $v(0)=3$, $v(t)=t^2+t+3$.

Therefore the distance traveled for $0 \le t \le 3$ is

$$\int_0^3 |t^2+t+3|dt = \frac{45}{2}.$$

5.5 The Substitution Rule

YouTube http://youtu.be/h7tmvmNOliU

The Substitution Rule replaces a relatively complicated integral by a simpler integral, by changing from the original variable x to a new variable u that is a function of x. The main challenge lies in finding an appropriate substitution. At first you can try choosing u to be some function in the integrand or some complicated part of the integrand.

For example, to evaluate integral

1
$$\int \sin^2 x \cos x \, dx,$$

introduce a new variable with $u = \sin x$. So $du = \cos x \, dx$. Now we can rewrite the given integral as a new integral in the new variable u as:

2
$$\int \sin^2 x \cos x \, dx = \int u^2 \, du = \frac{1}{3}u^3 + C = \frac{1}{3}\sin^3 x + C.$$

A simple check reveals that the above method of "substitution" works well:

$$\frac{d}{dx}\left(\frac{1}{3}\sin^3 x + C\right) = 3 \cdot \frac{1}{3}\sin^2 x \cos x = \sin^2 x \cos x.$$

In general, this method works whenever the given integral can be written in the form $\int f(g(x))g'(x)dx$. Since $F' = f$, and since by the Chain Rule,

$$\frac{d}{dx}[F(g(x))] = F'(g(x))g'(x)$$

then

3
$$\int F'(g(x))g'(x)dx = F(g(x)) + C.$$

With the "change of variable" or "substitution" $u = g(x)$, we have $du = g'(x)dx$. Then using the Chain Rule for differentiation 3 reduces to

$$\int F'(g(x))g'(x)dx = F(g(x)) + C = F(u) + C = \int F'(u)du$$

or, alternatively writing $F' = f$, we get $\int f(g(x))g'(x)dx = \int f(u)du$.

The Substitution Rule

If $u = g(x)$ is a differentiable function whose range is an interval I and f is continuous on I, then $\int f(g(x))g'(x)dx = \int f(u)du$.

Note At the final step, make sure to return to the original variable x.

EXAMPLE 1

Find $\int x\sin(x^2+1)dx$.

Solution The integrand suggests the substitution $u=x^2+1$. Then $du=2x\,dx$. Now using $x\,dx=du/2$ and the Substitution Rule, we have

$$\int x\sin(x^2+1)dx=\int \sin u\cdot\frac{1}{2}du=\frac{1}{2}\int \sin u\,du$$
$$=-\frac{1}{2}\cos u+C$$
$$=-\frac{1}{2}\cos(x^2+1)+C.$$

■

EXAMPLE 2

Find $\int x\sqrt{2x^2+1}\,dx$.

Solution 1. Put $u=2x^2+1$. Then $du=4x\,dx$, so $x\,dx=du/4$. Thus, the Substitution Rule gives

$$\int x\sqrt{2x^2+1}\,dx=\int\sqrt{u}\,\frac{du}{4}=\frac{1}{4}\int u^{\frac{1}{2}}du$$
$$=\frac{1}{4}\cdot\frac{2}{3}u^{\frac{3}{2}}+C=\frac{1}{6}u^{\frac{3}{2}}+C$$
$$=\frac{1}{6}(2x^2+1)^{\frac{3}{2}}+C.$$

2. Alternatively take $u=\sqrt{2x^2+1}$. Then

$$u^2=2x^2+1 \text{ so } u\,du=2x\,dx.$$

Therefore,

$$\int x\sqrt{2x^2+1}\,dx=\int u\cdot\frac{u}{2}du=\frac{1}{2}\int u^2\,du$$
$$=\frac{1}{2}\cdot\frac{1}{3}u^3+C=\frac{1}{6}(2x^2+1)^{\frac{3}{2}}+C.$$

■

EXAMPLE 3

Find $\int \frac{e^{1/x}}{x^2}dx$.

Solution Let $u=\frac{1}{x}$, then $du=-\frac{1}{x^2}dx$. Therefore,

$$\int \frac{e^{1/x}}{x^2}dx = \int e^{\frac{1}{x}}\left(\frac{1}{x^2}dx\right) = -\int e^u du$$
$$= -e^u + C = -e^{1/x} + C.$$ ■

EXAMPLE 4

Evaluate $\int \frac{1}{1+e^{-x}}dx$.

Solution If we multiply the numerator and denominator by e^x, then we obtain

$$\int \frac{1}{1+e^{-x}}dx = \int \frac{e^x}{1+e^x}dx.$$

Put $u = e^x + 1$, then $du = e^x dx$. Therefore,

$$\int \frac{1}{1+e^{-x}}dx = \int \frac{e^x}{1+e^x}dx = \int \frac{1}{u}du$$
$$= \ln|u| + C = \ln(e^x + 1) + C.$$ ■

EXAMPLE 5

Find $\int \tan x\, dx$.

Solution Since $\tan x = \sin x / \cos x$, this suggests the substitution $u = \cos x$. Then $du = -\sin x\, dx$

$$\int \tan x\, dx = \int \frac{\sin x}{\cos x}dx = -\int \frac{du}{u}$$
$$= -\ln|u| + C = -\ln|\cos x| + C.$$ ■

Since $-\ln|\cos x| = \ln(|\cos x|^{-1}) = \ln(1/|\cos x|) = \ln|\sec x|$, so this result can also be written as

4 $$\int \tan x\, dx = \ln|\sec x| + C$$

EXAMPLE 6

Find $\int \frac{(\tan^{-1} x)^3}{1+x^2}dx$.

Solution Put $u = \tan^{-1} x$, then $du = \frac{1}{1+x^2}dx$. Therefore,

$$\int \frac{(\tan^{-1} x)^3}{1+x^2}dx = \int (\tan^{-1} x)^3 \frac{1}{1+x^2}dx$$

$$= \int u^3\,du = \frac{1}{4}u^4 + C = \frac{1}{4}(\tan^{-1}x)^4 + C\,. \quad \blacksquare$$

EXAMPLE 7

Evaluate $\displaystyle\int_0^2 \left(3x^2 - 4x + \frac{4}{x^2+4}\right)dx$.

Solution Since $\displaystyle\int \frac{1}{x^2+4}\,dx = \frac{1}{2}\tan^{-1}\frac{x}{2} + C$, we have

$$\begin{aligned}\int_0^2 \left(3x^2 - 4x + \frac{4}{x^2+4}\right)dx &= \left[3\frac{x^3}{3} - 4\frac{x^2}{2} + 4\cdot\frac{1}{2}\tan^{-1}\frac{x}{2}\right]_0^2 \\ &= \left[x^3 - 2x^2 + 2\tan^{-1}\frac{x}{2}\right]_0^2 \\ &= 2^3 - 2(2^2) + 2\tan^{-1}1 - 0 = \frac{\pi}{2}. \quad \blacksquare\end{aligned}$$

Remark The substitution in the substitution rule is not unique. For example refer to example 2.

Using Substitution Rule for evaluating Definite Integrals

A definite integral can be evaluated by substitution, in two ways. One way is to evaluate the indefinite integral first and then use the FTC. Another way is to put everything in terms of the new variable u. This will change the integrand and the limits of integration. The new limits of integration are the values of u obtained from $x = a$ and $x = b$.

The Substitution Rule for Definite Integrals

If g' is continuous on $[a, b]$ and f is continuous on the range of $u = g(x)$, then

5 $$\int_{x=a}^{x=b} f(g(x))g'(x)\,dx = \int_{u=g(a)}^{u=g(b)} f(u)\,du.$$

Proof Let F be an antiderivative of f. Then, by 3, $F(g(x))$ is an antiderivative of $f(g(x))g'(x)$, so by Part 2 of the Fundamental Theorem, we have

$$\int_a^b f(g(x))g'(x)\,dx = [F(g(x))]_a^b = F(g(b)) - F(g(a))$$

But, applying the FTC 2 a second time, we also have

$$\int_{g(a)}^{g(b)} f(u)\,du = [F(u)]_{g(a)}^{g(b)} = F(g(b)) - F(g(a)). \quad \blacksquare$$

EXAMPLE 8

Find $\int_0^2 x\sqrt{2x^2+1}\,dx$.

Solution Put $u=2x^2+1$ and $du=4x\,dx$. Now when $x=0$, $u=1$ and when $x=2$, $u=9$ so the new limits of integration are $u=1$ to $u=9$. Therefore,

$$\begin{aligned}\int_0^2 x\sqrt{2x^2+1}\,dx &= \int_1^9 \frac{1}{4}\sqrt{u}\,du\\ &= \left[\frac{1}{4}\cdot\frac{2}{3}u^{\frac{3}{2}}\right]_1^9\\ &= \frac{1}{6}(9^{\frac{3}{2}}-1^{\frac{3}{2}})=\frac{13}{3}.\end{aligned}$$ ■

EXAMPLE 9

Calculate $\int_0^1 \frac{dx}{(2-3x)^2}$.

Solution Put $u=2-3x$. Then $du=-3dx$, so $dx=-du/3$. When $x=0$, $u=2$ and when $x=1$, $u=-1$. Thus

$$\begin{aligned}\int_0^1 \frac{dx}{(2-3x)^2} &= -\frac{1}{3}\int_2^{-1}\frac{du}{u^2}\\ &= -\frac{1}{3}\left[-\frac{1}{u}\right]_2^{-1}=\left[\frac{1}{3u}\right]_2^{-1}\\ &= \frac{1}{3}\left(-1-\frac{1}{2}\right)=-\frac{1}{2}.\end{aligned}$$ ■

EXAMPLE 10

Find $\int_1^e \frac{\ln x}{x}dx$.

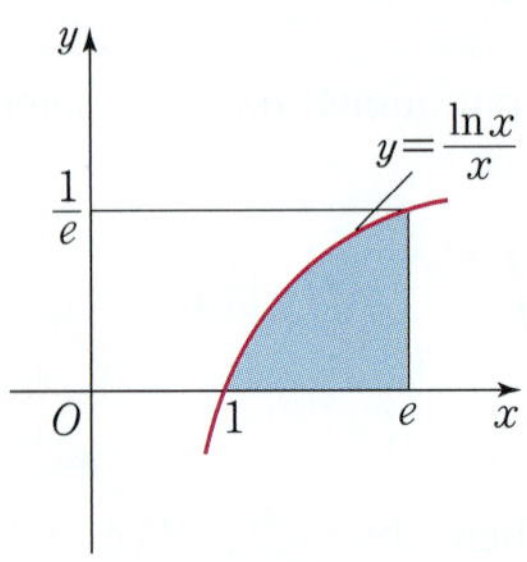

Figure 1

Solution The integrand suggests using $u=\ln x$, so then $du=dx/x$. Now when $x=1$, $u=\ln 1=0$; when $x=e$, $u=\ln e=1$. Thus

$$\int_1^e \frac{\ln x}{x}dx=\int_0^1 u\,du=\left[\frac{u^2}{2}\right]_0^1=\frac{1}{2}.$$ ■

Symmetry

When the graph of a function $y=f(x)$ is symmetric with respect to either the y-axis (even function) or the origin (odd function), then the definite integral of

f on a symmetric interval $[-a, a]$, that is $\int_{-a}^{a} f(x)\,dx$, can be evaluated by means of a shortcut.

Integrals of Symmetric Functions

(a) Even Function Rule
If f is an even integrable function on $[-a, a]$, then

$$\int_{-a}^{a} f(x)dx = 2\int_{0}^{a} f(x)dx.$$

(b) Odd Function Rule
If f is an odd integrable function on $[-a, a]$, then

$$\int_{-a}^{a} f(x)dx = 0.$$

Proof Split the given integral into two:

5 $$\int_{-a}^{a} f(x)dx = \int_{-a}^{0} f(x)dx + \int_{0}^{a} f(x)dx = -\int_{0}^{-a} f(x)dx + \int_{0}^{a} f(x)dx$$

In the first integral on the far right side put $u = -x$. Then $du = -dx$ and when $x = -a$, $u = a$. Therefore

$$-\int_{0}^{-a} f(x)dx = -\int_{0}^{a} f(-u)(-du) = \int_{0}^{a} f(-u)du.$$

So 5 becomes

6 $$\int_{-a}^{a} f(x)dx = \int_{0}^{a} f(-u)du + \int_{0}^{a} f(x)dx.$$

(a) If f is even, then $f(-u) = f(u)$ so Equation 6 gives

$$\int_{-a}^{a} f(x)dx = \int_{0}^{a} f(u)du + \int_{0}^{a} f(x)dx = 2\int_{0}^{a} f(x)dx$$

(b) If f is odd, then $f(-u) = -f(u)$ so Equation 6 gives

$$\int_{-a}^{a} f(x)dx = -\int_{0}^{a} f(u)du + \int_{0}^{a} f(x)dx = 0.$$ ■

For the case where f is positive and even, part (a) says that the area under $y = f(x)$ from $-a$ to a is twice the area from 0 to a because of symmetry.

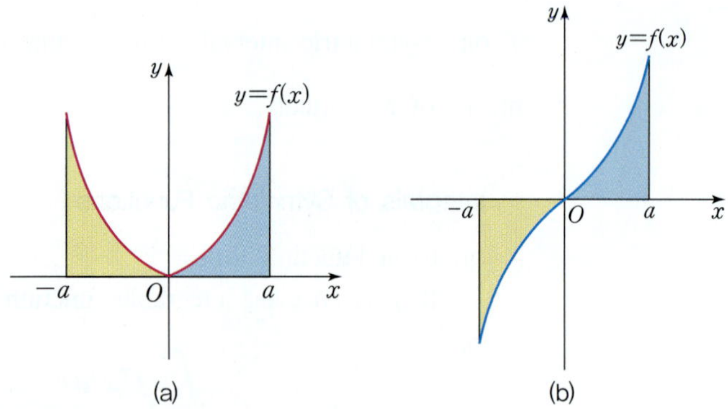

Figure 2

Recall that an integral $\int_a^b f(x)dx$ can be expressed as the area above the x-axis and below $y=f(x)$, minus the area below the x-axis and above $y=f(x)$. Thus, part (b) says the integral is 0 because the areas cancel. (See Figure 2.)

EXAMPLE 11

Evaluate $\int_{-\frac{\pi}{2}}^{\frac{\pi}{2}} \cos x\,dx$.

Solution In this case $f(x)=\cos x$ is an even function on the symmetric interval $\left[-\frac{\pi}{2}, \frac{\pi}{2}\right]$. (See Figure 2 (a).)

$$\int_{-\frac{\pi}{2}}^{\frac{\pi}{2}} \cos x\,dx = 2\int_0^{\frac{\pi}{2}} \cos x\,dx = 2[\sin x]_0^{\frac{\pi}{2}} = 2[1-0] = 2.$$ ■

EXAMPLE 12

Evaluate $\int_{-1}^{1} (x^3+x)dx$.

Solution In this case $f(x)=x^3+x$ is an odd function on the symmetric interval $[-1, 1]$. (See Figure 2 (b).)

$$\int_{-1}^{1} (x^3+x)dx = 0$$

```
var('x')
f=x^3 + x
plot(f)
integral(f, x, -1, 1)
```

■

5.5 EXERCISES (The Substitution Rule)

http://matrix.skku.ac.kr/Cal-Book/part1/CS-Sec-5-5-Sol.html YouTube http://youtu.be/0TMbpCPO4Uc

1-3. Evaluate the given indefinite integral using an appropriate substitution.

1. $\int 2x(x^2+3)^4 dx$

Solution $u = x^2+3 \Rightarrow du = 2x dx$,

$$\int 2x(x^2+3)^4 dx = \int u^4 du = \frac{1}{5}u^5 + c$$

$$= \frac{1}{5}(x^2+3)^5 + C$$

2. $\int \frac{1+4x}{\sqrt{1+x+2x^2}} dx$

Solution $1+x+2x^2 = u \Rightarrow (1+4x)dx = du$,

$$\int \frac{1+4x}{\sqrt{1+x+2x^2}} dx = \int \frac{1}{\sqrt{u}} du$$

$$= 2\sqrt{u} + C = 2\sqrt{1+x+2x^2} + C$$

3. $\int \frac{x}{(x^2+1)^2} dx$

Solution $x^2+1 = u \Rightarrow x dx = \frac{1}{2} du$,

$$\int \frac{x}{(x^2+1)^2} dx = \int \frac{1}{2u^2} du = -\frac{1}{2u} + C$$

$$= -\frac{1}{2(x^2+1)} + C$$

4-10. Find the indefinite integral.

CAS **4.** $\int \frac{x}{2x^2+1} dx$

http://matrix.skku.ac.kr/cal-lab/cal-5-5-4.html

Solution Let $t = x^2$ then $dt = 2x dx$.

$$\int \frac{x}{2x^2+1} dx = \int \frac{1}{2t+1}\frac{dt}{2} = \frac{1}{2}\int \frac{1}{2t+1} dt$$

$$= \frac{1}{4}\ln|2t+1| + C = \frac{1}{4}\ln|2x^2+1| + C$$

```
@interact
def _(c=((1,  10,  0.1)))
    f(x)=x/(2*x^2+1)
    F=diff(f(x),  x)+c
    P=plot(F)
    show(P)
```

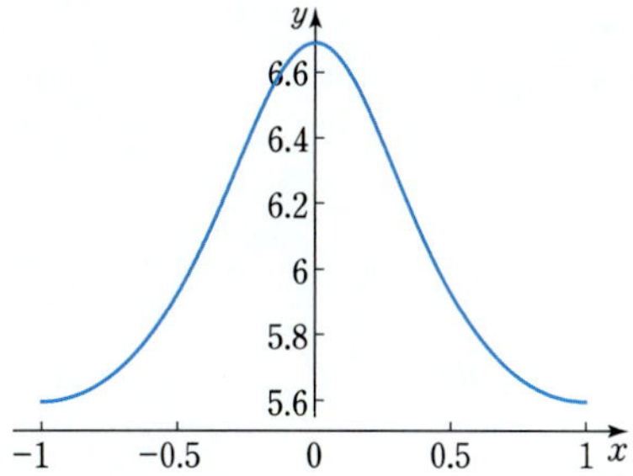

5. $\int \frac{ax+b}{\sqrt{ax^2+2bx+c}} dx$

Solution $\sqrt{ax^2+2bx+c} + C$

6. $\int \frac{dx}{x\ln x}$

Solution $\ln(\ln x) + C$

CAS **7.** $\int \sqrt[3]{x^3+1}\, x^5 dx$

Solution $\frac{1}{7}(x^3+1)^{7/3} - \frac{1}{4}(x^3+1)^{4/3} + C$

8. $\int x^a \sqrt{b+cx^{a+1}} dx \quad (c \neq 0,\ a \neq -1)$

Solution $b+cx^{a+1} = u \Rightarrow x^a dx = \frac{1}{(a+1)c} du$,

$$\int x^a\sqrt{b+cx^{a+1}} dx = \frac{1}{(a+1)c}\int \sqrt{u}\, du$$

$$= \frac{2}{3(a+1)c}u^{3/2} + C = \frac{2}{3(a+1)c}(b+cx^{a+1})^{3/2} + C$$

9. $\int \frac{x}{\sqrt[4]{x+2}} dx$

Solution $\frac{4}{7}(x+2)^{7/4} - \frac{8}{3}(x+2)^{3/4} + C$

CAS **10.** $\int \frac{x^2}{\sqrt{1-x}} dx$

Solution $-\frac{2}{5}(1-x)^{5/2} + \frac{4}{3}(1-x)^{3/2} - 2\sqrt{1-x} + C$

11-23. Evaluate the definite integral, if it exists.

11. $\int_0^2 (x-1)^{25}\,dx$

Solution $\int_0^2 (x-1)^{25}dx = \int_{-1}^1 u^{25}du = 0$

(Because x^{25} is an odd function)

12. $\int_0^7 \sqrt{4x+3}\,dx$

Solution $\frac{31}{6}\sqrt{31} - \frac{\sqrt{3}}{2}$

CAS **13.** $\int_0^1 x^2(1+2x^3)^5 dx$

Solution $\frac{182}{9}$

CAS **14.** $\int_0^{\sqrt{\pi}} x\cos(x^2)dx$

Solution 0

CAS **15.** $\int_0^{\pi} \sec^2(t/4)dt$

Solution 4

CAS **16.** $\int_{1/6}^{1/2} \csc \pi t \cot \pi t\, dt$

Solution $\pi\int_{\pi/6}^{\pi/2}\frac{1}{\sin t}\cdot\frac{\cos t}{\sin t}dt = \pi\int_{\pi/6}^{\pi/2}\frac{\cos t}{\sin^2 t}dt.$

(First of all we substitute $t=\pi t$.)

$\sin t = u \Rightarrow \cos t dt = du$

$$\pi\int_{\pi/6}^{\pi/2}\frac{\cos t}{\sin^2 t}dt = \pi\int_{1/2}^{1}\frac{1}{u^2}du$$
$$= \left[-\frac{\pi}{u}\right]_{1/2}^{1} = -\pi+\frac{\pi}{2} = -\frac{\pi}{2}$$

17. $\int_0^1 \frac{e^{3x}+2}{e^{3x}-2e^{2x}+2e^x}dx$

Solution $e^x = y \Rightarrow e^x dx = dy,$

$$\int_0^1 \frac{e^{3x}+2}{e^{3x}-2e^{2x}+2e^x}dx = \int_0^1 \frac{e^{3x}+2}{e^{4x}-2e^{3x}+2e^{2x}}e^x dx$$
$$= \int_1^e \frac{y^3+2}{y^4-2y^3+2y^2}dy = \int_1^e\left(\frac{1}{y}+\frac{1}{y^2}+\frac{1}{y^2-2y+2}\right)dy$$
$$= \left[\ln y - \frac{1}{y} + \tan^{-1}(y-1)\right]_1^e$$
$$= 2 - \frac{1}{e} + \tan^{-1}(e-1)$$

18. $\int_0^1 \frac{x^4}{x^3+1}dx$

Solution Since $\frac{x^4}{x^3+1} = x - \frac{x}{x^3+1}$

$= x + \frac{1}{3}\left(\frac{1}{x+1} - \frac{x+1}{x^2-x+1}\right),$

first we have

$$I := \int_0^1 \frac{x^4}{x^3+1}dx$$
$$= \int_0^1 x dx + \frac{1}{3}\int_0^1 \frac{1}{x+1}dx - \frac{1}{3}\int_0^1\frac{x+1}{x^2-x+1}dx$$
$$= \frac{1}{2} + \frac{\ln 2}{3} - \frac{1}{3}\int_0^1 \frac{x+1}{x^2-x+1}dx.$$

Then,

$$I_1 = \int_0^1 \frac{x+1}{x^2-x+1}dx$$
$$= \frac{1}{2}\left[\ln(x^2-x+1)\right]_0^1 + \frac{3}{2}\int_0^1 \frac{1}{\left(x-\frac{1}{2}\right)^2+\frac{3}{4}}dx$$
$$= 0 + \frac{3}{2}\int_0^1 \frac{1}{\left(x-\frac{1}{2}\right)^2+\frac{3}{4}}dx := \frac{3}{2}I_2,$$
$$I_2 = \int_0^1 \frac{1}{\left(x-\frac{1}{2}\right)^2+\frac{3}{4}}dx$$
$$= \int_{-\frac{1}{2}}^{\frac{1}{2}} \frac{1}{u^2+\left(\frac{\sqrt{3}}{2}\right)^2}du$$
$$= 2\frac{2}{\sqrt{3}}\tan^{-1}\left(\frac{2}{\sqrt{3}}u\right)\Bigg|_0^{1/2} = \frac{2\sqrt{3}}{9}\pi.$$

(Because $\frac{1}{u^2+\left(\frac{\sqrt{3}}{2}\right)^2}$ is an even function.)

Hence $I = \frac{1}{2} + \frac{\ln 2}{3} - \frac{1}{3}\left(\frac{3}{2}\cdot\frac{2\sqrt{3}}{9}\pi\right)$

$= \frac{1}{2} + \frac{\ln 2}{3} - \frac{\sqrt{3}}{9}\pi.$

CAS **19.** $\int_{-\pi/2}^{\pi/2} \frac{x^3}{1+x^6}dx$

Solution 0 because $\frac{x^3}{1+x^6}$ is odd function.

CAS **20.** $\int_0^{1/2} \frac{\sin^{-1}x}{\sqrt{1-x^2}}dx$

Solution $\frac{\pi^2}{72}$

CAS **21.** $\int_0^a x\sqrt{a^2-x^2}\,dx$

http://matrix.skku.ac.kr/cal-lab/cal-5-5-21.html

Solution $\dfrac{a^3}{3}$

```
var('x, a')
f(x)=x*sqrt(a^2 - x^2)
integral(f, x)
```

22. $\displaystyle\int_0^a x\sqrt{x^2+a^2}\,dx \quad (a>0)$

Solution $\dfrac{a^3(2\sqrt{2}-1)}{3}$

23. $\displaystyle\int_{-a}^a x\sqrt{x^2+a^2}\,dx$

Solution 0 because $x\sqrt{x^2+a^2}$ is odd function.

24-25. Find the exact area.

24. $y=\sqrt{2x+1}\,,\ 0\le x\le 1$

Solution $\sqrt{3}-\dfrac{1}{3}$

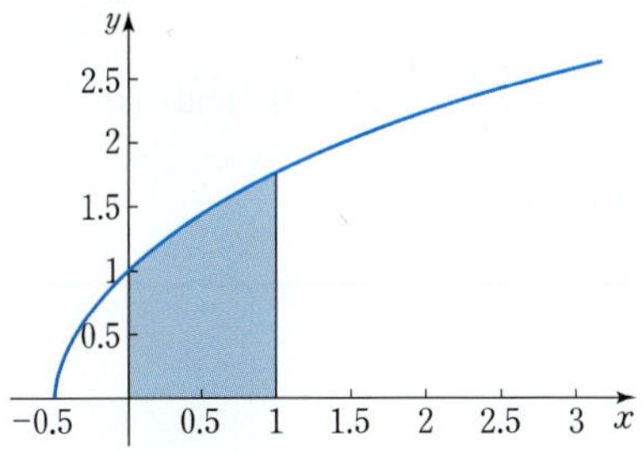

CAS **25.** $y=2\sin x-\sin 2x,\ 0\le x\le \pi$

Solution 4

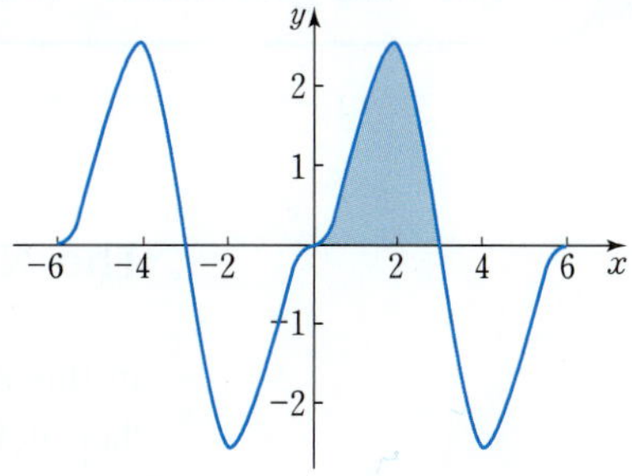

26. Let $f(x)$ be a function symmetrical with respect to $\left(\dfrac{1}{2},\ 0\right)$. Show $\displaystyle\int_0^1\left(x^2+\frac{1}{4}\right)f(x)dx=0$, when $\displaystyle\int_0^1 xf(x)dx=0$.

Solution Since $\displaystyle\int_0^1 xf(x)dx=0$

$$\int_0^1\left(x^2+\frac{1}{4}\right)f(x)dx=\int_0^1\left(x^2-x+\frac{1}{4}\right)f(x)dx.$$

Hence

$$\int_0^1\left(x-\frac{1}{2}\right)^2 f(x)dx=\int_{-\frac{1}{2}}^{\frac{1}{2}} y^2 f\left(y-\frac{1}{2}\right)dy=0$$

because $f(x)$ is a function symmetrical with respect to $\left(\dfrac{1}{2},\ 0\right)$.

5.6 The Logarithm Defined as an Integral

YouTube http://youtu.be/kD0Z9PqetsA

The Natural Logarithm

In this section, first of all we will use the FTC to define $\ln x$ as an integral and then define the exponential function as its inverse.

DEFINITION 1

The natural logarithmic function is the function defined by $\ln x = \int_1^x \frac{1}{t}dt$ $x > 0$. Note that it exists since $y = 1/t$ is a continuous function. If $x > 1$, then $\ln x$ can be interpreted geometrically as the area under the hyperbola $y = 1/t$ from $t = 1$ to $t = x$. (See Figure 1.) For $x = 1$, we have

$$\ln 1 = \int_1^1 \frac{1}{t}dt = 0$$

For $0 < x < 1$, $\ln x = \int_1^x \frac{1}{t}dt = -\int_x^1 \frac{1}{t}dt < 0$ and so $\ln x$ is the negative of the area shown in Figure 2.

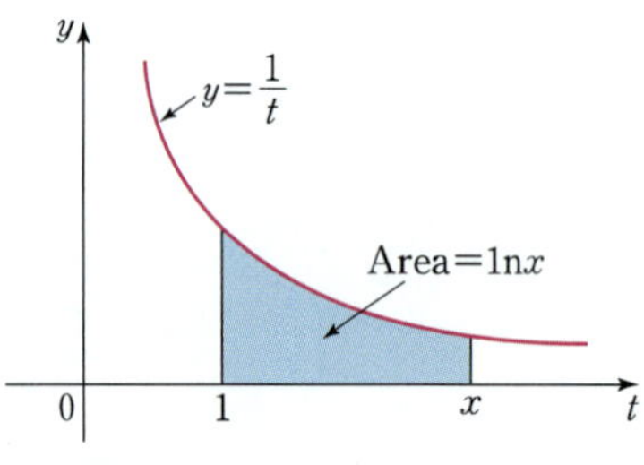

Figure 1

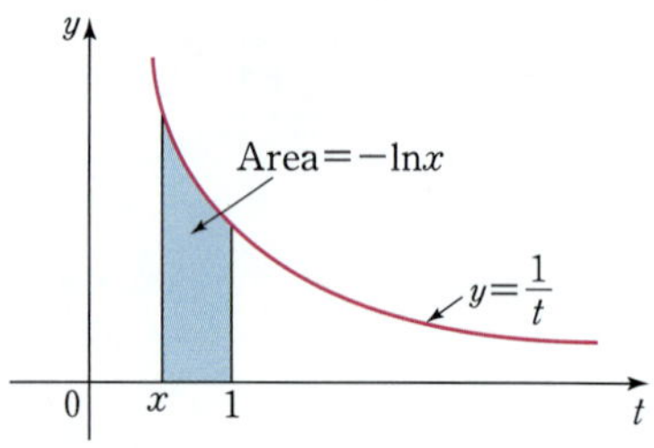

Figure 2

EXAMPLE 1

(a) By comparing areas, show that $\frac{1}{2} < \ln 2 < \frac{3}{4}$.

(b) Use the Midpoint Rule with $n = 10$ to estimate the value of $\ln 2$.

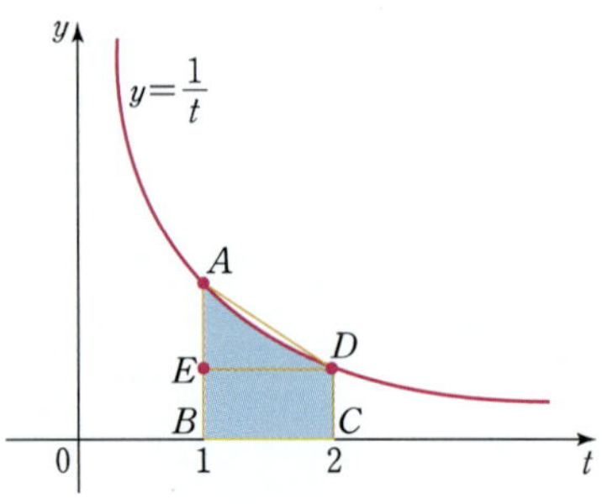

Figure 3

Solution (a) We can interpret $\ln 2$ as the area under the curve $y = 1/t$ from 1 to 2. From Figure 3 we see that this area is larger than the area of rectangle *BCDE* and smaller than the area of trapezoid *ABCD*. Thus, we have

$$\frac{1}{2} \cdot 1 < \ln 2 < 1 \cdot \frac{1}{2}\left(1 + \frac{1}{2}\right)$$

$$\frac{1}{2} < \ln 2 < \frac{3}{4}$$

(b) If we use the Midpoint Rule with $f(t) = 1/t$, $n = 10$, and $\Delta t = 0.1$, we get

$$\ln 2 = \int_1^2 \frac{1}{t} dt \approx (0.1)[f(1.05) + f(1.15) + \cdots + f(1.95)]$$

$$= (0.1)\left(\frac{1}{1.05} + \frac{1}{1.15} + \cdots + \frac{1}{1.95}\right) \approx 0.693.$$ ■

Using part 1 of the FTC we have

$$\frac{d}{dx}\int_1^x \frac{1}{t} dt = \frac{1}{x}$$

and so

1

$$\frac{d}{dx}(\ln x) = \frac{1}{x}$$

2 Laws of Logarithms

If x and y are positive numbers and r is a rational number.

1. $\ln(xy) = \ln x + \ln y$ 2. $\ln\left(\frac{x}{y}\right) = \ln x - \ln y$ 3. $\ln(x^r) = r \ln x$

Proof 1. Let $f(x) = \ln(ax)$, where a is a positive constant. Then, using 2 and the Chain Rule, we have

$$f'(x) = \frac{1}{ax}\frac{d}{dx}(ax) = \frac{1}{ax} \cdot a = \frac{1}{x}.$$

Therefore, $f(x)$ and $\ln x$ have the same derivative and so they must differ by a constant, that is, $\ln(ax) = \ln x + C$.
Putting $x = 1$ in this equation, we get $\ln a = \ln 1 + C = 0 + C = C$. Thus $\ln(ax) = \ln x + \ln a$.
If we now replace the constant a by any number y, then we have $\ln(xy) = \ln x + \ln y$.

2. Using Law 1 with $x = \frac{1}{y}$, we have $\ln 1/y + \ln y = \ln\left(\frac{1}{y} \cdot y\right)$ $= \ln 1 = 0$ and so $\ln\frac{1}{y} = -\ln y$.

Using Law 1 again, we have $\ln\left(\frac{x}{y}\right) = \ln\left(x \cdot \frac{1}{y}\right) = \ln x + \ln\frac{1}{y}$ $= \ln x - \ln y$.

The proof of Law 3 is left as an exercise. ■

In order to graph $y = \ln x$, we first determine its limits:

3 (a) $\lim_{x\to\infty} \ln x = \infty$ (b) $\lim_{x\to 0^+} \ln x = -\infty$

Proof (a) Using Law 3 with $x = 2$ and $r = n$ (where n is any positive integer), we have $\ln(2^n) = n\ln 2$. Now $\ln 2 > 0$, so this shows that $\ln(2^n) \to \infty$ as $n \to \infty$. $\ln x$ is an increasing function since its derivative is $1/x > 0$. Therefore, $\ln x \to \infty$ as $x \to \infty$.

(b) If we let $t = 1/x$, then $t \to \infty$ as $x \to 0^+$. Thus, using (a), we have

$$\lim_{x\to 0^+} \ln x = \lim_{t\to\infty} \ln\left(\frac{1}{t}\right) = \lim_{t\to\infty}(-\ln t) = -\infty.$$ ■

If $y = \ln x$, $x > 0$, then

$$\frac{dy}{dx} = \frac{1}{x} > 0 \text{ and } \frac{d^2y}{dx^2} = -\frac{1}{x^2} < 0$$

which shows that $\ln x$ is increasing and concave downward on $(0, \infty)$. With this information, we draw the graph of $y = \ln x$. (See Figure 4.)

Since $\ln 1 = 0$ and $\ln x$ is an increasing continuous function that takes on arbitrarily large values, the Intermediate Value Theorem shows that there is a number where ln x takes on the value 1. (See Figure 5.) This important number is denoted by e.

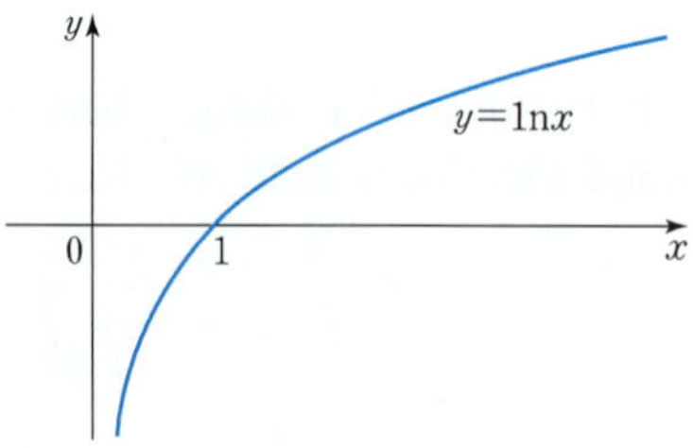

Figure 4

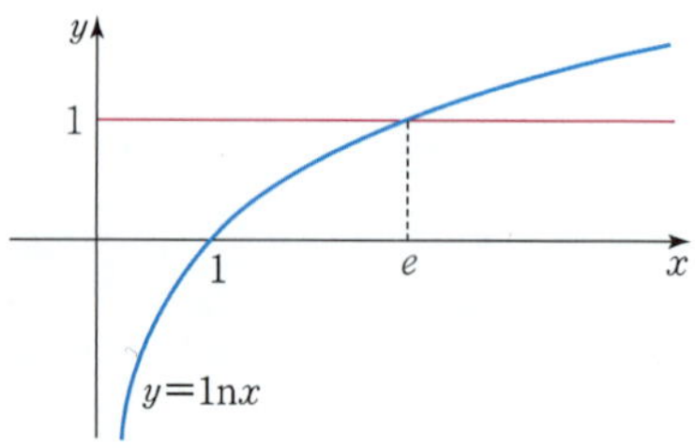

Figure 5

DEFINITION 2

e is the number such that $\ln e = 1$.

The Natural Exponential Function

The natural exponential function $f(x) = e^x$ is one of the most frequently occurring functions in calculus and its applications. We defined e^x before and can extend e^x continuously to all real value x.

Since the logarithm is an increasing function, it is one-to-one and therefore has an inverse function, which we denote by exp. Thus, according to the definition of an inverse function,

4 $$\exp(x) = y \Leftrightarrow \ln y = x$$

and the cancellation equations are

5 $$\exp(\ln x) = x \text{ and } \ln(\exp x) = x$$

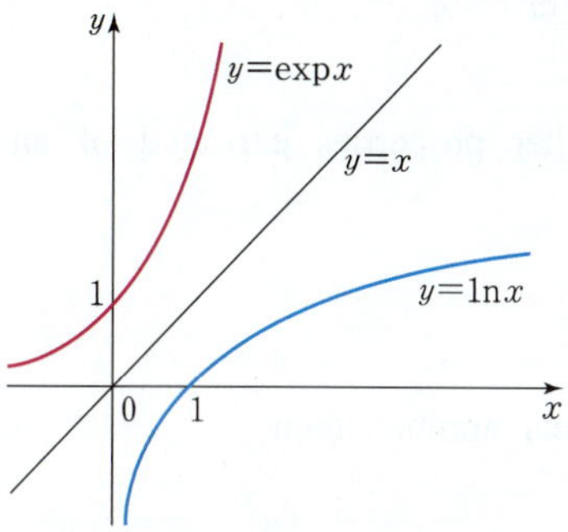

Figure 6 $f^{-1}(x) = y \Leftrightarrow f(y) = x$
$f^{-1}(f(x)) = x$
$f(f^{-1}(x)) = x$

In particular, we have

$$\exp(0) = 1 \quad \text{since} \quad \ln 1 = 0$$
$$\exp(1) = e \quad \text{since} \quad \ln e = 1$$

We obtain the graph of $y = \exp(x)$ by reflecting the graph of $y = \ln x$ about the line $y = x$. (See Figure 6.) The domain $D(\exp) = R(\log) = (-\infty, \infty)$ of exp is the range of the logarithm. $D(\exp) = R(\log) = (-\infty, \infty)$. The range of exp is the domain of log, $(0, \infty)$, which is the domain of the logarithm is the range of exp.

If r is any rational number, then the third law of logarithms gives i.e. $R(\exp) = D(\log) = (0, \infty)$

$$\ln(e^r) = r \ln e = r.$$

Therefore, by 4, $\exp(r) = e^r$.

Thus, $\exp(x) = e^x$ whenever x is a rational number. This leads us to write e^x, even for irrational values of x, by the equation

$$e^x = \exp(x)$$

In other words, for the reasons given, we define e^x to be the inverse of the function $\ln x$. In this notation 4 becomes

6 $$e^x = y \Leftrightarrow \ln y = x$$

and the cancellation equations 5 become

7 $$e^{\ln x} = x, \quad x > 0$$

$$\boxed{8} \quad \ln(e^x) = x \quad \text{for all } x$$

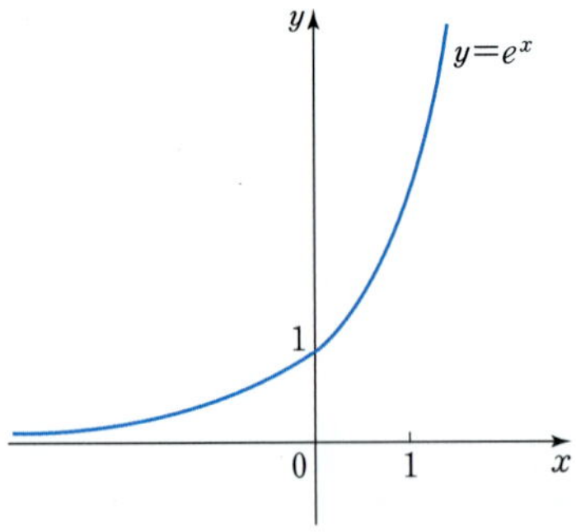

Figure 7 The natural exponential function

Properties of the Exponential Function

The exponential function $f(x) = e^x$ is an increasing continuous function with domain $\mathbb{R}$ and range $(0, \infty)$. Thus, $e^x > 0$ for all x. Also

$$\lim_{x \to -\infty} e^x = 0 \qquad \lim_{x \to \infty} e^x = \infty$$

So the x-axis is a horizontal asymptote of $f(x) = e^x$.

We now verify f where $f = e^x$ has the other properties expected of an exponential function.

9 Laws of Exponents

If x and y are real numbers and r is any real number, then

1. $e^{x+y} = e^x e^y$ 2. $e^{x-y} = \dfrac{e^x}{e^y}$ 3. $(e^x)^r = e^{rx}$

Proof 1. Using the first law of logarithms and Equation 8, we have

$$\ln(e^x e^y) = \ln(e^x) + \ln(e^y) = x + y = \ln(e^{x+y})$$

Since ln is a one-to-one function, it follows that $e^x e^y = e^{x+y}$. ■

THEOREM 3

Prove that

$$\frac{d}{dx}(e^x) = e^x$$

Proof The function $y = e^x$ is differentiable because it is the inverse function of $y = \ln x$. To find its derivative, we use the inverse function method. Let $y = e^x$. Then $\ln y = x$ and differentiating this latter equation implicitly with respect to x, we get

$$\frac{1}{y}\frac{dy}{dx} = 1$$

$$\frac{dy}{dx} = y = e^x.$$ ■

General Exponential Functions

If $a>0$ and r is any rational number, then by [7] and [9],

$$a^r = (e^{\ln a})^r = e^{r\ln a}.$$

Therefore, even for irrational numbers x, we define

[10]
$$a^x = e^{x\ln a}$$

Thus, for instance, $2^{\sqrt{3}} = e^{\sqrt{3}\ln 2} \approx e^{1.20} \approx 3.32$.

The function $f(x)=a^x$ is called the **exponential function with base** a. Notice that a^x is positive for all x because e^x is positive for all x.

Definition [10] allows us to extend one of the laws of logarithms. We already know that $\ln(a^r) = r\ln a$ where r is rational. But if we now let r be any real number we have, from Definition [10],

$$\ln a^r = \ln(e^{r\ln a}) = r\ln a$$

Thus

[11] $\ln a^r = r\ln a$ for any real number r

The general laws of exponents follow from Definition [10] together with the laws of exponents for e^x.

[12] Properties of Exponential Functions

If x and y are real numbers and $a,\ b>0$ then

1. $a^{x+y} = a^x a^y$
2. $a^{x-y} = a^x / a^y$
3. $(a^x)^y = a^{xy}$
4. $(ab)^x = a^x b^x$

Proof 1. Using Definition [10] and the properties of exponential functions e^x, we have

$$a^{x+y} = e^{(x+y)\ln a} = e^{x\ln a + y\ln a}$$
$$= e^{x\ln a}e^{y\ln a} = a^x a^y$$

3. Using Equation [11] we obtain

$$(a^x)^y = e^{y\ln(a^x)} = e^{yx\ln a}$$
$$= e^{xy\ln a} = a^{xy}.$$

■

The remaining proofs are left as exercises.[1]

The differentiation formula for exponential functions is also a consequence of

1) We can also define the number e from the exponent function. The number e is the number for which the tangent to a^x at $x=0$ is $y=x+1$. We can show this by means of an animation.
http://www.ima.umn.edu/~arnold//graphics.html

Definition [10] :

[13]

$$\frac{d}{dx}(a^x) = a^x \ln a$$

Proof $\frac{d}{dx}(a^x) = \frac{d}{dx}(e^{x \ln a}) = e^{x \ln a}\frac{d}{dx}(x \ln a)$

$= a^x \ln a$ ■

If $a > 1$, then $\ln a > 0$, so $(d/dx)a^x = a^x \ln a > 0$, which shows that $y = a^x$ is increasing (see Figure 8). If $0 < a < 1$, then $\ln a < 0$ and so $y = a^x$ is decreasing (see Figure 9).

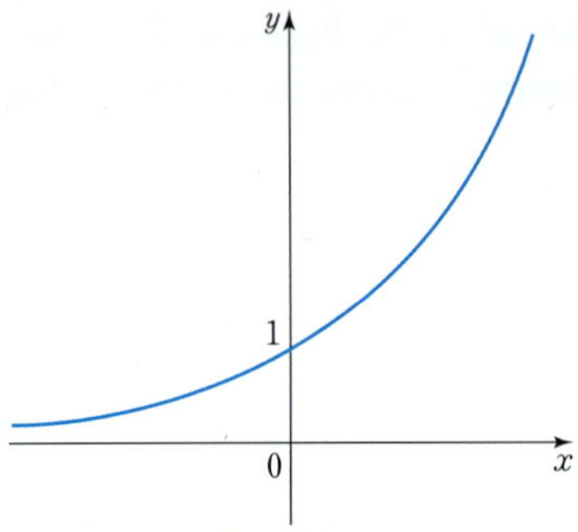

$\lim_{x \to -\infty} a^x = 0,\ \lim_{x \to \infty} a^x = \infty$

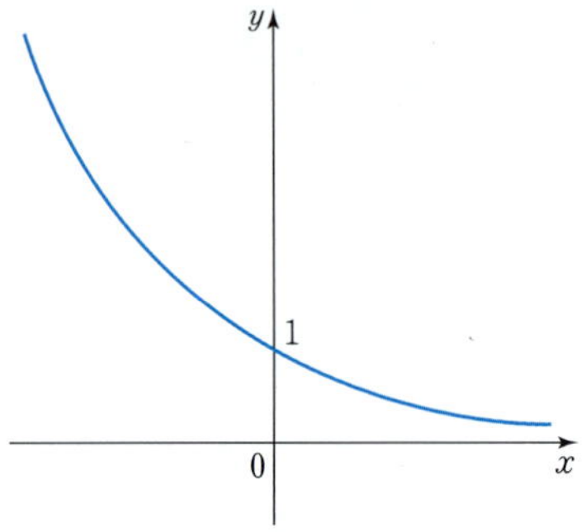

$\lim_{x \to -\infty} a^x = \infty,\ \lim_{x \to \infty} a^x = 0$

Figure 8　　Figure 9

http://matrix.skku.ac.kr/2011-BK-Coll/coll_20110512.htm

General Logarithmic Functions

If $a > 0$ and $a \neq 1$, then $f(x) = a^x$ is a one-to-one function. Its inverse function is called the **logarithmic function with base *a*** and is denoted by $\log_a$.

Thus

[14]

$$\log_a x = y \quad \Leftrightarrow \quad a^y = x$$

In particular, we see that $\log_e x = \ln x$.

The properties of logarithms are similar to those for the natural logarithm and can be deduced from the corresponding properties of exponential functions.

To differentiate $y = \log_a x$, we write the equation as $a^y = x$. From Equation [14] we have $y \ln a = \ln x$, so

$$\log_a x = y = \frac{\ln x}{\ln a}$$

Since $\ln a$ is a constant, we can differentiate as follows:

$$\frac{d}{dx}(\log_a x) = \frac{d}{dx}\frac{\ln x}{\ln a} = \frac{1}{\ln a}\frac{d}{dx}(\ln x) = \frac{1}{x \ln a}$$

15
$$\frac{d}{dx}(\log_a x) = \frac{1}{x \ln a}$$

The Number e Expressed as a Limit

In this section e is defined as the number such that $\ln e = 1$. It can also be defined as

16
$$e = \lim_{x \to 0}(1+x)^{1/x}$$

Proof Let $f(x) = \ln x$. Then $f'(x) = 1/x$, so $f'(1) = 1$. But, by the definition of derivative,

$$\begin{aligned} f'(1) &= \lim_{h \to 0}\frac{f(1+h)-f(1)}{h} = \lim_{x \to 0}\frac{f(1+x)-f(1)}{x} \\ &= \lim_{x \to 0}\frac{\ln(1+x)-\ln 1}{x} = \lim_{x \to 0}\frac{1}{x}\ln(1+x) \\ &= \lim_{x \to 0}\ln(1+x)^{1/x}. \end{aligned}$$

Since $f'(1) = 1$, we have $\lim_{x \to 0}\ln(1+x)^{1/x} = 1$.

Hence

$$e = e^1 = e^{\lim_{x \to 0}\ln(1+x)^{1/x}} = \lim_{x \to 0} e^{\ln(1+x)^{1/x}} = \lim_{x \to 0}(1+x)^{1/x}.$$ ■

5.6 EXERCISES (The Logarithm Defined as an Integral)

http://matrix.skku.ac.kr/Cal-Book/part1/CS-Sec-5-6-Sol.html

YouTube http://youtu.be/ymDImdIQ90c

CAS **1.** (a) By comparing areas, show that $\frac{1}{3} < \ln 1.5 < \frac{5}{12}$.

(b) Use the Midpoint Rule with $n = 10$ to estimate $\ln 1.5$.

http://matrix.skku.ac.kr/cal-lab/cal-RiemannSum.html

Solution

(a)

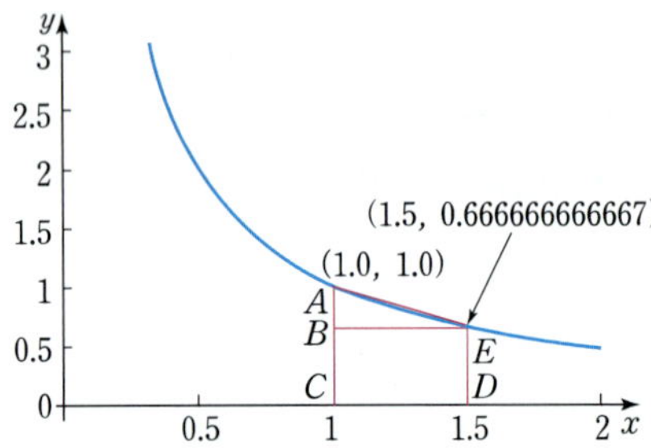

From above figure, we have

$$\square BCDE \le \int_1^{1.5} \frac{1}{x}dx \le \square ACDE.$$

Since $\square BCDE = \frac{1}{2} \cdot \frac{2}{3} = \frac{1}{3}$,

$$\square ACDE = \left(1 + \frac{2}{3}\right) \cdot \frac{1}{2} \cdot \frac{1}{2} = \frac{5}{12}$$

and $\int_1^{1.5} \frac{1}{x}dx = \ln 1.5$,

we have $\frac{1}{3} < \ln 1.5 < \frac{5}{12}$.

(b) Let $f(x) = \frac{1}{x}$. Then we get

$$\ln 1.5 = \int_1^{1.5} \frac{1}{x}dx$$
$$\approx (0.05)[f(1.025) + f(1.075) + \cdots + f(1.475)]$$
$$= (0.05)\left(\frac{1}{1.025} + \frac{1}{1.075} + \cdots + \frac{1}{1.475}\right) \approx 0.405.$$

2. By comparing areas, show that

$$\frac{1}{2} + \frac{1}{3} + \cdots + \frac{1}{n} < \ln n < 1 + \frac{1}{2} + \frac{1}{3} + \cdots + \frac{1}{n-1}$$

Solution Note that $\int_1^n \frac{1}{x}dx = \ln n$. Let $f(x) = \frac{1}{x}$. If we use the left endpoint rule with $n-1$ subintervals to estimate $\ln n$, then the ith height is $f(i) = \frac{1}{i}$. If we use the right endpoint rule with $n-1$ subinterval to estimate $\ln n$, the ith height is $f(i+1) = \frac{1}{i+1}$.

Since Δx for both rules is 1 and $\frac{1}{i+1} \le \frac{1}{i}$, we obtain the desired expressions.

3. (a) By comparing areas, show that $\ln 2 < 1 < \ln 3$.

(b) Deduce that $2 < e < 3$.

Solution (a) By similar approach as Example 1,

$$\ln 2 = \int_1^2 \frac{1}{x}dx \le \square ACEF = (1 + 0.5) \cdot \frac{1}{2} = \frac{3}{4}$$

and $\ln 3 = \int_1^3 \frac{1}{x}dx \ge \sum_{i=1}^{10} \frac{1}{5}\left(\frac{1}{1+0.2i}\right) \approx 1.035.$

Hence $\ln 2 < 1 < \ln 3$.

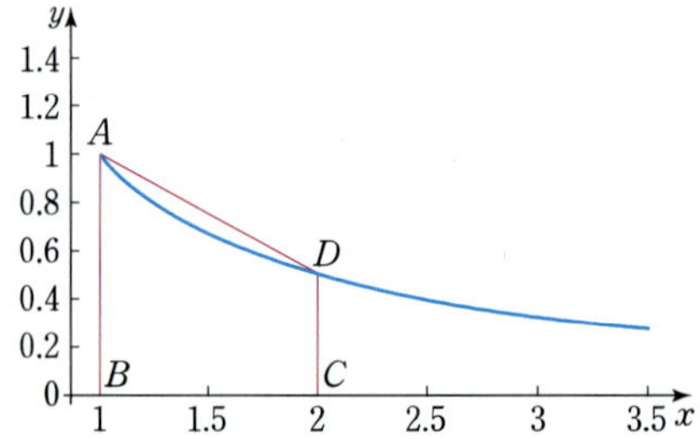

(b) Since $f(x) = e^x$ is an increasing function and $\ln 2 < 1 < \ln 3$, we have $e^{\ln 2} < e < e^{\ln 3}$. Therefore $2 < e < 3$.

4. Deduce the following laws of logarithms.

(a) $\log_a(xy) = \log_a x + \log_a y$

(b) $\log_a(x/y) = \log_a x - \log_a y$

(c) $\log_a(x^y) = y\log_a x$

Solution Let $\alpha = \log_a x$ and $\beta = \log_a y$.

Then $x = a^\alpha$ and $y = a^\beta$.

(a) $\log_a xy = \log_a a^\alpha a^\beta = \log_a a^{\alpha+\beta}$
$= \alpha + \beta = \log_a x + \log_a y$

(b) $\log_a x/y = \log_a a^\alpha / a^\beta = \log_a a^{\alpha-\beta}$
$= \alpha - \beta = \log_a x - \log_a y$

(c) $\log_a x^y = \log_a(a^\alpha)^y = \log_a a^{\alpha y}$
$= \alpha y = y\log_a x$

5. Show that $\lim_{n\to\infty}\sum_{i=1}^n \frac{1}{n} = \infty$ by using an integral.

Solution $\sum_{i=1}^n \frac{1}{n} \ge \int_1^n \frac{1}{x}dx = \ln n$ (See exercise 2.)

$$\Rightarrow \lim_{n\to\infty}\sum_{i=1}^n \frac{1}{n} \ge \lim_{n\to\infty}\int_1^n \frac{1}{x}dx = \lim_{n\to\infty}\ln n = \infty$$

6. Evaluate

$$\lim_{n\to\infty}\frac{1}{n}\left[\left(\frac{1}{n}\right)^9+\left(\frac{2}{n}\right)^9+\left(\frac{3}{n}\right)^9+\cdots+\left(\frac{n}{n}\right)^9\right].$$

Solution $\lim_{n\to\infty}\frac{1}{n}\left[\left(\frac{1}{n}\right)^9+\left(\frac{2}{n}\right)^9+\left(\frac{3}{n}\right)^9+\cdots+\left(\frac{n}{n}\right)^9\right]$

$$=\int_1^2 x^9dx=\left[\frac{1}{10}x^{10}\right]_1^2=102.4-0.1=102.3.$$

CAS **7.** Evaluate $\lim_{n\to\infty}\left(\frac{n!}{n^n}\right)^{1/n}$.

Solution Let $S_n=\left(\frac{n!}{n^n}\right)^{1/n}$.

Then $\ln S_n=\frac{1}{n}\left(\ln\frac{1}{n}+\ln\frac{2}{n}+\cdots+\ln\frac{n}{n}\right)$.

$$\lim_{n\to\infty}\ln S_n=\lim_{n\to 0}\frac{1}{n}\sum_{k=1}^{n}\left(\ln\frac{k}{n}\right)==-1$$

$$\int_0^1 \ln x dx=[x\ln x-x]_0^1$$

Hence we have $\lim_{n\to\infty}\left(\frac{n!}{n^n}\right)^{1/n}=e^{-1}$.

8. Evaluate $\lim_{n\to\infty}\sum_{i=0}^{n^2}\frac{\left(\ln\left(\frac{i}{n^2}\right)\right)^2}{\sqrt{\frac{i}{n^2}}}\frac{1}{n^2}$.

Solution The above limit can be written as a definite integral, namely $\int_0^1\frac{(\ln x)^2}{\sqrt{x}}dx$. To evaluate the integral, we first use the substitution $\ln x=y$. Then

$$\int_0^1\frac{(\ln x)^2}{\sqrt{x}}dx=\int_{-\infty}^0 y^2e^{\frac{y}{2}}dy=\lim_{k\to-\infty}\int_k^0 y^2e^{\frac{y}{2}}dy.$$

The last integral above is computed as follows:

$$\int_k^0 y^2e^{\frac{y}{2}}dy=\left[2y^2e^{\frac{y}{2}}\right]_k^0-4\int_k^0 ye^{\frac{y}{2}}dy$$

$$=-2k^2e^{\frac{k}{2}}-\left[8ye^{\frac{y}{2}}\right]_k^0+8\int_k^0 e^{\frac{y}{2}}dy$$

$$=(-2^2+8k-16)e^{\frac{k}{2}}+16.$$

Hence $\lim_{k\to-\infty}[(-2k^2+8k-16)e^{k/2}+16]=16$.

9. Find $\lim_{x\to3}\left(\frac{x}{x-3}\int_3^x\frac{\sin t}{t}dt\right)$.

Solution We know that $\frac{x}{x-3}\to\infty$ as $x\to3^+$ and $\frac{x}{x-3}\to-\infty$ as $x\to3^-$ while the second factor approaches $\int_3^3\frac{\sin t}{t}dt$, which is 0.

The integral $\int_3^x\frac{\sin t}{t}dt$ has x as its upper limit of integration (Part 1 the FTC) suggesting that differentiation might be involved. Then the denominator $(x-3)$ reminds us the definition of the derivative with $a=3$. So this becomes $F'(3)=\lim_{x\to3}\frac{F(x)-F(3)}{x-3}$. Thus define

$F(x)=\int_3^x\frac{\sin t}{t}dt$ then $F(3)=0$. Now

$$\lim_{x\to3}\left(\frac{x}{x-3}\int_3^x\frac{\sin t}{t}dt\right)$$

$$=\left(\lim_{x\to3}x\right)\cdot\left(\lim_{x\to3}\frac{\int_3^x\frac{\sin t}{t}dt}{x-3}\right)$$

$$=3\lim_{x\to3}\frac{F(x)-F(3)}{x-3}$$

$$=3F'(3)$$

$$=3\frac{\sin 3}{3}\text{ (By the FTC3)}$$

$$=\sin 3.$$

Note Alternatively you may use L'Hospital's Rule.
http://www.dnatube.com/video/19835/Lec-41--LHopitals-Rule-Example-3

Chapter 6

Applications of Integration

Seki KOWA

(関孝和) 1642-1708

Determinant

[Essentials of Mathematics]

A Japanese mathematician, Kowa created a new algebraic notation system and laid the foundation for the later development of early Japanese mathematics. He also had done some important work in calculus and integer indeterminate equations and is credited with the discovery of Bernoulli numbers, resultant and determinant.

YouTube http://youtu.be/o53phm5cqJE

The definite integral has numerous applications in the real life. For instance, it can be used to find the area bounded by curves, volumes of solids, the mass of a body with known density, the distance traveled by a particle (say, a car), work done by a force, average value of a function, just to name a few applications. In this chapter, we apply the definite integral to solve some of above mentioned problems.

6.1 Areas between Curves

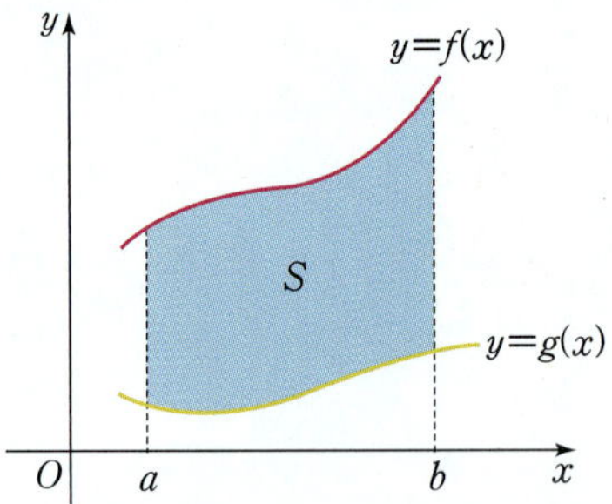

$f_1(x) \geq g(x)$ on $[a, b]$

Figure 1

$S = \{(x, y) | a \leq x \leq b, g(x) \leq y \leq f(x)\}$

As mentioned in Chapter 5, if $f(x)$ is a function on an integral $[a, b]$ such that $f(x) \geq 0$ for all x in $[a, b]$, then $\int_a^b f(x)dx$ is the area under the graph of $f(x)$.

Consider two continuous functions f and g defined on an interval $[a, b]$ such that $f(x) \geq g(x)$. Let S be the region that lies between the two given curves $y = f(x)$ and $y = g(x)$ and between the vertical lines $x = a$ and $x = b$. (See Figure 1.) We want to derive a formula for the area of S.

Suppose for a moment that $f(x) \geq g(x) \geq 0$ for all x in $[a, b]$. Then, if we denote by S_f the region under the graph of $f(x)$ and by S_g the region under the graph of $g(x)$, then we can easily see that

$$\begin{aligned} \text{Area of } S &= (\text{Area of } S_f) - (\text{Area of } S_g) \\ &= \int_a^b f(x)dx - \int_a^b g(x)dx = \int_a^b [f(x) - g(x)]\,dx. \end{aligned}$$

In the general case, when $f(x)$ and $g(x)$ are not necessarily non-negative, we can see that this formula remains true by shifting vertically both the graphs of $f(x)$ and of $g(x)$ by the same distance, so that both will be above the x -axis.

The area A of the region bounded by the curves $y = f(x)$ and $y = g(x)$ on $[a, b]$, and the lines $x = a$ and $x = b$ is

1 $$\begin{aligned} A &= [\text{area under } y = f(x)] - [\text{area under } y = g(x)] \\ &= \int_a^b f(x)dx - \int_a^b g(x)dx = \int_a^b [f(x) - g(x)]dx \end{aligned}$$

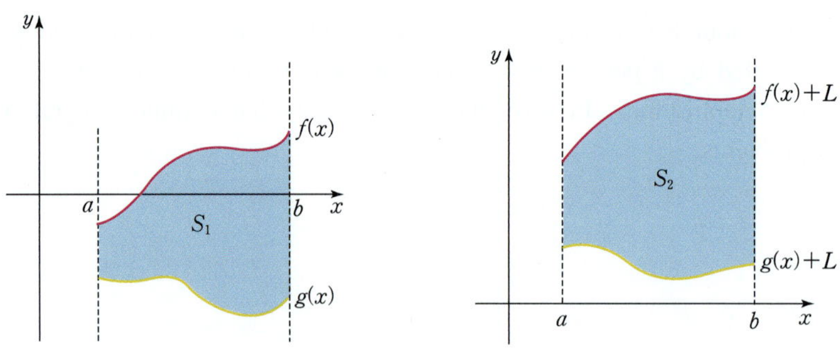

Figure 2

In the special case when $g(x) = 0$, S is the region under the graph of f(and above the x-axis). In general, if f and g are continuous functions with

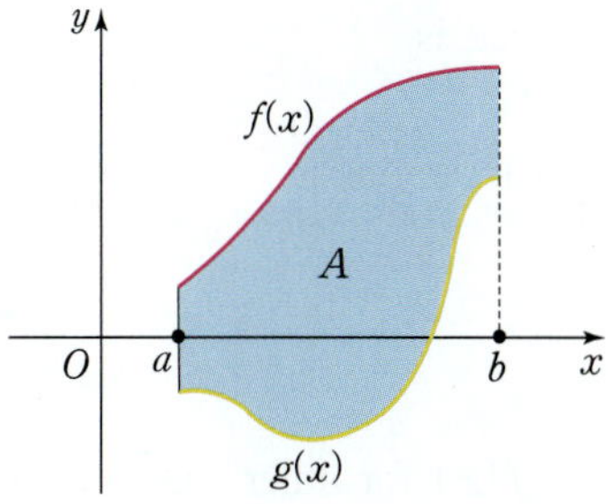

Figure 3 $A = \int_a^b f(x)dx - \int_a^b g(x)dx$

$f(x) \geq g(x)$ on the interval $[a, b]$ (see Figure 3), then the area A of the region bounded by the curves $y = f(x)$, $y = g(x)$ on $[a, b]$, the lines $x = a$ and $x = b$ is

$$A = \int_a^b [f(x) - g(x)]dx.$$

EXAMPLE 1

Determine the area of the region bounded above by $y = e^x$, bounded below by $y = \frac{1}{2}x$, and bounded on the sides by $x = 0$ and $x = 1$.

http://matrix.skku.ac.kr/cal-lab/cal-6-1-Exm-1.html

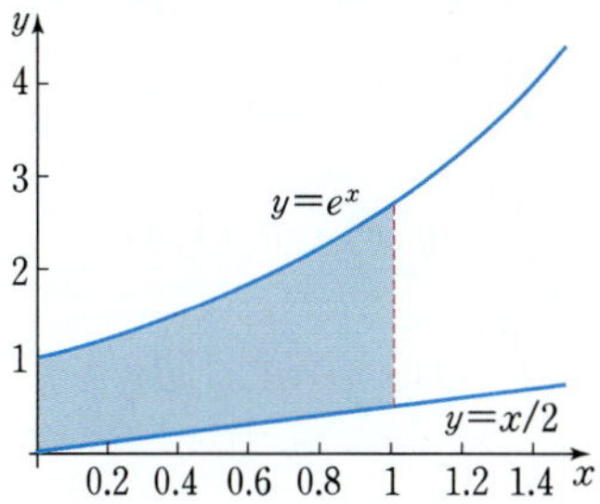

Figure 4

Solution Draw a typical approximating rectangle with width Δx (see Figure 4), with the upper boundary curve $y = f(x) = e^x$ and the lower boundary curve $y = g(x) = \frac{1}{2}x$, $a = 0$, and $b = 1$.

$$A = \int_0^1 \left(e^x - \frac{1}{2}x\right)dx = e^x - \frac{1}{4}x^2\Big]_0^1$$

$$= e - \frac{1}{4} - 1 = e - \frac{5}{4}.$$

```
f = exp(x)
p = plot(f, 0, 1, fill=x/2) + plot(f, 0, 1.5)+plot(x/2, 0, 1.5)
l1=line([(1, 1/2),(1, exp(1))], thickness=1.2, linestyle="--", color='red')
l2=line([(0, 0), (0, exp(1))], thickness=2, linestyle="--", color='red')
t1=text("$y=e^x$", (0.7, 3), fontsize=50, color='blue')
t2=text("$y= x/2$", (1.2, 0.3), fontsize=50, color='blue')
show(p+l1+l2+t1+t2)
integral(e^x - (1/2)*x, x, 0, 1)
```

Answer : e-5/4 ■

EXAMPLE 2

Calculate the area of the region enclosed by $y = x$ and $y = x^2$.

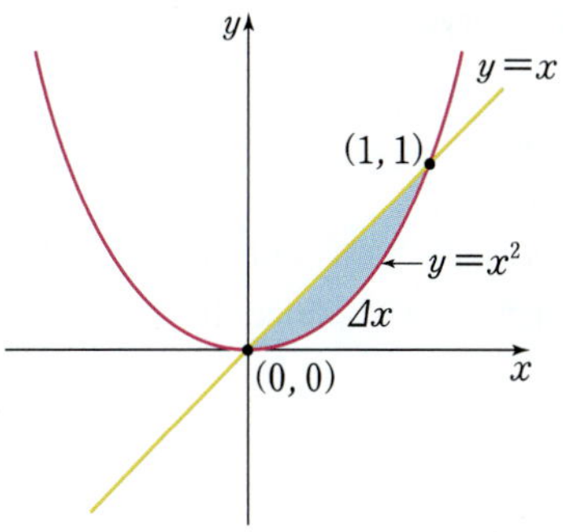

Figure 5

Solution First we need to find the intersection points of the line $y = x$ and parabola $y = x^2$, in order to know the bounds of the integral. The x coordinate of the intersection points satisfies both $y = x$ and $y = x^2$, hence it satisfies $x = x^2$, or $x^2 - x = 0$. Thus $x(x-1) = 0$, so $x = 0$ or 1. The points of intersection are $(0, 0)$ and $(1, 1)$. From Figure 5, the top and bottom boundaries are $y = x$ and $y = x^2$ respectively.

The area of a typical rectangle is $(x - x^2)\Delta x$ and the region lies between $x=0$ and $x=1$. So the total area is

$$A=\int_0^1 (x-x^2)dx=\left[\frac{1}{2}x^2-\frac{1}{3}x^3\right]_0^1$$
$$=\left(\frac{1}{2}-\frac{1}{3}\right)=\frac{1}{6}.$$ ■

If $f(x)\geq g(x)$ for some values of x but $g(x)\geq f(x)$ for other values of x, then the area between the curves $y=f(x)$ and $y=g(x)$ is obtained by splitting the given region S into several regions. In Figure 6, the region S is obtained by adding areas A_1, A_2, A_3 of S_1, S_2, S_3. We will use the following "word" formulas to ensure that we remember that the area is always the "larger" function minus the "smaller" function.

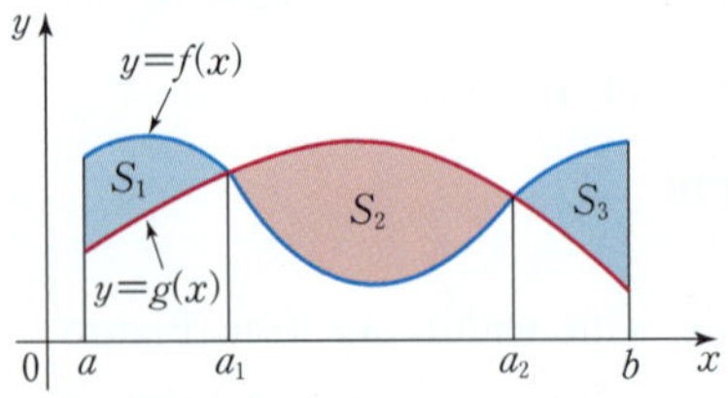

Figure 6

$$A_1=\int_a^{a_1}\left(\begin{matrix}\text{upper}\\ \text{function}\end{matrix}\right)-\left(\begin{matrix}\text{lower}\\ \text{function}\end{matrix}\right)dx=\int_a^{a_1}\{f(x)-g(x)\}dx$$
$$A_2=\int_{a_1}^{a_2}\left(\begin{matrix}\text{upper}\\ \text{function}\end{matrix}\right)-\left(\begin{matrix}\text{lower}\\ \text{function}\end{matrix}\right)dx=\int_{a_1}^{a_2}\{g(x)-f(x)\}dx$$
$$A_3=\int_{a_2}^{b}\left(\begin{matrix}\text{upper}\\ \text{function}\end{matrix}\right)-\left(\begin{matrix}\text{lower}\\ \text{function}\end{matrix}\right)dx=\int_{a_2}^{b}\{f(x)-g(x)\}dx$$

Then the area of the region S is defined to be the sum of the areas of the smaller regions S_1, S_2, S_3 that is, $A=A_1+A_2+A_3$. The area of the region S is defined to be the sum of the areas of the smaller regions S_1, S_2, $\cdots$, S_n that is, $A=A_1+A_2+\cdots+A_n$,

Since

2 $$|f(x)-g(x)|=\begin{cases} f(x)-g(x) & \text{when } f(x)\geq g(x)\\ g(x)-f(x) & \text{when } g(x)\geq f(x)\end{cases},$$

the area between the curves $y=f(x)$ and $y=g(x)$ and between $x=a$ and $x=b$ is given by

$$A=\int_a^b |f(x)-g(x)|dx.$$

EXAMPLE 3

Compute the area of the region bounded by the curves $y=x^2$, $y=(x-2)^2$, $x=0$ and $x=2$.

http://matrix.skku.ac.kr/cal-lab/cal-6-1-Exm-3.html

Solution The points of intersection of the curves $y=x^2$ and $y=(x-2)^2$ occur when $x^2=(x-2)^2$, that is, when $x=1$ (since $0\leq x\leq 1$). (See Figure 7.)

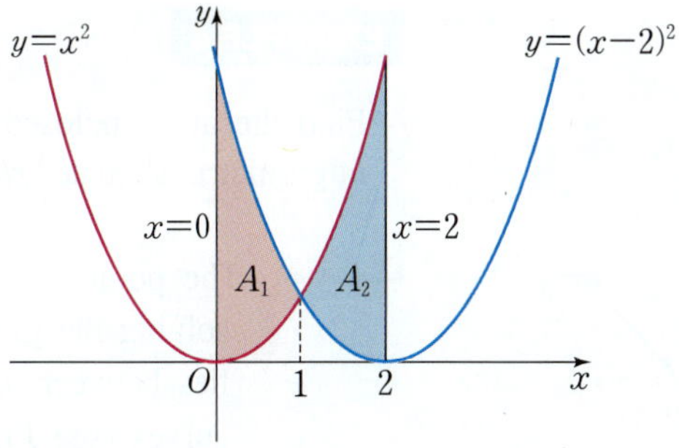

Figure 7

Note that $(x-2)^2 \geq x^2$ when $0 \leq x \leq 1$ but $x^2 \geq (x-2)^2$ when $1 \leq x \leq 2$. Therefore, the required area is

$$\begin{aligned} A = A_1 + A_2 &= \int_0^2 |(x-2)^2 - x^2| dx \\ &= \int_0^1 [(x-2)^2 - x^2] dx + \int_1^2 [x^2 - (x-2)^2] dx \\ &= [-2x^2 + 4x]_0^1 + [2x^2 - 4x]_1^2 \\ &= (-2+4+0-0) + (8-8-2+4) \\ &= 4. \end{aligned}$$

```
f = x^2
g = (x-2)^2
integral(abs( f- g), x, 0, 2)
```

Answer : 4

(The region is symmetric about $x = 1$,

so $A = 2A_1 = 2\int_0^1 [(x-2)^2 - x^2] dx$.)

■

Suppose that f and g are continuous functions with $f(y) \geq g(y)$ on the interval $[c, d]$. The area A of the region bounded by the curves $x = f(y)$, $x = g(y)$ on $[c, d]$, and the lines $y = c$, $y = d$ is

$$A = \int_c^d [f(y) - g(y)]\, dy. \text{ (See Figure 8 and 9.)}$$

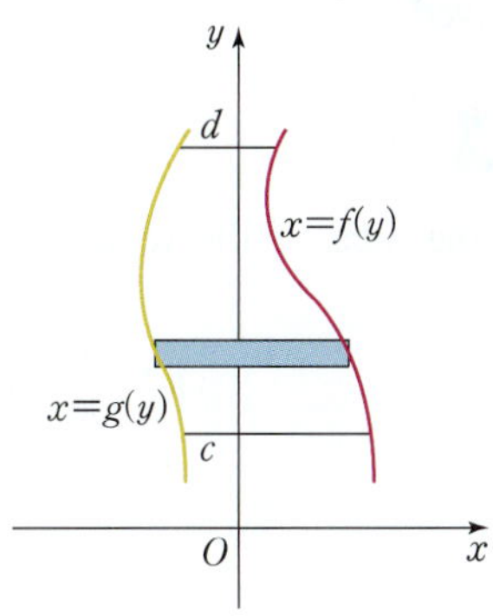

Figure 8

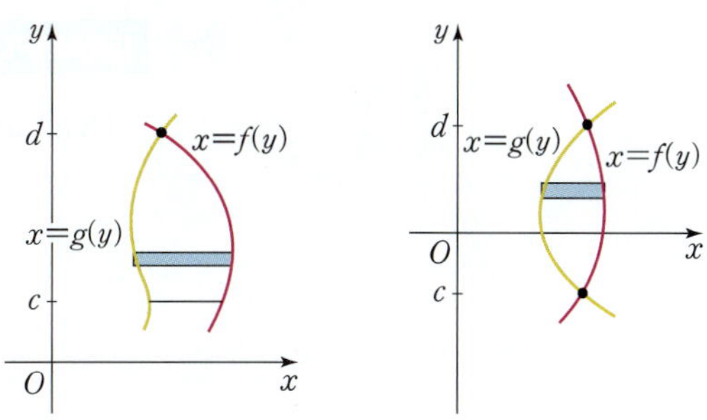

Figure 9

EXAMPLE 4

Find the area enclosed by the parabola $x=2y^2$ and the line $y=x-1$.
http://matrix.skku.ac.kr/cal-lab/cal-6-1-Exm-4.html

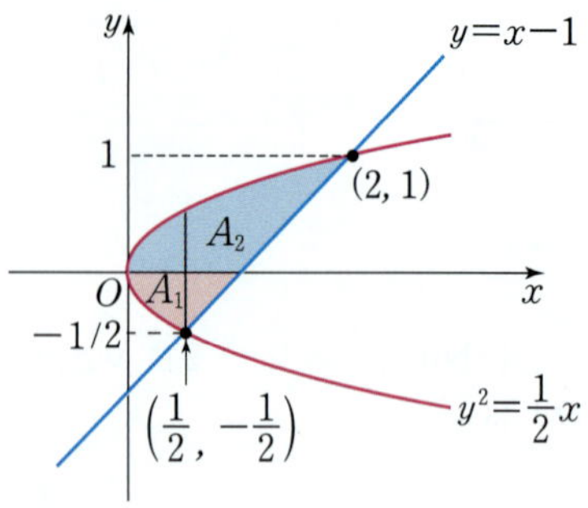

Figure 10

Solution The points of intersection $(1/2,\ -1/2)$ and $(2,\ 1)$ are obtained by solving the given two equation simultaneously. The region of interest lies between $y=-1/2$ and $y=1$ and with left and right boundary curves (see Figure 10) given by, respectively:

$$f(y)=2y^2,\quad g(y)=y+1.$$

Thus

$$\begin{aligned} A &= \int_{-\frac{1}{2}}^{1}(g(y)-f(y))\,dy \\ &= \int_{-\frac{1}{2}}^{1}(y+1-2y^2)dy \\ &= \left[-\frac{2}{3}y^3+\frac{1}{2}y^2+y\right]_{-\frac{1}{2}}^{1} \\ &= \left(-\frac{2}{3}(1)+\frac{1}{2}(1)+1\right)-\left(-\frac{2}{3}\left(-\frac{1}{8}\right)+\frac{1}{2}\left(\frac{1}{4}\right)+\left(-\frac{1}{2}\right)\right) \\ &= \frac{9}{8} \end{aligned}$$

```
var('x, y')
f(x) = 2*y^2
g(x) = y+1
solve([f==g], y)
integral(abs( f- g), y, -1/2, 1).n()
```

Answer : {9} over {8} =1.12500 ■

Caution In this example, if we try to find the area by integrating with respect to x instead of with respect to y, the region needs to be split into two areas labeled A_1 and A_2 (see Figure 10) with more calculations involved.

CAS **EXAMPLE 5**

Determine the area of the region bounded by the curves:

$$y=\frac{1}{x},\ y=x,\ x=2.$$

Solution

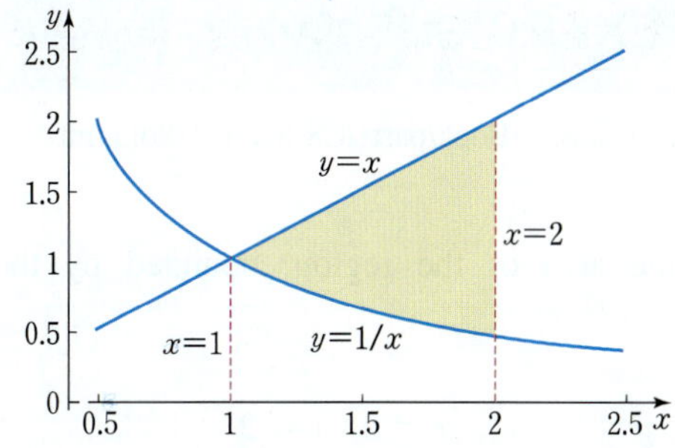

Figure 11

```
integral(x - 1/x, x, 1, 2)
```

Answer : -log(2) + 3/2 ■

CAS **EXAMPLE 6**

Find the area between the curves $y = 2x^2 + 10$ and $y = 4x + 16$ between $x = -2$ and $x = -5$.

http://matrix.skku.ac.kr/cal-lab/cal-6-1-example-6.html

Solution

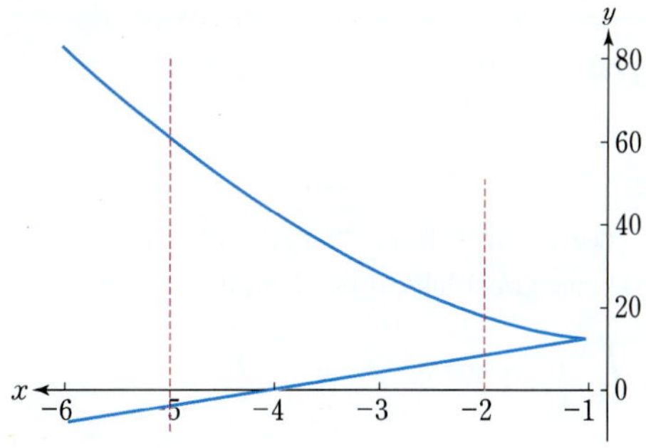

Figure 12

```
var('x')
f(x)= 2*x^2 +10
g(x)= 4*x +16
L1=line([(-2,0),(-2,50)], color="red", linestyle='--')
L2=line([(-5,-10),(-5,80)], color="red", linestyle='--')
P1=plot(f,(x,-6,-1))
P2=plot(g,(x,-6,-1), color="green")
show(P1+P2+L1+L2)
integral((2*x^2 +10) - (4*x +16), x, -5, -2)
```

Answer : 102 ■

6.1 EXERCISES (Areas between Curves)

http://matrix.skku.ac.kr/Cal-Book/part1/CS-Sec-6-1-Sol.html

1-21. Find the area of the region, bounded by the given curves.

1. $y = x - 1,\ y = 5 - x^2,\ x = 0,\ x = 2$
http://matrix.skku.ac.kr/cal-lab/cal-6-1-1.html

Solution $A = \int_0^2 [(5 - x^2) - (x - 1)]dx = \frac{22}{3}$

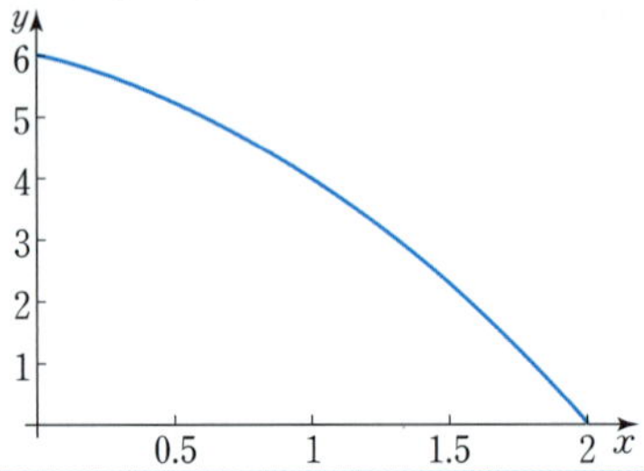


```
f(x)=abs((5-x^2)-(x-1))
plot(f(x),x,0,2)
```

```
print integral(f(x),x,0,2)
```

Answer : 22/3

2. $y = e^x,\ y = \cos x,\ x = 0,\ x = \pi/2$
http://matrix.skku.ac.kr/cal-lab/cal-6-1-2.html

Solution $A = \int_0^{\frac{\pi}{2}} (e^x - \cos x)dx = e^{\frac{\pi}{2}} - 2$

```
f(x)=exp(x)-cos(x)
plot(f(x),x,0,pi/2)
```

```
print integral(f(x),x,0,pi/2)
```

Answer : $e^{\pi/2} - 2$

3. $y = x,\ y^2 = x$
http://matrix.skku.ac.kr/cal-lab/cal-6-1-3.html

Solution $A = \int_0^1 (\sqrt{x} - x)dx = \frac{1}{6}$

```
f(x)=sqrt(x)-x
plot(f(x),x,0,1)
```

```
print integral(f(x),x,0,1)
```

Answer : 1/6

4. $y = x^3,\ y = x^5$

Solution $A = \int_1^2 (x^5 - x^3)dx = \frac{17}{12}$

5. $y = 1/x,\ y = 1/x^2,\ x = 3$

Solution $A = \int_1^3 \left(\frac{1}{x} - \frac{1}{x^2}\right)dx = -\frac{2}{3} + \log 3$

6. $y = \sqrt{x},\ y = (2 - x)/2$

Solution $A = \int_{4-2\sqrt{3}}^{4+2\sqrt{3}} \left[\left(1 - \frac{x}{2}\right) - \sqrt{x}\right]dx = \frac{40}{3} - 4\sqrt{3}$

7. $y = x,\ y = \sqrt[3]{x}$

Solution $A = \int_0^3 (\sqrt[3]{x} - x)dx = \frac{9}{4}\sqrt{3} - \frac{9}{2}$

8. $y = 4 - x^2,\ y = x^2 - 2$

Solution $A = \int_{-\sqrt{3}}^{\sqrt{3}} [(4 - x^2) - (x^2 - 2)]dx = 8\sqrt{3}$

9. $y = x^2 - x,\ y = 2x$

Solution $A = \int_0^3 [2x - (x^2 - x)]dx = \frac{9}{2}$

10. $y = \sqrt{x},\ y = (1/3)x,\ x = 12$

Solution $A = \int_0^9 \left(\sqrt{x} - \frac{1}{3}x\right)dx = \frac{9}{2}$

11. $y = 4 - x^2,\ y = x^2,\ x = -2,\ x = 2$

Solution $A = \int_{-2}^{-\sqrt{2}} [x^2 - (4 - x^2)]dx$

$+ \int_{-\sqrt{2}}^{\sqrt{2}} [(4 - x^2) - x^2]dx + \int_{\sqrt{2}}^{2} [x^2 - (4 - x^2)]dx$

$= \frac{8}{3}\sqrt{2} - \frac{16}{3}$

12. $x = y^2,\ x + y = 2$
http://matrix.skku.ac.kr/cal-lab/cal-6-1-12.html

Solution $A = \int_{-1}^{2} [(2 - y) - y^2]dy = \frac{9}{2}$

$(y - 2) - y^2 = 0$ has solutions $y = -2,\ y = 1$, so $(y - 2) - y^2$ is positive where $-2 \le y \le 1$.

```
var('x')
f(x)=sqrt(x)
g(x)=-sqrt(x)
h(x)=-x+2
P = plot(f,x, xmin=0, xmax=5, ymax=3, ymin=-3, color='blue')
Q = plot(g,x, xmin=0, xmax=5, ymax=3, ymin=-3, color='green')
R = plot(h,x, xmin=-1, xmax=5, ymax=3, ymin=-3, color='red', linestyle="--")
show(P+Q+R)
```

```
F(x)=(x+2)-x^2
print integral(F(x), x, -1,2)     # 9/2
plot(F(x), x, -1, 2)
```

13. $x+y^2=3,\ x=y$

Solution $A=\int_{\frac{-1-\sqrt{13}}{2}}^{\frac{-1+\sqrt{13}}{2}}[(3-y^2)-y]dy=\frac{\sqrt{13}}{6}$

14. $y=1-x^2,\ y=x^2-1$

Solution $A=\int_{-1}^{1}[(1-x^2)-(x^2-1)]dx=\frac{8}{3}$

15. $y=x\sin(\pi/4),\ y=x$

Solution $A=\int_{0}^{1}\left(x-\frac{\sqrt{2}}{2}x\right)dx=\frac{2-\sqrt{2}}{4}$

16. $y=\cos x,\ y=\cos 2x,\ x=0,\ x=\pi/2$

Solution $A=\int_{0}^{\frac{\pi}{2}}(\cos x-\cos 2x)dx=1$

17. $y=\sin x,\ y=\sin 2x,\ x=0,\ x=\pi/2$

Solution $A=\int_{0}^{\frac{\pi}{3}}(\sin 2x-\sin x)dx$

$+\int_{\frac{\pi}{3}}^{\frac{\pi}{2}}(\sin x-\sin 2x)dx=1$

18. $y=\cos x,\ y=1-(x/\pi)$

Solution $A=\int_{0}^{0.66}\left[\cos x-\left(1-\frac{x}{\pi}\right)\right]dx$

$=\frac{-0.0468831480266\pi+0.2178}{\pi}$

19. $y=|x|,\ y=x^2-1$

Solution $A=2\int_{0}^{\frac{1+\sqrt{5}}{2}}[x-(x^2-1)]dx=\sqrt{5}+2$

20. $y=x^2,\ y=1/(x^2+1)$

Solution $A=\int_{-0.79}^{0.79}\left(\frac{1}{x^2+1}-x^2\right)dx=1.00853446919$

21. $y=\cos(\pi x),\ y=x^2-x,\ x=3$

Solution $A=\int_{\frac{1}{2}}^{1}[\cos(\pi x)-(x^2-x)]dx=\frac{1}{24}-\frac{1}{\pi}$

22-23. Find the integral and interpret it as the area of a region.

22. $\int_{-1}^{1}|x^2-x|dx$

Solution $A=\int_{-1}^{0}(x^2-x)dx+\int_{0}^{1}[-(x^2-x)]dx=1$

23. $\int_{0}^{2}|\sqrt{x}-x|dx$

Solution $A=\int_{0}^{1}(\sqrt{x}-x)dx+\int_{1}^{2}(x-\sqrt{x})dx$

$=\frac{10}{3}-\frac{4}{3}\sqrt{2}$

24-25. Approximate the area of the region bounded by the given curves using the Midpoint Rule with $n=4$.
http://matrix.skku.ac.kr/cal-lab/cal-Midpoint.html

24. $y=\sin^2(\pi/2)x,\ y=\cos^2(\pi/2)x,\ -1\le x\le 1$

Solution $A=2\int_{0}^{1}\sin^2\left(\frac{\pi}{2}\right)x\,dx=1$

25. $y=\sqrt{9-x^2},\ y=x,\ x=0$
http://matrix.skku.ac.kr/cal-lab/cal-6-1-25.html

Solution $A=\int_{0}^{3}(x-\sqrt{9-x^2})dx=\frac{9}{2}-\frac{9}{4}\pi$

```
var('x')
f(x) = x - sqrt (9 - x^(2) )
integral(f, x, 0, 3)
```

Answer : 9/2 - 9/4*pi

26-29. Determine the area of the region bounded by the curves.

26. $y = x^3,\ y = 2\sin x$

Solution $A = \int_0^{1.24} (2\sin x - x^3)dx = 0.759353991122$

27. $y = x^2,\ y = x - x^3$

Solution $A = \int_0^{\frac{-1+\sqrt{5}}{2}} [(x - x^3) - x^2]dx = \frac{13}{24} - \frac{5}{24}\sqrt{5}$

28. $y = x\cos x,\ y = x^2$

Solution $A = \int_0^{0.74} (x\cos x - x^2)dx = 0.102367464813$

29. $y = e^x,\ y = 1 - x^2$

Solution $A = \int_{-0.71}^{0} [(1 - x^2) - e^x]dx = 0.08235$

---

30. Find the area of region defined by the inequalities $x \ge (y+2)^2,\ x \ge |y|$.

Solution $A = \int_{-4}^{-1} [(y+2)^2 - y]dy = \frac{21}{2}$

31. Find the area enclosed by the loop of the curve with equation $y^2 = x^2(x+1)$.
(Tschirnhausen's cubic.)
http://en.wikipedia.org/wiki/Tschirnhausen_cubic

Solution $A = 2\int_{-1}^{0} \sqrt{x^2(x+1)}\,dx = \frac{8}{15}$

32. Find the area of the region bounded by the curve $y = x^3$, the tangent line to this curve at $(2, 8)$, and the x-axis.
http://matrix.skku.ac.kr/cal-lab/cal-6-1-32.html

Solution $A = \int_0^{\frac{4}{3}} x^3 dx + \int_{\frac{4}{3}}^{2} [x^3 - (12x - 16)]dx = \frac{4}{3}$

```
var('x')
f(x)=12*x-16
g(x)=x^3
print integral(g(x), x, 0, 4/3)+ integral(g(x)-f(x), x, 4/3,2)
plot(f(x), 0,3, ymax=15)+plot(g(x), 0, 3, ymax=15, color='red')
```

Answer : 4/3

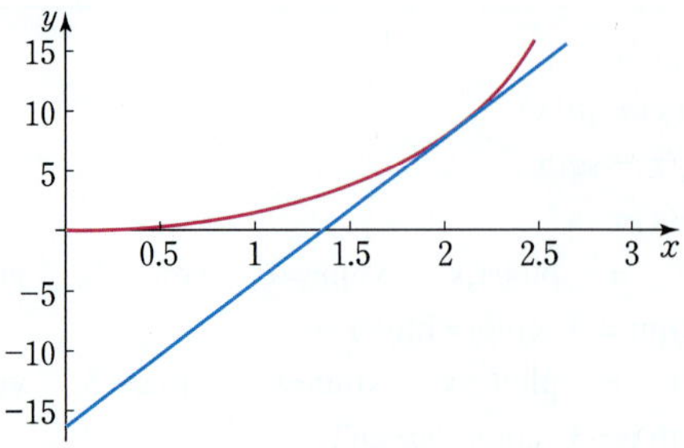

33. Find the number b such that the line $y = b$ divides the region bounded by the curves $y = x^2$ and $y = 3$ into two regions with equal area.

Solution $\int_0^b \sqrt{y}\,dy = \int_b^3 \sqrt{y}\,dy$, $b = \left(\frac{3}{2}\sqrt{3}\right)^{\frac{2}{3}}$

34. (a) Find the number a such that the line $x = a$ bisects the area under the curve $y = 1/x^2,\ 2 \le x \le 5$.

(b) Find the number b such that the line $y = b$ bisects the area in part (a).

Solution $\int_2^a \frac{1}{x^2}dx = \int_a^5 \frac{1}{x^2}dx,\ a = \frac{20}{7}$

$\int_b^1 \left(\frac{1}{\sqrt{y}} - 1\right)dy = \frac{3}{20},\ b = \frac{23}{20} - \frac{\sqrt{15}}{5}$

35. Find the values of c such that the area of the region enclosed by the parabolas $y = x^2 - c^2$ and $y = c^2 - x^2$ is 1944.

Solution $\int_{-c}^{c} (c^2 - x^2)dx - \int_{-c}^{c} (x^2 - c^2)dx = 1944$,

$c^3 = 729,\ c = 9$

For $y = x^2 - c^2$ and $y = c^2 - x^2$, $c = -9$ is another solution.

Therefore, $c = -9,\ 9$.

36-40. Find the area of the region bounded by the given curves.

36. $y = x^2 - x - 12,\ y = 0.$

Solution Since $y = x^2 - x - 12 = (x-4)(x+3)$,

$\int_{-3}^{4} (-x^2 + x + 12)dx = -\left[\frac{1}{3}x^3 - \frac{1}{2}x^2 - 12x\right]_{-3}^{4}$

$= \left(-9 - \frac{9}{2} + 36\right) - \left(\frac{64}{3} - 8 - 48\right)$

$= \frac{343}{6}.$

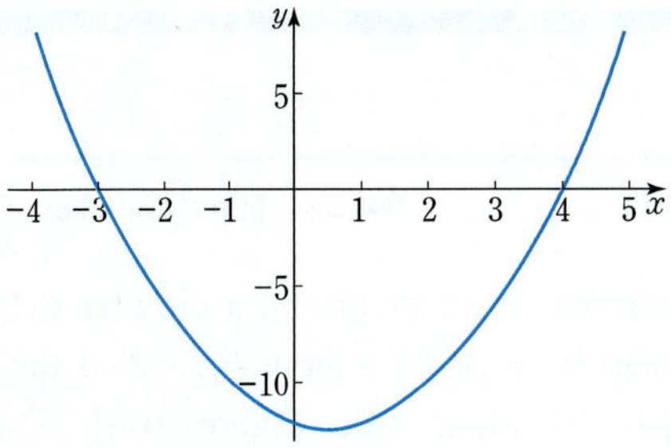

37. $y=12-x^2,\ y=x^2-8.$

Solution Since $x^2-8-12+x^2=0$

$$\Rightarrow 2x^2-20=0$$
$$\Rightarrow (x-\sqrt{10})(x+\sqrt{10})=0,$$

$$\int_{-\sqrt{10}}^{\sqrt{10}}(12-x^2-x^2+8)dx=2\int_0^{\sqrt{10}}(20-2x^2)dx$$
$$=4\left[10x-\frac{1}{3}x^3\right]_0^{\sqrt{10}}$$
$$=4\left(10\sqrt{10}-\frac{10\sqrt{10}}{3}\right)$$
$$=\frac{80\sqrt{10}}{3}.$$

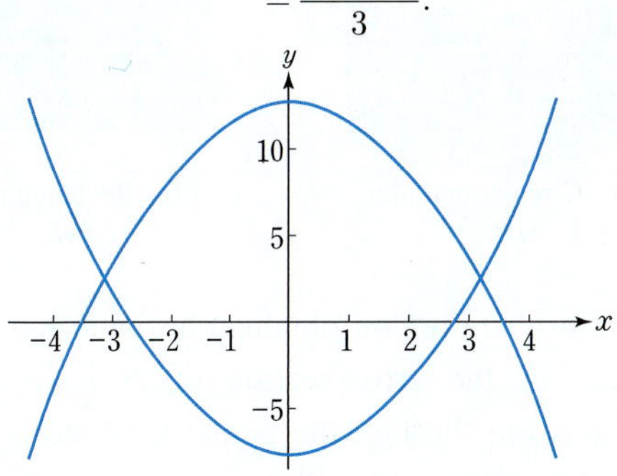

38. $x-2y=0,\ x=y^2+3y.$

Solution Since $y^2+3y=2y$

$$\Rightarrow y^2+y=y(y+1)$$
$$\Rightarrow y=-1 \text{ or } y=0,$$

$$\int_{-1}^{0}(2y-y^2-3y)dy=\int_{-1}^{0}(-y^2-y)dy$$
$$=-\left[\frac{1}{3}y^3+\frac{1}{2}y^2\right]_{-1}^{0}=\left(-\frac{1}{3}+\frac{1}{2}\right)=\frac{1}{6}.$$

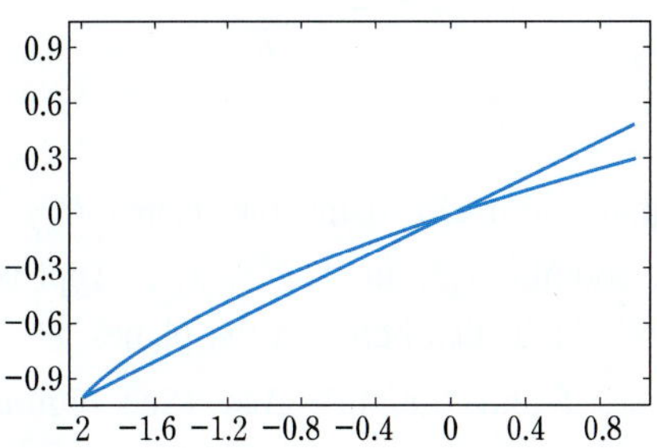

39. $y=\sin\frac{\pi x}{2},\ y=x^2-2x$

Solution $\displaystyle\int_0^2\left(\sin\frac{\pi x}{2}-x^2+2x\right)dx=\left[-\frac{2}{\pi}\cos\frac{\pi x}{2}\right]_0^2$

$$=\frac{2}{\pi}-\left(-\frac{2}{\pi}\right)=\frac{4}{\pi}.$$

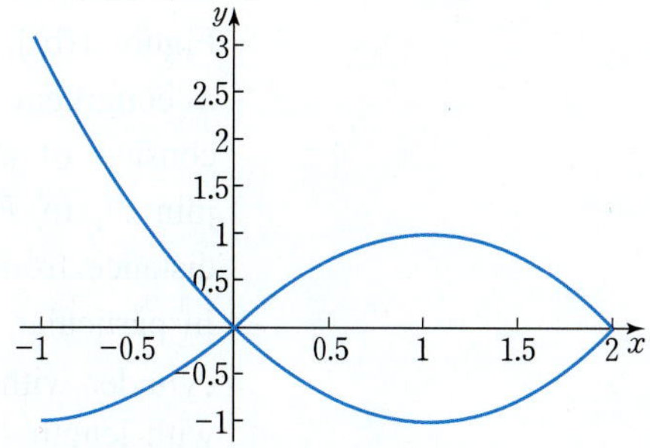

40. $y=\sqrt{2x},\ y=x^2,\ x=2$

http://matrix.skku.ac.kr/cal-lab/cal-6-1-40.html

Solution $\displaystyle\int_{2^{1/3}}^{2}(x^2-\sqrt{2x})dx=\left[\frac{1}{3}x^3-\frac{2\sqrt{2}}{3}x^{3/2}\right]_{2^{1/3}}^{2}$

$$=\left(\frac{8}{3}-\frac{8}{3}\right)-\left(\frac{2}{3}-\frac{4}{3}\right)$$
$$=\frac{2}{3}$$

```
var('x')
f(x) = sqrt (2*x)
g(x) = x^2
integral(g-f, x, 2^(1/3), 2)
```

Answer : 2/3

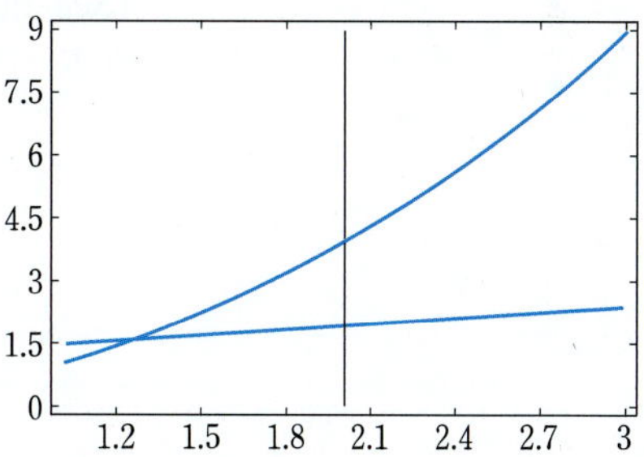

6.2 Volumes

YouTube http://youtu.be/4-ChOAFbJAs

Consider a solid called a **cylinder** (or, more precisely, a circular cylinder) [see Figure 1(b)]. A cylinder is bounded by a plane region B_1, called the **base**, and a congruent region B_2 in a parallel plane [see Figure 1(a)]. The cylinder consists of all points on line segments that are perpendicular to the base and join B_1 to B_2. If the area of the base is A and the height of the cylinder (the distance from B_1 to B_2) is h, then the volume V of the cylinder is $V = Ah$. In particular, if the base is a circle with radius r, then the cylinder is a circular cylinder with volume $V = \pi r^2 h$ [see Figure 1(b)], and if the base is a rectangle with length l and width w, then the cylinder is a rectangular box (also called a rectangular parallelepiped) with volume $V = lwh$ [see Figure 1(c)].

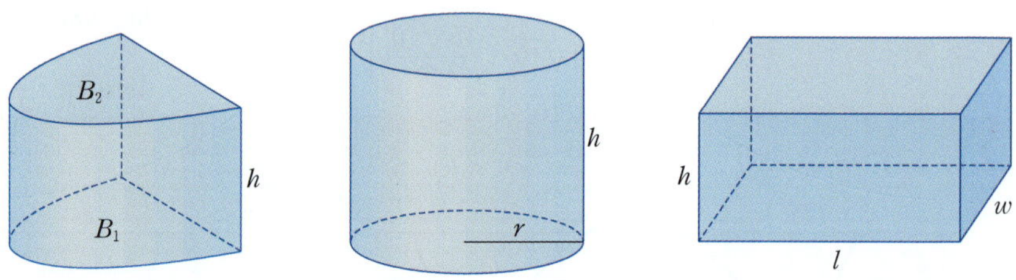

Figure 1 (a) Cylinder $V=Ah$ (b) Circular cylinder $V=\pi r^2 h$ (c) Rectangular box $V=lwh$

A **cross-section** of a solid S is a plane region obtained by intersecting S with a plane. Let $A(x)$ be the area of the cross-section of S in a plane P_x perpendicular to the x-axis and passing through the point x, where $a \le x \le b$. (See Figure 2.) The cross-sectional area $A(x)$ will vary as x increases from a to b.

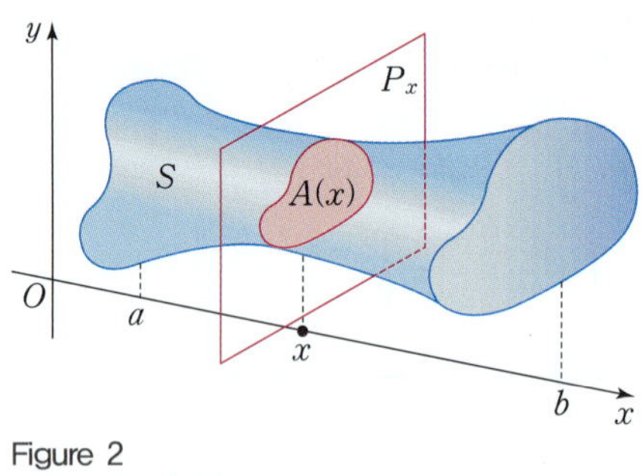

Figure 2

Divide S into n "slabs" of equal width Δx using the planes P_{x_1}, P_{x_2}, $\cdots$ to slice the solid. Choose sample points x_i^* in $[x_{i-1}, x_i]$. Approximate the volume of ith slab S_i (the part of S that lies between the planes $P_{x_{i-1}}$ and P_{x_i}) by a cylinder with base area $A(x_i^*)$ and "height" Δx. (See Figure 3.)

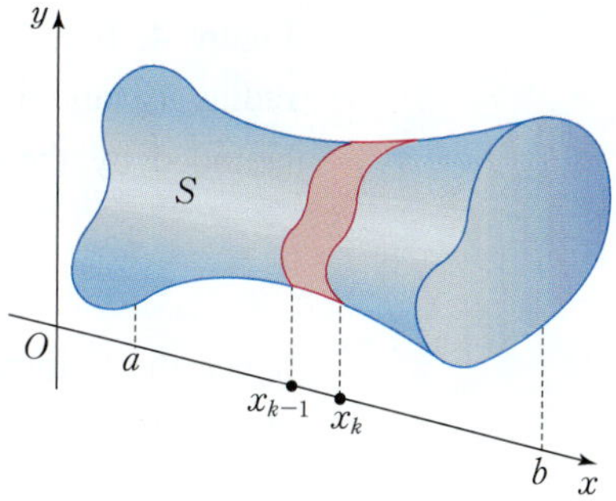

Figure 3

The volume of this cylinder is $A(x_i^*)\Delta x$. So an approximation of the volume of the ith slab S_i is $V(S_i) \approx A(x_i^*)\Delta x$. Adding the volumes of these slabs, we get an approximation to the total volume $V \approx \sum_{i=1}^{n} A(x_i^*)\Delta x$. As $n \to \infty$ the volume is the limit of a Riemann sum.

Definition of Volume

Let S be a solid that lies between $x=a$ and $x=b$. If the moving cross-sectional area of S in the plane P_x, through x and perpendicular to the x-axis, is $A(x)$, where A is a continuous function, then the **volume** of S is

$$V = \lim_{n\to\infty}\sum_{i=1}^{n} V(S_i) = \lim_{n\to\infty}\sum_{i=1}^{n} A(x_i^*)\Delta x = \int_a^b A(x)dx.$$

For a cylinder, the cross-sectional area is constant: $A(x)=A$ for all $x \in \mathrm{R}$. So, the definition of volume for a cylinder gives $V = \int_a^b A dx = A(b-a)$, which agrees with the formula $V = Ah$.

Solid of Revolution

A solid of revolution is a solid obtained by revolving a plane region about a line. (See Figure 4.) The volume of a solid of revolution is calculated using the formula if the cross-sections are perpendicular to the x-axis

$$V = \int_a^b A(x)dx$$

or if the cross-sections are perpendicular to the y-axis

$$V = \int_a^b A(y)dy.$$

Note that the cross-section $A(x)$ is either a disk or a ring, as shown in

Figure 4. If it is a disk with radius r, then $A=\pi r^2$. If it is a ring with outer radius r_1 and inner radius r_2, then $A=\pi r_1^2-\pi r_2^2$. Thus, if R is the region bounded by the curve $y=f(x)$, the lines $x=a$, $x=b$, and the x-axis, then

$$A(x)=\pi[f(x)]^2,$$

hence the volume V of the solid obtained by revolving the region R about the x-axis is

$$V=\int_a^b \pi[f(x)]^2\,dx. \text{ (See Figure 4 (a), (b).)}$$

If R is region bounded by $y=f(x)$, $y=g(x)$, and the lines $x=a$, $x=b$, with $f(x)\geq g(x)$ then the volume V of the solid obtained by revolving the region R about the x-axis is

$$V=\int_a^b \{\pi[f(x)]^2-\pi[g(x)]^2\}\,dx. \text{ (See Figure 4 (c).)}$$

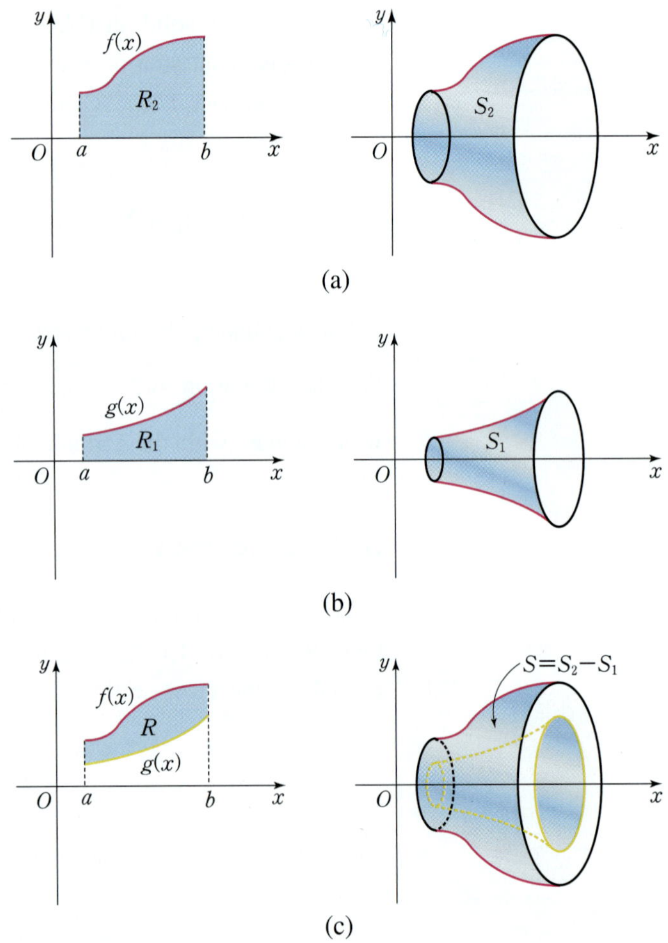

Figure 4 http://www.math.psu.edu/dlittle/java/calculus/volumedisks.html

EXAMPLE 1

Calculate the volume of the solid obtained by rotating the region bounded by the curves $y = e^x$ and $y = 0$ about the line $x = 1$ and $x = 2$ about the x -axis. [See Figure 4(b).]

Solution The cross-sectional area is $A(x) = \pi(e^x)^2$ and the volume of the solid is $V = \int_1^2 A(x)dx = \int_1^2 \pi(e^x)^2 dx = \frac{\pi}{2}(e^4 - e^2)$. ■

EXAMPLE 2

Find the volume of a sphere of radius r.

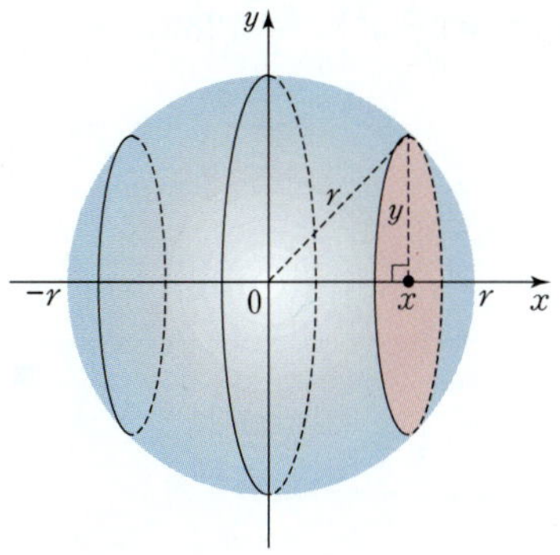

Figure 5

Solution Assume that the center of the sphere is at the origin. (See Figure 5.) Then the plane P_x intersects the sphere in a circle whose radius is $y = \sqrt{r^2 - x^2}$ So the cross-sectional area is

$$A(x) = \pi y^2 = \pi(r^2 - x^2).$$

From the definition of volume with $a = -r$ and $b = r$, we have

$$\begin{aligned} V &= \int_{-r}^{r} A(x)dx = 2\int_0^r \pi(r^2 - x^2)dx \\ &= 2\pi\int_0^r (r^2 - x^2)dx \\ &= 2\pi\left[r^2 x - \frac{x^3}{3}\right]_0^r = 2\pi\left(r^3 - \frac{r^3}{3}\right) \\ &= \frac{4}{3}\pi r^3. \end{aligned}$$

■

EXAMPLE 3

Compute the volume of the solid obtained by rotating the region bounded by $y = x^2$, $y = 16$, and $x = 0$ about the y-axis.
http://matrix.skku.ac.kr/cal-lab/cal-6-2-Exm-9.html

Solution The region bounded by $y = x^2$, $y = 16$, and $x = 0$ is shown in Figure 6(a). The resulting solid of revolution obtained by rotating this region about the y-axis is shown in Figure 6(b). Since the region is rotated about the y-axis, slice the solid perpendicular to the y-axis and integrate with respect to y. A slice at height y is a circular disk with radius x, where $x = \sqrt{y}$. So the area of a cross-section through y is $A(y) = \pi x^2 = \pi(\sqrt{y})^2 = \pi y$ and the volume of the approximating cylinder [shown in Figure 6(b)] is $A(y)\Delta y = \pi y \Delta y$. Since the solid lies between $y = 0$ and $y = 16$, its volume is

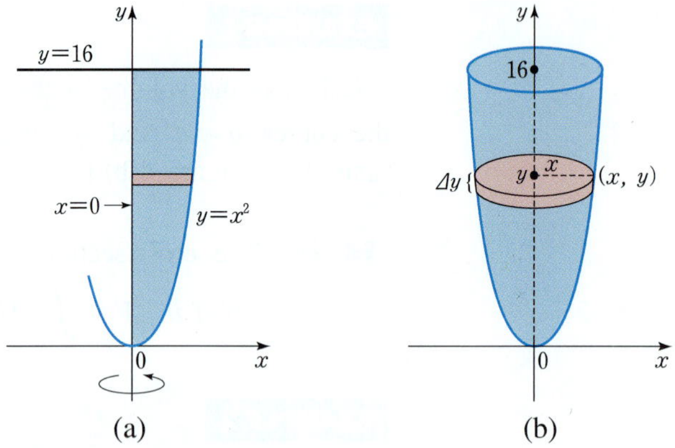

Figure 6

$$V = \int_0^{16} A(y)dy = \int_0^{16} \pi y dy = \pi\left[\frac{1}{2}y^2\right]_0^{16} = 128\pi.$$

```
r = var('r')
f(r) = pi* r^2
integral(f, r, 0, r)
```

Answer : 1/3*pi*r^3 ■

EXAMPLE 4

Calculate the volume of the solid obtained by rotating about the x-axis the region under the curve $y = \sqrt{x}$ from 0 to 2.

Solution The region under the curve $y = \sqrt{x}$ from 0 to 2 is shown in Figure 7(a). By rotating this region about the x-axis, the solid shown in Figure 7(b) is obtained. A slice through the point x, gives a disk with radius $\sqrt{x}$. Its area of cross-section is $A(x) = \pi(\sqrt{x})^2 = \pi x$ and its thickness is Δx.

The solid lies between $x = 0$ and $x = 2$, so its volume is

$$V = \int_0^2 A(x)dx = \int_0^2 \pi x dx = \pi \frac{x^2}{2}\Bigg]_0^2 = 2\pi.$$ ■

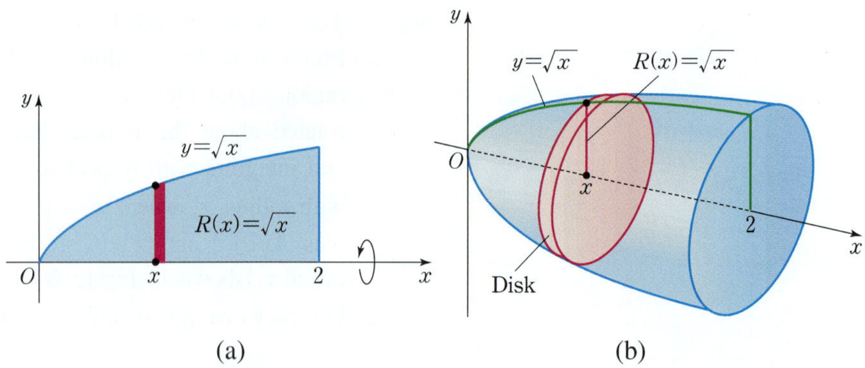

Figure 6

EXAMPLE 5

Find the volume of the solid of revolution obtained by revolving the region enclosed by the curves $y=\sqrt{x}$ and $y=x^2$ about the line $y=3$.

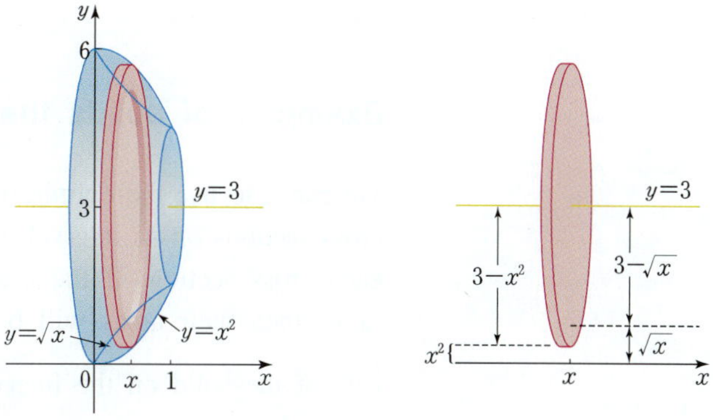

Figure 8

Solution The cross-section is a washer with the inner radius $3-\sqrt{x}$ and the outer radius $3-x^2$. (See Figure 8.) The cross-sectional area is

$$A(x)=\pi(3-x^2)^2-\pi(3-\sqrt{x})^2$$

and thus the volume of S is given by

$$\begin{aligned} V &= \int_0^1 A(x)dx \\ &= \pi\int_0^1 [(3-x^2)^2-(3-\sqrt{x})^2]dx \\ &= \pi\int_0^1 (x^4-6x^2-x+6\sqrt{x})dx \\ &= \pi\left[\frac{1}{5}x^5-2x^3-\frac{1}{2}x^2+4x^{\frac{3}{2}}\right]_0^1=\frac{17}{10}\pi . \end{aligned}$$

■

EXAMPLE 6

Find the volume of the solid of revolution obtained by rotating the region in Example 4 about the line $x=1$.

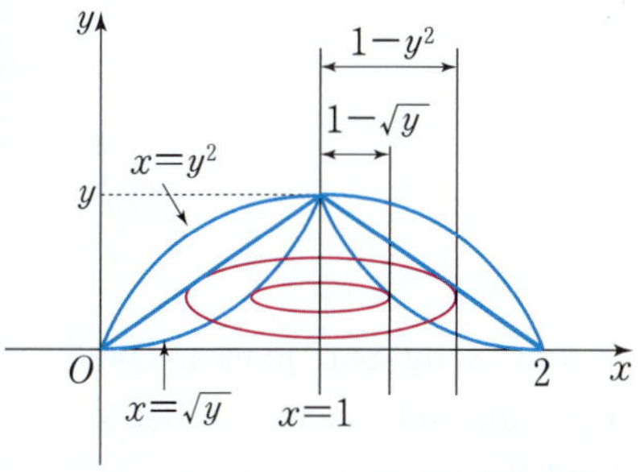

Figure 9

Solution A horizontal cross-section of a washer with inner radius $1-\sqrt{y}$ and outer radius $1-y^2$ is shown in Figure 9. Its cross-sectional area is

$$\begin{aligned} A(y) &= \pi(\text{outer radius})^2-\pi(\text{inner radius})^2 \\ &= \pi(1-y^2)^2-\pi(1-\sqrt{y})^2. \end{aligned}$$

The volume is

$$V=\int_0^1 A(y)dy=\pi\int_0^1[(1-y^2)^2-(1-\sqrt{y})^2]dy$$

$$= \pi \int_0^1 (y^4 - 2y^2 - y + 2\sqrt{y})dy$$

$$= \pi \left[\frac{1}{5}y^5 - \frac{2}{3}y^3 - \frac{1}{2}y^2 + \frac{4}{3}y^{\frac{3}{2}} \right]_0^1 = \frac{11}{30}\pi.$$ ■

Examples of solids that are not solids of revolution:

We can also use the definite integral to find the volume of a solid with specific cross sections on an interval, if we know a formula for the region determined by each cross section. If the cross sections generated are perpendicular to the x -axis, then their areas will be functions of x, denoted by $A(x)$. The volume (V) of the solid on the interval $[a, b]$ is $V = \int_a^b A(x)dx$.

EXAMPLE 7

Find the volume of a pyramid whose base is a square with side l and whose height is h.

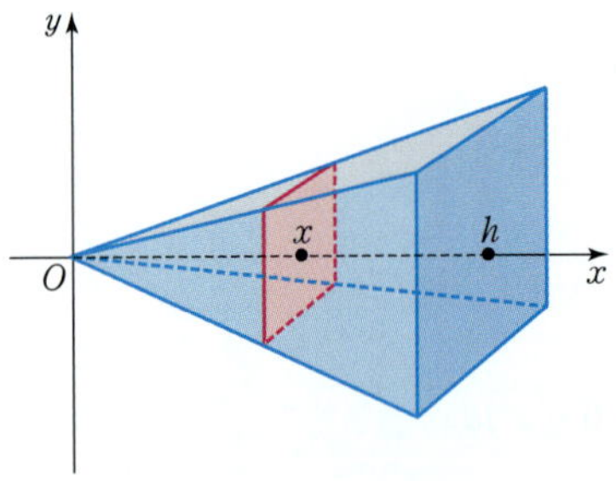

Figure 10

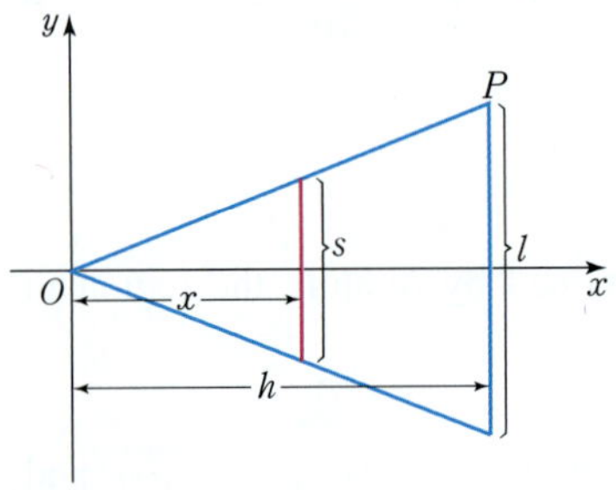

Figure 11

Solution Place the origin O at the vertex of the pyramid and the x-axis along its central axis as in Figure 10. Any plane P_x that passes through x and is perpendicular to the x-axis intersects the pyramid in a square with side of length s, say. Express s in terms of x by observing from the similar triangles in Figure 11 that

$$\frac{x}{h} = \frac{s/2}{l/2} = \frac{s}{l}$$

and so $s = lx/h$. [or observe that the line OP has slope $l/(2h)$ and so its equation is $y = lx/(2h)$.] Thus, the cross-sectional area is

$$A(x) = s^2 = \frac{l^2}{h^2}x^2.$$

The pyramid lies between $x = 0$ and $x = h$, so its volume is

$$V = \int_0^h A(x)dx = \int_0^h \frac{l^2}{h^2}x^2 dx$$

$$= \frac{l^2}{h^2}\frac{x^3}{3}\Big]_0^h = \frac{l^2 h}{3}.$$ ■

EXAMPLE 8

A wedge is cut out of a circular cylinder of radius 3 by two planes. One plane is perpendicular to the axis of the cylinder. The other intersects the first at an angle of 60° along a diameter of the cylinder. Find the volume of the wedge.

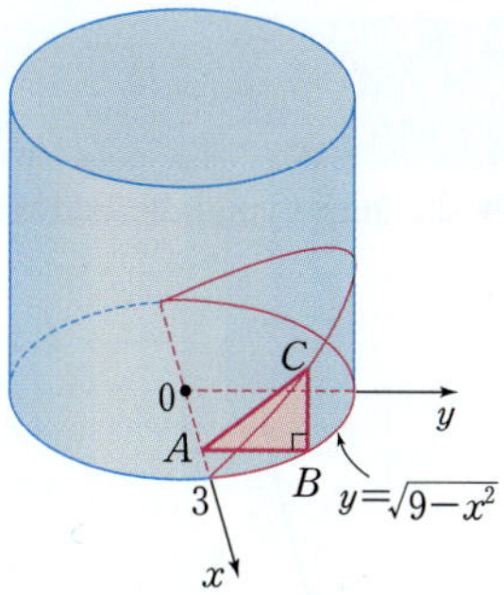

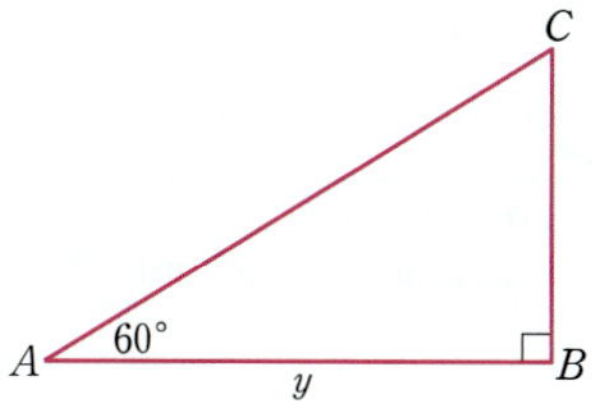

Figure 12

Solution Align the x-axis along the diameter where the planes meet. Then the base of the solid is a semicircle with equation $y=\sqrt{9-x^2}$, $-3 \le x \le 3$. A cross-section perpendicular to the x-axis at a distance x from the origin is a triangle ABC, as shown in Figure 12, whose base is $y=\sqrt{9-x^2}$ and whose height is $|BC| = y\tan 60^\circ = \sqrt{9-x^2}\sqrt{3}$. (See Figure 12.) Thus, the cross-sectional area is

$$A(x) = \frac{1}{2}\sqrt{9-x^2}\sqrt{3}\sqrt{9-x^2} = \frac{\sqrt{3}(9-x^2)}{2}$$

and the volume is

$$\begin{aligned} V &= \int_{-3}^{3} A(x)dx = \int_{-3}^{3} \sqrt{3}\,\frac{9-x^2}{2}dx \\ &= \sqrt{3}\int_{0}^{3}(9-x^2)dx = \sqrt{3}\left[9x - \frac{1}{3}x^3\right]_0^3 \\ &= 18\sqrt{3}. \end{aligned}$$

■

CAS **EXAMPLE 9**

Calculate the volume of the solid obtained by rotating the region bounded by the given curves about the specified line.

$$xy=4,\ x=2,\ x=4,\ y=0;\ \text{about the } x\text{-axis}$$

http://matrix.skku.ac.kr/cal-lab/cal-6-2-Exm-9.html

Solution

```
var('x, y')
p1 = parametric_plot((2, y), (y, -1, 3), color='blue')
p2 = parametric_plot((4, y), (y,-1, 3), color='green')
p3 = implicit_plot(x*y-4==0, (x, 0.5, 5), (y, -1, 3))
p4 = parametric_plot((x, 0), (x, 0.5, 5), color='red')
show(p1+p2+p3+p4)
integral(pi*(4/x)^2, x, 2, 4)
```

Answer : 4*pi

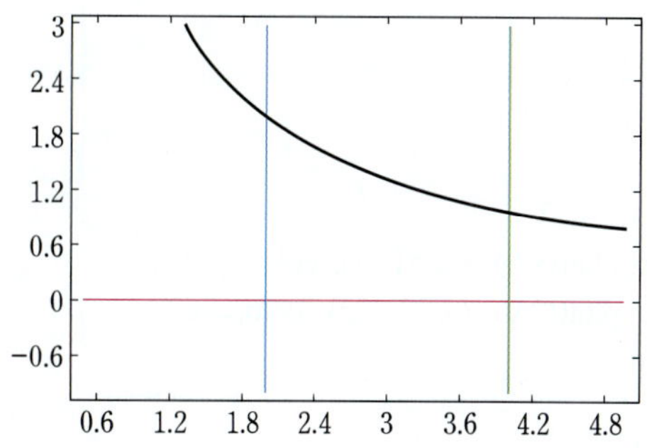

Figure 13

■

6.2 EXERCISES (Volumes)

http://matrix.skku.ac.kr/Cal-Book/part1/CS-Sec-6-2-Sol.html

YouTube http://youtu.be/Fd4Mguf2dbU

1-17. Calculate the volume of the solid obtained by rotating the region bounded by the given curves about the specified line.

1. $y=\dfrac{1}{x}$, $x=1$, $x=3$, $y=0$; about the x-axis

Solution $A(x)=\pi\left(\dfrac{1}{x}\right)^2$, $V=\displaystyle\int_1^3 A(x)dx=\dfrac{2}{3}\pi$

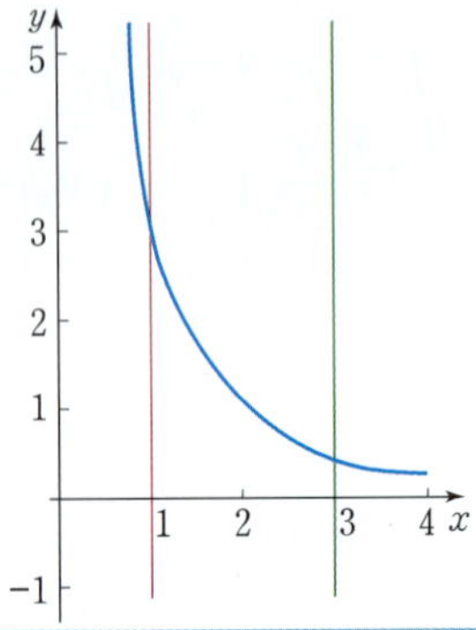

```
var('x, y')
f(x)=pi*(1/x)^2
print integral(f(x), x, 1, 3)          # 2/3 *pi
p0 = plot(f(x), x, 0, 2, ymax=30)
p1 = parametric_plot((1, y), (y, -1, 3), color='blue')
p2 = parametric_plot((3, y), (y,-1, 3), color='green')
show(p0+p1+p2)
```

Answer : 2/3*pi

2. $y=x^3$, $x=2$, $y=0$; about the x-axis
http://matrix.skku.ac.kr/cal-lab/cal-6-2-2.html

Solution $A(x)=\pi(x^3)^2$, $V=\displaystyle\int_0^2 A(x)\,dx=\dfrac{128}{7}\pi$

```
var('x, y')
f(x)=pi*(x^3)^2
print integral(f(x), x, 0, 2)
g=x-y^2
p1=plot(x^3, x, -1, 3)
p2=parametric_plot((x,0), (x, -1, 3), color='green')
p3=parametric_plot((2,y), (y, -1, 25), color='red')
show(p1+p2+p3)
```

Answer : 128/7*pi

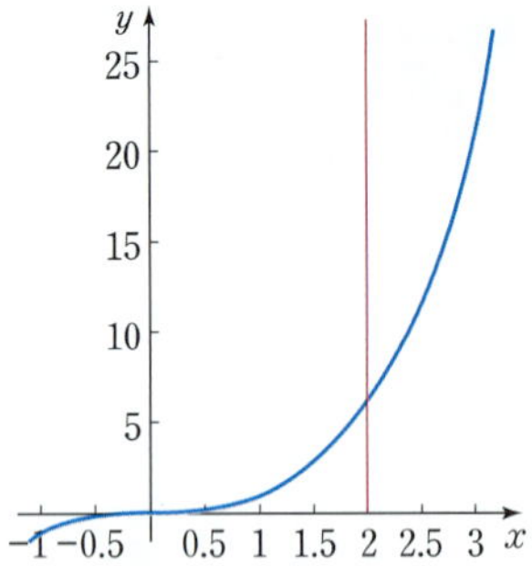

http://matrix.skku.ac.kr/cal-lab/cal-6-2-2-a.html

```
var('x, y, z')
implicit_plot3d(x^2 + z^2 == y^(6) , (x,-2, 2), (y,0, 8), (z,-2, 2), opacity=0.5, color="red")
```

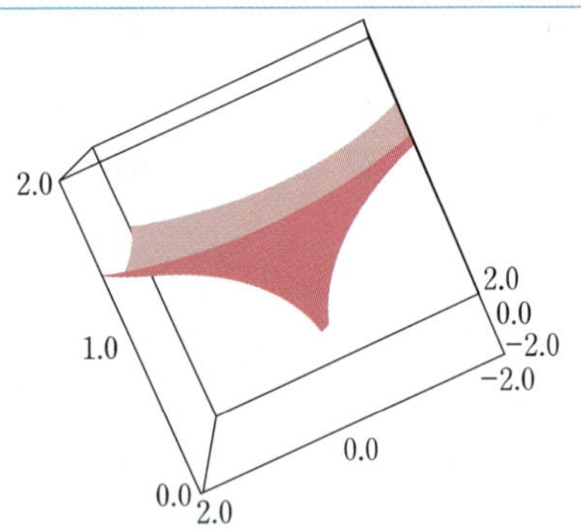

CAS **3.** $y=\dfrac{1}{x^2}$, $x=1$, $x=2$, $y=0$; about the x-axis
http://matrix.skku.ac.kr/cal-lab/cal-6-2-3.html

Solution $A(x)=\pi\left(\dfrac{1}{x^2}\right)^2$, $V=\displaystyle\int_1^2 A(x)dx=\dfrac{7}{24}\pi$

```
var('x, y')
f(x)=pi*(1/x^2)^2
print integral(f(x), x, 1, 2)
plot(f(x), x, 0, 2, ymax=50)
```

Answer : 7/24*pi

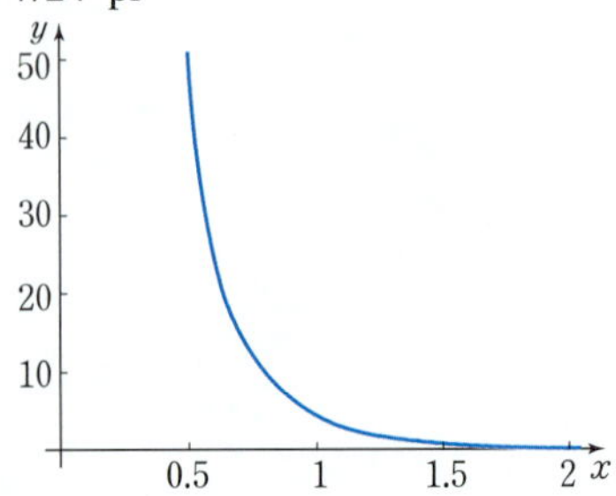

4. $y = \sqrt{x^2-1}$, $x=2$, $x=5$, $y=0$; about the x-axis
http://matrix.skku.ac.kr/cal-lab/cal-6-2-4.html

Solution $A(x) = \pi(\sqrt{x^2-1})^2$, $V = \int_2^5 A(x)dx = 36\pi$

```
var('x, y')
p1 = parametric_plot((2,y), (y, -1, 5), color='blue')
p2 = parametric_plot((5,y), (y,-1,5), color='green')
p3 = implicit_plot(y==sqrt(x^2-1), (x, 1, 6), (y, -1, 5))
p4 = parametric_plot((x,0), (x, 0.5, 6), color='red')
show(p1+p2+p3+p4)
print integral
```

Answer : 36*pi

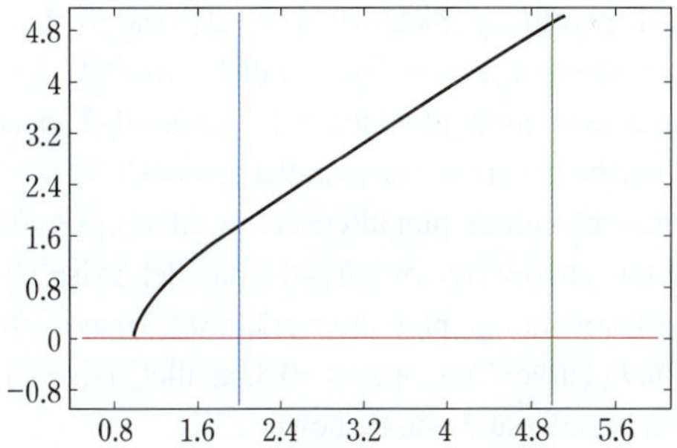

5. $y=x^2$, $0 \le x \le 3$, $y=9$, $x=0$; about the x-axis
http://matrix.skku.ac.kr/cal-lab/cal-6-2-5.html

Solution $A(x) = \pi(9)^2 - \pi(x^2)^2$,
$V = \int_0^3 A(x)dx = \frac{972}{5}\pi$

```
var('x, y')
f(x)=pi*(81-x^4)
print integral(f(x), x, 0, 3)
p1=plot(x^2, x, 0, 3)
p2=parametric_plot((x,9), (x, -1, 4), color='green')
p3=parametric_plot((0,y), (y, -1, 10), color='red')
show(p1+p2+p3)
```

Answer : 972/5*pi

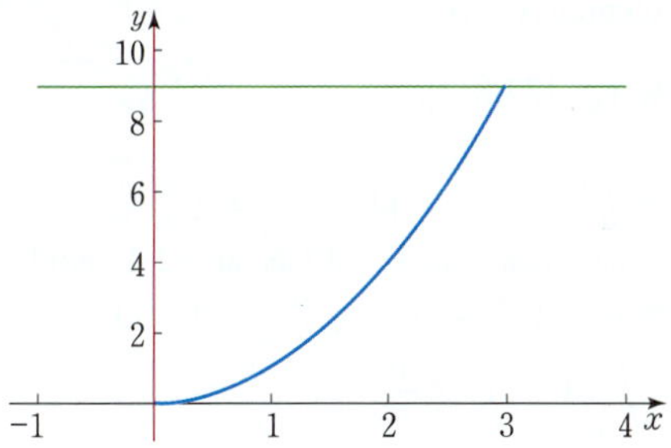

6. $x = 2y - y^2$, $x=0$; about the y-axis
http://matrix.skku.ac.kr/cal-lab/cal-6-2-6.html

Solution $A(y) = \pi x^2 = \pi(2y-y^2)^2$,
$V = \int_0^2 A(y)dy = \frac{16}{15}\pi$

```
f(y)=2*y-y^2
g(y)=pi*f^2
print integral(g(y), y, 0, 2)
plot(f(y), y, 0, 2)
```

Answer : 16/15*pi

7. $y=x^2$, $y^2=x$; about the x-axis

Solution $A(x) = \pi(\sqrt{x})^2 - \pi(x^2)^2$,
$V = \int_0^1 A(x)dx = \frac{3}{10}\pi$

8. $y^2 = x$, $x = y$; about the y-axis
http://matrix.skku.ac.kr/cal-lab/cal-6-2-8.html

Solution $A(y) = \pi(y^2 - y^4)$, $V = \int_0^1 A(y)dy = \frac{2}{15}\pi$

```
var('y')
f(y)=pi*(y^2-y^4)
print integral(f(y), y, 0, 1)
plot(f(y),0,1)
```

Answer : 2/15*pi

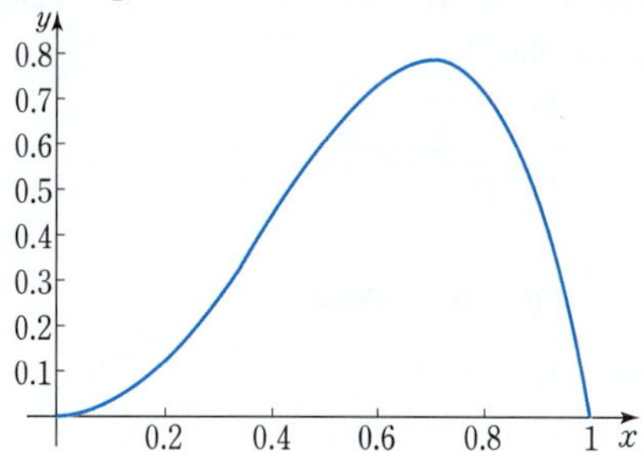

CAS **9.** $y = \sqrt[3]{x}$, $x=1$, $y=0$; about the x-axis
http://matrix.skku.ac.kr/cal-lab/cal-6-2-9.html

Solution $A(x) = \pi y^2 = \pi x^{\frac{2}{3}}$, $V = \int_0^1 A(x)dx = \frac{3}{5}\pi$

```
var('x')
f(x)=x^(1/3)
g(x)=pi*f^2
print integral(g(x), x, 0, 1)
```

Answer : 3/5*pi

10. $y=x$, $x=\sqrt{y}$; about $y=1$
http://matrix.skku.ac.kr/cal-lab/cal-6-2-10.html

Solution $A(x) = \pi(1-x)^2 - \pi(1-\sqrt{x})^2$,
$V = \int_0^1 A(x)dx = \frac{\pi}{6}$

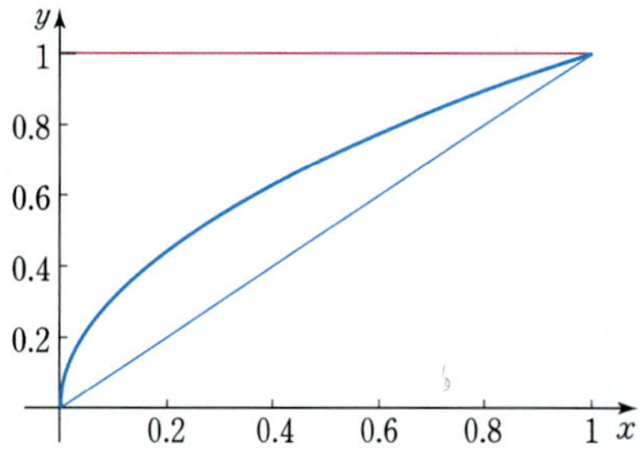


```
var('x, y')
f(x)=pi*(1-x)^2-pi*(1-sqrt(x))^2
print integral(f(x), x, 0, 1)
g=x-y^2
p1=plot(x, x, 0, 1)
p2=plot(sqrt(x), x, 0, 1)
p3=parametric_plot((x,1), (x, 0, 1), color='red')
show(p1+p2+p3)
```

```
var('x, y, z')
p1=implicit_plot3d( y^2+z^2==(1-x)^2 ,
(x,0, 1), (y,-1, 1), (z,-1, 1), opacity=0.5,
color="red")
p2=implicit_plot3d( y^2+z^2==(1-sqrt(x))^2
,(x,0,1), (y,-1, 1), (z,-1, 1), opacity=0.5,
color="blue")
show(p1+p2)
```

Answer : 1/6*pi

11. $y=x^2$, $y=9$; about $y=9$

Solution $A(x)=\pi(9-x)^2$, $V=\int_{-3}^{3} A(x)dx=5184$

CAS **12.** $y=x^3$, $y=1$; about $y=+1$

Solution $A(x)=\pi(1-2x^3+x^6)$, $V=\int_{0}^{1} A(x)dx=\frac{5}{14}$

13. $y=\frac{1}{x^2}$, $y=0$, $x=1$, $x=3$; about $y=-1$

http://matrix.skku.ac.kr/cal-lab/cal-6-2-13.html

Solution $A(x)=\pi\left(\frac{1}{x^2}-(-1)\right)^2-\pi(0-(-1))^2$,

$V=\int_{1}^{3} A(x)dx=\frac{134}{81}$

```
var('x, y')
f(x)=1/x^2
p1=plot(f(x), x, 0.5, 4)
p2=parametric_plot((1,y), (y, -2, 4), color='red')
p3=parametric_plot((3,y), (y, -2, 4), color='red')
p4=parametric_plot((x,-1), (y, -2, 4), color='red')
p5=parametric_plot((x,0), (x, -1, 4), color='red')
show(p1+p2+p3+p4+p5)
```

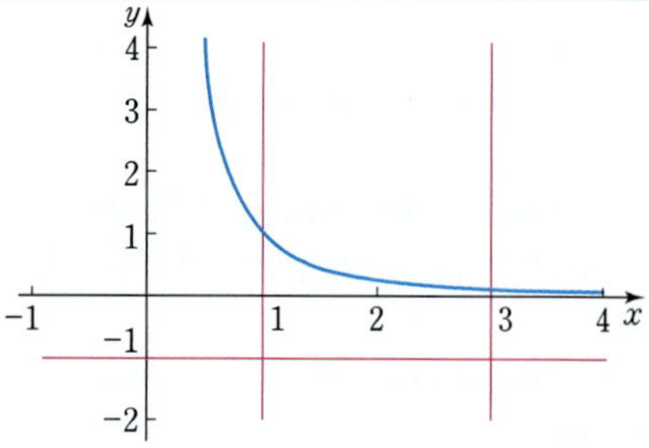


```
cover1=(1,x)
cover2=(3,x)
sur1=revolution_plot3d(f(x),(x,1,3),axis=(0,-1),opac
ity=0.8,show_curve=True,parallel_axis='x')
sur2=revolution_plot3d(0,(x,1,3),axis=(0,-1),opacity
=0.8,show_curve=True,parallel_axis='x')
sur3=revolution_plot3d(cover1,(x,0,f(1)),axis=(0,-1)
,show_curve=True,opacity=0.8,parallel_axis='x')
sur4=revolution_plot3d(cover2,(x,0,f(3)),axis=(0,-1)
,show_curve=True,opacity=0.8,parallel_axis='x')
(sur1+sur2+sur3+sur4).show()
```

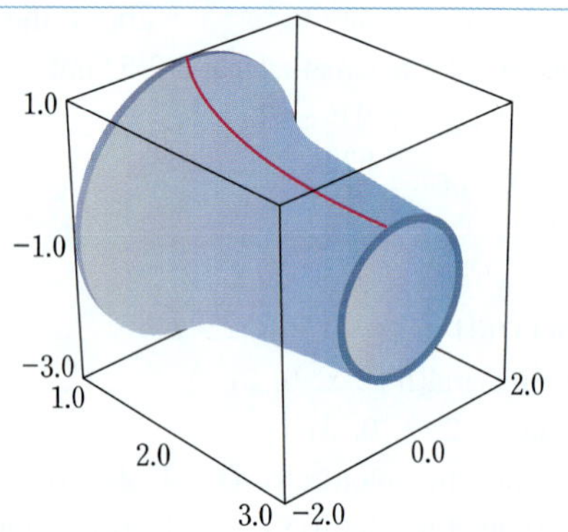


```
A(x)=((f(x)-(-1))^2-(0-(-1))^2)*pi
integral(A(x),x,1,3)
```

Answer : 134/81*pi

14. $x=y^2$, $x=1$; about $x=1$

http://matrix.skku.ac.kr/cal-lab/cal-6-2-14.html

Solution $A(y)=\pi(1-y^2)^2-\pi(1-1)^2$,

$V=\int_{0}^{1} A(y)dy=\frac{16}{15}\pi$

```
var('x, y')
f=x-y^2
p1=implicit_plot(f, (x, 0, 1.5), (y, -1.5, 1.5))
p2=parametric_plot((1,y), (y, -1.5, 1.5),
color='red')
show(p1+p2)
```

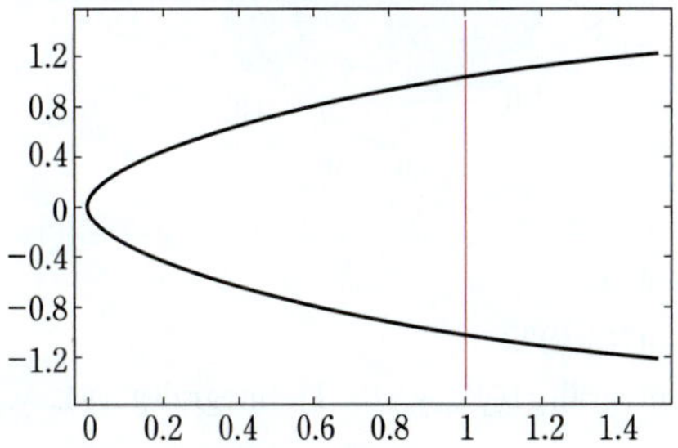

```
g1=(x^2,x)
g2=(x^2,-x)
sur1=revolution_plot3d(g1,(x,0,1),axis=(1,0),show_
curve=True,opacity=0.5)
sur2=revolution_plot3d(g2,(x,0,1),axis=(1,0),show_
curve=True,opacity=0.5)
(sur1+sur2).show()
```

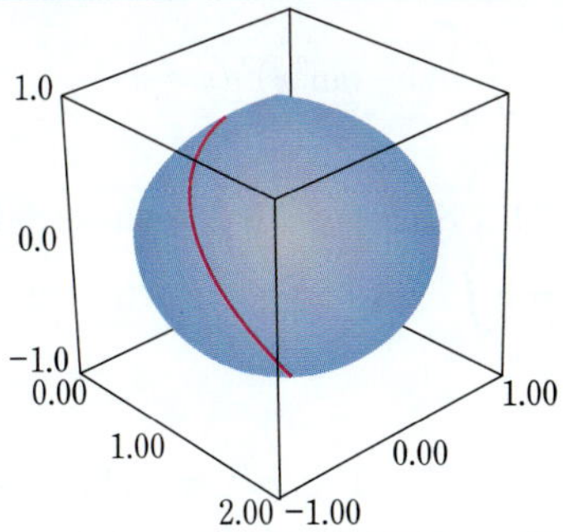

```
A(y)=(1-y^2)^2*pi
integral(A(y),y,-1,1)
```

Answer : 16/15*pi

CAS **15.** $y=x$, $y=\sqrt{x}$; about $x=1$

http://matrix.skku.ac.kr/cal-lab/cal-6-2-15.html

Solution $A(y)=\pi((1-y^2)^2-(1-y)^2)$,

$$V=\int_0^1 A(y)dy=\frac{\pi}{5}$$

```
var('x, y')
f(y)=pi*(1-y^2)^2-pi*(1-y)^2
print integral(f(y), y, 0, 1)
g(x)=x
p1=plot(g(x), x, -1, 2)
p2=plot(sqrt(x), x, 0, 2)
p3=parametric_plot((1,y), (y, -1, 2), color='red')
show(p1+p2+p3)
```

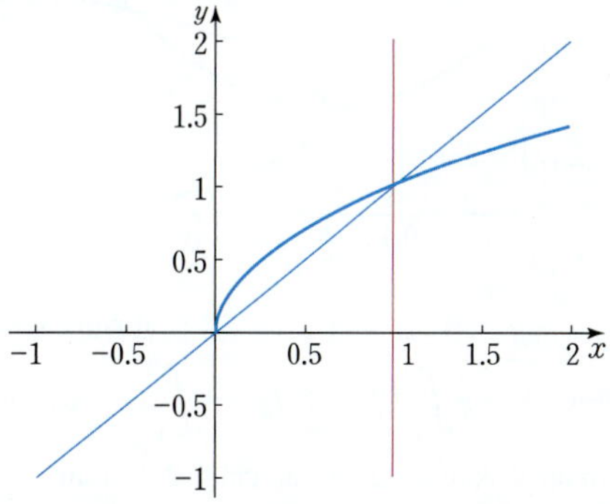

```
var('x, y, z')
p1=implicit_plot3d( (1-x)^2+z^2==y , (x,0, 2),
(y,0, 1), (z,-1, 1), opacity=0.5, color="red")
p2=implicit_plot3d( (1-x)^2+z^2==y^2 ,(x,0, 2),
(y,0, 1), (z,-1, 1), opacity=0.5, color="blue")
show(p1+p2)
```

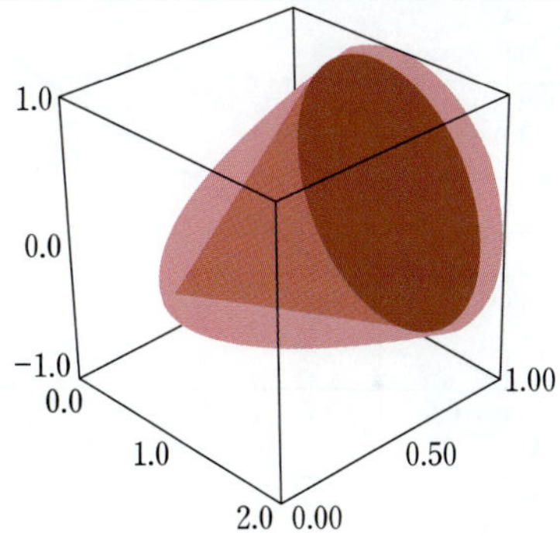

16. $y=x^2$, $x=y^2$; about $x=0$

http://matrix.skku.ac.kr/cal-lab/cal-6-2-16.html

Solution $A(y)=\pi(\sqrt{y})^2-\pi(y^2)^2$, $V=\int_0^1 A(y)dy=\frac{3}{10}\pi$

```
var('x, y')
f(y)=pi*y-pi*(y^2)^2
print integral(f(y), y, 0, 1)
g=x-y^2
p1=plot(x^2, x, -1, 1.5)
p2=implicit_plot(g, (x, -1, 2), (y, -1, 2))
p3=parametric_plot((0,y), (y, -1, 3), color='red')
show(p1+p2+p3)
```

Answer : 3/10*pi

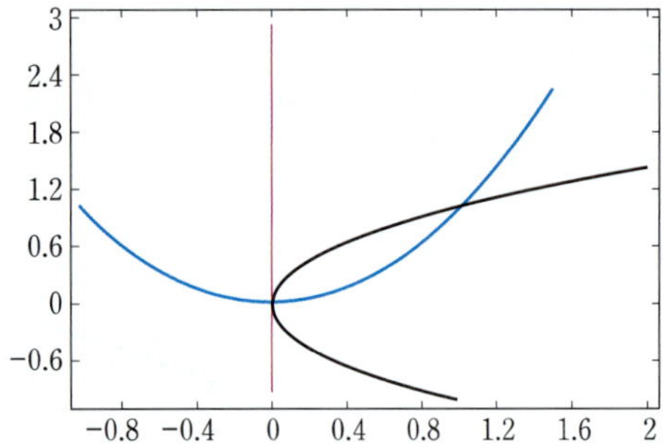

17. $y=x$, $y=0$, $x=1$, $x=3$; about $x=1$

Solution $V=\pi\int_0^1(3-1)^2dy+\pi\int_1^3(3-y)^2dy=\frac{20}{3}\pi$

http://matrix.skku.ac.kr/cal-lab/cal-6-2-17.html

```
var('x, y')
f(x)=x
p1=plot(f(x), x, 0, 4)
p2=parametric_plot((1,y), (y, 0, 4), color='red')
p3=parametric_plot((3,y), (y, 0, 4), color='red')
p4=parametric_plot((x,0), (x, 0, 4), color='red')
show(p1+p2+p3+p4)
```

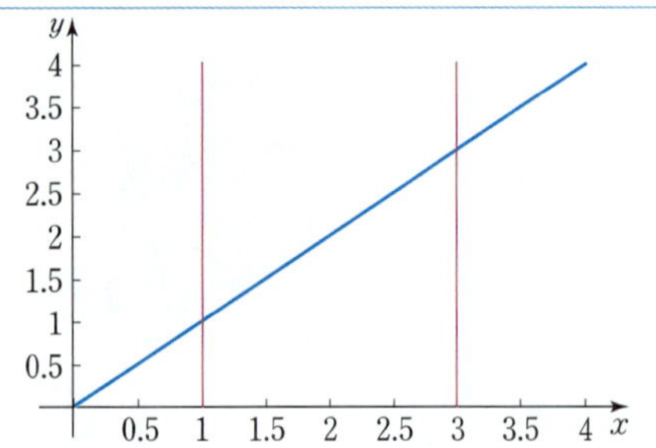

```
cover1=(1,y)
cover2=(3,y)
cover3=(x,0)
sur1=revolution_plot3d(f(x),(x,1,3),axis=(1,0),opaci
ty=0.7,show_curve=True,parallel_axis='z')
sur2=revolution_plot3d(cover1,(y,0,f(1)),axis=(1,0),
opacity=0.7,show_curve=True,parallel_axis='z')
sur3=revolution_plot3d(cover2,(y,0,f(3)),axis=(1,0),
opacity=0.7,show_curve=True,parallel_axis='z')
sur4=revolution_plot3d(cover3,(x,1,3),axis=(1,0),op
acity=0.7,show_curve=True,parallel_axis='z')
(sur1+sur2+sur3+sur4).show()
```

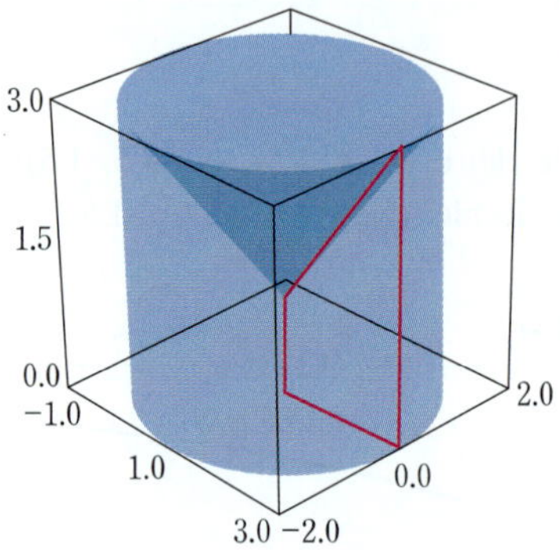

```
g1(y)=pi*4
g2(y)=pi*(3-y)^2
print integral(g1(y), y, 0, 1)+integral(g2(y), y, 1,
3)
```

Answer : 20/3*pi

18-23. Set up, but do not evaluate, an integral for the volume of the solid obtained by rotating the region bounded by the given curves about the specified line.

18. $y=\tan^2 x$, $y=1$, $x=0$; about $y=1$

Solution $V=\int_0^{\frac{\pi}{4}}(1-\tan^2 x)^2dx=\pi-\frac{8}{3}$

19. $y=(x-2)^2$, $8x-y=16$; about $x=6$

Solution $V=\pi\int_0^{64}[(6-(\frac{y}{8}+2))^2-(6-(2+\sqrt{y}))^2]dy$

$=3328\pi$

20. $y=0$, $y=\cos x$, $0\le x\le\pi$; about $y=1$

http://matrix.skku.ac.kr/cal-lab/cal-6-2-20.html

Solution $V=\int_0^{\frac{\pi}{2}}[\pi(1-0)^2-\pi(1-\cos x)^2]dx$

$+\int_{\frac{\pi}{2}}^{\pi}[\pi(1-\cos x)^2-\pi(1-0)^2]dx=4\pi$

```
var('x, y')
f(x)=pi-pi*(1-cos(x))^2
g(X)=pi*(1-cos(x))^2-pi
print integral(f(x), x, 0, pi/2)+integral(g(x), x,
pi/2, pi)
p1=plot(cos(x), 0, pi)
p2=parametric_plot((x,1), (x, -1, 4), color='red')
p3=parametric_plot((x,0), (x, -1, 4), color='green')
p4=parametric_plot((pi,y), (y, -1, 1), color='red')
show(p1+p2+p3+p4)
```

Answer : 4*pi

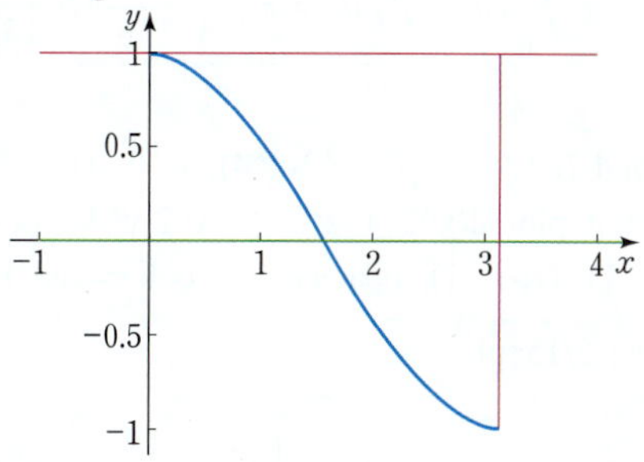

21. $y=0$, $y=\cos x$, $0 \le x \le \pi$; about $y=-2$

Solution $V=\pi[\int_0^{\pi}((\cos x-(-2))^2-(0-(-2))^2)]dx$

$= \dfrac{\pi^2}{2}$

CAS **22.** $x^2-y^2=1$, $x=2$; about $x=-1$

http://matrix.skku.ac.kr/cal-lab/cal-6-2-22.html

Solution $V=\pi\int_{-\sqrt{3}}^{\sqrt{3}}(-y^2+3)dy = 4\pi\sqrt{3}$

```
var('y')
f(y)= 3 - y^(2)
pi*integral(f(y), y, -3^(1/2), 3^(1/2))
```

Answer : 4*pi*sqrt(3)

23. $x+y=1$, $(y-1)^2=2-x$; about $x=-5$

http://matrix.skku.ac.kr/cal-lab/cal-6-2-23.html

Solution $V=\pi\int_0^3[(2-(y-1)^2+5)^2-(1-y+5)^2]dy$

$= \dfrac{243}{5}\pi$

```
var('x, y')
f(y)=pi*(2-(y-1)^2+5)^2-pi*(1-y+5)^2
print integral(f(y), y, 0, 3)
p1=plot(-x+1, x, -6, 3)
g=(y-1)^2-2+x
p2=implicit_plot(g, (x, -6, 3), (y, -4, 4))
p3=parametric_plot((-5,y), (y, -4, 6), color='red')
show(p1+p2+p3)
```

Answer : 243/5*pi

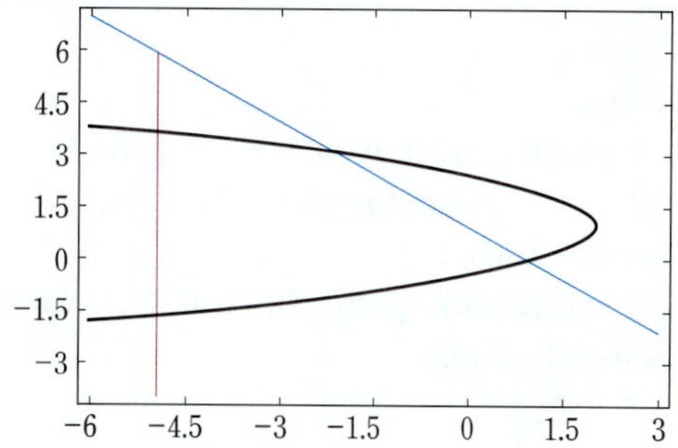

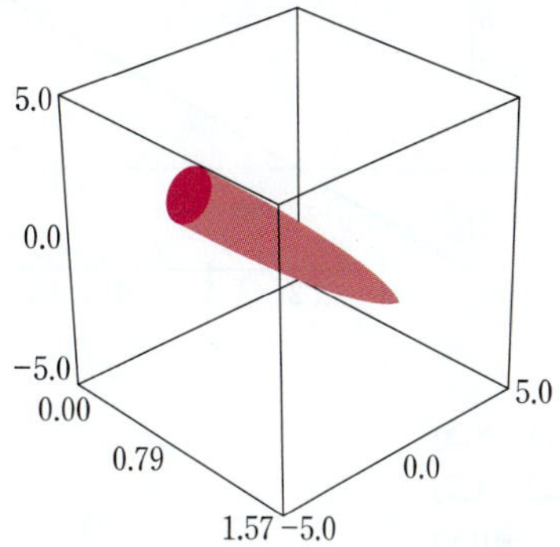

24-27. Describe the solid of revolution whose volume is represented by the integral.

24. $\pi\int_0^{\frac{\pi}{2}}\cos x\,dx$

http://matrix.skku.ac.kr/cal-lab/cal-6-2-24.html

Solution $R=\left\{(x,\ y)\,|\,0 \le x \le \dfrac{\pi}{2},\ 0 \le y \le \sqrt{\cos x}\right\}$;

about x-axis

$\int_0^{\pi/2}\cos x dx = \left(\sin\dfrac{\pi}{2}\right)-(\sin 0)=1$

$\pi\int_0^{\frac{\pi}{2}}\cos x\,dx = \pi$

```
var('x, y, z')
implicit_plot3d(y^2 + z^2 == cos(x) , (x,0, pi/2),
(y,-2, 2), (z,-2, 2), opacity=0.5, color="red")
integral (pi*cos(x), x, 0, pi/2)
```

Answer : pi

25. $\pi\int_2^5 y^2\,dy$

http://matrix.skku.ac.kr/cal-lab/cal-6-2-25.html

Solution $R=\{(x,\ y)\,|\,2 \le y \le 5 , 0 \le x \le y\}$;

about y-axis

```
var('x')
f(x)=x
p1=plot(f(x), x, 0, 6, ymin=-1, ymax= 6)
p2 = parametric_plot((x,2), (x, 0, 6), color='green')
p3 = parametric_plot((x,5), (x, 0, 6), color='red')
show(p1+p2+p3)
```

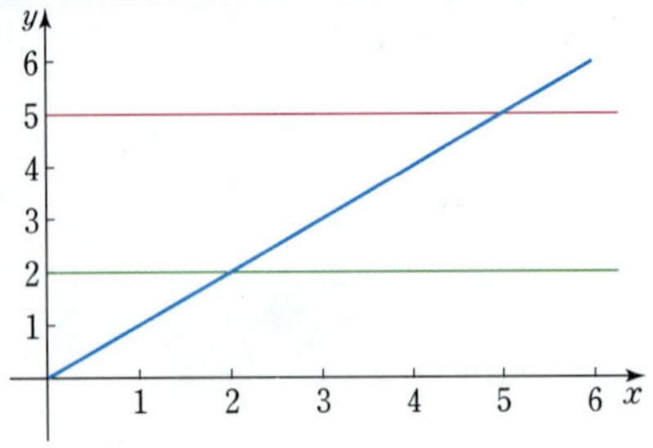

```
cover1=(x,5)
cover2=(x,2)
cover3=(0,x)
sur1=revolution_plot3d(f(x),(x,2,5),show_curve=True,opacity=0.7)
sur2=revolution_plot3d(cover1,(x,0,5),show_curve=True,opacity=0.7)
sur3=revolution_plot3d(cover2,(x,0,2),show_curve=True,opacity=0.7)
sur4=revolution_plot3d(cover3,(x,2,5),show_curve=True,opacity=0.7)
(sur1+sur2+sur3+sur4).show()
```

Answer : $R=\{(x,\ y)\,|\,2\le y\le 5\,,\,0\le x\le y\}$;
about y-axis

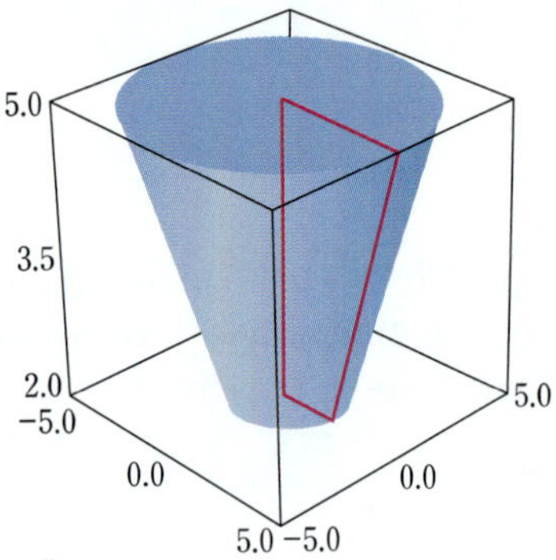

26. $\pi\int_0^1 (y^2-y^4)dy$

http://matrix.skku.ac.kr/cal-lab/cal-6-2-26.html

Solution $R=\{(x,\ y)\,|\,0\le y\le 1,\ y^2\le x\le y\}$;
about y-axis

$$\pi\int_0^1 (y^2-y^4)\,dy=\pi\left(\frac{1^3}{3}-\frac{1^5}{5}\right)=\frac{2}{15}\pi$$

```
var('x, y, z')
expand (integral (pi*(x^2-x^4), x, 0, 1))
implicit_plot3d(x^2 + z^2 == y^2-y^4 , (x,-1, 1), (y,0, 1), (z,-1, 1), opacity=0.5, color="red")
```

Answer : 2/15*pi

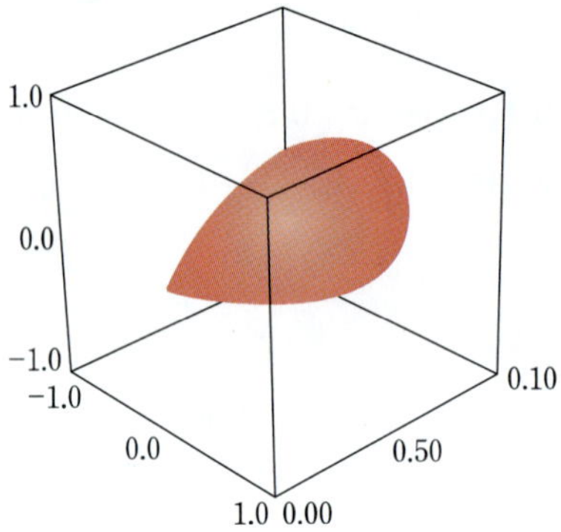

27. $\pi\int_0^{\frac{\pi}{2}} [(1-\cos x)^2-1^2]dx$

http://matrix.skku.ac.kr/cal-lab/cal-6-2-27.html

Solution $R=\left\{(x,y)\,|\,0\le x\le \frac{\pi}{2}\,,\,1\le y\le 1-\cos x\right\}$;
about x-axis

$$\pi\int_0^{\frac{\pi}{2}} [(1-\cos x)^2-1^2]dx=-\pi\int_0^{\frac{\pi}{2}} [1^2-(1-\cos x)^2]\,dx$$

```
var('x')
f(x)=1-cos(x)
p1=plot(f(x), x, 0, pi/2, ymin=-3, ymax= 5)
p2 = parametric_plot((x,1), (x, 0, pi/2), color='green')
show(p1+p2)
```

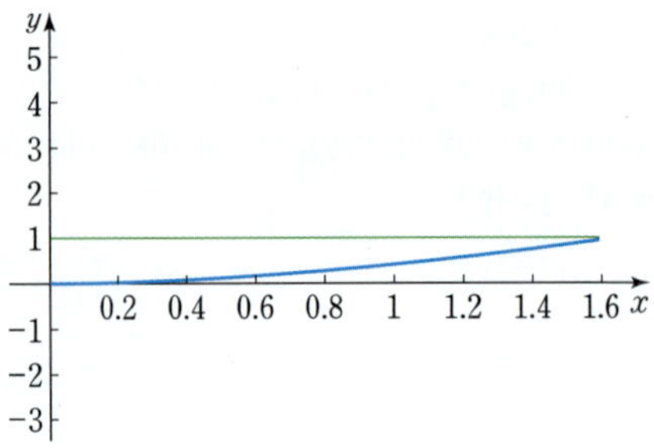

```
cover1=(x,1)
cover2=(0,x)
sur1=revolution_plot3d(f(x),(x,0,pi/2),show_curve=
"True",opacity=0.7,parallel_axis='x')
sur2=revolution_plot3d(cover1,(x,0,pi/2),show_curv
e="True",opacity=0.7,parallel_axis='x')
sur3=revolution_plot3d(cover2,(x,0,1),show_curve=
"True",opacity=0.7,parallel_axis='x')
(sur1+sur2+sur3).show()
```

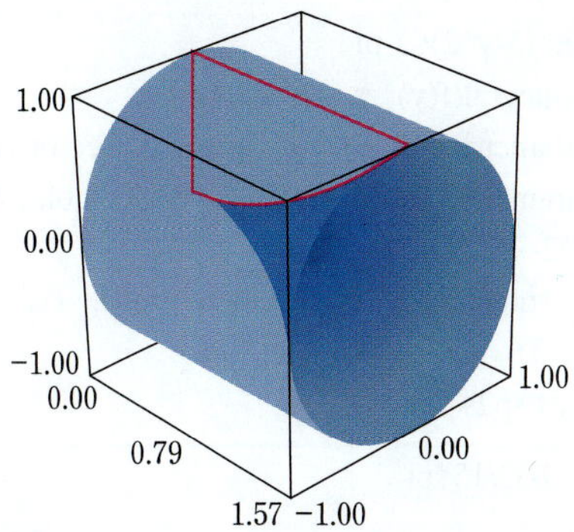

28-32. Find the volume of the described solid S.

28. A right circular cone with height h and base radius r.

Solution $V=\pi\displaystyle\int_0^h\left(\frac{r}{h}x\right)^2dx=\frac{1}{3}\pi r^2h$

29. A frustum of a right circular cone with height h, lower base radius R, and top radius r.

Solution $V=\pi\displaystyle\int_0^h\left(R-\frac{R-r}{h}y\right)^2dy=\frac{1}{3}\pi h(R^2+Rr+r^2)$

CAS **30.** A tetrahedron with three mutually perpendicular faces and three mutually perpendicular edges with lengths 6cm, 8cm and 10cm.

Solution $A(z)=\dfrac{1}{2}\cdot 6\left(\dfrac{10-z}{10}\right)\cdot 8\left(\dfrac{10-z}{10}\right)$

$=24\left(\dfrac{10-z}{10}\right)^2,\quad V=\displaystyle\int_0^{10}A(z)dz=80$

31. The base of S is an elliptical region with boundary curve $16x^2+9y^2=144$. Cross-sections perpendicular to the y-axis are isosceles right triangles with the hypotenuse in the base.

Solution $V=\displaystyle\int_{-3}^3 A(x)dx=2\int_0^3\frac{1}{4}(144-16x^2)dx$

$=144$

32. A bowl is shaped like a hemisphere with diameter 40 cm. A ball with diameter 20 cm is placed in the bowl and water is poured into the bowl to a depth of h centimeters. Find the volume of water in the bowl.

Solution $A(x)=20\pi x,\quad V(h)=\pi\displaystyle\int_0^h A(x)dx=10\pi h^2$.

CAS **33.** $(x-2)^2+y^2=4$; about the x-axis

http://matrix.skku.ac.kr/cal-lab/cal-6-3-18.html

http://matrix.skku.ac.kr/cal-lab/sage-grapher-imp.html

Solution

```
var('x, y')
x=sqrt(-y^2+4)+2
pi*integral(x, y, 0, 2)
```

Answer : (pi+4)*pi

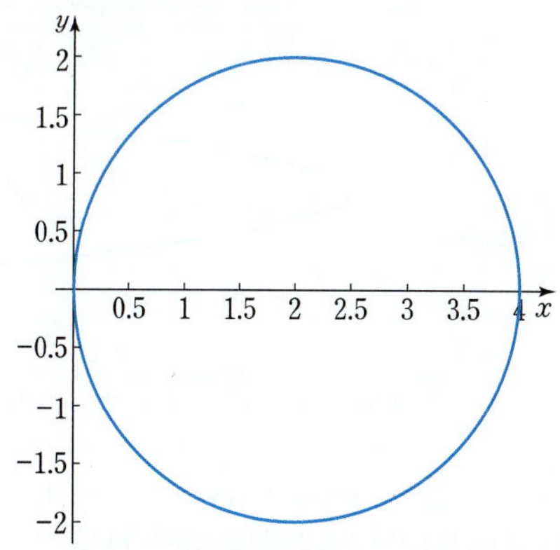

34-37. Find the volume of the solid obtained by rotating the region bounded by the given curves about the specified axis.

CAS **34.** $y=2x+4,\ y=2x^2$; about the x-axis

Solution $\pi\displaystyle\int_{-1}^2((2x+4)^2-(2x^2)^2)dx$

$=\pi\displaystyle\int_{-1}^2(4x^2+16x+16-4x^4)dx$

$=\pi\left[-\dfrac{1}{5}x^5+\dfrac{4}{3}x^3+8x^2+16x\right]_{-1}^2$

$=\pi\left(-\dfrac{32}{5}+\dfrac{32}{3}+32+32\right)-\pi\left(\dfrac{1}{5}-\dfrac{4}{3}+8-16\right)$

$= \dfrac{387}{5}\pi$

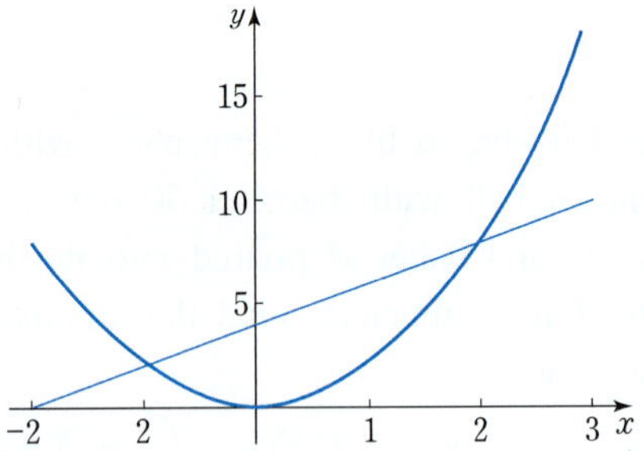

CAS **35.** $x = 2 + y^2$, $y = x - 4$; about y-axis

Solution $\pi \displaystyle\int_{-1}^{2} ((y+4)^2 - (2+y^2)^2)\,dy$

$= \pi \displaystyle\int_{-1}^{2} (y^2 + 8y + 16 - (y^4 + 4y^2 + 4))\,dy$

$= \pi \left[-\dfrac{1}{5}y^5 - y^3 + 4y^2 + 12y \right]_{-1}^{2}$

$= \pi \left(-\dfrac{32}{5} - 8 + 16 + 24 - \dfrac{1}{5} - 1 - 4 + 12 \right)$

$= \dfrac{162}{5}\pi$

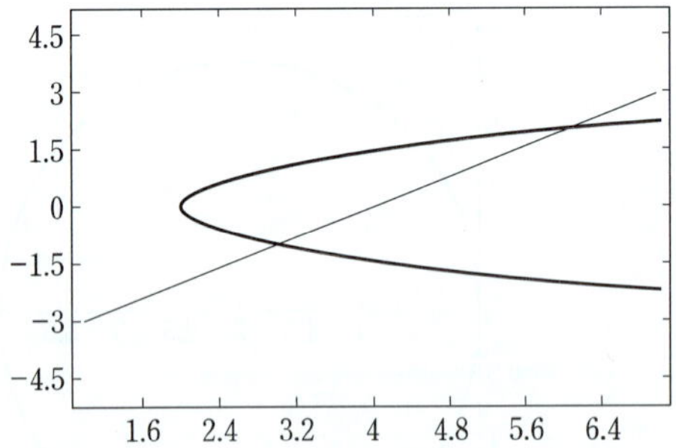

36. $x = 0$, $x = 4 - y^2$; about $x = -1$

http://math1.skku.ac.kr/home/pub/1142

Solution $\pi \displaystyle\int_{-2}^{2} ((4 - y^2 + 1)^2 - (1^2))\,dy$

$= \pi \displaystyle\int_{-2}^{2} (y^4 - 10y^2 + 25 - 1)\,dy$

$= 2\pi \displaystyle\int_{0}^{2} (y^4 - 10y^2 + 24)\,dy$

$= 2\pi \left[\dfrac{1}{5}y^5 - \dfrac{10}{3}y^3 + 24y \right]_0^2$

$= 2\pi \left(\dfrac{32}{5} - \dfrac{80}{3} + 48 \right) = \dfrac{832}{15}\pi$

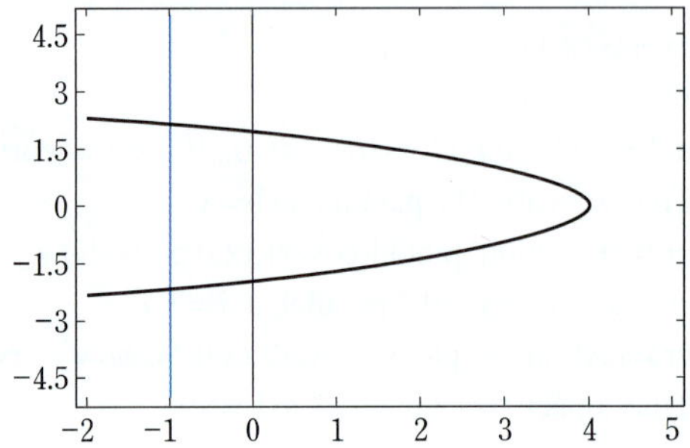


```
var('x, y')
f(y)=pi*(5-y^2)^2-pi
print integral(f(y), y, -2, 2)
p1=parametric_plot((0,y), (y, -3, 3), color='red')
p2=parametric_plot((-1,y), (y, -3, 3), color='blue')
f=4-y^2-x
832/15*pip3832/15*pi=implicit_plot(f, (x, -2, 5), (y, -3, 3))
show(p1+p2+p3)
```

Answer : 832/15*pi

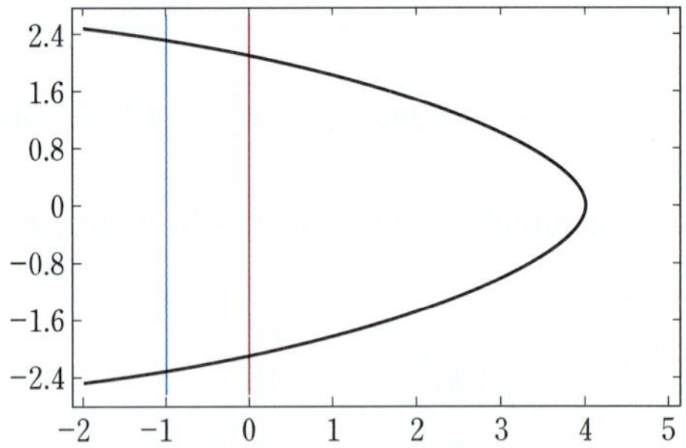

CAS **37.** $y = 2x^2 + 1$, $y = 13 - x^2$; about $y = -1$

http://matrix.skku.ac.kr/cal-lab/cal-6-2-37.html

Solution $\pi \displaystyle\int_{-2}^{2} ((13 - x^2 + 1)^2 - (2x^2 + 1 + 1)^2)\,dx$

$= \pi \displaystyle\int_{-2}^{2} (x^4 - 28x^2 + 196 - (4x^4 + 8x^2 + 4))\,dx$

$= 2\pi \displaystyle\int_{0}^{2} (-3x^4 - 36x^2 + 192)\,dx$

$= 2\pi \left[-\dfrac{3}{5}x^5 - 12x^3 + 192x \right]_0^2$

$= 2\pi \left(-\dfrac{96}{5} - 96 + 384 \right) = \dfrac{2688}{5}\pi$

```
var('x')
f(x)= -3*x^4 -36*x^2 + 192
g(x)= 2*pi*f
print integral(g(x), x, 0, 2)
plot(f(x), x, 0, 2)
```

Answer : 2688/5*pi

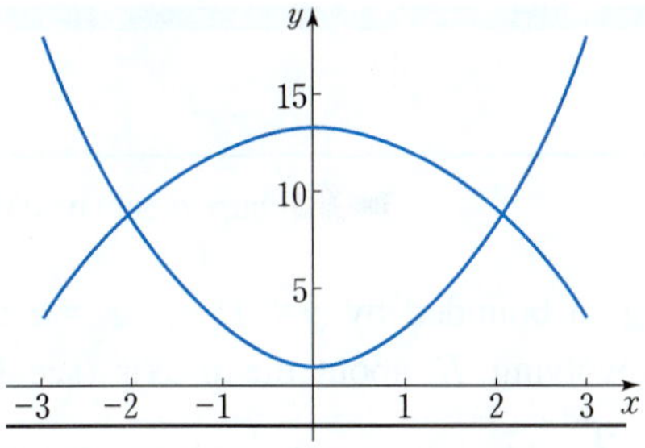

CAS **38.** Find the volume of the solids obtained by rotating the region bounded by the curves $y=x$ and $y=2x^2$ about the following lines:

(a) the x-axis

http://matrix.skku.ac.kr/cal-lab/cal-6-2-38.html

(b) the y-axis

(c) $y=2$

Solution (a) $\pi\int_0^{1/2}(x^2-(2x^2)^2)dx$

$$=\pi\int_0^{1/2}(x^2-4x^4)dx=\pi\left[\frac{1}{3}x^3-\frac{4}{5}x^5\right]_0^{1/2}$$

$$=\pi\left(\frac{1}{24}-\frac{1}{40}\right)=\frac{\pi}{60}$$

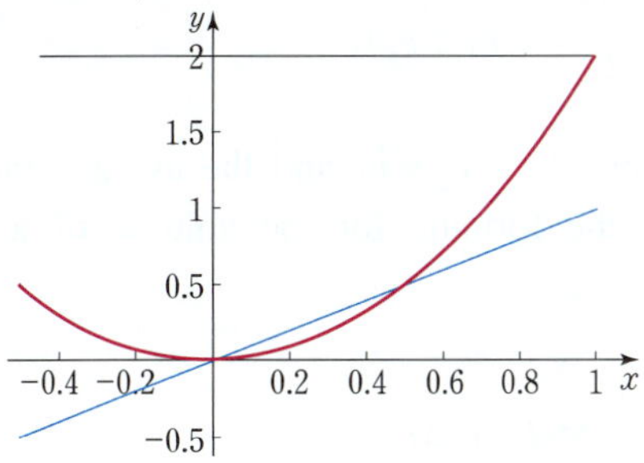


```
var('x, y')
f(x)= x^2 -4*x^4
g(x)= pi*f
print integral(g(x), x, 0, 1/2)
plot(f(x), x, 0, 1/2)
```

(b) $\pi\int_0^{1/2}\left(\frac{1}{2}y-y^2\right)dy=\pi\left[\frac{1}{4}y^2-\frac{1}{3}y^3\right]_0^{1/2}$

$$=\pi\left(\frac{1}{16}-\frac{1}{24}\right)=\frac{\pi}{48}$$

(c) $\pi\int_0^{1/2}\{(2-2x^2)^2-(2-x)^2\}dx$

$$=\pi\int_0^{1/2}\{4x^4-8x^2+4-(x^2-4x+4)\}dx$$

$$=\pi\left[\frac{4}{5}x^5-3x^3+2x^2\right]_0^{1/2}$$

$$=\pi\left(\frac{1}{40}-\frac{3}{8}+\frac{1}{2}\right)=\frac{3}{20}\pi$$

39. Let R be the region in the first quadrant bounded by the curves $y=x^2$ and $y=x-2x^3$. Calculate the following quantities.

(a) The area of R.

http://matrix.skku.ac.kr/cal-lab/cal-6-2-39.html

(b) The volume obtained by rotating R about the x-axis.

(c) The volume obtained by rotating R about the y-axis.

Solution (a) Answer= $\int_0^{1/2}[(x-2x^3)-(x^2)]dx$

which is

$$\left[\frac{1}{2}x^2-\frac{1}{2}x^4-\frac{1}{3}x^3\right]_0^{1/2}=\left(\frac{1}{8}-\frac{1}{32}-\frac{1}{24}\right)=\frac{5}{96}$$

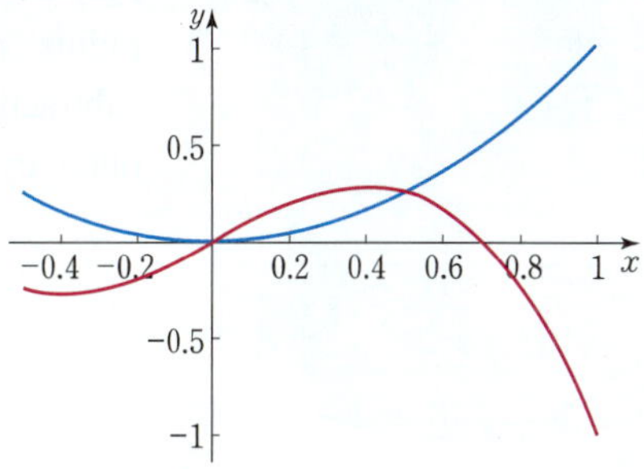


```
var('x, y')
f(x)= x -2*x^3 -x^2
print integral(f(x), x, 0, 1/2)
plot(f(x), x, 0, 1/2)          #    5/96
```

(b) $\pi\int_0^{1/2}((x-2x^3)^2-(x^2)^2)dx$

$$=\pi\int_0^{1/2}(4x^6-4x^4+x^2-x^4)dx$$

$$=\pi\left[\frac{4}{7}x^7-x^5+\frac{1}{3}x^3\right]_0^{1/2}$$

$$=\pi\left(\frac{1}{224}-\frac{1}{32}+\frac{1}{24}\right)$$

$$=\frac{5}{336}\pi$$

(c) $2\pi\int_0^{1/2}x(x-2x^3-x^2)dx$

$$=2\pi\int_0^{1/2}(x^2-2x^4-x^3)dx$$

$$=2\pi\left[-\frac{2}{5}x^5-\frac{1}{4}x^4+\frac{1}{3}x^3\right]_0^{1/2}$$

$$=2\pi\left(-\frac{1}{80}-\frac{1}{64}+\frac{1}{24}\right)$$

$$=\frac{13}{480}\pi$$

6.3 Volumes by Cylindrical Shells

YouTube http://youtu.be/qM1izf8qeX8

Suppose, for example, R is the region bounded by $y=f(x)$, $x=a$ and $x=b$ and V is true solid obtained by revolving R about the y-axis (see Figure 2). When we use the formula $V=\int_c^d A(y)dy$ for the volume, we must solve equation $y=f(x)$ for x to get an equation of the form $x=g(y)$. That task may be not easy, so in this section we are going to present another approach, one that does not require solving the equation $y=f(x)$ for x.

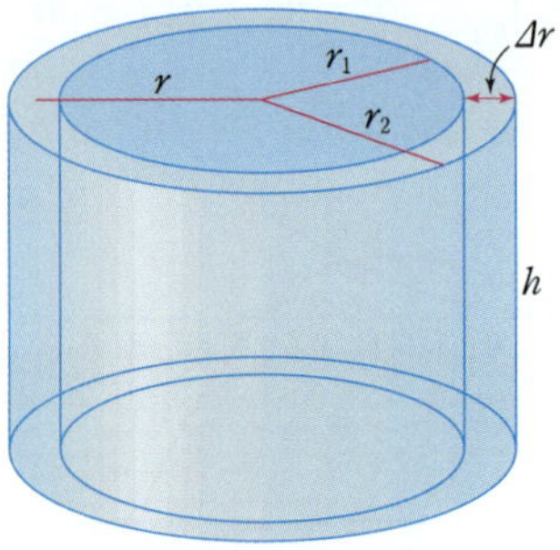

Figure 1

Method of cylindrical shells: Figure 1 shows a cylindrical shell with inner radius r_1, outer radius r_2, and height h. Its volume V is calculated by subtracting the volume V_1 of the inner cylinder from the volume V_2 of the outer cylinder:

$$\begin{aligned} V &= V_2 - V_1 \\ &= \pi r_2^2 h - \pi r_1^2 h = \pi(r_2^2 - r_1^2)h \\ &= \pi(r_2+r_1)(r_2-r_1)h \\ &= 2\pi\frac{r_2+r_1}{2}h(r_2-r_1). \end{aligned}$$

Let the thickness of the shell be $\Delta r = r_2 - r_1$ and the average radius of the shell be $r = 1/2(r_2+r_1)$. Then the formula for the volume of a cylindrical shell is

1

$$V = 2\pi rh\Delta r$$

$V=$ (circumference) · (height) · (thickness of shell)

Now let S be the solid obtained by rotating about the y-axis the region bounded by $y=f(x)$ (where $f(x)\geq 0$), $y=0$, $x=a$, and $x=b$, where $0\leq a<b$. (See Figure 2.)

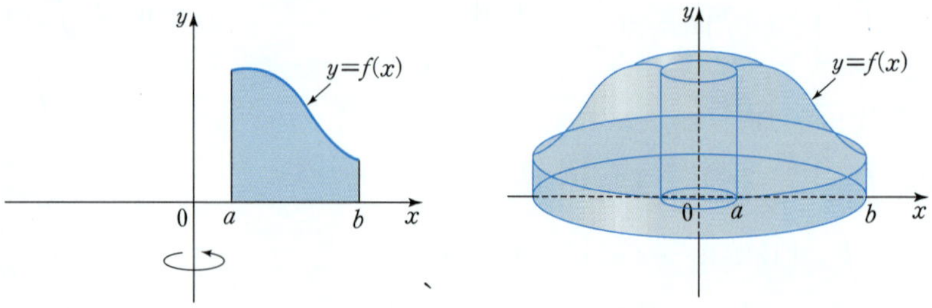

Figure 2

Divide the interval $[a, b]$ into n subintervals $[x_{i-1}, x_i]$ of equal width Δx and let $\overline{x}_i$ be the midpoint of the ith subinterval. If the rectangle with base $[x_{i-1}, x_i]$ and height $f(\overline{x}_i)$ is rotated about the y-axis, then the result is a cylindrical shell with average radius $\overline{x}_i$, height $f(\overline{x}_i)$, and thickness Δx (see Figure 3), so by Formula 1 its volume is

$$V_i = (2\pi\overline{x}_i)[f(\overline{x}_i)]\Delta x.$$

Therefore, an approximation to the volume V of S is given by the summation of the volumes of these shells:

$$V \approx \sum_{i=1}^{n} V_i = \sum_{i=1}^{n} 2\pi\overline{x}_i f(\overline{x}_i)\Delta x$$

As $n \to \infty$ we have

$$\lim_{n\to\infty}\sum_{i=1}^{n} 2\pi\overline{x}_i f(\overline{x}_i)\Delta x = \int_a^b 2\pi x f(x)dx.$$

Thus,

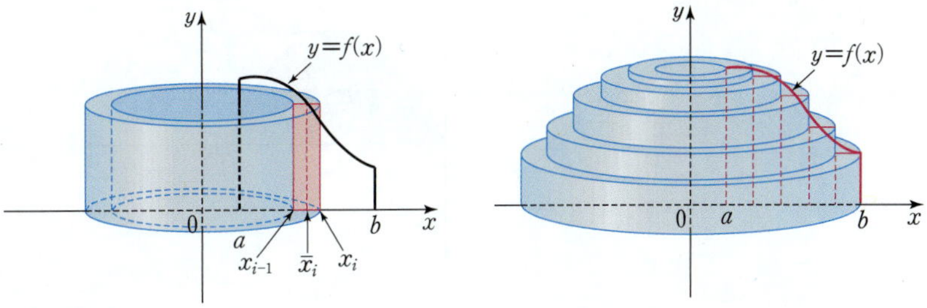

Figure 3

The volume of the solid in Figure 3, obtained by rotating about the y-axis the region under the curve $y = f(x)$ from a to b, is

2 $$V = \int_a^b 2\pi x f(x)dx \qquad \text{where} \qquad 0 \le a < b.$$

Remember Formula 2 by imagining a typical shell, cut and flattened as in Figure 4, with radius x, circumference $2\pi x$, height $f(x)$ and thickness Δx or dx:

$$\int_a^b \underbrace{(2\pi x)}\ \underbrace{[f(x)]}\ \underbrace{dx}$$

(circumference) · (height) · (thickness of shell).

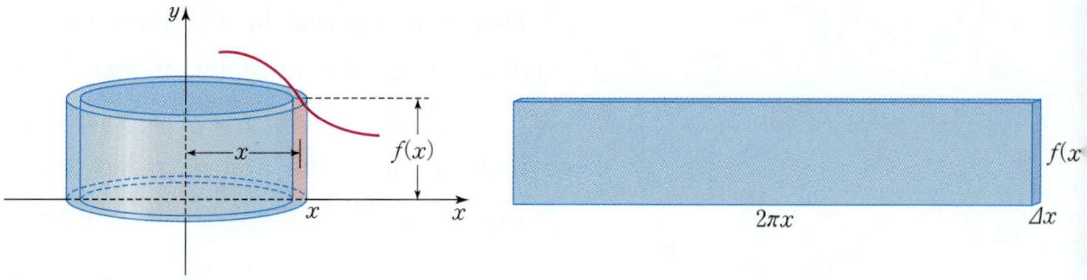

Figure 4

Note Draw a diagram to help identify the radius and height of a shell.[1)]

1)http://www.math.psu.edu/dlittle/java/calculus/volumeshells.html

EXAMPLE 1

The region R between the line $y=0$ and curve $y=3x-x^2$ rotated about the y-axis to form a solid of revolution S. Find the volume of the solid of revolution.

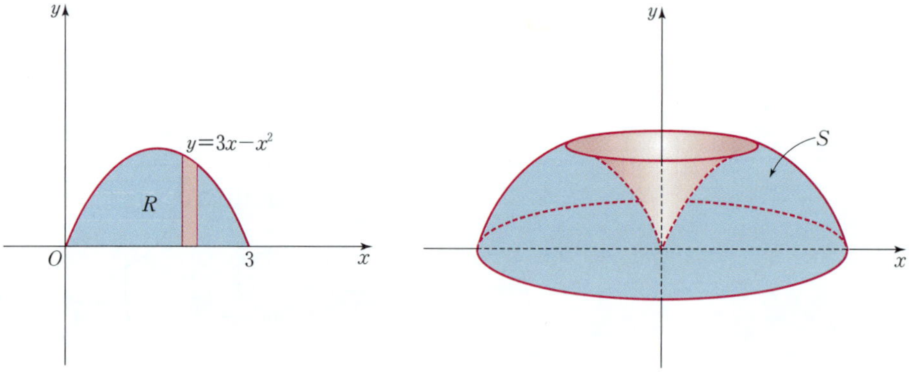

Figure 5

http://matrix.skku.ac.kr/cal-lab/cal-6-3-Exm-1.html

Solution We use the cylindrical shell method because y is the dependent variable. We see that curve crosses the x-axis at $x=0$ and $x=3$, and sketch the region in Figure 5. The volume is

$$V=\int_0^3 2\pi x\,(3x-x^2)\,dx = 2\pi\int_0^3 (3x^2-x^3)\,dx$$

$$=2\pi\left[x^3-\frac{1}{4}x^4\right]_0^3=\frac{27}{2}\pi.$$

```
show(plot(3*x-x^2, x, 0, 3, ticks=[[],[]], fill='axis', fillcolor='wheat') + plot(3*x-x^2, 1.9, 2, fill='axis'))
revolution_plot3d(3*x-x^2 , (x, 0, 3), show_curve= True, opacity=0.7).show(aspect_ratio=(1,1,1), parallel_axis='y')
```

■

EXAMPLE 2

The region R between the curves $y=x$ and $y=\sqrt{x}$ is rotated about the y -axis. Find the volume of the solid of revolution.

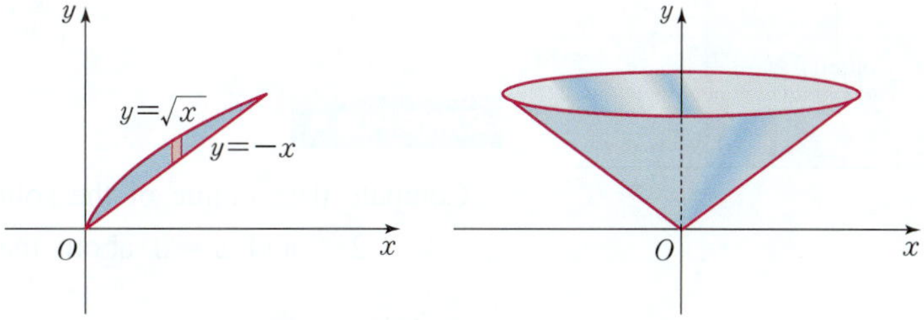

Figure 6

http://matrix.skku.ac.kr/cal-lab/cal-6-3-Exm-2.html

Solution We make a sketch in Figure 6 and find that the curves cross at $x=0$ and $x=1$. We take x for the independent variable and use the Cylindrical Shell Method.

$$V=\int_0^1 2\pi x(\sqrt{x}-x)\,dx = 2\pi\int_0^1 (x^{3/2}-x^2)\,dx$$

$$=2\pi\left[\frac{2}{5}x^{5/2}-\frac{1}{3}x^3\right]_0^1=\frac{2}{15}\pi$$

Some regions are more easily described by taking y as the independent variable, so that R is the region between $x=f(y)$ and $x=g(y)$ for $c\le y\le d$. The volumes of the solids of revolution are then computed by integrating with respect to y. Often we have a choice of either x or y as the independent variable.

```
var('u')
sur1=revolution_plot3d(sqrt(u), (u, 0, 1), opacity=0.5,
show_curve=True, parallel_axis='z')
sur2=revolution_plot3d(u, (u, 0, 1), opacity=0.5,
show_curve=True, parallel_axis='z')
(sur1+sur2).show(aspect_ratio=(1, 1, 1))
plot(sqrt(x), x, 0, 1, ticks=[[],[]], fill=x,fillcolor='wheat') +
plot(sqrt(x), x, 0.5, 0.55, ticks=[[],[]], fill=x,) + plot(x, 0, 1)
```

■

EXAMPLE 3

Determine the volume of the solid obtained by rotating about the x-axis the region under the curve $y=2\sqrt{x}$ from $x=0$ to $x=1$ using cylindrical shells.

Solution Rewrite the curve $y=2\sqrt{x}$ as $x=y^2/4$ in Figure 7. For rotation about the x-axis we see that a typical shell has radius y,

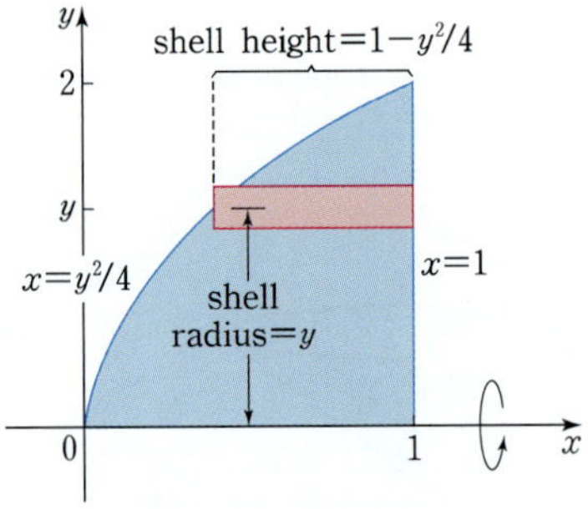

Figure 7

circumference $2\pi y$, and height $1-(y^2/4)$. So the volume is

$$V=\int_0^2 (2\pi y)\left(1-\frac{y^2}{4}\right)dy = 2\pi\int_0^2\left(y-\frac{y^3}{4}\right)dy$$
$$=2\pi\left[\frac{y^2}{2}-\frac{y^4}{16}\right]_0^2 = 2\pi.$$

■

EXAMPLE 4

Compute the volume of the solid obtained by rotating the region bounded by $y=4x-2x^2$ and $y=0$ about the line $x=2$.

Solution The region and a cylindrical shell formed by rotation about the line $x=2$ are shown in Figure 8. It has radius $2-x$, circumference $2\pi(2-x)$, and height $4x-2x^2$.

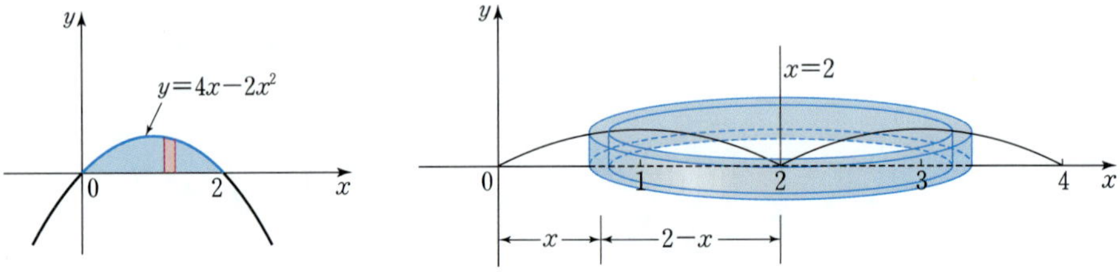

Figure 8

The volume of the given solid is

$$V=\int_0^2 2\pi(2-x)(4x-2x^2)dx = 2\pi\int_0^2 (2x^3-8x^2+8x)dx$$
$$=2\pi\left[\frac{1}{2}x^4-\frac{8}{3}x^3+4x^2\right]_0^2 = \frac{16}{3}\pi.$$

■

CAS **EXAMPLE 5**

Find the volume generated by rotating the region bounded by the given curves about the y–axis using the method of cylindrical shells.

$$y=-x+2\sin x,\ y=0,\ x=\frac{\pi}{3}$$

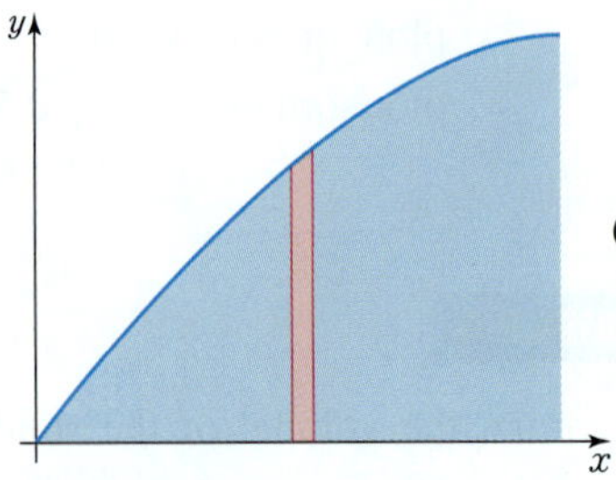
Figure 9

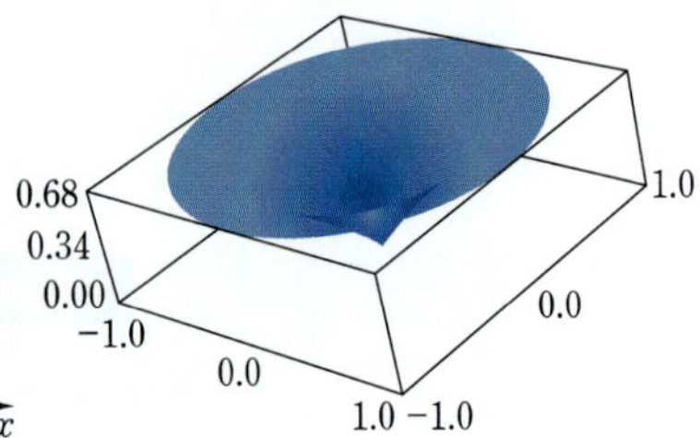
Figure 10

http://matrix.skku.ac.kr/cal-lab/cal-6-3-Exm-5.html

```
var('x, y')
p1 = plot(-x+2*sin(x), (x,0,pi/2), rgbcolor='red');
p2 = parametric_plot((x,0),(x,0,pi/2), rgbcolor='blue');
p3 = parametric_plot((0,y),(y,-1,1), rgbcolor='blue');
p4 = parametric_plot((pi/3,y),(y,-1,1), rgbcolor='blue');
show(p1+p2+p3+p4, aspect_ratio=1)
print integral(2*pi*x*(-x+2*sin(x)), x, 0, pi/3)
```

Answer : -2/81*(27*pi + pi^3 - 81*sqrt(3))*pi

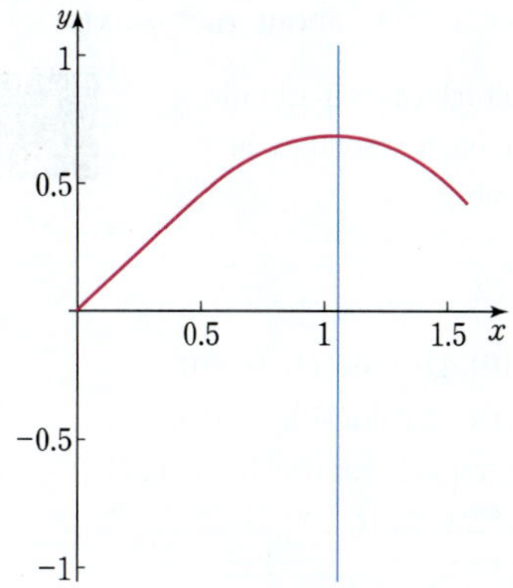

Figure 11

```
plot(-x+2*sin(x),x,0,pi/3,ticks=[[],[]],fill='axis',fillcolor='wheat')+plot(
-x+2*sin(x),x,0.5,0.55,fill='axis',ticks=[[],[]])+plot(-x+2*sin(x),0,pi/3)
```

```
revolution_plot3d(-x+2*sin(x),(x,0,pi/3),show_curve=False,opacity=0.
7).show(aspect_ratio=(1,1,1),parallel_axis='y')
```

■

6.3 EXERCISES (Volumes by Cylindrical Shells)

http://matrix.skku.ac.kr/Cal-Book/part1/CS-Sec-6-3-Sol.html

YouTube http://youtu.be/gNaKkA0UNHg

CAS **1-10.** Find the volume generated by rotating the region bounded by the given curves about the y-axis using the method of cylindrical shells.

1. $y = \dfrac{2}{x}$, $y = 0$, $x = 2$, $x = 3$; about the y-axis

http://matrix.skku.ac.kr/cal-lab/cal-6-3-1.html

[You may refer to other open resources in http://math1.skku.ac.kr/pub/]

Solution

```
var('x, y')
p1=plot(2/x, (x, 0, 10),rgbcolor=(1, 0, 0));
p2=plot(0, (y, -10, 10),rgbcolor=(0, 1, 0));
show(p1+p2, aspect_ratio=1, xmin=-1, ymax=10)
print integral(2*x*pi*2/x, x, 2, 3)
```

Answer : 4*pi

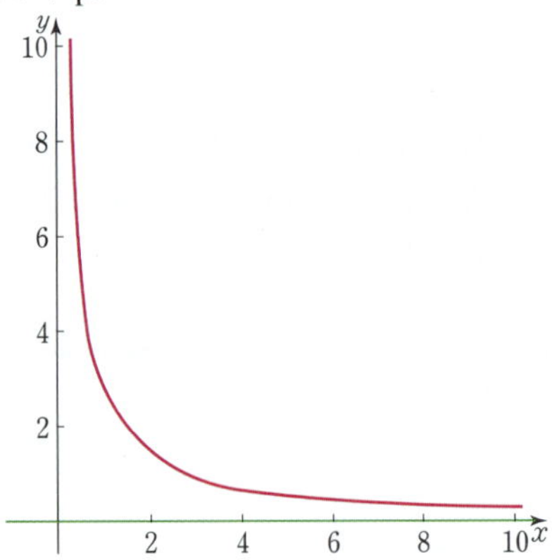

2. $y = -x^2 + 4$, $2x + y = 4$; about the y-axis

http://matrix.skku.ac.kr/cal-lab/cal-6-3-2.html

Solution

```
var('x,y')
p1=plot(-x^2+4 ,(x, -3, 3), rgbcolor=(1, 0, 0));
p2=plot(-2*x+4, (x, -3, 3), rgbcolor=(0, 1, 0));
show(p1+p2, aspect_ratio=1)
solve(-x^2+4==-2*x+4, x)
integral(2*pi*x(-x^2+4-(-2*x+4)), x, 0, 2)
```

Answer : 8/3*pi

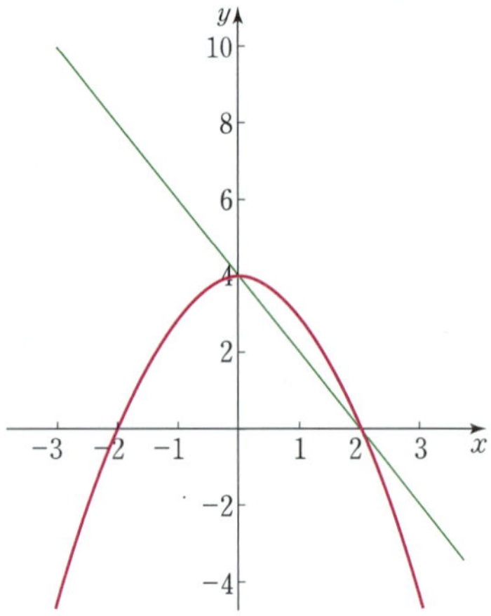

3. $y = 3(x-3)^2$, $y = x^2 - 6x + 11$; about the y-axis

http://matrix.skku.ac.kr/cal-lab/cal-6-3-3.html

Solution

```
var('x,y')
p1=plot(3*(x-3)^2, (x, 0, 5), color="red");
p2=plot(x^2-6*x+11, (x, 0, 5), color="green");
show(p1+p2, aspect_ratio=1)
solve(3*(x-3)^2==x^2-6*x+11, x)
integral(2*pi*x(x^2-6*x+11-(3*(x-3)^2)), x, 2, 4)
```

Answer : 16/3*pi

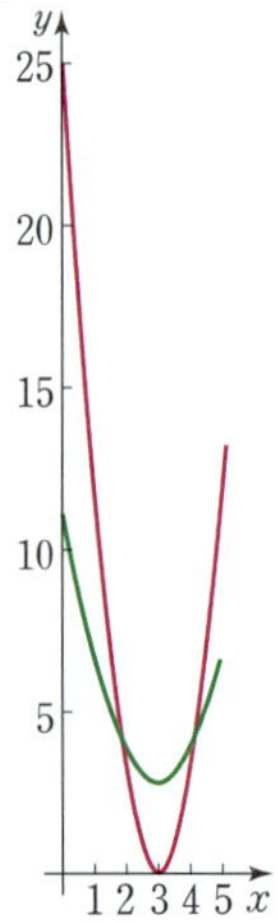

4. $x = 1 + 2y^2$, $x = 0$, $y = 1$, $y = 2$; about the x-axis

http://matrix.skku.ac.kr/cal-lab/cal-6-3-4.html

Solution

```
var('x,y')
p1=plot(1+2*y^2, (y, -3, 3), color="red");
show(p1, aspect_ratio=1)
integral(2*pi*y*(1+2*y^2), y, 1, 2)
```

Answer : 18*pi

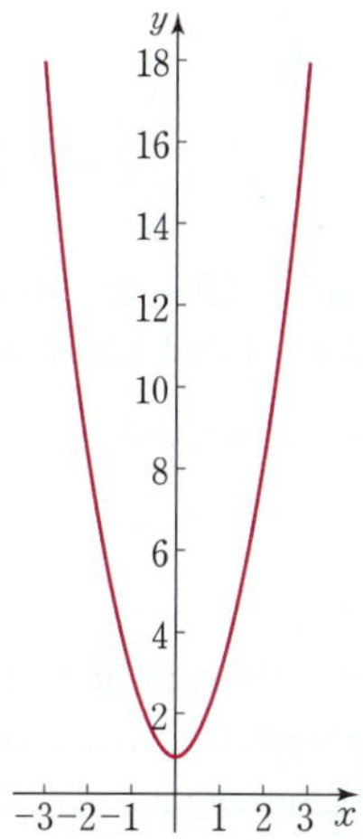

5. $y=2x^2$, $y=0$, $x=1$, $x=2$; about $x=1$

http://matrix.skku.ac.kr/cal-lab/cal-6-3-6.html

Solution

```
var('x')
diff(x^(15/14)+5*e^x,x)
var('x,y')
f(x)=2*x^2
g(x)=0
p1=plot(f(x),(x,-4,4),color="red");
p2=plot(g(x),(x,-4,4),color="green");
show(p1+p2,aspect_ratio=1)
integral(2*pi*x*(2*x^2),x,1,2)
```

Answer : 15*pi

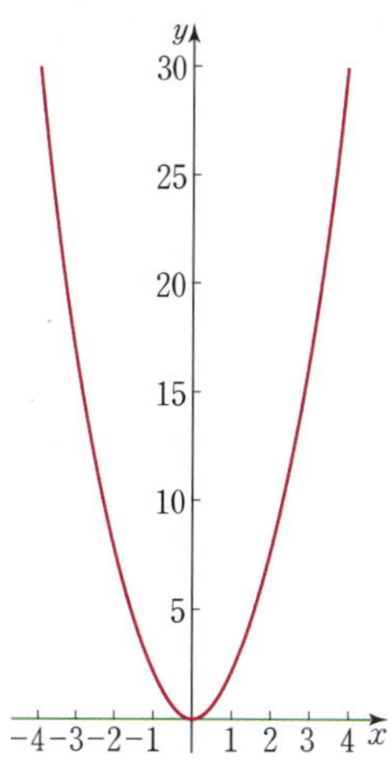

6. $y=2x-x^2$, $y=4x-2x^2$; about $x=-1$

http://matrix.skku.ac.kr/cal-lab/cal-6-3-7.html

Solution

```
var('x,y')
f(x)=2*x-x^2
g(x)=4*x-2*x^2
p1=plot(f(x),(x,-1,3),rgbcolor=(1,0,0));
p2=plot(g(x),(x,-1,3),rgbcolor=(0,1,0));
show(p1+p2,aspect_ratio=1)
solve(f(x)==g(x),x)
integral(2*pi*(x+1)*(g(x)-f(x)),x,0,2)
```

Answer : 16/3*pi

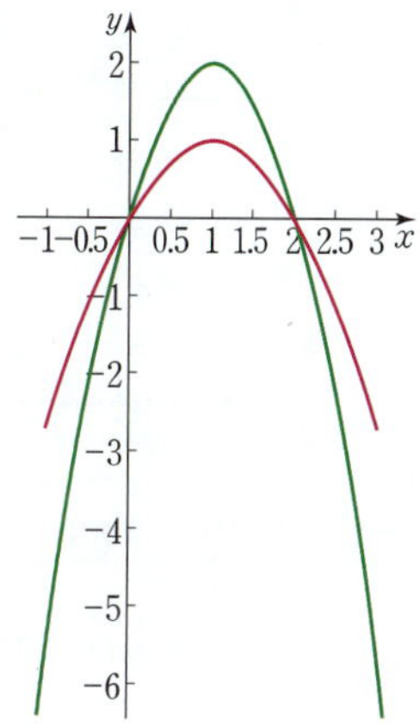

7. $y=\ln x$, $y=0$, $x=3$; about the y-axis

http://matrix.skku.ac.kr/cal-lab/cal-6-3-8.html

Solution

```
var('x,y')
f(x)=ln(x)
g(x)=0
p1=plot(f(x),(x,0,3),rgbcolor=(1,0,0));
p2=plot(g(x),(x,0,3),rgbcolor=(0,1,0));
show(p1+p2,aspect_ratio=1)
integral(2*pi*x*ln(x),x,1,3)
```

Answer : (9 ln3 - 4)*pi

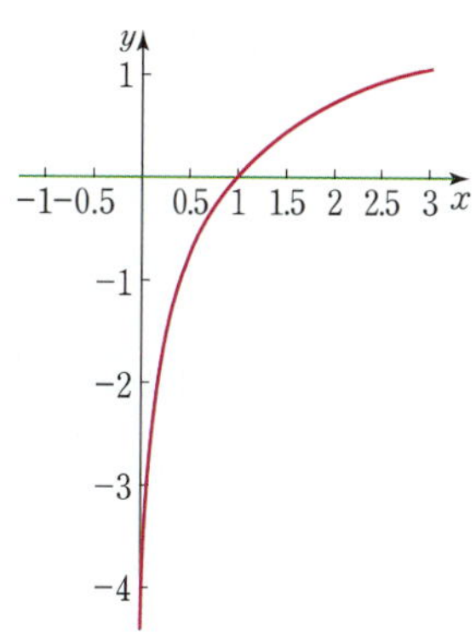

8. $y = x^4 - 1$, $y = \cos\frac{\pi}{2}$; about $x = -2$

http://matrix.skku.ac.kr/cal-lab/cal-6-3-9.html

Solution

```
var('x,y')
assume(x>=0)
f(x)=x^4-1
g(x)=cos(pi/2)
p1=plot(f(x),(x,-1,1),rgbcolor=(1,0,0));
p2=plot(g(x),(x,-1,1),rgbcolor=(0,1,0));
show(p1+p2,aspect_ratio=1)
integral(2*pi*(2+x)*(g(x)-f(x)),x,-1,1)
```

Answer : 32/5*pi

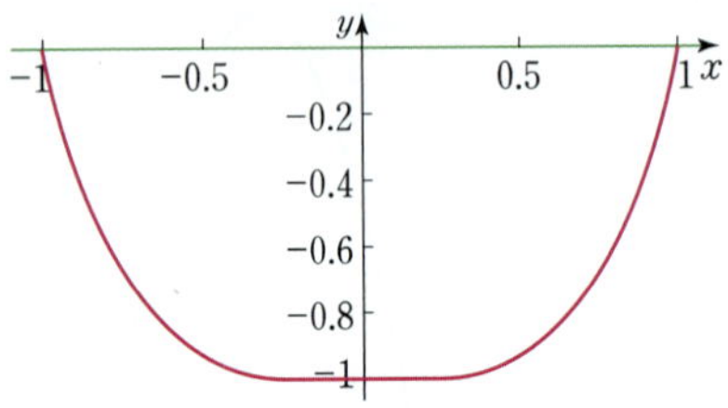

9. $y = x^2 - 5x + 6$, $y = 0$; about the y-axis

http://matrix.skku.ac.kr/cal-lab/cal-6-3-15.html

Solution

```
var('x')
f(x)=x^2-5*x+6
g(x)=0
p1=plot(f(x),(x,0,4),rgbcolor=(1,0,0));
p2=plot(g(x),(x,0,4),rgbcolor=(0,1,0));
show(p1+p2,aspect_ratio=1)
solve(f(x)==g(x),x)
integral(2*pi*x*abs(f(x)),x,2,3)
```

Answer : 5/6*pi

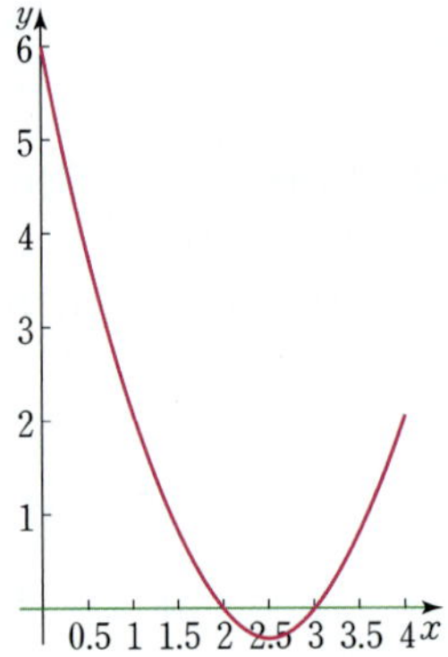

10. $y = 2$, $y = x^2 - 4x + 2$; about $x = -1$

http://matrix.skku.ac.kr/cal-lab/cal-6-3-16.html

Solution

```
var('x')
f(x)=2
g(x)=x^2-4*x+2
p1=plot(f(x),(x,0,4),rgbcolor=(1,0,0));
p2=plot(g(x),(x,0,4),rgbcolor=(0,1,0));
show(p1+p2,aspect_ratio=1)
solve(f(x)==g(x),x)
integral(2*pi*(x+1)*(f(x)-g(x)),x,0,4)
```

Answer : 64*pi

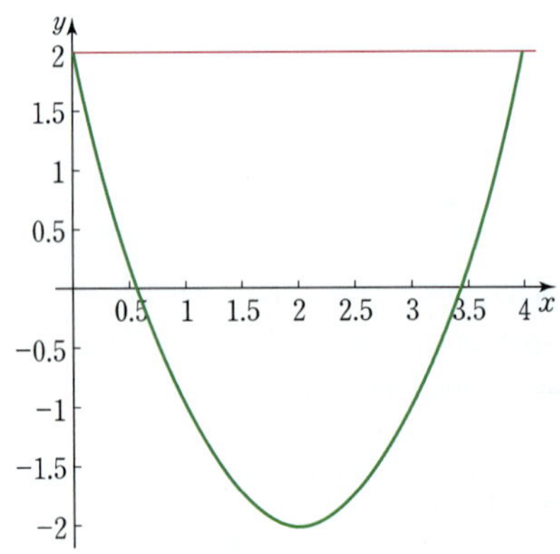

6.4 Work

YouTube http://youtu.be/u3ZaJWhKy6k

In this section we apply the definite integral to find the amount of work that is done by a force moving an object. Work is the total amount of effort required to perform a task. You may think of a force as describing a push or pull on an object. Consider an object moving along a straight line with position function $s(t)$. Then by Newton's Second Law of Motion the **force** F on the object (in the same direction) is

1 $$F = m\frac{d^2s}{dt^2}.$$

Thus, a force of 1newton (N = kg · m/s^2) acting on a mass of 1kg produces an acceleration of $1\ \mathrm{m/s^2}$.

In a first course in Physics you will typically look at the work done by a constant force, F. If the (net) force $F = ma$ acting on an object is constant, then the object's acceleration $a = \frac{d^2s}{dt^2}$ is constant (from 1 with constant mass m). In this case the work done is defined to be the product of the force F and the distance d that the object moves:

2 $$W = Fd \quad (\text{work} = \text{force} \times \text{distance})$$

The unit for W is a newton-meter, which is called a joule (J). Assume that the acceleration due to gravity is $g = 9.8\mathrm{m/s^2}$.

EXAMPLE 1

Find the work done in lifting a $1.4\ \mathrm{kg}$ book off the floor to put it on a desk that is $0.8\ \mathrm{m}$ high.

Solution From 1 the force exerted is equal and opposite to that exerted by gravity g, so

$$F = mg = (1.4)(9.8) = 13.72\ \mathrm{N}.$$

Then from 2 the work done is

$$W = Fd = (13.72)(0.8) \approx 10.976\ \mathrm{J}.$$

■

Variable Force

However, most forces are not constant and will depend the exact position where the force is acting. So, let us assume that the force at any x is given by $F(x)$. Then the **work done in moving the object from a to b** is

3 $$W=\int_a^b F(x)dx.$$

To see a justification of this formula see the Proof of Various Integral Properties[1] **by Paul Dawkins.** Notice that if the force is constant we get the same formula for a constant force $W=\int_a^b Fdx=Fx]_a^b=F(b-a)$ where $b-a$ is simply the distance moved, d. So, let us look at examples with non-constant forces.

1)http://tutorial.math.lamar.edu/Classes/CalcI/ProofIntProp.aspx#Extras_IntPf_Work'

EXAMPLE 2

Find the work done in moving a particle from $x=1$ to $x=5$ when it is located at a distance x meters from the origin, and a force of $\ln x$ Newtons acts on it.

Solution $W=\int_1^5(\ln x)dx=x\ln x-x\Big]_1^5=5\ln 5-4$

The work done is $5\ln 5-4$ (Joule). ■

Hooke's Law

The force, $F(x)$, required to maintain a spring stretched x units beyond its natural length is proportional to x, so $F(x)=kx$ where k is a positive constant (called the **spring constant**) provided that x is not too large. (See Figure 1.)

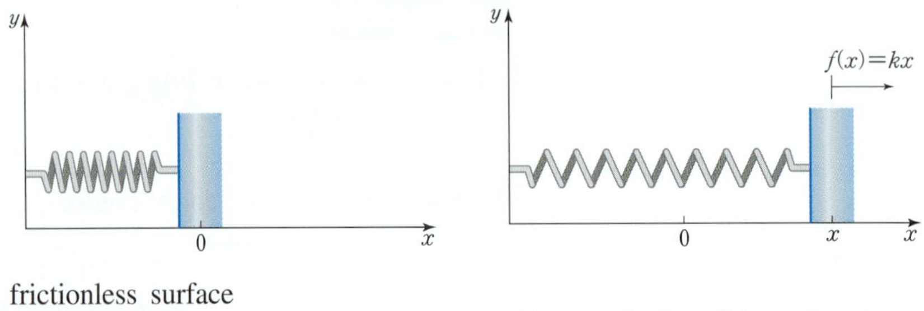

(a) Natural position of spring (b) Stretched position of spring

Figure 1 Hooke's Law

EXAMPLE 3

A spring has been stretched from its natural length of 20 cm to a length of 30 cm by a force of 40 newton(N). Find the work done in stretching the spring from 40 cm to 50 cm.

Solution This example will require Hooke's Law to determine the force. Hooke's Law tells us that the force required to stretch a spring a distance of x meters from its natural length is, $F(x) = kx$ where $k > 0$ is called the spring constant. The first thing that we need to do is determine the spring constant for this spring. We can do that using the initial information. A force of 40 N is required to stretch the spring 30 cm – 20 cm = 10 cm = 0.10 m from its natural length. Using Hooke's Law we have, $40 = 0.10k \Rightarrow k = 400$. Thus

$$F(x) = kx = 400x.$$

Now let us find the work done in stretching the spring from 40 cm to 50 cm. First we need to convert these distances from the natural length in centimeters, to meters, doing this gives x values of 0.2 (=0.40-0.20) m and 0.3 (=0.50-0.20) m. The work is then,

$$W = \int_{0.2}^{0.3} 400x\,dx = 200\,x^2\Big]_{0.2}^{0.3} = 10\,(\text{Joule}).$$ ■

EXAMPLE 4

We have a cable that weighs 2lbs attached to a bucket filled with coal that weighs 700lbs. The bucket is initially at the bottom of a 500ft mine shaft. Determine the amount of work required to lift the bucket all the way up the shaft.

Solution Before answering we first need to determine the force. In this case the force will be the weight of the bucket and cable at any point in the shaft. To determine a formula for this first we will need to set a convention for x, so for this problem, we will set x to be the amount of cable that has been pulled up. So at the bottom of the shaft $x = 0$, and at the top of the shaft $x = 500$. Also at any point in the shaft there is $500 - x$ feet of cable still in the shaft.
Then the force then for any x is nothing more than the weight of the cable and bucket at that point. This is,

$$\begin{aligned} F(x) &= \text{weight of cable + weight of bucket/coal} \\ &= 2(500 - x) + 700 = 1700 - 2x \end{aligned}$$

We can now answer the question. Then the total work done is

$$W = \int_0^{500} (1700 - 2x)\,dx = (1700x - x^2)\Big]_0^{500} = 600{,}000\,(\text{ft-lb}).$$ ■

EXAMPLE 5

A tank has the shape of an inverted circular cone with height 15 m and base radius 4 meters and is filled with water to a depth of 12 meters. Determine the amount of work needed to pump all of the water to the top of the tank. Assume that the density of the water is $1000\,\text{kg/m}^3$.[2)]

2) Example by Paul Dawkins

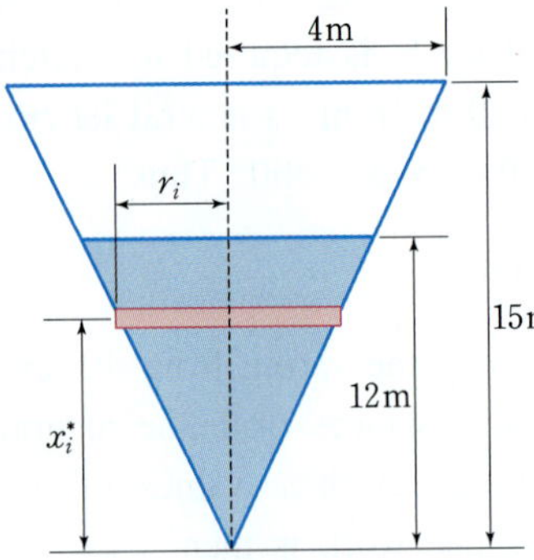

Figure 2

Solution In this case we cannot determine a force function, $F(x)$ simply. So, we are going to need to approach this from a different viewpoint. Let us first set $x=0$ to be the lower end of the tank/cone and $x=15$ to be the top of the tank/cone. With this definition of the x-values we can now see that the water in the tank will correspond to the interval [0, 12]. So, let us start by dividing [0, 12] into n subintervals each of width Δx and let x_i^* be any point from the ith subinterval where $i=1, 2, \cdots, n$. This divides the water into n layers. The ith layer is approximated by a circular cylinder with radius r_i and height Δx. Now, for each subinterval we will approximate the water in the tank corresponding to that interval as a cylinder of radius r_i and height Δx. Here is a quick sketch of the tank. (See Figure 2.)

The red strip in the sketch represents the "cylinder" of water in the ith subinterval. A quick application of similar triangles will allow us to relate r_i to x_i^* as follows. $\frac{r_i}{x_i^*}=\frac{4}{15}$, so $r_i=\frac{4}{15}(x_i^*)$. Now, the mass, m_i, of the volume of water, V_i, for the ith subinterval is simply,

$$m_i = \text{density} \times V_i.$$

We know the density of the water (it was given) and because we are approximating the water in the ith subinterval as a cylinder we can easily approximate the volume, $V_i \approx \pi(\text{radius})^2(\text{height})$, and hence the mass of the water in the ith subinterval. The mass for the ith subinterval is approximately,

$$\begin{aligned} m_i &= \text{density}\times \text{volume} = \text{density}\times V_i \\ &= 1000\pi(r_i)^2\Delta x = 1000\pi\left(\frac{4}{15}x_i^*\right)^2\Delta x \\ &= \frac{640}{9}\pi(x_i^*)^2\Delta x. \end{aligned}$$

To raise this volume of water we need to overcome the force of gravity that is acting on the volume and that is, $F_i = m_i g$, where $g=9.8\ \text{m/s}^2$ is the gravitational acceleration. The force to raise the volume of water in the ith subinterval is then approximately,

$$F_i = m_i g \approx (9.8)\frac{640}{9}\pi(x_i^*)^2\Delta x.$$

Next, in order to reach the top of the tank the water in the ith subinterval will need travel approximately $15-x_i^*$ m to reach the top of the tank. Because the volume of the water in the ith subinterval is constant the force needed to raise the water through any distance is also a constant force. Therefore the work to move the volume of water in the ith subinterval to the top of the tank, i.e. raise it a distance of $15 - x_i^*$, is then approximately,

$$W_i \approx F_i(15 - x_i^*) = (9.8)\frac{640}{9}\pi(x_i^*)^2 \Delta x\,(15 - x_i^*)$$

The total amount of work required to raise all the water to the top of the tank is then approximately the sum of each of the W_i for $i = 1, 2, \cdots, n$. Or,

$$W \approx \sum_{i=1}^{n} (9.8)\frac{640}{9}\pi(x_i^*)^2\,(15 - x_i^*)\,\Delta x$$

To get the actual amount of work we simply need to take $n \to \infty$. That is compute the following limit,

$$W = \lim_{n\to\infty}\sum_{i=1}^{n} (9.8)\frac{640}{9}\pi(x_i^*)^2\,(15 - x_i^*)\,\Delta x$$

This limit of a summation should look familiar to you. It has probably been some time, but recalling the definition of the definite integral we can see that this is nothing more than the following definite integral,

$$W = \int_0^{12} (9.8)\frac{640}{9}\pi x^2\,(15 - x)\,dx = (9.8)\frac{640}{9}\pi\int_0^{12}(15x^2 - x^3)\,dx$$
$$= (9.8)\frac{640}{9}\pi\left[5x^3 - \frac{1}{4}x^4\right]_0^{12} = 7{,}566{,}362.543\text{J}.$$ ■

As we have seen in the previous example we sometimes need to compute "incremental" work and then use that to determine the actual integral we need to compute. This idea does arise on occasion and we should not forget it.

CAS **EXAMPLE 6**

A particle is moved along the x-axis by a force that measures $3/(1 + \sin x)$ N at a point x meters from the origin. Find the work done in moving the particle from the origin to a distance of 3 m.

$$W = \int_0^3 \frac{3}{1 + \sin x}\,dx$$

http://matrix.skku.ac.kr/cal-lab/cal-6-4-exm-6.html

```
var('x')
f=3/(1+sin(x))
integral(f, x, 0, 3)
```

Answer : -6*cos(3)/(sin(3) + cos(3) + 1) - 6/(sin(3) + cos(3) + 1) + 6 ■

* An illustration of Work in a vector field: http://matrix.skku.ac.kr/cal-lab/cal-Work-flow-flux.html

6.4 EXERCISES (Work)

http://matrix.skku.ac.kr/Cal-Book/part1/CS-Sec-6-4-Sol.html

YouTube http://youtu.be/SmIo2yaxNsY

1. Find the work done in pushing a car a distance of 10m while exerting a constant force of 1000N.

Solution Since $F = 1000\text{N}$,

$W = Fd = 1000 \times 10 = 100000$. (in Joules)

2. How much work is done by a weightlifter in raising a 50kg barbell from the floor to a height of 3 meter?

Solution Since $m = 50\text{kg}$,

$F = mg = 50 \times 9.8 = 490$. (in Newtons)

So $W = Fd = 490 \times 3 = 1470$. (in Joules)

3. A particle is moved along the x-axis by a force that measures $5/(1+2x)^2$ pounds at a point x feet from the origin. Find the work done in moving the particle from the origin to a distance of 8ft.

http://matrix.skku.ac.kr/cal-lab/cal-6-4-3.html

Solution $W = \int_0^8 \frac{5}{(1+2x)^2} dx$. Use a change of variable $1+2x = t$, so $2dx = dt$. Then

$$W = \int_1^{17} \frac{5}{2t^2} dt = \frac{5}{2}\left[-\frac{1}{t}\right]_1^{17} = \frac{5}{2}\left(1 - \frac{1}{17}\right)$$

$$= \frac{40}{17}. \quad (\text{ft}-\text{lb})$$

```
var('x')
f(x)= 5/(2*x^2)
print integral(f, x, 1, 17)
```

4. Shown is the graph of a force function (in Newtons) that increases to its maximum value and then remains constant. How much work is done by the force in moving an object a distance of 8m?

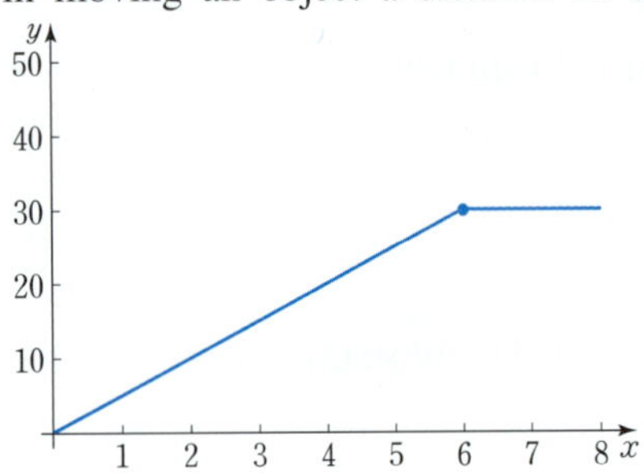

Solution $W = \frac{1}{2} \times 6 \times 30 + 2 \times 30 = 150$. (in Joules)

5. A force of 20 lb is required to hold a spring stretched 6 inch beyond its natural length. How much work is done in stretching it from its natural length to 12 inch beyond its natural length?

Solution Since 6 inch = 0.5 ft and 12 inch = 1 ft,

$k = \frac{20}{0.5} = 40$.

By Hooke's Law, $f(x) = 40x$.

So $W = \int_0^{0.5} 40x\,dx = \left[20x^2\right]_0^{0.5} = 5$ ft$-$lb.

6. A spring has a natural length of 10cm. If a 15N force is required to keep it stretched to a length of 20cm, how much work is required to stretch it from 10cm to 15cm?

Solution By Hooke's Law, $k = \frac{15}{0.1} = 150$,

so $f(x) = 150x$.

$W = \int_0^{0.05} 150x\,dx = \left[75x^2\right]_0^{0.05} = 0.1875$. (in Joules)

7. Suppose that 3J of work is needed to stretch a spring from its natural length of 20cm to a length 38cm, how much work is needed to stretch it from 25cm to 30cm.

Solution $3 = W = \int_0^{0.18} kx\,dx = \left[\frac{k}{2}x^2\right]_0^{0.18} = 0.0162k$.

That is $k = \frac{3}{0.0162} \approx 185.2$.

Hence $W = \int_0^{0.1} 185.2x\,dx = 0.926$. (in Joules)

8-14. Find the work done.

8. A heavy rope, 40ft long, weighs 0.5lb/ft and hangs over the edge of a building 150ft high. How much work is done in pulling the rope to the top of the building?

Solution $W = \int_0^{40} 0.5x\,dx = \left[\frac{1}{4}x^2\right]_0^{40} = 400$. (in Joules)

9. A chain lying on the ground is 15m long and its mass is 75kg. How much work is required to raise one end of the chain to a height of 8m?

Solution $k=\frac{75}{15}=5,\ f(x)=5x.$

So, $W=\int_0^8 5x\,dx=\left[\frac{5}{2}x^2\right]_0^8=160.$ (in Joules)

10. A cable that weight 3lb/ft is used to lift 700lb of coal up a mine shaft 400ft deep. Find the work done.

Solution Since the cable weight is 3lb/ft, the length of cable is $\frac{700}{3}$.

So $W=\int_0^{\frac{700}{3}} 3x\,dx+700\times\left(400-\frac{700}{3}\right)$

$=\left[\frac{3}{2}x^2\right]_0^{\frac{700}{3}}+\frac{350000}{3}=\frac{595000}{3}.$ (in Joules)

11. A bucket that weighs 5lb and a rope of negligible weight are used to draw water from a well that is 90ft deep. The bucket is filled with 30lb of water and is pulled up at a rate of 3ft/s, but water leaks out of a hole in the bucket at a rate of 0.4lb/s. Find the work done in pulling the bucket to the top of the well.

Solution Since $k=\frac{35-0.4t}{90}\times 3=\frac{35-0.4t}{30}$,

$W=\int_0^{90}\frac{35-0.4t}{30}\,t\,dt=1485.$ (in Joules)

12. An 8ft chain weighs 24lb and hangs from a ceiling. Find the work done in lifting the lower end of the chain to the ceiling so that it is level with the upper end.

Solution $k=\frac{24}{8}=3,\ f(x)=3x.$

So $W=\int_0^8 3x\,dx=\left[\frac{3}{2}x^2\right]_0^8=96.$ (in Joules)

13. An aquarium 3m long, 2m side and 2m deep is full of water. Find the work needed to pump half of the water out of the aquarium.

Solution $W=\int_0^{1.5}4x\,dx=\left[2x^2\right]_0^{1.5}=4.5.$ (in Joules)

14. A circular swimming pool has a diameter of 20ft, the sides are 4ft high and the depth of the water is 5ft. How much work is required to pump all of the water out over the side?

Solution $W=\int_0^4 100x\,dx=\left[50x^2\right]_0^4=800.$ (in Joules)

---

15. Newton's Law of Gravitation states that two bodies with masses m_1 and m_2 attract each other with a force

$$F=G\frac{m_1m_2}{r^2}$$

where r is the distance between the bodies and G is the gravitational constant. If one of the bodies is fixed, find the work needed to move the other from $r=a$ to $r=b$.

Solution $W=\int_a^b G\frac{m_1m_2}{r^2}r\,dr=-Gm_1m_2(-\ln b+\ln a)$
$=Gm_1m_2(\ln b-\ln a).$ (in Joules)

16. Use Newton's Law of Gravitation to compute the work required to launch a 1000kg satellite vertically to center.

Take the radius of Earth to be 6.37×10^6m and

$$G=6.67\times10^{-11}\mathrm{N}\cdot\mathrm{m}^2/\mathrm{kg}^2.$$

Solution Use the result of problem 15, then
$W=6.67\times10^{-11}\times1000.$ (in Joules)

6.5 Average Value of a Function

YouTube http://youtu.be/zmEeGmwQTB0

The average value of finitely many numbers y_1, y_2, ..., y_n is defined as

$$y_{\text{ave}} = \frac{y_1 + y_2 + \cdots + y_n}{n}.$$

To compute the average value of a function $y = f(x)$, $a \le x \le b$, divide the interval $[a, b]$ into n equal subintervals, each with length $\Delta x = \dfrac{b-a}{n}$. Take points x_i^*, ..., x_n^* in successive subintervals and calculate the average of the numbers $f(x_i^*)$, ..., $f(x_n^*)$:

Since $\Delta x = (b-a)/n$, so $n = (b-a)/\Delta x$ and the average value becomes

$$\frac{f(x_1^*) + \cdots + f(x_n^*)}{\dfrac{b-a}{\Delta x}} = \frac{1}{b-a}[f(x_1^*) + \cdots + f(x_n^*)]\Delta x.$$

As $n \to \infty$ the limiting value is

$$\lim_{n\to\infty} \frac{1}{b-a} \sum_{i=1}^{n} f(x_i^*)\Delta x = \frac{1}{b-a}\int_a^b f(x)dx.$$

Therefore, the **average value of *f*** on the interval $[a, b]$ is defined as

$$f_{\text{ave}} = \frac{1}{b-a}\int_a^b f(x)dx$$

EXAMPLE 1

Compute the average value of the function $f(x) = x + x^2$ on the interval $[-2, 1]$.

Solution Here $a = -2$ and $b = 1$ and $f(x) = x + x^2$. So

$$f_{\text{ave}} = \frac{1}{b-a}\int_a^b f(x)dx = \frac{1}{1-(-2)}\int_{-2}^{1}(x + x^2)dx$$
$$= \frac{1}{3}\left[\frac{x^2}{2} + \frac{x^3}{3}\right]_{-2}^{1} = \frac{1}{2}.$$

■

The following theorem (which is a consequence of the Mean Value Theorem for derivatives and the FTC) gives a number c at which the value of f is exactly equal to the average value of the function, that is, $f(c) = f_{\text{ave}}$.

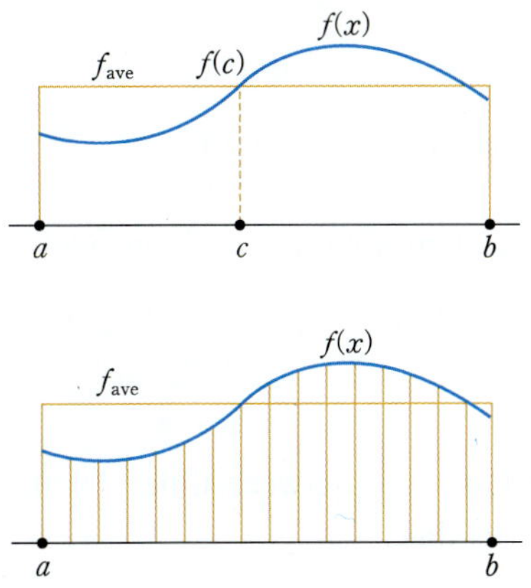

Figure 1

The Mean Value Theorem for Integrals

If f is continuous on $[a, b]$, then there exists a number c in $[a, b]$ such that

$$\int_a^b f(x)dx = f(c)(b-a).$$

Geometric interpretation : For positive functions f, there exists a number c such that the area of the rectangle with base $[a, b]$ and height $f(c)$ is same as the area of the region under the graph of f from a to b. (See Figure 1.)

EXAMPLE 2

Verify the Mean Value Theorem for Integrals for $f(x) = x + x^2$ on the interval $[-2, 1]$.

Solution Since $f(x) = x + x^2$ is continuous on the interval $[-2, 1]$, by the Mean Value Theorem for Integrals there exists a number c in $[-2, 1]$ such that

$$\int_{-2}^{1} (x + x^2)dx = f(c)[1-(-2)].$$

But from Example 1 above we know that $f_{ave} = 1/2$ so the value of c satisfies

$$f(c) = f_{ave} = \frac{1}{2}.$$

Therefore,

$$c + c^2 = \frac{1}{2}.$$

Thus, in this case there exists two numbers $c = 1/2(-1 \pm \sqrt{3})$ in the interval $[-2, 1]$ for which the condition of the Mean Value Theorem for Integrals is valid.
Examples 1 and 2 are illustrated by Figure 2. ■

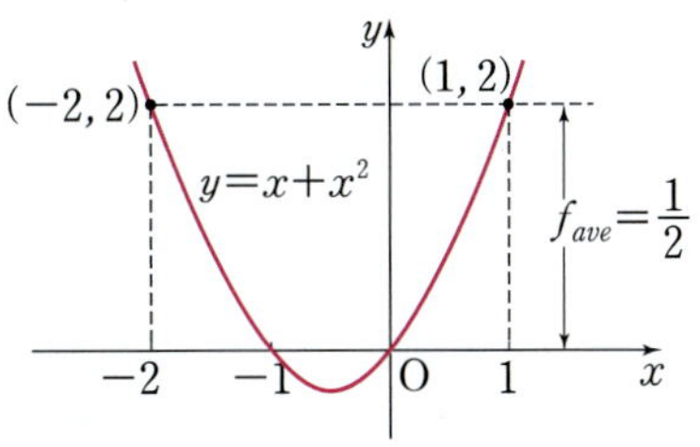

Figure 2

Average and Instantaneous Velocity

For an object moving in a straight line with position $s(t)$ at time t, the average velocity from time t to time $t + h$ is the average rate of change of position with respect to time:

$$v_{ave} = \frac{s(t+h) - s(t)}{h} = \frac{\Delta s}{\Delta t} \quad \text{(Average velocity)}$$
$$= \text{Average rate of change of position}$$

The instantaneous velocity at time t is

$$v = \lim_{h \to 0} \frac{s(t+h) - s(t)}{h} = \frac{ds}{dt} \quad \text{(Instantaneous velocity)}$$

In other words, instantaneous velocity is the derivative of position with respect to time.

EXAMPLE 3

A car travels with velocity $v = 4t + 10$, where t is time. Between times $t = 0$ and $t = 4$ find the average velocity with respect to time.

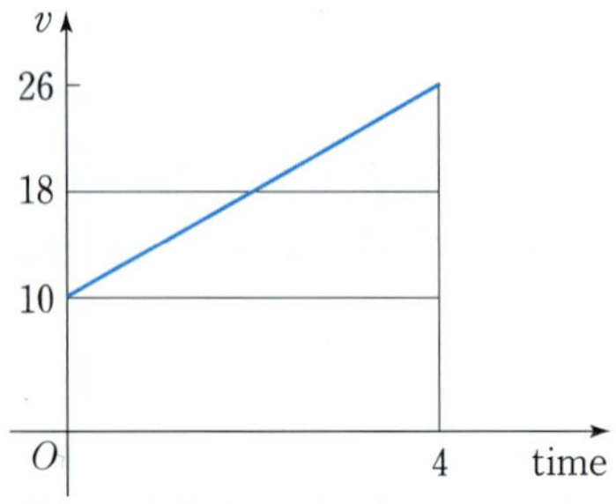

Figure 3

Solution Average velocity with respect to time is

$$v_{ave} = \frac{1}{4}\int_a^b v(t)dt = \frac{1}{4}\int_0^4 (4t+10)dt = \frac{2t^2 + 10t]_0^4}{4} = 18$$

which agrees with

$$v_{ave} = \frac{v(4) + v(0)}{2} = \frac{26 + 10}{2} = 18. \text{ (See Figure 3.)}$$

■

CAS **EXAMPLE 4**

Find the average value of the function on the given interval.

$$y = \sin\sqrt{x}, \ [1, 2]$$

http://matrix.skku.ac.kr/cal-lab/cal-6-5-exm-4.html

Solution

```
var('x, y')
f=sin(sqrt(x))
inc=2-1
fave=integral(f, x, 1, 2)/inc; fave
```

$v_{\text{ave}} = 2\sin\sqrt{2} - 2\sin(1) - 2\sqrt{2}\cos(\sqrt{2}) + 2\cos(1)$

```
p1 = plot(f, (x, 1, 2), rgbcolor='red')
p2 = parametric_plot((x, fave), (x, 1, 2), rgbcolor='blue')
p3 = parametric_plot((1, y), (y, 0.5, 1), rgbcolor='green')
p4 = parametric_plot((2, y), (y, 0.5, 1), rgbcolor='green')
show(p1+p2+p3+p4, aspect_ratio=1)
```

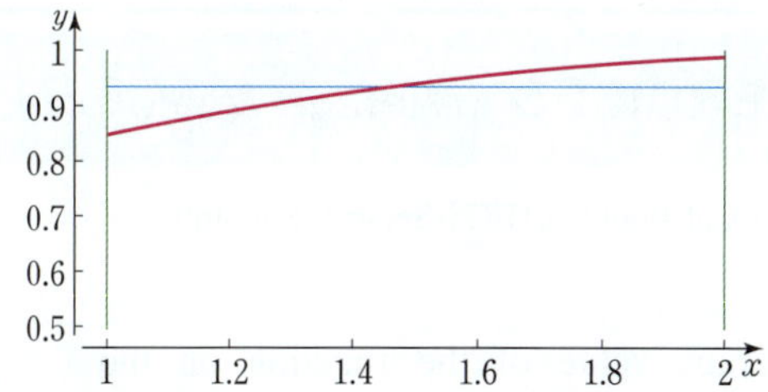

Figure 4

■

CAS **EXAMPLE 5**

Verify the Mean Value Theorem for $f(x) = e^{-x^2}\sin x$ in the interval $[0, 2]$.
http://matrix.skku.ac.kr/cal-lab/cal-6-5-exm-5.html

Solution By the Mean Value Theorem for Integrals, we have a number 0.21 in $[0, 2]$ such that $0.421 \approx \int_0^2 e^{-x^2}\sin x\, dx = f(0.21)(2-0) \approx 0.42$ as shown in the following graph.

```
f(x)=exp(-x^2)*sin(x)
a=0
b=2
f.plot(a, b, legend_label="$e^{-x^2}\sin(x)$").show(figsize=3)
avg=f.integrate(x, a, b)/(b-a).n()
avg
s=numerical_integral(f(x)/(b-a), a, b);s
avg=s[0];avg                    # 0.21
(plot(f, a, b, fill=True)+plot(avg, a, b)).show(figsize=3)
solve(f(x)==avg, x)
c1=find_root(f(x)==avg, 0, 1);c1
c2=find_root(f(x)==avg, 1, 2);c2
(plot(f, a, b)+plot(avg, a, b, fill=True) + line2d([(2, 0), (2,
f(c1))])).show(figsize=3)
```

■

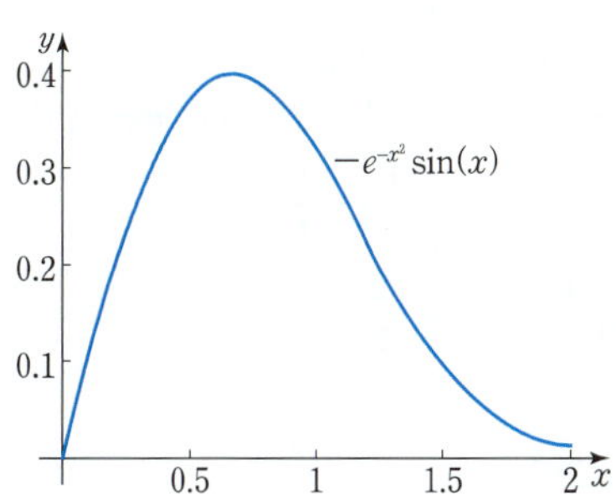

Figure 5

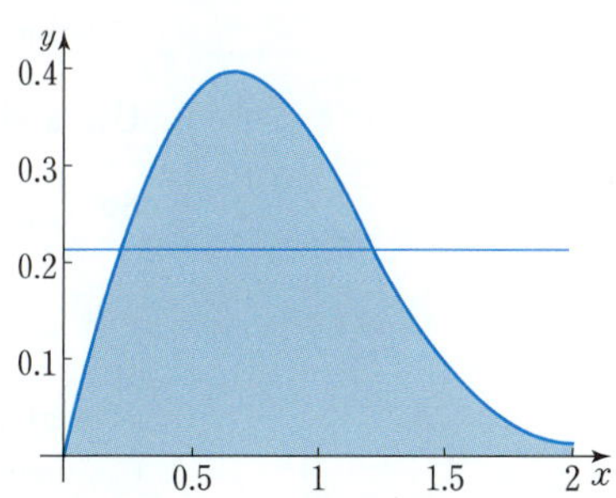

Figure 6

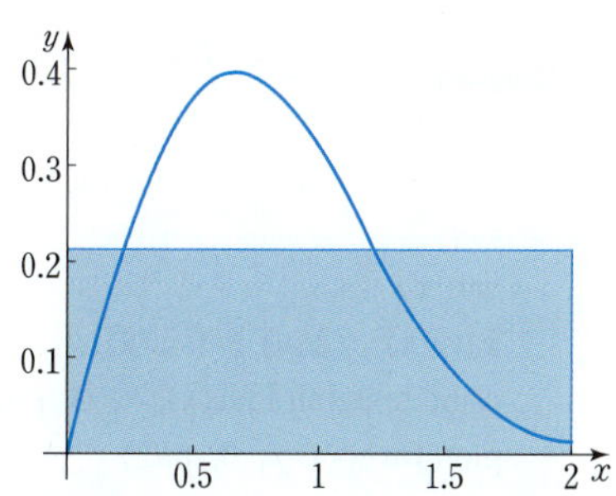

Figure 7

6.5 EXERCISES (Average Value of a Function)

http://matrix.skku.ac.kr/Cal-Book/part1/CS-Sec-6-5-Sol.html

YouTube http://youtu.be/BVahd-DJoe8

1-7. Find the average value of the function on the given interval.

1. $f(x)=x^3,\ [-1,\ 1]$

http://matrix.skku.ac.kr/cal-lab/cal-6-5-1.html

Solution $f_{ave}=\dfrac{1}{[1-(-1)]}\displaystyle\int_{-1}^{1}x^3dx=\dfrac{1}{2}\left[\dfrac{1}{4}x^4\right]_{-1}^{1}$

$=\dfrac{1}{2}\left(\dfrac{1}{4}-\dfrac{1}{4}\right)=0.$

```
var('x')
integral(x^3,x,-1,1)/(1-(-1))
```

Answer : 0

2. $f(x)=\ln x,\ [2,\ 4]$

http://matrix.skku.ac.kr/cal-lab/cal-6-5-2.html

Solution $f_{ave}=\dfrac{1}{(4-2)}\displaystyle\int_{2}^{4}\ln x dx$

$=\dfrac{1}{2}[x\ln x-x]_2^4$

$=\dfrac{1}{2}(4\ln 4-4-2\ln 2+2)$

$=\dfrac{1}{2}(6\ln 2-2)$

$=3\ln 2-1$

```
var('x')
fave(x)= (1/2)*ln(x)
print integral(fave(x), x, 2, 4)
plot(fave(x), x, 2, 4)      # 3*log(2) - 1
```

Answer : 3ln2-1

3. $f(x)=\sin x,\ [0,\ \pi/2]$

http://matrix.skku.ac.kr/cal-lab/cal-6-5-3.html

Solution $f_{ave}=\dfrac{2}{\pi}\displaystyle\int_{0}^{\frac{\pi}{2}}\sin x dx=\dfrac{2}{\pi}[-\cos x]_0^{\frac{\pi}{2}}$

$=\dfrac{2}{\pi}(0-(-1))=\dfrac{2}{\pi}$

```
var('x')
Tave(x)= (2/pi) * (sin(x))
print integral(Tave(x), x, 0, pi/2)
plot(Tave(x), x, 0, pi/2)
```

Answer : 2/pi

4. $f(x)=\dfrac{1}{\sqrt{1-x^2}},\ [0,\ 1]$

Solution $f_{ave}=\dfrac{1}{(1-0)}\displaystyle\int_{0}^{1}\dfrac{1}{\sqrt{1-x^2}}dx$

$=\displaystyle\int_{0}^{1}\dfrac{1}{\sqrt{1-x^2}}dx$

To calculate this integral, use a change of variable $x=\sin\theta$, then $dx=\cos\theta d\theta$. So,

$$f_{ave}=\int_{0}^{1}\frac{1}{\sqrt{1-x^2}}dx=\int_{0}^{\frac{\pi}{2}}\frac{\cos\theta}{\sqrt{1-\sin^2\theta}}d\theta$$

$$=\int_{0}^{\frac{\pi}{2}}1d\theta=\frac{\pi}{2}$$

5. $f(x)=2^x\ln 2,\ [0,\ 5]$

http://matrix.skku.ac.kr/cal-lab/cal-6-5-5.html

Solution $f_{ave}=\dfrac{1}{5-0}\displaystyle\int_{0}^{5}2^x\ln 2dx=\dfrac{1}{5}[2^x]_0^5$

$=\dfrac{1}{5}(32-1)=\dfrac{31}{5}$

```
var('x')
integral(2^(x)*ln(2),x,0,5)/(5-0)
```

6. $f(x)=-\csc x\cot x,\ [0,\ \pi]$

Solution $f_{ave}=\dfrac{1}{\pi-0}\displaystyle\int_{0}^{\pi}(-\csc x\cot x)dx$

$=\dfrac{1}{\pi}[\csc x]_0^{\pi}=-\infty$

7. $f(x)=\dfrac{2}{(1+2r)^2},\ [1,\ 6]$

http://matrix.skku.ac.kr/cal-lab/cal-6-5-7.html

Solution $f_{ave}=\dfrac{1}{6-1}\displaystyle\int_{1}^{6}\dfrac{2}{(1+2x)^2}dx.$

Use a change of variable $1+2x=t$, then $2dx=dt$.

So $f_{ave}=\dfrac{1}{5}\displaystyle\int_{3}^{13}\dfrac{1}{t^2}dt=\dfrac{1}{5}\left[-\dfrac{1}{t}\right]_3^{13}$

$=\dfrac{1}{5}\left(\dfrac{1}{3}-\dfrac{1}{13}\right)=\dfrac{2}{39}.$

```
var('x')
integral(2/(1+2*x)^2,x,1,6)/(6-1)
```

8-10. (a) Find the average value of f on the given interval.
(b) Find c such that $f_{ave} = f(c)$.
(c) Sketch the graph of f and a rectangle whose area is the same as the area under the graph of f.

8. $f(x) = 2(x-3)^2$, $[2, 5]$
http://matrix.skku.ac.kr/cal-lab/cal-6-5-8.html

Solution (a) $f_{\text{ave}} = \dfrac{1}{5-2}\displaystyle\int_2^5 2(x-3)^2 dx$

$= \dfrac{1}{3}\displaystyle\int_2^5 (2x^2 - 12x + 18)dx$

$= \dfrac{1}{3}\left[\dfrac{2}{3}x^3 - 6x^2 + 18x\right]_2^5$

$= \dfrac{1}{3}\left(\dfrac{250}{3} - 150 + 90 - \dfrac{16}{3} + 24 - 36\right)$

$= 2$

(b) $2 = f(c) = 2(c-3)^2$
$\Rightarrow (c-3)^2 = 1$
$\Rightarrow c - 3 = 1$ or $c - 3 = -1$
$\Rightarrow c = 4$ or $c = 2$

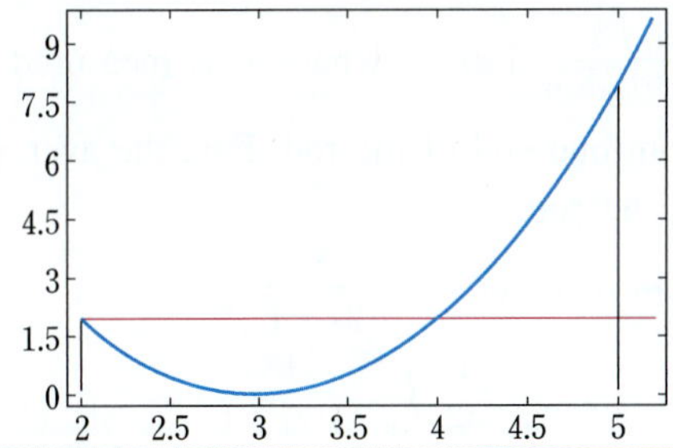

```
var('x')
f(x)=(2*(x-3)^2)
plot(f(x),x,2,5
```

9. $f(c) = \sqrt{2x}$, $[0, 4]$

Solution (a) $f_{\text{ave}} = \dfrac{1}{4}\displaystyle\int_0^4 \sqrt{2x}\,dx = \dfrac{1}{4}\left[\sqrt{2}\,\dfrac{2}{3}x^{\frac{3}{2}}\right]_0^4$

$= \dfrac{1}{4}\left(\dfrac{16\sqrt{2}}{3}\right) = \dfrac{4\sqrt{2}}{3}$

(b) $\dfrac{4\sqrt{2}}{3} = f(c) = \sqrt{2c}$

$\Rightarrow \sqrt{c} = \dfrac{4}{3} \Rightarrow c = \dfrac{16}{9}$

10. $f(x) = \dfrac{x}{(1+x^2)^2}$, $[0, 2]$.
http://matrix.skku.ac.kr/cal-lab/cal-6-5-10.html

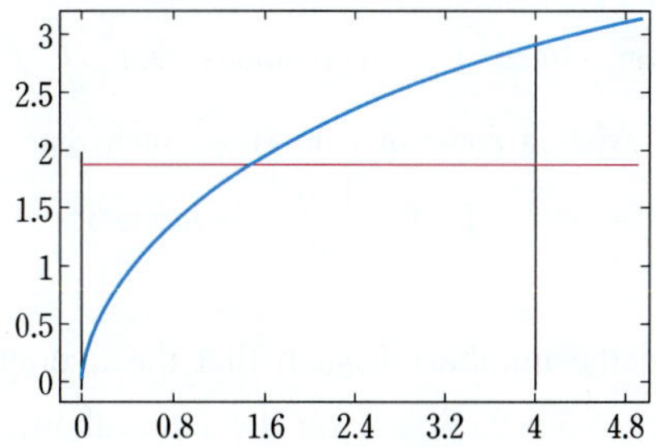

Solution (a) $f_{\text{ave}} = \dfrac{1}{2}\displaystyle\int_0^1 \dfrac{x}{(1+x^2)^2}dx$.

Use change of variable to $x = \tan\theta$, then $dx = \sec^2\theta$.

So $f_{\text{ave}} = \dfrac{1}{2}\displaystyle\int_0^{\frac{\pi}{4}} \dfrac{\tan\theta}{(1+\tan^2\theta)^2}\sec^2\theta d\theta$

$= \dfrac{1}{2}\displaystyle\int_0^{\frac{\pi}{4}} \dfrac{\tan\theta}{\sec^2\theta}d\theta$

$= \dfrac{1}{2}\displaystyle\int_0^{\frac{\pi}{4}} \dfrac{\sin\theta}{\cos^3\theta}d\theta$.

Also change of variable to $\cos\theta = t$, then $-\sin\theta d\theta = dt$.

$f_{\text{ave}} = -\dfrac{1}{2}\displaystyle\int_1^{\frac{\sqrt{2}}{2}} t^{-3}dt = \dfrac{1}{4}\left[t^{-2}\right]_1^{\frac{\sqrt{2}}{2}}$

$= \dfrac{1}{4}(2-1) = \dfrac{1}{4}$

```
var('x')
integral(x/((1+x^2)^2),x,0,2)/(2-0)
```

(b) $\dfrac{1}{5} = f(c) = \dfrac{c}{(1+c^2)^2}$

(c)

```
var('x')
f(x)=(x/(1+x^2)^2)
plot(f(x),x,0,2)
```

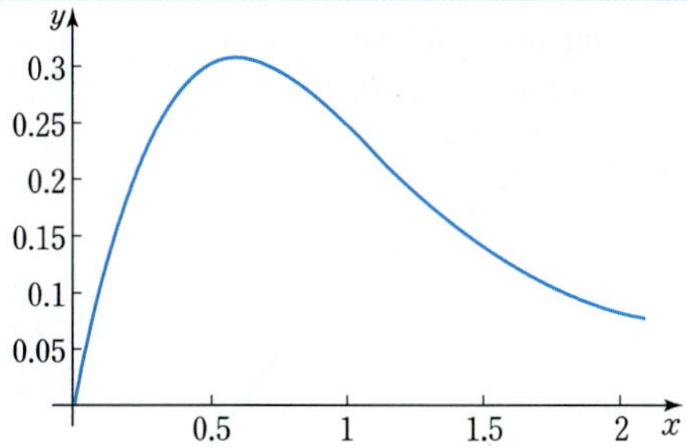

11. If f is continuous and $\displaystyle\int_1^5 f(x)dx = 12$, show that f takes on the value 3 at least once on the interval $[1, 5]$.

Solution Since f is continuous and $\int_1^5 f(x)dx = 12$, there exists a constant c in $[1, 5]$ such that

$$f(c) = \frac{1}{4}\int_1^5 f(x)dx = 3.$$

12. Find the numbers b such that the average value of $f(x) = 3 + 2x - 3x^2$ on the interval $[0, b]$ is equal to 2.

Solution $f_{ave} = \frac{1}{b}\int_0^b (3+2x-3x^2)dx$

$= \frac{1}{b}[3x+x^2-x^3]_0^b = \frac{1}{b}(3b+b^2-b^3)$

$= 3+b-b^2$

That is,

$$b^2-b-3=-2 \Rightarrow b^2-b-1=0 \Rightarrow b=\frac{1+\sqrt{5}}{2}.$$

13. In a certain city the temperature (in °F) t hours after 9 A.M. is modeled by the function

$$T(t) = 45 + 12\cos\frac{\pi t}{12}$$

Find the average temperature during the period from 9 A.M. to 9 P.M.

http://matrix.skku.ac.kr/cal-lab/cal-6-5-13.html

Solution $T_{\text{ave}} = \frac{1}{12}\int_0^{12}(45+12\cos\frac{\pi t}{12})dx$

$= \frac{1}{12}\left[45t + \frac{144}{\pi}\sin\frac{\pi t}{12}\right]_0^{12} = 45\,°\text{F}$

```
var('x')
Tave(x)= 1/12 * (45+12*cos(pi*x/12))
print integral(Tave(x), x, 0, 12)
plot(Tave(x), x, 0, 12)
```

Answer : 45

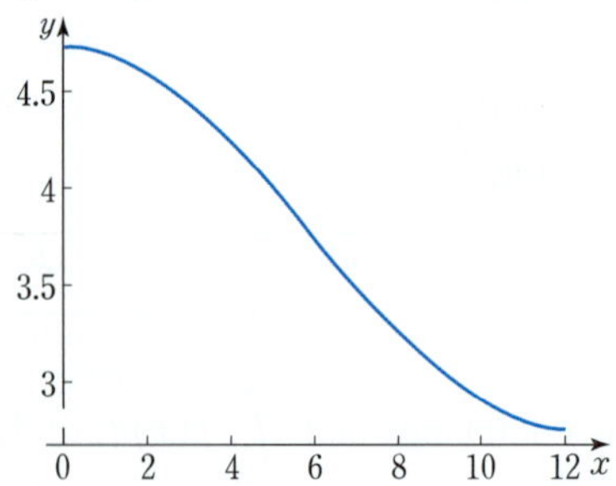

14. If a cup of coffee has temperature 90°C in a room where the temperature is 15°C then, according to Newton's Law of Cooling, the temperature of the coffee after t minutes is $T(t) = 25 + 80e^{-t/40}$. What is the average temperature of the coffee during the first half hour?

http://matrix.skku.ac.kr/cal-lab/cal-6-5-14.html

Solution $T_{ave} = \frac{1}{30}\int_0^{30}(25+80e^{-t/40})dt$

$= \frac{1}{30}[25t - 3200e^{-t/40}]_0^{30}$

$= \frac{1}{30}(750 - 3200e^{-3/4} + 3200)$

$= \frac{1}{30}(3275 - 3200e^{-3/4})$

```
var('x')
Tave(x)= 1/30 * (25+80*e^(-x/40))
print integral(Tave(x), x, 0, 30)
plot(Tave(x), x, 0, 30)
```

Answer : 5/3*(15*e^(3/4) - 64)*e^(-3/4) + 320/3

15. The linear density in a rod 12m long is $\frac{18}{\sqrt{2x+1}}$kg/m, where x is measured in meters from one end of the rod. Find the average density of the rod.

Solution Let $f(x) = \frac{18}{\sqrt{2x+1}}$.

Then $f_{\text{ave}} = \frac{1}{12}\int_0^{12}\frac{18}{\sqrt{2x+1}}dx$.

Change of variable to $2x+1=t$, then $2dx = dt$. So

$$f_{\text{ave}} = \frac{1}{12}\int_1^{25} 9t^{-1/2}dt = \frac{9}{12}[2t^{1/2}]_1^{25}$$
$$= \frac{3}{2}(5-1) = 6\text{kg/m}.$$

16. If $f_{\text{ave}}[a, b]$ denotes the average value of f on the interval $[a, b]$ and $a < c < b$, show that

$$f_{\text{ave}}[a, b] = \frac{c-a}{b-a}f_{\text{ave}}[a, c] + \frac{b-c}{b-a}f_{\text{ave}}[c, b].$$

Solution $\frac{c-a}{b-a}f_{\text{ave}}[a, c] + \frac{b-c}{b-a}f_{\text{ave}}[c, b]$

$= \frac{1}{b-a}\int_a^c f dx + \frac{1}{b-a}\int_c^b f dx$

$= \frac{1}{b-a}\left(\int_a^c f dx + \int_c^b f dx\right)$

$= \frac{1}{b-a}\left(\int_a^b f dx\right) = f_{\text{ave}}[a, b]$

The 12th International Congress on Mathematical Education
The 12th International Congress on Mathematical Educatio
July 8th to 15th, 2012 COEX, Seoul, Korea
KSICMI
ICME-12

ICME 12 logo

Chapter 7

Techniques of Integration

LIU, Hui

(劉徽) 220-280

[pi]

Nine Chapters on the Mathematical Art

Liu Hui was a Chinese mathematician who lived in the Wei Kingdom. In 263 AD, he published a book with solutions to mathematical problems presented in the famous Chinese book of mathematics known as The Nine Chapters on the Mathematical Art. He was a descendant of Marquis of Zixiang of the Han dynasty.

YouTube http://youtu.be/WX-6C9tCneE

Integration is not as straightforward as differentiation and in general requires more elaborate efforts. A definite integral can easily be evaluated via the Fundamental Theorem of Calculus (FTC) only when its antiderivative (or indefinite integral) is explicitly found. There are basic integration formulas for basic functions that we have encountered.

7.1 Integration by Parts

http://matrix.skku.ac.kr/cal-lab/sage-grapher-derivatives.html

There are many integrals that cannot be easily evaluated using basic formulas or integration by substitution. For example,

$$\int x \cos x \, dx$$

cannot be evaluated with what we know at this point. The integral can be solved by introducing an integration technique called **integration by parts**.

We have already observed that every differentiation rule gives rise to a corresponding integration rule. For example, the Substitution Rule for integration corresponds to the Chain Rule for differentiation. Integration by parts is one of the integration techniques that comes from the Product Rule for differentiation. The Product Rule for differentiation is given by

$$\frac{d}{dx}[f(x)g(x)] = f(x)g'(x) + f'(x)g(x).$$

Integrating with respect to x gives us

$$f(x)g(x) = \int \frac{d}{dx}[f(x)g(x)]dx = \int [f(x)g'(x) + f'(x)g(x)]dx$$

or

$$\int f(x)g'(x)dx + \int f'(x)g(x)dx = f(x)g(x).$$

Note that the constant of integration is not ignored! It is still there, we simply did not write it! Solving for the first integral on the left-hand side yields the **formula for integration by parts;**

1 $$\int f(x)g'(x)dx = f(x)g(x) - \int f'(x)g(x)$$

Integration by parts is thus a technique for evaluating integrals whose integrand is the product of two functions. If we choose $u = f(x)$ and $v = g(x)$, this formula then takes a simpler form:

2 $$\int u\,dv = uv - \int v\,du$$

To apply integration by parts, we need to make wise choices for u and dv. The integral on the right-hand side should be the one that is simpler than the original integral and is easier to evaluate. A rule of thumb is that $u = f(x)$ is chosen to be a function that becomes simpler when it is differentiated (or at least not more complicated) as long as $dv = g'(x)dx$ can be readily integrated to give v. (It is customary to write $\int 1dx$ as $\int dx$.)

EXAMPLE 1

Evaluate $\int xe^{2x}dx$.

http://matrix.skku.ac.kr/cal-lab/cal-7-1-Exm-1.html

Solution **1.** Take $f(x) = x$ and $g'(x) = e^{2x}$.

Then $f'(x) = 1$ and $g(x) = \frac{1}{2}e^{2x}$.

Using formula [1], we have

$$\int xe^{2x}dx = \frac{1}{2}xe^{2x} - \int \frac{1}{2}e^{2x}dx$$
$$= \frac{1}{2}xe^{2x} - \frac{1}{4}e^{2x} + C.$$

2. Let $u = x$ and $dv = e^{2x}dx$. Then $du = dx$ and $v = \frac{1}{2}e^{2x}$ and so

$$\int xe^{2x}dx = \frac{1}{2}xe^{2x} - \int \frac{1}{2}e^{2x}dx = \frac{1}{2}xe^{2x} - \frac{1}{4}e^{2x} + C. \quad \blacksquare$$

Remark If we choose $u = e^{2x}$ and $dv = xdx$. Then $du = 2e^{2x}$ and $v = \frac{1}{2}x^2$ and we have $\int xe^{2x}dx = x^2e^{2x} - \int x^2e^{2x}dx$. This is more complicated than the original one. That is why we need to make an appropriate choice for u and v. In particular, if $g'(x) = 1$ then $g(x) = x$. Thus we have,

$$\int f(x)dx = x \cdot f(x) - \int x \cdot f'(x)dx.$$

EXAMPLE 2

Find $\int \ln x\, dx$.

http://matrix.skku.ac.kr/cal-lab/cal-7-1-Exm-2.html

Solution Let $u = \ln x$ and $dv = dx$, then $du = \frac{1}{x}dx$ and $v = x$.

Integrating by parts, we get

$$\int \ln x\, dx = x\ln x - \int x\frac{1}{x}dx$$
$$= x\ln x - \int dx$$
$$= x\ln x - x + C. \quad \blacksquare$$

EXAMPLE 3

Evaluate $I=\int \sqrt{x^2+A}\,dx \quad (A>0)$.

http://matrix.skku.ac.kr/cal-lab/cal-7-1-Exm-3.html

Solution Let $u=\sqrt{x^2+A}$, $dv=dx$, then $du=\dfrac{x}{\sqrt{x^2+A}}dx$, $v=x$.

Thus $I=x\sqrt{x^2+A}-\int \dfrac{x^2}{\sqrt{x^2+A}}dx$. We have

$$\int \frac{x^2}{\sqrt{x^2+A}}dx=\int \sqrt{x^2+A}\,dx-A\int \frac{dx}{\sqrt{x^2+A}}$$

$$\left(\because \ \frac{x^2}{\sqrt{x^2+A}}=\frac{(x^2+A)-A}{\sqrt{x^2+A}}=\sqrt{x^2+A}-\frac{A}{\sqrt{x^2+A}}\right)$$

$$=I-A\int \frac{dx}{\sqrt{x^2+A}}$$

$$=I-A\ln\left|x+\sqrt{x^2+A}\right|.$$

$$I=x\sqrt{x^2+A}-\int \frac{x^2}{\sqrt{x^2+A}}dx$$

$$=x\sqrt{x^2+A}-\left(I-A\ln\left|x+\sqrt{x^2+A}\right|\right)+C_1$$

$$2I=x\sqrt{x^2+A}+A\ln\left|x+\sqrt{x^2+A}\right|+C_1$$

Hence, $I=\dfrac{1}{2}\left(x\sqrt{x^2+A}+A\ln\left|x+\sqrt{x^2+A}\right|\right)+C.\ (2C=C_1)$

```
var('x, A')
assume(A>0)
f(x)= sqrt(x^2 + A)
integral(f(x), x)
```

Answer : 1/2*A*arcsinh(x/sqrt(A)) + 1/2*sqrt(x^2 + A)*x ■

EXAMPLE 4

Evaluate $\int \dfrac{x^3}{\sqrt{1-x^2}}dx$.

http://matrix.skku.ac.kr/cal-lab/cal-7-1-Exmp-4.html

Solution Let $u=x^2$ and $dv=\dfrac{x}{\sqrt{1-x^2}}dx$. Then $du=2x\,dx$ and $v=-\sqrt{1-x^2}$. Integrating by parts gives

$$\int \frac{x^3}{\sqrt{1-x^2}}dx=-x^2\sqrt{1-x^2}+\int 2x\sqrt{1-x^2}\,dx$$

$$=-x^2\sqrt{1-x^2}-\frac{2}{3}\sqrt{(1-x^2)^3}+C.$$

```
var('x')
f(x)= (x^3)/((1- x^2)^(-1/2))
print integral(f(x), x)
```

Answer : -1/5*(-x^2 + 1)^(3/2)*x^2 - 2/15*(-x^2 + 1)^(3/2) ■

EXAMPLE 5

Evaluate $\int x\cos x\,dx$.

http://matrix.skku.ac.kr/cal-lab/cal-7-1-Exm-6.html

Solution Let $u = x$ and $dv = \cos x\,dx$, then $du = dx$ and $v = \sin x$. Hence,

$$\begin{aligned}\int x\cos x\,dx &= x\sin x - \int \sin x\,dx \\ &= x\sin x + \cos x + C.\end{aligned}$$

■

EXAMPLE 6

Evaluate $\int e^{ax}\cos bx\,dx,\ a \neq 0$. (See Figure 1.)

http://matrix.skku.ac.kr/cal-lab/cal-7-1-Exm-4.html

http://matrix.skku.ac.kr/cal-lab/sage-grapher-integral.html

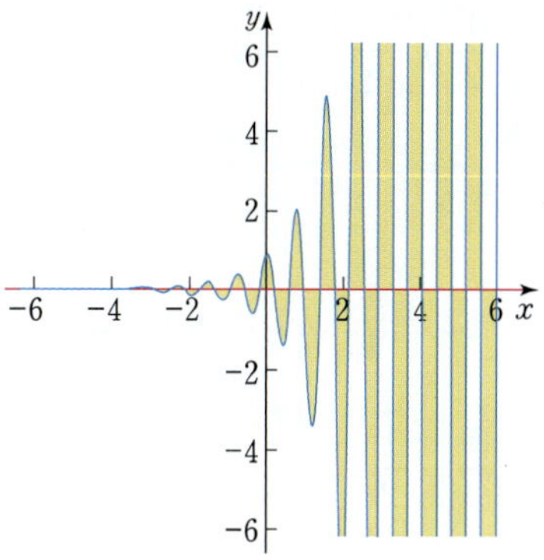

Figure 1
$f(x) = e^{ax}\cos bx\ (-10 < a < 10)$

Solution Let $u = \cos bx$ and $dv = e^{ax}dx$, then $du = -b\sin bx$, $v = \dfrac{e^{ax}}{a}$. Integrating by parts, we obtain

$$\int e^{ax}\cos bx dx = \frac{e^{ax}\cos bx}{a} + \frac{b}{a}\int e^{ax}\sin bx\,dx.$$

Using integration by parts to evaluate $\int e^{ax}\sin bx\,dx$ yields

$$\int e^{ax}\sin bx\,dx = \frac{e^{ax}\sin bx}{a} - \frac{b}{a}\int e^{ax}\cos bx\,dx,$$

$$\frac{b}{a}\int e^{ax}\sin bx\,dx = \frac{b}{a}\frac{e^{ax}\sin bx}{a} - \left(\frac{b}{a}\right)^2\int e^{ax}\cos bx\,dx.$$

Substituting this into the equation above, we get

$$\int e^{ax}\cos bx dx = \frac{e^{ax}\cos bx}{a} + \frac{b}{a}\frac{e^{ax}\sin bx}{a} - \left(\frac{b}{a}\right)^2\int e^{ax}\cos bx\,dx$$

which leads to the following.

$$\int e^{ax}\cos bx\,dx = \frac{e^{ax}}{a^2+b^2}(a\cos bx + b\sin bx) + C.$$

```
var('a, b, x')
f(x)= e^(a*x) * cos(b*x)
print integral(f, x)
```

Answer : (a*cos(b*x) + b*sin(b*x))*e^(a*x)/(a^2 + b^2) ■

Using FTC 2 and integration by parts, definite integrals can be evaluated as follows:

3
$$\int_a^b f(x)g'(x)dx = f(x)g(x)\Big]_a^b - \int_a^b g(x)f'(x)dx$$

In particular, the following is a special case of the above.

$$\int_a^b f(x)\,dx = [\,x \cdot f(x)\,]_a^b - \int_a^b x\, f'(x)\,dx.$$

EXAMPLE 7

Evaluate $\int_0^1 x\tan^{-1}x\,dx$.

Solution Let $u = \tan^{-1}x$, $dv = x\,dx$, then $du = \frac{1}{1+x^2}dx$, $v = \frac{1}{2}x^2$.

So, $\int_0^1 x\tan^{-1}x\,dx = \left[\frac{x^2}{2}\tan^{-1}x\right]_0^1 - \frac{1}{2}\int_0^1 \frac{x^2}{1+x^2}dx$

$= \frac{\pi}{8} - \frac{1}{2}\int_0^1 \frac{x^2}{1+x^2}dx.$

Since $\int_0^1 \frac{x^2}{1+x^2}dx = [x - \tan^{-1}x]_0^1 = 1 - \frac{\pi}{4}$, we finally get

$$\int_0^1 x\tan^{-1}x\,dx = \frac{\pi}{4} - \frac{1}{2}.$$ ■

Recall that

4
$$\int \sec x\,dx = \ln|\sec x + \tan x| + C$$

Proof Multiply and divide by $(\sec x + \tan x)$ and use the Substitution Rule

$$\int \sec x\,dx = \int \sec x \frac{\sec x + \tan x}{\sec x + \tan x}dx.$$

Set $u = \sec x + \tan x$, then we find

$$du = (\sec x \tan x + \sec^2 x)\, dx.$$

By substituting u and du, we have

$$\begin{aligned}\int \sec x \frac{\sec x + \tan x}{\sec x + \tan x} dx &= \int \frac{(\sec^2 x + \sec x \tan x)}{\sec x + \tan x} dx \\ &= \int \frac{du}{u} = \ln|u| + C \\ &= \ln|\sec x + \tan x| + C\end{aligned}$$

after substituting $u = \sec x + \tan x$. ■

CAS **EXAMPLE 8**

Evaluate $\int \sec^3 x\, dx$.

http://matrix.skku.ac.kr/cal-lab/cal-7-2-exm-9.html

Solution

```
f=sec(x)^3
integral(f, x)
```

Answer : -1/2*sin(x)/(sin(x)^2 - 1) - 1/4*log(sin(x) - 1) + 1/4*log(sin(x) + 1)

Another method: Let $u = \sec x$, $dv = \sec^2 x\, dx$, then $du = \sec x \tan x\, dx$, $v = \tan x$. Using integrating by parts gives

$$\begin{aligned}\int \sec^3 x\, dx = \int \sec^2 x \sec x\, dx &= \sec x \tan x - \int \tan^2 x \sec x\, dx \\ &= \sec x \tan x - \int (\sec^2 x - 1) \sec x\, dx \\ &= \sec x \tan x - \int (\sec^3 x - \sec x)\, dx \\ &= \sec x \tan x - \int \sec^3 x\, dx + \int \sec x\, dx \\ &= \sec x \tan x - \int \sec^3 x\, dx + \ln|\sec x + \tan x|.\end{aligned}$$

Solving for $\int \sec^3 x\, dx$ from this equation, we obtain

$$2\int \sec^3 x\, dx = \sec x \tan x + \ln|\sec x + \tan x| + C_1$$

$$\int \sec^3 x\, dx = \frac{1}{2}(\sec x \tan x + \ln|\sec x + \tan x|) + C. (2C = C_1)$$ ■

CAS **EXAMPLE 9**

Evaluate the integral using integration by parts

$$\int x\ln(x-1)dx.$$

http://matrix.skku.ac.kr/cal-lab/cal-7-1-exm-8.html

Solution integral(x*log(x-1),x)

```
var('x')
u(x)=x; V(x)=log(x-1)
v(x)=diff(V(x), x)
U(x)=integral(u(x), x)
I1=U(x)*V(x)-integral(U(x)*v(x), x)   # integration by parts
I2=integral(u(x)*V(x), x)
print I1
print I2
```

Answer : 1/2*x^2*log(x - 1) - 1/4*x^2 - 1/2*x - 1/2*log(x - 1)
1/2*x^2*log(x - 1) - 1/4*x^2 - 1/2*x - 1/2*log(x - 1)

```
bool(I1==I2)
```

Answer : True ■

7.1 EXERCISES (Integration by Parts)

http://matrix.skku.ac.kr/Cal-Book/part1/CS-Sec-7-1-Sol.html

YouTube http://youtu.be/jKCAGJ4HqvQ

1. $\int x(ax+b)^n dx$

http://matrix.skku.ac.kr/cal-lab/sage-grapher-integral.html

Solution $u = x$, $dv = (ax+b)^n dx$, $du = dx$,

$v = \frac{1}{a(n+1)}(ax+b)^{n+1}$.

$$\int x \cdot (ax+b)^n dx$$

$$= \left[\frac{x}{a(n+1)}(ax+b)^{n+1}\right] - \int \frac{1}{a(n+1)}(ax+b)^{n+1} dx$$

$$= \left[\frac{x}{a(n+1)}(ax+b)^{n+1}\right]$$

$$- \frac{1}{a^2(n+1)(n+2)}(ax+b)^{n+2} + C.$$

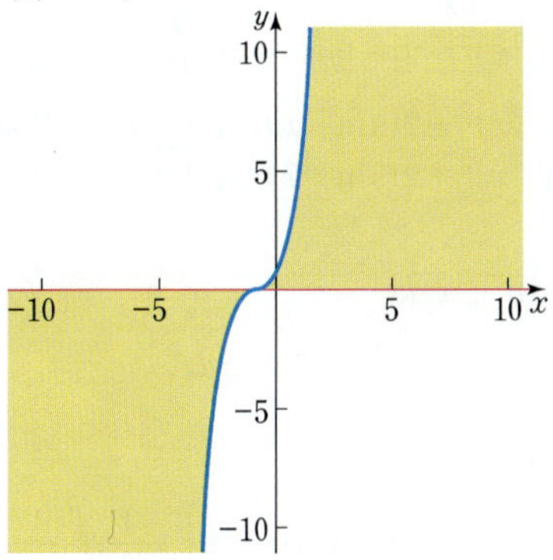

$a=1,\ b=1,\ C=3$

CAS **2.** $\int (x^2+2x+1)e^{7x} dx$

http://matrix.skku.ac.kr/cal-lab/cal-7-1-2.html

Solution $\int (x^2+2x+1)e^{7x} dx$

$$= \int x^2 e^{7x} dx + 2\int x e^{7x} dx + \int e^{7x} dx$$

$$= \frac{1}{343}(49x^2+84x-12)e^{7x}$$

```
integral((x^2+2*x+1)*e^(7*x), x)
```

Answer : 2/49*(7*x - 1)*e^(7*x) + 1/343*(49*x^2 - 14*x + 2)*e^(7*x) + 1/7*e^(7*x)

3. $\int_0^{\frac{\pi}{2}} x^2 \cos x\, dx$

Solution $\int_0^{\frac{\pi}{2}} x^2 \cos x\, dx = [x^2 \sin x]_0^{\frac{\pi}{2}} - \int_0^{\frac{\pi}{2}} 2x \sin x\, dx$

$$= \frac{\pi^2}{4} - 2\left[[-x\cos x]_0^{\frac{\pi}{2}} - \int_0^{\frac{\pi}{2}} -\cos x\, dx\right]$$

$$= \frac{\pi^2}{4} - 2[\sin x]_0^{\frac{\pi}{2}} = \frac{\pi^2}{4} - 2.$$

```
integral(x^2*cos(x),x,0,pi/2)
```

4. $\int x \sin x dx$

http://matrix.skku.ac.kr/cal-lab/cal-7-1-4.html

Solution $u = x$, $dv = \sin x\, dx$, $du = dx$, $v = -\cos x$

$$\int x \sin x\, dx = [-x\cos x] + \int \cos x\, dx$$

$$= -x\cos x + \sin x + C$$

```
integral(x*sin(x), x)
```

5. $\int x^2 \sin 4x dx$

Solution $\int x^2 \sin 4x dx = \left[\frac{-x^2}{4}\cos 4x\right]$

$$- \int \frac{-x}{2}\cos 4x\, dx$$

$$= \frac{-x^2}{4}\cos 4x + \frac{1}{2}\left[\left[\frac{x}{4}\sin 4x\right] - \int \frac{1}{4}\sin 4x\, dx\right]$$

$$= \frac{-x^2}{4}\cos 4x + \frac{x}{8}\sin 4x + \frac{x}{32}\cos 4x + C.$$

```
integral(x^2*sin(4*x),x)
```

CAS **6.** $\int x^3 \sin x^2 dx$

Solution

```
var('x')
integral(x^3 * sin(x^2), x)
```

Answer : -1/2*x^2*cos(x^2) + 1/2*sin(x^2) + C

7. $\int x^2 e^{-5x} dx$

Solution $\int x^2 e^{-5x} dx = -\frac{1}{5}x^2 e^{-5x} + \frac{2}{5}\int x e^{-5x} dx$

$$= -\frac{1}{5}x^2 e^{-5x} - \frac{2}{25}x e^{-5x} + \frac{2}{25}\int e^{-5x} dx$$

$$= -\frac{1}{5}x^2 e^{-5x} - \frac{2}{25}x e^{-5x} - \frac{2}{125}e^{-5x} + C.$$

CAS **8.** $\int \ln(1+\sqrt{x})dx$

http://matrix.skku.ac.kr/cal-lab/cal-7-1-21.html

Solution $u = \ln(1+\sqrt{x})dx$, $dv = 1$,

$$du=\frac{1}{2\sqrt{x}(1+\sqrt{x})},\ v=x-1$$

$$\int \ln(1+\sqrt{x})dx=(x-1)\ln(1+\sqrt{x})$$

$$-\int \frac{(x-1)}{2\sqrt{x}(1+\sqrt{x})}dx$$

$$=(x-1)\ln(1+\sqrt{x})-\frac{1}{2}x+\sqrt{x}+C.$$

```
integral(ln(1+sqrt(x)), x)
```

Answer : x*log(sqrt(x) + 1) - 1/2*x + sqrt(x) - log(sqrt(x) + 1)

CAS **9.** $\int (\ln x)^2 dx$

http://matrix.skku.ac.kr/cal-lab/cal-7-1-1.html

Solution $u=\ln x$, $dv=\ln x\,dx$, $du=\frac{1}{x}dx$, $v=x\ln(x)-x$

$$\int (\ln x)^2 dx=[\ln x(x\ln x-x)]-\int \frac{1}{x}(x\ln x-x)dx$$

$$=x(\ln x)^2-x\ln x-\int \ln x-1\,dx$$

$$=(\ln(x)^2-2\ln x+2)x+C.$$

```
integral(ln(x)^2, x)
```

Answer : (log(x)^2 - 2*log(x) + 2)*x

10. $\int x^6 \ln x\,dx$

Solution $u=\ln x$, $dv=x^6dx$, $du=\frac{1}{x}dx$, $v=\frac{1}{7}x^7$

$$\int x^6\ln x\,dx=\frac{1}{7}x^7(\ln x)-\int \frac{1}{7}x^6dx$$

$$=\frac{1}{7}x^7(\ln x)-\frac{1}{49}x^7+C.$$

11. $\int x^{-2}\ln x^4 dx$

Solution $\int x^{-2}\ln x^4 dx=\left[-\frac{1}{x}\ln x^4\right]+\int \frac{1}{x}\frac{4x^3}{x^4}dx$

$$=-\frac{1}{x}\ln x^4+4\int \frac{1}{x^2}dx$$

$$=-\frac{1}{x}\ln x^4-\frac{4}{x}+C.$$

12. $\int \frac{\ln x}{(1+x)^2}dx$

Solution

$$\int \frac{\ln x}{(1+x)^2}dx=\left[-\frac{\ln x}{(1+x)}\right]-\int -\frac{1}{x(1+x)}dx$$

$$=-\frac{\ln x}{(1+x)}+\int \frac{1}{x}-\frac{1}{1+x}dx$$

$$=-\frac{\ln x}{(1+x)}+\ln x-\ln(1+x)+C.$$

13. $\int \sin^{-1}x\,dx$

http://matrix.skku.ac.kr/cal-lab/cal-7-1-13.html

Solution $y=\sin^{-1}x$, $x=\sin y$, $dx=\cos y\,dy$, $\cos y=\sqrt{1-x^2}$

$$\int \sin^{-1}x\,dx=\int y\cos y\,dy$$

$$=[y\sin y]-\int \sin y\,dy$$

$$=x\sin^{-1}x+\sqrt{1-x^2}+C.$$

```
integral(x*sin(x), x)
```

14. $\int (\sin^{-1}x)^2dx$

Solution $y=\sin^{-1}x$, $x=\sin y$, $dx=\cos y\,dy$, $\cos y=\sqrt{1-x^2}$

$$\int (\sin^{-1}x)^2dx=\int y^2\cos y\,dy$$

$$=[y^2\sin y]-\int 2y\sin y\,dy$$

$$=x(\sin^{-1}x)^2-2[-[y\cos y]+\int \cos y\,dy]$$

$$=x(\sin^{-1}x)^2+2(\sin^{-1}x)(\sqrt{1-x^2})-2\sin y+C$$

$$=x(\sin^{-1}x)^2+2(\sin^{-1}x)(\sqrt{1-x^2})-2x+C.$$

15. $\int x\sin^{-1}x\,dx$

Solution $u=\sin^{-1}x$, $dv=x\,dx$, $du=\frac{1}{\sqrt{1-x^2}}dx$, $v=\frac{1}{2}x^2$

$$\int x\sin^{-1}x\,dx=\left[\frac{1}{2}x^2\sin^{-1}x\right]-\frac{1}{2}\int \frac{x^2}{\sqrt{1-x^2}}dx$$

$$=\frac{1}{2}x^2\sin^{-1}x-\frac{1}{4}\sin^{-1}x+\frac{x\sqrt{1-x^2}}{4}+C.$$

(by Example 1 Section 7.3)

16. $\int x^2\sin^{-1}x\,dx$

Solution $u=\sin^{-1}x$, $dv=x^2dx$, $du=\frac{1}{\sqrt{1-x^2}}dx$, $v=\frac{1}{3}x^3$

$$\int x^2\sin^{-1}x\,dx=\frac{1}{3}x^3\sin^{-1}x-\int \frac{x^3}{3\sqrt{1-x^2}}dx \cdots ①$$

Since $\int \frac{x^3}{\sqrt{1-x^2}}dx=\int \frac{-x(1-x^2)+x}{\sqrt{1-x^2}}dx$

$$=\int\left(-x\sqrt{1-x^2}+\frac{x}{\sqrt{1-x^2}}\right)dx$$

$$=-\frac{1}{3}x^2\sqrt{1-x^2}-\frac{2}{3}\sqrt{1-x^2}.$$

Therefore, from ①,

$$\int x^2\sin^{-1}x\,dx$$

$$= \frac{1}{3} x^3 \sin^{-1} x + \frac{1}{9} x^2 \sqrt{1-x^2} + \frac{2\sqrt{1-x^2}}{9}$$

$$= \frac{1}{3} x^3 \sin^{-1} x + \frac{\sqrt{1-x^2}}{9}(2+x^2) + C.$$

CAS **17.** $\int x \cos^{-1} x dx$

http://matrix.skku.ac.kr/cal-lab/cal-7-1-19.html

Solution $u = \cos^{-1} x$, $dv = x$, $du = -\frac{1}{\sqrt{1-x^2}}$, $v = \frac{1}{2}x^2$

$$\int x \cos^{-1} x dx = \frac{1}{2} x^2 \cos^{-1} x + \frac{1}{2} \int \frac{x^2}{\sqrt{1-x^2}} dx$$

$$= \frac{1}{2} x^2 \cos^{-1} x - \frac{x\sqrt{1-x^2}}{4} + \frac{1}{4} \sin^{-1} x + C.$$

```
integral(x*(arccos(x)), x)
```

Answer : 1/2*x^2*arccos(x) - 1/4*sqrt(-x^2 + 1)*x + 1/4*arcsin(x)

CAS **18.** $\int e^{-x} \sin^2 x dx$

http://matrix.skku.ac.kr/cal-lab/cal-7-1-20.html

Solution $\int e^{-x} \sin^2 x dx = -\frac{1}{10}(2x \sin 2x - \cos 2x + 5) e^{-x} + C.$

```
integral (e^(-x)*(sin(x))^2, x)
```

Answer : -1/10*(2*sin(2*x) - cos(2*x) + 5)*e^(-x)

CAS **19.** $\int_0^{\pi} x \cos\left(\frac{1}{2}x\right) dx$

http://matrix.skku.ac.kr/cal-lab/cal-7-1-24.html

Solution $\int_0^{\pi} x \cos\left(\frac{1}{2}x\right) dx = 2\pi - 4$

```
integral (x*cos(x/2), x, 0, pi)
```

Answer : 2*pi - 4

20. $\int \sqrt{a^2 - x^2}\, dx$

http://matrix.skku.ac.kr/cal-lab/cal-7-1-26.html

Solution $= \frac{1}{2}\left(x\sqrt{a^2-x^2} + a^2 \sin^{-1}\frac{x}{a}\right) + C.$

```
var('x, a');
integral(sqrt(a^2-x^2), x)
```

Answer : 1/2*a^2*arcsin(x/sqrt(a^2))+1/2*sqrt(a^2-x^2)*x

21. $\int_0^1 x e^x dx$

http://matrix.skku.ac.kr/cal-lab/cal-7-1-22.html

Solution $\int_0^1 x e^x dx = 1.$

```
integral (x*e^(x), x, 0, 1)
```

Ansewr : 1

CAS **22.** $\int_1^2 x^3 (\ln x)^2 dx$

http://matrix.skku.ac.kr/cal-lab/cal-7-1-25.html

Solution $\int_1^2 x^3 (\ln x)^2 dx = 4(\ln 2)^2 - 2\ln 2 + \frac{15}{32}.$

```
integral((x^3)*(ln(x)^2), x, 1, 2)
```

Answer : 4*log(2)^2-2*log(2)+15/32

23. $\int_0^1 \tan^{-1} x dx$

http://matrix.skku.ac.kr/cal-lab/cal-7-1-23.html

Solution $\int_0^1 \tan^{-1} x dx = \frac{\pi}{4} - \frac{1}{2}\ln 2.$

```
integral (arctan(x), x, 0, 1)
```

Answer : 1/4*pi - 1/2*log(2)

24. $\int_0^1 x^2 \tan^{-1} x dx$

Solution $\int_0^1 x^2 \tan^{-1} x dx$

$$= \left[\frac{1}{3} x^3 \tan^{-1} x\right]_0^1 - \frac{1}{3}\int_0^1 \frac{x^3}{1+x^2} dx$$

$$= \frac{\pi}{12} - \frac{1}{3}\int_0^1 \left(x - \frac{x}{1+x^2}\right) dx$$

$$= \frac{\pi}{12} - \frac{1}{3}\left[\frac{1}{2}x^2 - \frac{1}{2}\ln(1+x^2)\right]_0^1$$

$$= \frac{\pi}{12} - \frac{1}{6} + \frac{1}{6}\ln 2.$$

25. (a) Establish the reduction formula.

$$\int \sin^n x dx = -\frac{1}{n} \cos x \sin^{n-1} x + \frac{n-1}{n} \int \sin^{n-2} x\, dx \ (n \geq 1).$$

(b) Use part (a) to evaluate $\int \sin^2 x dx$.

(c) Use parts (a) and (b) to evaluate $\int \sin^4 x dx$.

Solution (a) $\int \sin^n x dx = -\frac{1}{n} \cos x \sin^{n-1} x + \frac{n-1}{n} \int \sin^{n-2} x\, dx \ (n \geq 1).$

$$\int \sin^n x\, dx = \int \sin x \sin^{n-1} x\, dx$$

$$= [-\cos x \sin^{n-1} x] + \int \cos x (n-1) \sin^{n-2} x \cos x\, dx$$

$=-\cos x\sin^{n-1}x+(n-1)\int \cos^2 x\sin^{n-2}x\,dx$

$=-\cos x\sin^{n-1}x+(n-1)$

$\int (1-\sin^2 x)\sin^{n-2}x\,dx=-\cos x\sin^{n-1}x+$

$(n-1)\int (\sin^{n-1}x-\sin^n x)\,dx.$

Therefore,

$$\int \sin^n x\,dx=-\frac{1}{n}\cos x\sin^{n-1}x+\frac{n-1}{n}\int \sin^{n-2}x\,dx.$$

(b) $\int \sin^2 x\,dx=-\frac{1}{2}\cos x\sin x+\frac{1}{2}\int 1\,dx$

$=-\frac{1}{2}\cos x\sin x+\frac{x}{2}+C.$

(c) $\int \sin^4 x\,dx=-\frac{1}{4}\cos x\sin^3 x+\frac{3}{4}\int \sin^2 dx$

$=-\frac{1}{4}\cos x\sin^3 x+\frac{3}{4}\left(-\frac{1}{2}\cos x\sin x+\frac{x}{2}\right)+C$

$=-\frac{1}{4}\cos x\sin^3 x-\frac{3}{8}\cos x\sin x+\frac{3x}{8}+C.$

26. (a) Establish the reduction formula.

$$\int x(\ln x)^n\,dx=\frac{1}{2}x^2(\ln x)^n-\frac{n}{2}\int x(\ln x)^{n-1}dx\ (n\ge 1)$$

(b) Find the integral explicitly for $n=3$.

Solution $u=(\ln x)^n,\ dv=dx,$

$du=n(\ln x)^{n-1}\frac{1}{x}dx,\ v=\frac{x^2}{2}.$

(a) $\int x(\ln x)^n dx$

$=\frac{1}{2}x^2(\ln x)^n-\int \frac{1}{2}x^2\frac{(\ln x)^{n-1}}{x}dx$

$=\frac{1}{2}x^2(\ln x)^n-\frac{n}{2}\int x(\ln x)^{n-1}dx\ (n\ge 1).$

(b) $\frac{x^2}{8}\{4(\ln x)^3-6(\ln x)^2+6\ln x-3\}+C.$

27. (a) Establish the reduction formula

$$\int \tan^n x\,dx=\frac{\tan^{n-1}x}{n-1}-\int \tan^{n-2}x\,dx\ (n\ne 1).$$

(b) Find the integral for $n=5$.

Solution (a) $\int \tan^n x\,dx=\frac{\tan^{n-1}x}{n-1}-\int \tan^{n-2}x\,dx$

$\int \tan^n x\,dx=\int \tan^2 x\tan^{n-2}x\,dx$

$=\int (\sec^2 x-1)\tan^{n-2}x\,dx$

$=\int \sec^2 x\tan^{n-2}x\,dx-\int \tan^{n-2}x\,dx.$

$\int \sec^2 x\tan^{n-2}x\,dx$

$=[\tan^{n-1}x]-\int (\tan x)(n-2)\tan^{n-3}x\sec^2 x\,dx$

$=\tan^{n-1}x-(n-2)\int \tan^{n-2}x\sec^2 x\,dx.$

$\int \sec^2 x\tan^{n-2}x\,dx=\frac{1}{n-1}\tan^{n-1}x+C_1.$

$\int \tan^n x\,dx=\frac{\tan^{n-1}x}{n-1}-\int \tan^{n-2}x\,dx.$

(b) $\int \tan^5 x\,dx=\frac{\tan^4 x}{4}-\int \tan^3 x\,dx$

$=\frac{\tan^4 x}{4}-\left\{\frac{\tan^2 x}{2}-\int \tan x\,dx\right\}$

$=\frac{\tan^4 x}{4}-\frac{\tan^2 x}{2}+\int \frac{\sin x}{\cos x}dx$

$=\frac{\tan^4 x}{4}-\frac{\tan^2 x}{2}-\ln|\cos x|+C.$

7.2 Trigonometric Integrals

You Tube http://youtu.be/sIR0zNGQbus

A trigonometric integral is an integral whose integrand contains powers of one or more trigonometric functions. This integral can be evaluated by making a clever substitution of trigonometric identities to turn the integrand into a new integrand that we can integrate.

There are guidelines to evaluate integrals of the form

$$\int \sin^m x \cos^n x \, dx,$$

where m and n are non-negative integers.

CASE Ⅰ : n is an odd positive integer.

If the power of cosine is odd $(n=2k+1)$, $k \in Z^+ \cup \{0\}$, factor out one cosine term and use $\cos^2 x = 1-\sin^2 x$ to express the remaining expression in terms of the sine:

$$\begin{aligned}\int \sin^m x \cos^{2k+1} x \, dx &= \int \sin^m x (\cos^2 x)^k \cos x \, dx \\ &= \int \sin^m x (1-\sin^2 x)^k \cos x dx\end{aligned}$$

Then use substitution with $u=\sin x$.

EXAMPLE 1

Find $\int \sin^2 x \cos^3 x \, dx$.

Solution Let $u=\sin x$, then we have $du = \cos x \, dx$ and so

$$\begin{aligned}\int \sin^2 x \cos^3 x \, dx &= \int \sin^2 x (1-\sin^2 x) \cos x \, dx \\ &= \int u^2 (1-u^2) \, du \\ &= \int (u^2 - u^4) \, du \\ &= \frac{1}{3}u^3 - \frac{1}{5}u^5 + C \\ &= \frac{1}{3}\sin^3 x - \frac{1}{5}\sin^5 x + C.\end{aligned}$$

■

CASE Ⅱ : m is an odd positive integer.

If the power of sine is odd $(m=2k+1)$, $k \in Z^+ \cup \{0\}$, factor out one sine term and use $\sin^2 x = 1-\cos^2 x$ to express the remaining factors in terms of cosine:

$$\int \sin^{2k+1}x \cos^n x\, dx = \int (\sin^2 x)^k \cos^n x \sin x\, dx$$
$$= \int (1-\cos^2 x)^k \cos^n x \sin x\, dx.$$

Then use substitution with $u = \cos x$.

EXAMPLE 2

Evaluate $\int \sin^3 x \cos^2 x\, dx$.

Solution Let $u = \cos x$, then we have $du = -\sin x\, dx$ and so

$$\int \sin^3 x \cos^2 x\, dx = \int (1-\cos^2 x)\cos^2 x \sin x\, dx$$
$$= \int (1-u^2)u^2(-du)$$
$$= \int (u^4 - u^2)\, du$$
$$= \frac{1}{5}u^5 - \frac{1}{3}u^3 + C$$
$$= \frac{1}{5}\cos^5 x - \frac{1}{3}\cos^3 x + C.$$

■

CASE III : m **and** n **are even positive integers.**

For the integrals whose the integrand contains even powers of both sine and cosine functions, we can apply the following half-angle formulas

$$\sin^2 x = \frac{1-\cos 2x}{2} \text{ and } \cos^2 x = \frac{1+\cos 2x}{2},$$

possibly several times, to reduce the powers in the integrand.

EXAMPLE 3

Find $\int_0^{\frac{\pi}{2}} \cos^2 x dx$.

Solution Rewriting the integrand gives

$$\int_0^{\frac{\pi}{2}} \cos^2 x dx = \frac{1}{2}\int_0^{\frac{\pi}{2}} (1+\cos 2x) dx$$
$$= \left[\frac{1}{2}\left(x + \frac{1}{2}\sin 2x\right)\right]_0^{\frac{\pi}{2}} = \frac{\pi}{4}.$$

■

EXAMPLE 4

Evaluate $\int \cos^4 x dx$.

Solution Rewrite $\cos^4 x = (\cos^2 x)^2$ and use the half-angle formula for cosine:

$$\begin{aligned}\int \cos^4 x\,dx &= \int\left(\frac{1+\cos 2x}{2}\right)^2 dx\\ &= \frac{1}{4}\int (1+2\cos 2x+\cos^2 2x)dx\\ &= \frac{1}{4}\int\left[1+2\cos 2x+\frac{1}{2}(1+\cos 4x)\right]dx\\ &= \frac{1}{4}\int\left(\frac{3}{2}+2\cos 2x+\frac{1}{2}\cos 4x\right)dx\\ &= \frac{1}{4}\left(\frac{3}{2}x+\sin 2x+\frac{1}{8}\sin 4x\right)+C.\end{aligned}$$ ■

There are also guidelines to evaluate the integrals of the form

$$\int \tan^m x\ \sec^n x\,dx.$$

CASE I : n is an even positive integer.

If the power of secant is even $(n=2k,\ k\geq 2),\ k\in N$ factor out a single $\sec^2 x$ and use the trigonometric identity $\sec^2 x=1+\tan^2 x$ to express the remaining factors in terms of $\tan x$:

$$\begin{aligned}\int \tan^m x\ \sec^{2k} x\,dx &= \int \tan^m x\ (\sec^2 x)^{k-1}\sec^2 x\,dx\\ &= \int \tan^m x\ (1+\tan^2 x)^{k-1}\sec^2 x\ dx.\end{aligned}$$

Then choose $u=\tan x$.

EXAMPLE 5

Find $\int \tan^3 x\sec^4 x\,dx$.

Solution Separate the $\sec^2 x$ term, and express the remaining $\sec^2 x$ factor in terms of tangent using the trigonometric identity $\sec^2 x=1+\tan^2 x$. Let $u=\tan x$, then we have $du=\sec^2 x\,dx$. We have

$$\begin{aligned}\int \tan^3 x\sec^4 x dx &= \int (\tan^3 x)(\tan^2 x+1)\sec^2 x\ dx\\ &= \int u^3(u^2+1)du\\ &= \frac{1}{6}u^6+\frac{1}{4}u^4+C\\ &= \frac{1}{6}\tan^6 x+\frac{1}{4}\tan^4 x+C.\end{aligned}$$ ■

CASE II : m **is an odd positive integer.**

If the power of tangent is odd $(m = 2k+1)$, $k \in Z^{+} \cup \{0\}$ in the integral $\int \tan^{m}x \sec^{n}x\,dx$, collect the term of $\sec x \tan x$ and use the trigonometric identity $\tan^2 x = \sec^2 x - 1$ to express the remaining factors in terms of $\sec x$:

$$\begin{aligned}\int \tan^{2k+1}x \sec^{n}x\,dx &= \int (\tan^2 x)^k \sec^{n-1}x \sec x \tan x\,dx \\ &= \int (\sec^2 x - 1)^k \sec^{n-1}x \sec x \tan x\,dx.\end{aligned}$$

Then choose $u = \sec x$.

EXAMPLE 6

Evaluate $\int \tan^3\theta \sec^3\theta\,d\theta$.

Solution Separate the $\sec\theta\tan\theta$ term, and convert the remaining power of tangent to an expression involving only the secant using the trigonometric identity $\tan^2\theta = \sec^2\theta - 1$. Let $u = \sec\theta$, so $du = \sec\theta\tan\theta\,d\theta$. We have

$$\begin{aligned}\int \tan^3\theta \sec^3\theta\,d\theta &= \int \tan^2\theta \sec^2\theta \sec\theta \tan\theta\,d\theta \\ &= \int (\sec^2\theta - 1)\sec^2\theta \sec\theta\tan\theta\,d\theta \\ &= \int (u^2-1)u^2\,du = \int (u^4 - u^2)\,du \\ &= \frac{1}{5}u^5 - \frac{1}{3}u^3 + C \\ &= \frac{1}{5}\sec^5\theta - \frac{1}{3}\sec^3\theta + C.\end{aligned}$$

■

Recall that

1 $$\int \tan x\,dx = \ln|\sec x| + C$$

EXAMPLE 7

Evaluate $\int \tan^5 x\,dx$.

Solution

$$\begin{aligned}\int \tan^5 x\,dx &= \int \tan^3 x(\sec^2 x - 1)\,dx \\ &= \int \tan^3 x \sec^2 x\,dx - \int \tan^3 x\,dx \\ &= \int \tan^3 x \sec^2 x\,dx - \int \tan x(\sec^2 x - 1)\,dx\end{aligned}$$

$$= \int \tan^3 x \sec^2 x\, dx - \int \tan x \sec^2 x\, dx + \int \tan x\, dx$$
$$= \frac{1}{4}\tan^4 x - \frac{1}{2}\tan^2 x - \ln|\cos x| + C.$$
■

CASE III : m is an even integer and n is an odd positive integer.

If the power of tangent is even $(n = 2k,\ k \geq 2)$ and the power of secant is odd $(m = 2k+1)$, $k \in Z^+ \cup \{0\}$ in the integral $\int \tan^m x \sec^n x\, dx$, collect the term of $\sec x \tan x$ and use the trigonometric identity $\tan^2 x = \sec^2 x - 1$ to express the remaining factors in terms of $\sec x$:

EXAMPLE 8

Evaluate $\int \tan^2 x \sec x\, dx$.

Solution $$\int \tan^2 x \sec x dx = \int (\sec^2 x - 1)\sec x\, dx$$
$$= \int \sec^3 x\, dx - \int \sec x\, dx$$
$$= \frac{1}{2}(\sec x \tan x + \ln|\sec x + \tan x|) - \ln|\sec x + \tan x|$$
$$= \frac{1}{2}\sec x \tan x - \frac{1}{2}\ln|\sec x + \tan x| + C.$$
■

Note If an even power of tangent appears with an odd power of secant, express the integrand completely in terms of $\sec x$ and use integration by parts.

Note Using the trigonometric identity $1 + \cot^2 x = \csc^2 x$, integrals of the form $\int \cot^m x \csc^n x\, dx$ can be evaluated using similar methods.

The integrals

(a) $\int \sin mx \cos nx\, dx$

(b) $\int \sin mx \sin nx\, dx$

or (c) $\int \cos mx \cos nx\, dx$

can be evaluated using the following trigonometric identities

(a) $\sin A \cos B = \frac{1}{2}[\sin(A-B) + \sin(A+B)]$

(b) $\sin A \sin B = \frac{1}{2}[\cos(A-B) - \cos(A+B)]$

(c) $\cos A \cos B = \frac{1}{2}[\cos(A-B) + \cos(A+B)]$.

EXAMPLE 9

Find $\int \sin x \sin 2x \sin 3x \, dx$.

Solution $\int \sin x \sin 2x \sin 3x \, dx$

$$
\begin{aligned}
&= \int \left(\frac{1}{2}\cos x - \frac{1}{2}\cos 3x\right)\sin 3x \, dx \\
&= \frac{1}{2}\int (\cos x \sin 3x - \cos 3x \sin 3x)dx \\
&= \frac{1}{2}\int \left[\frac{1}{2}(\sin 2x + \sin 4x) - \frac{1}{2}\sin 6x\right]dx \\
&= \frac{1}{4}\int (\sin 2x + \sin 4x - \sin 6x)dx \\
&= -\frac{1}{8}\cos 2x - \frac{1}{16}\cos 4x + \frac{1}{24}\cos 6x + C.
\end{aligned}
$$

■

<ICME 10, Felix Klein Medal>

7.2 EXERCISES (Trigonometric Integrals)

http://matrix.skku.ac.kr/Cal-Book/part1/CS-Sec-7-2-Sol.html

YouTube http://youtu.be/ytETYf1wLbs

CAS **1.** $\int \cos^3 x\, dx$

Solution $\int \cos^3 x\, dx = \int \cos^2 x \cos x\, dx$

$= \int (1-\sin^2 x)\cos x\, dx$

$= \sin x - \frac{1}{3}\sin^3 x + C.$

CAS **2.** $\int \sin^5 t\, dt$

http://matrix.skku.ac.kr/cal-lab/cal-7-2-9.html

Solution

```
var('t')
integral(sin(t)^5,t)
```

Answer : -1/5*cos(t)^5 + 2/3*cos(t)^3 - cos(t)

$$\int \sin^5 t\, dt = [-\cos t \sin^4 t] + \int \cos t\,(4\sin^3 t)\cos t\, dt$$
$$= -\cos t\sin^4 t + 4\int \cos^2 t \sin^3 t\, dt$$
$$= -\cos t\sin^4 t + 4\int (1-\sin^2 t)\sin^3 t\, dt$$
$$= -\cos t\sin^4 t + 4\int (\sin^3 t - \sin^5 t)\, dt.$$

Thus, $5\int \sin^5 t\, dt = -\cos t\sin^4 t + 4\int \sin^3 t\, dt.$

$$\int \sin^3 t\, dt = [-\cos t\sin^2 t] + \int \cos t\, 2\sin t\cos t\, dt$$
$$= -\cos t\sin^2 t + 2\int \cos^2 t\sin t\, dt$$
$$= -\cos t\sin^2 t + 2\int (1-\sin^2 t)\sin t\, dt$$
$$= -\cos t\sin^2 t + 2\int (\sin t - \sin^3 t)\, dt.$$

Thus, $3\int \sin^3 t\, dt = -\cos t\sin^2 t + 2\int \sin t\, dt.$

$$\int \sin^3 t\, dt = -\frac{1}{3}\cos t\sin^2 t - \frac{2}{3}\cos t.$$

Therefore,

$$\int \sin^5 t\, dt = -\frac{1}{5}\cos t\sin^4 t + \frac{4}{5}\left(-\frac{1}{3}\cos t\sin^2 t - \frac{2}{3}\cos t\right)$$
$$= -\frac{1}{5}\sin^4 t\cos t - \frac{4}{15}\sin^2 t\cos t - \frac{8}{15}\cos t + C.$$

CAS **3.** $\int \sin^5 x\cos^4 x\, dx$

http://matrix.skku.ac.kr/cal-lab/cal-7-2-2.html

Solution

```
integral(sin(x)^5*cos(x)^4,x)
```

Answer : -1/9*cos(x)^9 + 2/7*cos(x)^7 - 1/5*cos(x)^5

CAS **4.** $\int \sin^6 x\cos^3 x\, dx$

http://matrix.skku.ac.kr/cal-lab/cal-7-2-18.html

Solution

```
integral(sin(x)^6*cos(x)^3, x)x)
```

Answer : -1/9*sin(x)^9 + 1/7*sin(x)^7

Let $u = \sin x$, then $du = \cos x dx$.

$$\int \sin^6 x\cdot(1-\sin^2 x)\cdot\cos x dx = \int u^6(1-u^2)du$$
$$= \int (u^6 - u^8)du$$
$$= \frac{1}{7}u^7 - \frac{1}{9}u^9 + C$$
$$= \frac{1}{7}\sin^7 x - \frac{1}{9}\sin^9 u + C.$$

CAS **5.** $\int \frac{\sin^5 x}{\cos^6 x}dx$

http://matrix.skku.ac.kr/cal-lab/cal-7-2-4.html

Solution

```
integral(sin(x)^5/cos(x)^6,x)
```

Answer : 1/15*(15*cos(x)^4 - 10*cos(x)^2 + 3)/cos(x)^5

Let $u = \sec x$, then $du = \sec x\cdot\tan x dx$. Therefore,

$$\int \frac{\sin^5 x}{\cos^6 x}dx = \int (\sec^2 x - 1)^2\sec x\tan x dx$$
$$= \int (u^2-1)^2 du$$
$$= \int (u^4 - 2u^2 + 1)du$$
$$= \frac{1}{5}u^5 - \frac{2}{3}u^3 + u + C$$
$$= \frac{1}{5}\sec^5 x - \frac{2}{3}\sec^3 x + \sec x + C.$$

6. $\int \cos^2\theta d\theta$

Solution $\int \cos^2\theta d\theta = \int \frac{1+\cos 2\theta}{2}d\theta$

$= \frac{\theta}{2} + \int \frac{\cos 2\theta}{2}d\theta$

$$= \frac{\theta}{2} + \frac{\sin 2\theta}{4} + C.$$

CAS **7.** $\int \sin^2 x \cos^4 x\, dx$

http://matrix.skku.ac.kr/cal-lab/cal-7-2-5.html

Solution

```
integral(sin(x)^2*cos(x)^4,x)
```

Answer : 1/48*[sin(2*x)]^3 + 1/16*x - 1/64*sin(4*x)

CAS **8.** $\int \sin^7 3x \cos^2 3x\, dx$

http://matrix.skku.ac.kr/cal-lab/cal-7-2-10.html

Solution

```
integral(sin(3*x)^7*cos(3*x)^2,x)
```

Answer : 1/27*cos(3*x)^9 - 1/7*cos(3*x)^7 + 1/5*cos(3*x)^5 - 1/9*cos(3*x)^3

Let $u = \cos 3x$, then $du = -3\sin 3x\, dx$.

$$\int \sin^7 3x \cos^2 3x\, dx = \int \cos^2 3x (1-\cos^2 3x)^3 \sin 3x\, dx$$
$$= \int u^2 (1-u^2)^3 \left(-\frac{1}{3}\right) du$$
$$= -\frac{1}{3} \int (u^2 - 3u^4 + 3u^6 - u^8)\, du$$
$$= -\frac{1}{3}\left(\frac{1}{3}u^3 - \frac{3}{5}u^5 + \frac{3}{7}u^7 - \frac{1}{9}u^9\right) + C$$
$$= -\frac{1}{9}\cos^3 3x + \frac{1}{5}\cos^5 3x - \frac{1}{7}\cos^7 3x + \frac{1}{27}\cos^9 3x + C.$$

CAS **9.** $\int \sin^2(4x)\, dx$

http://matrix.skku.ac.kr/cal-lab/cal-7-2-3.html

Solution

```
integral(sin(4*x)^2,x)
```

Answer : 1/2*x − 1/16*sin(8*x)

Since $\sin^2 4x = \frac{1-\cos 8x}{2}$.

$$\int \sin^2 4x\, dx = \int \frac{1-\cos 8x}{2}\, dx$$
$$= \frac{1}{2}\int (1-\cos 8x)\, dx = \frac{1}{2}\left(x - \frac{1}{8}\sin 8x\right) + C$$
$$= \frac{1}{2}x - \frac{1}{16}\sin 8x + C.$$

CAS **10.** $\int \sin^4 2t \cos^4 2t\, dt$

http://matrix.skku.ac.kr/cal-lab/cal-7-2-11.html

Solution

```
var('t')
integral(sin(2*t)^4*cos(2*t)^4, t)
```

Answer : 3/128*t - 1/256*sin(8*t) + 1/2048*sin(16*t)

CAS **11.** $\int \sin 2x \cos 3x\, dx$

http://matrix.skku.ac.kr/cal-lab/cal-7-2-8.html

Solution

```
integral(sin(2*x)*cos(3*x), x)
```

Answer : -1/10*cos(5*x) + 1/2*cos(x)

CAS **12.** $\int \sin 4y \cos 5y\, dy$

http://matrix.skku.ac.kr/cal-lab/cal-7-2-15.html

Solution

```
var('y')
integral(sin(4*y)*cos(5*y), y)
```

Answer : -1/18*cos(9*y) + 1/2*cos(y)

CAS **13.** $\int_0^{\pi/2} \sin^3 x \cos^5 x\, dx$

http://matrix.skku.ac.kr/cal-lab/cal-7-2-17.html

Solution

```
integral(sin(x)^3*cos(x)^5, x, 0, pi/2)
```

Answer : 1/24

CAS **14.** $\int_{-\frac{\pi}{2}}^{\frac{\pi}{2}} \cos^3 x \sin^4 x\, dx$

http://matrix.skku.ac.kr/cal-lab/cal-7-2-21.html

Solution

```
integral(cos(x)^3*sin(x)^4, x, -pi/2, pi/2)
```

Answer : 4/35

CAS **15.** $\int x \cos^2 x \sin x\, dx$

http://matrix.skku.ac.kr/cal-lab/cal-7-5-16.html

Solution

```
integral(x*cos(x)^2*sin(x), x)
```

Answer : -1/12*x*cos(3*x) - 1/4*x*cos(x) + 1/36*sin(3*x) + 1/4*sin(x)

CAS **16.** $\int \cot^4 x\, dx$

http://matrix.skku.ac.kr/cal-lab/cal-7-2-6.html

Solution

```
integral(cot(x)^4, x)
```

Answer : 1/3*(3*tan(x)^2 - 1)/tan(x)^3 + arctan(tan(x))

CAS **17.** $\int \tan^3 x\,(\sec x)^{-1/2}\,dx$

http://matrix.skku.ac.kr/cal-lab/cal-7-2-7.html

Solution

```
integral(tan(x)^3*sec(x)^(-1/2), x)
```

Answer : 2*sqrt(cos(x)) + 2/3/cos(x)^(3/2)

CAS **18.** $\int \tan^3 3y\,\sec^3 3y\,dy$

http://matrix.skku.ac.kr/cal-lab/cal-7-2-12.html

Solution

```
var('y')
integral(tan(3*y)^3*sec(3*y)^3, y)
```

Answer : -1/45*(5*cos(3*y)^2 - 3)/cos(3*y)^5

CAS **19.** $\int (\tan t)^{-3}\sec^2 t\,dt$

http://matrix.skku.ac.kr/cal-lab/cal-7-2-14.html

Solution

```
var('t')
integral(tan(t)^(-3)*sec(t)^2, t)
```

Answer : -1/2/tan(t)^2

CAS **20.** $\int \cot^4 2x\,dx$

http://matrix.skku.ac.kr/cal-lab/cal-7-2-16.html

Solution

```
integral(cot(2*x)^4, x)
```

Answer : x + 1/6*(3*tan(2*x)^2 - 1)/tan(2*x)^3

CAS **21.** $\int_0^{\pi/4} \tan^3 x\sec^4 x\,dx$

http://matrix.skku.ac.kr/cal-lab/cal-7-2-19.html

Solution

```
integral(tan(x)^3*sec(x)^4, x, 0, pi/4)
```

Answer : 5/12

CAS **22.** $\int \cot x\csc^3 x\,dx$

http://matrix.skku.ac.kr/cal-lab/cal-7-2-13.html

Solution

```
integral(cot(x)*csc(x)^3, x)
```

Answer : -1/3/sin(x)^3

CAS **23.** $\int x\sec x\tan x\,dx$

http://matrix.skku.ac.kr/cal-lab/cal-7-2-20.html

Solution

```
integral(x*sec(x)*tan(x),x)
```

Answer : 1/2*(4*x*sin(2*x)*sin(x)+4*x*cos(2*x)*cos(x)+(sin(2*x)^2 +cos(2*x)^2 + 2*cos(2*x) + 1)*log(sin (x)^2 + cos(x)^2 - 2*sin(x) + 1) -(sin(2*x)^2+ cos(2*x)^2 + 2*cos(2*x) + 1)*log(sin(x)^2 + cos(x)^2 + 2*sin(x) + 1) + 4*x*cos(x))/(sin(2*x)^2 + cos(2*x)^2 + 2*cos(2*x) + 1)

7.3 Trigonometric Substitutions

YouTube http://youtu.be/avTqiEUi8u8

An integral that contains terms of the form $\sqrt{a^2-x^2}$, $\sqrt{a^2+x^2}$ or $\sqrt{x^2-a^2}$ can sometimes be evaluated by considering a new variable representing an angle, which may eliminate the square root function. We substitute a new variable for the original variable x and transform the integral into a trigonometric one (hence the name trigonometric substitution). Once the trigonometric integral has been found, we reverse the substitution in order to find the original integral in terms of the original variable. The three typical expressions, the corresponding substitution and trigonometric identities to use for each case are given in the following table.

Table of Trigonometric Substitutions

Expression	Substitution	Identity
$\sqrt{a^2-x^2}$	$x=a\sin\theta,\ -\frac{\pi}{2}\le\theta\le\frac{\pi}{2}$	$1-\sin^2\theta=\cos^2\theta$
$\sqrt{a^2+x^2}$	$x=a\tan\theta,\ -\frac{\pi}{2}<\theta<\frac{\pi}{2}$	$1+\tan^2\theta=\sec^2\theta$
$\sqrt{x^2-a^2}$	$x=a\sec\theta,\ 0\le\theta<\frac{\pi}{2}$ or $\pi\le\theta<\frac{3\pi}{2}$	$\sec^2\theta-1=\tan^2\theta$

EXAMPLE 1

Find $\displaystyle\int\frac{x^2}{\sqrt{16-x^2}}dx$.

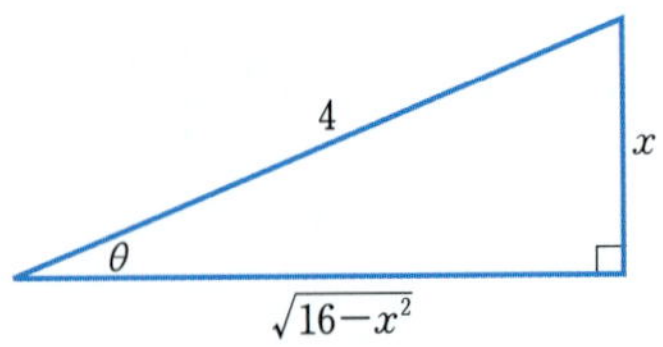

Figure 1

Solution Substitute $x=4\sin\theta$, where $-\pi/2\le\theta\le\pi/2$. (See Figure 1.) Then $dx=4\cos\theta\,d\theta$ and

$$\sqrt{16-x^2}=\sqrt{16-16\sin^2\theta}=\sqrt{16\cos^2\theta}=4|\cos\theta|=4\cos\theta.$$

Note that $\cos\theta\ge0$ because $-\pi/2\le\theta\le\pi/2$. Thus the Substitution Rule gives

$$\begin{aligned}\int\frac{x^2}{\sqrt{16-x^2}}dx&=\int\frac{16\sin^2\theta}{4\cos\theta}4\cos\theta\,d\theta\\&=16\int\sin^2\theta d\theta=8\int(1-\cos2\theta)d\theta\\&=8(\theta-\sin\theta\cos\theta)+C.\end{aligned}$$

Since $\sin\theta=\dfrac{x}{4}$, $\theta=\sin^{-1}\dfrac{x}{4}$, and $\cos\theta=\dfrac{\sqrt{16-x^2}}{4}$. We obtain

$$\int \frac{x^2}{\sqrt{16-x^2}}dx = 8\left(\sin^{-1}\frac{x}{4} - \frac{x}{4}\cdot\frac{\sqrt{16-x^2}}{4}\right) + C$$

$$= 8\sin^{-1}\frac{x}{4} - \frac{x\sqrt{16-x^2}}{2} + C.$$

■

EXAMPLE 2

Find $\int \sqrt{a^2-x^2}\,dx \quad (a>0)$.

http://matrix.skku.ac.kr/cal-lab/cal-7-8-36.html

Solution Substitute $x = a\sin\theta \ (-\frac{\pi}{2} \le \theta \le \frac{\pi}{2})$. Then $dx = a\cos\theta\, d\theta$. Since $\cos\theta \ge 0$, and

$$\sqrt{a^2-x^2} = \sqrt{a^2-a^2\sin^2\theta} = \sqrt{a^2\cos^2\theta} = a|\cos\theta| = a\cos\theta$$

$$\text{for } -\pi/2 \le \theta \le \pi/2,$$

we have

$$\int \sqrt{a^2-x^2}\,dx = \int (a\cos\theta)(a\cos\theta)d\theta$$
$$= a^2\int \cos^2\theta d\theta$$
$$= \frac{a^2}{2}\int (1+\cos 2\theta)d\theta$$
$$= \frac{a^2}{2}\left(\theta + \frac{1}{2}\sin 2\theta\right) + C$$
$$= \frac{a^2}{2}(\theta + \sin\theta\cos\theta) + C.$$

Since $\sin\theta = \frac{x}{a}$, $\theta = \sin^{-1}\frac{x}{a}$, and $\cos\theta = \frac{\sqrt{a^2-x^2}}{a}$, we obtain

$$\int \sqrt{a^2-x^2}\,dx = \frac{1}{2}\left(x\sqrt{a^2-x^2} + a^2\sin^{-1}\frac{x}{a}\right) + C.$$

```
var('x, a')
integral(sqrt(a^2-x^2), x)
```

Answer : 1/2*a^2*arcsin(x/sqrt(a^2)) + 1/2*sqrt(a^2 - x^2)*x ■

EXAMPLE 3

Find $\int \frac{x^2+3x+2}{\sqrt{5-4x-x^2}}dx$.

Solution Completing the square under the square root we have

$$5-4x-x^2 = 9-(x+2)^2.$$

This suggests that we make the substitution $u = x+2$. Then $du = dx$ and $x = u-2$, so

$$\int \frac{x^2+3x+2}{\sqrt{5-4x-x^2}}dx = \int \frac{u^2-u}{\sqrt{9-u^2}}du.$$

Let $u = 3\sin\theta$ where $-\pi/2 \le \theta \le \pi/2$. Then $du = 3\cos\theta\, d\theta$ and $\sqrt{9-u^2} = 3\cos\theta$, so

$$\begin{aligned}\int \frac{x^2+3x+2}{\sqrt{5-4x-x^2}}dx &= \int [9\sin^2\theta - 3\sin\theta]d\theta \\ &= \frac{9}{2}\theta - \frac{9}{2}\sin\theta\cos\theta + 3\cos\theta + C \\ &= \frac{9}{2}\sin^{-1}\left(\frac{u}{3}\right) + \left(1-\frac{1}{2}u\right)\sqrt{9-u^2} + C \\ &= \frac{9}{2}\sin^{-1}\left(\frac{x+2}{3}\right) - \frac{1}{2}x\sqrt{5-4x-x^2} + C.\end{aligned}$$

■

EXAMPLE 4

Evaluate $\displaystyle\int \frac{1}{\sqrt{4+x^2}}dx$.

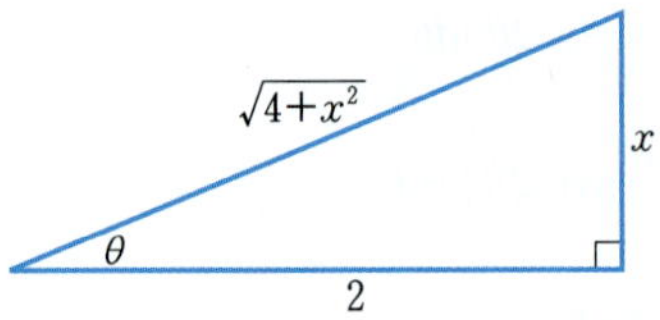

Figure 2

Solution Substitute $x = 2\tan\theta$ $(-\pi/2 < \theta < \pi/2)$. (See Figure 2.) Then $dx = 2\sec^2\theta\, d\theta$ and

$$\sqrt{x^2+4} = \sqrt{4(\tan^2\theta+1)} = \sqrt{4\sec^2\theta} = 2|\sec\theta| = 2\sec\theta.$$

Since $\sec\theta > 0$ for $-\pi/2 < \theta < \pi/2$

Thus, we have

$$\begin{aligned}\int \frac{dx}{\sqrt{4+x^2}} &= \int \frac{2\sec^2\theta}{2\sec\theta}d\theta \\ &= \int \sec\theta\, d\theta \\ &= \ln|\sec\theta + \tan\theta| + C.\end{aligned}$$

Drawing a right triangle such that $\tan\theta = \dfrac{x}{2}$:

we observe from this triangle that

$$\sec\theta = \frac{\sqrt{4+x^2}}{2}$$

so

$$\begin{aligned}\int \frac{dx}{\sqrt{4+x^2}} &= \ln\left|\frac{\sqrt{4+x^2}}{2} + \frac{x}{2}\right| + C \\ &= \ln\left|\sqrt{4+x^2} + x\right| - \ln 2 + C\end{aligned}$$

$$= \ln\left|\sqrt{4+x^2}+x\right| + K,$$

where $K = C - \ln 2$. ■

EXAMPLE 5

Evaluate $\displaystyle\int_0^{\sqrt{3}a} \frac{x^2}{(a^2+x^2)^{3/2}}\,dx$.

Solution We put $x = a\tan\theta$. Then $dx = a\sec^2\theta\, d\theta$. As a definite integral, we can convert the limits of x into values of θ. So,

$$\begin{aligned}\int_0^{\sqrt{3}a} \frac{x^2}{(a^2+x^2)^{3/2}}\,dx &= \int_0^{\frac{\pi}{3}} \frac{\sin^2\theta}{\cos\theta}\,d\theta \\ &= \left[-\sin\theta + \ln(\sec\theta+\tan\theta)\right]_0^{\frac{\pi}{3}} \\ &= -\sin\left(\frac{\pi}{3}\right) + \ln\left(\sec\left(\frac{\pi}{3}\right)+\tan\left(\frac{\pi}{3}\right)\right) \\ &= -\frac{\sqrt{3}}{2} + \ln(2+\sqrt{3}).\end{aligned}$$ ■

EXAMPLE 6

Find $\displaystyle\int \frac{dx}{x^3\sqrt{x^2-16}}$.

Solution Substitute $x = 4\sec\theta$, where $0 < \theta < \pi/2$ or $\pi < \theta < 3\pi/2$. Then $dx = 4\sec\theta\,\tan\theta\, d\theta$ and

$$\sqrt{x^2-16} = \sqrt{16(\sec^2\theta - 1)} = \sqrt{16\tan^2\theta} = 4|\tan\theta| = 4\tan\theta.$$

Since $\tan\theta > 0$ for $0 < \theta < \pi/2$ or $\pi < \theta < 3\pi/2$, we have

$$\begin{aligned}\int \frac{dx}{x^3\sqrt{x^2-16}} &= \frac{1}{64}\int \cos^2\theta\, d\theta \\ &= \frac{1}{128}\int (1+\cos 2\theta)\,d\theta \\ &= \frac{1}{128}\left(\theta + \frac{1}{2}\sin 2\theta\right) + C \\ &= \frac{1}{128}(\theta + \sin\theta\cos\theta) + C.\end{aligned}$$

A right triangle with $\sec\theta = \dfrac{x}{4}$ allows us to conclude that $\theta = \sec^{-1}\left(\dfrac{x}{4}\right)$, $\cos\theta = \dfrac{4}{x}$, and $\sin\theta = \dfrac{\sqrt{x^2-16}}{x}$, so we obtain

$$\int \frac{dx}{x^3\sqrt{x^2-16}} = \frac{1}{128}\left[\sec^{-1}\left(\frac{x}{4}\right) + \frac{4\sqrt{x^2-16}}{x^2}\right] + C.$$ ■

CAS **EXAMPLE 7**

Evaluate the integral using a trigonometric substitution.

$$\int \frac{1}{x\sqrt{x^2-1}}dx \ (x>0).$$

http://matrix.skku.ac.kr/cal-lab/cal-7-3-Exm-7.html

Solution Let $x = \csc t$, then

$$\frac{1}{x\sqrt{x^2-1}} = \frac{1}{\csc t\sqrt{\csc^2 t-1}} = \frac{1}{\csc t\cot t}$$

and $dx = -\csc t \cot t\, dt$.

Thus $\displaystyle\int \frac{1}{x\sqrt{x^2-1}}dx = \int \frac{1}{\csc t\cot t}(-\csc t\cot t)dt$

$$= -\int dt = -t + C$$

$$x = \csc t \ \Rightarrow \ \frac{1}{x} = \sin t \ \Rightarrow \ t = \sin^{-1}\frac{1}{x}$$

Hence $\displaystyle\int \frac{1}{x\sqrt{x^2-1}}dx = -\sin^{-1}\frac{1}{x} + C.$

```
var('x')
f=1/(x*sqrt(x^2-1))
assume(x>0)
integral(f, x)
```

Answer : -arcsin(1/x) ■

Note The following will apply in example 8 and 9. Let $\tan\frac{x}{2} = t$. Then,

$dx = \dfrac{2}{1+t^2}dt$ because $\dfrac{1}{2}\sec^2\dfrac{x}{2}\,dx = dt$.

$$\sin x = \sin 2\frac{x}{2} = 2\sin\frac{x}{2}\cos\frac{x}{2}$$

$$= \frac{2\sin\frac{x}{2}\cos\frac{x}{2}}{\cos^2\frac{x}{2}+\sin^2\frac{x}{2}} = \frac{2\tan\frac{x}{2}}{1+\tan^2\frac{x}{2}} = \frac{2t}{1+t^2},$$

$$\cos x = \cos 2\frac{x}{2} = \cos^2\frac{x}{2} - \sin^2\frac{x}{2}$$

$$= \frac{\cos^2\frac{x}{2}-\sin^2\frac{x}{2}}{\cos^2\frac{x}{2}+\sin^2\frac{x}{2}} = \frac{1-\tan^2\frac{x}{2}}{1+\tan^2\frac{x}{2}} = \frac{1-t^2}{1+t^2}$$

and $\tan x = \dfrac{\sin x}{\cos x} = \dfrac{\dfrac{2t}{1+t^2}}{\dfrac{1-t^2}{1+t^2}} = \dfrac{2t}{1-t^2}.$

EXAMPLE 8

Find $\displaystyle\int \frac{1}{\sin x}\,dx$.

Solution Let $\tan\dfrac{x}{2} = t$, then $\sin x = \dfrac{2t}{1+t^2}$ and $dx = \dfrac{2}{1+t^2}dt$.

$$\int \frac{1}{\sin x}\,dx = \int \frac{1}{\dfrac{2t}{1+t^2}} \cdot \frac{2}{1+t^2}\,dt$$
$$= \int \frac{1}{t}\,dt = \ln|t| = \ln\left|\tan\frac{x}{2}\right| + C.$$

■

EXAMPLE 9

Find $I = \displaystyle\int \frac{1}{\sin x + \cos x}\,dx$.

http://matrix.skku.ac.kr/cal-lab/cal-7-3-Exm-9.html

Solution Let $\tan\dfrac{x}{2} = t$, then $\sin x = \dfrac{2t}{1+t^2}$, $\cos x = \dfrac{1-t^2}{1+t^2}$ and $dx = \dfrac{2}{1+t^2}dt$.

$$I = \int \frac{1}{\dfrac{2t}{1+t^2} + \dfrac{1-t^2}{1+t^2}} \cdot \frac{2}{1+t^2}\,dt$$
$$= 2\int \frac{1}{1+2t-t^2}\,dt = 2\int \frac{1}{2-(t-1)^2}\,dt$$
$$= \frac{1}{\sqrt{2}}\ln\left|\frac{\sqrt{2}+(t-1)}{\sqrt{2}-(t-1)}\right| + C$$
$$= \frac{1}{\sqrt{2}}\ln\left|\frac{\sqrt{2}-1+\tan\dfrac{x}{2}}{\sqrt{2}+1-\tan\dfrac{x}{2}}\right| + C.$$

```
var('x')
integral(1/(sin(x) + cos(x)), x)
```

Answer : -1/2*sqrt(2)*log(-(2*sqrt(2) - 2*sin(x)/(cos(x) + 1)+ 2) / ((2*sqrt(2)) + 2*sin(x)/(cos(x) + 1) - 2))

■

7.3 EXERCISES (Trigonometric Substitutions)

http://matrix.skku.ac.kr/Cal-Book/part1/CS-Sec-7-3-Sol.html

YouTube http://youtu.be/utTQHIabTyI

CAS **1.** $\displaystyle\int \frac{\sqrt{1-x^2}}{x^2}dx = -\frac{\sqrt{1-x^2}}{x} - \sin^{-1}x + C$

http://matrix.skku.ac.kr/cal-lab/cal-7-3-1.html

Solution $x=\sin\theta,\ dx=\cos\theta\, d\theta$

$$\int \frac{\sqrt{1-x^2}}{x^2}dx = \int \frac{\sqrt{\cos^2\theta}}{\sin^2\theta}\cos\theta d\theta$$

$$= \int \frac{\cos^2\theta}{\sin^2\theta}d\theta = \int \cot^2\theta d\theta$$

$$= \int (\csc^2\theta - 1)d\theta = -\cot\theta - \theta + C$$

$$= \frac{\sqrt{1-x^2}}{x} - \sin^{-1}x + C.$$

```
integral((sqrt(1-x^2))/(x^2), x)
```

Answer : -sqrt(-x^2 + 1)/x - arcsin(x)

CAS **2.** $\displaystyle\int \frac{1}{(4-x^2)\sqrt{4+x^2}}dx$

http://matrix.skku.ac.kr/cal-lab/cal-7-3-2.html

Solution

```
integral(1/(sqrt(4+x^2)*(4-x^2)), x)
```

Answer : 1/16*sqrt(2)*arcsinh(2*x/abs(2*x + 4) - 4/abs(2*x + 4))+1/16*sqrt(2)*arcsinh(2*x/abs(2*x - 4) + 4/abs(2*x - 4))

CAS **3.** $\displaystyle\int \frac{1}{x^2\sqrt{9+x^2}}dx = -\frac{\sqrt{9+x^2}}{9x} + C$

http://matrix.skku.ac.kr/cal-lab/cal-7-3-3.html

Solution $x=3\tan\theta,\ dx=3\sec^2 x d\theta$

$$\int \frac{1}{x^2\sqrt{9+x^2}}dx = \int \frac{1}{9\tan^2\theta\, 3\sec\theta}3\sec^2\theta d\theta$$

$$= \frac{1}{9}\int \frac{1}{\sin\theta}\frac{\cos\theta}{\sin\theta}d\theta$$

$$= \frac{1}{9}\int \csc\theta\cot\theta\, d\theta = -\frac{1}{9}\csc\theta$$

$$= -\frac{\sqrt{9+x^2}}{9x} + C.$$

```
integral(1/((sqrt(9+x^2))*(x^2)), x)
```

Answer : -1/9*sqrt(x^2 + 9)/x

CAS **4.** $\displaystyle\int \frac{dx}{(1-x^2)^{3/2}} = \frac{x}{\sqrt{1-x^2}} + C$

http://matrix.skku.ac.kr/cal-lab/cal-7-3-4.html

Solution

```
integral(1/(-x^2 + 1)^(3/2), x)
```

Answer : x/sqrt(-x^2 + 1)

CAS **5.** $\displaystyle\int \frac{dx}{\sqrt{3x^2-2x-1}}$

http://matrix.skku.ac.kr/cal-lab/cal-7-3-5.html

Solution

```
integral(1/sqrt(3*x^2 -2*x-1), x)
```

Answer : 1/3*sqrt(3)*log(2*sqrt(3*x^2 - 2*x - 1)*sqrt(3)+6*x -2)

CAS **6.** $\displaystyle\int \frac{1}{3}x^2\sqrt{9-x^2}\,dx$

http://matrix.skku.ac.kr/cal-lab/cal-7-3-6.html

Solution

```
integral(1/3*x^2*sqrt(9-x^2), x)
```

Answer : -1/12*(-x^2 + 9)^(3/2)*x + 3/8*sqrt(-x^2 + 9)*x + 27/8*arcsin(1/3*x)

CAS **7.** $\displaystyle\int \frac{1}{(x^2+4x+13)^{3/2}}dx = \frac{x+2}{9\sqrt{x^2+4x+13}} + C$

http://matrix.skku.ac.kr/cal-lab/cal-7-3-7.html

Solution $x+2=3\tan u,\ dx=3\sec^2 u\,du$

$$\int \frac{1}{(x^2+4x+13)^{3/2}}dx = \int \frac{1}{((x+2)^2+9)^{3/2}}dx$$

$$= \int \frac{1}{(3\sec u)^3}3\sec^2 u\,du$$

$$= \int \frac{1}{9\sec u}du = \frac{1}{9}\int \cos u\,du = \frac{1}{9}\sin u$$

$$= \frac{1}{9}\frac{x+2}{\sqrt{x^2+4x+13}} + C$$

```
integral(1/(x^2+4*x+13)^(3/2), x)
```

Answer : 1/9*x/sqrt(x^2 + 4*x + 13) + 2/9/sqrt(x^2 + 4*x + 13)

CAS **8.** $\displaystyle\int \frac{x^2}{(4-x^2)^{3/2}}dx = \frac{x}{\sqrt{4-x^2}} - \sin^{-1}\left(\frac{x}{2}\right) + C$

http://matrix.skku.ac.kr/cal-lab/cal-7-3-8.html

Solution $x=2\sin\theta,\ dx=2\cos\theta d\theta$

$$\int \frac{x^2}{(4-x^2)^{3/2}}dx = \int \frac{4\sin^2\theta}{8\cos^3\theta}2\cos\theta d\theta$$
$$= \int \tan^2\theta d\theta$$
$$= \int (\sec^2\theta - 1)d\theta = \tan\theta - \theta$$
$$= \frac{x}{\sqrt{4-x^2}} - \sin^{-1}\left(\frac{x}{2}\right) + C$$

```
integral(x^2/(4-x^2)^(3/2), x)
```

Answer : x/sqrt(-x^2+4)-arcsin(1/2*x)

9. $\int \frac{1}{x\sqrt{x^2+9}}dx$

http://matrix.skku.ac.kr/cal-lab/cal-7-3-9.html

Solution

```
integral(1/(x*sqrt(9+x^2)), x)
```

Answer : -1/3*arcsinh(3/abs(x))

CAS **10.** $\int \frac{1}{x\sqrt{9-x^2}}dx$

http://matrix.skku.ac.kr/cal-lab/cal-7-3-10.html

Solution

```
integral(1/(x*sqrt(9-x^2)), x)
```

Answer : sqrt(-x^2+9)-3*log(6*sqrt(-x^2+9)/abs(x)+18/abs(x))

CAS **11.** $\int \frac{1}{\sqrt{x^2-a^2}}dx\,(a>0)$

http://matrix.skku.ac.kr/cal-lab/cal-7-3-11.html

Solution $x = a\sec\theta,\ dx = a\sec\theta\tan\theta\,d\theta,$

$\sqrt{x^2-a^2} = a\tan\theta.$

$$\int \frac{1}{\sqrt{x^2-a^2}}dx = \frac{1}{a}\int \sec\theta\,d\theta$$
$$= \frac{1}{a}\ln|\tan\theta + \sec\theta|$$
$$= \frac{1}{a}\ln\left|\frac{x+\sqrt{x^2-a^2}}{a}\right| + C.$$

```
var('x, a');
integral(1/sqrt(x^2-a^2), x)
```

Answer : -1/2*a^2*log(2*x + 2*sqrt(-a^2 + x^2)) + 1/2*sqrt(-a^2 + x^2)*x

CAS **12.** $\int \frac{\sqrt{2x^2-2}}{x^4}dx$

http://matrix.skku.ac.kr/cal-lab/cal-7-3-12.html

Solution

```
integral((sqrt(2*x^2-2)/x^4), x)
```

Answer : 1/6*(2*x^2 - 2)^(3/2)/x^3

CAS **13.** $\int \frac{x}{(1-x^2)^2}dx = -\frac{1}{2(x^2-1)} + C$

http://matrix.skku.ac.kr/cal-lab/cal-7-3-13.html

Solution

```
integral(x /((-x^2+1)^2), x)
```

Answer : -1/2/(x^2 - 1)

CAS **14.** $\int \frac{dx}{\sqrt{1-x^2}} = \sin^{-1}x + C$

http://matrix.skku.ac.kr/cal-lab/cal-7-3-14.html

Solution

```
integral(1/(sqrt(-x^2+1)), x)
```

Answer : arcsin(x)

CAS **15.** $\int \frac{dx}{\sqrt{x^2+3x+1}}$

http://matrix.skku.ac.kr/cal-lab/cal-7-3-15.html

Solution

```
integral(1/sqrt(x^2+3*x+1), x)
```

Answer : log(2*x + 2*sqrt(x^2 + 3*x + 1) + 3)

CAS **16.** $\int \frac{3x+1}{\sqrt{5-x-x^2}}dx$

http://matrix.skku.ac.kr/cal-lab/cal-7-3-16.html

Solution $x + \frac{1}{2} = \frac{\sqrt{21}}{2}\sin\theta,\ dx = \frac{\sqrt{21}}{2}\cos\theta\,d\theta$

$$\int \frac{3x+1}{\sqrt{5-x-x^2}}dx = \int \frac{3\left(x+\frac{1}{2}\right)-\frac{1}{2}}{\sqrt{\frac{21}{4}-\left(x+\frac{1}{2}\right)^2}}dx$$
$$= \int \frac{\frac{3\sqrt{21}}{2}\sin\theta - \frac{1}{2}}{\frac{\sqrt{21}}{2}\cos\theta}\frac{\sqrt{21}}{2}\cos\theta\,d\theta$$
$$= \int \left(\frac{3\sqrt{21}}{2}\sin\theta - \frac{1}{2}\right)d\theta = -\frac{3\sqrt{21}}{2}\cos\theta - \frac{\theta}{2}$$
$$= -3\sqrt{5-x-x^2} - \frac{1}{2}\sin^{-1}\frac{2x+1}{\sqrt{21}} + C.$$

```
integral((3*x+1)/sqrt(-x^2-x+5), x)
```

Answer : -3*sqrt(-x^2-x+5) +1/2*arcsin(-1/21*(2*x + 1)*sqrt(21))

CAS 17. $\displaystyle\int \frac{\sqrt{x^2-1}}{x}dx$

http://matrix.skku.ac.kr/cal-lab/cal-7-3-17.html

Solution

```
integral(sqrt(x^2-1)/x, x)
```

Answer : sqrt(x^2 - 1) + arcsin(1/abs(x))

CAS 18. $\displaystyle\int e^t\sqrt{9-e^{2t}}\,dt$

http://matrix.skku.ac.kr/cal-lab/cal-7-3-18.html

Solution $e^t = 3\sin\theta,\ e^t dt = 3\cos\theta\, d\theta$

$$\int e^t\sqrt{9-e^{2t}}\,dt = \int e^t\sqrt{9-(e^t)^2}\,dt$$

$$= \int 9\cos^2\theta d\theta = 9\int \frac{1+\cos 2\theta}{2}d\theta$$

$$= \frac{9}{2}\theta + \frac{9}{4}\sin 2\theta$$

$$= \frac{1}{2}e^t\sqrt{9-e^{2t}} + \frac{9}{2}\sin^{-1}\frac{e^t}{3} + C.$$

```
integral((e^x)*sqrt(9-e^(2*x)), x)
```

Answer : 1/2*sqrt(-e^(2*x) + 9)*e^x + 9/2*arcsin(1/3*e^x)

CAS 19. (a) Prove that

$$\int \frac{dx}{\sqrt{x^2+a^2}} = \ln(x+\sqrt{x^2+a^2}) + C.$$

(b) Use $x = a\sinh t$ to prove that

$$\int \frac{dx}{\sqrt{x^2+a^2}} = \sinh^{-1}\left(\frac{x}{a}\right) + C,\quad a \neq 0.$$

http://matrix.skku.ac.kr/cal-lab/cal-7-3-19.html

Solution (a) $x = a\tan\theta,\ dx = a\sec^2\theta\, d\theta$,

$\sqrt{x^2+a^2} = \sqrt{a^2(1+\tan^2\theta)} = a\sec\theta$.

$$\int \frac{dx}{\sqrt{x^2+a^2}} = \int \sec\theta\, d\theta = \frac{1}{a}\ln|\tan\theta + \sec\theta|$$

$$= \ln\left|\frac{x}{a} + \frac{\sqrt{x^2+a^2}}{a}\right| + c$$

$$= \ln(x+\sqrt{x^2+a^2}) + C.$$

```
var('x, a');
integral(sqrt(x^2+a^2), x)
```

Answer : 1/2*a^2*arcsinh(x/sqrt(a^2)) + 1/2*sqrt(a^2 + x^2)*x

(b) Use $x = a\sinh t$ to prove that $x = a\sinh\theta$,

$dx = a\cosh\theta\, d\theta,\ \theta = \sinh^{-1}\dfrac{x}{a}$.

$\sqrt{x^2+a^2} = \sqrt{a^2(1+\sinh^2\theta)} = a\cosh\theta$

$$\int \frac{dx}{\sqrt{x^2+a^2}} = \int 1\, d\theta = \theta + C$$

$$= \sinh^{-1}\left(\frac{x}{a}\right) + C.$$

7.4 Integration of Rational Functions by the Method of Partial Fractions and Computer Algebra System

YouTube http://youtu.be/KLTHp_7G4cI

It is not obvious how one should evaluate the integral of a complicated rational function (a ratio of polynomials), for instance,

$$\int \frac{x^2+3x-4}{3x^3-4x^2-5x+2}dx.$$

However, the integral can be evaluated by rewriting the rational function as a sum of simpler fractions, called *partial fractions,* which are easily integrable. The decomposition of a rational function into a sum of its partial fractions is called a partial fraction decomposition of the rational function.

To integrate a rational function f defined by

$$f(x)=\frac{P(x)}{Q(x)},$$

where $P(x)$ and $Q(x)$ are polynomials, by its partial fraction decomposition, it is required that the numerator of the rational function has degree less than the denominator. For example, the antiderivative $\int \frac{3x+2}{x^2-1}dx$ is in the correct form for one to proceed with the partial fraction decomposition. If it is not in the desired form, we divide $P(x)$ by $Q(x)$ using long division until a remainder $R(x)$ is obtained such that $\deg(R)<\deg(Q)$ to acquire the following form of f

1 $$f(x)=\frac{P(x)}{Q(x)}=S(x)+\frac{R(x)}{Q(x)}$$

where S and R are also polynomials. We now have

$$\int \frac{P(x)}{Q(x)}dx=\int \left[S(x)+\frac{R(x)}{Q(x)}\right]dx=\int S(x)dx+\int \frac{R(x)}{Q(x)}dx.$$

To evaluate the second integral, $\int \frac{R(x)}{Q(x)}dx$, on the right-hand side of the final expression, one must find the partial fraction decomposition of the integrand, which depends on the factored form of $Q(x)$, then integrate.

It is known that any polynomial $Q(x)$ can be factored as a product of linear factors (of the form $ax+b$) and irreducible quadratic factors (of the form ax^2+bx+c, where $b^2-4ac<0$). So the next step is to factor the denominator, $Q(x)$, completely. To rewrite the rational function $R(x)/Q(x)$ using **partial fractions,** we consider four possible cases:

CASE I : The denominator $Q(x)$ is a product of distinct linear factors.

In this case we can write

$$Q(x) = (a_1x + b_1)(a_2x + b_2)\cdots(a_kx + b_k)$$

where no factor is repeated (and no factor is a constant multiple of another).

In this case the partial fraction theorem states that there exist constants $A_1, A_2, \ldots, A_k$ such that

$$\frac{R(x)}{Q(x)} = \frac{A_1}{a_1x + b_1} + \frac{A_2}{a_2x + b_2} + \cdots + \frac{A_k}{a_kx + b_k}. \qquad \boxed{2}$$

EXAMPLE 1

Evaluate $\int \frac{x^3 + 1}{x + 2}\,dx$.

Solution Since the degree of the numerator is larger than the degree of the denominator, perform long division to obtain

$$\begin{array}{r|l} & x^2 - 2x + 4 \\ \hline x + 2 & x^3 \qquad\qquad + 1 \\ & x^3 + 2x^2 \\ \hline & -2x^2 \qquad + 1 \\ & -2x^2 - 4x \\ \hline & \qquad 4x + 1 \\ & \qquad 4x + 8 \\ \hline & \qquad\quad -7 \end{array}$$

$$\begin{aligned} \int \frac{x^3 + 1}{x + 2}\,dx &= \int \left(x^2 - 2x + 4 - \frac{7}{x + 2}\right)dx \\ &= \frac{x^3}{3} - x^2 + 4x - 7\ln|x + 2| + C. \end{aligned}$$

■

EXAMPLE 2

Find $\int \frac{x^2 + 3x - 4}{3x^3 - 4x^2 - 5x + 2}\,dx$.

http://matrix.skku.ac.kr/cal-lab/cal-7-4-Exm-2.html

Solution The degree of the numerator is smaller than the degree of the denominator, thus the long division is not needed. The denominator factors as follows:

$$3x^3 - 4x^2 - 5x + 2 = (x + 1)(x - 2)(3x - 1).$$

Since the denominator has three distinct linear factors, the integrand decomposes into three partial fractions of the form

3 $$\frac{x^2+3x-4}{(x+1)(x-2)(3x-1)}=\frac{A}{x+1}+\frac{B}{x-2}+\frac{C}{3x-1}.$$

To determine the constants A, B, and C, multiply both sides of this equation 3 by the product of the denominators, $(x+1)(x-2)(3x-1)$, resulting in

4 $$x^2+3x-4=A(x-2)(3x-1)+B(x+1)(3x-1)+C(x+1)(x-2).$$

Rewriting the right-hand side of 4 in the standard form for polynomials, we obtain

5 $$x^2+3x-4=(3A+3B+C)x^2+(-7A+2B-C)x+2A-B-2C.$$

Comparing the coefficients of powers of x on both sides of 5, we observe that the coefficient of x^2 on the right side, $3A+3B+C$, must equal the coefficient of x^2 on the left side namely, 1. Likewise, the coefficients of x are equal and the constant terms are equal. This gives the following system of equations for A, B, and C:

$$\begin{aligned} 3A+3B+C&=1\\ -7A+2B-C&=3\\ 2A-B-2C&=-4. \end{aligned}$$

Solving it, we get $A=-\frac{1}{2}$, $B=\frac{2}{5}$, $C=\frac{13}{10}$

$$\begin{aligned} &\int\frac{x^2+3x-4}{3x^3-4x^2-5x+2}dx\\ &=\int\left(-\frac{1}{2}\frac{1}{x+1}+\frac{2}{5}\frac{1}{x-2}+\frac{13}{10}\frac{1}{3x-1}\right)dx\\ &=-\frac{1}{2}\ln|x+1|+\frac{2}{5}\ln|x-2|+\frac{13}{30}\ln|3x-1|+C. \end{aligned}$$

```
var('x')
exp= (x^2 +3*x-4)/(3*x^3 -4*x^2 - 5*x +2)
print integral (exp, x)
exp.partial_fraction()
```

Answer : 2/5*log(x - 2) - 1/2*log(x + 1) + 13/30*log(3*x – 1)
2/5/(x - 2) - 1/2/(x + 1) + 13/10/(3*x – 1) ■

Note Alternatively the coefficients A, B, C in 4 can be determined by choosing values of x that simplify the equation 4. If we substitute $x=-1$ in 4, then the equation becomes $12A=-6$, $A=-\frac{1}{2}$.

Likewise, $x=2$ gives $15B=6$ and $x=\frac{1}{3}$ gives $-\frac{20}{9}C=-\frac{26}{9}$, so $B=\frac{2}{5}$ and $C=\frac{13}{10}$.

EXAMPLE 3

Find $\int \frac{dx}{x^2+(b-a)x-ab}$ where $a+b \neq 0$.

http://matrix.skku.ac.kr/cal-lab/cal-7-4-Exm-3.html

Solution The method of partial fractions gives

$$\frac{1}{x^2+(b-a)x-ab}=\frac{1}{(x-a)(x+b)}=\frac{A}{x-a}+\frac{B}{x+b}$$

and therefore $A(x+b)+B(x-a)=1$. Put $x=a$ in this equation and get $A(a+b)=1$, so $A=\frac{1}{a+b}$. If we put $x=-b$, we get $-B(a+b)=1$, so $B=-\frac{1}{a+b}$. Thus

$$\begin{aligned}\int \frac{dx}{x^2+(b-a)x-ab} &= \frac{1}{a+b}\int\left(\frac{1}{x-a}-\frac{1}{x+b}\right)dx \\ &= \frac{1}{a+b}(\ln|x-a|-\ln|x+b|)+C.\end{aligned}$$

```
var('a, b, x')
exp= 1/(x^2 + (b-a)*x - a*b)
print integral (exp, x)
exp.partial_fraction()
```

Answer : -log(b + x)/(a + b) + log(-a + x)/(a + b)
-1/((b + x)*(a - x)) ■

Example 3 provides us with an integration formula.

6 $$\int \frac{dx}{x^2+(b-a)x-ab}=\frac{1}{a+b}\ln\left|\frac{x-a}{x+b}\right|+C.$$

CASE II : $Q(x)$ is a product of linear factors, some of which are repeated.

Suppose the first linear factor (a_1x+b_1) is repeated r times; that is, $(a_1\ x+b_1)^r$ occurs in the factorization of $Q(x)$. Then instead of the single term $A_1/(a_1x+b_1)$ in Equation 2, we use

7 $$\frac{A_1}{a_1x+b_1}+\frac{A_2}{(a_1x+b_1)^2}+\cdots+\frac{A_r}{(a_1x+b_1)^r}$$

EXAMPLE 4

Evaluate $\int \frac{2x^4+x^3-6x^2-6x-2}{x^3-3x-2}dx.$

http://matrix.skku.ac.kr/cal-lab/cal-7-4-Exm-4.html

Solution Since the degree of the numerator is larger than the degree of the denominator, we perform the long division to obtain

$$\frac{2x^4+x^3-6x^2-6x-2}{x^3-3x-2}=2x+1+\frac{x}{x^3-3x-2}.$$

Next we factor, the denominator $Q(x)=x^3-3x-2$, which gives

$$x^3-3x-2=(x+1)^2(x-2).$$

Since the linear factor $x-1$ occurs twice, the partial fraction decomposition is

$$\frac{x}{(x+1)^2(x-2)}=\frac{A}{x+1}+\frac{B}{(x+1)^2}+\frac{C}{x-2}.$$

Multiplying by the least common denominator, $(x+1)^2(x-2)$, we get

8 $$\begin{aligned} x &= A(x+1)(x-2)+B(x-2)+C(x+1)^2 \\ &= (A+C)x^2+(-A+B+2C)x+(-2A-2B+C). \end{aligned}$$

Now we equate coefficients to obtain:

$$\begin{aligned} A \qquad\quad + C &= 0 \\ -A+\quad B+2C &= 1 \\ -2A-2B+\quad C &= 0 \end{aligned}$$

(or put $x=-1$ in 8: $B=\frac{1}{3}$, put $x=2$: $C=\frac{9}{2}$, put $x=0$: $A=-B+\frac{1}{2}C=-\frac{2}{9}$).

Solving, we obtain $A=-\frac{2}{9}$, $B=\frac{1}{3}$, $C=\frac{2}{9}$, so

$$\begin{aligned} \int \frac{2x^4+x^3-6x^2-6x-2}{x^3-3x-2}dx &= \int\left[2x+1-\frac{2}{9}\frac{1}{x+1}+\frac{1}{3}\frac{1}{(x+1)^2}+\frac{2}{9}\frac{1}{x-2}\right]dx \\ &= x^2+x-\frac{2}{9}\ln|x+1|-\frac{1}{3}\frac{1}{x+1}+\frac{2}{9}\ln|x-2|+C \\ &= x^2+x-\frac{1}{3}\frac{1}{x+1}+\frac{2}{9}\ln\left|\frac{x-2}{x+1}\right|+C. \end{aligned}$$

```
var('a, b, x')
exp= (2*x^4 + x^3 - 6*x^2 - 6*x -2)/(x^3 - 3*x - 2)
print exp.partial_fraction()        # partial fraction
print
print integral (exp, x)
```

Answer : 2*x + 2/9/(x - 2) - 2/9/(x + 1) + 1/3/(x + 1)^2 + 1
-1/3/(x + 1) + x^2 + x + 2/9*log(x - 2) - 2/9*log(x + 1) ■

CASE III : $Q(x)$ contains irreducible quadratic factors, none of which is repeated.

If $Q(x)$ has the factor ax^2+bx+c, where $b^2-4ac<0$, then, in addition to the partial fractions in Equations 2 and 7, the expression for $R(x)/Q(x)$ will have a term of the form

9 $$\frac{Ax+B}{ax^2+bx+c}$$

where A and B are constants to be determined. The term given in 9 can be integrated by completing the square and using the formula

10 $$\int \frac{dx}{x^2+a^2} = \frac{1}{a}\tan^{-1}\left(\frac{x}{a}\right)+C.$$ ■

EXAMPLE 5

Find $\int \frac{1}{x^4-1}dx$.

http://matrix.skku.ac.kr/cal-lab/cal-7-4-Exm-5.html

Solution Since $x^4-1=(x-1)(x+1)(x^2+1)$ cannot be factored further, we may write

$$\frac{1}{x^4-1} = \frac{A}{x-1}+\frac{B}{x+1}+\frac{Cx+D}{x^2+1}.$$

Multiplying by $(x-1)(x+1)(x^2+1)$, we have

$$\begin{aligned}1 &= A(x+1)(x^2+1)+B(x-1)(x^2+1)+(Cx+D)(x-1)(x+1)\\ &= (A+B+C)x^3+(A-B+D)x^2+(A+B-C)x+A-B-D.\end{aligned}$$

Equating coefficients, we obtain

$$A+B+C=0,\ A-B+D=0,\ A+B-C=0,\ A-B-D=1.$$

Thus $A=\frac{1}{4}$, $B=-\frac{1}{4}$, $C=0$, and $D=-\frac{1}{2}$ and so

$$\int \frac{1}{x^4-1}dx = \frac{1}{4}\int\left(\frac{1}{x-1}-\frac{1}{x+1}-\frac{2}{x^2+1}\right)dx$$

$$= \frac{1}{4}\ln\left|\frac{x-1}{x+1}\right| - \frac{1}{2}\tan^{-1}x + C \ .$$

```
var('a, b, x')
exp= 1/(x^4 - 1)
print integral (exp, x)
exp.partial_fraction()        # partial fraction
```

Answer : 1/4*log(x - 1) - 1/4*log(x + 1) - 1/2*arctan(x)
1/4/(x - 1) - 1/4/(x + 1) - 1/2/(x^2 + 1) ■

EXAMPLE 6

Find $\int \frac{1}{2x^3 - 3x^2 + 2x - 1} dx$.

http://matrix.skku.ac.kr/cal-lab/cal-7-4-Exm-6.html

Solution The partial fractions decomposition of the integrand, we will be of the form

$$\frac{1}{2x^3 - 3x^2 + 2x - 1} = \frac{1}{(x-1)(2x^2 - x + 1)}$$
$$= \frac{A}{x-1} + \frac{Bx + C}{2x^2 - x + 1}.$$

Note that $2x^2 - x + 1$ is an irreducible quadratic factor. Multiplying the above equation by $(x+1)(2x^2 - x + 1)$, we have

$$1 = A(2x^2 - x + 1) + (Bx + C)(x - 1)$$
$$= (2A + B)x^2 + (-A - B + C)x + A - C.$$

Equating coefficients, we obtain

$$2A + B = 0 \qquad -A - B + C = 0 \qquad A - C = 1.$$

Thus $A = \frac{1}{2}$, $B = -1$, and $C = -\frac{1}{2}$, and so

$$\int \frac{1}{2x^3 - 3x^2 + 2x - 1} dx$$
$$= \frac{1}{2}\int \frac{1}{x-1} dx - \frac{1}{2}\int \frac{2x+1}{2x^2 - x + 1} dx$$
$$= \frac{1}{2}\int \frac{1}{x-1} dx - \frac{1}{4}\int \frac{4x-1}{2x^2 - x + 1} dx - \frac{3}{4}\int \frac{1}{2x^2 - x + 1} dx$$
$$= \frac{1}{2}\ln|x-1| - \frac{1}{4}\ln(2x^2 - x + 1) - \frac{3}{8}\int \frac{1}{\left(x - \frac{1}{4}\right)^2 + \frac{7}{16}} dx$$
$$= \frac{1}{2}\ln|x-1| - \frac{1}{4}\ln(2x^2 - x + 1) - \frac{3\sqrt{7}}{14}\tan^{-1}\left(\frac{4x-1}{\sqrt{7}}\right) + C.$$

```
var('a, b, x')
exp= 1/(2*x^3 -3*x^2 + 2*x -1)
print integral (exp, x)
exp.partial_fraction()      # partial fraction
```

Answer : 1/2{ln}(x-1) - 1/4{ln}(2 x^2 - x + 1) - {3 sqrt7}/14 {tan^-1} (sqrt7/7 (4x - 1))-1/2*(2*x + 1)/(2*x^2 - x + 1) + 1/2/(x – 1) ■

Note The general procedure for integrating a partial fraction of the form

$$\frac{Ax+B}{ax^2+bx+c} \quad \text{where } b^2-4ac<0$$

is to complete the square of the denominator and then make a substitution that brings the integral into the form

$$\int \frac{Cu+D}{u^2+a^2}du = C\int \frac{u}{u^2+a^2}du + D\int \frac{1}{u^2+a^2}du.$$

Then the first integral is a logarithm and the second is expressed in terms of inverse tangent function.

CASE IV : $Q(x)$ contains a repeated irreducible quadratic factor.

If $Q(x)$ has the factor $(ax^2+bx+c)^r$, where $b^2-4ac<0$, then instead of the single partial fraction 9, the sum

11 $$\frac{A_1x+B_1}{ax^2+bx+c}+\frac{A_2x+B_2}{(ax^2+bx+c)^2}+\cdots+\frac{A_rx+B_r}{(ax^2+bx+c)^r}$$

occurs in the partial fraction decomposition of $R(x)/Q(x)$. Each of the terms in 11 can be integrated by first completing the square.

EXAMPLE 7

What is the form of the partial fraction decomposition of the function?

$$\frac{x^4+x^3+x^2+x+2}{(x+1)(x-2)(x^2-x+4)^4}$$

http://matrix.skku.ac.kr/cal-lab/cal-7-4-Exm-7.html

Solution

$$\frac{x^4+x^3+x^2+x+2}{(x+1)(x-2)(x^2-x+4)^4}$$
$$=\frac{A}{x+1}+\frac{B}{x-2}+\frac{Cx+D}{x^2-x+4}+\frac{Ex+F}{(x^2-x+4)^2}+\frac{Gx+H}{(x^2-x+4)^3}$$
$$+\frac{Ix+J}{(x^2-x+4)^4}.$$

Working carefully by hand, or by employing Sage, we may find the coefficients to be

$$A=-\frac{1}{1944},\ B=\frac{2}{243},\ C=-\frac{5}{648},\ D=-\frac{3}{324},$$
$$E=-\frac{5}{108},\ F=-\frac{3}{54},\ G=-\frac{5}{18},\ H=\frac{2}{3},\ I=\frac{4}{3},\ J=-1.$$

```
var('x')
exp=(x^4 + x^3 + x^2 + x + 2)/((x+1)*(x-2)*(x^2 - x+4)^4)
exp.partial_fraction()
```

Answer : -1/648*(5*x + 6)/(x^2 - x + 4) + 2/243/(x - 2) - 1/1944/(x + 1) - 1/108*(5*x + 6)/(x^2 - x + 4)^2 - 1/18*(5*x - 12)/(x^2 - x + 4)^3 + 1/3*(4*x - 3)/(x^2 - x + 4)^4 ■

EXAMPLE 8

Find $\int \frac{x^2+3}{(x+1)(x^2+2x+3)^2}dx$.

http://matrix.skku.ac.kr/cal-lab/cal-7-4-Exm-8.html

Solution The partial fraction decomposition of the integrand has the form

$$\frac{x^2+3}{(x+1)(x^2+2x+3)^2}=\frac{A}{x+1}+\frac{Bx+C}{x^2+2x+3}+\frac{Dx+E}{(x^2+2x+3)^2}.$$

Multiplying this equation by $(x+1)(x^2+2x+3)^2$, we have

$$\begin{aligned}x^2+3 &= A(x^2+2x+3)^2+(Bx+C)(x+1)(x^2+2x+3)\\ &\quad +(Dx+E)(x+1)\\ &= (A+B)x^4+(4A+3B+C)x^3\\ &\quad +(10A+5B+3C+D)x^2\\ &\quad +(12A+3B+5C+D+E)x+9A+E+3C.\end{aligned}$$

If we equate coefficients, we get the following system of equations

$$A+B=0 \quad 4A+3B+C=0 \quad 10A+5B+3C+D=1$$
$$12A+3B+5C+D+E=0 \quad 9A+E+3C=3,$$

which has the solution $A=1$, $B=-1$, $C=-1$, $D=-1$, and $E=-3$. Thus

$$\begin{aligned}&\int \frac{x^2+3}{(x+1)(x^2+2x+3)^2}dx\\ &= \int\left[\frac{1}{x+1}-\frac{x+1}{x^2+2x+3}-\frac{x+3}{(x^2+2x+3)^2}\right]dx\\ &= \ln|x+1|-\frac{1}{2}\ln(x^2+2x+3)-\int\frac{x+3}{(x^2+2x+3)^2}dx.\end{aligned}$$

Put $I=\int \frac{x+3}{(x^2+2x+3)^2}dx$.

Let $t=x+1$. Then $x^2+2x+3=t^2+2$. Thus

$$I=\int \frac{t}{(t^2+2)^2}dt+\int \frac{2}{(t^2+2)^2}dt.$$

Put $J_1=\int \frac{t}{(t^2+2)^2}dt,\ J_2=\int \frac{2}{(t^2+2)^2}dt.$

Let $w=t^2+2$. Then $dw=2tdt$. Thus

$$J_1=\frac{1}{2}\int w^{-2}dw=\frac{1}{2}\cdot(-1)\cdot w^{-1}=-\frac{1}{2w}=-\frac{1}{2}\cdot\frac{1}{t^2+2}.$$

Observe that

$$\int \frac{1}{(t^2+2)}dt=\int \frac{t^2+2}{(t^2+2)^2}=\int \frac{t^2}{(t^2+2)^2}dt+\int \frac{2}{(t^2+2)^2}dt.$$

$$\text{Thus } J_2=\int \frac{1}{(t^2+2)}dt-\int \frac{t^2}{(t^2+2)^2}dt$$
$$=\frac{1}{\sqrt{2}}\tan^{-1}\frac{t}{\sqrt{2}}-\int \frac{t^2}{(t^2+2)^2}dt.$$

Let $g'(t)=\frac{t}{(t^2+2)^2},\ f(t)=t \Rightarrow g(t)=J_1=-\frac{1}{2}\cdot\frac{1}{t^2+2},$
$f'(t)=1$. Thus,

$$\int \frac{t^2}{(t^2+2)^2}dt=-\frac{1}{2}\frac{t}{t^2+2}+\frac{1}{2}\int \frac{1}{t^2+2}dt$$
$$=-\frac{1}{2}\frac{t}{t^2+2}+\frac{1}{2\sqrt{2}}\tan^{-1}\frac{t}{\sqrt{2}}.$$

Therefore,

$$J_2=\frac{1}{2\sqrt{2}}\tan^{-1}\frac{t}{\sqrt{2}}+\frac{1}{2}\cdot\frac{t}{t^2+2}\quad\text{and}$$
$$I=-\frac{1}{2}\cdot\frac{1}{t^2+2}+\frac{1}{2\sqrt{2}}\tan^{-1}\frac{t}{\sqrt{2}}+\frac{1}{2}\cdot\frac{t}{t^2+2}.$$

Hence

$$\int \frac{x^2+3}{(x+1)(x^2+2x+3)^2}dx$$
$$=\ln|x+1|-\frac{1}{2}\ln(x^2+2x+3)-\frac{x}{2(x^2+2x+3)}$$
$$+\frac{1}{2\sqrt{2}}\tan^{-1}\frac{x+1}{\sqrt{2}}+C.$$

```
var('x')
exp= (x^2 + 3)/((x+1)*(x^2 + 2*x + 3)^2 )
print integral (exp, x)
exp.partial_fraction()
```

Answer : -1/4*sqrt(2)*arctan(1/2*(x + 1)*sqrt(2)) - 1/2*x/(x^2 + 2*x + 3) + log(x + 1) - 1/2*log(x^2 + 2*x + 3)-(x + 1)/(x^2 + 2*x + 3) + 1/(x + 1) - (x + 3)/(x^2 + 2*x + 3)^2 ■

Substitution Techniques

Some nonrational functions can be transformed into rational functions by means of appropriate substitutions. In particular, when an integrand contains an expression of the form $\sqrt[n]{g(x)}$, then the substitution of the form $u=\sqrt[n]{g(x)}$ may be effective.

EXAMPLE 9

Find $\int \frac{1}{x}\sqrt{\frac{x+1}{x-1}}\,dx$. $x\neq 0,\ x\neq 1$.

http://matrix.skku.ac.kr/cal-lab/cal-7-4-Exm-9.html

Solution Let $t=\sqrt{\frac{x+1}{x-1}}$. Then $x=\frac{t^2+1}{t^2-1}$ and $dx=\frac{-4t}{(t^2-1)^2}dt$.

Hence we have

$$\begin{aligned}\int \frac{1}{x}\sqrt{\frac{x+1}{x-1}}\,dx &= -4\int \frac{t^2}{(t^2+1)(t^2-1)}dt\\ &= -2\int\left(\frac{1}{t^2-1}+\frac{1}{t^2+1}\right)dt\\ &= -2\int \frac{1}{t^2-1}dt - 2\int \frac{1}{1+t^2}dt\\ &= -\ln\left|\frac{t-1}{t+1}\right| - 2\tan^{-1}t + C\\ &= \ln\frac{\sqrt{x+1}+\sqrt{x-1}}{\sqrt{x+1}-\sqrt{x-1}} - 2\tan^{-1}\left(\sqrt{\frac{x+1}{x-1}}\right) + C.\end{aligned}$$

```
var('x')
exp= (1/x)* ( ((x+1)/(x-1))^(-1/2) )
print integral (exp, x)
```

Answer : -log(sqrt((x + 1)/(x - 1)) - 1) + log(sqrt((x + 1)/(x - 1)) + 1) + 2*arctan(sqrt((x + 1)/(x – 1))) ■

CAS **EXAMPLE 10**

Evaluate the integral

$$\int \frac{x}{(x+1)(x+2)(x+3)}dx.$$

http://matrix.skku.ac.kr/cal-lab/cal-7-4-Exm-10.html
http://matrix.skku.ac.kr/cal-lab/sage-grapher-integral.html

Solution Let $\frac{x}{(x+1)(x+2)(x+3)} = \frac{A}{x+1} + \frac{B}{x+2} + \frac{C}{x+3}$ and use Sage to get A, B, C.

$$\frac{x}{(x+1)(x+2)(x+3)} = \frac{-1}{2(x+1)} + \frac{2}{x+2} + \frac{-3}{2(x+3)}$$

```
var('x')
f= x/((x+1)*(x+2)*(x+3))
print integral (f, x)
f.partial_fraction()
```

Answer : -1/2*log(x + 1) + 2*log(x + 2) - 3/2*log(x + 3),
-1/2/(x + 1) + 2/(x + 2) - 3/2/(x + 3) ■

7.4 EXERCISES (Integration of Rational Functions by the Method of Partial Fractions and Computer Algebra System)

http://matrix.skku.ac.kr/Cal-Book/part1/CS-Sec-7-4-Sol.html

YouTube http://youtu.be/SkNW_bax0YI

CAS **1.** $\int \frac{x^3+1}{x(x-1)^3}dx$

http://matrix.skku.ac.kr/cal-lab/cal-7-4-1.html

Solution $\frac{x^3+1}{x(x-1)^3}=\frac{A}{x}+\frac{B}{x-1}+\frac{C}{(x-1)^2}+\frac{D}{(x-1)^3}$

$\Rightarrow A=-1,\ B=2,\ C=1,\ D=2$

$$\int \frac{x^3+1}{x(x-1)^3}dx$$
$$=\int\left(\frac{-1}{x}+\frac{2}{x-1}+\frac{1}{(x-1)^2}+\frac{2}{(x-1)^3}\right)dx$$
$$=-\int\frac{1}{x}dx+\int\frac{2}{x-1}dx+\int\frac{1}{(x-1)^2}dx+\int\frac{2}{(x-1)^3}dx$$
$$=-\ln|x|+2\ln|x-1|-\frac{1}{x-1}-\frac{1}{(x-1)^2}+C.$$

```
integral((x^3+1)/(x*(x-1)^3), x)
```

Answer : -x/(x^2 - 2*x + 1) + 2*log(x - 1) - log(x)

CAS **2.** $\int \frac{dx}{x^3-x}$

http://matrix.skku.ac.kr/cal-lab/cal-7-4-2.html

Solution

```
integral(1/(x^3-x), x)
```

Answer : 1/2*log(x-1)+1/2*log(x+1)-log(x)

$$\frac{1}{x^3-x}=\frac{1}{x(x-1)(x+1)}$$
$$=\frac{a}{x}+\frac{b}{x-1}+\frac{c}{x+1},\quad \left(a=-1, b=\frac{1}{2}, c=\frac{1}{2}\right)$$
$$\therefore\ \int\frac{dx}{x^3-x}=\int\left(-\frac{1}{x}+\frac{1}{2(x-1)}+\frac{1}{2(x+1)}\right)dx$$
$$=-\ln|x|+\frac{1}{2}\ln|x-1|+\frac{1}{2}\ln|x+1|+C.$$

CAS **3.** $\int \frac{x-1}{x^2-4x-5}dx$

http://matrix.skku.ac.kr/cal-lab/cal-7-4-3.html

Solution

```
integral((x-1)/(x^2-4*x-5), x)
```

Answer : 2/3*log(x - 5) + 1/3*log(x + 1)

CAS **4.** $\int \frac{5x^3+2x^2-12x-8}{x^4-8x^2+16}dx$

http://matrix.skku.ac.kr/cal-lab/cal-7-4-4.html

Solution

```
integral((5*x^3+2*x^2-12*x-8)/(x^4-8*x^2+16),
```

Answer : -4/(x^2 - 4) + 3*log(x - 2) + 2*log(x + 2)

CAS **5.** $\int \frac{x^2}{(x^2+1)(x^2+4)}dx$

http://matrix.skku.ac.kr/cal-lab/cal-7-4-5.html

Solution

```
integral((x^2)/((x^2+1)*(x^2+4)), x)
```

Answer : 2/3*arctan(1/2*x) - 1/3*arctan(x)

CAS **6.** $\int \frac{x^3+2x^2}{x^2-1}dx$

http://matrix.skku.ac.kr/cal-lab/cal-7-4-6.html

Solution

```
integral((x^3+2*x^2)/(x^2-1), x)
```

Answer : 1/2*x^2 + 2*x + 3/2*log(x - 1) - 1/2*log(x + 1)

CAS **7.** $\int \frac{2x^2+1}{x(x^2+1)}dx$

http://matrix.skku.ac.kr/cal-lab/cal-7-4-7.html

Solution

```
integral((1+2*x^2)/(x*(x^2+1)), x)
```

Answer : 1/2*log(x^2 + 1) + log(x)

CAS **8.** $\int \frac{dx}{x^3+1}$

http://matrix.skku.ac.kr/cal-lab/cal-7-4-8.html

Solution $\frac{1}{x^3+1}=\frac{1}{(x+1)(x^2-x+1)}$

$$=\frac{A}{x+1}+\frac{Bx+C}{x^2-x+1}$$
$$\Rightarrow A=\frac{1}{3},\ B=-\frac{1}{3},\ C=\frac{2}{3}$$
$$\int\frac{1}{x^3+1}dx=\int\left(\frac{1}{3(x+1)}+\frac{-x+2}{3(x^2-x+1)}\right)dx$$
$$=\frac{1}{3}\int\frac{1}{x+1}dx-\frac{1}{3}\int\frac{x-2}{x^2-x+1}dx$$

$$= \frac{1}{3}\ln|x+1| - \frac{1}{6}\int \frac{2x-4}{x^2-x+1}dx$$

$$= \frac{1}{3}\ln|x+1| - \frac{1}{6}\int \frac{2x-1}{x^2-x+1}dx$$

$$+ \frac{1}{2}\int \frac{1}{x^2-x+1}dx$$

$$= \frac{1}{3}\ln|x+1| - \frac{1}{6}\ln|x^2-x+1|$$

$$+ \frac{1}{2}\int \frac{1}{\left(x-\frac{1}{2}\right)^2 + \left(\frac{\sqrt{3}}{2}\right)^{21}}dx$$

$$= \frac{1}{3}\ln|x+1| - \frac{1}{6}\ln|x^2-x+1|$$

$$+ \frac{1}{2}\cdot\frac{1}{\frac{\sqrt{3}}{2}}\tan^{-1}\frac{x-\frac{1}{2}}{\frac{\sqrt{3}}{2}}$$

$$= \frac{1}{3}\ln|x+1| - \frac{1}{6}\ln|x^2-x+1|$$

$$+ \frac{1}{\sqrt{3}}\tan^{-1}\frac{2x-1}{\sqrt{3}} + C.$$

```
integral(1/(x^3+1), x)
```

Answer : 1/3*sqrt(3)*arctan(1/3*(2*x-1)*sqrt(3))+1/3*log (x+1)-1/6 *log(x^2-x+1)

CAS **9.** $\int \frac{3x-4}{x^2-6x+9}dx$

http://matrix.skku.ac.kr/cal-lab/cal-7-4-9.html

Solution

```
integral((3*x-4)/(x^2-6*x+9), x)
```

Answer : -5/(x - 3) + 3*log(x - 3)

CAS **10.** $\int \frac{dx}{x^6-1}$

http://matrix.skku.ac.kr/cal-lab/cal-7-4-10.html

Solution

```
integral(1/(x^6-1), x)
```

Answer : -1/6*sqrt(3)*arctan(1/3*(2*x-1)*sqrt(3))-1/6* sqrt(3) *arctan (1/3*(2*x+1)*sqrt(3))+1/6*log(x-1)-1/6* log(x+1)+ 1/12*log (x^2-x+1)-1/12*log(x^2+x+1)

CAS **11.** $\int \frac{dx}{x^4+1}$

http://matrix.skku.ac.kr/cal-lab/cal-7-4-11.html

Solution

```
integral(1/(x^4+1), x)
```

Answer : -1/8*sqrt(2)*log(x^2-sqrt(2)*x+1)+1/8*sqrt(2)* log(x^2+ sqrt(2)*x+1) 1/4*sqrt(2)*arctan(1/2*(2*x-sqrt(2))*sqrt(2))+1/4* sqrt(2)*arctan(1/2*(2*x+sqrt(2))*sqrt(2))

CAS **12.** $\int \frac{x^2}{(x^2+4)(x^2+9)}dx$

http://matrix.skku.ac.kr/cal-lab/cal-7-4-12.html

Solution $\frac{x^2}{(x^2+4)(x^2+9)} = \frac{Ax+B}{x^2+4} + \frac{Cx+D}{x^2+9}$

$\Rightarrow A = C = 0,\ B = -4/5,\ D = 9/5$

$$\int \frac{x^2}{(x^2+4)(x^2+9)}dx$$

$$= -\frac{4}{5}\int \frac{1}{x^2+4}dx + \frac{9}{5}\int \frac{1}{x^2+9}dx$$

$$= -\frac{2}{5}\tan^{-1}\frac{x}{2} + \frac{3}{5}\tan^{-1}\frac{x}{3} + C.$$

```
integral((x^2)/((x^2+4)*(x^2+9)), x)
```

Answer : 3/5*arctan(1/3*x) - 2/5*arctan(1/2*x)

CAS **13.** $\int \frac{x^2}{(x^2+1)^2}dx$

http://matrix.skku.ac.kr/cal-lab/cal-7-4-13.html

Solution

```
integral(x^2/(x^2+1)^2, x)
```

Answer : -1/2*x/(x^2 + 1) + 1/2*arctan(x)

CAS **14.** $\int \frac{dx}{(x-1)(x^2+1)^2}$

http://matrix.skku.ac.kr/cal-lab/cal-7-4-14.html

Solution

```
integral(1/((x-1)*(x^2+1)^2), x)
```

Answer : -1/4*(x-1)/(x^2+1)+1/4*log(x-1)-1/8*log(x^2+1)- 1/2*arctan(x)

CAS **15.** $\int \frac{2x^2-x+1}{(x-1)^3(x-2)^2}dx$

Solution

```
integral((2*x^2-x+1)/((x-1)^3*(x-2)^2), x)
```

Answer : -(14*x^2 - 34*x + 19)/(x^3 - 4*x^2 + 5*x - 2) - 14*log(x - 2) + 14*log(x - 1)

CAS **16.** $\int \frac{5x+1}{2x^2+5x+3}dx$

http://matrix.skku.ac.kr/cal-lab/cal-7-4-16.html

Solution

```
integral((5*x+1)/(2*x^2+5*x+3), x)
```

Answer : -4*log(x + 1) + 13/2*log(2*x + 3)

CAS **17.** $\int \frac{\sqrt{x^2+9}}{x}dx$

http://matrix.skku.ac.kr/cal-lab/cal-7-4-17.html

Solution

```
integral((sqrt(x^2+9))/(x), x)
```

Answer : sqrt(x^2 + 9) - 3*arcsinh(3/abs(x))

CAS **18.** $\int \frac{dx}{x^3(1-x)^3}$

http://matrix.skku.ac.kr/cal-lab/cal-7-4-18.html

Solution

```
integral(1/((x^3)*(-x+1)^3), x)
```

Answer : -1/2*(12*x^3 - 18*x^2 + 4*x + 1)/(x^4 - 2*x^3 + x^2) - 6*log(x - 1)+ 6*log(x)

CAS **19.** $\int \frac{dx}{x(x^n+a)^3},\quad a \neq 0$

http://matrix.skku.ac.kr/cal-lab/cal-7-4-19.html

Solution

```
var('x, a, n');
integral(1/x*(x^n+a)^3, x)
```

Answer : a^3*log(x)+1/6*(2*x^(3*n)+18*x^n*a^2+9*a*x^(2*n))/n

$$\int \frac{dx}{x(x^n+a)^3}(a \neq 0) = \int \frac{x^{n-1}}{x^n(x^n+a)^3}dx = \frac{1}{n}\int \frac{1}{(t-a)t^3}dt.$$

$t = x^n + a$

$$\frac{At^2+Bt+C}{t^3} + \frac{D}{t-a} = \frac{1}{(t-a)t^3}$$

$(t-a)(At^2+Bt+C)+Dt^3 = 1$

$A+D=0,\ B-aA=0,\ C-aB=0,\ -aC=1$

$C=-\frac{1}{a},\ B=-\frac{1}{a^2},\ A=-\frac{1}{a^3},\ D=\frac{1}{a^3}$

$$\frac{At^2+Bt+C}{t^3} + \frac{D}{t-a} = \frac{1}{(t-a)t^3}$$

$(t-a)(At^2+Bt+C)+Dt^3 = 1$

$A+D=0,\ B-aA=0,\ C-aB=0,-aC=1$

$C=-\frac{1}{a},\ B=-\frac{1}{a^2},\ A=-\frac{1}{a^3},\ D=\frac{1}{a^3}$

$$\int \frac{1}{(t-a)t^3}dt = \int -\frac{1}{a^3 t} - \frac{1}{a^2t^2} - \frac{1}{at^3} + \frac{1}{a^3(t-a)}dt = -\frac{\ln t}{a^3} + \frac{1}{a^2 t} + \frac{1}{2at^2} + \frac{\ln(t-a)}{a^3} + C.$$

$$\int \frac{dx}{x(x^n+a)^3} = \frac{1}{n}\left[-\frac{\ln(x^n+a)}{a^3} + \frac{1}{a^2(x^n+a)} + \frac{1}{2a(x^n+a)^2} - \frac{\ln(x^n)}{a^3}\right] + C.$$

CAS **20.** $\int_0^1 \frac{x^4}{x^3+1}dx$

http://matrix.skku.ac.kr/cal-lab/cal-7-4-20.html

Solution

```
integral(x^4/(x^3+1), x, 0, 1)
```

Answer : -1/9*pi*sqrt(3) + 1/3*log(2) + 1/2

CAS **21.** $\int \frac{2x^2-x+4}{x^3+4x}dx$

http://matrix.skku.ac.kr/cal-lab/cal-7-4-21.html

Solution

```
integral((2*x^2-x+4)/(x^3+4*x), x)
```

Answer : 1/2*log(x^2 + 4) + log(x) - 1/2*arctan(1/2*x)

CAS **22.** $\int \frac{1}{(1+x^2)^2}dx$

http://matrix.skku.ac.kr/cal-lab/cal-7-4-22.html

Solution

```
integral(1/(1+x^2)^2, x)
```

Answer : 1/2*x/(x^2 + 1) + 1/2*arctan(x)

CAS **23.** $\int \frac{dx}{(x-1)\sqrt{x+2}}$

http://matrix.skku.ac.kr/cal-lab/cal-7-4-23.html

Solution

```
integral(1/((x-1)*sqrt(x+2)), x)
```

Answer : 1/3*sqrt(3)*log((sqrt(x + 2) - sqrt(3))/(sqrt(x + 2) + sqrt(3)))

24. $\int \frac{dx}{\sqrt{x}(1-\sqrt[3]{x})}$

http://matrix.skku.ac.kr/cal-lab/cal-7-4-25.html

Solution

```
integral(1/((sqrt(x))*(1-x^(1/3))), x)
```

Answer: -6*x^(1/6)-3*log(x^(1/6)-1)+3*log(x^(1/6) + 1)

25. $\int \frac{1}{\sqrt{x^2+A}}dx$

Solution $\sqrt{x^2+A} = u - x,\ \frac{2x}{2\sqrt{x^2+A}}dx = du - dx,$

$$du = \frac{x+\sqrt{x^2+A}}{\sqrt{x^2+A}}dx = \frac{u}{u-x}dx, \;\; dx = \frac{u-x}{u}du$$

$$\int \frac{1}{\sqrt{x^2+A}}dx = \int \frac{1}{u-x}\,\frac{u-x}{u}\,du = \int \frac{1}{u}du$$

$$= \ln|u| + C = \ln|x+\sqrt{x^2+A}| + C.$$

26. $\displaystyle\int \frac{dx}{x\sqrt{x^2+2x+2}}$

[Hint: Substitute $\sqrt{x^2+2x+2} = u - x$]

http://matrix.skku.ac.kr/cal-lab/cal-7-4-27.html

Solution

```
integral(1/(x*(sqrt(x^2+2*x+2))), x)
```

Answer : -1/2*sqrt(2)*arcsinh(x/abs(x) + 2/abs(x))

<ICM 2014 logo >

7.5 Formulas for Integration

YouTube http://youtu.be/Fgn8U4We60o

Most functions can be differentiated by applying the formulas for differentiation. However, integrating a given function may not be as obvious as finding the derivative of a function. We may find integrals without much difficulty when an appropriate integration technique is recognized and correctly employed. So far, the method of substitution, integration by parts, trigonometric substitution and the method of partial fraction decomposition have been applied and their effectiveness in dealing with a variety of integration problems has been shown. In this section a collection of miscellaneous integrals (there is no particular order), is given. No hard and fast rules can be given as to which technique applies in a given situation, but one must be aware of the following basic integration formulas.

Table of Integration Formulas [Constant of integration is omitted.]

1. $\int k\,dx = kx \quad (k \in R)$
2. $\int x^n dx = \dfrac{x^{n+1}}{n+1} \quad (n \neq -1)$
3. $\int \dfrac{1}{x} dx = \ln|x|,\ x \neq 0$
4. $\int e^x dx = e^x$
5. $\int a^x dx = \dfrac{a^x}{\ln a},\ a > 0$
6. $\int \sin x\,dx = -\cos x$
7. $\int \cos x\,dx = \sin x$
8. $\int \sec^2 x\,dx = \tan x$
9. $\int \csc^2 x\,dx = -\cot x$
10. $\int \sec x \tan x\,dx = \sec x$
11. $\int \csc x \cot x\,dx = -\csc x$
12. $\int \sec x\,dx = \ln|\sec x + \tan x|$
13. $\int \csc x\,dx = \ln|\csc x - \cot x|$
14. $\int \tan x\,dx = \ln|\sec x|$
15. $\int \cot x\,dx = \ln|\sin x|$
16. $\int \sinh x\,dx = \cosh x$
17. $\int \cosh x\,dx = \sinh x$
18. $\int \text{sech}^2 x\,dx = \tanh x$
19. $\int \text{csch}^2 x\,dx = -\coth x$
20. $\int \text{sech}\,x \tanh x\,dx = -\text{sech}\,x$
21. $\int \text{csch}\,x \coth x\,dx = -\text{csch}\,x$

22. $\displaystyle\int \frac{dx}{x^2+a^2} = \frac{1}{a}\tan^{-1}\left(\frac{x}{a}\right),\ a \neq 0$

23. $\displaystyle\int \frac{dx}{\sqrt{a^2-x^2}} = \sin^{-1}\left(\frac{x}{a}\right),\ a \neq 0$

24. $\displaystyle\int \frac{dx}{x^2-a^2} = \frac{1}{2a}\ln\left|\frac{x-a}{x+a}\right|,\ a \neq 0$

25. $\displaystyle\int \frac{1}{\sqrt{x^2+A}}\,dx = \ln\left|x+\sqrt{x^2+A}\right| + C,\ A \neq 0$

26. $\displaystyle\int \frac{1}{(x-a)^n}\,dx = \begin{cases} -\dfrac{1}{(n-1)(x-a)^{n-1}} + C & (n \neq 1) \\ \ln|x-a| + C & (n = 1) \end{cases}$

27. $\displaystyle\int \frac{x}{(x^2+a^2)^n}\,dx = \begin{cases} -\dfrac{1}{2(n-1)(x^2+a^2)^{n-1}} + C & (n \neq 1) \\ \dfrac{1}{2}\ln|x^2+a^2| + C & (n = 1) \end{cases}$

There is no universal set of rules that indicate which method to employ for every integral. A guide to which techniques you might try for a particular integrand is given below.

Strategy for integration

1. Use algebraic manipulation and/or trigonometric identities to simplify the integrand as much as possible.
2. Determine a suitable substitution $u = g(x)$ in the integrand whose differential $du = g'(x)dx$ also occurs, baring a constant factor.
3. Classify the integrand according to its form.
   (a) If $f(x)$ is a product of powers of $\sin x$ and $\cos x$, $\tan x$ and $\sec x$, $\cot x$ and $\csc x$, then try trigonometric integrations.
   (b) If $f(x)$ is a rational function, the method of partial fractions may be used.
   (c) If $f(x)$ is a product of a power of x (or a polynomial) and a transcendental function (such as a trigonometric, exponential, or logarithmic function), then consider integration by parts.
   (d) When certain radicals appear as part of the integrand, the following substitutions may by helpful.
      (i) If $\sqrt{\pm x^2 \pm a^2}$ occurs, use a trigonometric substitution.
      (ii) If $\sqrt[n]{ax+b}$ occurs, use the rationalizing substitution $u = \sqrt[n]{ax+b}$. Sometimes, this also works for $\sqrt[n]{g(x)}$.
4. If the first three cases do not lead to an answer:
   (a) When no substitution is obvious, try an appropriate substitution.
   (b) Integration by parts should performed as many times as it is necessary.

(c) Relate the problem to previous problems. Use a method on a given integral that is similar to a method you have already used by expressing the given integral in terms of a previous one.

(d) Use several methods, a combination of two or more methods may be required to evaluate an integral, involving several successive substitutions of different types, or it may combine integration by parts with one or more substitutions etc.

EXAMPLE 1

Find $\int \frac{\sin^4 x}{\cos^6 x} dx$.

Solution Rewrite the integral: $\int \frac{\sin^4 x}{\cos^6 x} dx = \int \tan^4 x \sec^2 x \, dx$.

The integral is now of the form $\int \tan^m x \sec^n x \, dx$ with n even. So,

$$\begin{aligned}\int \frac{\sin^4 x}{\cos^6 x} dx &= \int \tan^4 x \sec^2 x \, dx \\ &= \int u^4 \, du \\ &= \frac{1}{5} u^5 + C \\ &= \frac{1}{5} \tan^5 x + C.\end{aligned}$$

■

EXAMPLE 2

Find $\int \frac{dx}{(x+1)\sqrt{x-3}}$.

Solution Let $u = \sqrt{x-3}$. Then $x = u^2 + 3$, $dx = 2u \, du$, and

$$\int \frac{dx}{(x+1)\sqrt{x-3}} = 2\int \frac{du}{u^2+4} = \tan^{-1}\left(\frac{\sqrt{x-3}}{2}\right) + C.$$

■

Some of the following examples are not solved completely but some hints are given:

EXAMPLE 3

Find $\int \frac{x^5 + 2x + 1}{x^3 + x^2 - 12x} dx$.

Solution Since the degree of the numerator is greater than the degree of the denominator, we perform long division and apply method of partial fractions. ■

EXAMPLE 4

Find $\int x^2(\ln x)^2 dx$.

Solution Using integrate by parts.

$$\begin{aligned}\int x^2(\ln x)^2 dx &= \frac{1}{3}x^3(\ln x)^2 - \int \frac{1}{3}x^3 \cdot 2\ln x \cdot \frac{1}{x}dx \\ &= \frac{1}{3}x^3(\ln x)^2 - \frac{2}{3}\int x^2 \ln x\, dx \\ &= \frac{1}{3}x^3(\ln x)^2 - \frac{2}{3}\left[\frac{1}{3}x^3\ln x - \int \frac{1}{3}x^3 \cdot \frac{1}{x}dx\right] \\ &= \frac{1}{3}x^3(\ln x)^2 - \frac{2}{9}x^3\ln x + \frac{2}{9}\int x^2 dx \\ &= \frac{1}{3}x^3(\ln x)^2 - \frac{2}{9}x^3\ln x + \frac{2}{27}x^3 + C.\end{aligned}$$

■

EXAMPLE 5

Find $\int \sqrt{\frac{2+x}{2-x}}\,dx$.

Solution Multiplying the numerator and denominator by $\sqrt{2+x}$, we have

$$\begin{aligned}\int \sqrt{\frac{2+x}{2-x}}\,dx &= \int \frac{2+x}{\sqrt{2^2-x^2}}dx \\ &= \int \frac{2}{\sqrt{2^2-x^2}}dx + \int \frac{x}{\sqrt{2^2-x^2}}dx \\ &= 2\sin^{-1}\left(\frac{x}{2}\right) - \sqrt{4-x^2} + C.\end{aligned}$$

■

Note The rationalizing substitution

$$u = \sqrt{\frac{2+x}{2-x}}$$

may be used even if it makes the problem more cumbersome.

$$x = \frac{2(u^2-1)}{u^2+1},\quad dx = \frac{8u}{(u^2+1)^2}du.$$

So, we get

$$\begin{aligned}\int \sqrt{\frac{2+x}{2-x}}\,dx &= 8\int \frac{u^2}{(u^2+1)^2}du \\ &= 8\int \left(\frac{1}{u^2+1} - \frac{1}{(u^2+1)^2}\right)du\end{aligned}$$

$$= 8\left[\tan^{-1}u - \frac{1}{2}\left(\frac{u}{u^2+1} + \tan^{-1}u\right)\right] + C$$

$$= 4\tan^{-1}\left(\sqrt{\frac{2-x}{2+x}}\right) - \sqrt{4-x^2} + C.$$

Polynomials, rational functions, power functions (x^a), exponential functions (a^x), logarithmic functions, trigonometric and inverse trigonometric functions, hyperbolic and inverse hyperbolic functions and all functions that can be obtained from these by the five operations of addition, subtraction, multiplication, division and composition are called elementary functions and are the objects differentiated or integrated in calculus course. If f is an elementary function, then f' is also an elementary function but $\int f(x)dx$ need not to be an elementary function. For instance, the indefinite integral $\int e^{x^3}dx$ can not be expressed in terms of elementary functions. Similarly the following integrals cannot be expressed in terms of elementary functions.

$$\int \frac{e^x}{x}dx \qquad \int \sin(x^4)dx \qquad \int \sin(e^x)dx$$

$$\int \sqrt{x^5+1}\,dx \qquad \int \frac{1}{\ln x}dx \qquad \int \frac{\cos x}{x}dx$$

Remark The majority of elementary functions that one may encounter in various branches of science and engineering do not possess elementary antiderivatives.

CAS **EXAMPLE 6**

Evaluate the integral

$$\int \frac{1}{(x^2+1)(x^2+x)}dx.$$

Solution Let $\dfrac{1}{(x^2+1)(x^2+x)} = \dfrac{A}{x} + \dfrac{B}{x+1} + \dfrac{Cx+D}{x^2+1}$.

Use Sage to find A, B, C, D and the integral.

```
var('x')
f= 1/((x^2 +1)*(x^2 + x))
print integral (f, x)
f.partial_fraction()
```

Answer : -1/2*log(x + 1) - 1/4*log(x^2 + 1) + log(x) - 1/2*arctan(x)
-1/2*(x + 1)/(x^2 + 1)-1/2/(x + 1) + 1/x ■

7.5 EXERCISES (Formulas for Integration)

http://matrix.skku.ac.kr/Cal-Book/part1/CS-Sec-7-5-Sol.html

YouTube http://youtu.be/-N9Fe_Arp2c

CAS **1.** $\displaystyle\int \sqrt{\frac{a-x}{a+x}}\,dx$

http://matrix.skku.ac.kr/cal-lab/cal-7-5-1.html

Solution

```
var('x, a');
integral(sqrt(a-x/a+x),x)
```

Answer : -2/3*(a + x - x/a)^(3/2)/(1/a - 1)

CAS **2.** $\displaystyle\int \frac{1}{x}\sqrt{\frac{x+1}{x-1}}\,dx$

http://matrix.skku.ac.kr/cal-lab/cal-7-5-3.html

Solution $\displaystyle\int \frac{1}{x}\sqrt{\frac{x+1}{x-1}}\,dx=\int \frac{x+1}{x\sqrt{x^2-1}}dx$

$$=\int \frac{1}{\sqrt{x^2-1}}dx+\int \frac{1}{x\sqrt{x^2-1}}dx$$

$$\int \frac{1}{\sqrt{x^2-1}}dx=\ln\left|x+\sqrt{x^2-1}\right|+C$$

Let $t=\sqrt{\dfrac{x+1}{x-1}}$ .Then $x=\dfrac{t^2+1}{t^2-1}$ and $dx=\dfrac{-4t}{(t^2-1)^2}dt.$

Therefore we get:

$$\int \frac{1}{x\sqrt{x^2-1}}dx$$

$$=\int \frac{1}{\left(\frac{t^2+1}{t^2-1}\right)\sqrt{\left(\frac{t^2+1}{t^2-1}\right)^2-1}}\frac{-4t}{(t^2-1)^2}dt$$

$$=-\int \frac{4t}{(t^2+1)\sqrt{(t^2+1)^2-(t^2-1)^2}}dt$$

$$=-2\int \frac{1}{t^2+1}dt=-2\tan^{-1}t$$

$$=-2\tan^{-1}\left(\sqrt{\frac{x+1}{x-1}}\right)+C.$$

Therefore,

$$\int \frac{1}{x}\sqrt{\frac{x+1}{x-1}}\,dx=\ln\left|x+\sqrt{x^2-1}\right|-2\tan^{-1}\left(\sqrt{\frac{x+1}{x-1}}\right)+C.$$

```
integral((1/x)*sqrt((x+1)/(x-1)),x)
```

Answer : -log(sqrt((x + 1)/(x - 1)) - 1) + log(sqrt((x + 1)/(x - 1)) + 1) -2*arctan(sqrt((x + 1)/(x- 1)))

CAS **3.** $\displaystyle\int \frac{dx}{\sqrt{x}\,(1-\sqrt[3]{x}\,)}$

http://matrix.skku.ac.kr/cal-lab/cal-7-5-4.html

Solution Let $\sqrt[6]{x}=t$. Then $x=t^6$, so $dx=6t^5\,dt$.

$$\int \frac{1}{\sqrt{x}\,(1-\sqrt[3]{x}\,)}dx=\int \frac{1}{t^3(1-t^2)}\cdot 6t^5\,dt$$

$$=6\int \frac{t^2}{1-t^2}dt$$

$$=6\int \frac{1-1+t^2}{1-t^2}dt=6\int\left(\frac{1}{1-t^2}-1\right)dt$$

$$=3\int\left(\frac{1}{1+t}+\frac{1}{1-t}\right)dt-6\int dt$$

$$=3(\ln|1+t|-\ln|1-t|)-6t$$

$$=3\ln\left|\frac{1+t}{1-t}\right|-6t+C$$

$$=3\ln\left|\frac{1+\sqrt[6]{x}}{1-\sqrt[6]{x}}\right|-6\sqrt[6]{x}+C.$$

```
integral(1/(sqrt(x)*(1-x^(1/3))),x)
```

Answer : -6*x^(1/6)-3*log(x^(1/6) - 1) + 3*log(x^(1/6) + 1)

CAS **4.** $\displaystyle\int \frac{1}{1+\sqrt[3]{x+1}}dx$

http://matrix.skku.ac.kr/cal-lab/cal-7-5-7.html

Solution $1+\sqrt[3]{x+1}=t,\ \dfrac{1}{3}(x+1)^{-\frac{2}{3}}dx$

$$=dt \Rightarrow dx=3(t-1)^2dt$$

$$\int \frac{1}{1+\sqrt[3]{x+1}}dx=\int \frac{3(t-1)^2}{t}dt$$

$$=\int\left(3t-6+\frac{3}{t}\right)dt=\frac{3}{2}t^2-6t+3\ln|t|$$

$$=-3\sqrt[3]{x+1}+\frac{3}{2}(x+1)^{\frac{2}{3}}+3\ln\left|1+\sqrt[3]{x+1}\right|+C.$$

```
integral(1/(1+(x+1)^(1/3)), x)
```

Answer : 3/2*(x + 1)^(2/3) - 3*(x + 1)^(1/3) + 3*log((x + 1)^(1/3) + 1)

CAS **5.** $\displaystyle\int \frac{1}{\sqrt{x}\,(1+\sqrt[3]{x}\,)}dx$

http://matrix.skku.ac.kr/cal-lab/cal-7-5-8.html

Solution Let $\sqrt[6]{x}=t$. Then $x=t^6$, so $dx=6t^5\,dt$.

$$\int \frac{1}{\sqrt{x}\,(1+\sqrt[3]{x}\,)}dx=\int \frac{1}{t^3(1+t^2)}\cdot 6t^5\,dt$$

$$=6\int \frac{t^2}{1+t^2}dt$$

$$=6\int \frac{1-1+t^2}{1+t^2}dt=6\int\left(1-\frac{1}{1+t^2}\right)dt$$

$$=6(t-\tan^{-1}t)=6\left(\sqrt[6]{x}-\tan^{-1}(\sqrt[6]{x})\right)+C.$$

```
integral(1/(sqrt(x)*(1+x^(1/3))), x)
```

Answer : 6*x^(1/6) - 6*arctan(x^(1/6))

CAS **6.** $\int \frac{\sqrt[4]{x}}{1+x}dx$

http://matrix.skku.ac.kr/cal-lab/cal-7-5-9.html

Solution

```
integral((x^(1/4))/(1+x), x)
```

Answer : 1/2*sqrt(2)*log(-sqrt(2)*x^(1/4) + sqrt(x) + 1) -1/2*sqrt(2)*log(sqrt(2)*x^(1/4) + sqrt(x) + 1) -sqrt(2)*arctan(-1/2*(sqrt(2) - 2*x^(1/4))*sqrt(2)) -sqrt(2)*arctan(1/2*(sqrt(2) + 2*x^(1/4))*sqrt(2)) + 4*x^(1/4)

CAS **7.** $\int \frac{\sqrt{3+\ln x}}{x(\ln x)^2}dx$

http://matrix.skku.ac.kr/cal-lab/cal-7-5-13.html

Solution

```
integral(sqrt(3+ln(x))/(x*(ln(x))^2), x)
```

Answer : 1/6*sqrt(3)*log((sqrt(log(x) + 3) - sqrt(3))/(sqrt(log(x) + 3) +sqrt(3))) - sqrt(log(x)+3)/log(x)

CAS **8.** $\int \sqrt{3x-2-x^2}\,dx$

http://matrix.skku.ac.kr/cal-lab/cal-7-5-14.html

Solution

```
integral(sqrt(3*x-2-x^2), x)
```

Answer : 1/2*sqrt(-x^2 + 3*x - 2)*x - 3/4*sqrt(-x^2 + 3*x - 2) + 1/8*arcsin(2*x -3)

CAS **9.** $\int \frac{x^3}{\sqrt{1-x^2}}dx$

http://matrix.skku.ac.kr/cal-lab/cal-7-5-15.html

Solution $x=\sin\theta,\ dx=\cos\theta\, d\theta$

$$\begin{aligned}\int \frac{x^3}{\sqrt{1-x^2}}dx &= \int \frac{\sin^3\theta}{\sqrt{1-\sin^2\theta}}\cos\theta\, d\theta\\ &= \int \sin^3\theta\, d\theta\\ &= \int\left(\frac{3}{4}\sin\theta-\frac{1}{4}\sin 3\theta\right)d\theta\\ &= -\frac{3}{4}\cos\theta+\frac{1}{12}\cos 3\theta = \frac{1}{3}\cos^3\theta-\cos\theta\\ &= -\frac{1}{3}x^2\sqrt{1-x^2}-\frac{2}{3}\sqrt{1-x^2}+C.\end{aligned}$$

```
integral(x^3/sqrt(1-x^2), x)
```

Answer : -1/3*sqrt(-x^2 + 1)*x^2 - 2/3*sqrt(-x^2 + 1)

CAS **10.** $\int \frac{\tan^5 x}{\cos^5 x}dx$

http://matrix.skku.ac.kr/cal-lab/cal-7-5-5.html

Solution $\sec x = t \implies \sec x\tan x\, dx = dt$

$$\begin{aligned}\int \frac{\tan^5 x}{\cos^5 x}dx &= \int \tan^5 x\sec^5 x\, dx\\ &= \int \sec^4 x(\sec^2 x-1)^2\tan x\sec x\, dx\\ &= \int t^4(t^2-1)^2dt = \int (t^8-2t^6+t^4)dt\\ &= \frac{1}{9}t^9-\frac{2}{7}t^7+\frac{1}{5}t^5\\ &= \frac{1}{9}\sec^9 x-\frac{2}{7}\sec^7 x+\frac{1}{5}\sec^5 x+C.\end{aligned}$$

```
integral(tan(x)^5/cos(x)^5, x)
```

Answer : 1/315*(63*cos(x)^4 - 90*cos(x)^2 + 35)/cos(x)^9

CAS **11.** $\int \frac{\cos^2 x}{\sin^6 x}dx$

http://matrix.skku.ac.kr/cal-lab/cal-7-5-6.html

Solution $\cot x = t \implies \csc^2 x dx = -dt$

$$\begin{aligned}\int \frac{\cos^2 x}{\sin^6 x}dx &= \int \cot^2 x\csc^4 x dx\\ &= \int \cot^2 x(1+\cot^2 x)\csc^2 x dx\\ &= \int t^2(1+t^2)(-dt) = -\frac{1}{3}t^3-\frac{1}{5}t^5\\ &= -\frac{1}{3}\cot^3 x-\frac{1}{5}\cot^5 x+C.\end{aligned}$$

```
integral(cos(x)^2/sin(x)^6, x)
```

Answer : -1/15*(5*tan(x)^2 + 3)/tan(x)^5

CAS **12.** $\int x\sqrt[5]{x+1}\,dx$

http://matrix.skku.ac.kr/cal-lab/cal-7-5-17.html

Solution

```
integral(x*(x+1)^(1/5), x)
```

Answer : 5/11*(x + 1)^(11/5) - 5/6*(x + 1)^(6/5)

CAS **13.** $\int \frac{x^2}{\sqrt{x^2+1}}dx$

http://matrix.skku.ac.kr/cal-lab/cal-7-5-18.html

Solution

```
integral(x^2/sqrt(x^2+1), x)
```

Answer : 1/2*sqrt(x^2 + 1)*x - 1/2*arcsinh(x)

14. $\int \frac{x}{\sqrt{3-x^4}}dx$

http://matrix.skku.ac.kr/cal-lab/cal-7-5-19.html

Solution

```
integral(x/sqrt(3-x^4), x)
```

Answer : -1/2*arctan(sqrt(-x^4 + 3)/x^2)

15. $\int \frac{x^2-4x-2}{x^2-x-2}dx$

Solution $\int \frac{x^2-4x-2}{x^2-x-2}dx = \int \left(1-\frac{3x}{(x-2)(x+1)}\right)dx$

$= x+\int \left(\frac{1}{x+1}-\frac{1}{x-2}\right)dx$

$= x+\ln|x+1|-\ln|x-2|$

$= x+\ln\left|\frac{x+1}{x-2}\right|+C.$

16. $\int \frac{3x^2-x+6}{(x+1)^2(x^2+1)}dx$

http://matrix.skku.ac.kr/cal-lab/cal-7-5-11.html

Solution

```
integral((3*x^2-x+6)/((x+1)^2)*(x^2+1),x)
```

Answer : -20/(x+1) + x^3 - 7/2*x^2 + 20*x - 34*log(x + 1)

17. $\int e^{\sqrt{x+5}}dx$

http://matrix.skku.ac.kr/cal-lab/cal-7-5-2.html

Solution $\sqrt{x+5}=t \Rightarrow x+5=t^2 \Rightarrow dx=2t\,dt$

$\int e^{\sqrt{x+5}}dx = \int 2te^t\,dt = 2te^t - 2\int e^t\,dt = 2te^t - 2e^t$

$= 2(t-1)e^t = 2(\sqrt{x+5}-1)e^{\sqrt{x+5}}+C.$

```
integral(e^(sqrt(x+5)),x)
```

Answer : 2*(sqrt(x + 5) - 1)*e^(sqrt(x + 5))

18. $\int e^{\sqrt[4]{x}}dx$

http://matrix.skku.ac.kr/cal-lab/cal-7-5-12.html

Solution $\sqrt[4]{x}=t \Rightarrow x=t^4 \Rightarrow dx=4t^3\,dt$

$\int e^{\sqrt[4]{x}}dx = \int 4t^3e^t\,dt = 4t^3e^t - 12\int t^2e^t\,dt$

$= 4t^3e^t - 12t^2e^t + 24\int te^t\,dt$

$= 4t^3e^t - 12t^2e^t + 24te^t - 24\int e^t\,dt$

$= 4t^3e^t - 12t^2e^t + 24te^t - 24e^t$

$= 4e^{\sqrt[4]{x}}(\sqrt[4]{x^3} - 3\sqrt{x} + 6\sqrt[4]{x} - 6) + C.$

```
integral(e^(x^(1/4)),x)
```

Answer : 4*(x^(3/4)-3*sqrt(x) + 6*x^(1/4) - 6)*e^ (x^(1/4))

7.6 Integration Using Formulas

YouTube http://youtu.be/tn9jLkgTMp8

We may need to use an integral table to evaluate an integral. In this case we need to use substitution or algebraic manipulation to transform the given integral into one of the integrals listed in the table. (See Section 7.5.)

EXAMPLE 1

Find the volume of the solid obtained by rotating about the y-axis the region bounded by the curves $y=\sin^{-1}x$, $y=0$, and $x=\dfrac{1}{2}$. (See Figure 1.)

http://matrix.skku.ac.kr/cal-lab/cal-7-6-Exm-1.html

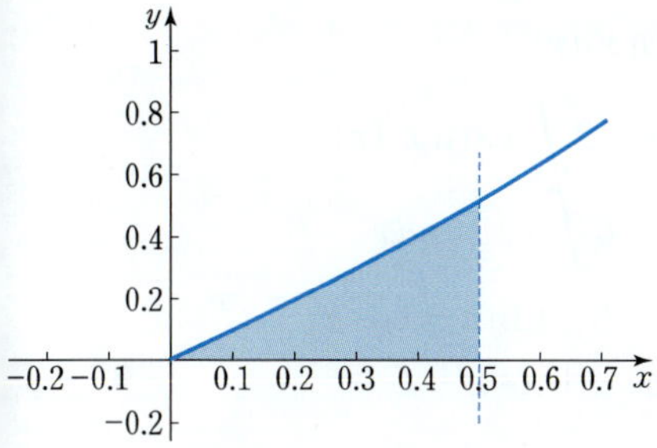

Figure 1

Solution The volume by the method of cylindrical shells is

$$V=\int_0^{1/2} 2\pi x \sin^{-1}x\,dx.$$

We know from the table

$$\int u\sin^{-1}u\,du=\frac{2u^2-1}{4}\sin^{-1}u+\frac{u\sqrt{1-u^2}}{4}+C.$$

Thus, the volume is

$$\begin{aligned} V&=2\pi\int_0^{1/2} x\sin^{-1}x\,dx \\ &=2\pi\left[\frac{2x^2-1}{2}\sin^{-1}x+\frac{x\sqrt{1-x^2}}{4}\right]_0^{1/2} \\ &=\frac{\sqrt{3}}{8}-\frac{\pi^2}{24}. \end{aligned}$$

■

EXAMPLE 2

Evaluate $\displaystyle\int \frac{x^2}{\sqrt{3x-2x^2}}\,dx.$

Solution We know from the table

$$\int \frac{u^2\,du}{\sqrt{2au-u^2}}=-\frac{(u+3a)}{2}\sqrt{2au-u^2}+\frac{3a^2}{2}\cos^{-1}\left(\frac{a-u}{a}\right)+C.$$

Thus,

$$\int \frac{x^2}{\sqrt{3x-2x^2}}\,dx=\frac{1}{\sqrt{2}}\int \frac{x^2}{\sqrt{(3/2)x-x^2}}\,dx$$

$$= \frac{1}{\sqrt{2}}\left[\frac{4x+9}{8}\sqrt{\frac{3}{2}x - x^2} + \frac{27}{32}\cos^{-1}\left(\frac{3-4x}{3}\right)\right] + C.$$ ■

EXAMPLE 3

Evaluate $\int x^3 \cos x\, dx$.

Solution We know from the table that

$$\int u^n \cos u du = u^n \sin u - n\int u^{n-1}\sin u\, du.$$

Using the reduction formula repeatedly, we have

$$\begin{aligned}\int x^3\cos x dx &= x^3 \sin x - 3\int x^2 \sin x\, dx \\ &= x^3\sin x - 3\left[x^2\sin x - 2\int x\sin x\, dx\right] \\ &= x^3\sin x - 3x^2\sin x + 6\int x\sin x\, dx \\ &= x^3\sin x - 3x^2\sin x + 6[x\sin x + \cos x] + C \\ &= (x^3 - 3x^2 + 6x)\sin x + 6\cos x + C.\end{aligned}$$ ■

EXAMPLE 4

Evaluate $\int (x^2 - 4)\sqrt{5 - 4x - x^2}\, dx$.

Solution First complete the square: $5 - 4x - x^2 = 9 - (x+2)^2$
If we make the substitution $u = x + 2$ (so $x = u - 2$), the integrand will involve the form $\sqrt{a^2 - u^2}$.

$$\begin{aligned}\int (x^2-4)\sqrt{5-4x-x^2}\, dx &= \int (u^2 - 4u)\sqrt{9-u^2}\, du \\ &= \int u^2\sqrt{9-u^2}\, du - 4\int u\sqrt{9-u^2}\, du \\ &= \int u^2\sqrt{9-u^2}\, du + \frac{4}{3}(9-u^2)^{3/2} + C.\end{aligned}$$

We know from the table that

$$\int u^2\sqrt{a^2-u^2}\, du = \frac{u}{8}(2u^2 - a^2)\sqrt{a^2-u^2} + \frac{a^4}{8}\sin^{-1}\left(\frac{u}{a}\right) + C.$$

Applying the formula with $a = 3$ gives

$$\begin{aligned}&\int (x^2-4)\sqrt{5-4x-x^2}\, dx \\ &= \frac{(x+2)}{8}(2x^2 + 8x - 1)\sqrt{5-4x-x^2} \\ &\quad + \frac{81}{8}\sin^{-1}\left(\frac{x+2}{3}\right) + \frac{4}{3}(5-4x-x^2)^{\frac{3}{2}} + C.\end{aligned}$$ ■

CAS **EXAMPLE 5**

Evaluate the integral

$$\int \frac{\tan^{-1}e^x}{e^x}dx.$$

Let $t=e^x$, $x=\ln t$, then $dx=\frac{1}{t}dt$.

$$\begin{aligned}
&\int \frac{\tan^{-1}e^x}{e^x}dx \\
&\qquad =\int \frac{\tan^{-1}t}{t}\cdot\frac{1}{t}dt=-\frac{1}{t}\cdot\tan^{-1}t+\int\frac{1}{t}\cdot\frac{1}{1+t^2}dt \\
&\qquad =-\frac{\tan^{-1}t}{t}+\int\left(\frac{1}{t}-\frac{t}{1+t^2}\right)dt \\
&\qquad =-\frac{\tan^{-1}t}{t}+\ln t-\frac{1}{2}\ln(1+t^2)+C \\
&\qquad =x-e^{-x}\tan^{-1}e^x-\frac{1}{2}\ln(1+e^{2x})+C.
\end{aligned}$$

```
var('x')
f=arctan(exp(x))/exp(x)
integral(f, x)
```

Answer : -e^(-x)*arctan(e^x) + x - 1/2*log(e^(2*x) + 1) ■

7.6 EXERCISES (Integration Using Formulas)

http://matrix.skku.ac.kr/Cal-Book/part1/CS-Sec-7-6-Sol.html

YouTube http://youtu.be/EnEQ9ZS3B_k

CAS **1.** $\int \frac{x^2}{\sqrt{7-3x^2}}dx$

http://matrix.skku.ac.kr/cal-lab/cal-7-6-1.html

Solution Let $x=\sqrt{\frac{7}{3}}\sin\theta$, then $dx=\sqrt{\frac{7}{3}}\cos\theta\, d\theta$.

$$\int \frac{x^2}{\sqrt{7-3x^2}}dx = \int \frac{\frac{7}{3}\sin^2\theta}{\sqrt{7-7\sin^2\theta}}dx$$

$$= \int \frac{\frac{7}{3}\sin^2\theta}{\sqrt{7}\cos\theta}\sqrt{\frac{7}{3}}\cos\theta\, d\theta$$

$$= \int \frac{7}{3\sqrt{3}}\sin^2\theta\, d\theta = \int \frac{7}{3\sqrt{3}}\frac{1-\cos 2\theta}{2}\, d\theta$$

$$= \frac{7}{6\sqrt{3}}\theta - \frac{7}{12\sqrt{3}}\sin 2\theta$$

Since $x=\sqrt{\frac{7}{3}}\sin\theta$, $\theta=\sin^{-1}\sqrt{\frac{3}{7}}x$.

Therefore, $\frac{7}{12\sqrt{3}}\sin 2\theta = \frac{1}{6}x\sqrt{7-3x^2}$

$$\int \frac{x^2}{\sqrt{7-3x^2}}dx = -\frac{1}{6}x\sqrt{7-3x^2} + \frac{7}{6\sqrt{3}}\sin^{-1}\sqrt{\frac{3}{7}}x + C.$$

```
integral(x^2/(sqrt(7-3*x^2)), x)
```

Answer : -1/6*sqrt(-3*x^2 + 7)*x + 7/18*sqrt(3)*arcsin(1/7*sqrt (21)*x)

CAS **2.** $\int \frac{1}{(2x-3)\sqrt{3x-2-x^2}}dx$

http://matrix.skku.ac.kr/cal-lab/cal-7-6-12.html

Solution

```
integral(1/((2*x-3)*(sqrt(-x^2+3*x-2))), x)
```

Answer : -log(sqrt(-x^2 + 3*x - 2)/abs(2*x - 3) + 1/2/abs(2*x - 3))

CAS **3.** $\int x\sqrt{x^2-4x+7}\, dx$

http://matrix.skku.ac.kr/cal-lab/cal-7-6-4.html

Solution

```
integral(x*(sqrt(x^2-4*x+7)), x)
```

Answer : 1/3*(x^2 - 4*x + 7)^(3/2) + sqrt(x^2 - 4*x + 7)*x - 2*sqrt(x^2 - 4*x + 7) + 3*arcsinh(1/3*(x − 2)*sqrt(3))

CAS **4.** $\int x^5 \sin x\, dx$

http://matrix.skku.ac.kr/cal-lab/cal-7-6-2.html

Solution

```
integral(x^5*(sin(x)), x)
```

Answer : 5*(x^4 - 12*x^2 + 24)*sin(x) - (x^5 - 20*x^3 + 120*x)*cos(x)

CAS **5.** $\int \sin^5 x \cos^4 x\, dx$

http://matrix.skku.ac.kr/cal-lab/cal-7-6-3.html

Solution

```
integral(sin(x)^5*cos(x)^4, x)
```

Answer : -1/9*cos(x)^9 + 2/7*cos(x)^7 - 1/5*cos(x)^5

CAS **6.** $\int \sec^7 x\, dx$

http://matrix.skku.ac.kr/cal-lab/cal-7-6-5.html

Solution

```
integral((sec(x))^7, x)
```

Answer : -1/48*(15*sin(x)^5 - 40*sin(x)^3 + 33*sin(x))/(sin(x)^6 - 3*sin(x)^4 +3*sin(x)^2 − 1) - 5/32* log(sin(x) - 1) + 5/32*log(sin(x) + 1)

CAS **7.** $\int x^3 \tan^{-1}(3x)dx$

http://matrix.skku.ac.kr/cal-lab/cal-7-6-6.html

Solution

```
integral(x^3*arctan(3*x), x)
```

Answer : 1/4*x^4*arctan(3*x) - 1/36*x^3 + 1/108*x - 1/324*arctan(3*x)

CAS **8.** $\int \frac{1}{e^x\sqrt{1+e^{2x}}}dx$

http://matrix.skku.ac.kr/cal-lab/cal-7-6-7.html

Solution

```
integral(1/(e^x*sqrt(1+e^(2*x))), x)
```

Answer : -sqrt(e^(2*x) + 1)*e^(-x)

CAS **9.** $\int e^x(2-3e^{2x})^{3/2}dx$

http://matrix.skku.ac.kr/cal-lab/cal-7-6-8.html

Solution

```
integral(e^x*(2-3*e^(2*x))^(3/2), x)
```

Answer : 1/4*(-3*e^(2*x) + 2)^(3/2)*e^x + 3/4*sqrt (-3*e^(2*x) + 2)*e^x+1/2*sqrt(3)*arcsin(1/2* sqrt(6)*e^x)

CAS **10.** $\displaystyle\int \frac{1}{x[(\ln x)^2+(\ln x)]}dx$

http://matrix.skku.ac.kr/cal-lab/cal-7-6-9.html

Solution $\ln x = t \Rightarrow \dfrac{1}{x}dx = dt$

$$\int \frac{dx}{x[(\ln x)^2+\ln x]} = \int \frac{dt}{t^2+t}$$
$$= \int \left(\frac{1}{t} - \frac{1}{t+1}\right)dt$$
$$= \ln|t| - \ln|t+1|$$
$$= \ln|\ln x| - \ln|\ln x+1| + C.$$

```
integral(1/(x*((ln(x))^2+(ln(x)))), x)
```

Answer : -log(log(x) + 1) + log(log(x))

CAS **11.** $\displaystyle\int \frac{1}{\sin^3 x\cos^3 x}dx$

http://matrix.skku.ac.kr/cal-lab/cal-7-6-10.html

Solution Let $\cos 2x = t$, then $-2\sin 2x\,dx = dt$.

$$\int \frac{1}{\sin^3 x\cos^3 x}dx = \int \frac{-4}{(1-t^2)}dt$$
$$= -2\int \left(\frac{1}{1-t} + \frac{1}{1+t}\right)dt$$
$$= 2\ln|1-t| - 2\ln|1+t|$$
$$= 2\ln(1-\cos 2x) - 2\ln(1+\cos 2x) + C.$$

```
integral(1/((sin(x))^3*(cos(x))^3), x)
```

Answer : -1/2*(2*sin(x)^2 - 1)/(sin(x)^4 - sin(x)^2) - log(sin(x)^2 - 1) +2*log(sin(x))

CAS **12.** $\displaystyle\int e^{2x}\cos(3x)dx$

http://matrix.skku.ac.kr/cal-lab/cal-7-6-11.html

Solution

```
integral(e^(2*x)*cos(3*x), x)
```

Answer : 1/13*(3*sin(3*x) + 2*cos(3*x))*e^(2*x)

<Counting Rods>

7.7 Approximate Integration and CAS

You Tube http://youtu.be/hg2pw1n1cZI

There are many definite integrals which are difficult or even impossible to evaluate since we are unable to find an antiderivative of the integrand or the function is given as a table of values of collected data. For instance, it is impossible to evaluate the integrals like $\int_0^{\pi/2} \cos(x^3)dx$ and $\int_{-1}^{1} \sqrt{1+x^7}\,dx$ exactly. In these cases we have no choice but to find approximate values of these definite integrals.

From the definition of the definite integral, we can obtain approximate integration formulas. Recall the definite integral is defined as a limit of Riemann sums, and any Riemann sum could be used as an approximation to the integral. If we divide the interval $[a, b]$ into n subintervals of equal length $\Delta x = (b-a)/n$, we then have

$$\int_a^b f(x)dx \approx \sum_{i=1}^{n} f(x_i^*)\Delta x$$

where x_i^* is any point in the ith subinterval $[x_{i-1}, x_i]$. A choice for x_i^* results in an approximate integration formula. Among all possible choices for x_i^*, the case where x_i^* is chosen to be the midpoint $\overline{x}_i$ of the subinterval $[x_{i-1}, x_i]$ is known as the 'midpoint approximation'.

1 Midpoint Rule

$$\int_a^b f(x)dx \approx M_n = \Delta x\left[f(\overline{x_1}) + f(\overline{x_2}) + \cdots + f(\overline{x_n})\right]$$

where $\Delta x = \dfrac{b-a}{n}$ and $\overline{x}_i = \dfrac{1}{2}(x_{i-1} + x_i)$ = midpoint of $[x_{i-1}, x_i]$.

http://matrix.skku.ac.kr/cal-lab/sage-grapher-integral2.html

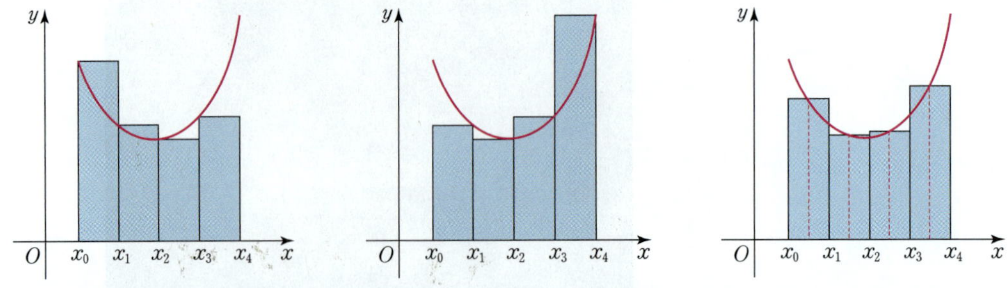

(a) Left endpoint approximation (b) Right endpoint approximation (c) Midpoint approximation

Figure 1

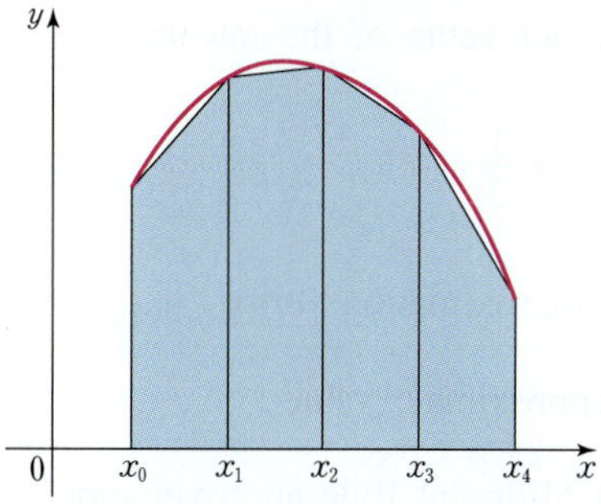

Figure 2 Trapezoidal approximation

We could see intuitively that using trapezoids instead of rectangles over each subinterval will give a better approximation to the area of the corresponding subregion. Based on this observation, we can construct another approximate integration formula

$$\int_a^b f(x)dx \approx \frac{\Delta x}{2}\left[\sum_{i=1}^{n}(f(x_{i-1})+f(x_i))\right]$$
$$= \frac{\Delta x}{2}[(f(x_0)+f(x_1))+(f(x_1)+f(x_2))+\cdots+(f(x_{n-1})+f(x_n))]$$
$$= \frac{\Delta x}{2}[f(x_0)+2f(x_1)+2f(x_2)+\cdots+2f(x_{n-1})+f(x_n)].$$

This approximation is called the Trapezoidal Rule.

2 The Trapezoidal Rule

$$\int_a^b f(x)dx \approx T_n = \frac{\Delta x}{2}[f(x_0)+2f(x_1)+2f(x_2)+\cdots+2f(x_{n-1})+f(x_n)]$$

where $\Delta x = (b-a)/n$ and $x_i = a + i\Delta x$. (See Figure 2.)

http://matrix.skku.ac.kr/cal-lab/Area-Sum.html

EXAMPLE 1

Approximate the integral $\int_1^2 \frac{1}{x^2}dx$ using (a) the Trapezoidal Rule and (b) the Midpoint Rule with $n=5$.

Solution (a) With $n=5$, $a=1$ and $b=2$, we have $\Delta x = (2-1)/5 = 0.2$, and so the Trapezoidal Rule gives

$$\int_1^2 \frac{1}{x^2}dx \approx T_5 = \frac{0.2}{2}[f(1)+2f(1.2)+2f(1.4)$$
$$+2f(1.6)+2f(1.8)+f(2)]$$
$$=0.1\left(\frac{1}{1^2}+\frac{2}{(1.2)^2}+\frac{2}{(1.4)^2}+\frac{2}{(1.6)^2}+\frac{2}{(1.8)^2}+\frac{1}{2^2}\right)$$
$$\approx 0.505783.$$

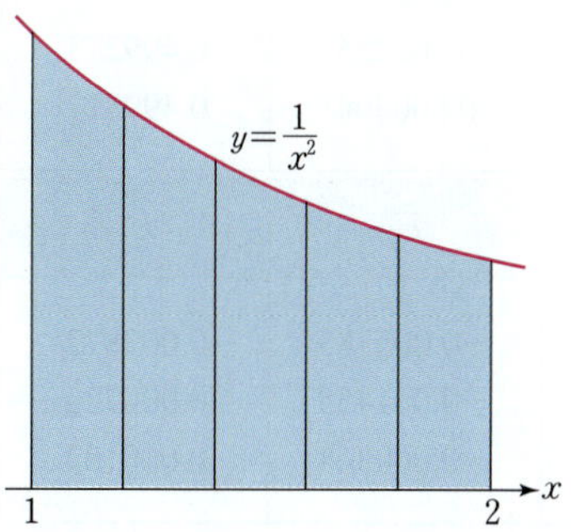

Figure 3

This approximation is illustrated in Figure 3.

(b) The midpoints of the five subintervals are 1.1, 1.3, 1.5, 1.7, and 1.9, so the Midpoint Rule gives

$$\int_1^2 \frac{1}{x^2}dx \approx \Delta x[f(1.1)+f(1.3)+f(1.5)+f(1.7)+f(1.9)]$$
$$=\frac{1}{5}\left(\frac{1}{(1.1)^2}+\frac{1}{(1.3)^2}+\frac{1}{(1.5)^2}+\frac{1}{(1.7)^2}+\frac{1}{(1.9)^2}\right)$$
$$\approx 0.497127.$$

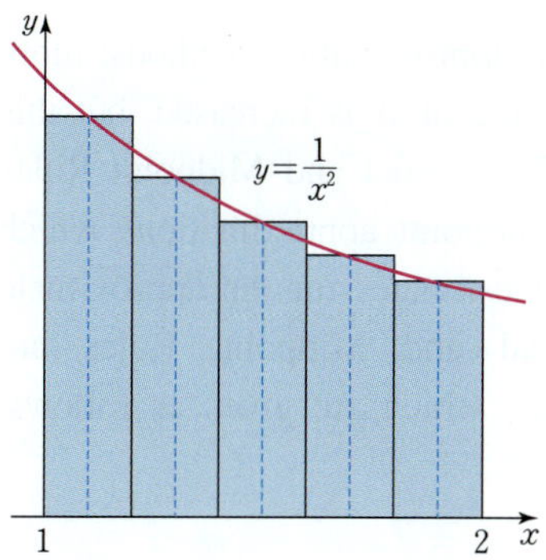

Figure 4

This approximation is illustrated in Figure 4.

By the FTC, compute explicitly the exact value of the integral:

$$\int_1^2 \frac{1}{x^2}dx = -\frac{1}{x}\Big]_1^2 = \frac{1}{2} = 0.5$$

$$\int_a^b f(x)dx = \text{value of the approximation} + \text{error}$$

$$\text{error} = \text{exact value} - \text{approximate value}$$

Now the errors in the Trapezoidal and Midpoint Rule approximations for $n = 5$ are

$$E_T \approx -0.005783 \text{ and } E_M \approx 0.002872.$$

In general, we have

$$E_T = \int_a^b f(x)dx - T_n \text{ and } E_M = \int_a^b f(x)dx - M_n.$$

The left end point and **right end point rules** may be found in http://en.wikipedia.org/wiki/Riemann_sum. The following tables show the results of calculations (similar to those in Example 1) for $n = 5,\ 10$ and 20 for the left, right endpoint approximations, Trapezoidal and Midpoint Rules.

n	L_n	R_n	T_n	M_n
5	0.580783	0.430783	0.505783	0.497127
10	0.538955	0.463955	0.501455	0.499273
20	0.519114	0.481614	0.500364	0.499817

n	E_L	E_N	E_T	E_M
5	−0.080783	0.069216	−0.005783	0.002872
10	−0.038955	0.036044	−0.001455	0.000726
20	−0.019114	0.018385	−0.000364	0.000182

http://sage.luther.edu:8443/home/pub/98/

http://matrix.skku.ac.kr/cal-lab/cal-RiemannSum.html ■

It should be mentioned that in all of the above approximation methods, more accurate approximations are obtained when the value of n is increased, but this results in an accumulated round-off error. The Trapezoidal and Midpoint Rules are known to be much more accurate than the endpoint approximations which can be observed from the tables above. The error estimates for the approximate integration values provided by the Trapezoidal and Midpoint Rules are established from text books on numerical analysis, which are given as follows:

3 Error Bounds

Suppose $|f^{(n)}(x)| \leq K$ for $a \leq x \leq b$. If E_T and E_M are the errors for the Trapezoidal and Midpoint Rules, respectively then

$$|E_T| \leq \frac{K(b-a)^3}{12n^2} \quad \text{and} \quad |E_M| \leq \frac{K(b-a)^3}{24n^2}.$$

Here, K can be any number larger than all the values of $|f^n(x)|$, but smaller values of K give better error bounds.

In the error estimates given above, observe that the Midpoint Rule is more accurate than the Trapezoidal Rule. Let us apply this error estimate, for the Trapezoidal Rule approximation in Example 1. If $f(x) = 1/x^2$ then $f'(x) = -2/x^3$ and $f''(x) = 6/x^4$. Since $1 \leq x \leq 2$, we have $1/x \leq 1$, so

$$|f''(x)| = \frac{6}{x^4} \leq \frac{6}{1^4} = 6.$$

Therefore, taking $K = 6$, $a = 1$, $b = 2$, and $n = 5$ in the error estimate 3, we find that

$$|E_T| \leq \frac{K(b-a)^3}{12n^2} = \frac{6(2-1)^3}{12(5)^2} = \frac{1}{50} = 0.02.$$

If we compare this error estimate of 0.02 with the actual error of around 0.005783, then we can see that the actual error is substantially less than the upper bound for the error provided by 3.

EXAMPLE 2

Determine how large should be n so that the Trapezoidal and Midpoint Rule approximations for $\int_1^2 (1/x^2)dx$ are accurate to within 0.0001.

Solution We know (from the above calculations) that $|f''(x)| \leq 6$ for $1 \leq x \leq 2$, so take $K = 6$, $a = 1$, and $b = 2$ in 3. Since the size of the error should be less than 0.0001, therefore choose n so that $\frac{6(1)^3}{12n^2} < 0.0001$.

Solving the inequality for n, we get $n^2 > \frac{1}{2(0.0001)}$ or $n > \frac{1}{\sqrt{0.0002}} \approx 70.7$.

Thus, $n = 71$ will ensure the desired accuracy. ■

Note A larger value for n would suffice, but 71 is the smallest value for which the error bound formula will guarantee us accuracy to within 0.0001.

For the same accuracy with the Midpoint Rule we choose n so that

$$|E_M| \le \frac{K(b-a)^3}{24n^2} = \frac{6(1)^3}{24n^2} < 0.0001,$$

which gives $n > \dfrac{1}{\sqrt{0.0004}} = 50.$

EXAMPLE 3

(a) Approximate the integral $\int_0^1 e^{x^3}\,dx$ using the Midpoint Rule with $n = 10$.

(b) Find an upper bound for the error involved in this approximation.

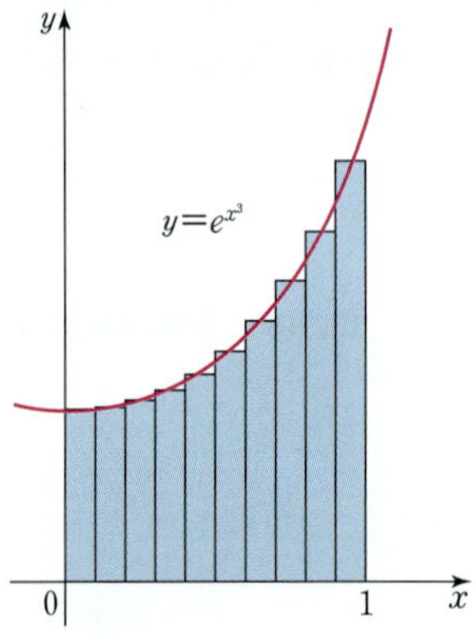

Figure 5

Solution (a) With $a = 0$, $b = 1$ and $n = 10$, the Midpoint Rule gives

$$\begin{aligned}\int_0^1 e^{x^3}\,dx &\approx \Delta x[f(0.05)+f(0.15)+\cdots+f(0.85)+f(0.95)]\\ &= 0.1[e^{0.000125}+e^{0.003375}+e^{0.015625}+e^{0.042875}+e^{0.091125}\\ &\quad +e^{0.166375}+e^{0.274615}+e^{0.421875}+e^{0.614125}+e^{0.857375}]\\ &\approx 1.338534.\end{aligned}$$

This approximation is illustrated in Figure 5.

(b) Since $f(x) = e^{x^3}$, we have $f'(x) = 3x^2e^{x^3}$ and

$$f''(x) = (6x + 9x^4)e^{x^3}.$$

Also, since $0 \le x \le 1$, we have $x^3 \le 1$ and so

$$0 \le f''(x) = (6x + 9x^4)e^{x^3} \le 15e. \qquad ■$$

Taking $K = 15e$, $a = 0$, $b = 1$, and $n = 10$ in the error estimate 3, an upper bound for the error is

$$\frac{15e(1)^3}{24(10^2)} = \frac{e}{160} \approx 0.0169.$$

Simpson's Rule–Approximation using Parabolas

By approximating a curve using parabolas instead of straight line segments that produce trapezoids, another well-known rule for approximate integration is obtained. As earlier, we make a partition of the interval $[a, b]$ into n subintervals of equal length $h = \Delta x = (b-a)/n$, but this time we require that n is an even number. We approximate the curve $y = f(x) \ge 0$ by a parabola on each consecutive pair of intervals as shown in Figure 6. If $y_i = f(x_i)$, then $P_i(x_i, y_i)$ is the point on the curve lying above x_i. A typical parabola passes through three consecutive points P_i, P_{i+1}, and P_{i+2}.

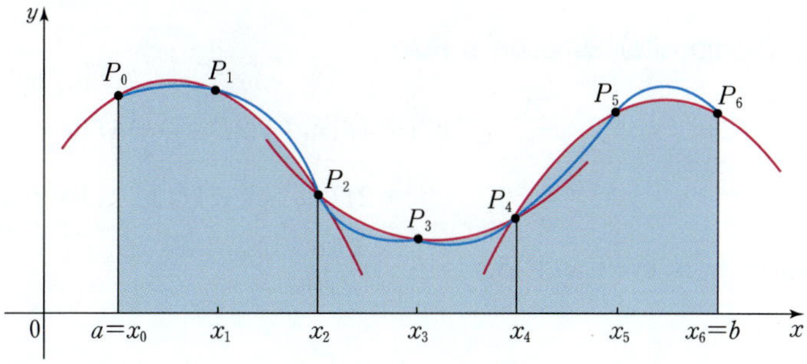

Figure 6

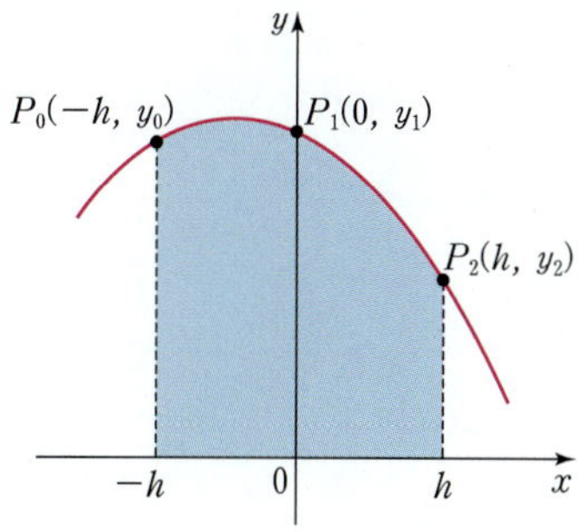

Figure 7

We consider the case where $x_0 = -h$, $x_1 = 0$, and $x_2 = h$. See Figure 7. The parabola through P_0, P_1, and P_2 has an equation of the form $y = Ax^2 + Bx + C$. So the area under the parabola from $x = -h$ to $x = h$ is

$$\int_{-h}^{h} (Ax^2 + Bx + C)dx = 2\int_0^h (Ax^2 + C)dx = 2\left[A\frac{x^3}{3} + Cx\right]_0^h$$
$$= 2\left(A\frac{h^3}{3} + Ch\right) = \frac{h}{3}(2Ah^2 + 6C).$$

Since the parabola passes through $P_0(-h,\ y_0)$, $P_1(0,\ y_1)$, and $P_2(h,\ y_2)$, we have

$$y_0 = A(-h)^2 + B(-h) + C = Ah^2 - Bh + C$$
$$y_1 = C$$
$$y_2 = Ah^2 + Bh + C$$

and therefore $y_0 + 4y_1 + y_2 = 2Ah^2 + 6C$.

Thus, the area under the parabola is

$$\frac{h}{3}(y_0 + 4y_1 + y_2).$$

Similarly, the area under the parabola through P_2, P_3, and P_4 from $x = x_2$ to $x = x_4$ is

$$\frac{h}{3}(y_2 + 4y_3 + y_4).$$

Computing the areas under all the parabolas in this manner and adding the results, we obtain

$$\int_a^b f(x)dx \approx \frac{h}{3}(y_0 + 4y_1 + y_2) + \frac{h}{3}(y_2 + 4y_3 + y_4) + \cdots$$
$$+ \frac{h}{3}(y_{n-2} + 4y_{n-1} + y_n)$$
$$= \frac{h}{3}(y_0 + 4y_1 + 2y_2 + 4y_3 + 2y_4 + \cdots + 2y_{n-2} + 4y_{n-1} + y_n)$$

Note the pattern of coefficients in Simpson' rule: 1, 4, 2, 4, 2, 4, 2, $\cdots$, 4, 2, 4, 1.

4 (Composite) Simpson's Rule

$$\int_a^b f(x)dx \approx S_n = \frac{\Delta x}{3}[f(x_0)+4f(x_1)+2f(x_2)+4f(x_3)+\cdots$$
$$+2f(x_{n-2})+4f(x_{n-1})+f(x_n)]$$

where n is even and $\Delta x = (b-a)/n$.

The approximations in Simpson's Rule are weighted averages of those in the Trapezoidal and Midpoint Rules:

$$S_{2n} = \frac{1}{3}T_n + \frac{2}{3}M_n.$$

http://matrix.skku.ac.kr/cal-lab/cal-7-8-Simpson.html
http://sage.luther.edu:8443/home/pub/98/

* Run the following Simpson's Rule code in http://sage2.skku.edu/

```
def integrate(y_vals, h):
    i=1
    total=y_vals[0]+y_vals[-1]
    for y in y_vals[1:-1]:
        if i%2 == 0:
            total+=2*y
        else:
            total+=4*y
        i+=1
    return total*(h/3.0)
y_values=[13, 45.3, 12, 1, 476, 0]
interval=1.2
area=integrate(y_values, interval)
print "The area is", area # The area is 469.680
```

EXAMPLE 4

Approximate $\int_1^2 (1/x^2)dx$ using Simpson's Rule with $n=10$.

Solution Here $f(x)=1/x^2$, $n=10$, and Simpson's Rule, we obtain

$$\int_1^2 \frac{1}{x^2}dx \approx S_{10} = \frac{\Delta x}{3}[f(1)+4f(1.1)+2f(1.2)+4f(1.3)$$
$$+\cdots+2f(1.8)+4f(1.9)+f(2)]$$
$$=\frac{0.1}{3}\left(\frac{1}{1^2}+\frac{4}{(1.1)^2}+\frac{2}{(1.2)^2}+\frac{4}{(1.3)^2}+\frac{2}{(1.4)^2}\right.$$
$$\left.+\frac{4}{(1.5)^2}+\frac{2}{(1.6)^2}+\frac{4}{(1.7)^2}+\frac{2}{(1.8)^2}+\frac{4}{(1.9)^2}+\frac{1}{2^2}\right)$$

$$\approx 0.5000124.$$ ■

Observation In Example 4, observe that Simpson's Rule gives a much better approximation $(S_{10} \approx 0.5000124)$ to the exact value of the integral $(1/2 = 0.5)$ than does the Trapezoidal Rule $(T_{10} \approx 0.501455)$ or the Midpoint Rule $(M_{10} = 0.499273)$.

EXAMPLE 5

Given the values of the function f at the following points,

x	1	1.2	1.4	1.6	1.8
$f(x)$	2.71828	2.30097	2.04272	1.86824	1.74290

approximate $\int_1^{1.8} f(x)dx$ using Simpson's Rule.

Solution With $n = 4$ and $\Delta x = 0.2$ the estimate of the integral by Simpson's Rule is given by

$$\int_1^{1.8} f(x)dx \approx \frac{\Delta x}{3}[f(1) + 4f(1.2) + 2f(1.4) + 4f(1.6) + f(1.8)]$$

$$= \frac{0.2}{3}[2.71828 + 4(2.30097) + 2(2.04272) + 4(1.86824) + 1.7429]$$

$$\approx 1.681564.$$ ■

5 Error Bound for Simpson's Rule

Assume that $|f^{(4)}(x)| \le K$ for $a \le x \le b$. Let E_S be the error involved in using Simpson's Rule, then

$$|E_S| \le \frac{K(b-a)^5}{180n^4}.$$

But it is often very difficult to compute the fourth derivatives and obtain a good upper bound K for $|f^{(4)}(x)|$ by hand. It is recommended to use CAS when it is needed.

EXAMPLE 6

Determine how large n should be so that the Simpson's Rule approximation for $\int_1^2 (1/x^2)dx$ is accurate to within 0.0001?

Solution Here $f(x) = 1/x^2$, then $f^{(4)}(x) = 120/x^6$. Since $x \ge 1$, we have $1/x \le 1$ and so

$$|f^{(4)}(x)| = \left|\frac{120}{x^6}\right| \le 120.$$

So take $K = 120$ in [5]. Thus, for an error less than 0.0001 we should choose n such that

$$\frac{120(1)^5}{180n^4} < 0.0001.$$

This gives $n^4 > \dfrac{120}{180(0.0001)}$ or $n > \dfrac{1}{\sqrt[4]{0.00015}} \approx 9.03.$

Therefore, $n = 10$ (n must be even) gives the desired accuracy. ■

Observation Compare this with Example 2, where we obtained $n = 71$ for the Trapezoidal Rule and $n = 50$ for the Midpoint Rule.

EXAMPLE 7

(a) Approximate the integral $\int_0^1 e^{x^3}dx$ using Simpson's Rule with $n = 10$.

(b) Determine the error involved using this approximation.

Solution (a) If $n = 10$, then $\Delta x = 0.1$ and Simpson's Rule gives

$$\begin{aligned}\int_0^1 e^{x^3}dx &\approx \frac{\Delta x}{3}[f(0) + 4f(0.1) + 2f(0.2) + \cdots + 2f(0.8) \\ &\qquad + 4f(0.9) + f(1)] \\ &= \frac{0.1}{3}[e^0 + 4e^{0.001} + 2e^{0.008} + 4e^{0.027} + 2e^{0.064} \\ &\qquad + 4e^{0.125} + 2e^{0.216} + 4e^{0.343} + 2e^{0.512} + 4e^{0.729} + e^1] \\ &\approx 1.342025.\end{aligned}$$

(b) The fourth derivative of $f(x) = e^{x^3}$ is

$$f^{(4)}(x) = (180x^2 + 324x^5 + 81x^8)e^{x^3}.$$

Since $0 \le x \le 1$, we have

$$0 \le f^{(4)}(x) \le (180 + 324 + 81)e^1 = 585e.$$

Therefore, taking $K = 585e$, $a = 0$, $b = 1$, and $n = 10$ in [4], we observe that the error is at most

$$\frac{585e(1)^5}{180(10)^4} \approx 0.000883.$$

Thus, correct to three decimal places, we have

$$\int_0^1 e^{x^3}dx \approx 1.342.$$

■

CAS **EXAMPLE 8**

Approximate the given integral with the specified n by the Trapezoidal Rule.

$$\int_{-1}^{2} e^{-x^2}dx,\ n=5.$$

Solution With $n=5$, $a=-1$ and $b=2$, we have $\Delta x=(2-(-1))/5=0.6$, and so the Trapezoidal Rule gives

$$\begin{aligned}\int_{-1}^{2} e^{-x^2}dx \approx T_5 &= \frac{0.6}{2}[f(-1)+2f(-0.4)+2f(0.2)\\ &\quad +2f(0.8)+2f(1.4)+f(2)]\\ &=0.3\left(e^{-1}+2e^{-0.4^2}+2e^{-0.2^2}+2e^{-0.8^2}+2e^{-1.4^2}+e^{-2^2}\right)\\ &\approx 1.60450.\end{aligned}$$

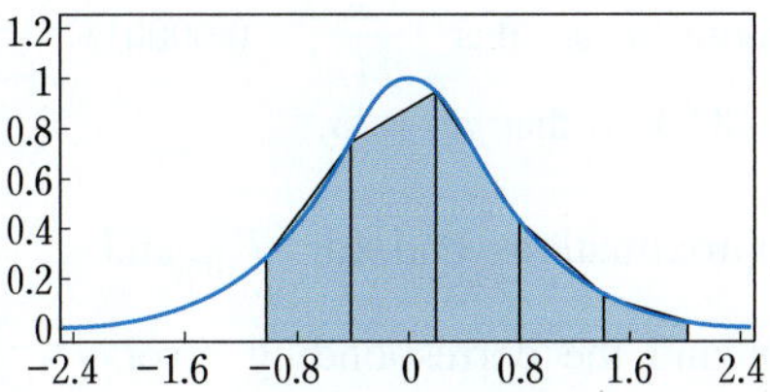

Figure 8

```
0.3*(exp(-1)+2*exp(-0.4^2)+2*exp(-0.2^2)+2*exp(-0.8^2)+2*ex
p(-1.4^2)+exp(-2^2)).n()
```

Answer : 1.60450896786763

http://matrix.skku.ac.kr/cal-lab/cal-7-7-Exm-8.html

```
var('x')
f=exp(-x^2)
integral(f, x, -1, 2).n()
```

Answer : 1.62890552357485

We see some differences in these results. ■

We may exercise numerical integrals with various rules in

http://matrix.skku.ac.kr/cal-lab/cal-Numerical-IntegrationMethods.html
http://matrix.skku.ac.kr/cal-lab/cal-Midpoint.html

Here procedures for left end point rule, right end point rule, mid point rule, trapezoidal rule and Simpson's rule are given.

Remark One can increase degree of the polynomial of approximation to get even better approximation rules. Approximation using 3rd degree polynomials is called Simpson's $\frac{3}{8}$ Rule for Numerical Integration[1]. The $n=5$ gives Boole's rule of integration and for $n=7$ gives called Weddle's rule of integration[2].

1) http://mathworld.wolfram.com/Simpsons38Rule.html
2) http://mathworld.wolfram.com/WeddlesRule.html

7.7 EXERCISES (Approximate Integration and CAS)

http://matrix.skku.ac.kr/Cal-Book/part1/CS-Sec-7-7-Sol.html

YouTube http://youtu.be/dfSkjvmSXYo

1-6. Find the integral using the midpoint, trapezoidal and Simpson's rule for the given n.

1. $\int_0^1 \sin x^3 dx,\ n=8$

http://matrix.skku.ac.kr/cal-lab/cal-Midpoint.html

http://matrix.skku.ac.kr/cal-lab/cal-7-8-Simpson.html

Solution Use sin(x^3) in http://matrix.skku.ac.kr/cal-lab/Area-Sum.html with midpoint and trapezoidal rules.

(a) 0.232771 with midpoint rule

(b) 0.2359771 with trapezoidal rule

(c) 0.233760

2. $\int_0^{\pi/4} e^{3x}\sin 2x\,dx,\ n=8\ (\approx 2.58862863250718)$

Solution Use e^(3*x)*sin(2*x) in http://matrix.skku.ac.kr/cal-lab/Area-Sum.html with midpoint and trapezoidal rules.

(a) 2.57670 with midpoint rule

(b) 2.612462 with trapezoidal rule

<Error = 0.0238343387259241>

(c) 2.588559

3. $\int_0^1 \frac{1}{1+x^8}dx,\ n=6$

Solution (a) 0.919952 (b) 0.927027 (c) 0.925237

4. $\int_0^{\pi/4} \ln(1+\tan x)dx,\ n=8$

Solution (a) 0.272198 (b) 0.272198 (c) 0.272198

5. $\int_0^1 \frac{x}{(1+x^5)^{2/5}}dx,\ n=10$

Solution (a) 0.457277 (b) 0.458528 (c) 0.458114

6. $\int_0^1 e^{x^7}dx,\ n=10$

Solution (a) 1.182973 (b) 1.160116 (c) 1.169130

---

7. (a) Determine the approximations T_{10} and M_{10} for $\int_0^1 \sin x^3 dx$.

(b) Find the errors involved in the approximations of part (a).

(c) Determine how large must n be so that the approximations T_n and M_n to the integral in part (a) are accurate to within 0.00001?

Solution (a) $T_{10}=0.235204,\ M_{10}=0.233162$

(b) $E_T=-0.001359,\ E_M=0.000683$

(c) $|f''(x)|\le|-9\sin(1)+6\cos(1)|=4.33142=K$

For T_n, we must choose n so that $\frac{K}{12n^2}<0.00001$ solving this, $n>189.9$ so that $n=190$.

For M_n, we must choose n so that $\frac{K}{24n^2}<0.00001$ solving this gives, $n>134.3$ so that $n=135$.

8. (a) Determine the approximations and for T_{10} and S_{10} for $\int_0^1 e^{x^4}dx$ and the corresponding errors E_T and E_S.

(b) Compare the actual errors in part (a) with the error estimates given by 3 and 4.

(c) Determine how large must we choose n so that the approximations T_n, M_n, and S_n for the integral in part (a) are accurate to within 0.00001?

Solution (a) $T_{10}=1.280262,\ S_{10}=1.271599,$
$E_T=-0.008975,\ E_S=-0.000312$

(b) Since $K=28e\approx 76.111891$ 3 and 4 gives

$$|E_T|\le\frac{76.111891}{12(10)^2}\approx 0.063426,$$

$$|E_S|\le\frac{76.111891}{180(10)^4}\approx 0.000042.$$

(c) For T_n, find n so that $\frac{K}{12n^2}<0.00001,\ n=797.$

For M_n, find n so that $\frac{K}{24n^2}<0.00001,\ n=564.$

For S_n, find n so that $\frac{K}{180n^4}<0.00001,\ n=15.$

9. Given the function f at the following values, approximate $\int_{1.8}^{2.4} f(x)dx$ using Simpson's Rule.

x	1.8	1.9	2.0	2.1	2.2	2.3	2.4
$f(x)$	0.028561	0.020813	0.015384	0.011525	0.008742	0.006709	0.004079

Solution Plot $f(x)$ and use Simpson's Rule

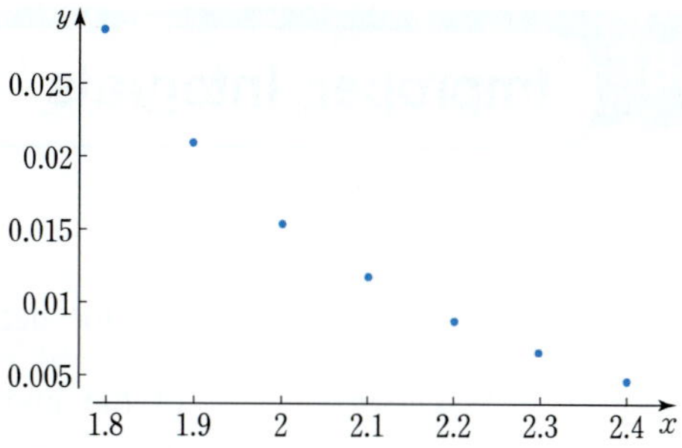

$\Delta x = (2.4 - 1.8)/6 = 0.1$

$$\Rightarrow \int_{1.8}^{2.4} f(x)dx \approx S_6$$

$$= \frac{0.1}{3}[f(1.8) + 4f(1.9) + 2f(2.0) + 4f(2.1) + 2f(2.2) + 4f(2.3) + f(2.4)]$$

$$\approx 0.007923.$$

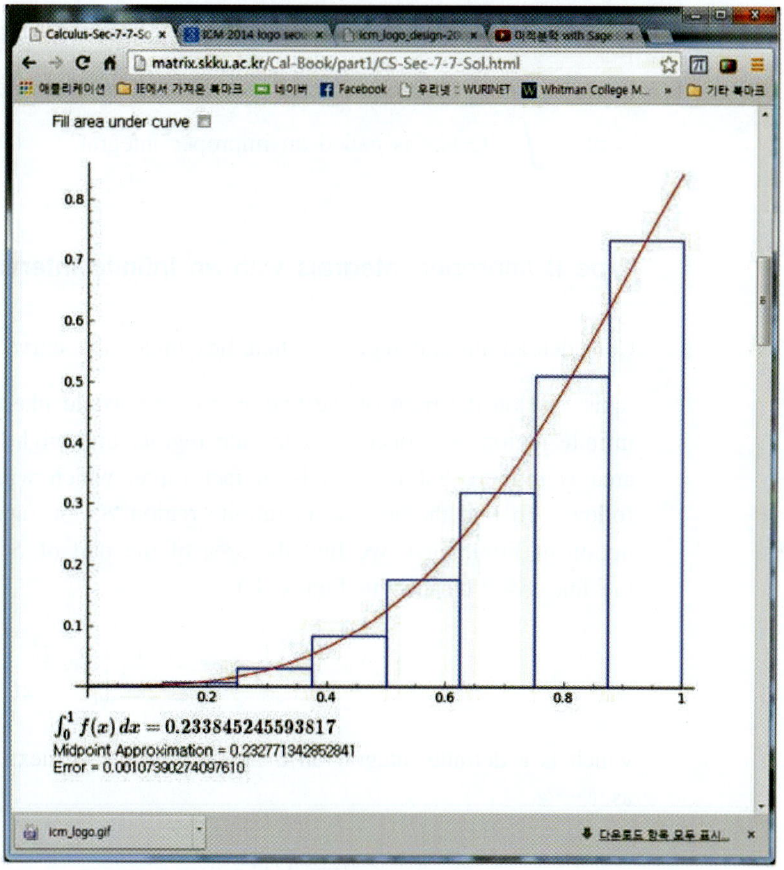

<integration tool>

7.8 Improper Integrals

You Tube http://youtu.be/rquxbYrC0Yc

For the definite integral, $\int_a^b f(x)dx$, (so far) we have assumed that the length of the integration interval $[a, b]$ is finite and that the range of the integrand $f(x)$ remains finite on this finite interval. Sometimes, however, we may be curious about what happens if either one or both of these assumptions is not satisfied. In fact, these situations occur frequently in applied science fields and probability theory. This question requires a new interpretation to the symbol $\int_a^b f(x)dx$. If either or both the limits of integration becomes infinite, or if the function $f(x)$ becomes infinite somewhere in the interval $[a, b]$, then the symbol $\int_a^b f(x)dx$ is called an improper integral.

Type I: Improper Integrals with an Infinite Interval

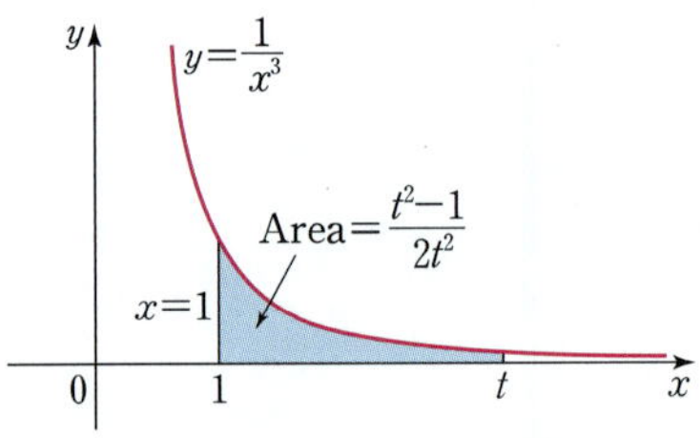

Figure 1

Consider an infinite region S that lies under the curve $y=\frac{1}{x^3}$ above the x-axis, and on the right of the line $x=1$. We would like to find the area of this infinite region S. Since S is infinite region, one might be tempted to say the area is infinite, but its area is in fact finite, which will be justified by what follows. To find the area of an infinite region S, our approach is based on the notion of limit. First, we find the area of the part of S that lies to the left of the line $x=t$ (shaded in Figure 1.):

$$A(t)=\int_1^t \frac{1}{x^3}dx=-\frac{1}{2x^2}\Big]_1^t=\frac{t^2-1}{2t^2},$$

which is a definite integral on a finite interval. We next find the limit of $A(t)$ as $t\to\infty$

$$\lim_{t\to\infty}A(t)=\lim_{t\to\infty}\frac{t^2-1}{2t^2}=\frac{1}{2}.$$

Since the limit exists, it is natural to assign this value to the area of the infinite region S and therefore we write

$$\text{Area of } S=\int_1^\infty \frac{1}{x^3}dx=\lim_{t\to\infty}\int_1^t \frac{1}{x^3}dx=\frac{1}{2}.$$

The discussion given above can be extended to any function $f(x)$. Therefore, we make the following definition.

DEFINITION 1 Definition of an Improper Integral of Type 1

(a) The symbol $\int_a^{\infty} f(x)dx$ is defined by the equation

$$\int_a^{\infty} f(x)dx = \lim_{t\to\infty}\int_a^t f(x)dx.$$

(b) The symbol $\int_{-\infty}^{b} f(x)dx$ is defined by the equation

$$\int_{-\infty}^{b} f(x)dx = \lim_{t\to-\infty}\int_t^b f(x)dx.$$

The improper integrals $\int_a^{\infty} f(x)dx$ and $\int_{-\infty}^{b} f(x)dx$ are called **convergent** if the corresponding limit exists; otherwise they are called **divergent**.

(c) If both $\int_a^{\infty} f(x)dx$ and $\int_{-\infty}^{a} f(x)dx$ are convergent, then we can define

$$\int_{-\infty}^{\infty} f(x)dx = \int_{-\infty}^{a} f(x)dx + \int_a^{\infty} f(x)dx.$$

EXAMPLE 1

Is the integral $\int_0^{\infty} \sin x\,dx$ convergent?

http://matrix.skku.ac.kr/cal-lab/cal-7-8-Exm-1.html

Solution We know that

$$\int_0^{\infty} \sin x\,dx = \lim_{t\to\infty}\int_0^t \sin x\,dx = \lim_{t\to\infty}[-\cos x]_0^t$$
$$= \lim_{t\to\infty}(1-\cos t).$$

```
p1=plot(1-cos(x), x, -10,10, color='blue')
print limit(1-cos(x), x=00)
show(p1, ymax=3, ymin=-0.5, aspect_ratio=1)
```

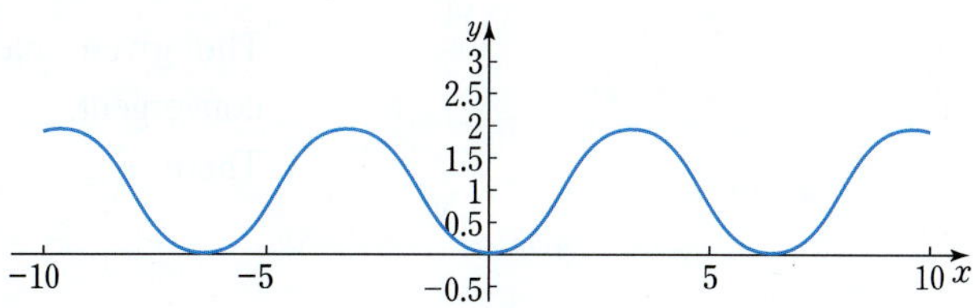

Figure 2

The improper integral $\int_0^{\infty} \sin x\,dx$ is divergent since the limit does not exist. ■

EXAMPLE 2

Find $\int_{-\infty}^{-1} xe^{-x^2}dx$.

Solution

$$\begin{aligned}\int_{-\infty}^{-1} xe^{-x^2}dx &= \lim_{t\to-\infty}\int_{t}^{-1} xe^{-x^2}dx \\ &= \lim_{t\to-\infty}\left[-\frac{1}{2}e^{-x^2}\right]_t^{-1} \\ &= \lim_{t\to-\infty}\left[-\frac{1}{2}e^{-1}+\frac{1}{2}e^{-t^2}\right]_t^{-1} \\ &= -\frac{1}{2e}\end{aligned}$$

■

EXAMPLE 3

Find $\int_{-\infty}^{\infty}\frac{dx}{(1+x^2)^{3/2}}$.

Solution We know that

$$\int_{-\infty}^{\infty}\frac{dx}{(1+x^2)^{3/2}}=\int_{-\infty}^{0}\frac{dx}{(1+x^2)^{3/2}}+\int_{0}^{\infty}\frac{dx}{(1+x^2)^{3/2}}.$$

Evaluating the integrals on the right-hand side separately:

$$\begin{aligned}\int_{0}^{\infty}\frac{dx}{(1+x^2)^{3/2}} &= \lim_{t\to\infty}\int_{0}^{t}\frac{dx}{(1+x^2)^{3/2}}=\lim_{t\to\infty}\left[\frac{x}{\sqrt{1+x^2}}\right]_0^t \\ &= \lim_{t\to\infty}\frac{t}{\sqrt{1+t^2}}=1\end{aligned}$$

and

$$\begin{aligned}\int_{-\infty}^{0}\frac{dx}{(1+x^2)^{3/2}} &= \lim_{t\to-\infty}\int_{t}^{0}\frac{dx}{(1+x^2)^{3/2}} \\ &= \lim_{t\to-\infty}\frac{x}{\sqrt{1+x^2}}\Bigg]_t^0=\lim_{t\to-\infty}\left(-\frac{t}{\sqrt{1+t^2}}\right)=1.\end{aligned}$$

The given integral is convergent since both of these integrals are convergent.
Therefore,

$$\int_{-\infty}^{\infty}\frac{dx}{(1+x^2)^{3/2}}=1+1=2.$$

■

EXAMPLE 4

Determine the values of p for which the integral $\int_1^\infty \frac{1}{x^p}dx$ is convergent.

Solution When $p=1$,

$$\int_1^\infty \frac{1}{x}dx = \lim_{t\to\infty}\int_1^t \frac{1}{x}dx = \lim_{t\to\infty}[\ln x]_1^t = \lim_{t\to\infty}(\ln t - \ln 1)$$

$$= \lim_{t\to\infty}\ln t = \infty.$$

When $p \neq 1$,

$$\int_1^\infty \frac{1}{x^p}dx = \lim_{t\to\infty}\int_1^t \frac{1}{x^p}dx = \lim_{t\to\infty}\left[\frac{1}{1-p}x^{1-p}\right]_1^t$$

$$= \lim_{t\to\infty}\frac{1}{1-p}(t^{1-p}-1)$$

If $p>1$, then $1-p<0$, so as $t\to\infty$ and $t^{1-p}\to 0$. Therefore

$$\int_1^\infty \frac{1}{x^p}dx = \frac{1}{p-1} \quad \text{if } p>1,$$

and so the integral converges.
But if $p<1$, then $1-p>0$ so as $t\to\infty$ and $t^{1-p}\to\infty$. Therefore

$$\int_1^\infty \frac{1}{x^p}dx = \infty \quad \text{if } p<1$$

and the integral diverges. Thus

$$\int_1^\infty \frac{1}{x^p}dx = \begin{cases} \dfrac{1}{p-1} & \text{convergent} \quad \text{if } p>1 \\ \infty & \text{divergent} \quad \text{if } p\le 1 \end{cases}.$$ ■

As a consequence of Example 4, we know

$$\int_1^\infty \frac{1}{x^2}dx \text{ converges, but } \int_1^\infty \frac{1}{x}dx \text{ diverges.}$$

Type 2: Improper Integrals for which the Integrand becomes Infinite

Consider the integral $\int_1^2 \frac{x}{\sqrt{x-1}}dx$. Observe that this is not a definite integral since the integrand is not continuous at $x=1$ and becomes infinite as x

approaches 1. Yet we can find the area under the graph of the integrand on the interval $[1, 2]$. How we do this is presented in the following definition.

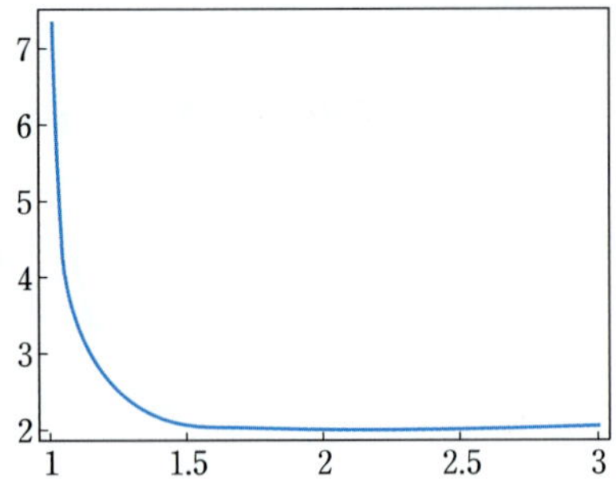

Figure 3 $f(x) = \dfrac{x}{\sqrt{x-1}}$

2 Definition of an Improper Integral of Type 2

(a) If $f(x)$ becomes infinite at $x = b$, then we define (See Figure 4.)

$$\int_a^b f(x)dx = \lim_{t \to b^-} \int_a^t f(x)dx.$$

(b) If $f(x)$ becomes infinite at $x = a$, then (See Figure 5.)

$$\int_a^b f(x)dx = \lim_{t \to a^+} \int_t^b f(x)dx.$$

The improper integral $\int_a^b f(x)dx$ is called **convergent** if the corresponding limit exists and **divergent** if the limit does not exist.

(c) If $f(x)$ becomes infinite for some value of $x = c$, where $a < c < b$, and both $\int_a^c f(x)dx$ and $\int_c^b f(x)dx$ are convergent, then we define (See Figure 6.)

$$\int_a^b f(x)dx = \int_a^c f(x)dx + \int_c^b f(x)dx$$

$$= \lim_{t \to c^-} \int_a^t f(x)dx + \lim_{t \to c^+} \int_t^b f(x)dx.$$

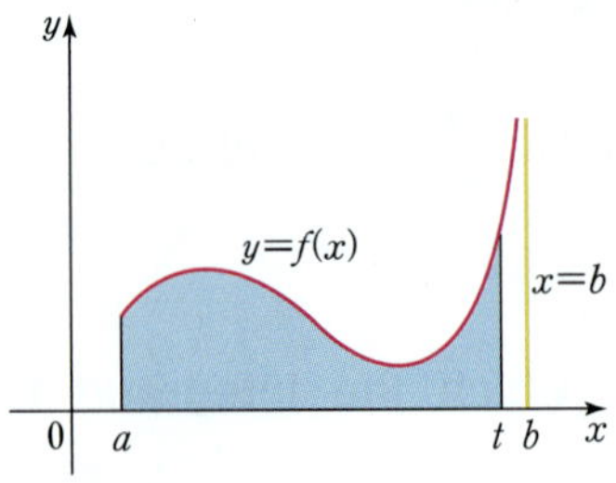

Figure 4

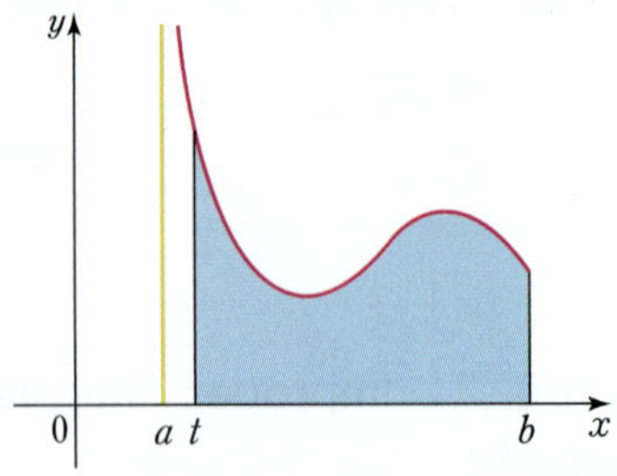

Figure 5

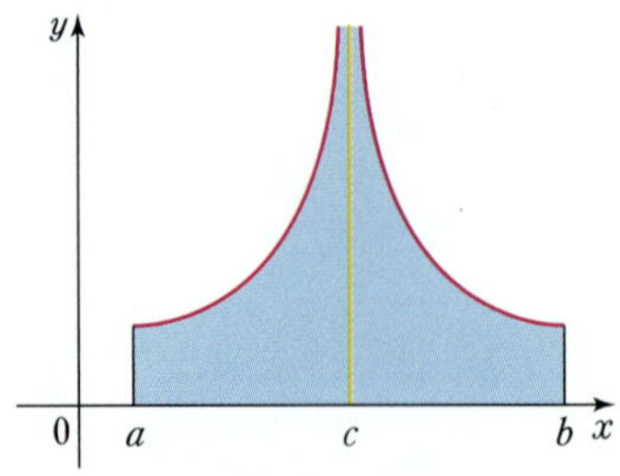

Figure 6

EXAMPLE 5

Evaluate $\displaystyle\int_1^2 \frac{x}{\sqrt{x-1}}\,dx$.

Solution Since $f(x)=\dfrac{x}{\sqrt{x-1}}$ has a vertical asymptote at $x=1$, the given integral is improper. We use part (b) of Definition 1 since the infinite discontinuity occurs at $x=1$. We have

$$\begin{aligned}\int_1^2 \frac{x}{\sqrt{x-1}}\,dx &= \lim_{t\to 1^+}\int_t^2 \frac{x}{\sqrt{x-1}}\,dx \\ &= \lim_{t\to 1^+}\left[2\sqrt{x-1}+\frac{2}{3}\sqrt{(x+1)^3}\right]_t^2 \\ &= \lim_{t\to 1^+}\left(\frac{8}{3}-2\sqrt{t-1}-\frac{2}{3}\sqrt{(t-1)^3}\right) \\ &= \frac{8}{3}.\end{aligned}$$

Thus, the given improper integral converges and the value is $\dfrac{8}{3}$. ■

EXAMPLE 6

Is $\displaystyle\int_0^2 \frac{dx}{\sqrt{4-x^2}}$ convergent or divergent?

Solution Since $\displaystyle\lim_{t\to 2^-}\frac{1}{\sqrt{4-x^2}}=\infty$, then the given integral is improper. But

$$\begin{aligned}\int_0^2 \frac{dx}{\sqrt{4-x^2}} &= \lim_{t\to 2^-}\int_0^t \frac{dx}{\sqrt{4-x^2}} \\ &= \lim_{t\to 2^-}\left[\sin^{-1}\left(\frac{x}{2}\right)\right]_0^t \\ &= \frac{\pi}{2}.\end{aligned}$$

Thus, the given improper integral is convergent. ■

EXAMPLE 7

Find $\displaystyle\int_0^3 \frac{1}{(x-1)^{2/3}}\,dx$. (See Figure 7.)

http://matrix.skku.ac.kr/cal-lab/cal-7-8-Exm-7.html

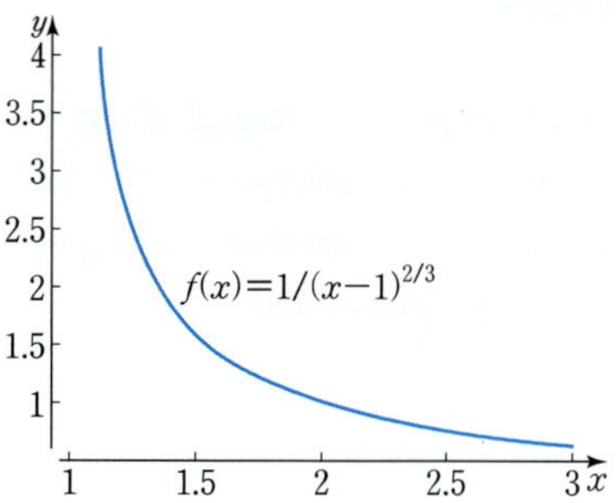

Figure 7

Solution The graph of the integrand has a vertical asymptote at $x=1$, which lies in the middle of the interval $[0,\ 3]$. Then

$$
\begin{aligned}
\int_0^3 \frac{dx}{(x-1)^{2/3}} &= \int_0^1 \frac{dx}{(x-1)^{2/3}} + \int_1^3 \frac{dx}{(x-1)^{2/3}} \\
&= \lim_{t\to1^-}\int_0^t \frac{dx}{(x-1)^{2/3}} + \lim_{s\to1^+}\int_s^3 \frac{dx}{(x-1)^{2/3}} \\
&= \lim_{t\to1^-}[3(x-1)^{1/3}]_0^t + \lim_{s\to1^+}[3(x-1)^{1/3}]_s^3 \\
&= 3\lim_{t\to1^-}[(t-1)^{1/3}+1] + 3\lim_{s\to1^+}[2^{1/3}-(s-1)^{1/3}] \\
&= 3+3\sqrt[3]{2}.
\end{aligned}
$$

■

EXAMPLE 8

Find $\int_0^1 (\ln x)^2 dx$. (See Figure 8.)

http://matrix.skku.ac.kr/cal-lab/cal-7-8-Exm-8.html

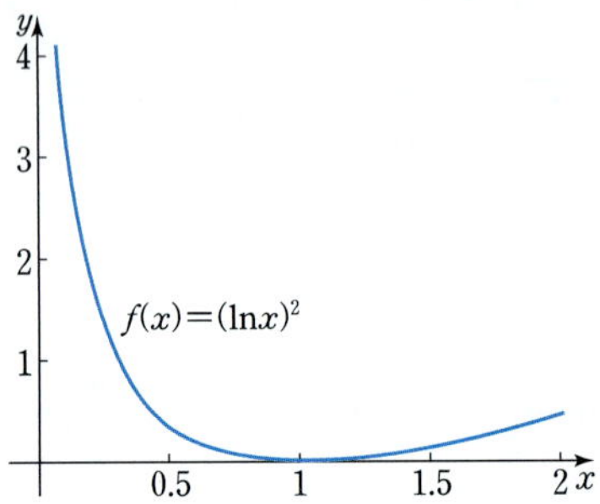

Figure 8

Solution Since $\lim_{x\to0^+}(\ln x)^2 = -\infty$, the function $f(x)=\ln x^2$ has a vertical asymptote at 0. Thus, $\int_0^1 (\ln x)^2 dx = \lim_{t\to0^+}\int_t^1 (\ln x)^2 dx$.

Integrating by parts gives

$$
\begin{aligned}
\int_t^1 (\ln x)^2 dx &= [x(\ln x)^2]_t^1 - 2\int_t^1 \ln x\, dx \\
&= -t(\ln t)^2 - 2[-t\ln t - 1 + t] \\
&= -t(\ln t)^2 + 2t\ln t + 2 - 2t.
\end{aligned}
$$

Using L'Hospital's Rule, we obtain $\lim_{t\to0^+} t(\ln t)^2 = \lim_{t\to0^+} t\ln t = 0$.

Therefore,

$$
\begin{aligned}
\int_0^1 (\ln x)^2 dx &= \lim_{t\to0^+}[-t(\ln t)^2 + 2t\ln t + 2 - 2t] \\
&= 0+0+2-0 = 2.
\end{aligned}
$$

■

A Comparison Test for Improper Integrals

Often it is not possible to find the exact value of an improper integral directly from the definition. Even in this case, the convergence or divergence of the improper integral may be determined by comparing with another improper integral whose convergence or divergence can be easily determined.

> **THEOREM 2 Comparison Theorem**
>
> Assume that f and g are continuous functions with $f(x) \geq g(x) \geq 0$ for $x \geq a$. Then
>
> (a) if $\int_a^{\infty} f(x)dx$ is convergent, then $\int_a^{\infty} g(x)dx$ is convergent.
>
> (b) if $\int_a^{\infty} g(x)dx$ is divergent, then $\int_a^{\infty} f(x)dx$ is divergent.

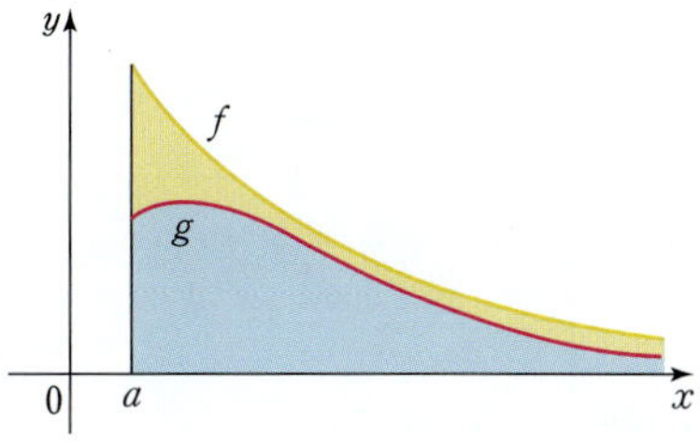

Figure 9

The Comparison Theorem makes sense if we interpret it in terms of area. If the area under the top curve $y=f(x)$ is finite, then so is the area under the bottom curve $y=g(x)$. Similarly if the area under $y=g(x)$ is infinite, then so is the area under $y=f(x)$ (see Figure 9). The integral $\int_1^{\infty} \frac{1+e^{-x^2}}{x^{1/2}}dx$ is divergent by the Comparison Theorem because $\frac{1+e^{-x^2}}{x^{1/2}} > \frac{1}{x^{1/2}}$ and $\int_1^{\infty} \frac{1}{x^{1/2}}dx$ is divergent by Example 4. However, the converse of the Comparison Theorem is not necessarily true: If $\int_a^{\infty} g(x)dx$ is convergent, $\int_a^{\infty} f(x)dx$ may or may not be convergent, and if $\int_a^{\infty} f(x)dx$ is divergent, $\int_a^{\infty} g(x)dx$ may or may not be divergent.

EXAMPLE 9

Prove that $\int_0^{\infty} \frac{e^{-x}}{x^{1/3}}dx$ is convergent.

Solution Because the integrand is unbounded near $x=0$ we decompose the integral into two parts:

$$\int_0^{\infty} \frac{e^{-x}}{x^{1/3}}dx = \int_0^{1} \frac{e^{-x}}{x^{1/3}}dx + \int_1^{\infty} \frac{e^{-x}}{x^{1/3}}dx.$$

In the first integral we have $e^{-x}/x^{1/3} \leq 1/x^{1/3}$, and the integral converges by the Comparison Theorem. In the second integral we have $e^{-x}/x^{1/3} \leq e^{-x}$, and once again the Comparison Theorem yields the result since $\int_1^{\infty} e^{-x}dx$ is convergent. Therefore, $\int_0^{\infty} \frac{e^{-x}}{x^{1/3}}dx$ is convergent. ■

CAS **EXAMPLE 10**

Determine whether the integral is convergent or divergent. Evaluate the integral if it is convergent.

$$\int_{-\infty}^{0} e^{3x} \sin 2x \, dx.$$

Solution

```
var('x')
f=e^(x*3)*sin(2*x)
plot(f,(x, -5, 0))
```

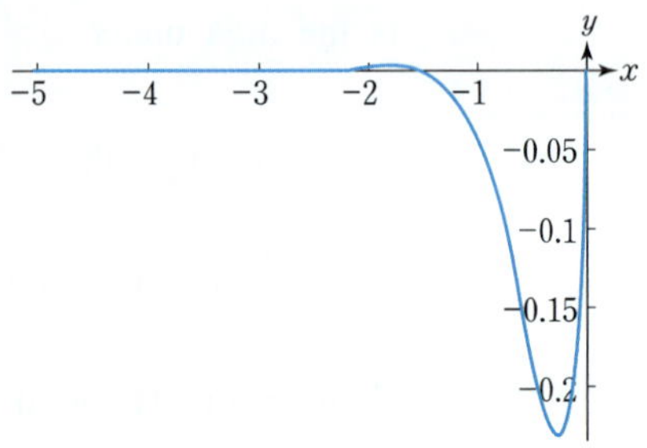

Figure 10

```
integrate(f, x, -infinity, 0)
```

Answer : -2/13 ■

7.8 EXERCISES (Improper Integrals)

http://matrix.skku.ac.kr/Cal-Book/part1/CS-Sec-7-8-Sol.html

YouTube http://youtu.be/rquxbYrC0Yc

1. $\displaystyle\int_e^{\infty} \frac{\ln x}{x} dx$

http://matrix.skku.ac.kr/cal-lab/cal-7-8-1.html

Solution $\displaystyle\lim_{t\to\infty}\int_e^t \frac{\ln x}{x}\, dx = \lim_{t\to\infty}\left(\frac{1}{2}(\ln t)^2 - \frac{1}{2}\right) = \infty.$

```
integral(ln(x)/x, x, e, infinity)
```

ValueError : Integral is divergent

http://sage2.skku.edu/?q=6bee9254-c31b-467b-a792-30317be25c93&lang=sage

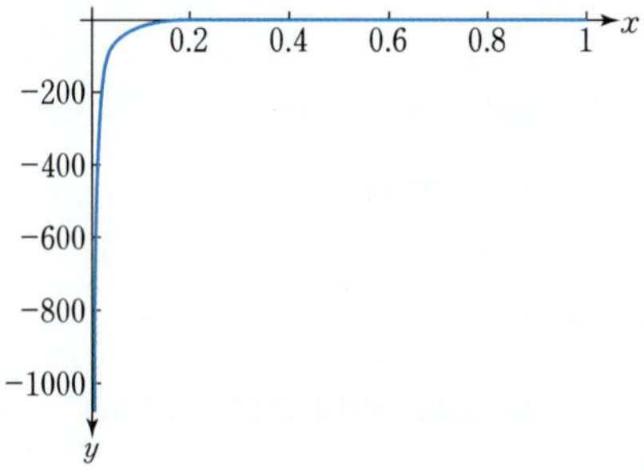

2. $\displaystyle\int_0^{\infty} xe^{-x} dx$

http://matrix.skku.ac.kr/cal-lab/cal-7-8-2.html

Solution

```
integral(x*e^(-x), x, 0, infinity)
```

http://sage2.skku.edu/?q=962bf6d6-05ca-4d1a-bb54-12c9e12b89d7&lang=sage

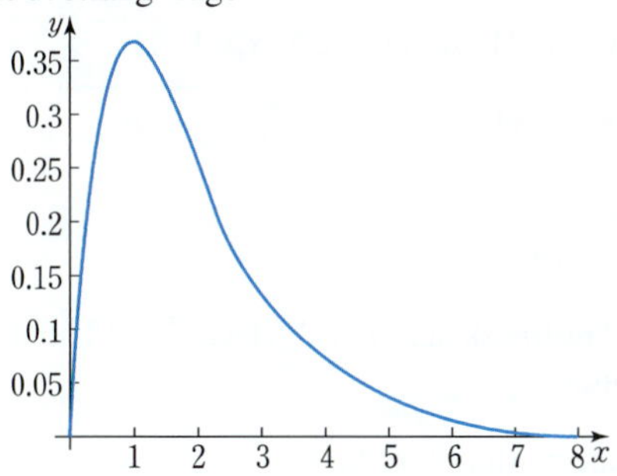

3. $\displaystyle\int_{-\infty}^{\infty} \frac{1}{1+x^2} dx$

http://matrix.skku.ac.kr/cal-lab/cal-7-8-3.html

Solution $\displaystyle\int_{-\infty}^{\infty} \frac{1}{1+x^2}\, dx = \int_0^{\infty} \frac{1}{1+x^2}\, dx + \int_{-\infty}^{0} \frac{1}{1+x^2}\, dx$

$$= \lim_{b\to\infty}\int_0^b \frac{1}{1+x^2}\, dx + \lim_{a\to-\infty}\int_a^0 \frac{1}{1+x^2}\, dx$$

$$= \lim_{b\to\infty}[\tan^{-1}x]_0^b + \lim_{a\to-\infty}[\tan^{-1}x]_a^0$$

$$= \lim_{b\to\infty}[\tan^{-1}b - \tan^{-1}0] + \lim_{a\to-\infty}[\tan^{-1}0 - \tan^{-1}a]$$

$$= \frac{\pi}{2} - \left(-\frac{\pi}{2}\right) = \pi.$$

```
integral(1/(x^2+1), x, -infinity, infinity)
```

Answer : pi

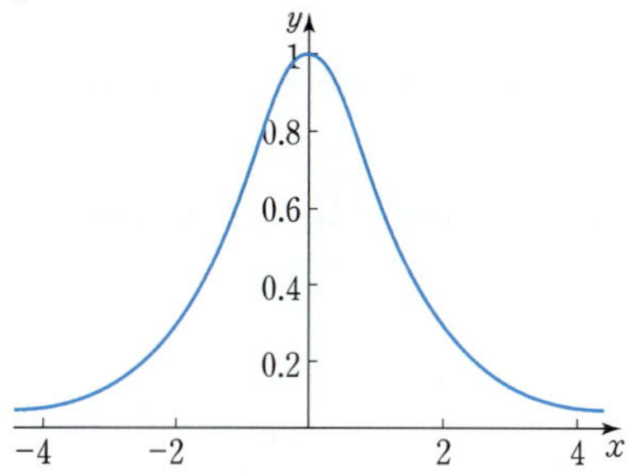

4. $\displaystyle\int_0^{\pi/2} \sec x\, dx$

http://matrix.skku.ac.kr/cal-lab/cal-7-8-4.html

Solution

```
integral(sec(x), x, 0, pi/2)
```

ValueError : Integral is divergent.

http://sage2.skku.edu/?q=302f1d18-5504-4b3a-9025-5993aff0a85c&lang=sage

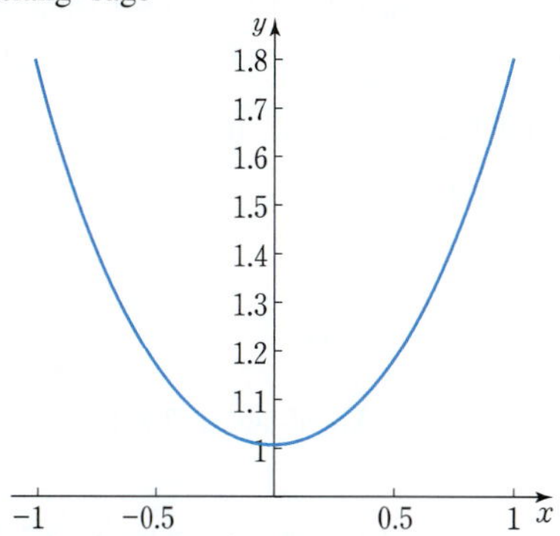

5. $\displaystyle\int_2^{\infty} \frac{1}{x(\ln x)^2} dx$

http://matrix.skku.ac.kr/cal-lab/cal-7-8-6.html

Solution

```
integral(1/(x*(ln(x))^2), x, 2, infinity)
```

Answer : 1/log(2)

http://sage2.skku.edu/?q=fdf0b67d-2fbc-457b-9cfa-b24c80fc42a3&lang=sage

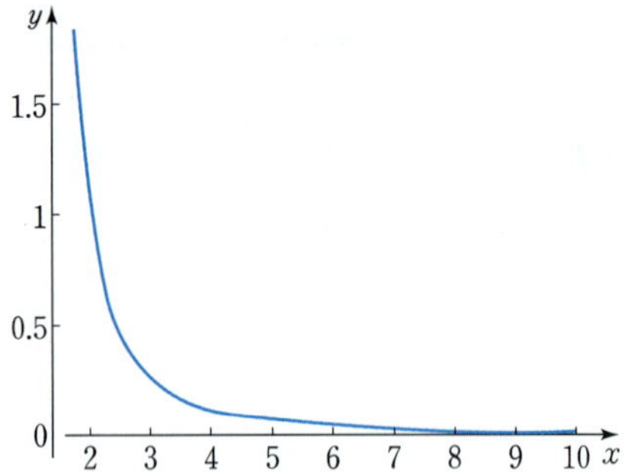

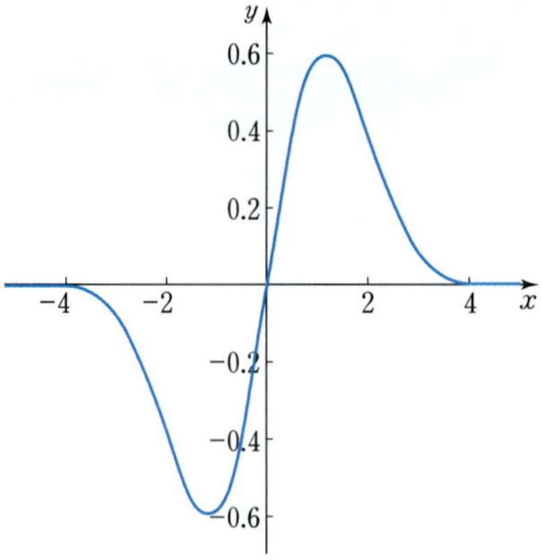

6. $\displaystyle\int_0^{\infty} e^{-x}\sin x\,dx$

http://matrix.skku.ac.kr/cal-lab/cal-7-8-7.html

Solution

```
integral(1/(x*(ln(x))^2), x, 2, infinity)
```

Answer : 1/2

7. $\displaystyle\int_0^{\infty} e^{-x}x^{n-1}dx = \Gamma(n) = (n-1)!$

http://matrix.skku.ac.kr/cal-lab/cal-7-8-9.html

Solution

```
var('x, n');
integral(e^(-x)*x^(n-1), x)
```

Answer : -gamma(n, x)

8. $\displaystyle\int_{-\infty}^{0} \frac{1}{(2x-1)^3}dx$

http://matrix.skku.ac.kr/cal-lab/cal-7-8-10.html

Solution

```
integral(1/(2*x-1)^3, x, -infinity, 0)
```

Answer : -1/4

9. $\displaystyle\frac{1}{\sqrt{2\pi}}\int_{-\infty}^{\infty} xe^{-\frac{x^2}{2}}dx$

http://matrix.skku.ac.kr/cal-lab/cal-7-8-11.html

Solution

```
integral((e^(-(x^2)/2))*x, x, -infinity, infinity)
```

Answer : 0

http://sage2.skku.edu/?q=3a3a14c1-be3d-4039-a1ff-458cadf8e86b&lang=sage

10. $\displaystyle\int_0^{1} x\sin x\,dx$

http://matrix.skku.ac.kr/cal-lab/cal-7-8-5.html

Solution

```
integral(x*sin(x), x, 0, 1)
```

Answer : sin(1) - cos(1)

11. $\displaystyle\int_0^{1} \frac{1}{\sqrt{x}}dx$

http://matrix.skku.ac.kr/cal-lab/cal-7-8-8.html

Solution

```
integral(1/sqrt(x), x, 0, 1)
```

Answer : 2

12. $\displaystyle\int_{-1}^{1} \frac{1}{\sqrt{1-x^2}}dx$

http://matrix.skku.ac.kr/cal-lab/cal-7-8-12.html

Solution

```
integral(1/(sqrt(1-x^2)), x, -1, 1)
```

Answer : pi

13. $\displaystyle\int_0^{1} \ln x\,dx$

http://matrix.skku.ac.kr/cal-lab/cal-7-8-13.html

Solution

```
integral(ln(x), x, 0, 1)
```

Answer : -1

14. $\displaystyle\int_1^{\infty} \frac{1}{x\sqrt{x^2-1}}dx$

http://matrix.skku.ac.kr/cal-lab/cal-7-8-14.html

Solution

```
integral(1/(x*(sqrt(x^2-1))), x, 1, infinity)
```

Answer : 1/2*pi

http://sage2.skku.edu/?q=0b2091ea-3973-4e61-ad0c-fb90ab700761&lang=sage

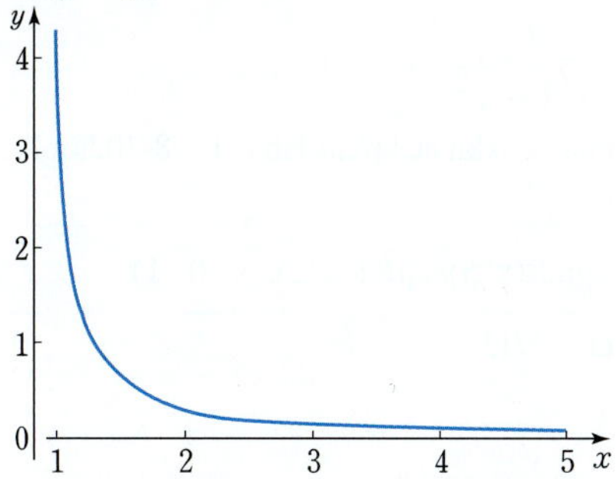

15. $\displaystyle\int_1^\infty \frac{1}{x(x^2+1)}dx$

http://matrix.skku.ac.kr/cal-lab/cal-7-8-17.html

Solution

```
integral(1/(x*(x^2+1)), x, 1, infinity)
```

Answer : 1/2*log(2)

16. $\displaystyle\int_1^\infty \frac{1}{x\sqrt{x^2-1}}dx$

http://matrix.skku.ac.kr/cal-lab/cal-7-8-21.html

Solution

```
integral(1/(x*sqrt(x^2-1)), x, 1, infinity)
```

Answer : 1/2*pi

17. $\displaystyle\int_0^\infty (\sqrt{x^2+1}-x)^2 dx$

http://matrix.skku.ac.kr/cal-lab/cal-7-8-23.html

Solution

```
integral((sqrt(x^2+1)-x)^2, x, 0, infinity)
```

Answer : 2/3

18. $\displaystyle\int_0^\infty \frac{1+x}{1+x^2}dx$

http://matrix.skku.ac.kr/cal-lab/cal-7-8-24.html

Solution

```
integral((1+x)/(1+x^2), x, 0, infinity)
```

Answer : ValueError : Integral is divergent

19. $\displaystyle\int_0^\infty x^{2n-1}e^{-x^2}dx$

http://matrix.skku.ac.kr/cal-lab/cal-7-8-27.html

Solution $\displaystyle\int_0^\infty x^{2n-1}e^{-x^2}dx = \frac{1}{2}(n-1)!.$

(n: positive integer)

```
var('x, n');
integral(x^(2*n-1)*e^(-x^2), x)
```

Answer : -1/2*x^(2*n)*gamma(n, x^2)/(x^2)^n

20. $\displaystyle\int_a^b \frac{1}{\sqrt{(a-x)(b-x)}}dx \ (a<b)$

Solution $a< b \Rightarrow \sqrt{a} < \sqrt{b}$

$$\int_a^b \frac{dx}{\sqrt{(a-x)(b-x)}} < \int_a^b \frac{dx}{\sqrt{(a-x)(a-x)}}$$
$$= \int_a^b \frac{dx}{\sqrt{(a-x)^2}}$$
$$= \int_a^b \frac{dx}{a-x}$$
$$= [\ln|a-x|]_a^b$$

convergent.

21. $\displaystyle\int_0^{\frac{\pi}{4}} \frac{1}{\sin x\cos x}dx$

http://matrix.skku.ac.kr/cal-lab/cal-7-8-16.html

Solution

```
integral(1/(sin(x)*cos(x)), x, 0, pi/4)
```

Answer : ValueError: Integral is divergent

22. $\displaystyle\int_0^{16} \frac{1}{\sqrt[4]{x}}dx$

http://matrix.skku.ac.kr/cal-lab/cal-7-8-18.html

Solution

```
integral(1/(x^(1/4)), x, 0, 16)
```

Answer : 32/3

23. $\displaystyle\int_0^1 x^n \ln x\, dx \ (n>-1)$

Solution $\displaystyle\int_0^1 x^n \ln x\, dx \ (n>-1) = -\frac{1}{(1+n)^2}$

24. $\displaystyle\int_0^a x^3\sqrt{\frac{x}{a-x}}\, dx \ (a>0)$

Solution $\displaystyle\int_0^a x^3\sqrt{\frac{x}{a-x}}\, dx \le a^3\sqrt{a}\int_0^a \frac{1}{\sqrt{(\sqrt{a})^2-x}}$

$$= a^3\sqrt{a}\left[\sin^{-1}\left(\frac{x}{a}\right)\right]_0^a$$
$$= a^3\sqrt{a}\sin^{-1}(1).$$

convergent.

25. $\int_0^{\frac{\pi}{2}} (\sec x - \tan x)dx$

http://matrix.skku.ac.kr/cal-lab/cal-7-8-22.html

Solution

```
integral(sec(x)-tan(x), x, 0, pi/2)
```

Answer : log(2)

26. $\int_0^1 \sin(\ln x)dx$

http://matrix.skku.ac.kr/cal-lab/cal-7-8-25.html

Solution

```
integral(sin(ln(x)), x, 0, 1)
```

Answer : -1/2

27. $\int_{-1}^1 \frac{(1+x)^3}{\sqrt{1-x^2}}dx$

http://matrix.skku.ac.kr/cal-lab/cal-7-8-26.html

Solution

```
integral((1+x)^3/sqrt(1-x^2), x, -1, 1)
```

Answer : 5/2*pi

28. $\int_0^\infty \frac{x^2}{1+x^2}dx$

http://matrix.skku.ac.kr/cal-lab/cal-7-8-28.html

Solution

```
integral(x^2/(1+x^2), x, 0, infinity)
```

Answer : ValueError: Integral is divergent

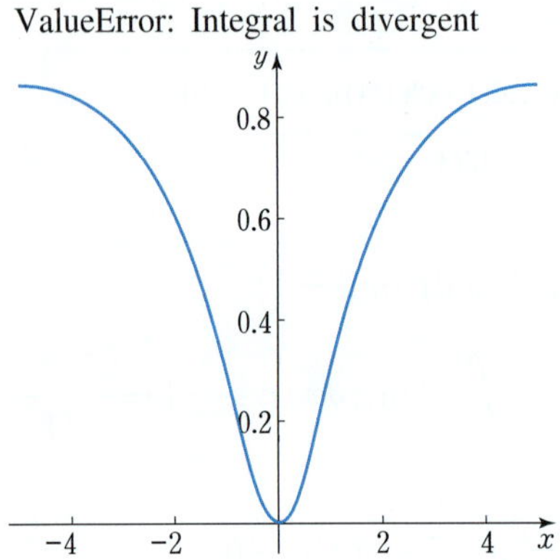

29. $\int_1^\infty \frac{x}{\sqrt{1+x^{98}}}dx$

http://matrix.skku.ac.kr/cal-lab/cal-7-8-29.html

Solution $\int_1^\infty \frac{x}{\sqrt{1+x^{98}}}dx \le \int_1^\infty \frac{x}{\sqrt{x^{98}}}dx$

$= \int_1^\infty \frac{1}{x^{48}}dx.$

```
integral(1/sqrt(x^48), x, 1, infinity)
```

Answer : 1/23

30. $\int_0^1 \frac{x^5}{\sqrt{1-x^2}}dx$

http://matrix.skku.ac.kr/cal-lab/cal-7-8-30.html

Solution

```
integral((x^5)/sqrt(1-x^2), x, 0, 1)
```

Answer : 8/15

31. $\int_1^\infty \frac{\sqrt{x}}{1+x^{3/2}}dx$

http://matrix.skku.ac.kr/cal-lab/cal-7-8-31.html

Solution $\int_1^\infty \frac{\sqrt{x}}{1+x^{3/2}}dx \ge \int_1^\infty \frac{\sqrt{x}}{x^{3/2}+x^{3/2}}dx$

$1+x^{3/2} \le x^{3/2}+x^{3/2} \Rightarrow \frac{\sqrt{x}}{1+x^{3/2}} \ge \frac{\sqrt{x}}{2x^{3/2}}$

$= \frac{1}{2}x^{\frac{1}{2}-\frac{3}{2}} = \frac{1}{2}x^{-1}$

Therefore, $\int_1^\infty \frac{\sqrt{x}}{1+x^{3/2}}dx \ge \frac{1}{2}\int_1^\infty \frac{1}{x}dx.$

```
integrate(1/x, x, 1, infinity)
```

Answer : ValueError: Integral is divergent.

32. $\int_0^1 \frac{\cos(1/x)}{\sqrt{x}}dx$

http://matrix.skku.ac.kr/cal-lab/cal-7-8-32.html

Solution $\int_0^1 \left|\frac{\cos(1/x)}{\sqrt{x}}\right|dx \le \int_0^1 \left|\frac{1}{\sqrt{x}}\right|.$

```
integrate(1/(sqrt(x)), x, 0, 1)
```

Answer : 2

33. $\int_0^\infty \frac{e^{-x}}{\sqrt{x}}dx$

http://matrix.skku.ac.kr/cal-lab/cal-7-8-33.html

Solution

```
integral(e^(-x)/sqrt(x),x,0,infinity)
```

Answer : sqrt(pi)

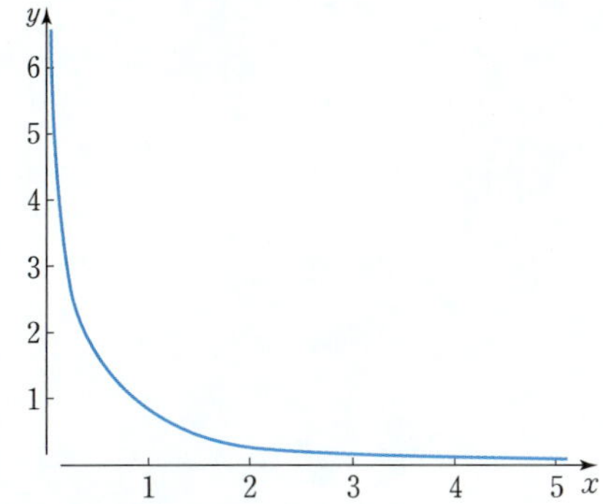

34. Show that $\displaystyle\int_0^{\infty} \frac{\ln x}{1+x^2}dx = 0.$

Proof $u=\dfrac{1}{x},\ 1+x^2=\dfrac{1}{u^2}+1,\ dx=-\dfrac{1}{u^2}du,$

$$\int_0^{\infty}\frac{\ln x}{1+x^2}dx = \int_{\infty}^{0}\frac{u^2}{1+u^2}(-\ln u)\left(-\frac{1}{u^2}du\right)$$
$$=-\int_0^{\infty}\frac{\ln u}{1+u^2}du$$

Therefore, $2\displaystyle\int_0^{\infty}\frac{\ln x}{1+x^2}dx=0.$

35. Prove that $\displaystyle\int_0^{\infty}\cos x^2 dx$ converges.

(Hint: let $u=x^2$)

Solution $u=x^2,\ 2xdx=du,\ dx=\dfrac{du}{2\sqrt{u}}$

$$\int_0^{\infty}\cos x^2dx=\int_0^{\infty}\frac{\cos u}{2\sqrt{u}}du$$
$$=\frac{1}{2}\int_0^1\frac{\cos u}{\sqrt{u}}du+\frac{1}{2}\int_1^{\infty}\frac{\cos u}{\sqrt{u}}du$$

$$\frac{1}{2}\int_0^1\frac{\cos u}{\sqrt{u}}du \le \frac{1}{2}\int_0^1\frac{1}{\sqrt{u}}du=1$$

$$\left|\int_1^{\infty}\frac{\cos u}{\sqrt{u}}du\right| = \left|\left[\frac{\sin u}{\sqrt{u}}\right]_1^{\infty}+\int_1^{\infty}\frac{\sin u}{u}du\right|$$
$$=\left|0+\left[\frac{-\cos u}{u}\right]_1^{\infty}-\int_1^{\infty}\frac{\cos u}{u^2}du\right|$$
$$\le |\cos(1)|+\left|\int_1^{\infty}\frac{1}{u^2}du\right|=\cos(1)+1.$$

<Calculators>

Chapter 8

Further Applications of Integration

8.1 Arc Length

8.2 Area of a Surface of Revolution

8.3 Center of Mass

*8.4 Differential Equations

Jean Baptiste Joseph Fourier

1768-1830

Fourier Series

[Heat Diffusion by Fourier Series]

A French mathematician and physicist, Fourier is credited as the discoverer of the greenhouse effect. His work included the Fourier series which dealt with problems of heat transfer. He was also at the forefront of the development of dimensional analysis. For a time he served as the Permanent Secretary of the French Academy of Sciences.

YouTube http://youtu.be/7OVqI20z_Bw

Applications of integrals to areas, volumes, work done and the average value have been considered in the previous chapters. In this chapter, we consider some more applications of integrals to the length of curves, the surface area of a surface of revolution, and center of mass. We also describe applications of integrals to simple differential equations.

8.1 Arc Length

We are well aware of the fact that the length of a polygon is found by adding the lengths of the line segments that form the polygon. A question arises: what about the length of a curve if it is not a polygon? To define the length of a general curve, we proceed as follows. First divide the curve into a large number of small parts, find the approximate lengths of the small parts and then add them up. This produces an approximation to the length of the curve. The larger the number of small parts, the better the approximation becomes. When, we take the limit as $n \to \infty$, where n is the number of parts.

Consider a curve C defined by the equation $y = f(x)$, $a \le x \le b$. Assume that C is continuously differentiable in the sense that the derivative $f'(x)$ is continuous for $a \le x \le b$. We divide the interval $[a, b]$ into n subintervals with endpoints x_i, $i = 1, 2, \ldots, n$ and, for the sake of convenience, of equal width Δx. If $y_i = f(x_i)$, then the point $P_i(x_i, y_i)$ lies on C. These points divide the curve into n smaller curves. Since the polygon with vertices P_1, P_2, ..., P_n approximates C (see Figure 1), we approximate the length L of C by the length of this polygon. As n increases, the approximation gets better.

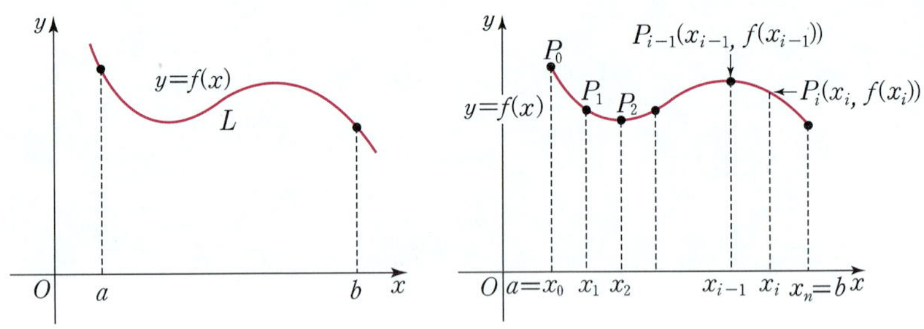

Figure 1

Therefore, the length L of the curve C is defined as follows:

1
$$L = \lim_{n\to\infty} \sum_{i=1}^{n} |P_{i-1}P_i|$$

provided the right hand limit exists.

We now try to deduce a more convenient expression for the computation of L. Let $\Delta y_i = y_i - y_{i-1}$ in Figure 2, then

$$|P_{i-1}P_i| = \sqrt{(\Delta x)^2 + (\Delta y_i)^2}\,.$$

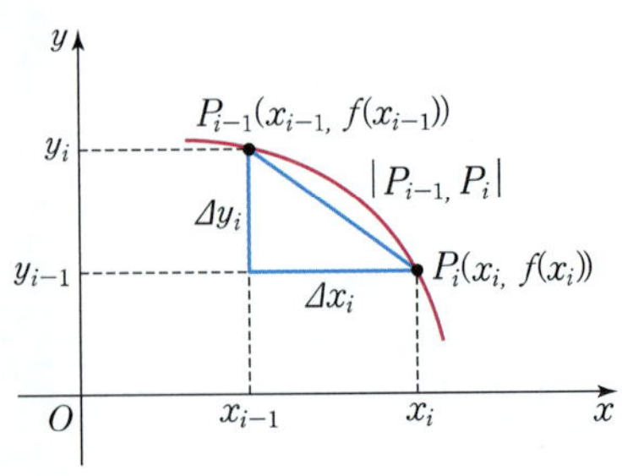

Figure 2

By the Mean Value Theorem applied to f on the interval $[x_{i-1}, x_i]$, there exists a number $x_i{}^*$ between x_{i-1} and x_i such that

$$\Delta y_i = f'(x_i{}^*)\Delta x$$

and so

$$
\begin{aligned}
|P_{i-1}P_i| &= \sqrt{(\Delta x)^2 + (\Delta y_i)^2} \\
&= \sqrt{(\Delta x)^2 + [f'(x_i^*)\Delta x]^2} \\
&= \sqrt{1 + [f'(x_i^*)]^2}\,\sqrt{(\Delta x)^2} \\
&= \sqrt{1 + [f'(x_i^*)]^2}\,\Delta x.
\end{aligned}
$$

Therefore,

$$L = \lim_{n\to\infty}\sum_{i=1}^{n}|P_{i-1}P_i| = \lim_{n\to\infty}\sum_{i=1}^{n}\sqrt{1+[f'(x_i^*)]^2}\,\Delta x.$$

Since the right side of this equation is the limit of a Riemann sum, we recognize it as being a definite integral and so we have the following formula.

The Arc Length Formula

If f' is continuous on $[a, b]$, then the length of the curve $y = f(x)$, $a \le x \le b$, is

2 $$L = \int_a^b \sqrt{1+[f'(x)]^2}\,dx = \int_a^b \sqrt{1+\left(\frac{dy}{dx}\right)^2}\,dx.$$

If $x = g(y)$, $c \le y \le d$ denotes the equation for a curve C, then the length of C is given by

3 $$L = \int_c^d \sqrt{1+[g'(y)]^2}\,dy = \int_c^d \sqrt{1+\left(\frac{dx}{dy}\right)^2}\,dy$$

which is obtained by interchanging the roles of x and y in 2.

EXAMPLE 1

For the curve $y = \frac{2}{3}(x-1)^{3/2}$, between the points $(1, 0)$ and $(4, 2\sqrt{3})$, determine the length of the curve.

http://matrix.skku.ac.kr/cal-lab/cal-8-1-Exm1.html

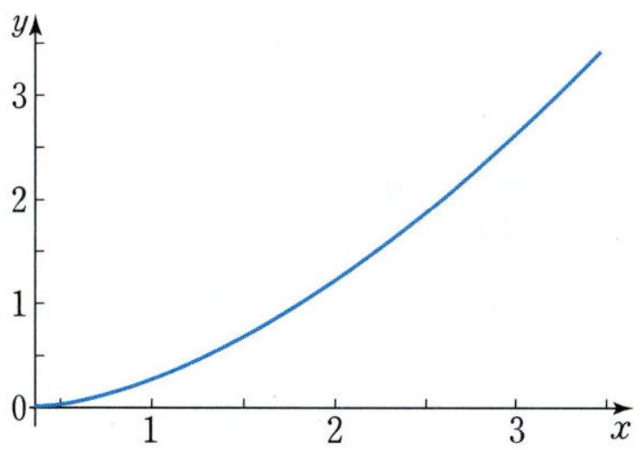

Figure 3 $y=\frac{2}{3}(x-1)^{3/2}$

Solution For the upper branch of the curve we have

$$\frac{dy}{dx} = \sqrt{x-1},$$

and so the arc length formula gives

$$L = \int_1^4 \sqrt{1+\left(\frac{dy}{dx}\right)^2}\,dx = \int_1^4 \sqrt{1+\left\{(x-1)^{\frac{1}{2}}\right\}^2}\,dx$$

$$= \int_1^4 \sqrt{1+x-1}\, dx = \int_1^4 \sqrt{x}\, dx = \left[\frac{2}{3} x^{\frac{3}{2}} \right]_1^4 = \frac{14}{3}.$$

```
var('x')
integral (sqrt(x), x, 1, 4)
```

Answer : 14/3 ■

EXAMPLE 2

Calculate the length of the curve $x = \frac{1}{4}y^4 + \frac{1}{8y^2}$ between $\left(\frac{3}{8}, 1\right)$ and $\left(\frac{33}{32}, 2\right)$.

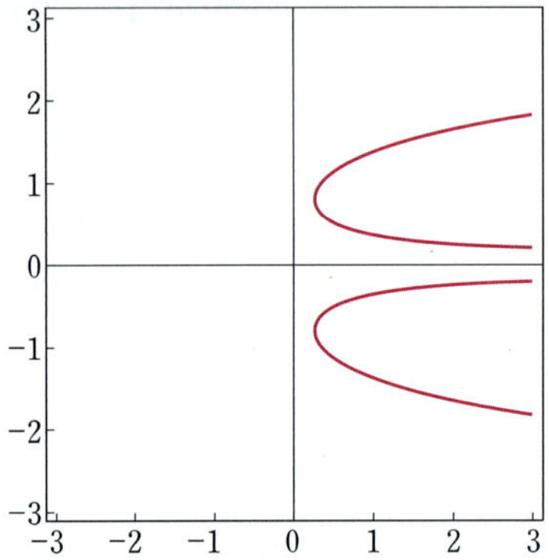

Figure 4 $x = \frac{1}{4}y^4 + \frac{1}{8y^2}$

Solution We have

$$\frac{dx}{dy} = y^3 - \frac{1}{4y^3}$$

and so, the length is

$$\begin{aligned} L &= \int_1^2 \sqrt{1+\left(\frac{dx}{dy}\right)^2}\, dy = \int_1^2 \sqrt{1+\left(y^3 - \frac{1}{4y^3}\right)^2}\, dy \\ &= \int_1^2 \sqrt{\left(y^3 + \frac{1}{4y^3}\right)^2}\, dy = \int_1^2 \left(y^3 + \frac{1}{4y^3}\right) dy \\ &= \left[\frac{1}{4}y^4 - \frac{1}{8y^2}\right]_1^2 \\ &= \frac{123}{32}. \end{aligned}$$

```
var('y')
integral (y^3 + 1/(4*y^3), y, 1, 2)
```

Answer : 123/32 ■

EXAMPLE 3

Determine the length of the curve $y^2 = 4x^3$, between the points $(0, 0)$ and $(2, 4\sqrt{2})$.

Solution For the curve we have

$$\frac{dy}{dx} = 3x^{\frac{1}{2}}.$$

The length of the curve on the interval $[0, 2]$ is

$$L = \int_0^2 \sqrt{1 + \left(3x^{\frac{1}{2}}\right)^2}\, dx = \int_0^2 \sqrt{1 + 9x}\, dx.$$

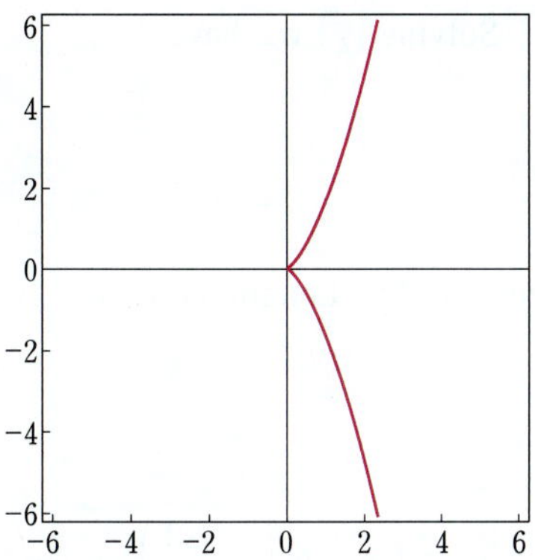

Figure 5 $y^2 = 4x^3$

Let $\sqrt{1+9x} = t$. Then $1 + 9x = t^2$ and $9dx = 2t\,dt$.
With the substitution, $x = 0$ becomes $t = 1$ and $x = 2$ becomes $t = \sqrt{19}$

$$L = \int_1^{\sqrt{19}} t \cdot \frac{2}{9} t\, dt = \frac{2}{9} \int_1^{\sqrt{19}} t^2\, dt$$
$$= \frac{2}{9}\left[\frac{1}{3}t^3\right]_1^{\sqrt{19}} = \frac{2}{27}(19\sqrt{19} - 1).$$

■

The Arc Length Function

Let $y = f(x)$, $a \le x \le b$ be the equation of a smooth curve C. Let $s(x)$ be the distance along C from the initial point $P_0(a, f(a))$ to the point $Q(x, f(x))$. Then the function $s(x)$ is called the **arc length function**, measuring the arc length of a part of the curve C, is given by

4 $$s(x) = \int_a^x \sqrt{1 + [f'(t)]^2}\, dt$$

Differentiating 4 using FTC 1, we have

5 $$\frac{ds}{dx} = \sqrt{1 + [f'(x)]^2} = \sqrt{1 + \left(\frac{dy}{dx}\right)^2}$$

From 5, the rate of change of s with respect to x is always at least 1 and is equal to 1 when $f'(x)$, the slope of the curve, is 0. The differential of the arc length is

6 $$ds = \sqrt{1 + \left(\frac{dy}{dx}\right)^2}\, dx$$

or in the symmetric form

7 $$(ds)^2 = (dx)^2 + (dy)^2$$

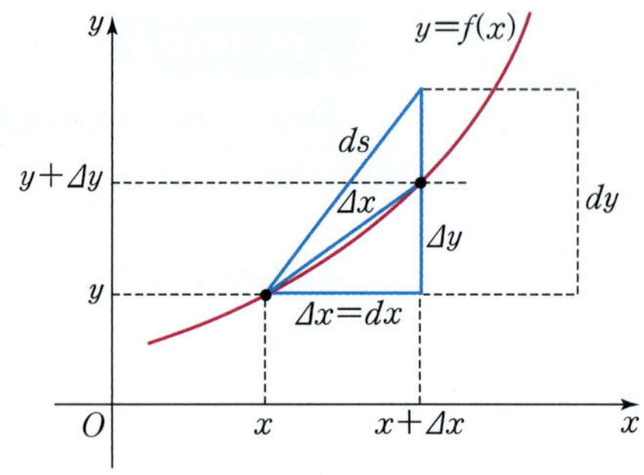

Figure 6

From the right triangle in Figure 6 we get [7]. Solving [7] we have

$$ds = \sqrt{1 + \left(\frac{dx}{dy}\right)^2}\, dy$$

Substituting the ds above in $L = \int ds$, we get the Arc Length Formula [2] or [3].

EXAMPLE 4

Determine the arc length function $y = \frac{4}{3}x^3 + \frac{1}{16x}$ taking $P_0\left(1, \frac{23}{16}\right)$ as the starting point.

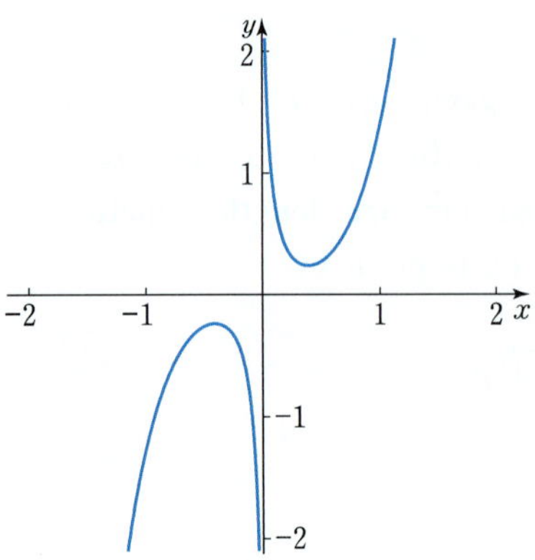

Figure 7 $y = \frac{4}{3}x^3 + \frac{1}{16x}$

Solution Here, $f(x) = \frac{4}{3}x^3 + \frac{1}{16x}$, then we have

$$f'(x) = 4x^2 - \frac{1}{16x^2},$$

$$1 + [f'(x)]^2 = 1 + \left(4x^2 - \frac{1}{16x^2}\right)^2 = \left(4x^2 + \frac{1}{16x^2}\right)^2,$$

$$\sqrt{1 + [f'(x)]^2} = 4x^2 + \frac{1}{16x^2}.$$

Thus, the arc length function is given by

$$\begin{aligned} s(x) &= \int_1^x \sqrt{1 + [f'(t)]^2}\, dt \\ &= \int_1^x \left(4t^2 + \frac{1}{16t^2}\right) dt = \left[\frac{4}{3}t^3 - \frac{1}{16t}\right]_1^x \\ &= \frac{4}{3}x^2 - \frac{1}{16x} - \frac{61}{48}. \end{aligned}$$

For example, the arc length along the curve from $\left(1, \frac{23}{16}\right)$ to $(3, f(3))$ is

$$s(3) = \frac{4}{3}(3)^2 - \frac{1}{16(3)} - \frac{61}{48} = \frac{257}{24}. \qquad \blacksquare$$

8.1 EXERCISES (Arc Length)

http://matrix.skku.ac.kr/Cal-Book/part1/CS-Sec-8-1-Sol.html

YouTube http://youtu.be/A8N-mDD0ja8

1-6. Find the length of the curve.

1. $y = x\sqrt{x}$, $0 \le x \le 4/3$

Solution Since the length of the curve is

$$L = \int_a^b \sqrt{1+[f'(x)]^2}\,dx = \int_a^b \sqrt{1+\left(\frac{dy}{dx}\right)^2}\,dx,$$

$$L = \int_0^{\frac{4}{3}} \sqrt{1+\left(\frac{3}{2}x^{\frac{1}{2}}\right)^2}\,dx \Rightarrow L = \int_0^{\frac{4}{3}} \sqrt{1+\frac{9}{4}x}\,dx.$$

Substitute $1+\frac{9}{4}x = a$ and $\frac{9}{4}dx = da$.

So, $L = \int_0^{\frac{4}{3}} \sqrt{1+\frac{9}{4}x}\,dx \Rightarrow L = \int_1^4 \frac{4}{9}\sqrt{a}\,da.$

$$L = \frac{4}{9}\left[\frac{2}{3}a^{\frac{3}{2}}\right]_1^4 \Rightarrow L = \frac{56}{27}$$

2. $y = \ln x - \frac{1}{8}x^2$, $1 \le x \le 2$

Solution

$$L = \int_1^2 \sqrt{1+\left(\frac{dy}{dx}\right)^2}\,dx = \int_1^2 \sqrt{1+\left(\frac{1}{x}-\frac{x}{4}\right)^2}\,dx$$

$$= \int_1^2 \sqrt{\left(\frac{1}{x}+\frac{x}{4}\right)^2}\,dx = \int_1^2 \left(\frac{1}{x}+\frac{x}{4}\right)dx$$

$$= \left[\ln x + \frac{x^2}{8}\right]_1^2 = \ln 2 + \frac{1}{2} - \left(\ln 1 + \frac{1}{8}\right)$$

$$= \ln 2 + \frac{3}{8}$$

3. $y = \cosh x$, $0 \le x \le \ln 2$

Solution $\frac{3}{4}$

4. $y^2 = x$ from $(0, 0)$ to $(1, 1)$

Solution Since $2y\,dy = dx$ implies $\frac{dx}{dy} = 2y$,

$$L = \int_0^1 \sqrt{1+\left(\frac{dx}{dy}\right)^2}\,dy = \int_0^1 \sqrt{1+(2y)^2}\,dy$$

$$= \int_0^1 \sqrt{1+4y^2}\,dy = 2\int_0^1 \sqrt{y^2+\frac{1}{4}}\,dy$$

Let $y = \frac{1}{2}\tan\theta$, then $dy = \frac{1}{2}\sec^2\theta\,d\theta$ and

$\sqrt{y^2+\frac{1}{4}} = \frac{1}{2}\sec\theta$.

$y \to 0 \Rightarrow \theta \to 0,\ y \to 1 \Rightarrow \theta \to \tan^{-1}2$

$\tan\theta = 2,\ \sec\theta = \sqrt{\tan^2\theta+1} = \sqrt{5}$

$$L = 2\int_0^1 \sqrt{y^2+\left(\frac{1}{2}\right)^2}\,dy$$

$$= 2\int_0^{\tan^{-1}2} \frac{1}{2}\sec\theta \cdot \frac{1}{2}\sec^2\theta\,d\theta$$

$$= \frac{1}{2}\int_0^{\tan^{-1}2} \sec^3\theta\,d\theta$$

$$= \frac{1}{4}\left[\sec\theta\tan\theta + \ln|\sec\theta+\tan\theta|\right]_0^{\tan^{-1}2}$$

$$= \frac{1}{4}\{2\sqrt{5}+\ln(\sqrt{5}+2)\}$$

$$= \frac{\sqrt{5}}{2} + \frac{\ln(\sqrt{5}+2)}{4}$$

5. $y^2 = (1+2x)^3$, $0 \le x \le 1$, $y > 0$

Solution $\frac{56\sqrt{7}-10\sqrt{10}}{27}$

6. $x = \ln(\sec y)$, $0 \le y \le \frac{\pi}{4}$

Solution Since $\frac{dx}{dy} = \tan y$,

$$L = \int_0^{\frac{\pi}{4}} \sqrt{1+\left(\frac{dx}{dy}\right)^2}\,dy$$

$$= \int_0^{\frac{\pi}{4}} \sqrt{1+\tan^2 y}\,dy = \int_0^{\frac{\pi}{4}} \sec y\,dy$$

$$= [\ln|\sec y + \tan y|]_0^{\frac{\pi}{4}} = \ln|\sqrt{2}+1|.$$

CAS **7.** Find the length of the curve as k goes to ∞.

$$y = \tanh(kx),\ 1 \le x \le 2$$

http://matrix.skku.ac.kr/cal-lab/cal-8-1-9.html

Solution

```
var('k, x')
@interact
def _(y = input_box(tanh(k*x), label="y="),
kappa=slider(0, 100, 0.01, default=1, label='k')):
plot(tanh(kappa*x), x, 0, 10,
color='purple').show(aspect_ratio=1, xmin=0,
xmax=2.5, ymin=0, ymax=1.2)
```

Answer : 1

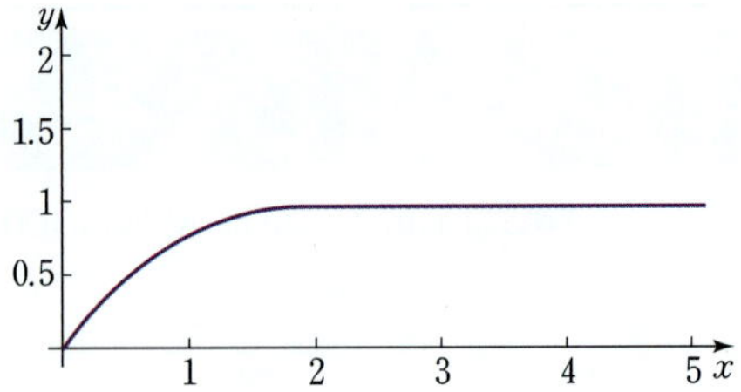

8. $y=\dfrac{1}{2}x\sqrt{x^2-2}-\ln\left(x+\sqrt{x^2-2}\right),\ 2\le x\le 3$

Solution $3\sqrt{2}-\sqrt{3}-\dfrac{1}{2}\ln(3+2\sqrt{2})+\dfrac{1}{2}\ln(2+\sqrt{3})$

CAS **9.** Find the length of the curve when k goes to 0.

$$y=\frac{k}{2}x^2,\ 0\le x\le 1$$

http://matrix.skku.ac.kr/cal-lab/cal-8-1-11.html

Solution

```
var('k, x')
@interact
def _(y = input_box(1/2*k*x^2, label="y="), kappa=slider(0, 100, 0.05, default=1,label='k')):
L= lim(integral(sqrt(1+(k*x)^2), x, 0, 1), k=kappa)
html('<p style="text-align: left;">The Arc Length is = ' + str(L)+'</p>')
plot(1/2*kappa*x^2, x,0,1).show(aspect_ratio=1, xmin=0, xmax=1, ymin=0, ymax=1)
```

Answer : 1

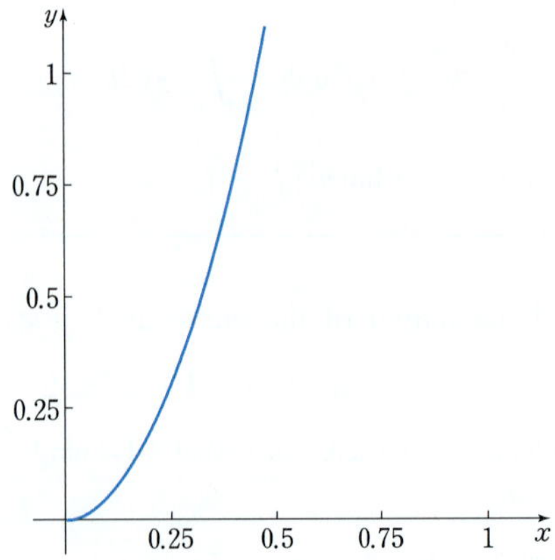

10. Find the length of the curve, $x=\cos^3 t,\ y=\sin^3 t,\ 0\le t\le 2\pi$.

http://matrix.skku.ac.kr/cal-lab/cal-8-1-12.html

Solution

```
var('t')
parametric_plot(((cos(t))^3, (sin(t))^3), 0, 2*pi, rgbcolor=hue(0.6))
```

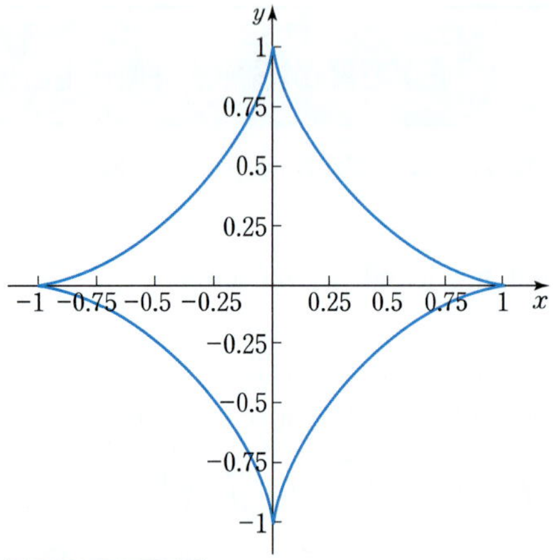

$$ds=\sqrt{\left(\frac{dx}{dt}\right)^2+\left(\frac{dy}{dt}\right)^2}\,dt=3|\sin t\cos t|$$

$$L=\int_0^{2\pi}ds=\int_0^{2\pi}3|\sin t\cos t|dt$$

$$=\int_0^{\frac{\pi}{2}}3\sin t\cos t\,dt-\int_{\frac{\pi}{2}}^{\pi}3\sin t\cos t\,dt$$

$$+\int_{\pi}^{\frac{3\pi}{2}}3\sin t\cos t\,dt-\int_{\frac{3\pi}{2}}^{2\pi}3\sin t\cos t\,dt$$

$$=4\int_0^{\frac{\pi}{2}}3\sin t\cos t\,dt$$

```
var('t')
4*integral(3*sin(t)*cos(t), t, 0, pi/2)
```

Answer : 6

CAS **11.** (a) Graph the curve $y=(1+x^2)^{2/3},\ 0\le x\le 4$.

(b) Compute the lengths of inscribed polygons with $n=1$, 2, and 4 sides. (Divide the interval into equal subintervals.) Illustrate by sketching these polygons.

(c) Set up an integral to find the length of the curve.

(d) Use any of your CAS tool to find the length of the curve to four decimal places. Compare with the approximations in part (b).

http://matrix.skku.ac.kr/cal-lab/cal-8-1-11-1.html

Solution

(a)

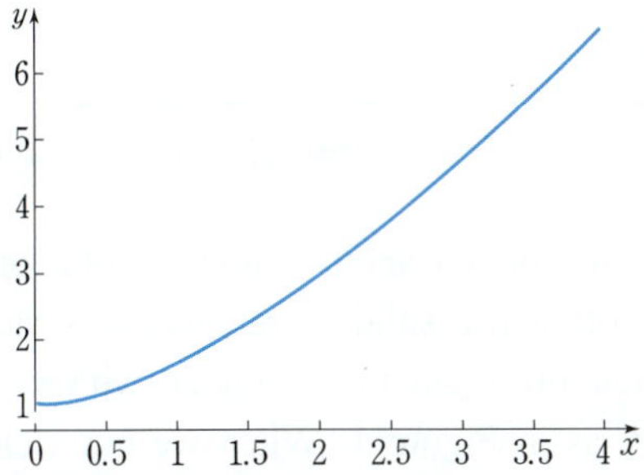

(b) 6.89 $(n=1)$, 6.97 $(n=2)$, 7.02 $(n=4)$

(c) $L=\frac{1}{3}\int_0^4 \sqrt{9+\frac{16x^2}{(1+x^2)^{2/3}}}\,dx$

```
var('x')
S = plot( (1+x^2)^(2/3), x, 0, 4)
L = integral((1/3)*((9 +(16*x^2)/((1 + x^2 )^(2/3) ) )^(1/2)) , x, 0, 4)
show (L)                # 7.06022221757
print numerical_integral((1/3)*((9 +(16*x^2)/((1 + x^2 )^(2/3) ) )^(1/2)) , 0, 4)[0]
#P = plot(f,x, xmax=2.2, xmin=1.5, ymax=10, ymin=-10, color='blue')
show (S)
```

(d) Simpson's Rule with $n=8$ gives $L\approx 7.0621$.

12. Repeat Exercise 11 for the curve $y=x^2\cos x$, $0\leq x\leq 2\pi$.

http://matrix.skku.ac.kr/cal-lab/cal-8-1-14.html
http://math1.skku.ac.kr/home/pub/1100/

Solution

(a)

```
var('x')
plot(x^2*cos(x), x, 0, 2*pi);
```

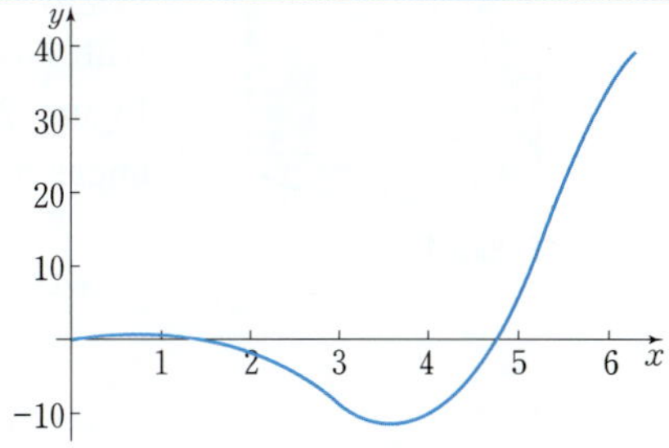

(b) 39.97 $(n=1)$, 59.80 $(n=2)$, 61.06 $(n=4)$

(c) $L=\int_0^{2\pi}\sqrt{1+(4x^2-x^4)\cos^2 x-4x^3\cos x\sin x+x^4}\,dx$

(d) Simpson's Rule with $n=8$ gives $L\approx 65.1036$.

<Abacus>

8.2 Area of a Surface of Revolution

http://youtu.be/Eq4i2A8eKxA

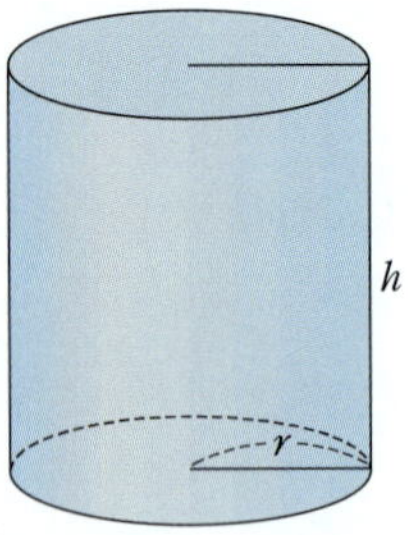

Figure 1

When a region bounded by curves is rotated about a line, a solid of revolution is generated. The lateral boundary of such a solid of revolution is the surface of revolution. The lateral surface area of a circular cylinder with radius r and height h is taken to be $A = 2\pi r h$ as in Figure 1. When we cut a circular cone with base radius r, arc length $2\pi r$ and slant height l along a dashed line as in Figure 2, and flatten it, it forms a sector of a circle with radius l and central angle $\theta = 2\pi r / l$. Therefore the lateral surface area A of this cone is

$$A = \frac{1}{2} l^2 \theta = \frac{1}{2} l^2 \left(\frac{2\pi r}{l} \right) = \pi r l.$$

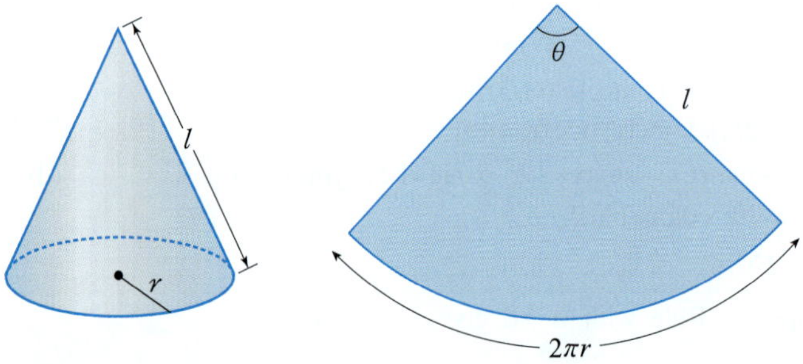

Figure 2

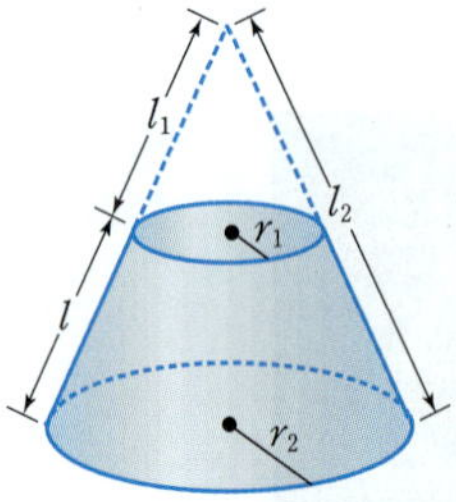

Figure 3

Observe that each of the lateral surfaces of a circular cylinder and a circular cone is generated when a simple polygon is rotated about a line. For later use in this section, let us find the lateral surface area of a portion of a circular cone with slant height l and upper and lower radii r_1 and r_2 as shown in Figure 3. It is found by subtracting the areas of smaller cone from the larger cone as shown in the Figure 3;

1 $$A = \pi r_2 l_2 - \pi r_1 l_1.$$

The slant heights in Figure 3 are proportional to the radii.

$$\frac{l_1}{r_1} = \frac{l_2}{r_2} \quad \text{so,} \quad r_1 l_2 = r_2 l_1.$$

The slant height l of the frustum is

$$l = l_2 - l_1.$$

Substituting this into 1, using the last two equations

$$\begin{aligned} A = \pi r_2 l_2 - \pi r_1 l_1 &= \pi (r_2 l_2 + r_1 l_2 - r_2 l_1 - r_1 l_1) \\ &= \pi (r_2 + r_1)(l_2 - l_1) \\ &= \pi (r_1 + r_2) l. \end{aligned}$$

We obtain the surface A as $A = \pi(r_1 + r_2)l$.

2 $$A = 2\pi rl$$

where $r = \frac{1}{2}(r_1 + r_2)$ is the average radius of the band.

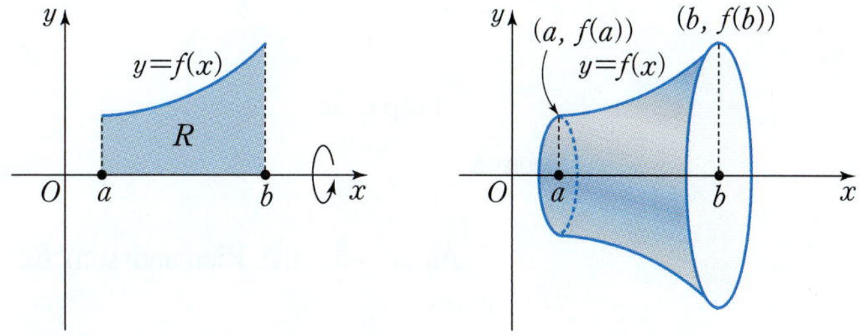

Figure 4

We now consider finding the area of the surface of revolution obtained when a smooth curve $y = f(x)$, $a \le x \le b$ is rotated about the x-axis. (See Figure 4.) Assume that f is positive and has a continuous derivative. We divide the interval $[a, b]$ into n subintervals with endpoints x_0, x_1, ..., x_n and equal width Δx. If $y_i = f(x_i)$, then the point $P_i(x_i, y_i)$ lies on the curve. The part of the surface between x_{i-1} and x_i is approximated by the band obtained by taking the line segment $P_{i-1}P_i$ and rotating it about the x-axis. As n increases, the approximation gets better. The band has slant height $l = |P_{i-1}P_i|$ and average radius $r = \frac{1}{2}(y_{i-1} + y_i)$. So, by 2, its surface area is

2 $$2\pi \frac{y_{i-1} + y_i}{2} |P_{i-1}P_i| (= 2\pi rl).$$

We have

$$|P_{i-1}P_i| = \sqrt{1 + [f'(x_i^*)]^2}\, \Delta x,$$

where x_i^* is some number in $[x_{i-1}, x_i]$. When Δx is small, we have $y_i = f(x_i) \approx f(x_i^*)$ and also $y_{i-1} = f(x_{i-1}) \approx f(x_i^*)$, since f is a continuous function. Therefore, we have

$$2\pi \frac{y_{i-1} + y_i}{2} |P_{i-1}P_i| \approx 2\pi f(x_i^*) \sqrt{1 + [f'(x_i^*)]^2}\, \Delta x.$$

So an approximation to the area of the total surface of revolution is

3 $$\sum_{i=1}^{n} 2\pi f(x_i^*) \sqrt{1 + [f'(x_i^*)]^2}\, \Delta x.$$

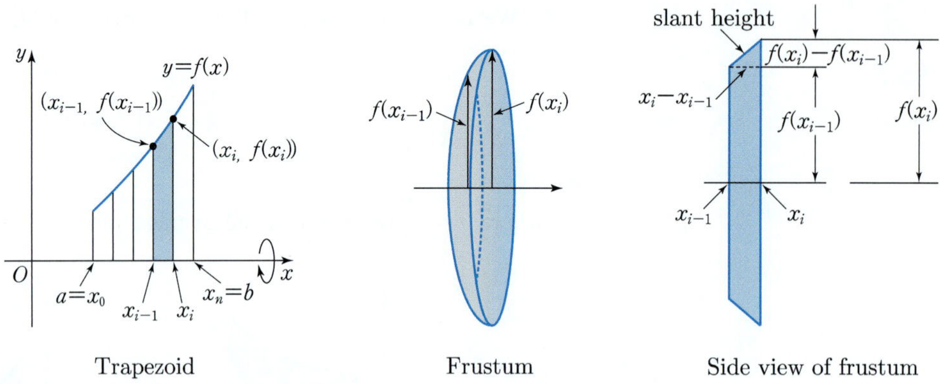

Figure 5

As $n\to\infty$ the Riemann sum for the function $g(x)=2\pi f(x)\sqrt{1+[f'(x)]^2}\,\Delta x$, gives

$$\lim_{n\to\infty}\sum_{i=1}^{n}2\pi f(x_i^*)\sqrt{1+[f'(x_i^*)]^2}\,\Delta x=\int_a^b 2\pi f(x)\sqrt{1+[f'(x)]^2}\,dx.$$

Therefore, the area of the surface of revolution obtained by rotating the curve $y=f(x)$, $a\le x\le b$, about the x-axis is given by

4 $$S=\int_a^b 2\pi f(x)\sqrt{1+[f'(x)]^2}\,dx$$

or

5 $$S=\int_a^b 2\pi y\sqrt{1+\left(\frac{dy}{dx}\right)^2}\,dx.$$

If the curve is given by the equation $x=g(y)$, $c\le y\le d$, then the area of the surface of revolution obtained by rotating the curve, about the y-axis becomes

6 $$S=\int_c^d 2\pi\,g(y)\sqrt{1+[g'(y)]^2}\,dy$$

or

7 $$S=\int_c^d 2\pi x\sqrt{1+\left(\frac{dx}{dy}\right)^2}\,dy.$$

Or symbolically, for the rotation about the x-axis the surface area is

8 $$S=\int 2\pi y\, ds$$

whereas, for the rotation about the y-axis, the surface area formula is

9 $$S=\int 2\pi x ds$$

where we use for ds either

$$ds=\sqrt{1+\left(\frac{dy}{dx}\right)^2}\,dx \text{ or } ds=\sqrt{1+\left(\frac{dx}{dy}\right)^2}\,dy.$$

Note that the circumference of a circle traced out by the point $(x,\ y)$ on the curve as it is rotated about the x-axis or y-axis, is $2\pi y$ or $2\pi x$ respectively. (See Figure 6.) Therefore, the surface area can be obtained by integrating $2\pi y$ of $2\pi x$.

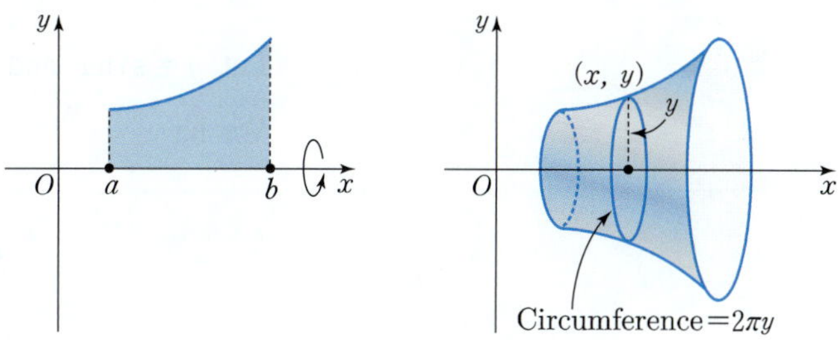

(a) Rotation about x-axis $S=\int 2\pi y\, ds$

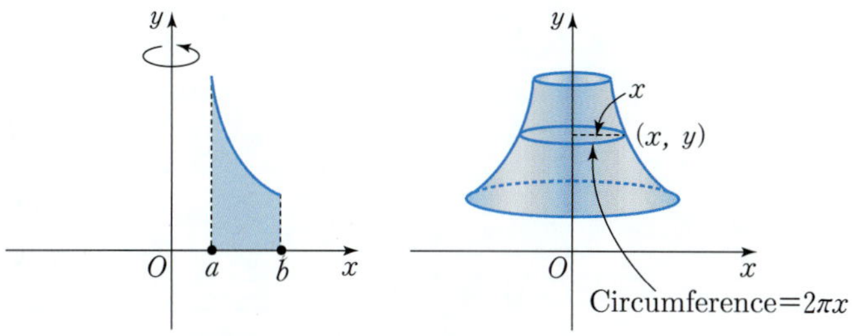

(b) Rotation about y-axis $S=\int 2\pi x\, ds$

Figure 6

EXAMPLE 1

Determine the area of the surface of revolution obtained by rotating about the x-axis the curve $y=\sin x,\ 0\le x\le \pi/2$.

Solution

```
var('u')
revolution_plot3d(sin(u), (u, 0, pi/2), opacity=0.5, color='blue',
show_curve=True, parallel_axis='x')
```

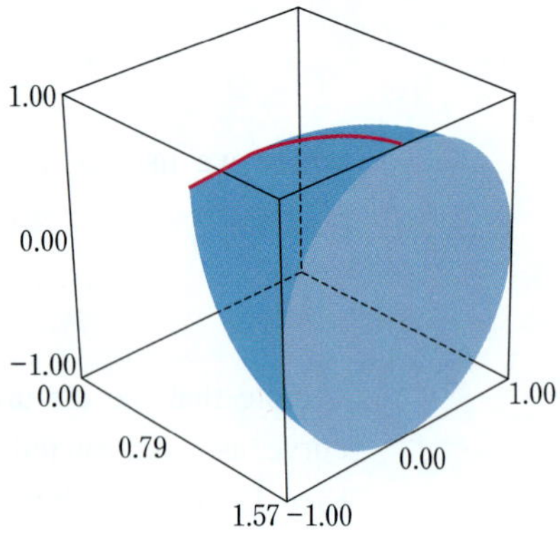

Figure 7

Let $y = \sin x$ and $\dfrac{dy}{dx} = \cos x$.

We have

$$
\begin{aligned}
S &= \int_0^{\pi/2} 2\pi y \sqrt{1 + \left(\frac{dy}{dx}\right)^2}\, dx \\
&= 2\pi \int_0^{\pi/2} \sin x \sqrt{1 + \cos^2 x}\, dx \\
&= 2\pi \int_0^1 \sqrt{1 + u^2}\, du \\
&= 2\pi \int_0^{\frac{\pi}{4}} \sec^3\theta\, d\theta \\
&= \pi \left[\sec\theta \tan\theta + \ln|\sec\theta + \tan\theta|\right]_0^{\pi/4} \\
&= \pi\left[\sqrt{2} + \ln(1 + \sqrt{2})\right].
\end{aligned}
$$

```
var('x')
integral ( 2*pi* (sec(x))^3 , x, 0, pi/4)
```

■

EXAMPLE 2

Determine the area of the surface of revolution obtained by rotating about the y-axis the arc of the parabola $y = x^3$ from $(0, 0)$ to $(2, 8)$.

http://matrix.skku.ac.kr/cal-lab/cal-8-2-Exm-2.html

Solution

```
var('u')
revolution_plot3d(u^3, (u, 0, 2), opacity=0.5, color='blue',
show_curve=True, parallel_axis='x')
```

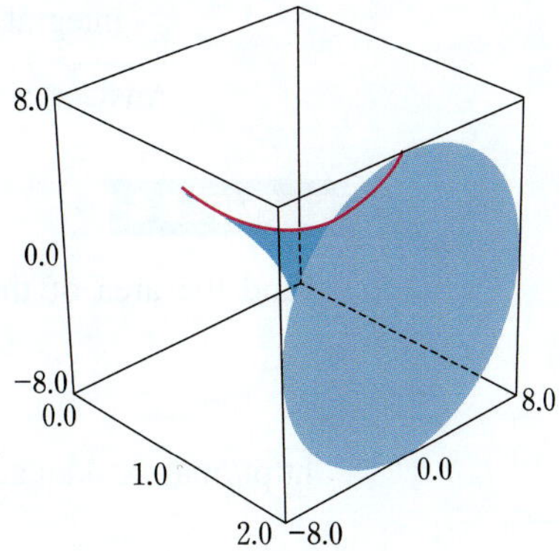

Figure 8

Since $y=x^3$ and $\dfrac{dy}{dx}=3x^2$ from (8) we have,

$$\begin{aligned} S &= \int 2\pi y\, ds \\ &= \int_0^2 2\pi x^3 \sqrt{1+\left(\frac{dy}{dx}\right)^2}\, dx \\ &= 2\pi \int_0^2 x^3 \sqrt{1+9x^4}\, dx \\ &= \frac{\pi}{9}\int_0^{12} u\sqrt{1+u^2}\, du \ \text{(where } u=3x^2\text{)} \end{aligned}$$

```
var('u')
pi/9*integral (u*(1+u^2)^1/2, u, 0, 12)
```

Answer : 1/27 (145*sqrt (145)-1)*pi ■

EXAMPLE 3

Find the area of the surface of a sphere of radius R, $R>0$.

Solution The equation of the upper circle with radius R and center at $(0,\ 0)$ is $y=\sqrt{R^2-x^2}$.
We have

$$\sqrt{1+\left(\frac{dy}{dx}\right)^2}=\frac{R}{\sqrt{R^2-x^2}}.$$

Therefore, the desired area of the surface is

$$S = 2\pi \int_{-R}^{R} \sqrt{R^2 - x^2} \cdot \frac{R}{\sqrt{R^2 - x^2}} dx$$

$$= 2\pi \int_{-R}^{R} R dx = 2\pi [Rx]_{-R}^{R} = 4\pi R^2.$$

```
var('x, R')
integral ( 2*pi * R , x, -R, R)     # 4*pi*R^2
```

Answer : 4*pi*R^2 ■

CAS **EXAMPLE 4**

Find the area of the surface obtained by rotating the curve about the x-axis.

$$y = x + \cos x, \; 0 \le x \le \pi$$

http://matrix.skku.ac.kr/cal-lab/cal-8-2-Exm-4.html

Solution

```
var('x, u')
y=x+cos(x)
P=plot(y, (x, 0, pi))
Q=revolution_plot3d(u+cos(u), (u, 0, pi), opacity=0.5, color='blue', show_curve=True, parallel_axis='x')
show(Q+P)
print numerical_integral(2*pi*y*sqrt(1+(y.diff(x))^2), 0, pi)[0]
```

Answer : 34.1375871765

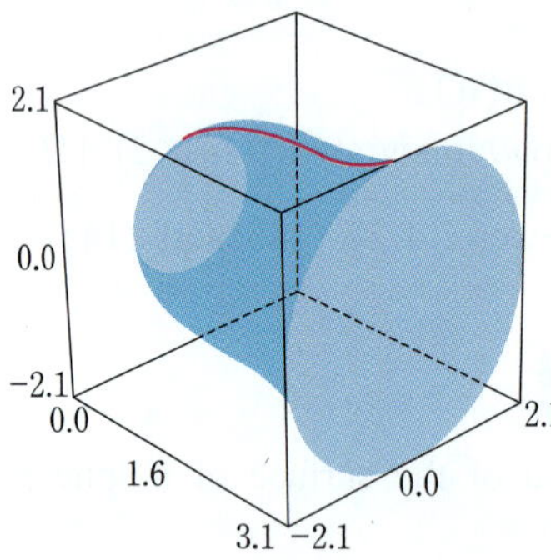

Figure 9

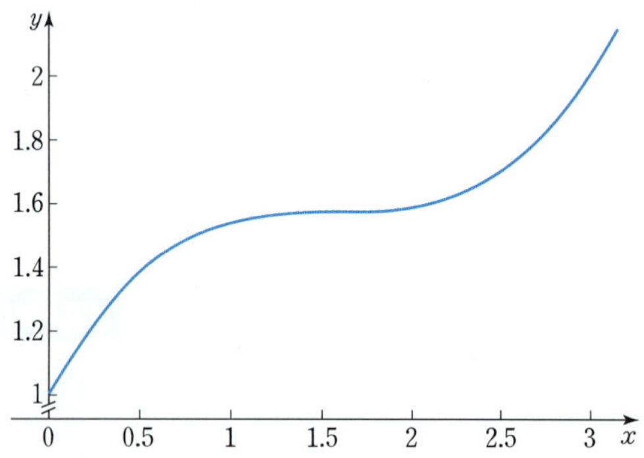

Figure 10 ■

8.2 EXERCISES (Area of a Surface of Revolution)

http://matrix.skku.ac.kr/Cal-Book/part1/CS-Sec-8-2-Sol.html

YouTube http://youtu.be/yZFJDJgTJfw

1-2. Set up, but do not evaluate, an integral for the area of the surface obtained by rotating the curve about the given axis.

1. $y = \cos^2 x,\ 0 \le x \le \pi/2;\ x$-axis

Solution $\displaystyle\int_0^{\frac{\pi}{2}} 2\pi\cos^2 x\sqrt{1+4\cos^2 x\sin^2 x}\,dx$

2. $4y = 2x^2 - \ln x,\ 1 \le x \le 4;\ x$-axis

http://matrix.skku.ac.kr/cal-lab/cal-8-2-2.html

Solution $\displaystyle S=\int_a^b 2\pi y\sqrt{1+\left(\frac{dy}{dx}\right)^2}\,dx$ (range from a to b)

Let, $y=\dfrac{2x^2-\ln x}{4}$

$$S=\int_a^b 2\pi y\sqrt{1+\left(\frac{dy}{dx}\right)^2}\,dx$$

$$\Rightarrow S=\int_1^4 2\pi\frac{2x^2-\ln x}{4}\sqrt{1+\left(x-\frac{1}{4x}\right)^2}\,dx$$

$$\Rightarrow S=\int_1^4 \pi\frac{2x^2-\ln x}{2}\sqrt{x^2+\frac{1}{16x^2}+\frac{1}{2}}\,dx$$

$$\Rightarrow S=\int_1^4 \pi\frac{2x^2-\ln x}{2}\sqrt{\left(x+\frac{1}{4x}\right)^2}\,dx$$

$$S=\frac{\pi}{2}\int_1^4 (2x-\ln x)\left(x+\frac{1}{4x}\right)dx$$

```
var('x,y')
implicit_plot(4*y==2*x^2-ln(x),(1,4),(0,8))
revolution_plot3d(y==0.5*x^2-0.25*ln(x),(x,1,4),(y,0,10),axis=(1,0,0),show_curve=False,opacity=0.7).show(aspect_ratio=(1,1,1))
```

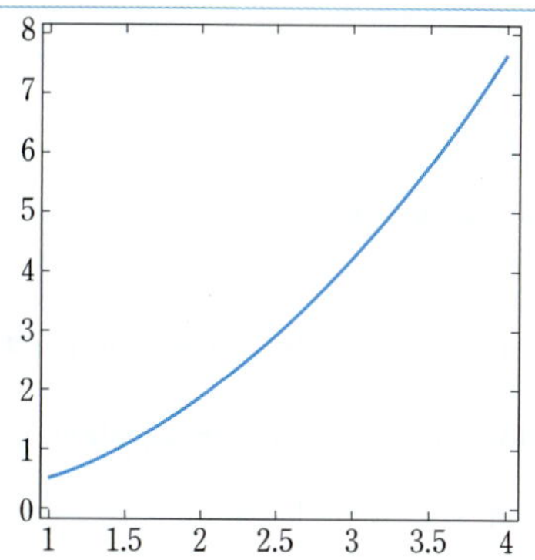

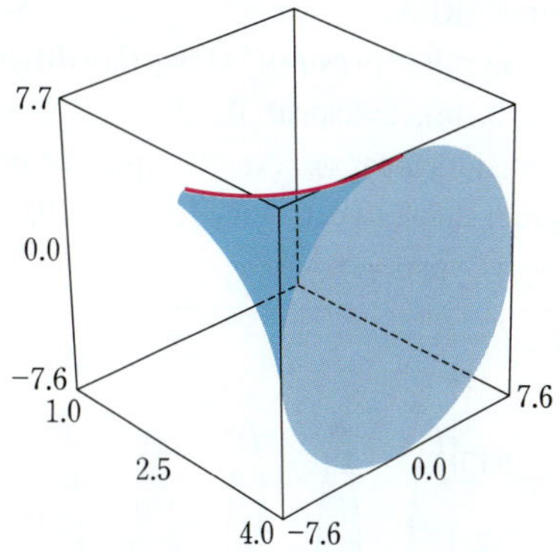

3-7. Find the area of the surface obtained by rotating the curve about the x-axis.

CAS **3.** Find the area of the surface when k goes to 0.

$$y=\frac{k}{2}x^2,\ 0 \le x \le 1$$

http://matrix.skku.ac.kr/cal-lab/cal-8-2-3.html

Solution

```
var('k, x')
@interact
def _(y = input_box(1/2*k*x^2, label="y="), kappa=slider(0, 1, 0.05, default=1, label='k')):
    html(AREA '$$=%s$$'%( lim(integral(pi*2*1/2*k*x^2*sqrt(1+(k*x)^2), x, 0, 1), k=kappa) ))
    plot(1/2*kappa*x^2, x, 0, 1, color='purple', fill=true).show(aspect_ratio=1, xmin=0, xmax=1, ymin=0, ymax=0.5)
```

Answer : 0

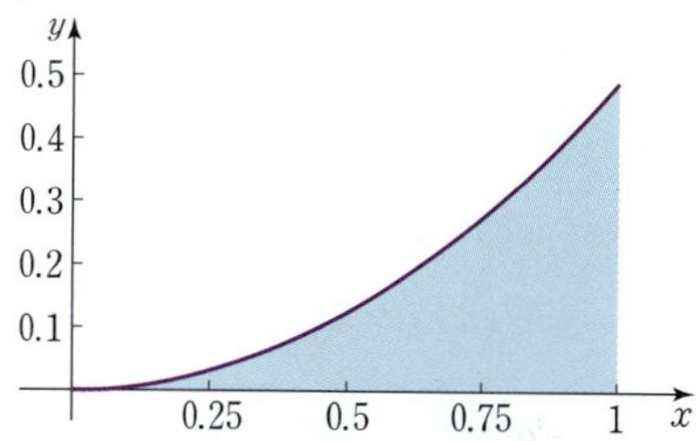

CAS **4.** Find the area of the surface when k is a positive even integer.

$$y=\sin(kx),\ 0 \le x \le \pi$$

http://matrix.skku.ac.kr/cal-lab/cal-8-2-4.html

Solution

```
k, x=var('k, x');
@interact
def _(y = input_box(sin(k*x), label="y="), kappa=slider(0, 100, 1, default=1, label=' k')):
    html(AREA '$$=%s$$'%( lim(integral(2*pi*sin(k*x)*sqrt(1+diff(sin(k*x), x)), x, 0, pi), k=kappa) ))
    plot(sin(kappa*x), x, 0, pi, color='purple', fill=true).show(aspect_ratio=1, xmin=0, xmax=pi, ymin=-1, ymax=1)
```

Answer : 0

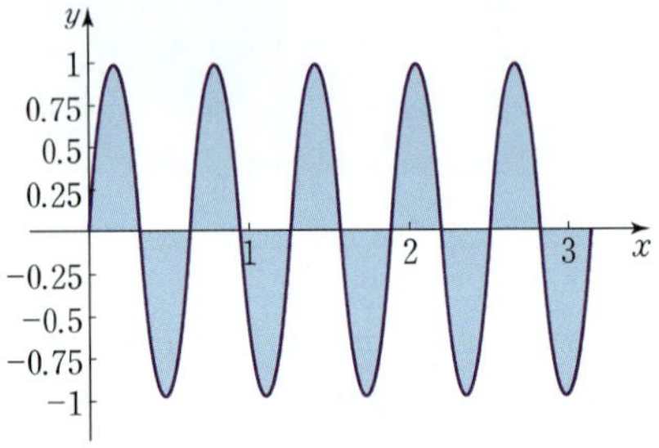

5. $y=\sqrt{9-x^2}$, $-3 \le x \le 3$

Solution $S=\displaystyle\int_{-3}^{3} 2\pi\sqrt{9-x^2}\sqrt{1+\left(-\frac{x}{\sqrt{9-x^2}}\right)^2}\,dx$

$=\displaystyle\int_{-3}^{3} 2\pi\sqrt{9-x^2}\sqrt{\frac{9}{9-x^2}}\,dx$

$=\displaystyle\int_{-3}^{3} 6\pi\,dx = 36\pi$

6. $y=e^x$, $0 \le x \le 1$

Solution The surface obtained by rotating the curve about the x-axis is $S=\displaystyle\int_a^b 2\pi y\sqrt{1+\left(\frac{dy}{dx}\right)^2}\,dx$.

Substitute $y=e^x$,

$S=\displaystyle\int_a^b 2\pi y\sqrt{1+\left(\frac{dy}{dx}\right)^2}\,dx$

$\Rightarrow S=\displaystyle\int_0^1 2\pi e^x\sqrt{1+(e^x)^2}\,dx = 2\pi\int_0^1 e^x\sqrt{1+e^{2x}}\,dx.$

Let $1+e^{2x}=\sec^2\theta$ and $e^x=\tan\theta\,(\because\ 1+\tan^2\theta=\sec^2\theta)$,

then $e^x dx=\sec^2\theta d\theta \Rightarrow dx=\dfrac{\sec^2\theta}{\tan\theta}d\theta.$

$S=2\pi\displaystyle\int_\alpha^\beta \sec^3\theta d\theta=\pi[\sec\theta\tan\theta+\ln|\sec\theta+\tan\theta|]_\alpha^\beta$

$(\because\ \tan\alpha=1,\ \tan\beta=e)$

Calculate and simplify to have

$S=\pi[e\sqrt{e^2-1}+\ln(e+\sqrt{1+e^2})-\sqrt{2}-\ln(\sqrt{2}+1)].$

7. Find the area of the surface obtained by rotating the curve about the x-axis.

$y=\max\left\{\sqrt{1-x^2}, -2x\log x\right\},\ 0 \le x \le 1$

http://matrix.skku.ac.kr/cal-lab/cal-8-2-7.html

Solution

```
p1=plot(sqrt(1-(x)^2), (0,1),
rgbcolor=hue(0.4))
p2=plot(-2*log(x)*x, (0, 1),
 rgbcolor=hue(0.6))
```

From the above figure, it is clear that the maximum is circle of unit radius in the first quadrant. When we rotate the unit circle in the first quadrant about the x-axis we obtain half unit sphere. Hence the required area is 2π.

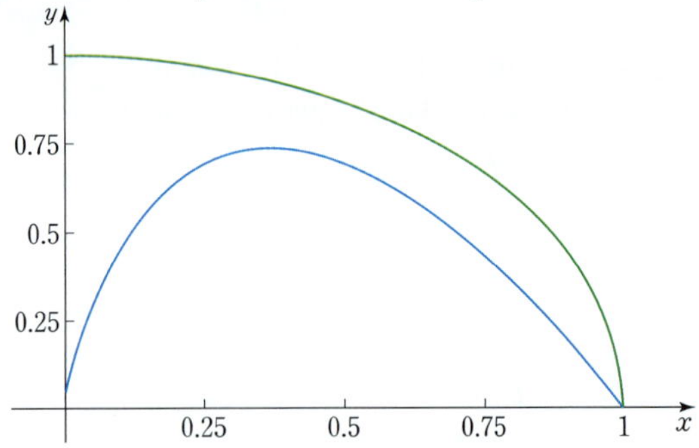

8-10. The given curve is rotated about the y-axis. Find the area of the resulting surface.

8. $x=\frac{1}{2}y^2$, $\sqrt{3} \le y \le 2\sqrt{2}$

Solution

$\pi\left[\frac{51}{4}\sqrt{2}-\frac{7\sqrt{3}}{4}+\frac{1}{8}\ln(3-2\sqrt{2})+\frac{1}{8}\ln(2+\sqrt{3})\right]$

9. $y=\ln x$, $0 \le x \le e$

Solution

$\pi[e\sqrt{1+e^2}+\ln(e+\sqrt{1+e^2})-\sqrt{2}-\ln(\sqrt{2}+1)]$

10. $y=x^2$, $1 \le x \le 3$

Solution The surface obtained by rotating the curve about the y-axis becomes

$S=\displaystyle\int_a^b 2\pi x\sqrt{1+\left(\frac{dx}{dy}\right)^2}\,dy$ (Range form a to b)

Then substitute $y=x^2$ to above equation.

$S=\displaystyle\int_1^9 2\pi\sqrt{y}\sqrt{1+\left(\frac{1}{2x}\right)^2}\,dy$

$=\displaystyle\int_1^9 2\pi\sqrt{y}\sqrt{1+\frac{1}{4x^2}}\,dy$

($y=x^2$, then $\sqrt{y}=x$ (because of $1 \le x \le 3$) and $1 \le x \le 3$ so $1 \le y \le 9$)

$$S=\int_1^9 2\pi\sqrt{y}\sqrt{1+\frac{1}{4y}}\,dy=\int_1^9 2\pi\sqrt{y+\frac{1}{4}}\,dy$$

$$=2\pi\int_1^9\sqrt{y+\frac{1}{4}}\,dy=2\pi\left[\frac{2}{3}\left(\frac{1}{4}+y\right)^{\frac{3}{2}}\right]_1^9$$

$$=\frac{37\sqrt{37}-5\sqrt{5}}{6}\pi$$

11. Find the area of the surface of the solid of revolution obtained by rotating about the y-axis the circle $(x-a)^2+(y-b)^2=R^2$, $a>R$.

Solution The right semicircle and left semicircle equations are given by $x_1=a+\sqrt{R^2-(y-b)^2}$ and $x_2=a-\sqrt{R^2-(y-b)^2}$, respectively.

$$S=2\pi\int_{b-R}^{b+R}x_1\sqrt{1+\left(\frac{dx_1}{dy}\right)^2}\,dy+2\pi\int_{b-R}^{b+R}x_2\sqrt{1+\left(\frac{dx_2}{dy}\right)^2}\,dy$$

$$=4\pi^2 aR.$$

CAS **12.** Use a CAS to find the area of the surface obtained by rotating the curve about the given axis. Use Simpson's Rule with $n=10$.

(a) $y=x^5$, $0 \le x \le 1$; x-axis

http://matrix.skku.ac.kr/cal-lab/cal-8-2-12.html

(b) $y=e^x$, $1 \le y \le 5$; y-axis

http://matrix.skku.ac.kr/cal-lab/cal-8-2-12-b.html

Solution

(a)

```
var('x');
y=x^5;
d=diff(y);
f(x)=y*(1+d^2)^(1/2);
q=n(2*pi*(0.1/3*(f(0)+4*f(0.1)+2*f(0.2)+4*f(0.3)+
2*f(0.4)+4*f(0.5)+2*f(0.6)+4*f(0.7)+2*f(0.8)+4*f(
0.9)+f(1))));
show(q)
```

Answer : 3.37004769810923

(b) 7.054993

```
var('y');
x=ln(y);
d=diff(x);
f(y)=x*(1+d^2)^(1/2);
k=4;
q=n(2*pi*(0.1*k/3*(f(1)+4*f(1+0.1*k)+2*f(1+0.2*
k)+4*f(1+0.3*k)+2*f(1+0.4*k)+4*f(1+0.5*k)+2*f(
1+0.6*k)+4*f(1+0.7*k)+2*f(1+0.8*k)+4*f(1+0.9*k
)+f(1+1*k))));
show(q)
```

Answer : 26.8537676835850

8.3 Center of Mass

YouTube http://youtu.be/1ZAJeP16pAQ

Center of Mass

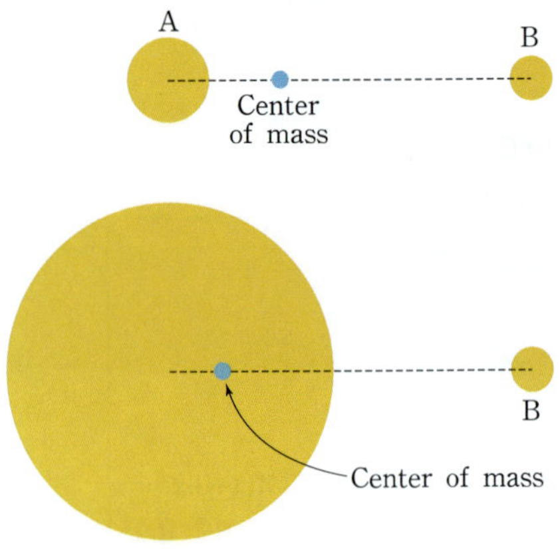

Figure 1 the center of mass r_c

In physics, the center of mass in space is the unique point where the weighted relative position of the distributed mass sums to zero. Calculations in mechanics are simplified when formulated with respect to the center of mass. In the case of a single rigid body, the center of mass is fixed in relation to the body, and if the body has uniform density, it will be located at the centroid. The center of mass may be located outside the physical body.

The center of mass is the mean location of all the mass in a system. We consider first two bodies, with masses m_1 and m_2, and positions r_1 and r_2 respectively. We denote by r_c the center of mass for the bodies. Then, due to the Law of the Lever (i.e., the rod is balanced at the center of mass), we note that $m_1(r_c - r_1) = m_2(r_2 - r_c)$.

Therefore, the center of mass in this case is given as

1
$$r_c = \frac{m_1 r_1 + m_2 r_2}{m_1 + m_2}.$$

In general, for N bodies, with masses $m_1, m_2, \ldots, m_N$ and their positions $r_1, r_2, \ldots, r_N$ from the origin, respectively, the center of mass, denoted by r_c, is located at

2
$$r_c = \frac{\sum_{k=1}^{N} m_k r_k}{m},$$

where $m = \sum_{k=1}^{N} m_k$ is the total mass of the bodies. The numerator in the formula 2, $\sum_{k=1}^{N} m_k r_k$ is called the moment of the masses.

EXAMPLE 1

Find the center of mass of an object that has masses 3, 4, and 8 at the points $(-2, 1)$, $(3, -1)$, and $(3, 2)$, respectively.

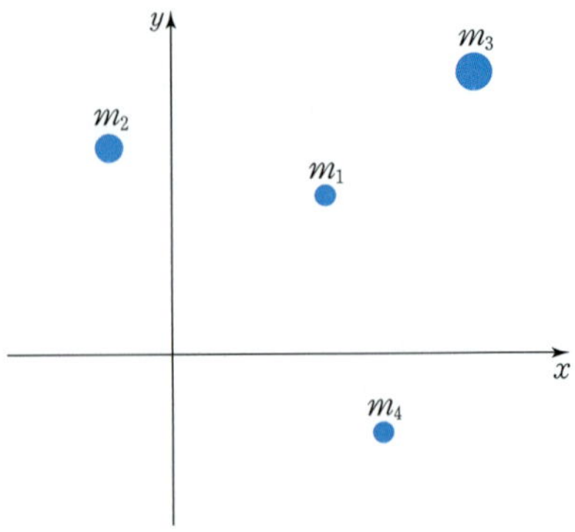

Figure 2 Discrete Masses

Solution We use the formula 2:

$$r_c = \frac{3(-2, 1) + 4(3, -1) + 8(3, 2)}{3 + 4 + 8} = \frac{(30, 15)}{15} = (2, 1).$$

Thus, the center of mass is $(2, 1)$. ■

Centroids

We consider a flat plate, denoted by D, with uniform density ρ in the plane.

The centroid of a plane region is geometrically the intersection of all straight lines that divide D into two parts of equal moment about the line. Roughly speaking, the centroid is the average of all points of D. The concept of the centroid can be extended to any domain R in $\mathbb{R}^n$ as follows: the centroid is the intersection of all hyperplanes that divide R into two parts of equal moment. The centroid, denoted by $C=(\overline{x_1}, \overline{x_2}, \cdots, \overline{x_n})$ of a domain R in $\mathbb{R}^n$ can be obtained as an integral

3 $$C=\frac{\int xI_R(x)dx}{\int I_R(x)dx},$$

where I_R is a characteristic function, which is 1 if $x\in\mathbb{R}$ and 0 otherwise. Another formula for each coordinate for the centroid is given as follows:

4 $$\overline{x_k}=\frac{\int zI_{R,k}(z)dz}{\int I_{R,k}(z)dz},$$

where $I_{R,k}(z)$ is the measure of the intersection of $\mathbb{R}$ and the hyperplane defined by the equation $x_k=z$. The denominator is nothing but the measure of R. For the plane, due to the formula 4, the coordinates of the centroid become

5 $$\overline{x}=\frac{\int xI_{D,y}(x)dx}{|D|},\quad \overline{y}=\frac{\int yI_{D,x}(y)dy}{|D|},$$

where $I_{D,y}(x)$ is the length of the intersection of the flat plate D and the line parallel to y-axis passing through $(x, 0)$ and $I_{D,x}(y)$ is similarly the length of the intersection of D and the line parallel to x-axis passing through $(0, y)$.

We remark that if the domain under consideration is symmetric with respect to a hyperplane, then the centroid lies on the hyperplane. Therefore, for a flat plane if D is symmetric about a line l, then the center of mass is on the line l. This is intuitively obvious, and is often referred as the "symmetric principle", from which a computation may become simpler, as long as symmetry is observed for the domains under consideration.

EXAMPLE 2

Consider a flat plate D, which is enclosed by $x=a$, $x=b$, x-axis, and the graph of f, where f is a continuous function. Suppose that the density is ρ, which is uniformly distributed on the plate. Show that the centroid (center of mass) of the plate is

6 $$\overline{x}=\frac{\int xf(x)dx}{|D|},\quad \overline{y}=\frac{\int \frac{1}{2}|f(x)|^2dx}{|D|}.$$

Solution $\overline{x}$ is given in the form of 6 directly from 5. It remains to show 6

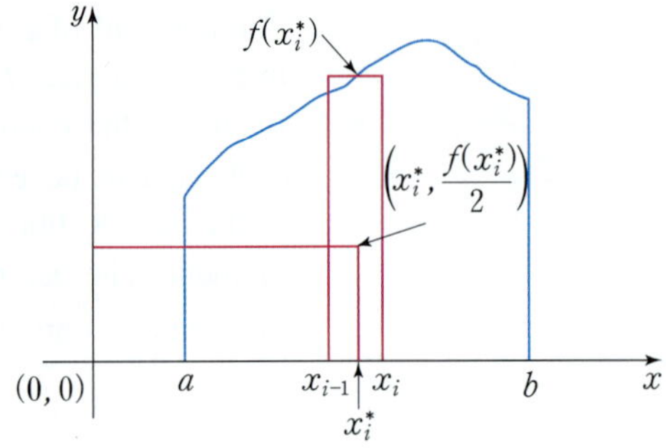

Figure 3

for the case of $\bar{y}$.

We divide the interval $[a, b]$ into n subintervals with endpoints x_0, x_1, x_2, ..., x_n and equal width $\Delta x = (b-a)/n$. We denote $x_i^* = (x_i + x_{i-1})/2$, which is the midpoint of ith subinterval. Note that the centroid of the ith approximating rectangle, denoted by D_i, is $(x_i^*, f(x_i^*)/2)$ and its mass is $\rho f(x_i^*)\Delta x$. Furthermore, the moment of R_i about the x-axis is $\rho f(x_i^*)\,\Delta x \cdot \dfrac{f(x_i^*)}{2} = \dfrac{\rho}{2}[f(x_i^*)]^2\,\Delta x$. The total moment of approximating the plate is the sum of each moment, that is, $\sum_{i=1}^{n}\rho\frac{1}{2}[f(x_i^*)]^2\Delta x$. Now passing it to the limit, total moment about x-axis becomes the following integral: $\dfrac{\rho}{2}\displaystyle\int_a^b |f(x)|^2\,dx$. Therefore, y-coordinate of centroid is just the above integral divided by the total mass. The formula 6 is deduced. ■

We remark that if a flat plate D is enclosed by $y=f(x)$ and $y=g(x)$, where $f(x) \geq g(x)$ for $a \leq x \leq b$, then the formula 6 becomes

7 $$\bar{x} = \frac{\int x(f(x)-g(x))dx}{|D|}, \quad \bar{y} = \frac{\int \frac{1}{2}(|f(x)|^2 - |g(x)|^2)dx}{|D|}.$$

Its verification is the same as the case of 6 and thus the details are omitted.

We close this section by recalling Pappus's centroid theorem (also known as the Guldinus theorem, Pappus–Guldinus theorem or Pappus' theorem), which shows a connection between the surface areas and volumes enclosed by surfaces and solids of revolution.

THEOREM 1 Pappus's Centroid Theorem (PART I)

The surface area A of a surface of revolution generated by rotating a plane curve C about an axis external to C and on the same plane is equal to the product of the arc length s of C and the distance d traveled by its geometric centroid.

$$A = sd.$$

THEOREM 2 Pappus's Centroid Theorem (PART II)

The volume V of a solid of revolution generated by rotating a plane figure F about an external axis is equal to the product of the area A of F and the distance d traveled by its geometric centroid.

$$V = Ad. \tag{8}$$

EXAMPLE 3

(a) Find the centroid of the region bounded by the line $y = 2x$ and the parabola $y = x^2$ (we assume that the density of the enclosed region is 1).

(b) Use Pappus's centroid theorem to find the volume of the solid obtained by rotating about the y-axis the region enclosed by the line $y = 2x$ and the parabola $y = x^2$.

Solution (a) The area of the enclosed region is

$$A = sd = \int_0^2 (2x - x^2)dx = \left[x^2 - \frac{x^3}{3}\right]_0^2 = 4 - \frac{8}{3} = \frac{4}{3}.$$

Due to formula 8, we have

$$\bar{x} = \frac{1}{A}\int_0^2 x(2x - x^2)dx = \frac{3}{4}\left[\frac{2x^3}{3} - \frac{x^4}{4}\right]_0^2 = 1,$$

$$\bar{y} = \frac{1}{A}\int_0^2 \frac{1}{2}(4x^2 - x^4)dx = \frac{3}{4}\left[\frac{2x^3}{3} - \frac{x^5}{10}\right]_0^2 = \frac{8}{5}.$$

So the centroid is $\left(1, \frac{8}{5}\right)$.

(b) Using result (a), we know that x-coordinate of the centroid for the enclosed region is $\bar{x} = 1$. Therefore, with the aid of Pappus' centroid theorem, the volume of rotating the region is $V = Ad$ $= A \cdot 2\pi\bar{x} = \frac{8\pi}{3}$. Since $\frac{1}{4 + y^2}dy = dx$ is separabl, using integration on both sides

$$\int \frac{1}{4 + y^2}dy = \int dx + \frac{C}{2}$$

$$\tan^{-1}\frac{y}{2} = \frac{x}{2} + C \quad \text{or} \quad y = 2\tan\left(\frac{x}{2} + C\right).$$

The solution is $y = 2\tan\left(\frac{x}{2} + C\right)$. ($C$ is an arbitrary constant.) ■

8.3 EXERCISES (Center of Mass)

http://matrix.skku.ac.kr/Cal-Book/part1/CS-Sec-8-3-Sol.html

YouTube http://youtu.be/

1. Find the center of mass of the region bounded by the cycloid and x-axis (we assume that the density of the enclosed region is 1).

$$x=(\theta-\sin\theta),\ y=(1-\cos\theta),\ 0\le\theta\le 2\pi$$

(Hint: $\int_0^{2\pi}\cos\theta d\theta=\int_0^{2\pi}\cos^3\theta d\theta = \int_0^{2\pi}\cos 2\theta d\theta = 0$)

Solution By the symmetry principle, the center of mass must lie on $x=\pi$, so $\bar{x}=\pi$. The area of the region is computed as follows:

$$A=\int_0^{2\pi} y dx=\int_0^{2\pi}(1-\cos\theta)^2 d\theta$$
$$=\int_0^{2\pi}(1-2\cos\theta+\cos^2\theta)d\theta$$
$$=\int_0^{2\pi}\left(1+\frac{1+\cos 2\theta}{2}\right)d\theta=3\pi.$$

Recall the formula for y-coordinate of the center of mass is given as follows:

$$\bar{y}=\frac{1}{A}\int_0^{2\pi}\frac{y^2}{2}dx=\frac{1}{6\pi}\int_0^{2\pi}(1-\cos\theta)^3 d\theta$$
$$=\frac{1}{6\pi}\int_0^{2\pi}(1+3\cos^2\theta)d\theta$$
$$=\frac{1}{6\pi}\int_0^{2\pi}\left(1+\frac{3}{2}(1+\sin 2\theta)\right)d\theta=\frac{5\pi}{6\pi}=\frac{5}{6}.$$

Summing up, the center of mass $(\bar{x},\ \bar{y})$ is $\left(\pi,\ \frac{5}{6}\right)$.

CAS **2.** By using graphical tools, find the approximate center of mass of the region which is bounded by the $x^2+y^2=2+\cos 2\theta,\ 0\le\theta\le 2\pi$.

http://matrix.skku.ac.kr/cal-lab/cal-8-3-3.html

Solution

```
theta=var('theta');
polar_plot(sqrt(2+cos(2*theta)), (0, 2*pi),
fill=True).show(aspect_ratio=1, xmin=-2,
xmax=2, ymin=-2, ymax=2)
```

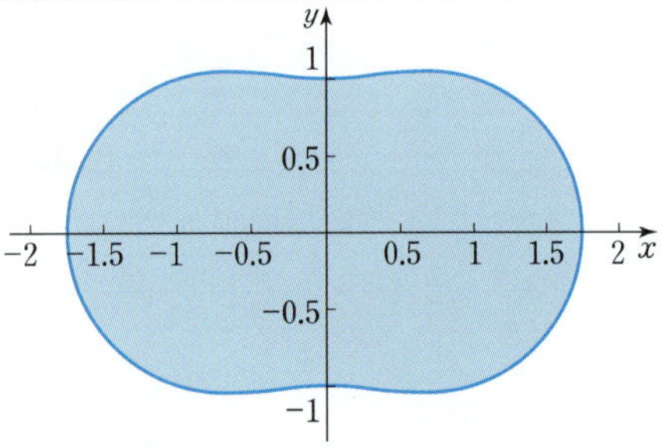

By looking at the figure it is clear that the center of mass of the region is approximately $(0,\ 0)$.

<Calculator>

*8.4 Differential Equations

YouTube http://youtu.be/uHfOjz8I4-s
http://www.youtube.com/watch?v=sca2RiBG3E0
http://www.youtube.com/watch?v=4wYqhJ0V76M

Differential equations is one of the most important applications of differentiation and integration in Calculus.[1] In this section we introduce several models of ordinary differential equations, which seem relatively simple but quite informative.[2] A method of finding solutions is also presented for a separable differential equation although in general, it is not an easy task to solve differential equations explicitly.

1)http://wstein.org/home/wstein/www/home/mhampton/ODE_6_2011/des-book-2011-09-05.pdf by Joyner and Hampton

2)http://www.sagemath.org/doc/reference/calculus/sage/calculus/desolvers.html

Models of Ordinary Differential Equations

One very simple model for population growth assumes that the rate of growth of a population is proportional to the size of the population. This assumption is ideal and may be observable in certain situations. For convenience, we denote by $P(t)$ the number of inhabitants in the population, where t indicates the time. The hypothesis is then written as the following equation:

1 $$\frac{dP}{dt} = kP,$$

where k is the proportionality constant. Suppose that the population size at a certain time (here without loss of generality such time is 0) can be measured, let us denote it as $P(0)$. Then the question is what the population is at time t. This type of question is called an initial-value problem. It turns out that the solution for 1 is $P(t) = P(0)e^{kt}$ (you can check this directly). In this case, finding the solution is not difficult and details of the procedure will be presented subsequently in example 2.

A modified version of the above **population model**[3] is the "logistic model", which is based on the following assumption: There is a number C, the carrying capacity, so that the population approaches C as time passes by. More precisely, if size of population is small, the rate of the growth of population is assumed to be proportional to the size of population as in the previous case. However, we suppose that $\frac{dP}{dt} < 0$ if $P > C$ (P decreases if it ever exceeds C). A typical example satisfying such assumptions would be

3)http://matrix.skku.ac.kr/2013-Album/GrowthModel-v1.html

2 $$\frac{dP}{dt} = kP\left(1 - \frac{P}{C}\right).$$

This seems more realistic compared to the previous case in the sense that it does not allow never-ending growth (but this does not mean that the logistic model reflects reality).

So far, we have considered some models for a single population, next we consider models for two populations that interact with one another. One simple example is predator-prey models, which can be written in the form of differential equations shown below: (**predator-prey models**[4] have a variety of descriptions but here we introduce only a simple case). In the predator-prey model, there are two species, one is the prey and the other is the predator. For convenience, we denote by $W(t)$ and $R(t)$ the number of the prey and the predator, respectively. In this case, the hypothesis is that the rate of predation on the prey is proportional to the rate at which the predators and the prey meet, and vice versa, possibly with a different proportional constant. The differential equations taking these assumptions into account is described as follows:

4)http://matrix.skku.ac.kr/2013-Album/Predator-Prey.html

$$\boxed{3} \qquad \frac{dR}{dt} = k_1 R - k_2 RW,$$

$$\boxed{4} \qquad \frac{dW}{dt} = -k_3 W + k_4 RW,$$

where k_1, k_2, k_3 and k_4 are positive constants. The system of this prey-predator model is sometimes referred to the **Lotka-Volterra equations.**[5] Compared to the previous two examples, the solutions are a pair of functions to a coupled system of differential equations.

5)http://matrix.skku.ac.kr/cal-lab/cal-8-4-Epidemic.html

Separable Differential Equations

For first-order ordinary differential equations, a separable equation is of the form

$$\frac{dy}{dx} = f(x)g(y).$$

If $g(y) \neq 0$, we separate variables in the following manner:

$$\frac{dy}{g(y)} = f(x)dx.$$

Integrating both sides of the equation we get

$$\int \frac{dy}{g(y)} = \int f(x)dx.$$

Summing up, if a differential equation is separable, it can be eventually separated into a pair of integrals.

EXAMPLE 1

Solve $\dfrac{dy}{dx} = -\dfrac{y}{x}$ with initial condition $y(1) = 1$.

http://matrix.skku.ac.kr/cal-lab/cal-8-4-Exm-1.html

Solution $\dfrac{1}{y}dy = -\dfrac{1}{x}dx$ is separable variable equation.

By integration on both sides of the equations

$$\int \frac{1}{y}\,dy = -\int \frac{1}{x}\,dx + C,$$

$$\ln|y| = -\ln|x| + \ln C \ \text{ or } \ y = \frac{C}{x}.$$

The general solution is $y = \frac{C}{x}$, where C is arbitrary constant.

From the given initial condition $y(1) = 1$, we obtain $C = 1$.

Thus the solution of the initial-value problem is $y = \frac{1}{x}$.

```
sage: x = var('x')
sage: y = function('y', x)
sage: de = lambda y: diff(y,x) + y/x
sage: desolve(de(y),[y,x],[1,1])     # 1/x
```

■

EXAMPLE 2

Find the general solution to $\frac{dy}{dx} = 4 + y^2$ either "by hand" or using Sage.

Solution The equation $\frac{1}{4+y^2}dy = dx$ is with separable variables. By integration of both sides we have

$$\int \frac{1}{4+y^2}dy = \int dx + \frac{C}{2},$$

$$\tan^{-1}\frac{y}{2} = \frac{x}{2} + C \ \text{ or } \ y = 2\tan\left(\frac{x}{2} + C\right).$$

The solution is $y = 2\tan\left(\frac{x}{2} + C\right)$. ($C$ is arbitrary constant.) ■

Next we recall the model of population growth [1], which can also be solved explicitly by the method of separable variables. We note that the solution is in fact growing if $k > 0$ and is decaying in case that $k < 0$.

EXAMPLE 3

Find the general solution of the equation [1], that is $\frac{dP}{dt} = kP$, where k is a constant.

Solution We can rewrite the equation as follows: $\frac{dP}{P} = kdt$, so integrating both sides,

$$\int \frac{dP}{P} = \int kdt$$

$$\ln|P(t)| = kt + c$$

$$|P(t)| = e^{kt+c} = e^c e^{kt}$$
$$P(t) = \pm e^c e^{kt} = Ce^{kt}$$

where C is a constant. Simplifying it, we obtain $P(t) = Ce^{kt}$. We obtained

$$P(0) = Ce^{k \cdot 0} = C.$$

Hence C is the initial value of the given equation. ■

The next example is for logistic models such as 2, which are again separable differential equations. This means that such equations are explicitly solvable (as long as we can find antiderivatives).

EXAMPLE 4

Solve $\dfrac{dP}{dt} = P\left(1 - \dfrac{P}{10}\right)$, with an initial condition $P(0) = 5$.

Solution Using the method of separable variables, we have

$$\int \frac{dP}{P(1 - P/10)} = \int dt.$$

Due to the method of partial fractions, the left side of the integrals can be rewritten and computed as follows:

$$\frac{1}{P(1-P/10)} = \frac{10}{P(10-P)} = \frac{1}{P} + \frac{1}{10-P}$$
$$\int \left(\frac{1}{P} + \frac{1}{10-P}\right) dP = \int dt$$
$$\ln|P| - \ln|10-P| = t + c$$
$$\ln\left|\frac{P}{10-P}\right| = t + c$$
$$\left|\frac{P}{10-P}\right| = e^{t+c} = e^c e^t$$
$$\frac{P}{10-P} = \pm e^c e^t = Ce^t \quad \text{where } C = \pm e^c$$
$$P(t) = \frac{10}{1 + Ce^{-t}}$$

Since $P(0) = \dfrac{10}{1+C} = 5 \Rightarrow C = 1$, the solution is

$$P(t) = \frac{10}{1 + e^{-t}}.$$ ■

EXAMPLE 5

Find the general solution to $xy' + y = e^x$ either "by hand" or using Sage.
http://matrix.skku.ac.kr/cal-lab/cal-8-4-Exm-8.html

Solution We rewrite this as $y' + \frac{1}{x}y = \frac{e^x}{x}$. Now compute $\mu = e^{\int \frac{1}{x}dx} = e^{\ln(x)} = x$ so the formula gives

```
sage: t = var('t')
sage: x = function('x', t)
sage: de = lambda y: diff(y,t) + (1/t)*y - exp(t)/t
sage: desolve(de(x),[x,t])     # (c + e^t)/t
```

Answer : (c + e^t)/t ■

Orthogonal Trajectory

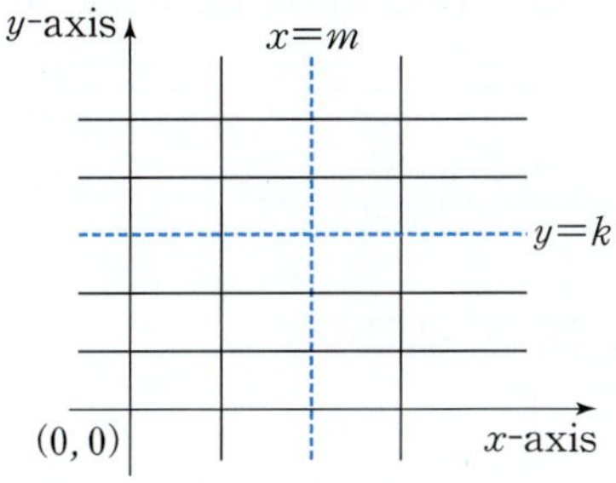

Figure 1 Orthogonal trajectories

We close this section by discussing orthogonal trajectories. Orthogonal trajectories are two families of curves in the plane that intersect a given family of curves at right angles. For example, the family of straight lines parallel to y-axis, that is $x = m$ for any $m \in \mathbb{R}$, is an orthogonal trajectory for the family of straight line parallel to x-axis, that is $y = k$ for any $k \in \mathbb{R}$. We say that two families of curves (curves including the straight lines) are orthogonal trajectories of each other.

```
var('y')
L=[]
for i in srange(-2, 2, 0.2):
    p=implicit_plot(x^2+y^2==i*x, (x, -2, 2), (y, -2, 2))
    L.append(p)
M=[]
for i in srange(-2, 2, 0.4):
    p=implicit_plot(x^2+y^2==i*y, (x, -2, 2), (y, -2, 2), cmap=['red'])
    M.append(p)
sum(L)+sum(M)
```

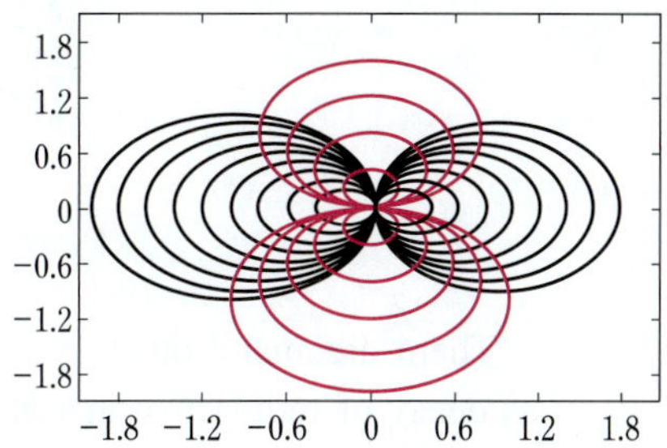

Figure 2

EXAMPLE 6

Find the orthogonal trajectories of the family of straight lines through the origin, that is $y = mx$ for any $m \in R$.

http://matrix.skku.ac.kr/cal-lab/cal-8-4-Exm-4.html

Solution If we differentiate $y = mx$, we get $\frac{dy}{dx} = m$. To eliminate m, we note that $m = y/x$ and so the differential equation can be rewritten as $\frac{dy}{dx} = y/x$. Since an orthogonal trajectory must have the slope of its tangent line that is negative reciprocal of this slope. Thus, the orthogonal trajectories must satisfy $\frac{dy}{dx} = -x/y$. Since it is separable, we compute

$$\int y dy = -\int x dx \Rightarrow y^2 = -x^2 + C,$$

where C is an arbitrary positive constant. Therefore, the orthogonal trajectory is the family of concentric circles centered at the origin.■

```
var('y')
L=[]
for i in srange(-2, 2, 0.4):
    p=implicit_plot(y==i*x, (x, -2, 2), (y, -2, 2))
    L.append(p)
M=[]
for i in srange(0, 2, 0.2):
    p=implicit_plot(x^2+y^2==i, (x, -2, 2), (y, -2, 2),
cmap=['red'])
    M.append(p)
sum(L)+sum(M)
```

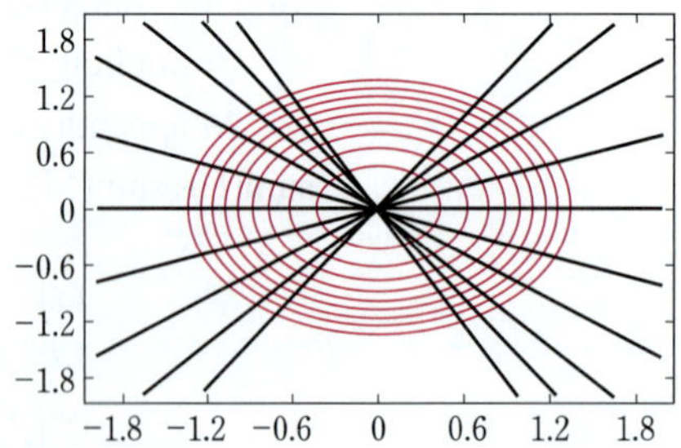

Figure 3

There are many other applications of first order differential equations, such as decay of radioactive materials, mixing problems involving two or more fluids, and so on (some examples are in the exercises).

8.4 EXERCISES (Differential Equations)

http://matrix.skku.ac.kr/Cal-Book/part1/CS-Sec-8-4-Sol.html

YouTube http://www.youtube.com/watch?v=flNqoAoZNLA

1. Find the solutions of the given initial-value problem $y' = 2xy\ln y,\ y(0) = e$.

Solution The given equation is separable, and therefore, it can be rewritten as

$$\frac{y'}{y\ln y} = 2x \Rightarrow \int \frac{dy}{y\ln y} = \int 2x dx$$

$$\Rightarrow \ln|\ln y| = x^2 + C \Rightarrow |\ln y| = Ce^{x^2}.$$

Using the initial condition, we know that $C=1$, and therefore, $y(t) = e^{e^{x^2}}$.

2. Find the orthogonal trajectories of the family of the following curves: $y = (x+k)^{-1}$.

Solution Since $y' = -(x+k)^{-2}$, the orthogonal trajectories must have their derivative perpendicular to the given family of curves, namely

$$y' = \frac{dy}{dx} = (x+k)^2 = y^{-2}. \qquad (1)$$

Note that the given curves never touch x-axis, which means that $y \neq 0$, and so the above differential equation (1) makes sense. Now noting that (1) is of separable type, we get

$$y^2 dy = dx \Rightarrow \int y^2 dy = \int dx \Rightarrow \frac{y^3}{3} = x + C$$

3. Solve the differential equation $\dfrac{dx}{dy} = x^3(1+y^2)$.

Solution Separation of variables gives $\dfrac{dy}{1+y^2} = x^3 dx$. By the integration, we have $\tan^{-1} y = \dfrac{1}{4}x^4 + C$ So, by simplification, we obtain $y = \tan\left(\dfrac{x^4}{4} + C\right)$.

4. Solve the differential equation $2dy/dt = -y(y-1)(y-2)$ that satisfies the initial condition $y(0) = 7$.

Solution We first note that differential equation is separable. Therefore, after splitting it into variables and using the integration in each variable, we have

$$\int \frac{2}{y(y-1)(y-2)} dy = -\int dt$$

$$\Rightarrow \int \left[\frac{1}{y} - \frac{2}{y-1} + \frac{1}{y-2}\right] dy = -t + C$$

$$\ln\left|\frac{y(y-2)}{(y-1)^2}\right| = -t + C \Rightarrow \frac{y(y-2)}{(y-1)^2} = Ce^{-t}.$$

Since $y(0) = 7$, we have $C = 35/36$. Summing up, we obtain $\dfrac{y(y-2)}{(y-1)^2} = \dfrac{35}{36}e^{-t}$.

5. A tank contains 1000 L of brine with 15kg of dissolved salt. Brine that contains 0.07kg of salt per liter of water enters the tank at a rate of 10L/min and another brine that contains 0.02kg of salt per liter of water enters the tank at a rate of 15L/min. The solution is kept thoroughly mixed and drains from the tank at a rate of 25L/min. How much salt is in the tank as t tends to ∞?

Solution Let $y(t)$ be the amount of salt (in kilograms) after t minutes. The rate at which enters the tank is

$$0.07\text{kg/L} \times 10\text{L/min} + 0.02\text{kg/L} \times 15\text{L/min} = 0.1\text{kg/min}.$$

The tank always contains 1000L of liquid and so the rate at which leaves the tank is

$$\frac{y(t)}{1000}\text{kg/L} \times 25\text{L/min} = \frac{y(t)}{40}\text{kg/min}.$$

Thus, the differential equation is

$$\frac{dy}{dt} = 0.1 - \frac{y(t)}{40} = \frac{10 - y(t)}{40},\ y(0) = 15.$$

Solving this separable differential equation, we obtain

$$\int \frac{dy}{10-y} = \int \frac{dt}{40} \Rightarrow -\ln|10-y| = \frac{t}{40} + C$$

$$\Rightarrow y(t) = 10 + Ce^{-\frac{t}{40}}$$

Since $y(0) = 15$, we have $y(t) = 10 + 5e^{-\frac{t}{40}}$. As t tends to ∞, y converges to 10kg.

6. Let $x(t)$ and $y(t)$ be the number of prey and predators at time t, respectively. We suppose that these two species satisfy the following system of differential equations:

$$\frac{dx}{dt} = 1000x - 100xy,\ \frac{dy}{dt} = -100y + xy.$$

(a) Find all equilibrium solutions and the expression for $\dfrac{dy}{dx}$.

(b) Solve $\dfrac{dy}{dx}$ with given initial conditions $x(0)$

$= 1000$ and $y(0) = 10$ by the method of separable variables.

Solution (a) The equilibrium are steady state solutions, and so they must satisfy $1000x - 100xy = 0$, $-100y + xy = 0 \Rightarrow (x, y) = (0, 0)$ or $(x, y) = (100, 10)$. Using the chain rule, we obtain

$$\frac{dy}{dx} = \frac{dy/dt}{dx/dt} = \frac{y(x-100)}{100(10-y)x}. \tag{1}$$

(b) Note that the initial condition $y = 10$ when $x = 1000$ and by solving the equation (1), we get

$$\frac{10-y}{y}dy = \frac{x-100}{100x}dx \Rightarrow 10\ln y - y = \frac{x}{100} - \ln x + C.$$

It remains to specify the constant C. Indeed, due to initial condition, we can see

$$C = 10\ln 10 - 10 - 10 + \ln 1000 = 13\ln 10 - 20.$$

Hence, the solution satisfies

$$10\ln y - y = \frac{x}{100} - \ln x + 13\ln 10 - 20.$$

7. The half-life of Cesium-137 is 30 years. Suppose that we have a 100mg sample. After how long will only 1mg remain?

Solution Let $m(t)$ be the mass of Cesium-137 (in milligrams) that remains after t years. Then $\frac{dm}{dt} = km$ and $y(0) = 100$, so we have $m(t) = m(0)e^{kt} = 100e^{kt}$. In order to determine the value of k, we use the fact that $y(30) = 50$. Thus we get

$$100e^{30k} = 50 \Rightarrow k = -\frac{\ln 2}{30} = 100e^{-\frac{\ln 2}{30}t}.$$

We want to find the value of t such that $m(t) = 1$. Solving the following equation for t and taking the natural logarithm of both sides, we have

$$100e^{-\frac{\ln 2}{30}t} = 1 \Rightarrow t = 30\frac{\ln 100}{\ln 2} \text{ years}.$$

8. Let $m > 0$ be a positive constant. We consider the following differential equation .

$$\frac{dy}{dt} = y^m(t),\ y(0) = 1,\ t \ge 0. \tag{1}$$

For what value of m does the solution of (1) exist for all time (global existence or blow up in a finite time)? Verify your answer.

Solution Since the equation is separable, we get $y'/y^m = 1$. We consider first the case $0 < m < 1$. Keeping in mind that $y(0) = 1$, we obtain

$$\frac{1}{1-m}(y^{1-m})' = 1 \Rightarrow y^{1-m}(t) = (1-m)t + 1$$

$$\Rightarrow y(t) = [(1-m)t+1]^{\frac{1}{1-m}}$$

In this case, a solution exists globally for all time. Next we consider the case $m = 1$. Since $y'/y = 1$ and $y(0) = 1$, we obtain $y(t) = e^t$, which again exists globally. Finally, in case that $m > 1$, as in the previous case of $m < 1$, we have

$$\frac{1}{1-m}(y^{1-m})' = 1 \Rightarrow \frac{1}{y^{m-1}(t)} = 1 - (m-1)t$$

$$\Rightarrow y(t) = \frac{1}{[1-(m-1)t]^{\frac{1}{m-1}}}$$

The above solution is well-defined as long as $t < \frac{1}{m-1}$ but its limit, goes to ∞ as t approaches $\frac{1}{m-1}$. Therefore, the solution blows up in a finite time for the case $m > 1$.

CAS **9.** Solve the differential equation $\frac{dy}{dx} = \frac{3x^2+2x+1}{y-1}$, with the initial condition $y(0) = -1$ and determine the interval in which the solution exists.

(Hint: To find the interval of definition, look for points where the integral curve has a vertical tangent)

(You may do it with Sage at http://math1.skku.ac.kr/.
Open resources in http://math1.skku.ac.kr/pub/)
http://matrix.skku.ac.kr/cal-lab/cal-8-4-9.html

Solution Since $\int (y-1)dy = \int (3x^2+2x+1)\,dx$ and $y(0) = -1$, we obtain the following integral curve.

```
f = lambda x,y: y^2-2*y-2*x^3-2*x^2-2*x-3
contour_plot(f, (-5, 5), (-5, 5), fill=0, contours=20,plot_points=200)
```

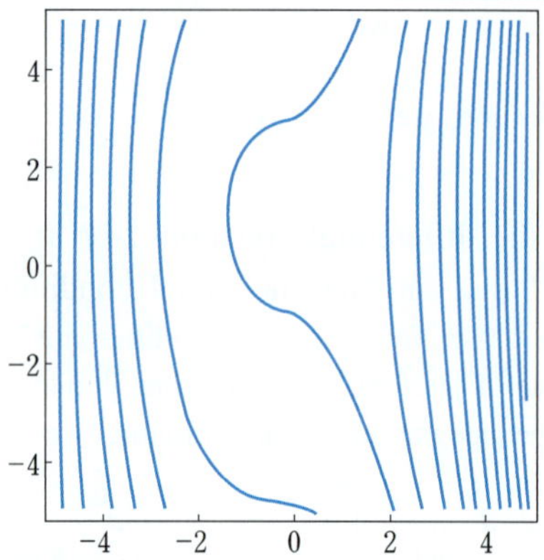

Let us enlarge it.

```
f = lambda x,y: y^2-2*y-2*x^3-2*x^2-2*x-3
contour_plot(f, (-5, 5), (-5, 5), fill=0, contours=2, plot_points=200)
```

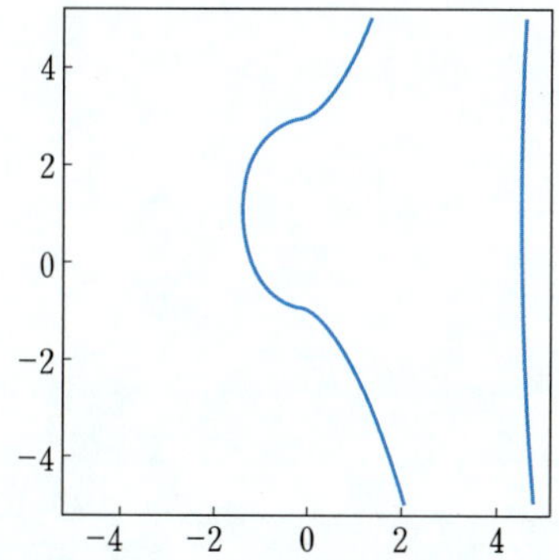

The integral curve has a vertical tangent around $(-2, 1)$. Therefore, the desired interval is $[-2, \infty]$.

Now, we can consider another solution. Since $x^3+2x^2+x=\frac{1}{2}y^2-y+c$ and $y(0)=-1$, we obtain $c=-\frac{3}{2}$. To find the interval of definition, look for points where the integral curve has a vertical tangent.

```
x,y=var('x, y')
p1=implicit_plot(y^2-2*y-2*x^3-2*x^2-2*x-3, (x,
-3, 3), (y, -3, 3))
show(p1)
```

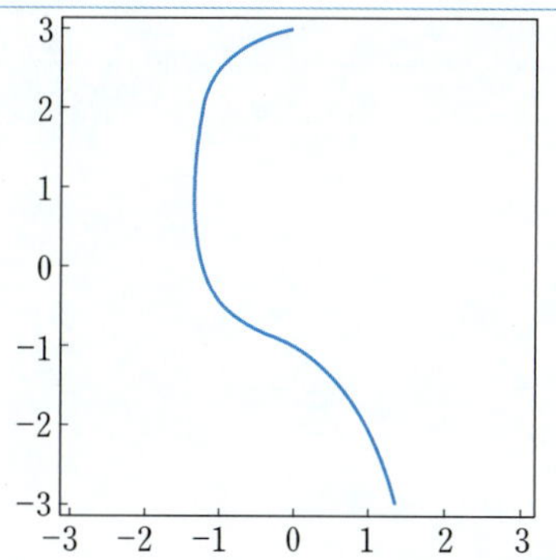

The integral curve has a vertical tangent around $(-2, 1)$. Therefore, the desired interval is $[-2, \infty]$.

CAS **10.** Solve the differential equation $\frac{dy}{dx}=\frac{3x}{6y+\sin 1y}$, with the initial condition $y(0)=-1$. Hence plot the solution curve. Also plot the solution curves for a range of initial conditions.

Solution

```
reset()
var('t')
y = function('y', x)
de = diff(y, x) == 3*x^2/(6*y+sin(y))
sol=desolve(de, y, [0,-1])
print "The required solution is given by"
show(sol)
s=sol.subs_expr(y(x)==t)
p=implicit_plot(s, (x,-3,3), (t,-3,3));
p
```

The required solution is given by

$$y(x)^2-\frac{1}{3}\cos(y(x))=\frac{1}{3}x^3-\frac{1}{3}\cos(1)+1.$$

We plot the solutions curves for a range of initial values.

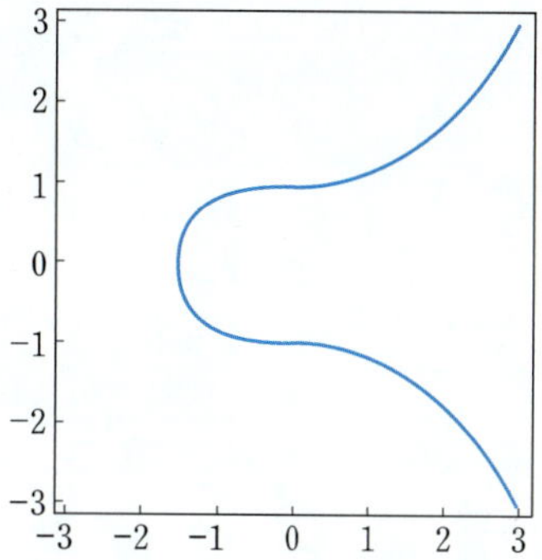

```
reset()
var('t')
y = function('y', x)
de = diff(y,x) == 3*x^2/(6*y+sin(y))
sol=desolve(de, y, [0,-1])
s=sol.subs_expr(y(x)==t)
p=implicit_plot(s, (x, -3, 3), (t, -3, 3))
x0 = srange(-1, 2, 0.2)
soln = [desolve(de, y, [0,float(c)]) for c in x0]
for i in range(len(soln)):
    g=soln[i].subs_expr(y(x)==t)
    p=p+implicit_plot(g, (x, -3, 3), (t, -3, 3))
p
```

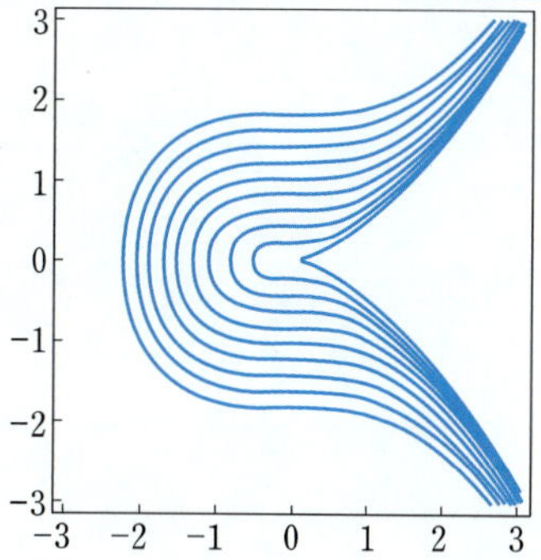

Chapter 9

Infinite Sequences and Infinite Series

Al-Khwarizmi

780-850

Algebra

Al-Khwarizmi was a Persian mathematician, astronomer and geographer. His Kitab al-Jabr wal-Muqabala presented the first systematic solution of linear and quadratic equations, along with Diophantus. In the twelfth century, Latin translations of his work on the Indian numerals introduced the decimal positional number system to the West.

YouTube http://youtu.be/

This chapter is divided into two parts: infinite sequence and infinite series. The first part deals with a brief introduction to infinite sequence, limit laws of sequences and some basic results. In the second part, we introduce infinite series, which are essentially the limit of sequences associated with each infinite series. Some of the standard tests of convergence of infinite series are also introduced in this chapter. We also study power series and their convergence. The chapter ends with Taylor's and Maclaurin's series expansions of functions, which are two of the very important tools for approximating functions. Infinite series have a wide range of applications and can be found in theoretical and practical problems in mathematics, physics and other scientific fields.

9.1 Sequences and Series

A list of numbers in a definite order, say, $a_1, a_2, a_3, \ldots, a_n, \ldots$, can be given precise mathematical meaning. A sequence is a function f whose domain is the set of positive integers. If $f(n) = a_n$, for the number n, it is customary to denote the sequence f by the notation $\{a_n\}_{n=1}^{\infty}$ or simply $\{a_n\}$ or sometimes by : $a_1, a_2, a_3, \ldots, a_n, \ldots$. The range of f, that is, the values a_n, are called the terms of the sequence. The number a_1 is called the first term of the sequence, a_2 the second term, and so on, and a_n, the nth term.

An example is the sequence $\left\{\dfrac{n-1}{n}\right\}_{n=1}^{\infty}$, in which $a_n = \dfrac{n-1}{n}$, or writing out its terms,

$$\left\{0, \frac{1}{2}, \frac{2}{3}, \frac{3}{4}, \ldots, \frac{n-1}{n}, \ldots\right\}.$$

Another example is the sequence $\left\{\sin\dfrac{n\pi}{3}\right\}_{n=0}^{\infty}$, in which $a_n = \sin\dfrac{n\pi}{3}$, or written in detail,

$$\left\{0, \frac{\sqrt{3}}{2}, \frac{\sqrt{3}}{2}, 0, \ldots, \sin\frac{n\pi}{3}, \ldots\right\}.$$

Notice that n does not have to start at 1.

Sequences are frequently specified by simply giving a formula for the nth term. For example, if $a_n = \sqrt{n+1}$, then the sequence $\{a_n\}$ is

$$\{\sqrt{2}, \sqrt{3}, \ldots, \sqrt{n+1}, \ldots\}.$$

Another way to specify a sequence is by means of a recursion formula. As an example, consider the sequence called the **Fibonacci sequence** $\{F_n\}$ whose terms satisfy the following conditions:

$$F_1 = 1, F_2 = 1 \text{ and } F_{n+1} = F_n + F_{n-1}, \quad n \geq 2.$$

Written in detail, the sequence is

$$\{1, 1, 2, 3, 5, 8, 13, 21, \ldots\}.$$

Decimal Expansion

The decimal expansion of a real number is its representation with base-10 (That is, the decimal system). For example, the decimal expansion of 25^2 is 625, of π is 3.14159⋯, and of $1/9$ is 0.1111⋯.

The decimal expansion of a real number can be found in Sage using the following commands: R = RealField(200); R(17/6); R(pi); R(e).

The decimal expansion of a number may terminate (in which case the number is called a **regular number** or finite decimal, for example, $1/2 = 0.5$), eventually become periodic (in which case the number is called a repeating decimal, for example, $1/3 = 0.\overline{3}$), or continue infinitely without repeating (in which case the number is called irrational).

The following table summarizes the decimal expansions of the first few unit fractions. As usual, the repeating portion of a decimal expansion is conventionally denoted as given in the following table.

fraction	decimal expansion	fraction	decimal expansion
1	1	1/11	$0.\overline{09}$
1/2	0.5	1/12	$0.08\overline{3}$
1/3	$0.\overline{3}$	1/13	$0.\overline{076923}$
1/4	0.25	1/14	$0.0\overline{714285}$
1/5	0.2	1/15	$0.0\overline{6}$
1/6	$0.1\overline{6}$	1/16	0.0625
1/7	$0.\overline{142857}$	1/17	$0.\overline{0588235294117647}$
1/8	0.125	1/18	$0.0\overline{5}$
1/9	$0.\overline{1}$	1/19	$0.\overline{052631578947368421}$
1/10	0.1	1/20	0.05

Limit of a Sequence

Suppose that $\{a_n\}$ is a given sequence. A real number L is said to be the **limit** of the sequence $\{a_n\}$, written as

$$\lim_{n\to\infty} a_n = L$$

or, alternatively, $a_n \to L$ as $n \to \infty$, if the terms a_n can be made as close to L as possible by taking n to be sufficiently large. The symbol $n \to \infty$ is read "n tends to infinity."

It is clear that if $\{a_n\}$ has a **limit,** then $\lim_{n\to\infty} a_{n+1} = \lim_{n\to\infty} a_n$.

Figure 1 illustrates the definition of a sequence by showing the graphs of two sequences that have the limit L.

The intuitive definition of limit of a sequence given above can be made more precise. We say that **a sequence $\{a_n\}$ has the limit L** and we write

$$\lim_{n\to\infty} a_n = L$$

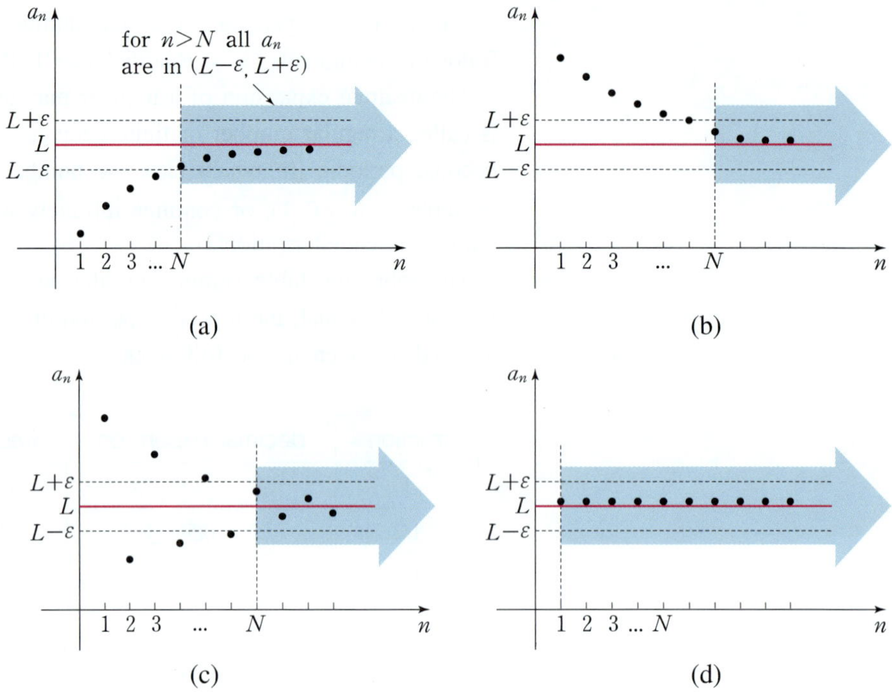

Figure 1 Graphs of two sequences with $\lim_{n\to\infty} a_n = L$

or, alternatively, $a_n \to L$ as $n \to \infty$, if for any given $\varepsilon > 0$, there is a corresponding integer N (which may depend on ε) such that

$$|a_n - L| < \varepsilon \quad \text{whenever } n \geq N.$$

If $\lim_{n\to\infty} a_n$ exists, we say that the sequence **converges** (or is **convergent**). Otherwise, the sequence is said to **diverge** (or is **divergent**).

Remember that $|a_n - L|$ represents the distance between a_n and L. The inequality $|a_n - L| < \varepsilon$ is equivalent to the double inequality $L - \varepsilon < a_n < L + \varepsilon$. This definition can, therefore, also be stated as follows: $\lim_{n\to\infty} a_n = L$ means that whatever interval $(L-\varepsilon,\ L+\varepsilon)$ is taken (with $\varepsilon > 0$), all the terms of the sequence except possibly the first $N-1$ (depending on ε) lie within this interval.

The sequence $\{a_n\}$ is said to **diverge** (or **diverges to** ∞) and we write

$$\lim_{n\to\infty} a_n = \infty$$

if, for any given positive number M, there is an integer N (depending on M) such that $a_n > M$ whenever $n \geq N$, in other words, a_n becomes large as n becomes large. Similarly, if for any given positive number M, there is an integer N (depending on M) such that $a_n < M$ whenever $n \geq N$, we write

$$\lim_{n\to\infty} a_n = -\infty.$$

Archimedean Property : Given any real number x, there exists a natural number N_x such that $N_x > x$.

EXAMPLE 1

Prove that

$$\lim_{n\to\infty}\frac{1}{n}=0.$$

Solution Consider a given $\varepsilon>0$. We have to determine N (depending on ε) so that

$$\left|\frac{1}{n}-0\right|=\frac{1}{n}<\varepsilon \text{ for every } n\geq N.$$

Now $\frac{1}{n}<\varepsilon$ if $n>\frac{1}{\varepsilon}$. Let N be any integer larger than $\frac{1}{\varepsilon}$ (follows from Archimedean Property). Then we have

$$\frac{1}{n}\leq\frac{1}{N}<\varepsilon \quad\text{if}\quad n\geq N,$$

as desired. ■

EXAMPLE 2

Prove that

$$\lim_{n\to\infty}\frac{2n-5}{3n}=\frac{2}{3}.$$

Solution Consider a given $\varepsilon>0$. We have to determine N (depending on ε) so that

$$\left|\frac{2n-5}{3n}-\frac{2}{3}\right|=\left|\frac{2n-5}{3n}-\frac{2n}{3n}\right|=\left|\frac{5}{3n}\right|<\varepsilon \text{ for every } n\geq N.$$

Now $\frac{5}{3n}<\varepsilon$ if $n>\frac{5}{3\,\varepsilon}$. Let N be any integer larger than $\frac{5}{3\,\varepsilon}$ (follows from Archimedean Property). Then we have

$$\frac{5}{3n}<\varepsilon \quad\text{if}\quad n\geq N,$$

as desired. ■

EXAMPLE 3

Prove that if $|r|<1$, then

$$\lim_{n\to\infty}r^n=0.$$

Solution Let $\varepsilon > 0$ be given. We have to determine N (depending on ε) such that

$$|r^n - 0| = |r^n| < \varepsilon \text{ for every } n \geq N.$$

We note that if $|r| < 1$, then $|r| = \dfrac{1}{1+s}$ for some $s > 0$. By the binomial theorem,

$$(1+s)^n = 1 + ns + \frac{n(n-1)}{2}s^2 + \cdots + s^n > ns$$

and therefore

$$|r^n - 0| = |r^n| = \frac{1}{(1+s)^n} < \frac{1}{ns} < \varepsilon \text{ if } n > \frac{1}{\varepsilon s}.$$

Let N be any integer larger than $\dfrac{1}{\varepsilon s}$. Then we have

$$|r^n| < \frac{1}{Ns} < \varepsilon \text{ if } n \geq N.$$

This completes the proof. ■

EXAMPLE 4

Prove that if $|r| > 1$, then

$$\lim_{n \to \infty} r^n = \infty.$$

Solution Let $M > 0$ be given. We have to determine N (depending on M) such that

$$r^n > M \quad \text{for every } n \geq N.$$

Since $|r| > 1$, we have $r = 1 + s$ for some $s > 0$. By the binomial theorem,

$$(1+s)^n > ns,$$

and therefore

$$r^n = (1+s)^n > ns > M \quad \text{if } n > \frac{M}{s}.$$

Let N be any integer larger than $\dfrac{M}{s}$. Then we have

$$r^n > M \quad \text{if } n \geq N.$$

This completes the proof. ■

The following theorem immediately follows from Example 2 and Example 3.

THEOREM 1 Limit of a geometric sequence $\{r^n\}$

If $-1 < r \le 1$, the sequence $\{r^n\}$ converges with

$$\lim_{n\to\infty} r^n = \begin{cases} 0, & -1 < r < 1 \\ 1, & r = 1 \end{cases}$$

and diverges for all other values of r. (See Figure 2.)

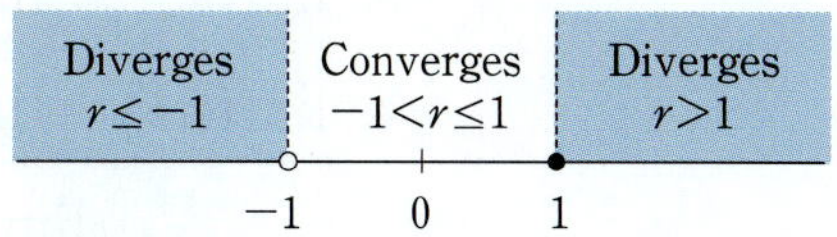

Figure 2

The definition of the limit of a sequence is very similar to that of a limit of a function at infinity. The only difference between $\lim_{n\to\infty} a_n = L$ and $\lim_{x\to\infty} f(x) = L$ is that n must be an integer. This gives rise to the following result, which proves to be useful for the evaluation of limits of sequences.

THEOREM 2

If $\lim_{x\to\infty} f(x) = L$ and $f(n) = a_n$ when n is an integer, then $\lim_{n\to\infty} a_n = L$. (See Figure 3.)

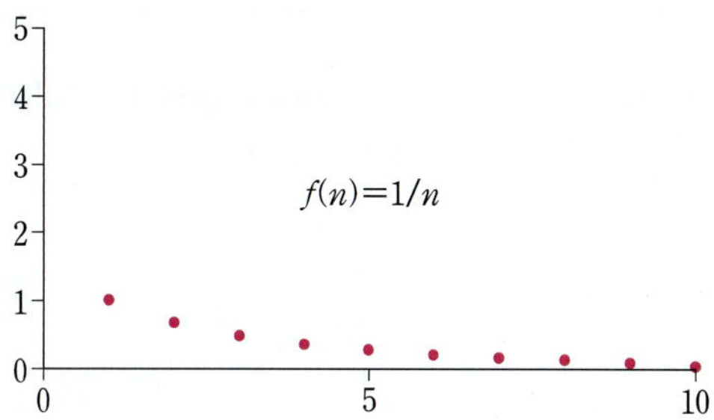

Figure 3

The fact that if $r > 0$, $\lim_{n\to\infty} \frac{1}{n^r} = 0$ can also be adapted from Theorem 1 and also holds $\lim_{x\to\infty} (1/x^r) = 0$.

The following limit laws of sequences, which are very similar to the limit laws of functions, can be deduced from Theorem 1 and are easily proved.

THEOREM 3 Limit Laws for Sequences

If $\{a_n\}$ and $\{b_n\}$ are convergent sequences and c is a constant, then

1. $\lim_{n\to\infty}(a_n + b_n) = \lim_{n\to\infty} a_n + \lim_{n\to\infty} b_n$
2. $\lim_{n\to\infty}(a_n - b_n) = \lim_{n\to\infty} a_n - \lim_{n\to\infty} b_n$
3. $\lim_{n\to\infty} ca_n = c\lim_{n\to\infty} a_n$ for any constant c
4. $\lim_{n\to\infty}(a_n b_n) = \lim_{n\to\infty} a_n \cdot \lim_{n\to\infty} b_n$
5. $\lim_{n\to\infty}\dfrac{a_n}{b_n} = \dfrac{\lim_{n\to\infty} a_n}{\lim_{n\to\infty} b_n}$ if $\lim_{n\to\infty} b_n \neq 0$
6. $\lim_{n\to\infty} a_n^p = [\lim_{n\to\infty} a_n]^p$ if $a_n > 0$ and $p > 0$

The Squeeze Theorem can be adapted for sequences.

THEOREM 4 Squeeze (Sandwich) Theorem for Sequences

If $a_n \leq b_n \leq c_n$ for $n \geq n_0$ and $\lim_{n\to\infty} a_n = \lim_{n\to\infty} c_n = L$, then

$$\lim_{n\to\infty} b_n = L.$$

Another useful result follows upon the application of the Squeeze Theorem for sequences.

If $\lim_{n\to\infty}|a_n| = 0$, then $\lim_{n\to\infty} a_n = 0$.

EXAMPLE 5

Find $\lim_{n\to\infty}\dfrac{n-1}{n}$.

Solution Dividing numerator and denominator by the highest power of n in the denominator, and then using the Limit Laws, we have

$$\lim_{n\to\infty}\frac{n-1}{n} = \lim_{n\to\infty}\frac{1-1/n}{1} = \frac{1-0}{1} = 1.$$ ■

EXAMPLE 6

Find $\lim_{n\to\infty}\frac{\ln n}{n^2}$.

Solution Even though both numerator and denominator approach infinity as $n\to\infty$, L'Hospital's Rule does not apply directly because numerator and denominator are not continuous. However, if we consider the related differentiable function $f(x)=\frac{\ln x}{x^2}$, then we can apply L'Hospital's Rule. In this case

$$\lim_{x\to\infty}\frac{\ln x}{x^2}=\lim_{x\to\infty}\frac{1/x}{2x}=0.$$

Consequently,

$$\lim_{n\to\infty}\frac{\ln n}{n^2}=0.$$

■

EXAMPLE 7

The sequence $\{(-1)^n\}$ diverges because its terms oscillate between 1 and -1 so that they do not approach any number as $n\to\infty$. ■

EXAMPLE 8

Evaluate $\lim_{n\to\infty}\frac{(-1)^n}{\sqrt{n}}$ if it exists.

Solution The limit of the absolute value is found to be

$$\lim_{n\to\infty}\left|\frac{(-1)^n}{\sqrt{n}}\right|=\lim_{n\to\infty}\frac{1}{\sqrt{n}}=0.$$

Therefore, by the application of the Squeeze Theorem for sequences

$$\lim_{n\to\infty}\frac{(-1)^n}{\sqrt{n}}=0.$$

■

Criterion for Convergence of Monotonic Sequences

We say that a sequence $\{a_n\}$ is

(a) strictly **increasing** if $a_n < a_{n+1}$ for all $n \geq 1$;
(b) strictly **decreasing** if $a_n > a_{n+1}$ for all $n \geq 1$.

A sequence $\{a_n\}$ is called **monotonic** if it is either increasing or decreasing.

EXAMPLE 9

Prove that the sequence $a_n = \dfrac{5n}{3n^2-1}$ is decreasing.

Solution 1. We must show that $a_{n+1} < a_n$, that is,

$$\frac{5(n+1)}{3(n+1)^2-1} < \frac{5n}{3n^2-1}.$$

By cross-multiplication, we get the numerator of $a_{n+1} - a_n$:

$$5(n+1)(3n^2-1) - 5n[3(n+1)^2-1] = -(3n^2+3n+1).$$

Since $3n^2+3n+1>0$ for all $n \ge 1$, $a_{n+1}-a_n<0$. Hence, $a_{n+1}<a_n$ for all $n \ge 1$. Therefore, $\{a_n\}$ is decreasing.

2. Let $a_n = f(n)$.

Consider the related function $f(x) = \dfrac{5x}{3x^2-1}$. Since

$$f'(x) = \frac{5(3x^2-1)-5x(6x)}{(3x^2-1)^2} = \frac{-5(3x^2+1)}{(3x^2-1)^2} < 0 \text{ when } x \ge 1,$$

$f(x)$ is decreasing on $[1, \infty)$ and so $f(n) > f(n+1)$ for all $n \ge 1$. Therefore, $\{a_n\}$ is decreasing. ■

One may find a positive number M such that the inequality $|a_n| \le M$ or $-M \le a_n \le M$ is satisfied for every n, however large. In such a case the sequence $\{a_n\}$ is said to be **bounded**. For example, $\{(-1)^n\}$ is a bounded sequence, but $\{n^3\}$ is not. The following theorem says that every convergent sequence is bounded.

THEOREM 5

Any convergent sequence is bounded.

Proof Suppose that $\lim_{n\to\infty} a_n = L$. Taking $\varepsilon = 1$ in the definition of limit, there exists an integer N such that $|a_n - L| < 1$ if $n \ge N$.
We have $|a_n| \le |a_n - L| + |L| \le 1 + |L|$ for $n \ge N$. Therefore, if we define $M = \max(|a_1|, |a_2|, \dots, |a_{N-1}|, 1+|L|)$, we have $|a_n| \le M$ for every n. ■

The converse of Theorem 5 is not true: there exists a sequence which is bounded but not convergent, for example, the sequence $\{(-1)^n\}$ is one such example. Thus not every bounded sequence is convergent. Also, not every monotonic sequence is convergent. For instance, consider $\{n\}$. However, if a sequence is both bounded and monotonic, it must be convergent. (See Figure 4.)

THEOREM 6 Monotonic Sequence Theorem

Every bounded, monotonic sequence is convergent.

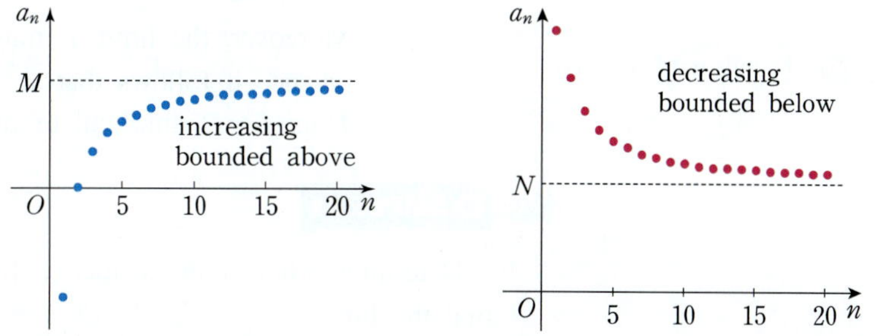

(a) increasing and bounded above (b) decreasing and bounded below

Figure 4

EXAMPLE 10

Determine whether the sequence $\{a_n\}$ defined by the recurrence relation

$$a_1 = \sqrt{2}, \quad a_{n+1} = \sqrt{2+a_n} \qquad \text{for } n = 1, 2, 3, \ldots$$

is convergent or divergent. If so, find the limit of the sequence.

Solution We first observe the first several terms:

$$a_1 = \sqrt{2} \simeq 1.41421$$
$$a_2 = \sqrt{2+\sqrt{2}} \simeq 1.84776$$
$$a_3 = \sqrt{2+\sqrt{2+\sqrt{2}}} \simeq 1.96157$$
$$a_4 = \sqrt{2+\sqrt{2+\sqrt{2+\sqrt{2}}}} \simeq 1.99037$$
$$a_5 = \sqrt{2+\sqrt{2+\sqrt{2+\sqrt{2+\sqrt{2}}}}} \simeq 1.99759$$
$$a_6 = \sqrt{2+\sqrt{2+\sqrt{2+\sqrt{2+\sqrt{2+\sqrt{2}}}}}} \simeq 1.9994.$$

The sequence $\{a_n\}$ and its terms seem to be increasing and approach 2. Mathematical induction will be used to establish $a_{n+1} > a_n$ for all $n \geq 1$. This is true for $n = 1$ because $a_2 = \sqrt{2+\sqrt{2}} > a_1 = \sqrt{2}$. Assuming that it is true for $n = k$, we have

$$a_{k+1} > a_k \text{ so } 2 + a_{k+1} > 2 + a_k \text{ and } \sqrt{2+a_{k+1}} > \sqrt{2+a_k},$$

which implies that $a_{n+1} > a_n$ is also true for $n = k+1$. Therefore the inequality is true for all n by induction.

We next prove that $\{a_n\}$ is bounded. Induction will also be used to verify that $a_n < 2$ for all n. This is true for $n = 1$ since $a_1 < 2$. Supposing that it is true for $n = k$, we have

$$a_k + 2 < 4 \text{ and } \sqrt{a_k + 2} < \sqrt{4} = 2,$$

so that $a_n < 2$ is also true for $n = k+1$. By induction, $a_n < 2$ for all n. Note that $a_n > 1$ for all n. Thus, $|a_n| < 2$ for all n. Since $\{a_n\}$ is increasing and bounded, $\lim_{n\to\infty} a_n = \lim_{n\to\infty} a_{n+1}$ and $L = \lim_{n\to\infty} a_n$ exists by the Monotonic Sequence Theorem.[1)] Moreover, the limit L must satisfy the relation $L = \sqrt{2+L}$ whence it follows that $L^2 - L - 2 = 0$ or $(L-2)(L+1) = 0$. Thus $L = 2$ since all terms a_n are positive. ■

1) Remark: The bound need not be equal to L.

CAS **EXAMPLE 11**

Determine whether the sequence $\{a_n\}$ converges or diverges. If it converges, find the limit.

$$a_n = \frac{1-2n^2}{2+5n^2}$$

http://matrix.skku.ac.kr/cal-lab/cal-10-1-Exm10.html

Solution

```
var('x i n')
p1=plot((3 -2*x)/(1 +5*x), (x, 1, 15), rgbcolor=(1,0,0))
p2=list_plot([(3 -2*x)/(1 + 5*x) for i in range(0, 16, 1)],
rgbcolor=(0, 0, 1))
show(p1+p2)
```

limit((1 - 2*n^2)/(2 + 5*n^2), n=+oo)

Answer : -2/5 ■

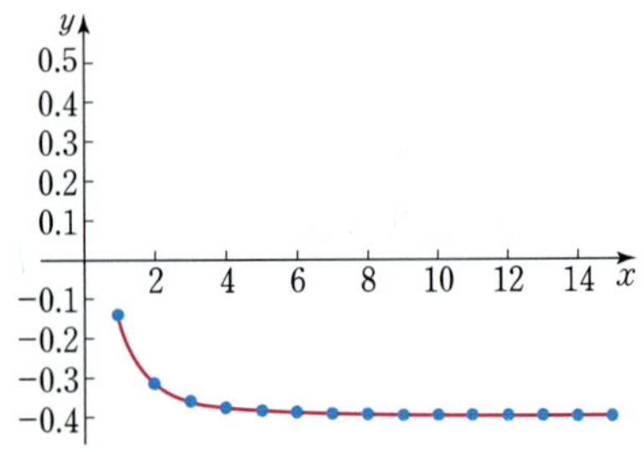

Figure 5

CAS **EXAMPLE 12**

Determine whether the sequence $\{a_n\}$ converges or diverges. If it converges, find the limit. $a_n = \left(3 - \frac{1}{3^n}\right)\left(2 + \frac{1}{2^n}\right)$.

http://matrix.skku.ac.kr/cal-lab/cal-10-1-Exm11.html

Solution

```
var('n')
limit((3 - 1/(3^n))* (2 + 1/(2^n)), n=+oo)
```

Answer : 3*2=6 ■

Series

Given a sequence of numbers, one might be led to add up the infinite number of its terms. Defining and implementing this summability in a mathematically precise sense requires the use of the limit.

Given a sequence $\{a_n\}_{n=1}^{\infty}$, we use the symbol

$$\sum_{n=1}^{\infty} a_n \equiv a_1 + a_2 + a_3 + \cdots + a_n + \cdots$$

or, more concisely when there is no possible ambiguity,

$$\sum a_n.$$

This symbol $\sum a_n$ is called an **infinite series,** or just a **series**. With $\{a_n\}_{n=1}^{\infty}$, we associate a new sequence $\{s_n\}$, where

$$\begin{aligned}
s_1 &= a_1 \\
s_2 &= a_1 + a_2 \\
s_3 &= a_1 + a_2 + a_3 \\
&\cdots \\
s_n &= a_1 + a_2 + a_3 + \cdots + a_n = \sum_{i=1}^{n} a_i. \\
&\cdots
\end{aligned}$$

The numbers s_n are called the **partial sums of the series** $\sum_{n=1}^{\infty} a_n$. If the sequence $\{s_n\}$ converges to s, we say that the series $\sum a_n$ converges (or is convergent), and write

$$\sum_{n=1}^{\infty} a_n = s.$$

The number s is called the **sum of the series** $\sum_{n=1}^{\infty} a_n$, in other words, by definition,

$$\sum_{n=1}^{\infty} a_n = \lim_{n \to \infty} \sum_{i=1}^{n} a_i$$

provided the limit exists. It should be understood that s is the limit of a sequence of partial sums, and is not simply obtained by addition.

If $\{s_n\}$ diverges, the series $\sum_{n=1}^{\infty} a_n$ is said to diverge (or is divergent).

For example, for the series

$$1 + 2 + 3 + 4 + 5 + \cdots + n + \cdots$$

the partial sums are as follows: 1, 3, 6, 10, 15, 21, $\cdots$. After the nth term we get $n(n+1)/2$, which becomes very large as n increases. Thus this series diverges. On the other hand, for the series

$$\frac{1}{2}+\frac{1}{4}+\frac{1}{8}+\frac{1}{16}+\frac{1}{32}+\frac{1}{64}+\cdots+\frac{1}{2^n}+\cdots$$

the sequence of its partial sums is given by

$$\frac{1}{2}, \frac{3}{4}, \frac{7}{8}, \frac{15}{16}, \frac{31}{32}, \frac{63}{64}, \cdots, \left(1-\frac{1}{2^n}\right), \cdots,$$

which converges to 1. Therefore, the series converges and its sum is

$$\frac{1}{2}+\frac{1}{4}+\frac{1}{8}+\frac{1}{16}+\cdots+\frac{1}{2^n}+\cdots=\sum_{n=1}^{\infty}\frac{1}{2^n}=1.$$

Let a_n and b_n be two sequences such that $a_n = b_{n+1} - b_n$, $n \geq 1$. In this case $\sum a_n$ is called a telescoping series. In mathematics, a telescoping series[2)] is a series whose partial sums eventually only have a fixed number of terms after cancellation. Such a technique is also known as the method of differences. The approach used to find a formula for the partial sums is to find the sum of the first N terms and then take the limit as N approaches infinity:

2)http://en.wikipedia.org/wiki/Teles-coping_series

EXAMPLE 13

Show that the series $\displaystyle\sum_{n=1}^{\infty}\frac{3}{(n+1)(n+2)}$ is convergent, and find its sum.

Solution Using the partial fraction decomposition

$$\frac{1}{(i+1)(i+2)}=\frac{1}{i+1}-\frac{1}{i+2}, \qquad \text{[Telescoping series]}$$

the nth partial sum can be written as

$$\begin{aligned} s_n &= 3\left(\frac{1}{2\cdot 3}+\frac{1}{3\cdot 4}+\cdots+\frac{1}{(n+1)(n+2)}\right) \\ &= 3\left[\left(\frac{1}{2}-\frac{1}{3}\right)+\left(\frac{1}{3}-\frac{1}{4}\right)+\left(\frac{1}{4}-\frac{1}{5}\right)+\cdots+\left(\frac{1}{n+1}-\frac{1}{n+2}\right)\right] \\ &= 3\left(\frac{1}{2}-\frac{1}{n+2}\right) \end{aligned}$$

and hence,

$$\lim_{n\to\infty} s_n = \lim_{n\to\infty} 3\left(\frac{1}{2}-\frac{1}{n+2}\right)=\frac{3}{2}.$$

Therefore, the series converges and its value is

$$\sum_{n=1}^{\infty}\frac{3}{(n+1)(n+2)}=\frac{3}{2}.$$

■

The simplest of all series is perhaps the geometric series $\sum_{n=0}^{\infty} ar^n$.

THEOREM 7 Geometric Series

If $|r|<1$, the series $\sum_{n=0}^{\infty} ar^n$ converges, and $\sum_{n=0}^{\infty} ar^n = \frac{a}{1-r}$.

If $|r| \geq 1$, the series diverges.

Proof If $r \neq 1$, the partial sums are $s_n = \frac{a(1-r^n)}{1-r} = \frac{a}{1-r} - \frac{ar^n}{1-r}$.

If $|r|<1$, $\lim_{n\to\infty} s_n = \frac{a}{1-r}$, and hence, the series converges and its sum is $\frac{a}{1-r}$.

If $|r| \geq 1$, then the series evidently diverges. For $r \geq 1$, $s_n \geq n$, so $\{s_n\}$ diverges and for $r \leq -1$, $\{r^n\}$ diverges and hence, $\lim_{n\to\infty} s_n$ does not exist. ■

EXAMPLE 14

Find the sum of the geometric series

$$3 - \frac{6}{5} + \frac{12}{25} - \frac{24}{125} + \cdots + 3\left(-\frac{2}{5}\right)^n + \cdots = 3\sum_{n=0}^{\infty}\left(-\frac{2}{5}\right)^n.$$

Solution The series is a geometric series with the first term $a=3$ and the common ratio $r=-\frac{2}{5}$. Since $|r|<1$, the series converges and its sum is $3/(1-(-2/5)) = 15/7$. ■

EXAMPLE 15

Determine whether the series $\sum_{n=1}^{\infty} 3^{2n} 5^{2-n}$ is convergent or divergent.

Solution The series can be rewritten in a form that we recognize.

$$\sum_{n=1}^{\infty} 3^{2n} 5^{2-n} = \sum_{n=1}^{\infty} \frac{9^n}{5^{n-2}} = \sum_{n=1}^{\infty} 25\left(\frac{9}{5}\right)^n.$$

This is a geometric series with $a=45$ and $r=9/5$. Since $r>1$, the series diverges. ■

EXAMPLE 16

Write the number $5.33\overline{27} = 5.33272727\cdots$ as a ratio of integers.

Solution We have

$$5.332727\cdots = 5.33 + \frac{27}{10^4} + \frac{27}{10^6} + \frac{27}{10^8} + \cdots.$$

After the first term, the right side is a geometric series with $a = 27/10^4$ and $r = 1/10^2$. Therefore,

$$5.33\overline{27} = 5.33 + \frac{\frac{27}{10^4}}{1 - \frac{1}{10^2}} = 5.33 + \frac{\frac{27}{10000}}{\frac{99}{100}}$$

$$= \frac{533}{100} + \frac{27}{9900} = \frac{2933}{550}.$$ ■

THEOREM 8 Necessary Condition for Convergence

If the series $\sum_{n=1}^{\infty} a_n$ converges, then $\lim_{n\to\infty} a_n = 0$.

Proof Note that $a_n = s_n - s_{n-1}$. If the series converges to s, then $\lim_{n\to\infty} s_n = s$ and $\lim_{n\to\infty} s_{n-1} = s$. Thus,

$$\lim_{n\to\infty} a_n = \lim_{n\to\infty} (s_n - s_{n-1}) = s - s = 0.$$ ■

Figure 6 shows $\lim_{n\to\infty} a_n = 0$ if $\sum_{n=1}^{\infty} a_n$ converges.

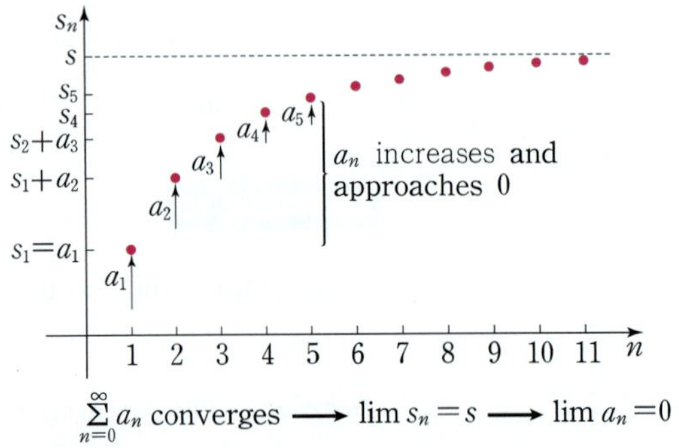

Figure 6

The condition $a_n \to 0$ is not, however, sufficient to ensure convergence of $\sum_{n=1}^{\infty} a_n$. That is, a series may diverge even though $\lim_{n\to\infty} a_n = 0$. A classical example illustrating this case is the series called the **harmonic series**

$$\sum_{n=1}^{\infty} \frac{1}{n} = 1 + \frac{1}{2} + \frac{1}{3} + \frac{1}{4} + \cdots.$$

This series diverges; for proof we refer to Theorem 2 in Section 9.2.
The contrapositive of Theorem 8 provides a useful test for divergence.

A Test for Divergence

If $\lim_{n\to\infty} a_n$ does not exist or if $\lim_{n\to\infty} a_n \neq 0$, then the series $\sum_{n=1}^{\infty} a_n$ is divergent.

EXAMPLE 17

Prove that the series $\sum_{n=1}^{\infty} \frac{3n^3+1}{7n^3+2n^2}$ diverges.

Solution Since $\lim_{n\to\infty} a_n = \lim_{n\to\infty} \frac{3n^3+1}{7n^3+2n^2} = \lim_{n\to\infty} \frac{3+\frac{1}{n^3}}{7+\frac{2}{n}} = \frac{3}{7} \neq 0$, the series diverges by the test for divergence. ■

We know that the sum of a convergent series is the limit of its partial sums. Using the properties of limits of sequences, the following properties of convergence can be easily established.

THEOREM 9 Sum of Two Convergent Series

Let c be a constant. If Σa_n and Σb_n are convergent series, then the series $\Sigma c a_n$, $\Sigma (a_n + b_n)$, and $\Sigma (a_n - b_n)$ are also convergent and we have

(i) $\sum_{n=1}^{\infty} c a_n = c \sum_{n=1}^{\infty} a_n$,

(ii) $\sum_{n=1}^{\infty} (a_n + b_n) = \sum_{n=1}^{\infty} a_n + \sum_{n=1}^{\infty} b_n$,

(iii) $\sum_{n=1}^{\infty} (a_n - b_n) = \sum_{n=1}^{\infty} a_n - \sum_{n=1}^{\infty} b_n$.

Thus two convergent series may be added or subtracted term by term, and the resulting series converges to the sum or difference of the two series.

EXAMPLE 18

Find the sum of the series $\sum_{n=1}^{\infty} \left(\frac{3}{(n+1)(n+2)} - \frac{2}{3^n} \right)$.

http://matrix.skku.ac.kr/cal-lab/cal-10-1-Exm17.html

Solution From Example 13 we have

$$\sum_{n=1}^{\infty} \frac{3}{(n+1)(n+2)} = \frac{3}{2}.$$

The second series $\sum_{n=1}^{\infty} 2/3^n$ is a geometric series with $a = 2/3$ and $r = 1/3$, and its sum is

$$\sum_{n=1}^{\infty} \frac{2}{3^n} = \frac{\frac{2}{3}}{1 - \frac{1}{3}} = 1.$$

Therefore, the series converges by Theorem 9 and it's sum is

$$\begin{aligned}\sum_{n=1}^{\infty} \left(\frac{3}{(n+1)(n+2)} - \frac{2}{3^n} \right) &= \sum_{n=1}^{\infty} \frac{3}{(n+1)(n+2)} - \sum_{n=1}^{\infty} \frac{2}{3^n} \\ &= \frac{3}{2} - 1 = \frac{1}{2}.\end{aligned}$$

```
fvar('n')
a(n) =  3/((n+1)*(n+2))
P=sum(a(n), n, 1, +oo)
b(n) =  2/(3^n)
Q=sum(b(n), n, 1, +oo)
P-Q
```

Answer : 1/2 ■

Remark The convergence or divergence of a series is not affected by adding to or subtracting from it a finite number of terms. This can be inferred from the relation

$$\sum_{n=1}^{\infty} a_n = \sum_{n=1}^{N} a_n + \sum_{n=N+1}^{\infty} a_n.$$

If the series $\sum_{n=N+1}^{\infty} a_n$ converges, then the full series

$$\sum_{n=1}^{\infty} a_n = \sum_{n=1}^{N} a_n + \sum_{n=N+1}^{\infty} a_n$$

also converges. A similar argument applies if the series diverges.

9.1 EXERCISES (Sequences and Series)

http://matrix.skku.ac.kr/Cal-Book/part1/CS-Sec-9-1-Sol.html

YouTube http://youtu.be/

1-3. Find a formula for the general term a_n of the sequence, assuming that the pattern of the first few terms continues.

1. $\left\{\frac{1}{2}, \frac{1}{5}, \frac{1}{8}, \frac{1}{11}, \cdots\right\}$

Solution $a_1 = \frac{1}{3 \cdot 0+2}$, $a_2 = \frac{1}{3 \cdot 1+2}$, $a_3 = \frac{1}{3 \cdot 2+2}$, $\cdots \Rightarrow a_n = \frac{1}{3(n-1)+2} = \frac{1}{3n-1}$.

Therefore, $a_n = \frac{1}{3n-1}$.

2. $\{3, 8, 13, 18, \ldots\}$

Solution $a_1 = 5 \cdot 0+3$, $a_2 = 5 \cdot 1+3$, $a_3 = 5 \cdot 2+3$, ... $\Rightarrow a_n = 5(n-1)+3 = 5n-2$.

Therefore, $a_n = 5n-2$.

3. $\left\{\frac{1}{2}, \frac{1}{5}, \frac{1}{10}, \frac{1}{17}, \cdots\right\}$

Solution $a_1 = \frac{1}{1^2+1}$, $a_2 = \frac{1}{2^2+1}$, $a_3 = \frac{1}{3^2+1}$, ...

Therefore, $a_n = \frac{1}{n^2+1}$.

4-15. Determine whether the sequence converges or diverges. If it converges, find the limit.

CAS **4.** $a_n = \frac{1-n^3}{2+3n^2}$

http://matrix.skku.ac.kr/cal-lab/cal-10-1-4.html

Solution

```
var('x, i, n')
p1 = plot((1-x^3)/(2+3*x^2), (x, 1, 15), rgbcolor=(1,0,0))
p2 = list_plot([(1 - i^2)/(2 + 3*i^2) for i in range(0, 16, 1)], rgbcolor=(0,0,1))
show(p1+p2)
```

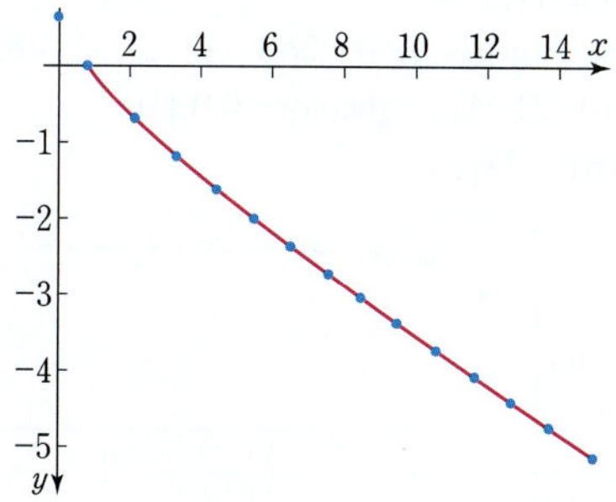


```
limit((1 - n^3)/(2 + 3*n^2), n=+oo)
```

Answer : - Infinity

CAS **5.** $a_n = \frac{n^2-2n+3}{2n^3+2}$

http://matrix.skku.ac.kr/cal-lab/cal-10-1-5.html

Solution

```
var('x, i, n')
p1 = plot((x^2 - 2*x + 3)/(2*x^3 + 2), (x, 1, 50), rgbcolor=(1,0,0))
p2 = list_plot([(i^2 - 2*i + 3)/(2*i^3 + 2) for i in range(0, 51, 1)], rgbcolor=(0,0,1))
show(p1+p2)
```

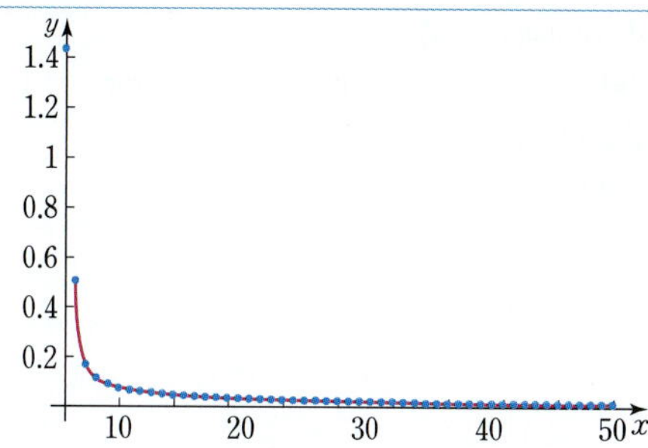


```
var('n')
limit((n^2 - 2*n + 3)/(2*n^3 + 2), n=+oo)
```

Answer : 0

CAS **6.** $a_n = \frac{(-1)^n n^2}{3+2n^2}$

http://matrix.skku.ac.kr/cal-lab/cal-10-1-6.html

Solution

```
var('a, b, i, n')
p1=plot(a^2/(3 + 2*a^2), (a, 1, 20), rgbcolor=(1,0,0))
p2=plot((-1)*b^2/(3 + 2*b^2), (b, 1, 20), rgbcolor=(1,0,0))
p3=list_plot([((-1))^i*i^2/(3 + 2*i^2) for i in range(0, 21, 1)], rgbcolor=(0,0,1))
show(p1+p2+p3)
```

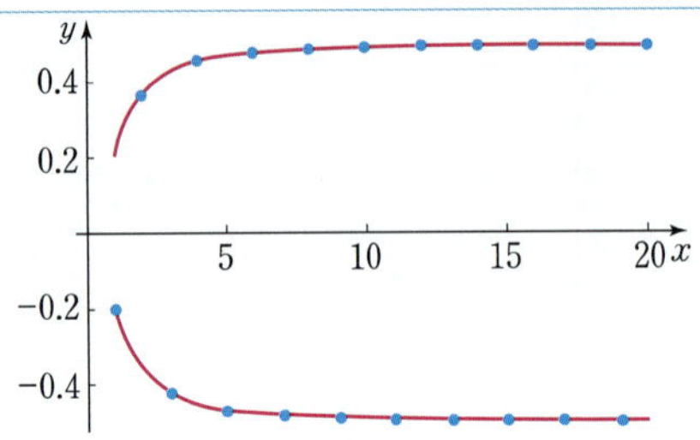


```
var('n')
limit(((-1))^n*n^2/(3 + 2*n^2), n=+oo)limit(ln(3*n)/ln(5*n), n=+oo)
```

Answer : ind (indefinite)

CAS **7.** $a_n = \dfrac{\sin^2 n}{3^n}$

http://matrix.skku.ac.kr/cal-lab/cal-10-1-7.html

Solution

```
var('x, i, n')
p1=plot((sin(x))^2/(3^x), (x, 1, 20), rgbcolor=(1,0,0))
p2=list_plot([(i,(sin(i))^2/(3^i)) for i in range(1,21,1)], rgbcolor=(0,0,1))
show(p1+p2)
```

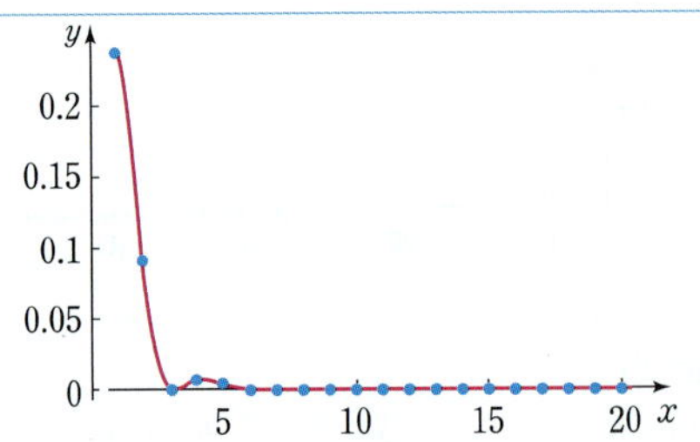


```
limit((sin(n))^2/(3^n), n=+oo)
```

Answer : 0

CAS **8.** $a_n = \dfrac{\ln 3n}{\ln 5n}$

http://matrix.skku.ac.kr/cal-lab/cal-10-1-8.html

Solution

```
var('x, i, n')
p1=plot(ln(3*(x))/ln(5*(x)), (x, 1, 40), rgbcolor=(1,0,0))
p2=list_plot([(i, ln(3*i) /ln(5*i)) for i in range(1,41,1)], rgbcolor=(0,0,1))
show(p1+p2)
```

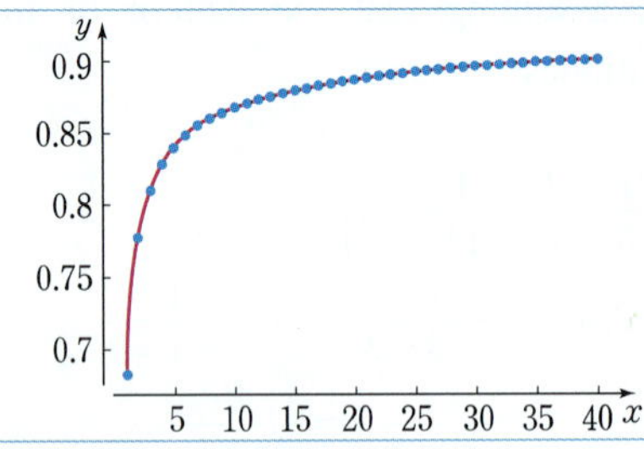


```
limit(ln(3*n)/ln(5*n), n=+oo)
```

Answer : 1

CAS **9.** $a_n = \left(1 - \dfrac{1}{n}\right)^n$

http://matrix.skku.ac.kr/cal-lab/cal-10-1-9.html

Solution

```
var('x, i, n')
p1=plot((1- 1/x)^x, (x, 1, 20), rgbcolor=(1,0,0))
p2=list_plot([(i,(1- 1/i)^i) for i in range(1,21,1)], rgbcolor=(0,0,1))
show(p1+p2)
```

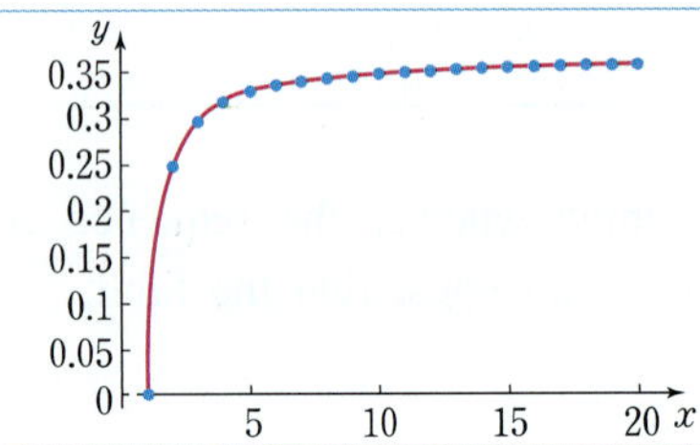


```
limit((1- 1/n)^n, n=+oo)
```

Answer : e^(-1)

CAS **10.** $a_n = \left(\dfrac{3}{n}\right)^{2/n}$

http://matrix.skku.ac.kr/cal-lab/cal-10-1-10.html

Solution

```
var('x, i, n')
p1=plot((3/x)^(2/x), (x, 1, 20), rgbcolor=(1,0,0))
p2=list_plot([(i,(3/i)^(2/i)) for i in range(1,21,1)], rgbcolor=(0,0,1))
show(p1+p2)
```

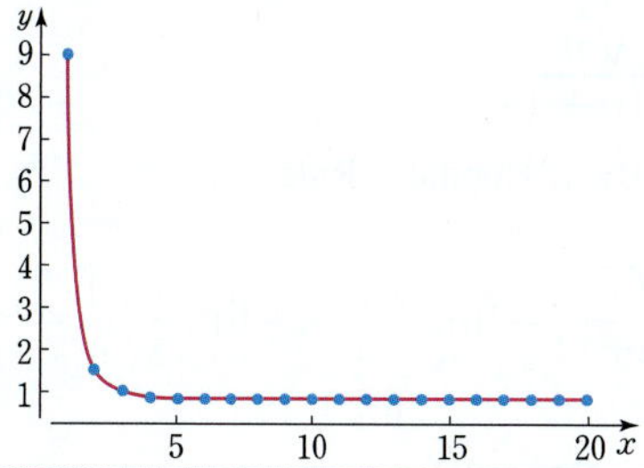


```
limit((3/n)^(2/n), n=+oo)
```

Answer : 1

11. $a_n = \left(\dfrac{3-n^2}{2+n^2}\right)^n$

Solution $\lim\limits_{n\to\infty} a_n = \lim\limits_{n\to\infty}\left(\dfrac{3-n^2}{2+n^2}\right)^n$

$$= \lim_{n\to\infty}\left(-1+\frac{5}{2+n^2}\right)^n$$

$$= \lim_{n\to\infty}(-1)^n\left(1-\frac{5}{2+n^2}\right)^{\left(-\frac{2+n^2}{5}\right)\left(-\frac{5}{2+n^2}\right)n}$$

$$= \lim_{n\to\infty}(-1)^n\left[\left(1-\frac{5}{2+n^2}\right)^{\left(-\frac{2+n^2}{5}\right)\left(-\frac{5}{2+n^2}\right)n^2}\right]^{\frac{1}{2}}$$

Here, $\lim\limits_{n\to\infty}\left(1-\dfrac{5}{2+n^2}\right)^{\left(-\frac{2+n^2}{5}\right)} = e$

and $\lim\limits_{n\to\infty}\left(-\dfrac{5}{2+n^2}\right)n^2 = -5$.

Thus, $\lim\limits_{n\to\infty} a_n = \lim\limits_{n\to\infty}(-1)^n e^{\left(-\frac{5}{2}\right)}$.

Hence, this sequence is divergent.

12. $a_n = (-1)^n(1+n^2)^{1/n}$

Solution $f(x) = (1+x^2)^{1/x} \Rightarrow \ln f(x) = \dfrac{1}{x}\ln(1+x^2)$

$\Rightarrow \lim\limits_{x\to\infty}\ln f(x) = \lim\limits_{x\to\infty}\dfrac{1}{x}\ln(1+x^2) = 0$

$\Rightarrow \lim\limits_{x\to\infty} f(x) = 1$

Thus, $\lim\limits_{n\to\infty} a_n = \lim\limits_{n\to\infty}(-1)^n(1+n^2)^{1/n}$

$= \lim\limits_{n\to\infty}(-1)^n \lim\limits_{n\to\infty}(1+n^2)^{1/n} = \lim\limits_{n\to\infty}(-1)^n \cdot 1.$

Hence, this sequence is divergent.

13. $a_n = n\sin\dfrac{2}{n}$

Solution $\lim\limits_{n\to\infty} a_n = \lim\limits_{n\to\infty} n\sin\dfrac{2}{n} = \lim\limits_{n\to\infty}\dfrac{2\sin\frac{2}{n}}{\frac{2}{n}} = 2$

Hence, this sequence is divergent.

14. $a_n = \dfrac{(\ln n)^2}{n}$

Solution $\lim\limits_{n\to\infty} a_n = \lim\limits_{n\to\infty}\dfrac{(\ln n)^2}{n}$

By L'Hopital's Rule,

$$\lim_{n\to\infty}\frac{(\ln n)^2}{n} = \lim_{n\to\infty}\frac{2(\ln n)\frac{1}{n}}{1} = \lim_{n\to\infty}\frac{2\ln n}{n} = \lim_{n\to\infty}\frac{\frac{2}{n}}{1} = 0.$$

Hence, this sequence is convergent.

15. $a_n = \dfrac{2^n+2}{e^n}$

Solution $\lim\limits_{n\to\infty} a_n = \lim\limits_{n\to\infty}\dfrac{2^n+2}{e^n} = \lim\limits_{n\to\infty}\left\{\left(\dfrac{2}{e}\right)^n + \left(\dfrac{2}{e^n}\right)\right\}$

Since $\dfrac{2}{e} < 1$, Therefore, $\lim\limits_{n\to\infty}\left\{\left(\dfrac{2}{e}\right)^n + \left(\dfrac{2}{e^n}\right)\right\} = 0.$

Hence, this sequence is convergent.

---

16. Investigate the sequence $\{a_n\}$ defined by the recurrence relation $a_1 = 2$, $a_{n+1} = \dfrac{1}{2}(a_n+6)$ for $n =$ 1, 2, 3,$\cdots$. In particular, show that

$$\lim_{n\to\infty} a_n = 6.$$

Solution $a_{n+1} = \dfrac{1}{2}(a_n+6) \Rightarrow a_{n+1} - 6 = \dfrac{1}{2}(a_n - 6)$

$\Rightarrow a_n - 6 = (-4)\left(\dfrac{1}{2}\right)^{n-1} \Rightarrow a_n = (-4)\left(\dfrac{1}{2}\right)^{n-1} + 6.$

Therefore, $\lim\limits_{n\to\infty} a_n = \lim\limits_{n\to\infty}\left((-4)\left(\dfrac{1}{2}\right)^{n-1} + 6\right) = 6.$

17. Let $a_n = \dfrac{3n-1}{5n+1}$

(a) Determine whether $\{a_n\}$ is convergent.

(b) Determine whether $\sum\limits_{n=1}^{\infty} a_n$ is convergent.

Solution

(a) $\lim\limits_{n\to\infty} a_n = \lim\limits_{n\to\infty}\dfrac{3n-1}{5n+1} = \dfrac{3}{5}$

Therefore, $\{a_n\}$ is convergent.

(b) $\lim\limits_{n\to\infty} a_n = \lim\limits_{n\to\infty}\dfrac{3n-1}{5n+1} = \dfrac{3}{5} \neq 0$

Therefore, $\sum\limits_{n=1}^{\infty} a_n$ is divergent.

18-24. Determine whether the following series is convergent or divergent. Find the sum if it is convergent.

18. $\sum_{n=1}^{\infty} \frac{(-1)^n n}{n+1}$

Solution $\lim_{n\to\infty}(-1)^{n+1}\frac{n}{n+1} = \lim_{n\to\infty}(-1)^{n+1}\lim_{n\to\infty}\frac{n}{n+1}$

$= \lim_{n\to\infty}(-1)^{n+1}1 \neq 0$. Thus, $\sum_{n=1}^{\infty}\frac{(-1)^n n}{n+1}$ is divergent.

```
var('n')
a(n) = ((-1)^n *n)/(n+1)
sum(a(n), n, 1, +oo)
```

Answer : Undefined

19. $\sum_{n=1}^{\infty}\left(\frac{e}{10}\right)^n$

http://matrix.skku.ac.kr/cal-lab/cal-10-1-19.html

Solution This is a geometric series and since $-1 < \frac{e}{10} < 1$, this series is convergent.

And, $\sum_{n=1}^{\infty}\left(\frac{e}{10}\right)^n = \frac{\frac{e}{10}}{1-\frac{e}{10}} = \frac{e}{10-e}$.

```
var('n')
a(n) = ((e)/10)^n
sum(a(n), n, 1, +oo)
```

Answer : -e/(e − 10)

20. $\sum_{n=1}^{\infty}\left(\frac{3}{n}-\frac{1}{3^n}\right)$

Solution $\sum_{n=1}^{\infty}\left(\frac{3}{n}-\frac{1}{3^n}\right) = \sum_{n=1}^{\infty}\left(\frac{3}{n}\right) - \sum_{n=1}^{\infty}\left(\frac{1}{3^n}\right)$

But, $\sum_{n=1}^{\infty}\left(\frac{3}{n}\right)$ is not convergent and $\sum_{n=1}^{\infty}\left(\frac{1}{3^n}\right) = \frac{\frac{1}{3}}{1-\frac{1}{3}} = \frac{1}{2}$ is convergent.

Thus, $\sum_{n=1}^{\infty}\left(\frac{3}{n}-\frac{1}{3^n}\right)$ is divergent.

21. $\sum_{n=1}^{\infty}\frac{1+3^n+5^n}{7^n}$

Solution $\sum_{n=1}^{\infty}\frac{1+3^n+5^n}{7^n} = \sum_{n=1}^{\infty}\left(\frac{1}{7}\right)^n + \sum_{n=1}^{\infty}\left(\frac{3}{7}\right)^n + \sum_{n=1}^{\infty}\left(\frac{5}{7}\right)^n$

$= \frac{\frac{1}{7}}{1-\frac{1}{7}} + \frac{\frac{3}{7}}{1-\frac{3}{7}} + \frac{\frac{5}{7}}{1-\frac{5}{7}} = \frac{1}{6}+\frac{3}{4}+\frac{5}{2} = \frac{41}{12}$.

Therefore, $\sum_{n=1}^{\infty}\frac{1+3^n+5^n}{7^n}$ is convergent to $\frac{41}{12}$.

22. $\sum_{n=1}^{\infty}\frac{\sqrt{n}}{\ln(n+1)}$

Solution By L'Hopital's Rule,

$$\lim_{n\to\infty}\frac{\sqrt{n}}{\ln(n+1)} = \lim_{n\to\infty}\frac{\frac{1}{2\sqrt{n}}}{\frac{1}{n+1}} = \lim_{n\to\infty}\frac{n+1}{2\sqrt{n}} = \infty.$$

Therefore, $\sum_{n=1}^{\infty}\frac{\sqrt{n}}{\ln(n+1)}$ is divergent.

23. $\sum_{n=1}^{\infty}\sin^n(2)$

Solution Since $-1 < \sin 2 < 1$, this geometric series with $r = \sin 2$ ($|r| < 1$) is convergent.

And, $\sum_{n=1}^{\infty}\sin^n(2) = \frac{\sin 2}{1-\sin 2} \approx 10.025$.

24. $\sum_{n=1}^{\infty}\ln\left(1+\frac{1}{n}\right)$

Solution $\sum_{n=1}^{\infty}\ln\left(1+\frac{1}{n}\right) = \sum_{n=1}^{\infty}\ln\left(\frac{n+1}{n}\right) = \sum_{n=1}^{\infty}(\ln(n+1)-\ln n)$

$\Rightarrow s_n = (\ln 2 - \ln 1) + (\ln 3 - \ln 2) + \cdots + (\ln(n+1) - \ln n)$

$= \ln(n+1) - \ln 1 = \ln(n+1)$

As $n\to\infty$, $s_n = \ln(n+1) \to \infty$.

Thus, $\sum_{n=1}^{\infty}\ln\left(1+\frac{1}{n}\right)$ is divergent.

25-26. Express the number as a ratio of integers.

25. $3.14\overline{159} = 3.14159159\cdots$

Solution $3.14\overline{159} = 3.14 + \frac{159}{10^5} + \frac{159}{10^8} + \cdots$

$= 3.14 + \frac{\frac{159}{10^5}}{1-\frac{1}{10^3}} = \frac{314}{100} + \frac{53}{33300} = \frac{20923}{6660}$

26. $0.\overline{83} = 0.838383\cdots$

http://matrix.skku.ac.kr/cal-lab/cal-10-1-26.html

Solution $0.\overline{83} = \frac{83}{10^2} + \frac{83}{10^4} + \cdots = \frac{\frac{83}{10^2}}{1-\frac{1}{10^2}} = \frac{83}{99}$

```
var('n')
a(n) = 83/(10^(2*n))
sum(a(n), n, 1, +oo)
```

Answer : 83/99

9.2 Tests for Convergence of Series

Finding the exact sum of a series is a very difficult task, except for series with simple forms such as for geometric series or telescoping series. It is often enough to know the convergence or divergence of a series. If we learn that a given series converges, we may obtain as close an estimate as we wish for its exact sum, which will often suffice for both theoretical and applied purposes. In the following sections, several tests will be introduced that enable us to determine whether a series converges or not.

Examples of series with known convergence may be obtained by applying the integral test due to Colin Maclaurin (1698-1746), which, as the name indicates, involves integration. As a motivating example, let us consider the series

$$\sum_{n=1}^{\infty}\frac{1}{n^4}=\frac{1}{1^4}+\frac{1}{2^4}+\frac{1}{3^4}+\frac{1}{4^4}+\frac{1}{5^4}+\cdots.$$

By geometric observation and calculation it can be shown that

$$s_n=\sum_{i=1}^{n}\frac{1}{i^4}=\frac{1}{1^4}+\frac{1}{2^4}+\cdots+\frac{1}{n^4}\le\frac{1}{1^4}+\int_1^{\infty}\frac{1}{x^4}dx=1+\frac{1}{3}=\frac{4}{3}$$

for all $n\ge 1$. (see Figure 1.) The sequence of partial sums $\{s_n\}$ is thus bounded above and is also increasing. Hence by the Monotonic Sequence Theorem, the partial sums of the sequence converge. Therefore the series $\sum_{n=1}^{\infty}\frac{1}{n^4}$ is convergent. Consider as another motivating example,

$$\sum_{n=1}^{\infty}\frac{1}{n^{1/3}}=\frac{1}{1^{1/3}}+\frac{1}{2^{1/3}}+\frac{1}{3^{1/3}}+\frac{1}{4^{1/3}}+\frac{1}{5^{1/3}}+\cdots.$$

By geometric observation (see Figure 2) and calculation it can also be shown that

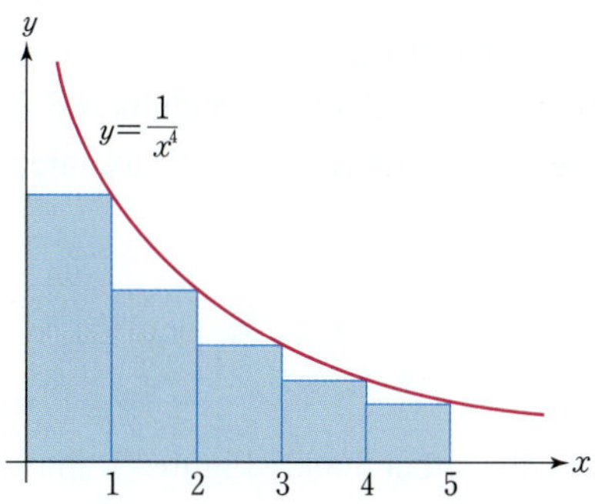

Figure 1 $s_n\le\frac{1}{1^4}+\int_1^{\infty}\frac{1}{x^4}dx$

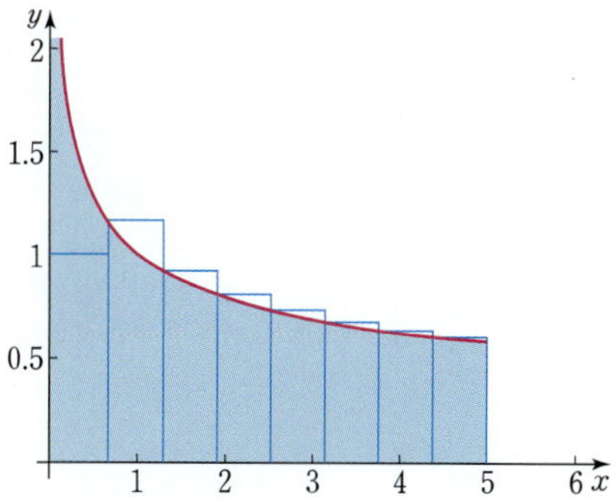

Figure 2 $s_n\ge\int_1^{n+1}\frac{1}{x^{1/3}}dx$

$$s_n = \sum_{i=1}^{n} \frac{1}{i^{1/3}} = \frac{1}{1^{1/3}} + \frac{1}{2^{1/3}} + \cdots + \frac{1}{n^{1/3}} \geq \int_1^{n+1} \frac{1}{x^{1/3}} dx = -\frac{3}{2} + \frac{3}{2}\sqrt[3]{(n+1)^2}.$$

Since the right side of the inequality goes to ∞ as $n \to \infty$, we get $s_n \to \infty$. Hence $\sum_{n=1}^{\infty} \frac{1}{n^{1/3}}$ is divergent.

It should be noted that the convergence and divergence of the series $\sum_{n=1}^{\infty} \frac{1}{n^4}$ and $\sum_{n=1}^{\infty} \frac{1}{n^{1/3}}$ were essentially determined by the convergence and divergence of the improper integrals $\int_1^{\infty} \frac{1}{x^4} dx$ and $\int_1^{\infty} \frac{1}{x^{1/3}} dx$, respectively. Also note that the functions $\frac{1}{x^4}$ and $\frac{1}{x^{1/3}}$, which are obtained from the n-terms $\frac{1}{n^4}$ and $\frac{1}{n^{1/3}}$ by replacing n with x, are continuous, positive and decreasing functions on $[1, \infty)$.

In general for series $\sum_{n=1}^{\infty} a_n$ where the function $f(x) = a_x (\approx f(n) = a_n)$ is continuous, positive and decreasing on $[1, \infty)$, the following powerful test for convergence applies.

THEOREM 1 The Integral Test

Suppose f is a continuous, positive, decreasing function on $[1, \infty)$ and let $a_n = f(n)$. Then the series $\sum_{n=1}^{\infty} a_n$ is convergent if and only if the improper integral $\int_1^{\infty} f(x)dx$ is convergent. In other words,

(i) If $\int_1^{\infty} f(x)dx$ is convergent, then $\sum_{n=1}^{\infty} a_n$ is convergent.

(ii) If $\int_1^{\infty} f(x)dx$ is divergent, then $\sum_{n=1}^{\infty} a_n$ is divergent.

Proof The area of the first shaded rectangle in Figure 3 is the value of f at the first right endpoint which is $f(2) = a_2$. So, by comparing the areas of the shaded rectangles with the area under $y = f(x)$ from 1 to n, we can write

$$\boxed{1} \qquad a_2 + a_3 + \cdots + a_n \leq \int_1^n f(x)dx.$$

On the other hand, Figure 3 shows that

$$\boxed{2} \qquad \int_1^n f(x)dx \leq a_1 + a_2 + \cdots + a_{n-1}.$$

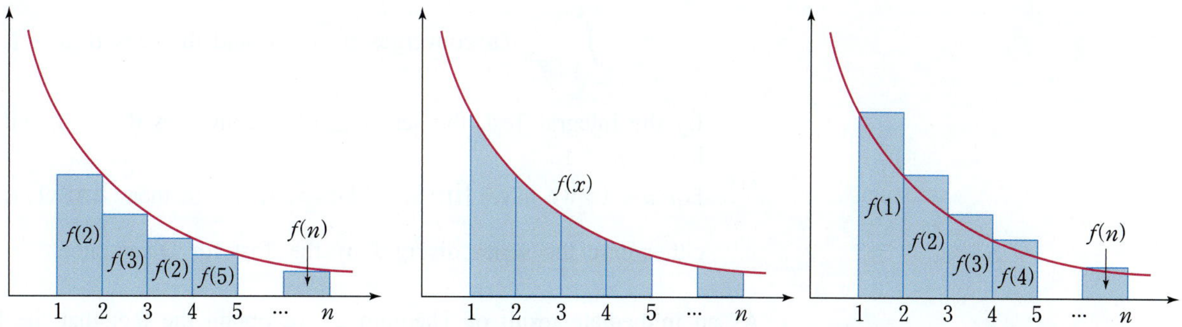

Figure 3 Subject to the conditions of the integral test, either the series $\sum_{n=1}^{\infty} a_n$ and the integral $\int_1^{\infty} f(x)dx$ both converge or diverge.

(i) If $\int_1^n f(x)dx$ is convergent, then 1 gives

$$\sum_{i=2}^{n} a_i \le \int_1^n f(x)dx \le \int_1^{\infty} f(x)dx$$

since $f(x) \ge 0$. Therefore,

$$s_n = a_1 + \sum_{i=2}^{n} a_i \le a_1 + \int_1^{\infty} f(x)dx = M.$$

Since this is true for all n, the sequence $\{s_n\}$ is bounded. Also $\{s_n\}$ is an increasing sequence since $s_{n+1} = s_n + a_{n+1} \ge s_n$, due to $a_{n+1} = f(n+1) \ge 0$. By the Monotonic Sequence Theorem, $\{s_n\}$ is convergent. Therefore, $\sum a_n$ is convergent.

(ii) If $\int_1^{\infty} f(x)dx$ is divergent, then $\int_1^n f(x)dx \to \infty$ as $n \to \infty$ because $f(x) \ge 0$. But 2 gives

$$\int_1^n f(x)dx \le \sum_{i=1}^{n-1} a_i = s_{n-1}$$

and so, $s_{n-1} \to \infty$. This implies that $s_n \to \infty$ and hence, $\sum a_n$ diverges. ■

We shall show how the Integral Test can be applied to the p-series $\sum \frac{1}{n^p}$.

THEOREM 2 p-series Test

The series $\sum \frac{1}{n^p}$ converges if $p > 1$ and diverges if $p \le 1$.

Proof If $p > 0$, then the function $f(x) = 1/x^p$ is clearly continuous, positive, and decreasing on $[1, \infty)$. We know that

$$\int_1^\infty \frac{1}{x^p}dx \text{ converges if } p>1 \text{ and diverges if } p\le 1.$$

By the Integral Test, the series $\Sigma 1/n^p$ converges if $p>1$ and diverges if $0<p\le 1$.

For $p<0$, we have $\lim_{n\to\infty}(1/n^p)=\infty$. If $p=0$, then $\lim_{n\to\infty}(1/n^p)=1$. In either case the series diverges by the Test for Divergence. ■

As an immediate result of Theorem 2, we obtain the fact that the harmonic series $\sum\frac{1}{n}$ diverges because it is a p-series with $p=1$.

EXAMPLE 1

Test the series $\sum_{n=1}^\infty \frac{1}{3n^2+1}$ for convergence or divergence.

Solution The function $f(x)=\frac{1}{3x^2+1}$ is continuous, positive, and decreasing on $[1,\ \infty)$ and we have

$$\begin{aligned}\int_1^\infty \frac{1}{3x^2+1}dx &= \lim_{t\to\infty}\int_1^t \frac{1}{3x^2+1}dx = \lim_{t\to\infty}\frac{1}{\sqrt{3}}\tan^{-1}\sqrt{3}\,x]_1^t \\ &= \lim_{t\to\infty}\frac{1}{\sqrt{3}}(\tan^{-1}\sqrt{3}\,t-\tan^{-1}\sqrt{3}) \\ &= \frac{1}{\sqrt{3}}\left(\frac{\pi}{2}-\frac{\pi}{3}\right)=\frac{\pi}{6\sqrt{3}}.\end{aligned}$$

Since the integral converges, by the Integral Test, the given series is convergent. ■

EXAMPLE 2

Test the series $\sum_{n=2}^\infty \frac{1}{n\ln n}$ for convergence or divergence.

Solution The function $f(x)=1/(x\ln x)$ is positive and continuous for $x>1$. Since $x\ln x$ is increasing, $f(x)$ is decreasing for $x>1$.
We have

$$\int_2^\infty \frac{1}{x\ln x}dx = \lim_{t\to\infty}\int_2^t \frac{1}{x\ln x}dx = \lim_{t\to\infty}\ln(\ln x)]_2^t=\infty,$$

which is divergent. The series $\sum 1/(n\ln n)$ is divergent by the Integral Test. ■

The next test shows that if the terms of a series with positive terms are dominated by the corresponding terms of a convergent series, then the first series is convergent. It also shows that if the terms of a series with positive terms dominate corresponding terms of a divergent series, then the first series is divergent.

THEOREM 3 The Comparison Test

Let Σa_n and Σb_n be series of *positive* terms with $a_n \le b_n$ for all $n \ge N$, where N is some fixed integer.

(i) If Σb_n is convergent, then Σa_n is also convergent.

(ii) If Σa_n is divergent, then Σb_n is also divergent.

Proof (i) Consider

$$s_n = \sum_{i=1}^{n} a_i, \qquad t_n = \sum_{i=1}^{n} b_i, \qquad t = \sum_{n=1}^{\infty} b_n.$$

Since both series have positive terms, the sequences $\{s_n\}$ and $\{t_n\}$ are increasing $(s_{n+1} = s_n + a_{n+1} \ge s_n)$. Also, $t_n \to t$, so $t_n \le t$ for all n. Since $a_i \le b_i$, we have $s_n \le t_n$. Thus, $s_n \le t$ for all n. Therefore $\{s_n\}$ is convergent by the Monotonic Sequence Theorem. Hence, Σa_n is convergent.

(ii) If Σa_n is divergent, then $s_n \to \infty$ since $\{s_n\}$ is increasing. But $a_i \le b_i$, so $s_n \le t_n$. Hence $t_n \to \infty$ as n increases. Therefore, Σb_n diverges. ■

The comparison test is very useful and is most frequently used to prove whether a series converges or diverges. To use it effectively, we must have at our disposal examples of series of positive terms where convergence or divergence of the series is known. Two of the most important series for comparison purposes are the geometric series and the p-series.

EXAMPLE 3

Test the series $\displaystyle\sum_{n=1}^{\infty} \frac{3}{n^3 + 2n^2 + 11}$ for convergence or divergence.

Solution The dominant term in the denominator is n^3 for large n. Note that

$$\frac{3}{n^3 + 2n^2 + 11} < \frac{3}{n^3} \quad \text{for } n \ge 1.$$

Since the p-series $\displaystyle\sum_{n=1}^{\infty} \frac{1}{n^3}$ converges, by the Comparison Test, the given series must also converge. ■

EXAMPLE 4

Use the Comparison Test to test for convergence or divergence of the series

$$\sum_{n=1}^{\infty} \frac{\ln(n+3)}{n+1}.$$

Solution Since $\ln(n+3) > 1$ for $n \geq 1$, we $\dfrac{\ln(n+3)}{n+1} > \dfrac{1}{n+1}$.

The series $\displaystyle\sum_{n=1}^{\infty} \frac{1}{n+1}$ is divergent by p-series with $p=1$.

The given series is divergent. ■

Note for the series $\displaystyle\sum_{n=2}^{\infty} \frac{1}{n^2-3}$, the inequality

$$\frac{1}{n^2-3} > \frac{1}{n^2} \quad \text{for } n \geq 2$$

is not useful for an application of the Comparison Test. In general, if the terms of the given series are larger than the terms of a convergent series or smaller than those of a divergent series, the Comparison Test may not be useful. In such cases the following test can be used. ■

THEOREM 4 The Limit Comparison Test

Let Σa_n and Σb_n be series of positive terms. Assume that

$\displaystyle\lim_{n\to\infty} \frac{a_n}{b_n} = L > 0$ exists. Then

Σa_n converges if and only if Σb_n converges.

Σa_n diverges if and only if Σb_n diverges.

Proof Let m and M be positive numbers such that $m < L < M$. Because a_n/b_n is close to L for large n, there is an integer N such that

$$m < \frac{a_n}{b_n} < M \quad \text{when } n > N$$

and so,

$$mb_n < a_n < Mb_n \quad \text{when } n > N.$$

If Σb_n converges, so does $\Sigma M b_n$. Thus, Σa_n converges by part (i) of the Comparison Test. If Σb_n diverges, so does $\Sigma m b_n$, and hence, part (ii) of the Comparison Test shows that Σa_n diverges. ■

EXAMPLE 5

Determine if the series $\displaystyle\sum_{n=1}^{\infty} \frac{3n-2}{n^3-2n^2+11}$ is convergent or divergent.

Solution The nth term $\dfrac{3n-2}{n^3-2n^2+11}$ behaves approximately like

$$\frac{3n}{n^3} = \frac{3}{n^2}, \text{ and let } a_n = \frac{3n-2}{n^3-2n^2+11} \text{ and } b_n = \frac{3}{n^2}.$$

Since

$$\lim_{n\to\infty}\frac{a_n}{b_n} = \lim_{n\to\infty}\frac{(3n-2)/(n^3-2n^2+11)}{3/n^2} = \lim_{n\to\infty}\frac{3n^3-2n^2}{3n^3-6n^2+33} = 1 > 0$$

and $\sum_{n=1}^{\infty} 1/n^2$ converges, by the Limit Comparison Test, the given series converges. ■

EXAMPLE 6

Test the series $\sum_{n=1}^{\infty}\frac{1}{\sqrt{n^2+19n}}$ for convergence or divergence.

Solution The nth term $\frac{1}{\sqrt{n^2+19n}}$ behaves approximately like $b_n = \frac{1}{n}$. Let $a_n = \frac{1}{\sqrt{n^2+19n}}$ and $b_n = \frac{1}{n}$. Since

$$\lim_{n\to\infty}\frac{a_n}{b_n} = \lim_{n\to\infty}\frac{1/\sqrt{n^2+19n}}{1/n} = \lim_{n\to\infty}\frac{n}{\sqrt{n^2+19n}} = 1 > 0$$

and $\Sigma\, b_n = \Sigma\, 1/n$ diverges, by the Limit Comparison Test, the given series diverges. ■

CAS **EXAMPLE 7**

Test for convergence or divergence of the series

$$\sum_{n=1}^{\infty} 2^n \sin\frac{\pi}{2^n}.$$

http://matrix.skku.ac.kr/cal-lab/cal-10-2-Exm7.html

Solution

```
var('n')
un=2^n*sin(pi/(2^n))
limit(un, n=+oo)          # pi
```

Answer : Since $u_n = 2^n \sin\frac{\pi}{2^n} \to \pi \neq 0$, the series diverges. ■

9.2 EXERCISES (Tests for Convergence of Series)

http://matrix.skku.ac.kr/Cal-Book/part1/CS-Sec-9-2-Sol.html

YouTube http://youtu.be/

1-5. Determine whether the series is convergent or divergent using the Integral Test.

1. $\sum_{n=1}^{\infty}\frac{1}{\sqrt[3]{n}}$

Solution The function $f(x)=\frac{1}{\sqrt[3]{x}}=x^{-\frac{1}{3}}$ is continuous, positive, and decreasing on $[1,\ \infty)$.

The Integral Test applies.

$$\int_1^{\infty}x^{-\frac{1}{3}}dx=\lim_{t\to\infty}\int_1^{t}x^{-\frac{1}{3}}dx=\lim_{t\to\infty}\left[\frac{3}{2}x^{\frac{2}{3}}\right]_1^t$$
$$=\lim_{t\to\infty}\left[\frac{3}{2}t^{\frac{2}{3}}-\frac{3}{2}\right]=\infty$$

Hence by the Integral Test, $\sum_{n=1}^{\infty}\frac{1}{\sqrt[3]{n}}$ diverges.

CAS **2.** $\sum_{n=1}^{\infty}ne^{-n}$

http://matrix.skku.ac.kr/cal-lab/cal-10-2-2.html

Solution

```
var('n')
a(n) = n*e^(-n)
sum(a(n), n, 1, +oo)
#integral(a(n), n, 1, +oo)  # gamma(2, 1)
```

Answer : -e/(2*e - e^2 - 1)

Thus, the series converges.

3. $\sum_{n=1}^{\infty}\frac{e^{1/n^2}}{n^2}$

Solution $\sum_{n=1}^{\infty}\frac{e^{1/n^2}}{n^2}\le\sum_{n=1}^{\infty}\frac{e}{n^2}$

$f(x)=\frac{e}{x^2}$ is continuous and positive on $[1,\ \infty)$, and also decreasing since $f'(x)=\frac{-2e}{x^3}<0$ for $x>1$.

Hence we can use the Integral Test.

$$\int_1^{\infty}\frac{e}{x^2}dx=\lim_{t\to\infty}\left[\frac{-e}{x}\right]_1^t=\lim_{t\to\infty}\left[\frac{-e}{t}+e\right]=e$$

Thus, $\sum_{n=1}^{\infty}\frac{e}{n^2}$ converges.

And by the Comparison Test, $\sum_{n=1}^{\infty}\frac{e^{1/n^2}}{n^2}$ is convergent.

```
var('n')
a(n) = exp(1/n^2)/n^2
sum(a(n), n, 1, +oo)
```

Answer : sum(e^(n^(-2))/n^2, n, 1, +Infinity)

4. $\sum_{n=2}^{\infty}\frac{\ln n}{n}$

Solution The function $f(x)=\frac{\ln x}{x}$ is continuous, positive on $[2,\ \infty)$, and also decreasing. hence by Integral test, since $f'(x)=\frac{1-\ln x}{x^2}<0$ for $x>2$, so we can use the Integral Test.

$$\int_2^{\infty}\frac{\ln x}{x}dx=\lim_{t\to\infty}\int_2^{t}\frac{\ln x}{x}dx=\lim_{t\to\infty}\left[\frac{1}{2}(\ln x)^2\right]_2^t$$
$$=\lim_{t\to\infty}\left[\frac{1}{2}(\ln t)^2-\frac{1}{2}(\ln 2)^2\right]=\infty$$

Hence by Integral Test $\sum_{n=2}^{\infty}\frac{\ln n}{n}$ diverges.

5. $\sum_{n=2}^{\infty}\frac{1}{n(\ln n)^2}$

Solution The function $f(x)=\frac{1}{x(\ln x)^2}$ is continuous, positive on $[2,\ \infty)$, and also decreasing, since $f'(x)=-\frac{2+\ln x}{x^2(\ln x)^3}<0$ if $x>e^{-2}$, so we can use the Integral Test and $x>2$.

$$\int_2^{\infty}\frac{1}{x(\ln x)^2}dx=\lim_{t\to\infty}\int_2^{t}\frac{1}{x(\ln x)^2}dx=\lim_{t\to\infty}\left[\frac{-1}{\ln x}\right]_2^t$$
$$=\lim_{t\to\infty}\left[\frac{-1}{\ln t}-\frac{-1}{\ln 2}\right]=\frac{1}{\ln 2}$$

Hence by the Integral Test, $\sum_{n=2}^{\infty}\frac{1}{n(\ln n)^2}$ converges.

6-7. Find the values of p for which the series is convergent.

6. $\sum_{n=3}^{\infty}\frac{2}{n\ln n(\ln(\ln n))^p}$

Solution When $p=1$, $f(x)=\dfrac{2}{x\ln x(\ln(\ln x))}$ is ontinuous and positive on $[3,\ \infty)$, and also decreasing since

$$f'(x)=-2\frac{(\ln x)(\ln(\ln x))+(\ln(\ln x))+1}{x^2(\ln x)^2(\ln(\ln x))^2}<0 \text{ for } x>3$$

($\because$ x, $\ln x$, $\ln(\ln x)>0$ for $x>3$). Hence we can use the Integral Test.

$$\int_3^\infty \frac{2}{x\ln x(\ln(\ln x))}dx=\lim_{t\to\infty}2[\ln(\ln(\ln x))]_3^t$$
$$=\lim_{t\to\infty}2[\ln(\ln(\ln t))-\ln(\ln(\ln 3))]=\infty$$

Thus, when $p=1$, $\displaystyle\sum_{n=3}^{\infty}\frac{2}{n\ln n(\ln(\ln n))^p}$ diverges.

When $p\neq 1$, $f(x)=\dfrac{2}{x\ln x(\ln(\ln x))^p}$ is continuous and positive on $[3,\ \infty)$, and also decreasing since

$$f'(x)=-2\frac{(\ln x)(\ln(\ln x))^p+(\ln(\ln x))^p+(\ln(\ln x))^{p-1}}{x^2(\ln x)^2(\ln(\ln x))^{2p}}<0$$

for $x>3$, so we can use the Integral Test.

$$\int_3^\infty \frac{2}{x\ln x(\ln(\ln x))^p}dx=\lim_{t\to\infty}\left[\frac{1}{(1-p)(\ln(\ln x))^{1-p}}\right]_3^t$$
$$=\lim_{t\to\infty}\left[\frac{1}{(1-p)(\ln(\ln t))^{1-p}}-\frac{1}{(1-p)(\ln(\ln 3))^{1-p}}\right]$$

This limit exists whenever $1-p<0 \Leftrightarrow p>1$, so the series converges for $p>1$.

7. $\displaystyle\sum_{n=2}^{\infty}\frac{\ln n}{7n^p}$

Solution We have already shown (in Exercise 4) that when $p=1$, the given series is divergent. Let us assume $p\neq 1$.

The function $f(x)=\dfrac{\ln x}{7x^p}$ is continuous, positive on $[2,\ \infty)$, and also decreasing since

$$f'(x)=\frac{7x^{p-1}(1-p\ln x)}{49x^{2p}}<0 \text{ if } x>e^{\frac{1}{p}},$$

so we can use the Integral Test and $x\geq 2$.

$$\int_2^\infty \frac{\ln x}{7x^p}dx=\lim_{t\to\infty}\int_2^t\frac{\ln x}{7x^p}dx$$
$$=\lim_{t\to\infty}\left(\left[\frac{x^{1-p}}{7(1-p)}\ln x\right]_3^t-\int_3^t\frac{1}{7x^{p+1}}dx\right)$$
$$=\lim_{t\to\infty}\left[\frac{t^{1-p}}{7(1-p)}\ln t+\frac{1}{t^p7p}-\frac{3^{1-p}}{7(1-p)}\ln 3+\frac{1}{3^p7p}\right]$$

Thus the series converges for $p>1$.

---

8. Find all positive values of δ for which the series $\displaystyle\sum_{n=1}^{\infty}3\delta^{\ln n}$ converges.

Solution The function $f(x)=3\delta^{\ln x}$ is continuous, positive on $[1,\ \infty)$, and also decreasing since

$$f'(x)=\frac{3\delta^{\ln x}\ln\delta}{x}<0 \text{ if } x>1,\ 0<\delta<1\ ,$$

so we can use the Integral Test:

$$\int_1^\infty 3\delta^{\ln x}dx=\lim_{t\to\infty}\int_1^t 3\delta^{\ln x}dx=\lim_{t\to\infty}\left[\frac{3x\delta^{\ln x}}{1+\ln\delta}\right]_1^t$$
$$=\lim_{t\to\infty}\left[\frac{3t\delta^{\ln t}}{1+\ln\delta}-\frac{3}{1+\ln\delta}\right]$$

Thus $\displaystyle\int_1^\infty 3\delta^{\ln x}dx<\infty$ for $0<\delta<1$. The series converges for $0<\delta<1$.

9-13. Test for convergence or divergence of the series.

9. $\displaystyle\sum_{n=1}^{\infty}\frac{\sin^2 n}{n^2+5}$

Solution $\dfrac{\sin^2 n}{n^2+5}\leq\dfrac{1}{n^2+5}\leq\dfrac{1}{n^2}$, so the series $\displaystyle\sum_{n=1}^{\infty}\frac{\sin^2 n}{n^2+5}$ converges by comparison with the p-series $\displaystyle\sum_{n=1}^{\infty}\frac{1}{n^2}$.

10. $\displaystyle\sum_{n=1}^{\infty}\frac{5^n+2^n}{8^n+3^n}$

Solution Use the Limit Comparison Test with $a_n=\dfrac{5^n+2^n}{8^n+3^n}$ and $b_n=\dfrac{5^n}{8^n}$.

$$\lim_{n\to\infty}\frac{a_n}{b_n}=\lim_{n\to\infty}\frac{\dfrac{5^n+2^n}{8^n+3^n}}{\dfrac{5^n}{8^n}}=\lim_{n\to\infty}\frac{1+\left(\dfrac{2}{5}\right)^n}{1+\left(\dfrac{3}{8}\right)^n}=1>0$$

Since $\displaystyle\sum_{n=1}^{\infty}b_n$ converges, $\displaystyle\sum_{n=1}^{\infty}\frac{5^n+2^n}{8^n+3^n}$ converges.

CAS **11.** $\displaystyle\sum_{n=1}^{\infty}\frac{2^n n!}{n^n}$

http://matrix.skku.ac.kr/cal-lab/cal-10-2-11.html

Solution

```
var('n')
u(n)=2^n*factorial(n)/n^n
limit(u(n+1)/u(n), n=+oo)   #   2*e^(-1)
bool(2*e^(-1)<1)            #   True
```

Answer : $2e^{-1}$

CAS **12.** $\sum_{n=1}^{\infty} \frac{n^2}{3^n}$

http://matrix.skku.ac.kr/cal-lab/cal-10-2-12.html

Solution

```
var('n')
u(n)=n^2/3^n
limit(u(n+1)/u(n), n=+oo)   #   1/3
```

Answer : 1/3

13. $\sum_{n=1}^{\infty}\left(1-\cos\frac{1}{n}\right)$

Solution Use the Limit Comparison Test with

$$a_n = 1-\cos\frac{1}{n} \text{ and } b_n = \frac{1}{4n^2}.$$

Then $\sum a_n$ and $\sum b_n$ are series with positive terms and

$$\lim_{n\to\infty}\frac{a_n}{b_n} = \lim_{n\to\infty}\frac{1-\cos\frac{1}{n}}{\frac{1}{4n^2}} = \lim_{n\to\infty}\frac{2\sin^2\frac{1}{2n}}{\frac{1}{4n^2}}$$

$$= \lim_{n\to\infty} 2\frac{\sin\frac{1}{2n}}{\frac{1}{2n}}\frac{\sin\frac{1}{2n}}{\frac{1}{2n}} = 2 > 0.$$

Since $\sum b_n$ converges, $\sum_{n=1}^{\infty}\left(1-\cos\frac{1}{n}\right)$ converges.

14. If $\sum a_n$ is a convergent series with non-negative terms, is it true that $\sum a_n^2$ is also convergent?

Solution Use the Comparison Test.

If $a_n < 1$, $a_n^2 < a_n$ for all n, then $\sum a_n^2$ is convergent. But, if $a_n > 1$, $a_n^2 > a_n$ for all n, then we cannot decide the convergence of $\sum a_n^2$.

15. If $\sum b_n$ is a convergent series with positive terms, is it true that $\sum \sin(b_n)$ is also convergent?

Solution Yes. Since $\sum b_n$ is a convergent series with positive terms, $\lim_{n\to\infty} b_n = 0$, and $\sum c_n = \sum \sin(b_n)$ is a series with positive terms (for large enough n).

We have $\lim_{n\to\infty}\frac{c_n}{b_n} = \lim_{n\to\infty}\frac{\sin(b_n)}{b_n} = 1 > 0$.

Thus, $\sum b_n$ is also convergent by the Limit Comparison Test.

<CHOI, Seok-Jeong>

9.3 Alternating Series and Absolute Convergence

Alternating Series

In this section will focus on tests for convergence for series with terms that are not necessarily positive.

There is a special type of series with terms that are alternately positive and negative, $\sum_{n=1}^{\infty}(-1)^{n-1}b_n$, the b_n are positive, it is called an **alternating series.**

The convergence of an alternating series is easily determined by the following result.

THEOREM 1 The Alternating Series Test

The alternating series

$$\sum_{n=1}^{\infty}(-1)^{n-1}b_n = b_1 - b_2 + b_3 - b_4 + b_5 - b_6 + \cdots, \text{ where } b_n > 0 \text{ for all } n.$$

is convergent if it satisfies

(i) $b_{n+1} \le b_n$ for all n,

(ii) $\lim_{n\to\infty} b_n = 0$.

Proof Let us first consider the partial sums of even orders

$$s_{2n} = b_1 - b_2 + b_3 - \cdots - a_{2n},$$
$$s_{2n+2} = s_{2n} + (b_{2n+1} - b_{2n+2}).$$

By hypothesis, the terms are decreasing, so that the quantity in the parenthesis is positive. Hence $s_{2n+2} > s_{2n}$, so that the partial sums of even orders form an increasing sequence. On the other hand,

$$s_{2n} = b_1 - (b_2 - b_3) - (b_4 - b_5) - \cdots - (b_{2n-2} - b_{2n-1}) - b_{2n}.$$

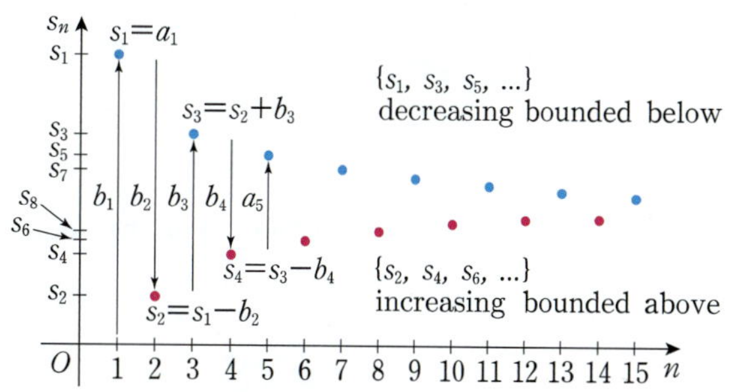

Figure 1 Alternating Series

Since the quantity in each parentheses is positive, $s_{2n} \le b_1$ for all n.

Hence the sequence $\{s_{2n}\}$ is increasing and bounded. By the Monotonic Sequence Theorem, it must converge to some value s.
Now,

$$s_{2n+1} = s_{2n} + b_{2n+1},$$

and

$$\lim_{n\to\infty} s_{2n+1} = \lim_{n\to\infty} s_{2n} + \lim_{n\to\infty} b_{2n+1}.$$

Since $\lim_{n\to\infty} b_{2n+1} = 0$, so the partial sums of even and odd orders tend to the same limit s. Thus the series is convergent. ■

Theorem 1 enables one to easily establish the convergence of an alternating series, since all that is necessary is to show that both (i) and (ii) hold as follows:

(i) $b_{n+1} \leq b_n$ for all n,

(ii) $\lim_{n\to\infty} b_n = 0$.

EXAMPLE 1

Test the convergence of the alternating harmonic series

$$1 - \frac{1}{2} + \frac{1}{3} - \frac{1}{4} + \cdots = \sum_{n=1}^{\infty} \frac{(-1)^{n-1}}{n}.$$

Solution The sequence $\{b_n\}$ where $b_n = \dfrac{1}{n}$ satisfies

(i) $b_{n+1} \leq b_n$ because $\dfrac{1}{n+1} < \dfrac{1}{n}$,

(ii) $\lim_{n\to\infty} b_n = \lim_{n\to\infty} \dfrac{1}{n} = 0$.

By the Alternating Series Test, the series is convergent. ■

EXAMPLE 2

Determine whether the serie is convergent or divergent

$$\sum_{n=1}^{\infty} (-1)^{n+1} \frac{3n+2}{5n+3}.$$

Solution For the sequence $\{b_n\}$ where $b_n = \dfrac{3n+2}{5n+3}$, we have

$$\lim_{n\to\infty} b_n = \lim_{n\to\infty} \frac{3n+2}{5n+3} = \frac{3}{5}.$$

So condition (ii) is not satisfied.

Thus this series is divergent by the Alternating Series Test. ■

Absolute Convergence, Conditional Convergence, and the Ratio and Root Tests

DEFINITION 2

A series $\sum a_n$ is **absolutely convergent** if $\sum |a_n|$ is convergent.
A series $\sum a_n$ is **conditionally convergent** if $\sum a_n$ converges, but $\sum |a_n|$ diverges.

For series with positive terms, there is no distinction between ordinary convergence and absolute convergence. We have the following relation.

THEOREM 3

If a series $\sum a_n$ is absolutely convergent, then the series is convergent.

The series $\sum_{n=1}^{\infty} \frac{(-1)^{n-1}}{n^3} = 1 - \frac{1}{2^3} + \frac{1}{3^3} - \frac{1}{4^3} + \cdots$ is absolutely convergent because

$$\sum_{n=1}^{\infty} \left| \frac{(-1)^{n-1}}{n^3} \right| = \sum_{n=1}^{\infty} \frac{1}{n^3} = 1 + \frac{1}{2^3} + \frac{1}{3^3} + \frac{1}{4^3} + \cdots$$

is a convergent p-series $(p=3)$. It is therefore convergent by Theorem 3.

The converse of Theorem 3 is false: a classical example is given by the alternating harmonic series

$$\sum_{n=1}^{\infty} \frac{(-1)^{n-1}}{n} = 1 - \frac{1}{2} + \frac{1}{3} - \frac{1}{4} + \cdots,$$

which converges by Theorem 1 (or see Example 1), but it does not converge absolutely because the corresponding series of absolute values

$$\sum_{n=1}^{\infty} \left| \frac{(-1)^{n-1}}{n} \right| = \sum_{n=1}^{\infty} \frac{1}{n} = 1 + \frac{1}{2} + \frac{1}{3} + \frac{1}{4} + \cdots$$

is divergent.

Absolute convergence is a stronger condition than ordinary convergence. A series $\sum a_n$ is said to be conditionally convergent if it is convergent but not absolutely convergent. Thus the series $\sum_{n=1}^{\infty} \frac{(-1)^{n-1}}{n}$ is conditionally convergent.

EXAMPLE 3

Is the series $\sum_{n=1}^{\infty} \frac{\sin n}{n^3} = \frac{\sin 1}{1^3} + \frac{\sin 2}{2^3} + \frac{\sin 3}{3^3} + \cdots$ absolutely convergent?

Solution The first three terms are positive, the next three are negative, and the following three are positive. The signs change irregularly. This series has both positive and negative terms, but it is not alternating. series. So, apply the Comparison Test to the series of absolute values

$$\sum_{n=1}^{\infty} \left| \frac{\sin n}{n^3} \right| = \sum_{n=1}^{\infty} \frac{|\sin n|}{n^3}.$$

Since $|\sin n| \leq 1$ for all n, we have $\frac{|\sin n|}{n^3} \leq \frac{1}{n^3}$. Since $\Sigma \frac{1}{n^3}$ is convergent (p-series with $p = 3$), by the Comparison Test, $\Sigma \frac{|\sin n|}{n^3}$ is convergent. Thus, the given series $\Sigma \frac{\sin n}{n^3}$ is absolutely convergent and therefore convergent by Theorem 3. ■

The ratio test, see below is very useful when determining whether a given series is absolutely convergent.

THEOREM 4 The Ratio Test

(i) If $\lim_{n\to\infty} \left| \frac{a_{n+1}}{a_n} \right| = L < 1$, then the series $\sum_{n=1}^{\infty} a_n$ is absolutely convergent (and therefore convergent).

(ii) If $\lim_{n\to\infty} \left| \frac{a_{n+1}}{a_n} \right| = L > 1$ or $\lim_{n\to\infty} \left| \frac{a_{n+1}}{a_n} \right| = \infty$, then the series $\sum_{n=1}^{\infty} a_n$ is divergent.

(iii) If $\lim_{n\to\infty} \left| \frac{a_{n+1}}{a_n} \right| = 1$, the Ratio Test is inconclusive; that is, no conclusion can be drawn on the convergence or divergence of Σa_n.

Proof (i) Since $L < 1$, there exists a number r such that $L < r < 1$. Since

$$\lim_{n\to\infty} \left| \frac{a_{n+1}}{a_n} \right| = L \quad \text{and} \quad L < r$$

the ratio $|a_{n+1}/a_n|$ will eventually be less than r; that is, there exists an integer N such that

$$|a_{n+1}/a_n| < r \quad \text{whenever } n \geq N$$

or, equivalently,

$\boxed{1}$ $$|a_{n+1}| < |a_n|\, r \quad \text{whenever } n \ge N.$$

Putting n successively equal to $N,\ N+1,\ N+2, \cdots$ in $\boxed{1}$, we obtain

$$\begin{aligned} |a_{N+1}| &< |a_N|\, r \\ |a_{N+2}| &< |a_{N+1}|\, r < |a_N|\, r^2 \\ |a_{N+3}| &< |a_{N+2}|\, r < |a_N|\, r^3 \end{aligned}$$

and, in general,

$\boxed{2}$ $$|a_{N+k}| < |a_N|\, r^k \quad \text{for all } \ k \ge 1.$$

Now, the series

$$\sum_{k=1}^{\infty} |a_N|\, r^k = |a_N|\, r + |a_N|\, r^2 + |a_N|\, r^3 + \cdots$$

is convergent because it is a geometric series with $0 < r < 1$. So, the inequality $\boxed{2}$, together with the Comparison Test, shows that the series

$$\sum_{n=N+1}^{\infty} |a_n| = \sum_{k=1}^{\infty} |a_{N+k}| = |a_{N+1}| + |a_{N+2}| + |a_{N+3}| + \cdots$$

is also convergent. It follows that the series $\Sigma_{n=1}^{\infty} |a_n|$ is convergent (Recall that a finite number of terms do not affect convergence.). Therefore, Σa_n is absolutely convergent.

(ii) If $|a_{n+1}/a_n| \to L > 1$ or $|a_{n+1}/a_n| \to \infty$, then the ratio $|a_{n+1}/a_n|$ will eventually be greater than 1; that is, there exists an integer N such that

$$\left| \frac{a_{n+1}}{a_n} \right| > 1 \quad \text{whenever } \ n \ge N.$$

This means that $|a_{n+1}| > |a_n|$ whenever $n \ge N$ and so $\lim_{n \to \infty} a_n \ne 0$.

Therefore, $\sum a_n$ diverges by the Ratio Test. ■

EXAMPLE 4

Investigate the convergence of the series $\displaystyle\sum_{n=1}^{\infty} (-1)^n \frac{n^2}{5^n}$.

Solution With $a_n = (-1)^n n^2 / 5^n$, we have

$$\left| \frac{a_{n+1}}{a_n} \right| = \left| \frac{\dfrac{(-1)^{n+1}(n+1)^2}{5^{n+1}}}{\dfrac{(-1)^n n^2}{5^n}} \right| = \frac{(n+1)^2}{5^{n+1}} \cdot \frac{5^n}{n^2}$$

$$= \frac{1}{5}\left(\frac{n+1}{n}\right)^2 = \frac{1}{5}\left(1+\frac{1}{n}\right)^2 \to \frac{1}{5} \quad \text{as } n \to \infty.$$

Since $1/5 < 1$, the given series is absolutely convergent by the Ratio Test and therefore the series is convergent. ■

EXAMPLE 5

Investigate the convergence of the series $\sum_{n=1}^{\infty} \frac{(-1)^{n-1}2^n}{n^2}$.

Solution Here $a_n = \frac{(-1)^{n-1}2^n}{n^2}$, we have

$$\lim_{n\to\infty}\left|\frac{a_{n+1}}{a_n}\right| = \lim_{n\to\infty}\left|\frac{(-1)^n}{(-1)^{n-1}} \cdot \frac{n^2}{(n+1)^2} \cdot \frac{2^{n+1}}{2^n}\right|$$
$$= \lim_{n\to\infty}\frac{2n^2}{(n+1)^2} = 2.$$

Since $2 > 1$, the given series diverges by the Ratio Test. ■

EXAMPLE 6

Investigate the convergence of the series $\sum_{n=1}^{\infty} \frac{1}{n^2}$.

Solution Here $a_n = \frac{1}{n^2}$. Note that the series $\sum_{n=1}^{\infty} \frac{1}{n^2}$ is convergent by the p -series test. But the Ratio Test is inconclusive.

$$\lim_{n\to\infty}\left|\frac{a_{n+1}}{a_n}\right| = \lim_{n\to\infty}\left|\frac{1}{(n+1)^2} \cdot \frac{n^2}{1}\right|$$
$$= \lim_{n\to\infty}\frac{n^2}{(n+1)^2} = 1$$

■

EXAMPLE 7

Investigate the convergence of the series $\sum_{n=1}^{\infty} \frac{1}{n}$.

Solution Here $a_n = \frac{1}{n}$. Then

$$\lim_{n\to\infty}\left|\frac{a_{n+1}}{a_n}\right| = \lim_{n\to\infty}\frac{n}{n+1} = 1$$

and the Ratio Test is inconclusive. By using the p-series test, the given series is shown to be divergent. ■

Examples 6 and 7 shows part (iii) of the Ratio Test: the condition $\lim_{n\to\infty}\left|\frac{a_{n+1}}{a_n}\right| = 1$ does not allow one to make a conclusion on the convergence or divergence of the series $\sum_{n=1}^{\infty} a_n$. In this case another test should be considered.

When nth powers occur, it is convenient to apply the following test:

THEOREM 5 The Root Test

(i) If $\lim_{n\to\infty} \sqrt[n]{|a_n|} = L < 1$, then the series $\sum_{n=1}^{\infty} a_n$ is absolutely convergent (and therefore convergent).

(ii) If $\lim_{n\to\infty} \sqrt[n]{|a_n|} = L > 1$ or $\lim_{n\to\infty} \sqrt[n]{|a_n|} = \infty$, then the series $\sum_{n=1}^{\infty} a_n$ is divergent.

(iii) If $\lim_{n\to\infty} \sqrt[n]{|a_n|} = 1$, the Root Test is inconclusive.

EXAMPLE 8

Is the series $\sum_{n=0}^{\infty} \frac{1}{2^{n+(-1)^n}}$ convergent?

Solution Let $a_n = \frac{1}{2^{n+(-1)^n}}$. Then

$$\lim_{n\to\infty} \sqrt[n]{|a_n|} = \lim_{n\to\infty} \left(\frac{1}{2^{n+(-1)^n}}\right)^{1/n} = \lim_{n\to\infty} \frac{1}{2}\left(\frac{1}{2^{(-1)^n/n}}\right) = \frac{1}{2}.$$

Since $\frac{1}{2} < 1$, the series converges by the Root Test. ■

EXAMPLE 9

Test for convergence or divergence, of the alternating series:

$$\sum_{n=2}^{\infty} (-1)^{n+1} \frac{1}{\ln n}$$

http://matrix.skku.ac.kr/cal-lab/cal-10-3-Exm-9.html

Solution

```
var('n')
u(n)=1/log(n)
assume(n>2)
bool(u(n+1)<u(n))  #  True
limit(u(n), n=+oo)
```

Answer : 0 ■

Thus $\sum_{n=2}^{\infty} (-1)^{n+1} \frac{1}{\ln n}$ is convergent.

9.3 EXERCISES (Alternating Series and Absolute Convergence)

http://matrix.skku.ac.kr/Cal-Book/part1/CS-Sec-9-3-Sol.html   YouTube http://youtu.be/

1-7. Test for convergence of the following alternating series:

1. $\dfrac{3}{\sqrt{1}}+\dfrac{3}{\sqrt{2}}+\dfrac{3}{\sqrt{3}}+\dfrac{3}{\sqrt{4}}+\cdots$

Solution Since $3\displaystyle\int_1^{\infty}\frac{1}{\sqrt{x}}dx$ is divergent, the series is divergent by the Integral Test.

2. $\dfrac{1}{3}-\dfrac{2}{5}+\dfrac{3}{7}-\dfrac{4}{9}+\dfrac{5}{11}-\cdots+(-1)^{n+1}\dfrac{n}{2n+1}+\cdots$

Solution Note that $|a_n|\to 0$ iff $a_n\to 0$,

$|a_n|=\dfrac{n}{2n+1}\to\dfrac{1}{2}$.

$$\lim_{n\to\infty}(-1)^{n+1}\frac{n}{2n+1}=\lim_{n\to\infty}(-1)^{n+1}\lim_{n\to\infty}\frac{n}{2n+1}$$
$$=\lim_{n\to\infty}(-1)^{n+1}\frac{1}{2}\neq 0.$$

By Alternating Series Test, the series is divergent.

3. $\dfrac{1}{\ln 2}-\dfrac{1}{\ln 3}+\dfrac{1}{\ln 4}-\dfrac{1}{\ln 5}+\cdots$

Solution $\displaystyle\sum_{n=2}^{\infty}a_n=\sum_{n=2}^{\infty}(-1)^n\frac{1}{\ln n},\ b_n=\frac{1}{\ln n}$.

It satisfies $b_{n+1}\le b_n$ because $\dfrac{1}{\ln(n+1)}\le\dfrac{1}{\ln n}$ and $\displaystyle\lim_{n\to\infty}b_n=\lim_{n\to\infty}\frac{1}{\ln n}=0$.

By the Alternating Series Test, the series is convergent.

CAS **4.** $\displaystyle\sum_{n=1}^{\infty}(-1)^{n+1}\frac{n}{n+1}$

http://matrix.skku.ac.kr/cal-lab/cal-10-3-4.html

Solution Note that $|a_n|\not\to 0$

```
var('n')
u(n)=(-1)^(n+1)*n/(n+1)
limit(abs(u(n)), n=+oo)
```

Answer : ind (indefinite)

5. $\displaystyle\sum_{n=1}^{\infty}(-1)^{n+1}\frac{\ln n}{n}$

Solution By the Integral Test,

$\displaystyle\int_1^{\infty}(-1)^{x+1}\frac{\ln x}{x}dx$ is divergent.

By Alternating Series Test, the series is divergent.

6. $\displaystyle\sum_{n=1}^{\infty}(-1)^{n+1}\frac{5^n}{n!}$

Solution $\displaystyle\sum_{n=1}^{\infty}a_n=\sum_{n=1}^{\infty}(-1)^{n+1}\frac{5^n}{n!}$,

$$\lim_{n\to\infty}\left|\frac{a_{n+1}}{a_n}\right|=\lim_{n\to\infty}\left|\frac{5}{n+1}\right|=0.$$

By the Ratio Test, this series is absolutely convergent.

Hence, $\displaystyle\sum_{n=1}^{\infty}(-1)^{n+1}\frac{5^n}{n!}$ is convergent.

7. $\displaystyle\sum_{n=1}^{\infty}(-1)^{n+1}\frac{3n-1}{2n+3}$

Solution $\displaystyle\lim_{n\to\infty}\frac{3n-1}{2n+3}=\lim_{n\to\infty}\frac{3}{2}\neq 0.$

By Alternating Series Test, the series is divergent.

---

8. For what values of p, is the series $\displaystyle\sum_{n=2}^{\infty}(-1)^{n+1}\frac{(\ln n)^p}{n}$ is convergent.

Solution By Exercise 5, when $p=1$, this series diverges.

Let $f(x)=\dfrac{(\ln x)^p}{x}$ and $p\neq 1$.

For$p<1$,

$$f'(x)=\frac{p(\ln x)^{p-1}-(\ln x)^p}{x^2}=\frac{(\ln x)^{p-1}(p-\ln x)}{x^2}<0$$

for all x.

And for $p<1$, $\displaystyle\lim_{n\to\infty}\frac{(\ln n)^p}{n}=0$.

Thus, by the Alternating Series Test, for $p<1$, this series is convergent. If $p>1$, this series is divergent.

9-14. Test whether the series is absolutely convergent, conditionally convergent, or divergent.

9. $\displaystyle\sum_{n=1}^{\infty}(-1)^n\frac{n^n}{n!}$

Solution $\displaystyle\sum_{n=1}^{\infty}a_n=\sum_{n=1}^{\infty}(-1)^n\frac{n^n}{n!}$,

$$\lim_{n\to\infty}\left|\frac{a_{n+1}}{a_n}\right| = \lim_{n\to\infty}\frac{\dfrac{(n+1)^{n+1}}{(n+1)!}}{\dfrac{n^n}{n!}} = \lim_{n\to\infty}\frac{(n+1)\left(\dfrac{n+1}{n}\right)^n}{n+1}$$

$$= \lim_{n\to\infty}\left(1+\frac{1}{n}\right)^n = e > 1.$$

Hence, the series is divergent.

CAS **10.** $\sum_{n=1}^{\infty} e^{-n} n!$

http://matrix.skku.ac.kr/cal-lab/cal-10-3-10.html

Solution Divergent by the Ratio test.

```
var('n')
u(n)=exp(-n)*factorial(n)
limit(u(n+1)/u(n), n=+oo)
```

Answer : +Infinity

11. $\sum_{n=1}^{\infty} \frac{\cos\frac{n\pi}{3}}{n!}$

Solution $\sum_{n=1}^{\infty}\left|\frac{\cos\frac{n\pi}{3}}{n!}\right| = \sum_{n=1}^{\infty}\frac{\left|\cos\frac{n\pi}{3}\right|}{n!}$.

Since $\left|\cos\frac{n\pi}{3}\right| \le 1$ for all n, we have $\frac{\left|\cos\frac{n\pi}{3}\right|}{n!} \le \frac{1}{n!}$.

Thus $\sum_{n=1}^{\infty}|a_n| \leqq \sum_{n=1}^{\infty}\frac{1}{n!}$. Let $\sum_{n=1}^{\infty}a_n = \sum_{n=1}^{\infty}\frac{1}{n!}$.

$$\lim_{n\to\infty}\left|\frac{a_{n+1}}{a_n}\right| = \lim_{n\to\infty}\frac{\dfrac{1}{(n+1)!}}{\dfrac{1}{n!}} = \lim_{n\to\infty}\frac{1}{n+1} = 0 < 1.$$

$\sum_{n=1}^{\infty}\frac{1}{n!}$ is absolutely convergent.

By the Comparison Test, $\sum_{n=1}^{\infty}\frac{\cos\frac{n\pi}{3}}{n!}$ is absolutely convergent, and hence convergent.

12. $\sum_{n=1}^{\infty}\left(\frac{n^2+3}{3n^2+2}\right)^n$

Solution $\sum_{n=1}^{\infty}a_n = \sum_{n=1}^{\infty}\left(\frac{n^2+3}{3n^2+2}\right)^n$.

Using the root test, $\lim_{n\to\infty}\sqrt[n]{|a_n|} = \lim_{n\to\infty}\frac{n^2+3}{3n^2+2} = \frac{1}{3} < 1$.

$\sum_{n=1}^{\infty}\left(\frac{n^2+3}{3n^2+2}\right)^n$ is absolutely convergent, and hence convergent.

13. $\sum_{n=1}^{\infty}(-1)^n\frac{3^n n!}{5\cdot 8\cdot 11\cdot \cdots \cdot(3n+2)}$

Solution $\sum_{n=1}^{\infty}a_n = \sum_{n=1}^{\infty}(-1)^n\frac{3^n n!}{5\cdot 8\cdot 11\cdot \cdots \cdot(3n+2)}$,

$$\lim_{n\to\infty}\left|\frac{a_{n+1}}{a_n}\right| = \lim_{n\to\infty}\frac{\dfrac{3^{n+1}(n+1)!}{5\cdot 8\cdot 11\cdot\cdots\cdot(3(n+1)+2)}}{\dfrac{3^n n!}{5\cdot 8\cdot 11\cdot\cdots\cdot(3n+2)}}$$

$$= \lim_{n\to\infty}\frac{3(n+1)}{3n+5} = 1.$$

We do not know if this series is absolutely convergent, when using the ratio test. Let us try another test.

$$\sum_{n=1}^{\infty}b_n = \sum_{n=1}^{\infty}\frac{3^n n!}{5\cdot 8\cdot 11\cdot\cdots\cdot(3n+2)}$$

Here, $b_{n+1} \le b_n$ for all n and

$$\lim_{n\to\infty}b_n = \lim_{n\to\infty}\frac{3^n n!}{5\cdot 8\cdot 11\cdot\cdots\cdot(3n+2)} = 0.$$

By the Alternating Series Test,

$$\sum_{n=1}^{\infty}(-1)^n\frac{3^n n!}{5\cdot 8\cdot 11\cdot\cdots\cdot(3n+2)}$$

is conditionally convergent.

14. $\sum_{n=2}^{\infty}(-1)^n\frac{1}{n\ln n}$

Solution $\sum_{n=2}^{\infty}a_n = \sum_{n=2}^{\infty}(-1)^n\frac{1}{n\ln n}$.

Then $\lim_{n\to\infty}\left|\frac{a_{n+1}}{a_n}\right| = \lim_{n\to\infty}\frac{\dfrac{1}{(n+1)\ln(n+1)}}{\dfrac{1}{n\ln n}} = 1.$

This means we cannot conclude convergence of this series using the Ratio Test.

Consider $\sum_{n=2}^{\infty}b_n = \sum_{n=2}^{\infty}\frac{1}{n\ln n}$.

Here, $b_{n+1} \le b_n$ for all n and $\lim_{n\to\infty}b_n = \lim_{n\to\infty}\frac{1}{n\ln n} = 0$.

By the Alternating Series Test, $\sum_{n=2}^{\infty}(-1)^n\frac{1}{n\ln n}$ is conditionally convergent.

9.4 Power Series

The most important series are power series. Any series of the form

$$\boxed{1} \qquad \sum_{n=0}^{\infty} c_n (x-a)^n = c_0 + c_1(x-a) + c_2(x-a)^2 + c_3(x-a)^3 + \cdots$$

is called a **power series with center** a or a **power series about** a. Here, the symbols a and c_0, c_1, c_2, ... denote real numbers called the center and coefficients of the series, respectively and x denotes a real variable. We are interested in criteria for the convergence of the power series ($\boxed{1}$ is related to Taylor Series in Section 9.5). In general, the series $\boxed{1}$ will converge or diverge, depending on the choice of x. The power series $\boxed{1}$ always converges when $x = a$, adopting the convention that $(x-a)^0 = 1$ even when $x = a$. In general, it may converge for some values of x and diverge for other values of x.

The geometric series

$$\sum_{n=0}^{\infty} x^n = 1 + x + x^2 + \cdots + x^n + \cdots$$

with the common ratio x is a power series with center 0. We know that this series converges if $|x| < 1$, and diverges if $|x| \geq 1$. It is very surprising that every power series turns out to exhibit a behavior similar to the geometric series. More specifically, for every power series there is an associated interval, the interval of convergence for the series $\boxed{1}$. This is an interval such that $\boxed{1}$ converges if x is inside of the interval and diverges if x is outside the interval. The behaviour at the endpoints of the interval of convergence varies, depending on the series.

THEOREM 1

For a given power series $\sum_{n=0}^{\infty} c_n (x-a)^n$ there are only three possibilities:

(i) The series converges only when $x = a$.
(ii) The series converges for all x.
(iii) The series converges if $|x-a| < R$, for some positive number R, and diverges if $|x-a| > R$. ($|x-a| = R$ one must check.)

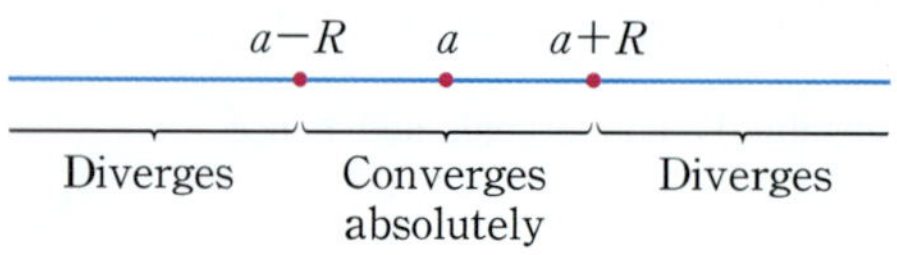

Figure 1

DEFINITION 2

For a given power series $\sum_{n=0}^{\infty} c_n (x-a)^n$, the series converges if $|x-a| < R$ for some positive number R, and diverges if $|x-a| > R$. The positive number R is called the **radius of convergence** of the power series. To cover all cases, we consider a point as an interval of $R=0$ (case (i)), and the set of real numbers as the inside of an interval of $R=\infty$ (case (ii)). The interval I that consists of all values of x for which the series converges is called the **interval of convergence** of the power series.

It should be noted that the case $|x-a|=R$ is not covered in Theorem 1. The reason for this is that some series may converge at $|x-a|=R$ and others diverge. For example, $\sum_{n=0}^{\infty} x^n$ may diverge for $|x|=1$, whereas $\sum_{n=0}^{\infty} \frac{x^n}{n^2}$ converges for $|x|=1$, which can be shown by using the same techniques as those in the examples. Therefore, the convergence or divergence of the series at the endpoints of the interval, that is, at $x=a \pm R$ should be investigated separately. In general, the radius of convergence R is determined by using the Ratio Test in which $R=\lim_{n\to\infty}\left|\frac{c_n}{c_{n+1}}\right|$ (and by using the Root Test in which $R=\frac{1}{\lim_{n\to\infty}\sqrt[n]{|c_n|}}$), provided this limit exists. At an endpoint of the interval of convergence, both the Ratio Test and Root Test always fail. Therefore, some other test should be used at the end points. (See Figure 2.)

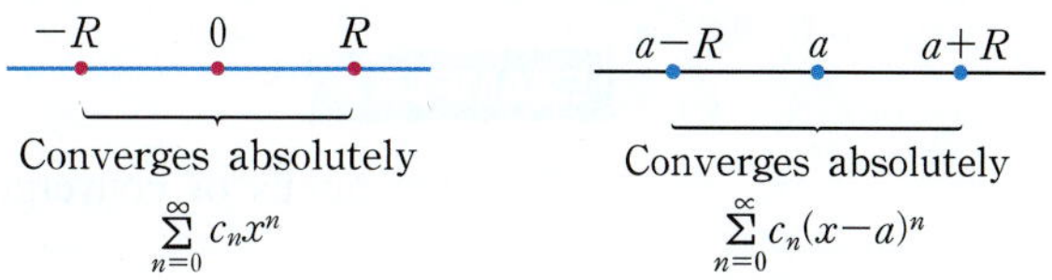

Figure 2

EXAMPLE 1

Find the radius of convergence and interval of convergence for the series $\sum_{n=0}^{\infty} \frac{x^n}{n!}$ where $0! = 1$.

Solution Let $a_n = x^n/n!$. Then

$$\lim_{n\to\infty}\left|\frac{a_{n+1}}{a_n}\right| = \lim_{n\to\infty}\left|\frac{x^{n+1}}{(n+1)!}\frac{n!}{x^n}\right| = \lim_{n\to\infty}\frac{|x|}{n+1} = 0 \text{ for all } x.$$

By the Ratio Test, the series converges for all x. Therefore, the radius of convergence is $R=\infty$, and hence the interval of convergence is $I = (-\infty, \infty)$. ■

EXAMPLE 2

Find the radius of convergence and interval of convergence for the series

$$\sum_{n=1}^{\infty}(-1)^n\frac{x^n}{n}.$$

Solution Let $a_n = (-1)^n x^n/n$. Then

$$\left|\frac{a_{n+1}}{a_n}\right| = \left|\frac{(-1)^{n+1}\dfrac{x^{n+1}}{n+1}}{(-1)^n\dfrac{x^n}{n}}\right| = \left|\frac{n}{n+1}x\right|$$

and hence,

$$\lim_{n\to\infty}\left|\frac{a_{n+1}}{a_n}\right| = |x|.$$

Therefore, the series converges for all $|x| < 1$ and diverges for $|x| > 1$. Thus the radius of convergence is $R = 1$. When $|x| = 1$, the Ratio Test is inconclusive, so we must consider $x = 1$ and $x = -1$, separately. When $x = -1$, the series becomes $\sum 1/n$, the harmonic series, which diverges. When $x = 1$, the series is $\sum(-1)^n/n$, which converges by the Alternating Series Test. Thus the given power series converges when $-1 < x \le 1$, so the interval of convergence is $I = (-1, 1]$. ■

EXAMPLE 3

Find the radius of convergence and interval of convergence for the series

$$\sum_{n=1}^{\infty}\frac{(-1)^n(x+2)^n}{\sqrt{n}\,3^n}.$$

Solution Let $a_n = \dfrac{(-1)^n(x+2)^n}{\sqrt{n}\,3^n}$. Then

$$\left|\frac{a_{n+1}}{a_n}\right| = \left|\frac{\dfrac{(-1)^{n+1}(x+2)^{n+1}}{\sqrt{n+1}\,3^{n+1}}}{\dfrac{(-1)^n(x+2)^n}{\sqrt{n}\,3^n}}\right| = \frac{1}{3}\sqrt{\frac{n}{n+1}}\,|x+2|$$

and hence,

$$\lim_{n\to\infty}\left|\frac{a_{n+1}}{a_n}\right| = \frac{1}{3}|x+2|.$$

Therefore, the series converges for all $|x+2|<3$ and diverges for all $|x+2|>3$. Thus the radius of convergence is $R=3$. When $|x+2|=3$ the Ratio Test is inconclusive, but substituting $x=-5$ in the series gives $\sum_{n=1}^{\infty}\frac{1}{\sqrt{n}}$, which is a divergent $p-$series with $p=1/2$. Setting $x=1$ gives $\sum_{n=1}^{\infty}\frac{(-1)^n}{\sqrt{n}}$, which is convergent by the Alternating Series Test. Therefore, the interval of convergence is $(-5, 1]$. ■

The function $f(x)=1/(1-x)$ can be represented as a power series,

2 $$\frac{1}{1-x} = \sum_{n=0}^{\infty} x^n, \qquad |x|<1$$

which is the geometric series with the first term $a=1$ and a common ratio $r=x$. Certain classes of functions which may be obtained by manipulating the function $f(x)=1/(1-x)$ may also be represented as a power series.

EXAMPLE 4

Represent the function $f(x)=1/(1+x^3)$ as a power series and find its interval of convergence.

Solution Replacing x by $-x^3$ in Equation 2, we have

$$\frac{1}{1+x^3} = \frac{1}{1-(-x^3)} = \sum_{n=0}^{\infty}(-x^3)^n$$
$$= \sum_{n=0}^{\infty}(-1)^n x^{3n} = 1-x^3+x^6-x^9+x^{12}-\cdots.$$

This converges when $|-x^3|<1$ that is, $x^3<1$ or $|x|<1$. Thus the interval of convergence is $(-1, 1)$. ■

EXAMPLE 5

Represent the function $f(x)=1/(x-5)$ as a power series and find its interval of convergence.

Solution In order to put this function in the form of the left side of Equation 2, we first factor -5 from the denominator. Then

$$\frac{1}{x-5} = \frac{1}{-5(1-x/5)} = \frac{1}{-5}\frac{1}{1-x/5}$$
$$= \left(-\frac{1}{5}\right)\sum_{n=0}^{\infty}\left(\frac{x}{5}\right)^n = \sum_{n=0}^{\infty}\frac{(-1)}{5^{n+1}}x^n.$$

This series converges when $|x/5| < 1$, that is, $|x| < 5$. Thus the interval of convergence is $(-5, 5)$. ■

EXAMPLE 6

Represent the function $f(x) = x^2/(x-5)$ as a power series and find its interval of convergence.

Solution By Example 5, $\frac{1}{x-5} = \sum_{n=0}^{\infty} \frac{(-1)}{5^{n+1}} x^n$, $|x| < 5$. Hence

$$\frac{x^2}{x-5} = x^2 \cdot \frac{1}{x-5} = x^2 \sum_{n=0}^{\infty} \frac{(-1)}{5^{n+1}} x^n = \sum_{n=0}^{\infty} \frac{(-1)}{5^{n+1}} x^{n+2}$$

$$= -\frac{1}{5}x^2 - \frac{1}{25}x^3 - \frac{1}{125}x^4 - \cdots, \ |x| < 5.$$

Thus the interval of convergence of the series is $(-5, 5)$. ■

The functions obtained by differentiating or integrating the functions with known power series, such as $f(x) = 1/(1-x)$ can be easily represented as a power series by using the technique called term-by-term differentiation and integration of the power series. Any power series can be differentiated or integrated term-by-term within its interval of convergence, similar to polynomials.

THEOREM 2

Suppose $f(x) = \sum c_n (x-a)^n$ has radius of convergence R. Then the function f is differentiable and integrable on the interval $(a-R, a+R)$ and

(i) $f'(x) = \sum_{n=1}^{\infty} n c_n (x-a)^{n-1}$,

(ii) $\int f(x)dx = C + \sum_{n=0}^{\infty} \frac{c_n}{n+1} (x-a)^{n+1}$,

or equivalently

(iii) $\frac{d}{dx}\left[\sum_{n=0}^{\infty} c_n (x-a)^n\right] = \sum_{n=0}^{\infty} \frac{d}{dx}[c_n (x-a)^n]$,

(iv) $\int \left[\sum_{n=0}^{\infty} c_n (x-a)^n\right] dx = \sum_{n=0}^{\infty} \int c_n (x-a)^n dx$.

Proof It should be remarked that the radii of convergence of the power series in Equations (i) and (ii) remain the same. But the interval of convergence I need not remain the same, that is, endpoints of the interval might change. ■

EXAMPLE 7

Find the radius of convergence of a power series representation of $f(x)=1/(1+x)^2$.

Solution Differentiating both sides of the equation:

$$\frac{1}{1+x}=1-x+x^2-x^3+\cdots=\sum_{n=0}^{\infty}(-1)^n x^n$$

we get

$$\frac{-1}{(1+x)^2}=-1+2x-3x^2+\cdots=\sum_{n=1}^{\infty}(-1)^n n x^{n-1}.$$

Hence, by multiplying both sides by -1, we have

$$\frac{1}{(1+x)^2}=1-2x+3x^2-\cdots=\sum_{n=1}^{\infty}(-1)^{n+1} n x^{n-1}.$$

Thus the radius of convergence of this power series representation is also $R=1$ by Theorem 2. ■

EXAMPLE 8

Find a power series representation of the function $f(x)=\ln(1-x)$ and determine its radius of convergence.

Solution The derivative of $\ln(1-x)$ is $-1/(1-x)$. We know that

$$-\frac{1}{1-x}=-\sum_{n=0}^{\infty}x^n=-1-x-x^2-\cdots, \quad \text{for } |x|<1.$$

Integrating both sides gives

$$\ln(1-x)=\int\left(-\frac{1}{1-x}\right)dx=\int(-1-x-x^2-\cdots)dx$$
$$=-\sum_{n=1}^{\infty}\frac{x^n}{n}+C, \quad |x|<1.$$

Putting $x=0$ in this equation to determine the value of C. Then, $\ln(1-0)=C$ or $C=0$. Thus,

$$\ln(1-x)=-\sum_{n=1}^{\infty}\frac{x^n}{n}, \quad |x|<1.$$

Here $R=1$ since the radius of convergence is the same as for the original series. (Interval of convergence) ■

EXAMPLE 9

Find a power series representation of $f(x)=\tan^{-1}x$ and its radius of convergence.

Solution Since $f'(x)=1/(1+x^2)$, integrating the power series for $1/(1+x^2)$ gives

$$\tan^{-1}x=\int\frac{1}{1+x^2}dx=\int(1-x^2+x^4-x^6+\cdots)dx$$
$$=C+x-\frac{x^3}{3}+\frac{x^5}{5}-\frac{x^7}{7}+\cdots,\quad \text{for } |x|<1.$$

Putting $x=0$, $C=\tan^{-1}0=0$. Therefore, for $|x|<1$

$$\tan^{-1}x=x-\frac{x^3}{3}+\frac{x^5}{5}-\frac{x^7}{7}+\cdots=\sum_{n=0}^{\infty}(-1)^n\frac{x^{2n+1}}{2n+1}.$$

The radius of convergence of this series for $\tan^{-1}x$ is 1. ■

CAS EXAMPLE 10

Determine the radius of convergence R of the following series:

$$\sum_{n=1}^{\infty}\frac{x^n}{n\cdot 2^n}$$

http://matrix.skku.ac.kr/cal-lab/cal-10-2-Exm10.html

Solution

```
var('n')
u(n)=1/(n*2^n)
rho=limit(abs(u(n+1)/u(n)), n=+oo)
R=1/rho; R
```

Answer : 2

We may employ CAS to test convergence at the end points. ■

9.4 EXERCISES (Power Series)

http://matrix.skku.ac.kr/Cal-Book/part1/CS-Sec-9-4-Sol.html

YouTube http://youtu.be/

1-10. Determine the radius of convergence R and interval of convergence I of the following series.

1. $\sum_{n=1}^{\infty}\frac{x^n}{n}$

Solution $\left|\frac{a_{n+1}}{a_n}\right|=\left|\frac{x^{n+1}}{n+1}\frac{n}{x^n}\right|=\frac{1}{1+1/n}|x|\to|x|$ as $n\to\infty$.

Using the Ratio Test, the given series is absolutely convergent and therefore convergent when $|x|<1$, and divergent when $|x|>1$.

If $x=1$, then the series becomes $\sum_{n=1}^{\infty}\frac{1}{n}$, which is divergent.

If $x=-1$, then the series becomes $\sum_{n=1}^{\infty}\frac{(-1)^n}{n}$, which converges by the Alternating Series Test.

Thus, the given power series converges for $-1\le x<1$. So, $R=1$ and $I=[-1, 1)$.

2. $\sum_{n=1}^{\infty}\frac{x^n}{n^2}$

Solution $\left|\frac{a_{n+1}}{a_n}\right|=\left|\frac{x^{n+1}}{(n+1)^2}\frac{n^2}{x^n}\right|=\left(\frac{1}{1+1/n}\right)^2|x|\to|x|$ as $n\to\infty$

Using the Ratio Test, the given series is absolutely convergent and therefore convergent when $|x|<1$, and divergent when $|x|>1$.

If $x=1$, then the series becomes $\sum_{n=1}^{\infty}\frac{1}{n^2}$, which converges by the Integral Test.

If $x=-1$, then the series becomes $\sum_{n=1}^{\infty}\frac{(-1)^n}{n^2}$, which converges by the Alternating Series Test.

Thus, the given power series converges for $-1\le x\le 1$. Hence, $R=1$ and $I=[-1, 1]$.

CAS **3.** $\sum_{n=0}^{\infty}\frac{x^n}{(2n)!}$

http://matrix.skku.ac.kr/cal-lab/cal-10-4-3.html

Solution

```
var('n')
u(n)=1/factorial(2*n)
rho=limit(abs(u(n+1)/u(n)), n=+oo)
rho
```

Answer : 0

4. $\sum_{n=0}^{\infty}\frac{(x-5)^n}{3^n}$

Solution $\left|\frac{a_{n+1}}{a_n}\right|=\left|\frac{(x-5)^{n+1}}{3^{n+1}}\frac{3^n}{(x-5)^n}\right|=\frac{1}{3}|x-5|$ as $n\to\infty$.

Using the Ratio Test, the given series is absolutely convergent and therefore convergent when $|x-5|<3$, and divergent when $|x-5|>3$.

If $x=8$, then the series becomes $\sum_{n=0}^{\infty}\frac{3^n}{3^n}$, which is divergent.

If $x=2$, then the series becomes $\sum_{n=1}^{\infty}\frac{(-3)^n}{3^n}$, which is not convergent.

Thus, the given power series converges for $2\le x\le 8$. So, $R=3$ and $I=[-4, 6]$.

5. $\sum_{n=1}^{\infty}\frac{(-1)^{n+1}x^n}{n}$

Solution $\left|\frac{a_{n+1}}{a_n}\right|=\left|\frac{x^{n+1}}{n+1}\frac{n}{x^n}\right|=\frac{1}{1+1/n}|x|\to|x|$ as $n\to\infty$.

Using the Ratio Test, the given series is absolutely convergent and therefore convergent when $|x|<1$, and divergent when $|x|>1$.

If $x=1$, then the series becomes $\sum_{n=1}^{\infty}\frac{(-1)^{n+1}}{n}$, which converges by the Alternating Series Test.

If $x=-1$, then the series becomes $\sum_{n=1}^{\infty}\frac{-1}{n}$ the negative series, which is divergent.

Thus, the given power series converges for $-1<x\le 1$. So, $R=1$ and $I=(-1, 1]$.

6. $\sum_{n=1}^{\infty}\frac{(x-3)^n}{n^2+2}$

Solution $\left|\dfrac{a_{n+1}}{a_n}\right| = \left|\dfrac{(x-3)^{n+1}}{(n+1)^2+2}\dfrac{n^2+2}{(x-3)^n}\right|$

$= \dfrac{n^2+2}{(n+1)^2+2}|x-3| \to |x-3|$ as $n\to\infty$.

Using the Ratio Test, the given series is absolutely convergent and therefore convergent when $|x-3|<1$, and divergent when $|x-3|>1$.

If $x=4$, then the series becomes $\sum_{n=1}^{\infty}\dfrac{1}{n^2+2}$. Since $\sum_{n=1}^{\infty}\dfrac{1}{n^2+2} \le \sum_{n=1}^{\infty}\dfrac{1}{n^2}$, by the Comparison test, $\sum_{n=1}^{\infty}\dfrac{1}{n^2+2}$ is convergent.

If $x=2$, then the series becomes $\sum_{n=1}^{\infty}\dfrac{(-1)^n}{n^2+2}$, which converges by the Alternating Series Test.

Thus, the given power series converges for $2\le x\le 4$. So, $R=1$ and $I=[2, 4]$.

7. $\sum_{n=1}^{\infty}\dfrac{(x-3)^n}{n^n}$

Solution By the Root test, $\lim_{n\to\infty}\sqrt[n]{|a_n|} = \lim_{n\to\infty}\dfrac{x-3}{n} = 0<1$ for all x.

Thus, the radius of convergence is $R=\infty$ and the interval of convergence is $I=[-\infty, \infty]$.

CAS **8.** $\dfrac{x}{1\cdot 2}+\dfrac{x^2}{2^2\cdot 2^2}+\dfrac{x^3}{3^2\cdot 2^3}+\dfrac{x^4}{4^2\cdot 2^4}+\cdots$

http://matrix.skku.ac.kr/cal-lab/cal-10-4-8.html

Solution

```
var('n')
u(n)=1/(n^2*2^n)
rho=limit(abs(u(n+1)/u(n)), n=+oo)
R=1/rho; R
```

Answer : 2

9. $\sum_{n=1}^{\infty}\dfrac{1\cdot 2\cdot 3\cdot\cdots\cdot n}{1\cdot 3\cdot 5\cdot\cdots\cdot(2n-1)}x^{2n+1}$

Solution $\left|\dfrac{a_{n+1}}{a_n}\right| = \left|\dfrac{n+1}{2n+1}x^2\right| \to \dfrac{1}{2}x^2$ as $n\to\infty$.

Using the Ratio Test, the given series is absolutely convergent and therefore convergent when $|x|<\sqrt{2}$, and divergent when $|x|>\sqrt{2}$. If $x=\sqrt{2}$, then the series becomes $\sum_{n=1}^{\infty}\sqrt{2}\dfrac{(2n)!}{(2n-1)!}$. Since

$\sum_{n=1}^{\infty}\sqrt{2}\dfrac{(2n)!}{(2n-1)!} \ge \sum_{n=1}^{\infty}\sqrt{2}\dfrac{(2n-1)!}{(2n-1)!} = \sum_{n=1}^{\infty}\sqrt{2}=\infty$,

by the Comparison test,

$\sum_{n=1}^{\infty}\sqrt{2}\dfrac{(2n)!}{(2n-1)!}$ is divergent.

If $x=-\sqrt{2}$, then the series becomes $\sum_{n=1}^{\infty}-\sqrt{2}\dfrac{(2n)!}{(2n-1)!}$.

Since $\sum_{n=1}^{\infty}\sqrt{2}\dfrac{(2n)!}{(2n-1)!}=\infty$, $\sum_{n=1}^{\infty}-\sqrt{2}\dfrac{(2n)!}{(2n-1)!}=-\infty$.

Thus, the given power series converges for $-\sqrt{2}<x<\sqrt{2}$. So, $R=\sqrt{2}$ and $I=(-\sqrt{2}, \sqrt{2})$.

10. $\sum_{n=1}^{\infty}\dfrac{n!}{n+1}(x+1)^{n+1}$

Solution $\left|\dfrac{a_{n+1}}{a_n}\right| = (n+1)\left(\dfrac{n+1}{n+2}\right)(x+1) \to \infty$ as $n\to\infty$.

Then, the given power series converges for $x=-1$. So, $R=0$ and $I=\{-1\}$.

11-13. Determine the interval of convergence of a power series representation for the function $f(x)$.

11. $f(x)=\dfrac{1}{1+2x}$

Solution $\dfrac{1}{1+2x}=\dfrac{1}{1-(-2x)}=\sum_{n=0}^{\infty}(-2x)^n$

$=\sum_{n=0}^{\infty}(-1)^n(2x)^n = 1-2x+4x^2-8x^3+\cdots$.

Since this is a geometric series, it converges when $|-2x|<1$, that is $|x|<\dfrac{1}{2}$. Therefore, the interval of convergence is $(-1/2, 1/2)$.

12. $f(x)=\dfrac{x}{5+x^2}$

Solution $\dfrac{x}{5+x^2}=x\cdot\dfrac{1}{5+x^2}=x\cdot\dfrac{1}{5-(-x^2)}$

$=x\sum_{n=0}^{\infty}(-x^2)^n=\sum_{n=0}^{\infty}(-1)^n x^{2n+1}$

$=x-x^3+x^5-x^7+\cdots$.

Since this is a geometric series, it converges when $|-x^2|<1$. Therefore, the interval of convergence is $(-1, 1)$.

13. $f(x)=\dfrac{2}{1-x^4}$

Solution $2\cdot\dfrac{1}{1-x^4}=2\sum_{n=0}^{\infty}(x^4)^n = 2+2x^4+2x^8+2x^{12}+\cdots$.

Since this is a geometric series, it converges when $x^4<1$. Therefore, the interval of convergence is $(-1, 1)$.

---

14. Express the function as the sum of a power series by first using partial fractions. Find the interval of convergence: $f(x)=\dfrac{3}{x^2+x-2}$

Solution $\dfrac{3}{x^2+x-2}=\dfrac{1}{x-1}-\dfrac{1}{x+2}$

$$=-\frac{1}{1-x}-\frac{1}{2}\frac{1}{1-(-x/2)}=-\sum_{n=0}^{\infty}x^n-\frac{1}{2}\sum_{n=0}^{\infty}\frac{(-1)^n x^n}{2^n}.$$

Since this is a geometric series, it converges when $|x|<1$ and $|x/2|<1$, respectively. Therefore, the interval of convergence is $(-1, 1)$.

15-16. Find a power series representation for the function and determine the radius of convergence.

15. $f(x)=\ln(3-x)$

Solution The derivative of $\ln(3-x)$ is $-1/(3-x)$. We have $-\dfrac{1}{3-x}=-\dfrac{1}{3}\dfrac{1}{1-x/3}=-\dfrac{1}{3}\displaystyle\sum_{n=0}^{\infty}\frac{x^n}{3^n}$

$=-\dfrac{1}{3}-\dfrac{x}{9}-\dfrac{x^2}{27}-\cdots$, for $\left|\dfrac{x}{3}\right|<1$.

Thus, $\ln(3-x)=\displaystyle\int\left(-\frac{1}{3-x}\right)dx$

$$=\int\left(-\frac{1}{3}-\frac{1}{9}x-\frac{1}{27}x^2-\cdots\right)dx$$

$$=-\frac{1}{3}\sum_{n=0}^{\infty}\frac{x^{n+1}}{(n+1)3^n}+C$$

$$=\sum_{n=1}^{\infty}\frac{(-1)x^n}{n3^n}+C,\ |x/3|<1.$$

We put $x=0$ in this equation to determine the value of C. That is, $\ln(3-0)=C$ or $C=\ln 3$. Thus,

$$\ln(3-x)=\sum_{n=1}^{\infty}\frac{(-1)x^n}{n3^n}+\ln 3,\quad |x/3|<1.$$

Here $R=3$ since the radius of convergence is the same as for the original series.

16. $f(x)=\tan^{-1}(x/2)$

Solution The derivative of $\tan^{-1}(x/2)$ is $2/(4+x^2)$. Now, we have

$$\frac{2}{4+x^2}=\frac{1}{2}\frac{1}{1-(-x^2/4)}=\frac{1}{2}\sum_{n=0}^{\infty}(-1)^n\frac{x^{2n}}{2^{2n}},\ \text{for } x^2<1.$$

Thus, $\tan^{-1}(x/2)=\displaystyle\int\frac{2}{4+x^2}dx$

$$=\sum_{n=0}^{\infty}(-1)^n\frac{x^{2n+1}}{(2n+1)2^{2n}}+C,\quad |x|<1.$$

We put $x=0$ in this equation to determine the value of C. Then, $\tan^{-1}(0)=C$ or $C=0$. Thus,

$$\tan^{-1}(x/2)=\sum_{n=0}^{\infty}(-1)^n\frac{x^{2n+1}}{(2n+1)2^{2n}},\quad |x|<1.$$

Here $R=1$ since the radius of convergence is the same as for the original series.

17. (a) Find a power series representation for $f(x)=x\ln(1-x)$. What is the radius of convergence?

(b) Use part (a) to find a power series for $f(x)=\ln(1-x^2)$.

Solution

(a) By Example 8 in Section 9.2,

$$x\ln(1-x)=x\left(-\sum_{n=1}^{\infty}\frac{x^n}{n}\right)=-\sum_{n=1}^{\infty}\frac{x^{n+1}}{n}.$$

(b) $\ln(1-x^2)=\ln(1+x)+\ln(1-x)$.

The power series for $\ln(1+x)$ is

$$\sum_{n=1}^{\infty}(-1)^{n-1}\frac{x^n}{n},\quad |x|<1.$$

By part (a), $\ln(1-x^2)=\displaystyle\sum_{n=1}^{\infty}(-1)^{n-1}\frac{x^n}{n}-\sum_{n=1}^{\infty}\frac{x^n}{n}$.

Thus, $\ln(1-x^2)=-\displaystyle\sum_{n=1}^{\infty}\frac{x^{2n}}{n}$, $|x|<1$.

18-19. Evaluate the indefinite integral as a power series and find the radius of convergence.

18. $\displaystyle\int\frac{t^2}{1-t^6}dt$

Solution $\dfrac{t^2}{1-t^6}=t^2\displaystyle\sum_{n=0}^{\infty}t^{6n}=\sum_{n=0}^{\infty}t^{6n+2}$, $|t|<1$.

Thus, the power series is

$$\int\frac{t^2}{1-t^6}dt=\sum_{n=0}^{\infty}\frac{t^{6n+3}}{6n+3},\quad |t|<1.$$

The above power series converges when $|t|<1$. Therefore, the interval of convergence is $(-1, 1)$.

19. $\displaystyle\int\frac{x-\tan^{-1}x}{x^2}dx$

Solution Since

$$\tan^{-1}x=x-\frac{x^3}{3}+\frac{x^5}{5}-\frac{x^7}{7}+\cdots=\sum_{n=0}^{\infty}(-1)^n\frac{x^{2n+1}}{2n+1},$$

$$x-\tan^{-1}x=\frac{x^3}{3}-\frac{x^5}{5}+\frac{x^7}{7}-\cdots=\sum_{n=1}^{\infty}(-1)^{n-1}\frac{x^{2n+1}}{2n+1}.$$

Hence,

$$\frac{x-\tan^{-1}x}{x^2}=\frac{x}{3}-\frac{x^3}{5}+\frac{x^5}{7}-\cdots=\sum_{n=1}^{\infty}(-1)^{n-1}\frac{x^{2n-1}}{2n+1}.$$

Therefore, $\displaystyle\int\frac{x-\tan^{-1}x}{x^2}dx=\sum_{n=1}^{\infty}(-1)^{n-1}\frac{x^{2n}}{2n(2n+1)}$.

9.5 Taylor, Maclaurin, and Binomial Series

Finding power series representations of functions, that is, the approximation of functions by polynomials is very useful in science and has many applications. In the preceding section we looked at a particular class of functions whose power series can be easily found by simple manipulations from a geometric series. This approach will not always work with other types of functions. In this section, we work with the class of functions that have a power series representations and determine methods to find such representations.

THEOREM 1

If f has a power series representation given by

$$f(x) = \sum_{n=0}^{\infty} c_n (x-a)^n, \quad |x-a| < R$$

then the coefficients c_n are given by

$$c_n = \frac{f^{(n)}(a)}{n!}.$$

Proof We start by supposing that f is any function that can be represented by a power series:

$$\boxed{1} \quad f(x) = c_0 + c_1(x-a) + c_2(x-a)^2 + c_3(x-a)^3 + \cdots, \quad |x-a| < R$$

Let us try to determine what the coefficients c_n must be in terms of f. To begin, notice that if we put $x = a$ in Equation $\boxed{1}$, then all terms after the first one are 0 and we get

$$f(a) = c_0.$$

We can differentiate the series in Equation 1 term by term:

$$\boxed{2} \quad f'(x) = c_1 + 2c_2(x-a) + 3c_3(x-a)^2 + \cdots, \quad |x-a| < R$$

and substitution of $x = a$ in Equation $\boxed{2}$ gives

$$f'(a) = c_1.$$

Now we differentiate both sides of Equation $\boxed{2}$ and obtain

$$\boxed{3} \quad f''(x) = 2c_2 + 2 \cdot 3c_3(x-a) + 3 \cdot 4c_4(x-a)^2 + \cdots, \quad |x-a| < R.$$

Again we put $x = a$ in Equation 3. The result is

$$f''(a) = c_2.$$

Let us apply the procedure one more time. Differentiation of series in Equation 3 gives

4 $$f'''(x) = 6c_3 + 2 \cdot 3 \cdot 4c_4(x-a) + 3 \cdot 4 \cdot 5c_5(x-a)^2 + \cdots, \quad |x-a| < R$$

and substitution of $x = a$ in Equation 4 gives

$$f'''(a) = 2 \cdot 3c_3 = 3!c_3$$

By now you can see the pattern. If we continue to differentiate and substitute $x = a$, we obtain

$$f^{(n)}(a) = 2 \cdot 3 \cdot 4 \cdot \cdots \cdot nc_n = n!c_n$$ ■

Solving this equation for the nth coefficient c_n, we get

$$c_n = \frac{f^{(n)}(a)}{n!}.$$

Theorem 1 tells us that if f can be represented as a power series about a, then it must be of the form

5 $$\begin{aligned} f(x) &= \sum_{n=0}^{\infty} \frac{f^{(n)}(a)}{n!}(x-a)^n \\ &= f(a) + \frac{f'(a)}{1!}(x-a) + \frac{f''(a)}{2!}(x-a)^2 + \frac{f'''(a)}{3!}(x-a)^3 + \cdots. \end{aligned}$$

Note that the sum of the power series is equal to f. The series on the right side is called the Taylor Series (after the English mathematician Brook Taylor, 1685–1731) of the function $\boldsymbol{f}$ at $\boldsymbol{a}$ (or about $\boldsymbol{a}$).

The Taylor series of f at $a = 0$ which is given by

6 $$f(x) = \sum_{n=0}^{\infty} \frac{f^{(n)}(0)}{n!}x^n = f(0) + \frac{f'(0)}{1!}x + \frac{f''(0)}{2!}x^2 + \cdots$$

is called the **Maclaurin Series** of f. (after the Scottish mathematician Colin Maclaurin, 1698–1746.)

If f can be represented as a power series about a, then the sum of its Taylor series is equal to f. In this case the value of the $f(x)$ can be determined around $x = a$ based on the knowledge of f and its derivatives at a.

The converse of the above Theorem 1 is false: there exists a function for which the sum of its Taylor series is not equal to the function. Consider the function

$$f(x) = \begin{cases} e^{-1/x}, & x > 0 \\ 0, & x \le 0 \end{cases}$$

$f^{(n)}(0) = 0$ for all n. Therefore, the sum of its Taylor series at $a = 0$ is

$$\sum_{n=0}^{\infty} \frac{f^{(n)}(0)}{n!} x^n = 0.$$

Hence the function f is not equal to its Taylor series. Therefore, the question arises: Under what circumstances is a function equal to the sum of its Taylor series? The following theorem and corollary answer this question.

THEOREM 2 Taylor's Theorem with Remainder

Let f be n times continuously differentiable on the interval $|x-a| < h$. Then for all x in this interval,

7
$$f(x) = \sum_{k=0}^{n-1} \frac{f^{(k)}(a)}{k!}(x-a)^k + R_n(x)$$

where $R_n(x) = \dfrac{f^{(n)}(\xi)}{n!}(x-a)^n$, for some point ξ lying between x and a.

The polynomial $T_n(x) = \sum_{k=0}^{n} \dfrac{f^{(k)}(a)}{k!}(x-a)^k$ is called the nth degree Taylor polynomial of f at a, and $R_n(x)$ the **remainder** of the Taylor series of f about a. As an immediate consequence of the above theorem, we obtain the following important result.

THEOREM 3

Let f be infinitely differentiable (smooth) on the interval $|x-a| < h$. Then f is equal to the sum of its Taylor series on the interval $|x-a| < R$ if and only if $\lim_{n\to\infty} R_n(x) = 0$, for $|x-a| < R$, where $R_n(x)$ is given in Theorem 2.

Proof We prove necessity only. The partial sums of Taylor series are

$$\begin{aligned} T_n(x) &= \sum_{i=0}^{n} \frac{f^{(i)}(a)}{i!}(x-a)^i \\ &= f(a) + \frac{f'(a)}{1!}(x-a) + \frac{f''(a)}{2!}(x-a)^2 + \cdots + \frac{f^{(n)}(a)}{n!}(x-a)^n. \end{aligned}$$

In general, $f(x)$ is the sum of its Taylor series if

$$f(x)=\lim_{n\to\infty} T_n(x).$$

Let $R_n(x)=f(x)-T_n(x)$, then $f(x)=T_n(x)+R_n(x)$. Thus

$$\lim_{n\to\infty} T_n(x)=\lim_{n\to\infty}[f(x)-R_n(x)]=f(x)-\lim_{n\to\infty} R_n(x)=f(x).$$

Therefore, $f(x)=\sum_{n=0}^{\infty}\frac{f^{(n)}(a)}{n!}(x-a)^n$ if $\lim_{n\to\infty} R_n(x)=0$. ■

Let us consider the series $\sum\frac{x^n}{n!}$. The convergence of the series $\sum\frac{x^n}{n!}$ gives the following fact

8 $$\lim_{n\to\infty}\frac{x^n}{n!}=0 \quad \text{for every real number } x$$

which is useful in proving that a function is equal to the sum of its Taylor series.

EXAMPLE 1

Prove that sin x is equal to its Maclaurin series.

Solution Let $f(x)=\sin x$. Then, taking derivatives for $f(x)$ and finding the each value at $x=0$,

$$\begin{aligned} f(x)&=\sin x & f(0)&=0\\ f'(x)&=\cos x & f'(0)&=1\\ f''(x)&=-\sin x & f''(0)&=0\\ f'''(x)&=-\cos x & f'''(0)&=-1\\ &\cdots \end{aligned}$$

In general, since $f^{(n)}(x)=\sin\left(x+\frac{n\pi}{2}\right)$,

$$f^{(n)}(0)=\sin\frac{n\pi}{2}=\begin{cases}0, & n \text{ is even}\\ +1, & n=1,\ 5,\ 9,\ \cdots\\ -1, & n=3,\ 7,\ 11,\ \cdots\end{cases}$$

By Theorem 2, we have

$$\sin x=x-\frac{x^3}{3!}+\frac{x^5}{5!}-\frac{x^7}{7!}+\cdots+R_n(x),$$

where $R_n(x)=\frac{x^n}{n!}\sin\left(\xi+\frac{n\pi}{2}\right)$, for some point ξ lying between x and 0.

Thus for the remainder $R_n(x)$ we have

$$|R_n(x)| \le \left|\frac{x^n}{n!}\right|$$

for all values of x. By 8, if R is any positive number, then for all values of x such that $|x| < R$,

$$\lim_{n\to\infty}\left|\frac{x^n}{n!}\right| = 0.$$

Consequently, $\lim_{n\to\infty} R_n(x) = 0$, for any finite interval $|x| < R$. Thus, by Corollary 3, sin x is equal to its Maclaurin series for all x, that is,

9
$$\begin{aligned}\sin x &= x - \frac{x^3}{3!} + \frac{x^5}{5!} - \frac{x^7}{7!} + \cdots + (-1)^n \frac{x^{2n+1}}{(2n+1)!} + \cdots \\ &= \sum_{n=0}^{\infty} (-1)^n \frac{x^{2n+1}}{(2n+1)!} \quad \text{for all } x.\end{aligned}$$

■

The graphs of sin x together with its Taylor polynomials,

$$T_1(x) = x$$
$$T_3(x) = x - \frac{x^3}{3!}$$
$$T_5(x) = x - \frac{x^3}{3!} + \frac{x^5}{5!}$$
$$T_7(x) = x - \frac{x^3}{3!} + \frac{x^5}{5!} - \frac{x^7}{7!}$$

are shown in Figure 1. Observe that $T_n(x)$ becomes a better approximation to sin x as n increases.

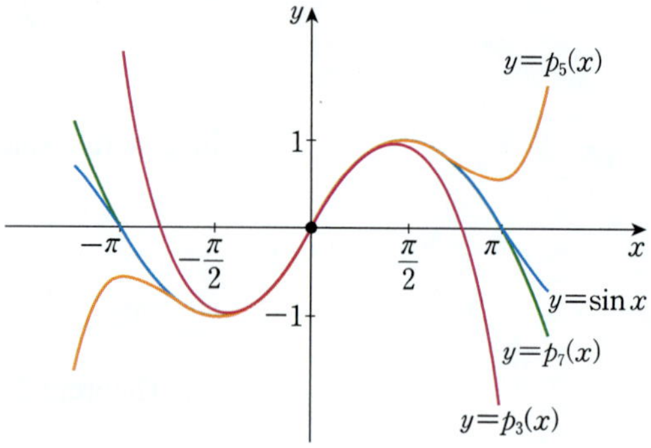

Figure 1

EXAMPLE 2

Find the Maclaurin series for e^x and prove that it is equal to e^x for all x.

Solution Let $f(x) = e^x$. Since $f^{(n)}(0) = 1$ for all n, the Maclaurin series is

$$\sum_{n=0}^{\infty} \frac{f^{(n)}(0)}{n!} x^n = \sum_{n=0}^{\infty} \frac{x^n}{n!}.$$

By Theorem 2, we have

$$e^x = \sum_{k=0}^{n-1} \frac{x^k}{k!} + R_n(x),$$

where $R_n(x) = \dfrac{e^{\xi}}{n!} x^n$, for some point ξ lying between x and 0. For each given x, therefore, we have

$$|R_n(x)| \le e^{|\xi|} \left| \frac{x^n}{n!} \right| \to 0 \qquad \text{as } n \to \infty.$$

which implies that $\lim_{n\to\infty} R_n(x) = 0$ for any x. Thus e^x is equal to the sum of its Maclaurin series for all x by Theorem 3, Hence we have

$$e^x = \sum_{n=0}^{\infty} \frac{x^n}{n!} = 1 + \frac{x}{1!} + \frac{x^2}{2!} + \frac{x^3}{3!} + \cdots \quad \text{for all } x.$$

■

In particular, when $x = 1$, we obtain another expression for e as a sum of an infinite series.

$$e = \sum_{n=0}^{\infty} \frac{1}{n!} = 1 + \frac{1}{1!} + \frac{1}{2!} + \frac{1}{3!} + \cdots.$$

We can easily find the Taylor series expansion of e^x about an arbitrary point a since

$$e^x = e^a e^{x-a} = \sum_{n=0}^{\infty} \frac{e^a}{n!} (x-a)^n \quad \text{for all } x.$$

The radius of convergence of the Maclaurin series of $f(x) = e^x$ given by $\sum_{n=0}^{\infty} \dfrac{x^n}{n!}$ may also be found by applying the Ratio Test. Let $a_n = x^n/n!$. For any real number x, we have

$$\left| \frac{a_{n+1}}{a_n} \right| = \left| \frac{x^{n+1}}{(n+1)!} \cdot \frac{n!}{x^n} \right| = \frac{|x|}{n+1} \to 0 < 1 \quad \text{as } n \to \infty.$$

Hence, the series converges for all x by the Ratio Test with its radius of convergence $R = \infty$.

EXAMPLE 3

Find the Maclaurin series for $\cos x$.

Solution Finding and then evaluating the derivatives of $f(x)$ at $x=0$, we get

$$\begin{aligned} f(x) &= \cos x & f(0) &= 1 \\ f'(x) &= -\sin x & f'(0) &= 0 \\ f''(x) &= -\cos x & f''(0) &= -1 \\ f'''(x) &= \sin x & f'''(0) &= 0 \\ f^{(4)}(x) &= \cos x & f^{(4)}(0) &= 1 \end{aligned}$$

The Maclaurin series of $\cos x$ is, therefore,

$$\begin{aligned} \cos x &= f(0) + \frac{f'(0)}{1!}x + \frac{f''(0)}{2!}x^2 + \frac{f'''(0)}{3!}x^3 + \cdots \\ &= 1 - \frac{x^2}{2!} + \frac{x^4}{4!} - \frac{x^6}{6!} + \cdots = \sum_{n=0}^{\infty}(-1)^n \frac{x^{2n}}{(2n)!}. \end{aligned}$$ ■

As in Example 1, it can be proved that $\cos x$ is equal to the sum of its Maclaurin series for all x. Thus,

$$\begin{aligned} \cos x &= 1 - \frac{x^2}{2!} + \frac{x^4}{4!} - \frac{x^6}{6!} + \cdots \\ &= \sum_{n=0}^{\infty}(-1)^n \frac{x^{2n}}{(2n)!} \quad \text{for all } x. \end{aligned}$$

EXAMPLE 4

Find the Maclaurin series for the function $f(x) = x\sin x$.

Solution Multiplying the Maclaurin series for $\sin x$ by x, we find

$$x\sin x = x\sum_{n=0}^{\infty}(-1)^n \frac{x^{2n+1}}{(2n+1)!} = \sum_{n=0}^{\infty}(-1)^n \frac{x^{2n+2}}{(2n+1)!}.$$ ■

EXAMPLE 5

Find the Taylor expansion about $x = \pi/3$ for the function $f(x) = \cos x$.

Solution The evaluation of the derivatives of $f(x)$ at $\pi/3$ gives

$$\begin{aligned} f(x) &= \cos x & f\left(\frac{\pi}{3}\right) &= \frac{1}{2} \\ f'(x) &= -\sin x & f'\left(\frac{\pi}{3}\right) &= -\frac{\sqrt{3}}{2} \\ f''(x) &= -\cos x & f''\left(\frac{\pi}{3}\right) &= -\frac{1}{2} \end{aligned}$$

$$f'''(x) = \sin x \qquad\qquad f'''\left(\frac{\pi}{3}\right) = \frac{\sqrt{3}}{2}$$

Therefore the Taylor series at $\frac{\pi}{3}$ is

$$\begin{aligned}\cos x &= f\left(\frac{\pi}{3}\right) + \frac{f'\left(\frac{\pi}{3}\right)}{1!}\left(x - \frac{\pi}{3}\right) + \frac{f''\left(\frac{\pi}{3}\right)}{2!}\left(x - \frac{\pi}{3}\right)^2 \\ &\quad + \frac{f'''\left(\frac{\pi}{3}\right)}{3!}\left(x - \frac{\pi}{3}\right)^3 + \cdots \\ &= \frac{1}{2} - \frac{\sqrt{3}}{2 \cdot 1!}\left(x - \frac{\pi}{3}\right) - \frac{1}{2 \cdot 2!}\left(x - \frac{\pi}{3}\right)^2 \\ &\quad + \frac{\sqrt{3}}{2 \cdot 3!}\left(x - \frac{\pi}{3}\right)^3 + \cdots.\end{aligned}$$

■

Some important Maclaurin series and their intervals of convergence are listed in the following table for future reference.

$\frac{1}{1-x} = \sum_{n=0}^{\infty} x^n = 1 + x + x^2 + x^3 + \cdots,$	$(-1, 1)$
$e^x = \sum_{n=0}^{\infty} \frac{x^n}{n!} = 1 + \frac{x}{1!} + \frac{x^2}{2!} + \frac{x^3}{3!} + \cdots,$	$(-\infty, \infty)$
$\sin x = \sum_{n=0}^{\infty} (-1)^n \frac{x^{2n+1}}{(2n+1)!} = x - \frac{x^3}{3!} + \frac{x^5}{5!} - \frac{x^7}{7!} + \cdots,$	$(-\infty, \infty)$
$\cos x = \sum_{n=0}^{\infty} (-1)^n \frac{x^{2n}}{(2n)!} = 1 - \frac{x^2}{2!} + \frac{x^4}{4!} - \frac{x^6}{6!} + \cdots,$	$(-\infty, \infty)$
$\tan^{-1} x = \sum_{n=0}^{\infty} (-1)^n \frac{x^{2n+1}}{2n+1} = x - \frac{x^3}{3} + \frac{x^5}{5} - \frac{x^7}{7} + \cdots,$	$[-1, 1]$

Taylor series are also useful for integrating functions that cannot be integrated explicitly using known integration techniques.

EXAMPLE 6

Evaluate $\int e^{-x^3}\,dx$ as an infinite series.

Solution The Maclaurin series for $f(x) = e^{-x^3}$ is obtained simply by replacing x with $-x^3$ in the Maclaurin series for e^x. Thus

$$e^{-x^3} = \sum_{n=0}^{\infty} \frac{(-x^3)^n}{n!} = \sum_{n=0}^{\infty} (-1)^n \frac{x^{3n}}{n!} = 1 - \frac{x^3}{1!} + \frac{x^6}{2!} - \frac{x^9}{3!} + \cdots$$

for all values of x. Integrate both sides term-by-term we obtained

$$\int e^{-x^3}dx = \int\left(1-\frac{x^3}{1!}+\frac{x^6}{2!}-\frac{x^9}{3!}+\cdots+(-1)^n\frac{x^{3n}}{n!}+\cdots\right)dx$$
$$= C+x-\frac{x^4}{4\cdot 1!}+\frac{x^7}{7\cdot 2!}-\frac{x^{10}}{10\cdot 3!}+\cdots$$
$$+(-1)^n\frac{x^{3n+1}}{(3n+1)n!}+\cdots.$$

This series converges for all x because the original series for e^{-x^3} converges for all x. ■

Limits can also be found using Taylor series.

EXAMPLE 7

Evaluate $\lim_{x\to 0}\frac{x-\sin x}{x^3}$.

Solution Using the Maclaurin series for $\sin x$, we have

$$\lim_{x\to 0}\frac{x-\sin x}{x^3} = \lim_{x\to 0}\frac{x-\left(x-\frac{x^3}{3!}+\frac{x^5}{5!}-\cdots\right)}{x^3}$$
$$= \lim_{x\to 0}\frac{\frac{x^3}{3!}-\frac{x^5}{5!}+\frac{x^7}{7!}-\cdots}{x^3}$$
$$= \lim_{x\to 0}\left(\frac{1}{3!}-\frac{x^2}{5!}+\frac{x^4}{7!}-\cdots\right)$$
$$= \frac{1}{6}.$$ ■

Power series can be added, subtracted, multiplied or divided like polynomials. However only the first few terms are calculated because the initial terms are more important and the calculations for the later terms become cumbersome.

EXAMPLE 8

Find the first three non-zero terms of the Maclaurin series for $e^x\cos x$.

Solution Multiplying the Maclaurin series for e^x and $\cos x$, then collecting like terms, we have

$$e^x\cos x = \left(1+\frac{x}{1!}+\frac{x^2}{2!}+\frac{x^3}{3!}+\cdots\right)\left(1-\frac{x^2}{2!}+\frac{x^4}{4!}-\frac{x^6}{6!}+\cdots\right)$$
$$= 1+x-\frac{1}{3}x^3-\cdots.$$ ■

The Binomial Series

If a and b are any real numbers and k is a positive integer, then by the Binomial Theorem, we have

$$\begin{aligned}(a+b)^k &= a^k + ka^{k-1}b + \frac{k(k-1)}{2!}a^{k-2}b^2 + \frac{k(k-1)(k-2)}{3!}a^{k-3}b^3 \\ &\quad + \cdots + \frac{k(k-1)(k-2)\cdots(k-n+1)}{n!}a^{k-n}b^n \\ &\quad + \cdots + kab^{k-1} + b^k.\end{aligned}$$

This can be written in the abbreviated form as

$$(a+b)^k = \sum_{n=0}^{k} \binom{k}{n} a^{k-n}b^n,$$

where

$$\binom{k}{0} = 1 \text{ and } \binom{k}{n} = \frac{k(k-1)(k-2)\cdots(k-n+1)}{n!}, \quad 1 \le n \le k.$$

These numbers are called the **binomial coefficients.**

For $a=1$ and $b=x$, we get

10 $$(1+x)^k = \sum_{n=0}^{k} \binom{k}{n} x^n.$$

We extend this formula to the case where k is no longer a positive integer. To attain this, we find the Maclaurin series for the function $f(x) = (1+x)^k$, which is found to be

11 $$\sum_{n=0}^{\infty} \frac{f^{(n)}(0)}{n!}x^n = \sum_{n=0}^{\infty} \frac{k(k-1)\cdots(k-n+1)}{n!}x^n = \sum_{n=0}^{\infty} \binom{k}{n} x^n.$$

This series is called the **binomial series**. The following theorem gives the conditions under which the binomial expansion holds, that is, $(1+x)^k$ is equal to the binomial series 11, for exponent not necessarily a positive integer.

THEOREM 4 The Binomial Series Expansion

If k is any real number and $|x| < 1$, then

$$\begin{aligned}(1+x)^k &= 1 + kx + \frac{k(k-1)}{2!}x^2 + \frac{k(k-1)(k-2)}{3!}x^3 + \cdots \\ &= \sum_{n=0}^{\infty} \binom{k}{n} x^n,\end{aligned}$$

where

$$\binom{k}{n} = \frac{k(k-1)\cdots(k-n+1)}{n!} \quad (n \ge 1) \quad \text{and} \quad \binom{k}{0} = 1.$$

It should be noted that if k is a positive integer and $n > k$, then the expression for $\binom{k}{n}$ contains a factor, $(k-k)$, so $\binom{k}{n}=0$ for $n>k$. In this case the series terminates and reduces to the ordinary binomial expansion.

EXAMPLE 9

Using the method of binomial series expansion, expand $\dfrac{1}{(1-x)^3}$ as a power series.

Solution We have $k=-3$, and the binomial coefficient is found to be

$$\begin{aligned}\binom{-3}{n} &= \frac{(-3)(-4)(-5)\cdots(-3-n+1)}{n!}\\ &= (-1)^n\frac{3\cdot4\cdot5\cdots(n+1)(n+2)}{n!}\\ &= (-1)^n\frac{(n+1)(n+2)}{2}.\end{aligned}$$

By Theorem 4, for $|x|<1$, we have

$$\begin{aligned}\frac{1}{(1-x)^3} &= (1-x)^{-3} = \sum_{n=0}^{\infty}\binom{-3}{n}(-x)^n\\ &= \sum_{n=0}^{\infty}\frac{(n+1)(n+2)}{2}x^n = 1+3x+6x^2+\cdots.\end{aligned}$$

■

EXAMPLE 10

Find the Maclaurin series for the function $f(x)=\dfrac{1}{\sqrt{1-x}}$ and find its radius of convergence.

Solution We begin by rewriting $f(x)$ to get it into the form that we can utilize.

$$\frac{1}{\sqrt{1-x}}=(1-x)^{-1/2}.$$

Now we can apply the binomial series with $k=-1/2$:

$$\begin{aligned}\frac{1}{\sqrt{1-x}} &= (1-x)^{-1/2} = \sum_{n=0}^{\infty}\binom{-1/2}{n}(-x)^n\\ &= 1+\left(-\frac{1}{2}\right)(-x)+\frac{\left(-\frac{1}{2}\right)\left(-\frac{3}{2}\right)}{2!}(-x)^2+\cdots\\ &\quad+\frac{1\cdot3\cdot5\cdots(2n-1)}{n!\cdot2^n}x^n+\cdots.\end{aligned}$$

This takes the following form,

$$\frac{1}{\sqrt{1-x}} = 1 + \sum_{n=1}^{\infty} \frac{1 \cdot 3 \cdot 5 \cdots (2n-1)}{2 \cdot 4 \cdot 6 \cdots (2n)} x^n$$

This series converges when $|-x| < 1$, that is, $|x| < 1$, so the radius of convergence is $R = 1$. ■

CAS **EXAMPLE 11**

Obtain the Taylor series for $f(x)$ about a, where $f(x) = \sin(x^2)$, $a = \frac{\pi}{4}$.

http://matrix.skku.ac.kr/cal-lab/cal-10-5-Exm-11.html

Solution

```
var('x')
f=sin(x^2)
g=f.taylor(x, pi/4, 4)
p1=plot(f, (x, -1, 2), rgbcolor=(0, 0, 1))
p2=plot(g, (x, -1, 2), rgbcolor=(1, 0, 0))
show(p1+p2)
```

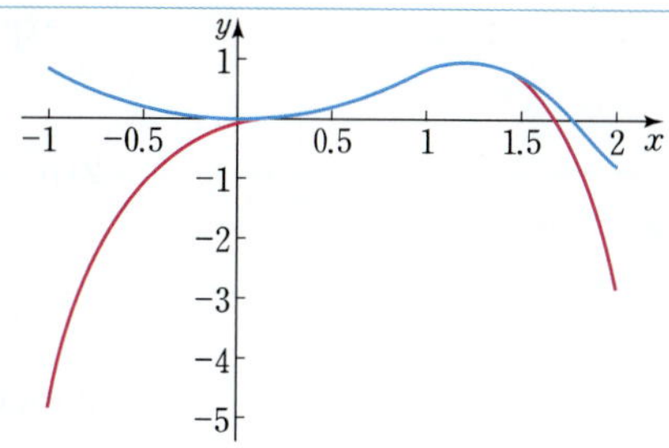

Figure 2

```
print g
```

Answer : 1/98304*(pi-4*x)^4 * (pi^4*sin(1/16*pi^2) - 48*pi^2 * cos(1/16*pi^2)- 192 * sin(1/16*pi^2)) + 1/3072 * (pi^3 * cos(1/16 * pi^2) + 24 * pi * sin(1/16 * pi^2)) * (pi - 4*x)^3 - 1/128*(pi^2 * sin (1/16*pi^2) - 8*cos(1/16*pi^2)) * (pi- 4*x)^2 - 1/8*(pi-4*x)*pi*cos (1/16*pi^2) + sin(1/16*pi^2) ■

Telescoping Series

Recall that $\sum a_n$ is called a telescoping series if a sequence $a_n = b_{n+1} - b_n$, $n \geq 1$.

We note that $s_1 = a_1 = b_2 - b_1$, $s_2 = a_1 + a_2 = (b_2 - b_1) + (b_3 - b_2) = b_3 - b_1$ and $s_n = a_1 + \cdots + a_n = (b_2 - b_1) + (b_3 - b_2) + \cdots + (b_{n+1} - b_n) = b_{n+1} - b_1$.

Thus we see that $\sum a_n$ converges if and only if $\lim b_n$ exists, in which case we have

$$\sum a_n = -b_1 + \lim b_n.$$

EXAMPLE 12

Show that the series $\sum \frac{n}{n^4+n^2+1}$ is a telescoping series and find its sum.
http://matrix.skku.ac.kr/cal-lab/cal-10-5-Exm13.html

Solution

```
var('n')
a(n)=n/(n^4+n^2+1)
an=n/(n^4+n^2+1)
s=a(n).partial_fraction()
show(s)
```

$$\frac{1}{2\,(n^2-n+1)} + \frac{-1}{2\,(n^2+n+1)}$$

```
b(n)=1/(2*(n^2-n+1))
expand(b(n+1))
```

$$\frac{1}{2\,(n^2+n+1)}$$

Thus if $b_n = \frac{1}{2\,(n^2-n+1)}$ then $b_{n+1} = \frac{1}{2\,(n^2+n+1)}$. This implies $a_n = b_n - b_{n+1}$. In particular it is a telescoping series.

```
var('k')
sum(a(n), n, 0, +oo)
```

Answer : 1/2 ■

Approximation using Taylor's Series

Recall that if $f(x)$ has a Taylor's series expansion in an interval $I=[a,\,b]$ about $x=c$ then we have

$$f(x) = T_n(x) + R_n(x)$$

where $R_{n(x)} = \frac{(x-c)^{n+1}}{(n+1)!} f^{(n+1)}(\zeta)$ is the error term for some ζ between x and c in $[a,\,b]$ and

$$T_n(x) = f(c) + \frac{(x-c)}{1!} f'(c) + \frac{(x-c)^2}{2!} f''(c) + \cdots + \frac{(x-c)^n}{n!} f^{(n)}(c)$$

is the nth degree Taylor polynomial.

Since ζ in $R_n(x)$ in the Taylor's theorem is not known, we cannot find the error term exactly for a given x in an interval I. However, we can obtain a bound on the error term as

$$|R_{n(x)}| = \left| \frac{(x-c)^{n+1}}{(n+1)!} f^{(n+1)}(c) \right| \le \max_{x \in I} \frac{|x-c|^{n+1}}{(n+1)!} \left[\max_{x \in I} |f^{(n+1)}(x)| \right]$$

Suppose $M := \max_{x \in I} |f^{(n+1)}(x)|$. Then we have

$$|R_{n(x)}| \le M.$$

CAS **EXAMPLE 13**

Approximate $\sqrt{1.1}$ using the first 10 Taylor polynomials.
http://matrix.skku.ac.kr/cal-lab/cal-10-5-Exm14.html

Solution

```
f(x)=sqrt(1+x)
```

```
for i in range(1, 10):
    T(x)=taylor(f(x), x, 0, i)
T(x=0.1)
```

1.05000000000000
1.04875000000000
1.04881250000000
1.04880859375000
1.04880886718750
1.04880884667969
1.04880884829102
1.04880884816010
Answer : 1.04880884817101 ∎

CAS **EXAMPLE 14**

Approximate $\sin 18^\circ$ using the first 10 Taylor polynomials. Note that $18^\circ = 18 \times \frac{\pi}{180} = \frac{\pi}{10}$ radians.
http://matrix.skku.ac.kr/cal-lab/cal-10-5-Exm15.html

Solution

```
f(x)=sin(x)
for i in range(1, 10):
    T(x)=taylor(f(x), x, 0, i)
T(x=pi/10).n()
```

0.314159265358979
0.314159265358979
0.308991552578929
0.308991552578929

```
0.309017054219328
0.309017054219328
0.309016994292875
0.309016994292875
```

Answer : 0.309016994375021 ∎

CAS **EXAMPLE 15**

Find the approximate value of $\cos 18°$ using the first 10 Taylor polynomials.

Solution

```
f(x)=cos(x)
for i in range(1, 10):
    T(x)=taylor(f(x), x, 0, i)
T(x=pi/10).n()
```

Answer : 0.951056516297732 ∎

CAS **EXAMPLE 16**

Find the approximate value of $\sin 0.2$.

Solution

```
sin(0.2).n()
```

Answer : 0.198669330795061 ∎

CAS **EXAMPLE 17**

Find the approximate value of $\ln 2.1$.

Solution

```
ln(2.1).n()
```

Answer : 0.741937344729377 ∎

9.5 EXERCISES (Taylor, Maclaurin, and Binomial Series)

http://matrix.skku.ac.kr/Cal-Book/part1/CS-Sec-9-5-Sol.html

1-4. Determine the radius of convergence of the Maclaurin series expansion for $f(x)$, where

1. $f(x) = \sin 2x$

Solution $f(x) = \sin 2x$, $f(0) = 0$,

$f'(x) = 2\cos 2x$, $f'(0) = 2 \cdot 1 = 2$,

$f''(x) = (-1)2^2 \sin 2x$, $f''(0) = 0$,

$f^{(3)}(x) = (-1)2^3 \cos 2x$, $f^{(3)}(0) = (-1)2^3 \cdot 1 = -8$,

$f^{(4)}(x) = (-1)2^4 \sin 2x$, $f^{(4)}(0) = 0$,

$\vdots \qquad \vdots$

$f^{(n-1)}(x) = (-1)^{n-3} 2^{n-1} \cos 2x$,

$f^{(n-1)}(0) = (-1)^{n-3} 2^{n-1}$,

$f^{(n)}(x) = (-1)^{n-2} 2^n \sin 2x$,

$f^{(n)}(0) = 0$.

Therefore, the function's Maclaurin series is

$$\sum_{n=0}^{\infty} \frac{f^{(n)}(0)}{n!} x^n = 2x + (-1)2^3 \frac{x^3}{3!} + \cdots + (-1)^{n-3} 2^{n-1} \frac{x^{n-1}}{(n-1)!} + \cdots$$

Here $a_{n-1} = \dfrac{x^{n-1}}{(n-1)!}$,

then $\left|\dfrac{a_{n-1}}{a_{n-3}}\right| = \left|\dfrac{x^{n-1}}{(n-1)!} \dfrac{(n-3)!}{x^{n-3}}\right| = \left|\dfrac{x^2}{(n-1)(n-2)}\right|$ $\to 0 < 1$ as $n \to \infty$.

Hence, by the Ratio Test, the series converges for all x, and the radius of convergence is $R = \infty$.

2. $f(x) = e^{-3x}$

Solution $f^{(n)}(x) = (-3)^n e^{-3x}$, $f^{(n)}(0) = (-3)^n$ for all n.

Therefore, Maclaurin series is

$$\sum_{n=0}^{\infty} \frac{f^{(n)}(0)}{n!} x^n = \sum_{n=0}^{\infty} \frac{(-3)^n}{n!} x^n$$

$$= 1 + \frac{(-3)}{1!} x + \frac{(-3)^2}{2!} x^2 + \cdots + \frac{(-3)^n}{n!} x^n + \cdots.$$

Let $a_n = \dfrac{(-3)^n}{n!} x^n$,

then $\left|\dfrac{a_{n+1}}{a_n}\right| = \left|\dfrac{(-3)^{n+1} x^{n+1}}{(n+1)!} \dfrac{n!}{(-3)^n x^n}\right| = \left|\dfrac{3x}{n+1}\right| \to 0 < 1$ as $n \to \infty$.

Hence by the Ratio Test, the series converges for all x, and the radius of convergence is $R = \infty$.

3. $f(x) = \ln(1+x)$

Solution $f(x) = \ln(1+x)$, $f(0) = 0$,

$f'(x) = (1+x)^{-1}$, $f'(0) = 1$,

$f''(x) = (-1)(1+x)^{-2}$, $f''(0) = -1$,

$f^{(3)}(x) = (-1)(-2)(1+x)^{-3}$, $f^{(3)}(0) = (-1)^2 2!$,

$f^{(4)}(x) = (-1)(-2)(-3)(1+x)^{-4}$, $f^{(4)}(0) = (-1)^3 3!$,

$\vdots \qquad \vdots$

$f^{(n)}(x) = (-1)(-2) \cdots (n-1)(1+x)^n$,

$f^{(n)}(0) = (-1)^{n-1}(n-1)!$.

Therefore, its Maclaurin series is

$$\sum_{n=0}^{\infty} \frac{f^{(n)}(0)}{n!} x^n = \sum_{n=1}^{\infty} \frac{(-1)^{n-1}}{n} x^n = x - \frac{x^2}{2} + \frac{x^3}{3} - \frac{x^4}{4} + \cdots.$$

Here $a_n = \dfrac{(-1)^{n-1} x^n}{n}$,

then $\left|\dfrac{a_{n+1}}{a_n}\right| = \left|\dfrac{x^{n+1}}{n+1} \dfrac{n}{x^n}\right| = \left|\dfrac{nx}{n+1}\right| \to |x|$ as $n \to \infty$.

Hence by the Ratio Test, the series converge when $|x| < 1$.

4. $f(x) = \sinh x$

Solution $f(x) = \sinh x$, $f(0) = 0$

$f'(x) = \cosh x$, $f'(0) = 1$

$f''(x) = \sinh x$, $f''(0) = 0$

$\vdots \qquad \vdots$

$f^{(n)}(x) = \sinh x$, $f^{(n)}(0) = 0$

$f^{(n+1)}(x) = \cosh x$, $f^{(n+1)}(0) = 1$.

Therefore, its Maclaurin series is

$$\sum_{n=0}^{\infty} \frac{f^{(n)}(0)}{n!} x^n = x + \frac{x^3}{3} + \frac{x^5}{5} + \cdots.$$

Let $a_n = \dfrac{x^n}{n!}$,

then $\left|\dfrac{a_{n+1}}{a_{n-1}}\right| = \left|\dfrac{x^{n+1}}{(n+1)!} \dfrac{(n-1)!}{x^{n-1}}\right| = \left|\dfrac{x^2}{n(n+1)}\right| \to 0 < 1$ as $n \to \infty$.

Hence by the Ratio Test, the series converges for all x, and the radius of convergence is $R = \infty$.

5-8. Obtain the Taylor series for $f(x)$ about a.

5. $f(x) = e^x$, $a = 2$

Solution $f(x) = \displaystyle\sum_{n=0}^{\infty} \frac{f^{(n)}(2)}{n!} (x-2)^n$

$$= e^2 + e^2(x-2) + \frac{e^2}{2!}(x-2)^2 + \frac{e^2}{3!}(x-2)^3 + \cdots.$$

6. $f(x)=x^5,\ a=-2$

Solution $f(x)=\sum_{n=0}^{\infty}\frac{f^{(n)}(-2)}{n!}(x+2)^n$

$$x^5=(x-a+a)^5$$
$$=(-a)^5+{}_5C_1(-a)^4(x+a)+{}_5C_2(-a)^3(x+a)^2$$
$$+{}_5C_3(-a)^2(x+a)^3+{}_5C_4(-a)(x+a)^4+{}_5C_5(x+a)^5$$
$$=-2^5+5\cdot(-2)^4(x+2)$$
$$+\frac{5\cdot4\cdot(-2)^3}{2!}(x+2)^2+\frac{5\cdot4\cdot3\cdot2^2}{3!}(x+2)^3$$
$$+\frac{5\cdot4\cdot3\cdot2\cdot(-2)}{4!}(x+2)^4+\frac{5\cdot4\cdot3\cdot2}{5!}(x+2)^5.$$

CAS **7.** $f(x)=\frac{1}{x^3},\ a=1$

http://matrix.skku.ac.kr/cal-lab/cal-10-5-7.html

Solution

```
var('x')
f(x)=1/(x^3)
g(x)=f.taylor(x, 1, 4)
p1=plot(f, (x, 0.5, 2))
p2=plot(g, (x, 0.5, 2), color="red", linestyle='--')
show(p1+p2)
print g
```

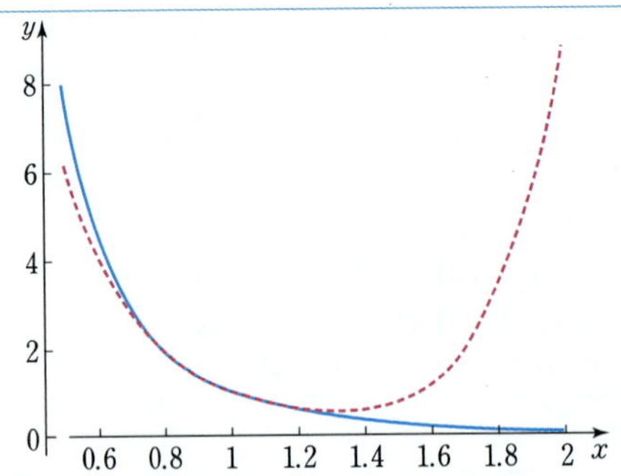

Answer : x |--> 15*(x - 1)^4 - 10*(x - 1)^3 + 6*(x - 1)^2 - 3*x + 4

CAS **8.** $f(x)=\ln\left(\frac{\sin x}{x}\right),\ a=2$

http://matrix.skku.ac.kr/cal-lab/cal-10-5-8.html

Solution

```
var('x')
f(x)=log(sin(x)/x)
g(x)=f.taylor(x, 2, 3)
p1=plot(f, (x, 0, 3))
p2=plot(g, (x, 0, 3), color="red", linestyle='--')
show(p1+p2)
print g
```

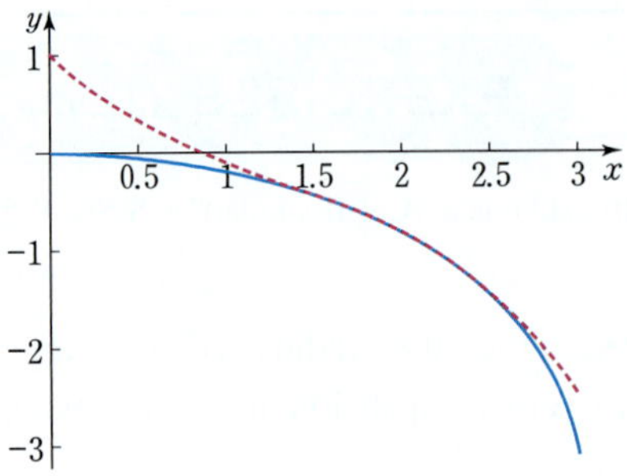

Answer : x |--> -1/24*(x - 2)^3*(sin(2)^3 - 8*sin(2)^2*cos(2) - 8*cos(2)^3)/sin(2)^3 - 1/8*(x - 2)^2*(3*sin(2)^2 + 4*cos(2)^2)/sin(2)^2 - 1/2*(x - 2)*(sin(2) - 2*cos(2))/sin(2) - log(2) + log(sin(2)).

9-11. Find the Maclaurin series for the given function.

9. $f(x)=\cos^2 x$ (Use $\cos^2 x=\frac{1}{2}(1+\cos 2x)$.)

Solution $f(x)=\frac{1}{2}(1+\cos 2x),\ f(0)=1$

$$f'(x)=\frac{1}{2}(-1)2\sin 2x=-\sin 2x,\ f'(0)=0$$
$$f''(x)=-2\cos 2x,\ f''(0)=-2$$
$$f^{(3)}(x)=2^2\sin 2x,\ f^{(3)}(0)=0$$
$$\vdots \qquad\qquad \vdots$$
$$f^{(n-1)}(x)=(-1)^{n-2}2^{n-2}\sin 2x,\ f^{(n-1)}(0)=0$$
$$f^{(n)}(x)=(-1)^{n-1}2^{n-1}\cos 2x,\ f^{(n-1)}(0)=(-1)^{n-1}2^{n-1}$$

Thus $f(x)=\sum_{n=0}^{\infty}\frac{f^{(n)}(0)}{n!}(x)^n$

$$=-2+4-\cdots+(-1)^{n-1}2^{n-1}+\cdots.$$

CAS **10.** $f(x)=x^2\tan^{-1}x$

http://matrix.skku.ac.kr/cal-lab/cal-10-5-10.html

Solution

```
var('x')
f=(x^2)*(arctan(x))
g=f.taylor(x, 0, 4)
p1=plot(f, (x,  0.5, 6), rgbcolor=(0, 0, 1))
p2=plot(g, (x, 0.5, 6), rgbcolor=(0, 1, 0))
show(p1+p2)
print g
```

Answer : x^3

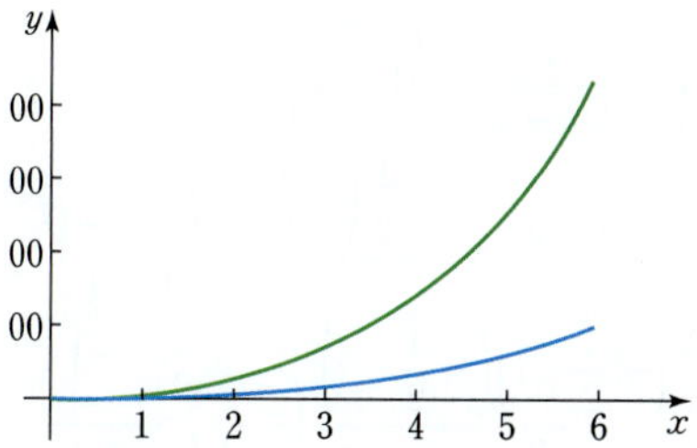

CAS **11.** $f(x)=e^{-x^2}$

http://matrix.skku.ac.kr/cal-lab/cal-10-5-11.html

Solution

```
var('x')
f=exp(-(x^2))
g=f.taylor(x, 0, 4)
p1=plot(f, (x, 0.5, 6), rgbcolor=(0, 0, 1))
p2=plot(g, (x, 0.5, 6), rgbcolor=(0, 1, 0))
show(p1+p2)
print g
```

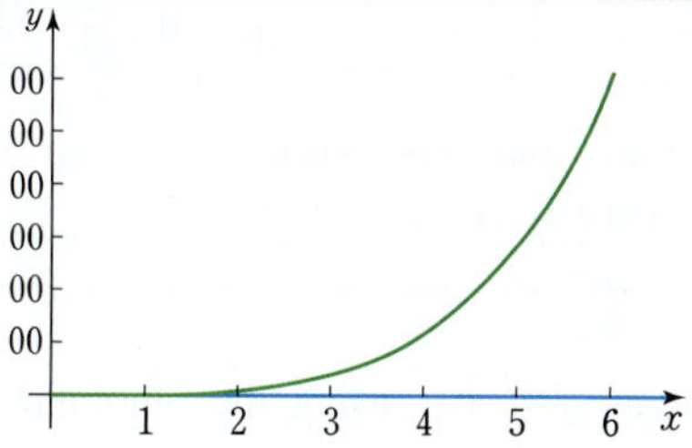

Answer : 1/2*x^4 - x^2 + 1

12-13. Evaluate the indefinite integral as an infinite series

12. $\int \frac{\sin t}{t}dt$

Solution $\sin t = t-\frac{t^3}{3!}+\frac{t^5}{5!}-\cdots$

$\frac{\sin t}{t}=1-\frac{t^2}{3!}+\frac{t^4}{5!}-\cdots$

Integrate both sides term by term:

$$\int \frac{\sin t}{t}dt=\int\left(1-\frac{t^2}{3!}+\frac{t^4}{5!}-\cdots\right)dt$$

$$=C+t-\frac{1}{3}\frac{t^3}{3!}+\frac{1}{5}\frac{t^5}{5!}-\cdots$$

$$+(-1)^n\frac{t^{2n+1}}{(2n+1)\cdot(2n+1)!}+\cdots.$$

CAS **13.** $\int \sqrt{1+t^2}\,dt$

http://matrix.skku.ac.kr/cal-lab/cal-10-5-13.html

Solution

```
var('t')
f=(1+(t^2))^(1/2)
g=f.taylor(t, 0, 4)
print g
```

Answer : -1/8*t^4 + 1/2*t^2 + 1

Thus $\sqrt{1+t^2}=-\frac{1}{8}t^4+\frac{1}{2}t^2+1.$

Integrate both sides term by term:

$$\int \sqrt{1+t^2}\,dt=\int\left(-\frac{1}{8}t^4+\frac{1}{2}t^2+1\right)dt$$

$$=-\frac{1}{40}t^5+\frac{1}{6}t^3+t+C$$

14-15. Evaluate the limit using a series:

CAS **14.** $\lim_{x\to 0}\frac{[\ln(1+x)]^2}{x\csc x-1}$

http://matrix.skku.ac.kr/cal-lab/cal-10-5-14.html

Solution

```
var('x')
f=log(1+x)
g=f.taylor(x, 0, 4)
h=1/(sin(x))
i=h.taylor(x, 0, 4)
j=(g^2)/(x*i-1)
p1=plot(j, (x, 0.5, 6), rgbcolor=(0, 1, 0))
show(p1)
print j
```

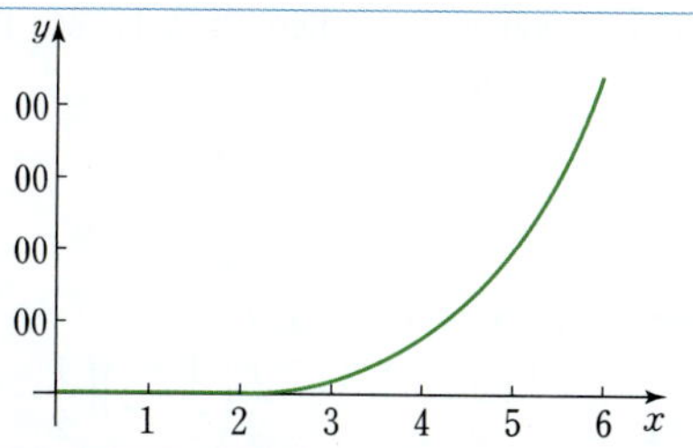

Answer: 5/2*(3*x^4 - 4*x^3 + 6*x^2 - 12*x)^2/((7*x^3 + 60*x + 360/x)*x - 360)

Thus $\lim_{x\to 0}\frac{[\ln(1+x)]^2}{x\csc x-1}=-360.$

15. $\lim_{x\to 0}\frac{\sin^2 x-x^2}{(e^{x^2}-1)^2}$

Solution $\sin^2 x=\left(x-\frac{x^3}{3!}+\frac{x^5}{5!}-\frac{x^7}{7!}+\cdots\right)^2$

$$=x^2-\frac{1}{3}x^4+\frac{2}{45}x^6-\frac{1}{315}x^8+\cdots,$$

$$e^{x^2}=1+\frac{x^2}{1!}+\frac{x^4}{2!}+\frac{x^6}{3!}+\cdots.$$

$$\lim_{x\to 0}\frac{\sin^2 x-x^2}{(e^{x^2}-1)^2}=\lim_{x\to 0}\frac{-\frac{1}{3}x^4+\frac{2}{45}x^6-\frac{1}{315}x^8+\cdots}{\left(x^2+\frac{x^4}{2!}+\frac{x^6}{3!}+\cdots\right)^2}=-\frac{1}{3}.$$

---

16. Deduce from the Maclaurin series for e^t that $\frac{1}{x^x}=\sum_{n=0}^{\infty}\frac{(-1)^n}{n!}(x\ln x)^n$.

Solution Since $\frac{1}{x^x}=(x^x)^{-1}=(e^{\ln x^x})^{-1}=e^{-\ln x^x}$,

$$e^t=\sum_{n=0}^{\infty}\frac{t^n}{n!}.$$

Put $t=-\ln x^x$,

then $\frac{1}{x^x}=\sum_{n=0}^{\infty}\frac{(-\ln x^x)^n}{n!}=\sum_{n=0}^{\infty}\frac{(-1)^n(x\ln x)^n}{n!}$.

17-20. Obtain the binomial series and radius of convergence of the function.

17. $\frac{x+x^2}{(1+x)^3}$

Solution $\frac{x+x^2}{(1+x)^3}=x(1+x)^{-2}$

$$=x\left(1-2x+\frac{-2(-2-1)}{2!}x^2+\frac{-2(-2-1)(-2-2)}{3!}x^3+\cdots\right)$$

$$=x-2x^2+\frac{-2(-2-1)}{2!}x^3+\frac{-2(-2-1)(-2-2)}{3!}x^4+\cdots,$$

this series converges when $|x|<1$, so the radius of convergence is $R=1$.

18. $\sqrt{1-x}$

Solution $\sqrt{1-x}=(1+(-x))^{\frac{1}{2}}$

$$=1-\frac{1}{2}x+\frac{\frac{1}{2}\left(\frac{1}{2}-1\right)}{2!}x^2-\frac{\frac{1}{2}\left(\frac{1}{2}-1\right)\left(\frac{1}{2}-2\right)}{3!}x^3+\cdots,$$

this series converges when $|-x|<1$. so the radius of convergence is $R=1$.

19. $\frac{1}{\sqrt{4-x}}$

Solution $\frac{1}{\sqrt{4-x}}=2(1-\frac{x}{4})^{-\frac{1}{2}}$

$$=2\left(1-\frac{1}{2}\left(-\frac{x}{4}\right)+\frac{-\frac{1}{2}\left(-\frac{1}{2}-1\right)}{2!}\left(-\frac{x}{4}\right)^2+\frac{-\frac{1}{2}\left(-\frac{1}{2}-1\right)\left(-\frac{1}{2}-2\right)}{3!}\left(-\frac{x}{4}\right)^3+\cdots\right)$$

$$=2\left(1+\frac{1}{8}x+\frac{1\cdot 3}{2!8^2}x^2+\frac{1\cdot 3\cdot 5}{3!8^3}x^3+\cdots\right),$$

this series converges when $\left|-\frac{x}{4}\right|<1$, so the radius of convergence is $R=4$.

20. $\frac{1}{\sqrt[3]{1+x}}$

Solution $\frac{1}{\sqrt[3]{1+x}}=(1+x)^{-\frac{1}{3}}=1+\left(-\frac{1}{3}\right)x+$

$$\frac{-\frac{1}{3}\left(-\frac{1}{3}-1\right)}{2!}x^2+\frac{-\frac{1}{3}\left(-\frac{1}{3}-1\right)\left(-\frac{1}{3}-2\right)}{3!}x^3+\cdots,$$

this series converges when $|x|<1$, so the radius of convergence is $R=1$.

---

21. Evaluate $f^{(12)}(0)$ using the binomial series where $f(x)=\frac{1}{\sqrt{1+2x^2}}$.

Solution Since

$$\sum_{n=0}^{\infty}\frac{f^{(n)}(0)}{n!}x^n=\sum_{n=0}^{\infty}\frac{k(k-1)\cdots(k-n+1)}{n!}x^n,$$

$$f(x)=\frac{1}{\sqrt{1+2x^2}}=(1+2x^2)^{-\frac{1}{2}}.$$

Thus $f^{(n)}(0)=-\frac{1}{2}\left(-\frac{1}{2}-1\right)\left(-\frac{1}{2}-2\right)\cdots\left(-\frac{1}{2}-n+1\right)$,

$$f^{(12)}(0)=-\frac{1}{2}\left(-\frac{1}{2}-1\right)\left(-\frac{1}{2}-2\right)\cdots\left(-\frac{1}{2}-12+1\right)$$

$$=-\frac{1}{2}\left(-\frac{3}{2}\right)\left(-\frac{5}{2}\right)\left(-\frac{7}{2}\right)\left(-\frac{9}{2}\right)\left(-\frac{11}{2}\right)\left(-\frac{13}{2}\right)\left(-\frac{15}{2}\right)\left(-\frac{17}{2}\right)\left(-\frac{19}{2}\right)\left(-\frac{21}{2}\right)\left(-\frac{23}{2}\right)$$

$$=\frac{3\cdot 5\cdot 7\cdot 9\cdot 11\cdot 13\cdot 15\cdot 17\cdot 19\cdot 21\cdot 23}{2^{12}}.$$

Fibonacci bus stop, Pisa

Chapter 10

Parametric Equations and Polar Coordinates

Teiji TAKAGI

(高木 貞治) 1875-1960

Founder of the Japanese School of Modern Mathematics

Teiji Takagi was a Japanese mathematician, best known for proving the Takagi existence theorem in class field theory. He learned mathematics from such European classic texts as Salmon's Algebra and Weber's Lehrbuch der Algeba. Aided by Hilbert, he then studied at Göttingen. He wrote a great number of Japanese textbooks on modern mathematics and geometry.

YouTube http://youtu.be/

In the previous chapters, we have been representing a plane curve by a single equation involving two variables x and y. In this chapter, we deal with two new ways of describing a plane curve: parametric representations and polar coordinates representations. Sometimes, it is useful to represent the variables x and y in terms of a third variable t, called a parameter. The parametric representation of a curve is useful for studying motion in a plane. The polar coordinate system is another convenient way of describing a curve which we deal with in this chapter. We also look at derivatives, arc length, area and curve sketching in parametric and polar coordinate systems. This chapter ends with conic sections.

10.1 Parametric Equations

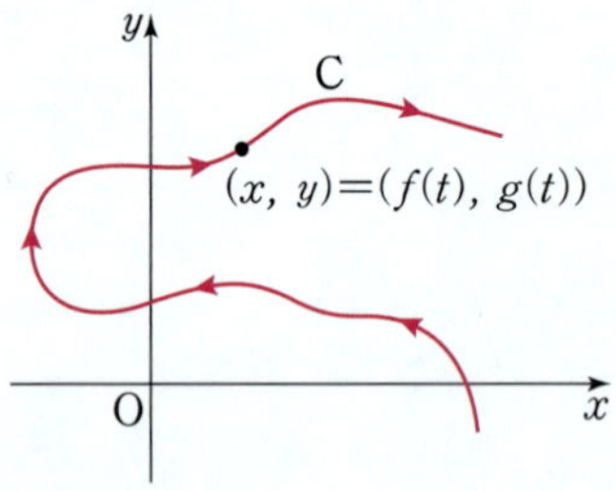

Figure 1

The motion of a particle along a curve C is shown in Figure 1. Since C fails the Vertical Line Test, the curve C cannot be described by $y=f(x)$. If there was such a relation it would not tell the whole story. Although such a function does tell us where the object has been, it does not tell us when the object was at a given point (x, y). Since the x and y-coordinates of the particle depend on time t, we can write $x=f(t)$ and $y=g(t)$. Such a pair of equations is often a convenient way of describing a curve.
If f and g are continuous functions of t on an interval I, then the set of ordered pairs $(f(t), g(t))$ is called a **plane curve** C. The equations $x=f(t)$ and $y=g(t)$ are called **parametric equations** for C, and t is called a **parameter.** (See Figure 1.)

Although in most applications the parameter t denotes time, it can also be represented by other letters and represent other quantities.

EXAMPLE 1

Sketch the plane curve defined by the parametric equations

$$x=t-1, \quad y=t^2-2t, \ -2 \le t \le 4.$$

http://matrix.skku.ac.kr/cal-lab/cal-9-1-exm-1.html

Solution Each value of t gives a point on the curve, as shown in the table. For instance, if $t=0$, then, $x=-1$, $y=0$ and so the corresponding point is $(-1, 0)$. In Figure 2 we plot the points (x, y) determined by several values of the parameter and we join them to produce a curve.

t	x	y
-2	-3	8
-1	-2	3
0	-1	0
1	0	-1
2	1	0
3	2	3
4	3	8

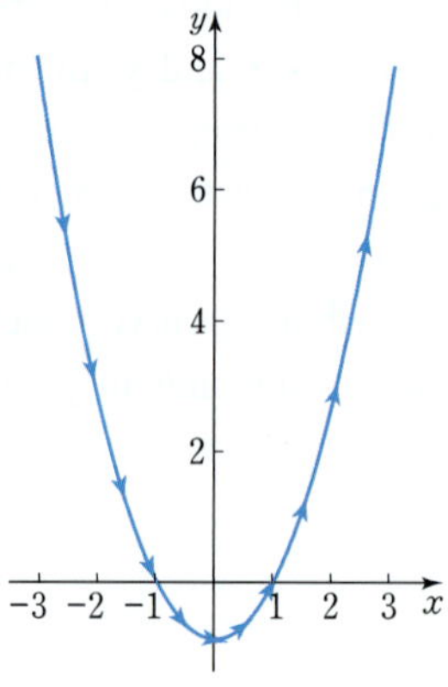

Figure 2

If the position of a particle is represented by parametric equations, then as t increases the particle moves along the curve in the direction of the arrows.

By eliminating the parameter t from both equations $x=t-1$ and $y=t^2-2t$, we have

$$y=t^2-2t=(x+1)^2-2(x+1)=x^2-1.$$

Thus, the curve represented by the given parametric equations is the parabola $y=x^2-1$. ■

It often happens that two different sets of parametric equations represent the same curve. For example, the parametric equations

$$x=t \quad \text{and} \quad y=t^2-1, \quad -3 \le t \le 3$$

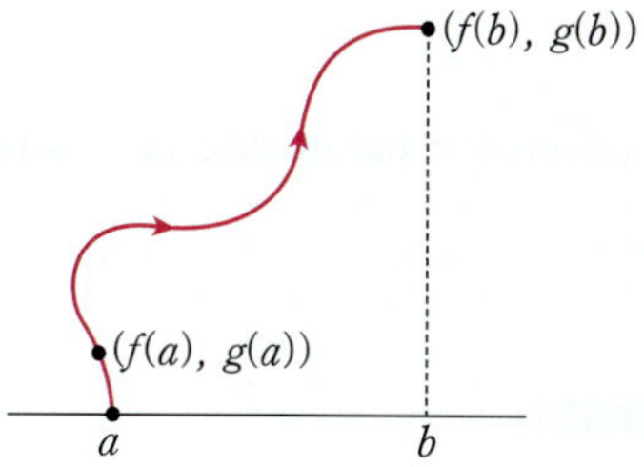

Figure 3

have the same curve as Example 1. However, comparing the value of t, we see that the above is traced out more rapidly (considering t as time) than Example 1. Thus, in applications, different parametric representations can be used to represent various speeds at which objects travel along a given path.

When $a \le t \le b$, the curve with parametric equations $x=f(t)$, $y=g(t)$ has an **initial point** $(f(a),\ g(a))$ and a terminal point $(f(b),\ g(b))$.

EXAMPLE 2

Find the curve represented by the parametric equations

$$x=2\cos\theta,\ y=3\sin\theta, \quad 0 \le \theta \le 2\pi.$$

http://matrix.skku.ac.kr/cal-lab/cal-9-1-exm-2.html

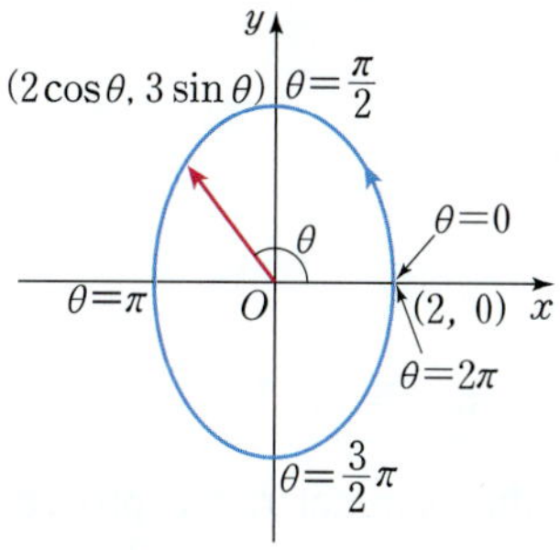

Figure 4

Solution First of all, we solve for $\sin\theta$ and $\cos\theta$ from the given equations.

$$\cos\theta=\frac{x}{2} \quad \text{and} \quad \sin\theta=\frac{y}{3}$$

Now we utilize the identity $\cos^2\theta+\sin^2\theta=1$ to form an equation involving only x and y.

$$\cos^2\theta+\sin^2\theta=\left(\frac{x}{2}\right)^2+\left(\frac{y}{3}\right)^2=1 \text{ or } \frac{x^2}{4}+\frac{y^2}{9}=1.$$

From this rectangular equation we see that the graph is an ellipse centered at (0, 0), with vertex (0, 3) and (0, -3). Note that the ellipse is traced out counterclockwise as θ varies from 0 to 2π. ■

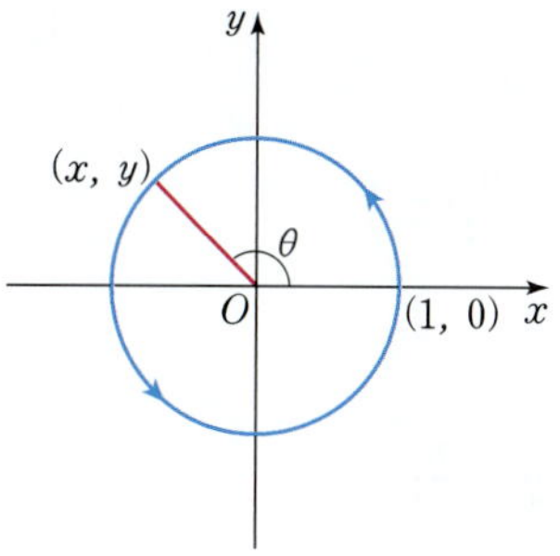

Figure 5

By eliminating θ from the above identity $\cos^2\theta+\sin^2\theta=1$ we have $x^2+y^2=1$, which is a unit circle with center at the origin and radius 1. Here the parameter θ is the angle (in radians) shown in Figure 5. As θ increases from 0 to 2π starting from the point $(1,\ 0)$, the point $(x,\ y)=(\cos\theta,\ \sin\theta)$ moves once around the circle in the counterclockwise direction.

Note that the parametric equations $x=\sin 2\theta$, $y=\cos 2\theta$, $0 \le \theta \le \pi$ also represent the above unit circle. Again different sets of parametric equations can represent the same curve.

In general, the graph of an equation of the form $x=g(y)$ can be represented by the parametric equations

$$x=g(t),\ y=t.$$

Notice that curves with equations $y=f(x)$ (the ones we are most familiar with-graphs of functions) can also be regarded as curves with parametric equations

$$x=t,\quad y=f(t).$$

Remark Note that we may not always be able to convert the parametric equations of a curve to a Cartesian equation.

Finding parametric equations for a graph

In the previous examples, we have been looking at techniques for sketching the graph represented by a set of parametric equations. Let us think about the reverse problem. How can we find a set of parametric equations for a given graph or a given physical description? We know that such a representation is not unique from the above discussion. In the following example, we will find two different parametric representations for a given graph.

EXAMPLE 3

Find a set of parametric equations to represent the graph of $y=x^2-1$, using the following parameters.

(a) $t=x$

(b) The slope $m=dy/dx$ at the point $(x,\ y)$.

Solution (a) Let $x=t$. Then we obtain

$$x=t \quad \text{and} \quad y=x^2-1=t^2-1 \text{ for all } t\in\mathbb{R}.$$

(b) To express x and y in terms of the parameter, we proceed as follows.

$$m=\frac{dy}{dx}=2x \quad\Rightarrow\quad x=\frac{m}{2}$$

$$y=x^2-1=\left(\frac{m}{2}\right)^2-1=\frac{m^2}{4}-1$$

Thus the parametric equation are

$$x=\frac{m}{2} \quad \text{and} \quad y=\frac{m^2}{4}-1. \qquad ■$$

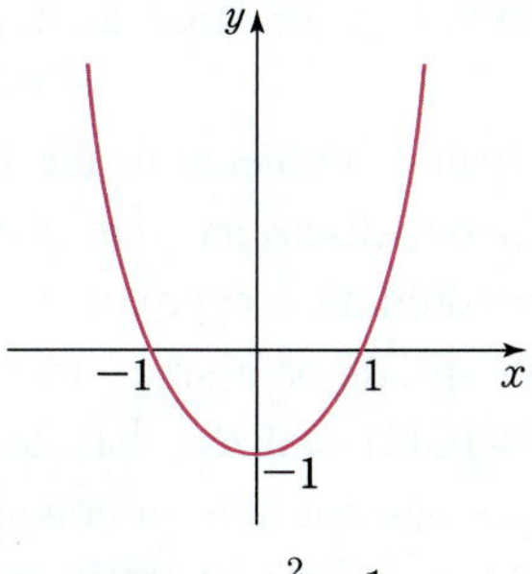

Figure 6 $y=x^2-1$

In Figure 6, the resulting curve has a left-to-right orientation as determined by the direction of increasing values of slope m.

The Cycloid

The **Cycloid** is the curve traced out by a point P on the circumference of a circle as the circle rolls along a straight line. (See Figure 7.)

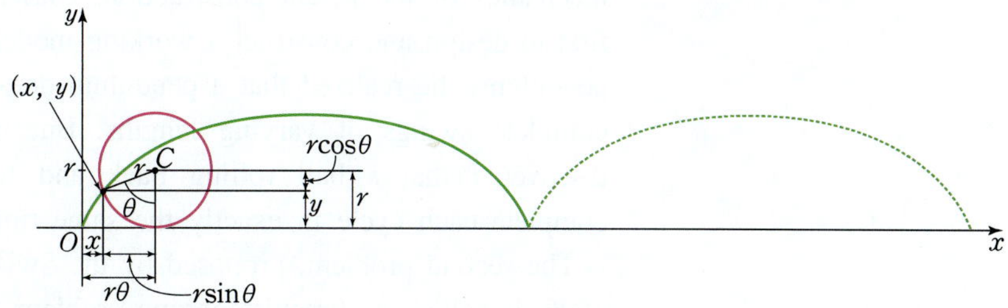

Figure 7

EXAMPLE 4

Determine the curves traced by a point P on the circumference of a circle of radius r rolling along the x-axis in a plane. One position of P is the origin.

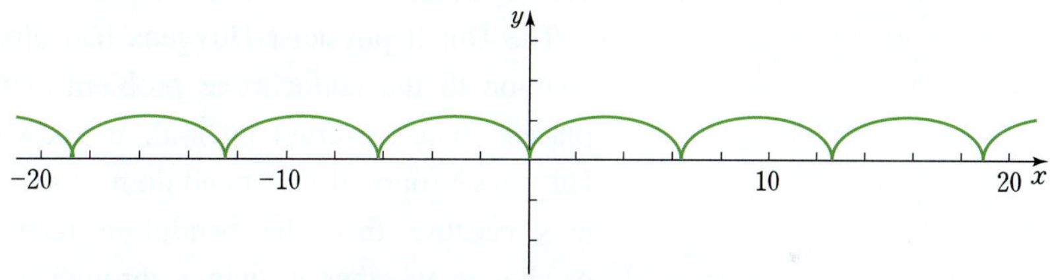

Figure 8 http://matrix.skku.ac.kr/cal-lab/sage-grapher-para.html

x(t) = a*((t) - sin(t)), y(t) = a*(1 - cos(t))

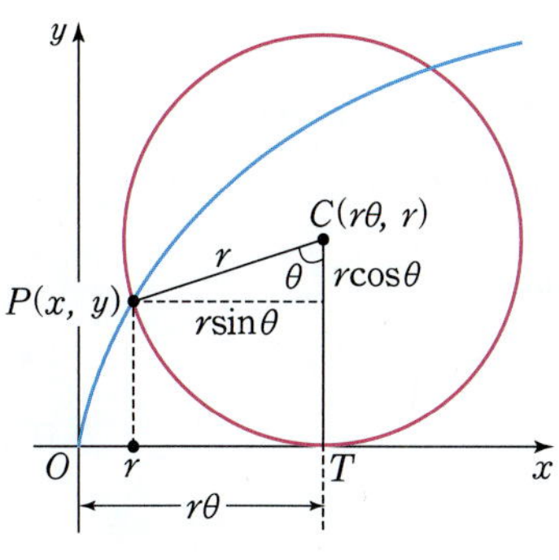

Figure 9

Solution Let the parameter θ be the measure of circle's rotation, and we let the point $P=(x,\ y)$ begin at the origin. When $\theta=0$, P is at the origin; when $\theta=\pi$, P is at a point $(\pi r,\ 2r)$; and when $\theta=2\pi$, P is back on x-axis at $(2\pi r,\ 0)$. From Figure 9 we see that the distance it has rolled from the origin is

$$|OT| = \text{arc } PT = r\,\theta.$$

Therefore, the center of the circle is $C(r\theta,\ r)$. Since $P(x,\ y)$ is any point on the cycloid, from Figure 9, we have

$$x=|OT|-|PQ|=r\,\theta-r\sin\theta=r(\theta-\sin\theta),$$
$$y=|TC|-|QC|=r-r\cos\theta=r(1-\cos\theta).$$

Thus, the parametric equations of the cycloid are

1 $$x=r(\theta-\sin\theta),\qquad y=r(1-\cos\theta),\qquad \theta\in\mathbb{R}.$$ ■

For one rotation of the circle described by $0 \le \theta \le 2\pi$, one arch of the cycloid is traced out.

Cycloid is related to one of the most famous (pairs) problems in the history of calculus. The first problem began with Galileo's discovery that the time required to complete a full swing of a given pendulum is approximately the same, whether it makes a large movement at high speeds or a small movement at a lower speeds. Late in his life, Galileo (1564-1642) realized that he could use this principle to construct a clock. However, he was not able to conquer the mechanics of the actual construction. Christian Huygens (1629-1695) was the first to design and construct a working model of such a clock. In his work with pendulums, he realized that a pendulum does not take exactly the same time to complete swings of varying lengths. But in studying the problem, Huygens discovered that a ball rolling back and forth on an inverted cycloid does complete each cycle in exactly the same time.

The second problem, proposed by the Swiss mathematician John Bernoulli, in 1696, is called the **brachistochrone problem** (in Greek brachys means short and chronos means time). In the brachistochrone problem, a curve is to be determined along which a particle slides in the shortest time (under the influence of gravity) from a point A to a lower point B not directly beneath A. He showed that among all possible curves that join A to B, the particle will take the least time sliding from A to B if the curve is part of an inverted arch of a cycloid.

The Dutch physicist Huygens had already shown that the cycloid is also the solution to the **tautochrone problem.** That is, no matter where a particle P is placed on an inverted cycloid, it takes the same time to slide to the bottom. Huygens proposed that pendulum clocks (which he invented) swing in cycloidal arcs because then the pendulum takes the same time to make a complete oscillation whether it swings through a wide or a small arc.

CAS **EXAMPLE 5**

Sketch the curve

$$x = t^3 - 1, \; y = 2t, \; -2 \le t \le 2.$$

http://matrix.skku.ac.kr/cal-lab/cal-9-1-exm-5.html

Solution

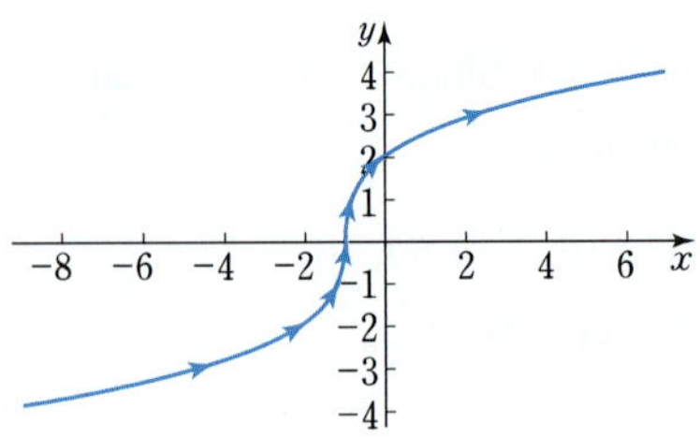

Figure 10

```
var('t,x,y')
x=t^3-1
y=2*t
a=-2
b=2
p=parametric_plot((x,y),(t,a,b))
small=0.001
step=0.5
```

```
n=(b-a)/step
arr=sum([arrow((x(t=a+i*step),
y(t=a+i*step)),(x(t=a+i*step+small), y(t= a+i*step+small))) for
i in range(1,n) ])
p+arr
```

■

CAS **EXAMPLE 6**

Sketch the curve

$$x = 2\cos t - \cos(2t),\ \ y = \sin(2t) - 2\sin t,\ \ 0 \le t \le 2\pi.$$

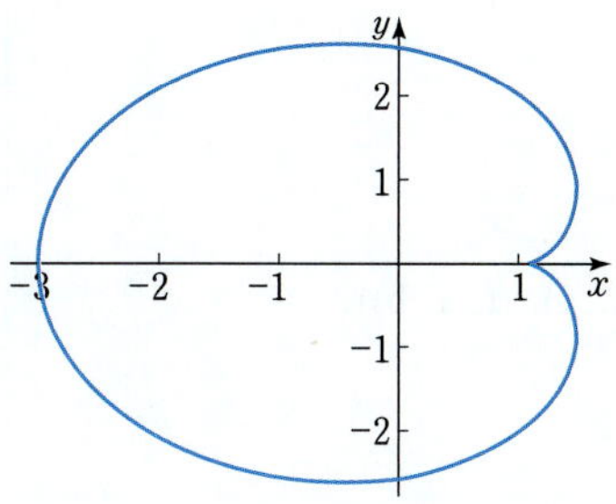

Figure 11

Solution

```
t=var('t')
parametric_plot((2*cos(t)-cos(2*t), sin(2*t)-2*sin(t)), (t, 0,
2*pi))
```

■

CAS **EXAMPLE 7**

Sketch the curve for $1 \le t \le 20$

$$x = 12\sin^3 t\,,\ \ y = 8\cos t - 5\cos 2t - 2\cos 3t - \cos 4t.$$

http://matrix.skku.ac.kr/cal-lab/sage-grapher-para.html

Solution

```
var('t, x, y')
x=12*sin(t)^3
y=8*cos(t)-5*cos(2*t)-2*cos(3*t)-cos(4*t)
p=parametric_plot((x,y), (t, 1, 20))
p
```

■

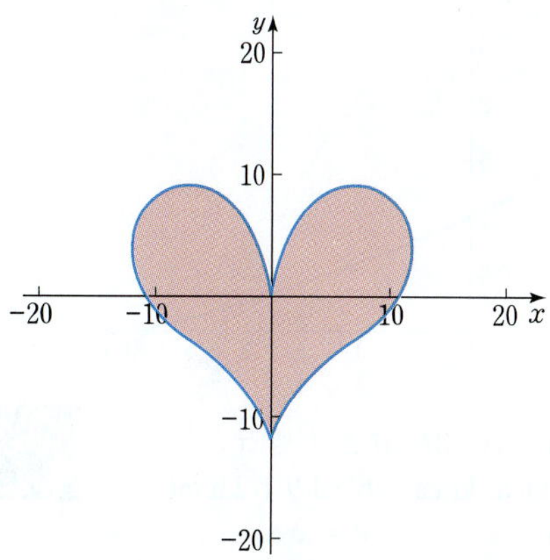

Figure 12

10.1 EXERCISES (Parametric Equations)

http://matrix.skku.ac.kr/Cal-Book/part1/CS-Sec-10-1-Sol.html

YouTube http://youtu.be/Ybs68e0iMZI

1-8. (a) Find the Cartesian equation of the curve.

CAS (b) Sketch the curve and indicate with an arrow the direction in which the curve is traced as the parameter increases.

1. $x = t^2 - 2t,\ y = 3 - t,\ 0 \le t \le 3$

http://matrix.skku.ac.kr/cal-lab/cal-9-1-1.html

Solution (a) From the second equation, we have $t = 3 - y$. Now from the first equation, we get

$$x = (3-y)^2 - 2(3-y) = y^2 - 4y + 3.$$

Notice that if $0 \le t \le 3$, we have $0 \le y \le 3$. Thus the Cartesian equation of the given curve is $x = y^2 - 4y + 3$, $0 \le y \le 3$.

(b)

```
var('t,x,y')
x=t^2-2*t
y=3-t
a=0
b=3
p=parametric_plot((x, y) , (t, a, b))
small=0.001
step=0.25
n=(b-a)/step
arr=sum([arrow((x(t=a+i*step),        y(t=a+i*step)),
(x(t=a+i*step+small),  y(t=a+i*step+small))) for i
in range(1,n) ])
p+arr
```

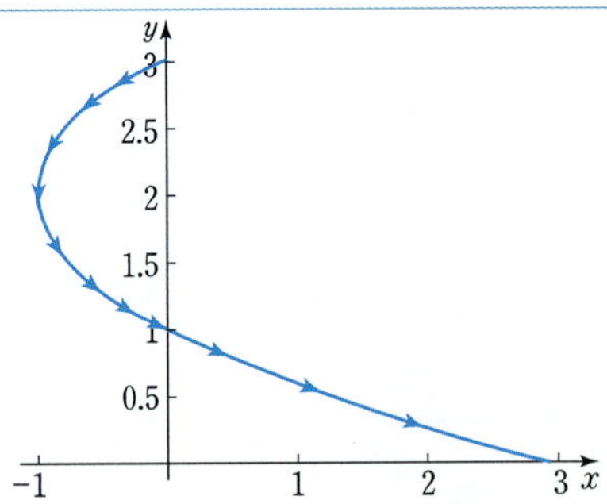

CAS **2.** $x = \sin^2 t,\ y = \cos 3t,\ 0 \le t \le \pi$

http://matrix.skku.ac.kr/cal-lab/cal-9-1-2.html

Solution (a) $y = \cos 3t \Rightarrow 3t = \arccos(y)$

$\Rightarrow t = \frac{1}{3}\arccos(y) \Rightarrow x = \sin^2\left(\frac{1}{3}\arccos(y)\right)$

$(-1 \le y \le 1)$

$\cos^2 t + x = 1 \Rightarrow \cos t = \pm\sqrt{1-x}$

$\therefore\ y = \pm(1-4x)\sqrt{1-x} \quad (\because \cos 3t = 4\cos^3 t - 3\cos t)$

(b)

```
var('t,x,y')
x=sin(t)^2
y=cos(3*t)
a=0
b=pi
p=parametric_plot((x, y), (t, a, b))
small=0.001
step=pi/8
n=(b-a)/step
arr=sum([arrow((x(t=a+i*step), y(t=a+i*step)), (x(
t= a+i*step + small), y(t=a+i*step+small))) for i
in range(1, n) ])
p+arr
```

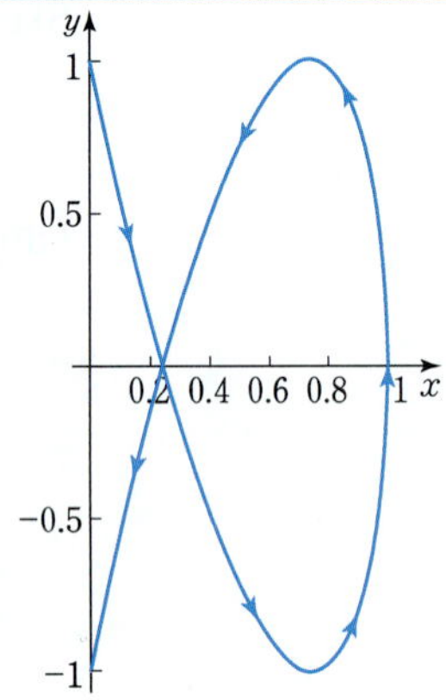

3. $x = \sin^2\theta,\ y = \cos^2\theta,\ 0 \le \theta \le 2\pi$

http://matrix.skku.ac.kr/cal-lab/cal-9-1-3.html

Solution (a) $x + y = 1 \Rightarrow y = -x + 1\ (0 \le x \le 1)$

(b)

```
var('t')
parametric_plot((sin(t)^2,cos(t)^2),(t,0,1/2)).show(a
spect_ratio=1,    xmin=-2,    xmax=2,    ymin=-1,
ymax=2)
parametric_plot((sin(t)^2,cos(t)^2),(t,0,1)).show(asp
ect_ratio=1, xmin=-2, xmax=2,
ymin=-1, ymax=2)
parametric_plot((sin(t)^2,cos(t)^2),(t,0,2)).show(asp
ect_ratio=1,    xmin=-2,    xmax=2,    ymin=-1,
ymax=2)
```

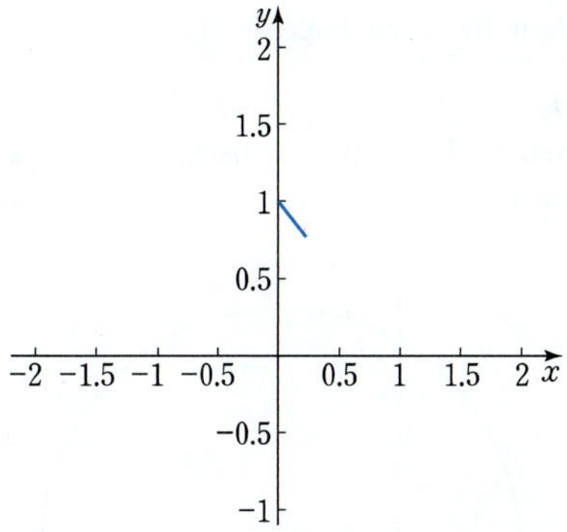

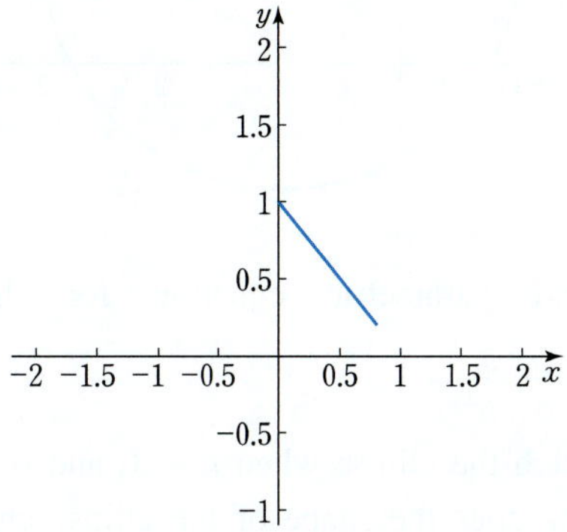

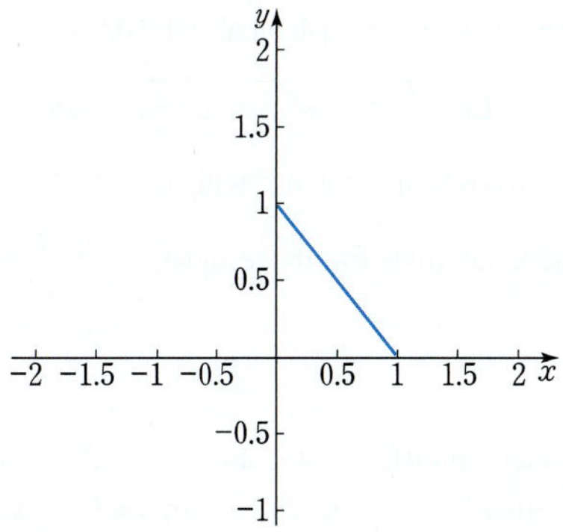

4. $x = \ln t,\ y = t - 1,\ t \geq 1.7$
http://matrix.skku.ac.kr/cal-lab/cal-9-1-4.html

Solution

(a) $y = t-1 \Rightarrow t = y+1 \Rightarrow x = \ln(y+1) \quad (y \geq 0.7)$

(b)

```
var('t')
parametric_plot((ln(t),t-1),(t,1,3/2)).show(aspect_ra
tio=1, xmin=-1, xmax=1, ymin=-1,
ymax=1.5)
parametric_plot((ln(t),t-1),(t,1,2)).show(aspect_rati
o=1, xmin=-1, xmax=1, ymin=-1, ymax=1.5)
parametric_plot((ln(t),t-1),(t,1,5/2)).show(aspect_ra
tio=1, xmin=-1, xmax=1, ymin=-1, ymax=1.5)
```

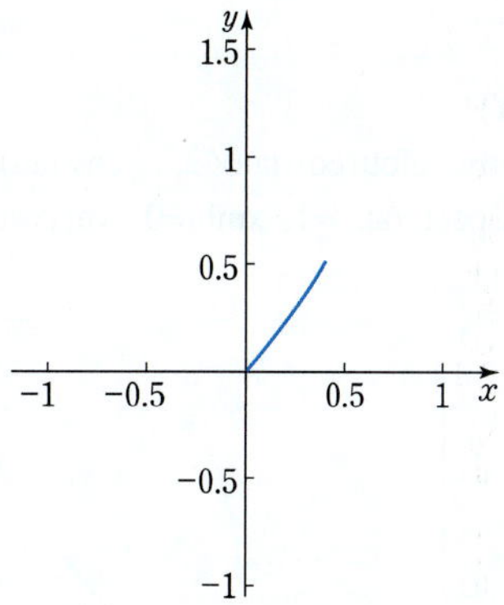

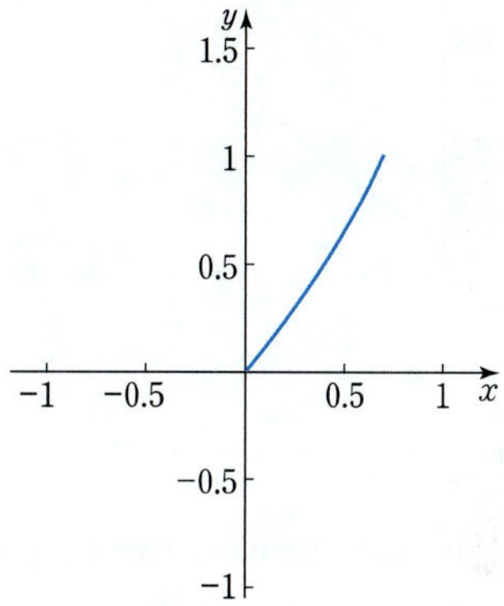

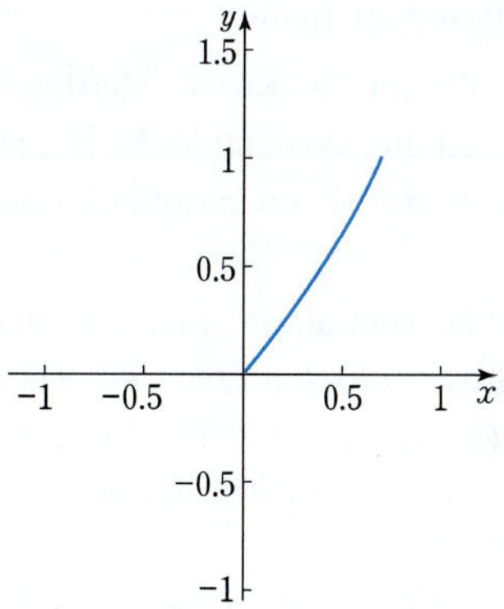

5. $x = \cosh t,\ y = \sinh t$

Solution $\cosh^2 t - \sinh^2 t = 1 \quad \Rightarrow \quad x^2 - y^2 = 1$

6. $x = 1 + \cos\theta,\ y = 2\cos\theta - 1,\ 0 \leq \theta \leq \pi$

Solution $x = 1 + \cos\theta \Rightarrow \cos\theta = x - 1,\ y = 2\cos\theta - 1$

$$\Rightarrow \cos\theta = \frac{(y+1)}{2}$$

$$x - 1 = \frac{(y+1)}{2} \Rightarrow 2x - 2 = y + 1 \Rightarrow y = 2x - 3$$

7. $x = \cos^2 t,\ y = \cos t,\ 0 \leq t \leq 4\pi$
http://matrix.skku.ac.kr/cal-lab/cal-9-1-7.html

Solution (a) Since $x = y^2$.

(b)

```
var('tau')
parametric_plot((cos(tau)^2,    cos(tau)),tau,0,4*pi).
show(aspect_ratio=1, xmin=0, xmax=0.7, ymin=0,
ymax=1)
```

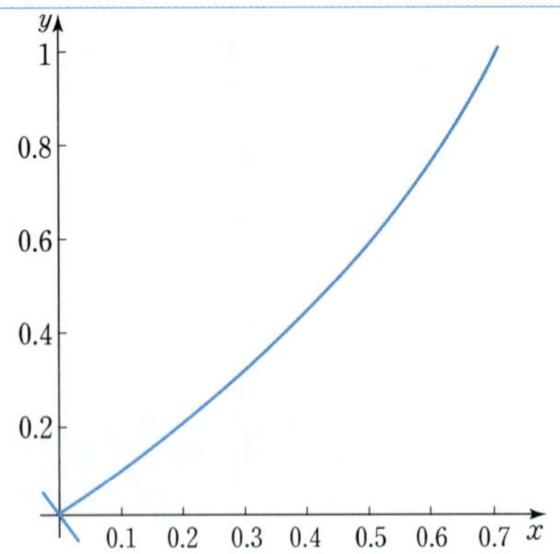

8. $x = e^{2t}$, $y = e^{-t}$

CAS **9.** Find a parametric equation for the path of a particle that moves along $(x-1)^2+(y-1)^2=4$ in the manner described below.

(a) Once around clockwise, starting at (3, 1).

(b) Twice around counterclockwise, starting at (3, 1).

(c) Halfway around counterclockwise, starting at (1, 3).

(d) Graph the semicircle traced by the particle.

http://matrix.skku.ac.kr/cal-lab/cal-9-1-9.html

Solution The circle $(x-1)^2+(y-1)^2=4$ has center (1, 1) and radius 2, so it can be represented by

$$x = 1+2\cos t,\ y = 1+2\sin t\ (0 \le t \le 2\pi).$$

This representation gives us the circle with a counterclockwise orientation starting (3, 1).

(a) To get a clockwise orientation, we can replace t by $2\pi - t$ in the equations to get,

$$x = 1+2\cos t,\ y = 1-2\sin t \quad (0 \le t \le 2\pi).$$

(b) To get twice around in the counterclockwise direction, we use the original equations

$$x = 1+2\cos t,\ y = 1+2\sin t$$

with the domain expanded to

$$0 \le t \le 4\pi.$$

(c) To start at (1, 3) using the original equations, we must have $x=1$; that is, $1+2\cos t = 1$ and $t = \frac{\pi}{2}$. We use the original equations $x = 1+2\cos t$, $y = 1+2\sin t$ with the domain $\frac{\pi}{2} \le t \le \frac{3\pi}{2}$ to be a counterclockwise orientation starting (1, 3).

(d) Graph it by using Sage.

```
var('t')
parametric_plot    ((1+2*cos(t),1    +    2*sin(t)),
(t,0,2*pi))
```

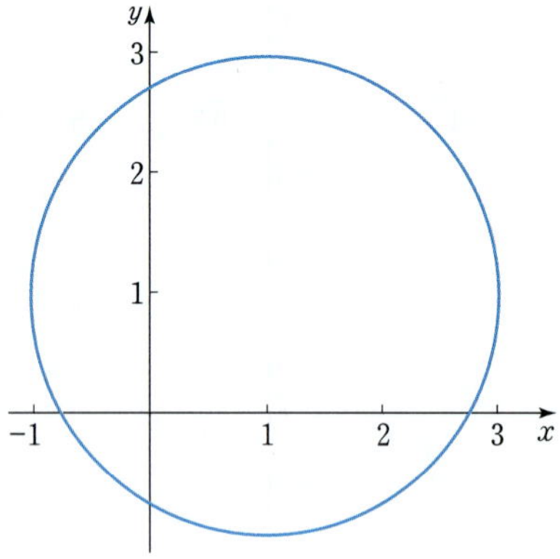

CAS **10.** (a) Find parametric equation for the ellipse $\frac{x^2}{a^2}+\frac{y^2}{b^2}=1$.

(b) Sketch the ellipse when $a=3$, and $b=1,\ 2,\ 4$.

(c) How does the shape of the ellipse change as a and b varies?

http://matrix.skku.ac.kr/cal-lab/cal-9-1-10.html

Solution (a) Let $\frac{x^2}{a^2}=\cos^2 t$, and $\frac{y^2}{b^2}=\sin^2 t$ to obtain $x = a\cos t$, $y = b\sin t$ with $0 \le t \le 2\pi$ as possible parametric equations for the ellipse $\frac{x^2}{a^2}+\frac{y^2}{b^2}=1$.

(b)

```
var('t')
parametric_plot((3*cos(t),sin(t)),(t,0,2*pi)).show(as
pect_ratio=1,    xmin=-3,    xmax=3,    ymin=-4,
ymax=4)
parametric_plot((3*cos(t),2*sin(t)),(t,0,2*pi)).show(
aspect_ratio=1,    xmin=-3,    xmax=3,    ymin=-4,
ymax=4)
parametric_plot((3*cos(t),4*sin(t)),(t,0,2*pi)).show(
aspect_ratio=1,    xmin=-3,    xmax=3,    ymin=-4,
ymax=4)
```

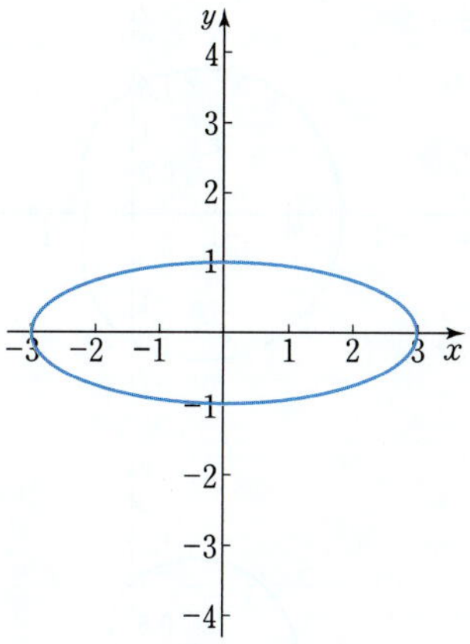

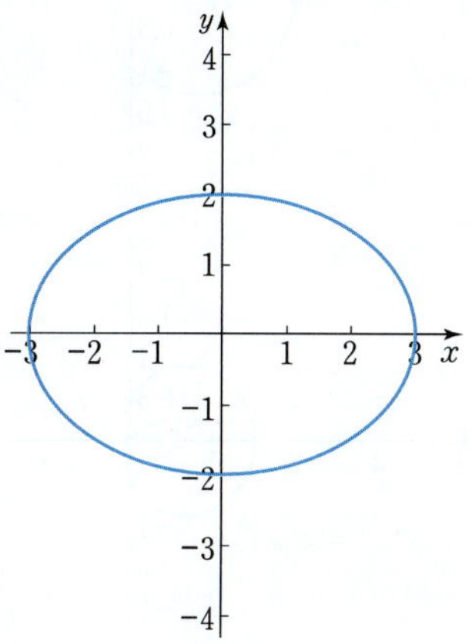

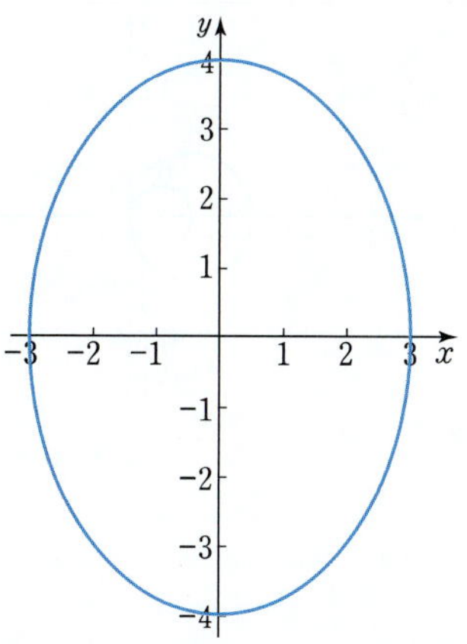

(c) As b increases, the ellipse stretches vertically.

11. Find three different sets of parametric equations to represent the curve $y=x^3$, $x\in\mathbb{R}$.

Solution i) $x=t,\ y=t^3,\quad t\in\mathbb{R}$,

ii) $x=t^{\frac{1}{3}},\ y=t,\quad t\in\mathbb{R}$,

iii) $x=t^3,\ y=t^9,\quad t\in\mathbb{R}$.

12. Suppose that the position of a particle at time t is given by $x_1=-2\sin t,\ y_1=3\cos t,\ 0\le t\le 2\pi$ and the position of the second particle is given by $x_2=-1+2\cos t,\ y_2=1+\sin t,\ 0\le t\le 2\pi$.

(a) Graph the paths of both particles. How many points of intersection are there?

(b) Are any of these points of intersection collision points? In other words, are the particles ever at the same place at same time? If so, find the collision points.

(c) Describe what happens if the path of the second particle is given by $x_2=2+\cos t$, $y_2=1+\sin t,\quad 0\le t\le 2\pi$.

http://matrix.skku.ac.kr/cal-lab/cal-9-1-12.html

Solution (a) There are 4-points of intersection :

```
var('t')
p1=parametric_plot((sin(t),3*cos(t)),(t,0,2*pi)).show(aspect_ratio=1, xmin=-5, xmax=3, ymin=-3, ymax=3)
p2=parametric_plot((-1+2*cos(t),1+sin(t)),(t,0,2*pi)).show(aspect_ratio=1, xmin=-5, xmax=3, ymin=-3, ymax=3)
```

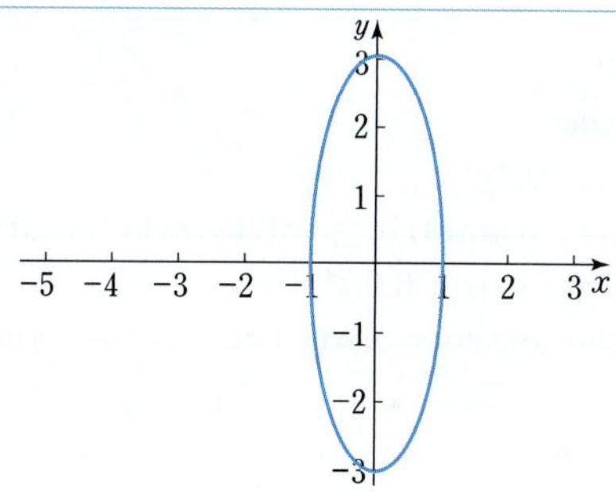

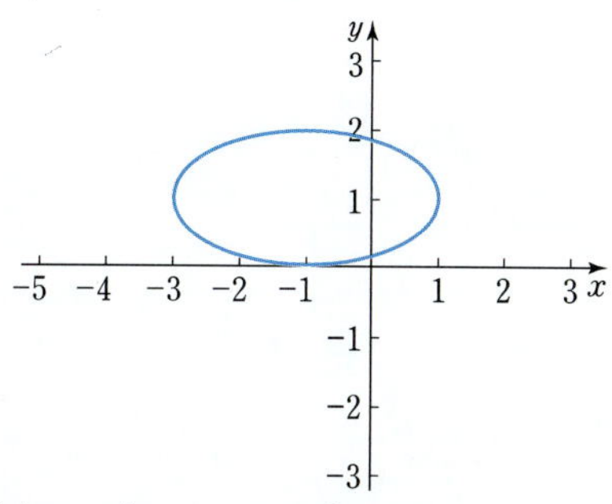

(b) A collision point occurs when $x_1=x_2,\ y_1=y_2$ for the same t. Thus we get the following equations :

$$\sin t=-1+2\cos t \qquad \cdots [1]$$

$$3\cos t=1+\sin t \qquad \cdots [2]$$

From [2], $\sin t=3\cos t-1$. Substituting into [1],

we get $3\cos t-1=-1+2\cos t \Rightarrow \cos t=0 \qquad \cdots [*]$

$$\Rightarrow t=\frac{\pi}{2} \text{ or } \frac{3\pi}{2}.$$

Note that $t=\frac{3\pi}{2}$ satisfies [1], [2], but $t=\frac{\pi}{2}$ does not.

So the only collision point occurs when $t=\dfrac{3\pi}{2}$, and this gives the point $(-2,\ 0)$.

[We could check our work by graphing x_1, x_2 together as functions of t and on another plot, y_1, y_2 as functions of t. If we do so, we see that the only value of t for which both pairs of graphs intersect is $t=\dfrac{3\pi}{2}$.]

(c) The equation $x_2=2+\cos t$, $y_2=1+\sin t$ is the circle centered at $(2,\ 1)$ instead of $(-2,\ 1)$. There are still 4 intersection points.

However there are no collision points, since (*) in part (b) becomes $\cos t=2>1$.

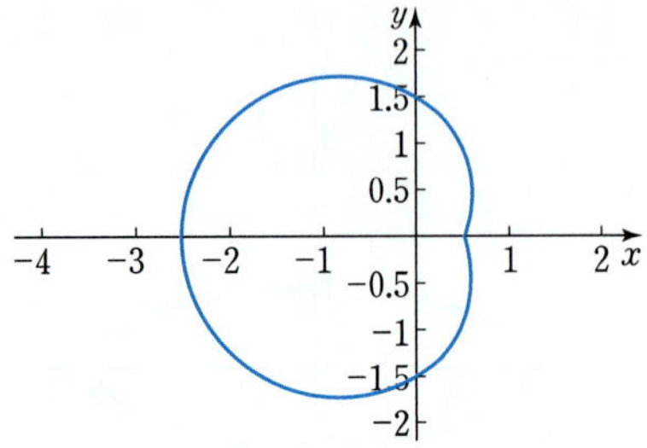

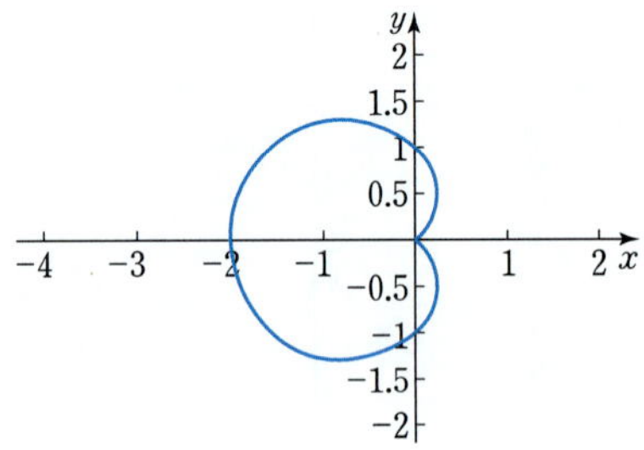

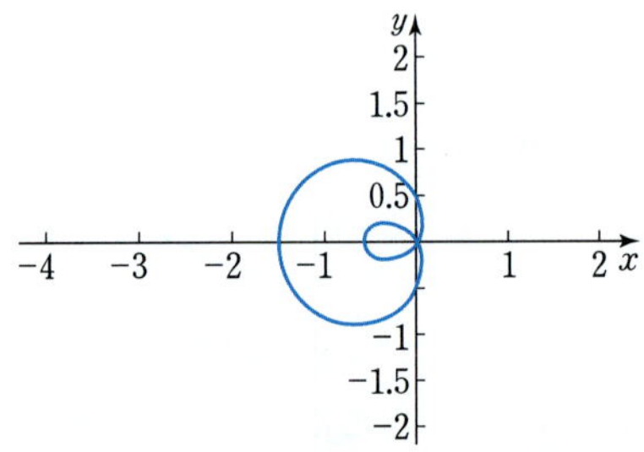

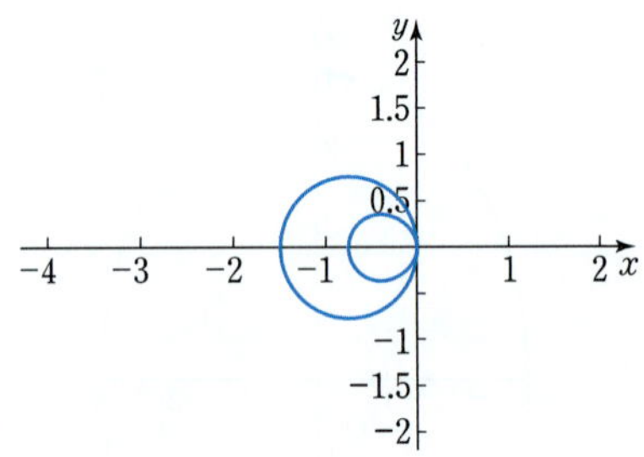

13. Investigate the family of curves defined by the parametric equations

$$x=\sin t(c-\sin t),\quad y=\cos t(c-\sin t).$$

How does the shape change as c changes? In particular, you should identify values of c for which the basic shape of the curve changes.

http://matrix.skku.ac.kr/cal-lab/cal-9-1-13.html

Solution

```
var('t')
@interact
def _(c=(0,2)):
    p=parametric_plot((sin(t)*(c-sin(t)),
cos(t)*(c-sin(t))), (t,0,2*pi))
    show(p,xmin=-3, xmax=2, ymin=-2, ymax=2)
```

10.2 Calculus with Parametric Curves

We know how to represent a curve by a set of parametric equations. It is natural to ask how to use calculus to study parametric curves. Therefore, we will study problems involving tangents, area, arc length, and the areas formed by revolutions of parametric curves.

Tangents

By eliminating the parameter t, curves defined by parametric equations $x=f(t)$ and $y=g(t)$ can also be expressed in the form $y=F(x)$. By substituting $x=f(t)$ and $y=g(t)$ in the equation $y=F(x)$, we get $g(t)=F(f(t))$. If g, F, and f are differentiable, then by using the chain rule, $g(t)$ can be differentiated with respect to t. Then we have

$$g'(t)=F'(f(t))f'(t)=F'(x)f'(t).$$

Solving for $F'(x)$, we get

1 $$F'(x)=\frac{g'(t)}{f'(t)} \quad \text{provided} \quad f'(t)\neq 0.$$

Thus, the tangent to a parametric curve (without having to eliminate the parameter t) can be calculated from 1. Since the slope of the tangent line to the curve
$y=F(x)$ at $(x,\ F(x))$ is $F'(x)$. Equation 1 can be written as

2 $$\frac{dy}{dx}=\frac{\frac{dy}{dt}}{\frac{dx}{dt}} \quad \text{if} \quad \frac{dx}{dt}\neq 0$$

If the motion of a particle is described by a parametric curve, then dy/dt and dx/dt are the vertical and horizontal velocities of the particle respectively. The dy/dx is the slope of the tangent, which is the ratio of the vertical and horizontal velocities.

From 2, note that the curve has a horizontal tangent when $dy/dt=0$ (provided that $dx/dt\neq 0$) and it has a vertical tangent when $dx/dt=0$ (provided that $dy/dt\neq 0$). This is useful in sketching parametric curves.

Differentiating 2 again with respect to x, we get the second derivative

$$\frac{d^2y}{dx^2}=\frac{d}{dx}\left(\frac{dy}{dx}\right)=\frac{d\left(\frac{dy}{dx}\right)/dt}{dx/dt}=\frac{\frac{d}{dt}\left(\frac{dy}{dx}\right)}{\frac{dx}{dt}}.$$

EXAMPLE 1

Find the slope and concavity at point $(2, 4)$ for the curve

$$x = \sqrt{t} \quad \text{and} \quad y = \frac{1}{2}(t^2 - 8).$$

Solution Since

$$\frac{dy}{dx} = \frac{dy/dt}{dx/dt} = \frac{t}{(1/2)t^{-1/2}} = 2t^{3/2}$$

and $$\frac{d^2y}{dx^2} = \frac{\frac{d}{dt}\left(\frac{dy}{dx}\right)}{dx/dt} = \frac{3t^{1/2}}{(1/2)t^{-1/2}} = 6t$$

and furthermore $(x, y) = (2, 4)$ is corresponding to $t = 4$. Therefore the slope is

$$\left.\frac{dy}{dx}\right|_{t=4} = 2(4)^{(3/2)} = 16.$$

Moreover, when $t = 4$, the second derivative is

$$\left.\frac{d^2y}{dx^2}\right|_{t=4} = 6(4) = 24 > 0.$$

Thus we conclude that the graph is concave upward at (2, 4), as shown in Figure 1. ■

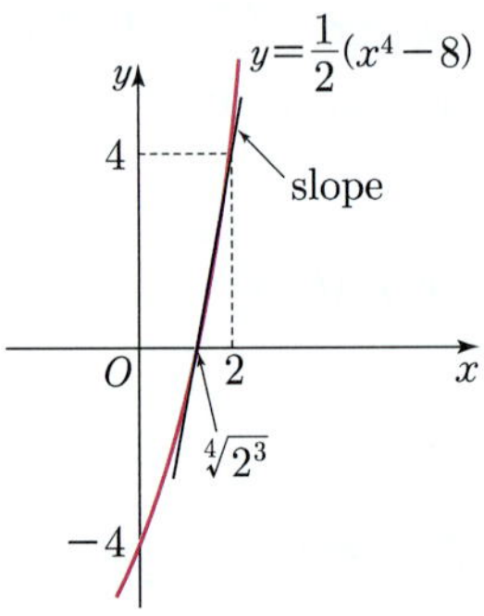

Figure 1

EXAMPLE 2

(a) Find the tangent to the circle $x = r\cos\theta$, $y = r\sin\theta$ $(0 \le \theta \le \pi)$ at the point where $\theta = \pi/3$.

(b) At what points is the tangent horizontal? When is it vertical?

Solution (a) The slope of the tangent line is

$$\frac{dy}{dx} = \frac{dy/d\theta}{dx/d\theta} = \frac{r\cos\theta}{-r\sin\theta} = -\cot\theta.$$

When $\theta = \pi/3$, we have

$$x = \frac{r}{2}, \quad y = \frac{\sqrt{3}}{2}r$$

and $$\left.\frac{dy}{dx}\right|_{\theta=\frac{\pi}{3}} = -\frac{1}{\sqrt{3}}.$$

Therefore, the slope of the tangent line at the point where to $\theta = \dfrac{\pi}{3}$ is $-\dfrac{1}{\sqrt{3}}$ and its equation is

$$y-\frac{\sqrt{3}}{2}r=-\frac{1}{\sqrt{3}}\left(x-\frac{r}{2}\right).$$

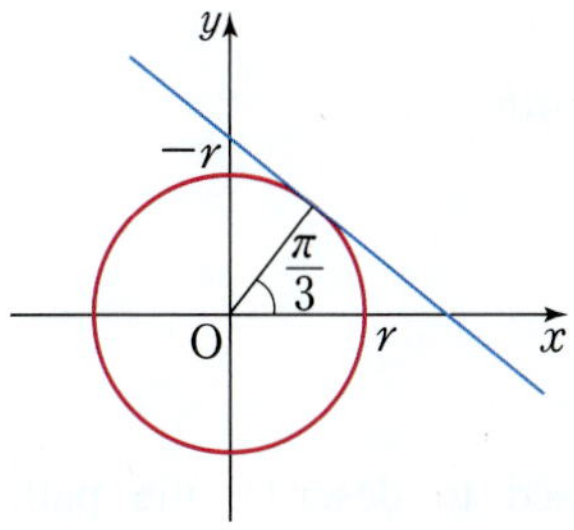

Figure 2

The tangent line is sketched in Figure 2.

(b) The tangent line is horizontal when $dy/dx=0$, which occurs when $\cos\theta=0$ and $\sin\theta\neq 0$. That is, $\theta=\frac{\pi}{2}$ or $\frac{3}{2}\pi$. The corresponding points on the circle are $(0,\ r)$ and $(0,\ -r)$.
The tangent line is vertical when $\frac{dx}{d\theta}=0$ (provide that $\frac{dy}{d\theta}\neq 0$), that is $\theta=0$ or π. The corresponding points on the circle are $(r,\ 0)$ and $(-r,\ 0)$. ■

Area

If $F(x)\geq 0$, then the **area** under a curve $y=F(x)$ from a to b is $A=\int_a^b F(x)\,dx$. When the curve is given by parametric equations $x=f(t)$, $y=g(t)$ and is traversed once as t increases from α to β, then

3 $$A=\int_a^b y\,dx=\int_\alpha^\beta g(t)\,f'(t)\,dt$$

$$\left[\text{ or } \int_\beta^\alpha g(t)f'(t)dt \text{ if } (f(\beta),\ g(\beta)) \text{ is the leftmost endpoint.}\right]$$

EXAMPLE 3

Find the area of the region enclosed by an ellipse

$$\frac{x^2}{a^2}+\frac{y^2}{b^2}=1 \text{ (for constants } a,\ b>0).$$

http://matrix.skku.ac.kr/cal-lab/sage-grapher-imp.html

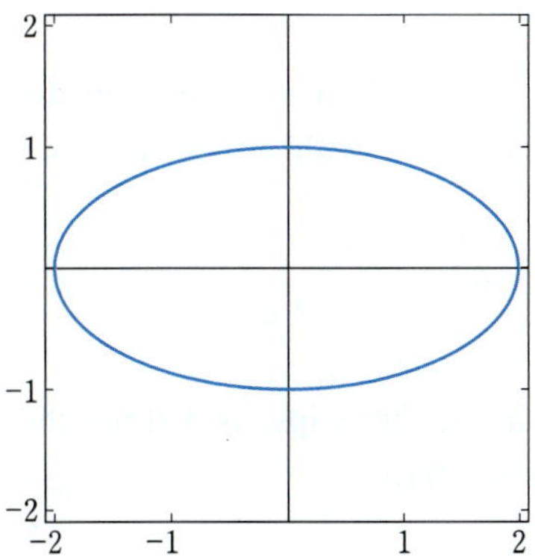

Figure 3

Solution The parametric equation of an ellipse is $x=a\cos\theta$, $y=b\sin\theta$ $(0\leq\theta\leq 2\pi)$. Due to symmetry, it is clear that area inside the ellipse is 4 times the area inside the first quadrant. Thus we have

$$\begin{aligned}A&=4\int_0^a y\,dx\\&=4\int_{\frac{\pi}{2}}^0 (b\sin\theta)(-a\sin\theta)\,d\theta\\&=4ab\int_0^{\frac{\pi}{2}}\sin^2\theta d\theta=4ab\int_0^{\frac{\pi}{2}}\frac{1-\cos 2\theta}{2}d\theta\end{aligned}$$

$$= 2ab\left[\theta - \frac{1}{2}\sin 2\theta\right]_0^{\frac{\pi}{2}} = \pi ab.$$ ■

Arc Length

We know how parametric equations can be used to describe the path of a particle moving in a plane. Now we develop a formula for determining the distance traveled by the particle along its path. Recall that the formula for the **arc length** of a curve C given by $y = h(x)$ over the interval $a \le x \le b$ is

4 $$L = \int_a^b \sqrt{1 + \left(\frac{dy}{dx}\right)^2}\, dx.$$

If C is represented by the parametric equations $x = f(t)$ and $y = g(t)$, $\alpha \le t \le \beta$, where $dx/dt = f'(t) > 0$ with C is traversed only once, as t increases from α to β and $f(\alpha) = a$, $f(\beta) = b$, we have

$$L = \int_a^b \sqrt{1 + \left(\frac{dy}{dx}\right)^2}\, dx = \int_\alpha^\beta \sqrt{1 + \left(\frac{dy/dt}{dx/dt}\right)^2}\, \frac{dx}{dt} dt$$

$$= \int_\alpha^\beta \sqrt{\left[1 + \left(\frac{dy/dt}{dx/dt}\right)^2\right]\left(\frac{dx}{dt}\right)^2}\, dt = \int_\alpha^\beta \sqrt{\left(\frac{dx}{dt}\right)^2 + \left(\frac{dy}{dt}\right)^2}\, dt$$

or

5 $$L = \int_\alpha^\beta \sqrt{\left(\frac{dx}{dt}\right)^2 + \left(\frac{dy}{dt}\right)^2}\, dt$$

This formula for arc length can be shown to be valid for any smooth curve C that does not intersect itself.

EXAMPLE 4

Find the arc length of the astroid $x = \cos^3 t$, $y = \sin^3 t$ $(0 \le t \le 2\pi)$.

Solution Looking at Figure 4, one arc is described by the parameter t in the interval $0 \le t \le \pi/2$. Differentiating x and y with respect to t we get

$$\frac{dx}{dt} = -3\cos^2 t \sin t \text{ and } \frac{dy}{dt} = 3\sin^2 t \cos t.$$

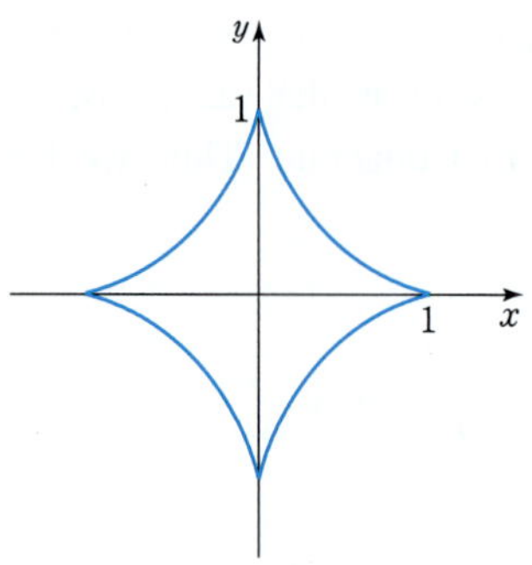

Figure 4

Due to symmetry, it is clear that area inside the ellipse is 4 times the area inside the first coordinates. Thus we have

$$L = 4\int_0^{\frac{\pi}{2}} \sqrt{\left(\frac{dx}{dt}\right)^2 + \left(\frac{dy}{dt}\right)^2}\, dt$$

$$= 4\int_0^{\frac{\pi}{2}} \sqrt{9\cos^4 t\ \sin^2 t + 9\cos^2 t\ \sin^4 t}\, dt$$

$$= 12\int_0^{\frac{\pi}{2}} \cos t\ \sin t\ dt$$

```
var('x, t')
x= (cos(t)*sin(t))
12*integral(x, t, 0, pi/2)
```

$$= 6\int_0^{\frac{\pi}{2}} \sin 2t\ dt = 6\left[-\frac{1}{2}\cos 2t\right]_0^{\frac{\pi}{2}} = 6. \quad \blacksquare$$

CAS **EXAMPLE 5**

Find the length of the curve

$$x = 3t^2 - 2,\ \ y = -t + 4, \quad -1 \le t \le 1.$$

Solution

```
var('x, y, t')
x=3*t^2-2;
y=-t+4;
parametric_plot((x, y), (t, -1, 1))
```

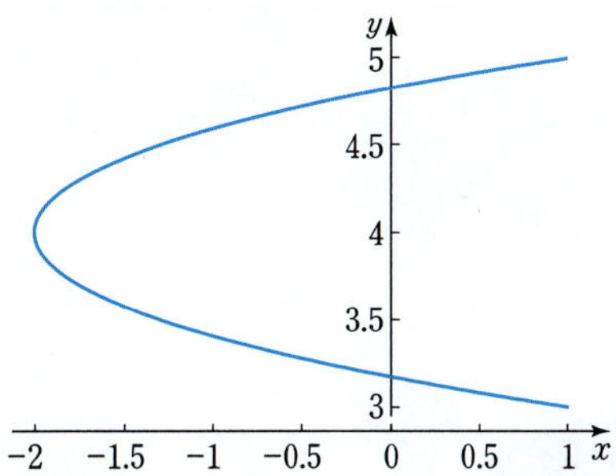

Figure 5

```
dxdt=diff(x, t)
dydt=diff(y, t)
integral(sqrt(dxdt^2+dydt^2), t, -1, 1)
```

Answer : sqrt(37) + 1/6*arcsinh(6) ■

Surface Area of a Surface of Revolution

As we did for arc length, we can use the formula for the surface area of a surface of revolution in rectangular form to develop a formula for a surface area in parametric form.

Suppose a curve C can be represented by the parametric equations $x=f(t)$, $y=g(t)$, $\alpha \leq t \leq \beta$. The general formula $S=\int 2\pi y\,ds$ and $S=\int 2\pi x\,ds$ can also be used for parametric curves with ds replaced by

$$ds=\sqrt{\left(\frac{dx}{dt}\right)^2+\left(\frac{dy}{dt}\right)^2}\,dt.$$

Assuming f' and g' are continuous and $g(t)\geq 0$, the area of the surface of revolution obtained by rotating the curve C about the x-axis, is given by

6
$$S=\int 2\pi y\,ds=\int_{\alpha}^{\beta} 2\pi y\sqrt{\left(\frac{dx}{dt}\right)^2+\left(\frac{dy}{dt}\right)^2}\,dt.$$

The surface area of the surface of revolution obtained by rotating the curve C about the y-axis, is given by

$$S=\int 2\pi x\,ds=\int_{\alpha}^{\beta} 2\pi x\sqrt{\left(\frac{dx}{dt}\right)^2+\left(\frac{dy}{dt}\right)^2}\,dt.$$

EXAMPLE 6

Find the surface area of the surface of revolution obtained by rotating $y=\sin x$ $(0\leq x\leq \pi/2)$ about the x-axis.

Solution Since $x=t$, $y=\sin t$, and $0\leq t\leq \pi/2$, we get

$$\begin{aligned}
S&=\int_0^{\frac{\pi}{2}} 2\pi\sin t\,\sqrt{1+\cos^2 t}\,dt\\
&=\int_0^1 2\pi\,\sqrt{1+u^2}\,du \quad (\text{where } u=\cos t)\\
&=\int_0^{\frac{\pi}{4}} 2\pi\sec^3\theta\,d\theta\\
&=2\pi\cdot\frac{1}{2}\left[\sec\theta\tan\theta+\ln|\sec\theta+\tan\theta|\right]_0^{\frac{\pi}{4}}\\
&=\pi(\sqrt{2}+\ln(1+\sqrt{2})).
\end{aligned}$$

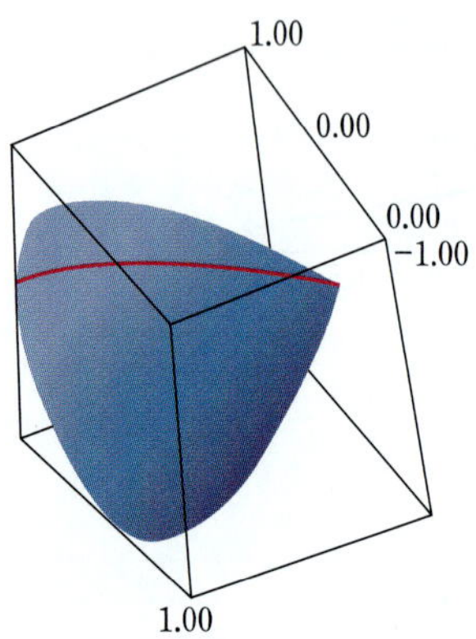

Figure 6

```
s=revolution_plot3d(sin(x), (x, 0, pi/2), parallel_axis='x',
show_curve=True, opacity=0.3, colour='green')
s.show(aspect_ratio=(1,1,1))
```

■

10.2 EXERCISES (Calculus with Parametric Curves)

http://matrix.skku.ac.kr/Cal-Book/part1/CS-Sec-10-2-Sol.html

YouTube http://youtu.be/yF5oZOQVnCE

1-2. Find $\dfrac{dy}{dx}$.

1. $x = t^3 + 2t^2$, $y = 1 - t$

Solution $\dfrac{dy}{dx} = \dfrac{\frac{dy}{dt}}{\frac{dx}{dt}} = \dfrac{\frac{d}{dt}(1-t)}{\frac{d}{dt}(t^3+2t^2)} = \dfrac{-1}{3t^2+4t}$

2. $x = \cos\theta + \sin 2\theta$, $y = \sin\theta + \cos 2\theta$

Solution $\dfrac{dy}{dx} = \dfrac{\frac{dy}{d\theta}}{\frac{dx}{d\theta}} = \dfrac{\frac{d}{d\theta}(\sin\theta + \cos 2\theta)}{\frac{d}{d\theta}(\cos\theta + \sin 2\theta)}$

$= \dfrac{\cos\theta - 2\sin 2\theta}{2\cos 2\theta - \sin\theta}$

3-6. Find an equation of the tangent to the curve at the point.

3. $x = t^4 + 3t^3$, $y = t - t^2$; $t = -1$

http://matrix.skku.ac.kr/cal-lab/cal-9-2-3.html

Solution $x = t^4 + 3t^3$, $\dfrac{dx}{dt} = 4t^3 + 9t^2$.

$y = t - t^2$, $\dfrac{dy}{dt} = 1 - 2t$.

At the point with parameter value t, the slope is

$$\frac{dy}{dx} = \frac{dy}{dt} \cdot \frac{dt}{dx} = \frac{1-2t}{4t^3+9t^2}.$$

Slope of the tangent at $t = -1$ is $\left.\dfrac{dy}{dx}\right|_{t=-1} = \dfrac{3}{5}$.

Hence, the equation of the tangent line is $y = \dfrac{3}{5}x - \dfrac{4}{5}$.

```
var('t')
x=t^4+3*t^3
y=t-t^2
T=diff(y)/diff(x)
print T (t=-1)
```

Answer : 3/5

4. $y = 2\sin 2t$, $x = 2\sin t$; $t = \dfrac{\pi}{6}$

Solution At the point with parameter value t, the slope is

$\dfrac{dy}{dx} = \dfrac{2\cos 2t}{\cos t}$.

Slope of the tangent at $t = \dfrac{\pi}{6}$ is $\left.\dfrac{dy}{dx}\right|_{t=\frac{\pi}{6}} = \dfrac{2\sqrt{3}}{3}$.

Hence, the equation of the tangent line is $y = \dfrac{2\sqrt{3}}{3}x - 1$.

5. $x = e^t$, $y = (t-1)^2$; $(1,\ 1)$

Solution At the point with parameter value t, the slope is $\dfrac{dy}{dx} = \dfrac{2(t-1)}{e^t}$.

The point $(1,\ 1)$ corresponds to the parameter value $t = 0$, so the slope of the tangent at the point is $\dfrac{dy}{dx_{t=0}} = -2$.

Hence, the equation of the tangent line is $y = -2x + 3$.

CAS **6.** $x = \sin t$, $y = \sin(t + t\sin t)$; $(0,\ 0)$

http://matrix.skku.ac.kr/cal-lab/cal-9-2-6.html

Solution At the point with parameter t, the slope is

$$\frac{dy}{dx} = \frac{dy}{dt} \cdot \frac{dt}{dx} = \frac{\frac{d(\sin t)}{dt}}{\frac{d(t+t\sin t)}{dt}} = \frac{\cos t}{1+\sin t + t\cos t}.$$

If $x = y = 0$, $t = 0$. The slope of the tangent at $t = 0$ is $\left.\dfrac{dy}{dx}\right|_{t=0} = 1$. Hence, the equation of the tangent line is $y = x$.

```
var('t')
x=sin(t)
y=sin(t+t*sin(t))
T=diff(y)/diff(x)
print T (t=0)
```

Answer : 1

```
p1=parametric_plot((sin(t),t+t*sin(t)),(t,0,2*pi))
p2=plot(t,color="red")
show(p1+p2)
```

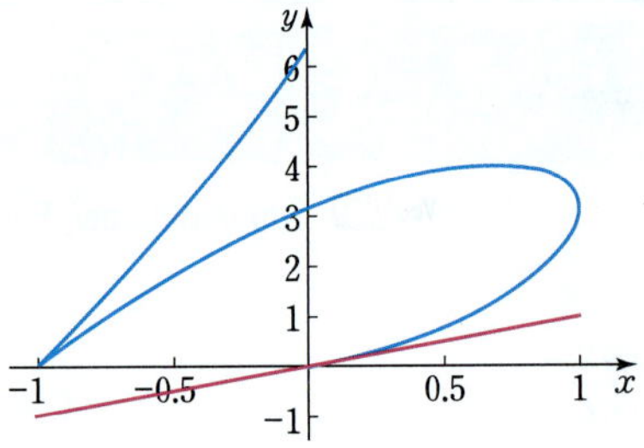

7-10. Find $\dfrac{d^2y}{dx^2}$. For what values of t is the curve concave upward or downward?

CAS **7.** $x = t + e^t,\ y = t - e^t$

http://matrix.skku.ac.kr/cal-lab/cal-9-2-7.html

Solution $\dfrac{dy}{dx} = \dfrac{1-e^t}{1+e^t},\ \dfrac{d^2y}{dx^2} = \dfrac{d}{dx}\left(\dfrac{dy}{dx}\right) = \dfrac{\dfrac{d}{dt}\left(\dfrac{dy}{dx}\right)}{\dfrac{dx}{dt}}$

$= \dfrac{-2e^t}{(1+e^t)^3}.$

$\Rightarrow$ Since $-2e^t < 0$ and $1+e^t > 0$, $\dfrac{d^2y}{dx^2} < 0$ for all t.

Hence, the curve is downward everywhere.

```
var('t')
x=t+exp(t)
y=t-exp(t)
parametric_plot([x, y], (t, -5, 2))
```

```
df1x=diff(x)
df1y=diff(y)
show(df1y/df1x)
```

Answer : (1-e^t)/(1+e^t)

```
df1=df1y/df1x
df2=diff(df1)
show((df2/df1x).factor())
```

Answer : -2*e^t/(1+e^t)^3

8. $x = t^3 + 2t^2,\ y = t^2 + 1$

Solution $\dfrac{dy}{dx} = \dfrac{2}{3t+4},\ \dfrac{d^2y}{dx^2} = \dfrac{-6}{t(3t+4)^3}.$

$\Rightarrow$ If $t(3t+4)^3 < 0$, $\dfrac{d^2y}{dx^2} > 0$. Thus if $t(3t+4)^3 > 0$, $\dfrac{d^2y}{dx^2} < 0$.

Hence, the curve is concave upward on $-\dfrac{3}{4} < t < 0$, and is concave downward on $t < \dfrac{4}{3}$ or $t > 0$.

CAS **9.** $x = 3\cos t,\ y = 2\sin t,\ 0 < t < 2\pi$

http://matrix.skku.ac.kr/cal-lab/cal-9-2-9.html

Solution $\dfrac{dy}{dx} = \dfrac{dy}{dt} \cdot \dfrac{dt}{dx} = \dfrac{\dfrac{d(2\sin t)}{dt}}{\dfrac{d(3\cos t)}{dt}} = \dfrac{2\cos t}{-3\sin t},$

$\dfrac{d^2y}{dx^2} = \dfrac{d}{dx}\left(\dfrac{dy}{dx}\right) = \dfrac{\dfrac{d}{dt}\left(\dfrac{dy}{dx}\right)}{\dfrac{dx}{dt}} = \dfrac{6}{(-3\sin t)^3}.$

If $\sin^3 t < 0$, $\dfrac{d^2y}{dx^2} > 0$. Thus if $\sin^3 t > 0$, $\dfrac{d^2y}{dx^2} < 0$.

Hence, the curve is upward on $\pi < t < 2\pi$, and the curve is downward on $0 < t < \pi$. Note that this curve is an ellipse.

```
var('t')
x= 3*cos(t)
y= 2*sin(t)
df1x=diff(x)
df1y=diff(y)
show(df1y/df1x)
df1=df1y/df1x
df2=diff(df1)
show(df2)
show((df2/df1x). simplify_trig())
```

Answer : $\dfrac{-2\cos(t)}{3\sin(t)},\ \dfrac{2\cos(t)^2}{3\sin(t)^2} + \dfrac{2}{3},\ \dfrac{-2}{9\sin(t)^3}$

10. $x = \sin 2\theta,\ y = \sin\theta,\ 0 < \theta < \pi$

11-12. Find the point on the curve where the tangent is horizontal or vertical.

11. $x = t^3 + t^2 - 8t,\ y = t^3 - t$

http://matrix.skku.ac.kr/cal-lab/cal-9-2-11.html

Solution $\dfrac{dx}{dt} = 3t^2 + 2t - 8$, so $\dfrac{dx}{dt} = 0 \Leftrightarrow t = \dfrac{4}{3},\ t = -2$

$\dfrac{dy}{dt} = 3t^2 - 1$, so $\dfrac{dy}{dt} = 0 \Leftrightarrow t = \dfrac{1}{\sqrt{3}},\ t = -\dfrac{1}{\sqrt{3}}$

The curve has horizontal tangents at

$\left(\dfrac{1}{3} - \dfrac{23}{3\sqrt{3}},\ -\dfrac{2}{3\sqrt{3}}\right)$, or $\left(\dfrac{1}{3} + \dfrac{23}{3\sqrt{3}},\ \dfrac{2}{3\sqrt{3}}\right)$,

and vertical tangents at $\left(-\dfrac{176}{27},\ \dfrac{28}{27}\right)$, or $(12, -6)$.

```
var('t')
x=t^3+t^2-8*t
y=t^3-t
print solve(diff(x)==0,t)
print solve(diff(y)==0,t)
print x (t=-2)
print y (t=-2)
print x (t=4/3)
print y (t=4/3)
print x (t=-1/3*sqrt(3))
print y (t=-1/3*sqrt(3))
print x (t=1/3*sqrt(3))
print y (t=1/3*sqrt(3))
parametric_plot([x, y], (t, -3, 3))
```

Answer : [t == -2,t == (4/3)]
[t == -1/3*sqrt(3), t == 1/3*sqrt(3)]
12
-6
-176/27
28/27
23/9*sqrt(3) + 1/3
2/9*sqrt(3)
-23/9*sqrt(3) + 1/3
-2/9*sqrt(3)

12. $x=\sin\theta,\ y=\cos 2\theta$

Solution $\dfrac{dy}{d\theta}=-2\sin 2\theta$, so $\dfrac{dy}{d\theta}=0 \Leftrightarrow \theta=\dfrac{n\pi}{2}$ for $n\in\mathbb{Z}$.

Points at which the tangent is horizontal and $\dfrac{dx}{d\theta}=\cos\theta$, so $\dfrac{dx}{d\theta}=0 \ \Leftrightarrow\ \theta=\dfrac{(2n+1)\pi}{2}$ for $n\in\mathbb{Z}$.

The curve has horizontal tangents at $(0,\ 1)$, or $(1,\ -1)$, or $(-1,\ -1)$, and vertical tangents at $(1,\ -1)$, or $(-1,\ -1)$.

13. Show that the curve $x=\cos t,\ y=\sin 2t$ has two tangents at $(0,\ 0)$ and find their equations.

Solution $$\frac{dy}{dx}=\frac{\frac{dy}{dt}}{\frac{dx}{dt}}=\frac{\frac{d}{dt}(\sin 2t)}{\frac{d}{dt}(\cos t)}=\frac{2\cos 2t}{-\sin t}.$$

Now $x=\cos t$ is 0 when $t=\dfrac{\pi}{2},\ t=\dfrac{3\pi}{2}$, so there are two tangents at the points $(0,\ 0)$ since both $t=\dfrac{\pi}{2},\ t=\dfrac{3\pi}{2}$ correspond to the origin. The tangent corresponding to $t=\dfrac{\pi}{2}$ has slope 2, and its equation is $y=2x$. The tangent corresponding to $t=\dfrac{3\pi}{2}$ has slope -2, and its equation is $y=-2x$.

CAS 14. At what point does the curve $x=1-2\cos^2 t$, $y=\tan t(1-2\cos^2 t)$ cross itself? Find the equations of both tangents at that point.
http://matrix.skku.ac.kr/cal-lab/cal-9-2-14.html

Solution

```
var('t')
parametric_plot(((1-2*cos(t)^2),
tan(t)*(1-2*cos(t)^2)), (t, -2, 2)).
show(aspect_ratio=1, xmin=-5, xmax=5, ymin=-7,
ymax=7)
```

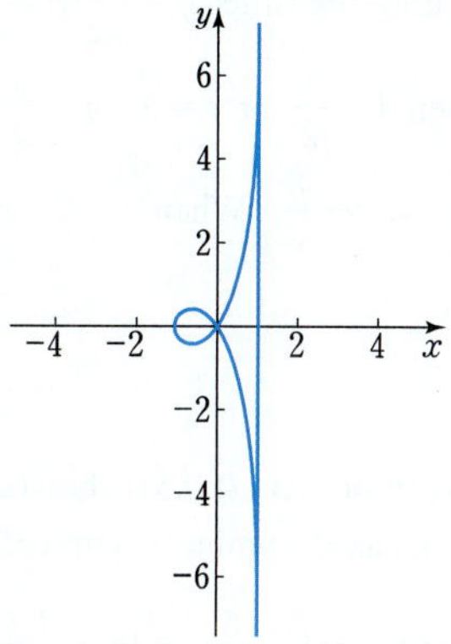

From the figure it is clear that at $(0,\ 0)$ the curve does cross itself. Moreover, $x=1-2\cos^2 t$ is 0 when $t=\dfrac{\pi}{4}$, $t=\dfrac{3\pi}{4},\ t=\dfrac{5\pi}{4},\ t=\dfrac{7\pi}{4}$.

The tangent corresponding to t has slope 1 or -1. Then the equations of tangents at $(0,\ 0)$ are $y=x,\ y=-x$.

15. At what points on the curve $x=t^3+4t,\ y=6t^2$ is the tangent parallel to the line with equations $x=-7t,\ y=12t-5$?
http://matrix.skku.ac.kr/cal-lab/cal-9-2-15.html

Solution Given latter function's slope is

$$\frac{dy}{dx}=\frac{dy}{dt}\cdot\frac{dt}{dx}=\frac{12}{-7},$$

so we should find t when the former function's slope is $-\dfrac{12}{7}$.

$$\frac{dy}{dx}=\frac{dy}{dt}\cdot\frac{dt}{dx}=\frac{12t}{3t^2+4},\quad \frac{dy}{dx}=-\frac{12}{7}.$$

We can get t is -1 or $-\dfrac{4}{3}$.

When t is -1 or $-\dfrac{4}{3}$, the former function's tangent parallel to the latter's.

```
var('t')
x1=-7*t
y1=12*t-5
x2=t^3+4*t
y2=6*t^2
d1=diff(y1)/diff(x1)
d2=diff(y2)/diff(x2)
print solve(d1==d2,t)
```

Answer : [t == (-4/3), t == -1]

16. Find the area bounded by the curve $x=t-\dfrac{1}{t}$, $y=t+\dfrac{1}{t}$ and the line $y=\dfrac{5}{2}$.

Solution When $t=\dfrac{1}{2}$ or $t=2$, $y=\dfrac{5}{2}$.

When $t=\dfrac{1}{2}$, $x=-\dfrac{3}{2}$. When $t=2$, $x=\dfrac{3}{2}$.

```
var('x,y,t')
x=t-1/t;
y=t+1/t;
parametric_plot((x,y),(t,0,5)).show(aspect_ratio=1,
xmin=-5, xmax=5, ymin=0, ymax=5)
```

$$\int_{-3/2}^{3/2}\left(\frac{5}{2}-y\right)dx=\int_{\frac{1}{2}}^{2}\left(\frac{5}{2}-y\right)\left(1+\frac{1}{t^2}\right)dt$$

$$=\int_{\frac{1}{2}}^{2}\left(\frac{5}{2}-t-\frac{1}{t}\right)\left(1+\frac{1}{t^2}\right)dt \qquad (\because dx=(1+\frac{1}{t^2})dt)$$

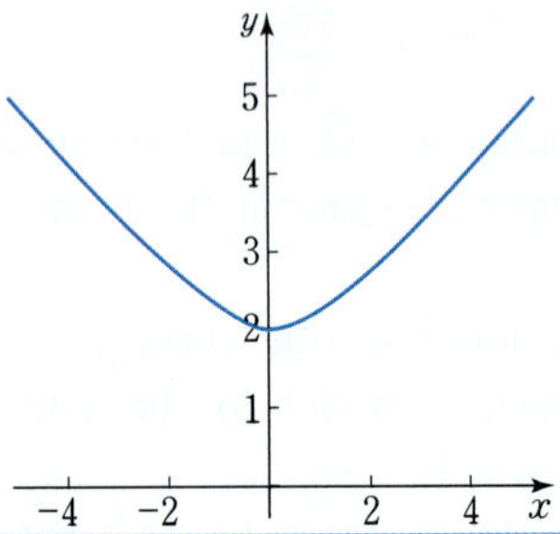


```
integral((5/2-y)*diff(x,t),t,1/2,2)
```

Answer : -4*log(2) + 15/4

CAS **17.** Find the area bounded by the curve $x=\cos t$, $y=e^t$, $0\le t\le\pi$, and the lines $y=9$ and $x=0$.

http://matrix.skku.ac.kr/cal-lab/cal-9-2-17.html

Solution The curve $x=\cos t$, $y=e^t$ $(0\le t\le\pi)$ intersects the y-axis when $t=\dfrac{\pi}{2}$. Then, the corresponding values of y are $e^{\pi/2}$. x and y are 0 and $e^{\pi/2}$ respectively.

```
var('t')
parametric_plot((cos(t), e^t), (t, 0,
pi)).show(aspect_ratio=1, xmin=-5, xmax=10,
ymin=-5, ymax=15)
```

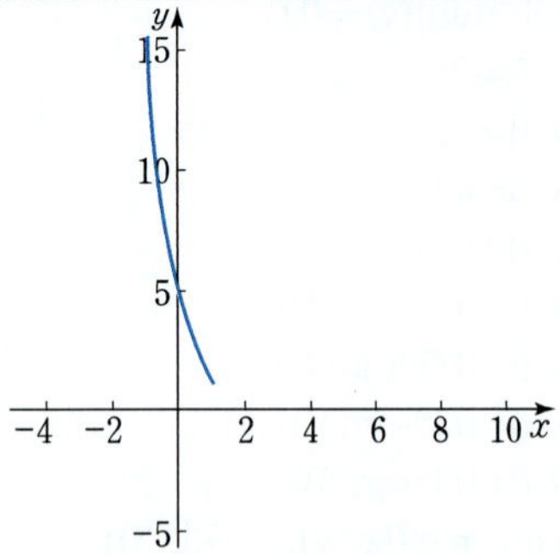


```
var('x, y, t')
x=cos(t)
y=e^t
integral((9-y)*diff(x, t), t, 0, pi)
```

Answer : 1/2*e^pi - 35/2

CAS **18.** Find the area of a region enclosed by the astroid $x=\cos^3\theta$, $y=\sin^3\theta$.

http://matrix.skku.ac.kr/cal-lab/cal-9-2-18.html

Solution

```
var('t')
parametric_plot((cos(t)^3, sin(t)^3), (t, 0,
2*pi)).show(aspect_ratio=1, xmin=-2, xmax=2,
ymin=-2, ymax=2)
```

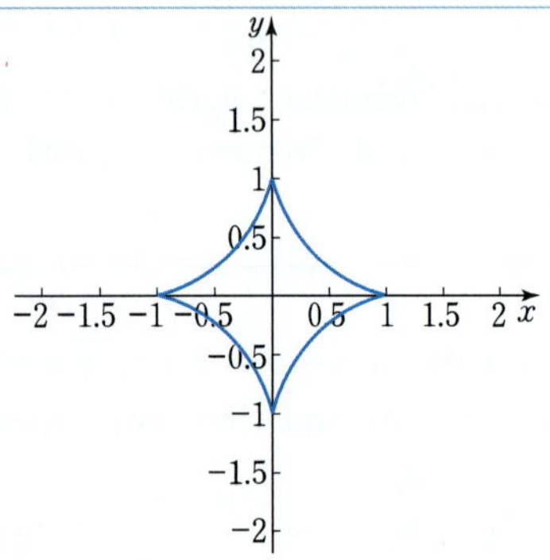

$$S=4\int_0^1 y\,dx=4\int_{\frac{\pi}{2}}^{0}3y(-\sin\theta)\cos^2\theta\,d\theta=\frac{3}{8}\pi$$

$\therefore$ $(dx=3(-\sin\theta)\cos^2\theta\,d\theta)$

```
x, y, t=var('x, y, t')
x=cos(t)^3
y=sin(t)^3
4*integral(y*diff(x, t), t, pi/2, 0)
```

Answer : 3/8*pi

19. Find the arc length of the curve defined by

$$x = t^3 - 3t,\ y = 3t^2;\ 0 \le t \le 1.$$

http://matrix.skku.ac.kr/cal-lab/sage-grapher-para.html

Solution $\dfrac{dx}{dt} = 3t^2 - 3,\ \dfrac{dy}{dt} = 6t.$

Then, $\left(\dfrac{dx}{dt}\right)^2 + \left(\dfrac{dy}{dt}\right)^2 = 9(t^2+1)^2.$

Hence, $L = \displaystyle\int_0^1 \sqrt{9(t^2+1)^2}\,dt = 3\int_0^1 (t^2+1)\,dt = 4.$

20. Find the arc length of the curve defined by

$$x = e^t \sin t,\ y = e^t \cos t\ ;\ 0 \le t \le \pi.$$

Solution $\dfrac{dx}{dt} = e^t \sin t + e^t \cos t,\ \dfrac{dy}{dt} = e^t \cos t - e^t \sin t$

$\left(\dfrac{dx}{dt}\right)^2 + \left(\dfrac{dy}{dt}\right)^2 = 2e^{2t}\sin^2 t + 2e^{2t}\cos^2 t$

$\displaystyle\int_0^\pi \sqrt{2e^{2t}\sin^2 t + 2e^{2t}\cos^2 t}\,dt$

```
var('x,y,t')
x=e^t*sin(t)
y=e^t*cos(t)
```

```
dxdt=diff(x,t)
dxdt
```

e^t*sin(t) + e^t*cos(t)

```
dydt=diff(y,t)
dydt
```

-e^t*sin(t) + e^t*cos(t)

```
L=integral(sqrt(dxdt^2+dydt^2),t,0,pi)
L
```

Answer : sqrt(2)*e^pi - sqrt(2)

21. Find the arc length of the curve defined by

$$x = 2\ln t,\ y = t + \frac{1}{t};\ 1 \le t \le 4$$

http://matrix.skku.ac.kr/cal-lab/cal-9-2-21.html

Solution

```
var('x,y,t')
x=2*ln(t);
y=t+(1/t);
parametric_plot((x,y),(t,1,4)).show(aspect_ratio=1,
xmin=-1, xmax=3, ymin=-2, ymax=5)
```

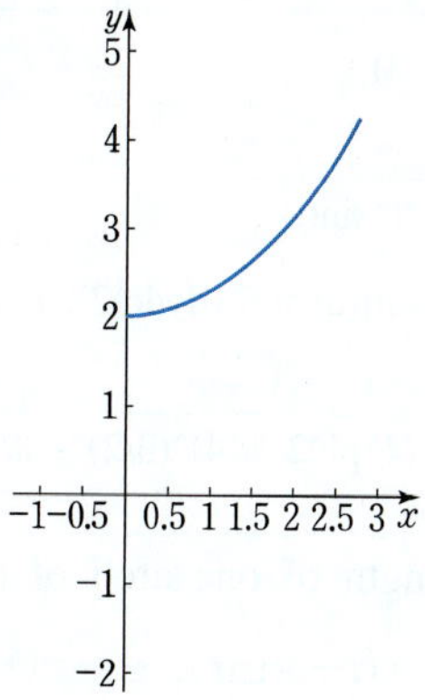

```
dxdt=diff(x,t)
dxdt
```

2/t

```
dydt=diff(y,t)
dydt
```

-1/t^2 + 1

```
L=integral(sqrt(dxdt^2+dydt^2),t,1,4)
L
```

Answer : 15/4

CAS **22.** Find the arc length of the curve defined by

$$x = t^2 \cos t,\ y = t^2 \sin t;\ 0 \le t \le 2\pi.$$

http://matrix.skku.ac.kr/cal-lab/cal-9-2-22.html

Solution

```
var('x, y, t')
x=t^2*cos(t);
y=t^2*sin(t);
parametric_plot((x,y),(t,0,2*pi)).show(aspect_ratio=
1, xmin=-20, xmax=40, ymin=-30, ymax=10)
```

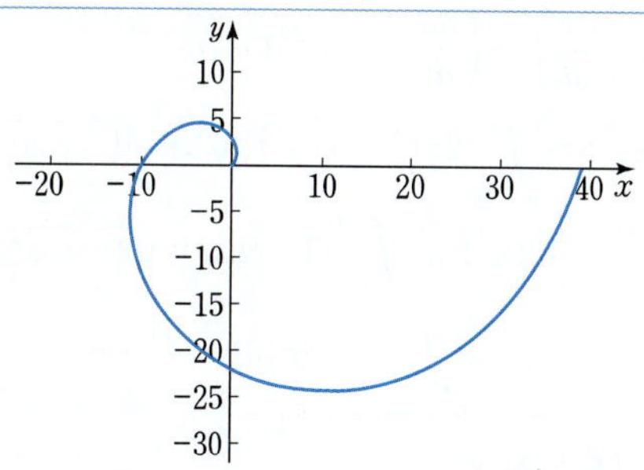

```
dxdt=diff(x, t)
dxdt
```

-t^2*sin(t) + 2*t*cos(t)

```
dydt=diff(y, t)
dydt
```

t^2*cos(t) + 2*t*sin(t)

```
L=integral(sqrt(dxdt^2+dydt^2), t, 0, 2*pi)
L
```

Answer : 1/3*(4*pi^2 + 4)^(3/2) - 8/3

23. Find the length of one arch of the cycloid

$$x = r(t - \sin t),\ y = r(1 - \cos t).$$

Solution $\dfrac{dx}{dt} = r(1-\cos t)$, $\dfrac{dy}{dt} = r\sin t$.

Then, $\left(\dfrac{dx}{dt}\right)^2 + \left(\dfrac{dy}{dt}\right)^2 = 2r^2(1-\cos t)$.

Hence, $L = \displaystyle\int_0^{2\pi} \sqrt{2}\, r\sqrt{1-\cos t}\, dt$

$= \sqrt{2}\, r \displaystyle\int_0^{2\pi} \sqrt{2}\sin\frac{t}{2} dt = 8r.$

24-26. Find the area of the surface obtained by rotating the given curve about the x-axis.

24. $x = a\cos t,\ y = a\sin t\ ;\ 0 \le t \le \pi$

Solution $\dfrac{dx}{dt} = -a\sin t$, $\dfrac{dy}{dt} = a\cos t$.

Then, $\left(\dfrac{dx}{dt}\right)^2 + \left(\dfrac{dy}{dt}\right)^2 = a^2$.

Hence, $S = \displaystyle\int_0^{\pi} 2\pi a \sin t(a) dt = 2\pi a^2 \int_0^{\pi} \sin t dt = 4\pi a^2.$

25. $x = r(t - \sin t),\ y = r(1 - \cos t)\ ;\ 0 \le t \le 2\pi$

http://matrix.skku.ac.kr/cal-lab/cal-10-2-25.html

Solution $\dfrac{dx}{dt} = r(1-\cos t)$, $\dfrac{dy}{dt} = r\sin t$.

Then, $\left(\dfrac{dx}{dt}\right)^2 + \left(\dfrac{dy}{dt}\right)^2 = 2r^2(1-\cos t)$.

Hence, $S = \displaystyle\int_0^{2\pi} 2\pi r(1-\cos t)(\sqrt{2}\, r\sqrt{1-\cos t})dt$

$= 2\sqrt{2}\,\pi r^2 \displaystyle\int_0^{2\pi} (1-\cos t)(\sqrt{1-\cos t})dt$

$= 16\pi r^2 \displaystyle\int_{-1}^{1} (1-k^2)dk = \frac{64}{3}\pi r^2.$

```
var('r, t, x, y, k')
x=r*(t-sin(t))
y=r*(1-cos(t))
dxdt=diff(x,t)
dxdt
```

-(cos(t) - 1)*r

```
dydt=diff(y,t)
dydt
```

r*sin(t)

```
L1=integral(16*pi*r^2*(1-k^2),k,-1,1)
L1
```

Answer : 64/3*pi*r^2

26. $x = 2\ln t,\ y = t^2\ ;\ 1 \le t \le \sqrt{3}$

27-29. Find the area of the surface generated by rotating the given curve about y-axis.

27. $x = t^2,\ y = t^3\ ;\ 0 \le t \le 1$

http://matrix.skku.ac.kr/cal-lab/cal-9-2-27.html

Solution $\dfrac{dx}{dt} = 2t$, $\dfrac{dy}{dt} = 3t^2$

$\left(\dfrac{dx}{dt}\right)^2 + \left(\dfrac{dy}{dt}\right)^2 = 4t^2 + 9t^4$

$\displaystyle\int_0^1 2\pi t^2 \sqrt{4t^2 + 9t^4}\, dt$

```
var('t, x, y')
x=t^2
y=r^3
dxdt=diff(x,t)
dydt=diff(y,t)
S=intergral(2*pi*x*sqrt(dxdt^2+dydt^2),t,0,1)
S
```

Answer : 32/1215*(sqrt(36*pi^2 + 4)*(243*pi^4 + 9*pi^2 - 2) + 4)*pi

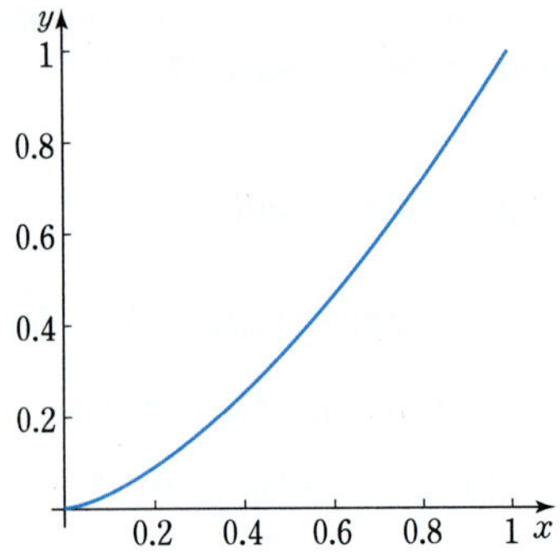

28. $x = e^t - t,\ y = 4e^{\frac{t}{2}};\ 0 \le t \le 1$

29. $x = \sin t \cos t,\ y = \sin^2 t;\ 0 \le t \le \dfrac{\pi}{2}$

10.3 Polar Coordinates

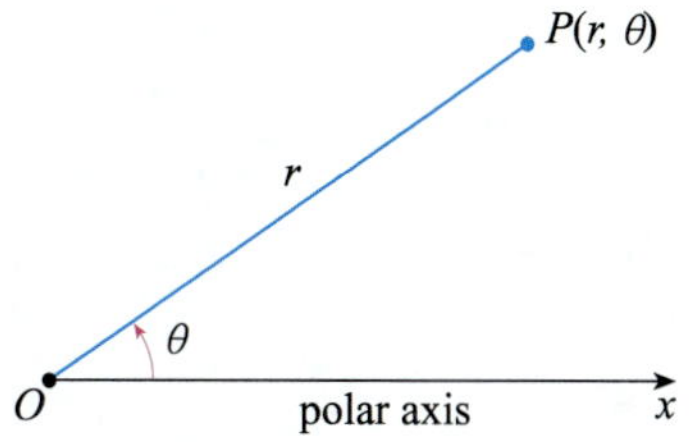

Figure 1

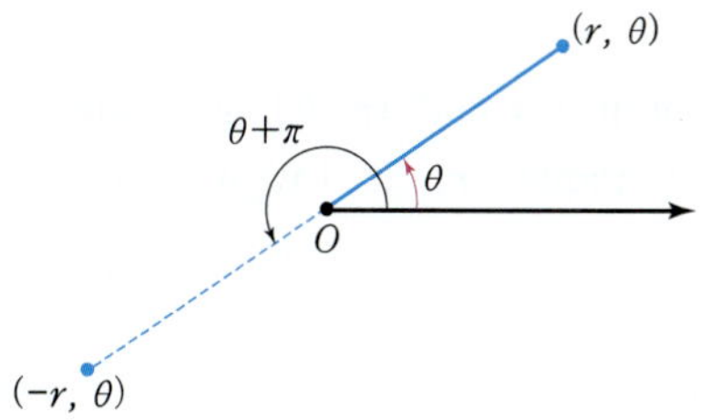

Figure 2

A coordinate system is a system which uses one or more numbers, or coordinates, to uniquely determine the position of a point in the coordinate plane. So far we have represented plane curves as collections of points (x, y) In the rectangular coordinate system (Cartesian coordinate system), where x and y represent the directed distances from the coordinate axes to the point (x, y). In some cases the **polar coordinate** system introduced by Newton is a more useful and convenient coordinate system.

Choose a point in the plane that is called the **pole** (or origin) and label it O. The **polar axis** is a ray (half-line) drawn starting at O. Usually the polar axis is drawn horizontally to the right and aligns with the positive x-axis in Cartesian coordinates.

For any point P in the plane, let r be the directed distance from O to P and let θ be the directed angle (usually measured in radians) between the polar axis and the line OP as in Figure 1. Then each point P in the plane can be assigned **polar coordinates** (r, θ). We assume that an angle is positive if measured in the counterclockwise direction from the polar axis and negative in the clockwise direction. If $P=O$, then $r=0$. Therefore, the pole for any value of θ can be represented by $(0, \theta)$.

When r is negative, the meaning of polar coordinates (r, θ) is extended as follows. Observe that the points $(-r, \theta)$ and (r, θ) (See Figure 2.) lie on the same line through O and at the same distance $|r|$ from O, but on opposite sides of O. If $r>0$, the point (r, θ) lies in the same quadrant as θ ; if $r<0$, it lies in the quadrant on the opposite side of the pole. Thus, $(-r, \theta)$ represents the same point as $(r, \theta+\pi)$. Since an angle 2π represents a complete counterclockwise rotation, the point represented by polar coordinates (r, θ) is also represented by

$$(r, \theta+2n\pi) \text{ and } (-r, \theta+(2n+1)\pi) \text{ for any integer } n.$$

EXAMPLE 1

Plot the points whose polar coordinates are given.

(a) $(1, \pi/3)$ (b) $(2, -\pi/4)$ (c) $(-1, 2\pi/3)$

Solution The points are plotted in Figure 3. In part (c), the point $(-1, 2\pi/3)$ is located three units from the pole in the fourth quadrant because the angle $2\pi/3$ is in the second quadrant and $r=-1$ is negative.

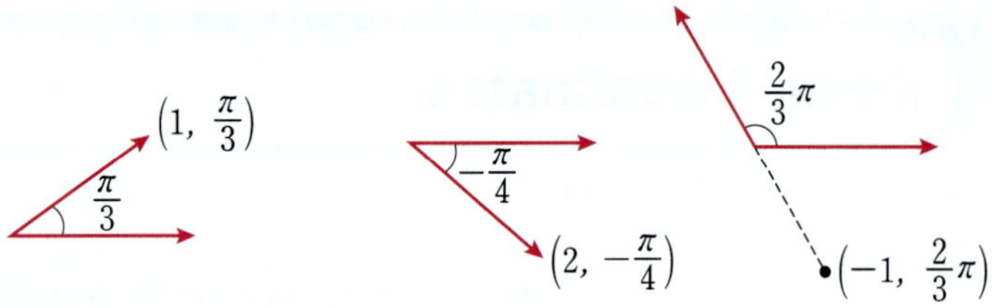

Figure 3

■

In the Cartesian coordinate system, every point has only one representation. This is not true with polar coordinates. For instance, the point $(1, \pi/3)$ in Example 1(a) could be written as $(1, -5/3\pi)$ or $(1, 7/3\pi)$ or $(-1, -2/3\pi)$ (See Figure 4.)

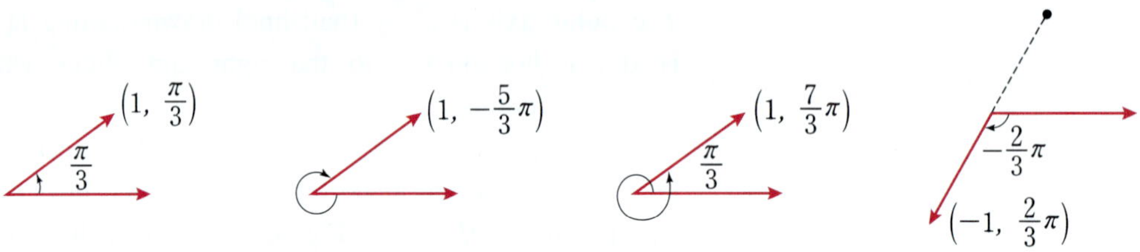

Figure 4

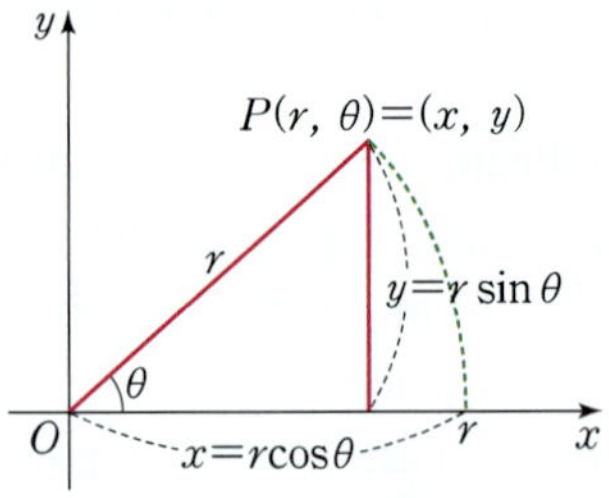

Figure 5

To establish the relationship between polar and rectangular coordinates, we let the polar axis coincide with the positive $x-$axis and the pole with the origin, as shown Figure 5.

Let the point P have (x, y) as Cartesian coordinates and (r, θ) as polar coordinates. Since (x, y) lies on a circle of radius r, it follows that $r^2 = x^2 + y^2$. Moreover, for $r > 0$, we have

$$\tan\theta = \frac{y}{x}, \quad \cos\theta = \frac{x}{r}, \quad \sin\theta = \frac{y}{r}.$$

Hence

1 $$x = r\cos\theta \quad \text{and} \quad y = r\sin\theta.$$

In case of $r < 0$, we can show that the same relationship holds. This relationship allows us to covert coordinates from one system to the other when we align polar axis with positive x-axis.

Note that for given values of x and y, as θ increases through the interval $0 \le \theta < 2\pi$, each value of $\tan\theta$ occurs twice. So in converting from Cartesian coordinates to polar coordinates, determine θ such that the point (r, θ) lies in the correct quadrant.

EXAMPLE 2

Find the Cartesian coordinates represent for polar coordinates $(1, \pi/6)$.

Solution Since $r=1$ and $\theta=\frac{\pi}{6}$, equation [1] gives

$$x=r\cos\theta=\cos\frac{\pi}{6}=\frac{\sqrt{3}}{2}, \text{ and}$$
$$y=r\sin\theta=\sin\frac{\pi}{6}=\frac{1}{2}.$$

Therefore, the point is $\left(\frac{\sqrt{3}}{2}, \frac{1}{2}\right)$ in the Cartesian coordinates. ■

EXAMPLE 3

Find the polar coordinates represent for Cartesian coordinates $(-5, 5\sqrt{3})$.

Solution If we choose r to be positive, then where [1] we have

$$r=\sqrt{x^2+y^2}=\sqrt{25+75}=10, \text{ and}$$
$$\tan\theta=\frac{y}{x}=-\sqrt{3}.$$

Since the point $(-5, 5\sqrt{3})$ lies in the second quadrant, we can choose $\theta=\frac{2\pi}{3}$. Thus, one possible answer is $\left(10, \frac{2}{3}\pi\right)$; another is $\left(-10, \frac{5}{3}\pi\right)$. ■

Polar Curves

A polar curve is represented by the **polar equation** $r=f(\theta)$, or more generally, $F(r, \theta)=0$. Its graph consists of all points P that have at least one polar representation (r, θ) satisfying the polar equation. Curve sketching in polar coordinates is similar to that of rectangular coordinates. We rely heavily on point-by-point plotting, aided by intercepts and symmetry. We can also convert to rectangular coordinates and then make the sketch, however, sometimes this can be difficult.

EXAMPLE 4

(a) Draw the curve with the polar equation $r=2\sin\theta$.
(b) Find a Cartesian equation for this curve.

Solution (a) In Figure 6, we find the values of r for some convenient values of θ and plot the corresponding points (r, θ). Then, we join these points to sketch the curve which appears to be a circle. We have used only values of θ between 0 and π. If we allow θ to increase beyond π, we obtain the same points again.

θ	$r=2\sin\theta$
0	0
$\pi/6$	1
$\pi/4$	$\sqrt{2}$
$\pi/3$	$\sqrt{3}$
$\pi/2$	2
$2\pi/3$	$\sqrt{3}$
$3\pi/4$	$\sqrt{2}$
$5\pi/6$	1
π	0
…	…

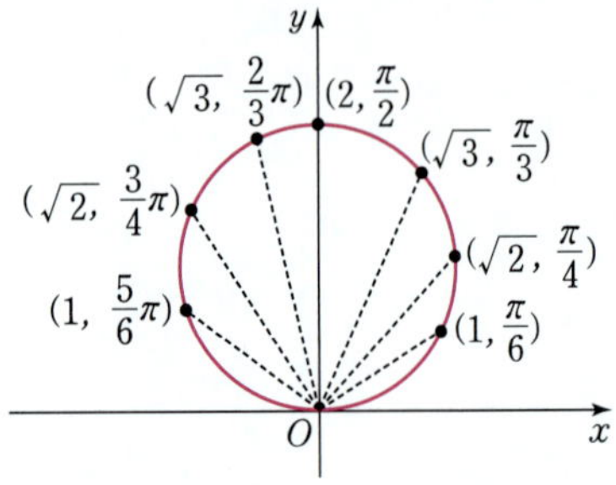

Figure 6 Table of values and graph of $r=2\sin\theta$

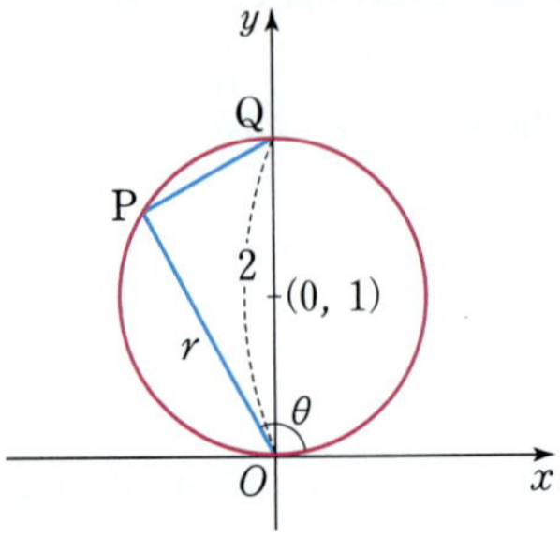

Figure 7

(b) We convert the given equation into the Cartesian equation. From $y=r\sin\theta$, we have $\sin\theta=y/r$. So the equation $r=2\sin\theta$ becomes $r=2y/r$, which gives

$$2y=r^2=x^2+y^2 \quad\text{or}\quad x^2+y^2-2y=0.$$

Completing the square, we obtain

$$x^2+(y-1)^2=1$$

which is an equation of a circle with center $(0,\ 1)$ and radius 1. ■

EXAMPLE 5

Sketch the curve $r=\cos 3\theta$, $0\le\theta\le\pi$.

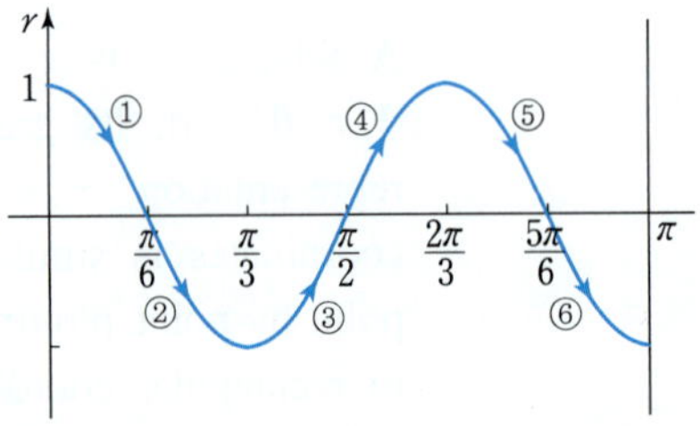

Figure 8 $r=\cos 3\theta$ in Cartesian coordinates

Solution As in Example 4, we first sketch $r=\cos 3\theta$, $0\le\theta\le\pi$, in the Cartesian coordinates in Figure 8. As θ increases from 0 to $\pi/6$, Figure 10 shows that r decreases from 1 to 0 and so we draw the corresponding portion of the polar curve in Figure 9 (indicated by ①). As θ increases from $\pi/6$ to $\pi/3$, r goes from 0 to -1. This means that the distance from O increases from 0 to 1, but instead of being in the first quadrant this portion of the polar curve (indicated by ②), it lies on the opposite side of the pole in the third quadrant. The remainder of the curve is drawn in a similar fashion, with the arrows and numbers indicating the order in which the portions are traced out. The resulting curve has three loops and is called a

three-leaved rose.

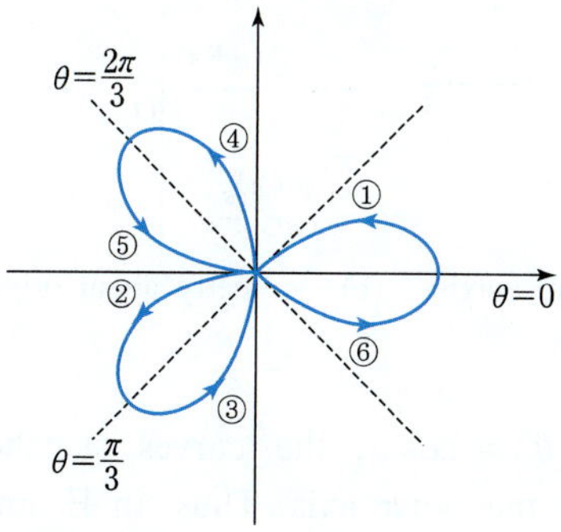

Figure 9 Three-leaved rose $r = \cos 3\theta$

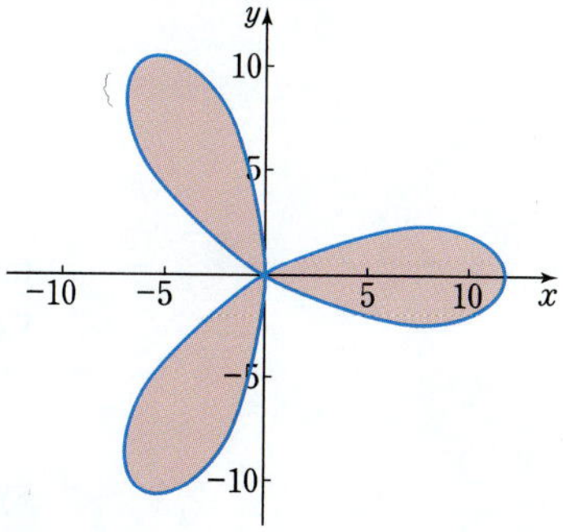

Figure 10
http://matrix.skku.ac.kr/cal-lab/sage-grapher-polar.html

■

Remark Shapes of the function $r = \sin(n\theta)$ and $r = \cos(n\theta)$

a cycle	$0 \le \theta \le \pi$, n: odd	$0 \le \theta \le 2\pi$, n: even
$r = \sin(n\theta)$	n leaves	$2n$ leaves (symmetric about y-axis)
$r = \cos(n\theta)$	n leaves	$2n$ leaves (symmetric about x-axis)

Symmetry

When making a table of points, it is often helpful to use at least one θ value from each of four quadrants. In the above Table, we see that no points are plotted in Quadrants III and IV because of r is negative for these values of θ. Thus, on the interval from $\theta = 0$ to $\theta = 2\pi$ the curve is traced twice. Moreover we also see that the curve is symmetric with respect to the line $\theta = \pi/2$. Therefore, if we had known about this symmetry, we could have plotted fewer points.

When we sketch polar curves, it is helpful to utilize the symmetry. The following three rules it will be useful in sketching polar curves.

(a) If $f(r, \theta) = f(r, -\theta)$ then the curve is **symmetric** about the polar axis(or x-axis).

(b) If $f(r, \theta) = f(-r, \theta)$ or $f(r, \theta) = f(r, \theta + \pi)$ then the curve is symmetric about the pole. (That is, the curve remains unchanged if we rotate it through 180° about the origin.)

(c) If $f(r, \theta) = f(r, \pi - \theta)$, then the curve is symmetric about the vertical line $\theta = \pi/2$(or y-axis). (See Figure 11.)

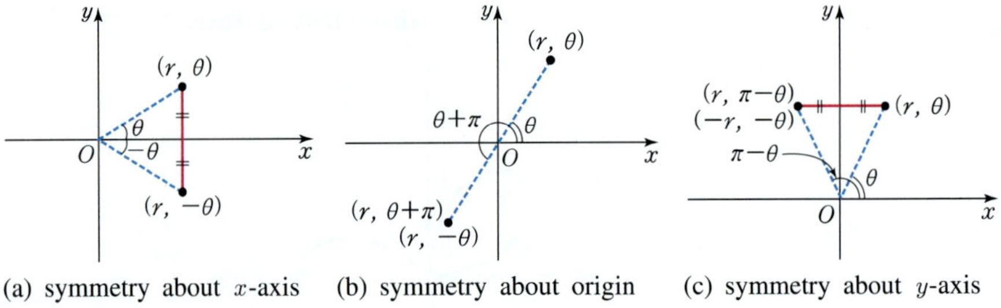

Figure 11

Since $\cos(-\theta)=\cos\theta$, the curves sketched in Examples 4 and 5 are symmetric about the polar axis. Thus, in Example 4, we should plot only the points for $0 \le \theta \le \pi/2$ and then reflect it about the polar axis to obtain the complete circle. The three-leaved rose is symmetric about the poler axis because $\cos(-2\theta)=\cos(2\theta)$. Since $\sin(\pi-\theta)=\sin\theta$, the curve $r=2\sin\theta$ is symmetric about $\theta=\pi/2$.

Tangents to Polar Curves

To find the slope of a tangent line to a polar curve, consider a differentiable function $r=f(\theta)$. Here we regard θ as a parameter, and write its parametric equations as

$$x=r\cos\theta=f(\theta)\cos\theta, \quad y=r\sin\theta=f(\theta)\sin\theta.$$

Then, using the parametric form of dy/dx given in Section 10.2, we have

3
$$\frac{dy}{dx}=\frac{\dfrac{dy}{d\theta}}{\dfrac{dx}{d\theta}}=\frac{f'(\theta)\sin\theta+f(\theta)\cos\theta}{f'(\theta)\cos\theta-f(\theta)\sin\theta}=\frac{\dfrac{dr}{d\theta}\sin\theta+r\cos\theta}{\dfrac{dr}{d\theta}\cos\theta-r\sin\theta}$$

From this formula, we can make the following observations:

1. Solutions to $dy/d\theta=0$ yield horizontal tangents. (provided $dx/d\theta \ne 0$)
2. Solutions to $dx/d\theta=0$ yield vertical tangents. (provided $dy/d\theta \ne 0$)

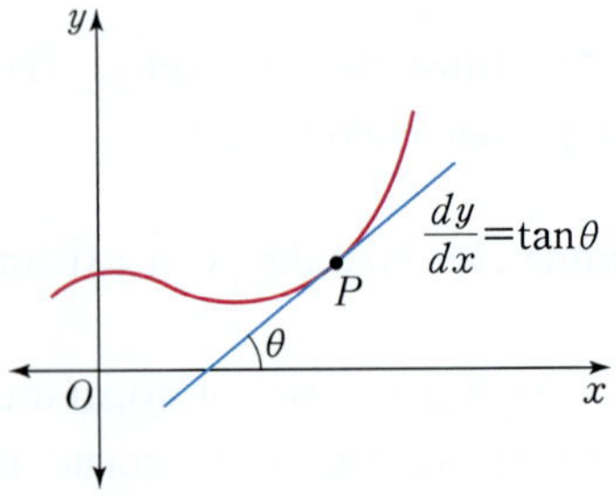

Figure 12

If $dy/d\theta$ and $dx/d\theta$ are simultaneously zero, then no conclusions can be drawn about tangent lines. Notice that if we are looking for tangent lines at the pole, then $r=0$, 3 simplifies to

$$\frac{dy}{dx}=\tan\theta \quad \text{if} \quad \frac{dr}{d\theta}\ne 0. \quad \text{(See Figure 12.)}$$

EXAMPLE 6

(a) For the cardioid $r=1-\sin\theta$, find the slope of the tangent line when $\theta=\pi/3$. (See Figure 13.)

(b) Find the points on the cardioid where the tangent line is horizontal or vertical.

http://matrix.skku.ac.kr/cal-lab/sage-grapher-polar2.html

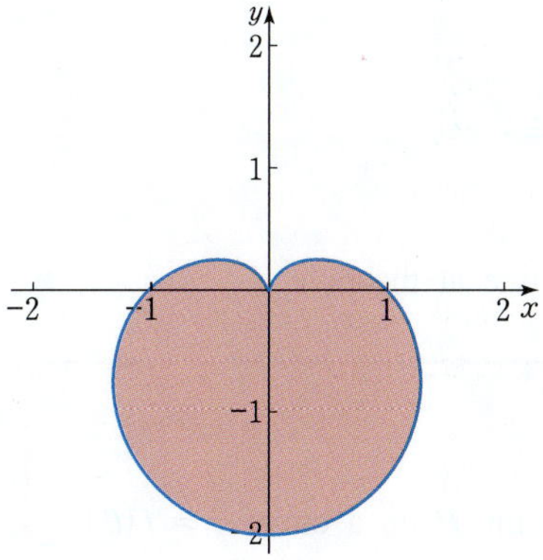

Figure 13

Solution Using 3 with $r = 1 - \sin\theta$, we have

$$\frac{dy}{dx} = \frac{\dfrac{dr}{d\theta}\sin\theta + r\cos\theta}{\dfrac{dr}{d\theta}\cos\theta - r\sin\theta} = \frac{-\cos\theta\sin\theta + (1-\sin\theta)\cos\theta}{-\cos\theta\cos\theta - (1-\sin\theta)\sin\theta}$$

$$= \frac{\cos\theta(1-2\sin\theta)}{-1+2\sin^2\theta - \sin\theta} = \frac{\cos\theta(1-2\sin\theta)}{-(1-\sin\theta)(1+2\sin\theta)}.$$

(a) The slope of the tangent at the point where $\theta = \pi/3$ is

$$\left.\frac{dy}{dx}\right|_{\theta=\pi/3} = \frac{\frac{1}{2}(1-\sqrt{3})}{-(1-\sqrt{3}/2)(1+\sqrt{3})}$$

$$= \frac{1-\sqrt{3}}{-(2-\sqrt{3})(1+\sqrt{3})} = 1.$$

(b) Observe that $\dfrac{dy}{d\theta} = \cos\theta(1-2\sin\theta) = 0$ when $\theta = \dfrac{\pi}{6}, \dfrac{\pi}{2}, \dfrac{5}{6}\pi, \dfrac{3}{2}\pi$.

```
x, t= var('x, t')
plot:  cos(t)* (1- 2*sin(t))
assume( 0<= t <= 2*pi)
p= cos(t)* (1- 2*sin(t))
solve(p==Integer(0), t)
```

$\dfrac{dx}{d\theta} = (-1+\sin\theta)(1+2\sin\theta) = 0$ when $\theta = \dfrac{\pi}{2}, \dfrac{7}{6}\pi, \dfrac{11}{6}\pi$.

Therefore, there are horizontal tangents at the points $\left(\dfrac{1}{2}, \dfrac{\pi}{6}\right)$ $\left(\dfrac{1}{2}, \dfrac{5}{6}\pi\right)$, $\left(2, \dfrac{3}{2}\pi\right)$ and vertical tangents at $\left(\dfrac{3}{2}, \dfrac{7}{6}\pi\right)$ and $\left(\dfrac{3}{2}, \dfrac{11}{6}\pi\right)$. When $\theta = \pi/2$, both $dy/d\theta$ and $dx/d\theta$ are 0, we must be careful. Using L'Hospital's Rule, we have

$$\lim_{\theta\to\left(\frac{\pi}{2}\right)-} \frac{dy}{dx} = \left(\lim_{\theta\to\left(\frac{\pi}{2}\right)-} \frac{1-2\sin\theta}{1+2\sin\theta}\right)\left(\lim_{\theta\to\left(\frac{\pi}{2}\right)-} \frac{\cos\theta}{1-\sin\theta}\right)$$

$$= \frac{1}{3}\lim_{\theta\to(\pi/2)-} \frac{\cos\theta}{1-\sin\theta}$$

$$= \frac{1}{3}\lim_{\theta\to(\pi/2)-} \tan\theta = \infty.$$

By symmetry,

$$\lim_{\theta \to \left(\frac{\pi}{2}\right)+} \frac{dy}{dx} = -\infty.$$

Thus, there is a vertical tangent line at the pole. ■

THEOREM 2

Let ψ be an angle between radius OP at point P on a curve $r = f(\theta)$ and tangent line at the point P. Then

[4] $$\tan\psi = \frac{r}{r'}$$

if $r' = \dfrac{dr}{d\theta} \neq 0$.

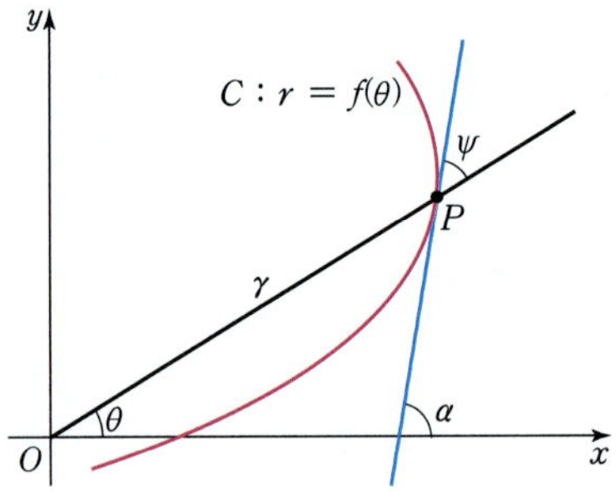

Figure 14

Proof From Figure 14, $\psi = \alpha - \theta$ implies that

[5] $$\tan\psi = \frac{\tan\alpha - \tan\theta}{1 + \tan\alpha\tan\theta}.$$

Since the slope of tangent line at the point P is $\tan\alpha$ and by [3],

$$\begin{aligned}\tan\alpha - \tan\theta &= \frac{r'\sin\theta + r\cos\theta}{r'\cos\theta - r\sin\theta} - \frac{\sin\theta}{\cos\theta} \\ &= \frac{r'\sin\theta\cos\theta + r\cos^2\theta - r'\sin\theta\cos\theta + r\sin^2\theta}{(r'\cos\theta - r\sin\theta)\cos\theta} \\ &= \frac{r}{(r'\cos\theta - r\sin\theta)\cos\theta}\end{aligned}$$

and

$$\begin{aligned}1 + \tan\alpha\tan\theta &= 1 + \frac{r'\sin\theta + r\cos\theta}{r'\cos\theta - r\sin\theta} \cdot \frac{\sin\theta}{\cos\theta} \\ &= \frac{r'}{(r'\cos\theta - r\sin\theta)\cos\theta}.\end{aligned}$$

Now substituting these values in [5] we get the required result in this theorem. ■

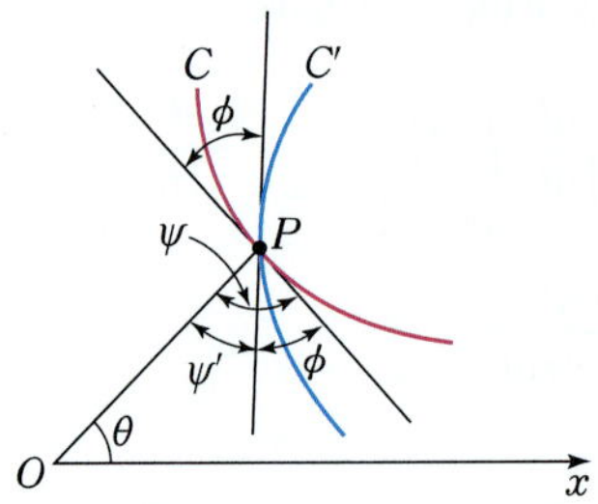

Figure 15

Let two curves C and C' meet at the point P. Say ψ and ψ' are angles between radius OP and curves C and C', respectively. Then the angle between two curves $\phi \in \left[0, \dfrac{\pi}{2}\right]$ is $|\psi - \psi'|$ and satisfies

[6] $$\tan\phi = \left|\frac{\tan\psi - \tan\psi'}{1 + \tan\psi\tan\psi'}\right|$$

where $\tan\psi$ and $\tan\psi'$ comes from [4]. (See Figure 15.)

EXAMPLE 7

Show two curves $r=2\cos\theta$ and $r=3\tan\theta$ meet at the point $(\sqrt{3},\ \pi/6)$, and find angle between the curves at the point.

Solution By solving $2\cos\theta=3\tan\theta$, we obtain $\theta=\dfrac{\pi}{6}$, then $2\cos\theta=2\left(\dfrac{\sqrt{3}}{2}\right)=\sqrt{3}$ and $3\tan\theta=3\left(\dfrac{\sqrt{3}}{3}\right)=\sqrt{3}$, hence the point $(\sqrt{3},\ \pi/6)$ satisfies two equations. Now we call ψ and ψ' be angles of the tangent lines of $r=2\cos\theta$ and $r=3\tan\theta$ at the point P, respectively. Then we have $\tan\psi=\dfrac{r}{r'}=\dfrac{2\cos(\pi/6)}{-2\sin(\pi/6)}=-\sqrt{3}$, $\tan\psi'=\dfrac{r}{r'}=\dfrac{3\tan(\pi/6)}{3\sec^2(\pi/6)}=\dfrac{\sqrt{3}}{4}$, respectively. Putting it in 6, the angle between two curves ϕ satisfies

$$\tan\phi=\left|\frac{\tan\psi-\tan\psi'}{1+\tan\psi\ \tan\psi'}\right|=\left|\frac{-\sqrt{3}-\dfrac{\sqrt{3}}{4}}{1+(-\sqrt{3})\dfrac{\sqrt{3}}{4}}\right|=5\sqrt{3}.$$

Hence the angle at the point that satisfies $\tan\phi=5\sqrt{3}$ is $\phi\in\left[0,\ \dfrac{\pi}{2}\right]$. ■

CAS **EXAMPLE 8**

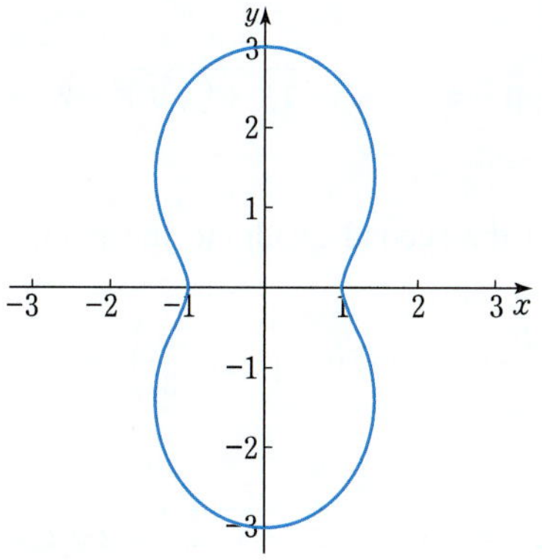

Figure 16

Sketch the curve with the given polar equation.

$$r=2-\cos 2\theta,\ 0\le\theta\le 2\pi$$

Solution

```
var('theta')
polar_plot(2-cos(2*theta),   (0,   2*pi)).show(aspect_ratio=1,
xmin=-3, xmax=3, ymin=-3, ymax=3)
```

■

10.3 EXERCISES (Polar Coordinates)

http://matrix.skku.ac.kr/Cal-Book/part1/CS-Sec-10-3-Sol.html

YouTube http://youtu.be/4hoVKvk8dq0

1-4. Plot the point whose polar coordinates are given. Then, find the Cartesian coordinates of the point.

CAS **1.** $(r,\ \theta)=(-1,\ \pi)$

http://matrix.skku.ac.kr/cal-lab/cal-9-3-1.html

http://matrix.skku.ac.kr/cal-lab/sage-grapher-polar.html

Solution Graph of polar functions

```
def Polar(r, theta):
    #converts Polar to Cartesian Coordinates
    CartC = ([r*cos(theta), r*sin(theta)]);
    return CartC;
```

```
pt=Polar(-1,pi);
vector(pt)
```

Answer : (x, y)=(1, 0)

```
list_plot([pt], aspect_ratio=1, xmin=-1, xmax=2,
ymin=-1, ymax=1)
```

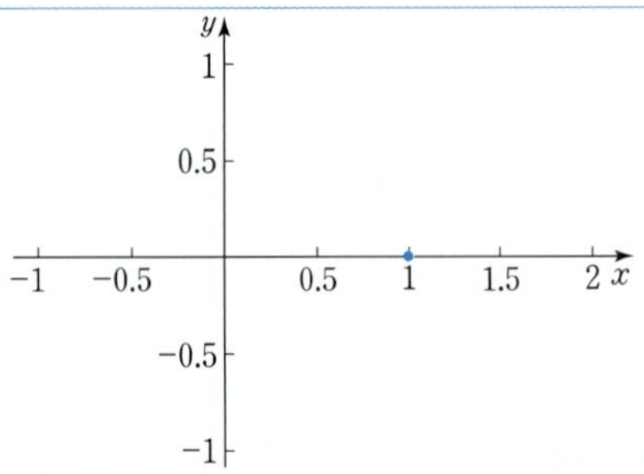

2. $(r,\ \theta)=\left(4,\ \dfrac{1}{4}\pi\right)$

http://matrix.skku.ac.kr/cal-lab/cal-10-3-2.html

Solution

```
def Polar(r, theta):
    #converts Polar to Cartesian Coordinates
    CartC = ([r*cos(theta), r*sin(theta)]);
    return CartC;
```

```
pt=Polar(4, pi/4);
vector(pt)
```

Answer : (x, y)=(2 sqrt {2}, 2 sqrt {2})

3. $(r,\ \theta)=\left(7,\ \dfrac{25}{6}\pi\right)$

Solution $\left(\dfrac{7}{2}\sqrt{3},\ \dfrac{7}{2}\right)$

4. $(r,\ \theta)=\left(2\sqrt{2},\ \dfrac{3}{4}\pi\right)$

Solution $(-2,\ 2)$

5-8. The Cartesian coordinates are given. Find two other pairs of polar coordinates of the point, one with $r>0$ and the other $r<0$.

5. $(\sqrt{3},\ -1)$

Solution $x=\sqrt{3},\ y=-1 \Rightarrow r=\sqrt{(\sqrt{3})^2+(-1)^2}=2$,

and $\theta=\tan^{-1}\left(-\dfrac{1}{\sqrt{3}}\right)=-\dfrac{\pi}{6}$.

Since $(\sqrt{3},\ -1)$ is in the fourth quadrant, the polar coordinates are

$$(1)\ r>0 : \left(2,\ \frac{11\pi}{6}\right),\ (2)\ r<0 : \left(-2,\ \frac{5\pi}{6}\right).$$

6. $(-1,\ \sqrt{3})$

Solution $x=-1,\ y=\sqrt{3} \Rightarrow r=\sqrt{(-1)^2+(\sqrt{3})^2}=2$,

and $\theta=\tan^{-1}(-\sqrt{3})=\dfrac{2\pi}{3}$.

Since $(-1,\ \sqrt{3})$ is in the second quadrant, the polar coordinates are

$$(1)\ r>0 : \left(2,\ \frac{2\pi}{3}\right),\ (2)\ r<0 : \left(-2,\ \frac{5\pi}{3}\right)$$

7. $(2,\ -2)$

Solution $x=2,\ y=-2 \Rightarrow r=\sqrt{(2)^2+(-2)^2}=2\sqrt{2}$,

and $\theta=\tan^{-1}(-1)=-\dfrac{\pi}{4}$.

Since $(2,\ -2)$ is in the fourth quadrant, the polar coordinates are

$$(1)\ r>0 : \left(2\sqrt{2},\ \frac{7\pi}{4}\right),\ (2)\ r<0 : \left(-2\sqrt{2},\ \frac{3\pi}{4}\right).$$

8. $(-4,\ 4)$

Solution $x=-4,\ y=4 \Rightarrow r=\sqrt{(-4)^2+4^2}=4\sqrt{2}$,

and $\theta=\tan^{-1}(-1)=\dfrac{3\pi}{4}$.

Since $(-4, 4)$ is in the second quadrant, the polar coordinates are

(1) $r > 0$: $\left(4\sqrt{2}, \dfrac{3\pi}{4}\right)$, (2) $r < 0$: $\left(-4\sqrt{2}, \dfrac{7\pi}{4}\right)$.

---

9. Find a formula for the distance between the points with polar coordinates (r_1, θ_1) and (r_2, θ_2).

Solution Let $(x_1, y_1) = (r_1\cos\theta_1, r_1\sin\theta_1)$, $(x_2, y_2) = (r_2\cos\theta_2, r_2\sin\theta_2)$

Let L be the distance between (x_1, y_1) and (x_2, y_2).

$$L = \sqrt{(x_2 - x_1)^2 + (y_2 - y_1)^2}$$
$$= \sqrt{r_1^2\cos^2\theta_1 + r_2^2\cos^2\theta_2 - 2r_1r_2\cos\theta_1\cos\theta_2 + r_1^2\sin^2\theta_1 + r_2^2\sin^2\theta_2 - 2r_1r_2\sin\theta_1\sin\theta_2}$$
$$= \sqrt{r_1^2 + r_2^2 - 2r_1r_2(\cos\theta_2\cos\theta_1 + \sin\theta_2\sin\theta_1)}$$
$$= \sqrt{r_1^2 + r_2^2 - 2r_1r_2\cos(\theta_2 - \theta_1)}$$

10-12. Find a polar equation for the curve represented by the given Cartesian equation.

10. $x + y = 0$

Solution $x + y = 0 \Leftrightarrow r\cos\theta + r\sin\theta = 0$
$\Leftrightarrow r(\cos\theta + \sin\theta) = 0$

11. $x^2 - y^3 = 4$

Solution $x^2 - y^3 = 4 \Leftrightarrow r^2\cos^2\theta - r^3\sin^3\theta = 4$
$\Leftrightarrow r^2(\cos^2\theta - r\sin^3\theta) = 4 \Leftrightarrow r^2 = \dfrac{4}{\cos^2\theta - r\sin^3\theta}$

12. $4x^2 - 24y - 5y^2 = 16$

Solution $4x^2 - 24y - 5y^2 = 16$
$\Leftrightarrow 4r^2\cos^2\theta - 24r\sin\theta - 5r^2\sin^2\theta = 16$

13-15. Find a Cartesian equation for the curve represented by the given polar equation.

13. $r = \cos\theta$

Solution $r = \cos\theta \Leftrightarrow r^2 = r\cos\theta \Leftrightarrow x^2 + y^2 = x$
$\Leftrightarrow \left(x - \dfrac{1}{2}\right)^2 + y^2 = \dfrac{1}{4}$, a circle of radius $\dfrac{1}{2}$ centered at $\left(\dfrac{1}{2}, 0\right)$. The first two equations are actually equivalent since $r^2 = r\cos\theta \Rightarrow r(r - \cos\theta) = 0 \Rightarrow r = 0$ or $r = \cos\theta$.
But $r = \cos\theta$ gives the point $r = 0$ when $\theta = \dfrac{\pi}{2}$.
Thus, the single equation $r = \cos\theta$ is equivalent to the compound condition ($r = 0$ or $r = \cos\theta$).

14. $r\cos\theta = 3$

Solution $r\cos\theta = 3 \Leftrightarrow x = 3$

15. $r = 1$

Solution $r = 1 \Leftrightarrow \sqrt{x^2 + y^2} = 1 \Leftrightarrow x^2 + y^2 = 1$

16-25. Sketch the curve with the given polar equation.

16. $\theta = \dfrac{\pi}{3}$

CAS **17.** $r = \sin\theta$

http://matrix.skku.ac.kr/cal-lab/cal-9-3-17.html

Solution

$x^2 + y^2 = r^2 = r\sin\theta = y$

$x^2 + \left(y - \dfrac{1}{2}\right)^2 = \dfrac{1}{4}$

```
var('theta')
polar_plot(sin(theta), (0, pi)).show(aspect_ratio=1,
xmin=-2, xmax=2, ymin=-2, ymax=2)
```

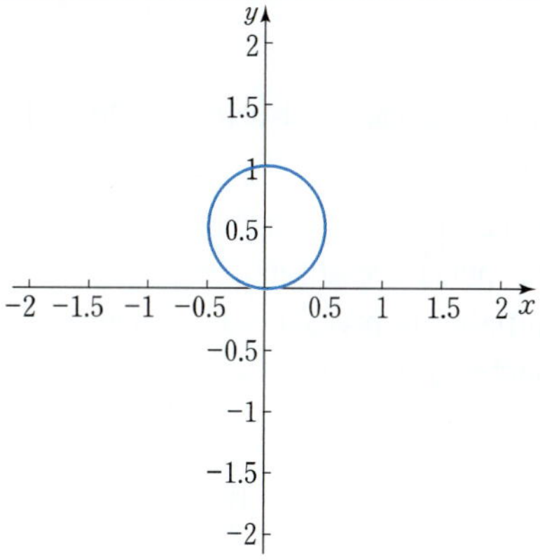

CAS **18.** $r^2 - 2r - 3 = 0$

http://matrix.skku.ac.kr/cal-lab/cal-9-3-18.html

Solution

```
r=var('r');
polar_plot(3,(0,        2*pi)).show(aspect_ratio=1,
xmin=-5, xmax=5, ymin=-5, ymax=5)
```

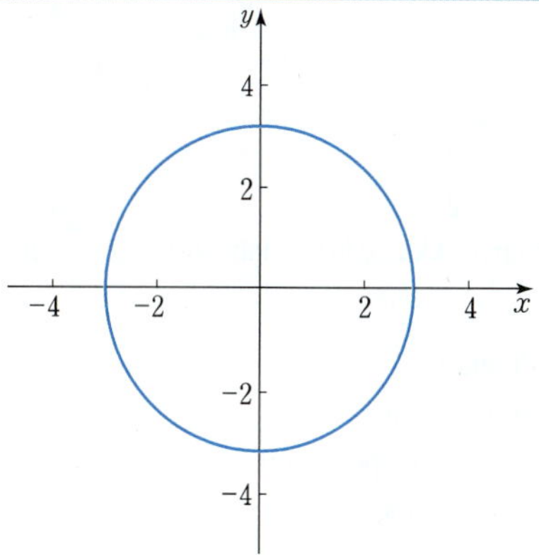

CAS **19.** $r = \theta$

http://matrix.skku.ac.kr/cal-lab/cal-9-3-19.html

Solution

```
theta=var('theta');
polar_plot(theta, (0, 2*pi)).show(aspect_ratio=1,
xmin=-10, xmax=10, ymin=-10, ymax=10)
```

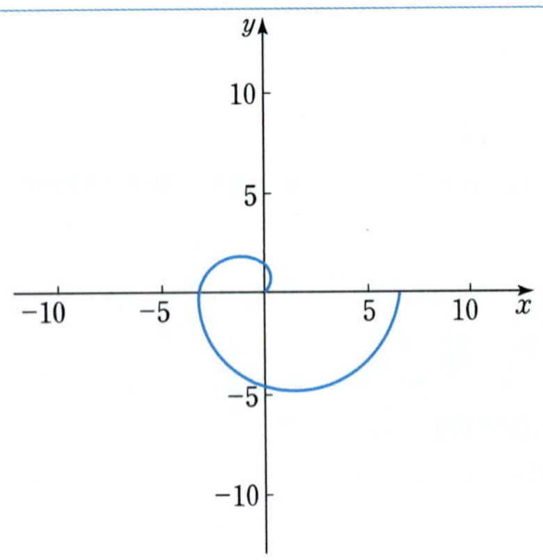

20. $r = 1 - 2\cos\theta$

http://matrix.skku.ac.kr/cal-lab/cal-9-3-20.html

Solution

```
var('theta')
polar_plot(1-2*cos(theta), (0,
2*pi)).show(aspect_ratio=1, xmin=-4, xmax=4,
ymin=-4, ymax=4)
```

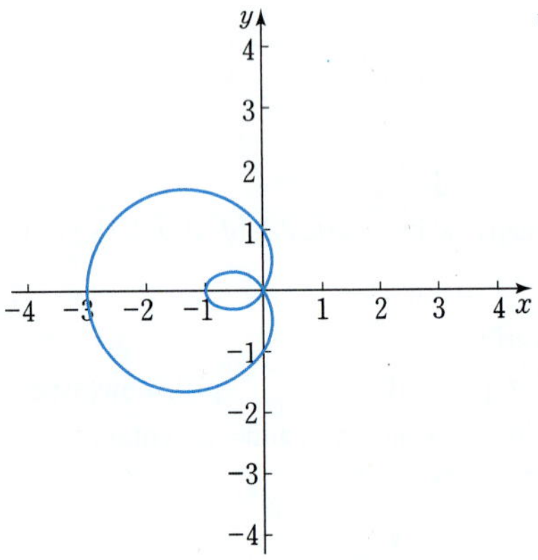

CAS **21.** $r = \sin\left(\dfrac{\theta}{2}\right)$

http://matrix.skku.ac.kr/cal-lab/cal-9-3-21.html

Solution

```
var('theta')
polar_plot(sin(theta/2), (0,
2*pi)).show(aspect_ratio=1, xmin=-2, xmax=2,
ymin=-2, ymax=2)
```

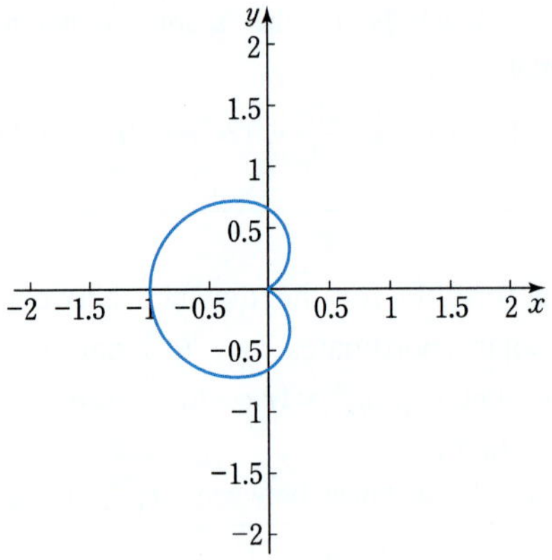

CAS **22.** $r = \sin(4\theta),\ (0,\ 2\pi)$

http://matrix.skku.ac.kr/cal-lab/cal-9-3-22.html

http://matrix.skku.ac.kr/cal-lab/sage-grapher-polar.html

Solution Graph of polar functions

```
var('theta')
polar_plot(sin(4*theta), (0,
2*pi)).show(aspect_ratio=1, xmin=-1, xmax=1,
ymin=-1, ymax=1)
```

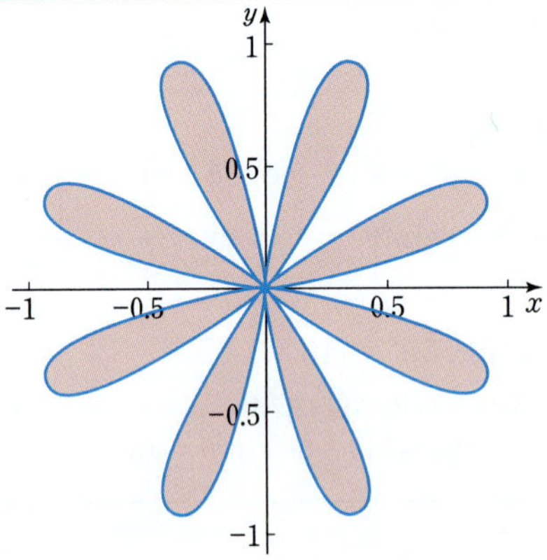

23. $r = 1 + 2\cos(2\theta),\ (0,\ 6\pi)$

Solution r(t) = 1+2*cos(2*t) in

Graph of polar functions

http://matrix.skku.ac.kr/cal-lab/sage-grapher-polar.html

```
var('theta')
polar_plot(1+2*cos(2*theta), (0,
6*pi)).show(aspect_ratio=1, xmin=-3, xmax=3,
ymin=-1, ymax=1)
```

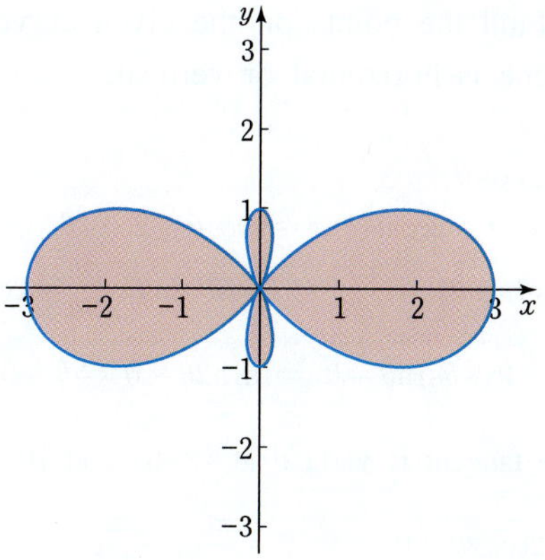

24. $r^2 = \cos(2\theta)$

25. $r = 1 + 2\sin\theta$

26. Show that the polar curve $r = 4 + 2\sec\theta$ (called a conchoid) has the line $x = 2$ as a vertical asymptote by showing that $\lim_{r\to\pm\infty} x = 2$. Use this fact to sketch the conchoid.
http://matrix.skku.ac.kr/cal-lab/cal-9-3-26.html

Solution $x = r\cos\theta = (4 + 2\sec\theta)\cos\theta = 4\cos\theta + 2$.

Now, $r \to \infty \Rightarrow (4 + 2\sec\theta) \to \infty$,

$\Rightarrow \theta \to \left(\frac{\pi}{2}\right)^- \text{ or } \theta \to \left(\frac{3\pi}{2}\right)^+$

so $\lim_{r\to\infty} x = \lim_{\theta\to\pi/2^-} (4\cos\theta + 2) = 2$.

Also, $r \to -\infty \Rightarrow (4 + 2\sec\theta) \to -\infty$

$\Rightarrow \theta \to \left(\frac{\pi}{2}\right)^+ \text{ or } \theta \to \left(\frac{3\pi}{2}\right)^-$,

so $\lim_{r\to-\infty} x = \lim_{\theta\to\pi/2^+} (4\cos\theta + 2) = 2$.

Therefore, $\lim_{r\to\pm\infty} x = 2 \Rightarrow x = 2$ is a vertical asymptote.

```
var('theta')
polar_plot(4+2*sec(theta), (0, 2*pi)).show(aspect_ratio=1, xmin=-5, xmax=15, ymin=-10, ymax=10)
```

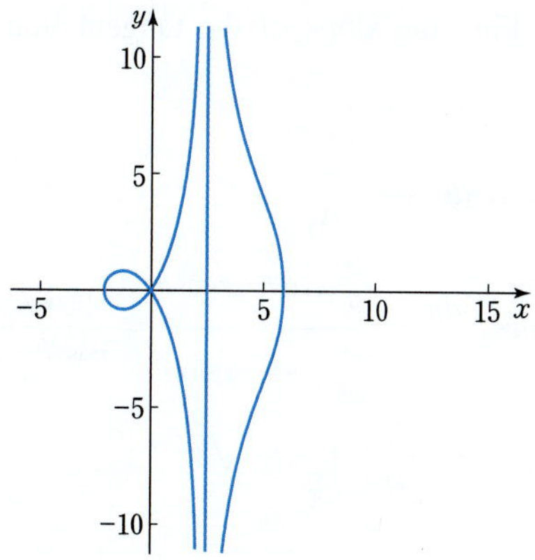

27. Show that the curve $r = 2 - \csc\theta$ (also a conchoid) has the line $y = -1$ as a horizontal asymptote by showing that $\lim_{r\to\pm\infty} y = -1$ Use this fact to help sketch conchoid.
http://matrix.skku.ac.kr/cal-lab/cal-9-3-27.html

Solution $y = r\sin\theta = (2 - \csc\theta)\sin\theta = 2\sin\theta - 1$.

Now, $r \to \infty \Rightarrow (2 - \csc\theta) \to \infty \Rightarrow \theta \to 0^+$ or $\theta \to \pi^-$,

so $\lim_{r\to\infty} y = \lim_{\theta\to 0^+} (2\sin\theta - 1) = -1$.

Also, $r \to -\infty \Rightarrow (2 - \csc\theta) \to -\infty$

$\Rightarrow \theta \to \pi^+$ or $\theta \to 2\pi^-$,

so $\lim_{r\to-\infty} y = \lim_{\theta\to\pi^+} (2\sin\theta - 1) = -1$.

Therefore,

$$\lim_{r\to\pm\infty} y = -1 \Rightarrow y = -1$$

is a horizontal asymptote.

```
var('theta')
polar_plot(2-csc(theta), (0, 2*pi)).show(aspect_ratio=1, xmin=-15, xmax=15, ymin=-10, ymax=10)
```

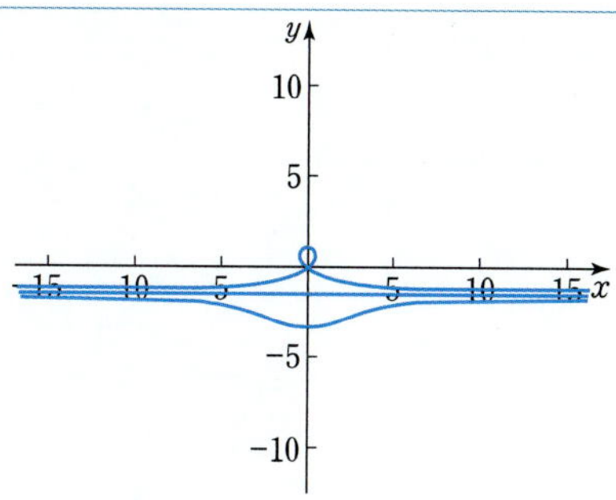

28. Show that the curve $r = \sin\theta\tan\theta$ (called a cissoid of Diocles) has the line $x = 1$ as a vertical asymptote. Show also that the curve lies entirely within the vertical strip $0 < x < 1$. Use these facts to sketch the cissoid.

29-32. Find the slope of the tangent line at the given point.

29. $r=\sin\theta,\ \theta=\dfrac{\pi}{3}$

Solution $\dfrac{dy}{dx}=\dfrac{\dfrac{dr}{d\theta}\sin\theta+r\cos\theta}{\dfrac{dr}{d\theta}\cos\theta-r\sin\theta}=\dfrac{2\cos\theta\sin\theta}{\cos^2\theta-\sin^2\theta}$

$\Rightarrow \left.\dfrac{dy}{dx}\right|_{\theta=\frac{\pi}{3}}=-\sqrt{3}$

30. $r=1-2\cos\theta,\ \theta=\pi$

Solution

$\dfrac{dy}{dx}=\dfrac{\dfrac{dr}{d\theta}\sin\theta+r\cos\theta}{\dfrac{dr}{d\theta}\cos\theta-r\sin\theta}=\dfrac{2(\sin^2\theta-\cos^2\theta)+\cos\theta}{4\cos\theta\sin\theta-\sin\theta}$

$\Rightarrow \left.\dfrac{dy}{dx}\right|_{\theta=\pi}=\infty$ (vertical tangent line)

31. $r=\sin3\theta,\ \theta=\dfrac{\pi}{2}$

Solution

$\dfrac{dy}{dx}=\dfrac{\dfrac{dr}{d\theta}\sin\theta+r\cos\theta}{\dfrac{dr}{d\theta}\cos\theta-r\sin\theta}=\dfrac{3\cos3\theta\sin\theta+\sin3\theta\cos\theta}{3\cos3\theta\cos\theta-\sin3\theta\sin\theta}$

$\Rightarrow \left.\dfrac{dy}{dx}\right|_{\theta=\frac{\pi}{2}}=0$

32. $r=\ln\theta,\ \theta=e\backslash$

33-36. Find the points on the given curve where the tangent line is horizontal or vertical.

33. $r=2\cos\theta$

Solution $r=2\cos\theta\Rightarrow x=r\cos\theta=2\cos^2\theta,\ y=r\sin\theta$

$=2\cos\theta\sin\theta \Rightarrow \dfrac{dx}{d\theta}=-4\cos\theta\sin\theta,\ \dfrac{dy}{d\theta}=2\cos2\theta.$

$\dfrac{dx}{d\theta}=-4\cos\theta\sin\theta=0 \Rightarrow \sin2\theta=0 \Leftrightarrow \theta=0$ or $\dfrac{\pi}{2}.$

So the tangent is vertical at $(2,\ 0)$ and $\left(0,\ \dfrac{\pi}{2}\right).$

$\dfrac{dy}{d\theta}=2\cos2\theta=0 \quad\Leftrightarrow\quad \theta=\dfrac{\pi}{4}$ or $\dfrac{3\pi}{4}.$

So the tangent is horizontal at $\left(\sqrt{2},\dfrac{\pi}{4}\right)$ and $\left(-\sqrt{2},\ \dfrac{3\pi}{4}\right).$

34. $r=\cos\theta+\sin\theta$

35. $r=1+\sin\theta$

36. $r=\cos(2\theta)$

37. Show that the polar equation $r=\sin\theta+\cos\theta$ represents a circle and find its center and radius.

Solution $r=\sin\theta+\cos\theta \Leftrightarrow r^2=r\sin\theta+r\cos\theta$

$\Leftrightarrow x^2+y^2=x+y \Leftrightarrow \left(x-\dfrac{1}{2}\right)^2+\left(y-\dfrac{1}{2}\right)^2=\dfrac{1}{2}.$

Then the polar equation $r=\sin\theta+\cos\theta$ represents a circle and its center : $\left(\dfrac{1}{2},\dfrac{1}{2}\right)$, radius : $\sqrt{\dfrac{1}{2}}.$

10.4 Areas and Lengths in Polar Coordinates

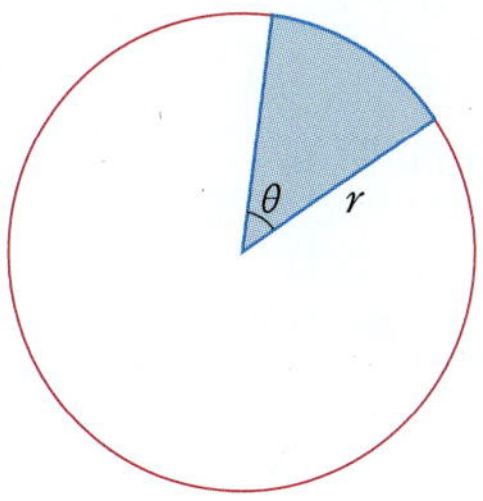

Figure 1

The formula for the area of a region whose boundary is given by a polar equation is considered in this section. The development of a formula for the area of a polar region parallels that for the area of regions in the rectangular coordinates system, but use sectors of a circle, instead of rectangles, as the basic element of area. We know that the area A of a sector of a circle with radius r and the radian measure θ of the central angle is

$$A = \frac{1}{2}r^2\theta \tag{1}$$

(See Figure 1.) since the area of a sector is proportional to its central angle: $2\pi : \theta = \pi r^2 : A$.

Consider the region R (See Figure 2.) bounded by the polar curve $r = f(\theta)$ and by the rays $\theta = a$ and $\theta = b$. Assume that f is a continuous and non-negative function and $0 < b - a \leq 2\pi$. We partition the interval $[a, b]$ into n equal subintervals

$$a = \theta_0 < \theta_1 < \theta_2 < \cdots < \theta_{n-1} < \theta_n = b$$

with equal width $\Delta\theta$. Then R is divided into n smaller regions by rays $\theta = \theta_i$ with central angle $\Delta\theta = \theta_i - \theta_{i-1}$. Choose θ_i^* in the ith subinterval $[\theta_{i-1}, \theta_i]$, then the area ΔA_i of the ith region is approximated by the area of the sector of a circle with central angle $\Delta\theta$ and radius $f(\theta_i^*)$. (See Figure 3.)

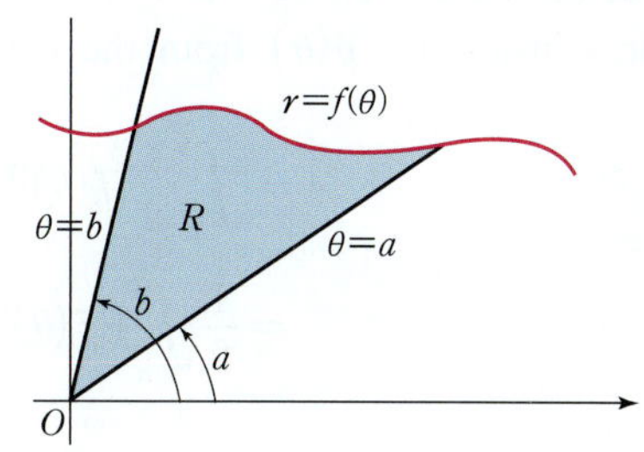

Figure 2

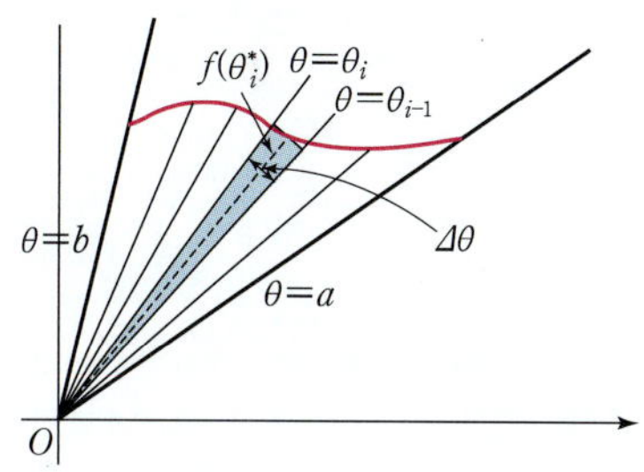

Figure 3

Using [1]

$$\Delta A_i \approx \frac{1}{2}[f(\theta_i^*)]^2\Delta\theta.$$

Thus an approximation to the total area A of R is

$$A \approx \sum_{i=1}^{n}\frac{1}{2}[f(\theta_i^*)]^2\Delta\theta. \tag{2}$$

As $n \to \infty$ the Riemann sums for the function $g(\theta) = \frac{1}{2}[f(\theta)]^2$, given by the sums in [2] converge to

$$\lim_{n\to\infty}\sum_{i=1}^{n}\frac{1}{2}[f(\theta_i^*)]^2\Delta\theta = \int_a^b \frac{1}{2}[f(\theta)]^2 d\theta.$$

Thus formula for the area A of the polar region R is

[3]
$$A = \int_a^b \frac{1}{2}[f(\theta)]^2 d\theta.$$

Since $r = f(\theta)$, Equation [3] may be written as

[4]
$$A = \int_a^b \frac{1}{2} r^2 d\theta.$$

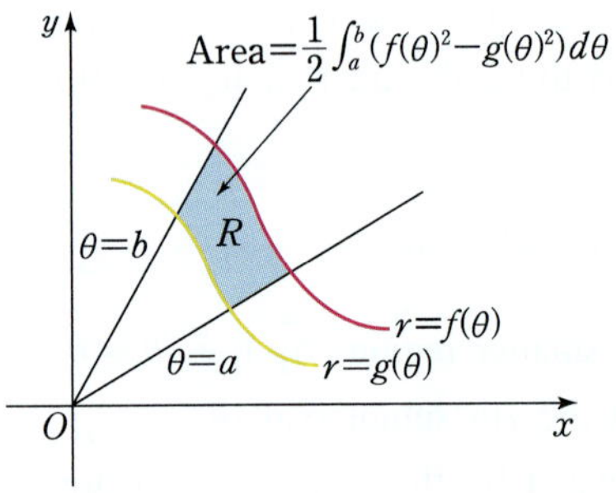

Figure 4

The area given by [3] or [4] may be thought of as being swept out by a rotating ray through O that starts with angle a and ends with angle b. We can use the same formula to find the area of a region bounded by the graph of a non-positive continuous function. However the formula is not necessarily valid if f takes on both positive and negative values in the interval $[a, b]$.

Sometimes the most difficult part of finding the area of a polar region is determining the limits of integration. A good sketch of the region helps.

Consider a region R bounded by two polar curves with polar equations $r = f(\theta)$, $r = g(\theta)$, $\theta = a$, and $\theta = b$. (See Figure 4.) We assume that $f(\theta)$ and $g(\theta)$ are continuous and $f(\theta) \geq g(\theta) \geq 0$ on $[a, b]$ and $0 < b - a \leq 2\pi$. The area of such a region R that is bounded by two curves is found by subtracting the area inside $r = g(\theta)$ from the area inside $r = f(\theta)$. Using [3], we have

$$\begin{aligned} A &= \int_a^b \frac{1}{2}[f(\theta)]^2 d\theta - \int_a^b \frac{1}{2}[g(\theta)]^2 d\theta \\ &= \frac{1}{2}\int_a^b ([f(\theta)]^2 - [g(\theta)]^2) d\theta. \end{aligned}$$

Note To find all points of intersection of two polar curves, it is recommended to draw the graphs of both curves, because a single point has many representations in polar coordinates.

EXAMPLE 1

Find the area of the region bounded by the limaçon with equation $r = 2 + 2\cos\theta$, $0 \leq \theta \leq 2\pi$.

Solution The curve $r = 2 + 2\cos\theta$ is sketched in Figure 5. From [4] we have

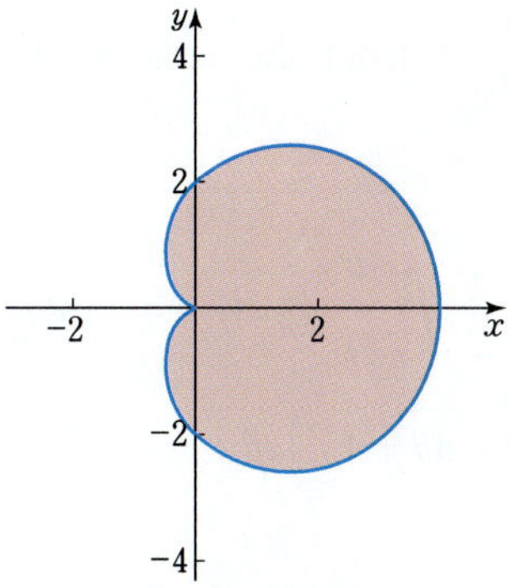
Figure 5

$$A = 2\int_0^{\pi} \frac{1}{2} r^2 \, d\theta = \int_0^{\pi} (2 + 2\cos\theta)^2 \, d\theta$$

```
var('x, t')
x= (2+ 2*cos(t))^2
integral(x, (t, 0, pi))
```

$$= \int_0^{\pi} 4(1 + 2\cos\theta + \cos^2\theta) d\theta = 4\int_0^{\pi} (2\cos\theta + \frac{3}{2} + \frac{1}{2}\cos 2\theta) \, d\theta$$

$$= 2\left[4\sin\theta + 3\theta + \frac{\sin 2\theta}{2}\right]_0^{\pi} = 6\pi.$$

■

EXAMPLE 2

Find the entire area of the lemniscate with equation $r^2 = a^2 \cos 2\theta$.

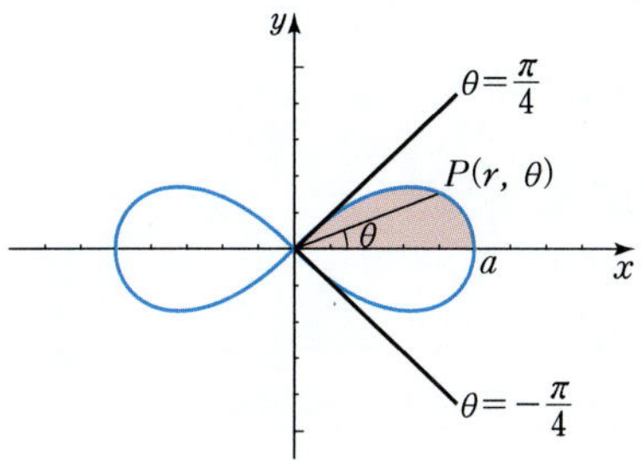

Figure 6

Solution The curve $r^2 = a^2 \cos 2\theta$ is sketched in Figure 6. Figure 6 is symmetric with respect to both x and y-axis, the whole area = 4× the area in the first quadrant. Since $r = 0$ when $\theta = \frac{\pi}{4}$. If θ varies from 0 to $\frac{\pi}{4}$, the radius vector OP sweeps over the area OAB. From 4 we have

$$A = 4\int_0^{\frac{\pi}{4}} \frac{1}{2} r^2 \, d\theta = 2\int_0^{\frac{\pi}{4}} a^2 \cos 2\theta \, d\theta = 2a^2 \int_0^{\frac{\pi}{4}} \cos 2\theta \, d\theta$$

$$= 2a^2 \left[\frac{1}{2}\sin 2\theta\right]_0^{\frac{\pi}{4}} = a^2.$$

```
var('x, a')
f(x)= 2*a^2* (cos(2*x))
print integral(f(x), x, 0, pi/4)          # a^2
```

■

EXAMPLE 3

Find the area of the region inside the circle $r = 4\sin\theta$ and outside the cardioid $r = 2\sin\theta - 1$.

http://matrix.skku.ac.kr/cal-lab/sage-grapher-polar2.html

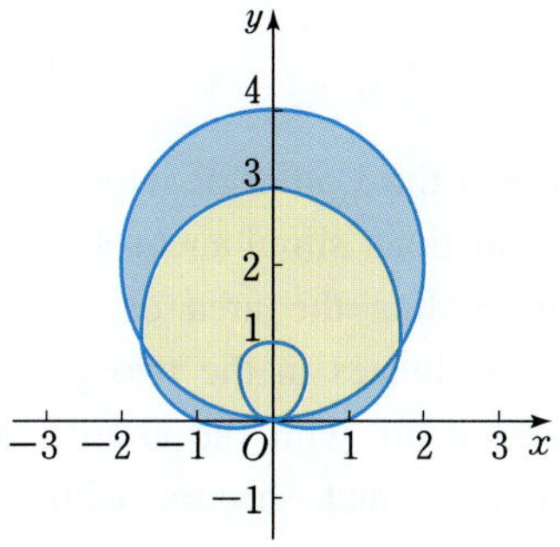
Figure 7

Solution The cardioid and the circle are shown in Figure 7 with the desired region shaded. The points of intersection of the two curves are given by $4\sin\theta = 2\sin\theta - 1$ which gives $\sin\theta = -\frac{1}{2}$, so $\theta = \frac{7}{6}\pi, \frac{11}{6}\pi$. The required area is obtained by subtracting the area inside the cardioid between $\theta = \frac{7}{6}\pi$ and $\theta = \frac{11}{6}\pi$ from the area inside the circle from $\theta = \frac{7}{6}\pi$ to $\frac{11}{6}\pi$. Thus

$$\begin{aligned} A &= \frac{1}{2}\int_{7\pi/6}^{11\pi/6}(4\sin\theta)^2 d\theta - \frac{1}{2}\int_{7\pi/6}^{11\pi/6}(2\sin\theta - 1)^2 d\theta \\ &= \frac{1}{2}\int_{7\pi/6}^{11\pi/6}[(4\sin\theta)^2 - (2\sin\theta - 1)^2]d\theta. \end{aligned}$$

Using symmetry of the region about the vertical axis $\theta = \frac{3}{2}\pi$, we can write

$$\begin{aligned} A &= \int_{7\pi/6}^{3\pi/2}[(4\sin\theta)^2 - (2\sin\theta - 1)^2]d\theta \\ &= \int_{7\pi/6}^{3\pi/2}[12\sin^2\theta + 4\sin\theta - 1]d\theta \\ &= \int_{7\pi/6}^{3\pi/2}(5 - 6\cos 2\theta + 4\sin\theta)\,d\theta \\ &= [5\theta - 3\sin 2\theta - 4\cos\theta]_{7\pi/6}^{3\pi/2} \\ &= \frac{5}{3}\pi - \frac{\sqrt{3}}{2}. \end{aligned}$$

```
var('x, a')
f(x)= 12 *((sin(x))^2) + 4*sin(x) -1
print integral(f(x), x, 7*pi/6, 3*pi/2)      # 5/3 π − √3/2
```

■

Because a point may be represented in different ways in polar coordinates, care must be taken in determining the points of intersection of two polar graphs. For example, consider the points of intersection of the graphs of $r = 1 - 2\cos\theta$ and $r = 1$. As with cartesian equations, let us solve two equation simultaneously to find the points of intersection. Then we would obtain the following.

$$1 = 1 - 2\cos\theta \Rightarrow \cos\theta = 0 \Rightarrow \theta = \frac{\pi}{2},\ \frac{3\pi}{2}$$

If you sketch the graphs, you will see that there is a third point of intersection that did not show up when we solved two polar equations simultaneously. This is one reason we stress sketching a graph when finding the area of a polar region. Because it does not occur with the same coordinates in the two graphs, we did not find the third point. The point which occurs with the coordinates $(1, \pi)$ on the graph of $r = 1$ is the same point which occurs with the coordinates $(-1, 0)$ on the graph of $r = 1 - 2\cos\theta$.

We can compare the problem of finding points of intersection of two polar graphs with that of finding collision points of two satellites in intersecting orbits about the earth. The satellites will not collide because they reach the points of intersection at different times (θ values). A collision will occur only at the points of intersection are a "simultaneous point"-those reached at the same time (θ value).

EXAMPLE 4

Find the points of intersection of the cardioid $r=1+\sin\theta$ and the cardioid $r=1-\cos\theta$.

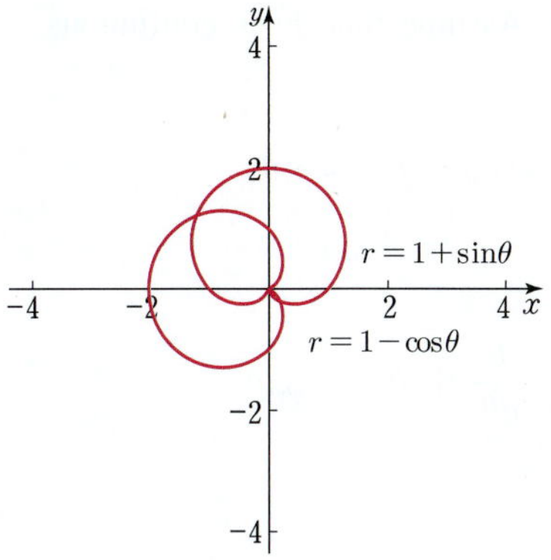

Figure 8

Solution By solving the equations $r=1+\sin\theta$ and $r=1-\cos\theta$, we get $\sin\theta+\cos\theta=0$ (See Figure 8.) and, therefore,

$$\sin\left(\theta+\frac{\pi}{4}\right)=0. \text{ This gives us } \theta+\frac{\pi}{4}=0 \text{ or } \pi.$$

```
var('x')
f(x)= sin (x + pi/4)
print solve (f == 0, x)
plot(f(x), x, 0, 2*pi)                    # θ = 3/4 π, 7/4 π
```

So the values of θ in $(0, 2\pi)$ satisfying both equations are $\theta=\frac{3}{4}\pi, \frac{7}{4}\pi$. Thus the two points of intersection are $\left(\frac{3}{2}, \frac{3}{4}\pi\right)$, $\left(\frac{1}{2}, \frac{7}{4}\pi\right)$. From Figure 8, we can see that the pole also is a point of intersection. ■

CAS **EXAMPLE 5**

Find the area of the region that is bounded by the given curve and lies in the specified sector.

$$r=\theta^2,\ 0\le\theta\le\pi$$

Solution

```
var('theta')
polar_plot(theta^2, (0, pi), fill=True).show(aspect_ratio=1,
xmin=-10, xmax=3, ymin=-2, ymax=5)
```

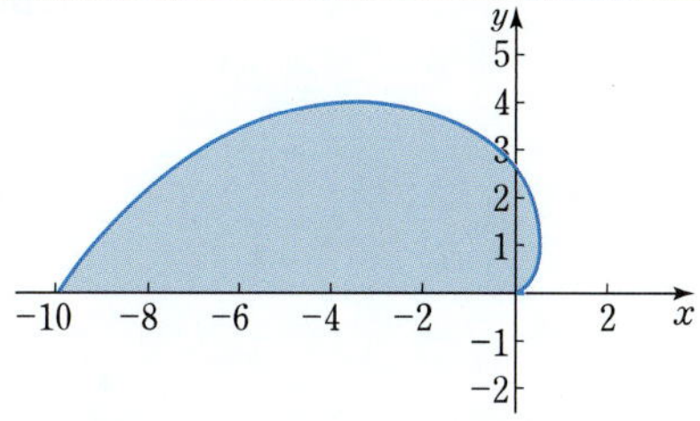
Figure 9

```
r=theta^2
integral(1/2*r^2, theta, 0, pi)
```

Answer : 1/10*pi^5 ■

Arc Length

The formula for the length of a polar arc can be obtained from arc length formula for a curve described by parametric equations. Consider θ as a parameter of a polar curve $r = f(\theta)$, $a \le \theta \le b$. Assume that f' is continuous. Then the parametric equations of the curve are

$$x = r\cos\theta = f(\theta)\cos\theta \quad \text{and} \quad y = r\sin\theta = f(\theta)\sin\theta.$$

Differentiating with respect to θ, we have

$$\frac{dx}{d\theta} = \frac{dr}{d\theta}\cos\theta - r\sin\theta \quad \text{and} \quad \frac{dy}{d\theta} = \frac{dr}{d\theta}\sin\theta + r\cos\theta.$$

Then

$$\begin{aligned}\left(\frac{dx}{d\theta}\right)^2 + \left(\frac{dy}{d\theta}\right)^2 &= \left(\frac{dr}{d\theta}\right)^2\cos^2\theta - 2r\frac{dr}{d\theta}\cos\theta\sin\theta + r^2\sin^2\theta \\ &\quad + \left(\frac{dr}{d\theta}\right)^2\sin^2\theta + 2r\frac{dr}{d\theta}\sin\theta\cos\theta + r^2\cos^2\theta \\ &= \left(\frac{dr}{d\theta}\right)^2 + r^2.\end{aligned}$$

Thus the arc length $L = \int_a^b \sqrt{\left(\frac{dx}{d\theta}\right)^2 + \left(\frac{dy}{d\theta}\right)^2}\, d\theta$ of the polar curve is

5
$$L = \int_a^b \sqrt{r^2 + \left(\frac{dr}{d\theta}\right)^2}\, d\theta.$$

EXAMPLE 6

Find the length of the cardioid $r = 1 + \cos\theta$.

Solution The full length of the cardioid (See Figure 10.) is given by the parameter interval $0 \le \theta \le 2\pi$. Using 5 we have

$$\begin{aligned}L &= \int_0^{2\pi} \sqrt{r^2 + \left(\frac{dr}{d\theta}\right)^2}\, d\theta = \int_0^{2\pi} \sqrt{(1+\cos\theta)^2 + (-\sin\theta)^2}\, d\theta \\ &= \int_0^{2\pi} \sqrt{2(1+\cos\theta)}\, d\theta = \int_0^{2\pi} \sqrt{4\cos^2\left(\frac{\theta}{2}\right)}\, d\theta,\end{aligned}$$

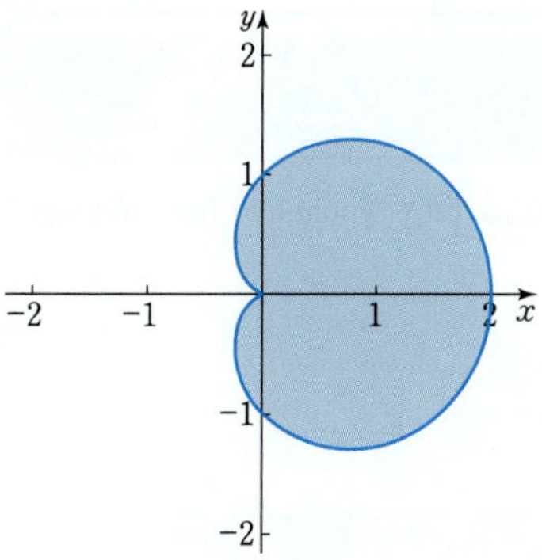

Figure 10 $r = 1 + \cos\theta$

$$= \int_0^{2\pi} \left| 2\cos\left(\frac{\theta}{2}\right) \right| d\theta \;= 2\int_0^{\pi} 2\cos\left(\frac{\theta}{2}\right) d\theta$$

$$= 8\left[\sin\left(\frac{\theta}{2}\right)\right]_0^{\pi} = 8.$$

```
var('x, t')
x= 2*(cos(t/2))
2*integral(x, t, 0, pi)
```

■

CAS **EXAMPLE 7**

Find the length of the curve.

$$r = \cos\theta, \; 0 \le \theta \le \frac{\pi}{3}$$

http://matrix.skku.ac.kr/cal-lab/cal-8-1-exm-4.html

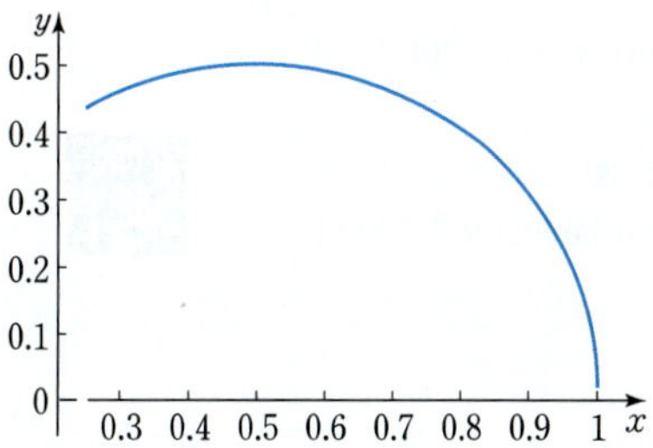

Figure 11

Solution

```
var('theta')
r=cos(theta)
polar_plot(r, (theta, 0, pi/3))
```

```
dr=diff(r, theta)
integral(sqrt(r^2+dr^2), theta, 0, pi/3)
```

Answer : 1/3*pi ■

CAS **EXAMPLE 8**

Find the approximated length of the curve $r = 1 - \cos\theta$, $0 \le \theta \le \frac{\pi}{3}$ with numerical approximation.

Solution

```
var('theta')
r= 1 - cos(theta)
dr=diff(r, theta)
h = integral(sqrt(r^2+dr^2), theta, 0, pi/3); h
h.n()
#print numerical_integral(sqrt(r^2+dr^2), theta, 0, pi/3)[0]
```

Answer : 0.535898384862246(-2*sqrt(3) + 4) ■

10.4 EXERCISES (Areas and Lengths in Polar Coordinates)

http://matrix.skku.ac.kr/Cal-Book/part1/CS-Sec-10-4-Sol.html

YouTube http://youtu.be/LRmasW9uqYY

1-4. Find the area of the region that is bounded by the given curve and lies in the specified sector.

CAS **1.** $r = \theta,\ 0 \le \theta \le \pi/2$

http://matrix.skku.ac.kr/cal-lab/cal-9-4-1.html

Solution

```
var('theta')
polar_plot(theta, (0, pi/2), fill=True).show(aspect_ratio=1, xmin=-3, xmax=3, ymin=-3, ymax=3)
```

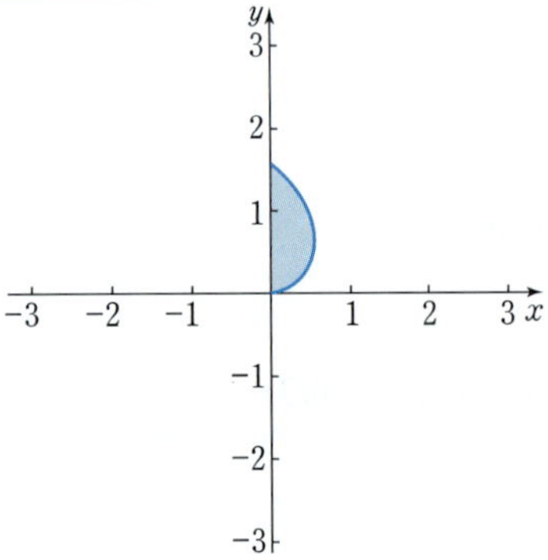


```
r=theta;
A=integral(1/2*r^2,theta,0,pi/2);
A
```

Answer : 1/48*pi^3

CAS **2.** $r = e^{\theta/3},\ \pi \le \theta \le 2\pi$

http://matrix.skku.ac.kr/cal-lab/cal-9-4-2.html

Solution $A = \displaystyle\int_{\pi}^{2\pi} \frac{1}{2}\left(e^{\frac{\theta}{3}}\right)^2 d\theta = \int_{\pi}^{2\pi} \frac{1}{2} e^{\frac{2\theta}{3}}\, d\theta$

$= \displaystyle\frac{1}{2}\left[\frac{3}{2}e^{\frac{2\theta}{3}}\right] = \frac{3}{4}\left(e^{\frac{4\pi}{3}} - e^{\frac{2\pi}{3}}\right)$

```
var('theta')
polar_plot(e^(theta/3), (pi, 2*pi), fill=True).show(aspect_ratio=1, xmin=-5, xmax=10, ymin=-5, ymax=5)
```

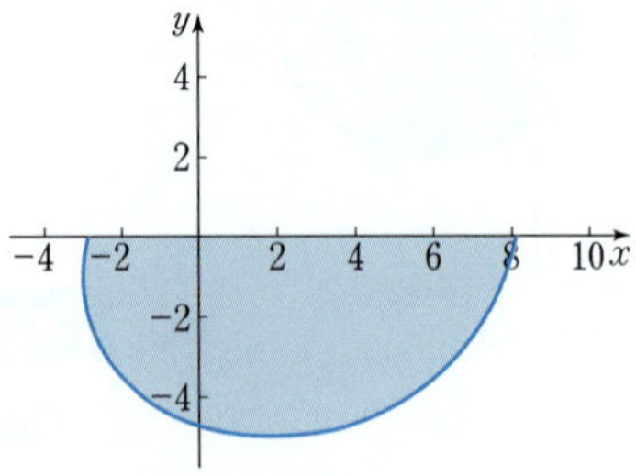


```
r=e^(theta/3);
A=integral(1/2*r^2,theta,pi,2*pi);
A
```

Answer : -3/4*e^(2/3*pi) + 3/4*e^(4/3*pi)

CAS **3.** $r = \sqrt{\theta},\ 0 \le \theta \le \pi/3$

http://matrix.skku.ac.kr/cal-lab/cal-9-4-3.html

Solution

```
var('theta')
polar_plot(sqrt(theta), (0, pi/3), fill=True).show(aspect_ratio=1, xmin=-2, xmax=2, ymin=-2, ymax=2)
```

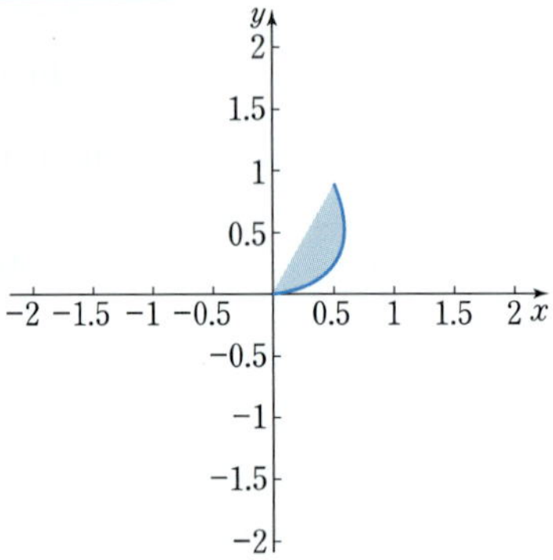


```
r=sqrt(theta);
A=integral(1/2*r^2,theta,0,pi/3);
A
```

Answer: 1/36*pi^2

4. $r = 3 + 2\sin\theta,\ 0 \le \theta \le 2\pi$

CAS **5-9.** Find the area bounded by one loop of the given curve.

5. $r = 2\cos 2\theta$

http://matrix.skku.ac.kr/cal-lab/cal-9-4-5.html

Solution

```
var('theta')
polar_plot(2*cos(2*theta), (0, 2*pi), fill=True).show(aspect_ratio=1, xmin=-3, xmax=3, ymin=-3, ymax=3)
```

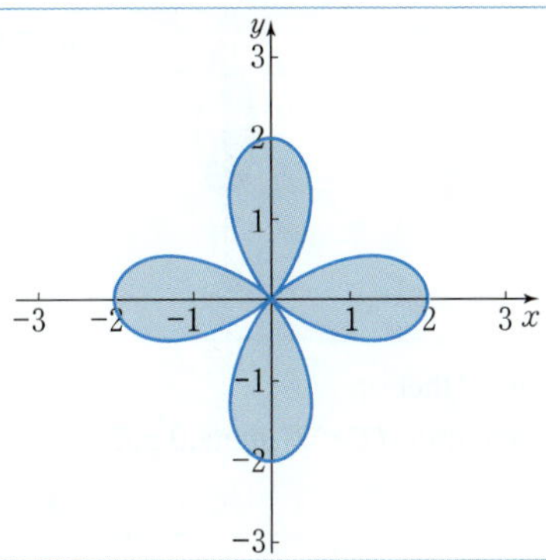

```
r=2*cos(2*theta);
A=integral(1/2*r^2,theta,pi/4,3*pi/4);
A
```

Answer : 1/2*pi

6. $r = 3\sin 3\theta$

http://matrix.skku.ac.kr/cal-lab/cal-9-4-6.html

Solution $A = \int_0^{\frac{\pi}{3}} \frac{1}{2}(3\sin 3\theta)^2 d\theta = \int_0^{\frac{\pi}{3}} \frac{9}{2}\sin^2 3\theta d\theta$

$= \frac{9}{2}\int_0^{\frac{\pi}{3}} \frac{1-2\cos 3\theta}{2} d\theta = \frac{9}{4}\left[\theta - \frac{2}{3}\sin 3\theta\right]$

```
var('theta')
polar_plot(3*sin(3*theta), (0, pi), fill=True).show(aspect_ratio=1, xmin=-3, xmax=3, ymin=-3, ymax=3)
```

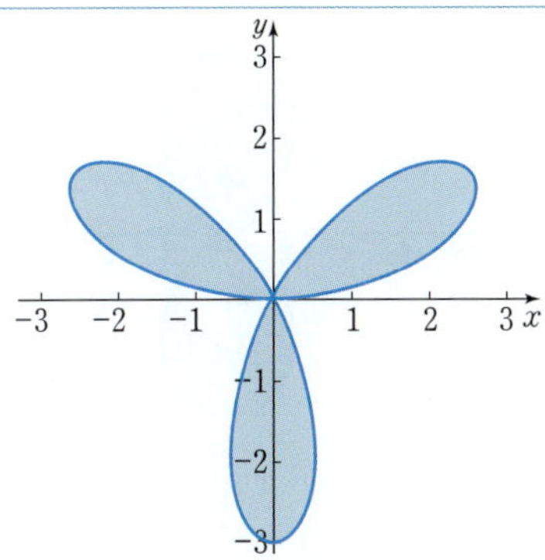

(3 leaves)

```
r=3*sin(3*theta);
A=integral(1/2*r^2,theta,0,pi/3);
A
```

Answer : 3/4*pi

7. $r = 2\cos 4\theta$

http://matrix.skku.ac.kr/cal-lab/cal-9-4-7.html

Solution

```
var('theta')
polar_plot(2*cos(4*theta), (0, 2*pi), fill=True).show(aspect_ratio=1, xmin=-3, xmax=3, ymin=-3, ymax=3)
```

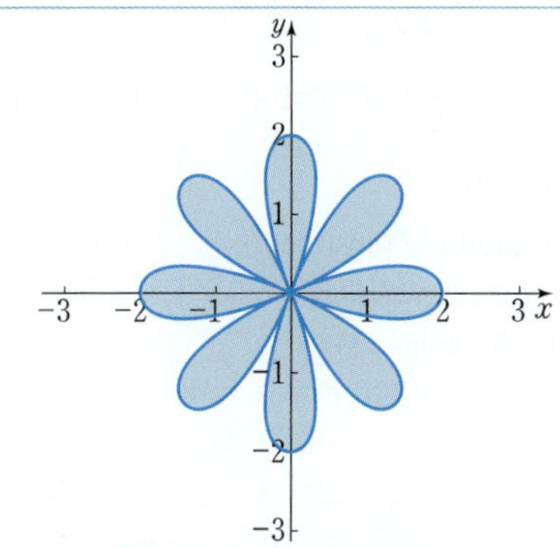

(8 leaves symmetric, about y-axis)

```
r=2*cos(4*theta);
A=integral(1/2*r^2,theta,pi/8,3*pi/8);
A
```

Answer : 1/4*pi

8. $r = 2 + \sin 2\theta$

http://matrix.skku.ac.kr/cal-lab/cal-9-4-8.html

Solution

```
var('theta')
polar_plot(2+sin(2*theta), (0, 2*pi), fill=True).show(aspect_ratio=1, xmin=-3, xmax=3, ymin=-3, ymax=3)
```

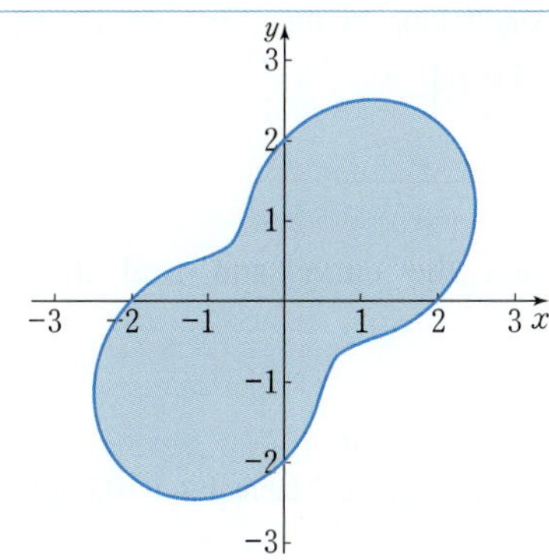

```
r=2+sin(2*theta);
A=integral(1/2*r^2,theta,7*pi/6,11*pi/6);
A
```

Answer : 3/2*pi + 1/16*sqrt(3)

9. Find the area between a large loop and the enclosed small loop of the curve $r=1+2\cos\theta$.
http://matrix.skku.ac.kr/cal-lab/cal-9-4-9.html

Solution $A=\int_0^{2\pi}\frac{1}{2}r^2d\theta=\int_0^{2\pi}\frac{1}{2}(1+2\cos\theta)^2d\theta$

$$=\frac{1}{2}\int_0^{2\pi}(4\cos^2\theta+4\cos\theta+1)d\theta$$

$$=\frac{1}{2}\int_0^{2\pi}(2\cos 2\theta+4\cos\theta+3)d\theta s$$

$$=\frac{1}{2}[3\theta+4\sin\theta+\sin 2\theta]=\frac{1}{2}(6\pi)=3\pi$$

```
var('theta')
polar_plot(1+2*cos(theta), (0, 2*pi), fill=True).show(aspect_ratio=1, xmin=-3, xmax=3, ymin=-3, ymax=3)
```

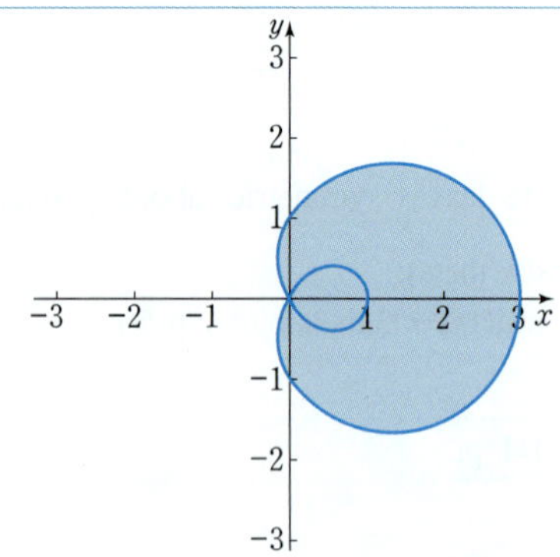


```
r=1+2*cos(theta);
B=integral(1/2*r^2,theta,0,2*pi);
B
```

Answer : 3*pi # 3π.

From figure, the area between a large loop and the enclosed small loop of the curve $r=1+2\cos\theta$ is $3\pi-\left(\pi-\frac{3\sqrt{3}}{2}\right)=2\pi+\frac{3\sqrt{3}}{2}$.

10-12. Sketch the curve and find the area that it encloses.

CAS **10.** $r=2\sin 3\theta,\ 0\le\theta\le\pi$
http://matrix.skku.ac.kr/cal-lab/cal-9-4-10.html

Solution

```
var('theta')
polar_plot(2*sin(3*theta), (0, pi), fill=True).show(aspect_ratio=1, xmin=-3, xmax=3, ymin=-3, ymax=3)
```

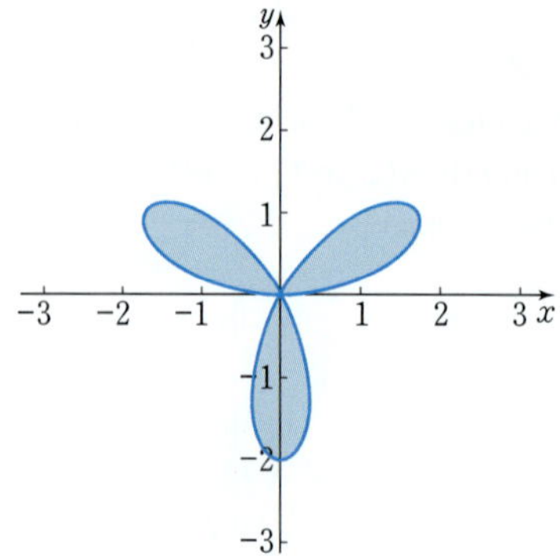


```
r=2*sin(3*theta);
A=3*integral(1/2*r^2 ,theta,0,pi/3);
A
```

Answer : pi

CAS **11.** $r=3(1-\cos 2\theta)$
http://matrix.skku.ac.kr/cal-lab/cal-9-4-11.html

Solution $A=\int_0^{2\pi}\frac{1}{2}r^2d\theta=\int_0^{2\pi}\frac{1}{2}[3(1-\cos 2\theta)]^2d\theta$

$$=\frac{9}{2}\int_0^{2\pi}(1-2\cos 2\theta+\cos^2 2\theta)d\theta$$

$$=\frac{9}{2}\int_0^{2\pi}(\frac{3}{2}-2\cos 2\theta+\frac{1}{2}\cos 4\theta)d\theta$$

$$=\frac{9}{2}\left[\frac{3}{2}\theta-\sin 2\theta+\frac{1}{8}sin4\theta\right]=\frac{9}{2}(3\pi)=\frac{27}{2}\pi$$

```
var('theta')
polar_plot(3*(1-cos(2*theta)), (0, 2*pi), fill=True).show(aspect_ratio=1, xmin=-8, xmax=8, ymin=-8, ymax=8)
```

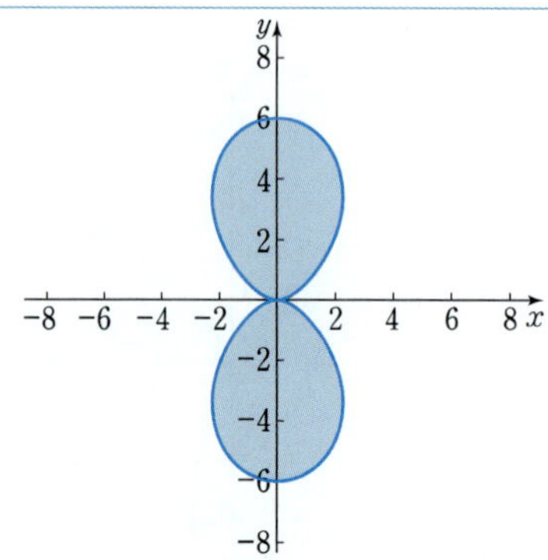

```
r=3*(1-cos(2*theta));
A=integral(1/2*r^2,theta,0,2*pi);
A
```

Answer : 27/2*pi

12. $r = 2\sin\theta + 2\sin 9\theta$

13-15. Find the area of the region that lies inside the first curve and outside the second curve.

13. $r = 2\sin\theta,\ r = 1$

http://matrix.skku.ac.kr/cal-lab/cal-9-4-13.html

Solution $r = 2\sin\theta = 1 \Rightarrow \theta = \dfrac{\pi}{6},\ \dfrac{5\pi}{6}$

$$A = 2\int_{\frac{\pi}{6}}^{\frac{\pi}{2}} \frac{1}{2}(r^2 - 1^2)\,d\theta = 2\int_{\frac{\pi}{6}}^{\frac{\pi}{2}} \frac{1}{2}[(2\sin\theta)^2 - 1]\,d\theta$$

$$= \int_{\frac{\pi}{6}}^{\frac{\pi}{2}} (4\sin^2\theta - 1)\,d\theta = \int_{\frac{\pi}{6}}^{\frac{\pi}{2}} (1 - 2\cos 2\theta)\,d\theta = [\theta - \sin 2\theta]$$

$$= \left(\frac{\pi}{2}\right) - \left(\frac{\pi}{6} - \frac{\sqrt{3}}{2}\right) = \frac{\pi}{3} + \frac{\sqrt{3}}{2}$$

```
var('theta')
p1=polar_plot(2*sin(theta), (0, 2*pi), fill=True).show(aspect_ratio=1, xmin=-2, xmax=2, ymin=-2, ymax=2)
p2=polar_plot(1, (0, 2*pi), fill=True).show(aspect_ratio=1, xmin=-2, xmax=2, ymin=-2, ymax=2)
```

```
r=2*sin(theta)
A=2*integral(1/2*(r^2-1), theta, pi/6, pi/2);
A
```

Answer : 1/3*pi + 1/2*sqrt(3)

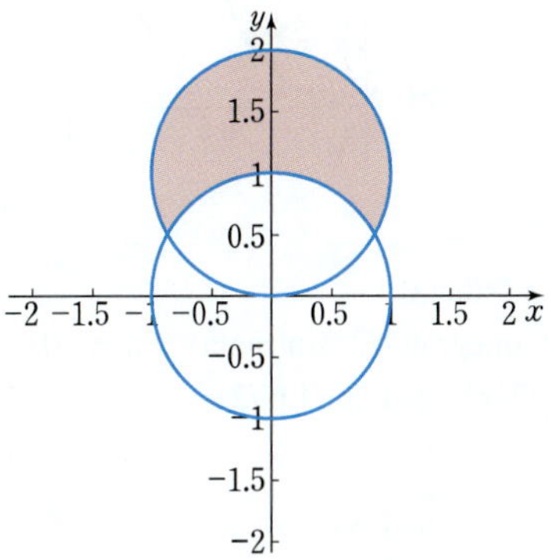

CAS 14. $r = 3\sin\theta,\ r = 1 + \cos\theta$

http://matrix.skku.ac.kr/cal-lab/cal-9-4-14.html

Solution

```
var('theta')
p1=polar_plot(3*sin(theta),(theta,0,2*pi), color="red", fill=True, fillcolor="red")
p2=polar_plot(1+cos(theta),(theta,0,2*pi), fill=True, fillcolor="blue")
show(p1+p2, aspect_ratio=1, ymin=-5, ymax=5, xmin=-5, xmax=5)
```

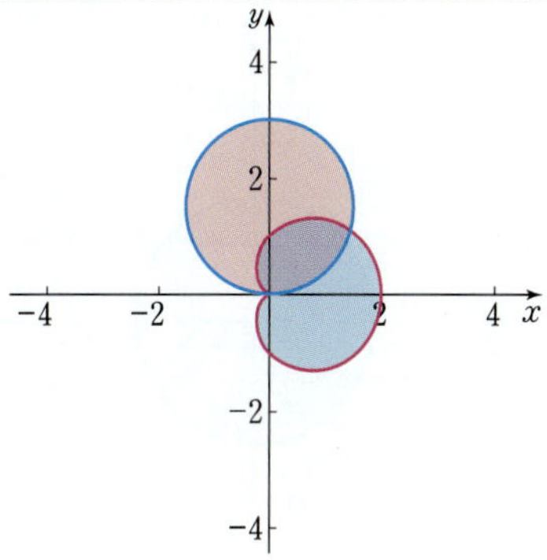


```
A=2*integral(1/2*((3*cos(theta))^2-(1+cos(theta))^2),theta,0,pi/3);
A
```

Answer : pi

CAS 15. $r^2 = 2\cos 2\theta,\ r = 1$

http://matrix.skku.ac.kr/cal-lab/cal-9-4-15.html

```
var('theta')
p1=polar_plot(sqrt(2*cos(2*(theta))), (- pi/4, pi/4), fill=True)
p2=polar_plot(sqrt(-2*cos(2*(theta))), ( pi/4, (3*pi)/4), fill=True)
p3=polar_plot(sqrt(-2*cos(2*(theta))), ( -3*pi/4, -pi/4), fill=True)
p4=polar_plot(sqrt(2*cos(2*(theta))), (3*pi/4, 5*pi/4), fill=True)
p5=polar_plot(1,(0, 2*pi), fill=True)
show(p1+p2+p3+p4+p5,aspect_ratio=1, xmin=-2, xmax=2, ymin=-2, ymax=2)
```

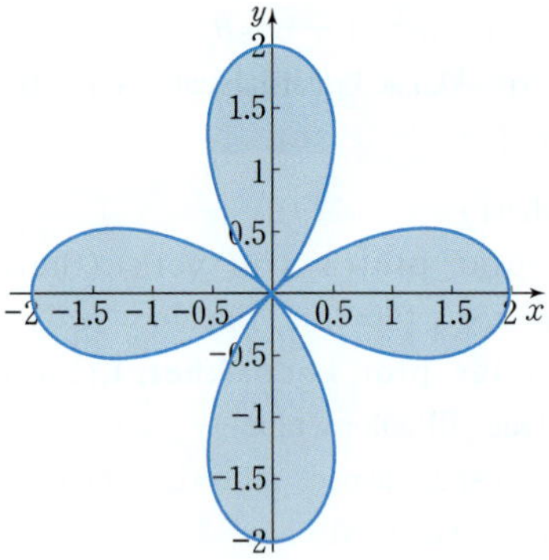

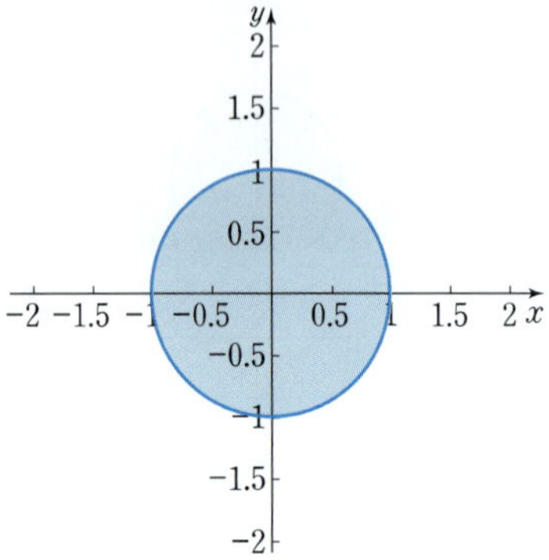

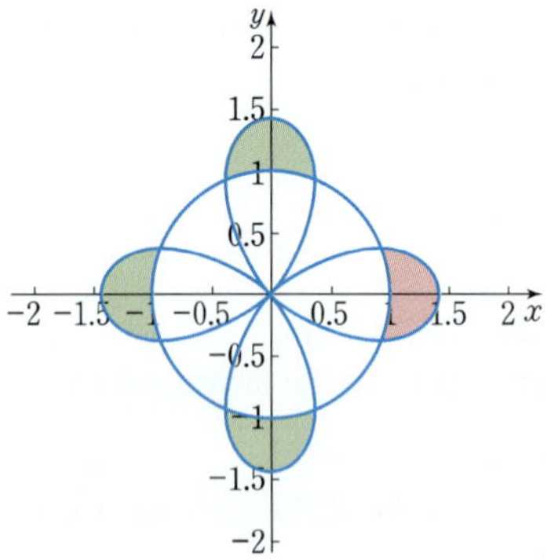


```
A=4*integral(1/2*(2*cos(2*theta)-1), theta, 2*pi/6, 4*pi/6);
A
```

Answer : -2/3*pi − 2*sqrt(3) # $2\sqrt{3}-\dfrac{2\pi}{3}$

16-17. Find all points of intersection of the given curves and find the area of the region that lies inside both curves.

16. $r=\sin 3\theta,\ r=\sin\theta$

http://matrix.skku.ac.kr/cal-lab/cal-9-4-16.html

Solution $\sin 3\theta=\sin\theta=3\sin\theta-4\sin^3\theta$

$4\sin^3\theta-2\sin\theta=0\quad 2\sin\theta(2\sin^2\theta-1)$

$\sin\theta=0$ or $\pm\dfrac{1}{2}$

$\theta=0,\ \pi,\ \dfrac{1}{6}\pi,\ \dfrac{5}{6}\pi,\ \dfrac{7}{6}\pi,\ \dfrac{11}{6}\pi$

$$A=2\int_0^{\frac{\pi}{4}}\frac{1}{2}\sin^2\theta d\theta+2\int_{\frac{\pi}{4}}^{\frac{\pi}{2}}\frac{1}{2}(\sin 3\theta)^2 d\theta$$

$$=\int_0^{\frac{\pi}{4}}\left(\frac{1}{2}-\frac{1}{2}\cos 2\theta\right)d\theta+\int_{\frac{\pi}{4}}^{\frac{\pi}{2}}\left(\frac{1}{2}-\frac{1}{2}\cos 6\theta\right)d\theta$$

$$=\left[\frac{1}{2}\theta-\frac{1}{4}\sin 2\theta\right]+\left[\frac{1}{2}\theta-\frac{1}{12}\sin 6\theta\right]$$

$$=\left(\frac{\pi}{8}-\frac{1}{4}\right)+\left[\frac{\pi}{4}-\left(\frac{\pi}{8}+\frac{1}{12}\right)\right]=\frac{\pi}{4}-\frac{1}{3}$$

```
var('theta')
p1=polar_plot(sin(3*(theta)), (0, pi), fill=True).show(aspect_ratio=1, xmin=-2, xmax=2, ymin=-2, ymax=2)
p2=polar_plot(sin(theta),(0, pi), fill=True).show(aspect_ratio=1, xmin=-2, xmax=2, ymin=-2, ymax=2)
```

The two curves intersect at $\theta=\pi/4$, and $3\pi/4$.

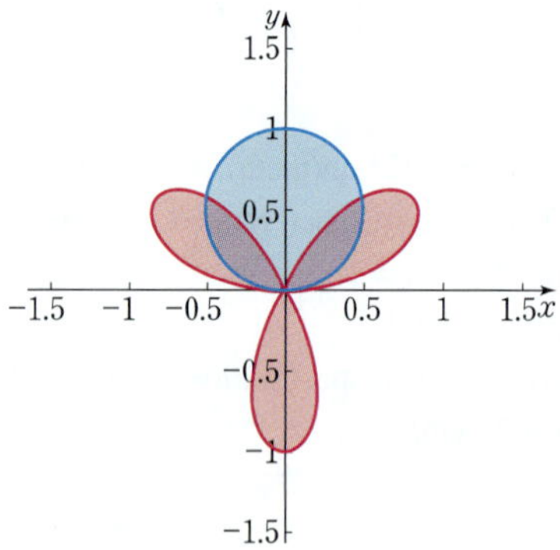

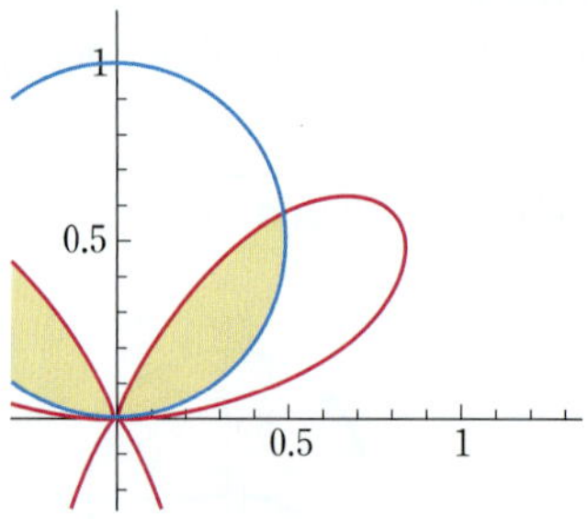

```
r=sin(3*theta);
A=2*integral(1/2*sin(theta)^2,theta,0,pi/4)+2*integral(1/2*r^2,theta,pi/4,pi/2);
A
```

Answer : 1/4*pi-1/3

CAS **17.** $r^2 = \cos 2\theta$, $r^2 = 4\sin 2\theta$

http://matrix.skku.ac.kr/cal-lab/cal-9-4-17.html

Solution

```
var('theta')
p1=polar_plot(cos(2*(theta)), (0, 2*pi), fill=True).show(aspect_ratio=1, xmin=-2, xmax=2, ymin=-2, ymax=2)
p2=polar_plot(4*sin(2*(theta)),(0, 2*pi), fill=True).show(aspect_ratio=1, xmin=-4, xmax=4, ymin=-4, ymax=4)
```

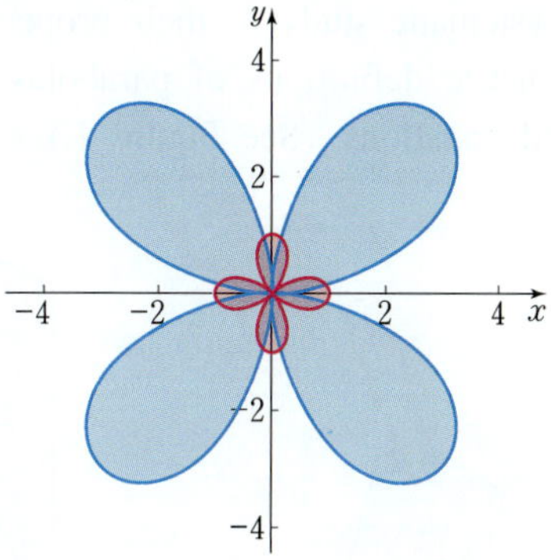

```
A=8*2*integral(1/2*sin(2*theta), theta, 0, pi/8);
A
```

Answer : -2*sqrt(2) + 4

18-21. Find the length of the polar curve.

18. $r = \theta^2$, $0 \le \theta \le \dfrac{3}{2}\pi$

19. $r = 2\theta$, $0 \le \theta \le \pi$

CAS **20.** Graph the curve $r = \cos^2\left(\dfrac{\theta}{3}\right)$ $0 \le \theta \le 6\pi$ and find its length.

http://matrix.skku.ac.kr/cal-lab/cal-9-4-20.html

Solution

$$L = 4\int_0^{\frac{3\pi}{2}} \sqrt{\left\{\cos^2\left(\frac{\theta}{3}\right)^2\right\}^2 + \left\{-\frac{2}{3}\cos\left(\frac{\theta}{3}\right)\sin\left(\frac{\theta}{3}\right)\right\}^2}\, d\theta$$

$$= 4\int_0^{\frac{3\pi}{2}} \cos\left(\frac{\theta}{3}\right)\sqrt{\cos^2\left(\frac{\theta}{3}\right) + \frac{4}{9}\sin^2\left(\frac{\theta}{3}\right)}\, d\theta$$

$$= \frac{18\sqrt{5}}{3}\sin^{-1}\left(\frac{\sqrt{5}}{3}\right) + 4$$

```
var('theta')
p1=polar_plot((cos(theta/3))^2, (0, 6*pi), fill=True).show(aspect_ratio=1, xmin=-2, xmax=2, ymin=-2, ymax=2)
```

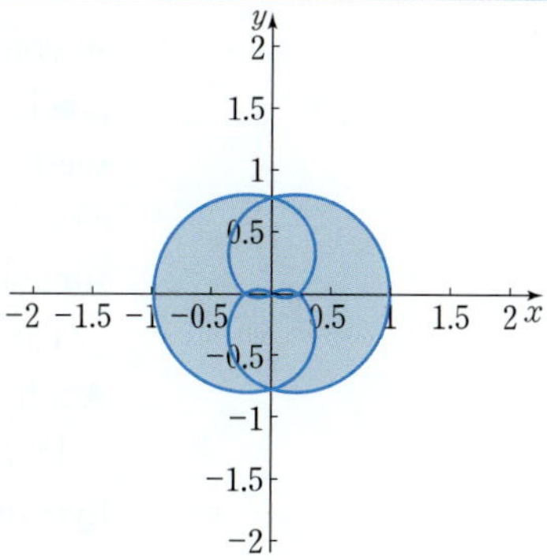

```
var('r')
r=(cos(theta/3))^2;
drd(theta)=diff(r,theta);
drd(theta)
```

-2/3*sin(1/3*theta)*cos(1/3*theta)

```
L=4*integral(cos(theta/3)*sqrt(cos(theta/3)^2+4/9*sin(theta/3)^2), theta, 0, 3*pi/2);
L
```

Answer : 18/3*sqrt(5)*arcsin(1/3*sqrt(5))+4

21. Show that the area of the surface generated by rotating the polar curve $r = f(\theta)$ $a \le \theta \le b$ with $0 \le a < b \le \pi$ about the polar axis is

$$S = \int_a^b 2\pi r\sin\theta\sqrt{r^2 + \left(\frac{dr}{d\theta}\right)^2}\, d\theta.$$

Solution From http://mathworld.wolfram.com/SurfaceofRevolution.html, we have the formula

$$S = \int_\alpha^\beta 2\pi y\sqrt{\left(\frac{dx}{dt}\right)^2 + \left(\frac{dy}{dt}\right)^2}\, dt.$$

Since $x = r\cos\theta = f(\theta)\cos\theta$, $y = r\sin\theta = f(\theta)\sin\theta$,

$$S = \int_\alpha^\beta 2\pi y\sqrt{\left(\frac{dx}{dt}\right)^2 + \left(\frac{dy}{dt}\right)^2}\, dt$$

$$= \int_a^b 2\pi r\sin\theta\sqrt{f(\theta)^2 + f'(\theta)^2}\, d\theta$$

$$= \int_a^b 2\pi r\sin\theta\sqrt{r^2 + \left(\frac{dr}{d\theta}\right)^2}\, d\theta.$$

10.5 Conic Sections

http://matrix.skku.ac.kr/2014-Album/Quadratic-form/index.htm

A conic section (or a conic) is a curve obtained by intersecting a cone (more precisely, a right circular conical surface) with a plane. The three types of **conic sections** are the hyperbola, the parabola, and the ellipse. The circle is a special case of the ellipse, and is of sufficient interest in its own right that it is sometimes called the fourth type of conic section.

The conic sections were named and studied as long ago as 200 BC, when Apollonius of Perga undertook a systematic study of their properties.

In this section we give the geometric definitions of parabolas, ellipses, and hyperbolas and derive their standard equations. (See Figure 1.)

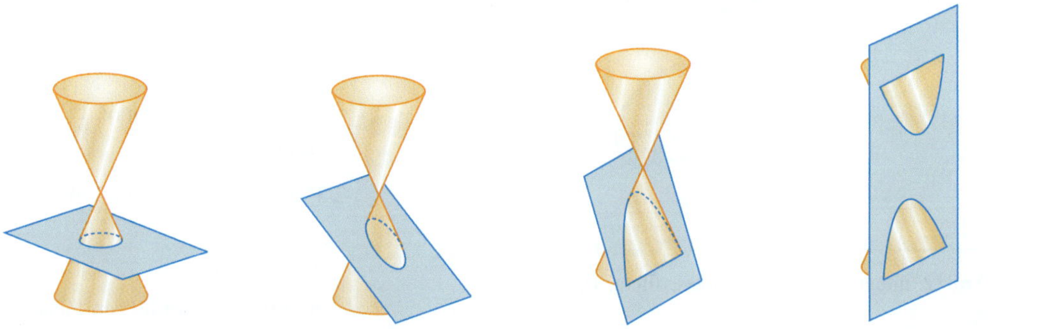

Figure 1 conics

Parabolas

The **parabola** is the set of all points (x, y) in a plane that are equidistant from a fixed point F (called the **focus**) and a fixed line (called the **directrix**). The focus does not lie on the directrix. The midpoint between the focus and the directrix is called the **vertex** and the line passing through the focus and the vertex is called the **axis** of the parabola.

In the 16th century Galileo showed that the path of a projectile that is shot into the air at an angle to the ground is a parabola. The parabola has many important applications, from automobile headlight reflectors to the design of ballistic missiles. They are frequently used in physics, engineering, and many other areas.

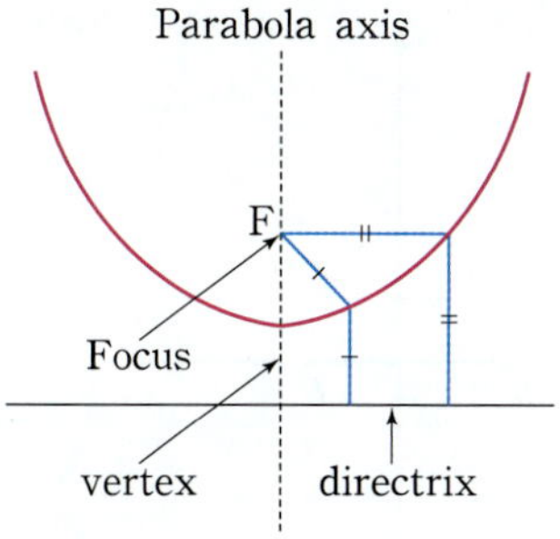

Figure 2

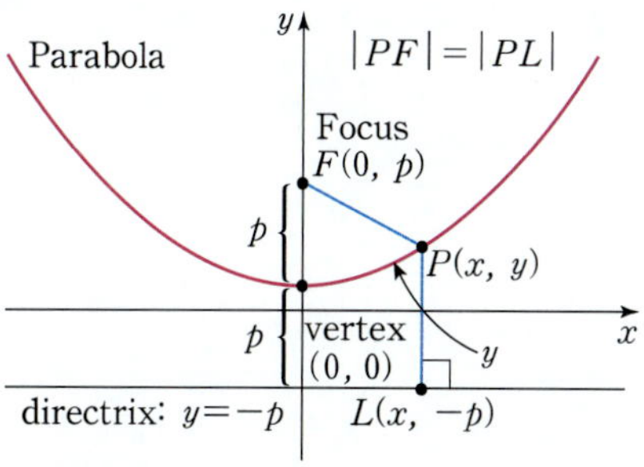

Figure 3

When vertex is at the origin O and its directrix is parallel to the x-axis (See Figure 3.), we obtain a particularly simple equation for a parabola. Suppose that the focus is the point $(0, p)$. Then the equation of the directrix is $y=-p$. For any point $P(x, y)$ on the parabola, the distance from P to the focus is

$$|PF| = \sqrt{x^2+(y-p)^2}$$

and the distance from P to the directrix is $|y+p|$. ($p>0$ in Figure 3.) According to the definition of a parabola, these distances are equal:

$$\sqrt{x^2+(y-p)^2} = |y+p|.$$

By squaring and simplifying the above equation we get

$$x^2+(y-p)^2 = |y+p|^2 = (y+p)^2$$
$$x^2+y^2-2py+p^2 = y^2+2py+p^2$$
$$x^2 = 4py.$$

Thus the equation of the parabola with focus $(0, p)$ and directrix $y=-p$ is

1 $$x^2 = 4py.$$

The standard equation of a parabola 1 is $y=ax^2$, where $a=1/(4p)$. The parabola opens upward if $p>0$ but downward if $p<0$.(See Figures 4(a) and 4(b).) Since 1 is a even function of x, the graph is symmetric with respect to the y-axis, which is the axis of the parabola.

An equation of the parabola with focus $(p, 0)$ and directrix $x=-p$ is

2 $$y^2 = 4px$$

which is obtained by interchanging x and y in 1.

The parabola 2 opens to the right if $p>0$ and to the left if $p<0$ (See Figures 4(c) and 4(d).). In both cases the graph is symmetric with respect to the x-axis.

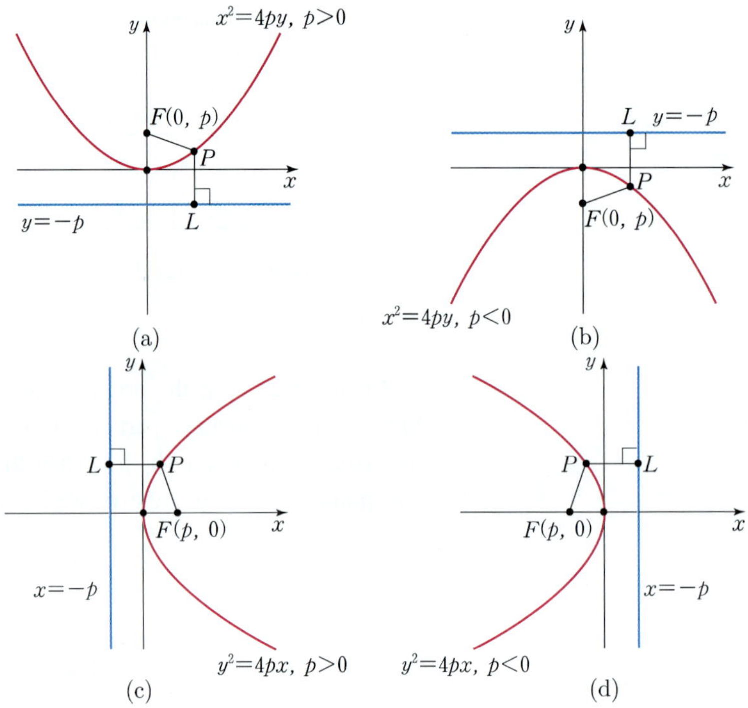

Figure 4

EXAMPLE 1

Draw the graph of the parabola $y^2+6x=0$ and find the focus and directrix.

Solution If we write the equation as $y^2=-6x$ and compare it with 2, we see that $4p=-6$, so $p=-\frac{3}{2}$. Thus, the focus is $(p,\ 0)=\left(-\frac{3}{2},\ 0\right)$ and the directrix is $x=\frac{3}{2}$. The sketch is shown in Figure 5. ■

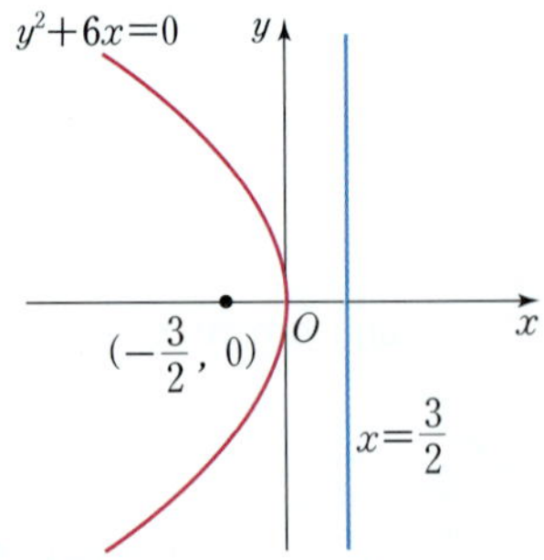

Figure 5

Ellipses

The ellipse was first studied by Menaechmus, investigated by Euclid, and named by Apollonius. The focus and conic section directrix of an ellipse were considered by Pappus. In 1602, Kepler believed that the orbit of Mars was oval; he later discovered that it was an ellipse with the Sun at one focus. In fact, Kepler introduced the word "focus" and published his discovery in 1609. In 1705, Halley showed that the comet now named after him moved in an elliptical orbit around the Sun. (MacTutor Archive.) An ellipse rotated about its minor axis gives an oblate spheroid, while an ellipse rotated about its major axis gives a prolate spheroid.

An is the set of points $(x,\ y)$ in a plane satisfying the sum of whose distances from two fixed points F_1 and F_2 is a constant. (See Figure 6.) These two fixed points are called the foci (plural of focus). One of Kepler's laws states that the orbits of the planets in the solar system are ellipses with the Sun at one focus. An interesting reflection property of the ellipses, similar to parabolas, is of practical use. For example, if a source of light or sound is placed at one focus of a surface with elliptical cross-sections, then all the light or sound is reflected off the surface to the other focus. In lithotripsy, a treatment for kidney stones, a reflector with elliptical cross-section is placed in such a way that the kidney stone is at one focus. High-intensity sound waves generated at the other focus are reflected to the stone to destroy it without damaging surrounding tissue.

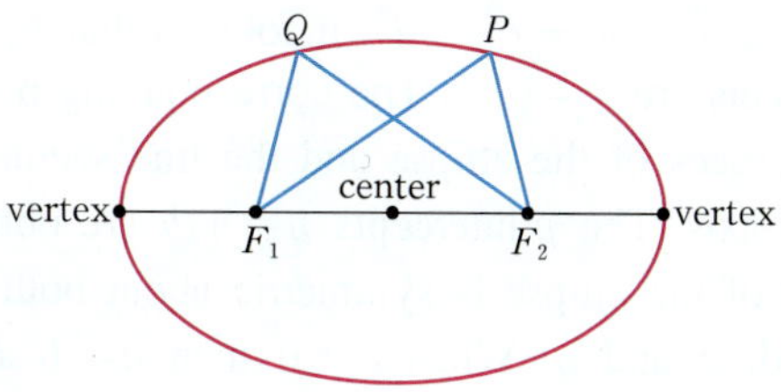

$|PF_1|+|PF_2|=|QF_1|+|QF_2|$
F_1 and F_2 are focuses
Figure 6

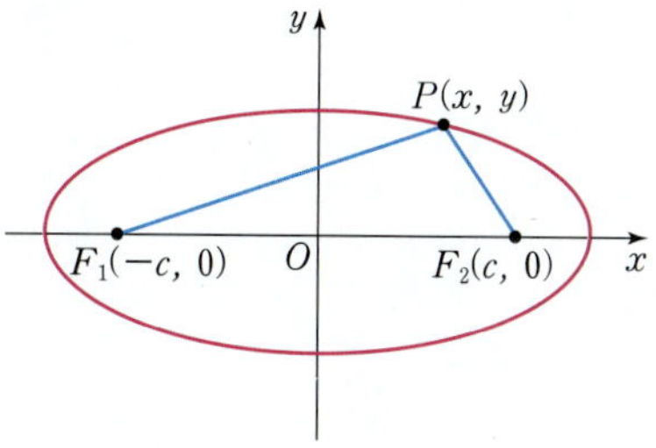

Figure 7

Suppose the foci are the points $(-c,\ 0)$ and $(c,\ 0)$ on the x-axis so that the origin is halfway between the foci. (See Figure 7.) Let $2a>0$ be the sum of the distances from a point on the ellipse to the foci. Let $P(x,\ y)$ be any point on the ellipse. Then

$$|PF_1|+|PF_2|=2a$$

That is

$$\sqrt{(x+c)^2+y^2}+\sqrt{(x-c)^2+y^2}=2a$$

or

$$\sqrt{(x-c)^2+y^2}=2a-\sqrt{(x+c)^2+y^2}\,.$$

By squaring on both sides, we get

$$x^2-2cx+c^2+y^2=4a^2-4a\sqrt{(x+c)^2+y^2}+x^2+2cx+c^2+y^2$$

Simplifying

$$a\sqrt{(x+c)^2+y^2}=a^2+cx\,.$$

By squaring again, we have

$$a^2(x^2+2cx+c^2+y^2)=a^4+2a^2cx+c^2x^2$$

Which is

$$(a^2-c^2)x^2+a^2y^2=a^2(a^2-c^2).$$

Note that $2a > 2c$. Since we have $2c < 2a$, so $c < a$ and, therefore, $a^2 - c^2 > 0$. Let $b^2 = a^2 - c^2$. Then the equation of the ellipse becomes

$$b^2x^2 + a^2y^2 = a^2b^2$$

or

3 $$\frac{x^2}{a^2} + \frac{y^2}{b^2} = 1.$$

Since $b^2 = a^2 - c^2 < a^2$, it follows that $b < a$. By putting $y = 0$ in 3, the x-intercepts are $x = \mp a$. The corresponding points $(a, 0)$ and $(-a, 0)$ are called the vertices of the ellipse and the line segment joining the vertices is called the major axis. The y-intercepts $y = \mp b$ are obtained by putting $x = 0$ in 3. The graph of the ellipse is symmetric about both x-axis and y-axis since 3 is even in both x and y. When $c = 0$ then $a = b$ and the foci coincide. Consequently the ellipse becomes a circle with radius $r = a = b$. Thus the ellipse

$$\frac{x^2}{a^2} + \frac{y^2}{b^2} = 1, \qquad a \geq b > 0$$

has foci at $(\pm c, 0)$, and vertices $(\pm a, 0)$. Here $c^2 = a^2 - b^2$. (See Figure 8.)

By interchanging x and y in 3, the ellipse with foci on the y-axis at $(0, \pm c)$, and vertices at $(0, \pm a)$ is

4 $$\frac{x^2}{b^2} + \frac{y^2}{a^2} = 1, \qquad a \geq b > 0.$$

Here $c^2 = a^2 - b^2$. (See Figure 9.)

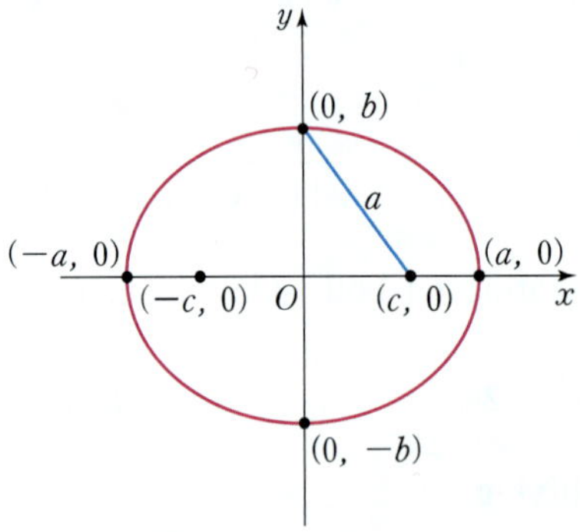

Major axis is horizontal
$\frac{x^2}{a^2} + \frac{y^2}{b^2} = 1, \; a \geq b > 0$

Figure 8

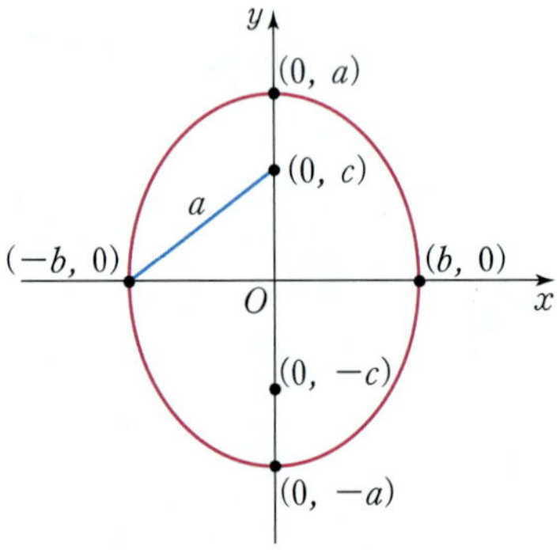

Major axis is vertical
$\frac{y^2}{a^2} + \frac{x^2}{b^2} = 1, \; b \geq a > 0$

Figure 9

EXAMPLE 2

Find the foci and sketch the graph of $4x^2 + 3y^2 = 12$.

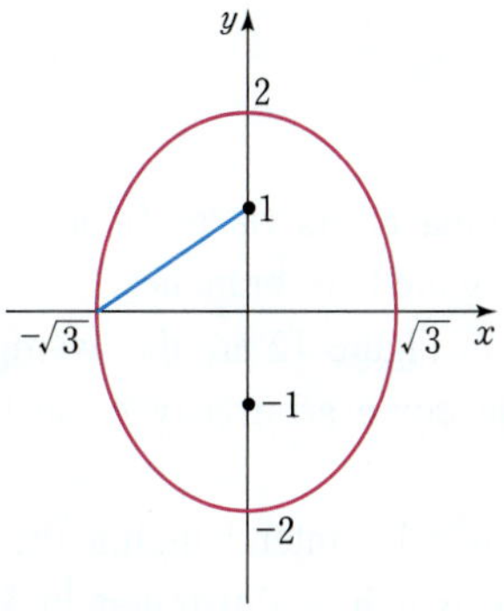

Figure 10

Solution The given equation becomes the standard form of an ellipse by dividing it by 12:

$$\frac{x^2}{3}+\frac{y^2}{4}=1$$

Thus $a^2=3$, $b^2=4$, $a=\sqrt{3}$, and $b=2$. Therefore the x-intercepts are $\pm\sqrt{3}$ and the y-intercepts are ± 2. Also, $c^2=b^2-a^2=1$, so $c=1$ and the foci are $(0, \pm 1)$. The graph is sketched in Figure 10. ■

EXAMPLE 3

Determine an equation of the ellipse with foci $(\pm 3, 0)$ and vertices $(\pm 5, 0)$.

Solution $c=3$ and $a=5$. Then $b^2=a^2-c^2=25-9=16$. Hence, the equation of the ellipse is

$$\frac{x^2}{25}+\frac{y^2}{16}=1 \text{ or } 16x^2+25y^2=400.$$ ■

Hyperbolas

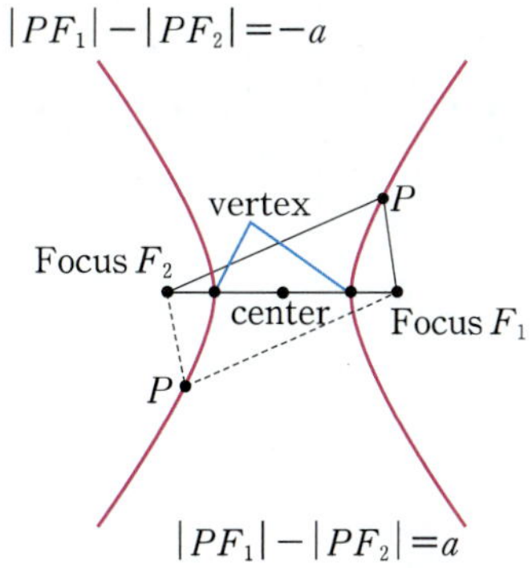

Figure 11

P is on the hyperbola when $|PF_1|-|PF_2|=\pm 2a$

The definition of a hyperbola is similar to that of an ellipse. A hyperbola is an open curve with two branches, the intersection of a plane with both halves of a double cone. A **hyperbola** is the set of all points in a plane the difference of whose distances from two fixed points F_1 and F_2 (the foci) is a constant. (See Figure 11.)

Hyperbola occur frequently as graphs of equations in chemistry, physics, biology, and economics (Boyle's Law, Ohm's law, supply and demand curves). A particularly significant application of hyperbolas was found in the navigation systems developed in World Wars I and II.

Notice that the derivation of the equation of a hyperbola is also similar to the one given earlier for an ellipse. You can show that, when the foci are on the x-axis at $(\pm c, 0)$ and the difference of distances is $|PF_1|-|PF_2|=\pm 2a$, the equation of the hyperbola is

5
$$\frac{x^2}{a^2}-\frac{y^2}{b^2}=1$$

where $c^2=a^2+b^2$. The x-intercepts are again $\pm a$ and the vertices are the points $(a, 0)$ and $(-a, 0)$. When we put $x=0$ in 5 we get $y^2=-b^2$, which means there is no real solution. Therefore, we don't have y-intercept.

Since 5 is even in both x and y, the graph of hyperbola is symmetric with respect to both x- and y-axes.

Rewriting 5 we have

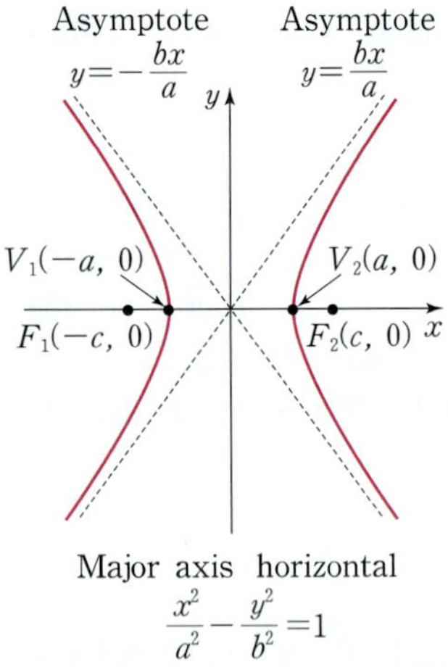

Figure 12

$$\boxed{6} \qquad \frac{x^2}{a^2}=1+\frac{y^2}{b^2}\geq 1$$

or $x^2 \geq a^2$, so $|x|=\sqrt{x^2}\geq a$. Thus the hyperbola exists (is real) only when $x\geq a$ or $x\leq -a$, and it consists of two parts, called its branches.

The dashed lines $y=(b/a)x$ and $y=-(b/a)x$ in Figure 12 are the **asymptotes** to the hyperbola. Both branches of the hyperbola come arbitrarily close to the asymptotes.

If the foci of a hyperbola are on the y-axis, then by interchanging the roles of x and y we obtain the following information, which is illustrated in Figure 13.

Similarly, the hyperbola

$$\boxed{7} \qquad \frac{y^2}{a^2}-\frac{x^2}{b^2}=1$$

has foci $(0,\ \pm c)$, on the y-axis, vertices at $(0,\ \pm a)$, and $y=\pm(a/b)x$ as asymptotes. Here $c^2=a^2+b^2$, ($\boxed{6}$ is obtained by interchanging x and y : see Figure 13.)

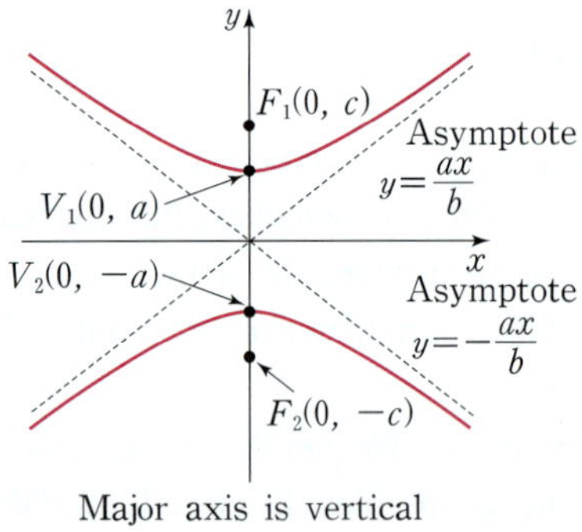

Figure 13 $\frac{y^2}{a^2}-\frac{x^2}{b^2}=1$

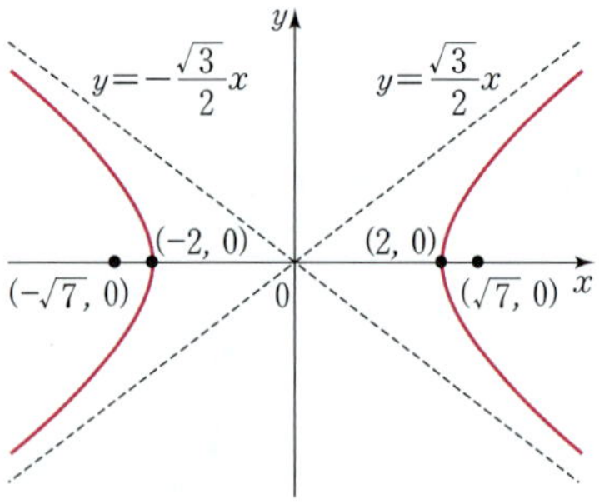

Figure 14 $3x^2-4y^2=12$

EXAMPLE 4

Sketch the graph and find the foci and asymptotes of the hyperbola $3x^2-4y^2=12$.

Solution Rewriting in form $\boxed{5}$, we have

$$\frac{x^2}{4}-\frac{y^2}{3}=1$$

with $a=2$ and $b=\sqrt{3}$. Since $c^2=4+3=7$, the foci are $(\pm\sqrt{7},\ 0)$. The asymptotes are the lines $y=\frac{\sqrt{3}}{2}x$ and $y=-\frac{\sqrt{3}}{2}x$. The graph is shown in Figure 14. ■

EXAMPLE 5

Determine the foci and equation of the hyperbola with vertices $(0, \pm 3)$ and asymptotes $y = \pm \dfrac{3}{2}x$.

Solution With reference to [7], we have $a = 3$ and $a/b = 3/2$. Thus, $b = 2$ and $c^2 = a^2 + b^2 = 13$. Then the foci are $(0, \pm\sqrt{13})$ and the equation of the hyperbola is $\dfrac{y^2}{9} - \dfrac{x^2}{4} = 1$. ■

Shifted Conics

By replacing x and y by $x - h$ and $y - k$, in the standard equations [1], [2], [4], [5] and [7], we obtain the shift conics.

http://www.youtube.com/watch?v=0KXOlU_7YU0

EXAMPLE 6

Find an equation of the ellipse with foci $(1, -3)$, $(3, -3)$ and vertices $(-1, -3)$, $(5, -3)$.

Solution The line segment joining the vertices $(-1, -3)$, $(5, -3)$ is the major axis and has length 6, so $a = 3$. The distance between the foci is 2, so $c = 1$. Thus, $b^2 = a^2 - c^2 = 8$. Since the center of the ellipse is $(2, -3)$, replace x by $x - 2$ and y by $y + 3$ in [4]. Thus the equation of the ellipse is

$$\frac{(x-2)^2}{9} + \frac{(y+3)^2}{8} = 1.$$

(See Figure 15.) ■

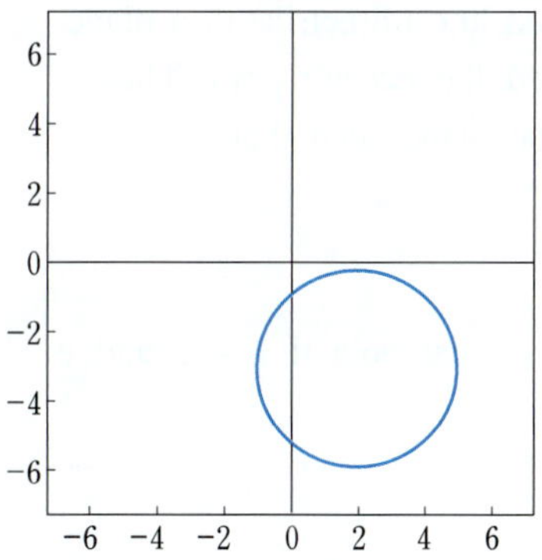

Figure 15

EXAMPLE 7

Find the foci and draw the conic $3x^2 - 5y^2 - 12x - 30y - 48 = 0$.

Solution Rearranging the terms and completing the squares, we get

$$3(x^2 - 4x) - 5(y^2 + 6y) = 48$$
$$3(x^2 - 4x + 4) - 5(y^2 + 6y + 9) = 15$$
$$\frac{(x-2)^2}{5} - \frac{(y+3)^2}{3} = 1.$$

With reference to [5], $a^2 = 5$, $b^2 = 3$, and $c^2 = 8$. The foci are $(2 + 2\sqrt{2}, -3)$ and $(2 - 2\sqrt{2}, -3)$ and the vertices are $(2 + \sqrt{5}, -3)$ and $(2 - \sqrt{5}, -3)$. The asymptotes are

$$y+3=\pm\frac{\sqrt{15}}{5}(x-2).$$ ■

Conic Sections in Polar Coordinates

We defined the parabola in terms of a focus and directrix, but we defined the ellipse and hyperbola in terms of two foci. Now we will give a unified treatment of all the three types of conic sections in terms of a focus and directrix. We know that the ellipse and hyperbolas take simple forms when the origin lies at their center. As it happens, there are many important applications of conics in which it is more convenient to use one of the foci as the reference point (the origin) for the coordinate system. Therefore, if we move one of the foci to the origin, a conic section has a simple polar equation which provides a convenient description of the motion of planets, satellites, and comets. In this section we will see that polar equations of conics have a simple form if one of the foci lies at the pole.

The following theorem use concepts of eccentricity to classify the three basic types of conics.

THEOREM 1

Let F be a fixed point (focus) and l be a fixed line (directrix) in a plane. And let e be a fixed positive number (called the eccentricity). Then a conic section is the set of all points P in the plane such that

$$\frac{|PF|}{|PL|}=e.$$

Furthermore the conic is an ellipse if $e<1$, a parabola if $e=1$ and a hyperbola if $e>1$. (See Figure 16.)

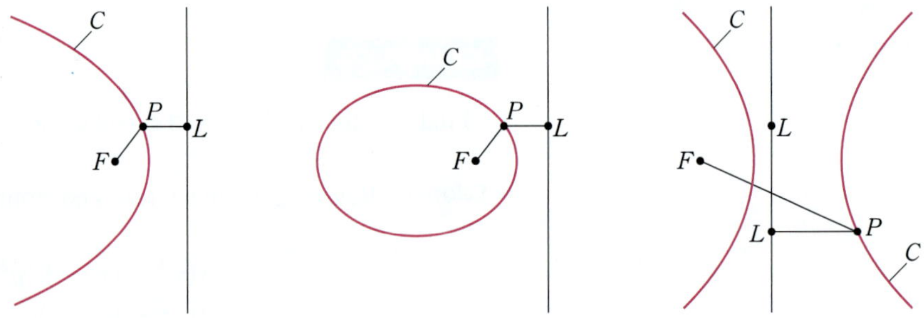

Parabola: $\frac{|PF|}{|PL|}=e=1$ Ellipse: $\frac{|PF|}{|PL|}=e,\ 0<e<1$ Hyperbola: $\frac{|PF|}{|PL|}=e>1$

Figure 16

Proof If $e=1$, then $|PF|=|PL|$. Therefore, the given condition simply becomes the definition of a parabola.

If $e\neq 1$, then we consider the focus F to lie at the origin $(0,\ 0)$ and the directrix $x = d$ to lie to the right of the origin, as shown in Figure 16. For the point $P = (r, \theta) = (x, y)$, we have

$$|PF| = r, \qquad |PL| = d - r\cos\theta.$$

Then

$$|PF|/|PL| = e, \text{ or } |PF| = e|PL|,$$

or

8 $$r = e(d - r\cos\theta).$$

By solving 8 for r, the polar equation of the conic (shown in Figure 16) is obtained as

$$r = \frac{ed}{1 + e\cos\theta}.$$

Squaring 8 on both sides and converting it to rectangular coordinates, we get

$$x^2 + y^2 = e^2(d - x)^2 = e^2(d^2 - 2dx + x^2)$$

or

$$(1 - e^2)x^2 + 2de^2x + y^2 = e^2d^2.$$

Rearranging the terms and completing the square, we get

9 $$\left(x + \frac{e^2d}{1 - e^2}\right)^2 + \frac{y^2}{1 - e^2} = \frac{e^2d^2}{(1 - e^2)^2}.$$

or

$$\frac{(x - h)^2}{a^2} + \frac{y^2}{b^2} = 1.$$

where

10 $$h = -\frac{e^2d}{1 - e^2} \quad a^2 = \frac{e^2d^2}{(1 - e^2)^2} \quad b^2 = \frac{e^2d^2}{1 - e^2}.$$

This is the equation of an ellipse if $e < 1$.
The foci of an ellipse are at a distance c from the center, where

11 $$c^2 = a^2 - b^2 = \frac{e^4d^2}{(1 - e^2)^2}.$$

This shows that

$$c = \frac{e^2d}{1 - e^2} = -h.$$

From 10 and 11, the eccentricity e is given by $e = c/a$.

If $e > 1$, then $1 - e^2 < 0$ then equation 9 can be rewritten in the form

$$\frac{(x-h)^2}{a^2} - \frac{y^2}{b^2} = 1$$

which represents a hyperbola with

$$e = \frac{c}{a} \text{ where } c^2 = a^2 + b^2.$$

■

THEOREM 2

A polar equation of the form

$$r = \frac{ed}{1 \pm e\cos\theta} \quad \text{or} \quad r = \frac{ed}{1 \pm e\sin\theta}$$

represents a conic section with eccentricity e, where $|d|$ is distance between the focus at the pole and its corresponding directrix.

In Figure 17(b) the directrix $x = -d$, is to the left of the focus. In Figure 17(c) and 17(d) the directrix is parallel to the polar axis as $y = \pm d$,

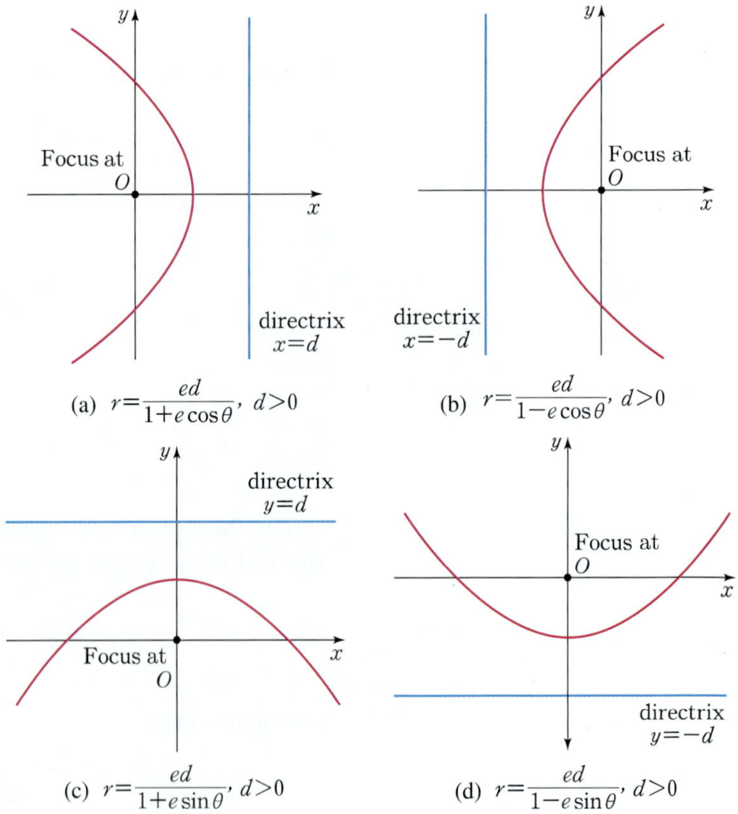

Figure 17

EXAMPLE 8

Find a polar equation for a parabola that has its focus at the origin and whose directrix is $y=3$.

Solution From Figure 17(c) and using Theorem 2 with $e=1$ and $d=3$ the equation of the parabola is

$$r=\frac{3}{1+\sin\theta}.$$

■

EXAMPLE 9

Find the eccentricity, locate the directrix, identify and sketch the conic given by the polar equation

$$r=\frac{16}{5-3\cos\theta}.$$

http://matrix.skku.ac.kr/cal-lab/sage-grapher-polar.html

Solution Rewriting

$$r=\frac{\frac{16}{5}}{1-\frac{3}{5}\cos\theta}\quad\left(e=\frac{3}{5}<1\right)$$

from Theorem 2 this represents an ellipse with $ed=\frac{16}{5}$. Since $ed=\frac{16}{5}$, we have

$$d=\frac{\frac{16}{5}}{\frac{3}{5}}=\frac{16}{3}$$

so the directrix has Cartesian equation $x=-\frac{16}{3}$. When $\theta=0$, $r=8$; when $\theta=\pi$, $r=2$. So the vertices have polar coordinates $(8,\ 0)$ and $(2,\ \pi)$. The ellipse is sketched in Figure 18.

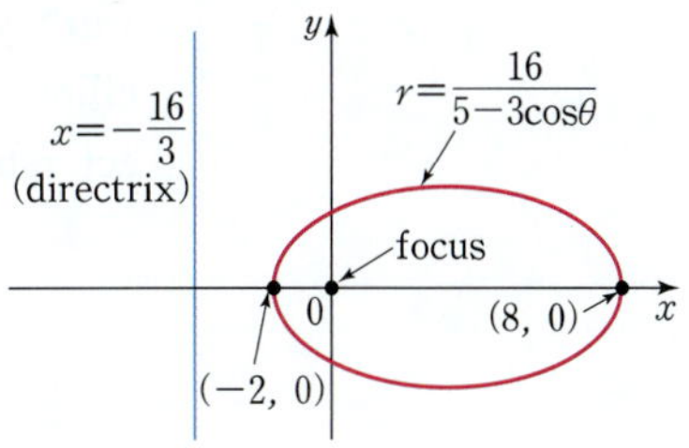

Figure 18

■

EXAMPLE 10

Sketch the conic $r=\dfrac{16}{4+8\sin\theta}$.

http://matrix.skku.ac.kr/cal-lab/sage-grapher-polar.html

Solution The rewritten equation

$$r=\frac{4}{1+2\sin\theta}$$

represents a hyperbola with the eccentricity $e=2$.
Since $ed=4$, $d=2$ and the directrix has equation $y=2$. The vertices occur when $\theta=\frac{\pi}{2}$ and $\frac{3}{2}\pi$, so they are $\left(\frac{4}{3},\ \frac{\pi}{2}\right)$ and $\left(-4,\ \frac{3\pi}{2}\right)=\left(4,\ \frac{\pi}{2}\right)$. When $\theta=0,\ \pi$ we get the x-intercepts with $r=4$ in both cases. As $r\to\pm\infty$ then $1+2\sin\theta\to 0^+$ or 0^- and $\sin\theta=-\frac{1}{2}$. Thus, the asymptotes are parallel to the rays $\theta=\frac{7}{6}\pi$ and $\theta=\frac{11}{6}\pi$. The hyperbola is sketched. (See Figure 19.)

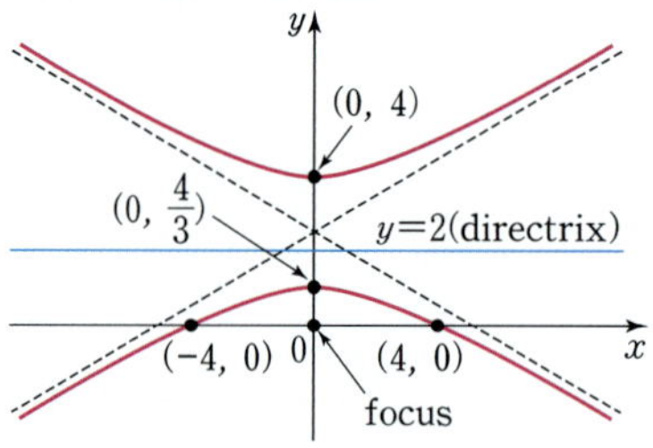

Figure 19 $r=\dfrac{16}{4+8\sin\theta}$

■

CAS **EXAMPLE 11**

Find the vertices and foci of the ellipse and sketch its graph.

$$\frac{x^2}{7}+\frac{y^2}{3}=1$$

Solution

```
var('x,y,a,b,c')
ellipse=implicit_plot(x^2/7+y^2/3==1,(x,-3,3),(y,-3,3)).show(asp
ect_ratio=1, xmin=-3, xmax=3, ymin=-3, ymax=3)
```

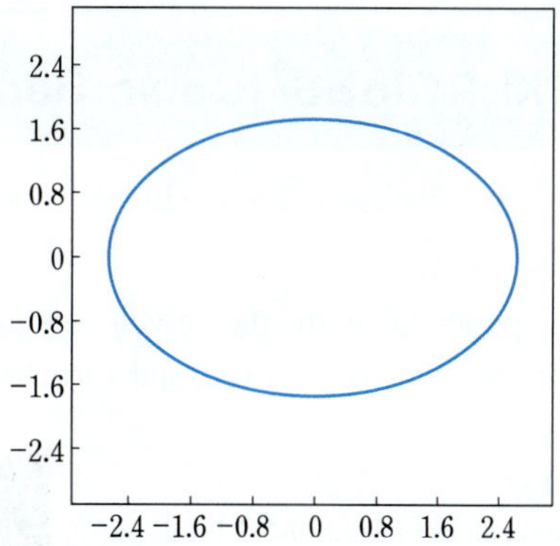

Figure 20

```
solve(a^2==7,a)
```

[a == -sqrt(7), a == sqrt(7)]

```
solve(b^2==3,b)
```

[b == -sqrt(3), b == sqrt(3)]

```
solve(c^2==7-3,c)
```

[c == -2, c == 2]

```
var('x, y')
ellipse=implicit_plot(x^2/7+y^2/3==1,(x,-3,3),(y,-3,3))
f1=point((-2, 0), pointsize=20, rgbcolor=(1,0,0));
f2=point((2, 0), pointsize=20, rgbcolor=(1,0,0));
v1=point((0,-sqrt(3)), pointsize=20, rgbcolor=(0,0,1));
v2=point((0,sqrt(3)), pointsize=20, rgbcolor=(0,0,1));
v3=point((-sqrt(7), 0), pointsize=20, rgbcolor=(0,0,1));
v4=point((sqrt(7), 0), pointsize=20, rgbcolor=(0,0,1));
show(ellipse+f1+f2+v1+v2+v3+v4,aspect_ratio=1,xmin=-3,xma
x=3, ymin=-3, ymax=3)
```

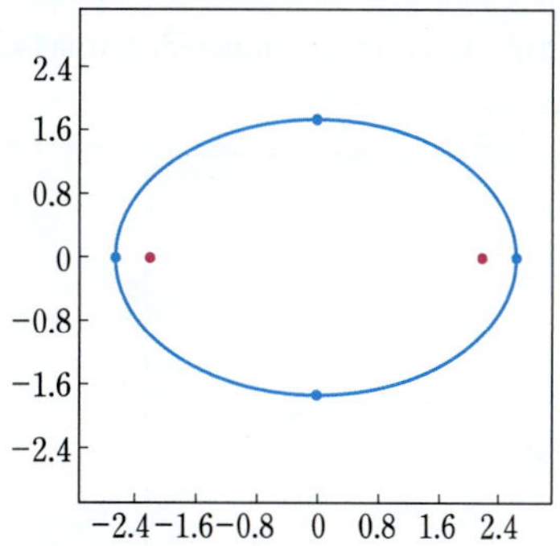

Figure 21

■

http://matrix.skku.ac.kr/Cal-Book/part1/CS-Sec-10-5-Sol.html

YouTube http://youtu.be/CZ9SHMtqVy4

1-4. Sketch the parabola with the given equation. Show and label its vertex, focus, axis, and directrix.

CAS **1.** $(y-2)^2 = 3(x-3)$

http://matrix.skku.ac.kr/cal-lab/cal-9-5-1.html

Solution

```
var('x,y')
implicit_plot((y-2)^2==3*(x-3), (x, -10, 10), (y, -10, 10)).show(aspect_ratio=1, xmin=0, xmax=12, ymin=-4, ymax=8)
```

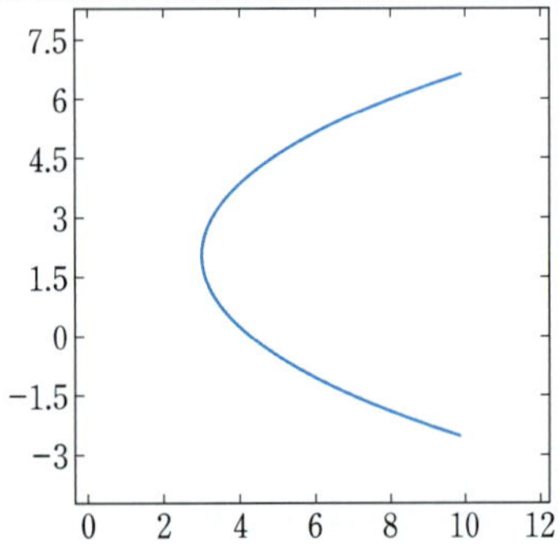

$\Rightarrow$ vertex : $(3,\ 2)$ / focus : $\left(\frac{15}{4},\ 2\right)$ / axis : $y=2$ /

directrix : $x = \frac{9}{4}$

CAS **2.** $6y + x^2 = 0$

http://matrix.skku.ac.kr/cal-lab/cal-9-5-2.html

Solution

```
var('x, y')
parabola=implicit_plot(6*y+x^2==0, (x, -3, 3), (y, -3, 3)).show(aspect_ratio=1, xmin=-3, xmax=3, ymin=-3, ymax=3)
```

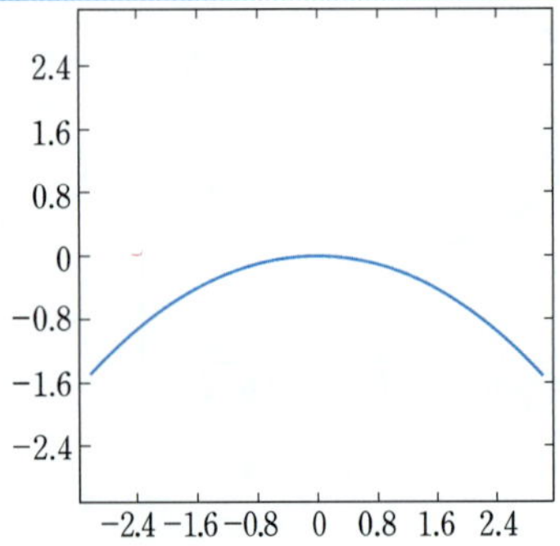

$\Rightarrow$ vertex : $(0,\ 0)$ / focus : $\left(0,\ -\frac{3}{2}\right)$ / axis : $x=0$ /

directrix : $y = \frac{3}{2}$.

CAS **3.** $(y-3)^2 + 2(x+1) = 0$

http://matrix.skku.ac.kr/cal-lab/cal-9-5-3.html

Solution

```
var('x, y')
parabola=implicit_plot((y-3)^2+2*(x+1)==0, (x, -5, 1), (y, 0, 7)).show(aspect_ratio=1, xmin=-5, xmax=1,
ymin=0, ymax=7)
```

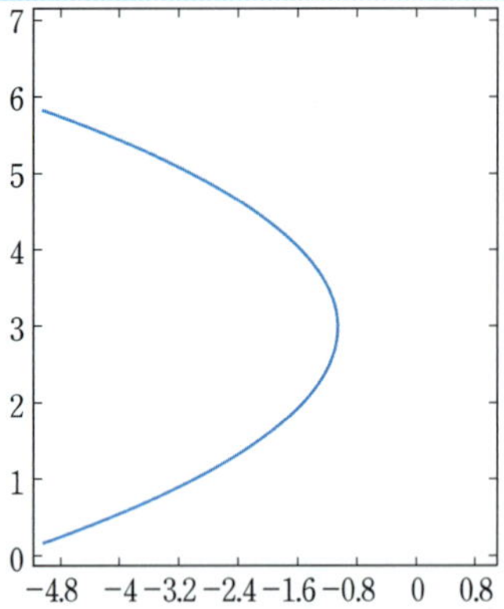

$\Rightarrow$ vertex : $(-1,\ 3)$ / focus : $\left(-\frac{3}{2},\ 3\right)$ / axis : $y=3$ /

directrix : $x = -\frac{1}{2}$

4. $4x^2 - 12x + 9y = 0$

5-8. Find the vertices and foci of the ellipse and sketch its graph.

CAS **5.** $\frac{x^2}{6} + \frac{y^2}{4} = 1$

http://matrix.skku.ac.kr/cal-lab/cal-9-5-5.html

Solution

```
var('x,y,a,b,c')
ellipse=implicit_plot(x^2/6+y^2/4==1,(x,-3,3),(y,-3,
3)).show(aspect_ratio=1,    xmin=-3,    xmax=3,
ymin=-3,
 ymax=3)
```

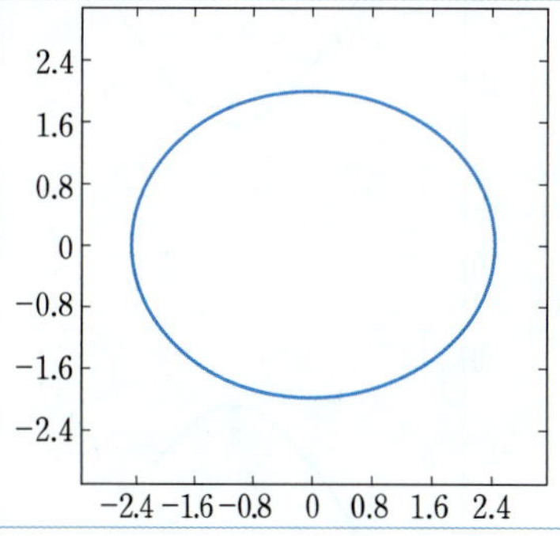

```
solve(a^2==6,a)
```

[a == -sqrt(6), a == sqrt(6)]

```
solve(b^2==4,b)
```

[b == -2, b == 2]

```
solve(c^2==6-4,c)
```

[c == -sqrt(2), c == sqrt(2)]

```
var('x,y,a,b,c')
ellipse=implicit_plot(x^2/6+y^2/4==1,(x,-3,3),(y,-3,3))
f1=point((-sqrt(2),0),                    pointsize=20,
rgbcolor=(1,0,0));
f2=point((sqrt(2),0),                     pointsize=20,
rgbcolor=(1,0,0));
v1=point((-sqrt(6),0),                    pointsize=20,
rgbcolor=(0,0,1));
v2=point((sqrt(6),0),                     pointsize=20,
rgbcolor=(0,0,1));
v3=point((0,-2), pointsize=20, rgbcolor=(0,0,1));
v4=point((0,2), pointsize=20, rgbcolor=(0,0,1));
show(ellipse+f1+f2+v1+v2+v3+v4,aspect_ratio=1,
xmin=-3, xmax=3, ymin=-3, ymax=3)
```

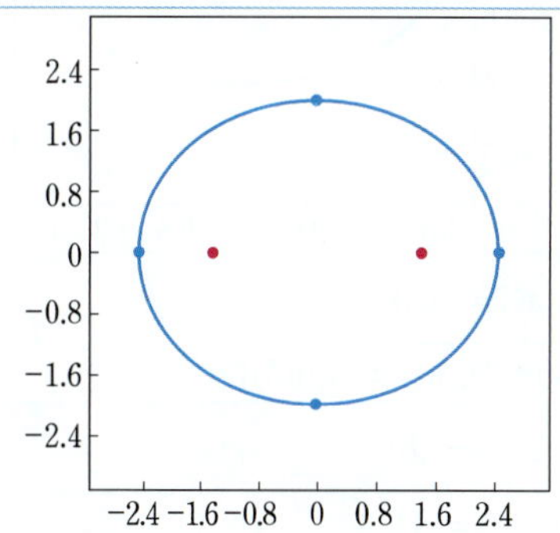

6. $4x^2 + 9y^2 = 144$

7. $9x^2 + 4y^2 - 32y + 28 = 0$

8. $7x^2 + 5y^2 - 42x + 20y + 48 = 0$

9-11. Find the vertices, foci, and asymptotes of the hyperbola and sketch its graph.

CAS **9.** $\dfrac{x^2}{36} - \dfrac{y^2}{16} = 1$

http://matrix.skku.ac.kr/cal-lab/cal-9-5-9.html

Solution

```
var('x,y,a,b,c')
hyperbola=implicit_plot(x^2/36-y^2/16==1,(x,-15,1
5),(y,-15,15)).show(aspect_ratio=1, xmin=-15,
xmax=15, ymin=-15, ymax=15)
```

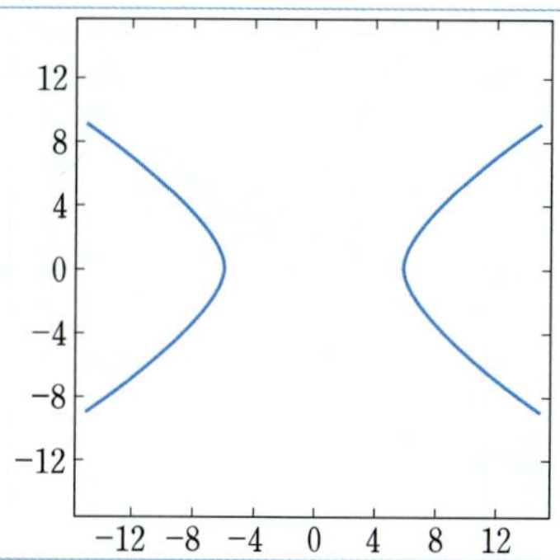

```
solve(a^2==36,a)
```

[a == -6, a == 6]

```
solve(b^2==16,b)
```

[b == -4, b == 4]

```
solve(c^2==36+16,c)
```

[c == -2*sqrt(13), c == 2*sqrt(13)]

```
var('x,y,a,b,c')
hyperbola=implicit_plot(x^2/36-y^2/16==1,(x,-15,1
5),(y,-15,15))
f1=point((-2*sqrt(13),0),              pointsize=20,
rgbcolor=(1,0,0));
f2=point((2*sqrt(13),0),               pointsize=20,
rgbcolor=(1,0,0));
v1=point((-6,0), pointsize=20, rgbcolor=(0,0,1));
v2=point((6,0), pointsize=20, rgbcolor=(0,0,1));
show(hyperbola+f1+f2+v1+v2,aspect_ratio=1,
xmin=-15, xmax=15, ymin=-15, ymax=15)
```

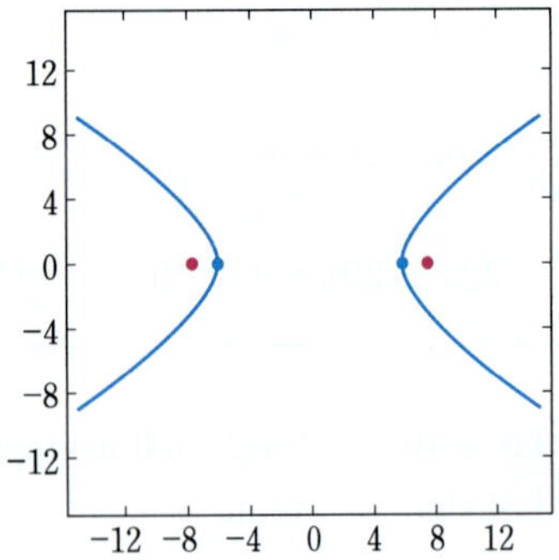

10. $\dfrac{y^2}{81} - \dfrac{x^2}{25} = 1$

http://matrix.skku.ac.kr/cal-lab/cal-9-5-10.html

Solution

```
var('x,y,a,b,c')
hyperbola=implicit_plot(y^2/81-x^2/25==1,(x,-15,1
5),(y,-20,20)).show(aspect_ratio=1, xmin=-15,
xmax=15, ymin=-20, ymax=20)
```

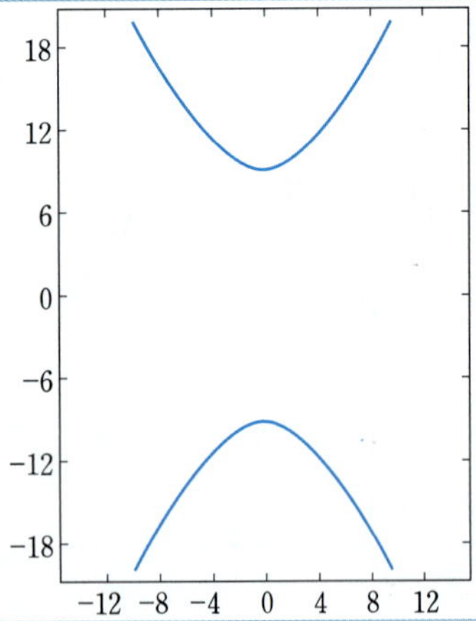


```
solve(a^2==81,a)
```

[a == -9, a == 9]

```
solve(b^2==25,b)
```

[b == -5, b == 5]

```
solve(c^2==81+25,c)
```

[c == -sqrt(106), c == sqrt(106)]

```
x,y,a,b,c=var('x,y,a,b,c')
hyperbola=implicit_plot(y^2/81-x^2/25==1,(x,-15,1
5),(y,-20,20))
f1=point((0,-sqrt(106)),              pointsize=20,
rgbcolor=(1,0,0));
f2=point((0,sqrt(106)), pointsize=20,
rgbcolor=(1,0,0));
v1=point((0,-9), pointsize=20, rgbcolor=(0,0,1));
v2=point((0,9), pointsize=20, rgbcolor=(0,0,1));
show(hyperbola+f1+f2+v1+v2,aspect_ratio=1,
xmin=-15, xmax=15, ymin=-20, ymax=20)
```

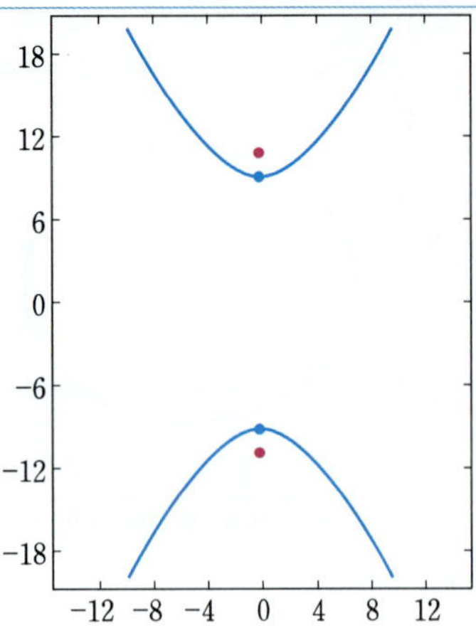

11. $9x^2 - 4y^2 + 18x + 8y = 31$

12-14. Identify the type of conic section whose equation is given and find the vertices and foci.

CAS **12.** $x^2 = 6y - 3y^2$

http://matrix.skku.ac.kr/cal-lab/cal-9-5-12.html

Solution $x^2 = 6y - 3y^2 \;\Rightarrow\; x^2 = -3(y^2 - 2y)$

$\Rightarrow\; x^2 + 3(y-1)^2 = 3 \;\Rightarrow\; \dfrac{x^2}{3} + \dfrac{(y-1)^2}{1} = 1$: Ellipse

```
var('x,y,a,b,c')
ellipse=implicit_plot(x^2/3+(y-1)^2/1==1,(x,-2,2),(y
,-2,4)).show(aspect_ratio=1,
xmin=-2, xmax=2, ymin=-2, ymax=4)
```

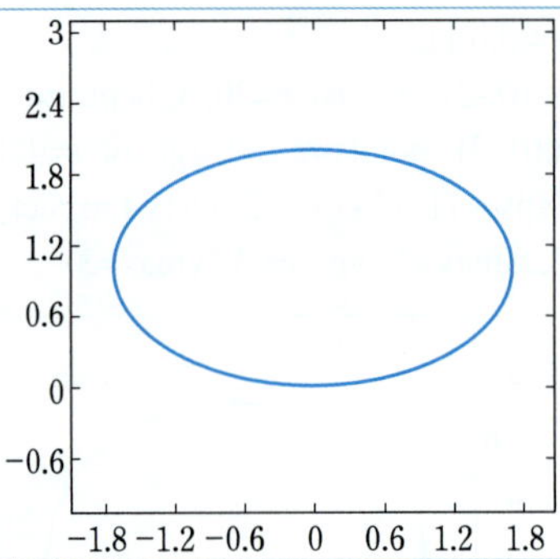


```
solve(a^2==3,a)
```

[a == -sqrt(3), a == sqrt(3)]

```
solve(b^2==1,b)
```

[b == -1, b == 1]

```
solve(c^2==3-1,c)
```

[c == -sqrt(2), c == sqrt(2)] : vertices $(\pm\sqrt{3},1)$ / focus $(\pm\sqrt{2},1)$

```
var('x,y')
ellipse=implicit_plot(x^2/3+(y-1)^2/1==1,(x,-2,2),(y,-2,4))
f1=point((-sqrt(2),1), pointsize=20, rgbcolor=(1,0,0));
f2=point((sqrt(2),1), pointsize=20, rgbcolor=(1,0,0));
v1=point((-sqrt(3),1), pointsize=20, rgbcolor=(0,0,1));
v2=point((sqrt(3),1), pointsize=20, rgbcolor=(0,0,1));
v3=point((0,0), pointsize=20, rgbcolor=(0,0,1));
v4=point((0,2), pointsize=20, rgbcolor=(0,0,1));
show(ellipse+f1+f2+v1+v2+v3+v4,aspect_ratio=1,xmin=-2, xmax=2, ymin=-1, ymax=3)
```

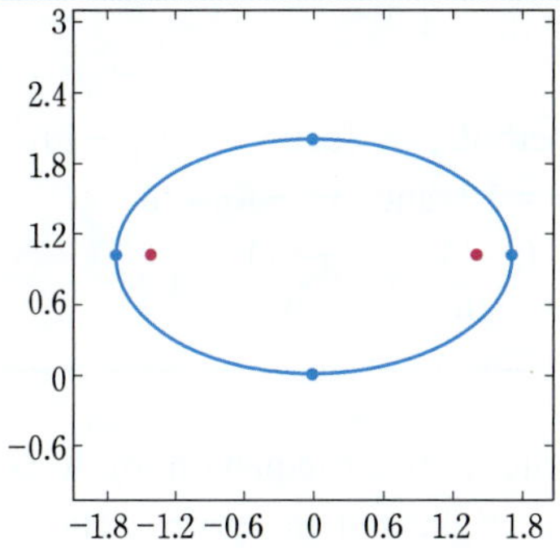

CAS **13.** $y^2 - 4y = 8x - 8$

http://matrix.skku.ac.kr/cal-lab/cal-9-5-13.html

Solution $(y-2)^2 = 8x-8 \Rightarrow (y-2)^2 = 8(x-1)$

: Parabola with vertices $(1,\ 2)$ and focus $(3,\ 2)$

```
var('x,y')
parabola=implicit_plot((y-2)^2==8*(x-1),(x,-1,8),(y,-4,8))
f1=point((1,2), pointsize=20, rgbcolor=(1,0,0));
v1=point((3,2), pointsize=20, rgbcolor=(0,0,1));
show(parabola+f1+v1,aspect_ratio=1, xmin=-1, xmax=8, ymin=-4, ymax=8)
```

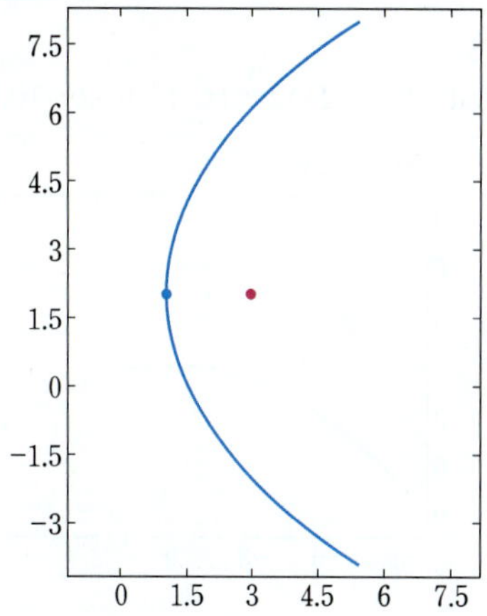

CAS **14.** $4y^2 - 5x^2 - 20x - 8y = 36$

http://matrix.skku.ac.kr/cal-lab/cal-9-5-14.html

Solution $4y^2 - 5x^2 - 20x - 8y = 36 \Rightarrow$

$\dfrac{(y-1)^2}{5} - \dfrac{(x+2)^2}{4} = 1$: Hyperbola with vertices $(-2,\ 1\pm\sqrt{5})$ and focus $(-2,\ 4)$, $(-2,\ -2)$

```
var('x,y')
hyperbola=implicit_plot((y-1)^2/5-(x+2)^2/4==1,(x,-10,87),(y,-8,10))
f1=point((-2,4), pointsize=20, rgbcolor=(1,0,0));
f2=point((-2,-2), pointsize=20, rgbcolor=(1,0,0));
v1=point((-2,1+sqrt Theorem), pointsize=20, rgbcolor=(0,0,1));
v2=point((-2,1-sqrt Theorem), pointsize=20, rgbcolor=(0,0,1));
show(hyperbola+f1+f2+v1+v2,aspect_ratio=1, xmin=-10, xmax=7, ymin=-8,ymax=10)
```

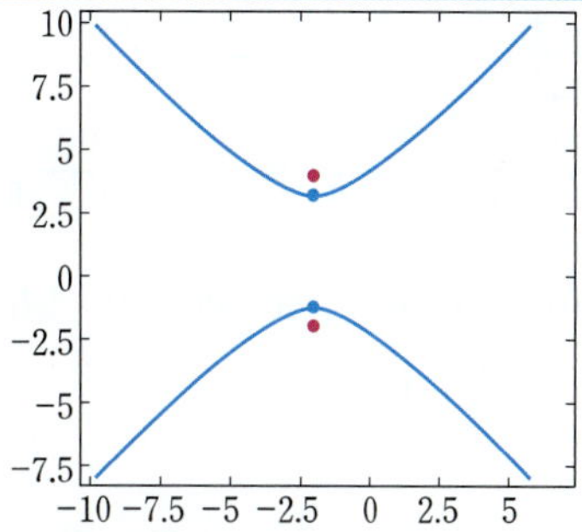

15-22. Find an equation for the conic that satisfies the given conditions.

15. Parabola, vertex $(0,\ 0)$, focus $(0,\ -3)$

http://matrix.skku.ac.kr/cal-lab/cal-10-5-15.html

Solution $x^2 = 4py = 4(-3)y \quad (\because p = -3)$

$= -12y$

```
var('x,y')
implicit_plot(x^2+12*y,(x,-10,10),(y,-10,10))
```

Answer : x^2 = -12y

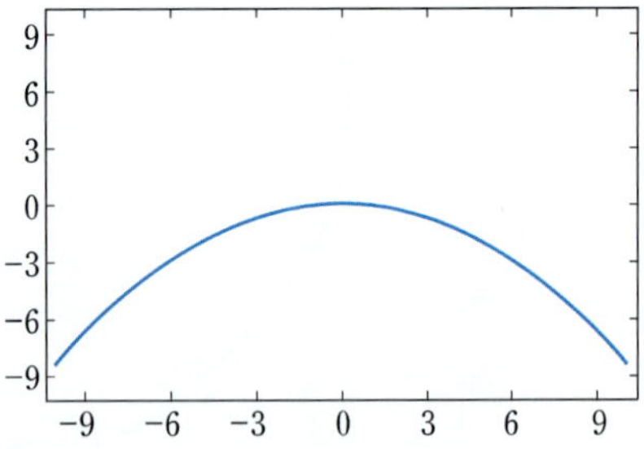

16. Parabola, vertex $(2,\ 1)$, focus $(-4,\ 1)$

Solution $(y-1)^2=-24(x-2)$

17. Ellipse, foci $(\pm 3,\ 0)$, vertices $(\pm 5,\ 0)$

http://matrix.skku.ac.kr/cal-lab/cal-10-5-17.html

Solution $\dfrac{x^2}{a^2}+\dfrac{y^2}{b^2}=1$

$$\frac{x^2}{a^2}+\frac{y^2}{a^2-c^2}=1$$

$$\frac{x^2}{25}+\frac{y^2}{16}=1 \quad (\because a=\pm 5, c=\pm 3)$$

```
var('x,y')
implicit_plot(x^2/25+y^2/16-1,(x,-10,10),(y,-10,10)
)
```

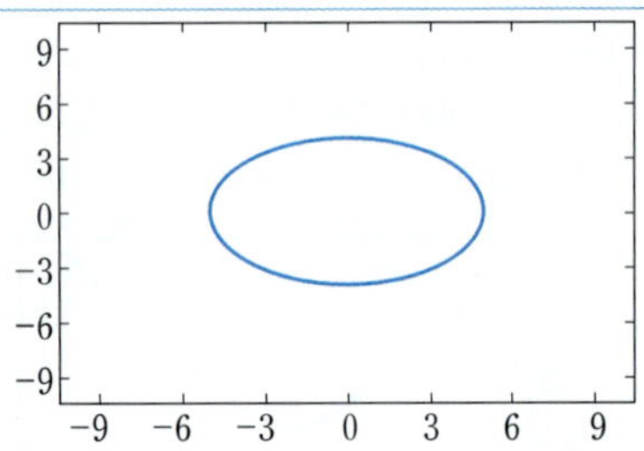

18. Ellipse, foci $(0,\ \pm 2)$, vertices $(0,\ \pm 5)$

Solution $\dfrac{x^2}{21}+\dfrac{y^2}{25}=1$

19. Ellipse, center $(1,\ -1)$, focus $(2,\ -1)$ vertex $(4,\ -1)$

Solution $\dfrac{(x-1)^2}{9}+\dfrac{(y+1)^2}{8}=1$

20. Hyperbola, foci $(0,\ \pm 3)$, vertices $(0,\ \pm 2)$

http://matrix.skku.ac.kr/cal-lab/cal-10-5-20.html

Solution $\dfrac{y^2}{a^2}-\dfrac{x^2}{b^2}=1$

$$\frac{y^2}{a^2}-\frac{x^2}{c^2-a^2}=1$$

$$\frac{y^2}{4}-\frac{x^2}{5}=1 \quad (\because a=\pm 2, c=\pm 3)$$

```
var('x,y')
implicit_plot(-1*x^2/5+y^2/4-1,(x,-10,10),(y,-10,10
))
```

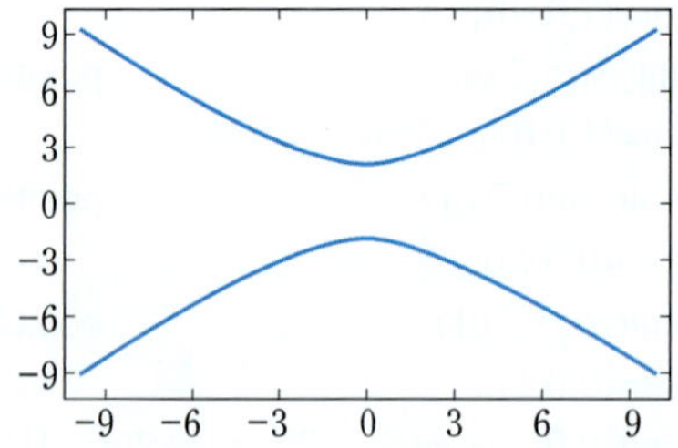

21. Hyperbola, foci $(\pm 5,\ 0)$, asymptotes $y=\pm\dfrac{3}{4}x$

Solution $\dfrac{x^2}{16}-\dfrac{y^2}{9}=1$

22. Hyperbola, focus $(8,\ -1)$, asymptotes $3x-4y=13$ and $3x+4y=5$

Solution $\dfrac{(x-3)^2}{16}-\dfrac{(y+1)^2}{9}=1$

23-28. Write a polar equation of a conic with the focus at the origin and the given data.

23. Hyperbola, eccentricity $\dfrac{8}{3}$, directrix $x=-\dfrac{3}{4}$

http://matrix.skku.ac.kr/cal-lab/cal-10-5-23.html

Solution
$$r=\frac{ed}{1-e\cos\theta}=\frac{\frac{8}{3}\times\frac{3}{4}}{1-\frac{8}{3}\cos\theta}\ \left(\because e=\frac{8}{3},\ d=\frac{3}{4}\right)=\frac{6}{3-8\cos\theta}$$

```
var('r,theta')
solve(3-8*cos(theta)==0,theta)
```

Answer : [theta == arccos(3/8)]

```
r=6/(3-8*cos(theta))
polar_plot
(r, theta,-1*arccos(3/8)+0.1,arccos(3/8)-0.1)
```

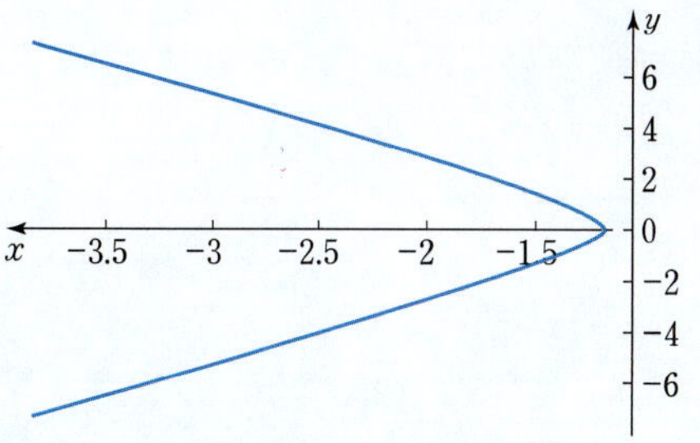

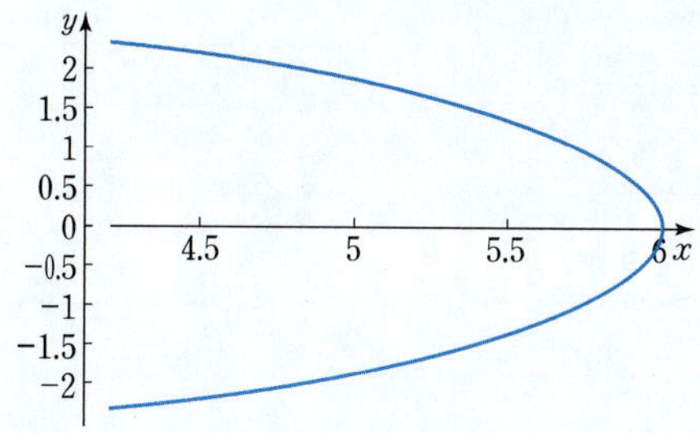

24. Hyperbola, eccentricity $\frac{5}{3}$, directrix $y=4$.

http://matrix.skku.ac.kr/cal-lab/cal-10-5-24.html

Solution $r=\dfrac{ed}{1+e\sin\theta}=\dfrac{\frac{5}{3}\cdot 4}{1+\frac{5}{3}\sin\theta}=\dfrac{20}{3+5\sin\theta}$

```
var('r,theta')
solve(3+5*sin(theta)==0,theta)
```

Answer : [theta == -arcsin(3/5)

```
r=20/(3+5*sin(theta))
polar_plot(r, theta,-0.5,2.5)
```

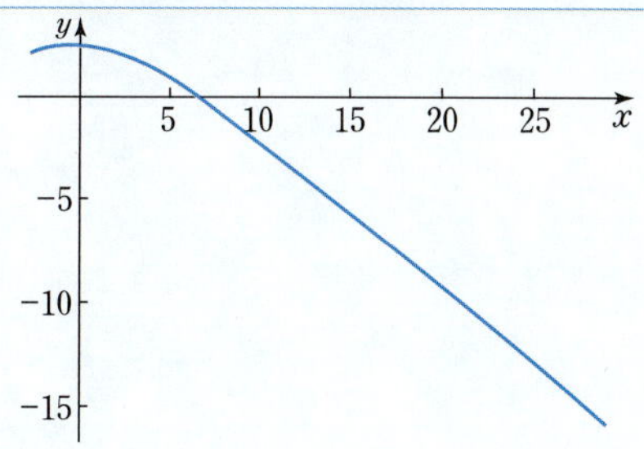

25. Ellipse, eccentricity $\frac{2}{3}$, directrix $x=-3$

http://matrix.skku.ac.kr/cal-lab/cal-10-5-25.html

Solution $r=\dfrac{ed}{1-e\cos\theta}=\dfrac{\frac{2}{3}\cdot 3}{1-\frac{2}{3}\cos\theta}=\dfrac{6}{3-2\cos\theta}$

```
var('r,theta')
solve(3-2*cos(theta)==0,theta)
```

Answer : [theta == arccos(3/2)]

```
r=6/(3-2*cos(theta))
polar_plot(r, theta,-0.5,0.5)
```

26. Ellipse, eccentricity $\frac{1}{2}$, directrix $y=\frac{1}{2}$

Solution $r=\dfrac{1}{4+2\sin(\theta)}$

27. Parabola, eccentricity 1, directrix $x=\frac{3}{5}$

Solution $r=\dfrac{1\cdot\frac{3}{5}}{1+1\cos\theta}=\dfrac{3}{5+5\cos\theta}$

28. Parabola, eccentricity 1, directrix $y=2$

Solution $r=\dfrac{1\cdot 2}{1+1\sin\theta}=\dfrac{2}{1+\sin\theta}$

29-32. Find the eccentricity, identify the conic, give an equation of the directrix, and sketch the conic.

29. $r=\dfrac{6}{1-\cos\theta}$

Solution $e=1$, $d=6$, Parabola, directrix : $x=-6$

30. $r=\dfrac{3}{1+\cos\theta}$

Solution $e=1$, $d=3$, Parabola, directrix : $x=3$

31. $r=\dfrac{6}{2+\sin\theta}$

Solution $e=\frac{1}{2}$, $d=6$, Ellipse, directrix : $y=6$

32. $r=\dfrac{12}{3-\sin\theta}$

Solution $e=\frac{1}{3}$, $d=12$, Ellipse, directrix : $y=-12$

Chapter 11

Vectors and the Geometry of Space

Euclid

BC330-BC275

Geometry

[Euclid's Elements]

Euclid is often referred to as the "Father of Geometry." His Elements is the most successful textbook and one of the most influential works in the history of mathematics, serving as the main textbook for teaching geometry until the early 20th century. In it, the principles of what is now called Euclidean geometry were deduced from a small set of axioms.

YouTube http://youtu.be/

In this chapter, we introduce vectors and coordinate systems for threedimensional space. We also deal with various concepts in vectors, such as the dot and the cross product of vectors, vector equations of lines, and planes in space. We will also discuss standard quadric surfaces.

11.1 Three-Dimensional Coordinate Systems

In a two-dimensional plane any point can be represented as an ordered pair (x, y) of real numbers, where x is the x-coordinate and y is the y-coordinate of the point. In three-dimensional space, a point is represented by three real numbers (x, y, z).

The **coordinate axes** labeled as the x-axis, y-axis and z-axis are three directed lines that are perpendicular to each other and passing through a chosen fixed point O (called the origin). This three-dimensional coordinate system (See Figure 1(a).) will enable us to represent a point in space. Normally, the x-axis and y-axis are taken as horizontal and the z-axis as being vertical forming a Right-Handed Coordinate System[1], as in Figure 1(b). The **three coordinate axes** determine three coordinate planes. Thus the xy-plane contains the x-axis and y-axis; the yz-plane contains the y-axis and z-axis; the xz-plane contains the x-axis and z-axis. These three coordinate planes divide the space into eight parts, called **octants**. The positive axes determine the **first octant**, in the foreground.(See Figure 2.)

1)http://mathworld.wolfram.com/Right-HandedCoordinateSystem.html

Figure 1 (a) Coordinate axes (b) Right-hand rule

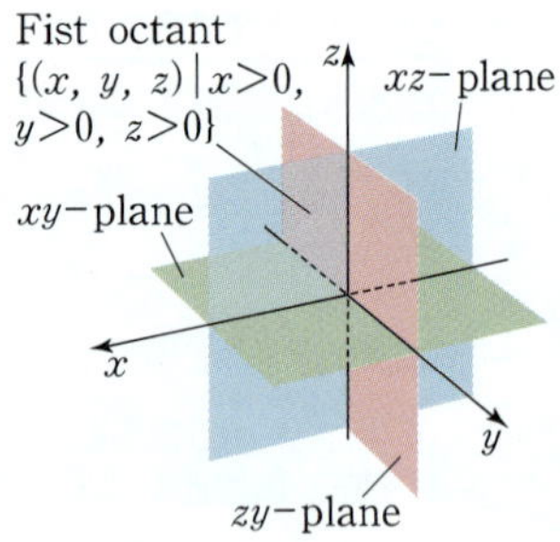

Figure 2 Octants

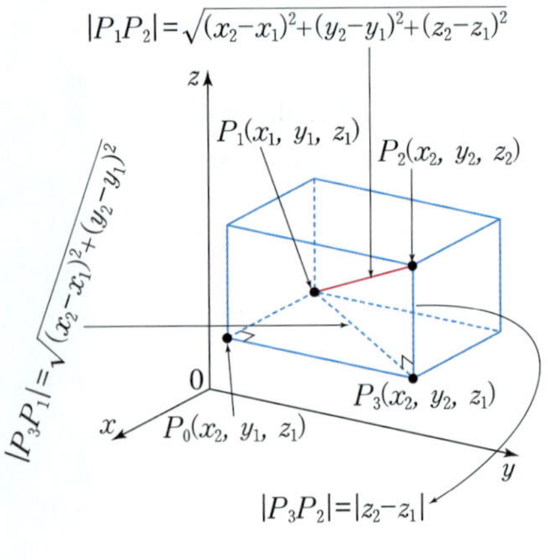

Figure 3

Let $P(a, b, c)$ be any point in space. Then, the real numbers a, b, c are the (directed) distances from P to the yz-plane, the xz-plane, and the xy-plane respectively. The point P is denoted by this ordered triple (a, b, c) and a, b and c are known as the **coordinates** of P; a is the x-coordinate, b is the y-coordinate, and c is the z-coordinate. Thus starting at the origin O, the point (a, b, c) is located by moving a units along the x-axis, then b units parallel to the y-axis, and then c units parallel to the z-axis as shown in Figure 3.

The three-dimensional rectangular coordinate system (Cartesian coordinate system) is the set of ordered triples given by the Cartesian product

$$\mathbb{R}^3 = \mathbb{R} \times \mathbb{R} \times \mathbb{R} = \{(x, y, z) \mid x, y, z \in \mathbb{R}\}.$$

Recall that an equation of x and y in two dimension represents a **curve.** Similarly, an equation of x, y and z in three dimension represents a **surface.** In general, if k is a constant, then $x = k$ represents a plane parallel to the yz-plane, $y = k$ is a plane parallel to the xz-plane, and $z = k$ is a plane parallel to the xy-plane in $\mathbb{R}^3$.

Distance Formula in Three Dimensions

The distance $|P_1P_2|$ between the points $P_1(x_1, y_1, z_1)$ and $P_2(x_2, y_2, z_2)$ is

$$|P_1P_2| = \sqrt{(x_2 - x_1)^2 + (y_2 - y_1)^2 + (z_2 - z_1)^2}.$$

EXAMPLE 1

Determine the equation of a sphere with radius r and center $C(a, b, c)$.

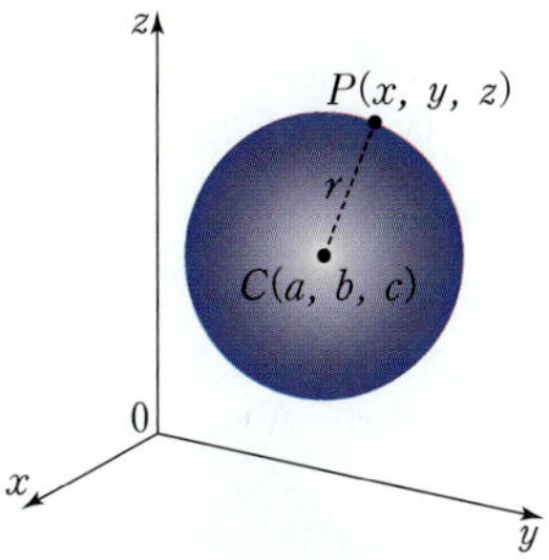

Figure 4

Solution A sphere is the set of all points $P(x, y, z)$ whose distance from the center C is r. (See Figure 4.) So, any point P on the sphere satisfies $|PC| = r$ or $|PC|^2 = r^2$.
Thus, the equation of the required sphere is

$$(x-a)^2 + (y-b)^2 + (z-c)^2 = r^2. \qquad \blacksquare$$

Note If the center is the origin O, then the equation of the sphere is

$$x^2 + y^2 + z^2 = r^2.$$

EXAMPLE 2

Show that $x^2 + y^2 + z^2 - 4x + 6y - 2z + 8 = 0$ is the equation of a sphere, and find its center and radius.

Solution We can rewrite the given equation in the form of an equation of a sphere if we complete the squares:

$$(x^2 - 4x + 4) + (y^2 + 6y + 9) + (z^2 - 2z + 1) = -8 + 4 + 9 + 1$$
$$(x-2)^2 + (y+3)^2 + (z-1)^2 = 6.$$

Comparing this equation with the standard form in the result of Example 1, we see that it is the equation of a sphere with center $(2, -3, 1)$ and radius $\sqrt{6}$. $\blacksquare$

CAS **EXAMPLE 3**

What region in $\mathbb{R}^3$ is represented by the following inequalities?

$$4 \le x^2 + y^2 + z^2 \le 9, \; z \ge 0.$$

http://matrix.skku.ac.kr/cal-lab/11-1-Example-3.html

Solution The inequalities $4 \le x^2 + y^2 + z^2 \le 9$ can be rewritten as

$2 \leq \sqrt{x^2+y^2+z^2} \leq 3$, so they represent the points (x, y, z) whose distance from the origin is at least 2 and at most 3. But we are also given that $z \geq 0$, so the points lie on or above the xy-plane. Thus, the given inequalities represent the region that lies between (or on) the spheres $x^2+y^2+z^2=4$ and $x^2+y^2+z^2=9$ and above (or on) the xy-plane. (See Figure 5.)

```
var('x, y, z')
p1=implicit_plot3d(x^2 + y^2 + z^2 == 4, (x,-5, 5), (y,-5, 5), (z,-5, 5), opacity=0.2, color="red")
p2=implicit_plot3d(x^2 + y^2 + z^2 == 9, (x,-5, 5), (y,-5, 5), (z,-5, 5), opacity=0.4)
p3=implicit_plot3d(z==0, (x,-5, 5), (y,-5, 5), (z,-5, 5), opacity=0.4, color="green")
show(p1+p2+p3)
```

■

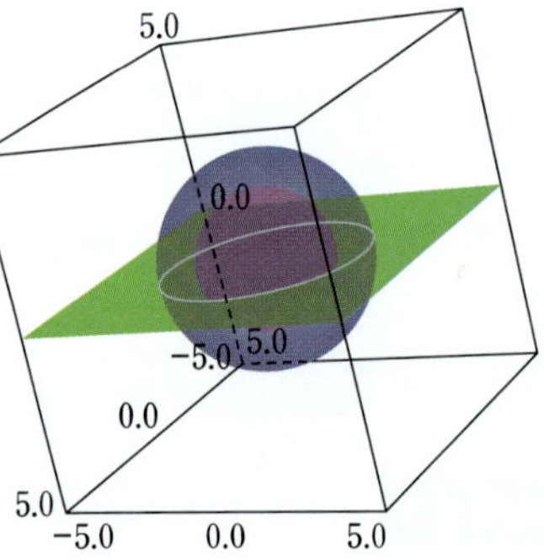

Figure 5

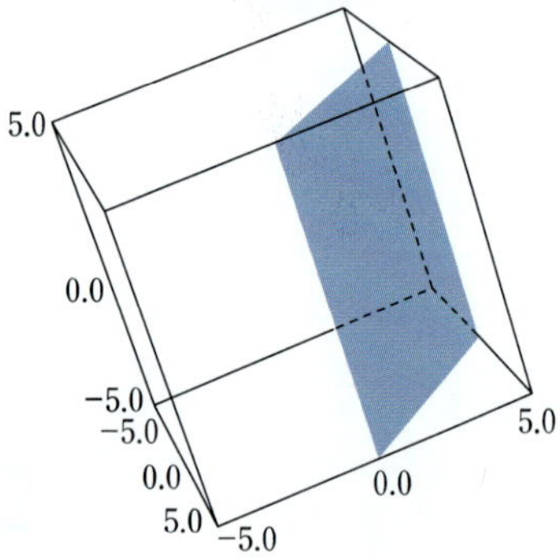

Figure 6

CAS **EXAMPLE 4**

Draw the surface $x+y=5$ in $\mathbb{R}^3$. (See Figure 6.)
http://matrix.skku.ac.kr/cal-lab/cal-11-1-Exm-4.html

Solution

```
var('x, y, z')
implicit_plot3d(x+y-5==0, (x, -5, 5), (y, -5, 5), (z, -5, 5), color='blue', opacity=0.5)
```

■

11.1 EXERCISES (Three-Dimensional Coordinate Systems)

http://matrix.skku.ac.kr/Cal-Book/part2/CS-Sec-11-1-Sol.html YouTube http://youtu.be/_s_2T1VVob8

CAS **1.** Draw the surface $2x-3y+z=1$ in $\mathbb{R}^3$.
http://matrix.skku.ac.kr/cal-lab/cal-11-1-1-a.html

Solution

```
var('x, y, z')
implicit_plot3d(2*x-3*y+z==1, (x, -3, 2), (y,-3,2), (z,-3,3))
```

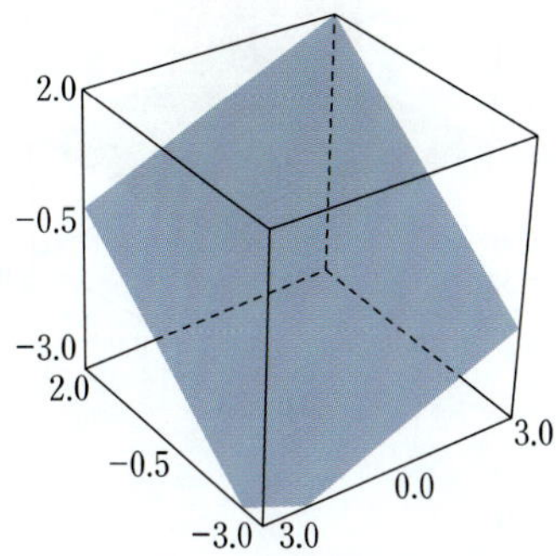

CAS **2.** Draw the surface $x^2-y^2=3$ in $\mathbb{R}^3$.
http://matrix.skku.ac.kr/cal-lab/cal-11-1-2.html

Solution

```
var('x, y, z')
s1=implicit_plot3d(x^2-y^2==3, (x,-3,3), (y,-3,3), (z,-2,0.5), color='red', opacity=0.3)
s1
```

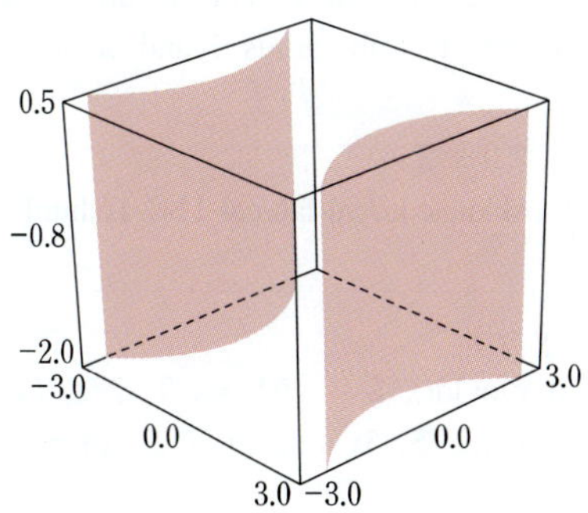

CAS **3.** Find the lengths of the sides of the triangle with vertices $A(1, 3, -2)$, $B(3, 1, -3)$ and $C(2, -1, -1)$. Is ABC a right triangle? Is it an isosceles triangle?
http://matrix.skku.ac.kr/cal-lab/cal-11-1-3.html

Solution Isosceles triangle
3
3
3*sqrt(2)
Answer : an isosceles triangle

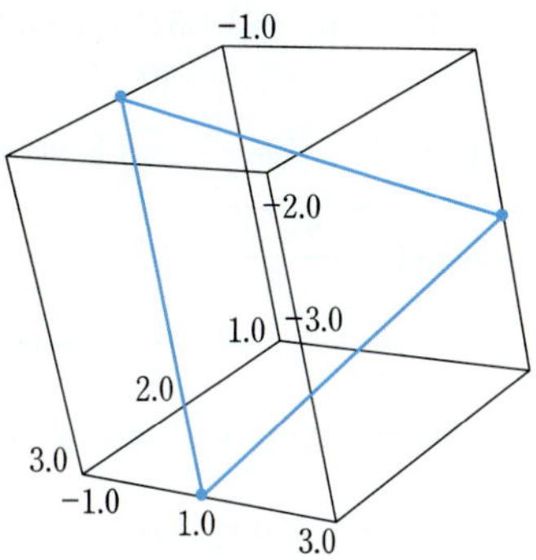

CAS **4.** Find the distance from $(3, -4, 5)$ to each of the following.

(a) The x-axis (b) The y-axis
(c) The z-axis (d) The xy-plane
(e) The yz-plane (f) The xz-plane

http://matrix.skku.ac.kr/cal-lab/cal-11-1-4.html

Solution Answer : (3, 4, 5, sqrt(41), sqrt(34), 5)

5. Find an equation of the sphere with center $(2, -4, 1)$ and radius 3. What is the intersection of this sphere with the yz-plane?

Solution An equation of the sphere :
$(x-2)^2+(y+4)^2+(z-1)^2=9$, and the intersection of this sphere with the yz-plane can be obtained by substituting $x=0$ in the equation. Hence

$$(y+4)^2+(z-1)^2=5.$$

6. Find an equation of the sphere that passes through the point $(-1, 4, 3)$ and has center $(5, 1, -1)$.

Solution The distance between $(-1,4,3)$ and $(5, 1, -1)$ is the radius of the sphere.

Hence, $r=\sqrt{(5-(-1))^2+(1-4)^2+(-1-3)^2}$
$=\sqrt{36+9+16}=\sqrt{61}$.

Thus, an equation of sphere is

$$(x-5)^2+(y-1)^2+(z+1)^2=61.$$

7-8. Show that the equation represents a sphere, and find its center and radius.

7. $x^2+y^2+z^2-8x+6y-12z=13$

Solution $x^2-8x+16+y^2+6y+9+z^2-12z+36$
$=13+16+9+36$
$\Rightarrow (x-4)^2+(y+3)^2+(z-6)^2=74.$
Hence center: $(4, -3, 6)$, radius: $\sqrt{74}$.

8. $3x^2+3y^2+3z^2-12x+18y=1$

Solution Completing squares in the equation gives :
$3(x^2-4x+4)+3(y^2+6y+9)+3z^2=1+12+27$
$\Rightarrow 3(x-2)^2+3(y+3)^2+3z^2=40$
$\Rightarrow (x-2)^2+(y+3)^2+z^2=\dfrac{40}{3}$, with the center $(2, -3, 0)$ and radius $\dfrac{2\sqrt{30}}{3}$.

---

9. (a) Prove that the midpoint of the line segment from $P_1(x_1, y_1, z_1)$ to $P_2(x_2, y_2, z_2)$ is $\left(\dfrac{x_1+x_2}{2}, \dfrac{y_1+y_2}{2}, \dfrac{z_1+z_2}{2}\right)$.

(b) Find the lengths of the medians of the triangle with vertices $A(1, 2, 3)$, $B(0, 3, -1)$ and $V(5, 1, -2)$.

10-16. Determine the region of $\mathbb{R}^3$ represented by the equation or inequality.

10. $x=8$

Solution The equation $x=8$ represents a plane parallel to the yz-plane and 8 units in front of it.

11. $z>2$

12. $0 \le y \le 7$

Solution The inequality $0 \le y \le 7$ represents all points on or between the horizontal planes $y=0$ (the zx-plane) and $y=7$. So the answer is all points on or between the horizontal plane $y=0$ (the zx-plane) and $y=7$.

CAS **13.** $x^2+y^2=3,\ z=-3$
http://matrix.skku.ac.kr/cal-lab/cal-11-1- 13.html

Solution

```
var('x, y, z')
s1=implicit_plot3d(x^2+y^2==3, (x,-3,3), (y,-3,3), (z,-2,0.5), color='red', opacity=0.3)
s2=implicit_plot3d(z==-1, (x,-3,3), (y,-3,3), (z,-2,0.5), color='green', opacity=0.5)
s3=implicit_plot3d(x^2+y^2==3, (x,-3,3), (y,-3,3), (z,-0.99,-1.01), color='blue')
s1+s2+s3
```

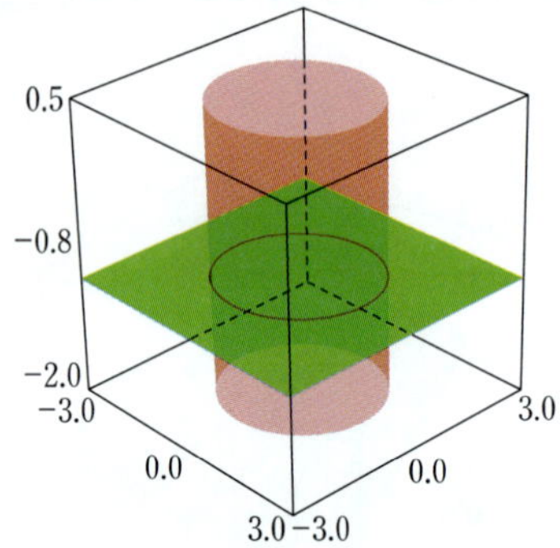

14. $x^2+y^2=9$

Solution The set of all points in $\mathbb{R}^3$ whose distance from the z-axis is 3. This is a cylinder of radius 3 and axis along z-axis.

15. $x^2+y^2+z^2>2x$

Solution The inequality
$$x^2+y^2+z^2>2x \Leftrightarrow (x-1)^2+y^2+z^2>1$$
is equivalent to $\sqrt{(x-1)^2+y^2+z^2}>1$. So the region consists of those points whose distance from the point $(1, 0, 0)$ is greater than 1. This is the set of all points outside the sphere with radius 1 and center $(1, 0, 0)$.

CAS **16.** $x^2+z^2 \le 9-2z$
http://matrix.skku.ac.kr/cal-lab/cal-11-1-16.html

Solution

```
var('x, y, z')
implicit_plot3d(x^2 + z^2 == 9 -2*z, (x, -5, 5), (y, -5, 5), (z, -5, 5), opacity=0.2, color="red")
```

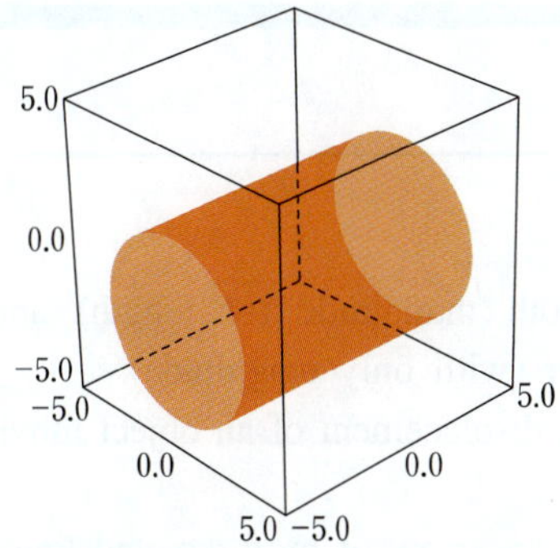

17-18. Describe the given region by an inequality.

17. The half-space consisting of all points to the left of the yz-plane.

Solution This describes all points with positive x-coordinates, that is, $x > 0$.

CAS **18.** The solid rectangular box in the first octant bounded by the planes $x = 1$, $y = 3$, and $z = 2$.
http://matrix.skku.ac.kr/cal-lab/cal-11-1-18.html

Solution

```
var('x, y, z')
p1=implicit_plot3d(x == 1, (x, -5, 5), (y, -5, 5),
(z, -5, 5), opacity=0.2, color="red")
p2=implicit_plot3d(y == 3, (x, -5, 5), (y, -5, 5),
(z, -5, 5), opacity=0.2, color="blue")
p3=implicit_plot3d(z == 2, (x, -5, 5), (y, -5, 5),
(z, -5, 5), opacity=0.2, color="green")
show(p1+p2+p3)
```

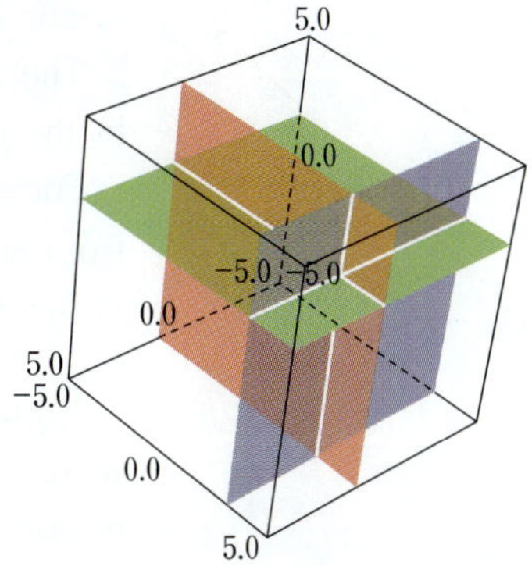

11.2 Vectors

Vectors are quantities that have both magnitude (or length) and direction. Recall that a **scalar** is a real number (with only magnitude).

For example, velocity, acceleration, displacement of an object moving in space and a force are all vector quantities.

A vector is denoted graphically by an arrow or by a directed line segment. A vector is printed as a boldface letter or by putting an arrow above the letter.

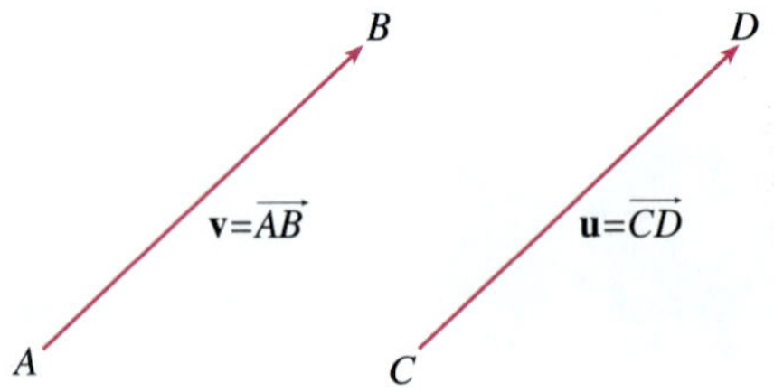

Figure 1

The magnitude of the vector is the length of the arrow and the arrow points in the direction of the vector.

For example in Figure 1, the displacement of a particle along a line segment from point A to point B is represented by $\mathbf{v} = \overrightarrow{AB}$. This **displacement vector v**, has **initial point** A (the tail) and **terminal point** B (the tip).

Equality of vectors: Two vectors **u** and **v** are **equal** (or **equivalent**) and write **u**=**v** if they have the same length and same direction (or parallel) but not necessarily the same initial point. The **zero vector**, denoted by **0**: it has length 0 and no direction.

Addition of vectors: The sum of two vectors **u** and **v** denoted by **u**+**v** is the vector from the initial point of **u** to the terminal point of **v** as shown in the triangle law in Figure 2.

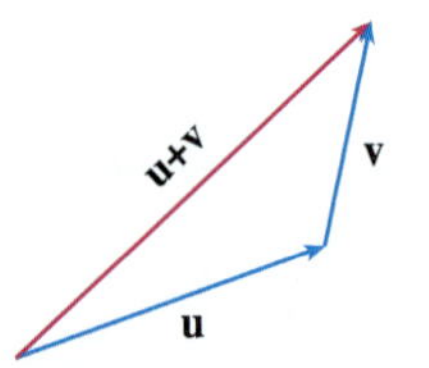

Figure 2 The Triangle Law

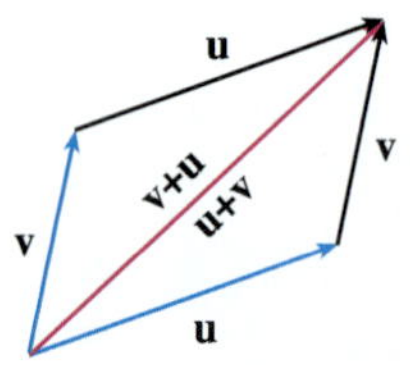

Figure 3 The Parallelogram Law

By placing **v** with the same initial point as **u**, we can construct a parallelogram with **u** and **v** as adjacent sides. Then, $\mathbf{u}+\mathbf{v}$ is the diagonal of the parallelogram starting at the same initial point. (This is called the **Parallelogram Law**: See Figure 3.) Note that **u**+**v**=**v**+**u**. (*i.e.* vector addition is commutative.) For instance, if several forces are acting on an object, the **resultant force** experienced by the object is the vector sum of these forces.

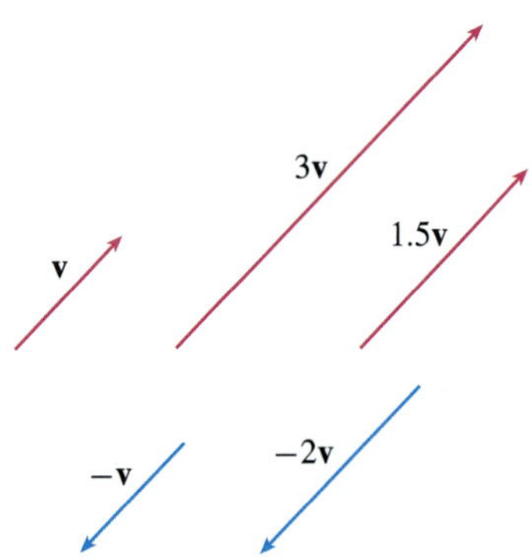

Figure 4

Scalar Multiplication: For any scalar c and a vector **v**, the **scalar multiple** $c\mathbf{v}$ is the vector whose length is $|c|$ times the length of **v** and whose direction is the same as **v** if $c > 0$ and is opposite to **v** if $c < 0$. If $c = 0$, then $0\mathbf{v} = \mathbf{0}$ (the zero vector). Scalar multiplication amounts to scaling the vectors, which can be elongation or contraction.

For example $-\mathbf{v} = (-1)\mathbf{v}$, known as **negative** of **v**, has the same length as **v** but points in the opposite direction. The **difference** of two vectors **u** and **v** is defined by

$$\mathbf{u}-\mathbf{v}=\mathbf{u}+(-\mathbf{v}).$$

Also $c\mathbf{v}=\mathbf{0}$ if $c=0$ or $\mathbf{v}=\mathbf{0}$.

Two non zero vectors are **parallel** if they are scalar multiples of one another.

Component Form of a Vector

Vectors can be treated algebraically by introducing a coordinate system. The **position vector** of a point $P(a_1, a_2, a_3)$ is the vector $\overrightarrow{OP}$ with the initial point at the origin $O(0, 0, 0)$ and terminal point at P. The entries a_1, a_2, a_3 are called the **components** of $\overrightarrow{OP}$ and the vector $\overrightarrow{OP}$ was written as $\langle a_1, a_2, a_3\rangle$. The ordered triple $\langle a_1, a_2, a_3\rangle$ refers to a vector while the ordered triple (a_1, a_2, a_3) refers to a point in the space.

For any vector $\overrightarrow{AB}$ with the initial point $A(x_1, y_1, z_1)$ and the terminal point $B(x_2, y_2, z_2)$, the vector representation of $\overrightarrow{AB}$ is $\langle x_2-x_1, y_2-y_1, z_2-z_1\rangle$ by $\overrightarrow{AB}=\overrightarrow{OB}-\overrightarrow{OA}$.

For example, for the vector represented by the directed line segment with the initial point $A(1, -7, 5)$ and terminal point $B(-6, 3, 1)$, is

$$\overrightarrow{AB}=\langle -6-1, 3+7, 1-5\rangle=\langle -7, 10, -4\rangle.$$

The **magnitude** (or **length**) of the vector $\mathbf{v}$ is denoted by the symbol $|\mathbf{v}|$ or $\|\mathbf{v}\|$. (read as norm $\mathbf{v}$)

The length (or magnitude) of the position vector $\mathbf{a}=\langle a_1, a_2, a_3\rangle$ is $|\mathbf{a}|=\sqrt{a_1^2+a_2^2+a_3^2}$.

A **unit vector** is a vector whose length is 1. In general, if $\mathbf{a}\neq 0$, then the unit vector that has the same direction as $\mathbf{a}$ is

1 $$\mathbf{u}=\frac{1}{|\mathbf{a}|}\mathbf{a}=\frac{\mathbf{a}}{|\mathbf{a}|}.$$

The vector $\mathbf{u}=\frac{1}{|\mathbf{a}|}\mathbf{a}$ is the normalization of the vector $\mathbf{a}$, $\mathbf{u}$ has the same direction as $\mathbf{a}$ and

$$|\mathbf{u}|=\frac{1}{|\mathbf{a}|}|\mathbf{a}|=1.$$

In component form, vectors are added by just adding their corresponding components. Similarly, we can subtract vectors by subtracting the corresponding components and multiple of a vector by a scalar c is done by multiplying each component by the scalar c. Thus,

$$\langle a_1, a_2, a_3\rangle+\langle b_1, b_2, b_3\rangle=\langle a_1+b_1, a_2+b_2, a_3+b_3\rangle,$$

$$\langle a_1, a_2, a_3\rangle - \langle b_1, b_2, b_3\rangle = \langle a_1 - b_1, a_2 - b_2, a_3 - b_3\rangle,$$
$$c\langle a_1, a_2, a_3\rangle = \langle ca_1, ca_2, ca_3\rangle.$$

We denote by $\mathbb{R}^3$ the set of all three dimensional vectors. More generally, $\mathbb{R}^n$ denotes the set of all n-dimensional vectors. An n-dimensional vector is an ordered n-tuple: $\mathbf{a} = \langle a_1, a_2, \ldots, a_n\rangle$ having n components $a_1, a_2, \ldots, a_n$. For n-dimensional vectors, addition and scalar multiplication is defined in terms of its components just as for the cases $n = 2$ and $n = 3$.

Algebra of Vectors

If $\mathbf{a}$, $\mathbf{b}$ and $\mathbf{c}$ are vectors in $\mathbb{R}^n$ and c and d are scalars, then

1. $\mathbf{a}+\mathbf{b} = \mathbf{b}+\mathbf{a}$
2. $\mathbf{a}+(\mathbf{b}+\mathbf{c}) = (\mathbf{a}+\mathbf{b})+\mathbf{c}$
3. $\mathbf{a}+\mathbf{0} = \mathbf{a}$
4. $\mathbf{a}+(-\mathbf{a}) = \mathbf{0}$
5. $c(\mathbf{a}+\mathbf{b}) = c\mathbf{a}+c\mathbf{b}$
6. $(c+d)\mathbf{a} = c\mathbf{a}+d\mathbf{a}$
7. $(cd)\mathbf{a} = c(d\mathbf{a})$
8. $1\mathbf{a} = \mathbf{a}$
9. $0\mathbf{a} = \mathbf{0}$
10. $c\mathbf{0} = \mathbf{0}$

Equality of Vectors: Two vectors $\mathbf{u}$ and $\mathbf{v}$ are equal if and only if their corresponding components are equal.

Let $\mathbf{i}$, $\mathbf{j}$ and $\mathbf{k}$ be unit vectors in the directions of the positive x-, y-, and z-axes. Thus unit vectors are write as follow:

$$\mathbf{i} = \langle 1, 0, 0\rangle \quad \mathbf{j} = \langle 0, 1, 0\rangle \quad \mathbf{k} = \langle 0, 0, 1\rangle$$

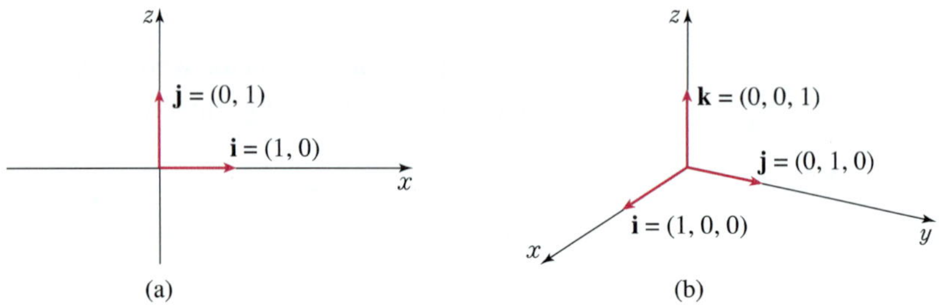

Figure 5 Standard basis vectors in $\mathbb{R}^2$ and $\mathbb{R}^3$

Any vector in $\mathbb{R}^3$ can be expressed in terms of the **standard basis vectors** (or unit vectors along axes) $\mathbf{i}$, $\mathbf{j}$ and $\mathbf{k}$. For $\mathbf{a} = \langle a_1, a_2, a_3\rangle$, we can write

2
$$\begin{aligned} \mathbf{a} &= \langle a_1, a_2, a_3\rangle = \langle a_1, 0, 0\rangle + \langle 0, a_2, 0\rangle + \langle 0, 0, a_3\rangle \\ &= a_1\langle 1, 0, 0\rangle + a_2\langle 0, 1, 0\rangle + a_3\langle 0, 0, 1\rangle \\ &= a_1\mathbf{i} + a_2\mathbf{j} + a_3\mathbf{k}. \end{aligned}$$

The above representation is unique. For example, $\langle 3, 5, -7\rangle = 3\mathbf{i}+5\mathbf{j}-7\mathbf{k}$.

For example, if $\mathbf{a} = \mathbf{i}-\mathbf{j}+3\mathbf{k}$ and $\mathbf{b} = 5\mathbf{i}+4\mathbf{j}-2\mathbf{k}$, then $5\mathbf{a}-3\mathbf{b} = -10\mathbf{i}-17\mathbf{j}+21\mathbf{k}$.

For example, the unit vector in the direction of the vector $4\mathbf{i}+\mathbf{j}-3\mathbf{k}$ is

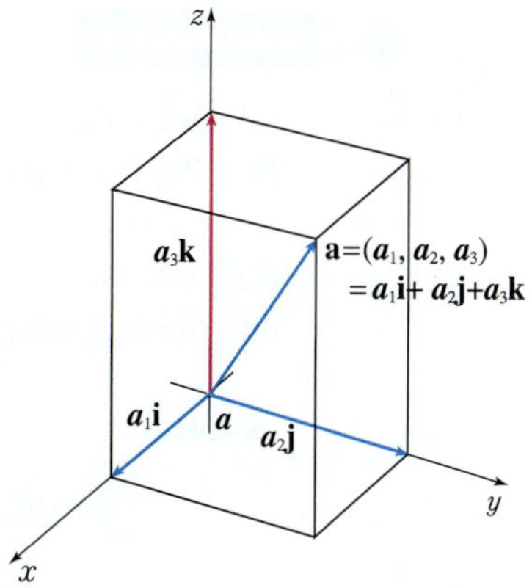

Figure 6

$\dfrac{4\mathbf{i}+\mathbf{j}-3\mathbf{k}}{\sqrt{4^2+1^2+3^2}}=\dfrac{4\mathbf{i}+\mathbf{j}-3\mathbf{k}}{\sqrt{26}}$. Note that the unit vector in the opposite direction of the vector $4\mathbf{i}+\mathbf{j}-3\mathbf{k}$ is $-(4\mathbf{i}+\mathbf{j}-3\mathbf{k})/\sqrt{26}$.

Application

Two forces $\mathbf{F}_1$ and $\mathbf{F}_2$ with magnitudes 10 lb and 12 lb act on an object at a point P as shown in Figure 7. Find the resultant force $\mathbf{F}$ acting at P as well as its magnitude and direction. (Indicate the direction by finding the angle θ shown in Figure 7.)

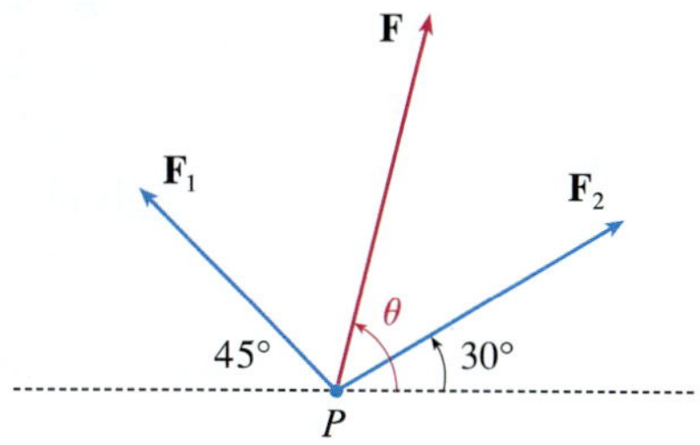

Figure 7

We want to find $\mathbf{F}=\mathbf{F}_1+\mathbf{F}_2$. First we can find as

$$\mathbf{F}_1 = 10\cos 30^\circ\,\mathbf{i} + 10\sin 30^\circ\,\mathbf{j} = 10\times\frac{\sqrt{3}}{2}\,\mathbf{i} + 10\times\frac{1}{2}\,\mathbf{j}$$
$$= 5\sqrt{3}\,\mathbf{i} + 5\,\mathbf{j}$$
$$\mathbf{F}_2 = 12\cos 135^\circ\,\mathbf{i} + 12\sin 135^\circ\,\mathbf{j} = 12\times\left(-\frac{1}{\sqrt{2}}\right)\mathbf{i} + 12\times\frac{1}{\sqrt{2}}\,\mathbf{j}$$
$$= -6\sqrt{2}\,\mathbf{i} + 6\sqrt{2}\,\mathbf{j}$$

Hence, $\mathbf{F}=\mathbf{F}_1+\mathbf{F}_2=(5\sqrt{3}-6\sqrt{2})\mathbf{i}+(5+6\sqrt{2})\mathbf{j}$.

CAS **EXAMPLE 1**

(a) Draw the vectors $\mathbf{a} = \langle -2, 3 \rangle$, $\mathbf{b} = \langle -2, 5 \rangle$ and $\mathbf{c} = \langle 7, 0 \rangle$.

(b) Show by means of a sketch, that there are scalars s and t such that $\mathbf{c} = s\mathbf{a} + t\mathbf{b}$.

(c) Find the exact values of s and t.

http://matrix.skku.ac.kr/cal-lab/cal-11-2-Exm-1.html

Solution

```
z=(0, 0); a=(-2, 3); b=(-2, -5); c=(7, 0)
  s, t=var('s, t')
  Ab = matrix(QQ, 2, 3, [a[0], b[0], c[0], a[1], b[1], c[1]])
  T = Ab.echelon_form()
  s=T[0, 2];t=T[1, 2]
  A=arrow(z, a, color=(2, 1, 1))
  A=A+text("a", (a[0]+0.3, a[1]+0.3))
  A=A+arrow(z, (s*a[0], s*a[1]), color='blue')
  A=A+text("sa", (s*a[0]+0.3, s*a[1]+0.3))
  A=A+arrow(z, b, color=(3, 1, 0))
  A=A+text("b", (b[0]+0.3, b[1]+0.3))
  A=A+arrow(z, (t*b[0], t*b[1]), color='green')
  A=A+arrow((t*b[0], t*b[1]), c, color='red')
 A=A+text("tb", (t*b[0]+0.3, t*b[1]+0.3))
  A=A+arrow((s*a[0], s*a[1]), c, color='red')
  A=A+arrow(z, c, color='black')
  A=A+text("c", (c[0]+0.3, c[1]+0.3))
  A=A+point((a, b, c, (s*a[0], s*a[1]), (t*b[0], t*b[1])),
rgbcolor='brown', size=30)
  show(A)
```

Answer : -35/16, -21/16

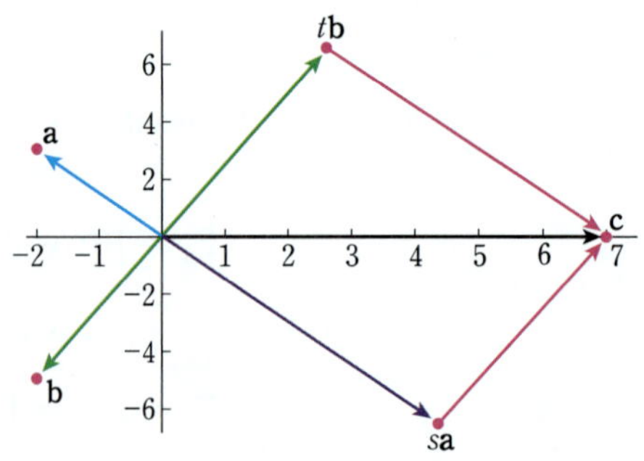

Figure 10

■

CAS **EXAMPLE 2**

For given vectors $\mathbf{a} = \langle 1, -1, 2 \rangle$, $\mathbf{b} = \langle 3, -1, 2 \rangle$, $\mathbf{c} = \langle 0 1, 2 \rangle$ and $\mathbf{d} = \langle 2, 1, 4 \rangle$. Find scalars r, s and t such that $\mathbf{d} = r\mathbf{a} + s\mathbf{b} + t\mathbf{c}$.

http://matrix.skku.ac.kr/cal-lab/cal-11-2-Example-2.html

Solution

```
z=vector([0, 0,0]); a=vector([1, -1, 2]); b=vector([3, -1, 2]);
c=vector([0,1,2]); d=vector([2,1,4])
Ad= matrix(QQ, [a, b, c, d]).transpose()
print Ad
T = Ad.echelon_form()
print T
print -1/4*a + 3/4*b + 3/2*c
p=plot(a,color='red')+plot(b,color='green')+plot(c)+plot(d,color='
goldenrod')
p
var('r,s,t')
n=3
eq=[r*a[i]+s*b[i]+t*c[i]==d[i] for i in range(3)]
solve(eq,r,s,t)
```

```
[ 1  3  0  2]
[ 1  3  0  2]
[-1 -1  1  1]
[ 2  2  2  4]
[   1    0    0 -1/4]
[   0    1    0  3/4]
[   0    0    1  3/2]
```

(2, 1, 4) # vector $\mathrm{d} = < 2, 1, 4 >$

[[r == (-1/4), s == (3/4), t == (3/2)]]

Answer : $(r, s, t) = \left(-\frac{1}{4}, \frac{3}{4}, \frac{3}{2}\right)$.

d = -1/4*a + 3/4*b + 3/2*c. ■

11.2 EXERCISES (Vectors)

http://matrix.skku.ac.kr/Cal-Book/part2/CS-Sec-11-2-Sol.html

YouTube http://youtu.be/BFgh6irMqsc

1-2. Determine $\overrightarrow{AB}$.

CAS **1.** (a) $A(2,\ 5),\ B(-3,\ 1)$

(b) $A(5,\ -3),\ B(10,\ 7)$

Solution (a) $\overrightarrow{AB}=\langle -3-2,\ 1-5\rangle=\langle -5,\ -4\rangle$

(b) $\overrightarrow{AB}=\langle 10-5,\ 7-(-3)\rangle=\langle 5,\ 10\rangle$

```
v1=vector([5, -3])
v2=vector([10, 7])
v2- v1
```

Answer : (5, 10)

2. $A(-1,\ 2,\ 3),\ B(-2,\ 1,-2)$

Solution $\overrightarrow{AB}=\langle -1+2,\ 2-1,\ 3-2\rangle=\langle 1,\ 1,\ 1\rangle$.

3-4. Find the sum of the given vectors.

3. $\langle -1,\ 4\rangle,\ \langle 3,\ -1\rangle$

Solution $\langle -1,\ 4\rangle+\langle 3,\ -1\rangle=\langle 2,\ 3\rangle$.

CAS **4.** $\langle 0,\ 1,\ -4\rangle,\ \langle 0,\ 2,\ 0\rangle$

http://matrix.skku.ac.kr/cal-lab/cal-11-2-4.html

Solution

```
v1=vector([0, 1, -4])
v2=vector([0, 2, 0])
v1+v2
```

Answer : (0, 3, -4)

5-8. Compute $|\mathbf{a}|$, $\mathbf{a}+\mathbf{b}$, $2\mathbf{a}-3\mathbf{b}$ and $|\mathbf{a}-\mathbf{b}|$.

5. $\mathbf{a}=\langle 5,\ 3\rangle,\ \mathbf{b}=\langle -3,\ 2\rangle$

Solution $|\mathbf{a}|=\sqrt{5^2+3^2}=\sqrt{34}$,

$\mathbf{a}+\mathbf{b}=\langle 5+(-3),\ 3+2\rangle=\langle 2,\ 5\rangle$,

$2\mathbf{a}-3\mathbf{b}=\langle 10,\ 6\rangle-\langle 9,\ 6\rangle$

$=\langle 10-(-9),\ -6\rangle=\langle 19,\ 0\rangle$,

$|\mathbf{a}-\mathbf{b}|=|\langle 5-(-3),\ 3-2\rangle|=|\langle 8,\ 1\rangle|$

$=\sqrt{8^2+1^2}=\sqrt{65}$.

CAS **6.** $\mathbf{a}=\langle 5,\ -1,\ 3\rangle,\ \mathbf{b}=\langle -1,\ 3,\ -2\rangle$.

http://matrix.skku.ac.kr/cal-lab/cal-11-2-6.html

Solution

```
a=vector([5, -1, 3])
b=vector([-1, 3, -2])
a.norm()
```

Answer : sqrt(35)

```
a+b
```

Answer : (4, 2, 1)

```
2*a-3*b
```

Answer : (13, -11, 12)

```
(a-b).norm()
```

Answer : sqrt(77)

7. $\mathbf{a}=\mathbf{i}-3\mathbf{j}+\mathbf{k},\ \mathbf{b}=\mathbf{j}+3\mathbf{k}$

Solution $|\mathbf{a}|=\sqrt{1^2+(-3)^2+1^2}=\sqrt{11}$,

$\mathbf{a}+\mathbf{b}=\mathbf{i}+(-3+1)\mathbf{j}+(1+3)\mathbf{k}=\mathbf{i}-2\mathbf{j}+4\mathbf{k}$,

$2\mathbf{a}-3\mathbf{b}=2(\mathbf{i}-3\mathbf{j}+\mathbf{k})-3(\mathbf{j}+3\mathbf{k})=2\mathbf{i}-9\mathbf{j}-7\mathbf{k}$,

$|\mathbf{a}-\mathbf{b}|=|(\mathbf{i}-3\mathbf{j}+\mathbf{k})-(\mathbf{j}+3\mathbf{k})|$

$=|\mathbf{i}-4\mathbf{j}-2\mathbf{k}|=\sqrt{1^2+(-4)^2+(-2)^2}=\sqrt{21}$.

CAS **8.** $\mathbf{a}=3\mathbf{i}-4\mathbf{j},\ \mathbf{b}=\mathbf{i}-\mathbf{j}+\mathbf{k}$

http://matrix.skku.ac.kr/cal-lab/cal-11-2-8.html

Solution

```
a=vector([3, -4, 0])
b=vector([1, -1, 1])
print a.norm()
print a+b
print 2*a-3*b
print (a-b).norm()
```

Answer : 5

(4, -5, 1)

(3, -5, -3)

sqrt(14)

9. Determine a unit vector that has the same direction as $7\mathbf{i}-3\mathbf{j}+\mathbf{k}$.

Solution The vector $7\mathbf{i}-3\mathbf{j}+\mathbf{k}$ has length $|7\mathbf{i}-3\mathbf{j}+\mathbf{k}|$ $=\sqrt{7^2+(-3)^2+1^2}=\sqrt{59}$, so the unit vector with the same direction is

$$\frac{1}{\sqrt{59}}(7\mathrm{i}-3\mathrm{j}+\mathrm{k})=\frac{7}{\sqrt{59}}\mathrm{i}-\frac{3}{\sqrt{59}}\mathrm{j}+\frac{1}{\sqrt{59}}\mathrm{k}.$$

CAS **10.** Find a vector that has the same direction as $\langle 2, -4, -2\rangle$ but has length 6.
http://matrix.skku.ac.kr/cal-lab/cal-11-2-10.html

Solution

```
v1=vector([2,-4,-2])
v2=6/v1.norm()*v1
print v2
v2.norm()
```

Answer : (sqrt(6), -2*sqrt(6), -sqrt(6))
6

11. A clothesline is tied between two poles, 6m apart. The line is quite taut and has negligible sag. When a wet shirt with a mass of 0.8kg is hung at the middle of the line, the midpoint is pulled down 6cm. Find the tension in each half of the clothesline.

Solution Let T_1 and T_2 represent the tension vectors in each side of the clothe line as shown in the figure. Then T_1 and T_2 have equal vertical components and opposite horizontal components, so

$$T_1=-a\mathrm{i}+b\mathrm{j} \text{ and } T_2=a\mathrm{i}+b\mathrm{j}\ (a,\ b>0).$$

By similar triangles, $\frac{b}{a}=\frac{0.06}{3}\Rightarrow a=50b$.

The force due to gravity acting on the shirt has magnitude $0.8g\approx(0.8)(9.8)=7.84\text{N}$, hence we have $\mathrm{w}=-7.84\mathrm{j}$. The resultant T_1+T_2 of the tensile forces counterbalances w, so

$T_1+T_2=-\mathrm{w}\Rightarrow(-a\mathrm{i}+b\mathrm{j})+(a\mathrm{i}+b\mathrm{j})=7.84\mathrm{j}$
$\Rightarrow(-50b\mathrm{i}+b\mathrm{j})+(50b\mathrm{i}+b\mathrm{j})=2b\mathrm{j}=7.84\mathrm{j}$
$\Rightarrow b=\frac{7.84}{2}=3.92$
and $a=50b=196$.

Thus, the tensions are

$$T_1=-196\mathrm{i}+3.92\mathrm{j},\ T_2=196\mathrm{i}+3.92\mathrm{j}.$$

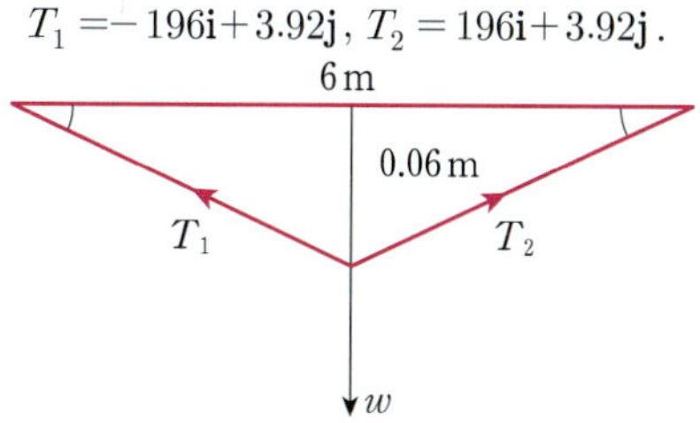

12. The tension T at each end of the chain has magnitude 50N. What is the weight of the chain?

CAS **13.** (a) Draw the vectors $\mathbf{a}=\langle 2, -3\rangle$, $\mathbf{b}=\langle -2, -1\rangle$ and $\mathbf{c}=\langle 6, -5\rangle$.
http://matrix.skku.ac.kr/cal-lab/cal-11-2-13.html

(b) Show, by means of a sketch, that there are scalars s and t such that $\mathbf{c}=s\mathbf{a}+t\mathbf{b}$.

(c) Find the exact values of s and t.

Solution (a)

```
a=vector([2, -3])
b=vector([-2, -1])
c=vector([6, -5])
p=plot(a, color='red')+plot(b, color='green')+plot(c)
t = text("a",
(a[0]+0.3,a[1]+0.3))+text("b",(b[0]-0.3,b[1]+0.3))+t
ext("c",(c[0]+0.3,c[1]+0.3))
show(p+t, figsize=4)
```

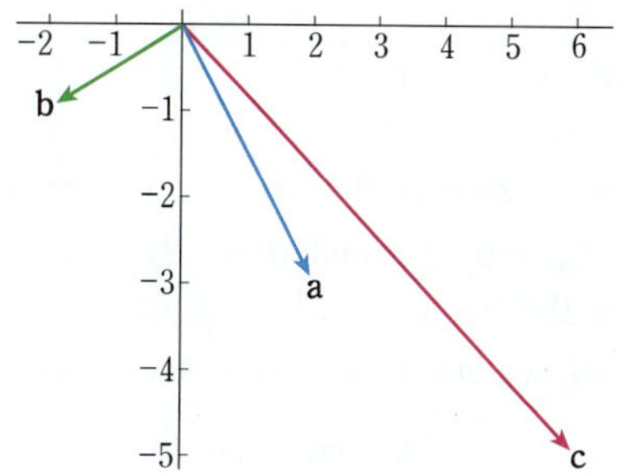

(b)

```
j=2
k=-1
z=(0, 0)
a=(2, -3)
b=(-2, -1)
c=(6, -5)
A=arrow(z, a, color=(2, 1, 1))+arrow(z, b,
color=(3, 1, 0))+arrow(z, c, color='black')
A=A+arrow(z, (j*a[0], j*a[1]),
color='blue')+arrow(z, (k*b[0], k*b[1]),
color='green')
A=A+arrow((k*b[0], k*b[1]), c,
color='red')+arrow((j*a[0], j*a[1]), c, color='red')
A=A+text("a", (a[0]+0.3, a[1]+0.3))+text("ja",
(j*a[0]+0.3, j*a[1]+0.3))+text("b", (b[0]+0.3,
b[1]+0.3))
A=A+text("kb", (k*b[0]+0.3,
k*b[1]+0.3))+text("c", (c[0]+0.3, c[1]+0.3))
A=A+point((a, b, c, (j*a[0], j*a[1]), (k*b[0],
k*b[1])), rgbcolor='brown', size=30)
show(A)
```

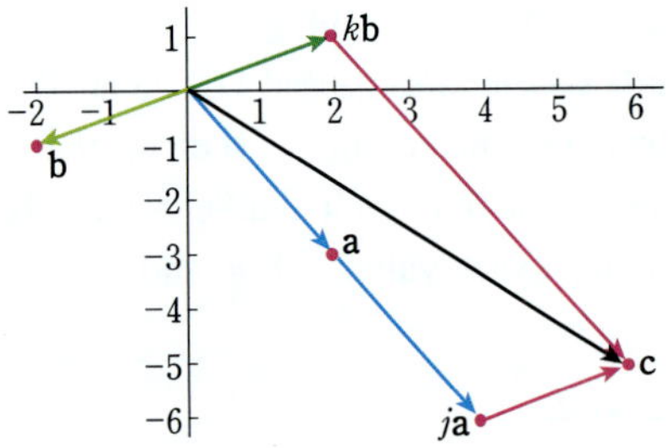


```
var('s,t')
a=vector([2,-3])
b=vector([-2,-1])
c=vector([6,-5])
n=3
eq=[s*a[i]+t*b[i]==c[i] for i in range(2)]
solve(eq,s,t)
```

Answer : $s=2,\ \ t=-1$

CAS **14.** Let $\mathbf{a}=\langle 2,\ -3,\ 1\rangle$, $\mathbf{b}=\langle -2,\ -2,\ 1\rangle$, $\mathbf{c}=\langle 6,\ -5,\ 2\rangle$ and $\mathbf{d}=\langle 2,\ 1,\ 2\rangle$ in $\mathbb{R}^3$.

(i) Plot the vectors $\mathbf{a}$, $\mathbf{b}$, $\mathbf{c}$ and $\mathbf{d}$.

(ii) Find scalars r, s and t such that

$$\mathbf{d}=r\mathbf{a}+s\mathbf{b}+t\mathbf{c}.$$

Solution

```
a=vector([2,-3,1])
b=vector([-2,-2,1])
c=vector([6,-5,2])
d=vector([2,1,2])
p=plot(a,color='red')+plot(b,color='green')+plot(c)+p
lot(d,color='goldenrod')
p
```

```
var('r,s,t')
n=3
eq=[r*a[i]+s*b[i]+t*c[i]==d[i] for i in range(3)]
solve(eq,r,s,t)
```

Answer : [[r == (-28/3), s == (8/3), t == (13/3)]]

15. If $\mathbf{r}=\langle x,\ y,\ z\rangle$ and $\mathbf{r}_0=\langle x_0,\ y_0,\ z_0\rangle$, describe the set of all points $(x,\ y,\ z)$ such that $|\mathbf{r}-\mathbf{r}_0|=3$.

Solution $|\langle x,\ y,\ z\rangle-\langle x_0,\ y_0,\ z_0\rangle|=3$

$\Rightarrow |\langle x-x_0,\ y-y_0,\ z-z_0\rangle|=3$

$\Rightarrow (x-x_0)^2+(y-y_0)^2+(z-z_0)^2=9$

Therefore the surface of a sphere with a center $(x_0,\ y_0,\ z_0)$ and a radius 3.

11.3 The Dot Product

The product of two vectors which is known as a scalar product or dot product or even as inner product, is defined as follows:

The **dot product** of $\mathbf{a} = \langle a_1, a_2, a_3 \rangle$ and $\mathbf{b} = \langle b_1, b_2, b_3 \rangle$ is the real number (scalar) $\mathbf{a} \cdot \mathbf{b}$ given by

$$\mathbf{a} \cdot \mathbf{b} = a_1 b_1 + a_2 b_2 + a_3 b_3 .$$

The dot product of $\mathbf{a}$ and $\mathbf{b}$ is obtained by multiplying corresponding components and then adding the individual products.

For example,

$$\langle 2, 3, 4 \rangle \cdot \langle 1, 5, -3 \rangle = 2(1) + 3(5) + 4(-3) = 5,$$
$$(2\mathbf{i} - 3\mathbf{j} + 5\mathbf{k}) \cdot (\mathbf{i} + 2\mathbf{j} - 6\mathbf{k}) = 2(1) + (-3)(2) + 5(-6) = -34.$$

Properties of the Dot Product

If $\mathbf{a}$, $\mathbf{b}$ and $\mathbf{c}$ are vectors in $\mathbb{R}^3$ and α is a scalar, then

(i) $\mathbf{a} \cdot \mathbf{a} = |\mathbf{a}|^2$

(ii) $\mathbf{a} \cdot \mathbf{b} = \mathbf{b} \cdot \mathbf{a}$

(iii) $\mathbf{a} \cdot (\mathbf{b} + \mathbf{c}) = \mathbf{a} \cdot \mathbf{b} + \mathbf{a} \cdot \mathbf{c}$

(iv) $(\alpha \mathbf{a}) \cdot \mathbf{b} = \alpha(\mathbf{a} \cdot \mathbf{b}) = \mathbf{a} \cdot (\alpha \mathbf{b})$

(v) $\mathbf{0} \cdot \mathbf{a} = 0$

Proof Let $\mathbf{a} = \langle a_1, a_2, a_3 \rangle$, $\mathbf{b} = \langle b_1, b_2, b_3 \rangle$ and $\mathbf{c} = \langle c_1, c_2, c_3 \rangle$. We shall prove few of them and remaining as exercise.

(i) $\mathbf{a} \cdot \mathbf{a} = a_1^2 + a_2^2 + a_3^2 = |\mathbf{a}|^2$

(iii)
$$\begin{aligned} \mathbf{a} \cdot (\mathbf{b} + \mathbf{c}) &= \langle a_1, a_2, a_3 \rangle \cdot \langle b_1 + c_1, b_2 + c_2, b_3 + c_3 \rangle \\ &= a_1(b_1 + c_1) + a_2(b_2 + c_2) + a_3(b_3 + c_3) \\ &= a_1 b_1 + a_1 c_1 + a_2 b_2 + a_2 c_2 + a_3 b_3 + a_3 c_3 \\ &= (a_1 b_1 + a_2 b_2 + a_3 b_3) + (a_1 c_1 + a_2 c_2 + a_3 c_3) \\ &= \mathbf{a} \cdot \mathbf{b} + \mathbf{a} \cdot \mathbf{c} \end{aligned}$$

■

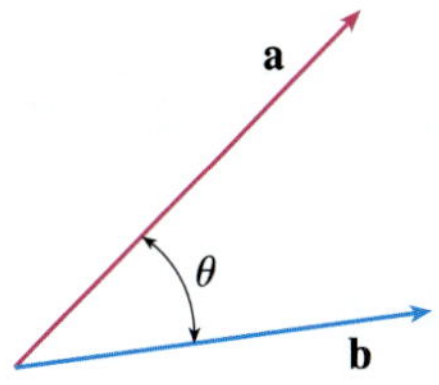

$\mathbf{a} \cdot \mathbf{b} = |\mathbf{a}|\,|\mathbf{b}| \cos(\theta)$

Figure 1

Geometric interpretation: The dot product enables us to find the angle between two nonzero vectors. Let θ be **angle between** $\mathbf{a}$ and $\mathbf{b}$ where $0 \leq \theta \leq \pi$ in Figure 1.

Then the dot product of two nonzero vectors $\mathbf{a}$ and $\mathbf{b}$ is $\mathbf{a} \cdot \mathbf{b} = |\mathbf{a}||\mathbf{b}| \cos\theta$. Thus, the formula of the angle between two nonzero vectors $\mathbf{a}$ and $\mathbf{b}$ is given by

$$\cos\theta = \frac{\mathbf{a}\cdot\mathbf{b}}{|\mathbf{a}||\mathbf{b}|}.$$

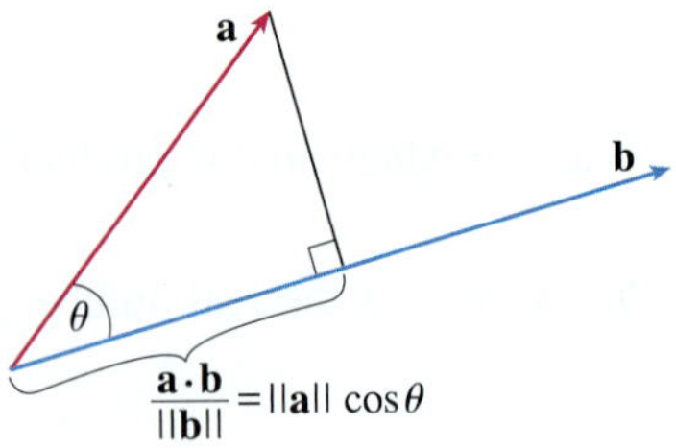

Figure 2

http://www.youtube.com/watch?v=g55dfkmlTHE&feature=youtu.be

The dot product of two vectors is itself a scalar. Two special cases immediately arise:

(i) a and b are **perpendicular** $(\theta=\pi/2)$ if and only if $\mathbf{a}\cdot\mathbf{b}=0$.
(ii) a and b are **parallel** $(\theta=0 \text{ or } \pi)$ if and only if $\mathbf{a}\cdot\mathbf{b}=\pm|\mathbf{a}||\mathbf{b}|$.

Note The zero vector 0 is perpendicular to all vectors.
Specially, for the unit vectors in $\mathbb{R}^3$,

$$\mathbf{i}\cdot\mathbf{i}=1,\quad \mathbf{j}\cdot\mathbf{j}=1,\quad \mathbf{k}\cdot\mathbf{k}=1,\quad \mathbf{i}\cdot\mathbf{j}=0,\quad \mathbf{j}\cdot\mathbf{k}=0,\quad \mathbf{k}\cdot\mathbf{i}=0.$$

For example, consider the vectors a and b having lengths 5 and 8 with an angle $\pi/6$ between a and b.
As a result, we have the dot product of a and b is $\mathbf{a}\cdot\mathbf{b}=|\mathbf{a}||\mathbf{b}|\cos\pi/6 = 5\cdot 8(\sqrt{3}/2)=20\sqrt{3}$.

CAS **EXAMPLE 1**

$\mathbf{a}=\langle 2, -3\rangle$, $\mathbf{b}=\langle 1, -7\rangle$. Find $\mathbf{a}\cdot\mathbf{b}$.
http://matrix.skku.ac.kr/cal-lab/cal-11-3-Exs-6.html

Solution

```
a=vector(QQ, [2, -3]);b=vector(QQ, [1, -7])
ab=a.dot_product(b)
```

Answer : $\mathbf{a}\cdot\mathbf{b}=23$ ■

CAS **EXAMPLE 2**

Find the angle between the vectors $\mathbf{a}=\langle 2, 4, -4\rangle$ and $\mathbf{b}=\langle 2, 2, 0\rangle$.
http://matrix.skku.ac.kr/cal-lab/cal-11-3-Example-2.html

Solution Since $|\mathbf{a}|=\sqrt{2^2+4^2+(-4)^2}=6$, $|\mathbf{b}|=\sqrt{2^2+2^2+0^2}=2\sqrt{2}$ and

$$\mathbf{a}\cdot\mathbf{b}=2(2)+4(2)+(-4)(0)=12.$$

We have

$$\cos\theta=\frac{\mathbf{a}\cdot\mathbf{b}}{|\mathbf{a}||\mathbf{b}|}=\frac{12}{12\sqrt{2}}=\frac{1}{\sqrt{2}}$$

That is, the angle between a and b is $\theta=45^\circ$

```
def anglebetween(a,b)
    return arccos(a.dot_product(b)/(a.norm()*b.norm()))
a=vector([2, 4, -4])
b=vector([2, 2, 0])
anglebetween(a, b)
```

Answer : 45° ■

EXAMPLE 3

Show that $3\mathbf{i}-5\mathbf{j}+2\mathbf{k}$ is perpendicular to $6\mathbf{i}+2\mathbf{j}-4\mathbf{k}$.

Solution Since $(3\mathbf{i}-5\mathbf{j}+2\mathbf{k})\cdot(6\mathbf{i}+2\mathbf{j}-4\mathbf{k})=3(6)+(-5)(2)+2(-4)=0$, these vectors are perpendicular. ■

Since $\cos\theta>0$ if $0\le\theta<\pi/2$ and $\cos\theta<0$ if $\pi/2<\theta\le\pi$, it follows that $\mathbf{a}\cdot\mathbf{b}$ is positive for $\theta<\pi/2$ and negative for $\theta>\pi/2$. The dot product $\mathbf{a}\cdot\mathbf{b}$ measures the extent to which $\mathbf{a}$ and $\mathbf{b}$ point in the same direction.

If $\mathbf{a}\cdot\mathbf{b}>0$, then $\mathbf{a}$ and $\mathbf{b}$ point in the same general direction, if $\mathbf{a}\cdot\mathbf{b}=0$, then they are perpendicular, and if $\mathbf{a}\cdot\mathbf{b}<0$, then they point in generally opposite directions. (See Figure 3.) When $\mathbf{a}$ and $\mathbf{b}$ point in exactly the same direction, we have $\theta=0$, so $\cos\theta=1$ and

$$\mathbf{a}\cdot\mathbf{b}=|\mathbf{a}||\mathbf{b}|.$$

If $\mathbf{a}$ and $\mathbf{b}$ point in exactly opposite directions, then $\theta=\pi$ and so $\cos\theta=-1$ and

$$\mathbf{a}\cdot\mathbf{b}=-|\mathbf{a}||\mathbf{b}|.$$

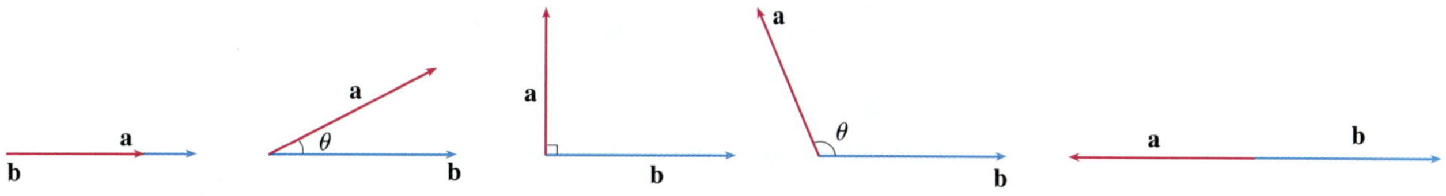

Figure 3

Proposition: Let $\mathbf{a}$, $\mathbf{b}$ and $\mathbf{c}$ be vectors. If $\mathbf{a}$ and $\mathbf{b}$ are both perpendicular to $\mathbf{c}$, then every linear combination $\alpha\mathbf{a}+\beta\mathbf{b}$ is perpendicular to $\mathbf{c}$.

Proof Suppose that $\mathbf{a}$ and $\mathbf{b}$ are both perpendicular to $\mathbf{c}$. Then

$$\mathbf{a}\cdot\mathbf{c}=0 \quad \text{and} \quad \mathbf{b}\cdot\mathbf{c}=0.$$

It follows that

$$(\alpha\mathbf{a}+\beta\mathbf{b})\cdot\mathbf{c}=\alpha(\mathbf{a}\cdot\mathbf{c})+\beta(\mathbf{b}\cdot\mathbf{c})=0$$

and therefore $\alpha\mathbf{a}+\beta\mathbf{b}$ is perpendicular to $\mathbf{c}$. ■

Direction Angles and Direction Cosines

The angles α, β, and γ which a nonzero vector **a** makes with the positive x-axis, y-axis and z-axis are known as the **direction angles** of **a**. These angles lie in the interval $[0, \pi]$.

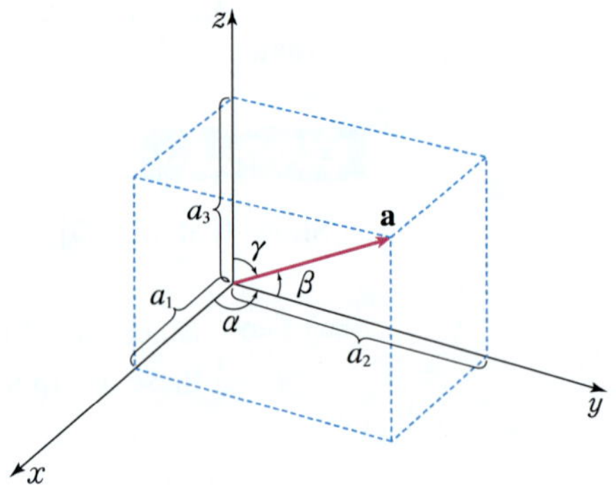

Figure 4

The **direction cosines** of a vector **a** are the cosines of these direction angles α, β and γ that are $\cos\alpha$, $\cos\beta$ and $\cos\gamma$, respectively.
Thus we have

1 $$\cos\alpha = \frac{\mathbf{a}\cdot\mathbf{i}}{|\mathbf{a}||\mathbf{i}|} = \frac{a_1}{|\mathbf{a}|},\quad \cos\beta = \frac{\mathbf{a}\cdot\mathbf{j}}{|\mathbf{a}||\mathbf{j}|} = \frac{a_2}{|\mathbf{a}|},\quad \cos\gamma = \frac{\mathbf{a}\cdot\mathbf{k}}{|\mathbf{a}||\mathbf{k}|} = \frac{a_3}{|\mathbf{a}|}.$$

Consequently,

2 $$\cos^2\alpha + \cos^2\beta + \cos^2\gamma = 1.$$

Also,

$$\begin{aligned}\mathbf{a} &= \langle a_1,\ a_2,\ a_3\rangle = \langle |\mathbf{a}|\cos\alpha,\ |\mathbf{a}|\cos\beta,\ |\mathbf{a}|\cos\gamma\rangle \\ &= |\mathbf{a}|\,\langle \cos\alpha,\ \cos\beta,\ \cos\gamma\rangle.\end{aligned}$$

Therefore, dividing by $|\mathbf{a}|$ both sides,

3 $$\frac{1}{|\mathbf{a}|}\mathbf{a} = \langle \cos\alpha,\ \cos\beta,\ \cos\gamma\rangle.$$

Thus, the direction cosines of **a** are the components of the unit vector in the direction of **a**.

EXAMPLE 4

Find the direction angles of the vector $\mathbf{a} = \langle 1, -2, 3\rangle$.

Solution Since $|\mathbf{a}| = \sqrt{1^2 + (-2)^2 + 3^2} = \sqrt{14}$, 1 gives

$$\cos\alpha = \frac{1}{\sqrt{14}}, \quad \cos\beta = -\frac{2}{\sqrt{14}}, \quad \cos\gamma = \frac{3}{\sqrt{14}}.$$

Hence

$$\alpha = \cos^{-1}\left(\frac{1}{\sqrt{14}}\right) \approx 74°, \quad \beta = \cos^{-1}\left(\frac{-2}{\sqrt{14}}\right) \approx 122°,$$
$$\gamma = \cos^{-1}\left(\frac{3}{\sqrt{14}}\right) \approx 37°.$$

■

Projections and Components

Consider two vectors a and b with the same initial point P and with θ as the angle between them, represented by $\overrightarrow{PQ}$ and $\overrightarrow{PR}$. Let S be the foot of the perpendicular from R to the line containing $\overrightarrow{PQ}$. Then $\overrightarrow{PS}$ is called the **vector projection** of b onto a and is denoted by $\mathrm{proj}_{\mathbf{a}}\mathbf{b}$.

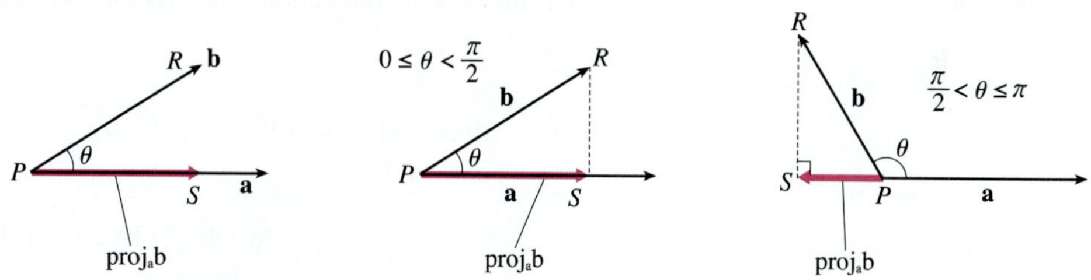

Figure 5 Vector projections

The **component of b along a** (or the **scalar projection** of b onto a) is the signed magnitude of the vector projection, which is the number $|\mathbf{b}|\cos\theta$. We denote this by $\mathrm{comp}_{\mathbf{a}}\mathbf{b}$. Observe that it is negative if $\pi/2 < \theta \le \pi$. Since

$$\mathbf{a} \cdot \mathbf{b} = |\mathbf{a}||\mathbf{b}|\cos\theta = |\mathbf{a}|(|\mathbf{b}|\cos\theta) = |\mathbf{a}|\mathrm{comp}_{\mathbf{a}}\,\mathbf{b}$$

the dot product of a and b is the product of the magnitude of a and the scalar projection of **b** on to a. Also

$$|\mathbf{b}|\cos\theta = \frac{\mathbf{a}\cdot\mathbf{b}}{|\mathbf{a}|} = \frac{\mathbf{a}}{|\mathbf{a}|}\cdot\mathbf{b}$$

the component of b along a is obtained by the dot product of b with the unit vector in the direction of a. Thus, as a summary, the followings are true:

Scalar projection of b onto a: $\mathrm{comp}_{\mathbf{a}}\mathbf{b} = \dfrac{\mathbf{a}\cdot\mathbf{b}}{|\mathbf{a}|}$,

Vector projection of b onto a: $\mathrm{proj}_{\mathbf{a}}\mathbf{b} = \left(\dfrac{\mathbf{a}\cdot\mathbf{b}}{|\mathbf{a}|}\right)\dfrac{\mathbf{a}}{|\mathbf{a}|} = \dfrac{\mathbf{a}\cdot\mathbf{b}}{|\mathbf{a}|^2}\mathbf{a}$

Note The vector projection is the scalar projection times the unit vector in the direction of **a**.

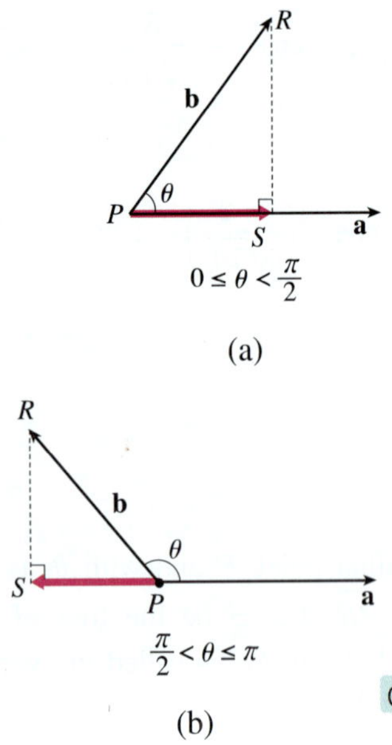

Figure 6

For each vector $\mathbf{a} = a_1\mathbf{i} + a_2\mathbf{j} + a_3\mathbf{k}$,

$$\text{comp}_{\mathbf{i}}\,\mathbf{a} = \mathbf{a}\cdot\mathbf{i} = a_1,\ \ \text{comp}_{\mathbf{j}}\,\mathbf{a} = \mathbf{a}\cdot\mathbf{j} = a_2,\ \ \text{comp}_{\mathbf{k}}\,\mathbf{a} = \mathbf{a}\cdot\mathbf{k} = a_3.$$

This agrees with our previous use of the term "component" and gives the identity

$$\mathbf{a} = a_1\mathbf{i} + a_2\mathbf{j} + a_3\mathbf{k} = (\mathbf{a}\cdot\mathbf{i})\mathbf{i} + (\mathbf{a}\cdot\mathbf{j})\mathbf{j} + (\mathbf{a}\cdot\mathbf{k})\mathbf{k}.$$

The vector $\mathbf{b} - \text{proj}_{\mathbf{a}}\mathbf{b}$ is orthogonal to the projection vector $\text{proj}_{\mathbf{a}}\mathbf{b}$. Hence the equation

$$\mathbf{b} = \text{proj}_{\mathbf{a}}\mathbf{b} + (\mathbf{b} - \text{proj}_{\mathbf{a}}\mathbf{b}) = \left(\frac{\mathbf{a}\cdot\mathbf{b}}{|\mathbf{a}|^2}\right)\mathbf{a} + \left(\mathbf{b} - \left(\frac{\mathbf{a}\cdot\mathbf{b}}{|\mathbf{a}|^2}\right)\mathbf{a}\right).$$

Note $\mathbf{b} - \text{proj}_{\mathbf{a}}\mathbf{b}$ is orthogonal to vector $\mathbf{a}$.

CAS **EXAMPLE 5**

Find the scalar projection and vector projection of $\mathbf{b} = \langle 2, 3, -1\rangle$ onto $\mathbf{a} = \langle 1, 5, -3\rangle$.

Solution Since $|\mathbf{a}| = \sqrt{(1)^2 + 5^2 + (-3)^2} = \sqrt{35}$, the scalar projection of $\mathbf{b}$ onto $\mathbf{a}$ is

$$\text{comp}_{\mathbf{a}}\mathbf{b} = \frac{\mathbf{a}\cdot\mathbf{b}}{|\mathbf{a}|} = \frac{(1)(2) + 5(3) + (-3)(-1)}{\sqrt{35}} = \frac{20}{\sqrt{35}}.$$

The vector projection is this scalar projection times the unit vector in the direction of $\mathbf{a}$. That is,

$$\text{proj}_{\mathbf{a}}\mathbf{b} = \frac{20}{\sqrt{35}}\frac{\mathbf{a}}{|\mathbf{a}|} = \frac{20}{35}\mathbf{a} = \left\langle \frac{20}{35}, \frac{100}{35}, -\frac{60}{35}\right\rangle$$
$$= \left\langle \frac{4}{7}, \frac{20}{7}, -\frac{12}{7}\right\rangle.$$

http://www.youtube.com/watch?v=YB976T1w0kE&feature=youtu.be

```
a=vector([1,5,-3])
b=vector([2,3,-1])
(a.dot_product(b))/(a.norm()^2)*a
```

Answer : (4/7, 20/7, -12/7) ■

Some Applications

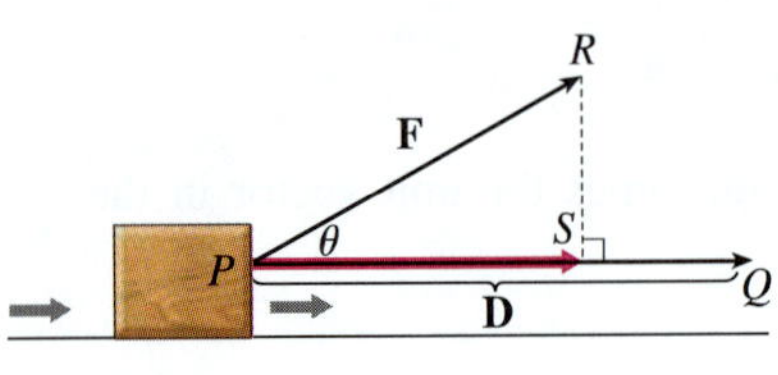

Figure 7

When a force $\mathbf{F}$ is directed along the line of motion of the object, then the work done by $\mathbf{F}$ in moving the object through a distance d is $W = \mathbf{F}d$. Suppose that the constant force $\mathbf{F} = \overrightarrow{PR}$, pointing in some other direction (See Figure 7.), moves the object from P to Q, then the **displacement vector** is $\mathbf{D} = \overrightarrow{PQ}$. Then, the **work** done by this force is defined to be the product of the component of the force along D and the distance moved. If θ is the angle between F and D, then

$$W = (|\mathbf{F}|\cos\theta)|\mathbf{D}| = \mathbf{F}\cdot\mathbf{D}.$$

If F causes a displacement D of a body, then the work done is

$$W = \mathbf{F}\cdot\mathbf{D}.$$

EXAMPLE 6

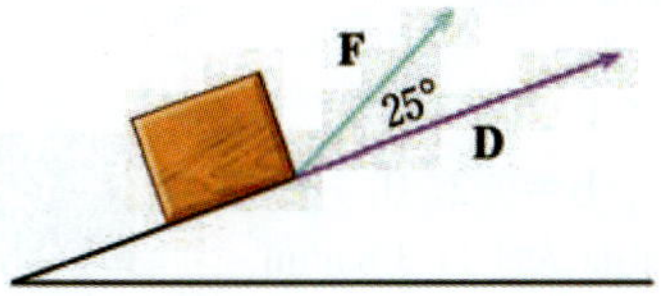

Figure 8

A crate is hauled 10m up a ramp under a constant force of 160N applied at an angle of 25° to the ramp. Find the work done.

Solution If F and D are the force and displacement vectors, as pictured in Figure 8, then the work done is

$$\begin{aligned} W &= \mathbf{F}\cdot\mathbf{D} = |\mathbf{F}||\mathbf{D}|\cos 25^\circ \\ &= (160)(10)\cos 25^\circ \approx 1450\,\mathrm{N}\cdot\mathrm{m} = 1450\,\mathrm{J}. \end{aligned}$$

■

A constant force with vector representation F moves an object along a straight line from a point to anthor point. The work done when we find in this book the distance is measured in meters and the magnitude of the force is measured in newtons.

CAS **EXAMPLE 7**

Find a work done when a force $\mathbf{F} = \langle 2, -3\rangle$ and the **displacement vector** is $\mathbf{D} = \langle 1, -7\rangle$.

http://matrix.skku.ac.kr/cal-lab/cal-11-3-Exs-6.html

Solution

```
def proj_ab(a,b):
return (a.dot_product(b))/(a.norm()^2)*a
a=vector([1,5,-3])
b=vector([2,3,-1])
proj_ab(a,b)
```

```
a=vector(QQ, [2, -3]);b=vector(QQ, [1, -7])
ab=a.dot_product(b)
show(ab)
```

Answer : W=F·D=23 ■

11.3 EXERCISES (The Dot Product)

http://matrix.skku.ac.kr/Cal-Book/part2/CS-Sec-11-3-Sol.html

YouTube http://youtu.be/qKTqcazFEhw

1. Determine the dot product of two vectors if their lengths are 8 and $1/4$ and the angle between them is $\pi/3$.

Solution Let the vectors be $\mathbf{a}$ and $\mathbf{b}$. Then,

$$\mathbf{a}\cdot\mathbf{b}=|\mathbf{a}||\mathbf{b}|\cos\theta=(8)\left(\frac{1}{4}\right)\cos\frac{\pi}{3}=1$$

by definition of the dot product.

2-6. Find the dot product of $\mathbf{a}$ and $\mathbf{b}$.

CAS **2.** $\mathbf{a}=\langle 5,\ -3\rangle,\ \mathbf{b}=\langle 4,\ 6\rangle.$

http://matrix.skku.ac.kr/cal-lab/cal-11-3-2. html

Solution $\mathbf{a}\cdot\mathbf{b}=5\times4-3\times6=2$

```
a=vector([5, -3]);
b=vector([4, 6]);
a.dot_product(b);
```

Answer : 2

3. $\mathbf{a}=\langle s,\ -2s,\ 4s\rangle,\ \mathbf{b}=\langle -t,\ 3t,\ 2t\rangle$

Solution

$\mathbf{a}\cdot\mathbf{b}=\langle s, -2s, 4s\rangle\cdot\langle -t, 3t, 2t\rangle$
$=s(-t)+(-2s)(3t)+(4s)(2t)=-st-6st+8st=st.$

4. $\mathbf{a}=\mathbf{i}-2\mathbf{j}+5\mathbf{k}\ ,\ \mathbf{b}=4\mathbf{i}+2\mathbf{k}$

Solution 14

5. $\mathbf{a}=\langle 3,\ 0,\ -2\rangle,\ \mathbf{b}=\langle 5,\ -1,\ 9\rangle$

Solution $\mathbf{a}\cdot\mathbf{b}=\langle 3, 0, -2\rangle\cdot\langle 5, -1, 9\rangle$
$=(3)(5)+(0)(-1)+(-2)(9)=-3$

6. $|\mathbf{a}|=12,\ |\mathbf{b}|=8,$ and the angle between $\mathbf{a}$ and $\mathbf{b}$ is $\pi/6$.

Solution $\mathbf{a}\cdot\mathbf{b}=|\mathbf{a}||\mathbf{b}|\cos\theta=12\times8\times\frac{\sqrt{3}}{2}=48\sqrt{3}.$

7-9. Compute the angle between the vectors.

7. $\mathbf{a}=\langle 3,\ -2\rangle,\ \mathbf{b}=\langle 4,\ 6\rangle$

Solution $|\mathbf{a}|=\sqrt{3^2+(-2)^2}=\sqrt{13},$

$|\mathbf{b}|=\sqrt{4^2+6^2}=\sqrt{52}$ and

$\mathbf{a}\cdot\mathbf{b}=(3)(4)+(-2)(6)=0.$

From the definition of the dot product, we have

$$\cos\theta=\frac{\mathbf{a}\cdot\mathbf{b}}{|\mathbf{a}||\mathbf{b}|}=\frac{0}{\sqrt{13}\sqrt{52}}=0.$$

Hence, the angle between $\mathbf{a}$ and $\mathbf{b}$ is

$$\theta=\cos^{-1}(0)=\frac{\pi}{2}=90^\circ.$$

That is, $\mathbf{a}$ and $\mathbf{b}$ are orthogonal.

CAS **8.** (a) $\mathbf{a}=\langle 6,\ 3\rangle,\ \mathbf{b}=\langle 4,\ 2\rangle$

(b) $\mathbf{a}=\langle 6,\ -3,\ 4\rangle\ ,\ \mathbf{b}=\langle 2,\ 0,\ -3\rangle$

http://matrix.skku.ac.kr/cal-lab/cal-11-3-8.html

Solution (a)

```
def anglebetween(a,b):
    return arccos(a.dot_product(b)
        /(a.norm()*b.norm()))
a=vector([6,3])
b=vector([4,2])
anglebetween(a,b)
```

Answer : [t == 0]

(b)

```
a=vector([6, -3, 4]);
b=vector([2, 0, -3]);
anglebetween(a,b)
```

Answer : [t == 1/2*pi]

9. $\mathbf{a}=2\mathbf{i}-\mathbf{j}+3\mathbf{k}\ ,\ \mathbf{b}=3\mathbf{i}+\mathbf{j}-5\mathbf{k}.$

http://matrix.skku.ac.kr/cal-lab/11-3-9.html

Solution $|\mathbf{a}|=\sqrt{2^2+(-1)^2+3^2}=\sqrt{14},$

$|\mathbf{b}|=\sqrt{3^2+(-1)^2+(-5)^2}=\sqrt{35},$ and

$\mathbf{a}\cdot\mathbf{b}=(2)(3)+(-1)(1)+(3)(-5)=-10.$

From the definition of the dot product, we have

$$\cos\theta=\frac{\mathbf{a}\cdot\mathbf{b}}{|\mathbf{a}||\mathbf{b}|}=\frac{-10}{\sqrt{14}\sqrt{35}}=\frac{-10}{7\sqrt{10}}=-\frac{\sqrt{10}}{7}$$ and

$$\theta=\cos^{-1}\left(-\frac{\sqrt{10}}{7}\right)\approx 116^\circ.$$

```
a=vector([2, -1, 3])
b=vector([3, 1, -5])
print "The angle between a and b is",
anglebetween(a,b).n(), "radians"
print "The angle between a and b is",
anglebetween(a,b).n()*180/pi.n(), "degrees"
```

The angle between a and b is 2.03952668853669 radians
The angle between a and b is 116.856271457445 degrees

CAS **10.** Verify whether the given vectors are orthogonal, parallel, or neither.

(a) $\mathbf{a} = \langle -5, 7, 10 \rangle$, $\mathbf{b} = \langle 4, 0, 2 \rangle$

(b) $\mathbf{a} = \langle 6, 3 \rangle$, $\mathbf{b} = \langle 4, 2 \rangle$

(c) $\mathbf{a} = -3\mathbf{i} - 9\mathbf{j} + 6\mathbf{k}$, $\mathbf{b} = 2\mathbf{i} + 6\mathbf{j} - 4\mathbf{k}$

http://matrix.skku.ac.kr/cal-lab/cal-11-3-10.html

Solution (a) Since $\mathbf{a} \cdot \mathbf{b} = 0$, $\mathbf{a}$ and $\mathbf{b}$ are orthogonal.

(b)

```
def anglebetween(a,b):
    return
arccos(a.dot_product(b)/(a.norm()*b.norm()))
var('t')
a=vector([6, 3]);
b=vector([4, 2]);
solve(a.dot_product(b)/(a.norm()*b.norm())==cos(t)
, t)
anglebetween(a,b)
```

Answer : [t == 0]

(c) Parallel.

CAS **11.** Determine t such that the vectors $\langle 2, -6, t \rangle$ and $\langle t, t, t^2 \rangle$ are orthogonal.

http://matrix.skku.ac.kr/cal-lab/cal-11-3-11.html

Solution

```
var('t')
a=vector([2, -6, t])
b=vector([t, t, t^2])
x=a.dot_product(b)
solve([x==0], t)
```

Answer : [t == -2, t == 2, t == 0]

12. Find a unit vector that is orthogonal to both $\langle 1, 1, 0 \rangle$ and $\langle 1, 0, 1 \rangle$.

Solution $\left\langle \frac{1}{\sqrt{3}}, -\frac{1}{\sqrt{3}}, -\frac{1}{\sqrt{3}} \right\rangle$.

13-14. Find the direction cosines and direction angles of the vector. (Give the direction angles correct to the nearest degree.)

13. $\mathbf{a} = \langle 4, 5, 3 \rangle$

Solution Since $|\mathbf{a}| = \sqrt{4^2 + 5^2 + 3^2} = \sqrt{50} = 5\sqrt{2}$, the direction cosines of the vector are

$$\cos\alpha = \frac{4}{5\sqrt{2}}, \quad \cos\beta = \frac{5}{5\sqrt{2}}, \quad \cos\gamma = \frac{3}{5\sqrt{2}}.$$

Hence, $\alpha = \cos^{-1}\left(\frac{4}{5\sqrt{2}}\right) \approx 55°$,

$$\beta = \cos^{-1}\left(\frac{5}{5\sqrt{2}}\right) \approx 45°, \quad \gamma = \cos^{-1}\left(\frac{3}{5\sqrt{2}}\right) \approx 64°.$$

14. $\mathbf{a} = -6\mathbf{i} + 2\mathbf{j} + 3\mathbf{k}$

Solution direction cosine: $\left\langle -\frac{6}{7}, \frac{2}{7}, \frac{3}{7} \right\rangle$,

direction angle: $\alpha = \cos^{-1}\left(-\frac{6}{7}\right)$, $\beta = \cos^{-1}\left(\frac{2}{7}\right)$,

$$\gamma = \cos^{-1}\left(\frac{3}{7}\right).$$

---

15. Prove that the vector $\mathrm{orth}_{\mathbf{a}}\mathbf{b} = \mathbf{b} - \mathrm{proj}_{\mathbf{a}}\mathbf{b}$ known as **orthogonal projection** of $\mathbf{b}$, is orthogonal to $\mathbf{a}$.

Solution
$$\begin{aligned}(\mathrm{orth}_{\mathbf{a}}\mathbf{b}) \cdot \mathbf{a} &= (\mathbf{b} - \mathrm{proj}_{\mathbf{a}}\mathbf{b}) \cdot \mathbf{a} \\ &= \mathbf{b} \cdot \mathbf{a} - (\mathrm{proj}_{\mathbf{a}}\mathbf{b}) \cdot \mathbf{a} \\ &= \mathbf{b} \cdot \mathbf{a} - \frac{\mathbf{a} \cdot \mathbf{b}}{|\mathbf{a}|^2}\mathbf{a} \cdot \mathbf{a} \\ &= \mathbf{b} \cdot \mathbf{a} - \frac{\mathbf{a} \cdot \mathbf{b}}{|\mathbf{a}|^2}|\mathbf{a}|^2 \\ &= \mathbf{b} \cdot \mathbf{a} - \mathbf{a} \cdot \mathbf{b} = 0.\end{aligned}$$

This proves the result.

16-19. Find the scalar and vector projections of b onto a and **orthogonal projection** of b, and a.

16. $\mathbf{a} = \langle 3, 0 \rangle$, $\mathbf{b} = \langle 5, -4 \rangle$

Solution scalar projection: 5,

vector projection: $\mathrm{proj}_{\mathbf{a}}\mathbf{b} == \frac{5}{3}\langle 3, 0 \rangle = \langle 5, 0 \rangle$,

orthogonal projection: $\langle 0, -4 \rangle$.

CAS **17.** $\mathbf{a} = \langle 2, -1, -2 \rangle$, $\mathbf{b} = \langle 4, 3, 3 \rangle$

http://matrix.skku.ac.kr/cal-lab/cal-11-3-17.html

Solution

```
a=vector(QQ, [2, -1, -2])
b=vector(QQ, [4, 3, 3])
ab=a.dot_product(b)
an=a.norm()
scal_proj=ab/an
scal_proj
```

Answer : -1/3

```
vec_proj=scal_proj/an*a
vec_proj
```

Answer : (-2/9, 1/9, 2/9)

```
orth_proj=b-vec_proj
orth_proj
```

Answer : (38/9, 26/9, 25/9)

```
vec_proj.dot_product(orth_proj)
```

Answer : 0

18. a = i − j, b = i + k

Solution scalar projection: $\frac{1}{\sqrt{2}}$,

vector projection: $\left\langle \frac{1}{2}, -\frac{1}{2}, 0 \right\rangle$,

orthogonal projection: $\left\langle \frac{1}{2}, \frac{1}{2}, 1 \right\rangle$.

19. a = 2i + 6j − 2k, b = i − 3j + k

Solution $|\mathbf{a}| = \sqrt{2^2 + 6^2 + (-2)^2} = \sqrt{44}$,

so $\text{comp}_{\mathbf{a}}\mathbf{b} = \frac{\mathbf{a} \cdot \mathbf{b}}{|\mathbf{a}|} = \frac{(2)(1)+(6)(-3)+(-2)(1)}{\sqrt{44}}$

$= -\frac{18}{\sqrt{44}}$

and $\text{proj}_{\mathbf{a}}\mathbf{b} = -\frac{18}{\sqrt{44}}\frac{1}{\sqrt{44}}(2\mathbf{i}+6\mathbf{j}-2\mathbf{k})$

$= -\frac{9}{11}\mathbf{i} - \frac{27}{11}\mathbf{j} + \frac{9}{11}\mathbf{k}.$

And $\text{orth}_{\mathbf{a}}\mathbf{b} = (\mathbf{i} - 3\mathbf{j} + \mathbf{k}) + \left(\frac{9}{11}\mathbf{i} + \frac{27}{11}\mathbf{j} - \frac{9}{11}\mathbf{k}\right)$

$= \frac{20}{11}\mathbf{i} - \frac{6}{11}\mathbf{j} + \frac{2}{11}\mathbf{k}.$

---

20. Prove that the distance from a point $P_1(x_1, y_1)$ to the line $ax + by + c = 0$ is $\frac{|ax_1 + by_1 + c|}{\sqrt{a^2 + b^2}}$.

Hence, find the distance from the point $(-2, 3)$ to the line $x - 4y + 5 = 0$.

21. Prove that the Cauchy-Schwarz Inequality :

$$|\mathbf{a} \cdot \mathbf{b}| \le |\mathbf{a}||\mathbf{b}|.$$

Solution Since $|\cos\theta| \le 1$, $|\mathbf{a}||\mathbf{b}|\cos\theta \le |\mathbf{a}||\mathbf{b}|$,

$|\mathbf{a} \cdot \mathbf{b}| = \left||\mathbf{a}||\mathbf{b}|\cos\theta\right| = |\mathbf{a}||\mathbf{b}|\cos\theta.$

22. Prove the Triangle Inequality :

$$|\mathbf{a} + \mathbf{b}| \le |\mathbf{a}| + |\mathbf{b}|.$$

Solution Note that it is enough to prove

$|\mathbf{a} + \mathbf{b}|^2 \le (|\mathbf{a}| + |\mathbf{b}|)^2.$

Consider the L·H·S ,

$|\mathbf{a}+\mathbf{b}|^2 = (\mathbf{a}+\mathbf{b}) \cdot (\mathbf{a}+\mathbf{b}) = |\mathbf{a}|^2 + 2\mathbf{a} \cdot \mathbf{b} + |\mathbf{b}|^2$

$\le |\mathbf{a}|^2 + 2|\mathbf{a}||\mathbf{b}| + |\mathbf{b}|^2 = (|\mathbf{a}| + |\mathbf{b}|)^2$

Hence we have $|\mathbf{a} + \mathbf{b}|^2 \le (|\mathbf{a}| + |\mathbf{b}|)^2$

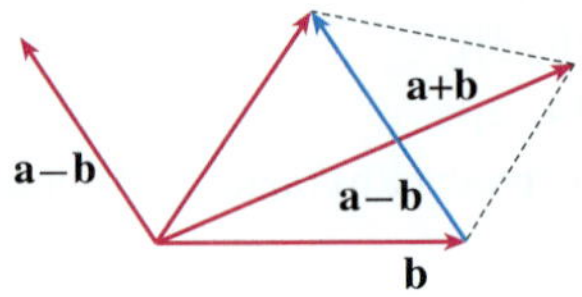

23. Prove that

$$|\mathbf{a} + \mathbf{b}|^2 + |\mathbf{a} - \mathbf{b}|^2 = 2|\mathbf{a}|^2 + 2|\mathbf{b}|^2.$$

Solution $|\mathbf{a}+\mathbf{b}|^2 = (\mathbf{a}+\mathbf{b}) \cdot (\mathbf{a}+\mathbf{b})$

$= |\mathbf{a}|^2 + 2(\mathbf{a} \cdot \mathbf{b}) + |\mathbf{b}|^2$

and $|\mathbf{a}-\mathbf{b}|^2 = (\mathbf{a}-\mathbf{b}) \cdot (\mathbf{a}-\mathbf{b})$

$= |\mathbf{a}|^2 - 2(\mathbf{a} \cdot \mathbf{b}) + |\mathbf{b}|^2.$

Adding these two equations gives

$|\mathbf{a}+\mathbf{b}|^2 + |\mathbf{a}-\mathbf{b}|^2 = 2|\mathbf{a}|^2 + 2|\mathbf{b}|^2.$

24. Show that vectors a and b are orthogonal if and only if $|\mathbf{a} - \mathbf{b}|^2 = |\mathbf{a}|^2 + |\mathbf{b}|^2$.

25. Show that $|\mathbf{a}| = |\mathbf{b}|$ if and only if a + b is orthogonal to a − b.

26. Give geometric interpretation of the above two problems.

11.4 The Cross Product

The cross product is only defined for vectors in $\mathbb{R}^3$ and results in another vector in $\mathbb{R}^3$.

Component Form of the Cross Product

We define the cross product of two vectors $\mathbf{a}$ and $\mathbf{b}$ in the terms of the components of the vector.

THEOREM 1 Cross Product of Two Vectors

The cross product of two vectors $\mathbf{a} = \langle a_1, a_2, a_3 \rangle$ and $\mathbf{b} = \langle b_1, b_2, b_3 \rangle$ is the vector

$$\mathbf{a} \times \mathbf{b} = \langle a_2b_3 - a_3b_2,\ a_3b_1 - a_1b_3,\ a_1b_2 - a_2b_1 \rangle.$$

The cross product of the vectors $\mathbf{a} = a_1\mathbf{i} + a_2\mathbf{j} + a_3\mathbf{k}$ and $\mathbf{b} = b_1\mathbf{i} + b_2\mathbf{j} + b_3\mathbf{k}$ can also be obtained as follows

1 $$\mathbf{a} \times \mathbf{b} = \begin{vmatrix} \mathbf{i} & \mathbf{j} & \mathbf{k} \\ a_1 & a_2 & a_3 \\ b_1 & b_2 & b_3 \end{vmatrix} = \begin{vmatrix} a_2 & a_3 \\ b_2 & b_3 \end{vmatrix}\mathbf{i} - \begin{vmatrix} a_1 & a_3 \\ b_1 & b_3 \end{vmatrix}\mathbf{j} + \begin{vmatrix} a_1 & a_2 \\ b_1 & b_2 \end{vmatrix}\mathbf{k}$$

or in the determinant form

2 $$\mathbf{a} \times \mathbf{b} = \begin{vmatrix} \mathbf{i} & \mathbf{j} & \mathbf{k} \\ a_1 & a_2 & a_3 \\ b_1 & b_2 & b_3 \end{vmatrix} \begin{matrix} \leftarrow \text{Unit vectors} \\ \leftarrow \text{Components of } \mathbf{a}. \\ \leftarrow \text{Components of } \mathbf{b} \end{matrix}$$

For example, we consider two vectors $\mathbf{a} = \langle 2, 4, 5 \rangle$ and $\mathbf{b} = \langle 3, 1, -2 \rangle$. Then

$$\mathbf{a} \times \mathbf{b} = \begin{vmatrix} \mathbf{i} & \mathbf{j} & \mathbf{k} \\ 2 & 4 & 5 \\ 3 & 1 & -2 \end{vmatrix} = \begin{vmatrix} 4 & 5 \\ 1 & -2 \end{vmatrix}\mathbf{i} - \begin{vmatrix} 2 & 5 \\ 3 & -2 \end{vmatrix}\mathbf{j} + \begin{vmatrix} 2 & 4 \\ 3 & 1 \end{vmatrix}\mathbf{k}$$
$$= \langle -13, 19, -10 \rangle.$$

EXAMPLE 1

Find a vector perpendicular to the plane that passes through the points $P(2, 1, 3)$, $Q(-1, 5, 2)$ and $R(1, -2, -1)$.
http://matrix.skku.ac.kr/cal-lab/11-4-Example-1.html

Solution The vector $\overrightarrow{PQ}\times\overrightarrow{PR}$ is perpendicular to both $\overrightarrow{PQ}$ and $\overrightarrow{PR}$ and is therefore perpendicular to the plane through P, Q and R. We know that

$$\overrightarrow{PQ}=(-1-2)\mathbf{i}+(5-1)\mathbf{j}+(2-3)\mathbf{k}=-3\mathbf{i}+4\mathbf{j}-\mathbf{k},$$
$$\overrightarrow{PR}=(1-2)\mathbf{i}+(-2-1)\mathbf{j}+(-1-3)\mathbf{k}=-\mathbf{i}-3\mathbf{j}-4\mathbf{k}.$$

We compute the cross product of these vectors:

$$\overrightarrow{PQ}\times\overrightarrow{PR}=\begin{vmatrix} \mathbf{i} & \mathbf{j} & \mathbf{k} \\ -3 & 4 & -1 \\ -1 & -3 & -4 \end{vmatrix}$$
$$=(-16-3)\mathbf{i}-(12-1)\mathbf{j}+(9+4)\mathbf{k}$$
$$=-19\mathbf{i}-11\mathbf{j}+13\mathbf{k}.$$

Hence, the vector $\langle -19, -11, 13\rangle$ is perpendicular to the given plane. Any nonzero scalar multiple of this vector, such as $\langle -38, -22, 26\rangle$ is also perpendicular to the Plane Equation. ■

Remark 1. Note that $\mathbf{a}\cdot(\mathbf{a}\times\mathbf{b})=0$ and $\mathbf{b}\cdot(\mathbf{a}\times\mathbf{b})=0$. (use determinant form for simplicity) Thus $\mathbf{a}\times\mathbf{b}$ is a vector which is orthogonal to both $\mathbf{a}$ and $\mathbf{b}$.

2. Suppose $\mathbf{c}=\langle c_1, c_2, c_3\rangle$ be a vector which is orthogonal to vectors $\mathbf{a}=\langle a_1, a_2, a_3\rangle$ and $\mathbf{b}=\langle b_1, b_2, b_3\rangle$. Then components of $\mathbf{a}\times\mathbf{b}$ can be obtained by solving two equations $\mathbf{a}\cdot\mathbf{c}=0$ and $\mathbf{b}\cdot\mathbf{c}=0$.

3. $\mathbf{a}\times\mathbf{b}\neq\mathbf{b}\times\mathbf{a}$, that is the vector product is not commutative. (In fact, $\mathbf{a}\times\mathbf{b}=-\mathbf{b}\times\mathbf{a}$)

Right-Hand Rule

An alternative characterization of the cross product uses the right-hand rule. As seen in Figure 1, if the fingers of the right hand point along the vector $\mathbf{a}$ and then curl toward the vector $\mathbf{b}$, the thumb will give the direction of $\mathbf{a}\times\mathbf{b}$. In Figure 1, the right-hand rule shows the direction of $\mathbf{b}\times\mathbf{a}$.

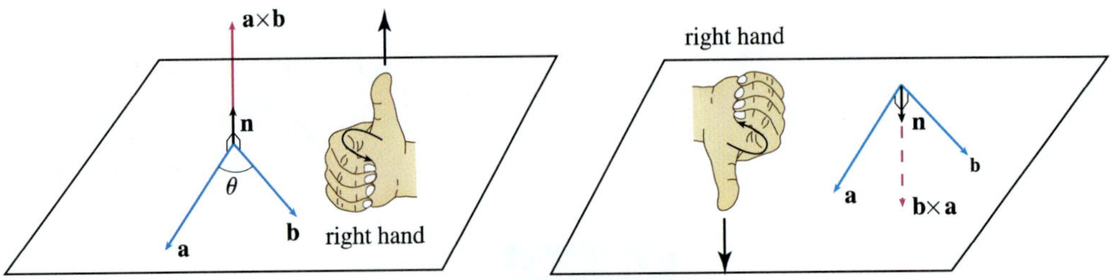

Figure 1

THEOREM 2

Let $\mathbf{a}$ and $\mathbf{b}$ be two vectors in $\mathbb{R}^3$. Then the cross product $\mathbf{a}\times\mathbf{b}$ is a vector with magnitude

$$|\mathbf{a}\times\mathbf{b}|=|\mathbf{a}||\mathbf{b}|\sin\theta$$

where $0\le\theta\le\pi$ is the angle between $\mathbf{a}$ and $\mathbf{b}$. The direction of $\mathbf{a}\times\mathbf{b}$ is the direction given by the right-hand rule. (See Figure 1.) When $\mathbf{a}\times\mathbf{b}=0$, the direction of $\mathbf{a}\times\mathbf{b}$ is undefined.

Proof If $\mathbf{a}=\langle a_1, a_2, a_3\rangle$ and $\mathbf{b}=\langle b_1, b_2, b_3\rangle$, then

$$|\mathbf{a}\times\mathbf{b}|^2=(\mathbf{a}\times\mathbf{b})\cdot(\mathbf{a}\times\mathbf{b}).$$

It is easy to see that

$$\begin{aligned}|\mathbf{a}\times\mathbf{b}|^2 &= (a_1^2+a_2^2+a_3^2)^2(b_1^2+b_2^2+b_3^2)^2-(a_1b_1+a_2b_2+a_3b_3)^2\\ &=|\mathbf{a}|^2|\mathbf{b}|^2-(\mathbf{a}\cdot\mathbf{b})^2=|\mathbf{a}|^2|\mathbf{b}|^2-|\mathbf{a}|^2|\mathbf{b}|^2\cos^2\theta\\ &=|\mathbf{a}|^2|\mathbf{b}|^2\sin^2\theta\end{aligned}$$

Taking square root both sides and observing that

$$\sin\theta\ge 0 \text{ for } 0\le\theta\le\frac{\pi}{2},$$

we get the result. ■

Area

Let two nonzero vectors $\mathbf{a}$ and $\mathbf{b}$ be two sides of a parallelogram, then the area of the parallelogram is

$$\begin{aligned}A&=(\text{base})(\text{altitude})\\ &=|\mathbf{a}|(|\mathbf{b}|\sin\theta)=|\mathbf{a}||\mathbf{b}|\sin\theta=|\mathbf{a}\times\mathbf{b}|.\end{aligned}$$

Likewise, we see that the area of a triangle with sides $\mathbf{a}$ and $\mathbf{b}$ is

$$A=\frac{1}{2}|\mathbf{a}\times\mathbf{b}|.$$

Thus, the length of the cross product $\mathbf{a}\times\mathbf{b}$ is equal to the area of the parallelogram (See Figure 2.) determined by $\mathbf{a}$ and $\mathbf{b}$.

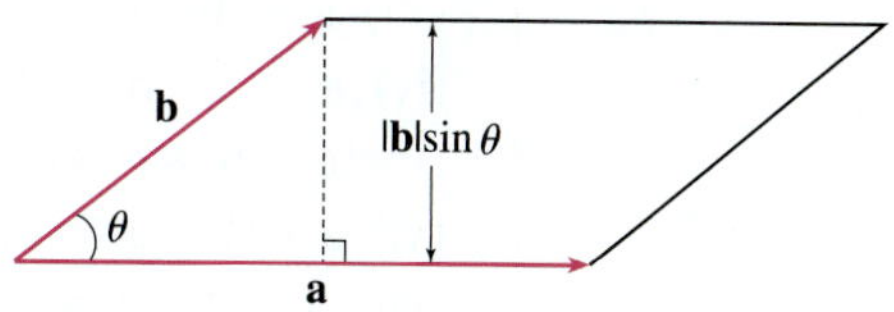

Figure 2

EXAMPLE 2

Find the area of the triangle with vertices $P(2,\ 1,\ 3)$, $Q(-1,\ 5,\ 2)$ and $R(1,\ -2,\ -1)$.

Solution In Example 1, we computed $\overrightarrow{PQ}\times\overrightarrow{PR}=\langle -19,\ -11,\ 13\rangle$. The area of the parallelogram with adjacent sides PQ and PR is the length of this cross product:

$$\begin{aligned}|\overrightarrow{PQ}\times\overrightarrow{PR}| &= \sqrt{(-19)^2+(-11)^2+13^2}\\ &= \sqrt{361+121+169}\\ &= \sqrt{651}.\end{aligned}$$

The area A of the triangle POR is half the area of this parallelogram. Hence the required area of its triangle POR is $\frac{1}{2}\sqrt{651}$. ■

The following theorem is a consequence of the definition of the cross product.

THEOREM 3 Parallel Vector

Two nonzero vectors $\mathbf{a}$ and $\mathbf{b}$ are parallel ($\theta=0$ or π) if and only if $\mathbf{a}\times\mathbf{b}=\mathbf{0}$.

Using $\theta=\pi/2$, then it follows that for the standard basis vectors $\mathbf{i}$, $\mathbf{j}$, and $\mathbf{k}$, we have

$$\begin{array}{lll}\mathbf{i}\times\mathbf{j}=\mathbf{k}, & \mathbf{j}\times\mathbf{k}=\mathbf{i}, & \mathbf{k}\times\mathbf{i}=\mathbf{j},\\ \mathbf{j}\times\mathbf{i}=-\mathbf{k}, & \mathbf{k}\times\mathbf{j}=-\mathbf{i}, & \mathbf{i}\times\mathbf{k}=-\mathbf{j}.\end{array}$$

Observe that $\mathbf{i}\times\mathbf{j}\neq\mathbf{j}\times\mathbf{i}$.

In general, $(\mathbf{a}\times\mathbf{b})\times\mathbf{c}\neq\mathbf{a}\times(\mathbf{b}\times\mathbf{c})$. Hence, the associative law for multiplication usually does not hold. For example $\mathbf{i}\times(\mathbf{i}\times\mathbf{j})=\mathbf{i}\times\mathbf{k}=-\mathbf{j}$ whereas $(\mathbf{i}\times\mathbf{i})\times\mathbf{j}=\mathbf{0}\times\mathbf{j}=\mathbf{0}$.

For vector products, the following usual laws of algebra hold.

THEOREM 4 Properties of the Cross Product

If $\mathbf{a}$, $\mathbf{b}$ and $\mathbf{c}$ are vectors and α is a scalar, then

(i) $\mathbf{a}\times\mathbf{b}=-\mathbf{b}\times\mathbf{a}$

(ii) $(\alpha\mathbf{a})\times\mathbf{b}=\alpha(\mathbf{a}\times\mathbf{b})=\mathbf{a}\times(\alpha\mathbf{b})$

(iii) $\mathbf{a}\times(\mathbf{b}+\mathbf{c})=\mathbf{a}\times\mathbf{b}+\mathbf{a}\times\mathbf{c}$

(iv) $\mathbf{a}\cdot(\mathbf{a}\times\mathbf{b})=0$ or $\mathbf{b}\cdot(\mathbf{a}\times\mathbf{b})=0$

(v) $(\mathbf{a}+\mathbf{b})\times\mathbf{c}=\mathbf{a}\times\mathbf{c}+\mathbf{b}\times\mathbf{c}$

(vi) $\mathbf{a}\cdot(\mathbf{b}\times\mathbf{c})=(\mathbf{a}\times\mathbf{b})\cdot\mathbf{c}$

(vii) $\mathbf{a}\times(\mathbf{b}\times\mathbf{c})=(\mathbf{a}\cdot\mathbf{c})\mathbf{b}-(\mathbf{a}\cdot\mathbf{b})\mathbf{c}$

Readers are encouraged to verify the above properties by using components of vectors.

Triple Scalar Product or Box Product

If $\mathbf{a} = \langle a_1, a_2, a_3 \rangle$, $\mathbf{b} = \langle b_1, b_2, b_3 \rangle$ and $\mathbf{c} = \langle c_1, c_2, c_3 \rangle$, then

3

$$\mathbf{a} \cdot (\mathbf{b} \times \mathbf{c}) = \begin{vmatrix} a_1 & a_2 & a_3 \\ b_1 & b_2 & b_3 \\ c_1 & c_2 & c_3 \end{vmatrix}.$$

The product $\mathbf{a} \cdot (\mathbf{b} \times \mathbf{c})$ is called the triple scalar product of $\mathbf{a}$, $\mathbf{b}$ and $\mathbf{c}$.

Note that $\mathbf{a} \cdot (\mathbf{b} \times \mathbf{c}) = (\mathbf{a} \times \mathbf{b}) \cdot \mathbf{c}$ follows from the properties of determinants.

http://matrix.skku.ac.kr/2012-LAwithSage/interact/1.html

Volume of a Parallelepiped

Consider a parallelepiped with coterminous edges determined by the vectors $\mathbf{a}$, $\mathbf{b}$, and $\mathbf{c}$. (See Figure 3.)

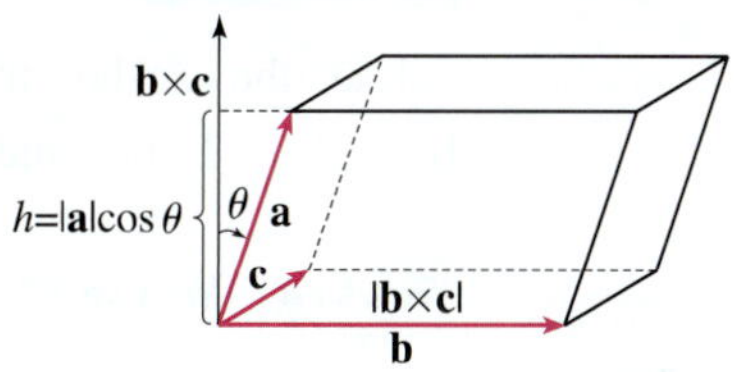

Figure 3

The area of the base parallelogram is $A = |\mathbf{b} \times \mathbf{c}|$. If θ is the angle between $\mathbf{a}$ and $\mathbf{b} \times \mathbf{c}$, then the height h of the parallelepiped is $h = |\mathbf{a}||\cos\theta|$. Therefore, the volume of the parallelepiped determined by the coterminous edge vectors $\mathbf{a}$, $\mathbf{b}$ and $\mathbf{c}$ is

$$\begin{aligned} V &= (\text{area of base})(\text{height}) = Ah = |\mathbf{b} \times \mathbf{c}||\mathbf{a}||\cos\theta| \\ &= |\mathbf{a} \cdot (\mathbf{b} \times \mathbf{c})| \\ &= \text{the magnitude of the scalar triple product.} \end{aligned}$$

CAS **EXAMPLE 3**

Find the volume of the box (parallelepiped) determined by $\mathbf{a} = 3\mathbf{i} + \mathbf{j} + \mathbf{k}$, $\mathbf{b} = \mathbf{i} + 4\mathbf{j} + \mathbf{k}$ and $\mathbf{c} = \mathbf{i} + \mathbf{j} + 5\mathbf{k}$.

http://matrix.skku.ac.kr/2012-LAwithSage/interact/2/2.html

Solution The volume of the parallelepiped is given by

$$\mathbf{a}\cdot(\mathbf{b}\times\mathbf{c})=\begin{vmatrix}3&1&1\\1&4&1\\1&1&5\end{vmatrix}$$

$$=3\begin{vmatrix}4&1\\1&5\end{vmatrix}-\begin{vmatrix}1&1\\1&5\end{vmatrix}+\begin{vmatrix}1&4\\1&1\end{vmatrix}=50.$$

Therefore, the volume is $|\mathbf{a}\cdot(\mathbf{b}\times\mathbf{c})|=50$ units cubed.

```
a=vector([3, 1, 1])
b=vector([1, 4, 1])
c=vector([1, 1, 5])
a.dot_product(b.cross_product(c))
```

Answer : 50 ■

THEOREM 5

If the three vectors $\mathbf{a}$, $\mathbf{b}$, $\mathbf{c}$ are **coplanar** (lie in the same plane) then the volume of the parallelepiped is zero, and consequently $\mathbf{a}\cdot(\mathbf{b}\times\mathbf{c})=0$.

EXAMPLE 4

Use the scalar triple product to show that the vectors $\mathbf{a}=\langle 2,3,1\rangle$, $\mathbf{b}=\langle 1,-1,0\rangle$ and $\mathbf{c}=\langle 7,3,2\rangle$ are coplanar.

Solution We use [3] to compute their scalar triple product:

$$\mathbf{a}\cdot(\mathbf{b}\times\mathbf{c})=\begin{vmatrix}2&3&1\\1&-1&0\\7&3&2\end{vmatrix}$$

$$=2\begin{vmatrix}-1&0\\3&2\end{vmatrix}-3\begin{vmatrix}1&0\\7&2\end{vmatrix}+1\begin{vmatrix}1&-1\\7&3\end{vmatrix}$$

$$=2(-2-0)-3(2-0)+1(3+7)$$

$$=-4-6+10=0.$$

Therefore, by Theorem 5 the volume of the parallelepiped determined by $\mathbf{a}$, $\mathbf{b}$ and $\mathbf{c}$ is 0. This means that $\mathbf{a}$, $\mathbf{b}$ and $\mathbf{c}$ are coplanar. ■

Torque

Suppose a force $\mathbf{F}$ is acting on a rigid body at a point given by a position vector $\mathbf{r}$. Let θ be the angle between the position and force vectors. The **torque vector** τ with reference to the origin is $\tau=\mathbf{r}\times\mathbf{F}$.

It measures the tendency of the body to rotate about the origin. Its direction indicates the axis of rotation. Its magnitude $|\tau|=|\mathbf{r}\times\mathbf{F}|=|\mathbf{r}||\mathbf{F}|\sin\theta$ is equal to the area of the parallelogram determined by $\mathbf{r}$ and $\mathbf{F}$.

For example, if a bolt is tightened by applying a force to a wrench (See Figure 4.), it produces, a turning effect. Note that the only component of F that can cause a rotation is the one perpendicular to $\mathbf{r}$, that is, $|\mathbf{F}|\sin\theta$.

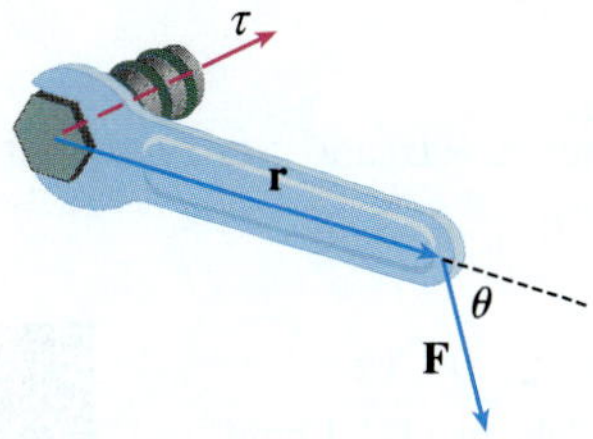

Figure 4

EXAMPLE 5

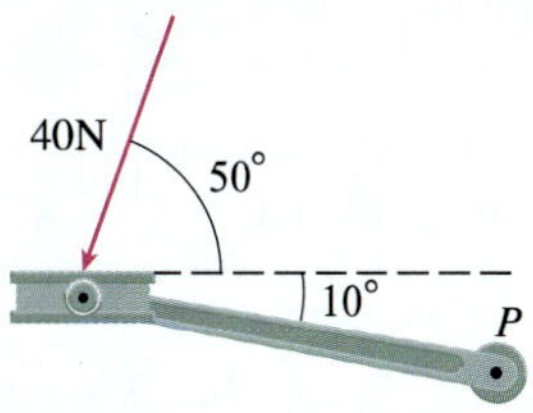

Figure 5

A bicycle pedal is pushed by a foot with a 40N force as shown in Figure 5. The shaft of the pedal is 20cm long. Find the magnitude of the torque about P.

Solution $|\tau| = |\mathbf{r}\times\mathbf{F}| = |\mathbf{r}||\mathbf{F}|\sin 60°|\mathbf{n}|$

$= 20 \cdot 40\sin 60°$

$= 400\sqrt{3}\,\mathrm{N}\cdot\mathrm{cm} = 4\sqrt{3}\ \mathrm{N}\cdot\mathrm{m}.$ ■

CAS **EXAMPLE 6**

Find the cross product $\mathbf{a}\times\mathbf{b}$ and verify that it is orthogonal to both $\mathbf{a}$ and $\mathbf{b}$: $\mathbf{a} = \langle 1,\ 1,\ -2\rangle$ and $\mathbf{b} = \langle 1,\ 0,\ -1\rangle$

http://matrix.skku.ac.kr/cal-lab/cal-11-4-Exs-6.html

Solution

```
a=vector(QQ, [1, 1, -2]);b=vector(QQ, [1, 0, -1])
c=a.cross_product(b)
show(c)
show(e.dot_product(a))
show(e.dot_product(b))
```

```
Answer : (-1, -1, -1)  # a.cross_product(b)
         0             # (e.dot_product(a))
         0             # (e.dot_product(b))
```
■

11.4 EXERCISES (The Cross Product)

http://matrix.skku.ac.kr/Cal-Book/part2/CS-Sec-11-4-Sol.html

YouTube http://youtu.be/0gtL8BqjU2c

1-5. Find the cross product $\mathbf{a}\times\mathbf{b}$ and verify that it is orthogonal to both $\mathbf{a}$ and $\mathbf{b}$.

CAS **1.** $\mathbf{a}=\langle 1, -1, 1\rangle$, $\mathbf{b}=\langle 2, 0, 3\rangle$

http://matrix.skku.ac.kr/cal-lab/cal-11-4-1.html

Solution

```
a=vector(QQ, [1, -1, 1])
b=vector(QQ, [2, 0, 3])
c=a.cross_product(b)
show(c)
show(e.dot_product(a))
show(e.dot_product(b))
```

Answer : $(-3, -1, 2)$

0

0

2. $\mathbf{a}=\mathbf{i}-2\mathbf{j}+3\mathbf{k}$, $\mathbf{b}=5\mathbf{i}+\mathbf{j}-\mathbf{k}$

Solution

$$\mathbf{a}\times\mathbf{b}=\begin{vmatrix}\mathbf{i} & \mathbf{j} & \mathbf{k}\\ 1 & -2 & 3\\ 5 & 1 & -1\end{vmatrix}=\begin{vmatrix}-2 & 3\\ 1 & -1\end{vmatrix}\mathbf{i}-\begin{vmatrix}1 & 3\\ 5 & -1\end{vmatrix}\mathbf{j}+\begin{vmatrix}1 & -2\\ 5 & 1\end{vmatrix}\mathbf{k}$$

$$=-\mathbf{i}+16\mathbf{j}+11\mathbf{k}$$

Now, $(\mathbf{a}\times\mathbf{b})\cdot\mathbf{a}=\langle 1, 16, 11\rangle\cdot\langle 1, -2, 3\rangle$

$=-1-32+33=0$ and

$(\mathbf{a}\times\mathbf{b})\cdot\mathbf{b}=\langle 1, 16, 11\rangle\cdot\langle 5, 1, -1\rangle$

$=-5+16-11=0$

So, $\mathbf{a}\times\mathbf{b}$ is orthogonal to both $\mathbf{a}$ and $\mathbf{b}$.

3. $\mathbf{a}=\langle 3, 1, -1\rangle$, $\mathbf{b}=\langle -2, 1, 1\rangle$

Solution $\langle 2, -1, 5\rangle$.

4. $\mathbf{a}=\mathbf{i}-e^{t}\mathbf{j}+e^{-t}\mathbf{k}$, $\mathbf{b}=e^{t}\mathbf{j}-e^{-t}\mathbf{k}$.

Solution $\langle 0, e^{-t}, e^{t}\rangle$.

5. $\mathbf{a}=\langle 1, t, t^4\rangle$, $\mathbf{b}=\langle 2t, 3t, t^2\rangle$

Solution $\langle t^3-3t^5, 2t^5-t^2, 3t-2t^2\rangle$.

---

6. If $\mathbf{a}=\langle 3, 1, -2\rangle$ and $\mathbf{b}=\langle 0, 2, -1\rangle$, find $\mathbf{a}\times\mathbf{b}$ and $\mathbf{b}\times\mathbf{a}$.

Solution $\mathbf{a}\times\mathbf{b}=\langle 3, 3, 6\rangle$, $\mathbf{b}\times\mathbf{a}=\langle -3, -3, -6\rangle$.

7. If $\mathbf{a}=\langle 1, 2, 1\rangle$, $\mathbf{b}=\langle -1, -1, 0\rangle$ and $\mathbf{c}=\langle 4, 0, -4\rangle$, show that $\mathbf{a}\times(\mathbf{b}\times\mathbf{c})\neq(\mathbf{a}\times\mathbf{b})\times\mathbf{c}$.

Solution

(i) $$\mathbf{a}\times\mathbf{b}=\begin{vmatrix}\mathbf{i} & \mathbf{j} & \mathbf{k}\\ 1 & 2 & 1\\ -1 & -1 & 0\end{vmatrix}=\begin{vmatrix}2 & 1\\ -1 & 0\end{vmatrix}\mathbf{i}-\begin{vmatrix}1 & 1\\ -1 & 0\end{vmatrix}\mathbf{j}+\begin{vmatrix}1 & 2\\ -1 & -1\end{vmatrix}\mathbf{k}$$

$$=\mathbf{i}+\mathbf{j}+\mathbf{k},$$

$$(\mathbf{a}\times\mathbf{b})\times\mathbf{c}=\begin{vmatrix}\mathbf{i} & \mathbf{j} & \mathbf{k}\\ 1 & 1 & 1\\ 4 & 0 & -4\end{vmatrix}=\begin{vmatrix}1 & 1\\ 0 & -4\end{vmatrix}\mathbf{i}-\begin{vmatrix}1 & 1\\ 4 & -4\end{vmatrix}\mathbf{j}+\begin{vmatrix}1 & 1\\ 4 & 0\end{vmatrix}\mathbf{k}$$

$$=-4\mathbf{i}+8\mathbf{j}-4\mathbf{k}.$$

(ii)

$$\mathbf{b}\times\mathbf{c}=\begin{vmatrix}\mathbf{i} & \mathbf{j} & \mathbf{k}\\ -1 & -1 & 0\\ 4 & 0 & -4\end{vmatrix}=\begin{vmatrix}-1 & 0\\ 0 & -4\end{vmatrix}\mathbf{i}-\begin{vmatrix}-1 & 0\\ 4 & -4\end{vmatrix}\mathbf{j}+\begin{vmatrix}-1 & -1\\ 4 & 0\end{vmatrix}\mathbf{k}$$

$$=4\mathbf{i}-4\mathbf{j}+4\mathbf{k},$$

$$\mathbf{a}\times(\mathbf{b}\times\mathbf{c})=\begin{vmatrix}\mathbf{i} & \mathbf{j} & \mathbf{k}\\ 1 & 2 & 1\\ 4 & -4 & 4\end{vmatrix}=\begin{vmatrix}2 & 1\\ -4 & 4\end{vmatrix}\mathbf{i}-\begin{vmatrix}1 & 1\\ 4 & 4\end{vmatrix}\mathbf{j}+\begin{vmatrix}1 & 2\\ 4 & -4\end{vmatrix}\mathbf{k}$$

$$=12\mathbf{i}-12\mathbf{k}.$$

Hence, $(\mathbf{a}\times\mathbf{b})\times\mathbf{c}\neq\mathbf{a}\times(\mathbf{b}\times\mathbf{c})$.

CAS **8.** Find two unit vectors orthogonal to both $\langle 1, -1, 2\rangle$ and $\langle 3, 0, 1\rangle$.

http://matrix.skku.ac.kr/cal-lab/cal-11-4-8.html

Solution

```
a=vector(QQ, [1, -1, 2])
b=vector(QQ, [3, 0, 1])
c=a.cross_product(b)
cn=c.norm()
e=c/cn
show(e)
show(-e)
```

Answer : (−1/35*sqrt(35), 1/7*sqrt(35), 3/35*sqrt(35))
(1/35*sqrt(35), −1/7*sqrt(35), −3/35*sqrt(35))

$\begin{vmatrix}\mathbf{i} & \mathbf{j} & \mathbf{k}\\ 1 & -1 & 2\\ 3 & 0 & 1\end{vmatrix}=\langle -1, 5, 3\rangle$. Thus, two unit vectors orthogonal to both are

$$\pm\frac{\langle -1, 5, 3\rangle}{\sqrt{1+25+9}}=\pm\left\langle -\frac{1}{\sqrt{35}}, \frac{5}{\sqrt{35}}, \frac{3}{\sqrt{35}}\right\rangle,$$

that is, $\left\langle -\frac{1}{\sqrt{35}}, \frac{5}{\sqrt{35}}, \frac{3}{\sqrt{35}}\right\rangle$ and

$\left\langle \frac{1}{\sqrt{35}}, -\frac{5}{\sqrt{35}}, -\frac{3}{\sqrt{35}}\right\rangle$.

9. Find two unit vectors orthogonal to both $-\mathbf{i}+\mathbf{j}+2\mathbf{k}$ and $2\mathbf{i}-2\mathbf{j}+\mathbf{k}$.

Solution

$$\begin{vmatrix} \mathbf{i} & \mathbf{j} & \mathbf{k} \\ -1 & 1 & 2 \\ 2 & -2 & 1 \end{vmatrix} = \begin{vmatrix} 1 & 2 \\ -2 & 1 \end{vmatrix}\mathbf{i} - \begin{vmatrix} -1 & 2 \\ 2 & 1 \end{vmatrix}\mathbf{j} + \begin{vmatrix} -1 & 1 \\ 2 & -2 \end{vmatrix}\mathbf{k} = 5\mathbf{i}+5\mathbf{j}+0\mathbf{k}.$$

Thus, two unit vectors orthogonal to both are $\pm\frac{1}{5\sqrt{2}}\langle 5, 5, 0\rangle$, that is, $\left\langle \frac{1}{\sqrt{2}}, \frac{1}{\sqrt{2}}, 0 \right\rangle$ and $\left\langle -\frac{1}{\sqrt{2}}, -\frac{1}{\sqrt{2}}, 0 \right\rangle$.

CAS **10.** Find the area of the parallelogram with vertices $A(0, 1)$, $B(1, 4)$, $C(2, 1)$ and $D(1, -2)$.

http://matrix.skku.ac.kr/cal-lab/cal-11-4-10.html

Solution We may think of these points in xy-plane in the space.

```
A=vector([0, 1, 0])
B=vector([2, 1, 0])
C=vector([1, 4, 0])
D=vector([1, -2, 0])
AB=B-A; AC=C-A
Area= abs(AB.cross_product(AC))
show(Area)
```

Answer : 6

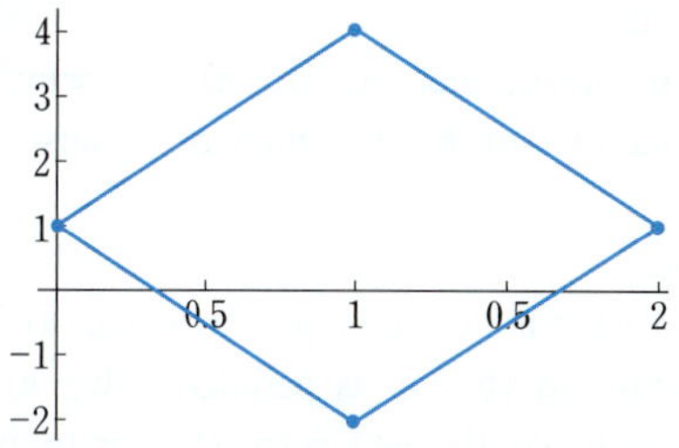

11. Find the area of the parallelogram with vertices $A(1, 1, 2)$, $B(3, -1, 0)$, $C(-1, 2, 1)$ and $D(1, 0, -1)$.

Solution The parallelogram is determined by the vectors $\overrightarrow{AB}=\langle 2, -2, -2\rangle$ and $\overrightarrow{AC}=\langle -2, 1, -1\rangle$, so the area of parallelogram $ABCD$ is

$$\overrightarrow{AB}\times\overrightarrow{AC} = \begin{vmatrix} \mathbf{i} & \mathbf{j} & \mathbf{k} \\ 2 & -2 & -2 \\ -2 & 1 & -1 \end{vmatrix} = \begin{vmatrix} -2 & -2 \\ 1 & -1 \end{vmatrix}\mathbf{i} - \begin{vmatrix} 2 & -2 \\ -2 & -1 \end{vmatrix}\mathbf{j} + \begin{vmatrix} 2 & -2 \\ -2 & 1 \end{vmatrix}\mathbf{k}$$
$$= 4\mathbf{i}+6\mathbf{j}-2\mathbf{k}$$

Then $|\overrightarrow{AB}\times\overrightarrow{AC}| = 2\sqrt{14}$.

12-13. Find a vector perpendicular to the plane through the points P, Q and R.

12. $P(1, 0, 0)$, $Q(4, 1, -1)$, $R(2, -1, -2)$

http://matrix.skku.ac.kr/cal-lab/cal-11-4-12.html

Solution

```
P=vector([1, 0, 0]);
Q=vector([4, 1, -1]);
R=vector([2, -1, -2]);
PQ=Q-P;
PR=R-P;
PQ.cross_product(PR)
```

Answer : (-3, 5, -4)

13. $P(2, 0, -1)$, $Q(3, 0, 6)$, $R(-1, 3, 4)$

Solution $\overrightarrow{PQ} = <1,0,7>$ and $\overrightarrow{PR} = <-3,3,5>$, so a vector orthogonal to the plane through P, Q, and R is

$$\overrightarrow{PQ}\times\overrightarrow{PR} = \begin{vmatrix} \mathbf{i} & \mathbf{j} & \mathbf{k} \\ 1 & 0 & 7 \\ -3 & 3 & 5 \end{vmatrix} = \begin{vmatrix} 0 & 7 \\ 3 & 5 \end{vmatrix}\mathbf{i} - \begin{vmatrix} 1 & 7 \\ -3 & 5 \end{vmatrix}\mathbf{j} + \begin{vmatrix} 1 & 0 \\ -3 & 3 \end{vmatrix}\mathbf{k}$$
$$= -21\mathbf{i}-26\mathbf{j}+3\mathbf{k}.$$

That is, $< -21, -26, 3 >$ is orthogonal to the plane through P, Q, and R.

14-15. Find the area of triangle PQR.

CAS **14.** $P(1, 0, 0)$, $Q(4, 1, -1)$, $R(2, -1, -2)$

http://matrix.skku.ac.kr/cal-lab/cal-11-4-14.html

Solution

```
P=vector([1, 0, 0]);
Q=vector([4, 1, -1]);
R=vector([2, -1, -2]);
PQ=Q-P;
PR=R-P;
CP=PQ.cross_product(PR);
1/2*CP.norm()
```

Answer : 5*sqrt(2)

15. $P(2, 0, -1)$, $Q(3, 1, 6)$, $R(-1, 3, 4)$.

Solution $11\sqrt{2}$.

16-17. Find the volume of the parallelepiped with adjacent edges PQ, PR, and PS.

CAS **16.** $P(2, 0, 1)$, $Q(4, 2, 0)$, $R(3, 1, -1)$, $S(1, 1, 0)$

http://matrix.skku.ac.kr/cal-lab/cal-11-4-16.html

Solution

```
P=vector([2, 0, 1]);
Q=vector([4, 2, 0]);
R=vector([3, 1, -1]);
S=vector([1, 1, 0]);
PQ=Q-P;
PR=R-P;
PS=S-P;
CP=PQ.cross_product(PR);
PS.inner_product(CP).abs()
```

Answer : 6

17. $P(0, 1, -2)$, $Q(1, -1, 4)$, $R(5, 0, 1)$, $S(0, 2, 3)$.

Solution $\mathbf{a} = \overrightarrow{PQ} = \langle 1, -2, 6 \rangle$, $\mathbf{b} = \overrightarrow{PR} = \langle 5, -1, 3 \rangle$ and $\mathbf{c} = \overrightarrow{PS} = \langle 0, 1, 5 \rangle$.

$$\mathbf{a} \cdot (\mathbf{b} \times \mathbf{c}) = \begin{vmatrix} 1 & -2 & 6 \\ 5 & -1 & 3 \\ 0 & 1 & 5 \end{vmatrix} = \begin{vmatrix} -1 & 3 \\ 1 & 5 \end{vmatrix} 1 - \begin{vmatrix} 5 & 3 \\ 0 & 5 \end{vmatrix} (-2) + \begin{vmatrix} 5 & -1 \\ 0 & 1 \end{vmatrix} 6$$
$$= -8 + 50 + 30 = 72.$$

So, the volume of the parallelepiped is 72 cubic units.

CAS **18.** Show that the vectors $\mathbf{a} = \langle 1, 1, 0 \rangle$, $\mathbf{b} = \langle 2, -1, 4 \rangle$ and $\mathbf{c} = \langle 2, 1, 4 \rangle$ are not coplanar.

http://matrix.skku.ac.kr/cal-lab/cal-11-4-18.html

Solution Not coplanar.

```
a=vector(QQ, [1, 1, 0])
b=vector(QQ, [2, -1, 4])
c=vector(QQ, [2, 1, 4])
M=matrix(QQ, [a, b, c]);M
print M
print M.det()
```

Answer : [1 1 0]
[2 -1 4]
[2 1 4]
-8

19. Determine whether the points $P(1, 1, 0)$, $Q(3, 4, 1)$, $R(2, 0, 0)$ and $S(8, 4, 2)$ lie in the same plane.

Solution $\mathbf{a} = \overrightarrow{PQ} = \langle 2, 3, 1 \rangle$, $\mathbf{b} = \overrightarrow{PR} = \langle 1, -1, 0 \rangle$ and $\mathbf{c} = \overrightarrow{PS} = \langle 7, 3, 2 \rangle$.

$$\mathbf{a} \cdot (\mathbf{b} \times \mathbf{c}) = \begin{vmatrix} 2 & 3 & 1 \\ 1 & -1 & 0 \\ 7 & 3 & 2 \end{vmatrix} = \begin{vmatrix} -1 & 0 \\ 3 & 2 \end{vmatrix} 2 - \begin{vmatrix} 1 & 0 \\ 7 & 2 \end{vmatrix} 3 + \begin{vmatrix} 1 & -1 \\ 7 & 3 \end{vmatrix} 1$$
$$= -4 - 6 + 10 = 0.$$

Thus, the volume of the parallelepiped determined by $\mathbf{a}$, $\mathbf{b}$ and $\mathbf{c}$ is 0. This says that these vectors lie in the same plane. Therefore, their initial and terminal points P, Q, R and S also lie in the same plane.

20. A wrench 40cm long lies along the positive y-axis and grips a bolt at the origin. A force is applied in the direction $\langle 0, 4, -3 \rangle$ at the end of the wrench. Find the magnitude of the force needed to supply 100N of torque to the bolt.

21. Suppose that $\mathbf{a} \neq 0$. Prove or disprove the following statements.

(a) If $\mathbf{a} \cdot \mathbf{b} = \mathbf{a} \cdot \mathbf{c}$, then $\mathbf{b} = \mathbf{c}$?

(b) If $\mathbf{a} \times \mathbf{b} = \mathbf{a} \times \mathbf{c}$, then $\mathbf{b} = \mathbf{c}$?

(c) If $\mathbf{a} \cdot \mathbf{b} = \mathbf{a} \cdot \mathbf{c}$ and $\mathbf{a} \times \mathbf{b} = \mathbf{a} \times \mathbf{c}$, then $\mathbf{b} = c$?

Solution (a) False.

If $\mathbf{a} \cdot \mathbf{b} = \mathbf{a} \cdot \mathbf{c}$, then $\mathbf{a} \cdot (\mathbf{b} - \mathbf{c}) = 0$, hence $\mathbf{a}$ is perpendicular to $\mathbf{b} - \mathbf{c}$.

This can happen if $\mathbf{b} \neq \mathbf{c}$.

For example, let $\mathbf{a} = \langle 1, 1, 1 \rangle$, $\mathbf{b} = \langle 1, 0, 0 \rangle$ and $\mathbf{c} = \langle 0, 1, 0 \rangle$, then $\mathbf{a} \cdot \mathbf{b} = \mathbf{a} \cdot \mathbf{c}$.

(b) False.

If $\mathbf{a} \times \mathbf{b} = \mathbf{a} \times \mathbf{c}$, then $\mathbf{a} \times (\mathbf{b} - \mathbf{c}) = 0$, which implies that $\mathbf{a}$ is parallel to $\mathbf{b} - \mathbf{c}$, which of course can happen if $\mathbf{b} \neq \mathbf{c}$.

(c) True.

Since $\mathbf{a} \cdot \mathbf{b} = \mathbf{a} \cdot \mathbf{c}$, $\mathbf{a}$ is perpendicular to $\mathbf{b} - \mathbf{c}$, by part (a). From part (b), $\mathbf{a}$ is parallel to $\mathbf{b} - \mathbf{c}$. Since $\mathbf{a} \neq 0$, and is both parallel and perpendicular to $\mathbf{b} - \mathbf{c}$, we have $\mathbf{b} - \mathbf{c} = 0$. Hence $\mathbf{b} = \mathbf{c}$.

22. Show that $\mathbf{a} \times (\mathbf{b} \times \mathbf{c}) = (\mathbf{a} \cdot \mathbf{c})\mathbf{b} - (\mathbf{a} \cdot \mathbf{b})\mathbf{c}$.

23. If $\mathbf{a} = \langle 1, 0, 1 \rangle$, $|\mathbf{b}| = 1$, $\mathbf{c} = \langle 0, 1, 0 \rangle$ and $\mathbf{a} \times \mathbf{b} = \mathbf{c}$, then find $\mathbf{b}$.

11.5 Equations of Lines and Planes

YouTube http://youtu.be/YB976T1w0kE

In this section, we will derive vector equations of lines and planes in $\mathbb{R}^2$ and $\mathbb{R}^3$, and we will deal with shortest distance problems related to these equations.

Lines

In the plane $\mathbb{R}^2$, an equation of the line can be uniquely determined when a slope and a specified point on the line are given. In general it can be generally written as follows:

$$ax+by+c=0$$

where a, b and c are real numbers and a and b are not both zero.

Let's find equation of a line in $\mathbb{R}^3$. If a line passes through the point $P_0(x_0, y_0, z_0)$ and is parallel to $\mathbf{v}=a\mathbf{i}+b\mathbf{j}+c\mathbf{k}$, then the vector $\overrightarrow{P_0P}$ is parallel to v (See Figure 1.), where $P(x, y, z)$ is any point on the line.

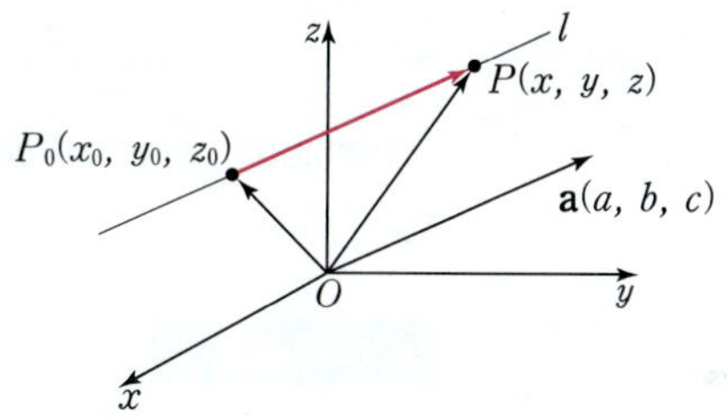

Figure 1

That is, the line is a set of all points $P(x, y, z)$ that satisfies the following equation 1 :

1 $$\overrightarrow{P_0P}=t\mathbf{v} \quad (t\in\mathbb{R})$$

Thus, if $\mathbf{r}$ is any point on the line through $\mathbf{r}_0$ that is parallel to $\mathbf{r}$, then the vector $\mathbf{r}-\mathbf{r}_0$ is parallel to $\mathbf{v}$, so $\mathbf{r}-\mathbf{r}_0=t\mathbf{v}$ for some $t\in\mathbb{R}$. This line can be represented as the equation 2.

2 $$\mathbf{r}=\mathbf{r}_0+t\mathbf{v} \; (\mathbf{r}=\overrightarrow{OP},\ \mathbf{r}_0=\overrightarrow{OP_o})$$

We call this a **vector equation** of the line through $\mathbf{r}_0$ that is parallel to $\mathbf{r}$.

A vector equation of a line can be split into a set of scalar equations by equating corresponding components; these are called **parametric equations** 3 of the line. Thus we have

$$(x, y, z) = (x_0, y_0, z_0) + t(a, b, c) = (x_0 + ta, y_0 + tb, z_0 + tc), \ (t \in \mathbb{R})$$

which implies

3
$$\begin{aligned} x &= x_0 + ta \ (t \in \mathbb{R}) \\ y &= y_0 + tb \\ z &= z_0 + tc \end{aligned}$$

From 3, symmetric equations of the line can be written defined as the following:

4
$$\frac{x - x_0}{a} = \frac{y - y_0}{b} = \frac{z - z_0}{c}$$

where a, b and c are nonzero constants.

If P_0 and P_1 are distinct points with position vectors $\mathbf{r}_0$ and $\mathbf{r}_1$ in $\mathbb{R}^2$ **or** $\mathbb{R}^3$, then the line determined by these points is parallel to the vector $\mathbf{v} = \mathbf{r}_1 - \mathbf{r}_0$, so it follows from 2 that the line can be expressed in vector form as

5
$$\mathbf{r} = \mathbf{r}_0 + t(\mathbf{r}_1 - \mathbf{r}_0) = (1 - t)\mathbf{r}_0 + t\mathbf{r}_1, \ (t \in \mathbb{R}).$$

Equation 5 is called the two-point vector equation of the line through $\mathbf{r}_0$ and $\mathbf{r}_1$.

EXAMPLE 1

(a) Find a vector, parametric and symmetric equations of the line that pass through the point $P_0(2, 3, 5)$ and is parallel to the vector $\mathbf{v} = -\mathbf{i} + 3\mathbf{j} - 5\mathbf{k} = \langle -1, 3, -5 \rangle$.

(b) Find two other points on the line.

Solution (a) Here $\mathbf{r}_0 = \langle 2, 3, 5 \rangle$ and $\mathbf{v} = \langle -1, 3, -5 \rangle$.
The vector equation is

$$\begin{aligned} \langle x, y, z \rangle &= \langle 2, 3, 5 \rangle + t\langle -1, 3, -5 \rangle \\ &= \langle 2 - t, 3 + 3t, 5 - 5t \rangle. \end{aligned}$$

Parametric equations are $x = 2 - t$, $y = 3 + 3t$, $z = 5 - 5t$.
The symmetric equation is

$$\frac{x - 2}{-1} = \frac{y - 3}{3} = \frac{z - 5}{-5}.$$

(b) Choosing the parameter value $t = 1$ gives $x = 1$, $y = 6$, and $z = 0$, so $(1, 6, 0)$ is a point on the line. Similarly, $t = 2$ gives the point $(0, 9, -5)$. ■

EXAMPLE 2

Find a vector, parametric and symmetric equations for the line that pass through the points $A(1,\ 1,\ -2)$ and $B(4,\ -1,\ 0)$.

Solution Two points $A(1,\ 1,\ -2)$ and $B(4,\ -1,\ 0)$ with position vectors $\mathbf{r}_0$ and $\mathbf{r}_1$ forms a vector

$$\overrightarrow{AB} = \mathbf{v} = \langle 4-1,\ -1-1,\ 0-(-2)\rangle = \langle 3,\ -2,\ 2\rangle$$

and the vector equation $\mathbf{r} = \mathbf{r}_0 + t(\mathbf{r}_1 - \mathbf{r}_0) = (1-t)\mathbf{r}_0 + t\mathbf{r}_1$ can be written as

$$\begin{aligned}\langle x,\ y,\ z\rangle &= \langle 1,\ 1,\ -2\rangle + t\langle 3,\ -2,\ 2\rangle \\ &= \langle 1+3t,\ 1-2t,\ -2+2t\rangle,\ (t \in \mathbb{R}).\end{aligned}$$

Thus, the parametric equations are $x = 1+3t$, $y = 1-2t$, $z = -2+2t$ and symmetric equations of the line are

$$\frac{x-1}{3} = \frac{y-1}{-2} = \frac{z+2}{2}.$$

■

Point-Normal Equation of Planes

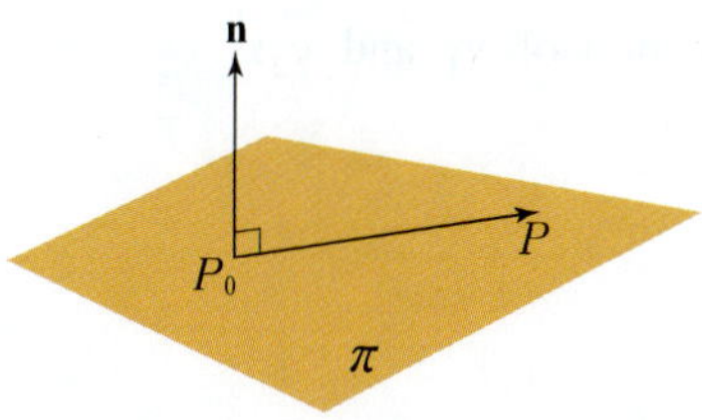

Figure 2 π: $ax+by+cz+d=0$

A plane in $\mathbb{R}^3$ can be uniquely obtained by specifying a point $P_0(x_0, y_0, z_0)$ in the plane and a **nonzero vector** $\mathbf{n}$ that is perpendicular to the plane. (See Figure 2) The vector $\mathbf{n}$ is called the **normal vector** to the plane. If $P(x, y, z)$ is any point in this plane, then the $\overrightarrow{P_0P} = \mathbf{r} - \mathbf{r}_0$ is orthogonal to $\mathbf{n}$, by the property of the dot (inner) product. (See Figure 2.) So

5 $$\mathbf{n} \cdot \overrightarrow{P_0P} = \mathbf{n} \cdot (\mathbf{r} - \mathbf{r}_0) = 0$$

From 5,

$$(a,\ b,\ c) \cdot (x - x_0,\ y - y_0,\ -z_0) = 0$$

or

6 $$a(x - x_0) + b(y - y_0) + c(z - z_0) = 0$$

where a, b and c are not all zero.

We call 6 a **point-normal equation** of the plane through $P_0(x_0,\ y_0,\ z_0)$ with normal $\mathbf{n} = (a, b, c)$.

For convenience, we simplify the left terms of 6 as follows

7 $$ax + by + cz + d = 0$$

where a, b and c are not all zero and $d = -(ax_0 + by_0 + cz_0)$.
We call [7] the **general equation** of the plane.

EXAMPLE 3

Find a point-normal equation and a general equation of the plane that passes through $(2, -3, 4)$ with normal $\mathbf{n} = \langle 2, -5, 3 \rangle$.

Solution From [6], a point-normal equation of the plane is

$$2(x-2) - 5(y+3) + 3(z-4) = 0.$$

Multiplying out and taking the constants to the right side yields the general equation

$$2x - 5y + 4z - 31 = 0.$$ ■

Vector and Parametric Equations of Planes

A plane W can be uniquely obtained by passing through a point $P_0(x_0, y_0, z_0)$ in W and two nonzero vectors $\mathbf{v}_1$ and $\mathbf{v}_2$ that are parallel to W and are not scalar multiples of one another. (See Figure 3.) That is, if $P(x, y, z)$ is any point in the plane W and $\mathbf{v}_1$ and $\mathbf{v}_2$ are positioned with their initial points at W_0, then $\mathbf{x} - \mathbf{x}_0$ is expressed as a linear combination of $\mathbf{v}_1$ and $\mathbf{v}_2$;

$$\mathbf{x} - \mathbf{x}_0 = \mathrm{t}_1\mathbf{v}_1 + \mathrm{t}_2\mathbf{v}_2$$

or

[8] $$\mathbf{x} = \mathbf{x}_0 + \mathrm{t}_1\mathbf{v}_1 + \mathrm{t}_2\mathbf{v}_2$$

where $\overrightarrow{OP} = \mathbf{x}$ and $\overrightarrow{OP_0} = \mathbf{x}_0$ and t_1 and t_2, called **parameters**, are in $\mathbb{R}$. We call [8] a **vector equation of the plane** through P_0 that is parallel to $\mathbf{v}_1$ and $\mathbf{v}_2$. Let $\mathbf{x} = (x, y, z)$ be any point in the plane through $\mathbf{x}_0 = (x_0, y_0, z_0)$ that is parallel to the vectors $\mathbf{v}_1 = \langle a_1, b_1, c_1 \rangle$ and $\mathbf{v}_2 = \langle a_2, b_2, c_2 \rangle$. Then [8] can be expressed in the component form as

$$\langle x, y, z \rangle = \langle x_0, y_0, z_0 \rangle + t_1 \langle a_1, b_1, c_1 \rangle + t_2 \langle a_2, b_2, c_2 \rangle$$

or

[9] $$\begin{aligned} x &= x_0 + a_1t_1 + a_2t_2 \\ y &= y_0 + b_1t_1 + b_2t_2, \ (t_1, t_2 \in \mathbb{R}) \\ z &= z_0 + c_1t_1 + c_2t_2 \end{aligned}$$

We call [9] **parametric equations** of the plane.

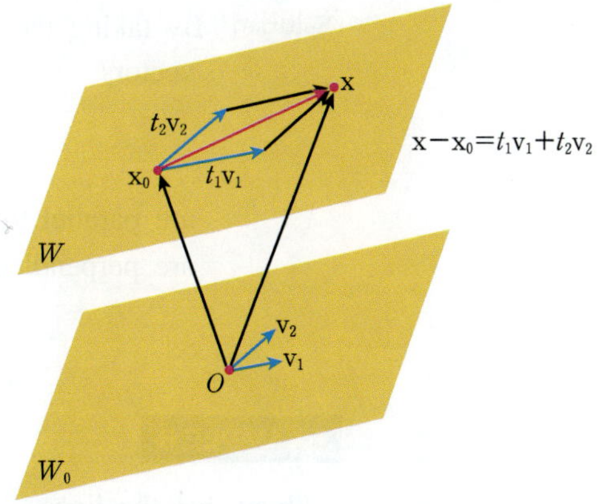

Figure 3
$W_0 = \{\mathbf{x} \mid \mathbf{x} = t_1\mathbf{v}_1 + t_2\mathbf{v}_2,\ t_1, t_2 \in R\}$
$W = \mathbf{x}_0 + W_0 = \{\mathbf{x} \mid \mathbf{x} = \mathbf{x}_0 + t_1\mathbf{v}_1 + t_2\mathbf{v}_2,\ t_1, t_2 \in R\}$

EXAMPLE 4

Find vector and parametric equations of the plane that passes through three points: $A(4, -3, 1)$, $B(6, -4, 7)$, $C(1, 2, 2)$.

Solution Let $\mathbf{r} = \langle x, y, z\rangle$, $\mathbf{r}_0 = \langle 4, -3, 1\rangle$, $\mathbf{r}_1 = \langle 6, -4, 7\rangle$ and $\mathbf{r}_2 = \langle 1, 2, 2\rangle$. Then we have two vectors that parallel to the plane as

$$\overrightarrow{AB} = \mathbf{r}_1 - \mathbf{r}_0 = \langle 2, -1, 6\rangle,\ \overrightarrow{AC} = \mathbf{r}_2 - \mathbf{r}_0 = \langle -3, 5, 1\rangle.$$

From 8, a vector equation of the plane is

$$\langle x, y, z\rangle = \langle 4, -3, 1\rangle + t_1\langle 2, -1, 6\rangle + t_2\langle -3, 5, 1\rangle.$$

Also we have parametric equations as

$$x = 4 + 2t_1 - 3t_2, y = -3 - t_1 + 5t_2, z = 1 + 6t_1 + t_2.$$ ■

Perpendicular and Parallel Lines

Two lines L_1 and L_2 with direction vectors $\mathbf{v}_1$ and $\mathbf{v}_2$, respectively, are

(i) perpendicular if $\mathbf{v}_1 \cdot \mathbf{v}_2 = 0$,
(ii) parallel if $\mathbf{v}_2 = k\mathbf{v}_1$ for some nonzero scalar k,
(iii) skew if neither (i) nor (ii).

EXAMPLE 5

Show that the lines

$$L_1 : x = 2t,\ y = -4t,\ z = 5t,$$
$$L_2 : x = 1 + s,\ y = 3s,\ z = 3 + 2s$$

are perpendicular.

Solution By taking the value for the parameters $t=1$ and $s=1$, we get two vectors

$$\mathbf{v}_1 = 2\mathbf{i} - 4\mathbf{j} + 5\mathbf{k} \quad \text{and} \quad \mathbf{v}_2 = \mathbf{i} + 3\mathbf{j} + 2\mathbf{k}$$

are parallel to L_1 and L_2, respectively. We can verify that two lines are perpendicular to each other by

$$\mathbf{v}_1 \cdot \mathbf{v}_2 = 2 - 12 + 10 = 0.$$

■

EXAMPLE 6

Show that the lines

$$L_1: \ x = 3 + 4t, \ y = 5 + 2t, \ z = 1 - 8t,$$
$$L_2: \ \mathbf{r} = \langle 1, 1, 4 \rangle + s\langle 2, 1, -4 \rangle$$

are parallel.

Solution The coefficient of the parameters t and s, we see that

$$\mathbf{v}_1 = 4\mathbf{i} + 2\mathbf{j} - 8\mathbf{k} \quad \text{and} \quad \mathbf{v}_2 = 2\mathbf{i} + \mathbf{j} - 4\mathbf{k}$$

are the direction vectors for L_1 and L_2, respectively. Since $\mathbf{v}_1 = 2\mathbf{v}_2$, the two lines are parallel. ■

EXAMPLE 7

Show that the lines L_1 and L_2 with parametric equations

$$L_1: \ \frac{x}{1} = \frac{y-1}{2} = \frac{z-2}{3},$$
$$L_2: \ \frac{x-3}{-4} = \frac{y-2}{-3} = \frac{z-1}{2}$$

are **skew lines**; that is, they do not intersect and are not parallel (and therefore do not lie in the same plane).

http://matrix.skku.ac.kr/cal-lab/11-5-Exmaple-7.html

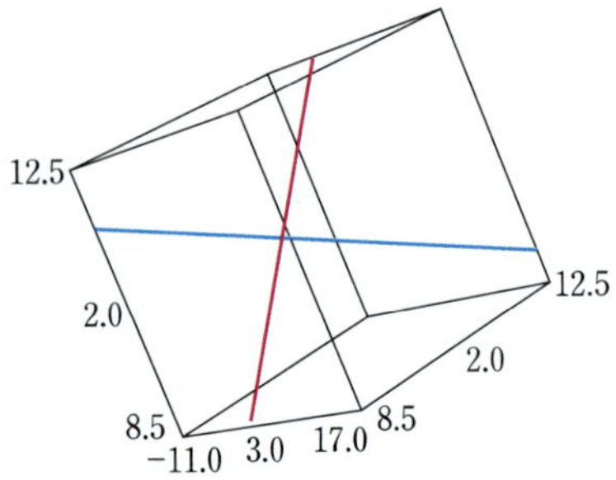

Figure 4

Solution The lines are not parallel because the corresponding direction vectors $\langle 1, 2, 3 \rangle$ and $\langle -4, -3, 2 \rangle$ are not parallel. (Their components are not proportional.) If L_1 and L_2 have a point of intersection, there would be values of t and s such that

$$t = 3 - 4s,$$
$$2t + 1 = 2 - 3s,$$
$$3t + 2 = 2s + 1.$$

If we solve the first two equations, we get $t = -1$ and $s = 1$, and these values do not satisfy the third equation. Therefore, no values of t and s satisfy all three equations. Thus L_1 and L_2 do not intersect.

Hence, L_1 and L_2 are skew lines.

```
var ('t, s')
L1=parametric_plot3d([t,2*t+1,3*t        +2],       (t,-3.5,3.5),
color='blue', thickness=5)
L2=parametric_plot3d([3-4*s,2-3*s,1+2*s],       (s,-3.5,3.5),
color='red', thickness=3)
show(L1+L2)
```

■

The Distance from a Point to a Line in Space

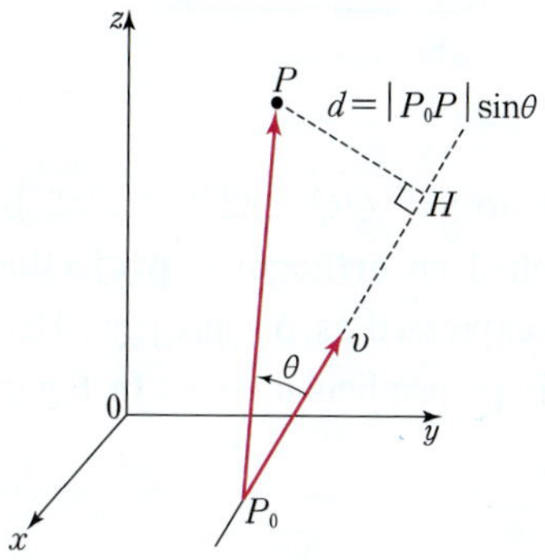

Figure 5

To find the distance from a point P to a line that passes through a point P_0 parallel to a vector $\mathbf{v}$, we can find the relationship of the distance d and the length $|\overrightarrow{P_0P}|$ as follows: (See Figure 5.)

$$d=|\overrightarrow{P_0P}|\sin\theta \quad \text{where } \theta \text{ is the angle between } \overrightarrow{P_0P} \text{ and } \mathbf{v}.$$

From Theorem 2 in section 11.4,

$$|\mathbf{v}\times\overrightarrow{P_0P_1}|=|\mathbf{v}||\overrightarrow{P_0P_1}|\sin\theta \quad \text{which implies} \quad \frac{|\mathbf{v}\times\overrightarrow{P_0P_1}|}{|\mathbf{v}|}=|\overrightarrow{P_0P_1}|\sin\theta .$$

Then the distance from a point P to a line L is $d=\dfrac{|\mathbf{v}\times\overrightarrow{P_0P}|}{|\mathbf{v}|}$.

EXAMPLE 8

Find the distance from the point $P(2,1,-1)$ to the line

$$L : x=3t,\ y=1+2t,\ z=-5-t .$$

Solution We see from the equations for L that it passes through $P_0(0,1,-5)$ and is parallel to $\mathbf{v}=3\mathbf{i}+2\mathbf{j}-\mathbf{k}$. We obtain

$$\overrightarrow{P_0P}=(2-0)\mathbf{i}+(1-1)\mathbf{j}+(-1+5)\mathbf{k}=2\mathbf{i}+4\mathbf{k}$$

and

$$\begin{aligned}\mathbf{v}\times\overrightarrow{P_0P}&=(3\mathbf{i}+2\mathbf{j}-\mathbf{k})\times(2\mathbf{i}+4\mathbf{k})\\ &=\begin{vmatrix}\mathbf{i}&\mathbf{j}&\mathbf{k}\\3&2&-1\\2&0&4\end{vmatrix}=8\mathbf{i}-14\mathbf{j}-4\mathbf{k}.\\ |\mathbf{v}\times\overrightarrow{P_0P}|&=|8\mathbf{i}-14\mathbf{j}-4\mathbf{k}|\\ &=\sqrt{8^2+(-14)^2+(-4)^2}=\sqrt{276}\\ |\mathbf{v}|&=|3\mathbf{i}+2\mathbf{j}-\mathbf{k}|=\sqrt{3^2+2^2+(-1)^2}=\sqrt{14}\\ d&=|\overrightarrow{P_0P_1}|\sin\theta=\frac{|\mathbf{v}\times\overrightarrow{P_0P_1}|}{|\mathbf{v}|}\end{aligned}$$

$$= \frac{\sqrt{276}}{\sqrt{14}} = \frac{\sqrt{138}}{\sqrt{7}}. \quad \blacksquare$$

Projection of Vectors

To find the distance from a point P to a plane in $\mathbb{R}^3$, we introduce the concept of the orthogonal projection of vectors.

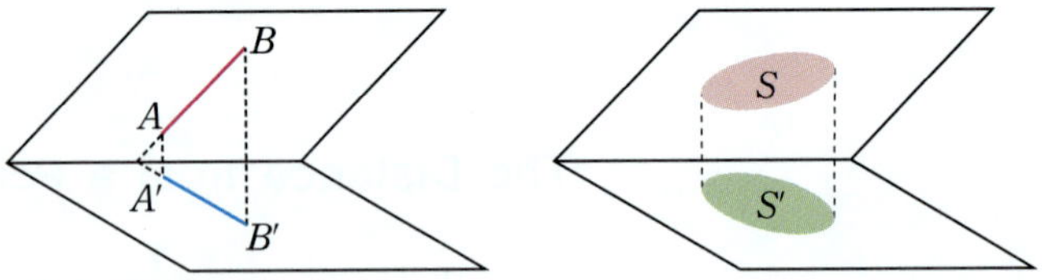

(a) (b)

Figure 6

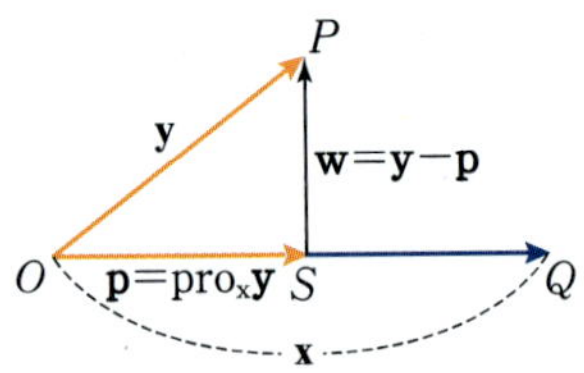

Figure 7

Let S be the foot of the perpendicular from P to $\mathbf{x} = \overrightarrow{OQ}$ and $\mathbf{y} = \overrightarrow{OP}$ be vectors in $\mathbb{R}^3$ with $\mathbf{x} \neq \mathbf{0}$. Then $\mathbf{p} = \overrightarrow{OS}$ is called an **orthogonal projection** of $\mathbf{y} = \overrightarrow{OP}$ onto $\mathbf{x} = \overrightarrow{OQ}$. The vector $\mathbf{p}$ can be expressed as $\mathbf{p} = \mathrm{proj}_{\mathbf{x}}\mathbf{y}$. Here $\mathbf{w} = \overrightarrow{SP}$ is called a vector component of $\mathbf{y}$ that is perpendicular to $\mathbf{x}$. In Figure 7, a vector $\mathbf{y}$ is sum of $\mathbf{p}$ and $\mathbf{w}$;

$$\mathbf{y} = \mathbf{p} + \mathbf{w}.$$

The following theorem give expression for the orthogonal projection $\mathbf{p}$.

THEOREM 1

If $\mathbf{x} \neq \mathbf{0}$ and $\mathbf{y}$ are in $\mathbb{R}^3$, then we have the following:

(a) $\mathrm{proj}_{\mathbf{x}}\mathbf{y} = \dfrac{\mathbf{y} \cdot \mathbf{x}}{\mathbf{x} \cdot \mathbf{x}}\mathbf{x}$ (b) $|\mathrm{proj}_{\mathbf{x}}\mathbf{y}| = \dfrac{|\mathbf{y} \cdot \mathbf{x}|}{|\mathbf{x}|}$

Proof (a) Since $\mathrm{proj}_{\mathbf{x}}\mathbf{y}$ is parallel to $\mathbf{x}$, we have

$$\mathrm{proj}_{\mathbf{x}}\mathbf{y} = t\mathbf{x} \quad (t \in \mathbb{R}).$$

Since $\mathbf{y} = t\mathbf{x} + \mathbf{w}$ and $\mathbf{w} \cdot \mathbf{x} = 0$ because $\mathbf{w}$ is perpendicular to $\mathbf{x}$ (See Figure 6), we can get the value t as

$$t = \frac{\mathbf{y} \cdot \mathbf{x}}{|\mathbf{x}|^2}$$

that satisfies $\mathbf{y} \cdot \mathbf{x} = (t\mathbf{x} + \mathbf{w}) \cdot \mathbf{x} = t|\mathbf{x}|^2 + \mathbf{w} \cdot \mathbf{x} = t|\mathbf{x}|^2 = t\mathbf{x} \cdot \mathbf{x}$.

Therefore $\mathrm{proj}_{\mathbf{x}}\mathbf{y} = \dfrac{\mathbf{y} \cdot \mathbf{x}}{\mathbf{x} \cdot \mathbf{x}}\mathbf{x}$.

(b) $|\mathrm{proj}_{\mathbf{x}}\mathbf{y}| = \left| \dfrac{\mathbf{y} \cdot \mathbf{x}}{|\mathbf{x}|^2}\mathbf{x} \right| = \dfrac{|\mathbf{y} \cdot \mathbf{x}|}{|\mathbf{x}|^2}|\mathbf{x}| = \dfrac{|\mathbf{y} \cdot \mathbf{x}|}{|\mathbf{x}|}$. $\blacksquare$

EXAMPLE 9

Find the projection $\text{proj}_{\mathbf{x}}\mathbf{y}$ of $\mathbf{y}$ onto $\mathbf{x}$ and a vector component of $\mathbf{y}$ perpendicular to $\mathbf{x}$ for $\mathbf{x}=\langle 4,\ -1,\ 3\rangle$ and $\mathbf{y}=\langle 3,\ 1,\ -2\rangle$.

Solution Since $\mathbf{y}\cdot\mathbf{x}=5$ and $|\mathbf{x}|=\sqrt{26}$, the projection of $\mathbf{y}$ onto $\mathbf{x}$ is

$$\text{proj}_{\mathbf{x}}\mathbf{y}=\frac{\mathbf{y}\cdot\mathbf{x}}{\mathbf{x}\cdot\mathbf{x}}\mathbf{x}=\frac{5}{26}\langle 4,\ -1,\ 3\rangle=\left\langle\frac{10}{13},\ -\frac{5}{26},\ \frac{15}{26}\right\rangle$$

and a vector component of $\mathbf{y}$ perpendicular to $\mathbf{x}$ is

$$\begin{aligned}\mathbf{w}=\mathbf{y}-\text{proj}_{\mathbf{x}}\mathbf{y}&=\langle 3,\ 1,\ -2\rangle-\left\langle\frac{10}{13},\ -\frac{5}{26},\ \frac{15}{26}\right\rangle\\&=\left\langle\frac{29}{13},\ \frac{31}{26},\ \frac{67}{26}\right\rangle.\end{aligned}$$

■

The Distance from a Point to a Plane in Space

We find a way to determine the **distance** D from a point $P_1(x_1,\ y_1,\ z_1)$ to the plane $ax+by+cz+d=0$. Note that $\mathbf{n}=\langle a,\ b,\ c\rangle$ is the normal to the plane.

Let $P_0(x_0,\ y_0,\ z_0)$ be any point in the given plane and let $\mathbf{b}=\overrightarrow{P_0P_1}$. Then, $\mathbf{b}=\langle x_1-x_0,\ y_1-y_0,\ z_1-z_0\rangle$.

The distance D from P_1 to the plane is equal to the absolute value of the scalar projection of $\mathbf{b}$ onto the normal vector $\mathbf{n}=\langle a,\ b,\ c\rangle$. (See Figure 8.) Thus,

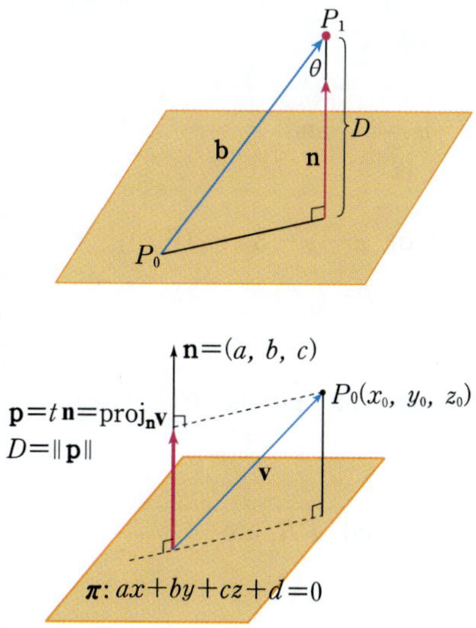

Figure 8 $\text{proj}_{\mathbf{n}}\mathbf{v}=t\mathbf{n}$

$$\begin{aligned}D=|\text{proj}_{\mathbf{n}}\mathbf{b}|&=\frac{|\mathbf{n}\cdot\mathbf{b}|}{|\mathbf{n}|}\\&=\frac{|a(x_1-x_0)+b(y_1-y_0)+c(z_1-z_0)|}{\sqrt{a^2+b^2+c^2}}\\&=\frac{|(ax_1+by_1+cz_1)-(ax_0+by_0+cz)|}{\sqrt{a^2+b^2+c^2}}.\end{aligned}$$

Since P_0 lies in the plane, its coordinates satisfy the equation of the plane, so we have $ax_0+by_0+cz_0+d=0$. Thus,

10

$$D=\frac{|ax_1+by_1+cz_1+d|}{\sqrt{a^2+b^2+c^2}}.$$

EXAMPLE 10

Find the distance D from a point $P(3,-1,2)$ to the plane $x+3y-2z-5=0$.

Solution Since a normal vector is $(1,3,-2)$,

$$D = \frac{|3+3\times(-1)-2\times 2-5|}{\sqrt{1^2+3^2+(-2)^2}} = \frac{9}{\sqrt{14}}.$$ ■

Similarly, in $\mathbb{R}^2$ the distance from $P_0(x_0, y_0)$ to a line $ax+by+c=0$ is

$$D = \frac{|ax_0+by_0+c|}{\sqrt{a^2+b^2}}.$$

EXAMPLE 11

Find the distance D from a point $P(3,-1)$ to the plane $x+3y-5=0$.

Solution Since a normal vector is $(1, 3)$,

$$D = \frac{|3+3\times(-1)-5|}{\sqrt{1^2+3^2}} = \frac{5}{\sqrt{10}}.$$ ■

Now we have some examples for properties of lines and planes in $\mathbb{R}^3$.

EXAMPLE 12

Find an equation of the plane through the point $(2, 1, -5)$ with the normal vector $\mathbf{n} = \langle -1, 1, 3\rangle$.

Solution Putting $a=-1$, $b=1$, $c=3$, $x_0=2$, $y_0=1$ and $z_0=-5$, from [6], we see that an equation of the plane is

$$-1(x-2)+1(y-1)+3(z+5)=0 \text{ or } x-y-3z=16.$$ ■

EXAMPLE 13

Find an equation of the plane that passes through the points $P(1,1,2)$, $Q(3,-1,2)$ and $R(1,2,4)$.

Solution The vectors $\mathbf{a}$ and $\mathbf{b}$ corresponding to $\overrightarrow{PQ}$ and $\overrightarrow{PR}$ are

$$\mathbf{a} = \langle 2, -2, 0\rangle \quad \mathbf{b} = \langle 0, 1, 2\rangle.$$

Since both $\mathbf{a}$ and $\mathbf{b}$ lie in the plane, their cross product $\mathbf{a}\times\mathbf{b}$ is orthogonal to the plane and can be taken as the normal vector. Thus,

$$\mathbf{n} = \mathbf{a}\times\mathbf{b} = \begin{vmatrix} \mathbf{i} & \mathbf{j} & \mathbf{k} \\ 2 & -2 & 0 \\ 0 & 1 & 2 \end{vmatrix} = -4\mathbf{i}-4\mathbf{j}+2\mathbf{k}.$$

With the point $P(1,1,2)$ and the normal vector $\mathbf{n}$, the equation of the plane is

$$-4(x-1)-4(y-1)+2(z-2)=0 \text{ or } 2x+2y-z=2.$$ ■

EXAMPLE 14

Find the point at which the line $x=2+t$, $y=-2t$, $z=3+2t$ intersects the plane $3x+2y+4z=20$.

Solution We substitute the expressions for x, y and z from the parametric equations into the equation of the plane:

$$3(2+t)+2(-2t)+4(3+2t)=20$$

This simplifies to $7t=2$, so $t=\frac{2}{7}$. Therefore, the point of intersection occurs when the parameter value is $t=\frac{2}{7}$. Thus, $x=2+\frac{2}{7}=\frac{16}{7}$, $y=-2\cdot\frac{2}{7}=-\frac{4}{7}$, $z=3+2\cdot\frac{2}{7}=\frac{25}{7}$. Hence the point of intersection is $\left(\frac{16}{7}, -\frac{4}{7}, \frac{25}{7}\right)$. ■

EXAMPLE 15

(a) Find the angle between the planes $x+2y-z=1$ and $x+y+3z=1$.

(b) Find symmetric equations for the line of intersection L of these two planes.

Solution (a) The normal vectors of these planes are $\mathbf{n}_1=\langle 1, 2, -1\rangle$ and $\mathbf{n}_2=\langle 1, 1, 3\rangle$, respectively.

Let θ be the angle between the two planes, which is called a dihedral angle. This dihedral angle is equal to the angle between two normal vectors $\mathbf{n}_1$, $\mathbf{n}_2$. By the property of the dot product,

$$\cos\theta=\frac{\mathbf{n}_1\cdot\mathbf{n}_2}{|\mathbf{n}_1||\mathbf{n}_2|}=\frac{1(1)+2(1)+(-1)(3)}{\sqrt{1+4+1}\,\sqrt{1+1+9}}=0,$$

and $\theta=\cos^{-1}(0)=\pi/2\approx 90°$. Therefore two planes are orthogonal to each other.

(b) We first need to find a point on L. For instance, we can find the point where the line intersects the xy-plane by setting $z=0$ in the equations of both planes. This gives the equations $x+2y=1$ and $x+y=1$, whose solution is $x=1$, $y=0$. Hence, the point $(1, 0, 0)$ lies on L.

Now, we observe that, since L lies in both planes, it is perpendicular to both of the normal vectors. Thus, a vector v parallel to L is given by the cross product

$$\mathbf{v}=\mathbf{n}_1\times\mathbf{n}_2=\begin{vmatrix}\mathbf{i} & \mathbf{j} & \mathbf{k}\\ 1 & 2 & -1\\ 1 & 1 & 3\end{vmatrix}=7\mathbf{i}-4\mathbf{j}-\mathbf{k}.$$

Hence, the symmetric equations of L can be written as

$$\frac{x-1}{7} = \frac{y}{4} = \frac{z}{-1}. \quad \blacksquare$$

Note The line of intersection can also be obtained by solving the equations of the planes for two of the variables in terms of the third, which can be taken as the parameter.
For instance, the line L was given as the line of intersection of the planes $x+y+z=1$ and $x-2y+3z=1$. The symmetric equations that we found for L could be written as

$$\frac{x-1}{5} = \frac{y}{-2} \text{ and } \frac{y}{-2} = \frac{z}{-3}$$

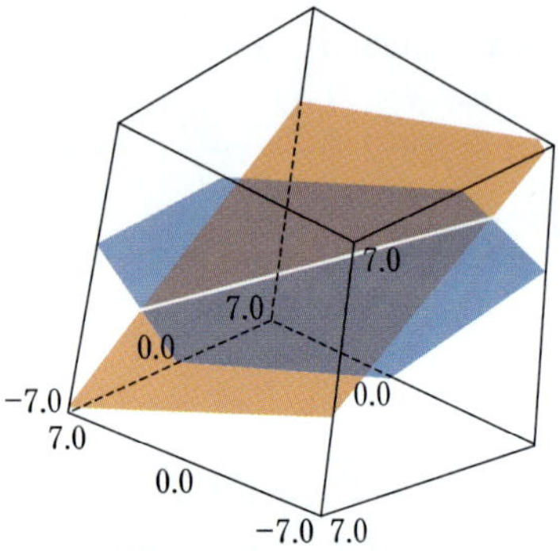

Figure 9 Shows how the line L in Example 17 can also be regarded as the line of intersection of planes derived from its symmetric equations (Sage).

which is again a pair of linear equations. They exhibit L as the line of intersection of the planes $(x-1)/5 = y/(-2)$ and $y/(-2) = z/(-3)$. (See Figure 9.)
http://matrix.skku.ac.kr/cal-lab/11-5-Exmaple-14.html

```
var ('x, y, z')
P1=implicit_plot3d(x + y + z == 1,(x,-7,7),(y,-7,7),(z,-7,7),color='blue',opacity=0.3)
P2=implicit_plot3d(x - 2*y + 3*z == 1,(x,-7,7),(y,-7,7),(z,-7,7),color='red',opacity=0.3)
show(P1+P2)
```

Perpendicular and Parallel Planes

Two planes α_1 : $a_1x+b_1y+c_1z=d_1$ and α_2 : $a_2x+b_2y+c_2z=d_2$ with normal vectors $\mathbf{n}_1 = \langle a_1, b_1, c_1 \rangle$ and $\mathbf{n}_2 = \langle a_2, b_2, c_2 \rangle$, respectively, are

(i) perpendicular : if $\mathbf{n}_1 \cdot \mathbf{n}_2 = 0$, and
(ii) parallel if $\mathbf{n}_2 = k\mathbf{n}_1$ for some nonzero scalar k.

EXAMPLE 16

Show that the planes are parallel and find the distance between the parallel planes $3x+y-2z=1$ and $6x+2y-4z=3$.

Solution First, we note that the planes are parallel because their normal vectors $\langle 3, 1, -2 \rangle$ and $\langle 6, 2, -4 \rangle$ are parallel. To find the distance d between the planes, we choose any point on one plane and calculate its distance to the other plane. In particular, if we put $y=z=0$ in the equation of the first plane, we get $3x=1$, so $\left(\frac{1}{3}, 0, 0\right)$ is a point in this plane. Then, the distance between $\left(\frac{1}{3}, 0, 0\right)$ and the plane $6x+2y-4z-3=0$ is

$$D=\frac{\left|6\cdot\frac{1}{3}+0+0-3\right|}{\sqrt{36+4+16}}=\frac{1}{\sqrt{56}}=\frac{\sqrt{14}}{28}.$$

Hence, the distance between the planes is $\frac{\sqrt{14}}{28}$. ■

EXAMPLE 17

In Example 7, we showed that the lines

$$L_1:\ \frac{x}{1}=\frac{y-1}{2}=\frac{z-2}{3},$$
$$L_2:\ \frac{x-3}{-4}=\frac{y-2}{-3}=\frac{z-1}{2}.$$

are skew. Find the distance between them.

Solution Since the two lines L_1 and L_2 are skew, they can be viewed as lying on two parallel planes P_1 and P_2. The distance between L_1 and L_2 is the same as the distance between P_1 and P_2 which can be computed as in Example 16. The common normal vector to both planes must be orthogonal to both $\mathbf{v}_1=\langle 1, 2, 3\rangle$ (the direction of L_1) and $\mathbf{v}_2=\langle -4, -3, 2\rangle$ (the direction of L_2). Hence, a normal vector is

$$\mathbf{n}=\mathbf{v}_1\times\mathbf{v}_2=\begin{vmatrix}\mathbf{i} & \mathbf{j} & \mathbf{k}\\ 1 & 2 & 3\\ -4 & -3 & 2\end{vmatrix}=13\mathbf{i}-14\mathbf{j}+5\mathbf{k}.$$

If we put $x=3$ in the equations of L_2 we get the point $(3, 2, 1)$ on L_2, so, an equation for P_2 is

$$13(x-3)-14(y-2)+5(z-1)=0 \text{ or}$$
$$13x-14y+5z-16=0.$$

If we now set $x=0$ in the equations for L_1, we get the point $(0, 1, 2)$ on P_1. So the distance between L_1 and L_2 is the same as the distance from $(0, 1, 2)$ to $13x-14y+5z-16=0$. Then, this distance is

$$D=\frac{|-14+10-16|}{\sqrt{(13)^2+(-14)^2+(5)^2}}=\frac{20}{\sqrt{390}}.$$ ■

CAS **EXAMPLE 18**

Find the distance from the point $(-3, 1, 5)$ to the plane $2x-4y+3z=2$.

Solution Use 8 to get the distance.

```
a=vector(QQ, [2, -4, 3]);
d= -2
p=vector(QQ, [-3, 1, 5])
dis=abs(a.dot_product(p)+d)/a.norm()
show(dis)
```

Answer : $\frac{3}{29}\sqrt{29}$ ∎

CAS **EXAMPLE 19**

Plot the two planes $x+y+z=1$ and $2x-y+z=2$. Find the line of intersection of the two planes and plot it along with the two planes. (See Figure 10.)

http://matrix.skku.ac.kr/cal-lab/cal-11-5-19.html

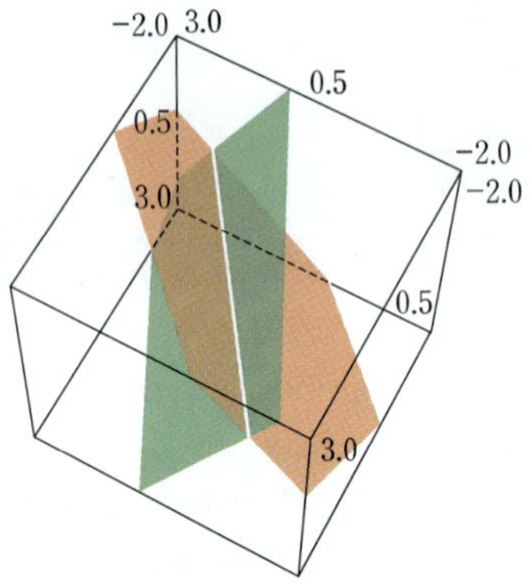

Figure 10

Solution

```
var ('x,y,z')
P1=implicit_plot3d(x    +    y    +    z    ==
1,(x,-2,3),(y,-2,3),(z,-2,3),color='orange',opacity=0.3)
P2=implicit_plot3d(2*x    -    y    +    z    ==
2,(x,-2,3),(y,-2,3),(z,-2,3),color='green',opacity=0.3)
show(P1+P2)
```

We may also get the parametric equation of the line of intersection of the two planes in Sage.

```
var('x, y, z')
solve([x+y+z==1, 2*x-y+z==2], [x, y, z])
```

Answer : $x=-\frac{2}{3}r+1,\ y=-\frac{1}{3}r,\ z=r$ ∎

11.5 EXERCISES (Equations of Lines and Planes)

http://matrix.skku.ac.kr/Cal-Book/part2/CS-Sec-11-5-Sol.html

YouTube http://youtu.be/lxuGE_Erthg

1-7. Find a vector equation, parametric equations and symmetric equations for the line.

1. Through the point $(1,\ 0,\ 2)$ and parallel to the vector $\mathbf{i}-3\mathbf{j}+2\mathbf{k}=(1,\ 0,\ 2)$.

Solution For this line, we have $\mathbf{r}_0=\mathbf{i}+2\mathbf{k}$ and $\mathbf{v}=\mathbf{i}-3\mathbf{j}+2\mathbf{k}$. Hence a vector equation is
$\mathbf{r}=\mathbf{r}_0+t\mathbf{v}=(\mathbf{i}+2\mathbf{k})+t(\mathbf{i}-3\mathbf{j}+2\mathbf{k})$
$=(1+t)\mathbf{i}-3t\,\mathbf{j}+(2+2t)\mathbf{k}$
and parametric equations are

$$x=1+t,\ y=-3t,\ z=2+2t.$$

The symmetric equations are $\dfrac{x-1}{1}=\dfrac{y}{-3}=\dfrac{z-2}{2}$.

CAS **2.** Through the point $(3,\ 5,\ 1)$ and parallel to the vector $(4,\ 1,\ -1)$.

http://matrix.skku.ac.kr/cal-lab/cal-11-5-2.html

Solution

```
var('t')
r0=vector([3, 5, 1])
pv=vector([4, 1, -1])
r0+t*pv
```

Answer : (4*t + 3, t + 5, -t + 1)

parametric equation: $x=4t+3,\ y=t+5,\ z=-t+1$,

symmetric equation: $\dfrac{x-3}{4}=\dfrac{y-5}{1}=\dfrac{z-1}{-1}$.

3. Through the origin and parallel to the line $x=3t$, $y=1+t,\ z=3+2t$.

Solution This line has the same direction as the vector, $\mathbf{v}=3\mathbf{i}+\mathbf{j}+2\mathbf{k}$.
Here $\mathbf{r}_0=0\mathbf{i}+0\mathbf{j}+0\mathbf{k}$, so a vector equation is $\mathbf{r}=3t\mathbf{i}+t\mathbf{j}+2t\mathbf{k}$ and parametric equations are $x=3t,\ y=t,\ z=2t$. The symmetric equations are $\dfrac{x}{3}=\dfrac{y}{1}=\dfrac{z}{2}$.

4. Through the point $(2,\ 0,\ -3)$ and perpendicular to the plane $2x-y-2z=3$.

Solution This line is passing through $(2,0,-3)$ and along the normal vector $(2,-1,2)$ to the given plane.
$\mathbf{r}(t)=(2,0,-3)+t(2,-1,-2)\quad (t\in\mathbb{R})\ \Rightarrow$
vector equation: $\mathbf{r}(t)=(2t+2)\mathbf{i}+(-t)\mathbf{j}+(-2t-3)\mathbf{k}$,
parametric equation: $x=2t+2,\ y=-t,\ z=-2t-3$,
symmetric equation: $\dfrac{x-2}{2}=\dfrac{y}{-1}=\dfrac{z+3}{-2}$.

5. Through the origin and the point $(4,\ 1,\ 0)$.

Solution For this line, we have $\mathbf{r}_0=0\mathbf{i}+0\mathbf{j}+0\mathbf{k}$ and $\mathbf{v}=4\mathbf{i}+\mathbf{j}+0\mathbf{k}$. Hence a vector equation is $\mathbf{r}=\mathbf{r}_0+t\mathbf{v}=4t\mathbf{i}+t\mathbf{j}$ and parametric equations are $x=4t,\ y=t,\ z=0$.

The symmetric equations are $\dfrac{x}{4}=\dfrac{y}{1},\ z=0$.

CAS **6.** Through the points $(3,\ 5,\ -3)$ and $(-1,\ 0,\ 5)$.

http://matrix.skku.ac.kr/cal-lab/cal-11-5-6.html

Solution parametric equation: $x=-4t+3,\ y=-5t+5$, $z=8t-3$,

symmetric equation: $\dfrac{x-3}{-4}=\dfrac{y-5}{-5}=\dfrac{z+3}{8}$.

```
var('t')
A=vector([3, 5, -3])
B=vector([-1, 0, 5])
v=B-A
A+t*v
```

Answer : (-4*t + 3, -5*t + 5, 8*t - 3)

7. Through $(1,\ 2,\ 0)$ and perpendicular to both $\mathbf{i}-\mathbf{j}$ and $2\mathbf{j}+\mathbf{k}$.

Solution A line perpendicular to the given two vectors has the same direction as a cross product of the two vectors. That is,

$$\mathbf{v}=\begin{vmatrix}\mathbf{i}&\mathbf{j}&\mathbf{k}\\1&-1&0\\0&2&1\end{vmatrix}=\begin{vmatrix}-1&0\\2&1\end{vmatrix}\mathbf{i}-\begin{vmatrix}1&0\\0&1\end{vmatrix}\mathbf{j}+\begin{vmatrix}1&-1\\0&2\end{vmatrix}\mathbf{k}=-\mathbf{i}-\mathbf{j}+2\mathbf{k}.$$

Here, $\mathbf{r}_0=1\mathbf{i}+2\mathbf{j}+0\mathbf{k}$, so a vector equation is $\mathbf{r}=(1-t)\mathbf{i}+(2-t)\mathbf{j}+2t\mathbf{k}$ and parametric equations are $x=1-t,\ y=2-t,\ z=2t$. The symmetric equations are $\dfrac{x-1}{-1}=\dfrac{y-2}{-1}=\dfrac{z}{2}$.

CAS **8.** Is the line through $(3,\ 4,\ 5)$ and $(-2,\ 0,\ 1)$ parallel to the line through $(2,\ 1,\ 4)$ and $(-3,\ -3,\ -3)$?

http://matrix.skku.ac.kr/cal-lab/cal-11-5-8.html

Solution The lines are not parallel because the corresponding vectors $<-5,-4,-4>$, $<-5,-4,-7>$ are not parallel.

```
var('t, s')
A=vector([3, 4, 5])
B=vector([-2, 0, 1])
v=B-A
print A+t*v
C=vector([2, 1, 4])
D=vector([-3, -3, -3])
w=D-C
print C+s*w
A=vector([3,4,5])
B=vector([-2,0,1])
C=vector([2,1,4])
D=vector([-3,-3,-3])
L1=line3d([A,B])
L2=line3d([C,D],color='red')
show(L1+L2)
```

Answer : (-5*t + 3, -4*t + 4, -4*t + 5)
(-5*s + 2, -4*s + 1, -7*s + 4)

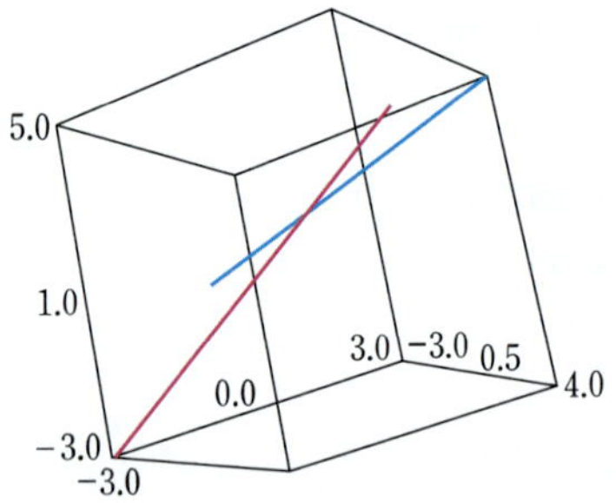

9. Is the line through $(3, 2, -5)$ and $(1, 2, 3)$ perpendicular to the line through $(1, -2, 3)$ and $(5, 2, 0)$?

Solution Direction vectors of the lines are

$$\mathbf{v}_1 = \langle -2, 0, 8\rangle \text{ and } \mathbf{v}_2 = \langle 4, 4, -3\rangle.$$

Since $\mathbf{v}_1 \cdot \mathbf{v}_2 = -8+0-24 \neq 0$, the vectors and the lines are not perpendicular.

10-13. Determine whether the lines L_1 and L_2 are parallel, skew, or intersecting. If they intersect, find the point of intersection.

CAS **10.** L_1: $x = 25$, $y = 1+3t$, $z = -4t$,
L_2: $x = 3+s$, $y = 4-2s$, $z = s$.
http://matrix.skku.ac.kr/cal-lab/cal-11-5-10.html

Solution It is apparent that the lines are skew in the following figure.

```
var('t, s');
L1=parametric_plot3d((25, 1+3*t, -4*t), (t, -50, 50))
L2=parametric_plot3d((3+s, 4-2*s, s), (s, -50, 50), color="red")
show(L1+L2)
```

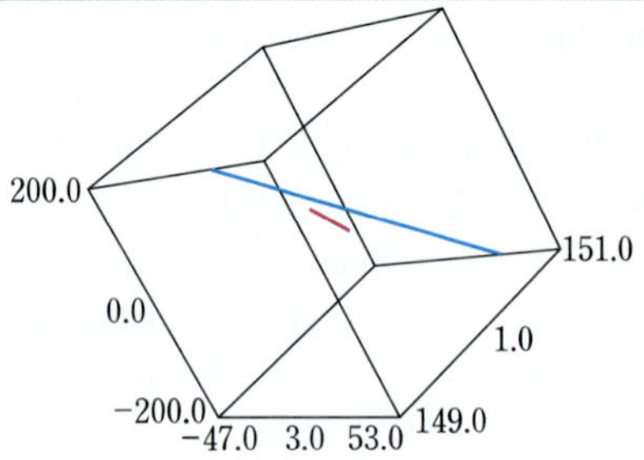

11. L_1: $x = 1+3t$, $y = 4-t$, $z = 1-3t$, $t \in \mathbb{R}$.
L_2: $x = -2-6s$, $y = 2+2s$, $z = 4+6s$, $s \in \mathbb{R}$.

Solution Since the direction vectors are $\mathbf{v}_1 = \langle 3, -1, -3\rangle$ and $\mathbf{v}_2 = \langle -6, 2, 6\rangle$, we have $\mathbf{v}_1 = -\frac{1}{2}\mathbf{v}_2$. Hence the lines are parallel.

12. L_1: $\dfrac{x-1}{2} = \dfrac{y-4}{3} = \dfrac{z+1}{-1}$,
L_2: $\dfrac{x+1}{1} = \dfrac{y}{-2} = \dfrac{z-1}{-4}$.

Solution The lines are not parallel because the corresponding vectors $\langle -2, 3, -1\rangle$, $\langle 1, -2, -4\rangle$ are not parallel. If L_1 and L_2 have a point of intersection, there would be values of t and s such that

$2t+1 = s-1$,
$3t+4 = -2s$,
$-t-1 = -4s+1$
$\Rightarrow t = -\frac{8}{7}$, $s = -\frac{2}{7}$.

L_1 and L_2 are skew.

CAS **13.** L_1: $\dfrac{x-4}{3} = y-3 = \dfrac{z-7}{4}$,
L_2: $\dfrac{x-3}{-2} = \dfrac{y}{2} = z-2$.
http://matrix.skku.ac.kr/cal-lab/cal-11-5-13.html

Solution From the figure it is clear that the lines are intersecting.

```
var('x');
diff(x^(15/14)+5*e^x,x)
var('s, t')
A=vector([3,1,4])
B=vector([-2,2,1])
C=vector([4,3,7])
D=vector([3,0,2])
Lt=A+t*B
Ls=C+s*D
print "Clearly the two lines are intersecting"
parametric_plot3d(Lt,(t,-3,3))+parametric_plot3d(Ls,(s,-3,3),color='red')
```

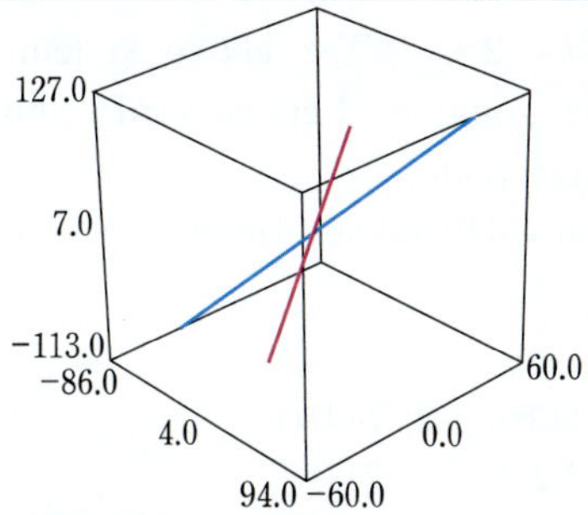


```
sol=solve([Lt[0]==Ls[0],Lt[1]==Lt[1],Lt[2]==Ls[2]], s, t, solution_dict=True)
show(sol)
```

Answer : [{t:1,s:−1}] They intersect.

14-15. Find an equation of the plane.

14. Through the point $(6,\ 3,\ 6)$ and perpendicular to the vector $\langle 2,\ -1,\ 5\rangle$.

Solution $2(x-6)-1(y-3)+5(z-6)=0$
$\Rightarrow 2x-y+5z=39$.

15. Through the point $(0,\ 4,\ -1)$ and with normal vector $\mathbf{i}+2\mathbf{j}-3\mathbf{k}$.

Solution $\mathbf{i}+2\mathbf{j}-3\mathbf{k}=\langle 1,\ 2,\ -3\rangle$ is a normal vector to the plane and $(0,4,-1)$ is a point of the plane.
Then $1(x-0)+2(y-4)-3(z+1)=0$
or $x+2y-3z=11$ to be the equation of the plane.

16. Which of the following four planes are parallel?

$P_1:\ 2x-3y+5z=3,\quad P_2:\ x-2y+2z=4,$
$P_3:\ 4x+2y-2z=10,\quad P_4:\ z=2x+y-5.$

Solution P_3 and P_4 are parallel.

17. Which of the following four lines are parallel?

$L_1:\ x=1+2t,\ y=t,\ z=3-4t,$
$L_2:\ x=1+4t,\ y=2t,\ z=13-t,$
$L_3:\ x+1=2y=1-2z,$
$L_4:\ \mathbf{r}=\langle 1,\ 1,\ 4\rangle+t\langle 4,\ 2,\ -8\rangle.$

Solution L_1 and L_4 are parallel.

18-19. Find an equation of the plane through the given point with the normal vector which is the direction of the line with the given parametric equations.

18. $(1,\ 2,\ 4);\ x=1+t,\ y=3-2t,\ z=4t.$

Solution $\mathbf{i}-2\mathbf{j}+4\mathbf{k}=\langle 1,\ -2,\ 4\rangle$ is a normal vector to the plane and $(1,2,4)$ is a point of the plane.
Then $1(x-1)-2(y-2)+4(z-4)=0$
or $x-2y+4z=13$ to be the equation of the plane.

19. $(1,\ 1,\ 0);\ x=5-3t,\ y=t,\ z=2+t.$

Solution $-3\mathbf{i}+\mathbf{j}+\mathbf{k}=\langle -3,\ 1,\ 1\rangle$ is a normal vector to the plane and $(1,1,0)$ is a point of the plane.
Then $-3(x-1)+1(y-1)+1(z-0)=0$
or $3x-y-z=2$ is the equation of the plane.

20-21. Find the distance from the point to the given plane.

CAS **20.** $(3,\ 1,\ 5);\ 2x-y+3z=4.$

http://matrix.skku.ac.kr/cal-lab/cal-11-5-20.html

Solution The normal vector to the plane is $\mathbf{n}=\langle 2,\ -1,\ 3\rangle$

```
n=vector(QQ, [2, -1, 3])
d= -4
p=vector(QQ, [3, 1, 5])
dis=abs(n.dot_product(p)+d)/a.norm()
dis
```

Answer : 8/7*sqrt(14)

21. $(1,\ 4,\ -2);\ 4x-4y+4z=1.$

Solution The distance

$$D=\frac{|4-16-8-1|}{\sqrt{16+16+16}}=\frac{21}{4\sqrt{3}}=\frac{21\sqrt{3}}{12}=\frac{7\sqrt{3}}{4}.$$

22-23. Find the distance between the given parallel planes.

22. $z=2x+y+3,\ 2x+y-z=2.$

Solution P_1: $2x+y-z=-3$, P_2: $2x+y-z=2$. Note that $(-1, 0, 1)$ is a point of the first plane.
Since the planes are parallel, the distance d between the two planes is the distance from $(-1, 0, 1)$ to the second plane. Then, the distance between $(-1, 0, 1)$ and the plane $2x+y-z=2$ is

$$D=\frac{|2\times -1+0-1\times 1-2|}{\sqrt{4+1+1}}=\frac{5}{\sqrt{6}}=\frac{5}{6}\sqrt{6}.$$

23. $3x+2y-4z=4,\ 6x+4y-8z=2.$

Solution Put $x=y=0$ in the equation of the first plane to get the point $(0,0,-1)$ on the plane.
Since the planes are parallel, the distance d between the two planes is the distance from $(0,0,-1)$ to the second plane. Hence

$$D=\frac{|8-2|}{\sqrt{36+16+64}}=\frac{6}{\sqrt{116}}=\frac{3}{\sqrt{29}}.$$

---

24. Find the distance between the two skew lines

$$\frac{x-1}{1}=\frac{y-1}{6}=\frac{z}{2}\ \text{and}\ \frac{x-1}{2}=\frac{y-5}{15}=\frac{z+2}{6}.$$

25. Prove that the distance between the parallel planes $ax+by+cz+d_1=0$ and $ax+by+cz+d_2=0$ is

$$D=\frac{|d_1-d_2|}{\sqrt{a^2+b^2+c^2}}.$$

26. (Line of intersection of two planes) Plot the two planes $x+y+z=1$ and $2x-y+z=2$. Find the line of intersection of two planes and hence plot this.

http://matrix.skku.ac.kr/cal-lab/cal-11-5-26.html

Solution

```
var('x,y,z,t')
solve([x + y + z == 1,2*x - y + z == 2],x,y,z)
```

Answer : x == -2/3*r1 + 1, y == -1/3*r1, z == r1

Clearly the line of intersection of the is

$$x=1-\frac{2}{3}t,\ y=-\frac{1}{3}t,\ z=t.$$

```
P1=implicit_plot3d(x + y + z == 1, (x,-3,3),
(y,-3,3), (z,-3,3), color='blue', opacity=0.3)
P2=implicit_plot3d(2*x - y + z == 2, (x,-3,3),
(y,-3,3), (z,-3,3), color='red', opacity=0.3)
L=parametric_plot3d([1-2/3*t,-1/3*t,t], (t,-3.5,3.5),
color='green',thickness=3)
show(P1+P2+L)
```

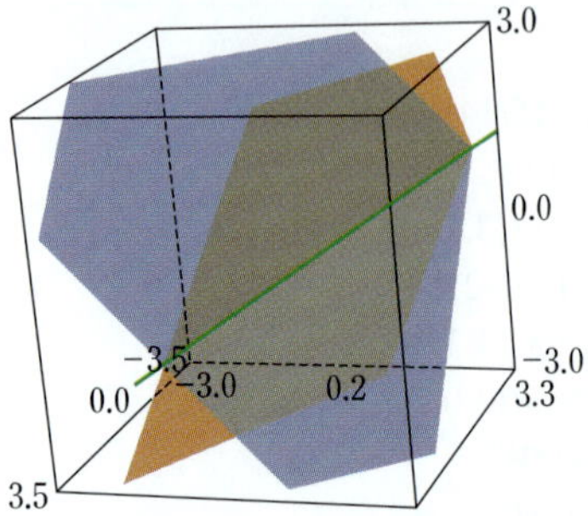

CAS 27. (Line of intersection of two planes) If the two lines have a point in common then there exists t and s such that $4+3t=3-2s$, $3+t=2s$, $7+4t=2+s$. The above system must have a unique solution. Let is verify this and find a common point.

http://matrix.skku.ac.kr/cal-lab/cal-11-5-27.html

Solution

```
var('s, t')
Lt=(4+3*t, 3+t, 7+4*t)
Ls=(3-2*s, 2*s, 2+s)
sol=solve([Lt[0]==Ls[0],Lt[1]==Ls[1],Lt[2]==Ls[2]
], s, t, solution_dict=True)
show(sol)
```

Clearly, $t=1$, $s=-1$ gives a unique solution and point $(1, 3, 5)$ is the common point.

```
def axes(xmin=-1,xmax=1,ymin=-1, ymax=1,
zmin=-1,zmax=1,**kwds):
    ex = vector((1,0,0))
    ey = vector((0,1,0))
    ez = vector((0,0,1))
    G =
arrow3d(xmin*ex,xmax*ex,radius=0.01)
    G =
G+arrow3d(ymin*ey,ymax*ey,radius=0.01)
    G =
G+arrow3d(zmin*ez,zmax*ez,radius=0.01);
    labels=text3d('x', (xmax+0.2,0,0)) +
text3d("y", (0, ymax+0.20, 0)) + text3d("z",
(0, 0, zmax+0.2))
    return G+labels
A=axes(xmin=-3, xmax=3, ymin=-3,
ymax=3, zmin=-3, zmax=3, color='red',
thickness=2)
A
```

11.6 Cylinders and Quadric Surfaces

Cylinder and Traces

A **cylinder** is one of the most basic curvilinear geometric shapes, the surface formed by the points at a fixed distance from a given line segment, the axis of the cylinder. The solid enclosed by this surface and by two planes perpendicular to the axis is also called a cylinder. The surface area and the volume of a cylinder have been known since deep antiquity.

The cylinder is parallel to one of the coordinate axes.

DEFINITION 1 Cylinder

Given a curve C in a plane P and a line L not in P, a **cylinder** is the surface consisting of all lines parallel to given line L that passes through a given plane curve C.

DEFINITION 1 Trace

A trace of a surface is the set of points at which the surface intersects a plane that is parallel to one of the coordinate planes. The traces in the coordinate planes are called xy-trace, yz-trace and zx-trace.

EXAMPLE 1

Draw a graph of the surface $x = z^2$.
http://matrix.skku.ac.kr/cal-lab/cal-11-6-Exam-1.html

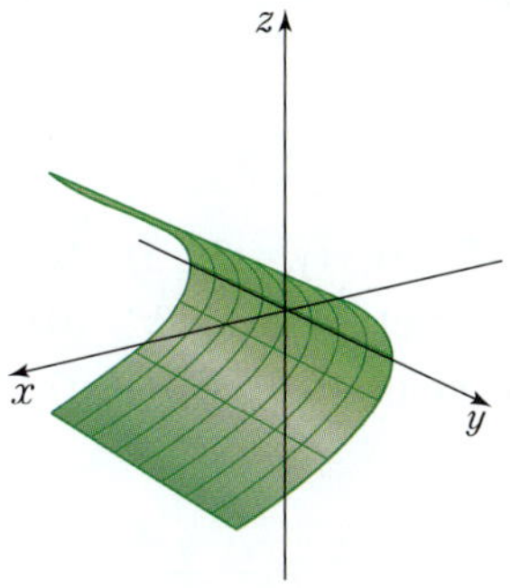

Figure 1 The surface $x = z^2$ is a parabolic cylinder.

Solution Since there are no y's in the equation, the trace of the graph in the plane $y = k$ is the same for every k. The trace of the cylinder in every plane parallel to the zx-plane is parabola $x = z^2$ in the zx-plane and moving it in the direction of the y-axis. The graph is a surface, called a **parabolic cylinder and it is** made up of infinitely many shifted copies of the same parabola. Here the rulings of the cylinder are parallel to the y-axis.

```
var('x, y, z, v, u')
S=implicit_plot3d(x==z^2, (x,-2,2), (y,-2,2), (z,-2,2),
plot_points=100, smooth=True, color='goldenrod', opacity=0.3)
p1=sum([parametric_plot3d((i^2 , v, i), (v,-2,2), color='red')
for i in srange(-sqrt(2), sqrt(2), 0.5)])
p2=sum([parametric_plot3d((u^2 ,i, u), (u,- sqrt(2) , sqrt(2)),
color='red') for i in srange(-2,2,0.5)])
show(S+p1+p2, frame=False)
```

■

CAS **EXAMPLE 2**

Sketch the graph of the surfaces $z = \cos x$ in $\mathbb{R}^3$.
http://matrix.skku.ac.kr/cal-lab/cal-11-6-Exam-2.html

Solution An equation in $\mathbb{R}^3$, $z = \cos x$ is missing the variable y. For all real values y, the graph is a cylinder consisting of lies parallel to the y -axis passing through the curve $z = \cos x$ in the zx-plane.
Graph the curve $z = \cos x$ in the zx-plane, which in the zx-trace of the surface.
Draw a second trace in a plane parallel to the zx-plane.
Draw lines parallel to the y-axis passing through the two traces.
The result is a cylinder, running parallel to the y-axis, consisting of copies of the curve $z = \cos x$.

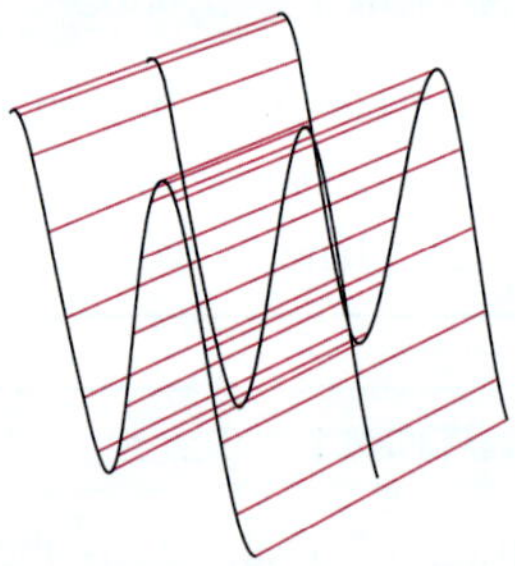

Figure 2

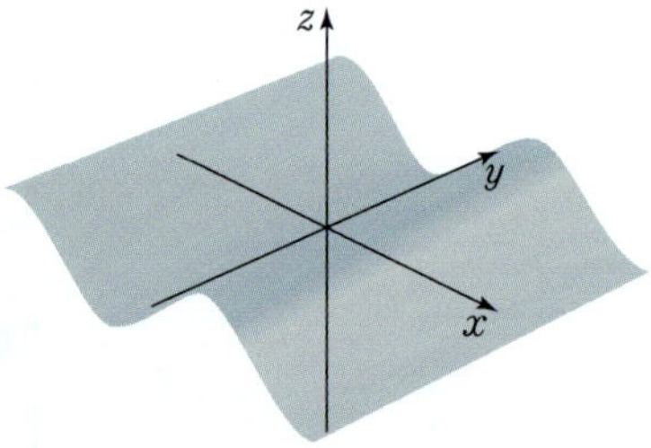

Figure 3

```
var('i,u,x,y,z')
p1=parametric_plot3d((i, 0, sin(i)), (i, -5, 5), opacity=0.5) + parametric_ plot3d((i, -5, sin(i)), (i, -5, 5), opacity=0.5)+parametric_plot3d((i, 5, sin(i)), (i, -5, 5), opacity=0.5)
p2=sum([parametric_plot3d((i, u, sin(i)), (u, - 5 , 5), color=' red ') for i in srange(-5, 5, 0.5)])
show(p1+p2, frame=False)
var('x,y,z')
p =implicit_plot3d(z==cos(x), (x, -5, 5), (y, -5, 5), (z, -5, 5),opacity=0.5, smooth=True, axes=True) + text3d('z', (0,0,5), color=(0.5,0,0))+text3d("y", (0,5,0), color=(0,0.5,0))+text3d('x', (5,0,0), color=(0,0.5,0))
show(p, frame=False)
```

■

Quadric Surfaces

The general second-degree equation in x, y, z,

$$Ax^2 + By^2 + Cz^2 + Dxy + Eyz + Fxz + Gx + Hy + Iz + J = 0$$

is a **quadric surface**. Here A, B, C, ..., J are constants. This equation can be brought into one of the two standard forms

$$Ax^2 + By^2 + Cz^2 + J = 0 \text{ or } Ax^2 + By^2 + Iz = 0.$$

Quadric surfaces are the three dimensional counterparts of the conic sections in the plane.

EXAMPLE 3

Sketch the quadric surface defined by the equation

$$\frac{x^2}{25} + \frac{y^2}{16} + \frac{z^2}{9} = 1$$

is an ellipsoid.

http://matrix.skku.ac.kr/cal-lab/cal-11-6-Exam-3.html

Solution By substituting $z = 0$, we find that the trace in the xy-plane is $x^2/25 + y^2/16 = 1$ which we recognize as an equation of an ellipse. In general, the horizontal trace in the plane $z = k$ is

$$\frac{x^2}{25} + \frac{y^2}{16} = 1 - \frac{k^2}{9}, \qquad z = k,$$

which is an ellipse, provided that $k^2 < 9$ that is, $-3 < k < 3$. The largest ellipse parallel to the xy-plane occurs with $k = 0$; it is the xy-plane, which the ellipse $\frac{x^2}{25} + \frac{y^2}{16} = 1$ with axes of length 10 and 8. Similarly, the vertical traces are also ellipses:

$$\frac{y^2}{16} + \frac{z^2}{9} = 1 - \frac{k^2}{25}, \qquad x = k, \quad (\text{if } -5 < k < 5)$$

$$\frac{x^2}{25} + \frac{z^2}{9} = 1 - \frac{k^2}{16}, \qquad y = k, \quad (\text{if } -4 < k < 4).$$

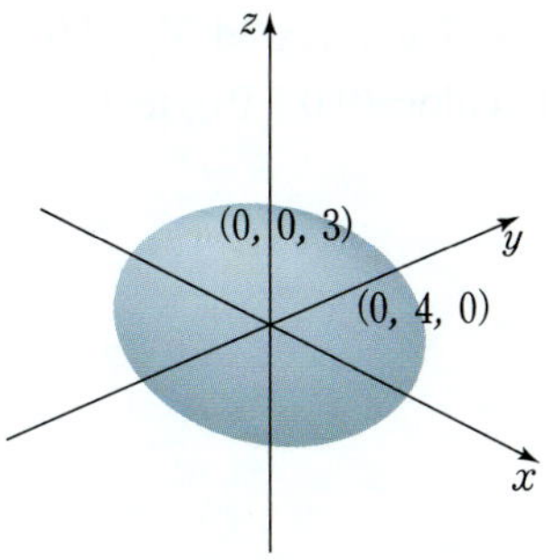

Figure 4

The ellipsoid $\frac{x^2}{25} + \frac{y^2}{16} + \frac{z^2}{9} = 1$

Figure 4 shows how drawing some traces indicates the shape of the surface. It's called an **ellipsoid** because all of its traces are ellipses. Notice that it is symmetric with respect to each coordinate plane; this is a reflection of the fact that its equation involves only even powers of x, y and z.

```
var('x,y,z')
p = implicit_plot3d((x^2)/25+(y^2)/16+(z^2)/9==1, (x, -10, 10), (y, -10, 10), (z, -10, 10), opacity=0.5, smooth=True, axes=True)+text3d("(0, 0, 3)", (0,0,3), color=(0.5,0,0)) + text3d("(0, 4, 0)", (0,4,0), color=(0,0.5,0)) + text3d('z', (0,0,10), color=(0.5,0,0)) + text3d("y", (0,10,0), color = (0,0.5,0)) + text3d('x', (10, 0, 0), color=(0, 0.5, 0))
show(p, frame=False)
```

■

CAS **EXAMPLE 4**

Sketch the quadric surface defined by the equation

$$x = \frac{y^2}{9} + \frac{z^2}{16}$$

is an ellipsoid paraboloid.

http://matrix.skku.ac.kr/cal-lab/cal-11-6-Exam-4.html

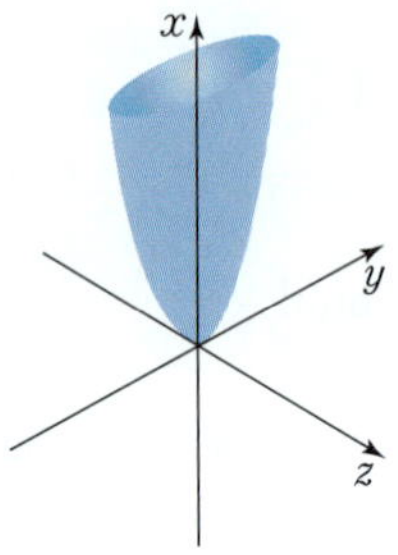

Figure 5

Solution If we put $y = 0$, we get $x = z^2/16$, so the zx-plane intersects the surface in a parabola. If we put $y = k$ (a constant), we get $x = \frac{k^2}{9} + \frac{z^2}{16}$. This means that if we slice the graph with any plane parallel to the zx-plane, we obtain a parabola that opens leftward. Similarly, if $z = k$, the trace is, which is again a parabola that opens leftward. If we put $x = k$, we get the horizontal traces $\frac{y^2}{9} + \frac{z^2}{16} = k$, which we recognize as a family of ellipses. Knowing the shapes of the traces, we can sketch the graph in Figure 5. Because of the elliptical and parabolic traces, the quadric surface $x = \frac{y^2}{9} + \frac{z^2}{16}$ is called an **elliptic paraboloid**.

```
var('x,y,z')
p =implicit_plot3d((x^2)*5+y^2==z, (x, -5, 5), (y, -5, 5),(z, -5, 5),opacity=0.5,smooth=True, axes=True)+text3d('z', (0,0,5), color=(0.5,0,0))+text3d("y", (0,5,0), color=(0,0.5,0))+text3d('x', (5,0,0), color=(0,0.5,0))
show(p,frame=False)
```

■

EXAMPLE 5

Sketch the surface $z = x^2 - y^2$.

Solution The traces in the vertical planes $x = k$ are the parabolas $z = k^2 - y^2$ which open upward. The traces in $y = k$ are the parabolas $z = x^2 - k^2$, which open downward. The horizontal traces are $x^2 - y^2 = k$, a family of hyperbolas. We draw the families of traces in Figure 6, and we show how the traces appear when placed in their correct planes in Figure 7.

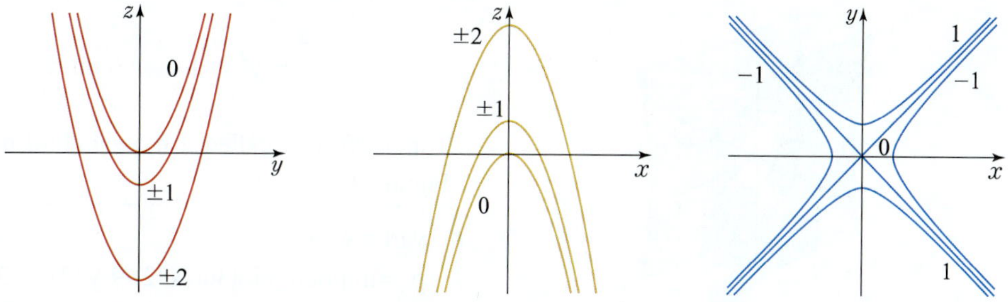

Traces in $y=k$ are $z=x^2-k^2$ Traces in $x=k$ are $z=-y^2+k^2$ Traces in $z=k$ are $x^2-y^2=k$

Figure 6 Vertical traces are parabolas; horizontal traces are hyperbolas. All traces are labeled with the value of k.

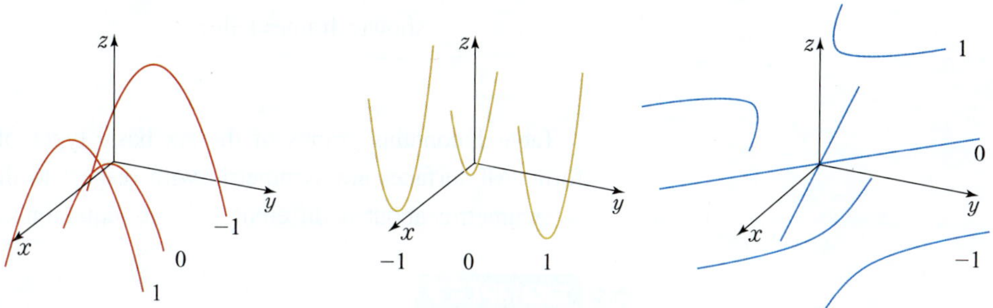

$x=k$ $y=k$ $z=k$

Figure 7 Traces moved to their correct planes.

In Figure 8, we fit together the traces from Figure 7 to form the surface $z=y^2-x^2$ a **hyperbolic paraboloid**. Notice that the shape of the surface near the origin resembles that of a saddle.

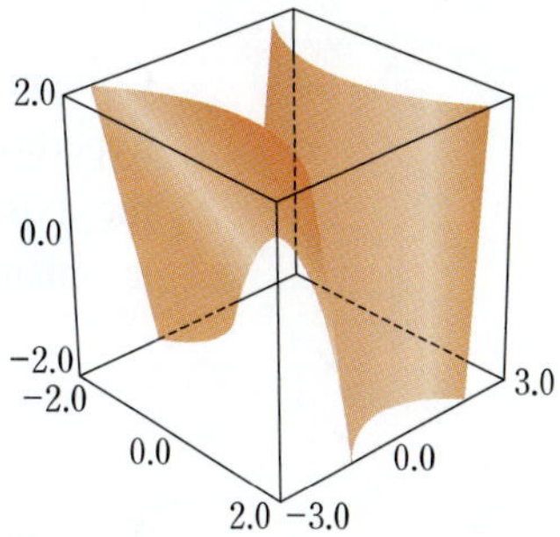

Figure 8 The surface $z=x^2-y^2$ is a hyperbolic paraboloid.

■

CAS **EXAMPLE 6**

Sketch the surface $\frac{x^2}{9}-y^2+\frac{z^2}{9}=1$.

http://matrix.skku.ac.kr/cal-lab/cal-11-6-Exam-6.html

Solution The trace in any horizontal plane $y=k$ is the ellipse

$$\frac{x^2}{9}+\frac{z^2}{9}=1+k^2,\ \ y=k,$$

but the traces in the xy- and yz-planes are the hyperbolas

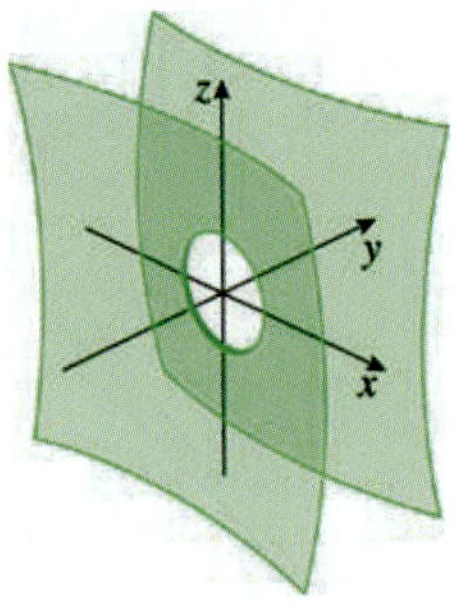

Figure 9

$$\frac{x^2}{9} - y^2 = 1 \quad z = 0 \quad \text{and} \quad -y^2 + \frac{z^2}{9} = 1 \quad x = 0.$$

This surface is called a **hyperboloid of one sheet** and is sketched in Figure 9.

```
var('x,y,z')
p =implicit_plot3d((x^2)/9-y^2+(z^2)/9==1, (x, -10, 10), (y, -10, 10),(z, -10, 10),opacity=0.5,smooth=True, axes=True)+text3d('z', (0,0,10), color=(0.5,0,0))+ text3d("y", (0,10,0), color=(0,0.5,0))+text3d('x', (10,0,0), color=(0,0.5,0))
show(p,frame=False)
```

■

Table 1 contains graphs of the six basic types of quadric surfaces in standard form. All surfaces are symmetric with respect to the z-axis. If a quadric surface is symmetric about a different axis, its equation changes accordingly.

CAS **EXAMPLE 7**

Draw a graph of the surface $x^2 - 4y^2 - 4z^2 = 4$.
http://matrix.skku.ac.kr/cal-lab/cal-11-6-Exam-7.html

Solution Dividing by 4, we first put the equation in standard form:

$$\frac{x^2}{4} - y^2 - \frac{z^2}{4} = 1.$$

Comparing this equation with Table 1, we see that it represents a hyperboloid of two sheets, the only difference being that in this case the axis of the hyperboloid is the x-axis. The traces in the xy- and xz-planes are the hyperbolas

$$\frac{x^2}{4} - y^2 = 1 \quad z = 0 \quad \text{and} \quad \frac{x^2}{4} - \frac{z^2}{4} = 1 \quad y = 0.$$

The surface has no trace in the yz-plane, but traces in the vertical planes $x = k$ for $|k| > 2$ are the ellipses

$$y^2 + \frac{z^2}{4} = \frac{k^2}{4} - 1, \qquad x = k.$$

which can be written as

$$\frac{y^2}{\frac{k^2}{4} - 1} + \frac{z^2}{4\left(\frac{k^2}{4} - 1\right)} = 1, \qquad x = k.$$

These traces are used to make the sketch in Figure 10.

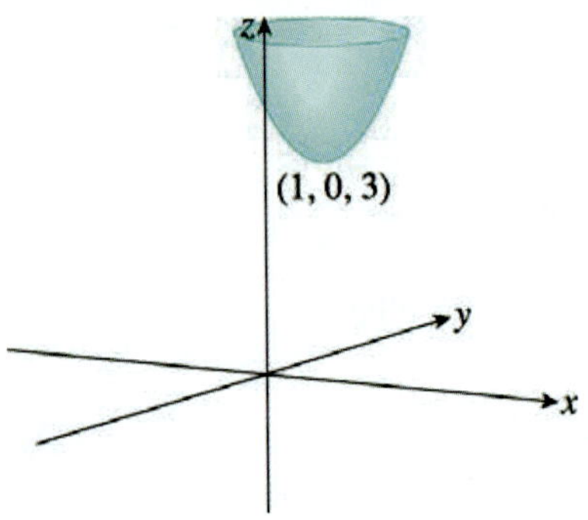

Figure 11 $x^2 + 2y^2 - 2x - z + 4 = 0$

```
var('x,y,z')
p =implicit_plot3d((x^2)*5-y^2+2*z^2+5==0, (x, -5, 5), (y, -5, 5), (z, -5, 5), opacity=0.5, smooth=True, axes = True) + text3d('z', (0,0,5), color = (0.5,0,0)) + text3d("y", (0,5,0), color=(0,0.5,0)) + text3d('x', (5,0,0), color=(0, 0.5, 0)) + text3d("(0, -2, 0)", (0,-2,0), color = (0.5,0,0)) + text3d("(0, 2, 0)", (0, 2, 0), color=(0, 0.5, 0))
show(p,frame=False)
```

■

TABLE 1 Graphs of quadric surfaces

Surface	Equation & Traces	Surface	Equation & Traces
Ellipsoid 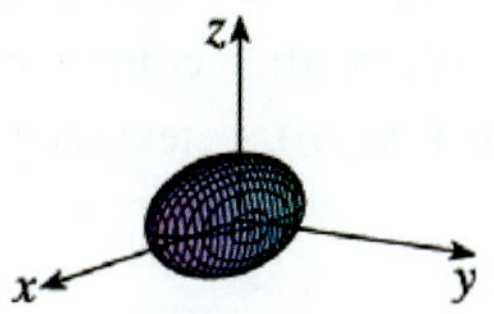	$\frac{x^2}{a^2}+\frac{y^2}{b^2}+\frac{z^2}{c^2}=1$ All traces are ellipses. If $a=b=c$, the ellipsoid is a sphere.	**Cone** 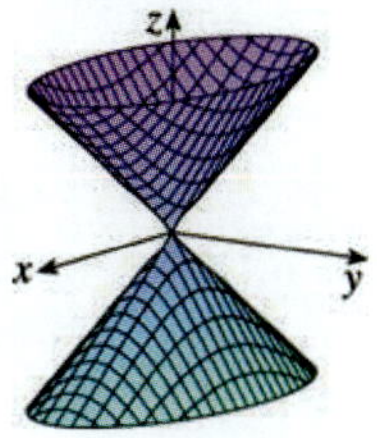	$\frac{z^2}{c^2}=\frac{x^2}{a^2}+\frac{y^2}{b^2}$ Horizontal traces are ellipses. Vertical traces in the planes $x=k$ and $y=k$ are hyperbolas if $k\neq 0$ but are pairs of lines if $k=0$.
Elliptic Paraboloid 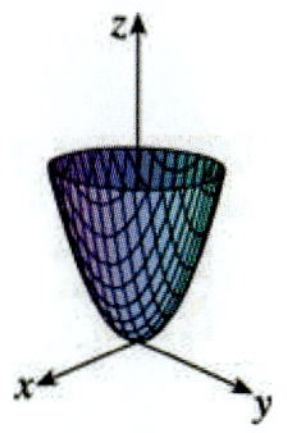	$\frac{z}{c}=\frac{x^2}{a^2}+\frac{y^2}{b^2}$ Horizontal traces are ellipses. Vertical traces are parabolas. The variable raised to the first power indicates the axis of the paraboloid.	**Hyperboloid of One Sheet** 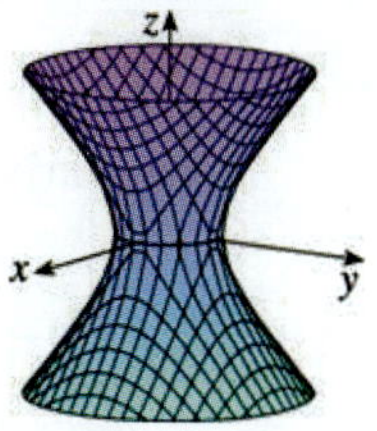	$\frac{x^2}{a^2}+\frac{y^2}{b^2}-\frac{z^2}{c^2}=1$ Horizontal traces are ellipses. Vertical traces are hyperbolas. The axis of symmetry corresponds to the variable whose coefficient is negative.
Hyperbolic Paraboloid 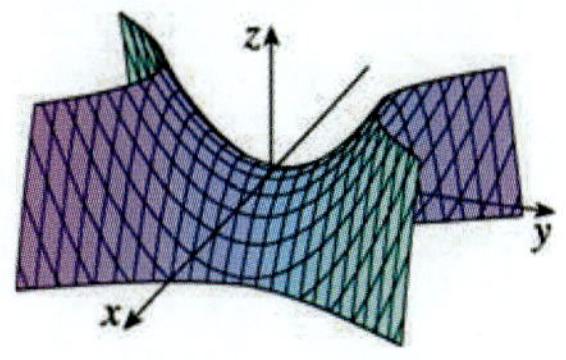	$\frac{z}{c}=\frac{x^2}{a^2}-\frac{y^2}{b^2}$ Horizontal traces are hyperbolas. Vertical traces are parabolas. The case where $c>0$ is illustrated.	**Hyperboloid of Two Sheets** 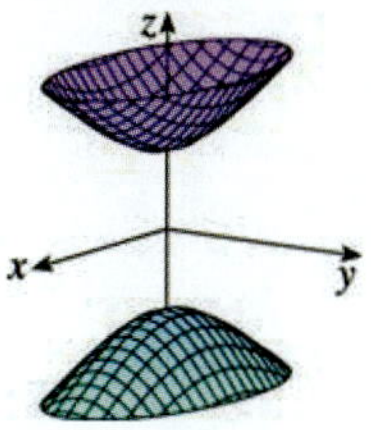	$-\frac{x^2}{a^2}-\frac{y^2}{b^2}+\frac{z^2}{c^2}=1$ Horizontal traces in $z=k$ are ellipses if $k>c$ or $k<-c$. Vertical traces are hyperbolas. The two minus signs indicate two sheets.

CAS **EXAMPLE 8**

Classify the quadric surface $x^2+2y^2-2x-z+4=0$.
http://matrix.skku.ac.kr/cal-lab/cal-11-6-Exam-8.html

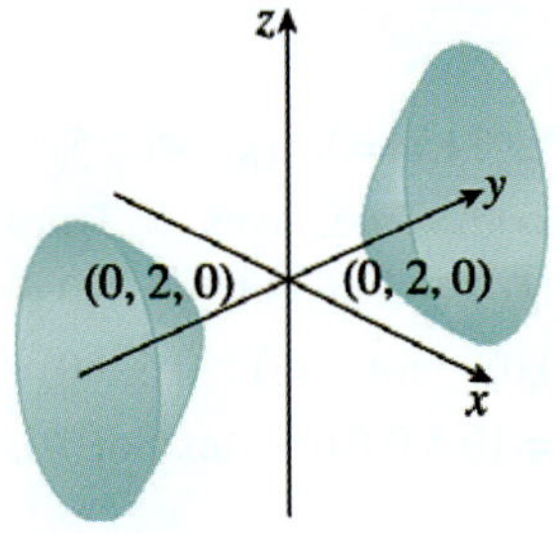

Figure 10 $x^2 - 4y^2 - 4z^2 = 4$

Solution By completing the square we rewrite the equation as

$$z - 3 = (x-1)^2 + 2y^2$$

Comparing this equation with Table 1, we see that it represents an elliptic paraboloid. Here, however, the axis of the paraboloid is parallel to the xy-axis, and it has been shifted so that its vertex is the point $(1, 0, 3)$. The traces in the plane $z = k(k > 3)$ are the ellipses

$$(x-1)^2 + 2y^2 = k - 3, \qquad z = k.$$

The trace in the xz-plane is the parabola with equation $z = 3 + (x-1)^2$, $y = 0$. The paraboloid is sketched in Figure 11.

```
var('x,y,z')
p = implicit_plot3d(x^2+2*y^2-2*x-z+4 ==0, (x, -5, 5), (y, -5, 5), (z, -5, 5), opacity=0.5, axes=True) + text3d('z', (0, 0, 5), color=(0.5, 0, 0)) + text3d("y", (0, 5, 0), color=(0, 0.5, 0)) + text3d('x', (5, 0, 0), color=(0, 0.5, 0)) + text3d("(1, 0, 3)", (1, 0, 3), color=(0, 0.5, 0))
show(p, frame=False)
```

■

CAS **EXAMPLE 9**

Sketch the equation

$$x^2 = 3y^2 + 2z^2.$$

http://matrix.skku.ac.kr/cal-lab/cal-11-6-Exm-9.html

Solution

```
var('x, y, z')
implicit_plot3d(x^2-3*y^2-2*z^2==0, (x, -1.5, 1.5), (y, -1.5, 1.5), (z, -1.5, 1.5), color='red', plot_points=80, smooth=True)
```

■

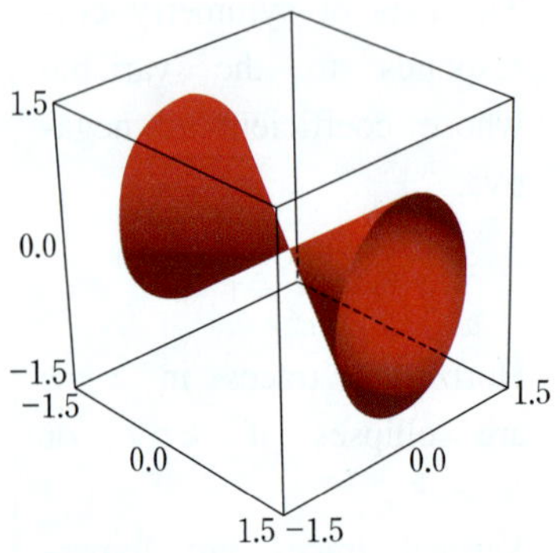

Figure 12

11.6 EXERCISES (Cylinders and Quadric Surfaces)

http://matrix.skku.ac.kr/Cal-Book/part2/CS-Sec-11-6-Sol.html

YouTube http://youtu.be/

CAS **1.** (a) What does the equation $y = 2x^2$ represent as a curve in $\mathbb{R}^2$?

(b) What does it represent as a surface in $\mathbb{R}^3$?

(c) What does the equation $z = 2y^2$ represent?

http://matrix.skku.ac.kr/cal-lab/cal-11-6-Exs-1.html

Solution (a) Equation $y = 2x^2$ represents a parabola of slope 2 passing through origin in $\mathbb{R}^2$.

(b) The equation of the graph is $y = 2x^2$, which doesn't involve z in $\mathbb{R}^3$. This means that any vertical plane with equation $z = k$ (parallel to the xy-plane) intersects the graph in a curve with $y = 2x^2$, that is, a parabola. Below figure shows how the graph is formed by taking the parabola $y = 2x^2$ in the xy-plane and moving it in the direction of the z-axis. So the graph is a surface, called a parabolic cylinder, made up of infinitely many shifted copies of the same parabola.

```
S=implicit_plot3d(z==y^2,(x,-2,2),(y,-2,2),(z,-2,2),p
lot_points=100,smooth=True,color='goldenrod',
opacity=0.3)
p1=sum([parametric_plot3d([i,y,y^2],(y,-sqrt(2),sqrt
(2)),color='red') for i in srange(-2,2,0.5)])
show(S+A+p1,frame=False)
```

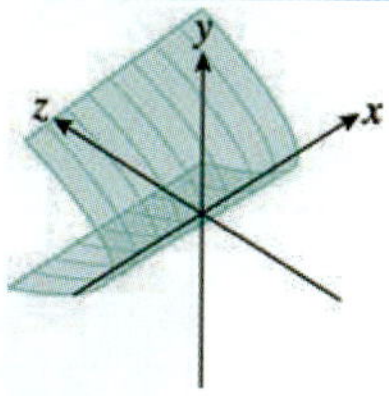

(c) $z = 2y^2$ also represents a parabolic cylinder, this time with axis the x-axis in $\mathbb{R}^3$.

CAS **2.** (a) Sketch the graph of $y = e^{2x}$ as a curve in $\mathbb{R}^2$.

http://matrix.skku.ac.kr/cal-lab/cal-11-6-2.html

(b) Sketch the graph of $y = e^{2x}$ as a surface in $\mathbb{R}^3$.

(c) Describe and sketch the surface $z = e^{2y}$.

Solution (a)

```
plot(exp(2*x), (x, -1, 1))
```

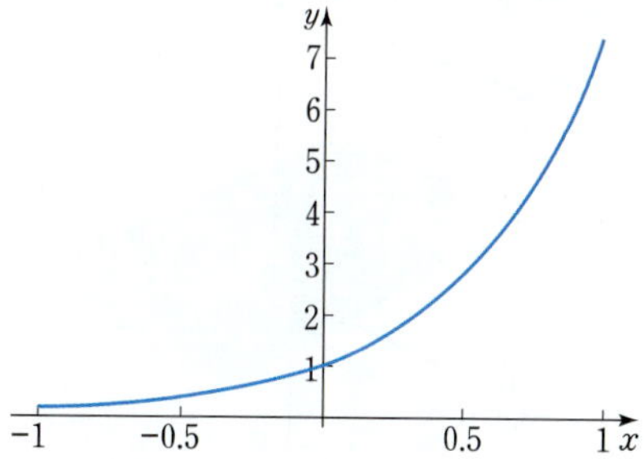

(b)

```
var('x, y, z')
implicit_plot3d(y==exp(2*x), (x, -1, 1), (y, -1, 1),
(z, -1, 1))
```

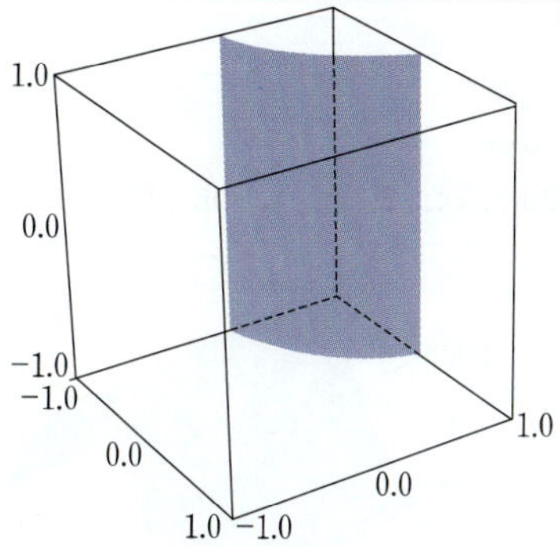

(c)

```
var('x, y, z')
implicit_plot3d(z==exp(2*y), (x, -1, 1), (y, -1, 1),
(z, -1, 1))
```

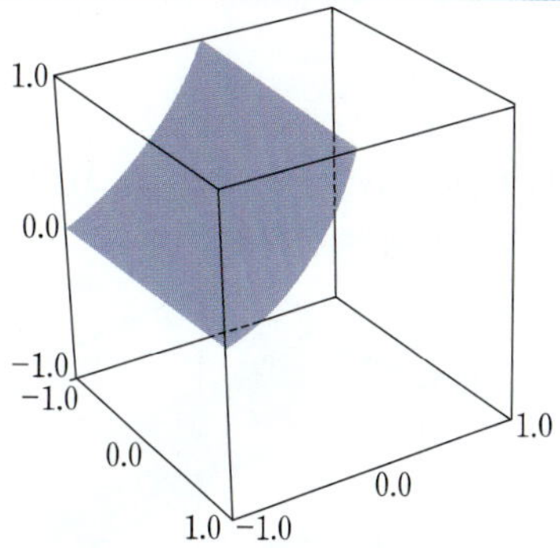

3-5. Describe and sketch the surface.

CAS **3.** $y^2 + 5z^2 = 5$

http://matrix.skku.ac.kr/cal-lab/cal-11-6-3.html

Solution

```
var('x, y, z')
implicit_plot3d(y^2+5*z^2-5==0, (x, -3, 3), (y, -3, 3), (z, -3, 3), opacity=0.5)
```

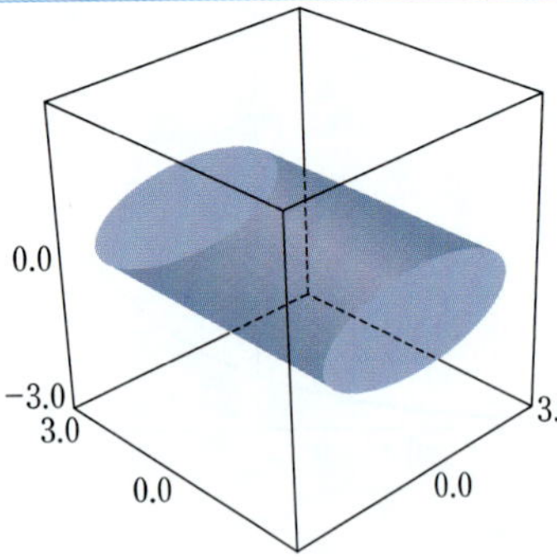

CAS **4.** $z = \cos y$

http://matrix.skku.ac.kr/cal-lab/cal-11-6-4.html

Solution

```
var('x, y')
z=cos(y)
plot3d(z, (x, -2, 2), (y, -pi, pi))
```

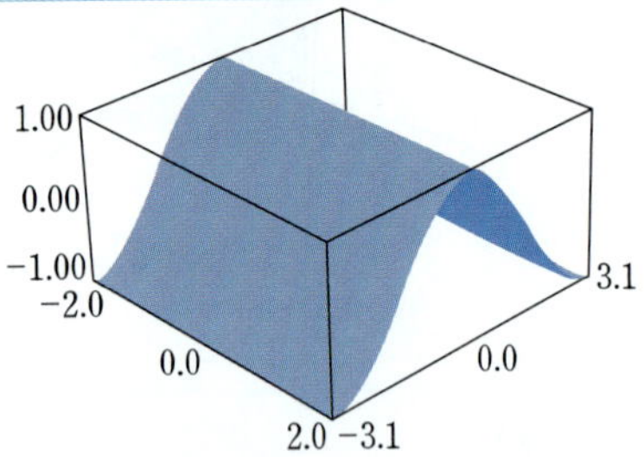

CAS **5.** $yz = 3$

http://matrix.skku.ac.kr/cal-lab/cal-11-6-5.html

Solution

```
var('x, y, z')
implicit_plot3d(y*z==3, (x, -3, 3), (y, -10, 10), (z, -10, 10), opacity=0.5)
```

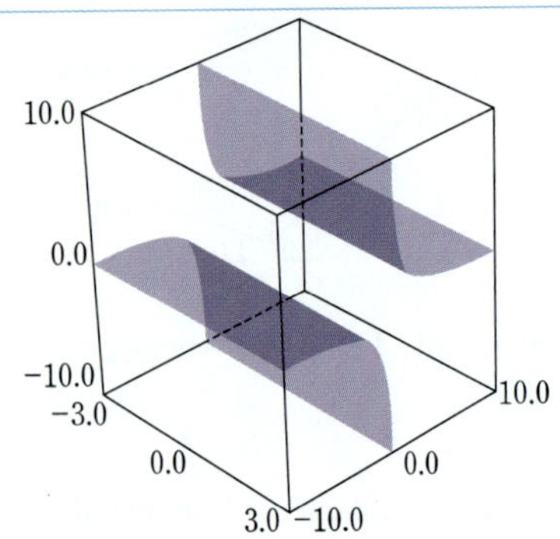

6-10. Find the traces of the given surface in $x = k$, $y = k$, $z = k$. Then, identify the surface and sketch it.

CAS **6.** $x^2 + 9y^2 + 9z^2 = 9$

http://matrix.skku.ac.kr/cal-lab/cal-11-6-5.html

Solution

```
var('x, y, z')
implicit_plot3d(x^2+9*y^2+9*z^2==9, (x, -3, 3), (y, -3, 3), (z, -3, 3), opacity=0.5)
```

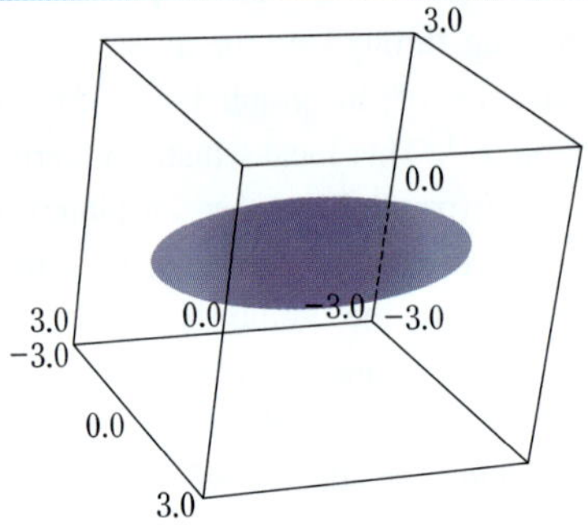

7. $x = y^2 + 5z^2$

Solution The trace in $x = k$ are ellipses of the form $y^2 + 5z^2 = k$, $k \geq 0$, the trace in $y = k$ are parabolas of the form $x = k^2 + 5z^2$, and the trace in $z = k$ are parabolas of the form $x = y^2 + 5k^2$.
Combining these traces we form the graph.

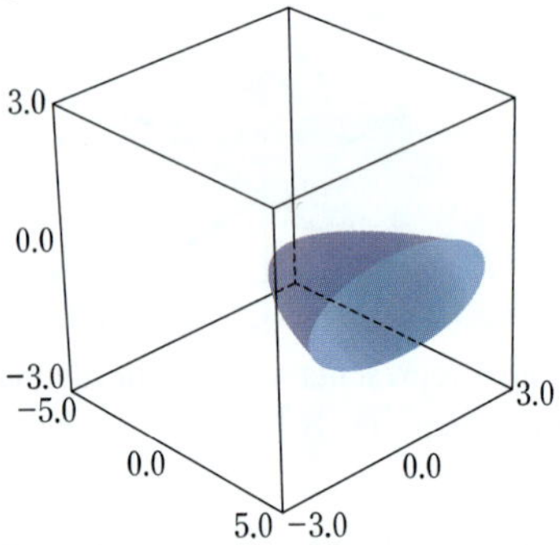

CAS **8.** $25x^2 + z^2 = 100 + 4y^2$

http://matrix.skku.ac.kr/cal-lab/cal-11-6-8.html

Solution

```
var('x, y, z')
implicit_plot3d(25*x^2+z^2==100+4*y^2, (x, -3, 3), (y, -3, 3), (z, -3, 3), opacity=0.5)
```

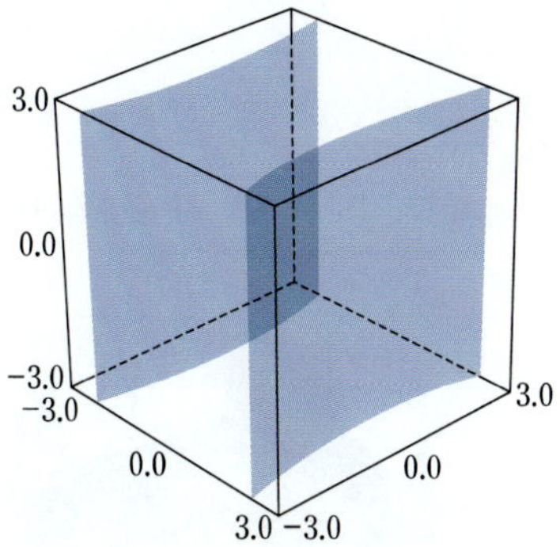

9. $-x^2+5y^2-z^2=5$

Solution The trace in $x=k$ are hyperbolas of the form $y^2-\dfrac{z^2}{5}=1+\dfrac{k^2}{5}$, the trace in $y=k$ are circles of the form $x^2+z^2=5k^2-5$, $k\geq 1$, and the trace in $z=k$ are hyperbolas of the form $-\dfrac{x^2}{5}+y^2=1+\dfrac{k^2}{5}$. Combining these traces we form the graph.

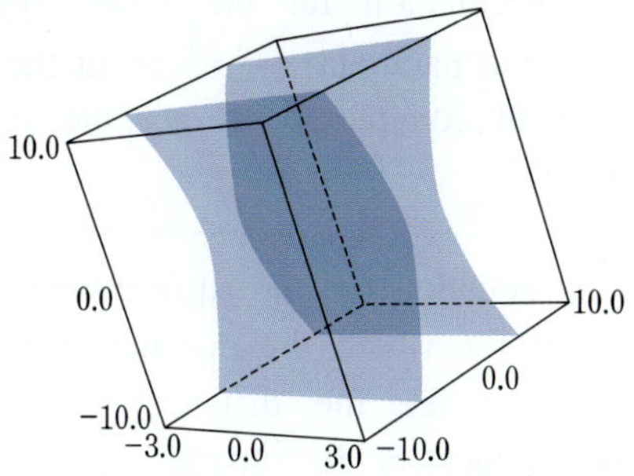

CAS **10.** $y^2+4z^2-x=0$

http://matrix.skku.ac.kr/cal-lab/cal-11-6-10.html

Solution

```
var('x, y, z')
implicit_plot3d(y^2+4*z^2-x==0, (x, -3, 3), (y, -3, 3), (z, -3, 3), opacity=0.5)
```

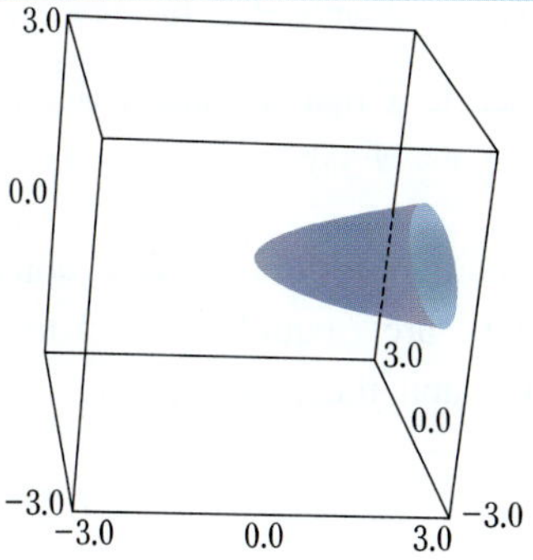

11-14. Reduce the equation to one of the standard forms, classify the surface, and sketch it.

11. $x=3y^2+5z^2$

Solution Dividing both sides by 15 gives $\dfrac{x}{15}=\dfrac{y^2}{5}+\dfrac{z^2}{3}$, an elliptic paraboloid with vertex $(0,0,0)$ and axis the horizontal line $y=0$, $z=0$.

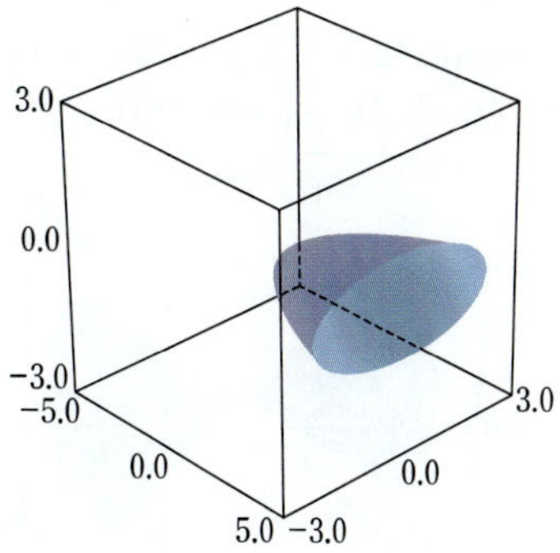

CAS **12.** $z^2=9x^2+4y^2-5$

http://matrix.skku.ac.kr/cal-lab/cal-11-6-12.html

Solution

```
var('x, y, z')
implicit_plot3d(z^2 == 9*x^2+4*y^2-5, (x, -3, 3), (y, -3, 3), (z, -3, 3), opacity=0.5)
```

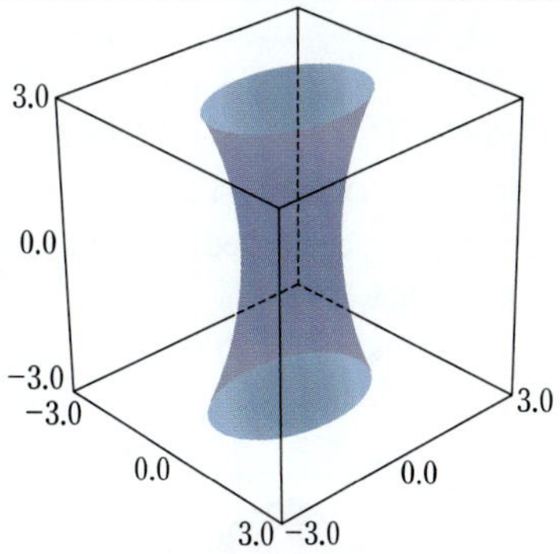

13. $x^2-y^2+z^2-4x+4y+8z+12=0$

Solution Completing squares in x, y, and z gives $(x-2)^2-(y-2)^2+(z+4)^2=4$ or $\dfrac{(x-2)^2}{4}-\dfrac{(y-2)^2}{4}+\dfrac{(z+4)^2}{4}=1$, a hyperboloid of one sheet.

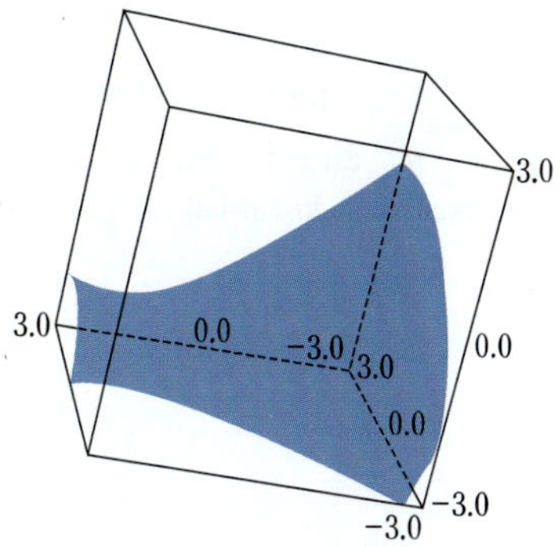

CAS 14. $x^2 = 3y^2 + 2z^2$

http://matrix.skku.ac.kr/cal-lab/cal-11-6-14.html

Solution

```
var('x, y, z')
implicit_plot3d(x^2 == 3*y^2+2*z^2, (x, -3, 3),
(y, -3, 3), (z, -3, 3), opacity=0.5)
```

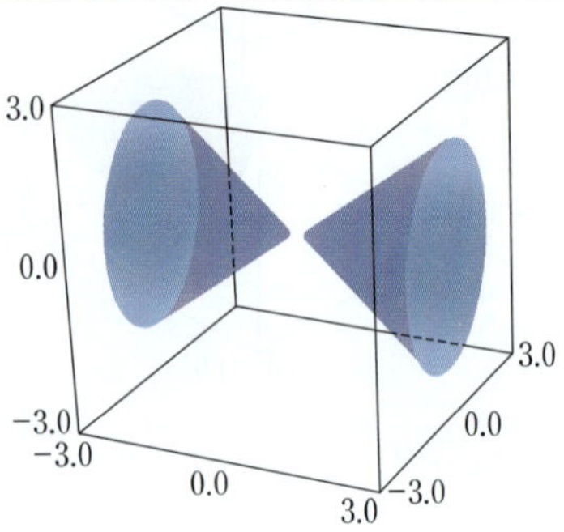

CAS 15. $(x^2 + \frac{9}{4}y^2 + z^2 - 1)^3 = x^2z^3 + \frac{9}{80}y^2z^3$

http://matrix.skku.ac.kr/cal-lab/cal-11-6-15.html

Solution

```
var('x, y, z')
implicit_plot3d((x^2+9/4*y^2+z^2-1)^3-x^2*z^3-9/
80 *y^2*z^3==0, (x, -1.5, 1.5), (y, -1.5, 1.5), (z,
-1.5, 1.5), color='red', plot_points=80,
smooth=True).show()
```

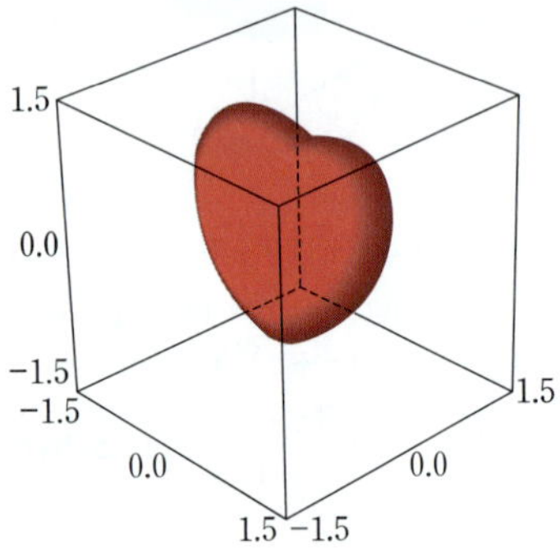

16. Sketch the region bounded by the surfaces $z - \sqrt{x^2 + y^2}$ and $x^2 + y^2 = 4$ for $2 \le z \le 4$.

http://matrix.skku.ac.kr/cal-lab/cal-11-6-16.html

Solution

```
var('x, y, z')
z=(x^2+y^2)^(1/2)
plot3d(z, (x, -2, 2), (y, -2, 2), (z, 2, 4))
```

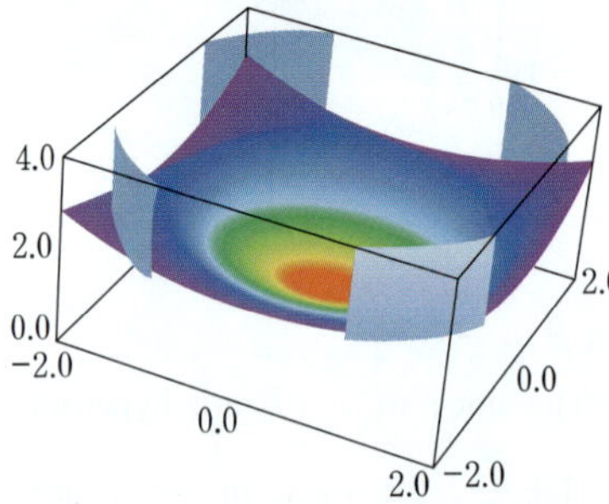

Ajit add: May be we can add set of example indicating surface of revolution of different curves about different axes.

17. Find an equation for the surface obtained by rotating the parabola $z = x^2$ about the z-axis. (Use revolution_plot3d to get the plot of this surface.)

18. Find an equation for the surface consisting of all points P for which the distance from P to the y-axis is twice the distance from P to the xz-plane. Identify the surface.

Solution Let $P = (x, y, z)$ be an arbitrary point whose distance from th y-axis is twice its distance from the xz-plane. The distance from P to the y-axis is $\sqrt{x^2 + (y-y)^2 + z^2} = \sqrt{x^2 + z^2}$ and the distance from P to the xz-plane$(y=0)$ is $\frac{|y|}{1} = |y|$.

Thus

$$\sqrt{x^2 + z^2} = 2|y| \iff x^2 + z^2 = 4y^2 \iff y^2 = \frac{x^2}{2^2} + \frac{z^2}{2^2}.$$

So, the surface is a right circular cone with vertex the origin and axis the y-axis.

19. Find an equation for the surface consisting of all points that are equidistant from the point $(0, -1, 0)$ and the plane $y = 1$.

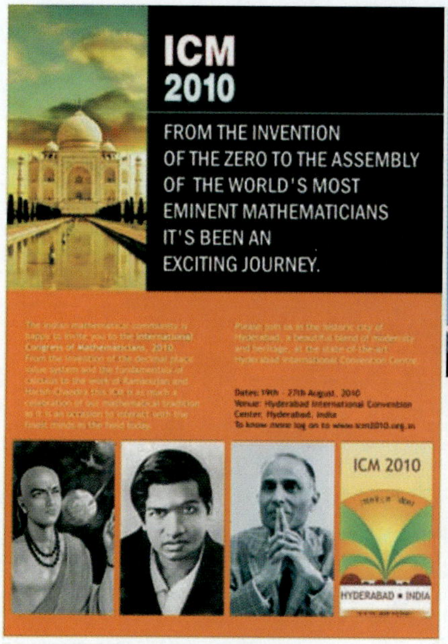

SEOUL 2014
ICM
INTERNATIONAL
CONGRESS OF
MATHEMATICIANS

ICM 2010 and 2014

Chapter 12

Vector Valued Functions

Leonhard Euler
1707-1783

Analysis

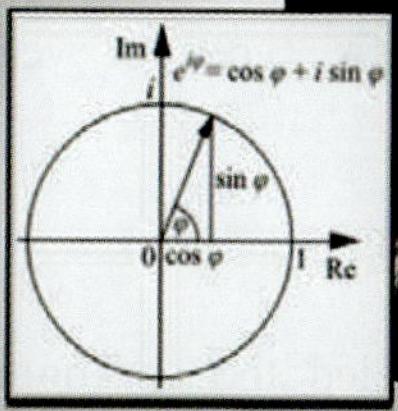

[Euler's Formula]

Euler was a pioneering Swiss mathematician who spent most of his life in Russia and Germany. Euler made important discoveries in fields as diverse as infinitesimal calculus, graph theory, mechanics, fluid dynamics, optics, and astronomy. He also introduced much of the modern mathematical terminology and notation, particularly for mathematical analysis.

YouTube http://youtu.be/0pvywjBjsQw

So far, we have been dealing with real-valued functions of real-variables. In this chapter, we will study functions whose values are vectors in a plane or in space. Such functions are used to describe plane and space curves and surfaces in space. The chapter also discusses limits, continuity and derivative of vector-valued functions, curvature of curves and motion along space curves.

12.1 Vector-Valued Functions and Space Curves

A **vector-valued function**, or simply a **vector function,** assigns to each element in the set of real numbers an element in the set of vectors. We will consider vector functions $\mathbf{r}$ in three-dimensional vector space. This means that a vector function on $\mathbb{R}$ is a function $\mathbf{r}: \mathbb{R} \to \mathbb{R}^3$. These specific types of vector functions are denoted by expressions such as

$$\mathbf{r}(t) = \langle f(t),\ g(t),\ h(t)\rangle = f(t)\mathbf{i} + g(t)\mathbf{j} + h(t)\mathbf{k}$$

where $f(t)$, $g(t)$ and $h(t)$ are real-valued functions called the **component functions** of the vector $\mathbf{r}(t)$. The domain of a vector function $\mathbf{r}(t)$ is the intersection of the domain of the component functions $f(t)$, $g(t)$ and $h(t)$. Often we imagine t to be the time and $\mathbf{r}(t)$ to be the position of a moving particle at time t.

EXAMPLE 1

If $\mathbf{r}(t) = \langle \sin t,\ \ln(5-t),\ \sqrt{3+t}\,\rangle$, then the component functions are $f(t) = \sin t$, $g(t) = \ln(5-t)$ and $h(t) = \sqrt{3+t}$, the domain of $\mathbf{r}$ is the interval $[-3,\ 5)$. ■

DEFINITION 1 Limit of a Vector Function

A vector function $\mathbf{r}$ approaches to the limit L as t approaches to a, written $\lim_{t\to a} \mathbf{r}(t) = L$, provided $\lim_{t\to a} |\mathbf{r}(t) - L| = 0$.

Another way of defining the limit is as follows:

$\lim_{t\to a} \mathbf{r}(t) = \boldsymbol{L}$ if and only if for every $\varepsilon > 0$ there is a number $\delta > 0$ such that $|\mathbf{r}(t) - \boldsymbol{L}| < \varepsilon$ whenever $0 < |t-a| < \delta$.

THEOREM 2

Let $\mathbf{r}(t) = \langle f(t),\ g(t),\ h(t)\rangle$ be a vector function. Then a vector $\boldsymbol{L} = \langle L_1,\ L_2,\ L_3\rangle \in \mathbb{R}^3$ is the limit of $\mathbf{r}(t)$ as $t \to a$ if and only if

$$\lim_{t\to a} f(t) = L_1,\ \lim_{t\to a} g(t) = L_2 \text{ and } \lim_{t\to a} h(t) = L_3.$$

The **limit** of a vector function $\mathbf{r}(t) = \langle f(t),\ g(t),\ h(t)\rangle$. Suppose that

$$\lim_{t\to a} f(t) = L_1,\ \lim_{t\to a} g(t) = L_2,\ \lim_{t\to a} h(t) = L_3 \text{ iff } \lim_{t\to a} \mathbf{r}(t) = \boldsymbol{L}.$$

Proof $\lim_{t \to a} \mathbf{r}(t) = \boldsymbol{L}$ iff $\lim_{t \to a} |\mathbf{r}(t) - \mathbf{L}| = 0$

$$\text{iff } \lim_{t \to a} \sqrt{[f(t) - L_1]^2 + [g(t) - L_2]^2 + [h(t) - L_3]^2} = 0$$

$$\text{iff } \lim_{t \to a} f(t) = L_1,\ \lim_{t \to a} g(t) = L_2,\ \lim_{t \to a} h(t) = L_3.$$

In particular, $\lim_{t \to a} \mathbf{r}(t) = \left\langle \lim_{t \to a} f(t), \lim_{t \to a} g(t), \lim_{t \to a} h(t) \right\rangle = \langle L_1, L_2, L_3 \rangle = \boldsymbol{L}$. The **limit** of a vector function $\mathbf{r}(t)$ is determined by computing the limits of its components. ■

Remark $\lim_{t \to a} \mathbf{r}(t)$ exists iff limits of component functions exist.

EXAMPLE 2

If $\mathbf{r}(t) = (2+t)^3\,\mathbf{i} - (4-t)e^{-t}\,\mathbf{j} + \cos t\,\mathbf{k}$, then find $\lim_{t \to 0} \mathbf{r}(t)$.

Solution
$$\begin{aligned}\lim_{t \to 0} \mathbf{r}(t) &= \lim_{t \to 0}(2+t)^3\,\mathbf{i} - \lim_{t \to 0}(4-t)e^{-t}\,\mathbf{j} + \lim_{t \to 0}(\cos t)\,\mathbf{k} \\ &= 8\,\mathbf{i} - 4\,\mathbf{j} + \mathbf{k}.\end{aligned}$$
■

Rules of limits of vector functions are the same as rules of limits of real-valued functions. We mentioned the following properties.

THEOREM 2 Rules of Limits

Let $\mathbf{u}$ and $\mathbf{v}$ be vector functions and c be a constant. Let f and g be real-functions. Suppose that $\lim_{t \to a} \mathbf{u}(t)$, $\lim_{t \to a} \mathbf{v}(t)$ and $\lim_{t \to a} f(t)$ exist and $\lim_{s \to s_0} g(s) = a$. Then we have the following properties of limits:

(i) $\lim_{t \to a} c\mathbf{u}(t) = c \lim_{t \to a} \mathbf{u}(t)$

(ii) $\lim_{t \to a} [\mathbf{u}(t) + \mathbf{v}(t)] = \lim_{t \to a} \mathbf{u}(t) + \lim_{t \to a} \mathbf{v}(t)$

(iii) $\lim_{t \to a} [f(t)\mathbf{u}(t)] = \lim_{t \to a} f(t) \lim_{t \to a} \mathbf{u}(t)$

(iv) $\lim_{t \to a} [\mathbf{u}(t) \cdot \mathbf{v}(t)] = \lim_{t \to a} \mathbf{u}(t) \cdot \lim_{t \to a} \mathbf{v}(t)$

(v) $\lim_{t \to a} [\mathbf{u}(t) \times \mathbf{v}(t)] = \lim_{t \to a} \mathbf{u}(t) \times \lim_{t \to a} \mathbf{v}(t)$

(vi) $\lim_{s \to s_0} \mathbf{u}(g(s)) = \lim_{t \to a} \mathbf{u}(t)$

All these properties can be easily proof ed by taking the limit of the corresponding functions.

DEFINITION 4 Continuity

A vector function $\mathbf{r}$ is said to **continuous** at a if
(i) $\mathbf{r}(a)$ is defined
(ii) $\lim_{t\to a}\mathbf{r}(t)$ exists
(iii) $\lim_{t\to a}\mathbf{r}(t)=\mathbf{r}(a)$.

In view of Definition 1, one can easily show that $\mathbf{r}$ is continuous at a if and only if its component functions f, g, and h are continuous at a.

DEFINITION 5 Continuity

Let f, g and h be continuous real-valued functions on an interval I. A **space curve** is the set C of all points (x, y, z) in space, where

$$x=f(t),\ y=g(t),\ z=h(t)$$

and t varies throughout the interval I. These equations are called **parametric equations** of C and t is called a **parameter.**

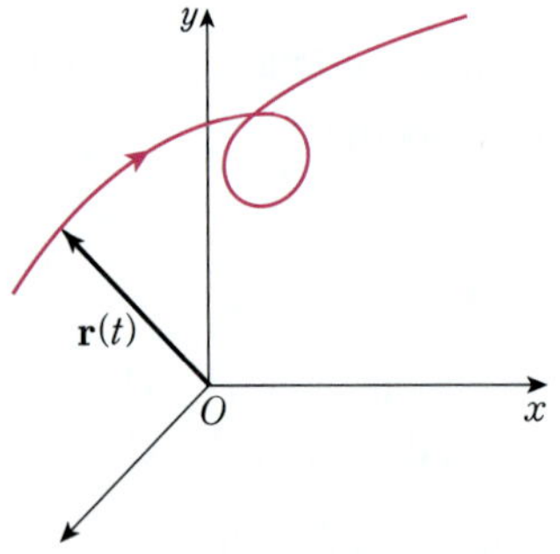

Figure 1

We think of C as being traced out by a moving particle whose position at time t is $(f(t), g(t), h(t))$. The vector

$$\mathbf{r}(t)=\langle f(t),\ g(t),\ h(t)\rangle=f(t)\,\mathbf{i}+g(t)\,\mathbf{j}+h(t)\,\mathbf{k}$$

from the origin to the particle's position $P(f(t), g(t), h(t))$ at time t is the particle's **position vector**. Thus, any continuous vector function $\mathbf{r}$ defines a space curve C that is traced out by the tip of the moving vector $\mathbf{r}(t)$, as shown in Figure 1.

EXAMPLE 3

Describe the curve defined by the vector function

$$\mathbf{r}(t)=\langle 2+t,\ 4-5t,\ -3+4t\rangle.$$

Solution The corresponding parametric equations are

$$x=2+t,\quad y=4-5t,\quad z=-3+4t.$$

These equations are parametric equations of a line passing through the point $(2, 4, -3)$ and parallel to the vector $\langle 1, -5, 4\rangle$. Alternatively, we could observe that the function $\mathbf{r}(t)$ can be written as $\mathbf{r}(t)=\mathbf{r}_0+t\mathbf{v}$, where $\mathbf{r}_0=\langle 2, 4, -3\rangle$ and $\mathbf{v}=\langle 1, -5, 4\rangle$.■

Plane curves can also be described in vector notation. For instance, the curve given by the parametric equations $x=t^2+2t$ and $y=t-1$ can also be described by the vector equation

$$\mathbf{r}(t)=\langle t^2+2t,\ t-1\rangle=(t^2+2t)\,\mathbf{i}+(t-1)\,\mathbf{j}$$

where $\mathbf{i}=\langle 1,0\rangle$ and $\mathbf{j}=\langle 0,1\rangle$. We recognize this as the equation of a parabola with vertex at $(-1,-2)$.

EXAMPLE 4

Sketch the curve whose vector equation is

$$\mathbf{r}(t)=\cos t\,\mathbf{i}+\sin t\,\mathbf{j}+2t\,\mathbf{k}.$$

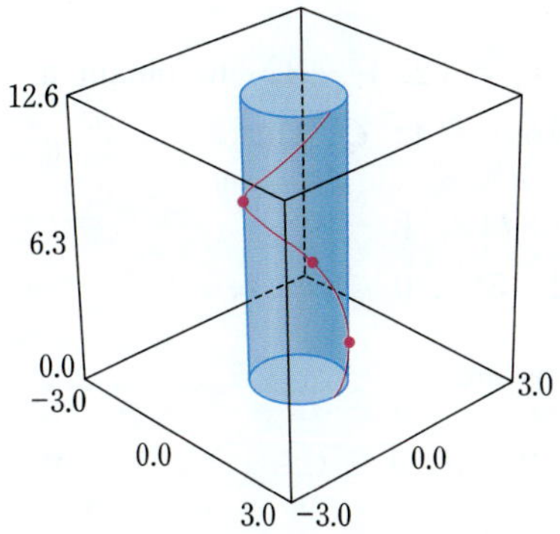

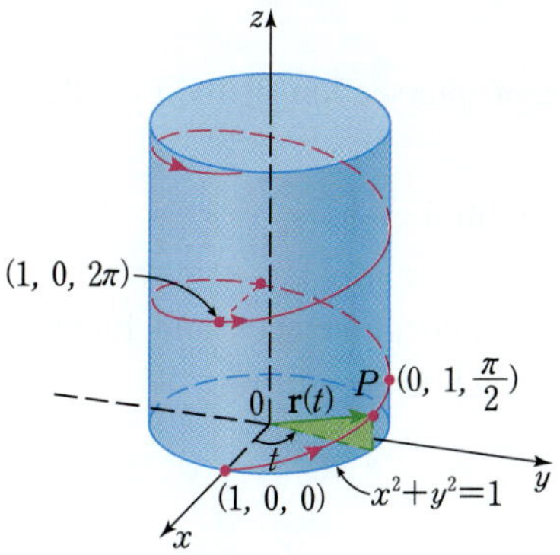

Figure 2

Solution The parametric equations for this curve are

$$x=\cos t,\quad y=\sin t,\quad z=2t.$$

The curve must lie on the circular cylinder $x^2+y^2=1$, because the x-coordinate and y-coordinate of the tip of $\mathbf{r}$ satisfy the cylinder's equation

$$x^2+y^2=\cos^2 t+\sin^2 t=1.$$

The point (x, y, z) lies directly above the point $(x, y, 0)$, which moves counterclockwise around the circle $x^2+y^2=1$ in the xy-plane. The curve spirals upward around the cylinder as $z=2t$ increases. That is, each time t increases by 2π, the curve completes one turn around the cylinder. The curve, shown in Figure 2, is called a **helix**. ■

http://matrix.skku.ac.kr/cal-lab/cal-12-1-Exmaple-4.html

```
var('x,y,z,t')
r=vector([cos(t),sin(t),2*t])
s=implicit_plot3d(x^2+y^2==1,(x,-3,3),(y,-3,3),(z,0,4*pi),color='goldenro
d',opacity=0.2)
c=parametric_plot3d([cos(t),sin(t),2*t],(t,0,2*pi),color='green')
t1=pi/4
t2=2*pi/3
t3=4*pi/3
t4=5*pi/2
Dt=0.1
Ar1=arrow3d(r(t=t1),r(t=t1+Dt),color='green')
Ar2=arrow3d(r(t=t2),r(t=t2+Dt),color='green')
Ar3=arrow3d(r(t=t3),r(t=t3+Dt),color='green')
Ar4=arrow3d(r(t=t4),r(t=t4+Dt),color='green')
s+c+Ar1+Ar2+Ar3+Ar4
```

EXAMPLE 5

Find a vector equation and parametric equations for the line segment that joins the point $P(-1, 3, 2)$ to the point $Q(2, 4, -3)$.

Solution We know that the vector equation for the line segment joining the tips of the vectors $\mathbf{r}_0$ and $\mathbf{r}_1$ and oriented from $\mathbf{r}_0$ to $\mathbf{r}_1$ is given by

$$\mathbf{r}(t) = (1-t)\,\mathbf{r}_0 + t\,\mathbf{r}_1,\ 0 \le t \le 1.$$

Here we take $\mathbf{r}_0 = \langle -1, 3, 2\rangle$ and $\mathbf{r}_1 = \langle 2, 4, -3\rangle$ to obtain a vector equation of the line segment from P to Q. Thus

$$\begin{aligned}\mathbf{r}(t) &= (1-t)\langle -1,\ 3,\ 2\rangle + t\langle 2,\ 4,\ -3\rangle \\ &= \langle -1+3t,\ 3+t,\ 2-5t\rangle,\ 0 \le t \le 1.\end{aligned}$$

The corresponding parametric equations are

$$x = -1+3t,\ y = 3+t,\ z = 2-5t,\ 0 \le t \le 1.$$

■

EXAMPLE 6

Find a vector equation that represents the curve of intersection of the cylinder $x^2+y^2=1$ and the plane $-y+z=2$.
http://matrix.skku.ac.kr/cal-lab/cal-12-1-Exmaple-6.html

Solution Figure 3 shows how the plane and the cylinder intersect, and Figure 4 shows the curve of intersection C, which is an ellipse.

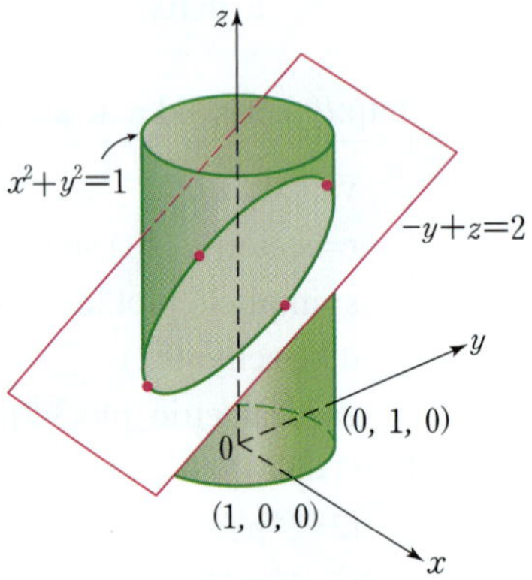

Figure 3

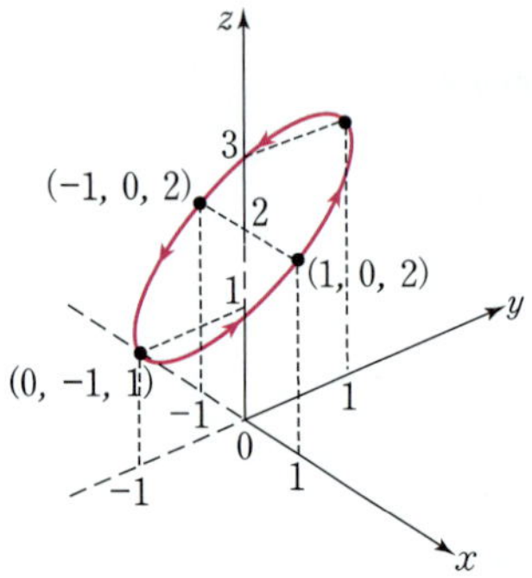

Figure 4

The projection of C onto the xy-plane is the circle $x^2+y^2=1$, $z=0$. So we can write

$$x = \cos t,\ y = \sin t,\ 0 \le t \le 2\pi.$$

From the equation of the plane, we have

$$z = 2+y = 2+\sin t.$$

So we can write parametric equations for C as

$$x=\cos t,\ y=\sin t,\ z=2+\sin t,\ 0\le t\le 2\pi.$$

The corresponding vector equation is

$$\mathbf{r}(t)=\cos t\,\mathbf{i}+\sin t\,\mathbf{j}+(2+\sin t)\,\mathbf{k},\ 0\le t\le 2\pi.$$

This equation is called a parameterization of the curve C. The arrows in Figure 3 indicate the direction in which C is traced as the parameter t increases. ■

CAS **EXAMPLE 7**

Find a vector function that represents the curve of intersection of the cylinder $x^2+y^2=5$ and the surface $z=xy$.
http://matrix.skku.ac.kr/cal-lab/cal-13-1-Exm7.html

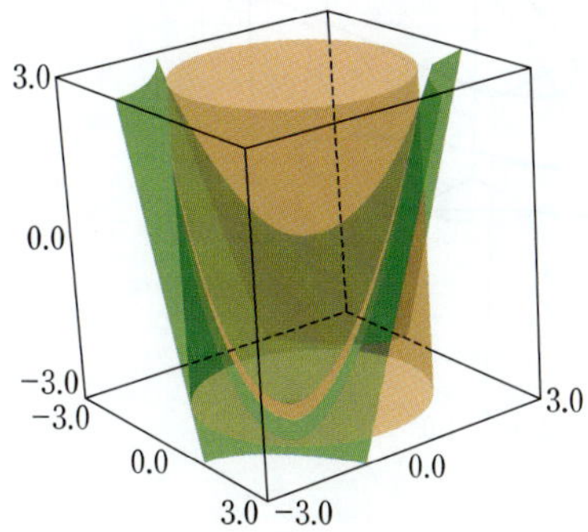

Figure 5

Solution Let $x=\sqrt{5}\cos t,\ y=\sqrt{5}\sin t$. Then $z=xy=5\cos t\sin t$
The answer is $\mathbf{r}(t)=\langle\sqrt{5}\cos t,\ \sqrt{5}\sin t,\ 5\sin t\cos t\rangle$.
Figure 5 illustrates the Vector Function.

```
x, y, z=var('x, y, z')
p1=implicit_plot3d(x^2+y^2-5==0, (x, -3, 3), (y, -3, 3), (z,
-3, 3),
rgbcolor='orange', opacity=0.5)
p2=implicit_plot3d(x*y-z==0, (x, -3, 3), (y, -3, 3),
(z, -3, 3), rgbcolor='green', opacity=0.5)
show(p1+p2)
```

A vector function that represents the curve of their intersection is $\mathbf{r}(t)=\langle\sqrt{5}\cos t,\ \sqrt{5}\sin t,\ 5\sin t\cos t\rangle$.

```
var('t')
x1=sqrt(5) *cos(t)
y1=sqrt(5) *sin(t)
z1=5*sin(t) *cos(t)
p3=parametric_plot3d((x1, y1, z1), (t, 0, 2*pi),
rgbcolor='blue')
show(p3)
```

■

Many surfaces are obtained by revolving a plane curve about various axes. For example if the upper semi-circle x^2 + y^2 = 1 with −1 <= x <= 1 and y > = 0 is rotated about the x-axis, we get the sphere. We can easily write a parametrization of such surfaces. If we take a curve y = f(x) on an interval [a, b]. Then we think of this as a curve in xz-plane.

http://matrix.skku.ac.kr/cal-lab/cal-12-1-Rotations.html
http://matrix.skku.ac.kr/cal-lab/cal-12-1-Rotations-A.html
http://matrix.skku.ac.kr/cal-lab/cal-12-1-Rotations-B.html

12.1 EXERCISES (Vector-Valued Functions and Space Curves)

http://matrix.skku.ac.kr/Cal-Book/part2/CS-Sec-12-1-Sol.html

YouTube http://youtu.be/jvMI6OzdR_I

1-2. Find the domain of the vector functions.

1. $\mathbf{r}(t) = \langle t^3,\ \sqrt{3-t},\ \sqrt{t+1}\,\rangle$.

Solution $-1 \le t \le 3$.

2. $\mathbf{r}(t) = \ln(4-t^2)\mathbf{i} + \dfrac{t+1}{t-1}\mathbf{j} + \cos t\mathbf{k}$.

Solution $4-t^2 > 0$ and $t \ne 1$.

Thus $-2 < t < 2$ and $t \ne 1$.

3-5. Find the limit.

CAS **3.** $\lim_{t\to 0}\left\langle \dfrac{\sin t}{t},\ \cos t,\ t\ln t\right\rangle = \langle 1,\ 1,\ 0\rangle$.

http://matrix.skku.ac.kr/cal-lab/cal-13-1-3.html

Solution

```
var('t')
(limit(sin(t)/t, t=0), limit(cos(t), t=0), limit(t*ln(t), t=0))
```

Answer : (1, 1, 0)

4. $\lim_{t\to 2}\left(\dfrac{t-2}{t^2-4}\mathbf{i} + \sqrt{t+1}\,\mathbf{j} + \dfrac{\sin t}{(t-1)\cos t}\mathbf{k}\right) = \dfrac{1}{4}\mathbf{i} + \sqrt{3}\,\mathbf{j} + \tan 2\mathbf{k}$.

5. $\lim_{t\to\infty}\left\langle \dfrac{\sin 2t}{t}, \dfrac{1-\cos^2 2t}{t}, \dfrac{\ln t}{t}\right\rangle = \langle 0,\ 0,\ 0\rangle$.

CAS **6-9.** Sketch curves with the given vector equations. Indicate with an arrow the direction in which t increases.

6. $\mathbf{r}(t) = \langle \cos 2t, \sin 3t, t\rangle$.

http://matrix.skku.ac.kr/cal-lab/cal-13-1-Exs-6.html

Solution

```
var('t')
r=vector([cos(2*t),sin(3*t),t])
a=0
b=2*pi
C=parametric_plot3d(r,(t,a,b),color='green',thickness=2)
t1=pi/10
Dt=0.1
t2=2*pi/3
t3=5*pi/3
Ar1=arrow3d(r(t=t1),r(t=t1+Dt),color='green')
Ar2=arrow3d(r(t=t2),r(t=t2+0.05),color='green')
Ar3=arrow3d(r(t=t3),r(t=t3+Dt),color='green')
C+Ar1+Ar2+Ar3
```

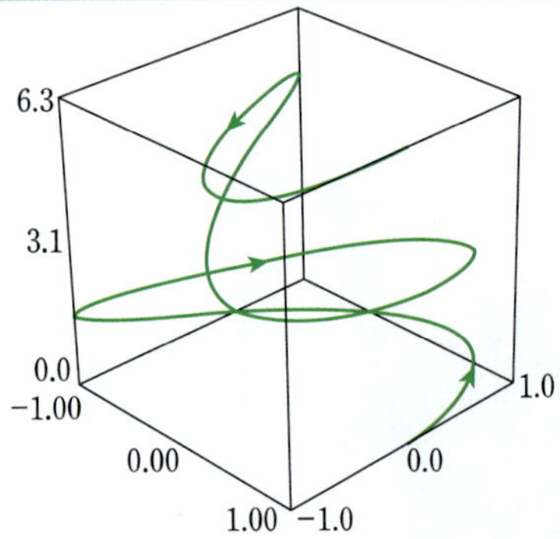

7. $\mathbf{r}(t) = \langle t^2,\ t,\ t^3\rangle$.

http://matrix.skku.ac.kr/cal-lab/cal-13-1-Exs-7.html

Solution

```
var('t')
r=vector([t^2,t,t^3])
a=0
b=2*pi
C=parametric_plot3d(r,(t,a,b),color='green',thickness=2)
t1=pi/10
Dt=0.1
t2=2*pi/3
t3=5*pi/3
Ar1=arrow3d(r(t=t1),r(t=t1+Dt),color='green',thickness=4)
Ar2=arrow3d(r(t=t2),r(t=t2+0.05),color='green',thickness=4)
Ar3=arrow3d(r(t=t3),r(t=t3+Dt),color='green',thickness=4)
C+Ar1+Ar2+Ar3
```

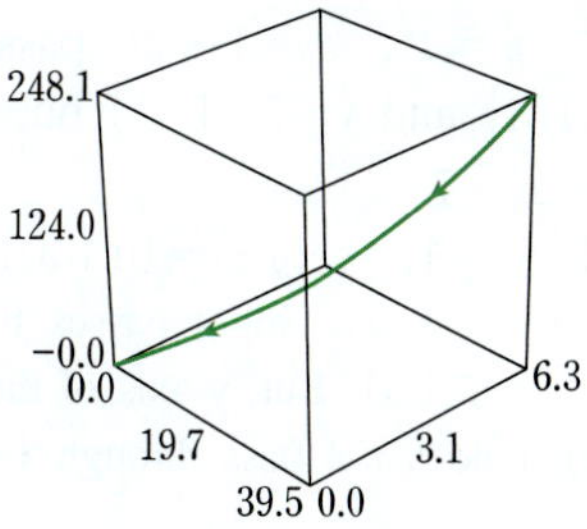

8. $\mathbf{r}(t) = t\mathbf{i} + t\mathbf{j} + \sin t\mathbf{k}$.

http://matrix.skku.ac.kr/cal-lab/cal-13-1-Exs-8.html

Solution

```
var('t')
r=vector([t,t,sin(t)])
a=0
b=2*pi
C=parametric_plot3d(r,(t,a,b),color='green',thickness=2)
t1=pi/10
Dt=0.1
t2=2*pi/3
t3=5*pi/3
Ar1=arrow3d(r(t=t1),r(t=t1+Dt),color='green',thickness=4)
Ar2=arrow3d(r(t=t2),r(t=t2+0.05),color='green',thickness=4)
Ar3=arrow3d(r(t=t3),r(t=t3+Dt),color='green',thickness=4)
C+Ar1+Ar2+Ar3
```

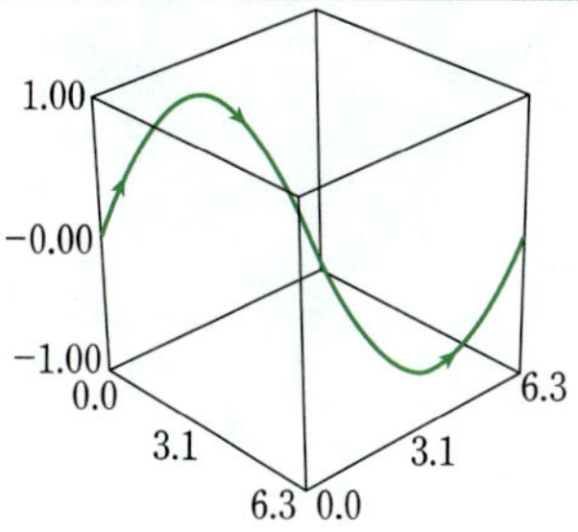

9. $\mathbf{r}(t) = t^2\cos t\mathbf{i} + t^2\sin t\mathbf{j} + \sin t\mathbf{k}$.

http://matrix.skku.ac.kr/cal-lab/cal-13-1-Exs-9.html

Solution

```
var('t')
x(t)=t^2*cos(t);y(t)=t^2*sin(t);z(t)=sin(t);r(t)=(x(t), y(t), z(t))
dr(t)=(diff(x(t), t), diff(y(t), t, ), diff(z(t), t));
t0=2
p1=r(t0)
p2=dr(t0)/abs(dr(t0))+r(t0)
p=parametric_plot3d((x(t), y(t), z(t)), (t, 1, 5))
utv=line3d([p1, p2], rgbcolor=(1, 0, 0), arrow_head=True, thickness=3)
show(p+utv)
```

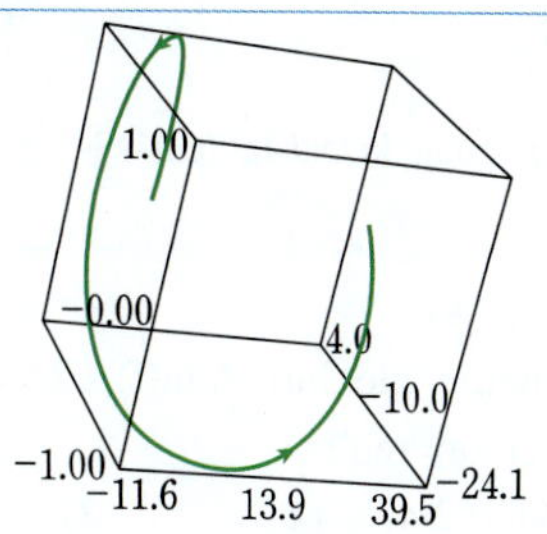

10-11. Find vector equations and parametric equations for the line segment P to Q.

10. $P(0,0,0)$, $Q(2,1,3)$.

Solution $\mathbf{r}(t) = (1-t)\overrightarrow{OP} + t\overrightarrow{OQ}$

$= (1-t)\langle 0,\ 0,\ 0\rangle + t\langle 2,\ 1,\ 3\rangle = \langle 2t,\ t,\ 3t\rangle$,

$\mathbf{r}(t) = 2t\mathbf{i} + t\mathbf{j} + 3t\mathbf{k}$.

11. $P(-1,1,2)$, $Q(1,4,7)$.

Solution

$\mathbf{r}(t) = (1-t)\overrightarrow{OP} + t\overrightarrow{OQ} = (1-t)\langle -1,\ 1,\ 2\rangle + t\langle 1,\ 4,\ 7\rangle$

$= \langle 2t-1,\ 3t+1,\ 5t+2\rangle$,

$\mathbf{r}(t) = (2t-1)\mathbf{i} + (3t+1)\mathbf{j} + (5t+2)\mathbf{k}$.

CAS **12.** Show that the curve with parametric equation $x = t\cos t$, $y = t\sin t$, $z = t$ lies on the cone $z^2 = x^2 + y^2$.

http://matrix.skku.ac.kr/cal-lab/cal-13-1-12.html

Solution We have $x^2 + y^2 = t^2\sin^2 t + t^2\cos^2 t = t^2 = z^2$.

```
var('x, y, z, t')
p1=parametric_plot3d((t^2*cos(t), t^2*sin(t), t^2), (t, -pi, pi), color='red')
p2=implicit_plot3d(z^2==x^2+y^2, (x, -10, 10), (y, -10, 10), (z, 0, 30), color='yellow', opacity=0.5)
show(p1+p2)
```

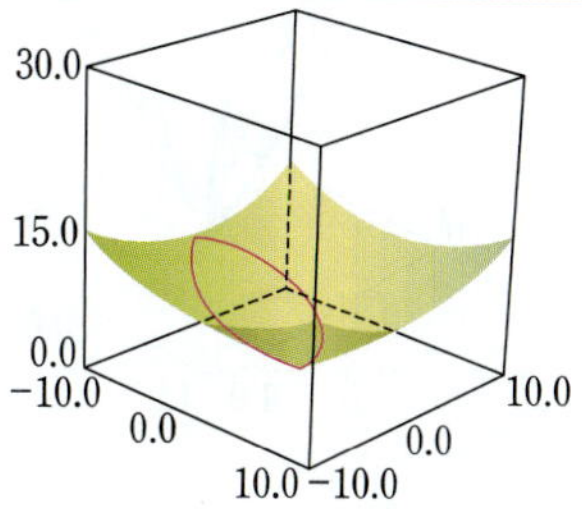

We can see that the curve (red) lies on the cone (yellow).

CAS 13. Show that the curve with parametric equations $x=2\sin t,\quad y=2\cos t,\quad z=\sin^2 t$ is the curve of intersection of the surfaces $z=\frac{1}{4}x^2$ and $x^2+y^2=4$.

http://matrix.skku.ac.kr/cal-lab/cal-13-1-13.html

Solution

```
var('x, y, z, t')
p1=parametric_plot3d((2*sin(t), 2*cos(t), t^2), (t, -pi, pi), color='red');
p2=plot3d(x^2/4, (x, -2, 2), (y, -2, 2), color='green', opacity=0.5)
p3=implicit_plot3d(x^2+y^2==4, (x, -2, 2), (y, -2, 2), (z, -2, 2), color='yellow', opacity=0.5)
show(p1+p2+p3)
```

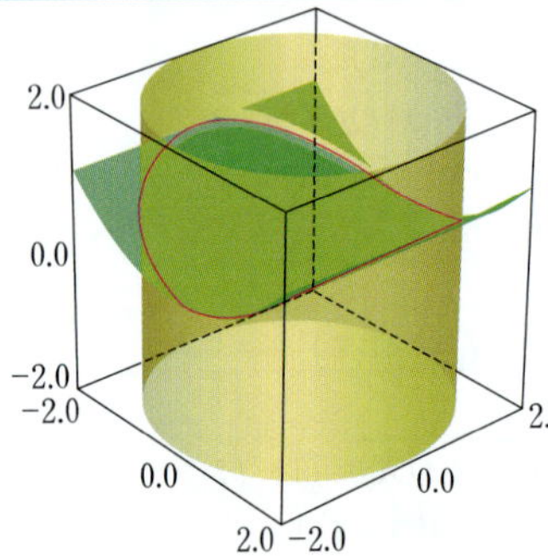

We can see that the curve (red) is intersection of two surfaces (yellow, green).

CAS 14. Graph the curve with parametric equations

$$x=(1+\sin 4t)\cos t,$$
$$y=(1+\sin 4t)\sin t,$$
$$z=1+\sin 4t.$$

http://matrix.skku.ac.kr/cal-lab/cal-13-1-14.html

Solution

```
var('t')
parametric_plot3d(((1+sin(4*t))*cos(t), (1+sin(4*t))*sin(t), (1+sin(4*t))), (t, -2*pi, 2*pi))
```

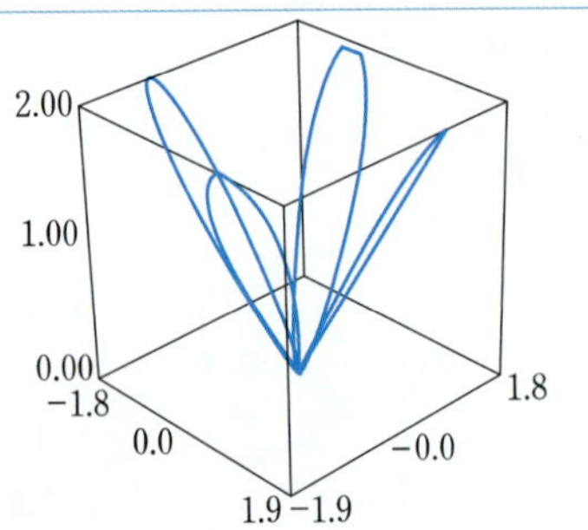

15. Show that the curve with parametric equations $x=1-t^3,\ y=t^2,\ z=1+2t$ passes through the points $(0,1,3)$ and $(-7,4,5)$ but not through the point $(8,-2,-7)$.

Solution If $t=1$, $(x,y,z)=(0,1,3)$, and if $t=2$, $(x,y,z)=(-7,4,5)$. So, it passes through the points $(0,1,3)$ and $(-7,4,5)$. But, y-axis of the curve cannot be negative, so it does not pass through $(8,-2,-7)$.

CAS 16-18. Find vector functions that represent the curves of intersection of the two surfaces.

16. The cylinder $x^2+y^2=7$ and the surface $z=2xy$.

http://matrix.skku.ac.kr/cal-lab/cal-13-1-16.html

Solution Let $x=\sqrt{7}\cos t,\ y=\sqrt{7}\sin t$.

Then $z=2xy=14\cos t\sin t=7\sin 2t$.

So, $\langle x, y, z\rangle=\langle \sqrt{7}\cos t, \sqrt{7}\sin t, 7\sin 2t\rangle$, $0\le t\le 2\pi$ is the vector function of intersection of two surfaces.

```
var('x, y, z, t')
p1=parametric_plot3d((sqrt(7)*cos(t), sqrt(7)*sin(t), 7*sin(2*t)), (t, -pi, pi), color='red');
p2=plot3d(2*x*y, (x, -3, 3), (y, -3, 3), color='green', opacity=0.5);
p3=implicit_plot3d(x^2+y^2==7, (x, -3, 3), (y, -3, 3), (z, -3, 30), color='yellow', opacity=0.5);
show(p1+p2+p3);
```

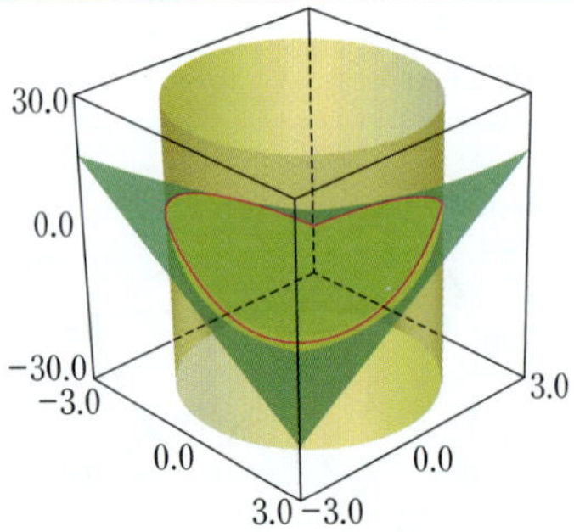

: See the graph.

17. The planes $3x-6y-2z=15$ and $2x+y-2z=5$.

http://matrix.skku.ac.kr/cal-lab/cal-13-1-17.html

Solution Let $x=t$. Then $-6y-2z=15-3t$, $y-2z=5-2t$.

Subtracting two equations, $-7y=10-t$, so we have $y=\frac{t-10}{7}$.

Substituting it to first equation,

$$-2z=\frac{6(t-10)}{7}+15-3t=\frac{-15t+45}{7},$$

so $z=\frac{15}{14}(t-3)$.

So, $\langle x, y, z \rangle = \left\langle t, \dfrac{t-10}{7}, \dfrac{15}{14}(t-3) \right\rangle$ is the vector function of intersection of two planes.

Graph

```
var('x, y, z, t')
p1=parametric_plot3d((t, (t-10)/7, 15*(t-3)/14), (t, -3, 3), color='red');
p2=implicit_plot3d(3*x-6*y-2*z==15, (x, -3, 3), (y, -3, 3), (z, -21, 6), color='green', opacity=0.5)
p3=implicit_plot3d(2*x+y-2*z==5, (x, -3, 3), (y, -3, 3), (z, -21, 6), color='yellow', opacity=0.5)
show(p1+p2+p3)
```

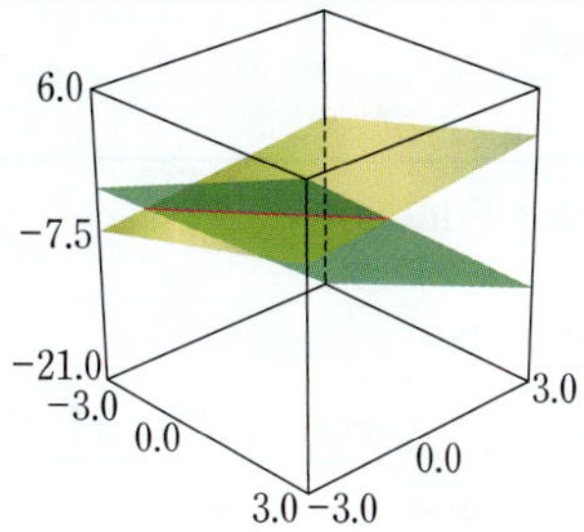

: See the graph.

18. The paraboloid $z = x^2 + 4y^2$ and the parabolic cylinder $x = y^2$.

http://matrix.skku.ac.kr/cal-lab/cal-13-1-18.html

Solution Put $y = t$. Then $x = y^2 = t^2$, and $z = x^2 + 4y^2 = t^4 + 4t^2$.

So, $\langle x, y, z \rangle = \langle t^2, t, t^4 + 4t^2 \rangle$ is the vector function of intersection of two surfaces.

We can see the answer is correct by drawing the graph of surfaces and vector function.

```
var('x, y, z, t')
p1=parametric_plot3d((t^2, t, t^4+4*t^2), (t, -sqrt(3), sqrt(3)), color='red')
p2=plot3d(x^2+4*y^2, (x, -3, 3), (y, -3, 3), color='green', opacity=0.5)
p3=implicit_plot3d(x-y^2==0, (x, -3, 3), (y, -3, 3), (z, -3, 45), color='yellow', opacity=0.5)
show(p1+p2+p3)
```

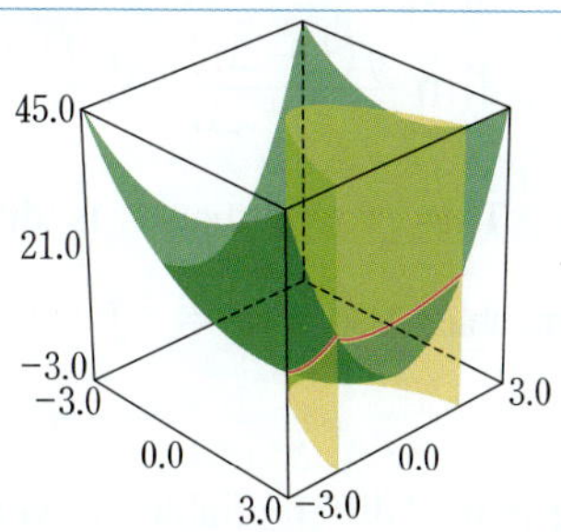

19. If two objects travel through space along two different curves, it is important to know whether they will collide (Will a missile hit its moving target? Will two aircraft collide?) The curves might intersect, but we need to know whether the objects are in the same position at the same time. Suppose the trajectories of two particles are given by the vector function.

$\mathbf{r}_1(t) = \langle t^2, 5t-9, t^2 \rangle$,

$\mathbf{r}_2(t) = \langle 3t-2, t^2, 4t-5 \rangle$ for $t > 0$.

Do the particles collide?

Solution Two particles collide means $\mathbf{r}_1(t) = \mathbf{r}_2(t)$ has a solution.

But for the second coordinate, $t^2 = 5t - 9$. Hence $t^2 - 5t + 9 = 0$ has no real root, and hence there would be no collision.

CAS **20.** Two particles travel along the space curves

$\mathbf{r}_1(t) = \langle t, t^2, t^3 \rangle$,

$\mathbf{r}_2(t) = \langle 1+t, 2+5t, 3+12t \rangle$.

Do the particles collide? Do their paths intersect?

http://matrix.skku.ac.kr/cal-lab/cal-13-1-20.html

Solution The first coordinate of the two particles cannot be the same, and hence there would be no collision.

To see if there is an intersection, we draw a graph of $\mathbf{r}_1$ and $\mathbf{r}_2$.

```
var('t')
p1=parametric_plot3d((t, t^2, t^3), (t, -5, 5), color='red')
p2=parametric_plot3d((1+t,2+5*t,3+12*t),(t,-5, 5));
show(p1+p2)
```

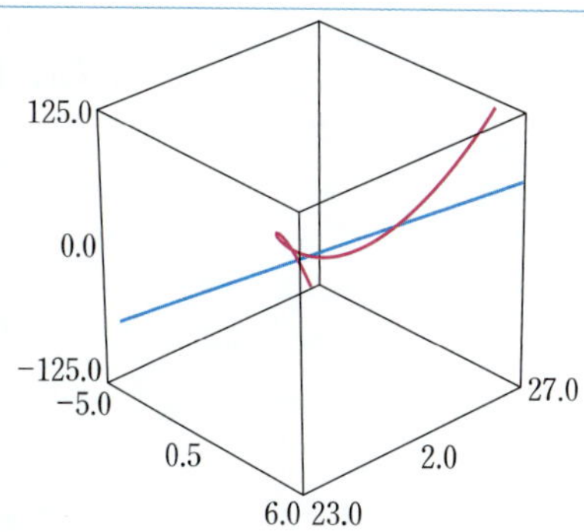

By rotating the resultant graph, we can see one intersection.

12.2 Calculus of Vector Functions

In this section, we introduce differentiation and integration of vector functions.

Derivatives

The derivative of a vector function $\mathbf{r}(t)$ at a point t can be defined in exactly same way as for one variable functions.

DEFINITION 1 Derivative of a Vector Function

The derivative of a vector function $\mathbf{r}$ is

$$\frac{d\mathbf{r}}{dt}=\mathbf{r}'(t)=\lim_{\Delta t\to 0}\frac{\Delta \mathbf{r}}{\Delta t}=\lim_{\Delta t\to 0}\frac{\mathbf{r}(t+\Delta t)-\mathbf{r}(t)}{\Delta t}$$

for all t for which the limit exists.

The next theorem gives us a way for computing the derivative of a vector function $\mathbf{r}$.

THEOREM 2 Differentiation

If $\mathbf{r}(t)=\langle f(t),\ g(t),\ h(t)\rangle=f(t)\mathbf{i}+g(t)\mathbf{j}+h(t)\mathbf{k}$, then $\mathbf{r}$ is differentiable at t, if f, g and h are differentiable at t. Moreover,

$$\mathbf{r}'(t)=\langle f'(t),\ g'(t),\ h'(t)\rangle=f'(t)\mathbf{i}+g'(t)\mathbf{j}+h'(t)\mathbf{k}.$$

Proof Note that

$$\frac{\mathbf{r}(t+\Delta t)-\mathbf{r}(t)}{\Delta t}=\left\langle \frac{f(t+\Delta t)-f(t)}{\Delta t}, \frac{g(t+\Delta t)-g(t)}{\Delta t}, \frac{h(t+\Delta t)-h(t)}{\Delta t}\right\rangle.$$

By Theorem 2 of section 12.1, $\lim_{\Delta t\to 0}\frac{\mathbf{r}(t+\Delta t)-\mathbf{r}(t)}{\Delta t}$ exists if and only if $\lim_{\Delta t\to 0}\frac{f(t+\Delta t)-f(t)}{\Delta t}$, $\lim_{\Delta t\to 0}\frac{g(t+\Delta t)-g(t)}{\Delta t}$ and $\lim_{\Delta t\to 0}\frac{h(t+\Delta t)-h(t)}{\Delta t}$ exists. This proves that $\mathbf{r}$ is differentiable if and only if f, g and h are differentiable. Also $\mathbf{r}'(t)=\langle f'(t), g'(t), h'(t)\rangle$. ■

A vector function $\mathbf{r}$ is **differentiable** if it is differentiable at every point of its domain. The **second derivative** of a vector function $\mathbf{r}$ is the derivative of $\mathbf{r}'$, that is, $\mathbf{r}'' = (\mathbf{r}')'$. Higher order derivatives of $\mathbf{r}$ may be defined in a similar way.

EXAMPLE 1

Find the second derivative of $\mathbf{r}(t) = \sin^2 t\,\mathbf{i} + e^{-t}\,\mathbf{j} + \cos 2t\,\mathbf{k}$.

Solution The first derivative of $\mathbf{r}(t)$ is

$$\begin{aligned}\mathbf{r}'(t) &= 2\sin t\cos t\,\mathbf{i} - e^{-t}\,\mathbf{j} - 2\sin 2t\,\mathbf{k}\\ &= \sin 2t\,\mathbf{i} - e^{-t}\,\mathbf{j} - 2\sin 2t\,\mathbf{k}.\end{aligned}$$

The second derivative of $\mathbf{r}(t)$ is

$$\mathbf{r}''(t) = 2\cos 2t\,\mathbf{i} + e^{-t}\,\mathbf{j} - 4\cos 2t\,\mathbf{k}.$$ ■

The curve traced by $\mathbf{r}$ is **smooth** if r' is continuous and never 0; that is, if f, g and h have continuous first derivatives that are not simultaneously 0. A curve that is made up of a finite number of smooth curves is called **piecewise smooth**.

EXAMPLE 2

Determine whether the semicubical parabola $\mathbf{r}(t) = \langle 1+t^3,\ t^2\rangle$ is smooth.

Solution Since $\mathbf{r}'(t) = \langle 3t^2,\ 2t\rangle$, we have $\mathbf{r}'(0) = \langle 0,\ 0\rangle = \mathbf{0}$ and therefore, the curve is not smooth. The point that corresponds to $t = 0$ is $(1,\ 0)$, and we see from the graph in Figure 1 that there is a sharp corner at (1, 0). Any curve with this type of behavior, an abrupt change in direction, is not smooth.

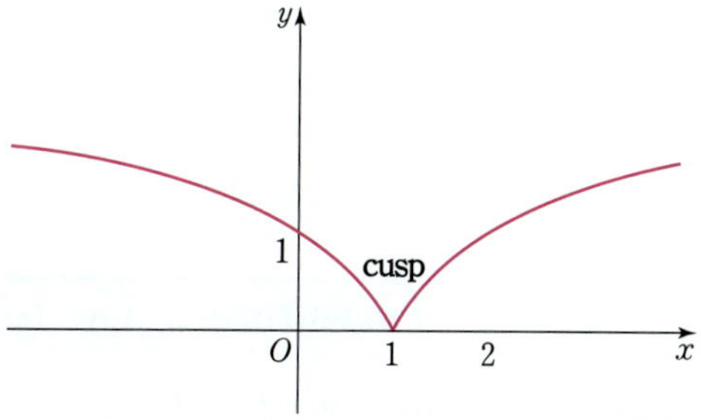

Figure 1 The curve $\mathbf{r}(t) = \langle 1+t^3,\ t^2\rangle$ is not smooth. ■

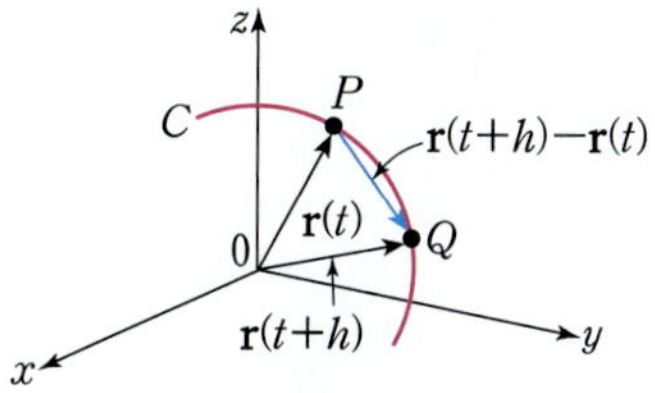

(a) The secant vector

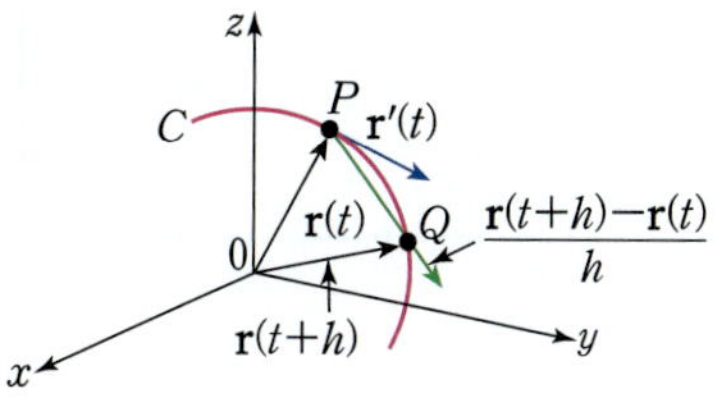

(b) The tangent vector

Figure 2

The geometric significance of the definition of derivative is shown in Figure 2. The points P and Q have position vectors $\mathbf{r}(t)$ and $\mathbf{r}(t+h)$, and the vector $\overrightarrow{PQ}$ is represented by the vector $\Delta\mathbf{r} = \mathbf{r}(t+h) - \mathbf{r}(t)$. For $h > 0$, the scalar multiple $\Delta\mathbf{r}/h$ has the same direction as the vector $\overrightarrow{PQ}$. As $h \to 0$ the vector $\Delta\mathbf{r}/h$ approaches a vector that is tangent to the curve at P. The vector

$\mathbf{r}'(t)$ is called the **tangent vector** to the curve defined by $\mathbf{r}$ at the point P, provided that $\mathbf{r}'(t)$ exists and $\mathbf{r}'(t) \neq 0$. The **tangent line** to C at P is defined to be the line through P parallel to the tangent vector $\mathbf{r}'(t)$. Look at the sage worksheet which gives a geometric meaning of $\mathbf{r}'(t)$ as the limit of the difference quotient

$$\frac{\mathbf{r}(t+\Delta t)-\mathbf{r}(t)}{\Delta t}.$$

For a vector $\mathbf{r}(t)=\cos t\mathbf{i}+\sin 2t\mathbf{j}+t\mathbf{k}$, we want to find a tangent vector at $t=0.6$. (See Figure 3.) It is a behavior of tangent vectors as t values near $t=0.6$. ■

http://matrix.skku.ac.kr/cal-lab/cal-12-2-Exm-2.html

```
t=var('t')
r=vector((cos(t),sin(2*t),t))
rprime=diff(r,t)
a=0.6
p = parametric_plot3d(r,(t,0,2),color='red',thickness=3)+point3d(r(t=a),size=15)
p = p + sum([arrow3d(r(t=a),r(t=a)+(r(t=a+h)-r(t=a))/h,width=0.3) for h in [0.01..1,step=0.1]])
p = p + arrow3d(r(t=a),r(t=a)+rprime(t=a),color='green',width=0.5)
show(p)
```

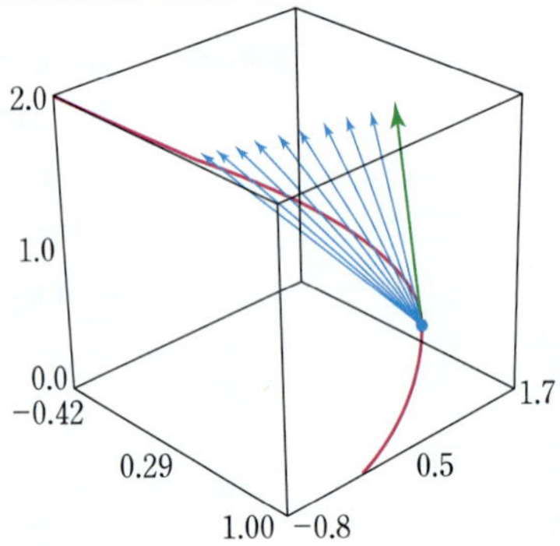

Figure 3

DEFINITION 3 Unit Tangent Vector

Let $\mathbf{r}(t)=f(t)\mathbf{i}+g(t)\mathbf{j}+h(t)\mathbf{k}$ be a smooth parameterized curve. If $\mathbf{r}(t)\neq 0$, then the unit tangent vector of $\mathbf{r}(t)$ is

$$\mathbf{T}(t)=\frac{\mathbf{r}'(t)}{|\mathbf{r}'(t)|}.$$

EXAMPLE 3

Find the unit tangent vector of $\mathbf{r}(t)=\sin t\,\mathbf{i}+\cos t\,\mathbf{j}+e^{-t}\mathbf{k}$ at the point $t=0$.

Solution According to Theorem 2, we differentiate each component of $\mathbf{r}$:

$$\mathbf{r}'(t) = \cos t\,\mathbf{i} - \sin t\,\mathbf{j} - e^{-t}\mathbf{k}.$$

Since $\mathbf{r}(0) = \mathbf{j} + \mathbf{k}$ and $\mathbf{r}'(0) = \mathbf{i} - \mathbf{k}$, the unit tangent vector at the point $(0, 1, 1)$ is

$$\mathbf{T}(0) = \frac{\mathbf{r}'(0)}{|\mathbf{r}'(0)|} = \frac{\mathbf{i} - \mathbf{k}}{\sqrt{1+1}} = \frac{1}{\sqrt{2}}\mathbf{i} - \frac{1}{\sqrt{2}}\mathbf{k}.$$ ■

EXAMPLE 4

For the curve $\mathbf{r}(t) = (t-1)\mathbf{i} + (t^2+2)\mathbf{j}$, find $\mathbf{r}'(t)$ and sketch the position vector $\mathbf{r}(1)$ and the tangent vector $\mathbf{r}'(1)$.

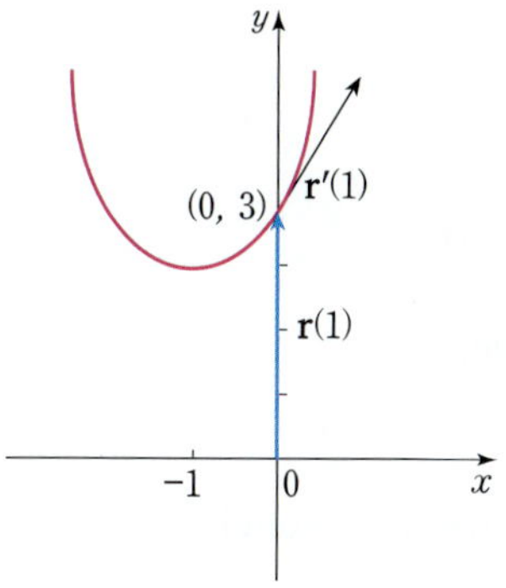

Figure 4

Solution We have

$$\mathbf{r}'(t) = \mathbf{i} + 2t\,\mathbf{j} \quad \text{and} \quad \mathbf{r}'(1) = \mathbf{i} + 2\mathbf{j}.$$

The curve is a plane curve and elimination of the parameter from the equations $x = t - 1$, $y = t^2 + 2$ gives $y = x^2 + 2x + 3$. In Figure 4, we draw the position vector $\mathbf{r}(1) = 3\mathbf{j}$ starting at the origin and the tangent vector $\mathbf{r}'(1) = \mathbf{i} + 2\mathbf{j}$ starting at the corresponding point $(0, 3)$. ■

EXAMPLE 5

Find parametric equations for the tangent line to the helix with parametric equations

$$x = \cos t, \quad y = 2\sin t, \quad z = t$$

at $t = \pi/3$.

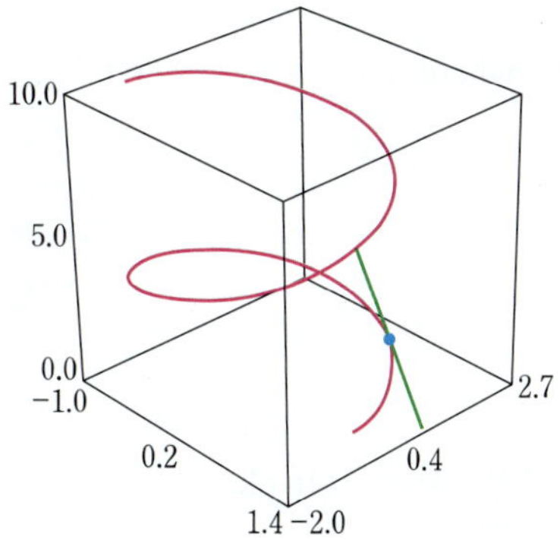

Figure 5

Solution The vector equation of the helix is

$$\mathbf{r}(t) = \langle \cos t,\ 2\sin t,\ t \rangle.$$

Hence $\mathbf{r}'(t) = \langle -\sin t, 2\cos t, 1 \rangle$. At $t = \pi/3$, the corresponding point is $(1/2, \sqrt{3}, \pi/3)$, so the tangent vector, at $t = \pi/3$ is $\mathbf{r}'(\pi/3) = \langle -\sqrt{3}/2, 1, 1 \rangle$. The tangent line is the line through $(1/2, \sqrt{3}, \pi/3)$ parallel to the vector $\langle -\sqrt{3}/2, 1, 1 \rangle$. Hence its parametric equations are

$$x = -\frac{\sqrt{3}}{2}t + \frac{1}{2}, \quad y = t + \sqrt{3}, \quad z = t + \frac{\pi}{3}.$$ ■

The following are the standard properties of derivatives of vector function.

THEOREM 4 Rules of Differentiation

Suppose that $\mathbf{u}$ and $\mathbf{v}$ are differentiable vector functions, c is a scalar, and f is a differentiable real-valued function. Then

(i) $\dfrac{d}{dt}[\mathbf{u}(t)+\mathbf{v}(t)]=\mathbf{u}'(t)+\mathbf{v}'(t)$

(ii) $\dfrac{d}{dt}[c\mathbf{u}(t)]=c\mathbf{u}'(t)$

(iii) $\dfrac{d}{dt}[f(t)\mathbf{u}(t)]=f'(t)\mathbf{u}(t)+f(t)\mathbf{u}'(t)$

(iv) $\dfrac{d}{dt}[\mathbf{u}(t)\cdot\mathbf{v}(t)]=\mathbf{u}'(t)\cdot\mathbf{v}(t)+\mathbf{u}(t)\cdot\mathbf{v}'(t)$

(v) $\dfrac{d}{dt}[\mathbf{u}(t)\times\mathbf{v}(t)]=\mathbf{u}'(t)\times\mathbf{v}(t)+\mathbf{u}(t)\times\mathbf{v}'(t)$

(vi) $\dfrac{d}{dt}\left[\mathbf{u}\left(f(t)\right)\right]=f'(t)\mathbf{u}'(f(t))$ (Chain Rule).

We will prove (iv) and leave the remaining proofs as exercises.

Proof Let $\mathbf{u}(t)=u_1(t)\,\mathbf{i}+u_2(t)\,\mathbf{j}+u_3(t)\,\mathbf{k}$ and $\mathbf{v}(t)=v_1(t)\,\mathbf{i}+v_2(t)\,\mathbf{j}+v_3(t)\,\mathbf{k}$.
Then, by the product rule

$$\begin{aligned}\frac{d}{dt}\left(\mathbf{u}(t)\cdot\mathbf{v}(t)\right)&=\frac{d}{dt}\left(u_1(t)v_1(t)+u_2(t)v_2(t)+u_3(t)v_3(t)\right)\\&=u_1'(t)v_1(t)+u_2'(t)v_2(t)+u_3'(t)v_3(t)\\&\quad+u_1(t)v_1'(t)+u_2(t)v_2'(t)+u_3(t)v_3'(t)\\&=\mathbf{u}'(t)\cdot\mathbf{v}(t)+\mathbf{u}(t)\cdot\mathbf{v}'(t).\end{aligned}$$

■

EXAMPLE 6

Find $\dfrac{d}{dt}(\mathbf{u}(t)\cdot\mathbf{v}(t))$ and $\dfrac{d}{dt}(\mathbf{u}(t)\times\mathbf{v}(t))$, where $\mathbf{u}(t)=\tan^{-1}t\,\mathbf{i}+5\mathbf{k}$, $\mathbf{v}(t)=\mathbf{i}+\ln t\,\mathbf{j}-2t\mathbf{k}$.

Solution Since $\mathbf{u}(t)\cdot\mathbf{v}(t)=\tan^{-1}t-10t$,

$$\frac{d}{dt}(\mathbf{u}(t)\cdot\mathbf{v}(t))=\frac{1}{1+t^2}-10.$$

And since

$\mathbf{u}(t)\times\mathbf{v}(t)=-5\ln t\,\mathbf{i}+(2t\tan^{-1}t+5)\mathbf{j}+\tan^{-1}t\ln t\,\mathbf{k}$,

$$\frac{d}{dt}(\mathbf{u}(t)\times\mathbf{v}(t))=-\frac{5}{t}\mathbf{i}+\left(2\tan^{-1}t+\frac{2t}{1+t^2}\right)\mathbf{j}+\left(\frac{\ln t}{1+t^2}+\frac{\tan^{-1}t}{t}\right)\mathbf{k}.$$

■

THEOREM 5

If $\mathbf{r}(t)$ is a vector function and $|\mathbf{r}(t)|=c$, where c is a constant for all t, then

$$\mathbf{r}'(t)\cdot\mathbf{r}(t)=0,$$

that is, $\mathbf{r}'(t)$ and $\mathbf{r}(t)$ are orthogonal vectors for all t.

Proof We recall that $|\mathbf{r}(t)|^2=\mathbf{r}(t)\cdot\mathbf{r}(t)=c^2$. It follows from Formula (iv) of Theorem 4 that

$$0=\frac{d}{dt}[\mathbf{r}(t)\cdot\mathbf{r}(t)]=\mathbf{r}'(t)\cdot\mathbf{r}(t)+\mathbf{r}(t)\cdot\mathbf{r}'(t)=2\left(\mathbf{r}'(t)\cdot\mathbf{r}(t)\right).$$

Then $\mathbf{r}'(t)\cdot\mathbf{r}(t)=0$, which says that $\mathbf{r}'(t)$ is orthogonal to $\mathbf{r}(t)$. ■

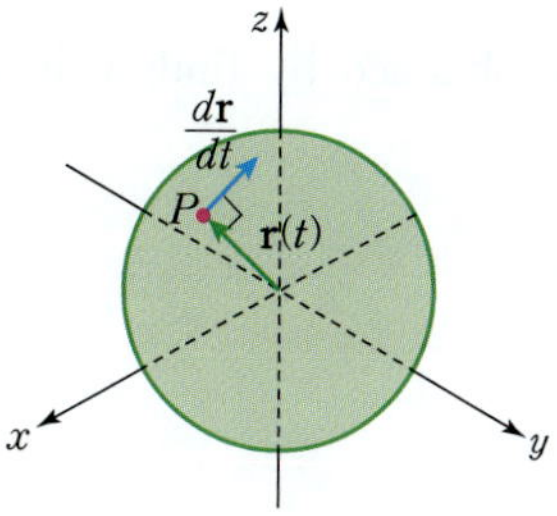

Figure 6

Remark Geometrically, Theorem 5 says that if a curve lies on a sphere with center at origin, then the tangent vector $\mathbf{r}'(t)$ is always perpendicular to the radial vector $\mathbf{r}(t)$. (See Figure 6.)

CAS **EXAMPLE 7**

Find the derivative of the vector function.

$$\mathbf{r}(t)=\cos^{-1}t\mathbf{i}+t^2\mathbf{j}+(1/\sqrt{1+t^2})\mathbf{k}.$$

http://matrix.skku.ac.kr/cal-lab/cal-13-2-Exm6.html

Solution

```
t=var('t')
x(t)=arccos(t)
y(t)=t^2
z(t)=1/sqrt(1+t^2)
(diff(x(t), t), diff(y(t), t), diff(z(t), t))
```

Answer : (-1/sqrt(-t^2 + 1), 2*t, -t/(t^2 + 1)^(3/2)) ■

Integrals of Vector Functions

A differentiable vector function $\mathbf{R}(t)$ is antiderivative of a vector function $\mathbf{r}(t)$ on an interval I if $\mathbf{R}'(t)=\mathbf{r}(t)$. If $\mathbf{r}(t)=f(t)\mathbf{i}+g(t)\mathbf{j}+h(t)\mathbf{k}$ then an antiderivative of $\mathbf{r}(t)$ is $\mathbf{R}(t)=F(t)\mathbf{i}+G(t)\mathbf{j}+H(t)\mathbf{k}$ where F, G and H are antiderivatives of f, g and h, respectively. Every antiderivative of $\mathbf{r}(t)$ on I has the form $\mathbf{R}(t)+C$ where C is an arbitrary constant vector. The set of all antiderivative of $\mathbf{r}(t)$ on I is the **indefinite integral** of $\mathbf{r}(t)$ on I.

DEFINITION 6 Indefinite Integral of a Vector Function

Let $\mathbf{r}(t)=f(t)\mathbf{i}+g(t)\mathbf{j}+h(t)\mathbf{k}$ be a vector function and $\mathbf{R}(t)=F(t)\mathbf{i}+G(t)\mathbf{j}+H(t)\mathbf{k}$, where F, G and H are anti- derivatives of f, g and h, respectively. Then the indefinite integral of $\mathbf{r}(t)$ is

$$\int \mathbf{r}(t)\,dt=\mathbf{R}(t)+\mathbf{C},$$

where $\mathbf{C}$ is an arbitrary constant vector.

Note that F, G and H in Definition 6 can be obtained by finding indefinite integrals of f, g and h, respectively.

EXAMPLE 8

If $\mathbf{r}(t)=\sec^2 t\,\mathbf{i}-2\cos t\,\mathbf{j}-\dfrac{1}{\sqrt{1-t^2}}\mathbf{k}$, then

$$\begin{aligned}\int \mathbf{r}(t)\,dt &= \left(\int \sec^2 t\,dt\right)\mathbf{i}-\left(\int 2\cos t\,dt\right)\mathbf{j}-\left(\int \frac{1}{\sqrt{1-t^2}}\,dt\right)\mathbf{k}\\ &= \tan t\,\mathbf{i}-2\sin t\,\mathbf{j}-\sin^{-1}t\,\mathbf{k}+\mathbf{C},\end{aligned}$$

where $\mathbf{C}$ is a vector constant of integration. ■

Definite integrals of a vector function are evaluated by applying the Fundamental Theorem of Calculus to each component of a vector function.

DEFINITION 7 Definite Integral of a Vector Function

If $\mathbf{r}(t)=f(t)\mathbf{i}+g(t)\mathbf{j}+h(t)\mathbf{k}$ is integrable (that is, it has anti-derivative) on $[a, b]$, then the definite integral of $\mathbf{r}(t)$ from a to b is given by

$$\int_a^b \mathbf{r}(t)\,dt=\left(\int_a^b f(t)\,dt\right)\mathbf{i}+\left(\int_a^b g(t)\,dt\right)\mathbf{j}+\left(\int_a^b h(t)\,dt\right)\mathbf{k}.$$

This means that an integral of a vector function is obtained by integrating each component function.

Remark The Fundamental Theorem of Calculus can be extended to continuous vector functions as follows :

$$\int_a^b \mathbf{r}(t)\,dt=[\mathbf{R}(t)]_a^b=\mathbf{R}(b)-\mathbf{R}(a)$$

where $\mathbf{R}$ is an anti-derivative of $\mathbf{r}$, that is, $\mathbf{R}'(t)=\mathbf{r}(t)$. We use the notation $\int \mathbf{r}(t)\,dt$ for indefinite integral of $\mathbf{r}(t)$, that is, the set of all anti-derivatives of $\mathbf{r}(t)$.

EXAMPLE 9

If $\mathbf{r}(t)=\sec^2 t\,\mathbf{i}-2\cos t\,\mathbf{j}-\mathbf{k}$, then

$$\begin{aligned}\int_0^{\frac{\pi}{4}}\mathbf{r}(t)\,dt &= \int_0^{\frac{\pi}{4}}(\sec^2 t\,\mathbf{i}-2\cos t\,\mathbf{j}-\mathbf{k})dt \\ &= [\tan t\,\mathbf{i}-2\sin t\,\mathbf{j}-t\,\mathbf{k}]_0^{\pi/4} \\ &= \mathbf{i}-\sqrt{2}\mathbf{j}-\frac{\pi}{4}\mathbf{k}.\end{aligned}$$

■

Properties of the Integrals of Vector Functions

Let $\mathbf{r}_1$ and $\mathbf{r}_2$ be continuous vector functions on $[a,b]$, c a scalar. Then the following hold:

(a) $\int_a^b c\mathbf{r}_1(t)dt = c\int_a^b \mathbf{r}_1(t)dt$,

(b) $\int_a^b [\mathbf{r}_1(t)+\mathbf{r}_2(t)]\,dt = \int_a^b \mathbf{r}_1(t)dt + \int_a^b \mathbf{r}_2(t)dt$,

(c) $\left|\int_a^b \mathbf{r}_1(t)dt\right| \le \int_a^b |r_1(t)|dt.$

<Nicolas Bourbaki>

12.2 EXERCISES (Calculus of Vector Functions)

http://matrix.skku.ac.kr/Cal-Book/part2/CS-Sec-12-2-Sol.html

YouTube http://youtu.be/VS5rPyOjP2I

CAS **1-5.** Find $\mathbf{r}'(t)$ and draw the position vector $\mathbf{r}(t)$ and the tangent vector $\mathbf{r}'(t)$ for the given value of t.

1. $\mathbf{r}(t)=\langle \cos t,\ t\rangle,\ t=\dfrac{\pi}{4};\ \mathbf{r}=\left(\dfrac{\pi}{4}\right)=\left\langle \dfrac{1}{\sqrt{2}},\ \dfrac{\pi}{4}\right\rangle$

http://matrix.skku.ac.kr/cal-lab/cal-13-2-1.html

Solution (-sin(t), 1)

So, we can see that $\mathbf{r}'(t)=(-\sin t, 1)$; $\mathbf{r}'\left(\dfrac{\pi}{4}\right)=\left\langle -\dfrac{1}{\sqrt{2}},\ 1\right\rangle$

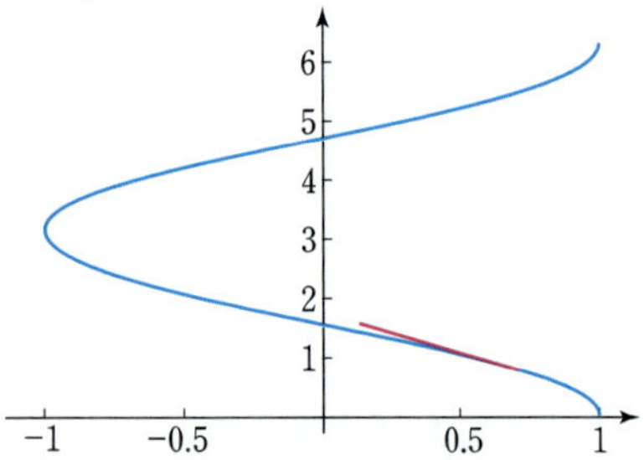

2. $\mathbf{r}(t)=\langle \sin t,\ t\rangle,\ t=\dfrac{\pi}{6};\ \mathbf{r}=\left(\dfrac{\pi}{6}\right)=\left\langle \dfrac{1}{2},\ \dfrac{\pi}{6}\right\rangle$

http://matrix.skku.ac.kr/cal-lab/cal-13-2-2.html

Solution

```
var('t')
x(t)=sin(t);y(t)=t
r(t)=(x(t), y(t))
dr(t)=(diff(x(t), t), diff(y(t), t));dr(t)
```

Answer : (cos(t), 1)

So, we can see that $\mathbf{r}'(t)=\langle \cos t, 1\rangle$; $\mathbf{r}'\left(\dfrac{\pi}{6}\right)=\left\langle \dfrac{\sqrt{3}}{2},\ 1\right\rangle$

```
t0=0
p1=r(t0)
p2=dr(t0)/abs(dr(t0))+r(t0)
p=parametric_plot((x(t), y(t)), (t, 0, 2*pi))
utv=line([p1, p2], rgbcolor=(1, 0, 0), thickness=2)
show(p+utv)
```

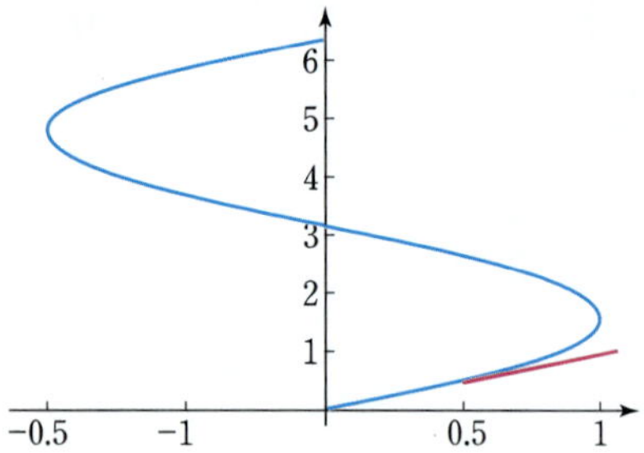

3. $\mathbf{r}(t)=t^3\mathbf{i}+t^2\mathbf{j},\ t=1;\ \mathbf{r}(1)=\langle 1,\ 1\rangle$

http://matrix.skku.ac.kr/cal-lab/cal-13-2-3.html

Solution

```
var('t')
x(t)=t^3;y(t)=t^2
r(t)=(x(t), y(t))
dr(t)=(diff(x(t), t), diff(y(t), t));dr(t)
```

Answer : (3*t^2, 2*t)

So, we can see that $\mathbf{r}'(t)=3t^2\mathbf{i}+2t\mathbf{j}$. $\mathbf{r}'(1)=\langle 3,\ 2\rangle$

```
t0=1
p1=r(t0)
p2=dr(t0)/abs(dr(t0))+r(t0)
p=parametric_plot((x(t), y(t)), (t, 0, 2))
utv=line([p1, p2], rgbcolor=(1, 0, 0), thickness=2)
show(p+utv)
```

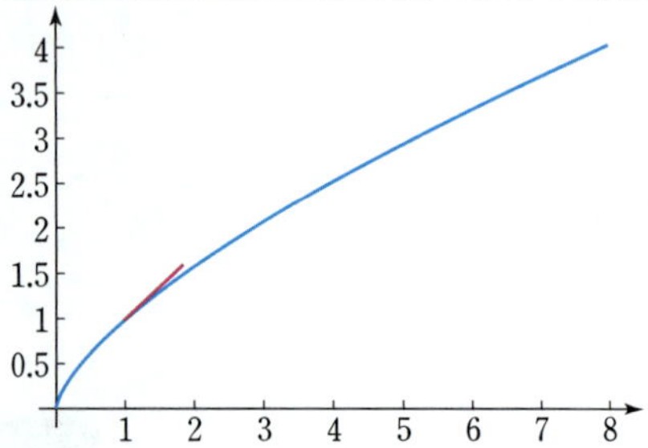

4. $\mathbf{r}(t)=e^{2t}\mathbf{i}+e^{-3t}\mathbf{j},\ t=0;\ \mathbf{r}(0)=\langle 1,\ 1\rangle$

http://matrix.skku.ac.kr/cal-lab/cal-13-2-4.html

Solution

```
var('t')
x(t)=exp(2*t)
y(t)=exp(-3*t)
r(t)=(x(t), y(t))
dr(t)=(diff(x(t), t), diff(y(t), t))
dr(t)
```

Answer : (2*exp(2*t), -3*exp(-3*t))

So, we can see that $\mathbf{r}'(t) = 2e^{2t}\mathbf{i} - 3e^{-3t}\mathbf{j}$;
$\mathbf{r}'(0) = \langle 2, -3 \rangle$

```
t0=0
p1=r(t0)
p2=dr(t0)/abs(dr(t0))+r(t0)
p=parametric_plot((x(t), y(t)), (t, -2, 2))
utv=line([p1, p2], rgbcolor=(1, 0, 0), thickness=2)
show(p+utv)
```

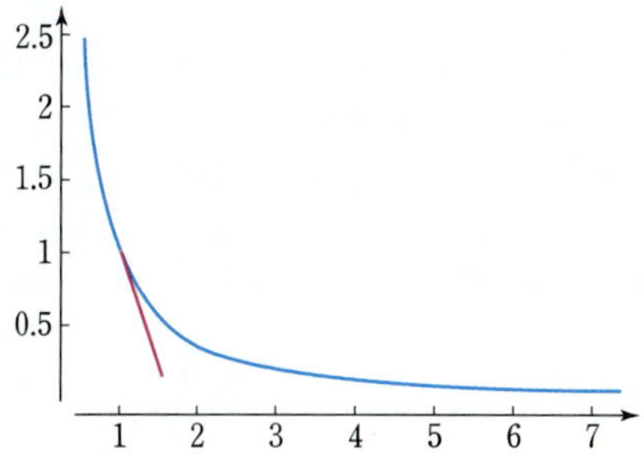

5. $\mathbf{r}(t) = 3\sin t\mathbf{i} + 2\cos t\mathbf{j}$, $t = \frac{\pi}{3}$;

$\mathbf{r}\left(\frac{\pi}{3}\right) = \left\langle \frac{3\sqrt{3}}{2}, 1 \right\rangle$

http://matrix.skku.ac.kr/cal-lab/cal-13-2-5.html

Solution

```
var('t')
x(t)=3*sin(t)
y(t)=2*cos(t)
r(t)=(x(t), y(t))
dr(t)=(diff(x(t), t), diff(y(t), t))
dr(t)
```

Answer : (3*cos(t), -2*sin(t))

So, we can see that $\mathbf{r}'(t) = 3\cos t\mathbf{i} - 2\sin t\mathbf{j}$;

$\mathbf{r}'\left(\frac{\pi}{3}\right) = \left\langle \frac{3}{2}, -\sqrt{3} \right\rangle$

```
t0=pi/3
p1=r(t0)
p2=dr(t0)/abs(dr(t0))+r(t0)
p=parametric_plot((x(t), y(t)), (t, -2*pi, 2*pi))
utv=line([p1, p2], rgbcolor=(1, 0, 0), thickness=2)
show(p+utv)
```

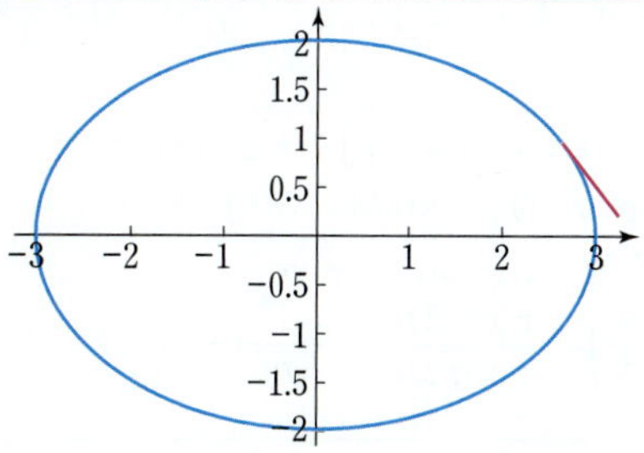

6-10. Find the derivatives of the vector functions.

6. $\mathbf{r}(t) = \langle 2+t, t^3, \sqrt{t} \rangle$.

Solution $\mathbf{r}'(t) = \left\langle 1, 3t^2, \frac{1}{2\sqrt{t}} \right\rangle$.

7. $\mathbf{r}(t) = \langle t, \cos 2t, \sin 2t \rangle$.

Solution $\mathbf{r}'(t) = \langle 1, -2\sin 2t, 2\cos 2t \rangle$.

8. $\mathbf{r}(t) = \tan^{-1}t\mathbf{i} + \mathbf{j} + \sqrt{1+t^2}\mathbf{k}$.

Solution $\mathbf{r}'(t) = \frac{1}{1+t^2}\mathbf{i} + \frac{t}{\sqrt{1+t^2}}\mathbf{k}$.

9. $\mathbf{r}(t) = e^{t^2}\mathbf{i} - e^{-t}\mathbf{j} + \ln(1+t)\mathbf{k}$.

Solution $\mathbf{r}'(t) = 2te^{t^2}\mathbf{i} + e^{-t}\mathbf{j} + \frac{1}{1+t}\mathbf{k}$.

10. $\mathbf{r}(t) = at\sin 3t\mathbf{i} + b\sin^3 t\mathbf{j} + c\cos^3 t\mathbf{k}$.

Solution $\mathbf{r}'(t) = (3at\cos 3t + a\sin 3t)\mathbf{i} + 3b\sin^2 t\cos t\mathbf{j}$
$-3c\sin t\cos^2 t\mathbf{k}$.

11-13. Find the unit tangent vector $\mathbf{T}(t)$ at the point with the given value of the parameter t.

11. $\mathbf{r}(t) = \left\langle \frac{1}{3}t^3, \frac{1}{2}t^2, t \right\rangle$, $t = 0$.

Solution $\mathbf{r}'(t) = \langle t^2, t, 1 \rangle$,

$|\mathbf{r}'(t)| = \sqrt{t^4 + t^2 + 1}$,

Hence $\mathbf{T}(0) = \frac{\mathbf{r}(0)}{|\mathbf{r}'(0)|} = \langle 0, 0, 1 \rangle$.

12. $\mathbf{r}(t) = \sqrt{2}\ln t\,\mathbf{i} + \frac{1}{t}\mathbf{j} + t\,\mathbf{k},\ t=1.$

Solution $\mathbf{r}'(t) = \frac{\sqrt{2}}{t}\mathbf{i} - \frac{1}{t^2}\mathbf{j} + \mathbf{k}.$

$|\mathbf{r}'(t)| = \sqrt{\frac{2}{t^2} + \frac{1}{t^4} + 1}.$

Therefore, $\mathbf{T}(1) = \frac{\mathbf{r}'(1)}{|\mathbf{r}'(1)|} = \frac{\sqrt{2}}{2}\mathbf{i} - \frac{1}{2}\mathbf{j} + \frac{1}{2}\mathbf{k}.$

13. $\mathbf{r}(t) = \cos t\,\mathbf{i} + \sin t\,\mathbf{j} + t\,\mathbf{k},\ t = \pi/2.$

Solution $\mathbf{r}'(t) = -\sin t\,\mathbf{i} + \cos t\,\mathbf{j} + \mathbf{k},$

$|\mathbf{r}'(t)| = \sqrt{(-\sin t)^2 + \cos^2 t + 1} = \sqrt{2},$

$\mathbf{T}\left(\frac{\pi}{2}\right) = \frac{\mathbf{r}'(\pi/2)}{|\mathbf{r}'(\pi/2)|} = -\frac{1}{\sqrt{2}}\mathbf{i} + \frac{1}{\sqrt{2}}\mathbf{k}.$

---

14. If $\mathbf{r}(t) = \langle t^2,\ t,\ t^3\rangle$, find $\mathbf{r}'(t)$, $\mathbf{T}(1)$, $\mathbf{r}''(t)$, and $\mathbf{r}'(t)\times\mathbf{r}''(t)$.

Solution $\mathbf{r}'(t) = \langle 2t,\ 1,\ 3t^2\rangle$, $|\mathbf{r}'(t)| = \sqrt{9t^4 + 4t^2 + 1}.$

Therefore, $\mathbf{T}(1) = \frac{\mathbf{r}'(1)}{|\mathbf{r}'(1)|} = \frac{\langle 2,\ 1,\ 3\rangle}{\sqrt{14}}$

$= \left\langle \frac{2}{\sqrt{14}},\ \frac{1}{\sqrt{14}},\ \frac{3}{\sqrt{14}} \right\rangle$, $\mathbf{r}''(t) = \langle 2\ ,0,\ 6t\rangle.$

$\mathbf{r}'(t)\times\mathbf{r}''(t) = \begin{vmatrix} \mathbf{i} & \mathbf{j} & \mathbf{k} \\ 2t & 1 & 3t^2 \\ 2 & 0 & 6t \end{vmatrix} = 6t\mathbf{i} - 6t^2\mathbf{j} - 2\mathbf{k}$

$= \langle 6t,\ -6t^2,\ -2\rangle.$

CAS **15.** If $\mathbf{r}(t) = \langle e^t\cos t,\ e^t\sin t,\ e^t\rangle$, find $\mathbf{T}(0)$, $\mathbf{r}''(0)$ and $\mathbf{r}'(t)\cdot\mathbf{r}''(t)$.

Solution $\mathbf{r}'(t) = \langle e^t(\cos t - \sin t),\ e^t(\sin t + \cos t),\ e^t\rangle.$

$|\mathbf{r}'(t)| = \sqrt{e^{2t}[(\cos t - \sin t)^2 + (\cos t + \sin t)^2] + e^{2t}}$

$= \sqrt{3}\,e^t.$

$\mathbf{T}(0) = \frac{\mathbf{r}'(0)}{|\mathbf{r}'(0)|} = \frac{\langle 1,\ 1,\ 1\rangle}{\sqrt{3}} = \left\langle \frac{1}{\sqrt{3}},\ \frac{1}{\sqrt{3}},\ \frac{1}{\sqrt{3}} \right\rangle.$

$\mathbf{r}''(t) = \langle -2e^t\sin t,\ 2e^t\cos t,\ e^t\rangle.$

$\mathbf{r}''(0) = \langle 0,\ 2,\ 1\rangle.$

$\mathbf{r}'(t)\cdot\mathbf{r}''(t) = 3e^{2t}.$

http://matrix.skku.ac.kr/cal-lab/cal-12-2-Exs-15.html

```
var('t')
r=vector([e^t*cos(t),e^t*sin(t),e^t])
rprime=diff(r,t)
show(rprime)
show(rprime.norm())
T=r/rprime.norm()
 show(T(t=0))
rrprime=diff(r,t,2)
show(rrprime(t=0))
S=rprime.dot_product(rrprime)
show(S.simplify_trig())
```

Answer : $(e^t(\cos t - \sin t),\ e^t(\sin t + \cos t),\ e^t)$

$\sqrt{e^{2t}[(\cos t - \sin t)^2 + (\cos t + \sin t)^2] + e^{2t}}$

$\left(\frac{1}{\sqrt{3}},\ \frac{1}{\sqrt{3}},\ \frac{1}{\sqrt{3}}\right)$

$(0,\ 2,\ 1)$

$3e^{2t}$

16-19. Find parametric equations for the tangent line to the curve with the given parametric equations at the specified point.

16. $x = t^2,\ y = t^3,\ z = t^4;\ (1,\ 1,\ 1).$

Solution We can see that $t = 1$.

For $\mathbf{r}(t) = \langle t^2,\ t^3,\ t^4\rangle$, $\mathbf{r}'(t) = \langle 2t,\ 3t^2,\ 4t^3\rangle.$

So, $\mathbf{r}'(1) = \langle 2,\ 3,\ 4\rangle.$

Hence $T(s) = \langle 1,\ 1,\ 1\rangle + s\langle 2,\ 3,\ 4\rangle$

$= \langle 1+2s,\ 1+3s,\ 1+4s\rangle.$

17. $x = e^{-t},\ y = e^{-t}\cos t,\ z = e^{-t}\sin t;\ (1,\ 1,\ 0).$

Solution We can see that $t = 0$.

For $\mathbf{r}(t) = \langle e^{-t},\ e^{-t}\cos t,\ e^{-t}\sin t\rangle,$

$\mathbf{r}'(t) = \langle -e^{-t},\ -e^{-t}(\cos t + \sin t),\ e^{-t}(\cos t - \sin t)\rangle.$

Therefore, $\mathbf{r}'(0) = \langle -1,\ -1,\ 1\rangle$

Hence

$T(s) = \langle 1,\ 1, 0\rangle + s\langle -1,\ -1,\ 1\rangle = \langle 1-s,\ 1-s, s\rangle.$

18. $x = t^2,\ y = -\frac{1}{t+1},\ z = 4 - t^2;\ (4,1,0).$

Solution We can see that $t = -2$ by solving $t^2 = 4$, $-\frac{1}{1+t} = 0,\ 4 - t^2 = 0.$

For $\mathbf{r}(t) = \left\langle t^2,\ -\frac{1}{t+1},\ 4 - t^2 \right\rangle,$

$\mathbf{r}'(t) = \left\langle 2t,\ \frac{1}{(t+1)^2},\ -2t \right\rangle.$

So, $\mathbf{r}'(-2) = \langle -4,\ 1,\ 4\rangle.$

Hence

$T(s) = \langle 4,\ 1,\ 0\rangle + s\langle -4,\ 1,\ 4\rangle = \langle 4-4s,\ 1+s,\ 4s\rangle.$

19. $x = \ln t,\ y = 4t,\ z = t^2;\ (0,\ 4,\ 1).$

Solution We can see that $t = 1$.

For $\mathbf{r}(t) = \langle \ln t,\ 4t,\ t^2\rangle$, $\mathbf{r}'(t) = \left\langle \frac{1}{t},\ 4,\ 2t \right\rangle.$

So, $\mathbf{r}'(1) = \langle 1,\ 4,\ 2\rangle.$

Hence

$T(s) = \langle 0,\ 4,\ 1\rangle + s\langle 1,\ 4,\ 2\rangle = \langle s,\ 4+4s,\ 1+2s\rangle.$

CAS **20-21.** Find parametric equations for the tangent line to the curve with the given parametric equations at the specified point. Illustrate by graphing both the curve and the tangent line together.

20. $x=t,\ y=2\cos t,\ z=2\sin t;$ $(\pi/3,\ 1,\ \sqrt{3})$.
http://matrix.skku.ac.kr/cal-lab/cal-13-2-20.html

Solution We can see that $t=\dfrac{\pi}{3}$.

```
var('t,s')
x(t)=t; y(t)=2*cos(t); z(t)=2*sin(t);
r(t)=(x(t), y(t), z(t));
dr(t)=(diff(x(t), t), diff(y(t), t), diff(z(t), t));
t0=pi/3;
pe=r(t0)+dr(t0)*s;
print pe
```

(1/3*pi + s, -sqrt(3)*s + 1, s + sqrt(3))
So, parametric equation is
$(x,\ y,\ z)=(\pi/3+s,\ 1-\sqrt{3}s,\ \sqrt{3}+s)$.

```
p1=parametric_plot3d(r(t), (t, -pi, pi));
p2=parametric_plot3d(pe, (s, -pi, pi), color='red');
show(p1+p2);
```

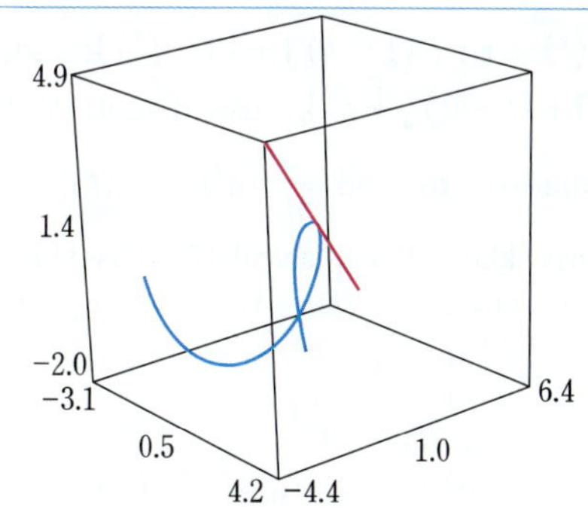

21. $x=e^t,\ y=\sin t,\ z=2e^{-3t};$ $(1,0,2)$.
http://matrix.skku.ac.kr/cal-lab/cal-13-2-21.html

Solution We can see that $t=0$.

```
var('t,s')
x(t)=exp(t); y(t)=sin(t); z(t)=2*exp(-3*t)
r(t)=(x(t), y(t), z(t));
dr(t)=(diff(x(t), t), diff(y(t), t), diff(z(t), t))
t0=0
pe=r(t0)+dr(t0)*s; print pe
```

(1 + s, s, 2 -6*s)
So, parametric equation is
$(x,y,z)=(s+1,\ s,\ -6s+2)$.

```
p1=parametric_plot3d(r(t), (t, -2, 2));
p2=parametric_plot3d(pe, (s, -2, 2), color='red');
show(p1+p2);
```

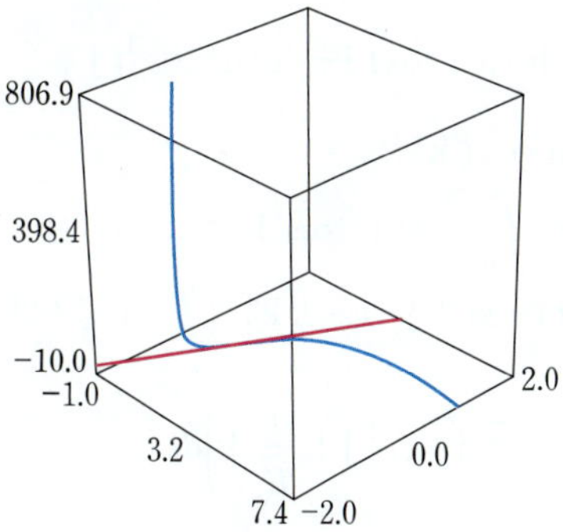

22-24. Determine whether the following curves are smooth.

22. $\mathbf{r}(t)=\langle t^2,\ t^3,\ t^4\rangle$.

Solution $\mathbf{r}'(t)=\langle 2t,3t^2,4t^3\rangle$. Since $\mathbf{r}'(0)=0$, this curve is not smooth.

23. $\mathbf{r}(t)=\langle t^3+t,\ t^2,\ t^3\rangle$.

Solution $\mathbf{r}'(t)=\langle 3t^2+1,2t,3t^2\rangle$.
First coordinate of $\mathbf{r}'(t)$ cannot be zero for any real number t, so this curve is smooth.

24. $\mathbf{r}(t)=\langle \cos t,\ e^t\rangle$.

Solution $\mathbf{r}'(t)=\langle -\sin t,\ e^t\rangle$.
Second coordinate of $\mathbf{r}'(t)$ cannot be zero for any real number t, so this curve is smooth.

25-30. Evaluate the integrals.

25. $\displaystyle\int_0^1 (3t^2\,\mathbf{i}+4t^3\,\mathbf{j}+5t^4\,\mathbf{k})dt$.
http://matrix.skku.ac.kr/cal-lab/cal-12-2-Exs-25.html

Solution $\displaystyle\int_0^1 (3t^2\,\mathbf{i}+4t^3\,\mathbf{j}+5t^4\,\mathbf{k})dt=t^3\,\mathbf{i}+t^4\,\mathbf{j}+t^5\,\mathbf{k}\Big|_{t=0}^{t=1}$
$=\mathbf{i}+\mathbf{j}+\mathbf{k}$.

26. $\displaystyle\int_0^1\left(\frac{t}{4+t^2}\,\mathbf{i}+\frac{1}{4+t^2}\,\mathbf{j}\right)dt$.

Solution $\displaystyle\int_0^1\left(\frac{t}{4+t^2}\,\mathbf{i}+\frac{1}{4+t^2}\,\mathbf{j}\right)dt$
$\displaystyle=\frac{1}{2}\ln(4+t^2)\,\mathbf{i}+\frac{1}{2}\tan^{-1}\frac{t}{2}\,\mathbf{j}\Big|_{t=0}^{t=1}$
$\displaystyle=\frac{\ln(5/4)}{2}\,\mathbf{i}+\frac{\tan^{-1}(1/2)}{2}\,\mathbf{j}$.

27. $\int_0^{\pi/2} (4\sin^3 t\cos t\,\mathbf{i} + 4\sin t\cos^3 t\,\mathbf{j} + 4\sin t\cos t\,\mathbf{k})dt$.

Solution $\int_0^{\pi/2} (4\sin^3 t\cos t\mathbf{i} + 4\sin t\cos^3 t\mathbf{j} + 4\sin t\cos t\mathbf{k})dt$

$= [\sin^4 t\,\mathbf{i} - \cos^4 t\,\mathbf{j} - \cos 2t\,\mathbf{k}]_{t=0}^{t=\pi/2} = \mathbf{i}+\mathbf{j}+2\mathbf{k}$.

28. $\int \left(\frac{3}{2}\sqrt{t}\,\mathbf{i} + te^t\,\mathbf{j} + \frac{1}{t^2}\,\mathbf{k}\right)dt$.

Solution

$\int \left(\frac{3}{2}\sqrt{t}\,\mathbf{i} + te^t\mathbf{j} + \frac{1}{t^2}\,\mathbf{k}\right)dt = t^{3/2}\mathbf{i} + (t-1)e^t\mathbf{j} - \frac{1}{t}\mathbf{k} + \mathbf{C}$

where $\mathbf{C}$ is a constant vector.

29. $\int (3t^2\,\mathbf{i} + 2e^{2t}\,\mathbf{j} + \ln t\,\mathbf{k})dt$.

Solution

$\int (3t^2\,\mathbf{i} + 2e^{2t}\mathbf{j} + \ln t\,\mathbf{k})dt = t^3\mathbf{i} + e^{2t}\,\mathbf{j} + t(\ln t - 1)\,\mathbf{k} + \mathbf{C}$

where $\mathbf{C}$ is a constant vector.

30. $\int_0^{\pi} (\mathbf{i} + \cos t\,\mathbf{j} - 2t\,\mathbf{k})dt$.

Solution

$\int_0^{\pi} (\mathbf{i} + \cos t\mathbf{j} - 2t\mathbf{k})dt = t\mathbf{i} + \sin t\mathbf{j} - t^2\mathbf{k}\Big|_{t=0}^{t=\pi} = \pi\mathbf{i} - \pi^2\mathbf{k}$.

---

31. Find $\mathbf{r}(t)$ if $\mathbf{r}'(t) = 16t^3\,\mathbf{i} - 9t^2\,\mathbf{j} + 25t^4\,\mathbf{k}$ and $\mathbf{r}(0) = 3\mathbf{j}$.

Solution

$\int (16t^3\,\mathbf{i} - 9t^2\,\mathbf{j} + 25t^4\,\mathbf{k})dt = 4t^4\,\mathbf{i} - 3t^3\,\mathbf{j} + 5t^5\,\mathbf{k} + \mathbf{C} = \mathbf{r}(t)$,

where $\mathbf{C}$ is a constant vector.

Since $3\mathbf{j} = \mathbf{r}(0) = \mathbf{C}$, we finally have

$$\mathbf{r}(t) = 4t^4\,\mathbf{i} - 3(t^3-1)\,\mathbf{j} + 5t^5\,\mathbf{k}.$$

32. Find $\mathbf{r}(t)$ if $\mathbf{r}'(t) = \pi\cos\pi t\,\mathbf{i} + \pi\sin\pi t\,\mathbf{j} + 2t\,\mathbf{k}$ and $\mathbf{r}(0) = \mathbf{i} - \mathbf{j} + 2\mathbf{k}$.

http://matrix.skku.ac.kr/cal-lab/cal-12-2-Exs-33.html

Solution

$\int \pi\cos\pi t\mathbf{i} + \pi\sin\pi t\mathbf{j} + 2t\mathbf{k} = \sin\pi t\mathbf{i} - \cos\pi t\mathbf{j} + t^2\mathbf{k} + \mathbf{C}$

$= \mathbf{r}(t)$, where $\mathbf{C}$ is a constant vector.

Since $\mathbf{r}(0) = -\mathbf{j} + \mathbf{C} = \mathbf{i} - \mathbf{j} + 2\mathbf{k}$, $\mathbf{C} = \mathbf{i} + 2\mathbf{k}$.

So, we finally have

$\mathbf{r}(t) = (\sin\pi t + 1)\mathbf{i} - \cos\pi t\mathbf{j} + (t^2+2)\mathbf{k}$.

CAS **33.** If $\mathbf{u}(t) = t\mathbf{i} + t^2\mathbf{j} + t^3\mathbf{k}$ and $\mathbf{v}(t) = \sin t\mathbf{i} + \sin 2t\mathbf{j} + t\mathbf{k}$, use Theorem 4(Rules of Differentiation) to find $\frac{d}{dt}[\mathbf{u}(t) \cdot \mathbf{v}(t)]$.

Solution $\mathbf{u}'(t) = \mathbf{i} + 2t\,\mathbf{j} + 3t^2\,\mathbf{k}$,

```
int _{} ^{} {} ( pi cos pi t bold{\i} + pi sin pi
t bold{₩j} +2t bold{₩k})dt =sin pi t bold{₩i}
-cos pi t bold{₩j} +t ^{2} bold{₩k}
+{rmbold C}= bold{₩r} (t)
```

$\mathbf{v}'(t) = \cos t\,\mathbf{i} + 2\cos 2t\,\mathbf{j} + \mathbf{k}$.

$\mathbf{u}'(t) \cdot \mathbf{v}(t) = \sin t + 2t\sin 2t + 3t^3$,

$\mathbf{u}(t) \cdot \mathbf{v}'(t) = t\cos t + 2t^2\cos 2t + t^3$.

So, $\frac{d}{dt}[\mathbf{u}(t) \cdot \mathbf{v}(t)] = \mathbf{u}'(t) \cdot \mathbf{v}(t) + \mathbf{u}(t) \cdot \mathbf{v}'(t)$

$= \sin t + 2t\sin 2t + t\cos t + 2t^2\cos 2t + 4t^3$.

```
var('t')
u=vector([t,t^2,t^3])
v=vector([sin(t),sin(2*t),t])
r=u.dot_product(v)
rprime=diff(r,t)
show(rprime);
```

Answer : $\sin t + 2t\sin 2t + t\cos t + 2t^2\cos 2t + 4t^3$

CAS **34.** If $\mathbf{u}(t) = t\,\mathbf{i} + (1-t)\,\mathbf{j} + (3+t^2)\,\mathbf{k}$ and $\mathbf{v}(t) = (3-t)\,\mathbf{i} + (t-2)\,\mathbf{j} + t^2\,\mathbf{k}$, use Theorem 2(Rules of differentiation) to find $\frac{d}{dt}[\mathbf{u}(t) \times \mathbf{v}(t)]$.

http://matrix.skku.ac.kr/cal-lab/cal-12-2-Exs-34.html

Solution $\mathbf{u}'(t) = \mathbf{i} - \mathbf{j} + 2t\mathbf{k}$, $\mathbf{v}'(t) = -\mathbf{i} + \mathbf{j} + 2t\mathbf{k}$,

$\mathbf{u}'(t) \times \mathbf{v}(t) = \begin{vmatrix} \mathbf{i} & \mathbf{j} & \mathbf{k} \\ 1 & -1 & 2t \\ 3-t & t-2 & t^2 \end{vmatrix}$

$= (-3t^2+4t)\,\mathbf{i} + (6t-3t^2)\,\mathbf{j} + \mathbf{k}$,

$\mathbf{u}(t) \times \mathbf{v}'(t) = \begin{vmatrix} \mathbf{i} & \mathbf{j} & \mathbf{k} \\ t & 1-t & 3+t^2 \\ -1 & 1 & 2t \end{vmatrix}$

$= (-3+2t-3t^2)\,\mathbf{i} - (3+3t^2)\,\mathbf{j} + \mathbf{k}$,

$\frac{d}{dt}[\mathbf{u}(t) \times \mathbf{v}(t)] = (\mathbf{u}'(t) \times \mathbf{v}(t)) + (\mathbf{u}(t) \times \mathbf{v}'(t))$

$= (-6t^2+6t-3)\,\mathbf{i} + (-6t^2+6t-3)\,\mathbf{j} + 2\mathbf{k}$.

```
var('t')
u=vector([t,1-t,3+t^2])
v=vector([3-t,t-2,t^2])
r=u.cross_product(v)
rprime=diff(r,t);
show((expand(rprime[0]),expand(rprime[1]),expand
(rprime[2])))
```

Answer : $(-6t^2+6t-3,\ -6t^2+6t-3,\ 2)$

35. Show that if $\mathbf{r}(t)$ is a vector function such that

$\mathbf{r}''(t)$ exists, then $\frac{d}{dt}[\mathbf{r}(t)\times\mathbf{r}'(t)] = \mathbf{r}(t)\times\mathbf{r}''(t)$.

Proof Using Theorem 4 (Rules of Differentiation),

$$\frac{d}{dt}[\mathbf{r}(t)\times\mathbf{r}'(t)] = \mathbf{r}'(t)\times\mathbf{r}'(t)+\mathbf{r}(t)\times\mathbf{r}''(t)$$
$$=\mathbf{r}(t)\times\mathbf{r}''(t).$$

36. If $\mathbf{r}(t)\neq 0$, show that

$$\frac{d}{dt}|\mathbf{r}(t)| = \frac{1}{|\mathbf{r}(t)|}(\mathbf{r}(t)\cdot\mathbf{r}'(t)).$$

Proof We already know that $|\mathbf{r}(t)|^2 = \mathbf{r}(t)\cdot\mathbf{r}(t)$. Differentiating both sides,

$$2|\mathbf{r}(t)|\frac{d}{dt}|\mathbf{r}(t)| = \mathbf{r}'(t)\cdot\mathbf{r}(t)+\mathbf{r}(t)\cdot\mathbf{r}'(t) = 2\mathbf{r}(t)\cdot\mathbf{r}'(t).$$

So, we finally have $\frac{d}{dt}|\mathbf{r}(t)| = \frac{1}{|\mathbf{r}(t)|}(\mathbf{r}(t)\cdot\mathbf{r}'(t))$.

CAS **37.** Find the unit tangent vector to the curve $x=4\sin(-t)$, $y=2\cos 2t$, $z=5t$ at any time $t>0$.

http://matrix.skku.ac.kr/cal-lab/cal-14-1-19.html

Solution

```
x,y,z,t,u=var('x,y,z,t,u')
x(t)=4*sin(-t)
y(t)=2*cos(2*t)
z(t)=5*t
r(t)=(x(t),y(t),z(t))
r(t)
```

Answer : (4*sin(-t), 2*cos(2*t), 5*t)

```
dr(t)=(diff(x(t),t),diff(y(t),t),diff(z(t),t)); # tangent vector(dr/dt)
dr(t)
```

(-4*cos(-t), -4*sin(2*t), 5)

```
dr(t)/abs(dr(t))+r(t)    # unit tangent vector((dr/dt)/abs(dr/dt))
```

Answer : (-4*cos(-t)/sqrt(16*sin(2*t)^2 + 16*cos(-t)^2 + 25) + 4*sin(-t), + 16*cos(-t)^2 + 25) + 2*cos(2*t), 5*t + 5/sqrt(16*sin(2*t)^2 + 16*cos(-t)^2 + 25))

And we can draw a curve with given unit tangent vector.

```
t0=2
p1=r(t0)
p2=dr(t0)/abs(dr(t0))+r(t0)
p=parametric_plot3d((x(t),y(t),z(t)),(t,0,2*pi))
utv=line3d([p1,p2],rgbcolor=(1,0,0),arrow_head=True, thickness=5)
show(p+utv)
```

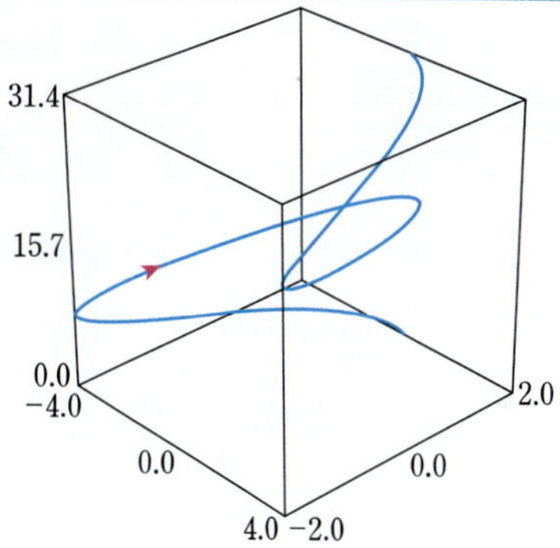

12.3 Arc Length and Curvature

The length of a space curve is defined in exactly the same way as a plane curve. Suppose that a curve C has the vector equation $\mathbf{r}(t) = \langle f(t), g(t), h(t) \rangle$, $a \le t \le b$, or, equivalently, the parametric equations

$$x = f(t),\ y = g(t),\ z = h(t)$$

where f', g' and h' are continuous. If the curve C is traced exactly once as t increases from a to b, then its length of C is defined as

1
$$L = \int_a^b \sqrt{[f'(t)]^2 + [g'(t)]^2 + [h'(t)]^2}\, dt = \int_a^b \sqrt{\left(\frac{dx}{dt}\right)^2 + \left(\frac{dy}{dt}\right)^2 + \left(\frac{dz}{dt}\right)^2}\, dt.$$

Notice that for space curves $\mathbf{r}(t) = f(t)\,\mathbf{i} + g(t)\,\mathbf{j} + h(t)\,\mathbf{k}$,

$$|\mathbf{r}'(t)| = |f'(t)\,\mathbf{i} + g'(t)\,\mathbf{j} + h'(t)\,\mathbf{k}| = \sqrt{[f'(t)]^2 + [g'(t)]^2 + [h'(t)]^2}$$

and hence the arc length formula can be also be written as

2
$$L = \int_a^b |\mathbf{r}'(t)|\, dt.$$

EXAMPLE 1

Find the length of the arc of the circular helix with vector equation $\mathbf{r}(t) = 2\cos t\,\mathbf{i} + 2\sin t\,\mathbf{j} + t\,\mathbf{k}$ from the point $(2, 0, 0)$ to the point $(2, 0, 2\pi)$.
http://matrix.skku.ac.kr/cal-lab/cal-12-3-Exm-1.html

Solution The arc from $(2, 0, 0)$ to $(2, 0, 2\pi)$ is described by the parameter interval $0 \le t \le 2\pi$. Hence from Equation 2, we have

$$L = \int_0^{2\pi} |\mathbf{r}'(t)|\, dt = \int_0^{2\pi} \sqrt{(-2\sin t)^2 + (2\cos t)^2 + 1^2}\, dt = \int_0^{2\pi} \sqrt{5}\, dt = 2\sqrt{5}\,\pi.$$

```
var('t')
r=vector([2*cos(t),2*sin(t),t])
a=0
b=2*pi
rprime=diff(r,t)
L=integral(rprime.norm(),t,a,b)
show(L)
```

Answer : $2\sqrt{5}\,\pi$ ■

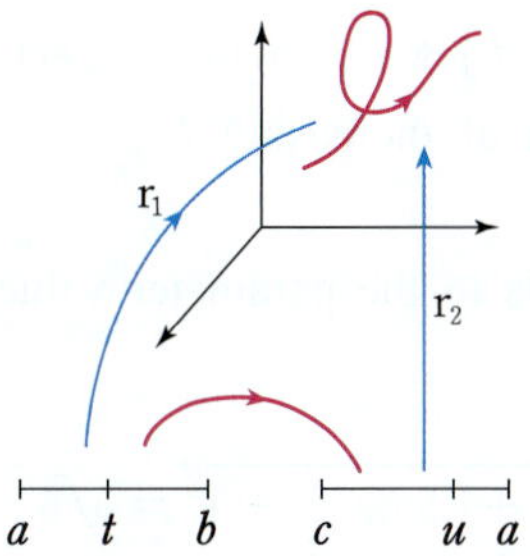

Figure 1

Remark A single curve C can be represented by more than one vector function. Fortunately, we have the same arc length no matter which parameterization of the curve is used in 2. We can easily show this as follows. Suppose that

$$\mathbf{r}=\mathbf{r}_1(t),\quad t\in[a,\ b]\ \text{and also}\ \ \mathbf{r}=\mathbf{r}_2(u),\quad u\in[c,\ d]$$

where $t=h(u)$ is such that $a=h(c)$ and $b=h(d)$, and h is continuous. Then we have

$$\mathbf{r}=\mathbf{r}_1(t)=\mathbf{r}_1(h(u)).$$

Thus from Theorem 4 (vi) in Section 12.2,

$$\frac{d\mathbf{r}}{du}=\frac{d\mathbf{r}_1}{dt}\frac{dt}{du}.$$

The technique of change of variables from calculus gives us

$$\int_c^d\left|\frac{d\mathbf{r}}{du}\right|du=\int_{h(c)}^{h(d)}\left|\frac{d\mathbf{r}_1}{dt}\frac{dt}{du}\right|du=\int_a^b\left|\frac{d\mathbf{r}}{dt}\right|dt.$$

Now let C be a curve given by a vector function

$$\mathbf{r}(t)=\langle f(t),\ g(t),\ h(t)\rangle,\ a\le t\le b,$$

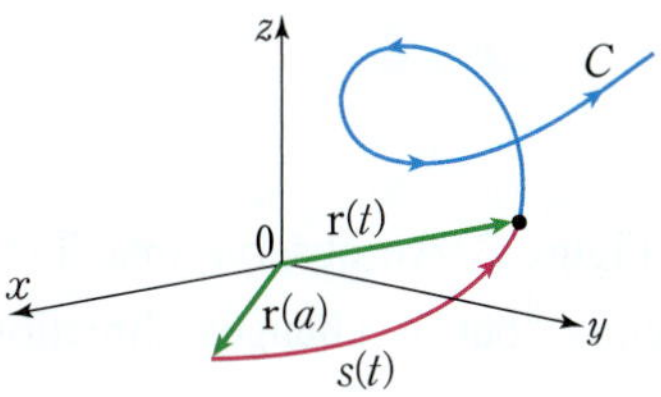

Figure 2

where $\mathbf{r}'$ is continuous and C is traced exactly once as t increases from a to b. The **arc length function** is the real valued function $s:[a,\ b]\to\mathbb{R}$ defined by

3 $$s(t)=\int_a^t|\mathbf{r}'(u)|du=\int_a^t\sqrt{\left(\frac{dx}{du}\right)^2+\left(\frac{dy}{du}\right)^2+\left(\frac{dz}{du}\right)^2}\,du.$$

Thus $s(t)$ is the length of the part of C between $\mathbf{r}(a)$ and $\mathbf{r}(t)$. (See Figure 2.) Differentiating both sides of 3 using the Fundamental Theorem of Calculus, we obtain

4 $$\frac{ds}{dt}=|\mathbf{r}'(t)|.$$

Since arc length is independent of a particular coordinate system parameterizion and arises from the shape of the curve, **parameterizion** of a curve with respect to arc length is often useful. If a curve $\mathbf{r}(t)$ is already given in terms of a parameter t and $s(t)$ is the arc length function given by 3, then we may be able to solve for t as a function of $s : t = t(s)$. Hence the curve can be reparametrized in terms of s by substituting for $t : \mathbf{r} = \mathbf{r}(t(s))$. It is the arc length parameterization.

EXAMPLE 2

Reparametrize the helix $\mathbf{r}(t) = 2\cos t\,\mathbf{i} + 2\sin t\,\mathbf{j} + t\,\mathbf{k}$ with respect to arc length measured from $(1,\ 0,\ 0)$ in the direction of increasing t.

Solution The initial point $(1,\ 0,\ 0)$ corresponds to the parameter value $t = 0$. From Example 1, we have

$$\frac{ds}{dt} = |\mathbf{r}'(t)| = \sqrt{(-2\sin t)^2 + (2\cos t)^2 + 1^2} = \sqrt{5}$$

and so

$$s = s(t) = \int_0^t |\mathbf{r}'(u)|\, du = \int_0^t \sqrt{5}\, du = \sqrt{5}\, t.$$

Solving this equation for t gives $t = s/\sqrt{5}$. Therefore, the required reparameterization is obtained by substituting for t:

$$\mathbf{r}\left(t(s)\right) = 2\cos\left(\frac{s}{\sqrt{5}}\right)\mathbf{i} + 2\sin\left(\frac{s}{\sqrt{5}}\right)\mathbf{j} + \left(\frac{s}{\sqrt{5}}\right)\mathbf{k}.$$ ■

Curvature of Curves

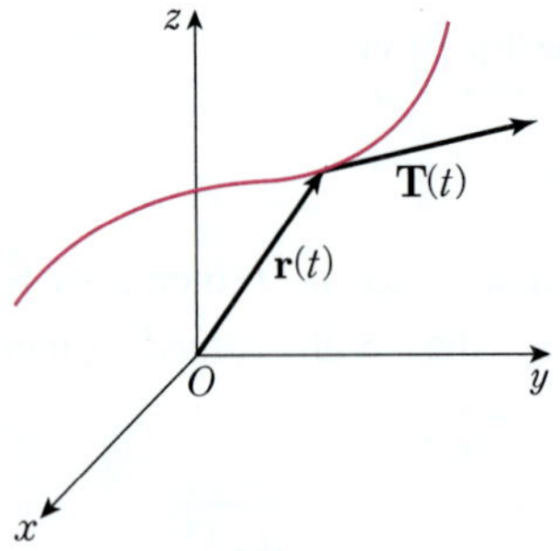

Figure 3

Let C be a smooth curve defined by the vector function $\mathbf{r}$. Recall that the unit tangent vector $\mathbf{T}(t)$ is given by

$$\mathbf{T}(t) = \frac{\mathbf{r}'(t)}{|\mathbf{r}'(t)|}$$

and points in the direction of the curve C. From Figure 3, we observe that $\mathbf{T}(t)$ changes direction gradually when C bends slightly, but it changes direction rapidly when C bends more sharply.

DEFINITION 1

The **curvature** κ of a curve $\mathbf{r}$ is the magnitude of the rate of change of the unit tangent vector with respect to arc length. That is,

$$\kappa = \left| \frac{d\mathbf{T}}{ds} \right|$$

where $\mathbf{T}$ is the unit tangent vector.

Remark If a smooth curve $\mathbf{r}(t)$ is given in terms of parameter t rather than the arc length parameter s, we can compute the curvature as follows. We use Theorem 4 (vi) in Section 12.2 to write

$$\frac{d\mathbf{T}}{dt} = \frac{d\mathbf{T}}{ds}\frac{ds}{dt} \qquad \text{and } \kappa = \left| \frac{d\mathbf{T}}{ds} \right| = \left| \frac{\frac{d\mathbf{T}}{dt}}{\frac{ds}{dt}} \right|.$$

However $ds/dt = |\mathbf{r}'(t)|$ from 4, thus we have the following formula for calculating curvature

5 $$\kappa(t) = \frac{|\mathbf{T}'(t)|}{|\mathbf{r}'(t)|}.$$

EXAMPLE 3

Find the curvature of a circle of radius a.

Solution We can take a circle of radius a, centered at the origin, and then a parameterization is

$$\mathbf{r}(t) = a\cos t\,\mathbf{i} + a\sin t\,\mathbf{j}.$$

Then

$$\mathbf{r}'(t) = -a\sin t\,\mathbf{i} + a\cos t\,\mathbf{j} \quad \text{and } |\mathbf{r}'(t)| = a.$$

From this, we find the unit tangent vector

$$\mathbf{T}(t) = \frac{\mathbf{r}'(t)}{|\mathbf{r}'(t)|} = -\sin t\,\mathbf{i} + \cos t\,\mathbf{j} \quad \text{and } \mathbf{T}'(t) = -\cos t\,\mathbf{i} - \sin t\,\mathbf{j}.$$

Hence by 5, the curvature is

$$\kappa(t) = \frac{|\mathbf{T}'(t)|}{|\mathbf{r}'(t)|} = \frac{1}{a} \quad \text{since } |\mathbf{T}'(t)| = 1.$$ ■

We see from Example 3 that the curvature of a circle is constant, which is the reciprocal of its radius. So small circles have large curvature and large circles have small curvature. We can see directly from the definition of curvature that the curvature of a straight line is 0 because the unit tangent vector $\mathbf{T}$ points in the same direction, that is, $\mathbf{T}$ is constant.

The following theorem provides another formula for curvature in terms of a general parameter t.

THEOREM 2

If C is smooth curve defined by the vector function $\mathbf{r}$, then the curvature κ can be expressed as

$$\kappa(t)=\frac{|\mathbf{r}'(t)\times\mathbf{r}''(t)|}{|\mathbf{r}'(t)|^3}.$$

Proof It follows from definition of the unit tangent vector $\mathbf{T}$ and 4 $\dfrac{ds}{dt}=|\mathbf{r}'(t)|$ that

$$\mathbf{r}'=|\mathbf{r}'(t)|\,\mathbf{T}=\frac{ds}{dt}\mathbf{T}.$$

Differentiating with respect to t, we get

$$\mathbf{r}''=\frac{d^2s}{dt^2}\mathbf{T}+\frac{ds}{dt}\mathbf{T}'.$$

However the cross product of a vector with itself is a zero vector, so we have

$$\begin{aligned}\mathbf{r}'\times\mathbf{r}''&=\left(\frac{ds}{dt}\mathbf{T}\right)\times\left(\left(\frac{d^2s}{dt^2}\right)\mathbf{T}\right)+\left(\frac{ds}{dt}\mathbf{T}\right)\times\left(\frac{ds}{dt}\mathbf{T}'\right)\\&=\left(\frac{ds}{dt}\right)\left(\frac{d^2s}{dt^2}\right)(\mathbf{T}\times\mathbf{T})+\left(\frac{ds}{dt}\right)^2(\mathbf{T}\times\mathbf{T}').\\&=0+\left(\frac{ds}{dt}\right)^2(\mathbf{T}\times\mathbf{T}')=\left(\frac{ds}{dt}\right)^2(\mathbf{T}\times\mathbf{T}').\end{aligned}$$

Since $|\mathbf{T}(t)|=1$ for all t so $\mathbf{T}$ and $\mathbf{T}'$ are orthogonal by Theorem 5 in Section 12.2. Hence, by the definition of cross product and $|\mathbf{T}\times\mathbf{T}'|=|\mathbf{T}||\mathbf{T}'|\sin\frac{\pi}{2}=|\mathbf{T}||\mathbf{T}'|$.

$$|\mathbf{r}'\times\mathbf{r}''|=\left(\frac{ds}{dt}\right)^2|\mathbf{T}\times\mathbf{T}'|=\left(\frac{ds}{dt}\right)^2|\mathbf{T}|\,|\mathbf{T}'|=\left(\frac{ds}{dt}\right)^2|\mathbf{T}'|.$$

Then

$$|\mathbf{T}'|=\frac{|\mathbf{r}'\times\mathbf{r}''|}{(ds/dt)^2}=\frac{|\mathbf{r}'\times\mathbf{r}''|}{|\mathbf{r}'|^2}.$$

Thus

$$\kappa = \frac{|\mathbf{T}'|}{|\mathbf{r}'|} = \frac{|\mathbf{r}' \times \mathbf{r}''|}{|\mathbf{r}'|^3}.$$

EXAMPLE 4

Find the curvature of the twisted cubic $\mathbf{r}(t) = \langle a\sin t, a\cos t, bt \rangle$ where a and b are positive.

Solution We have

$$\mathbf{r}'(t) = \langle a\cos t, -a\sin t, b \rangle \quad \mathbf{r}''(t) = \langle -a\sin t, -a\cos t, 0 \rangle.$$

Hence

$$\begin{aligned} \mathbf{r}'(t) \times \mathbf{r}''(t) &= \begin{vmatrix} \mathbf{i} & \mathbf{j} & \mathbf{k} \\ a\cos t & -a\sin t & b \\ -a\sin t & -a\cos t & 0 \end{vmatrix} \\ &= ab\cos t\,\mathbf{i} - ab\sin t\,\mathbf{j} - a^2\,\mathbf{k}. \end{aligned}$$

Therefore

$$|\mathbf{r}'(t)| = \sqrt{a^2 + b^2}$$

and

$$|\mathbf{r}'(t) \times \mathbf{r}''(t)| = \sqrt{(ab)^2 + a^4} = a\sqrt{a^2 + b^2}.$$

By Theorem 2,

$$\kappa(t) = \frac{|\mathbf{r}'(t) \times \mathbf{r}''(t)|}{|\mathbf{r}'(t)|^3} = \frac{a\sqrt{a^2+b^2}}{(\sqrt{a^2+b^2})^3} = \frac{a}{a^2+b^2}.$$

Remark For a plane curve with equation $y = f(x)$, we take x as the parameter, and then a parameterization is

$$\mathbf{r}(x) = x\,\mathbf{i} + f(x)\,\mathbf{j}.$$

Then

$$\mathbf{r}'(x) = \mathbf{i} + f'(x)\,\mathbf{j} \quad \text{and} \quad \mathbf{r}''(x) = f''(x)\,\mathbf{j}$$

and hence $\mathbf{r}'(x) \times \mathbf{r}''(x) = f''(x)\mathbf{k}$, and $|\mathbf{r}'(x)| = \sqrt{1 + [f'(x)]^2}$. Thus by Theorem 2, we have

6
$$\kappa(x) = \frac{|f''(x)|}{[1 + (f'(x))^2]^{3/2}}.$$

Let us look at this in another form.

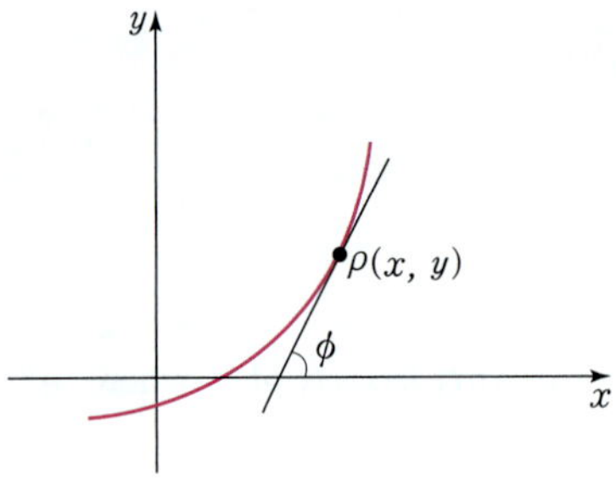

Figure 4

If $\dfrac{dy}{dx} = \tan\phi$ as $\phi = \tan^{-1}\dfrac{dy}{dx}$, then $\dfrac{d\phi}{dx} = \dfrac{\dfrac{d^2y}{dx^2}}{1+\left(\dfrac{dy}{dx}\right)^2}$.

The arc length function is

$$s(x) = \int_a^x \sqrt{\left(\frac{dx}{dt}\right)^2 + \left(\frac{dy}{dt}\right)^2}\,dt = \int_a^x \sqrt{1+\left(\frac{dy}{dx}\right)^2}\,dx.\ \text{Then,}$$

$$\frac{ds}{dx} = \sqrt{1+\left(\frac{dy}{dx}\right)^2} \quad \text{(by 1st FTC)}$$

$$\Rightarrow \frac{d\phi}{ds} = \frac{d\phi}{dx}\cdot\frac{dx}{ds} = \frac{d\phi}{dx}\cdot\frac{1}{\dfrac{ds}{dx}}$$

$$= \frac{\dfrac{d^2y}{dx^2}}{1+\left(\dfrac{dy}{dx}\right)^2}\cdot\frac{1}{\sqrt{1+\left(\dfrac{dy}{dx}\right)^2}} = \frac{\dfrac{d^2y}{dx^2}}{\left\{1+\left(\dfrac{dy}{dx}\right)^2\right\}^{3/2}}.$$

Thus $\kappa = \left|\dfrac{d\phi}{ds}\right|$, which means the curvature κ at any point $P(x, y)$ on a curve $y = f(x)$, is $\left|\dfrac{d\phi}{ds}\right|$.

Curvature in Parametric Form

Remark Using Theorem 2, we can show that the curvature of a plane parametric curve $x = f(t),\ y = g(t)$ is

$$\kappa = \frac{\left|\dfrac{dx}{dt}\dfrac{d^2y}{dt^2} - \dfrac{dy}{dt}\dfrac{d^2x}{dt^2}\right|}{\left[\left(\dfrac{dx}{dt}\right)^2 + \left(\dfrac{dy}{dt}\right)^2\right]^{3/2}}.$$

Remark Using Theorem 2, we can show that the curvature of a plane polar function $r = f(\theta)$ is

$$\kappa = \frac{|r^2 + 2r(r')^2 - rr''|}{\{r^2 + (r')^2\}^{3/2}}.$$

A Circle of Curvature

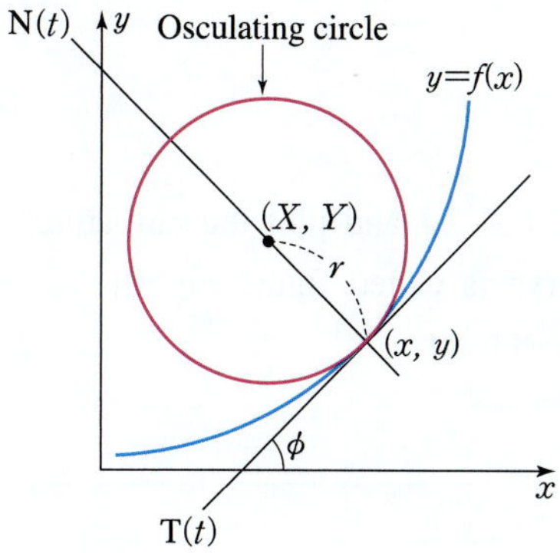

Figure 5

In Figure 5 a osculating circle on $y=f(x)$ at a point (x,y) is called a circle of curvature. By Example 3, a **radius of a circle of curvature** is

$$\rho=\frac{1}{\kappa} \text{ where } \kappa=\frac{|y''|}{\{1+(y')^2\}^{3/2}}.$$

If $f'(x)>0$ and $f''(x)>0$ (concave downward), then the coordinates of center of a circle of curvature is $(X=x-\rho\sin\phi,\ Y=y+\rho\cos\phi)$. Also since

$$\tan\phi=y',\ \sin\phi=\frac{y'}{\sqrt{1+(y')^2}},\ \cos\phi=\frac{1}{\sqrt{1+(y')^2}}.$$

Then the **center of the circle of curvature** (X, Y) can be obtained by

$$X=x-\frac{y'\{1+(y')^2\}}{|y''|},\quad Y=y-\frac{1+(y')^2}{|y''|}.$$

EXAMPLE 5

Find the radius of curvature for the curve $y=\ln\sec x$.

Solution Since $y'=\tan x$ and $y''=\sec^2 x$,

$$\rho=\frac{1}{\kappa}=\frac{(1+\tan^2 x)^{3/2}}{\sec^2 x}=\sec x.$$ ■

EXAMPLE 6

Find the center of curvature for the parabola $y=2x-x^2$ at $(1, 1)$.

Solution Since $y'=2-2x$ and $y''=-2$,
$X=1$ and $Y=1/2$. Therefore the center is $(1, 1/2)$. ■

EXAMPLE 7

Find the curvature of $y=\dfrac{x^4}{4}$ and sketch the graph of its curvature function $\kappa(x)$.

Solution Since $y'=x^3$ and $y''=3x^2$, 6 gives

$$\kappa(x)=\frac{|y''|}{[1+(y')^2]^{3/2}}=\frac{3x^2}{(1+x^6)^{3/2}}.$$

The curvature at $(0, 0)$ is $\kappa(0)=0$. At the point $\left(1, \dfrac{1}{4}\right)$, it is $\kappa(1)=\dfrac{3}{2\sqrt{2}}$. Observe from the expression for $\kappa(x)$ or the graph of κ in Figure 6 that $\kappa(x)\to 0$ as $x\to\pm\infty$. This corresponds to

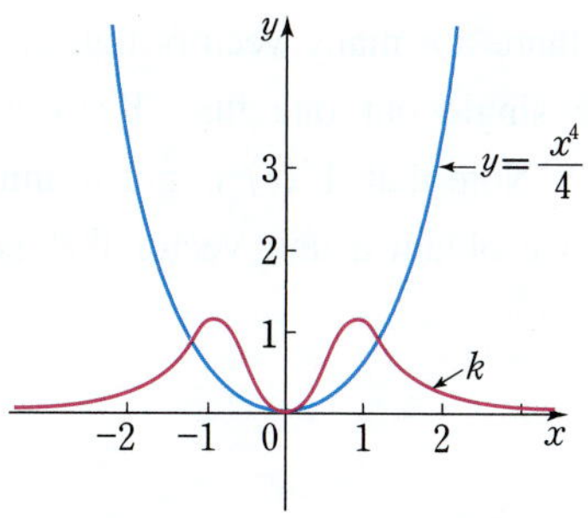

Figure 6

the fact that the graph appears to become flatter as $x \to \pm\infty$. ■

http://interact.sagemath.org/taxonomy/term/13

EXAMPLE 8

Sketch the graph of $\mathbf{r}(t) = (\cos t, \sin 2t)$, $0 \leq t \leq 2\pi$ and plot the curvature function of $\mathbf{r}(t)$ along with this curve. [This curve is called figure eight.]
http://matrix.skku.ac.kr/cal-lab/cal-12-3-Example-8.html

Solution

```
var('t')
x=sin(t)
y=sin(2*t)
x1=diff(x,t)
y1=diff(y,t)
x2=diff(x1,t)
y2=diff(y1,t)
curvature=(x1(t)*y2(t)-x2(t)*y1(t))/((x1(t)^2+y1(t)^2)^(3/2))
plot(curvature,(t,0,2*pi),color='red',dpi=120)+parametric_plot((x
,y),(t,0,2*pi),dpi=120)
```

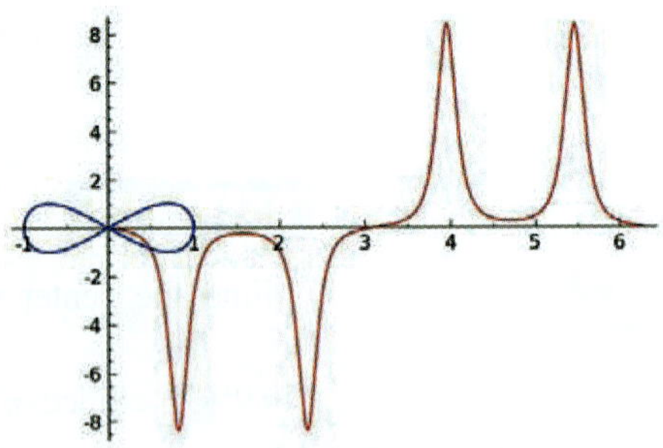

Figure 7

■

The Normal Vector and Binormal Vector

At a given point on a smooth space curve $\mathbf{r}(t)$, there are many vectors that are orthogonal to the unit tangent vector $\mathbf{T}(t)$. We single out one that $\mathbf{T}'(t)$ is orthogonal to $\mathbf{T}(t)$ by Theorem 5 in Section 12.2. Note that $\mathbf{T}'(t)$ is not a unit vector. If we divide $\mathbf{T}'(t)$ by its length $|\mathbf{T}'(t)|$, we obtain a unit vector that is orthogonal to $\mathbf{T}(t)$.

DEFINITION 3

The **principal unit normal vector** $\mathbf{N}(t)$ (or simply **unit normal**) for a smooth curve in space is

$$\mathbf{N}(t) = \frac{\mathbf{T}'(t)}{|\mathbf{T}'(t)|}.$$

DEFINITION 4

The **binormal vector** $\mathbf{B}(t)$ of a smooth curve in a space is given by

$$\mathbf{B}(t) = \mathbf{T}(t) \times \mathbf{N}(t).$$

Remark Since $\mathbf{T}$ and $\mathbf{N}$ are both unit vectors, the binormal vector $\mathbf{B}$ is also a unit vector and is perpendicular to both $\mathbf{T}$ and $\mathbf{N}$. (See Figure 8) The vectors $\mathbf{T}$, $\mathbf{N}$, and $\mathbf{B}$ determine three mutually perpendicular planes that move along the curve as t varies. Three mutually perpendicular unit vectors in a space are called the $\mathbf{TNB}$ frame. It has important applications in differential geometry and in the study of the motion of spacecraft.

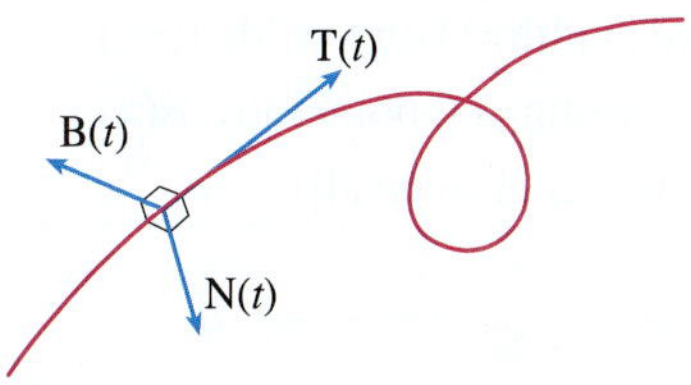

Figure 8

CAS **EXAMPLE 9**

Find the principal unit normal and binormal vectors for the circular helix

$$x = a\cos t,\ y = a\sin t,\ z = bt.$$

http://matrix.skku.ac.kr/cal-lab/cal-12-3-Example-9.html

Solution The vector function for the helix is

$$\mathbf{r}(t) = a\cos t\,\mathbf{i} + a\sin t\,\mathbf{j} + b\,t\,\mathbf{k}.$$

Then we have

$$\mathbf{r}'(t) = -a\sin t\,\mathbf{i} + a\cos t\,\mathbf{j} + b\mathbf{k},\ |\mathbf{r}'(t)| = \sqrt{a^2 + b^2}$$

$$\mathbf{T}(t) = \frac{\mathbf{r}'(t)}{|\mathbf{r}'(t)|} = \frac{1}{\sqrt{a^2+b^2}}(-a\sin t\,\mathbf{i} + a\cos t\,\mathbf{j} + b\mathbf{k}),$$

$$\mathbf{T}'(t) = \frac{1}{\sqrt{a^2+b^2}}(-a\cos t\,\mathbf{i} - a\sin t\,\mathbf{j}),$$

$$|\mathbf{T}'(t)| = \frac{a}{\sqrt{a^2+b^2}},$$

$$\mathbf{N}(t) = \frac{\mathbf{T}'(t)}{|\mathbf{T}'(t)|} = -\cos t\,\mathbf{i} - \sin t\,\mathbf{j}$$
$$= \langle -\cos t,\ -\sin t,\ 0 \rangle.$$

This shows that the normal vector at a point on the helix is horizontal and points toward the z-axis. The binormal vector is

$$\mathbf{B}(t)=\mathbf{T}(t)\times\mathbf{N}(t)=\frac{1}{\sqrt{a^2+b^2}}\begin{vmatrix}\mathbf{i} & \mathbf{j} & \mathbf{k}\\ -a\sin t & a\cos t & b\\ -\cos t & -\sin t & 0\end{vmatrix}$$

$$=\frac{1}{\sqrt{a^2+b^2}}\langle b\sin t,\ -b\cos t,\ a\rangle .$$

Plot the T, N and B vector for helix

```
t=var('t')
r=vector((cos(t),sin(t),t))
T=diff(r,t)/norm(diff(r,t))
N=diff(T,t)/norm(diff(T,t))
B=T.cross_product(N)
p = parametric_plot(r,(t,0,2*pi),color='goldenrod',thickness=2)
p = p+sum([arrow3d(r(t=x), r(t=x)+T(t=x),width=1)+arrow3d(
    r(t=x), r(t=x)+N(t=x),color='red',width=1)+arrow3d(r(t=x),
r(     t=x)+B(t=x), color='green', width=1)+point3d((cos(x),
sin(x), x),     pointsize=20) for x in [-3..3,step=1]])
show(p)
```

■

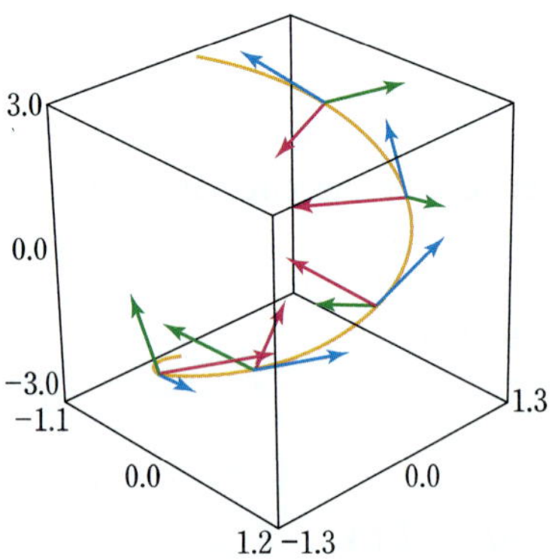

Figure 9

The three unit vectors, $\mathbf{T}$, $\mathbf{N}$ and $\mathbf{B}$ form a right-handed set of mutually orthogonal vectors called the **moving trihedral**. The plane of $\mathbf{T}$ and $\mathbf{N}$ is called the **osculating plane** at a point P on a curve C, the plane of $\mathbf{N}$ and $\mathbf{B}$ is called the **normal plane** at a point P on a curve C, the plane of $\mathbf{T}$ and $\mathbf{B}$ is called **rectifying plane** at a point P on a curve C.

The three mutually orthogonal unit vectors $\mathbf{T}$, $\mathbf{N}$ and $\mathbf{B}$ can be thought of as a movable right-handed coordinate system. (See Figure 9.)

$$\mathbf{B}(t)=\mathbf{T}(t)\times\mathbf{N}(t),\quad \mathbf{N}(t)=\mathbf{B}(t)\times\mathbf{T}(t),\quad \mathbf{T}(t)=\mathbf{N}(t)\times\mathbf{B}(t).$$

This movable coordinate system is referred to as the $\mathbf{TNB}$-frame.

The **osculating circle** or the **circle of curvature** of C at P is the circle with radius $\rho=1/\kappa$ that lies in the osculating plane of C at P, has the same tangent as C at P, lies on the concave side of C towards which $\mathbf{N}$ points. (See Figure 8) The behavior of the curve C near P is best described by osculating circle. Also it shares the same tangent, normal, and curvature at P. The circle at P is called circle of curvature and its center is the center of curvature. The circle of curvature has same tangent line at P as the curve C, and its center lies on the concave side of C.

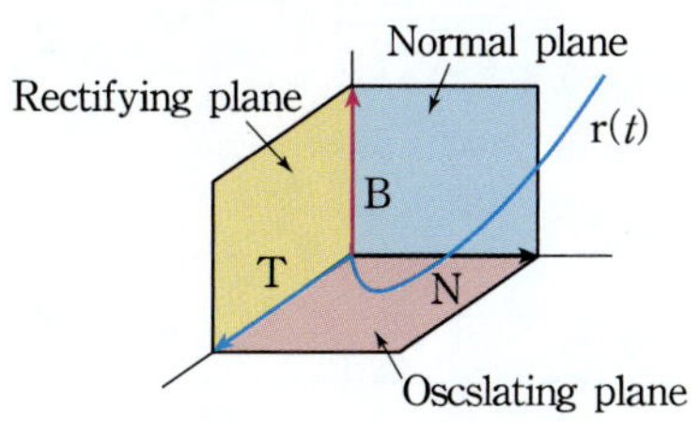

Figure 10

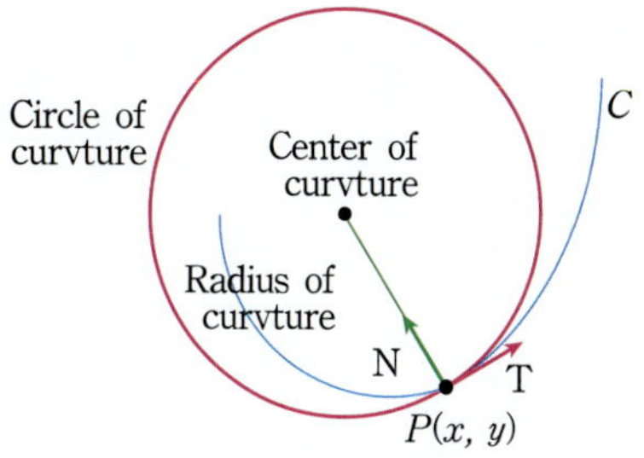

Figure 11

CAS **EXAMPLE 10**

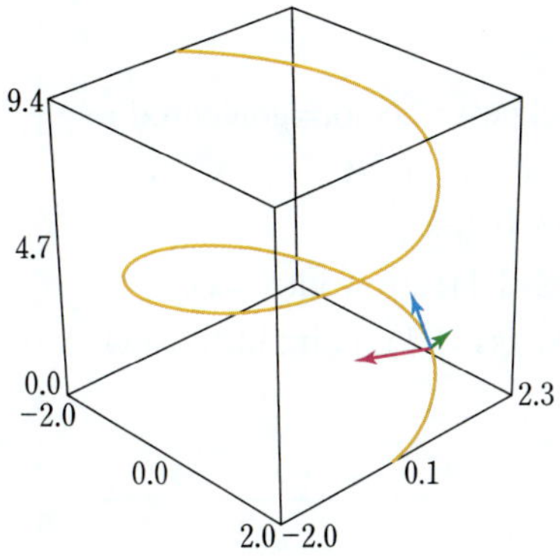

Figure 12

Find the equations of the normal plane, osculating plane and rectifying plane of the helix

$$\mathbf{r}(t) = 2\cos t\,\mathbf{i} + 2\sin t\,\mathbf{j} + t\,\mathbf{k}$$

at the point $P\left(1,\ \sqrt{3},\ \frac{\pi}{3}\right)$. (See Figure 12.)

http://matrix.skku.ac.kr/cal-lab/cal-12-3-Example-10.html

Solution The normal plane at P, the tangent vector is $\mathbf{r}'(\pi/3) = \langle -\sqrt{3},\ 1,\ 1\rangle$. Hence the equation of the normal plane is

$$-\sqrt{3}\,(x-1) + 1(y-\sqrt{3}) + 1\left(z - \frac{\pi}{3}\right) = 0 \quad \text{or}$$

$$-\sqrt{3}\,x + y + z = \frac{\pi}{3}.$$

The osculating plane at P is determined by the vectors $\mathbf{T}$ and $\mathbf{N}$, hence its normal vector is $\mathbf{T} \times \mathbf{N} = \mathbf{B}$.

$$\mathbf{B}(t) = \frac{1}{\sqrt{13}}\langle 3\sin t,\ -3\cos t,\ 2\rangle,$$

$$\mathbf{B}\left(\frac{\pi}{3}\right) = \frac{1}{\sqrt{13}}\left\langle \frac{3\sqrt{3}}{2},\ -\frac{3}{2},\ 2\right\rangle.$$

Hence an equation of the osculating plane is

$$\frac{3\sqrt{3}}{2\sqrt{13}}(x-1) - \frac{3}{2\sqrt{13}}(y-\sqrt{3}) + \frac{2}{\sqrt{13}}\left(z - \frac{\pi}{3}\right) = 0$$

or

$$\sqrt{3}\,x - y + \frac{4}{3}z = \frac{2\pi}{9}.$$

The rectifying plane at P has normal vector

$$\mathbf{N}\left(\frac{\pi}{3}\right) = -\frac{1}{2}\mathbf{i} - \frac{\sqrt{3}}{2}\mathbf{j}.$$

Hence an equation of the rectifying plane is

$$-\frac{1}{2}(x-1)+\frac{\sqrt{3}}{2}(y-\sqrt{3})+0\left(z-\frac{\pi}{3}\right)=0$$

or

$$x-\sqrt{3}\,y=4.$$

```
t=var('t')
r=vector((2*cos(t),2*sin(t),t))
T=diff(r,t)/norm(diff(r,t))
N=diff(T,t)/norm(diff(T,t))
B=T.cross_product(N)
x=pi/3
C=parametric_plot3d(r,(t,0,3*pi),thickness=2,color='goldenrod')
A = arrow3d(r(t=x),
r(t=x)+T(t=x),thickness=1.1)+arrow3d(r(t=x),
  r(t=x)+N(t=x),color='red',thickness=1.1)+arrow3d(r(t=x),
  r(t=x)+B(t=x),color='green',thickness=1.1)+point3d((cos(x),
    sin(x),x),pointsize=20)
C+A
```

■

CAS **EXAMPLE 11**

Find the vectors **T**, **N**, and **B** for $\mathbf{r}(t)=\langle t^3, t-2, t^2\rangle$ at the given point $P(1, -1, 1)$.

http://matrix.skku.ac.kr/cal-lab/cal-12-3-Example-11.html

Solution For point $(1, -1, 1)$, the corresponding value t is 1.

```
t=var('t')
x(t)=t^3
y(t)=t-2
z(t)=t^2
r(t)=(x(t), y(t), z(t))
dr(t)=(diff(x(t), t),  diff(y(t), t),  diff(z(t), t))  # r'(t)
ndr=dr(t).norm()  #  |r'(t)|
T=dr(t)/ndr  #  T(t)
dT=diff(T, t)  #  T'(t)
ndT=dT.norm()  #  |T'(t)|
N=dT/ndT  #  N(t)
print T(1)
print N(1)
print T(1).cross_product(N(1))
```

(3/14*sqrt(14), 1/14*sqrt(14), 1/7*sqrt(14))

(9/266 * sqrt(14) * sqrt(19), -11/266 * sqrt(14) * sqrt(19), -4/133 *

```
sqrt(14) * sqrt(19))
(1/19*sqrt(19), 3/19*sqrt(19), -3/19*sqrt(19)
```

```
t=var('t')
r=vector((t^3, t-2, t^2))
T=diff(r,t)/norm(diff(r,t))
N=diff(T,t)/norm(diff(T,t))
B=T.cross_product(N)
x=1
C=parametric_plot3d(r,(t,0,pi/2),thickness=2,color='goldenrod')
A=arrow3d(r(t=x), r(t=x)+T(t=x), thickness=1.1)+arrow3d
 (r(t=x), r(t=x)+N(t=x) ,color='red' ,thickness=1.1)+arrow3d
 (r(t=x), r(t=x) +B(t=x), color='green', thickness=1.1) +
 point3d((cos(x), sin(x), x), pointsize=20)
C+A
```

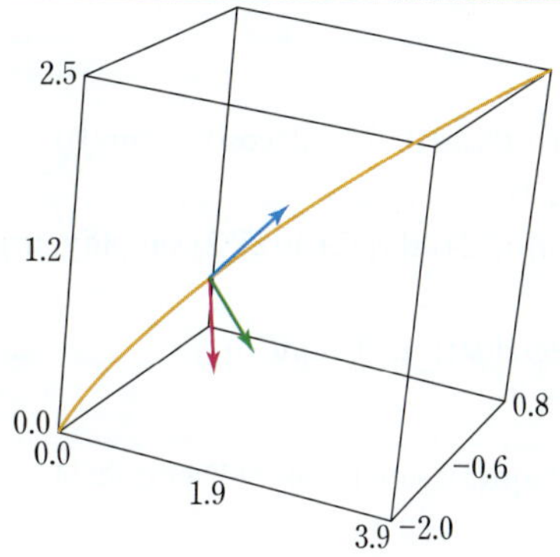

Figure 13 ■

<YANG, Hui>

12.3 EXERCISES (Arc Length and Curvature)

http://matrix.skku.ac.kr/Cal-Book/part2/CS-Sec-12-3-Sol.html

YouTube http://youtu.be/w-R4ez7vg24

1-5. Find the length of the curve.

1. $\mathbf{r}(t)=\langle 4t,\ 3\cos t,\ 3\sin t\rangle,\ -5\le t\le 5.$

Solution $\mathbf{r}'(t)=\langle 4,\ -3\sin t,\ 3\cos t\rangle.$

$|\mathbf{r}'(t)|=\sqrt{4^2+(-3\sin t)^2+(3\cos t)^2}=\sqrt{25}=5.$

$L=\int_{-5}^{5}5dt=50.$

CAS **2.** $\mathbf{r}(t)=\langle 2\sin t-t\cos t,\ 2t,\ 2\cos t+t\sin t\rangle,\ 0\le t\le\sqrt{5}.$

http://matrix.skku.ac.kr/cal-lab/cal-13-3-2.html

Solution

```
var('t')
r(t)=(2*sin(t)-t*cos(t), 2*t, 2*cos(t)+t*sin(t))
dr=diff(r(t), t);
s(t)=sqrt((dr[0]^2+dr[1]^2+dr[2]^2).simplify_trig())
;
length=integral(s(t), t, 0, sqrt(5))
print length
```

Answer : 1/2*sqrt(5)*sqrt(10) + 5/2*arcsinh(1)

3. $\mathbf{r}(t)=e^{2t}\mathbf{i}+2\sqrt{2}t\mathbf{j}+e^{-2t}\mathbf{k},\ 0\le t\le\ln 2.$

Solution $\mathbf{r}'(t)=2e^{2t}\mathbf{i}+2\sqrt{2}\,\mathbf{j}-2e^{-2t}\mathbf{k}.$

$|\mathbf{r}'(t)|=\sqrt{(2e^{2t})^2+(2\sqrt{2})^2+(-2e^{-2t})^2}$

$=\sqrt{4e^{4t}+8+4e^{-4t}}=\sqrt{4(e^{2t}+e^{-2t})^2}=2e^{2t}+2e^{-2t}.$

$L=\int_0^{\ln 2}2e^{2t}+2e^{-2t}dt=e^{2t}-e^{-2t}\Big|_0^{\ln 2}=4-\frac{1}{4}=\frac{15}{4}.$

CAS **4.** $\mathbf{r}(t)=t\mathbf{i}+2\mathbf{j}+\ln t\mathbf{k},\ 1\le t\le\sqrt{3}.$

http://matrix.skku.ac.kr/cal-lab/cal-13-3-4.html

Solution

```
var('t')
x(t)=t
y(t)=2
z(t)=ln(t)
dx(t)=diff(x(t), t)
dy(t)=diff(y(t), t)
dz(t)=diff(z(t), t)
s(t)=sqrt(dx(t)^2+dy(t)^2+dz(t)^2)
length=integral(s(t), t, 1, sqrt(3))
print length
```

Answer : -sqrt(2) - 1/2*log(sqrt(2) - 1) + 1/2*log(sqrt(2) + 1) - 1/2*log(3) + 2

5. $\mathbf{r}(t)=3t^2\mathbf{i}+t^3\mathbf{j}+5\mathbf{k},\ 0\le t\le 1.$

Solution $\mathbf{r}'(t)=6t\mathbf{i}+3t^2\mathbf{j},$

$|\mathbf{r}'(t)|=\sqrt{(6t)^2+(3t^2)^2}=\sqrt{36t^2+9t^4}=3t\sqrt{4+t^2},$

$L=\int_0^1 3t\sqrt{4+t^2}\,dt\underset{x=t^2+4}{=}\int_4^5\frac{3\sqrt{x}}{2}dx$

$=x\sqrt{x}\,\Big|_4^5=5\sqrt{5}-8.$

6-8. Reparametrize the curve with respect to arc length measured from the point $t=0$ in the direction of increasing t.

6. $\mathbf{r}(t)=(2-5t)\mathbf{i}+4t\mathbf{j}-(1+3t)\mathbf{k}.$

Solution Since $\mathbf{r}'(t)=-5\mathbf{i}+4\mathbf{j}-3\mathbf{k},$

$|\mathbf{r}'(t)|=\sqrt{(-5)^2+4^2+(-3)^2}=\sqrt{50}=5\sqrt{2}.$

Thus $s=s(t)=\int_0^t 5\sqrt{2}\,dt=5\sqrt{2}\,t,$ this implies $t=\frac{s}{5\sqrt{2}}$. Substituting $t=\frac{s}{5\sqrt{2}}$ in $\mathbf{r}(t)$, we have

$\mathbf{r}(t(s))=\left(2-\frac{s}{\sqrt{2}}\right)\mathbf{i}+\frac{2\sqrt{2}}{5}s\mathbf{j}-\left(1+\frac{3}{5\sqrt{2}}s\right)\mathbf{k}.$

7. $\mathbf{r}(t)=e^t\sin 3t\mathbf{i}+e^t\cos 3t\mathbf{j}+\mathbf{k}$

Solution Since

$\mathbf{r}'(t)=e^t(\sin 3t+3\cos 3t)\mathbf{i}+e^t(\cos 3t-3\sin 3t)\mathbf{j},$

$|\mathbf{r}'(t)|=\sqrt{10e^{2t}}=\sqrt{10}\,e^t.$

Thus $s=s(t)=\int_0^t\sqrt{10}\,e^t dt=\sqrt{10}\,e^t\ \Rightarrow\ t=\ln\frac{s}{\sqrt{10}}.$

Substituting $t=\ln\frac{s}{\sqrt{10}}$ in $\mathbf{r}(t)$, we have

$\mathbf{r}(t(s))=\frac{s}{\sqrt{10}}\sin\left(3\ln\frac{s}{\sqrt{10}}\right)\mathbf{i}+\frac{s}{\sqrt{10}}\cos\left(3\ln\frac{s}{\sqrt{10}}\right)\mathbf{j}+\mathbf{k}.$

CAS **8.** $\mathbf{r}(t)=3t\sin t\mathbf{i}+3t\cos t\mathbf{j}-2\sqrt{2}t^{3/2}\mathbf{k}.$

http://matrix.skku.ac.kr/cal-lab/cal-13-3-8.html

Solution Reparametrization

```
var('t')
x(t)=3*t*sin(t)
y(t)=3*t*cos(t)
var('t')
```

```
x(t)=3*t*sin(t)
y(t)=3*t*cos(t)
z(t)=--2*sqrt(2)*t^(3/2)
dx(t)=diff(x(t), t)
dy(t)=diff(y(t), t)
dz(t)=diff(z(t), t)
s(t)=sqrt(dx(t)^2+dy(t)^2+dz(t)^2)
length=integral(s(t), t, 1, sqrt(3))
print length
```

Answer : 1/2*(t+1)*sqrt19*t^2+18*t+9)-3/2

9-10. Find the unit tangent $\mathbf{T}(t)$, unit normal vectors $\mathbf{N}(t)$ and the curvature κ.

9. $\mathbf{r}(t)=\langle 5\cos t,\ 5\sin t,\ 4t\rangle$.

Solution $\mathbf{r}'(t)=\langle -5\sin t,\ 5\cos t,\ 4\rangle$,

$|\mathbf{r}'(t)|=\sqrt{(-5\sin t)^2+(5\cos t)^2+4^2}=\sqrt{41}$.

Hence

$$\mathbf{T}(t)=\frac{\mathbf{r}'(t)}{|\mathbf{r}'(t)|}=\left\langle -\frac{5}{\sqrt{41}}\sin t,\ \frac{5}{\sqrt{41}}\cos t,\ \frac{4}{\sqrt{41}}\right\rangle,$$

$$\mathbf{T}'(t)=\left\langle -\frac{5}{\sqrt{41}}\cos t,\ -\frac{5}{\sqrt{41}}\sin t,\ 0\right\rangle,$$

$$|\mathbf{T}'(t)|=\frac{5}{\sqrt{41}}.$$

Hence $\mathbf{N}(t)=\dfrac{\mathbf{T}'(t)}{|\mathbf{T}'(t)|}=\langle -\cos t,\ -\sin t,\ 0\rangle$.

$$\kappa=\frac{|\mathbf{T}'(t)|}{|\mathbf{r}'(t)|}=\frac{5/\sqrt{41}}{\sqrt{41}}=\frac{5}{41}.$$

CAS **10.** $\mathbf{r}(t)=\langle 2\sin t-t\cos t,\ 2t,\ 2\cos t+t\sin t\rangle$, $t>0$.

11-13. Use Theorem 2 to find the curvature.

11. $\mathbf{r}(t)=\langle 3t,\ 3+t,\ 3-t^2\rangle$.

http://matrix.skku.ac.kr/cal-lab/cal-12-3-Exercise-11.html

Solution $\mathbf{r}'(t)=\langle 3,\ 1,\ -2t\rangle$, $\mathbf{r}''(t)=\langle 0,\ 0,\ -2\rangle$,

$|\mathbf{r}'(t)|=\sqrt{10+4t^2}$, $\mathbf{r}'(t)\times\mathbf{r}''(t)=\langle -2,\ 6,\ 0\rangle$,

$|\mathbf{r}'(t)\times\mathbf{r}''(t)|=\sqrt{40}=2\sqrt{10}$,

$$\kappa(t)=\frac{|\mathbf{r}'(t)\times\mathbf{r}''(t)|}{|\mathbf{r}'(t)|^3}=\frac{2\sqrt{10}}{(10+4t^2)^{3/2}}=\frac{\sqrt{5}}{(5+2t^2)^{3/2}}.$$

```
t=var('t')
r=vector((3*t,3+t,3-t^2))
rprime=diff(r,t)
rrprime=diff(rprime,t)
curvature = abs(rprime.cross_product
(rrprime))/abs(rprime)^3
k=curvature.full_simplify()
show(k)
```

Answer : $\dfrac{\sqrt{5}}{(5+2t^2)^{3/2}}$

CAS **12.** $\mathbf{r}(t)=(3+4t^{3/2})\mathbf{i}+6t\mathbf{j}+\dfrac{3}{2}t^2\mathbf{k}$.

http://matrix.skku.ac.kr/cal-lab/cal-13-3-12.html

Solution

```
var('t')
r(t)=(3+4*t^(3/2), 6*t,3/2*t^2)
dr=diff(r(t), t)
ddr=diff(r(t), t, 2)
ANSWER=(dr.cross_product(ddr)).norm()/(dr.norm(
))^3
print ANSWER
```

Answer : 1/3(t^(5/2) +4*t^(3/2) + 4*sqrt(t))

13. $\mathbf{r}(t)=\cos t\,\mathbf{i}+\sin t\,\mathbf{j}+n(\cos t)\,\mathbf{k}$.

Solution $\mathbf{r}'(t)=-\sin t\,\mathbf{i}+\cos t\,\mathbf{j}-n(\sin t)\,\mathbf{k}$,

$\mathbf{r}''(t)=-\cos t\,\mathbf{i}-\sin t\,\mathbf{j}-n(\cos t)\,\mathbf{k}$,

$|\mathbf{r}'(t)|=\sqrt{1+n^2\sin^2 t}$,

$\mathbf{r}'(t)\times\mathbf{r}''(t)=-n\,\mathbf{i}+\mathbf{k}$, $|\mathbf{r}'(t)\times\mathbf{r}''(t)|=\sqrt{n^2+1}$.

$$\kappa(t)=\frac{|\mathbf{r}'(t)\times\mathbf{r}''(t)|}{|\mathbf{r}'(t)|^3}=\frac{\sqrt{n^2+1}}{(1+n^2\sin^2 t)^{3/2}}.$$

---

14. Find the curvature of $\mathbf{r}(t)=\langle e^t\cos 2t,\ e^t\sin 2t,\ 6\rangle$ at the point $(1,0,6)$.

Solution

$\mathbf{r}'(t)=\langle e^t(\cos 2t-2\sin 2t), e^t(\sin 2t+2\cos 2t), 0\rangle$,

$\mathbf{r}''(t)=\langle -e^t(3\cos 2t-4\sin 2t),\ e^t(-3\sin 2t+4\cos 2t), 0\rangle$

$|\mathbf{r}'(t)|=\sqrt{e^{2t}[(\cos 2t-2\sin 2t)^2+(\sin 2t+2\cos 2t)^2]}$

$=\sqrt{5e^{2t}}=e^t\sqrt{5}$,

$\mathbf{r}'(t)\times\mathbf{r}''(t)$

$=\langle 0,\ 0,\ e^{2t}[(\cos 2t-2\sin 2t)(-3\sin 2t+4\cos 2t)$

$+(\sin 2t+2\cos 2t)(3\cos 2t+4\sin 2t)\rangle$

$=\langle 0,\ 0,\ 10e^{2t}\rangle$,

$$\kappa(t)=\frac{|\mathbf{r}'(t)\times\mathbf{r}''(t)|}{|\mathbf{r}'(t)|^3}=\frac{10e^t}{5\sqrt{5}e^{3t}}=\frac{\sqrt{2}}{\sqrt{5}e^{2t}},$$

$$\kappa(1)=\frac{\sqrt{2}}{5e^2},\ \kappa(0)=\frac{2}{\sqrt{5}}.$$

15. Find the curvature of $\mathbf{r}(t)=\langle 6t,\ 3t^2,\ t^3\rangle$ at the point $(6,\ 3,\ 1)$.

Solution

$\mathbf{r}'(t)=\langle 6,\ 6t,\ 3t^2\rangle$, $\mathbf{r}''(t)=\langle 0,\ 6,\ 6t\rangle$,

$\mathbf{r}'(t)\times\mathbf{r}''(t)=\langle 18t^2,\ -36t,\ 36\rangle$,

$|\mathbf{r}'(t)|=\sqrt{6^2+(6t)^2+(3t^2)^2}=\sqrt{36+36t^2+9t^4}$

$=\sqrt{9(t^2+2)^2}=3(t^2+2)$,

$$|\mathbf{r}'(t)\times\mathbf{r}''(t)|=18\sqrt{t^4+4t^2+4}=18(t^2+2),$$
$$\kappa(t)=\frac{|\mathbf{r}'(t)\times\mathbf{r}''(t)|}{|\mathbf{r}'(t)|^3}=\frac{18(t^2+2)}{27(t^2+2)^3}=\frac{2}{3(t^2+2)^2},$$
$$\kappa(1)=\frac{2}{27}.$$

16-18. Use Formula 6 to find the curvature.

16. $y=x^2+1$.

Solution $y'=2x,\ y''=2,$
$$\kappa(x)=\frac{|y''|}{(1+y'^2)^{3/2}}=\frac{2}{(1+4x^2)^{3/2}}.$$

17. $y=\cos x+\sin x$.

Solution $y'=-\sin x+\cos x,\ y''=-\cos x-\sin x,$
$$\kappa(x)=\frac{|y''|}{(1+y'^2)^{3/2}}=\frac{|\cos x+\sin x|}{(2-\sin 2x)^{3/2}}.$$

18. $y=4x^{5/2}$.

Solution $y'=10x^{3/2},\ y''=15x^{1/2},$
$$\kappa(x)=\frac{|y''|}{(1+y'^2)^{3/2}}=\frac{15\sqrt{x}}{(1+100x^3)^{3/2}}.$$

CAS **19-20.** Find the vectors T, N and B at the given point, and plot them.

19. $\mathbf{r}(t)=\langle t+1,\ 2t,\ t^2\rangle,\ (2,2,1)$.

http://matrix.skku.ac.kr/cal-lab/cal-13-3-19.html

Solution

```
var('t')
r(t)=(t+1, 2*t, t^2)
dr=diff(r(t), t)
T=dr/dr.norm()
dT=diff(T, t)
N=dT/dT.norm()
B=T.cross_product(N)
print T.subs(t=1)
print N.subs(t=1)
print B.subs(t=1)
```

Answer : (1/3, 2/3, 2/3)
(-2/15*sqrt(5), -4/15*sqrt(5), 1/3*sqrt(5))
(2/5*sqrt(5), -1/5*sqrt(5), 0)

```
t=var('t')
r=vector((t+1, 2*t, t^2))
T=diff(r,t)/norm(diff(r,t))
N=diff(T,t)/norm(diff(T,t))
B=T.cross_product(N)
x=1
C=parametric_plot3d(r,(t,0,pi),thickness=2,color='goldenrod')
A = arrow3d(r(t=x),r(t=x)+T(t=x),thickness=1.1)+arrow3d(r(t=x),r(t=x)+N(t=x),color='red',thickness=1.1)+arrow3d(r(t=x),r(t=x)+B(t=x),color='green',thickness=1.1)+point3d((cos(x),sin(x),x),pointsize=20)
C+A
```

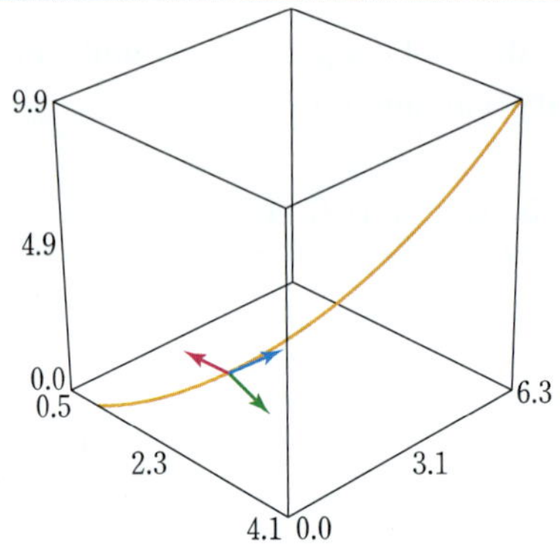

20. $\mathbf{r}(t)=\langle e^t\cos t,\ e^t\sin t,\ \sqrt{2}e^t\rangle,$
$(1,\ 0,\ \sqrt{2})$.

http://matrix.skku.ac.kr/cal-lab/cal-13-3-20.html

Solution

```
var('t')
r(t)=(exp(t)*cos(t), exp(t)*sin(t), sqrt(2)*exp(t))
dr=diff(r(t), t)
T=dr/dr.norm()
dT=diff(T, t)
N=dT/dT.norm()
B=T.cross_product(N)
print T.subs(t=0)
print N.subs(t=0)
print B.subs(t=0)
```

Answer : (1/2, 1/2, 1/2*sqrt(2))
(-sqrt(1/2), sqrt(1/2), 0)
(-1/2*sqrt(1/2)*sqrt(2),-1/2*sqrt(1/2)*sqrt(2), qrt(1/2))

```
t=var('t')
r=vector([exp(t)*cos(t), exp(t)*sin(t), sqrt(2)*exp(t)])
T=diff(r,t)/norm(diff(r,t))
N=diff(T,t)/norm(diff(T,t))
B=T.cross_product(N)
x=0
```

```
C=parametric_plot3d(r,(t,-pi/2,pi/2),thickness=2,color='goldenrod')
A = arrow3d(r(t=x), r(t=x)+T(t=x),thickness=1.1)+arrow3d(r(t=x), r(t=x)+N(t=x), color='red', thickness=1.1) + arrow3d(r(t=x), r(t=x) + B(t=x), color='green', thickness=1.1) + point3d((cos(x), sin(x), x), pointsize=20)
C+A
```

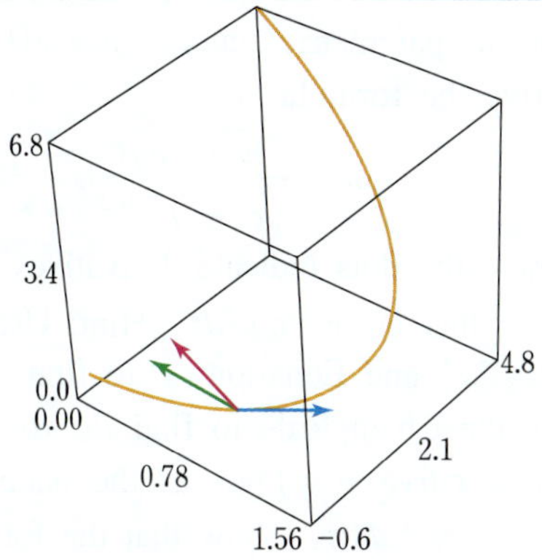

CAS 21-22. Find equations of the normal plane and osculating plane of the curve at the given point, and plot the graphs.

21. $x = 2\sin t,\ y = 5t,\ z = 2\cos t$; $(0,\ 5\pi,\ -2)$.

http://matrix.skku.ac.kr/cal-lab/cal-13-3-21.html

Solution Normal plane

```
var('x, y, z, t')
r(t)=(2*sin(t), 5*t, 2*cos(t))
dr=diff(r(t), t)
T=dr/dr.norm()
dT=diff(T, t)
N=dT/dT.norm()
B=T.cross_product(N)
N=N.subs(t=pi)
expand(N[0]*(x-0)+N[1]*(y-5*pi)+N[2]*(z+2)==0)
```

Answer : sqrt(1/29)*sqrt(29)*z + 2*sqrt(1/29)*sqrt(29)==0

```
Osculating plane
B=B.subs(t=pi)
expand(B[0]*(x-0)+B[1]*(y-5*pi)+B[2]*(z+2)==0)
```

Answer : -10*pi*sqrt(1/29) + 5*sqrt(1/29)*x + 2*sqrt(1/29)*y == 0

```
Graph
t=var('t')
r=vector([2*sin(t), 5*t, 2*cos(t)])
T=diff(r,t)/norm(diff(r,t))
N=diff(T,t)/norm(diff(T,t))
B=T.cross_product(N)
x=pi
C=parametric_plot3d(r,(t,0,2*pi),thickness=2,color='goldenrod')
A=arrow3d(r(t=x), r(t=x) + T(t=x), thickness=1.1) + arrow3d(r(t=x), r(t=x) + N(t=x), color='red', thickness=1.1) + arrow3d(r(t=x), r(t=x) + B(t=x), color='green', thickness=1.1) + point3d((cos(x), sin(x), x), pointsize=20)
C+A
```

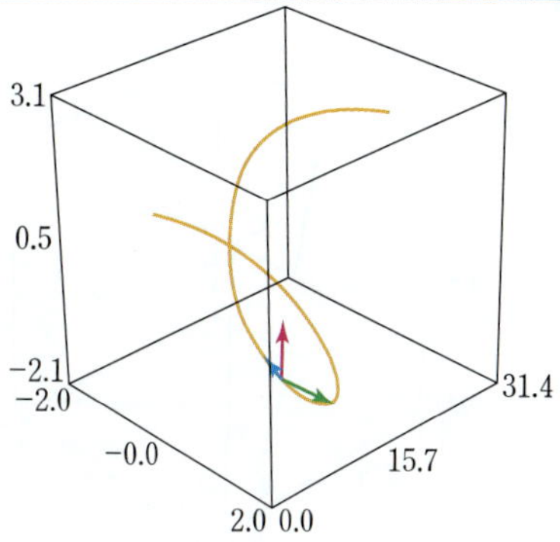

22. $x = t^2,\ y = \dfrac{2}{3}t^3,\ z = t$; $\left(1,\ \dfrac{2}{3},\ 1\right)$.

http://matrix.skku.ac.kr/cal-lab/cal-13-3-22.html

Solution Normal plane

```
var('x, y, z, t')
r(t)=(t^2, 2/3*t^3, t)
dr=diff(r(t), t)
T=dr/dr.norm()
dT=diff(T, t)
N=dT/dT.norm()
B=T.cross_product(N)
N=N.subs(t=1)
expand(N[0]*(x-1)+N[1]*(y-2/3)+N[2]*(z-1)==0)
```

Answer : -1/3*x + 2/3*y - 2/3*z + 5/9 == 0

```
Osculating plane
B=B.subs(t=1)
expand(B[0]*(x-1)+B[1]*(y-2/3)+B[2]*(z-1)==0)
```

Answer : -2/3*x + 1/3*y + 2/3*z - 2/9 == 0

```
Graph
t=var('t')
r=vector([t^2, 2/3*t^3, t])
T=diff(r,t)/norm(diff(r,t))
N=diff(T,t)/norm(diff(T,t))
B=T.cross_product(N)
x=1
C=parametric_plot3d(r,(t,0,pi/2),thickness=2,color='
goldenrod')
A=arrow3d(r(t=x), r(t=x) + T(t=x), thickness=1.1)
+ arrow3d(r(t=x), r(t=x) + N(t=x),
  color='red', thickness=1.1) + arrow3d(r(t=x),
r(t=x) + B(t=x), color='green',
  thickness=1.1) + point3d((cos(x), sin(x), x),
pointsize=20)
C+A
```

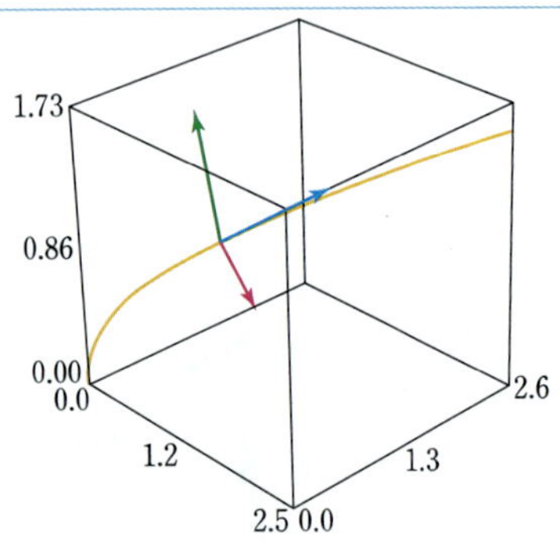

CAS **23.** At what point on the curve $x=t^4$, $y=3t$, $z=t^2$ is the tangent plane parallel to the plane $4x+3y+2z=1$?

http://matrix.skku.ac.kr/cal-lab/cal-13-3-23.html

Solution

```
var('x, y, z, t')
r(t)=(t^4, 3*t, t^2)
dr=diff(r(t), t)
T=dr/dr.norm()
T=T.subs(t=1)
expand(T[0]*(x-r(1)[0])+T[1]*(y-r(1)[1])+T[2]*(z-r
(1)[2])==0)
```

Answer : 4/29*sqrt(29)*x + 3/29*sqrt(29)*y + 2/29*sqrt(29)*z - 15/29*sqrt(29) == 0.

24. The curvature at a point P of a curve is defined as $\kappa=\left|\dfrac{d\phi}{ds}\right|$, where ϕ is the angle of inclination of the tangent line at P, as shown in the figure. Thus, the curvature is the absolute value of the rate of change of ϕ with respect to arc length. It can be regarded as a measure of the rate of change of direction of the curve at P.

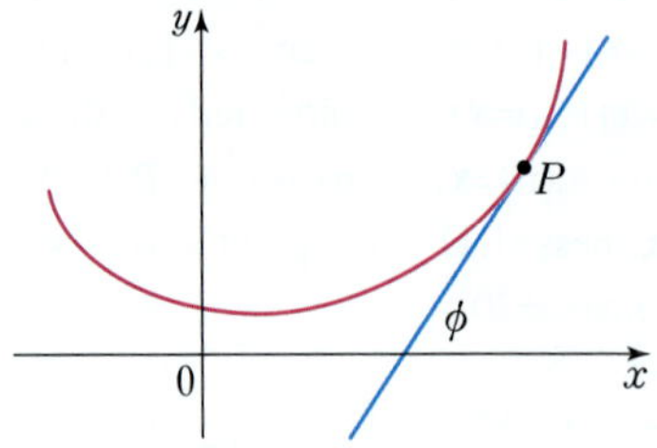

(a) For a parametric curve $x=x(t)$, $y=y(t)$ derive the formula

$$\kappa=\frac{|\dot{x}\ddot{y}-\ddot{x}\dot{y}|}{[\dot{x}^2+\dot{y}^2]^{3/2}}.$$

where the dots indicate derivatives with respect to t; that is, $\dot{x}=dx/dt$. [Hint: Use $\phi=\tan^{-1}(dy/dx)$ and Equation 2 to find $d\phi/dt$. Then use the Chain Rule to find $d\phi/ds$.]

(b) For a curve $y=f(x)$ as the parametric curve $x=x$, $y=f(x)$, show that the formula in part (a) becomes $\kappa=\dfrac{d^2y/dx^2}{[1+(dy/dx)^2]^{3/2}}$.

Solution

(a) Here
$\mathbf{r}(t)=\langle f(t),\ g(t)\rangle$, $\mathbf{r}'(t)=\langle f'(t),\ g'(t)\rangle$,
$\mathbf{r}''(t)=\langle f''(t),\ g''(t)\rangle$,
$|\mathbf{r}'(t)|^3=\left|\sqrt{(f'(t))^2+(g'(t))^2}\right|^3=[(f'(t))^2+(g'(t))^2]^{3/2}$
$=(\dot{x}^2+\dot{y}^2)^{3/2}$, and
$|\mathbf{r}'(t)\times\mathbf{r}''(t)|=|\langle 0,\ 0,\ f'(t)g''(t)-f''(t)g'(t)\rangle|$
$=[(\dot{x}\ddot{y}-\ddot{x}\dot{y})^2]^{1/2}=|\dot{x}\ddot{y}-\ddot{x}\dot{y}|$. Thus $\kappa=\dfrac{|\dot{x}\ddot{y}-\ddot{x}\dot{y}|}{[\dot{x}^2+\dot{y}^2]^{3/2}}$.

(b) Here $\mathbf{r}(x)=\langle x,\ f(x)\rangle$, $\mathbf{r}'(x)=\langle 1,\ f'(t)\rangle$,
$\mathbf{r}''(x)=\langle 0,\ f''(t)\rangle$.
By applying the result of (a) to this case, we obtain that

$$\kappa=\frac{|f''(x)|}{[1+(f'(x))^2]^{3/2}}=\frac{|d^2y/dx^2|}{[1+(dy/dx)^2]^{3/2}}.$$

25. (a) Show that the curvature at each point of a straight line is $\kappa=0$.

(b) Show that the curvature at each point of a circle of radius r is $\kappa=1/r$.

Solution (a) For a straight line, we parametrize $x=x$, $y=f(x)$. Since $f(x)$ is a straight line, $f'(x)$ is a constant, and hence $f''(x)$ is zero. $\kappa=0$.
(b) $\mathbf{r}(t)=\langle r\cos t,\ r\sin t\rangle$, $\mathbf{r}'(t)=\langle -r\sin t,\ r\cos t\rangle$,
$\mathbf{r}''(t)=\langle -r\cos t,\ -r\sin t\rangle$
Then, $|\mathbf{r}'(t)|^3=\left|\sqrt{(-r\sin t)^2+(r\cos t)^2}\right|^3=r^3$
and $|\mathbf{r}'(t)\times\mathbf{r}''(t)|=r^2$. $\kappa=1/r$.

*12.4 Motion Along A Space Curve: Velocity and Acceleration

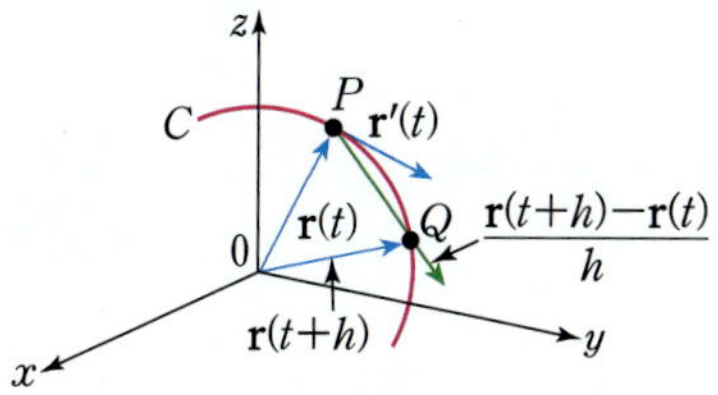

Figure 1

In this section, we will show how tangent and normal vectors can be used in physics to study the motion of a particle moving along a space curve, including its velocity and acceleration.

Suppose that a particle P is moving through space along a smooth curve C defined by the vector function $\mathbf{r}$. From Figure 1, for small values of h, consider the vector

$$\boxed{1} \qquad \frac{\mathbf{r}(t+h)-\mathbf{r}(t)}{h}.$$

The vector $\boxed{1}$ gives the average velocity over a time interval of length h. Then the **velocity vector** $\mathbf{v}(t)$ at time t is defined by

$$\boxed{2} \qquad \mathbf{v}(t)=\lim_{h\to 0}\frac{\mathbf{r}(t+h)-\mathbf{r}(t)}{h}=\mathbf{r}'(t).$$

Hence the velocity vector is also the tangent vector and points in the direction of the tangent line. We have also defined the **speed** at time t to be the magnitude of the velocity vector, that is, $|\mathbf{v}(t)|$. Notice that from $\boxed{3}$ and $\boxed{4}$ in Section 12.3, we have

$$|\mathbf{v}(t)|=|\mathbf{r}'(t)|=\frac{ds}{dt}= \text{rate of change of distance with respect to time.}$$

The **acceleration vector** $\mathbf{a}(t)$ at time t is defined as the derivative of the velocity:

$$\mathbf{a}(t)=\mathbf{v}'(t)=\mathbf{r}''(t).$$

EXAMPLE 1

The position vector of an object moving in a plane is given by $\mathbf{r}(t)=(4\cos t)\,\mathbf{i}+(\sqrt{2}\sin t)\,\mathbf{j}$. Find its velocity, speed, and acceleration when $t=\pi/4$ and illustrate geometrically.

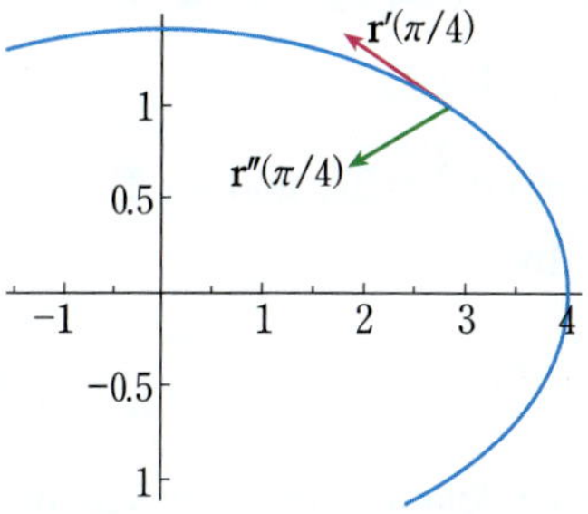

Figure 2

Solution The components of $\mathbf{r}$ give the parametric equations

$$x=4\cos t \quad \text{and} \quad y=\sqrt{2}\sin t.$$

Hence $\dfrac{x^2}{16}+\dfrac{y^2}{2}=1.$

In particular, the object is moving along the ellipse. (See Figure 2.)

The velocity and acceleration at time t are

$$\mathbf{v}(t) = \mathbf{r}'(t) = (-4\sin t)\,\mathbf{i} + (\sqrt{2}\cos t)\,\mathbf{j},$$
$$\mathbf{a}(t) = \mathbf{v}'(t) = (-4\cos t)\,\mathbf{i} + (-\sqrt{2}\sin t)\,\mathbf{j},$$

and the speed is

$$|\mathbf{v}(t)| = \sqrt{16\sin^2 t + 2\cos^2 t}.$$

When $t = \frac{\pi}{4}$, we have

$$\mathbf{v}\left(\frac{\pi}{4}\right) = -2\sqrt{2}\,\mathbf{i} + \mathbf{j}, \quad \mathbf{a}\left(\frac{\pi}{4}\right) = -2\sqrt{2}\,\mathbf{i} - \mathbf{j}, \quad \left|\mathbf{v}\left(\frac{\pi}{4}\right)\right| = 3.$$

These velocity and acceleration vectors are shown in Figure 2. ■

Since the velocity vector $\mathbf{v}(t)$ of a particle moving in space can be obtained by differentiating the position vector $\mathbf{r}(t)$, it follows that $\mathbf{r}(t)$ can be obtained by integrating $\mathbf{v}(t)$. Similarly, $\mathbf{v}(t)$ can be obtained by integrating $\mathbf{a}(t)$.

EXAMPLE 2

A moving particle starts at an initial position $\mathbf{r}(0) = \langle 3, 0, 0\rangle$ with initial velocity $\mathbf{v}(0) = 3\,\mathbf{j}$. Its acceleration is $\mathbf{a}(t) = -3\cos t\,\mathbf{i} - 3\sin t\,\mathbf{j} + 4\,\mathbf{k}$. Find its position at time t.

Solution Since $\mathbf{a}(t) = \mathbf{v}'(t)$ we have

$$\begin{aligned}\mathbf{v}(t) &= \int \mathbf{a}(t)\,dt = \int (-3\cos t\,\mathbf{i} - 3\sin t\,\mathbf{j} + 4\,\mathbf{k})\,dt \\ &= (-3\sin t)\,\mathbf{i} + (3\cos t)\,\mathbf{j} + 4t\,\mathbf{k} + \mathbf{C}\ .\end{aligned}$$

To find the value of the constant vector $\mathbf{C}$, we use the fact that $\mathbf{v}(0) = 3\,\mathbf{j}$. The preceding equation gives $3\,\mathbf{j} = \mathbf{v}(0) = 3\,\mathbf{j} + \mathbf{C}$, and hence $\mathbf{C} = \mathbf{0}$. Thus

$$\mathbf{v}(t) = (-3\sin t)\,\mathbf{i} + (3\cos t)\,\mathbf{j} + 4t\,\mathbf{k}\ .$$

Since $\mathbf{v}(t) = \mathbf{r}'(t)$ we have

$$\begin{aligned}\mathbf{r}(t) &= \int \mathbf{v}(t)\,dt = \int (-3\sin t\,\mathbf{i} + 3\cos t\,\mathbf{j} + 4t\,\mathbf{k})\,dt \\ &= (3\cos t)\,\mathbf{i} + (3\sin t)\,\mathbf{j} + 2t^2\,\mathbf{k} + \mathbf{D}\end{aligned}$$

To find $\mathbf{D}$, we use the initial condition $\mathbf{r}(0) = 3\mathrm{i}$, and so $\mathbf{D} = \mathbf{0}$. Hence the position at time t is given by

$$\mathbf{r}(t) = (3\cos t)\,\mathbf{i} + (3\sin t)\,\mathbf{j} + 2t^2\mathbf{k}.$$ ■

Remark The Newton's Second Law gives us an exact relationship between force, mass and acceleration. It can be expressed as a mathematical vector version:

$$\mathbf{F} = m\mathbf{a}$$

where $\mathbf{F}$ is a force, m a mass and $\mathbf{a}$ an acceleration.

A projectile is some object thrown in air or space. The curved path along which the projectile travels is what is known as trajectory. Projectile motion is the free fall motion of any body in a horizontal path with constant velocity.

EXAMPLE 3

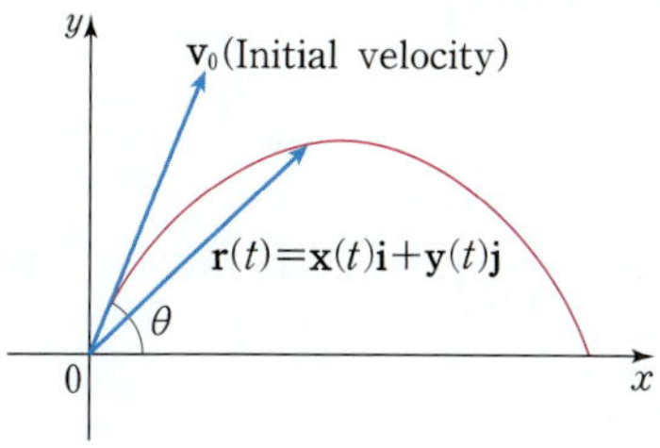

Figure 3

Suppose that a projectile is fired from the origin at time $t=0$ at angle of θ above the horizontal and initial velocity $\mathbf{v}_0$. (See Figure 3.) Assume that air resistance is negligible and the only external force is due to the gravity. What is the position vector $\mathbf{r}(t)$ of the projectile?

Solution By Newton's Second Law and the gravity, the force is

$$\mathbf{F} = m\mathbf{a} = -mg\mathbf{j}$$

where $g = |\mathbf{a}| \approx 9.8\,\mathrm{m/s^2}$. Integrating $\mathbf{v}'(t) = \mathbf{a} = -g\mathbf{j}$, $\mathbf{v} = -gt\mathbf{j} + \mathbf{C}$. Since $\mathbf{C} = \mathbf{v}(0) = \mathbf{v}_0$, $\mathbf{r}'(t) = \mathbf{v}(t) = -gt\mathbf{j} + \mathbf{v}_0$.

Integrating $\mathbf{r}'(t)$ again, the position vector is

$$\mathbf{r}(t) = -\frac{1}{2}gt^2\mathbf{j} + \mathbf{v}_0 t + \mathbf{D}.$$

Since $\mathbf{r}(0) = \mathbf{0}$, $\mathbf{D} = \mathbf{0}$. Hence

$$\boxed{3} \qquad \mathbf{r}(t) = -\frac{1}{2}gt^2\mathbf{j} + \mathbf{v}_0 t.$$

Let $|\mathbf{v}_0| = v_0$ be the initial speed of the projectile. Then

$$\mathbf{v}_0 = v_0\cos\theta\,\mathbf{i} + v_0\sin\theta\,\mathbf{j}.$$

Substituting $\mathbf{v}_0$ from $\boxed{3}$, the position vector of the projectile is

$$\mathbf{r}(t) = (v_0\cos\theta)t\,\mathbf{i} + \left((v_0\sin\theta)t - \frac{1}{2}gt^2\right)\mathbf{j}. \qquad \blacksquare$$

EXAMPLE 4

A foot ball player kicks a football at an angle of $30°$ above the horizontal in the field. The initial speed of the ball is 18 m/s. How far away from the kick off point will the football land, and what speed on the ground?

Solution With $v_0 = 18\,\mathrm{m/s}$, $\theta = 30°$ and $g = 9.8\,\mathrm{m/s^2}$, we get the components of position vector $\mathbf{r}(t)$ at time t

$$x = 18\cos 30°\, t = 9\sqrt{3}\,t \text{ and } y = 18\sin 30°\, t - 4.9t^2 = 9t - 4.9t^2.$$

Impact occurs when $y=0$ on the ground, that is, $9t-4.9t^2=0$. Solving this equation, we get $t=0, \dfrac{9}{4.9}$.

Then $x=9\sqrt{3}\times\dfrac{9}{4.9}\approx 28.63$ which means the football hits the field about 28.63m away.

The position vector of this football at time t is

$$\mathbf{r}(t)=9\sqrt{3}\,t\mathbf{i}+(9t-4.8t^2)\mathbf{j}.$$

Then the velocity of this football at time t is

$$\mathbf{v}(t)=9\sqrt{3}\,\mathbf{i}+(9-9.8t)\mathbf{j}.$$

Hence its speed on the ground is

$$\left|\mathbf{v}\left(\frac{9}{4.9}\right)\right|=\sqrt{(9\sqrt{3})^2+\left(9-9.8\times\frac{9}{4.9}\right)^2}=17.17\,\mathrm{m/s}. \qquad \blacksquare$$

Tangential and Normal components of Acceleration

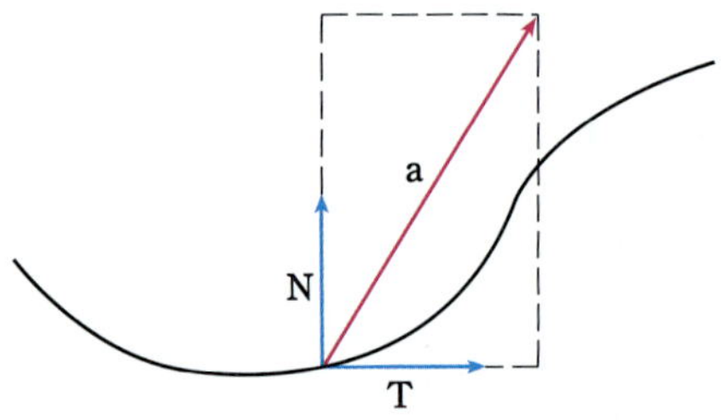

Figure 4

We are going to think about acceleration of the motion of an object along a curve as the second derivative of position, and we can further break down acceleration into two components called the tangential and normal components. (See Figure 4.)

The **tangential acceleration** a_{T} allows us to know how much of the acceleration acts in the direction of motion. The **normal acceleration** a_{N} is how much of the acceleration is orthogonal to the tangential acceleration. The tangential acceleration is a measure of the rate of change in the magnitude of the velocity vector, that is, speed, and the normal acceleration is a measure of the rate of change of the direction of the velocity vector.

Let $\mathbf{r}(t)$ be a position vector along a curve at time t. We can find the tangential acceleration by the Chain Rule to rewrite the velocity vector

$$\mathbf{v}=\frac{d\mathbf{r}}{dt}=\frac{d\mathbf{r}}{ds}\frac{ds}{dt}=\mathbf{T}\frac{ds}{dt}$$

using properties of Section 12.3.

Now, since acceleration is simply the derivative of velocity, we can find acceleration

$$\begin{aligned}\mathbf{a}&=\frac{d\mathbf{v}}{dt}=\frac{d}{dt}\left(\mathbf{T}\frac{ds}{dt}\right)=\frac{d^2s}{dt^2}\mathbf{T}+\frac{ds}{dt}\frac{d\mathbf{T}}{dt}\\&=\frac{d^2s}{dt^2}\mathbf{T}+\frac{ds}{dt}\left(\frac{d\mathbf{T}}{ds}\frac{ds}{dt}\right)=\frac{d^2s}{dt^2}\mathbf{T}+\frac{ds}{dt}\left(\kappa\mathbf{N}\frac{ds}{dt}\right)\\&=\frac{d^2s}{dt^2}\mathbf{T}+\kappa\left(\frac{ds}{dt}\right)^2\mathbf{N}.\end{aligned}$$

Note that $\frac{d\mathbf{T}}{ds} = \frac{d\mathbf{T}/dt}{ds/dt} = \frac{\mathbf{T}'}{|\mathbf{T}'|}\frac{|\mathbf{T}'|}{|\mathbf{r}'|} = \kappa\mathbf{N}$.

Let the speed of the object be $v = |\mathbf{v}| = |\mathbf{r}'| = \frac{ds}{dt}$. Since $\frac{d^2s}{dt^2} = \frac{d}{dt}v = v'$ and $\kappa\left(\frac{ds}{dt}\right)^2 = kv^2$, we define the tangential acceleration and normal acceleration of acceleration by

$$a_{\mathrm{T}} = v' \text{ and } a_{\mathrm{N}} = \kappa v^2, \tag{4}$$

respectively.

Then the acceleration of the motion of an object is

$$\mathbf{a} = a_{\mathrm{T}}\mathbf{T} + a_{\mathrm{N}}\mathbf{N}. \tag{5}$$

Note Since $\mathbf{T}$ and $\mathbf{N}$ are unit vectors and orthogonal to each other, $|\mathbf{a}|^2 = \mathbf{a}\cdot\mathbf{a} = (a_{\mathrm{T}}\mathbf{T} + a_{\mathrm{N}}\mathbf{N})\cdot(a_{\mathrm{T}}\mathbf{T} + a_{\mathrm{N}}\mathbf{N}) = a_{\mathrm{T}}^2 + a_{\mathrm{N}}^2$. The magnitude of the acceleration is $|\mathbf{a}| = \sqrt{a_{\mathrm{T}}^2 + a_{\mathrm{N}}^2}$.

Although 4 provides useful expressions for the tangential and normal components of acceleration, it is often desirable to have expressions in terms of $\mathbf{r}$, $\mathbf{r}'$ and $\mathbf{r}''$. To get this, we take the dot product of $\mathbf{v}$ and $\mathbf{a}$ as given by 5 :

$$\begin{aligned}\mathbf{v}\cdot\mathbf{a} &= v\mathbf{T}\cdot(v'\mathbf{T} + \kappa v^2\mathbf{N})\\ &= vv'\mathbf{T}\cdot\mathbf{T} + \kappa v^3\mathbf{T}\cdot\mathbf{N}\\ &= vv' \text{ (since } \mathbf{T}\cdot\mathbf{T} = 1 \text{ and } \mathbf{T}\cdot\mathbf{N} = 0).\end{aligned}$$

Thus

$$a_{\mathrm{T}} = v' = \frac{\mathbf{v}\cdot\mathbf{a}}{v} = \frac{\mathbf{r}'(t)\cdot\mathbf{r}''(t)}{|\mathbf{r}'(t)|}. \tag{6}$$

By Theorem 2 in Section 12.3, we have

$$a_{\mathrm{N}} = \kappa v^2 = \frac{|\mathbf{r}'(t)\times\mathbf{r}''(t)|}{|\mathbf{r}'(t)|^3}|\mathbf{r}'(t)|^2 = \frac{|\mathbf{r}'(t)\times\mathbf{r}''(t)|}{|\mathbf{r}'(t)|}. \tag{7}$$

EXAMPLE 5

A particle moves with position function $\mathbf{r}(t) = t\mathbf{i} + \frac{1}{2}t^2\mathbf{j} + \frac{1}{3}t^3\mathbf{k}$. Find the tangential and normal components of acceleration.

http://buzzard.pugetsound.edu/talks/beezer-2012-kms-math-ed.pdf

Solution We have

$$\mathbf{r}(t) = t\,\mathbf{i} + \frac{1}{2}t^2\,\mathbf{j} + \frac{1}{3}t^3\,\mathbf{k},$$

$$\mathbf{r}'(t) = 1\,\mathbf{i} + t\,\mathbf{j} + t^2\,\mathbf{k}\,,$$
$$\mathbf{r}''(t) = \mathbf{j} + 2t\,\mathbf{k}\,,$$
$$|\mathbf{r}'(t)| = \sqrt{1+t^2+t^4}\,.$$

Hence by 6, the tangential component is

$$a_{\mathrm{T}} = \frac{\mathbf{r}'(t)\cdot\mathbf{r}''(t)}{|\mathbf{r}'(t)|} = \frac{t+2t^3}{\sqrt{1+t^2+t^4}}\,.$$

Since

$$\mathbf{r}'(t)\times\mathbf{r}''(t) = \begin{vmatrix} \mathbf{i} & \mathbf{j} & \mathbf{k} \\ 1 & t & t^2 \\ 0 & 1 & 2t \end{vmatrix} = t^2\,\mathbf{i} - 2t\,\mathbf{j} + \mathbf{k}\,.$$

By 7, the normal component is

$$a_{\mathrm{N}} = \frac{|\mathbf{r}'(t)\times\mathbf{r}''(t)|}{|\mathbf{r}'(t)|} = \frac{\sqrt{1+4t^2+t^4}}{\sqrt{1+t^2+t^4}}\,.$$

■

CAS **EXAMPLE 6**

Find the velocity, acceleration, and speed of a particle with the given position function:

$$\mathbf{r}(t) = \langle \cos t,\ t^2,\ 3\sin t \rangle$$

http://matrix.skku.ac.kr/cal-lab/cal-13-4-Exm6.html

Solution

```
t=var('t')
x(t)=cos(t)
y(t)=t^2
z(t)=3*sin(t)
r(t)=(x(t), y(t), z(t))
v(t)=(diff(x(t), t), diff(y(t), t), diff(z(t), t)) # velocity
a(t)=(diff(x(t), t, 2), diff(y(t), t, 2), diff(z(t), t, 2)) #
acceleration
s(t)=v(t).norm() # speed
v(t)
```

Answer : (-sin(t), 2*t, 3*cos(t)) # velocity
a(t)
Answer : (-cos(t), 2, -3*sin(t)) # acceleration
s(t)
Answer : sqrt(abs(2*t)^2 + abs(-sin(t))^2 + abs(3*cos(t))^2) ■

12.4 EXERCISES (Motion Along A Space Curve: Velocity and Acceleration)

http://matrix.skku.ac.kr/Cal-Book/part2/CS-Sec-12-4-Sol.html

YouTube http://youtu.be/JlgB5XN3LI4

CAS **1-5.** Find the velocity, acceleration, and speed of a particle with the given position function. Sketch the path of the particle and draw the velocity and acceleration vectors for specified value of t.

1. $\mathbf{r}(t) = \langle t^3+1,\ t\rangle,\ t=2.$

http://matrix.skku.ac.kr/cal-lab/cal-13-4-1.html

Solution

```
var('t')
r(t)=(t^3+1, t)
v=diff(r(t), t)
a=diff(v, t)
s=v.norm()
print v, a, s
v2=v.subs(t=2)
a2=a.subs(t=2)
print v2, a2
```

Answer : (3*t^2, 1) (6*t, 0) sqrt(abs(3*t^2)^2 + 1) (12, 1) (12, 0)

```
p1=parametric_plot(r(t), (t, 0, 3))
p2=line([r(2), r(2)+v2],  color='red')
p3=line([r(2), r(2)+a2], color='green')
show(p1+p2+p3)
```

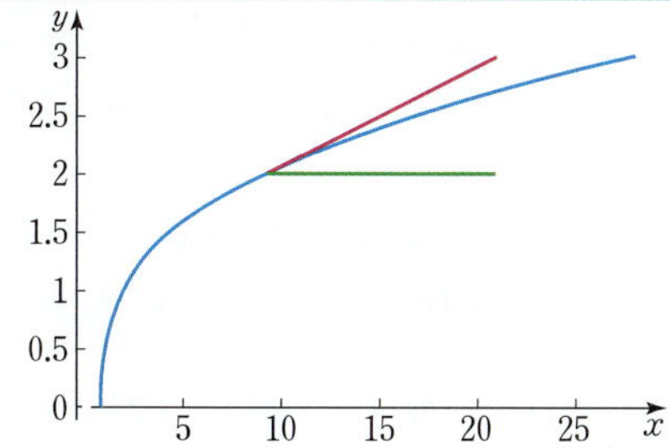

2. $\mathbf{r}(t) = \langle 2+2t,\ 4\sqrt{t}\rangle,\ t=1.$

http://matrix.skku.ac.kr/cal-lab/cal-13-4-2.html

Solution

```
var('t')
r(t)=(2+2*t, 4*sqrt(t))
v=diff(r(t), t)
a=diff(v, t)
v1=v.subs(t=1)
a1=a.subs(t=1)
s=v.norm()
print v, a, s
```

Answer : (2, 2/sqrt(t)) (0, -1/t^(3/2)) sqrt(abs(2/sqrt(t))^2 + 4)
(2, 2), (0, -1)

```
p1=parametric_plot(r(t), (t, 0, 3))
p2=line([r(1), r(1)+v1],  color='red')
p3=line([r(1), r(1)+a1], color='green')
show(p1+p2+p3)
```

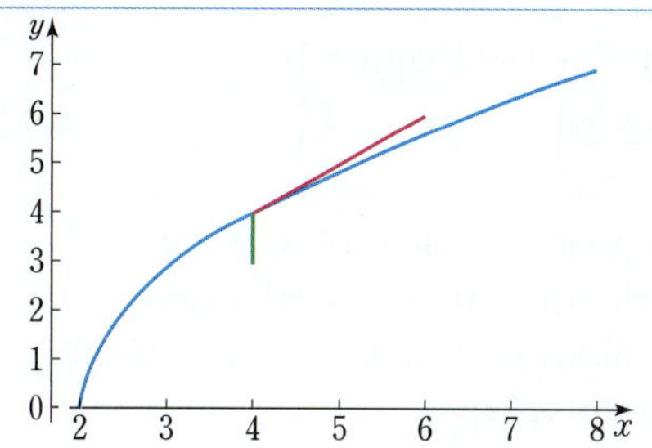

3. $\mathbf{r}(t) = e^{2t}\mathbf{i} + e^{-t}\mathbf{j},\ t=0.$

http://matrix.skku.ac.kr/cal-lab/cal-13-4-3.html

Solution

```
var('t')
r(t)=(exp(2*t), exp(-t))
v=diff(r(t), t)
a=diff(v, t)
v0=v.subs(t=0)
a0=a.subs(t=0)
s=v.norm()
print v, a, s
```

Answer : (2*e^(2*t), -e^(-t)) (4*e^(2*t), e^(-t)) sqrt(abs(-e^(-t))^2 + abs(2*e^(2*t))^2)
(2, -1), (4, 1)

```
p1=parametric_plot(r(t), (t, -1, 1))
p2=line([r(0), r(0)+v0], color='red')
p3=line([r(0), r(0)+a0], color='green')
show(p1+p2+p3)
```

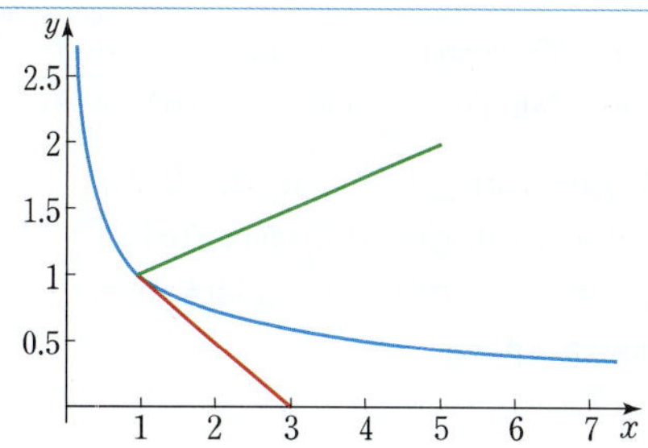

4. $\mathbf{r}(t) = 2\sin t\mathbf{i} + \cos t\mathbf{j}$, $t = \pi/3$.

http://matrix.skku.ac.kr/cal-lab/cal-13-4-4.html

Solution

```
var('t')
r(t)=(2*sin(t), cos(t))
v=diff(r(t), t)
a=diff(v, t)
vt=v.subs(t=pi/3)
at=a.subs(t=pi/3)
s=v.norm()
print v, a, s
```

Answer : (2*cos(t), -sin(t)) (-2*sin(t), -cos(t)) sqrt(abs(-sin(t))^2 + abs(2*cos(t))^2)

$\left(1, -\frac{\sqrt{3}}{2}\right), \left(-\sqrt{3}, -\frac{1}{2}\right)$

```
p1=parametric_plot(r(t), (t, 0, 3))
p2=line([r(pi/3), r(pi/3)+vt], color='red')
p3=line([r(pi/3), r(pi/3)+at], color='green')
show(p1+p2+p3)
```

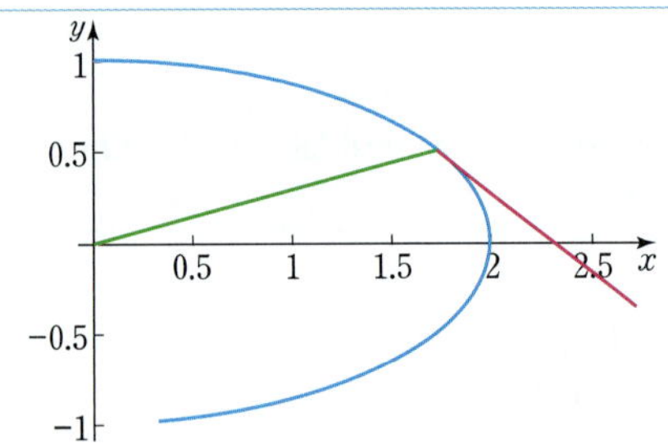

5. $\mathbf{r}(t) = 3\sin t\mathbf{i} + t\mathbf{j} + 3\cos t\mathbf{k}$, $t = 0$.

http://matrix.skku.ac.kr/cal-lab/cal-13-4-5.html

Solution

```
var('t')
r(t)=(3*sin(t), t, 3*cos(t))
v=diff(r(t), t)
a=diff(v, t)
v0=v.subs(t=0)
a0=a.subs(t=0)
s=v.norm()
print v, a, s
```

Answer : (3*cos(t), 1, -3*sin(t)) (-3*sin(t), 0, -3*cos(t)) sqrt(abs(-3*sin(t))^2 + abs(3*cos(t))^2 + 1)

```
p1=parametric_plot3d(r(t), (t, -1, 2))
p2=line([r(0), r(0)+v0], color='red')
p3=line([r(0), r(0)+a0],  color='green')
show(p1+p2+p3)
```

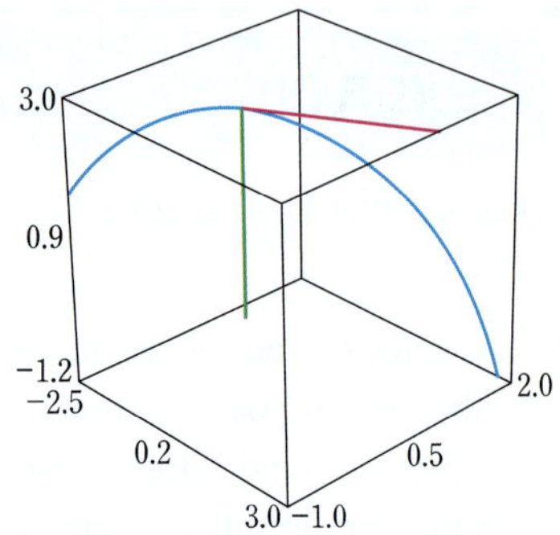

6-10. Find the velocity, acceleration and speed of a particle with the given position function.

6. $\mathbf{r}(t) = \langle 2t^2+1,\ t^3,\ 2t^2-1 \rangle$.

Solution $\mathbf{v}(t) = \mathbf{r}'(t) = \langle 4t,\ 3t^2,\ 4t \rangle$,

$\mathbf{a}(t) = \mathbf{r}''(t) = \langle 4,\ 6t,\ 4 \rangle$,

$s(t) = |\mathbf{v}(t)| = \sqrt{(4t)^2+(3t^2)^2+(4t)^2} = \sqrt{9t^4+32t^2}$.

7. $\mathbf{r}(t) = \langle 3\cos t,\ 2t,\ 3\sin t \rangle$.

Solution $\mathbf{v}(t) = \mathbf{r}'(t) = \langle -3\sin t,\ 2,\ 3\cos t \rangle$,

$\mathbf{a}(t) = \mathbf{r}''(t) = \langle -3\cos t,\ 0,\ -3\sin t \rangle$,

$s(t) = |\mathbf{v}(t)| = \sqrt{(-3\sin t)^2+2^2+(3\cos t)^2} = \sqrt{13}$.

8. $\mathbf{r}(t) = t^2\,\mathbf{i} + \ln t\,\mathbf{j} + t^3\,\mathbf{k}$.

Solution $\mathbf{v}(t) = \mathbf{r}'(t) = 2t\,\mathbf{i} + \frac{1}{t}\,\mathbf{j} + 3t^2\,\mathbf{k}$,

$\mathbf{a}(t) = \mathbf{r}''(t) = 2\,\mathbf{i} - \frac{1}{t^2}\,\mathbf{j} + 6t\,\mathbf{k}$,

$s(t) = |\mathbf{v}(t)| = \sqrt{(2t)^2+\frac{1}{t^2}+(3t^2)^2} = \sqrt{4t^2+9t^4+\frac{1}{t^2}}$.

9. $\mathbf{r}(t) = e^{2t}\cos t\,\mathbf{i} + e^{2t}\sin t\,\mathbf{j} + t\,\mathbf{k}$.

Solution $\mathbf{v}(t) = \mathbf{r}'(t) = e^{2t}(2\cos t - \sin t)\,\mathbf{i} + e^{2t}(2\sin t + \cos t)\,\mathbf{j} + \mathbf{k}$,

$\mathbf{a}(t) = \mathbf{r}''(t) = e^{2t}(3\cos t - 4\sin t)\,\mathbf{i} + e^{2t}(3\sin t + 4\cos t)\,\mathbf{j}$,

$s(t) = |\mathbf{v}(t)|$

$= \sqrt{e^{4t}(2\cos t - \sin t)^2 + e^{4t}(2\sin t + \cos t)^2 + 1}$

$= \sqrt{5e^{4t}+1}$.

10. $\mathbf{r}(t) = 2t\sin t\,\mathbf{i} + 2t\cos t\,\mathbf{j} + 2t^2\,\mathbf{k}$.

Solution $\mathbf{v}(t) = \mathbf{r}'(t) = (2t\cos t + 2\sin t)\,\mathbf{i} + (-2t\sin t + 2\cos t)\,\mathbf{j} + 4t\,\mathbf{k}$,

$\mathbf{a}(t) = \mathbf{r}''(t) = (-2t\sin t + 4\cos t)\,\mathbf{i} + (-2t\cos t - 4\sin t)\,\mathbf{j} + 4\,\mathbf{k}$,

$s(t) = |\mathbf{v}(t)|$

$= \sqrt{(2t\cos t + 2\sin t)^2 + (-2t\sin t + 2\cos t)^2 + (4t)^2}$

$= \sqrt{20t^2+4} = 2\sqrt{5t^2+1}$.

11-12. Find the velocity and position vectors of a particle that has the given acceleration and the given initial velocity and position.

11. $\mathbf{a}(t)=2\mathbf{k},\ \mathbf{v}(0)=\mathbf{i}-\mathbf{j},\ \mathbf{r}(0)=0.$

Solution $\mathbf{v}(t)=\int \mathbf{a}(t)dt=\int 2\,\mathbf{k}dt=2t\,\mathbf{k}+\mathbf{C}.$

Since $\mathbf{C}=\mathbf{v}(0)=\mathbf{i}-\mathbf{j}$, we have $\mathbf{v}(t)=\mathbf{i}-\mathbf{j}+2t\,\mathbf{k}$.

$\mathbf{r}(t)=\int \mathbf{v}(t)dt=\int \mathbf{i}-\mathbf{j}+2t\,\mathbf{k}dt=t\,\mathbf{i}-t\,\mathbf{j}+t^2\,\mathbf{k}+\mathbf{D}.$

Since $\mathbf{D}=\mathbf{r}(0)=0$, we have $\mathbf{r}(t)=t\,\mathbf{i}-t\,\mathbf{j}+t^2\,\mathbf{k}$.

12. $\mathbf{a}(t)=-4\mathbf{k},\ \mathbf{v}(0)=\mathbf{i}+\mathbf{j}-\mathbf{k},\ \mathbf{r}(0)=\mathbf{i}+2\mathbf{j}.$

Solution $\mathbf{v}(t)=\int \mathbf{a}(t)dt=\int -4\mathbf{k}\,dt=-4t\mathbf{k}+\mathbf{C}.$

Since $\mathbf{C}=\mathbf{v}(0)=\mathbf{i}+\mathbf{j}-\mathbf{k}$, we have

$\mathbf{v}(t)=\mathbf{i}+\mathbf{j}-(4t+1)\mathbf{k}.$

$$\mathbf{r}(t)=\int \mathbf{v}(t)dt$$
$$=\int \mathbf{i}+\mathbf{j}-(4t+1)\mathbf{k}dt=t\mathbf{i}+t\mathbf{j}-(2t^2+t)\mathbf{k}+\mathbf{D}.$$

Since $\mathbf{D}=\mathbf{r}(0)=\mathbf{i}+2\mathbf{j}$, we have

$\mathbf{r}(t)=(t+1)\mathbf{i}+(t+2)\mathbf{j}-(2t^2+t)\mathbf{k}.$

13-14. Find the position vector of a particle that has the given acceleration and the specified initial velocity and position.

13. $\mathbf{a}(t)=2\mathbf{i}+2\mathbf{j}+6t\,\mathbf{k},\ \mathbf{v}(0)=0,\ \mathbf{r}(0)=\mathbf{i}+\mathbf{k}.$

Solution $\mathbf{v}(t)=\int \mathbf{a}(t)dt=\int 2\mathbf{i}+2\mathbf{j}+6t\,\mathbf{k}dt$

$=2t\,\mathbf{i}+2t\,\mathbf{j}+3t^2\,\mathbf{k}+\mathbf{C}$. Since $\mathbf{C}=\mathbf{v}(0)=0$,

we have $\mathbf{v}(t)=2t\,\mathbf{i}+2t\,\mathbf{j}+3t^2\,\mathbf{k}$.

$\mathbf{r}(t)=\int \mathbf{v}(t)dt=\int 2t\mathbf{i}+2t\mathbf{j}+3t^2\mathbf{k}+\mathbf{C}\ dt$

$=t^2\mathbf{i}+t^2\mathbf{j}+t^3\mathbf{k}+\mathbf{D}$. Since $\mathbf{D}=\mathbf{r}(0)=\mathbf{i}+\mathbf{k}$, we have

$\mathbf{r}(t)=(t^2+1)\mathbf{i}+t^2\,\mathbf{j}+(t^3+1)\mathbf{k}.$

14. $\mathbf{a}(t)=4t\,\mathbf{i}+3t^2\,\mathbf{j}+\cos 2t\,\mathbf{k},\ \mathbf{v}(0)=\mathbf{i}+2\mathbf{j},$
$\mathbf{r}(0)=\mathbf{j}-4\mathbf{k}.$

Solution $\mathbf{v}(t)=\int \mathbf{a}(t)dt=\int 4t\,\mathbf{i}+3t^2\,\mathbf{j}+\cos 2t\,\mathbf{k}dt=$

$2t^2\,\mathbf{i}+t^3\,\mathbf{j}+\dfrac{\sin 2t}{2}\mathbf{k}+\mathbf{C}.$

Since $\mathbf{C}=\mathbf{v}(0)=\mathbf{i}+2j$, we have

$\mathbf{v}(t)=(2t^2+1)\mathbf{i}+(t^3+2)\mathbf{j}+\dfrac{\sin 2t}{2}\mathbf{k}.$

$$\mathbf{r}(t)=\int \mathbf{v}(t)dt=\int (2t^2+1)\mathbf{i}+(t^3+2)\,\mathbf{j}+\frac{\sin 2t}{2}\mathbf{k}\,dt$$
$$=\left(\frac{2}{3}t^3+t\right)\mathbf{i}+\left(\frac{1}{4}t^4+2t\right)\mathbf{j}-\frac{1}{4}\cos 2t\mathbf{k}+\mathbf{D}$$

Since $\mathbf{r}(0)=-\dfrac{1}{4}\mathbf{k}+\mathbf{D}=\mathbf{j}-4\,\mathbf{k}$, we have

$\mathbf{D}=\mathbf{j}-\dfrac{15}{4}\mathbf{k}$. Hence,

$$\mathbf{r}(t)=\left(\frac{2}{3}t^3+t\right)\mathbf{i}+\left(\frac{1}{4}t^4+2t\right)\mathbf{j}-\frac{1}{4}\cos 2t\,\mathbf{k}+\mathbf{j}-\frac{15}{4}\mathbf{k}$$

15. The position function of a particle is given by $r(t)=\langle 2t,\ t,\ t^2-4t\rangle$. When is the speed a minimum?

http://matrix.skku.ac.kr/cal-lab/cal-13-4-15.html

Solution

```
var('t')
r(t)=(t^2, t, t^2-4*t)
v=diff(r(t), t)
s=v.norm()
s0=diff(s, t)
solve(s0==0, t)
```

Answer : t=1

```
plot(s, (t, 0, 3))
```

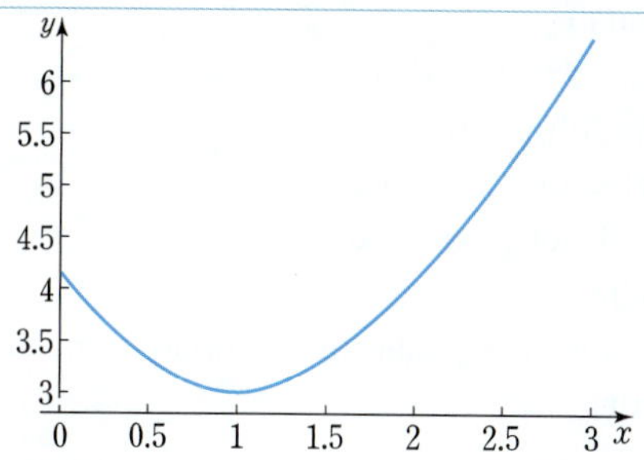

CAS **16-20.** Find the tangential and normal components of the acceleration vector.

16. $\mathbf{r}(t)=(t-t^3)\,\mathbf{i}+t^2\,\mathbf{j}.$

http://matrix.skku.ac.kr/cal-lab/cal-13-4-16.html

Solution

```
var('t')
r(t)=(t-t^3, t^2, 0)
dr=diff(r(t), t)
ddr=diff(r(t), t, 2)
T=dr.dot_product(ddr)/dr.norm()
print T
N=(dr.cross_product(ddr)).norm() / dr.norm()
print N
```

Answer : 2*(3*(3*t^2 - 1)*t + 2*t)/sqrt(abs(-3*t^2 + 1)^2 + abs(2*t)^2) sqrt(abs(6*t^2 + 2)^2)/sqrt(abs(-3*t^2 + 1)^2 + abs(2*t)^2)

17. $\mathbf{r}(t)=\cos t\mathbf{i}+\sin t\mathbf{j}+2t\mathbf{k}.$

http://matrix.skku.ac.kr/cal-lab/cal-13-4-17.html

Solution

```
var('t')
r(t)=(cos(t), sin(t), 2*t)
dr=diff(r(t), t)
ddr=diff(r(t), t, 2)
T=dr.dot_product(ddr)/dr.norm()
print T
N=(dr.cross_product(ddr)).norm() / dr.norm()
print N
```

Answer : 0 sqrt(abs(sin(t)^2 + cos(t)^2)^2 + abs(2*sin (t))^2 + abs(-2*cos(t))^2)/sqrt(abs(-sin(t))^2 + abs(cos(t))^2 + 4)

18. $\mathbf{r}(t)=2t\mathbf{i}+t^2\mathbf{j}+t^3\mathbf{k}.$

http://matrix.skku.ac.kr/cal-lab/cal-13-4-18.html

Solution

```
var('t')
r(t)=(2*t, t^2, t^3)
dr=diff(r(t), t)
ddr=diff(r(t), t, 2)
T=dr.dot_product(ddr)/dr.norm()
print T
N=(dr.cross_product(ddr)).norm() / dr.norm()
print N
```

Answer : 2*(9*t^3 + 2*t)/sqrt(abs(3*t^2)^2 + abs(2*t)^2 + 4) sqrt(abs(6*t^2)^2 + abs(-12*t)^2 + 16)/sqrt (abs(3*t^2)^2 + abs(2*t)^2 + 4)

19. $\mathbf{r}(t)=e^{2t}\,\mathbf{i}+t\,\mathbf{j}+e^{-2t}\,\mathbf{k}.$

http://matrix.skku.ac.kr/cal-lab/cal-13-4-19.html

Solution

```
var('t')
r(t)=(exp(2*t), t, exp(-2*t))
dr=diff(r(t), t)
ddr=diff(r(t), t, 2)
T=dr.dot_product(ddr)/dr.norm()
print T
N=(dr.cross_product(ddr)).norm() / dr.norm()
print N
```

Answer : -8*(e^(-4*t) - e^(4*t))/sqrt(abs(-2*e^(-2*t))^2 + abs(2*e^(2*t))^2 + 1) sqrt(abs(4*e^(-2*t))^2 + abs(-4*e^(2*t))^2 + 256)/sqrt(abs(-2*e^(-2*t))^2 + abs(2*e^(2*t))^2 + 1)

20. $\mathbf{r}(t)=t\,\mathbf{i}+\dfrac{1}{2}\cos^2 t\,\mathbf{j}+\dfrac{1}{2}\sin^2 t\,\mathbf{k}.$

http://matrix.skku.ac.kr/cal-lab/cal-13-4-20.html

Solution

```
var('t')
r(t)=(t, 1/2*(cos(t))^2, 1/2*(sin(t))^2)
dr=diff(r(t), t)
ddr=diff(r(t), t, 2)
T=dr.dot_product(ddr)/dr.norm()
print T
N=(dr.cross_product(ddr)).norm() / dr.norm()
print N
```

Answer : -2*(sin(t)^2 - cos(t)^2)*sin(t)*cos(t)/sqrt(abs(-sin (t)*cos(t))^2 + abs(sin(t)*cos(t))^2 + 1) sqrt(2)*sqrt(abs(sin(t)^2 - cos(t)^2)^2)/sqrt(abs (-sin(t)*cos (t))^2 + abs(sin(t)*cos(t))^2 + 1)

Torus and ICME 11 logo

Chapter 13

Partial Derivatives

Pierre de Fermat

1601-1665

Fermat's Last Theorem

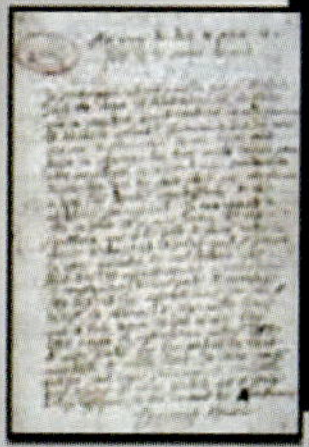

[Note's of Fermat]

A lawyer by profession, Fermat was an amateur mathematician who is credited with work that led to the development of calculus. He also made notable contributions in the areas of analytic geometry, probability and optics as well as the theory of numbers. He is best known for Fermat's Last Theorem, which led to tremendous advances in number theory .

YouTube http://youtu.be/

In this chapter, we consider functions of several variables. Such functions arise naturally in real life. We extend some of the standard notions of single variable calculus such as limit, continuity, derivative, tangent line to functions of several variables. One of the main concepts which we deal with, is partial derivatives and its applications to local maxima and local minima.

13.1 Multivariate Functions

Multivariate Functions

So far we have discussed calculus of functions of one variable. In this section we deal with functions of two or more variables. However in real life a function depends upon many dependent variables. Such functions, in general are called multi-variable functions or functions of several variables. In this section, we primarily deal with functions of two variables. All the concepts can be analogously extended to functions of two or more variables. Many functions in real life depend upon more than one variables. In chapters 3 and 5, we discussed how to obtain the derivatives and integrals of functions with one independent variable. In this section, we will consider functions with two or more independent variables.

Let us look at some examples: (i) The area A of a rectangle depends upon the length l and width b of the rectangle. Therefore, we can think of area as function of two variables, $A(l, b)$. (ii) The temperature T of particular place on earth depends on the longitude x and the latitude y of that point. So we can think of temperature as a function of two variables, say $T(x, y)$. (iii) The cost of a mobile hand-set depends upon many factors.

In this section we deal with real valued functions of two real variables. The domain of such functions is a set of points in $\mathbb{R}^2 = \{(x, y) \mid x, y \in \mathbb{R}\}$. We will assume that $\mathbb{R}^2$ is endowed with Euclidean distance, that is distance between any two points, (x_1, y_1) and (x_2, y_2) is given by

$$d((x_1, y_1), (x_2, y_2)) = \sqrt{(x_2 - x_1)^2 + (y_2 - y_1)^2}.$$

A point in $\mathbb{R}^2$ can also be referred to as a vector, which is the position vector of that point.

A real valued function of two real variables in general can be defined as follows:

DEFINITION 1

Let $D \subset \mathbb{R}^2$. A function f of two variables is a rule that assigns to each point $(a, b) \in D$ a unique real number denoted by $f(x, y)$. The set D is called the domain of f and its range is the set of values that f takes.

Generally, we write $z = f(x, y)$ to make explicit the value taken on by f at the point (x, y). Here x and y are called **independent variables** and z is called the **dependent variable**.

Most often the domain of the function is not mentioned explicitly, however it can be obtained from the function's expression by simply looking at set of points on which the function is well defined. Let us look at some examples.

EXAMPLE 1

Find the domain D of the following function:

(a) $f(x, y) = \dfrac{xy}{x^2 + y^2}$

(b) $f(x, y) = \dfrac{xy}{x^2 - y^2}$

(c) $f(x, y) = \sqrt{6 - 2x + 3y}$

(d) $f(x, y) = \ln[(16 - x^2 - y^2)(4 - x^2 - y^2)]$

(e) $f(x, y) = \sin^{-1}\left[\dfrac{2x - y}{x + y}\right]$.

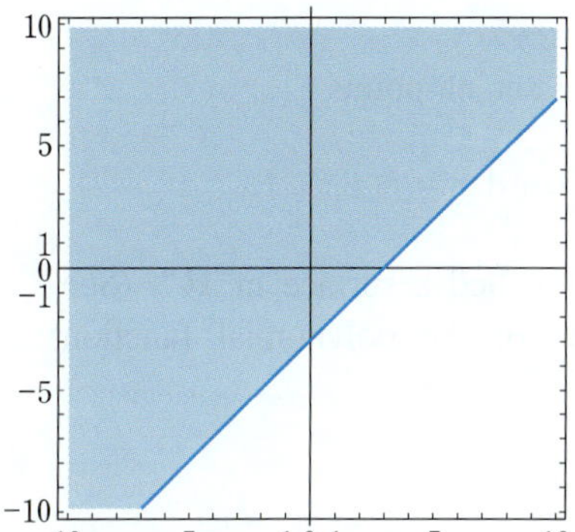

Figure 1

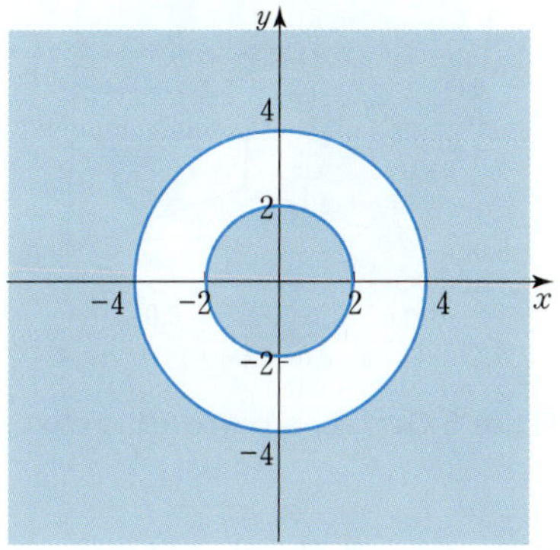

Figure 2

Solution (a) Note that the right handside of the expression works for all values except when the denominator is zero. That is, the domain of this function is

$$D = \{(x, y) \in \mathbb{R}^2 \mid x^2 + y^2 \neq 0\} = \mathbb{R}^2 - \{(0, 0)\}.$$

(b) The domain of this function is set of points (x, y) such that $x^2 - y^2 \neq 0$. That is, $x = \pm y$. Thus the domain D of f is $D = \mathbb{R}^2 - \{(x, y) \in \mathbb{R}^2 \mid x \neq \pm y\}$.

(c) The domain of f is a set of points $(x, y) \in \mathbb{R}^2$ such that $6 - 2x + 3y \geq 0$. See Figure 1. That is

$$D = \left\{(x, y) \in \mathbb{R}^2 \mid \frac{x}{3} - \frac{y}{2} \leq 1\right\}.$$

(d) We know that $\ln$ is defined only for positive reals. That is $(16 - x^2 - y^2)(4 - x^2 - y^2) > 0$. This happens when (i) $16 - x^2 - y^2 > 0$, $4 - x^2 - y^2 > 0$ or (ii) $16 - x^2 - y^2 < 0$, $4 - x^2 - y^2 < 0$. For (i), $D_1 = \{(x, y) \in \mathbb{R}^2 | x^2 + y^2 < 4\}$. For (ii), $D_2 = \{(x, y) \in \mathbb{R}^2 | x^2 + y^2 > 16\}$.

Then the domain D of f is $D_1 \cup D_2$. Thus $D = \mathbb{R}^2 - \{(x, y) \in \mathbb{R}^2 | 4 \leq x^2 + y^2 \leq 16\}$. See Figure 2.

(e) The domain of this function is when right handside make sense. That when $x + y \neq 0$ and $\left|\dfrac{2x - y}{x + y}\right| \leq 1$. The domain is plotted in Figure 3.

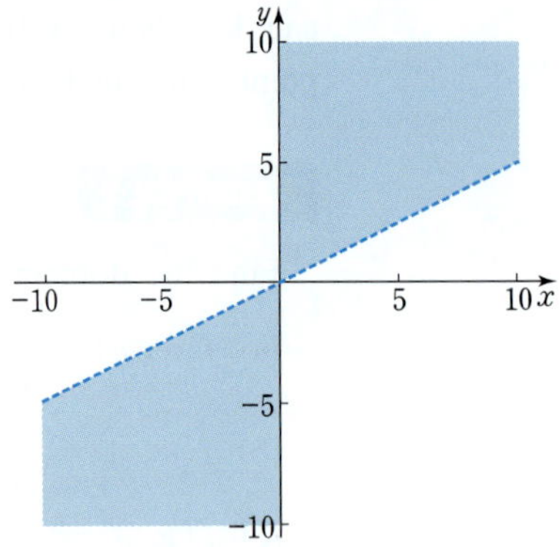

Figure 3

Graph of Multivariate Function

Let $f : D \subseteq \mathbb{R}^2 \to \mathbb{R}$. Then the graph of f is a set of points

$$\text{graph } f = \{(x, y, z) \in \mathbb{R}^3 \mid (x, y) \in D \text{ and } z = f(x, y)\}.$$

Usually, the graph of a function $z = f(x, y)$ is called a surface in $\mathbb{R}^3$. (See Figure 4.) In Figure 5 the surface is the graph of the polynomial function $z = 2x^2 - 2y^2 + 2$ is plotted.

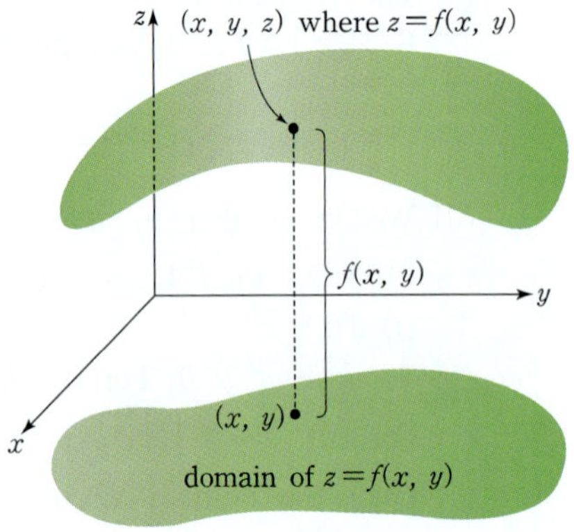

Figure 4 Graph of a function of x and y is a surface

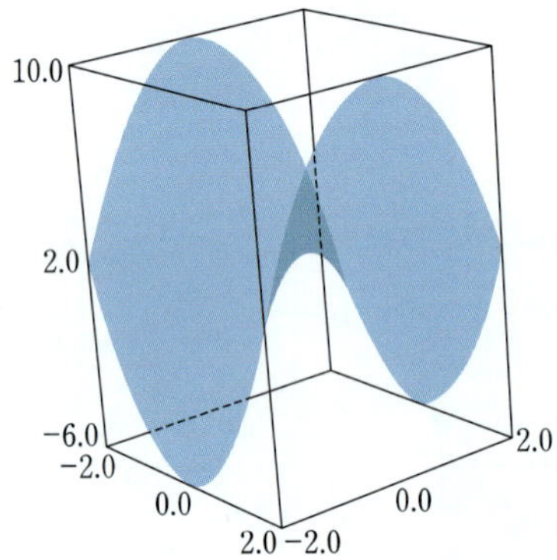

Figure 5 Graph of a polynomial function

Unlike one variable function $y = f(x)$, plotting the graph of a function $z = f(x, y)$ in $\mathbb{R}^3$ is not an easy task. One way to plot is to look at set of points $(x, y, f(x, y) = c)$ for various values of c, which is the intersection of the surface and the plane $z = c$. The intersection is called the level curve at the level $z = c$. If we know these level curves for all values of c, then we can visualize the surface $z = f(x, y)$.

Note that level curve $f(x, y) = c$ is the set of all points in the domain of f at which f takes value c. Level curves are also called contour lines. Thus to plot the graph of a surface $z = f(x, y)$ all we need is to plot level curves (contour lines) for various values of c and then put these level curves at the corresponding level $z = c$ planes. (See Figure 6(a).)

A contour map of a function $f(x, y)$ is a 2-dimensional graph showing several level curves corresponding to several values of c. A contour map of $f(x, y)$ lies in the xy-plane. (See Figure 6(b).)

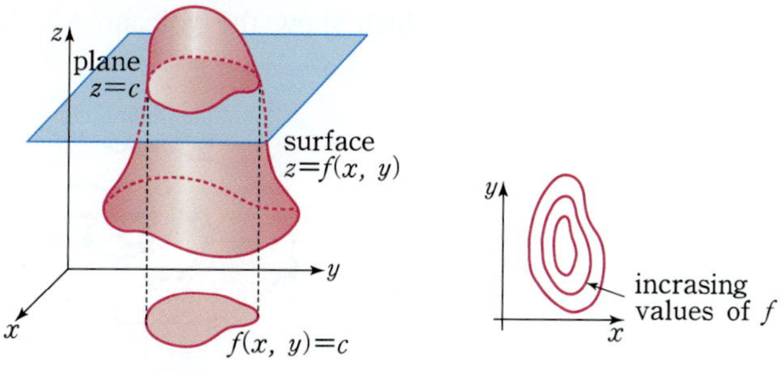

(a) (b)

Figure 6 Surface in (a) and level curves in (b)

EXAMPLE 2

Plot the surface and level curves of $z = x^2 - y^2$.

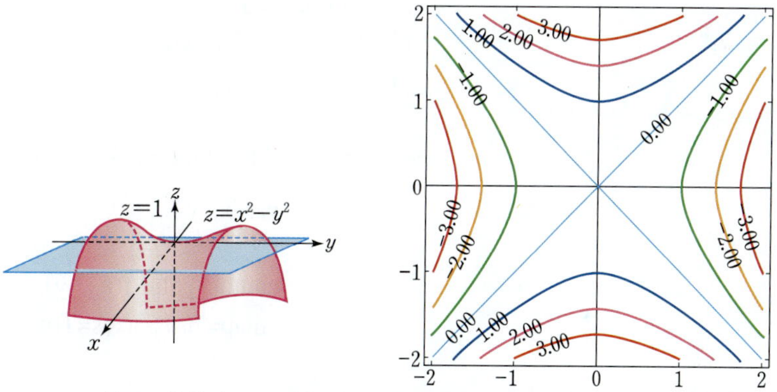

Figure 7 Surface and level curves in Example 2

In most instances the task of graphing level curves of a function of two variables $z = f(x, y)$ is formidable. A CAS was used to generate the surfaces and corresponding level curves in Figure 8.

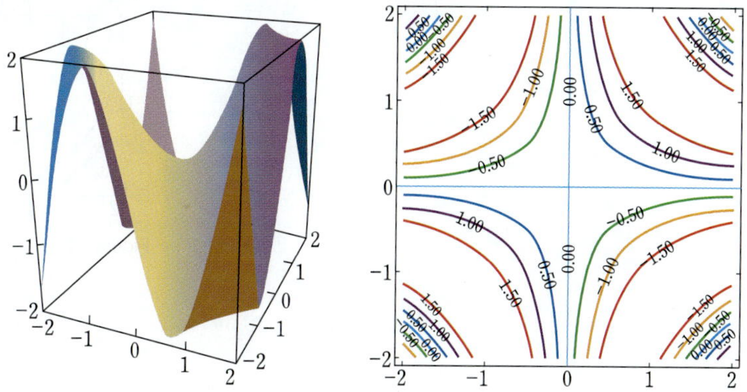

(a) (b)

Figure 8 Graph of $f(x, y) = 2\sin xy$ in (a); level curves in (b)

The level curves of a function f are also called **contour lines**. On a practical level, **contour maps** are often used to display curves of equal elevation. In Figure 9, we see that a contour map illustrates the various segments of a hill that have a given altitude. This shows heights of the mountain in in Figure 10, which show the volcano Mt. Halla, in the state of Jeju, Korea.

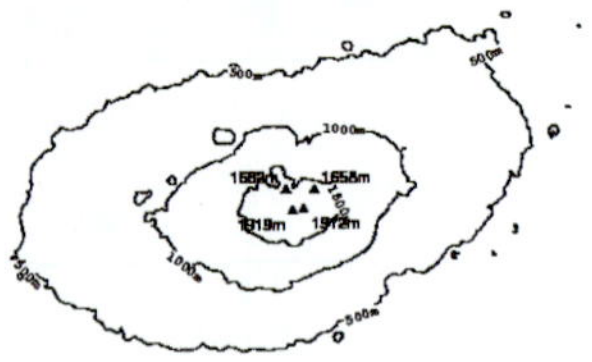

Figure 9 Contour map of Mt. Halla

Figure 10 Map of Mt. Halla in the state of Jeju, South Korea

CAS **EXAMPLE 3**

Graph the level curves of the following functions along with the surface:

(1) $f(x, y) = x^2 + y^2$

(2) $f(x, y) = x^2 - y^2$

(3) $f(x, y) = \sin x \cos y$

(4) $f(x, y) = 4xye^{-x^2-y^2}$

Solution (1) http://matrix.skku.ac.kr/cal-lab/m-Sec13-1-Exm-3.html

```
var('x, y')
f(x, y)=x^2+y^2
contour_plot(f(x, y), (x, -2, 2), (y, -2, 2), fill=False, cmap='hsv', axes=True, labels=True)
```

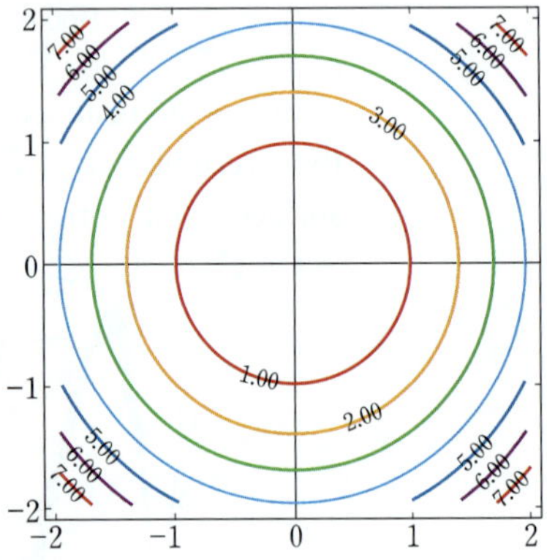

Figure 11

(2)

```
var('x, y')
f(x, y)=x^2-y^2
contour_plot(f(x, y), (x, -2, 2), (y, -2, 2), fill=False,
cmap='hsv', axes=True, labels=True)
```

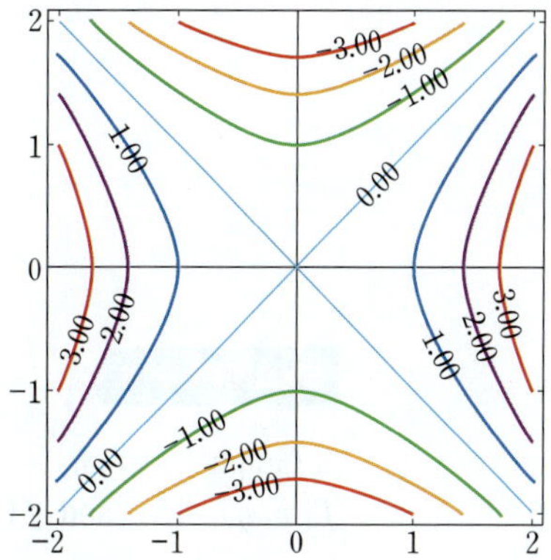

Figure 12

(3)

```
var('x, y')
f(x, y)=sin(x)*cos(y)
contour_plot(f(x, y), (x, -2, 2), (y, -2, 2), fill=False,
cmap='hsv', axes=True, labels=True)
```

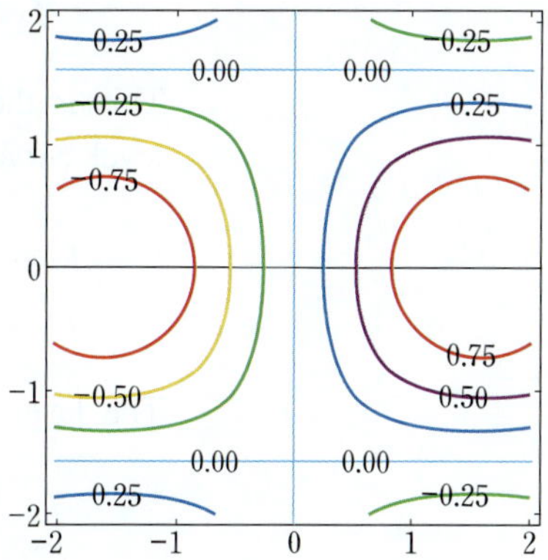

Figure 13

(4) Level curves along with the surface:

```
var('x, y')
f(x, y)=4*x*y*exp(-x^2-y^2)
contour_plot(f(x, y), (x, -2, 2), (y, -2, 2), fill=False,
cmap='hsv', axes=True, labels=True)
```

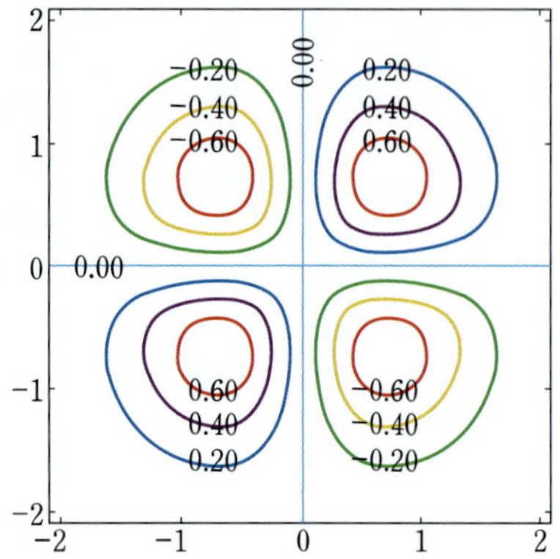

Figure 14

■

EXAMPLE 4

Graph $f(x,y)=100-x^2-y^2$ and plot the level curves $f(x,y)=0$, $f(x,y)=51$ and $f(x,y)=75$ in the domain of f in the plane.

Solution The domain of f is the entire xy-plane, and the range of f is the set of real numbers less than or equal to 100. The graph is the paraboloid $z=100-x^2-y^2$, a portion of which is shown in Figure 15.

In Figure 16, The level curve $f(x,y)=0$ is the set of points in the $xy-$plane such that

$$f(x,y)=100-x^2-y^2=0 \quad \text{or} \quad x^2+y^2=100.$$

This is the circle of radius 10 centered at the origin. Similarly, the level curves $f(x,\ y)=51$ and $f(x,\ y)=75$ are the circles

$$f(x,\ y)=100-x^2-y^2=51 \quad \text{or} \quad x^2+y^2=49$$
$$f(x,\ y)=100-x^2-y^2=75 \quad \text{or} \quad x^2+y^2=25.$$

The level curve $f(x,\ y)=100$ consists of the origin alone. ■

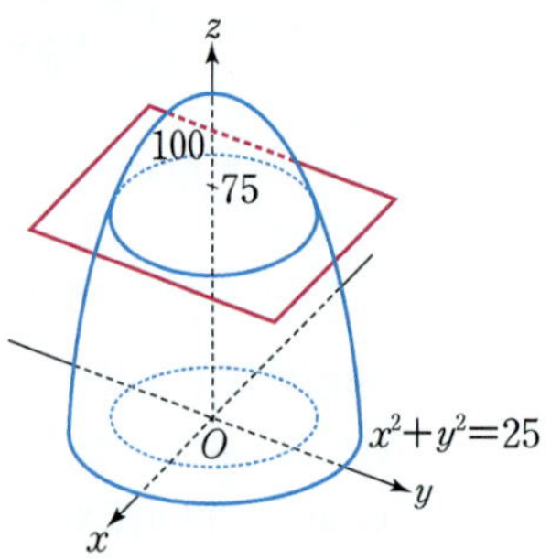

Figure 15

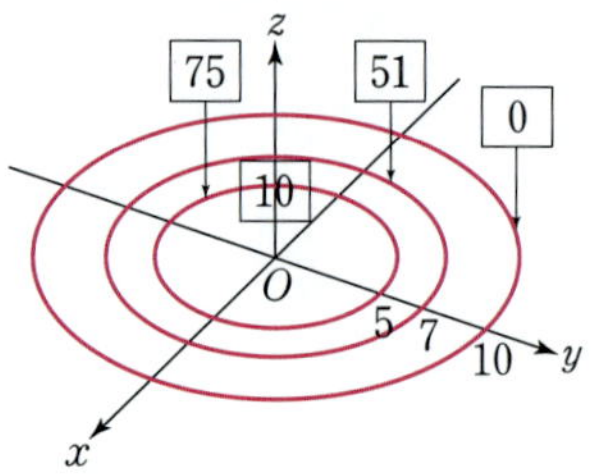

Figure 16

Functions of three variables

The definition of functions of three or more variables are simply generalizations of the definition of one variable. For example, a function of three variables is a rule of correspondence that assigns to each ordered triple of real number (x, y, z) in a subset of $\mathbb{R}^3$, one and only one number w in the set R of real numbers. A function of three variables is usually denoted by $w = f(x, y, z)$ or $w = F(x, y, z)$.

For example, the volume V and surface area S of a rectangular box are polynomial functions of three variables:

$$V = xyz \text{ and } S = 2xy + 2yz + 2zx.$$

The set of points (x, y, z) in space where a function of three independent variables has a value $f(x, y, z) = c$ is called a level surface of f.

EXAMPLE 5

Describe the level surfaces of the function $w = \sqrt{x^2 + y^2 + z^2}$.

Solution The value of w is the distance from the origin to the point (x, y, z). Each level surface

$$\sqrt{x^2 + y^2 + z^2} = a, \quad a > 0$$

is a half-sphere of radius a centered at the origin. The level surface $\sqrt{x^2 + y^2 + z^2} = 0$ consists of the origin alone. ■

Note We are not graphing the function in this example. We are looking at level surfaces in the domain of the function. The level surfaces show how the function values change as we move through its domain D. If we remain on a sphere of radius a centered at the origin, the function maintains a constant value, namely a. If we move from a point on one sphere to a point on another, the function value increases if we move away from the origin and decreases if we move toward the origin. The manner in which the values change depend on the direction chosen; the relationship between the function values and the direction travelled is important.

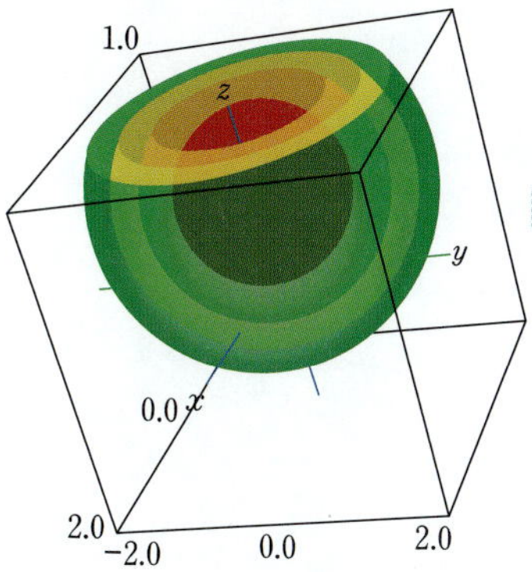

Figure 17 Level surfaces in (b) of Example 6

CAS **EXAMPLE 6**

(a) The level surfaces of the polynomial $f(x, y, z) = x - y + 2z$ are a family of parallel planes defined by $x - y + 2z = c$. (See Figure 18.)

(b) The level surfaces of the polynomial $f(x, y, z) = x^2 + y^2 + z^2$ are a family of concentric spheres defined by $x^2 + y^2 + z^2 = c$, $c > 0$. See Figure 17.

(c) The level surfaces of the rational function $f(x, y, z) = \dfrac{x^2 + z^2}{y}$ are given by $\dfrac{x^2+z^2}{y}=c$ or $x^2+z^2=cy$. A few members of this family of paraboloids are given in Figure 19.

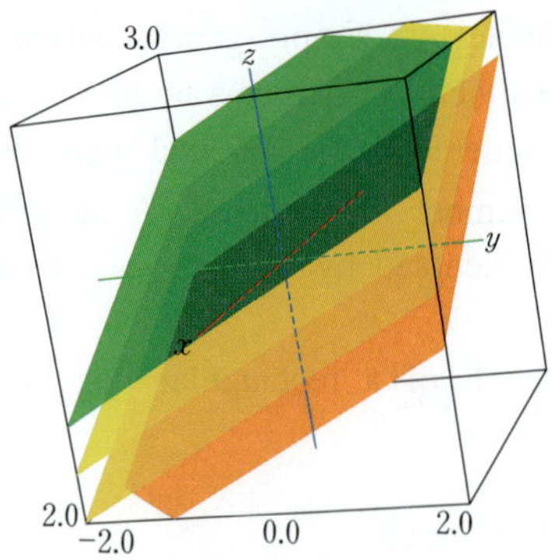

Figure 18 Level surfaces in (a) of Example 6

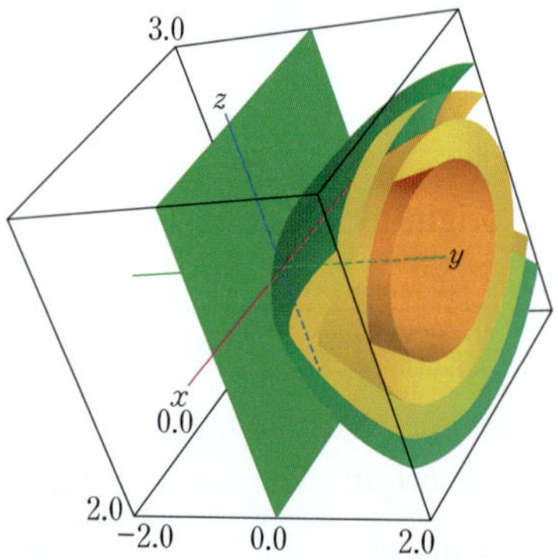

Figure 19 Level surfaces in (c) of Example 6

Solution (a) http://matrix.skku.ac.kr/cal-lab/m-Sec13-1-Exm-6.html

```
var('x,y,z')
f(x,y,z)=x-y+2*z
p=Graphics()
for k, col in [(1, 'red'),(2, 'orange'),(3, 'yellow'),(4, 'green')]:
p+=implicit_plot3d(f(x,y,z)==k,(x,-2,2),(y,-2,2),(z,-1,3),opacity=
0.7, color=col)
p
```

(b)

```
var('x,y,z')
f(x,y,z)=x^2+y^2+z^2
p=Graphics()
for k, col in [(1, 'red'),(2, 'orange'),(3, 'yellow'),(4, 'green')]:
p+=implicit_plot3d(f(x,y,z)==k,(x,-2,2),(y,-2,2),(z,-3,1),opacity=
0.7, color=col)
p                          # Figure 17
```

(c)

```
var('x,y,z')
f(x,y,z)=(x^2+z^2)/y
p=Graphics()
for k, col in [(1, 'red'),(2, 'orange'),(3, 'yellow'),(4, 'green')]:
p+=implicit_plot3d(f(x,y,z)==k,(x,-2,2),(y,-2,2),(z,-3,3),opacity=
0.7, color=col)
p
```

■

13.1 EXERCISES (Multivariate Functions)

http://matrix.skku.ac.kr/Cal-Book/part2/CS-Sec-13-1-Sol.html YouTube http://youtu.be/As_0AYApHIM

1. If $f(x, y) = x^3 - 3xy + y^3$, find (a) $f(-3, 3)$, (b) $f\left(\frac{1}{x}, \frac{3}{y}\right)$, (c) $\frac{f(x+k, y) - f(x,y)}{k}$.

Solution (a) $f(-3,3) = (-3)^3 - 3(-3)3 + 3^3 = 27.$

(b) $f\left(\frac{1}{x}, \frac{3}{y}\right) = \left(\frac{1}{x}\right)^3 - 3\left(\frac{1}{x}\right)\left(\frac{3}{y}\right) + \left(\frac{3}{y}\right)^3$

$= \frac{1}{x^3} - \frac{9}{xy} + \frac{27}{y^3}.$

(c) $\frac{f(x+k, y) - f(x,y)}{k}$

$= \frac{1}{k}\{(x+k)^3 - 3(x+k)y + y^3 - [x^3 - 3xy + y^3]\}$

$= 3x^2 + 3xk + k^2 - 3y.$

2. $f(x, y, z) = \sqrt{16 - 4x^2 - y^2 + 9z^2}$, find (a) $f(1, -2, 3)$, (b) $f\left(\frac{1}{x}, \frac{3}{y}, \frac{2}{z}\right)$.

Solution (a) $f(1, -2, 3) = \sqrt{16 - 4 \cdot 1^2 - (-2)^2 + 9 \cdot 3^2}$ $= \sqrt{89}$.

(b) $f\left(\frac{1}{x}, \frac{3}{y}, \frac{2}{z}\right) = \sqrt{16 - 4\left(\frac{1}{x}\right)^2 - \left(\frac{3}{y}\right)^2 + 9\left(\frac{2}{z}\right)^2}$

$= \sqrt{16 - \frac{4}{x^2} - \frac{9}{y^2} + \frac{36}{z^2}}.$

3-4. Find the domain of the given function.

3. $f(x, y) = \dfrac{1}{(x^2 - 4y^2)^2}$

4. $f(x, y, z) = \dfrac{\sqrt{16 - x^2 - z^2}}{y - 4}$

5-6. Find the range of the given function.

5. $f(x, y) = 1 + e^{xy}$

Solution

```
var('x,y,z')
implicit_plot3d(1+exp(x*y)==z,(x,-2.5,2.5),(y,-2.5,2
.5),(z,0,3),opacity=0.3)
```

Then the range is $R = \{f(x, y) \mid f(x, y) > 1\}$.

6. $f(x, y, z) = \cos(x + y + 2z)$

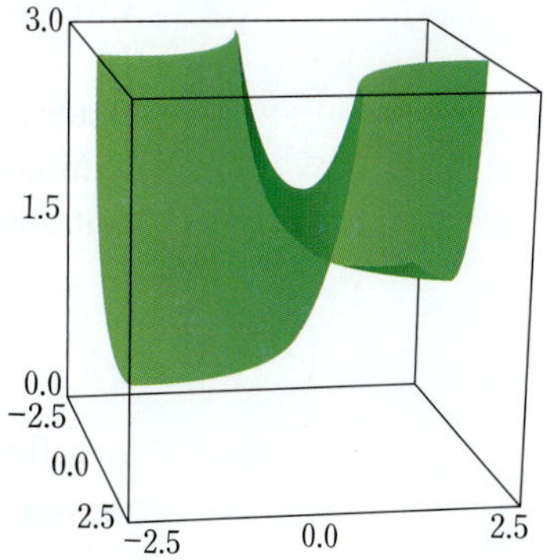

7-8. Sketch typical level surfaces of the function.

7. $f(x, y, z) = y + z$

Solution

```
var('x,y,z')
f(x,y,z)=y+z
p=Graphics()
for k, col in [(1,'red'), (2,'orange'), (3, 'yellow'),
(4,'green')]:
    p+=implicit_plot3d(f(x,y,z)==k,(x,-2,2),
(y,-2,2),(z,-1,3),opacity=0.7,color=col)
p
```

8. $f(x, y, z) = \dfrac{x^2}{4} + \dfrac{y^2}{16} + \dfrac{z^2}{9}$

9. $f(x, y, z) = \ln\left(\dfrac{x^2}{4} + \dfrac{y^2}{16} + \dfrac{z^2}{9}\right)$

10. Given that

$$f(x, y, z) = \frac{x^2}{4} + \frac{y^2}{16} + \frac{z^2}{9}$$

find the $x-$, $y-$ and $z-$intercepts of the level surface that passes through $(2, -4, 3)$.

13.2 Limits and Continuity of Multivariate Functions

Limits and Continuity

Before discussing limits and continuity, we need to introduce some terminology about sets that will be useful in the sections and chapters that follow. The set

$$\{(x, y) \mid (x-x_0)^2+(y-y_0)^2 < \delta^2\}$$

is called an **open disk** centered at (x_0, y_0) with radius δ. On the other hand, the set

$$\{(x, y) \mid (x-x_0)^2+(y-y_0)^2 \le \delta^2\}$$

is a closed disk. A closed disk contains all points interior to and on a circle. See Figure 1 (a). Let R be some region of the xy-plane. Then a point (a, b) is said to be an **interior point** of R, if there is at least one open disk centered at (a, b) that contains only points of R. We say that (a, b) is a **boundary point** of R if the interior of each open disk centered at (a, b) contains points from both R and R^c. The region R is said to be **open** if it contains no boundary points and **closed** if it contains all its boundary points. See Figure 1. A region R is said to be **bounded** if it can be contained in a sufficiently large rectangle or disk in the plane. These concepts can be carry over naturally to $\mathbb{R}^3$. For example, an open ball consists of all points interior to, but not on, a sphere with center (x_0, y_0) and radius $\delta > 0$;

$$\{(x, y, z) \mid (x-x_0)^2+(y-y_0)^2+(z-z_0)^2 < \delta^2\}.$$

A region in $\mathbb{R}^3$ is bounded if it can be contained in a sufficiently large disk.

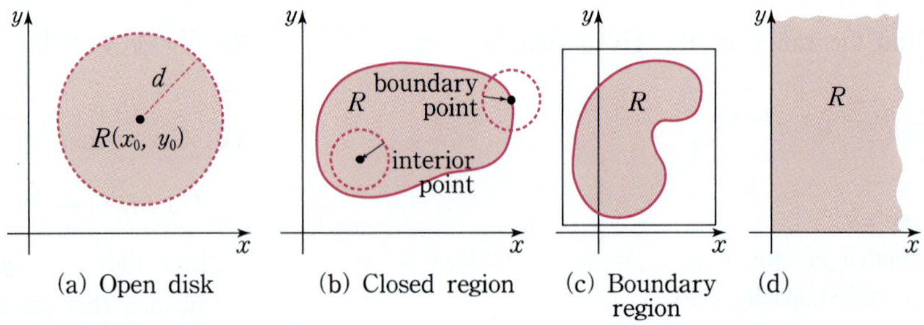

Figure 1 Various regions in $\mathbb{R}^2$

Let $z=f(x, y)$ be a function defined on a region $D \subseteq \mathbb{R}^2$.

$$\lim_{(x, y) \to (a, b)} f(x, y) = l$$

which means $f(x, y)$ is getting close to the number l as (x, y) close to (a, b). A precise definition can be given as follows;

DEFINITION 1

Let $f: D \subseteq \mathbb{R}^2 \to \mathbb{R}$ and $(a, b) \in \mathbb{R}^2$ (The point (a, b) need not necessarily be in D). Then we say that the limit of $f(x, y)$ as (x, y) approaches to (a, b) is l, if for every $\epsilon > 0$, there exists $\delta > 0$ such that such that $|f(x, y) - l| < \epsilon$ when $0 < \sqrt{(x-a)^2 + (y-b)^2} < \delta$ for $(x, y) \in D$.

This can be also expressed as

$$\lim_{(x,y) \to (a,b)} f(x,y) = l \text{ and } f(x, y) \to l \text{ as } (x, y) \to (a, b).$$

EXAMPLE 1

Let $f(x, y) = 5$. Let us check if $\lim_{(x, y) \to (1, -2)} f(x, y)$ exists.

Solution It is straightforward to guess that the limit exists and it is 5. Let $\epsilon > 0$ be given. To prove that $\lim_{(x, y) \to (1, -2)} f(x, y) = 5$, we need to find $\delta > 0$ such that when $0 < \sqrt{(x-1)^2 + (y+2)^2} < \delta$, we have $|f(x, y) - 5| < \epsilon$. Note that $|f(x, y) - 5| = 0$ for all (x, y). Hence any choice of δ will work. Thus

$$\lim_{(x, y) \to (1, -2)} f(x, y) = 5.$$ ■

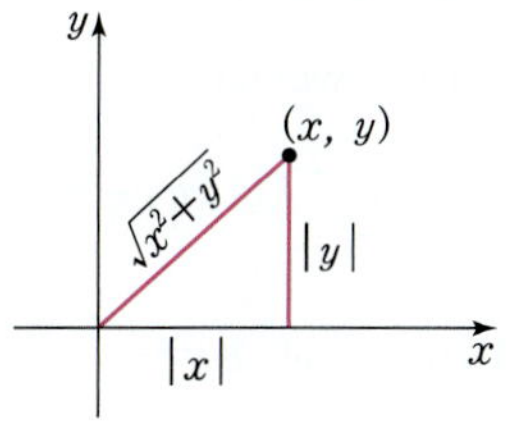

Figure 2

EXAMPLE 2

Let $f(x, y) = x + y$. Let us check if $\lim_{(x, y) \to (0, 0)} f(x, y)$ exists.

Solution It is not hard to guess that when $(x, y) \to (0, 0)$, $x + y \to 0$. Therefore, the limit must be 0. Let us prove this. Let $\epsilon > 0$. We need to find $\delta > 0$ such that when

$$0 < \sqrt{(x-0)^2 + (y-0)^2} = \sqrt{x^2 + y^2} < \delta,$$

we have $|f(x, y) - 0| = |x + y| < \epsilon$.

For any $(x, y) \in \mathbb{R}^2$, it is easy to that (See Figure 2.) $|x| < \sqrt{x^2 + y^2}$ and $|y| < \sqrt{x^2 + y^2}$. Thus we have

$$|x + y| \leq |x| + |y| \qquad \text{(Triangle Law of Modulus)}$$

$$\leq \sqrt{x^2+y^2}+\sqrt{x^2+y^2}=2\sqrt{x^2+y^2}.$$

Therefore,

$$\boxed{1} \qquad |f(x,y)-0|=|x+y|\leq |x|+|y|\leq 2\sqrt{x^2+y^2}.$$

This suggests that we may choose $\delta=\epsilon/2$. Now if $0<\sqrt{x^2+y^2}<\delta=\epsilon/2$, from $\boxed{1}$ we have $|f(x,y)-0|<\epsilon$. Hence, $\lim_{(x,y)\to(0,0)} f(x,y)$ exists and it is 0. ■

EXAMPLE 3

(1) Let $f(x,y)=x$. Then show that $\lim_{(x,y)\to(a,b)} f(x,y)=a$.

(2) Let $f(x,y)=y$. Then show that $\lim_{(x,y)\to(a,b)} f(x,y)=b$.

Solution (1) Let $\epsilon>0$. We need to find $\delta>0$ such that when

$$0<\sqrt{(x-a)^2+(y-b)^2}<\delta,$$

we have $|f(x,y)-a|=|x-a|<\epsilon$.
For any $(x,y)\in\mathbb{R}^2$, it is easy to that

$$\boxed{2} \qquad |x-a|\leq\sqrt{(x-a)^2+(y-b)^2}.$$

This suggests that we may choose $\delta=\epsilon$. Now if $0<\sqrt{(x-a)^2+(y-b)^2}<\delta=\epsilon$, from $\boxed{2}$ we have

$$|f(x,y)-a|<\epsilon. \text{ Hence, } \lim_{(x,y)\to(a,b)} f(x,y)=a.$$

(2) Similar to (1), we can prove $\lim_{(x,y)\to(a,b)} f(x,y)=b$. ■

Limit Along a Path

The notion of (x,y) "approaching" a point (a,b) is not as simple as it is for functions of one variable. In the xy-plane, there are an infinite number of ways of approaching a point (a,b), as shown in Figure 3. In order that $\lim_{(x,y)\to(a,b)} f(x,y)$ exist, we require that f approach to l along every possible curve or path through (a,b). In particular,

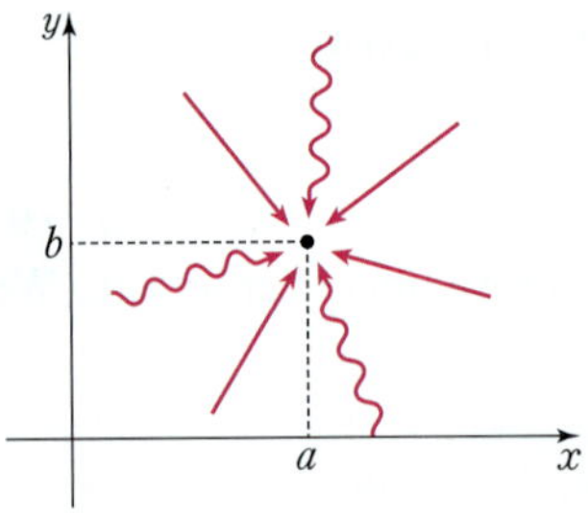

Figure 3

(i) If $f(x,y)$ does not approach the same number l for two different paths to (a,b), then $\lim_{(x,y)\to(a,b)} f(x,y)$ does not exist.

(ii) If the limit of (x,y) does not exists along some path through (a,b), then limit $f(x,y)$ at (a,b) does not exist.

In the discussion of $\lim_{(x,y)\to(a,b)} f(x,y)$ that follows we shall assume that the

function f is defined at every point (x, y) in an open disk centered at (a, b) but not necessarily at (a, b) itself.

EXAMPLE 4

Investigate the limit of

$$f(x,y)=\begin{cases}\dfrac{x^2-y^2}{x^2+y^2}, & (x,y)\neq(0,0)\\ 0, & (x,y)=(0,0)\end{cases} \quad \text{at } (0,0).$$

Solution Let $x \to 0$ and $y \to 0$ in such a way that $y = mx$. Then along this line

$$\lim_{(x,y)\to(0,0)}\frac{x^2-y^2}{x^2+y^2}=\lim_{x\to 0}\frac{x^2-m^2x^2}{x^2+m^2x^2}=\lim_{x\to 0}\frac{x^2(1-m^2)}{x^2(1+m^2)}=\frac{1-m^2}{1+m^2}.$$

Since the limit of the function depends on the manner in which (x, y) approach to $(0, 0)$, the function $f(x, y)$ does not have a limit at $(0, 0)$. For example, for $m = 0$, $f(x, y) \to 1$ and for $m = 1$, $f(x, y) \to 0$. ■

EXAMPLE 5

Find the limit, if it exists, of the following function as $(x, y) \to (0, 0)$.

(a) $f(x,y)=(x+y)\sin\dfrac{1}{x}\sin\dfrac{1}{y} \quad (x>0,\ y>0)$

(b) $f(x,y)=\dfrac{x^2+y^2}{xy+(x-y)^2} \quad (x>0,\ y>0)$

Solution (a) Since $\left|\sin\dfrac{1}{x}\right|\le 1$ and $\left|\sin\dfrac{1}{y}\right|\le 1$,

we have $\left|(x+y)\sin\dfrac{1}{x}\sin\dfrac{1}{y}\right|\le |x+y|\le |x|+|y|$.

Note that $(x, y)\to(0, 0)$ iff $|x|\to 0$ and $|y|\to 0$.
Thus, $|x+y|\to 0$ as $(x, y)\to(0, 0)$, and so

$$(x+y)\sin\frac{1}{x}\sin\frac{1}{y}\to 0.$$

That is,

$$\lim_{(x,y)\to(0,0)}(x+y)\sin\frac{1}{x}\sin\frac{1}{y}=0.$$

(b) Let $x \to 0$ and $y \to 0$ in such a way that $y = mx$. Then along this line

$$\lim_{(x,y)\to(0,0)} \frac{x^2+y^2}{xy+(x-y)^2} = \lim_{x\to 0}\frac{x^2+m^2x^2}{mx^2+(1-m)^2x^2}$$
$$= \lim_{x\to 0}\frac{x^2(1+m^2)}{x^2(m+(1-m)^2)} = \frac{1+m^2}{1+m^2-m}.$$

Since the limit of the function depends on the manner in which (x, y) approach to $(0,0)$, the function $f(x,y)$ does not have limit at $(0,0)$. For example, for $m=0$, $f(x,y)\to 1$ and for $m=1$, $f(x,y)\to 2$. ■

Finding the limit of a function directly from the definition is not an easy task. The following are properties of limit of two variable functions which are similar to properties of one variable functions. These properties give a tool to compute limit of various functions.

THEOREM 2 The Power Rule

Let c and d be real numbers and suppose that $\lim_{(x,y)\to(a,b)} f(x,y)=L$, $\lim_{(x,y)\to(a,b)} g(x,y)=M$. Then

1. $\lim_{(x,y)\to(a,b)} cf(x,y)=cL,$
2. $\lim_{(x,y)\to(a,b)} \{f(x,y)+g(x,y)\}=L+M$
3. $\lim_{(x,y)\to(a,b)} \{f(x,y)-g(x,y)\}=L-M$
4. $\lim_{(x,y)\to(a,b)} \{f(x,y)g(x,y)\}=LM$
5. $\lim_{(x,y)\to(a,b)} \frac{f(x,y)}{g(x,y)}=\frac{L}{M}$, if $M\neq 0$
6. $\lim_{(x,y)\to(a,b)} \{f(x,y)\}^n=L^n$
7. If m and n are integers, m and n have no common factors and $n\neq 0$, then
$$\lim_{(x,y)\to(a,b)} \{f(x,y)\}^{m/n}=L^{m/n},$$
where we have $L>0$ if n is even and m and n are integers.

EXAMPLE 6

Using the laws of limit in Theorem 2, find the limit of

(1) $f(x,y)=3x+y^2+4xy$ at $(1,1)$.

(2) $\lim_{(x,y)\to(1,-2)} \frac{x+y}{x^2+y^2}$.

Solution (1) From Theorem 2, we have we have

$$\lim_{(x,y)\to(1,1)} f(x,y) = \lim_{(x,y)\to(1,1)} (3x+y^2+4xy)$$
$$=3+1+4=8.$$

(2) From Theorem 2, we have

$$\lim_{(x,y)\to(1,-2)} (x+y) = 1-2 = -1$$
$$\lim_{(x,y)\to(1,-2)} (x^2+y^2) = 1+4 = 5 \neq 0.$$

Therefore, $\displaystyle\lim_{(x,y)\to(1,-2)} f(x,y) = \lim_{(x,y)\to(1,-2)} \frac{x+y}{x^2+y^2} = -\frac{1}{5}.$ ■

EXAMPLE 7

Find the limit of the following function as $(x,y)\to(0,0)$.

(a) $z = \dfrac{3x^2+5y^2}{\sqrt{x^2+y^2}}$ (b) $z = \dfrac{x-y}{x+y}$

Solution (a) Let $x = r\cos\theta$, $y = r\sin\theta$. Then one can show that $r\to 0$ if and only if $(x,y)\to(0,0)$. We have $z = r(3\cos^2\theta + 5\sin^2\theta)$. Therefore, $\displaystyle\lim_{(x,y)\to(0,0)} z = \lim_{r\to 0} r(3\cos^2\theta + 5\sin^2\theta) = 0$.

(b) When the point (x,y) is on the straight line $y = mx$, $z = \dfrac{x-y}{x+y} = \dfrac{1-m}{1+m}$. Therefore, the limit of z is $\dfrac{1-m}{1+m}$ as (x,y) approaches $(0,0)$ along line $y = mx$. This limit changes as m changes. Hence, z does not have a limit as $(x,y)\to(0,0)$. ■

EXAMPLE 8

Find the limit of the following functions as $(x,y)\to(0,0)$.

(a) $(x+y)\sin\dfrac{y}{x}$ (b) $\dfrac{2xy^2}{x^2+y^4}$

Solution (a) Since, $|\sin\frac{y}{x}| \leq 1$, we have

$$\left|(x+y)\sin\frac{y}{x}\right| \leq |x+y| \leq |x| + |y|$$

Note that $(x,y)\to(0,0)$ iff $|x|\to 0$ and $|y|\to 0$
Thus, $|x+y|\to 0$ as $(x,y)\to(0,0)$, and so

$$(x+y)\sin\frac{y}{x}\to 0.$$

That is,

$$\lim_{(x,y)\to(0,0)} (x+y)\sin\frac{y}{x} = 0.$$

(b) Letting $(x,y)\to(0,0)$ along any non-vertical line through the origin. Let $(x,y)\to(0,0)$ along the $y^2 = ax$, we have

$$\frac{2xy^2}{x^2+y^4} = \frac{2a}{(1+a^2)}.$$

Therefore, depending on the value of a, different paths will lead to different values of the limit as $(x, y) \to (0, 0)$. Hence the limit does not exist as $(x, y) \to (0, 0)$. ■

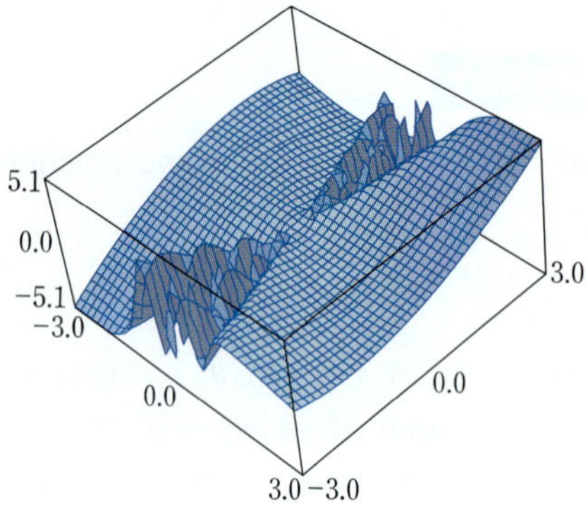

Figure 4

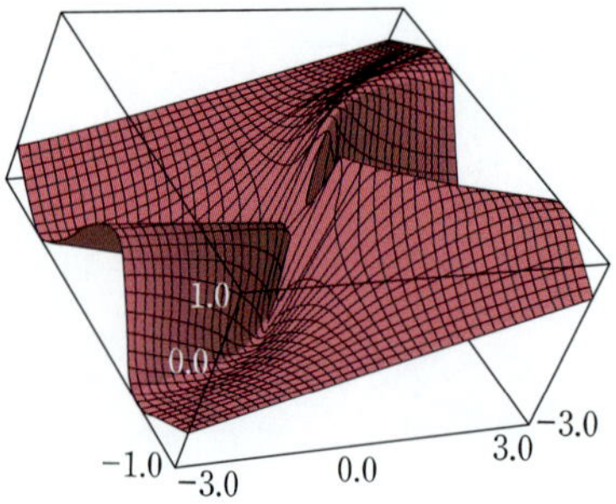

Figure 5

CAS **EXAMPLE 9**

Find $\displaystyle\lim_{(x,y)\to(1,2)} \frac{x-2y}{x+3y}$.

http://matrix.skku.ac.kr/cal-lab/m-Sec13-2-Exm-9.html

Solution

```
var('x,y')
f(x,y)=(x-2*y)/(x+3*y)
limit(limit(f,x=1),y=2)
```

Answer : (x, y) |--> -3/7

```
var('x,y')
f(x,y)=(x-2*y)/(x+3*y)
limit(limit(f,y=2),x=1)
```

Answer : -3/7 ■

THEOREM 3

For a function $f(x, y)$ in neighborhood of point (a, b), suppose that

$$g(y) = \lim_{x\to a} f(x, y), \quad h(x) = \lim_{y\to b} f(x, y)$$

exist. Let

3 $$\lim_{(x,y)\to(a,b)} f(x,y) = l\,.$$

Then $\lim_{y\to b} g(y)$ and $\lim_{x\to a} h(x)$ converges to l. Furthermore,

4 $$\lim_{y\to b}\left(\lim_{x\to a} f(x,y)\right) = \lim_{x\to a}\left(\lim_{y\to b} f(x,y)\right) = l$$

Caution 3 and 4 can be distinguished by denoting the former as the double limit while the latter as the successive limit. Since 4 is practically a limit of single variable function, we just repeats the limit process twice.

Proof Since $g(y) = \lim_{x\to a} f(x,y)$, for any given $\epsilon > 0$, there exists a number $\delta_1 > 0$ such that

$$\text{if } 0 < |x-a| < \delta_1 \quad \text{then} \quad |f(x,y) - g(y)| < \frac{\epsilon}{2}.$$

With the same argument, since $h(x) = \lim_{y\to b} f(x,y)$, for any given $\epsilon > 0$, there exists a number $\delta_2 > 0$ such that

$$\text{if } 0 < |y-b| < \delta_2 \quad \text{then} \quad |f(x,y) - h(x)| < \frac{\epsilon}{2}.$$

Since $\lim_{(x,y)\to(a,b)} f(x,y) = l$, for any given $\epsilon > 0$, there exists a number $\delta_1 > 0$ and $\delta_2 > 0$ such that

$$\text{if } 0 < |x-a| < \delta_1,\ 0 < |y-b| < \delta_2 \quad \text{then} \quad |f(x,y) - l| < \frac{\epsilon}{2}.$$

Thus, if $\min(\delta_1, \delta_2) = \delta$, then $0 < |x-a| < \delta$ and simultaneously if $0 < |y-b| < \delta$, then

$$|g(y) - l| \le |g(y) - f(x,y)| + |f(x,y) - l| < \frac{\epsilon}{2} + \frac{\epsilon}{2} = \epsilon.$$

This implies

$$\lim_{y\to b}\left(\lim_{x\to a} f(x,y)\right) = \lim_{y\to b} g(y) = l\,.$$

Likewise, this may be proven in a similar manner for limits taken in varying orders, that is, first x, then y and first y, then x. ■

Remark In Example 7 (b) $\lim_{x\to 0}\left[\lim_{y\to 0} z\right] = 1$ and $\lim_{y\to 0}\left[\lim_{x\to 0} z\right] = -1$. Thus $\lim_{x\to a}\left[\lim_{y\to b} f(x,y)\right]$ and $\lim_{y\to b}\left[\lim_{x\to a} f(x,y)\right]$ exist but two limits are not equal. Then $\lim_{(x,y)\to(a,b)} f(x,y)$ does not exists.

The function $f(x, y) = \dfrac{xy}{x^2+y^2}$ is one of examples where the converse of Theorem 3 is false. For this function,

$$\text{if } y \neq 0, \text{ then } g(y) = \lim_{x\to 0} f(x, y) = \lim_{x\to 0}\frac{xy}{x^2+y^2} = 0 \text{ and}$$

$$\text{if } x \neq 0, \text{ then } h(x) = \lim_{y\to 0} f(x, y) = \lim_{y\to 0}\frac{xy}{x^2+y^2} = 0.$$

Therefore,

$$\lim_{y\to 0}\left(\lim_{x\to 0} f(x, y)\right) = \lim_{x\to 0}\left(\lim_{y\to 0} f(x, y)\right) = 0.$$

However, when $x = \pm y \neq 0$, respectively $f(x, y) = \dfrac{xy}{x^2+y^2} = \pm\dfrac{1}{2}$. Thus in any domain near $(0, 0)$, no matter how small, there are points that are assigned $\dfrac{1}{2}$ and points that are assigned $-\dfrac{1}{2}$ by the function $\dfrac{xy}{x^2+y^2}$. This means that the limit of the function at $(0, 0)$ does not exists. The $\displaystyle\lim_{(x, y)\to(0, 0)} f(x, y)$ may not exist, even if 4 is satisfied.

EXAMPLE 10

Using Sage, find the limit of

(1) $f(x, y) = 3x + y^2 + 4xy$ at $(1, 1)$.

(2) $\displaystyle\lim_{(x, y)\to(0,0)} \frac{x^2-4y+1}{x+2y+3}$.

(3) Find the example of $\displaystyle\lim_{x\to a}\left(\lim_{y\to b} f(x, y)\right) \neq \lim_{y\to b}\left(\lim_{x\to a} f(x, y)\right)$.

Solution (1) http://matrix.skku.ac.kr/cal-lab/m-Sec13-2-Exm-10.html

```
var('x, y')
f(x, y)=(3*x+y^2+4*x*y)
limit(f(x, y), x=1)
```

y^2 + 4*y + 3

```
limit(f(1, y), y=1)
```

Answer : 8

(2)

```
var('x, y')
f(x, y)=(x^2-4*y+1)/(x+2*y+3)
limit(f(x, y), x=0)
```

-(4*y - 1)/(2*y + 3)

```
limit(f(0, y), y=0)
```

Answer : 1/3

(3) Let $f(x,y)=\dfrac{2x^2+y^2}{x^2-2y^2}$. We have

$$\lim_{x\to 0}\left(\lim_{y\to 0} f(x,y)\right)=\lim_{x\to 0}\left(\lim_{y\to 0}\frac{2x^2+y^2}{x^2-2y^2}\right)=\lim_{x\to 0}\frac{2x^2}{x^2}=2,$$

$$\lim_{y\to 0}\left(\lim_{x\to 0} f(x,y)\right)=\lim_{y\to 0}\left(\lim_{x\to 0}\frac{2x^2+y^2}{x^2-2y^2}\right)=\lim_{y\to 0}\frac{y^2}{-2y^2}=-\frac{1}{2}.$$

Then since $\lim_{x\to 0}\left(\lim_{y\to 0} f(x,y)\right)\neq \lim_{y\to 0}\left(\lim_{x\to 0} f(x,y)\right)$, a function $f(x,y)=\dfrac{2x^2+y^2}{x^2-2y^2}$ is an example of (3). ■

Continuity of Functions of Two Variables

The following definition of continuity for function of two variables is analogous to the definition of continuity for a one variable function.

DEFINITION 4

Let $f: D\subseteq \mathbb{R}^2\to\mathbb{R}$ and $(a,b)\in D$. Then f is said to be continuous at (a,b) if $\lim_{(x,y)\to(a,b)} f(x,y)=f(a,b)$. That is, $\lim_{(x,y)\to(a,b)} f(x,y)$ exists and it is equal to the value of the function. If f is continuous at every point in D, then we say that f is continuous on D.

EXAMPLE 11

It is easy to see that functions defined in Examples 1, 2 and 3 are continuous on $\mathbb{R}^2$. The proof is similar to the solutions given in Examples 1, 2 and 3. ■

EXAMPLE 12

Determine the points at which the following function is continuous.

$$f(x,y)=\begin{cases}\dfrac{2xy^2}{x^2+y^4} & \text{if } (x,y)\neq(0,0)\\ 0 & \text{if } (x,y)=(0,0).\end{cases}$$

http://matrix.skku.ac.kr/cal-lab/m-Sec13-2-Exm-12.html

Solution The function $\dfrac{2xy^2}{x^2+y^4}$ is a rational function, so it is continuous at all points of its domain, which consists of all points of $\mathbb{R}^2$ except $(0,0)$. In order for f to be continuous at $(0,0)$, we must show that

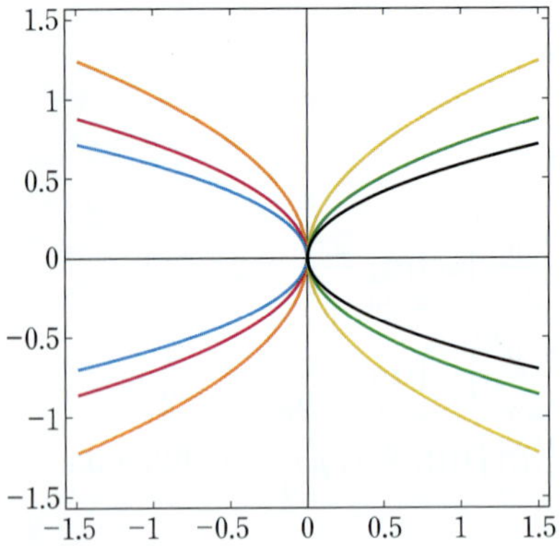

Figure 6

For the choice of $x = my^2$, notice that if x is replaced by my^2 in f, the result involves the same power of y (in this case, y^4) in the numerator and denominator were canceled.

$$\lim_{(x,\,y)\to(0,\,0)} \frac{2xy^2}{x^2+y^4} = f(0,0) = 0.$$

You can verify that as (x, y) approaches $(0, 0)$ along paths of the form $y = mx$, where m is any constant, the function values approach $f(0,0)=0$. Now consider parabolic paths of the form $x = my^2$, where m is a nonzero constant. (See Figure 6.) This time we substitute $x = my^2$ and note that $x \to 0$ as $y \to 0$;

$$\lim_{(x,\,y)\to(0,\,0)} \frac{2xy^2}{x^2+y^4} = \lim_{y\to 0} \frac{2(my^2)y^2}{(my^2)^2+y^4} = \lim_{y\to 0} \frac{2my^4}{m^2y^4+y^4}$$
$$= \lim_{y\to 0} \frac{2m}{m^2+1} = \frac{2m}{m^2+1}.$$

We see that along parabolic paths, the limit depends on the approach path. For example, with $m = -1$, along the path $x = y^2$, the function values approach -1. (See Figure 7.) Because function values approach two different numbers along two different paths, the limit at $(0,0)$ does not exist, and f is not continuous at $(0,0)$. ■

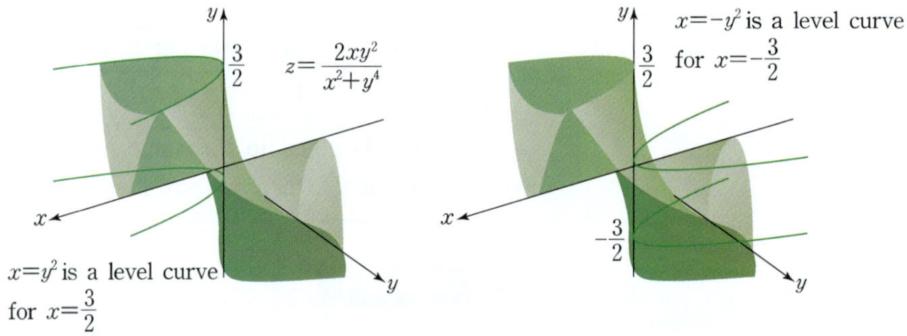

Figure 7

EXAMPLE 13

Investigate the continuity of

$$f(x,y) = \begin{cases} xy\sin^{-1}\left(\dfrac{x^2-y^2}{x^2+y^2}\right), & (x,y)\neq(0,0) \\ 0, & (x,y)=(0,0) \end{cases} \quad \text{at } (0,0).$$

Solution

$$-\frac{\pi}{2} \le \sin^{-1}\left(\frac{x^2-y^2}{x^2+y^2}\right) \le \frac{\pi}{2},$$

$$-\frac{\pi}{2}|xy| \le xy\sin^{-1}\left(\frac{x^2-y^2}{x^2+y^2}\right) \le \frac{\pi}{2}|xy|$$

$$\Rightarrow \lim_{(x,y)\to(0,0)} xy\sin^{-1}\left(\frac{x^2-y^2}{x^2+y^2}\right) = 0 = f(0,0)$$

Hence it is continuous at $(0,0)$. ■

Properties of Continuous Functions

Using the properties of limits in Theorem 2, one can show that sums, differences, products and quotients of continuous functions are continuous on their domains under appropriate conditions. For example:

(i) Any polynomial function of two variables x and y is continuous on $\mathbb{R}^2$. Note that a polynomial of x and y is a function which is obtained by adding terms of the form $cx^i y^j$ where i and j are non negative integers.

(ii) A rational function $f(x, y) = \dfrac{x+y}{1+x^2+y^2}$ is continuous on $\mathbb{R}^2$ as numerator and denominator are continuous on $\mathbb{R}^2$ and the denominator satisfies $1+x^2+y^2 \neq 0$ for all $x,\ y \in \mathbb{R}$.

THEOREM 5 Continuity of Composite Functions

If $u = g(x, y)$ is continuous at (a, b) and $z = f(u)$ is continuous at $g(a, b)$, then the composite function $z = f(g(x, y))$ is continuous at (a, b).

EXAMPLE 14

Investigate the continuity of

$$f(x, y) = e^{x+y},\ g(x, y) = \sin(x^2+y^2).$$

Solution Since $z = f(u) = e^u$ is continuous for all u and $u = h(x, y) = x + y$ is continuous for all (x, y), $f(x, y) = e^{x+y}$ is also continuous for all (x, y) by using Theorem 5. Similarly, since $z = g(u) = \sin u$ is continuous for all u and $u = s(x, y) = x^2 + y^2$ is continuous for all (x, y), $g(x, y) = \sin(x^2+y^2)$ is also continuous for all (x, y) by using Theorem 5. ■

The following theorem is an important concept which is used in quite often while dealing with continuous functions. It says that if $f(x,\ y)$ is a continuous function and has a non zero value at a point (a, b), then it continues to be non zero near $(a,\ b)$ in the domain of f.

THEOREM 6

If $f(x, y)$ is a continuous function at the point (x_1, y_1) and $f(x_1, y_1) \neq 0$, then $f(x, y)$ has the same sign as $f(x_1, y_1)$ near the point (x_1, y_1).

Continuity of Functions of Three Variables

The notions of limit and continuity for functions of three or more variables are natural extensions of those just considered. For example, a function $w = f(x, y, z)$ is continuous at (a, b, c) if

$$\lim_{(x,y,z)\to(a,b,c)} f(x, y, z) = f(a, b, c).$$

The polynomial function $f(x, y, z) = xy^2z^3$ is continuous throughout $\mathbb{R}^3$. The rational function

$$f(x, y, z) = \frac{xy^2}{x^2 + y^2 + (z-1)^2}$$

is continuous except at the single point $(0, 0, 1)$. The rational function

$$f(x, y, z) = \frac{x + 3y}{2x + 5y + z}$$

is continuous except at points (x, y, z) on the plane $2x + 5y + z = 0$.

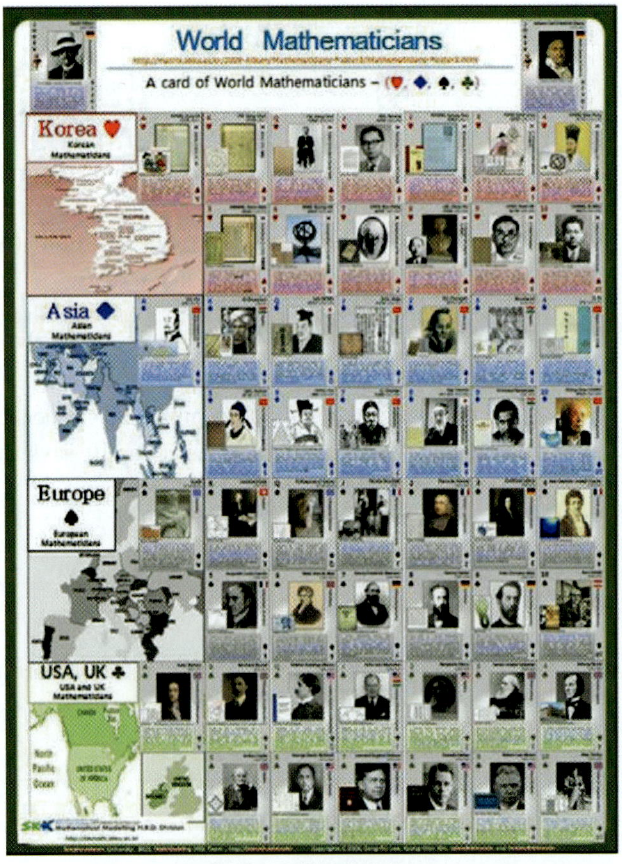

<World Mathematicians>
http://filedir.com/company/skku-matrix-lab/

13.2 EXERCISES (Limits and Continuity of Multivariate Functions)

http://matrix.skku.ac.kr/Cal-Book/part2/CS-Sec-13-2-Sol.html

1-7. Find the limit, if it exists, or show that the limit does not exist.

1. $\lim_{(x,y)\to(0,0)} \frac{y^4}{x^4+3y^4}$

Solution Let $x \to 0$ and $y \to 0$ along the line $y = mx$. Then $\lim_{(x,y)\to(0,0)} \frac{y^4}{x^4+3y^4} = \lim_{x\to 0} \frac{m^4x^4}{(1+3m^4)x^4}$ $= \frac{m^4}{1+3m^4}$. For example, for $m=0$, $f(x,y) \to 0$ and $f(x,y) = \frac{y^4}{y^4+3y^4} \to 1/4$ for $m=1$. Therefore the limit does not exists.

CAS **2.** $\lim_{(x,y)\to(0,0)} \frac{x^3}{2x^2+6y^4}$.

http://matrix.skku.ac.kr/cal-lab/cal-12-1-1.html

Solution

$$\frac{r^3\cos^3\theta}{r^2(2\cos^2\theta+6r^2\sin^4\theta)} = \frac{r\cos^3\theta}{2\cos^2\theta+6r^2\sin^4\theta} \to 0.$$

```
var('x,y')
f(x,y)=x^3/(2*x^2+6*y^4)
limit(limit(f(x,y),x=0),y=0)
```

Answer : 0

```
limit(limit(f(x,y),y=0),x=0)
```

Answer : 0

The limit exists and it is 0.

CAS **3.** $\lim_{(x,y)\to(0,0)} \frac{x^2-y^2}{x^2+y^2}$

http://matrix.skku.ac.kr/cal-lab/cal-12-1-Exs-3.html

Solution

```
var('x, y')
f(x, y)=(x^2-y^2)/(x^2+y^2)
limit(f(x, 0), x=0)
```

1

```
limit(f(0, y), y=0)
```

-1

Answer : The limit does not exist.

CAS **4.** $\lim_{(x,y)\to(0,0)} \frac{\sin(x^3+y^3)}{x^2+y^2}$.

http://matrix.skku.ac.kr/cal-lab/cal-12-1-3.html

Solution After plotting the function, we know it converges. So the order in x and y does not effect.

```
var('x, y')
f(x, y)=(sin(x^3 + y^3))/(x^2+y^2)
limit(f(x, y), x=0)
```

sin(y^3)/y^2

```
limit(f(0, y), y=0)
```

Answer : 0

5. $\lim_{(x,y)\to(0,0)} \frac{xy+y^3}{x^2+y^2}$.

http://matrix.skku.ac.kr/cal-lab/cal-12-1-4.html

Solution Let $x \to 0$ and $y \to 0$ along the line $y = mx$.

```
var('x,y')
f(x, y)=(sin(x^3 + y^3))/(x^2+y^2)
limit(limit(f(x,y),x=0),y=0)
```

y

```
limit(limit(f(x,y),x=0),y=0)
```

0

Answer : The limit exists and it is 0.

6. $\lim_{(x,y)\to(0,0)} \frac{xy^2}{x^2+y^2}$.

Solution After plotting the function, we know it converges. Let $x \to 0$ and $y \to 0$ along the line $y = mx$.

Answer : 0

7. $\lim_{(x,y,z)\to(0,0,0)} \frac{xyz}{xy+yz+zx}$.

http://matrix.skku.ac.kr/cal-lab/cal-12-1-6.html

Solution

```
var('x, y, z')
f(x, y)=x*y*z/(x*y+y*z+z*x)
limit(f(x, y, z), x=0)
```

0

So $\lim_{z\to 0}\left[\lim_{y\to 0}\left(\lim_{x\to 0}\frac{xyz}{xy+yz+zx}\right)\right]$

$$= \lim_{y\to0}\left[\lim_{z\to0}\left(\lim_{x\to0}\frac{xyz}{xy+yz+zx}\right)\right]$$
$$= \lim_{y\to0}\left(\lim_{z\to0}0\right) = \lim_{z\to0}\left(\lim_{y\to0}0\right) = 0$$

```
show(limit(f(0, y, z), y=0))
show(limit(f(0, y, z), z=0))
```

0
0
Answer : The limit exists and it is 0.

---

8. What value of $f(0,0)$ for
$f(x,y)=(x+y)\sin\dfrac{y}{x^2+y^2}$ $((x,y)\neq 0)$ will make the function continuous at $(0,0)$.

Solution

$$\left|\sin\frac{y}{x^2+y^2}\right| \le 1 \Rightarrow -|x+y| \le (x+y)\sin\frac{y}{x} \le |x+y|$$

Since $\lim_{(x,y)\to(0,0)}|x+y|=0$, then $\lim_{(x,y)\to(0,0)}(x+y)\sin\dfrac{y}{x}=0$ by Squeeze Theorem.

Hence, if we define $f(0,0)=0$, then $f(x,y)$ will be continuous at $(0,0)$.

9-10. Let each of the following functions have the value 0 at the origin. Which of them are continuous at the origin? Explain your answer.

9. $f(x,y)=\dfrac{x^2y^2}{x^2+y^4}$.

http://matrix.skku.ac.kr/cal-lab/cal-12-1-7.html

Solution

```
var('x, y')
f(x, y)=(x*2*y^2)/(x^2+y^4)
limit(f(x, y), x=0)
```

0

```
limit(f(0, y), y=0)
```

0

The limit is 0. Hence, if we define $f(0,0)=0$, then $f(x,y)$ is continuous at the origin.

10. $f(x,\ y,\ z,\ w)=\dfrac{x^2+y^4+z^2w^2}{x^4+y^2+z^2+w^2}$

http://matrix.skku.ac.kr/cal-lab/cal-12-1-9.html

Solution

```
var('x, y, z, w')
f(x,                    y,                    z,
w)=(x^2+y^4+z^2*w^2)/(x^4+y^2+z^2+w^2)
limit(f(x, 0, 0, 0), x=0)
```

+Infinity
The limit does not exist.

11-12. Find $h(x,y)=g(f(x,y))$ and the set on which h is continuous.

11. $g(t)=t^2+\sqrt{t}$, $f(x,y)=2x+3y-6$.

Solution Since $f(x,y)$ is continuous for all (x,y) and $g(t)$ is continuous for all $t\ge0$, $h(x,y)=g(f(x,y))$ is continuous on a set $\{(x,y)\in R^3\,|\,2x+3y-6\ge0\}$ by using Theorem 5 in Section 13.2.

12. $g(t)=t+\ln t$, $f(x,y)=\dfrac{1-xy}{1+x^2y^2}$.

Solution Since $f(x,y)$ is continuous for all (x,y) and $g(t)$ is continuous for all $t>0$, $h(x,y)=g(f(x,y))$ is continuous on a set $\{(x,y)\in R^2\,|\,xy<1\}$ by using Theorem 5 in Section 13.2.

13.3 Partial Derivatives

YouTube http://youtu.be/LR89Ct3cEDY

In this section, we will deal with partial derivatives, which is an important concept in multivariate calculus. Partial derivatives are (essentially) derivative of one variable functions. Consider $f(x, y)$ defined on a domain, $D \subseteq \mathbb{R}^2$. Let us fix a point $(a, b) \in D$. Suppose we fix one of the variables, say $y = b$, then $f(x, y) = f(x, b)$ becomes a function of one variable x. Let $f(x, b) = g(x)$. If g is differentiable at $x = a$, that is,

$$\lim_{h \to 0} \frac{g(x+h) - g(x)}{h} = \lim_{h \to 0} \frac{f(x+h, b) - f(x, b)}{h}$$

exists, then we say that f has a partial derivative at (a, b) with respect to x. Similarly we can define the partial derivative of f with respect to y at (a, b). See Definition 1 for a generalization of this idea.

First-Order Partial Derivatives

DEFINITION 1

Let $f(x, y)$ be a function of two variables x and y.
The partial derivative of f with respect to x is defined as

$$f_x(x, y) = \lim_{h \to 0} \frac{f(x+h, y) - f(x, y)}{h}, \text{ provided this limit exists.}$$

The partial derivative of f with respect to y is defined as

$$f_y(x, y) = \lim_{h \to 0} \frac{f(x, y+h) - f(x, y)}{h}, \text{ provided this limit exists.}$$

The partial derivative of $z = f(x, y)$ with respect to x may be denoted by (any of the following)

$$\frac{\partial}{\partial x} f(x, y), \ \frac{\partial f}{\partial x}, \ \frac{\partial z}{\partial x}, \ f_x(x, y), \ z_x.$$

Similarly, the partial derivative of $z = f(x, y)$ with respect to y may be denoted by

$$\frac{\partial}{\partial y} f(x, y), \ \frac{\partial f}{\partial y}, \ \frac{\partial z}{\partial y}, \ f_y(x, y), \ z_y.$$

Geometric Meaning of the Partial Derivative

Let us consider the function $z = f(x, y)$. Let $P(x_1, y_1, z_1)$ be a point on the surface where $z=f(x,y)$. We wish to look at the geometric meaning of $f_x(x_1,y_1)$ and $f_y(x_1,y_1)$. To find $f_x(x_1,y_1)$, we look at the curve of intersection of the surface $z=f(x,y)$ and the plane $y=y_1$, parallel to the xz−plane. This is the same as saying fix $y=y_1$ in $z=f(x,y)$. Then $f_x(x_1,y_1)$ is the slope of the tangent to the curve of intersection at P as the rate change of z for x at this point. (See Figure 1.)

Similarly, $f_y(x_1,y_1)$ is the slope of the tangent of the curve at the point P as the rate change of z for y at this point. (See Figure 2.)

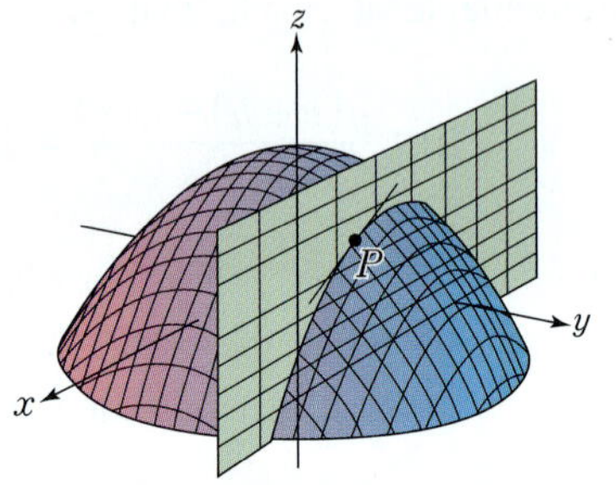

Figure 1 Intersection of the surface $z=f(x,y)$ with the plane $y=y_1$

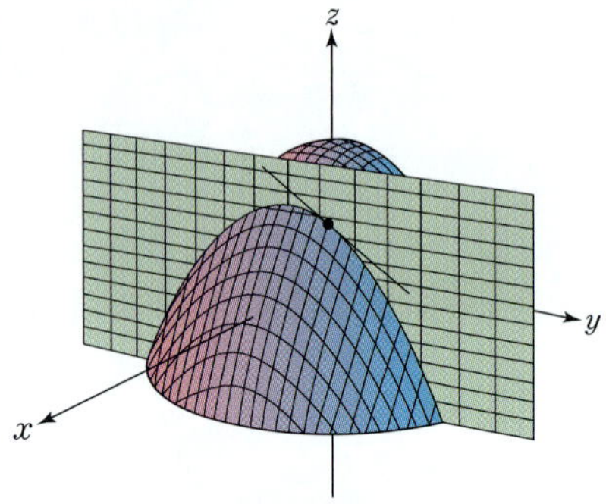

Figure 2 Intersection of the surface $z=f(x,y)$ with the plane $x=x_1$

EXAMPLE 1

Find partial derivatives with respect to x and y for the following functions at the given point.

(a) $z=x^3+y^3+3xy$, $(1,-1)$.

(b) $z=\log_x y$, $(2,e)$.

(c) $z=\sin^{-1}\dfrac{x}{\sqrt{x^2+y^2}}$, $(1,1)$.

Solution (a) $\dfrac{\partial z}{\partial x}=3(x^2+y)$, $\dfrac{\partial z}{\partial y}=3(y^2+x)$.

Thus $\left.\dfrac{\partial z}{\partial x}\right|_{(1,-1)}=0$, $\left.\dfrac{\partial z}{\partial y}\right|_{(1,-1)}=6$.

(b) The partial derivatives are

$$\frac{\partial z}{\partial x}=\ln y\frac{\partial}{\partial x}\left(\frac{1}{\ln x}\right)=(\ln y)\frac{-1/x}{(\ln x)^2}=-\frac{\ln y}{x(\ln x)^2},$$

$$\frac{\partial z}{\partial y}=\frac{1}{\ln x}\frac{\partial}{\partial y}(\ln y)=\frac{1}{\ln x}\cdot\frac{1}{y}=\frac{1}{y\ln x}.$$

Then $\left.\dfrac{\partial z}{\partial x}\right|_{(2,e)}=-\dfrac{1}{2(\ln 2)^2}$, $\left.\dfrac{\partial z}{\partial x}\right|_{(2,e)}=\dfrac{1}{e\ln 2}$.

(c) Let $u=\dfrac{x}{\sqrt{x^2+y^2}}$. Then

$$\frac{\partial z}{\partial x}=\frac{dz}{du}\frac{\partial u}{\partial x}=\left(\frac{d}{du}\sin^{-1}u\right)\frac{\partial u}{\partial x}=\frac{1}{\sqrt{1-u^2}}\frac{\partial}{\partial x}\left(\frac{x}{\sqrt{x^2+y^2}}\right)$$

$$= \frac{\sqrt{x^2+y^2}}{\sqrt{y^2}} \cdot \frac{y^2}{(x^2+y^2)^{3/2}} = \frac{y^2}{|y|(x^2+y^2)} = \frac{|y|}{(x^2+y^2)}$$

$$\frac{\partial z}{\partial y} = \left(\frac{d}{du}\sin^{-1}u\right)\frac{\partial u}{\partial y} = -\frac{xy}{|y|(x^2+y^2)}$$

Then $\left.\frac{\partial z}{\partial x}\right|_{(1,-1)} = \frac{1}{2}$ and $\left.\frac{\partial z}{\partial y}\right|_{(1,-1)} = \frac{1}{2}$. ■

EXAMPLE 2

Show that $x\frac{\partial z}{\partial x} + y\frac{\partial z}{\partial y} = z$ for $z = \frac{xy}{y+x}$.

Proof The partial derivatives are

$$\frac{\partial z}{\partial x} = \frac{y(x+y)-xy}{(x+y)^2} = \frac{y^2}{(x+y)^2} \text{ and}$$

$$\frac{\partial z}{\partial y} = \frac{x(x+y)-xy}{(x+y)^2} = \frac{x^2}{(x+y)^2}.$$

Therefore, $x\frac{\partial z}{\partial x} + y\frac{\partial z}{\partial y} = \frac{xy^2}{(x+y)^2} + \frac{x^2y}{(x+y)^2} = \frac{xy}{x+y} = z$. ■

Functions of Three Variables

Let $w = f(x, y, z)$ be a function of three variables x, y and z. The partial derivatives of f with respect to x, with respect to y, and with respect to z are defined as follows

$$\frac{\partial w}{\partial x} = \lim_{\Delta x \to 0} \frac{f(x+\Delta x, y, z) - f(x, y, z)}{\Delta x}$$

$$\frac{\partial w}{\partial y} = \lim_{\Delta y \to 0} \frac{f(x, y+\Delta y, z) - f(x, y, z)}{\Delta y}$$

$$\frac{\partial w}{\partial z} = \lim_{\Delta z \to 0} \frac{f(x, y, z+\Delta z) - f(x, y, z)}{\Delta z}$$

provided the limits on the right hand side exist.

In general, if $u = (x_1, x_2, \cdots, x_n)$ is a function of n variables, then the partial derivative of f with respect to the ith variable, x_i $(i = 1, 2, \cdots, n)$, is defined to be

$$\frac{\partial u}{\partial x_i} = \lim_{\Delta x \to 0} \frac{f(x_1, x_2, \cdots, x_i + \Delta x, \cdots, x_n) - f(x_1, x_2, \cdots, x_n)}{\Delta x}.$$

To compute $\partial u / \partial x_i$, we differentiate with respect to x_i while holding the remaining $n-1$ variables fixed.

EXAMPLE 3

Find $\frac{\partial f}{\partial x}$, $\frac{\partial f}{\partial y}$ and $\frac{\partial f}{\partial z}$ if $f(x, y, z) = e^{xy}\ln z$.

Solution $\frac{\partial f}{\partial x} = ye^{xy}\ln z$, $\frac{\partial f}{\partial y} = xe^{xy}\ln z$ and $\frac{\partial f}{\partial z} = \frac{e^{xy}}{z}$.

```
var('x,y,z')
f=exp(x*y)*log(z)
fx=diff(f,x);fy=diff(f,y);fz=diff(f,z)
print fx
print fy
print fz
```

Answer : y*e^(x*y)*log(z), x*e^(x*y)*log(z), e^(x*y)/z. ■

Second and Higher Order Partial Derivatives

Suppose that $z = f(x, y)$ has first order partial derivatives $\frac{\partial z}{\partial x}$ and $\frac{\partial z}{\partial y}$. Note that $\frac{\partial z}{\partial x}$ and $\frac{\partial z}{\partial y}$ are also function of x and y. Therefore, we can talk about their partial derivatives if they exist. If $\frac{\partial z}{\partial x}$ and $\frac{\partial z}{\partial y}$ have partial derivatives with respect to x and y, then the following four possibilities exist:

$$\frac{\partial}{\partial x}\left(\frac{\partial z}{\partial x}\right) = \frac{\partial^2 z}{\partial x^2} = z_{xx} = \frac{\partial^2 f}{\partial x^2} = f_{xx}(x, y),$$

$$\frac{\partial}{\partial x}\left(\frac{\partial z}{\partial y}\right) = \frac{\partial^2 z}{\partial x \partial y} = z_{yx} = \frac{\partial^2 f}{\partial x \partial y} = f_{yx}(x y),$$

$$\frac{\partial}{\partial y}\left(\frac{\partial z}{\partial x}\right) = \frac{\partial^2 z}{\partial y \partial x} = z_{xy} = \frac{\partial^2 f}{\partial y \partial x} = f_{xy}(x, y),$$

$$\frac{\partial}{\partial y}\left(\frac{\partial z}{\partial y}\right) = \frac{\partial^2 z}{\partial y^2} = z_{yy} = \frac{\partial^2 f}{\partial y^2} = f_{yy}(x, y).$$

These are called second order partial derivatives of z. $\frac{\partial^2 z}{\partial x \partial y}$ and $\frac{\partial^2 z}{\partial y \partial x}$ are called mixed second order partial derivatives. In general, the second order mixed partial derivatives need not be equal.

Third order partial derivatives may also be defined, for instance

$$\frac{\partial}{\partial x}\left(\frac{\partial^2 z}{\partial x^2}\right) = \frac{\partial^3 z}{\partial x \partial x^2} = \frac{\partial^3 z}{\partial x^3} = z_{xxx} = \frac{\partial^3 f}{\partial x^3} = f_{xxx}(x, y).$$

$$\frac{\partial}{\partial x}\left(\frac{\partial^2 z}{\partial x \partial y}\right) = \frac{\partial^3 z}{\partial x \partial x \partial y} = \frac{\partial^3 z}{\partial x^2 \partial y} = z_{yxx} = \frac{\partial^3 f}{\partial x^2 \partial y} = f_{yxx}(x, y).$$

$$\frac{\partial}{\partial x}\left(\frac{\partial^2 z}{\partial y \partial x}\right) = \frac{\partial^3 z}{\partial x \partial y \partial x} = z_{xyx} = \frac{\partial^3 f}{\partial x \partial y \partial x} = f_{xyx}(x, y).$$

The higher partial derivative may be defined similarly if they exist.

EXAMPLE 4

Let

$$f(x, y) = \begin{cases} \frac{x^3 y - x y^3}{x^2 + y^2}, & (x, y) \neq (0,0) \\ 0, & (x,y) = (0,0) \end{cases}$$

be a function of two variables x, y.

Verify that $f_{xy}(x, y)$ and $f_{yx}(x, y)$ are not define at $(0, 0)$.

Solution

```
var('x,y')
f=(x^3*y-x*y^3)/(x^2+y^2)
limit(limit(f,x=0),y=0)
```

Answer : 0

```
var('x,y')
f=(x^3*y-x*y^3)/(x^2+y^2)
limit(limit(f,y=0),x=0)
```

Answer : 0

Then $f(x, y)$ is continuous on $\mathbb{R}^2$.

```
var('x,y')
f(x,y)=(x^3*y-x*y^3)/(x^2+y^2)
f_xy=diff(diff(f,x),y)
fxy=limit(limit(f_xy,x=0),y=0)
print fxy
fyx=limit(limit(f_xy,y=0),x=0)
print fyx
```

Answer : -1, 1

Since two limits are different, $f_{xy}(x,y)$ is not define at $(0, 0)$.

Actually $f_{xy}(x, y) = f_{yx}(x, y)$ on $\mathbb{R}^2$ except $(0, 0)$.

```
var('x,y')
f(x,y)=(x^3*y-x*y^3)/(x^2+y^2)
f_xy=diff(diff(f,x),y)
f_yx=diff(diff(f,y),x)
print bool(f_xy==f_yx) # except the origin
```

Answer : True

Then it is easy to show that f_x, f_y, f_{xy}, f_{yx} exist. However $f_{xy}(0, 0) \neq f_{yx}(0, 0)$ because $\frac{\partial^2 f}{\partial y\, \partial x}(0, 0) = 1$ and

$$\frac{\partial^2 f}{\partial x \partial y}(0,0) = -1.$$
■

This example shows second order partial derivatives f_{xy} and f_{yx} may be different, so the order of differentiation cannot be interchanged arbitrarily. The following theorem provides a sufficient condition for changing the order of differentiation.

THEOREM 2 Equality of Mixed Partial Derivatives

Suppose $f(x, y)$ and it's derivatives f_x, f_y, f_{xy} and f_{yx} are defined in a domain containing a point (a, b) and all are continuous at (a, b). Then

$$f_{xy}(a, b) = f_{yx}(a, b).$$

Caution As a consequence, if f_{xy} and f_{yx} both exist and are continuous, then $f_{xy} = f_{yx}$. If third partial derivatives exist and are continuous, then $f_{xxy}(x,y) = f_{xyx}(x,y) = f_{yxx}(x,y)$ is true as well. These results holds for all higher order partial derivatives.

EXAMPLE 5

For $f(x, y) = \sin(x-y) + e^{-xy}$, find the second order partial derivatives.
http://matrix.skku.ac.kr/cal-lab/m-Sec13-3-Exm-5.html

Solution

$$\frac{\partial f}{\partial x} = \cos(x-y) - ye^{-xy}, \quad \frac{\partial f}{\partial y} = -\cos(x-y) - xe^{-xy},$$

$$\frac{\partial^2 f}{\partial x^2} = -\sin(x-y) + y^2 e^{-xy}, \quad \frac{\partial^2 f}{\partial y^2} = -\sin(x-y) + x^2 e^{-xy},$$

$$\frac{\partial^2 f}{\partial y \partial x} = \sin(x-y) - e^{-xy} + xye^{-xy},$$

$$\frac{\partial^2 f}{\partial x \partial y} = \sin(x-y) - e^{-xy} + xye^{-xy}.$$

```
var('x,y')
f=sin(x-y)+exp(-x*y)
fxx=diff(f,x,2)
fyy=diff(f,y,2)
fxy=diff(f,x,y)
fyx=diff(f,y,x)
print fxx
print fyy
print fxy
print fyx
```

Answer : y^2*e^(-x*y) - sin(x - y)

x^2*e^(-x*y) - sin(x - y)
x*y*e^(-x*y) - e^(-x*y) + sin(x - y)
x*y*e^(-x*y) - e^(-x*y) + sin(x – y) ■

EXAMPLE 6

Show that the partial derivatives $f_x(0,0)$ and $f_y(0,0)$ both exist, but the function $f(x,y)$ is not continuous at $(0,0)$.

$$f(x,y) = \begin{cases} \dfrac{2xy}{x^2+y^2}, & \text{if } (x,y) \neq (0,0) \\ 0, & \text{if } (x,y) = (0,0) \end{cases}$$

Solution Since the definition of partial derivatives, we have

$$f_x(0,0) = \lim_{h\to 0}\frac{f(0+h,0)-f(0,0)}{h} = \lim_{h\to 0}\frac{0-0}{h} = 0$$

$$f_y(0,0) = \lim_{h\to 0}\frac{f(0,0+h)-f(0,0)}{h} = \lim_{h\to 0}\frac{0-0}{h} = 0.$$

However, $f(x,y)$ is not continuous at $(0,0)$ because the limit of $f(x,y)$ does not exists as follows;
letting $(x,y)\to(0,0)$ along the line $y=mx$,

$$\lim_{(x,y)\to(0,0)}\frac{2xy}{x^2+y^2} = \lim_{x\to 0}\frac{2mx^2}{(1+m^2)x^2} = \frac{2m}{1+m^2}.$$ ■

<Father of Modern Korean Math Education>

http://matrix.skku.ac.kr/Cal-Book/part2/CS-Sec-13-3-Sol.html

YouTube http://youtu.be/rSYLp1mSMXY

CAS **1.** Find partial derivatives with respect to x and y for the function $f(x,y)=24xy-6x^2y$.

Solution $f_x(x,y)=24y-12xy,\ f_y(x,y)=24x-6x^2$

```
var('x,y')
f=24*x*y-6*x^2*y
fx=diff(f,x);fy=diff(f,y)
print fx
print fy
```

Answer : -12*x*y+24*y, -6*x^2 + 24*x

2. Find partial derivatives with respect to x and y for the function $f(x,y)=\dfrac{x^3y-xy^3}{x^2+y^2}$.

Solution $f_x(x,y)=\dfrac{x^4y+4x^2y^3-y^5}{(x^2+y^2)^2}$,

$f_y(x,y)=\dfrac{x^5-4x^3y^2-xy^4}{(x^2+y^2)^2}$

3. Find all second order partial derivatives of the function $f(x,y)=e^{xy^2}$.

Solution $f_x(x,y)=y^2e^{xy^2},\ f_y(x,y)=2xye^{xy^2}$

$f_{xx}(x,y)=y^4e^{xy^2},\ f_{xy}(x,y)=2ye^{xy^2}+2xy^3e^{xy^2}$

$f_{yx}(x,y)=2ye^{xy^2}+2xy^3e^{xy^2},\ f_{yy}(x,y)=2xe^{xy^2}+4x^2y^2e^{xy^2}$

4. Find all second order partial derivatives of the function $f(x,\ y)=\sin(xy^2)$.

Solution

$f_x(x,y)=y^2\cos xy^2,\ f_y(x,y)=2xy\cos xy^2$

$f_{xx}(x,y)=-y^4\sin xy^2,\ f_{xy}(x,y)=2y\cos xy^2-2xy^2\sin xy^2$

$f_{yx}(x,y)=2y\cos xy^2-2xy^2\sin xy^2,\ f_{yy}(x,y)$

$=2x\cos xy^2-4x^2y^4\sin xy^2$

CAS **5.** Take an example of the solution for the $U_{xx}+U_{yy}=0$.

Solution

http://matrix.skku.ac.kr/cal-lab/cal-12-2-6.html

```
var('x, y')
f(x, y)=x*y+x^3+y^3
solve([derivative(f,x,2)+derivative(f,y,2)==0], x, y)
```

([x == -y], [1])

Answer : $f(x,y)=xy+x^3+y^3$.

6-8. Laplace's equation A classical equation of mathematics is Laplace's equation, which arises in both theory and applications. It governs ideal fluid flow, electrostatic potentials, and the steady-state distribution of heat in a conducting medium. In two dimensions, Laplace's equation is

$$\frac{\partial^2 u}{\partial x^2}+\frac{\partial^2 u}{\partial y^2}=0$$

Show that the following functions are harmonic; that is, they satisfy Laplace's equation.

6. $u(x,y)=x(x^2-3y^2)$

http://matrix.skku.ac.kr/cal-lab/m-Sec13-3-Exs-6.html

Solution

```
var('x, y')
u(x, y)=x*(x^2-3*y^2)
u_xx=diff(u(x,y), x, 2)
u_yy=diff(u(x,y), y, 2)
bool(u_xx+u_yy==0)
```

True

7. $u(x,y)=e^{ax}\cos ay$ for any real number a.

Solution True

8. $u(x,y)=\tan^{-1}\left(\dfrac{y}{x-1}\right)-\tan^{-1}\left(\dfrac{y}{x+1}\right)$

Solution

```
var('x, y')
u(x,y)= arctan(y/(x-1)) - arctan(y/(x+1))
u_xx=diff(u(x,y), x, 2)
u_yy=diff(u(x,y), y, 2)
bool(u_xx+u_yy==0)
```

True

9. The volume of a right circular cone of radius r and height h is $V=\dfrac{1}{3}\pi r^2h$. Show that if the height remains constant while the radius changes the volume satisfies $\dfrac{\partial V}{\partial r}=\dfrac{2v}{r}$.

10. Let $z=x^2+y^2$. Show that

$$\left(\frac{\partial z}{\partial x}-\frac{\partial z}{\partial y}\right)^2=4\left(1-\frac{\partial z}{\partial x}-\frac{\partial z}{\partial y}\right).$$

13.4 Differentiability and Total Differential

Differentiability

Once again we try to generalize, the definition of differentiability of one variable function $y = f(x)$ at $x = a$. Recall that f is differentiable at $x = a$, with its derivative $f'(a)$, if $\lim_{\Delta x \to 0} \frac{f(a+\Delta x) - f(a)}{\Delta x} = f'(a)$. Using the $\epsilon - \delta$ definition of limit, this means the following: Given any $\epsilon > 0$, $\left| \frac{f(a+\Delta x) - f(a)}{\Delta x} - f'(a) \right| < \epsilon$ for sufficiently small Δx. That is,

$$\Delta y = f(a+\Delta x) - f(a) \approx f'(a)\Delta x + \epsilon \Delta x$$

with the condition that $\epsilon \to 0$ as $\Delta x \to 0$. We can generalize this definition to a function $z = f(x, y)$. In the sequel, we assume that f is defined on some open subset of $\mathbb{R}^2$.

The analogous requirement with several variables is the definition of differentiability for functions or two (or more) variables.

DEFINITION 1

The function $z = f(x, y)$ is differentiable at (a, b) provided $f_x(a, b)$ and $f_y(a, b)$ exist and the change $\Delta z = f(a+\Delta x, b+\Delta y) - f(a, b)$ equals

$$\Delta z = f_x(a, b)\Delta x + f_y(a, b)\Delta y + \epsilon_1 \Delta x + \epsilon_2 \Delta y$$

where for fixed a and b, ϵ_1 and ϵ_2 are functions that depend only on Δx and Δy, with $(\epsilon_1, \epsilon_2) \to (0, 0)$ as $(\Delta x, \Delta y) \to (0, 0)$. A function is differentiable on an open region R if it is differentiable at each point of R.

Several observations are needed here. First, the definition extends to functions of more than two variables. Second, we will show how differentiability is related to linear approximation and the existence of a tangent plane in Section 13.7. Finally, the conditions of the definition are generally difficult to verify. The following theorem may be useful in checking differentiablity.

THEOREM 2 Conditions for Differentiability

Suppose the function f has partial derivatives f_x and f_y defined on an open region R and f_x and f_y are continuous on R. Then f is differentiable at every point of R.

This theorem states that existence of f_x and f_y at (a, b) is not enough to ensure differentiability of f at (a, b); we also need their continuity. Polynomials and rational functions are differentiable at all points of their domains, as are compositions of exponential, logarithmic and trigonometric functions with other differentiable functions.

We close with the analog of Theorem 3 in Section 3.1, which states that differentiability implies continuity.

THEOREM 3 Differentiability Implies Continuity

If a function f is differentiable at (a, b), then it is continuous at (a, b).

Proof By the definition of differentiability,

$$\Delta z = f_x(a, b)\Delta x + f_y(a, b)\Delta y + \epsilon_1 \Delta x + \epsilon_2 \Delta y,$$

where $(\epsilon_1, \epsilon_2) \to (0, 0)$ as $(\Delta x, \Delta y) \to (0, 0)$. Because f is assumed to be differentiable, as Δx and Δy approach 0, we see that

$$\lim_{(\Delta x, \Delta y) \to (0, 0)} \Delta z = 0.$$

Also, because $\Delta z = f(a + \Delta x, b + \Delta y) - f(a, b)$, it follows that

$$\lim_{(\Delta x, \Delta y) \to (0, 0)} f(a + \Delta x, b + \Delta y) = f(a, b),$$

which implies continuity of f at (a, b). ■

EXAMPLE 1

Discuss the differentiability and continuity of the function

$$f(x, y) = \begin{cases} \dfrac{xy}{x^2 + y^2}, & (x, y) \neq (0, 0) \\ 0, & (x, y) = (0, 0). \end{cases}$$

Solution As the definition of a rational function, f is continuous and differentiable at all points $(x, y) \neq (0, 0)$. The interesting behavior occurs at the origin. It can be shown that if the origin is approached along the line $y = mx$, then

$$\lim_{(x, y) \to (0, 0)} \frac{xy}{x^2 + y^2} = \frac{m}{m^2 + 1}.$$

Therefore, the value of the limit depends on the direction of approach, which implies that the limit does not exist, and f is not continuous at $(0, 0)$. By Theorem 3, f is not differentiable at $(0, 0)$. Let's look at the first partial derivatives of f at $(0, 0)$. A short calculation shows that

$$f_x(0,0)=\lim_{h\to 0}\frac{f(0+h,0)-f(0,0)}{h}=\lim_{h\to 0}\frac{0-0}{h}=0$$

$$f_y(0,0)=\lim_{h\to 0}\frac{f(0,y+h)-f(0,0)}{h}=\lim_{h\to 0}\frac{0-0}{h}=0.$$

That is, the partial derivatives exist at $(0,0)$.
Despite the fact that f is not differentiable at $(0,0)$, its first partial derivatives exist at $(0,0)$. Existence of first partial derivatives at a point is not enough to ensure differentiability at that point. As expressed in Theorem 2, continuity of first partial derivatives is required for differentiability. It can be shown that f_x and f_y are not continuous at $(0,0)$. ■

Total Differential

The partial derivative introduced in the previous section tells us how a function changes when we vary one variable and hold the others fixed. Sometimes we wish to know how a function changes when more than one variable is changed. In section 3.4, we dealt with differential of a function $y=f(x)$ at x. Recall that

$$df=dy=f'(x)\,dx.$$

In this section we wish to generalize this for the function of two variables. Let $z=f(x,\,y)$ be a function of two variables. Assume that f has first order partial derivatives $\frac{\partial z}{\partial x}$ and $\frac{\partial z}{\partial y}$ at $(x,\,y)$ in some neighborhood of $(x,\,y)$. Let us also assume for convenience that $\frac{\partial z}{\partial x}$ and $\frac{\partial z}{\partial y}$ are also continuous.

Let Δx and Δy represent infinitesimal changes in x and y, respectively, at $(x,\,y)$. Let Δz be the **increment** of the function $z=f(x,y)$. That is,

1 $$\Delta z=f(x+\Delta x,\,y+\Delta y)-f(x,\,y).$$

We add and subtract $f(x,\,y+\Delta y)$ in 1, and get

2 $$\Delta z=[f(x+\Delta x,\,y+\Delta y)-f(x,\,y+\Delta y)]+[f(x,\,y+\Delta y)-f(x,\,y)]$$

Using the mean value theorem in the interval $[x,\,x+\Delta x]$, there exists x_1 between x and $x+\Delta x$ such that

3 $$\frac{f(x+\Delta x,\,y+\Delta y)-f(x,\,y+\Delta y)}{\Delta x}=\frac{\partial f}{\partial x}(x_1,\,y+\Delta y).$$

Similarly using the mean value theorem in the interval $[y, y+\Delta y]$, there exists y_1 between y and $y+\Delta y$ such that

4
$$\frac{f(x, y+\Delta y)-f(x, y)}{\Delta y}=\frac{\partial f}{\partial y}(x, y_1).$$

Using 3 and 4, in 2, we get

5
$$\Delta z=\frac{\partial f}{\partial x}(x_1, y+\Delta y)\Delta x+\frac{\partial f}{\partial y}(x, y_1)\Delta y.$$

The above equation can be rewritten as follows: there exist $\theta_1, \theta_2 \in (0, 1)$ such that

6
$$\Delta z=\frac{\partial f}{\partial x}(x+\theta_1 \Delta x, y+\Delta y)\Delta x+\frac{\partial f}{\partial y}(x, y+\theta_2 \Delta y)\Delta y.$$

Note that as $\Delta x \to 0$ andy $\Delta y \to 0$, we have $x_1 \to x$ and $y_1 \to y$. Thus under the limiting case we have

7
$$dz=df=\frac{\partial f}{\partial x}dx+\frac{\partial f}{\partial y}dy.$$

The 7 is called the differential or total differential of f. At time in infinitesimal case we write $dx \approx \Delta x$, $dy \approx \Delta y$ and $dz \approx \Delta z$, then the differential can be written as

8
$$\Delta z \approx \frac{\partial f}{\partial x}\Delta x+\frac{\partial f}{\partial y}\Delta y.$$

In case, $u=f(x, y, z)$ is a function of three variables then the total differential is given by

9
$$du=df=\frac{\partial f}{\partial x}dx+\frac{\partial f}{\partial y}dy+\frac{\partial f}{\partial z}dz.$$

EXAMPLE 2

If $z=a\left(\frac{y}{x}\right)^{1/2}$, find the total differential of z.

Solution By definitions of the total differential, we have

$$dz=\frac{\partial z}{\partial x}dx+\frac{\partial z}{\partial y}dy.$$

The first order partial derivatives of z with respect to x and y are as follows :

$$\frac{\partial z}{\partial x}=\frac{a}{2}\left(\frac{y}{x}\right)^{-1/2}\left(\frac{-y}{x^2}\right)=-\frac{a}{2}\left(\frac{y}{x}\right)^{1/2}\left(\frac{1}{x}\right)=-\frac{z}{2}\left(\frac{1}{x}\right),$$

$$\frac{\partial z}{\partial y}=\frac{a}{2}\left(\frac{y}{x}\right)^{-1/2}\left(\frac{1}{x}\right)=\frac{a}{2}\left(\frac{y}{x}\right)^{1/2}\left(\frac{1}{y}\right)=\frac{z}{2}\left(\frac{1}{y}\right)$$

Hence, we obtain

$$dz=-\frac{z}{2}\frac{dx}{x}+\frac{z}{2}\frac{dy}{y}=\frac{z}{2}\left(\frac{dy}{y}-\frac{dx}{x}\right).$$ ■

EXAMPLE 3

The base diameter and height of a cylinder are measured as 5cm and 8cm, respectively, with a possible error in measurement of as much as $\pm 0.1\text{cm}$ for each. Estimate maximum relative error and percentage error in the calculated volume of the cylinder.

Solution The volume V of a cylinder with base diameter $r$cm and height h cm is $V=\frac{1}{4}\pi r^2 h$. Taking natural logarithms on both sides of this equation, we get

$$\ln V=\ln\frac{\pi}{4}+2\ln r+\ln h.$$

Now we take the differentials on both sides to have

$$\frac{dV}{V}=\frac{2dr}{r}+\frac{dh}{h}.$$

$\frac{dV}{V}$ is a maximum when dr and dh are positive. Therefore, we take $dr=0.1$ and $dh=0.1$ along with $r=5,\ h=8$. This gives

$$\frac{dV}{V}=\frac{2(0.1)}{5}+\frac{0.1}{8}=0.0525.$$

The maximum relative error in the calculated volume is 0.0525. Thus the percentage error in the calculated volume is 5.25%. ■

EXAMPLE 4

The dimensions of a rectangular cuboid are measured to be 70 cm, 60 cm and 50 cm, with a possible error in measurement as much 0.3 cm in each. Calculate differential and increment to estimate the maximum error in the volume of the rectangular cuboid.

Solution Let x, y and z be the dimensions of the rectangular cuboid. Then the volume is $V=xyz$. By 9, since

$$dV=\frac{\partial V}{\partial x}dx+\frac{\partial V}{\partial y}dy+\frac{\partial V}{\partial z}dz=yz\,dx+xz\,dy+xy\,dz,$$

so $dV=60\times 50\times 0.3+70\times 50\times 0.3+70\times 60\times 0.3=3210$ if $x=70,\ y=60,\ z=50$ and $dx=dy=dz=0.3$.

Next we will find the increment of V as follows

$$\Delta z = V(70.3, 60.3, 50.3) - V(70, 60, 50) = 3226.227 \ . \qquad \blacksquare$$

Given $z = f(x, y)$, we have found the total differential dz. If we have the total differential $dz = \frac{\partial z}{\partial x}dx + \frac{\partial z}{\partial y}dy$, can we find z?

EXAMPLE 5

If $dz = (x^2 + y^2)dx + (2xy + 1)dy$, find z.

Solution By [7], $\frac{\partial z}{\partial x} = x^2 + y^2$, $\frac{\partial z}{\partial y} = 2xy + 1$.

Integrating $\frac{\partial z}{\partial x} = x^2 + y^2$ with respect to x, we obtain

$$z = f(x,y) = \int (x^2 + y^2)dx = \frac{x^3}{3} + y^2 x + k(y) \ , \qquad [10]$$

where $k(y)$ is a constant with respect to x, that is a function of y. Next we differentiate both sides of [10] with respect to y, and we have

$$\frac{\partial z}{\partial y} = \frac{\partial}{\partial y}\left(\frac{x^3}{3} + y^2 x + k(y)\right) = 2xy + 1 \ .$$

Then $2xy + k'(y) = 2xy + 1$. Thus $k'(y) = 1$.
Integrating $k'(y) = 1$ with respect to y, we have

$$k(y) = y + C$$

where C is a constant. Putting this in [10], we have

$$z = f(x, y) = \frac{x^3}{3} + xy^2 + y + C \ . \qquad \blacksquare$$

13.4 EXERCISES (Differentiability and Total Differential)

http://matrix.skku.ac.kr/Cal-Book/part2/CS-Sec-13-4-Sol.html

YouTube http://youtu.be/qDmCWBiXbIA

1. Find the total differential of $z=f(x,y)$ when $f(x,y)=\ln(xy)$.

Solution $f_x(x,y)=\dfrac{1}{x}$, $f_y(x,y)=\dfrac{1}{y}$

$$dz=\frac{1}{x}dx+\frac{1}{y}dy$$

2. If $f(x,y)=xy^2+y\sin x$, find the total differential of $z=f(x,y)$.

Solution $f_x(x,y)=y^2+y\cos x$,

$f_y(x,y)=2xy+\sin x$

$$dz=(y^2+y\cos x)dx+(2xy+\sin x)dy$$

3. The base diameter and height of a right circular cone are measured as 10cm and 25cm, respectively, with a possible error in measurement of as much as ± 0.1cm for each. Estimate maximum relative error and percentage error in the calculated volume of the right circular cone.

Solution Since the volume of right circular cone is

$$V=\frac{\pi r^2 h}{3},\ dV=\frac{2\pi rh}{3}dr+\frac{\pi r^2}{2}dh.$$

Let $dr=0.1$, $dh=0.1$, $r=10$, $h=25$, then

$$dV=\frac{500\pi}{3}(0.1)+\frac{100\pi}{3}(0.1)=20\pi.$$

The maximum relative error is

$\dfrac{dV}{V}=\dfrac{20\pi}{V}=\dfrac{20\pi}{25^2\pi}=\dfrac{4}{125}$ and the percentage error is 3.2%.

4. Find an approximation using total differential.

(1) Find an approximation of $f(1.1, 2.9)$ when $f(x,y)=\dfrac{1}{1+x-y}$.

(2) Find an approximation of $\sqrt{2.98^2+4.01^2}$.

Solution

(1) Since $f_x(x,y)=\dfrac{-1}{(1+x-y)^2}$ and

$f_x(x,y)=\dfrac{1}{(1+x-y)^2}$, the linear approximation is

$$f(x,y)\approx f(1,3)+f_x(1,3)(x-1)+f_y(1,3)(y-3)$$
$$=-1-(x-1)+(y-3)$$

Then $f(1.1, 2.9)\approx -1-0.1-0.1=-1.2$.

(2) Let $f(x,y)=\sqrt{x^2+y^2}$. Since

$f_x(x,y)=\dfrac{x}{\sqrt{x^2+y^2}}$ and $f_y(x,y)=\dfrac{y}{\sqrt{x^2+y^2}}$,

the linear approximation is

$$f(x,y)\approx f(3,4)+f_x(3,4)(x-3)+f_y(3,4)(y-4)$$
$$=5-0.6(x-3)+0.8(y-4)$$
$$f(2.98, 4.01)\approx 5-0.6(2.98-3)+0.8(4.01-4)$$
$$=5.02.$$

Bhāskara II

13.5 The Chain Rule

YouTube http://youtu.be/r3dGYL1vkEU

The Chain Rule

Let $z = f(x,y)$. If x and y depend on other variables, we would like to find derivatives or partial derivatives of z with respect to these independent variables. There are several variations that can occur. For example, x and y can be function of another variable, say t, or x and y can be functions of two variables u and v. The Chain Rule for multivariable function helps to find derivative, and partial derivatives of compositions of functions. Let us consider the first case, where x and y are functions of t.

THEOREM 1 Chain Rule (One Independent Variable)

If $z = f(x,y)$ is a function of x and y, and $x = x(t)$ and $y = y(t)$ are differentiable functions of t, then the composition $z = f(x(t), y(t))$ is a differentiable function of t and

1
$$\frac{dz}{dt} = \frac{\partial z}{\partial x}\frac{dx}{dt} + \frac{\partial z}{\partial y}\frac{dy}{dt}.$$

Proof If $z = f(x,y)$ and $x = g(t)$ and $y = h(t)$ are both functions of the parameter of t, then t is the independent variable. Using 6 in Section 13.4, we have

$$\frac{\Delta z}{\Delta t} = f_x(x+\theta_1\Delta x, y+\Delta y)\frac{\Delta x}{\Delta t} + f_y(x, y+\theta_2\Delta y)\frac{\Delta y}{\Delta t}.$$

When $\Delta t \to 0$, $\Delta x \to 0$ and $\Delta y \to 0$. By hypothesis $f_x(x,y)$ and $f_y(x,y)$ are continuous

$$\lim_{\Delta t\to 0}\frac{\Delta z}{\Delta t} = \lim_{\Delta t\to 0} f_x(x+\theta_1\Delta x, y+\Delta y)\frac{\Delta x}{\Delta t} + \lim_{\Delta t\to 0} f_y(x, y+\theta_2\Delta y)\frac{\Delta y}{\Delta t}$$

$$\frac{dz}{dt} = f_x(x,y)\frac{dx}{dt} + f_y(x,y)\frac{dy}{dt}$$

or, $$\frac{dz}{dt} = \frac{\partial z}{\partial x}\frac{dx}{dt} + \frac{\partial z}{\partial y}\frac{dy}{dt}.$$ (See Figure 1.) ■

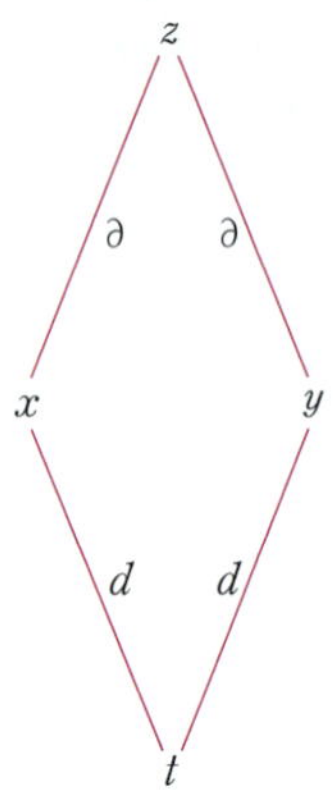

Figure 1

THEOREM 2

Let $z=f(x,y)$ have continuous first order partial derivatives and $y=g(x)$ be a differentiable function of x. Then the composition $z=f(x,g(x))$ is a differentiable function of x and

$$\frac{dz}{dx}=\frac{\partial z}{\partial x}+\frac{\partial z}{\partial y}\frac{dy}{dx}.$$

Proof Using [6] in Section 13.4, we have

$$\frac{\Delta z}{\Delta x}=f_x(x+\theta_1\Delta x,y+\Delta y)\frac{\Delta x}{\Delta x}+f_y(x,y+\theta_2\Delta y)\frac{\Delta y}{\Delta x}.$$

Taking the limit of the above equation as $\Delta x\to 0$ gives.
In particular

$$\lim_{\Delta x\to 0}\frac{\Delta z}{\Delta x}=\frac{dz}{dx} \text{ and } \lim_{\Delta x\to 0}\frac{\Delta y}{\Delta x}=\frac{dy}{dx}.$$

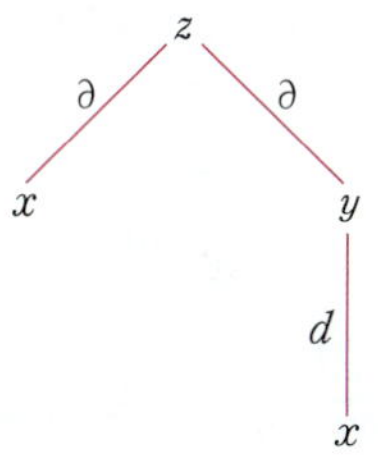

Figure 2

Taking the limit of the above equation as $\Delta x\to 0$ gives

$$\frac{dz}{dx}=f_x(x,y)+f_y(x,y)\frac{dy}{dx}$$

or,

$$\frac{dz}{dx}=\frac{\partial z}{\partial x}+\frac{\partial z}{\partial y}\frac{dy}{dx}. \text{ (See Figure 2.)}$$ ■

THEOREM 3 Chain Rule (Two Independent Variables)

Let $z=f(x,y)$ have continuous first order partial derivatives with respect to x and y. Let $x=\phi(u,v)$ and $y=\psi(u,v)$ also have continuous first order partial derivatives with respect to u and v. Then

[3]
$$\frac{\partial z}{\partial u}=\frac{\partial z}{\partial x}\frac{\partial x}{\partial u}+\frac{\partial z}{\partial y}\frac{\partial y}{\partial u}$$

$$\frac{\partial z}{\partial v}=\frac{\partial z}{\partial x}\frac{\partial x}{\partial v}+\frac{\partial z}{\partial y}\frac{\partial y}{\partial v}.$$

Proof Using [6] in Section 13.4, we have

$$\frac{\Delta z}{\Delta u}=f_x(x+\theta_1\Delta x,y+\Delta y)\frac{\Delta x}{\Delta u}+f_y(x,y+\theta_2\Delta y)\frac{\Delta y}{\Delta u}$$

for some $0<\theta_1,\theta_2<1$. When $\Delta x,\Delta y\to 0$ as $\Delta u\to 0$.
Now taking the limit of the above equation as $\Delta u\to 0$ gives

$$\lim_{\Delta u\to 0}\frac{\Delta x}{\Delta u}=\frac{\partial x}{\partial u},\quad \lim_{\Delta u\to 0}\frac{\Delta y}{\Delta u}=\frac{\partial y}{\partial u}$$

$$\frac{\partial z}{\partial u}=f_x(x,y)\frac{\partial x}{\partial u}+f_y(x,y)\frac{\partial y}{\partial u}\ .$$

In other words,

$$\frac{\partial z}{\partial u}=\frac{\partial z}{\partial x}\frac{\partial x}{\partial u}+\frac{\partial z}{\partial y}\frac{\partial y}{\partial u}\ .$$

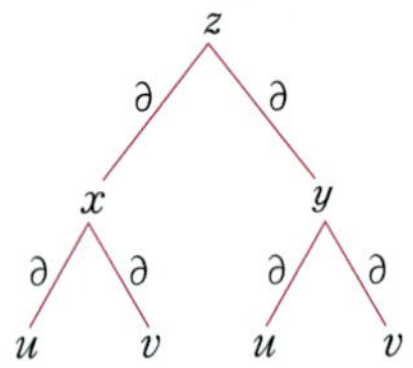

Figure 3

It may be shown, similarly that

$$\frac{\partial z}{\partial v}=\frac{\partial z}{\partial x}\frac{\partial x}{\partial v}+\frac{\partial z}{\partial y}\frac{\partial y}{\partial v}\ .$$ (See Figure 3.) ■

THEOREM 4

Let $w=f(x,y,z)$ have continuous first order partial derivatives with respect to x, y, z. Let x, y and z be functions of independent variables u and v. Also assume that x, y and z have continuous first order partial derivatives with respect to u and v. Then

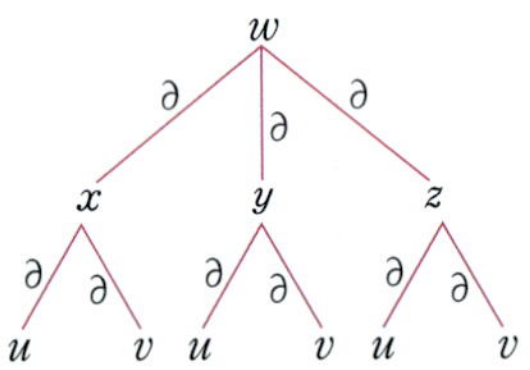

Figure 4

4

$$\frac{\partial w}{\partial u}=\frac{\partial f}{\partial u}=\frac{\partial f}{\partial x}\frac{\partial x}{\partial u}+\frac{\partial f}{\partial y}\frac{\partial y}{\partial u}+\frac{\partial f}{\partial z}\frac{\partial z}{\partial u}$$ and

$$\frac{\partial w}{\partial v}=\frac{\partial f}{\partial v}=\frac{\partial f}{\partial x}\frac{\partial x}{\partial v}+\frac{\partial f}{\partial y}\frac{\partial y}{\partial v}+\frac{\partial f}{\partial z}\frac{\partial z}{\partial v}\ .$$ (See Figure 4.)

THEOREM 5

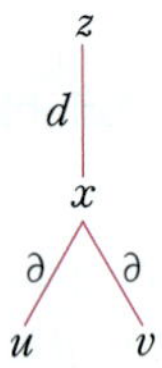

Figure 5

If $z=f(x)$ and $x=g(u,v)$ then

$$\frac{\partial z}{\partial u}=\frac{dz}{dx}\frac{\partial x}{\partial u} \quad \text{and} \quad \frac{\partial z}{\partial v}=\frac{dz}{dx}\frac{\partial x}{\partial v}.$$ (See Figure 5.)

EXAMPLE 1

If $z=\sin\dfrac{x}{y}$ where $x=e^t$ and $y=t^2$, find $\dfrac{dz}{dt}$.

Solution $\dfrac{\partial z}{\partial x}=\dfrac{1}{y}\cos\dfrac{x}{y},\ \dfrac{\partial z}{\partial y}=-\dfrac{x}{y^2}\cos\dfrac{x}{y},\ \dfrac{dx}{dt}=e^t,\ \dfrac{dy}{dt}=2t\,.$

Using the chain rule.

$$\begin{aligned}\frac{dz}{dt}&=\frac{dz}{dx}\frac{dx}{dt}+\frac{dz}{dy}\frac{dy}{dt}=\left(\frac{1}{y}\cos\frac{x}{y}\right)e^t+\left(-\frac{x}{y^2}\cos\frac{x}{y}\right)2t\\&=\left(\frac{e^t}{y}-\frac{2xt}{y^2}\right)\cos\frac{x}{y}=\frac{(t-2)e^t}{t^3}\cos\frac{e^t}{t^2}.\end{aligned}$$ ■

CAS **EXAMPLE 2**

If $u = \ln(x^2 + y^2 + z^2)$ where $y = x\sin x$ and $z = x\cos x$, find $\dfrac{du}{dx}$.

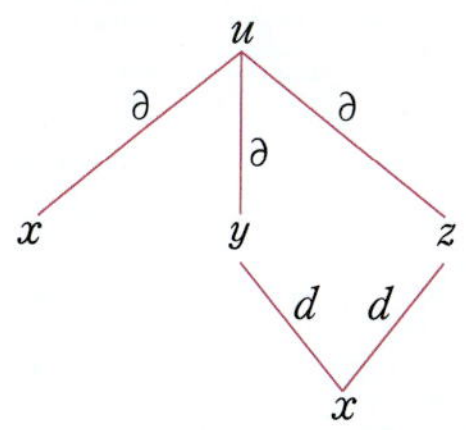

Figure 6

Solution Let us look at the tree diagram in Figure 6. Using the chain rule gives

$$\frac{du}{dx} = \frac{\partial u}{\partial x} + \frac{\partial u}{\partial y}\frac{dy}{dx} + \frac{\partial u}{\partial z}\frac{dz}{dx}$$

Now

$$\frac{\partial u}{\partial x} = \frac{2x}{x^2+y^2+z^2},\ \frac{\partial u}{\partial y} = \frac{2y}{x^2+y^2+z^2},\ \frac{\partial u}{\partial z} = \frac{2z}{x^2+y^2+z^2}.$$

$$\frac{dy}{dx} = \sin x + x\cos x,\ \frac{dz}{dx} = \cos x - x\sin x.$$

Therefore

$$\begin{aligned}\frac{du}{dx} &= \frac{2x}{x^2+y^2+z^2} + \frac{2y}{x^2+y^2+z^2}(\sin x + x\cos x) \\ &\quad + \frac{2z}{x^2+y^2+z^2}(\cos x - x\sin x) \\ &= \frac{2}{x^2+y^2+z^2}[x + y(\sin x + x\cos x) + z(\cos x - x\sin x)] \\ &= \frac{2}{x^2+y^2+z^2}[x + x\sin x(\sin x + x\cos x) \\ &\qquad + x\cos x(\cos x - x\sin x)] \\ &= \frac{2}{2x^2}(x+x) = \frac{2}{x}.\end{aligned}$$

http://matrix.skku.ac.kr/cal-lab/cal-12-2-Exm6.html

```
var('u, x, y, z')
u=ln(x^2+y^2+z^2)
u
```

log(x^2 + y^2 + z^2)

```
dudx=diff(u, x).factor()
dudx
```

2*x/(x^2 + y^2 + z^2)

```
dudy=diff(u, y).factor()
dudy
```

2*y/(x^2 + y^2 + z^2)

```
dudz=diff(u, z).factor()
dudz
```

2*z/(x^2 + y^2 + z^2)

```
y=x*sin(x)
dydx=diff(y, x).factor()
dydx
```

x*cos(x) + sin(x)

```
z=x*cos(x)
dzdx=diff(z, x).factor()
dzdx
```

-x*sin(x) + cos(x)

```
(dudx+dudy*dydx+dudz*dzdx).subs(y=x*sin(x),
z=x*cos(x)).trig_simplify()
```

Answer : 2/x ■

EXAMPLE 3

If $z=\sin^{-1}\left(\dfrac{y}{x}\right)$ where $x=u^2+v^2$, $y=2uv$, $u^2>v^2$, find $\dfrac{\partial z}{\partial u}$.

Solution Using the chain rule, we have

$$\frac{\partial z}{\partial u}=\frac{\partial z}{\partial x}\frac{\partial x}{\partial u}+\frac{\partial z}{\partial y}\frac{\partial y}{\partial u}.$$

Since $x>0$,

$$\frac{\partial z}{\partial x}=\frac{-\dfrac{y}{x^2}}{\sqrt{1-\left(\dfrac{y}{x}\right)^2}}=\frac{-y}{x\sqrt{x^2-y^2}},\quad \frac{\partial x}{\partial u}=2u,$$

$$\frac{\partial z}{\partial y}=\frac{\dfrac{1}{x}}{\sqrt{1-\left(\dfrac{y}{x}\right)^2}}=\frac{1}{\sqrt{x^2-y^2}},\quad \frac{\partial y}{\partial u}=2v.$$

Therefore, we have

$$\frac{\partial z}{\partial u}=\frac{-y}{x\sqrt{x^2-y^2}}2u+\frac{1}{\sqrt{x^2-y^2}}2v=\frac{-2uy+2vx}{x\sqrt{x^2-y^2}},$$

$$x^2-y^2=(u^2+v^2)^2-4u^2v^2=(u^2-v^2)^2,$$

Since $u^2 > v^2$, $\sqrt{x^2 - y^2} = u^2 - v^2$.

Therefore

$$\frac{\partial z}{\partial u} = \frac{-2u \cdot 2uv + 2v(u^2+v^2)}{(u^2+v^2)(u^2-v^2)} = \frac{-2v(u^2-v^2)}{(u^2+v^2)(u^2-v^2)}$$
$$= \frac{-2v}{u^2+v^2}.$$

■

http://matrix.skku.ac.kr/cal-lab/m-Sec13-5-Exm-3.html

EXAMPLE 4

If $z = r^2 e^{-\theta}$ where $x = r\cos\theta$ and $y = r\sin\theta$, find $\frac{\partial r}{\partial x}$, $\frac{\partial \theta}{\partial x}$, $\frac{\partial r}{\partial y}$, $\frac{\partial \theta}{\partial y}$ and $\frac{\partial^2 z}{\partial x \partial y}$.

Solution Taking the first order partial derivatives with respect to x and y of $x = r\cos\theta$ and $y = r\sin\theta$, we obtain

$$1 = \cos\theta \frac{\partial r}{\partial x} - r\sin\theta \frac{\partial \theta}{\partial x} \qquad (1)$$
$$0 = \sin\theta \frac{\partial r}{\partial x} + r\cos\theta \frac{\partial \theta}{\partial x} \qquad (2)$$
$$0 = \cos\theta \frac{\partial r}{\partial y} - r\sin\theta \frac{\partial \theta}{\partial y} \qquad (3)$$
$$1 = \sin\theta \frac{\partial r}{\partial y} + r\cos\theta \frac{\partial \theta}{\partial y} \qquad (4)$$

Solving (1) and (2), we have

$$\frac{\partial r}{\partial x} = \cos\theta \text{ and } \frac{\partial \theta}{\partial x} = -\frac{\sin\theta}{r}.$$

Solving (3) and (4), we have

$$\frac{\partial r}{\partial y} = \sin\theta \text{ and } \frac{\partial \theta}{\partial y} = \frac{\cos\theta}{r}.$$

To find $\frac{\partial^2 z}{\partial x \partial y}$, first we can get the following

$$\frac{\partial z}{\partial y} = \frac{\partial z}{\partial r}\frac{\partial r}{\partial y} + \frac{\partial z}{\partial \theta}\frac{\partial \theta}{\partial y} = 2re^{-\theta}\frac{\partial r}{\partial y} - r^2 e^{-\theta}\frac{\partial \theta}{\partial y}$$
$$= re^{-\theta}(2\sin\theta - \cos\theta).$$

Thus we have

$$\frac{\partial^2 z}{\partial x \partial y} = \frac{\partial}{\partial x}\left(\frac{\partial z}{\partial y}\right) = \frac{\partial}{\partial r}\left(\frac{\partial z}{\partial y}\right)\frac{\partial r}{\partial x} + \frac{\partial}{\partial \theta}\left(\frac{\partial z}{\partial y}\right)\frac{\partial \theta}{\partial x}$$

$$= \frac{\partial}{\partial r}\left[re^{-\theta}(2\sin\theta - \cos\theta)\right]\frac{\partial r}{\partial x} + \frac{\partial}{\partial\theta}\left[re^{-\theta}(2\sin\theta - \cos\theta)\right]\frac{\partial\theta}{\partial x}$$

$$= e^{-\theta}(2\sin\theta - \cos\theta)\frac{\partial r}{\partial x} + \left[-re^{-\theta}(2\sin\theta - \cos\theta) + re^{-\theta}(2\cos\theta + \sin\theta)\right]\frac{\partial\theta}{\partial x}$$

$$= e^{-\theta}(2\sin\theta\cos\theta - \cos^2\theta) + re^{-\theta}(3\cos\theta - \sin\theta)\left(\frac{-\sin\theta}{r}\right)$$

$$= e^{-\theta}\left[2\sin\theta\cos\theta - \cos^2\theta - 3\sin\theta\cos\theta + \sin^2\theta\right].$$

Therefore $\dfrac{\partial^2 z}{\partial x\partial y} = e^{-\theta}\left(\sin^2\theta - \sin\theta\cos\theta - \cos^2\theta\right)$.

http://matrix.skku.ac.kr/cal-lab/m-cal-13-5-Exm-4.html

```
var('r,t,x,y,r_x,t_x,r_y,t_y')
z=r^2*exp(-t)
x=r*cos(t);y=r*sin(t)
solve([1==cos(t)*r_x-r*sin(t)*t_x,0==sin(t)*r_x+r*cos(t)*t_x],r_
x,t_x)
```

Answer : r_x = cos(t)/(sin(t)^2 + cos(t)^2),
t_x =-sin(t)/(r*sin(t)^2 +r*cos(t)^2)]]

```
r_x=cos(t)/(sin(t)^2 + cos(t)^2)
t_x=-sin(t)/(r*sin(t)^2+r*cos(t)^2)
print r_x.simplify_trig()
print t_x.simplify_trig()
```

Answer : cos(t), -sin(t)/r

```
solve([0==cos(t)*r_y-r*sin(t)*t_y,1==sin(t)*r_y+r*cos(t)*t_y],r_
y,t_y)
```

Answer : r_y = sin(t)/(sin(t)^2 + cos(t)^2),
t_y = cos(t)/(r*sin(t)^2 +r*cos(t)^2)

```
r_y = sin(t)/(sin(t)^2 + cos(t)^2)
t_y = cos(t)/(r*sin(t)^2+r*cos(t)^2)
print r_y.simplify_trig()
print t_y.simplify_trig()
```

Answer : sin(t), cos(t)/r

```
z_y=diff(z,r)*r_y+diff(z,t)*t_y
z_y.simplify_trig()
```

Answer : (2*r*sin(t) - r*cos(t))*e^(-t)

```
z_yx=diff(z_y,r)*r_x+diff(z_y,t)*t_x
z_yx.simplify_trig()
```

Answer : -(sin(t)*cos(t) + 2*cos(t)^2 – 1)*e^(-t). ■

Implicit Differentiation[1]

1)http://archives.math.utk.edu/visual.calculus/3/implicit.7/index.html

(I) For the relationship between two variables x and y, we say that y is an implicit function of x if we are given an equation

13 $$f(x, y) = 0$$

Here y can be thought of as a single-valued function in a small neighborhood of x. If the relationship where $y = f(x)$ exists, then y is an explicit function of x. The following theorem holds for implicit function $f(x,y)=0$, if f has continuous first order partial derivatives of f with respect to x in the neighborhood at the point where $f_y(x,y) \neq 0$.

THEOREM 6 Implicit Differentiation

Let f have continuous first order derivatives on its domain. Suppose that $f(x,y)=0$ defines y as a differentiable function of x. Then

14 $$\frac{dy}{dx} = -\frac{\frac{\partial f}{\partial x}}{\frac{\partial f}{\partial y}}, \quad \text{provided} \quad \frac{\partial f}{\partial y} \neq 0. \text{ (See Figure 7.)}$$

Proof We can apply the chain rule to differentiate both sides of the equation $f(x,y)=0$ with respect to x. Since both x and y are functions of x, we get

$$\frac{\partial f}{\partial x} + \frac{\partial f}{\partial y}\frac{dy}{dx} = 0.$$

Then, $\dfrac{dy}{dx} = -\dfrac{\frac{\partial f}{\partial x}}{\frac{\partial f}{\partial y}}$, if $\dfrac{\partial f}{\partial y} \neq 0$. ■

(II) Now we suppose that z is given implicity as a function

15 $$z = \phi(x, y)$$

by an equation of the form $f(x,y,z)=0$. This means that $f(x,y,\phi(x,y))=0$ for all (x,y) in the domain of ϕ. If f and ϕ are differentiable, then we can apply Chain Rule to differentiate the equation $f(x,y,z)=0$ as follows:

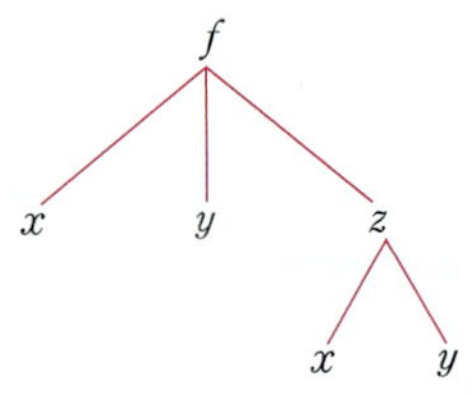

Figure 8

THEOREM 7 Implicit Differentiation

Let f be differentiable on its domain and suppose that $f(x, y, z) = 0$ defines z as an differentiable function of x and y. Then

16 $$\frac{\partial z}{\partial x} = -\frac{\frac{\partial f}{\partial x}}{\frac{\partial f}{\partial z}}, \quad \frac{\partial z}{\partial y} = -\frac{\frac{\partial f}{\partial y}}{\frac{\partial f}{\partial z}}, \quad \text{where } \frac{\partial f}{\partial z} \neq 0. \text{(See Figure 8.)}$$

Proof We assumed that $f(x, y, z) = 0$ defines z implicitly as a function of x and y. Then $\frac{\partial f}{\partial x}\frac{\partial x}{\partial x} + \frac{\partial f}{\partial y}\frac{\partial y}{\partial x} + \frac{\partial f}{\partial z}\frac{\partial z}{\partial x} = 0$.
So this equation becomes

17 $$\frac{\partial f}{\partial x} + \frac{\partial f}{\partial z}\frac{\partial z}{\partial x} = 0 \text{ because } \frac{\partial}{\partial x}(x) = 1 \text{ and } \frac{\partial}{\partial x}(y) = 0.$$

Similarly,

18 $$\frac{\partial f}{\partial y} + \frac{\partial f}{\partial z}\frac{\partial z}{\partial y} = 0 \text{ because } \frac{\partial}{\partial y}(x) = 0 \text{ and } \frac{\partial}{\partial y}(y) = 1.$$

If $\frac{\partial f}{\partial z} \neq 0$, then we have

$$\frac{\partial z}{\partial x} = -\frac{\frac{\partial f}{\partial x}}{\frac{\partial f}{\partial z}}, \quad \frac{\partial z}{\partial y} = -\frac{\frac{\partial f}{\partial y}}{\frac{\partial f}{\partial z}}.$$ ■

(Ⅲ) When y and z are functions of x

19 $$f(x,y,z) = 0, \; g(x,y,z) = 0,$$

we obtain the derivative with respect to x

$$\frac{\partial f}{\partial x} + \frac{\partial f}{\partial y}\frac{dy}{dx} + \frac{\partial f}{\partial z}\frac{dz}{dx} = 0,$$
$$\frac{\partial g}{\partial x} + \frac{\partial g}{\partial y}\frac{dy}{dx} + \frac{\partial g}{\partial z}\frac{dz}{dx} = 0$$

by solving the system of equations above, we find $\frac{dy}{dx}, \frac{dz}{dx}$. Let us assume that the determinant

$$\Delta = \begin{vmatrix} \frac{\partial f}{\partial y} & \frac{\partial f}{\partial z} \\ \frac{\partial g}{\partial y} & \frac{\partial g}{\partial z} \end{vmatrix} = \frac{\partial f}{\partial y}\frac{\partial g}{\partial z} - \frac{\partial f}{\partial z}\frac{\partial g}{\partial y} \neq 0,$$

then

$$\boxed{20} \qquad \frac{dy}{dx} = -\frac{1}{\Delta}\begin{vmatrix} \frac{\partial f}{\partial x} & \frac{\partial f}{\partial z} \\ \frac{\partial g}{\partial x} & \frac{\partial g}{\partial z} \end{vmatrix}, \quad \frac{dz}{dx} = -\frac{1}{\Delta}\begin{vmatrix} \frac{\partial f}{\partial y} & \frac{\partial f}{\partial x} \\ \frac{\partial g}{\partial y} & \frac{\partial g}{\partial x} \end{vmatrix}.$$

EXAMPLE 5

If $x\sin y + y\cos x = 0$, find $\frac{dy}{dx}$.

Solution Let $f(x, y) = x\sin y + y\cos x$. Then

$$f_x = \sin y - y\sin x, \ f_y = x\cos y + \cos x.$$

Therefore, from formula 14, we obtain

$$\frac{dy}{dx} = -\frac{\sin y - y\sin x}{x\cos y + \cos x}. \qquad \blacksquare$$

```
var('x,y')
f=x*sin(y)+y*cos(x)
f_x=diff(f,x);f_y=diff(f,y)
dydx=-f_x/f_y
dydx
```

Answer : (y*sin(x) - sin(y))/(x*cos(y) + cos(x))

EXAMPLE 6

Suppose z is an implicit function of x and y on any rectangular region that satisfy $x^3e^{y+z} - y\sin(x-z) = 0$. Find $\frac{\partial z}{\partial x}$ and $\frac{\partial z}{\partial y}$.

Solution Let $f(x, y, z) = x^3e^{y+z} - y\sin(x-z)$. Since

$$\frac{\partial f}{\partial x} = 3x^2e^{y+z} - y\cos(x-z), \ \frac{\partial f}{\partial y} = x^3e^{y+z} - \sin(x-z)$$

and $\frac{\partial f}{\partial z} = x^3e^{y+z} + y\cos(x-z)$,

from 16, we obtain

$$\frac{\partial z}{\partial x} = -\frac{3x^2e^{y+z} - y\cos(x-z)}{x^3e^{y+z} + y\cos(x-z)}$$

$$\frac{\partial z}{\partial y} = -\frac{x^3e^{y+z} - \sin(x-z)}{x^3e^{y+z} + y\cos(x-z)}. \qquad \blacksquare$$

```
var('x,y,z')
f=x^3*exp(y+z)-y*sin(x-z)
f_x=diff(f,x);f_y=diff(f,y);f_z=diff(f,z)
dzdx=-f_x/f_z;dzdy=-f_y/f_z
print dzdx
print dzdy
```

Answer : -(3*x^2*e^(y+z)-y*cos(x-z))/(x^3*e^(y+z)+y*cos(x-z))
-(x^3*e^(y+z)-sin(x-z))/(x^3*e^(y+z)+y*cos(x-z))

EXAMPLE 7

If equation $z^2 = xy - x\ln y$, find $\frac{\partial z}{\partial x}$ and $\frac{\partial z}{\partial y}$.

Solution Let $f(x,y,z) = xy - x\ln y - z^2$. Then

$$\frac{\partial f}{\partial x} = y - \ln y, \quad \frac{\partial f}{\partial y} = x - \frac{x}{y} \quad \text{and} \quad \frac{\partial f}{\partial z} = 2z.$$

Therefore, from formula 16, we obtain

$$\frac{\partial z}{\partial x} = \frac{\ln y - y}{2z} \quad \text{and} \quad \frac{\partial z}{\partial y} = \frac{-\left(x - \frac{x}{y}\right)}{2z} = \frac{x - xy}{2yz}.$$

■

13.5 EXERCISES (The Chain Rule)

http://matrix.skku.ac.kr/Cal-Book/part2/CS-Sec-13-5-Sol.html

YouTube http://youtu.be/vzN5By6qzvM

1. If $f(x, y, z) = x + 2y + z^2$ where $x = \frac{r}{s}$, $y = r^2 + \ln s$ and $z = 2r$, find $\frac{\partial f}{\partial r}$ and $\frac{\partial f}{\partial s}$ at $r = 1$ and $s = 1$.

Solution $\frac{\partial f}{\partial r} = \frac{\partial f}{\partial x}\frac{\partial x}{\partial r} + \frac{\partial f}{\partial y}\frac{\partial y}{\partial r} + \frac{\partial f}{\partial z}\frac{\partial z}{\partial r}$

$= (1)\left(\frac{1}{s}\right) + (2)(2r) + (2z)(2) = \frac{1}{s} + 12r$

$\Rightarrow \left.\frac{\partial f}{\partial r}\right|_{(r,s)=(1,1)} = 13$

$\frac{\partial f}{\partial s} = \frac{\partial f}{\partial x}\frac{\partial x}{\partial s} + \frac{\partial f}{\partial y}\frac{\partial y}{\partial s} + \frac{\partial f}{\partial z}\frac{\partial z}{\partial s}$

$= (1)\left(-\frac{r}{s^2}\right) + (2)\left(\frac{1}{s}\right) + (2z)(0) = -\frac{r}{s^2} + \frac{2}{s}$

$\Rightarrow \left.\frac{\partial f}{\partial s}\right|_{(r,s)=(1,1)} = 1.$

2. Find $\frac{\partial w}{\partial t}$ and $\frac{\partial^2 w}{\partial s \partial t}$ if $w = e^x y + (x + y)\tan^{-1}(z^2 + 1)$, where $x = s + t$, $y = s - t$, $z = s + 8$.

Solution (1) $\frac{\partial x}{\partial t} = 1$, $\frac{\partial y}{\partial t} = -1$, $\frac{\partial z}{\partial t} = 0$,

$\frac{\partial w}{\partial t} = f_x \frac{\partial x}{\partial t} + f_y \frac{\partial y}{\partial t} + f_z \frac{\partial z}{\partial t} = f_x - f_y$

$= [e^x\ y + \tan^{-1}(z^2+1)] - [e^x + \tan^{-1}(z^2+1)]$

$= e^x\ (y-1)$ or $e^{s+t}(s-t-1)$.

(2) $\frac{\partial^2 w}{\partial s \partial t} = \frac{\partial}{\partial s}\left(\frac{\partial w}{\partial t}\right)$

$= \left[\frac{\partial}{\partial x}\left(\frac{\partial w}{\partial t}\right)\right]\frac{\partial x}{\partial s} + \left[\frac{\partial}{\partial y}\left(\frac{\partial w}{\partial t}\right)\right]\frac{\partial y}{\partial s}$

$= e^x\ (y-1) \times 1 + e^x \times 1$

$= e^x\ y$ or $e^{(s+t)}(s-t)$.

3. Find $\frac{\partial z}{\partial x}$, $\frac{\partial z}{\partial y}$ of $x^2 + y^2 + z^2 = 1$.

Solution $f(x,y,z) = x^2 + y^2 + z^2 - 1 = 0$

$f_x(x,y,z) = 2x$, $f_y(x,y,z) = 2y$, $f_z(x,y,z) = 2z$

$\frac{\partial z}{\partial x} = -\frac{f_x(x,y,z)}{f_z(x,y,z)} = -\frac{2x}{2z} = -\frac{x}{z}$,

$\frac{\partial z}{\partial y} = -\frac{f_y(x,y,z)}{f_z(x,y,z)} = -\frac{2y}{2z} = -\frac{y}{z}$

4. Find $\frac{\partial z}{\partial x}$ of $xyz = \cos(x + y + z)$.

Solution Take a partial derivative of both sides with respect to x, then we have

$yz + xy\frac{\partial z}{\partial x} = -\sin(x+y+z)\left(1 + \frac{\partial z}{\partial x}\right)$.

$\Rightarrow \frac{\partial z}{\partial x} = -\frac{yz + \sin(x+y+z)}{xy + \sin(x+y+z)}$.

5. Find $\frac{\partial z}{\partial x}$ and $\frac{\partial z}{\partial y}$ when z is an implicitly defined function of x and y in
$f(x, y, z) = x^3 e^{y+z} - y\sin(x - z) = 0$.

Solution If $f(x,y,z) = x^3 e^{y+z} - y\sin(x-z)$

$f_x(x,y,z) = 3x^2 e^{y+z} - y\cos(x-z)$,

$f_y(x,y,z) = x^3 e^{y+z} - \sin(x-z)$

$f_z(x,y,z) = x^3 e^{y+z} + y\cos(x-z)$

$\frac{\partial z}{\partial x} = -\frac{3x^2 e^{y+z} - y\cos(x-z)}{x^3 e^{y+z} + y\cos(x-z)}$,

$\frac{\partial z}{\partial y} = -\frac{x^3 e^{y+z} - \sin(x-z)}{x^3 e^{y+z} + y\cos(x-z)}$

6. Find $\frac{\partial^2 z}{\partial x^2}$, $\frac{\partial^2 z}{\partial x \partial y}$ of $x^3 + y^3 + z^3 = 1$.

Solution $3x^2 + 3z^2\frac{\partial z}{\partial x} = 0 \Rightarrow \frac{\partial z}{\partial x} = -\frac{x^2}{z^2}$ because

$x^2 + z^2\frac{\partial z}{\partial x} = 0$.

$2x + 2z\left(\frac{\partial z}{\partial x}\right)^2 + z^2\frac{\partial^2 z}{\partial x^2} = 0$

$\Rightarrow \frac{\partial^2 z}{\partial x^2} = -\frac{2x + 2z\left(\frac{\partial z}{\partial x}\right)^2}{z^2} = -\frac{2x + 2z\left(-\frac{x^2}{z^2}\right)^2}{z^2}$

$= -\frac{2x(x^3 + z^3)}{z^5}$

$3y^2 + 3z^2\frac{\partial z}{\partial y} = 0 \Rightarrow \frac{\partial z}{\partial y} = -\frac{y^2}{z^2}$ because $y^2 + z^2\frac{\partial z}{\partial y} = 0$.

$2z\frac{\partial z}{\partial x} \cdot \frac{\partial z}{\partial y} + z^2\frac{\partial^2 z}{\partial x \partial y} = 0 \Rightarrow$

$\frac{\partial^2 z}{\partial x \partial y} = -\frac{2\frac{\partial z}{\partial x} \cdot \frac{\partial z}{\partial y}}{z} = -2\frac{\left(-\frac{x^2}{z^2}\right)\left(-\frac{y^2}{z^2}\right)}{z} = -\frac{2x^2y^2}{z^5}$

7. Find $\frac{dy}{dx}$, $\frac{dz}{dx}$ when $\begin{cases} x + y + z = 1 \\ x^2 + y^2 + z^2 = 4 \end{cases}$.

Solution $f(x,y,z) = x + y + z - 1 = 0 \Rightarrow$

$f_x(x,y,z) = 1$, $f_y(x,y,z) = 1$, $f_z(x,y,z) = 1$

$g(x,y,z) = x^2 + y^2 + z^2 - 4 = 0 \Rightarrow$

$g_x(x,y,z) = 2x$, $g_y(x,y,z) = 2y$, $g_z(x,y,z) = 2z$

$$\Delta = \begin{vmatrix} 1 & 1 \\ 2y & 2z \end{vmatrix} = 2z - 2y = 2(z-y)$$

$$\frac{dy}{dx} = -\frac{1}{2(z-y)}\begin{vmatrix} 1 & 1 \\ 2x & 2z \end{vmatrix} = -\frac{2(z-x)}{2(z-y)} = -\frac{z-x}{z-y}$$

$$\frac{dz}{dx} = -\frac{1}{2(z-y)}\begin{vmatrix} 1 & 1 \\ 2y & 2x \end{vmatrix} = -\frac{2(x-y)}{2(z-y)} = -\frac{x-y}{z-y}$$

CAS **8.** Let $f(x, y) = \dfrac{xy}{x^2+y^2}$ with $f(0,0)=0$.

(a) Show that f_x and f_y exist everywhere.

(b) Are f_x and f_y continuous at the origin? Justify your answer.

http://matrix.skku.ac.kr/cal-lab/cal-12-2-5.html

Solution (a) If $x=0$ and $y=0$, $f_x=0$, $f_y=0$.

Else

```
var('x, y')
f(x, y)=x*y/(x^2+y^2)
fx(x, y)=limit((f(x+h, y)-f(x, y))/h, h=0)
fy(x, y)=limit((f(x, y+h)-f(x, y))/h, h=0)
print(fx(x, y))
print((fy(x, y))
```

```
-(x^2*y - y^3)/(x^4 + 2*x^2*y^2 + y^4)
(x^3 - x*y^2)/(x^4 + 2*x^2*y^2 + y^4)
```

Thus f_x and f_y exist everywhere.

(b)

```
print(limit(fx(x, 0), x=0))
print(limit(fx(0, y), y=0))
```

```
0
Infinity
```

f_x is not continuous at the origin.

```
print(limit(fy(x, 0), x=0))
print(limit(fy(0, y), y=0))
```

```
Infinity
0
```

f_y is not continuous at the origin.

9. A function $f(x, y)$ is called a homogeneous function of degree n if all the terms in $f(x, y)$ are of degree n. In other words, $f(tx, ty) = t^n f(x, y)$ for any parameter t. If $f(x, y)$ is a homogeneous function of degree n then show that

$$x\frac{\partial f}{\partial x} + y\frac{\partial f}{\partial y} = nf(x, y)$$

This is also called Euler's theorem for homogeneous function.

10. Verify the Euler's theorem for the following:

(i) $u = \cos\left(\dfrac{xy+yz+zx}{x^2+y^2+z^2}\right)$.

(ii) $u = x^2\tan^{-1}\dfrac{y}{x} + y^2\tan^{-1}\dfrac{x}{y}$.

(iii) $u = \dfrac{x^3y^3z^3}{x^3+y^3+z^3}$.

(iv) $u = \left(\dfrac{x}{y} + \dfrac{y}{z} + \dfrac{z}{x}\right)^n$.

(v) $u = x^3\sin^{-1}\left(\dfrac{\sqrt{y}+\sqrt{x}}{\sqrt{y}-\sqrt{x}}\right)$.

(vi) $u = \left(\dfrac{x}{y}\right)^{\frac{y}{x}}$.

11. If $u = f(x, y)$ is a homogeneous function of degree n in x, y and z then show that

$$x^2\frac{\partial^2 u}{\partial x^2} + 2xy\frac{\partial^2 u}{\partial x\,\partial y} + y^2\frac{\partial^2 u}{\partial y^2} = 2n(n-1)u\frac{\partial^2 u}{\partial x^2}.$$

12. Let $u = \dfrac{x^4+y^4}{x^2y^2} + x^6\tan^{-1}\left(\dfrac{x^2+y^2}{x^2+2xy}\right)$. Find $x^2\dfrac{\partial^2 u}{\partial x^2} + 2xy\dfrac{\partial^2 u}{\partial x\,\partial y} + y^2\dfrac{\partial^2 u}{\partial y^2} + x\dfrac{\partial u}{\partial x} + y\dfrac{\partial u}{\partial y}$ at $x=1$, $y=1$.

13. Let z is a homogeneous function of degree n in x and y. If $z = f(u)$ then show that

$$x\frac{\partial u}{\partial x} + y\frac{\partial u}{\partial y} = n\frac{f(u)}{f'(u)}.$$

14. Let $u = \sin^{-1}\left(\dfrac{x^{1/4}+y^{1/4}}{x^{1/5}+y^{1/5}}\right)$. Show that

$$x\frac{\partial u}{\partial x} + y\frac{\partial u}{\partial y} = \frac{1}{20}\tan u.$$

15. Let $u = \tan^{-1}\left(\dfrac{x^4+y^4}{x^2+y^2}\right)$. Show that

$$x\frac{\partial u}{\partial x} + y\frac{\partial u}{\partial y} = \sin 2u.$$

16. Show that $z = f(x+at) - \phi(x-at)$ is a solution of $a^2\dfrac{\partial^2 z}{\partial x^2} = \dfrac{\partial^2 z}{\partial t^2}$ for all t and ϕ assuming that a is a constant.

17. If $u=a^{ax}\sin(bt+x)$ is the solution of the equation $\dfrac{\partial u}{\partial t}=b\dfrac{\partial^2 u}{\partial x^2}$ with the condition that $u \to 0$ as $x \to 0$, find the values of a and b.

18. Let $u=x^2+y^2+z^2$, $x=e^t$, $y=e^t\sin t$ and $z=e^t\cos t$. Find $\dfrac{du}{dt}$.

19. Let $z=e^{xy^2}$, $x=t\cos t$, $y=t\sin t$. Find $\dfrac{dz}{dt}$ at $t=\dfrac{\pi}{2}$.

<3D Printing>

13.6 Directional Derivatives and Gradient

YouTube http://youtu.be/o8L_ShRANjo

The partial derivatives $f_x(x_0, y_0)$ and $f_y(x_0, y_0)$ are the rates of change of $z = f(x, y)$ at (x_0, y_0) in the positive x- and y-directions. Rates of change in other directions are given by directional derivatives. We open this section by defi ning directional derivatives and then use the Chain Rule to derive a formula for their values in terms of x- and y-derivatives. Then we study gradient vectors and show how they are used to determine how directional derivatives at a point change as the direction changes, and, in particular, how they can be used to find the maximum and minimum directional derivatives at a point.

In this section we introduce a type of derivative, called a directional derivative, that enables us to find the rate of change of a function of two or more variables in any direction.

Directional Derivatives

DEFINITION 1 Directional Derivatives

Let $\mathbf{u} = (u_1, u_2)$ be a unit vector. The **directional derivative** of f at (a, b) in the direction of $\mathbf{u}$ is defined as

$$D_{\mathbf{u}} f(a, b) = \lim_{h \to 0} \frac{f(a + hu_1, b + hu_2) - f(a, b)}{h}$$

if the limit exist.

Remark 1. Note that if $\mathbf{u} = (1, 0)$, then $D_{\mathbf{u}}(a, b) = f_x(a, b)$ and if $\mathbf{v} = (0, 1)$, then $D_{\mathbf{v}}(a, b) = f_y(a, b)$.

2. Consider a curve $\gamma(h) = (a, b) + h(u_1, u_2)$ in the domain of f. Note that $\gamma(h)$ is the line passing through (a, b) in the direction of $\mathbf{u} = (u_1, u_2)$ furthermore, $\gamma(0) = (a, b)$ and $\gamma'(0) = (u_1, u_2)$. Then it is easy to see that $D_{\mathbf{u}} f(a, b) = \frac{d}{dh}[f(\gamma(h))]_{h=0}$.

EXAMPLE 1

Find the directional derivative of $f(x, y) = 4x^2 y + 3xy + 6x + 2$ at $(1, -1)$ in direction of $(1, 0)$.

http://matrix.skku.ac.kr/cal-lab/m-Sec13-6-Exm-1.html

Solution

```
var('x,y,h')
f(x, y)=4*x^2*y+3*x*y+6*x+2
limit((f(1+h,-1+h*0) - f(1,-1))/h, h=0)
```

Answer : -5 ■

THEOREM 2

Suppose that f is differentiable at (a, b) and $\mathbf{u} = \langle u_1, u_2 \rangle$ is any unit vector. Then

$$D_{\mathbf{u}} f(a, b) = f_x(a, b) u_1 + f_y(a, b) u_2$$
$$= \langle f_x(a, b), f_y(a, b) \rangle \cdot \langle u_1, u_2 \rangle$$

Proof $D_{\mathbf{u}} f(x_0, y_0) = \lim_{h \to 0} \frac{f(x_0 + hu_1, y_0 + hu_2) - f(x_0, y_0)}{h}$

$$= \lim_{h \to 0} \frac{f(x_0 + hu_1, y_0 + hu_2) - f(x_0, y_0 + hu_2) + f(x_0, y_0 + hu_2) - f(x_0, y_0)}{h}$$

$$= \lim_{h \to 0} \frac{f(x_0 + hu_1, y_0 + hu_2) - f(x_0, y_0 + hu_2)}{h} + \lim_{h \to 0} \frac{f(x_0, y_0 + hu_2) - f(x_0, y_0)}{h}$$

$$= f_x(x_0, y_0) u_1 + f_y(x_0, y_0) u_2 \qquad \text{(by the Chain rule)}$$

$$= \langle f_x(x_0, y_0), f_y(x_0, y_0) \rangle \cdot \langle u_1, u_2 \rangle.$$ ■

EXAMPLE 2

Find the directional derivative $D_{\mathbf{u}} f(1, 2)$ if

$$f(x, y) = x^3 - 3xy + 4y^2$$

and $\mathbf{u}$ is the unit vector given by angle $\theta = \pi/6$.

http://matrix.skku.ac.kr/cal-lab/m-Sec13-6-Exm-2.html

Solution When $\theta = \pi/6$, the unit vector is

$$\mathbf{u} = \left\langle \cos\frac{\pi}{6}, \sin\frac{\pi}{6} \right\rangle = \left\langle \frac{\sqrt{3}}{2}, \frac{1}{2} \right\rangle.$$

Since $f_x(x, y) = 3x^2 - 3y$ and $f_y(x, y) = -3x + 8y$,

$$f_x(1, 2) = -3 \text{ and } f_y(1, 2) = 13.$$

Then $D_{\mathbf{u}} f(1, 2) = \frac{13}{2} - \frac{3\sqrt{3}}{2}$. ■

```
var('x,y')
f(x,y)=x^3-3*x*y+4*y^2
delf=vector([diff(f(x,y),x),diff(f(x,y),y)])
delfp=delf.subs(x=1,y=2)
u=vector([cos(pi/6),sin(pi/6)])
dirdr=delfp.dot_product(u)
print dirdr
```

Answer : -3/2*sqrt(3) + 13/2

Gradient Vectors

For convenience, we define the **gradient** of a function to be the vector-valued function whose components are the first-order partial derivatives of f, as specified in Definition 3. We denote the gradient of a function f by **grad** f or ∇f.

DEFINITION 3

The gradient of $f(x, y)$ is the vector-valued function

$$\nabla f(x, y) = \left\langle \frac{\partial f}{\partial x}, \frac{\partial f}{\partial y} \right\rangle = \frac{\partial f}{\partial x}\mathbf{i} + \frac{\partial f}{\partial y}\mathbf{j},$$

provided both first order partial derivatives exist.

Remark 3. In view of Theorem 2, we have

$$\begin{aligned} D_{\mathbf{u}} f(a, b) &= f_x(a, b)u_1 + f_y(a, b)u_2 \\ &= \nabla f(a, b) \cdot \langle u_1, u_2 \rangle = \nabla f \cdot \mathbf{u}. \end{aligned}$$

EXAMPLE 3

Find the gradients of

(i) $f(x, y) = 4x^2 + 3y^2 - 5xy$,

(ii) $f(x, y) = \dfrac{xy}{x^2 + y^2 + 1}$.

Solution (i) http://matrix.skku.ac.kr/cal-lab/m-Sec13-6-Exm-3.html

```
var('x,y')
f(x, y)=4*x^2+3*y^2-5*x*y
f.gradient()
```

Answer : (x, y) |--> (8*x - 5*y, -5*x + 6*y)

$$\nabla f(x, y) = (8x - 5y)\mathbf{i} + (6y - 5x)\mathbf{j}$$

(ii)

```
var('x,y')
f(x, y)=x*y/(x^2+y^2+1)
f.gradient()
```

Answer : (x, y) |--> (-2*x^2*y/(x^2 + y^2 + 1)^2 + y/(x^2 + y^2 + 1), -2*x*y^2/(x^2 + y^2 + 1)^2 + x/(x^2 + y^2 + 1))

$$\nabla f(x, y) = \left(\frac{-2x^2y}{(x^2+y^2+1)^2} + \frac{y}{x^2+y^2+1} \right)\mathbf{i} + \left(\frac{-2xy^2}{(x^2+y^2+1)^2} + \frac{x}{x^2+y^2+1} \right)\mathbf{j}.$$

■

Interpretations of the Gradient

The gradient is not limited to calculating partial derivatives; it plays important roles in multivariable calculus. Our present goal is to develop some intuition about the meaning of the gradient.

From Remark 3, $D_{\mathbf{u}}f(a, b) = \nabla f(a, b) \cdot \mathbf{u}$. Using properties of the dot product, we have

$$\begin{aligned} D_{\mathbf{u}}f(a, b) &= \nabla f(a, b) \cdot \mathbf{u} \\ &= |\nabla f(a, b)| \cdot |\mathbf{u}| \cos\theta \\ &= |\nabla f(a, b)| \cos\theta, \quad |\mathbf{u}| = 1 \end{aligned}$$

where θ is the angle between $\nabla f(a, b)$ and $\mathbf{u}$. It follows that $D_{\mathbf{u}}f(a, b)$ has its maximum value when $\cos\theta = 1$, which corresponds to $\theta = 0$. Therefore, $D_{\mathbf{u}}f(a, b)$ has its maximum value and f has its greatest rate of increase when $\nabla f(a, b)$ and $\mathbf{u}$ point in the same direction. Notice that when $\cos\theta = 1$, the actual rate of increase is $D_{\mathbf{u}}f(a, b) = |\nabla f(a, b)|$.

Similarly, when $\theta = \pi$, we have $\cos\theta = -1$, and f has its greatest rate of decrease when $\nabla f(a, b)$ and $\mathbf{u}$ point in opposite directions. The actual rate of decrease is $D_{\mathbf{u}}f(a, b) = -|\nabla f(a, b)|$. These observations are summarized as follows:

Write these as three observations:

Observation 1. The gradient $f(a, b)$ points in the direction of steepest ascent at (a, b).

2. The negative of the gradient $f(a, b)$ points is the direction of steepest descent.

3. Notice that $D_{\mathbf{u}}f(a, b) = 0$ when the angle between $\nabla f(a, b)$ and $\mathbf{u}$ is $\pi/2$, which means $\nabla f(a, b)$ and $\mathbf{u}$ are orthogonal.

EXAMPLE 4

(a) Find the derivative of $f(x, y) = 100 - x^2 - y^2$ at the point $P_0(3,4)$ in the direction of the unit vector $\mathbf{u} = u_1\mathbf{i} + u_2\mathbf{j}$.

(b) What is the direction in which the function f increase most rapidly at P_0 in xy-plane region?

(c) Determine the direction of zero change in f at (3, 4).

Solution (a) Since $f_x(3,4) = -2x\,|_{(3,4)} = -6$,

$f_y(3,4) = -2y\,|_{(3,4)} = -8$,

then the directional derivative is

$D_{\mathbf{u}}f = \nabla f \cdot \mathbf{u} = \langle -6, -8\rangle \cdot \langle u_1, u_2\rangle = -6u_1 - 8u_2$.

(b) The function increases most rapidly in the direction of the gradient. The gradient is

$$|\nabla f| = \sqrt{(-6)^2 + (-8)^2} = \sqrt{36+64} = 10.$$

Its direction is

$$\frac{\nabla f}{|\nabla f|} = -\frac{6}{10}\mathbf{i} - \frac{8}{10}\mathbf{j} = -\frac{3}{5}\mathbf{i} - \frac{4}{5}\mathbf{j}.$$

(c) The directions of zero change in f at (3, 4) are the directions orthogonal to ∇f.

$$\mathbf{n} = \frac{4}{5}\mathbf{i} - \frac{3}{5}\mathbf{j} \text{ and } -\mathbf{n} = -\frac{4}{5}\mathbf{i} + \frac{3}{5}\mathbf{j}.$$

Computing $\nabla f \cdot \mathbf{n}$ shows that they are perpendicular.

$$\nabla f \cdot \mathbf{n} = (-6\mathbf{i} - 8\mathbf{j}) \cdot \left(\frac{4}{5}\mathbf{i} - \frac{3}{5}\mathbf{j}\right)$$
$$= -\frac{24}{5} + \frac{24}{5} = 0.$$

■

The Gradient and Level Curves

Note that Observation 3 says that in any direction orthogonal to the gradient $\nabla f(a,b)$, the function f does not change at (a, b). Recall the curve $f(x, y) = z_0$, where z_0 is a constant, is a **level curve**, on which function values is constant. Combing these two observations, we conclude that gradient $\nabla f(a,b)$ is orthogonal to the line tangent to the level curve through (a, b).

THEOREM 4 **The Gradient and Level Curves**

Given a function f differentiable at (a, b), the line tangent to the level curve of f at (a, b) is orthogonal to the gradient $\nabla f(a, b)$.

Proof A level curve of the function $z = f(x, y)$ is a curve in the xy-plane of the form $f(x, y) = z_0$, where z_0 is a constant. By Theorem 6 in section 13.5, the slope of the line tangent to the level curve is $y'(x) = -f_x/f_y$. It follows that any vector that points in the direction of the tangent line at the point (a, b) is a scalar multiple of the vector

$$\mathbf{t} = \langle -f_y(a, b), f_x(a, b) \rangle.$$

At that same point, the gradient points in the direction

$$\nabla f(a, b) = \langle f_x(a, b), f_y(a, b) \rangle.$$

The dot product of $\mathbf{t}$ and $\nabla f(a, b)$ is

$$\begin{aligned} \mathbf{t} \cdot \nabla f(a,b) &= \langle -f_y, f_x \rangle_{(a,b)} \cdot \langle f_x, f_y \rangle_{(a,b)} \\ &= -f_y(a,b)f_x(a,b) + f_x(a,b)f_y(a,b) = 0 \end{aligned}$$

which implies that $\mathbf{t}$ and $\nabla f(a, b)$ are orthogonal to each other. ■

An immediate consequence of Theorem 4 is an alternative equation of the tangent line. The curve described by $f(x, y) = z_0$ can be viewed as a level curve for a surface. By Theorem 4, the line tangent to the curve at (a, b) is orthogonal to $\nabla f(a, b)$. Therefore, if (x, y) is a point on the tangent line, then $\nabla f(a,b) \cdot \langle x-a, y-b \rangle = 0$, which, when simplified, gives an equation of the line tangent to the curve $f(x, y) = z_0$.

EXAMPLE 5

Consider the upper sheet $z = f(x, y) = \sqrt{1+2x^2+y^2}$ of a hyperboloid of two sheets.

(a) Verify that the gradient at $(1, 1)$ is orthogonal to the corresponding level curve at that point.

(b) Find an equation of the tangent line to the level curve at $(1, 1)$.

Solution (a) You can verify that $(1, 1, 2)$ is on the surface; therefore, $(1, 1)$ is on the level curve corresponding to $z = 2$. Setting $z = 2$ in the equation of the surface and squaring both sides, the equation of the level curve is $4 = 1 + 2x^2 + y^2$ or $2x^2 + y^2 = 3$, which is the equation of an ellipse. Differentiating $2x^2 + y^2 = 3$ with respect to x gives $4x + 2yy'(x) = 0$, which implies that the slope of the

level curve is $y'(x) = -\dfrac{2x}{y}$. Therefore, at the point $(1, 1)$, the slope of the tangent line is -2. Any vector proportional to $\mathbf{t} = \langle 1, -2 \rangle$ has slope -2 and points in the direction of the tangent line.
We now compute the gradient:

$$\nabla f(x, y) = \langle f_x, f_y \rangle = \left\langle \frac{2x}{\sqrt{1+2x^2+y^2}}, \frac{y}{\sqrt{1+2x^2+y^2}} \right\rangle.$$

It follows that $\nabla f(1, 1) = \left\langle 1, \dfrac{1}{2} \right\rangle$. The tangent vector $\mathbf{t}$ and the gradient are orthogonal because

$$\mathbf{t} \cdot \nabla f(1, 1) = \langle 1, -2 \rangle \left\langle 1, \frac{1}{2} \right\rangle = 0.$$

(b) An equation of the line tangent to the level curve at $(1, 1)$ is

$$f_x(1, 1)(x-1) + f_y(1, 1)(y-1) = 0.$$

Then the tangent line is $x + y - 2 = 0$. ■

Functions of Three Variables

Let $f(x, y, z)$ be a function of three variables defined on some domain D in $\mathbb{R}^3$. The notion of directional derivative and gradient of $f(x, y, z)$ can be defi ned just by introducing extra component z in definition of two variable function $f(x, y)$. For the sake of completion, let us write these definitions:

DEFINITION 5

Let $f(x, y, z)$ be a function of three variables defined in a domain D. Let $(a, b, c) \in D$ and $\mathbf{u} = \langle u_1, u_2, u_3 \rangle$ be an unit vector in $\mathbb{R}^3$. Then the directional derivative of f at (a, b, c) in the direction of $\mathbf{u}$ is given by

$$D_{\mathbf{u}}(a, b, c) = \lim_{h \to 0} \frac{f(a+hu_1, b+hu_2, c+hu_3) - f(a, b, c)}{h}$$

if the right hand side limit exists.
Again using chain rule it turns out that

$$D_{\mathbf{u}}(a, b, c) = f_x(a, b, c)u_1 + f_y(a, b, c)u_2 + f_z(a, b, c)u_3.$$

DEFINITION 6

Let $f(x,y,z)$ be a function of three variables defined in a domain D. Let $(a,b,c) \in D$ and f possesses continuous first order partial derivative at (a,b,c). Then the gradient of f at (a,b,c) is defined as

$$\nabla f(a,b,c) = f_x(a,b,c)\mathbf{i} + f_y(a,b,c)\mathbf{j} + f_z(a,b,c)\mathbf{k}$$

EXAMPLE 6

Find the directional derivative of the function $f(x,y,z) = xy^2 - 4x^2y + z^2$ at $(1,-1,2)$ in direction of $\mathbf{v} = 6\mathbf{i} + 2\mathbf{j} + 3\mathbf{k}$.

Solution We have $\partial f/\partial x = y^2 - 8xy$, $\partial f/\partial y = 2xy - 4x^2$, and $\partial f/\partial z = 2z$ so that

$$\nabla f(x,y,z) = (y^2 - 8xy)\mathbf{i} - (2xy - x^2)\mathbf{j} + 2z\mathbf{k}$$

$$\nabla f(1,-1,2) = 9\mathbf{i} - 6\mathbf{j} + 4\mathbf{k}.$$

Since $|\mathbf{v}| = |6\mathbf{i} + 2\mathbf{j} + 3\mathbf{k}|$ and $\mathbf{u} = \frac{1}{|\mathbf{v}|}\mathbf{v} = \left|\frac{6}{7}\mathbf{i} + \frac{2}{7}\mathbf{j} + \frac{3}{7}\mathbf{k}\right|$ is a unit vector in the indicated direction. From last definition we obtain

$$D_{\mathbf{u}} f(1,-1,2) = (9\mathbf{i} - 6\mathbf{j} + 4\mathbf{k}) \cdot \left(\frac{6}{7}\mathbf{i} + \frac{2}{7}\mathbf{j} + \frac{3}{7}\mathbf{k}\right) = \frac{54}{7}. \quad \blacksquare$$

Suppose $f(x,y,z)$ has continuous first order First order partial derivatives. Then the surface

$$S: f(x,y,z) = c$$

is one of the level surface of f. Let

$$x = x(t),\ y = y(t),\ z = z(t)$$

be differentiable curves passing through the point $P_0(x_0, y_0, z_0)$ on S where $x_0 = x(t_0)$, $y_0 = y(t_0)$ and $z_0 = z(t_0)$. For all t, we obtain

$$f(x(t), y(t), z(t)) = c.$$

Then using the chain rule, we have

$$\frac{d}{dt} f(x(t), y(t), z(t)) = \frac{d}{dt}(c) = 0.$$

$$\Rightarrow \frac{\partial f}{\partial x}\frac{dx}{dt} + \frac{\partial f}{\partial y}\frac{dy}{dt} + \frac{\partial f}{\partial z}\frac{dz}{dt} = 0.$$

Therefore

$$\nabla f \cdot \mathbf{v} = 0, \tag{1}$$

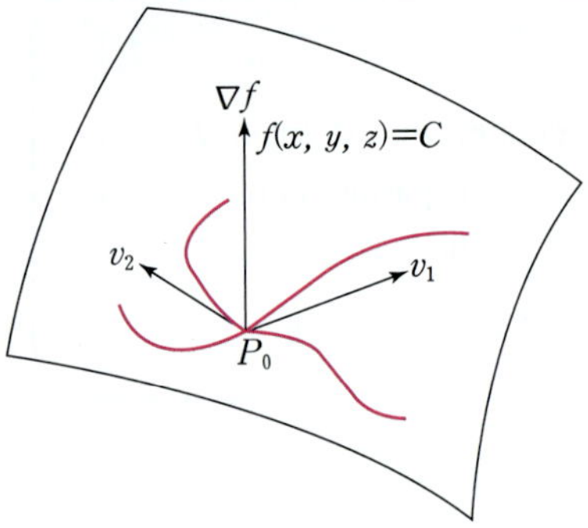

Figure 1 ∇f is perpendicular to tangent vector of differentiable function at P_0. So tangent vector is on the surface.

where $\mathbf{v}=(x'(t), y'(t), z'(t))$ is a tangent vector of the curve. What we conclude from this is that the gradient passing through P_0 is perpendicular to the tangent vector of all differentiable curves along S. (See Figure 1.) Thus ∇f is perpendicular to the tangent vector of all differentiable curves passing through P_0 on S. Therefore, these lines are perpendicular to ∇f and in the plane passing through P_0.

EXAMPLE 7

Find the directional derivative of the function $f(x, y, z)=x^3-xy^2-z$ at the point $P_0(1,1,0)$ in the direction of the vector $\mathbf{A}=2\mathbf{i}-3\mathbf{j}+6\mathbf{k}$.

Solution The unit vector of $\mathbf{A}$ is

$$\mathbf{u}=\frac{\mathbf{A}}{|\mathbf{A}|}=\frac{2}{7}\mathbf{i}-\frac{3}{7}\mathbf{j}+\frac{6}{7}\mathbf{k}$$

with $|\mathbf{A}|=\sqrt{(2)^2+(-3)^2+(6)^2}=7$.

The partial derivatives of f at P_0 are

$$f_x=3x^2-y^2\big|_{(1,1,0)}=2,$$
$$f_y=-2xy\big|_{(1,1,0)}=-2,$$
$$f_z=-1\big|_{(1,1,0)}=-1.$$

The gradient of f at P_0 is

$$\nabla f(1,\ 1,\ 0)=2\mathbf{i}-2\mathbf{j}-\mathbf{k}.$$

Hence the directional derivative of f at P_0 in the direction of A is

$$D_{\mathbf{u}}f(1,\ 1,\ 0)=\nabla f(1,\ 1,\ 0)\cdot\mathbf{u}$$
$$=(2\mathbf{i}-2\mathbf{j}-\mathbf{k})\cdot\left(\frac{2}{7}\mathbf{i}-\frac{3}{7}\mathbf{j}+\frac{6}{7}\mathbf{k}\right)=\frac{4}{7}.$$ ■

EXAMPLE 8

Estimate how much the value of

$$f(x, y, z)=xe^y+yz$$

will change if the point $P(x, y, z)$ moves $\Delta s=0.1$ unit from $P_0(2,0,0)$ straight toward $P_1(4,1,-2)$.

Solution We first find the directional derivative of f at P_0 in the direction of the vector $\overrightarrow{P_0P_1}=2\mathbf{i}+\mathbf{j}-2\mathbf{k}$. The unit vector of $\overrightarrow{P_0P_1}$ is

$$\mathbf{u}=\frac{\overrightarrow{P_0P_1}}{|\overrightarrow{P_0P_1}|}=\frac{\overrightarrow{P_0P_1}}{3}=\frac{2}{3}\mathbf{i}+\frac{1}{3}\mathbf{j}-\frac{2}{3}\mathbf{k}.$$

The gradient of f at P_0 is

$$\nabla f\,|_{(2,0,0)} = (e^y\mathbf{i} + (xe^y + z)\mathbf{j} + y\mathbf{k})\,\big|_{(2,0,0)} = \mathbf{i} + 2\mathbf{j}\ .$$

Therefore,

$$(D_{\mathbf{u}}f)_{P_0} = (\mathbf{i} + 2\mathbf{j}) \cdot \left(\frac{2}{3}\mathbf{i} + \frac{1}{3}\mathbf{j} - \frac{2}{3}\mathbf{k}\right) = \frac{2}{3} + \frac{2}{3} = \frac{4}{3}\ .$$

The change Δf in f that results from moving $\Delta s = 0.1$ unit away from P_0 in the direction of u is approximately

$$\Delta f \approx (D_{\mathbf{u}}f)\Delta s = \left(\frac{4}{3}\right)(0.1) \approx 0.13\,\text{unit.}$$ ■

EXAMPLE 9

Plot the contour plot and level surface of

$$f(x, y) = 2\sin x\ \cos y\,.$$

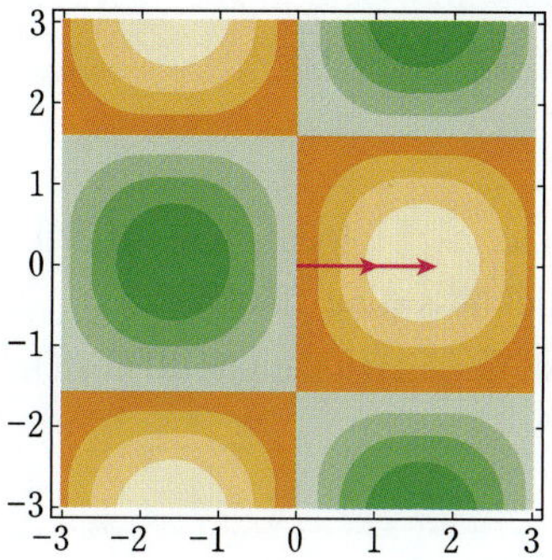

Figure 2

http://matrix.skku.ac.kr/cal-lab/cal-DirectionalDerivative.html

```
html("<i>  <b> Directional derivative <p></p> </b>"   )
var('x,y,t,z')
f(x,y)=2*sin(x)*cos(y)
line_thickness=3
surface_color='blue'
plane_color='purple'
line_color='red'
tangent_color='green'
gradient_color='orange'
@interact
def myfun(location=input_grid(1, 2, default=[0,0], label = "Location (x,y)", width=2), angle=slider(0,2*pi, label = "Angle"), show_surface=("Show surface", True), show_contour=("Show Contour Plot", True)):
    location3d = vector(location[0]+[0])
    location = location3d[0:2]
    direction3d = vector(RDF, [cos(angle), sin(angle), 0])
    direction=direction3d[0:2]
    cos_angle = math.cos(angle)
    sin_angle = math.sin(angle)
    df = f.gradient()
    direction_vector=line3d([location3d, location3d+direction3d], arrow_head=True, rgbcolor=line_color, thickness=line_thickness)
    curve_point = (location+t*direction).list()
    curve = parametric_plot(curve_point+[f(*curve_point)], (t,-3,3),color=line_color,thickness=line_thickness)
```

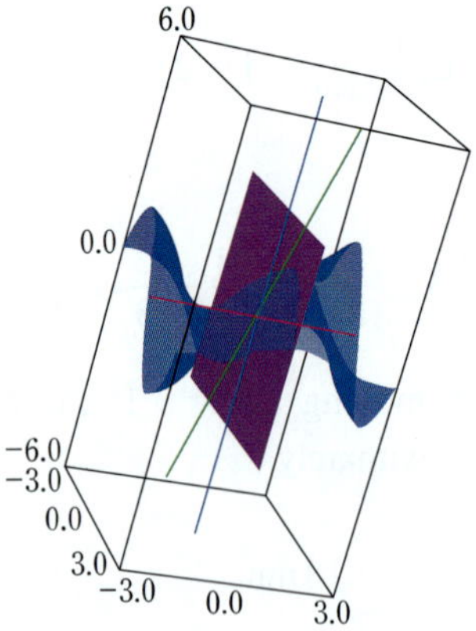

Figure 3

```
    plane = parametric_plot((cos_angle*x+location[0],sin_angle*x+location[1],t), (x, -3,3), (t,-3,3),opacity=0.8, color=plane_color)
    pt = point3d(location3d.list(),color='green', size=10)
 tangent_line = parametric_plot((location[0]+t*cos_angle, location[1]+t*sin_angle, f(*location)+t*df(*location)*(direction)), (t, -3,3), thickness=line_thickness, color=tangent_color)
    picture3d = direction_vector+curve+plane+pt+tangent_line
    picture2d = contour_plot(f(x,y), (x,-3,3),(y,-3,3), plot_points=100)
    picture2d += arrow(location.list(), (location+direction).list(),color="red")
    picture2d += point(location.list(),rgbcolor='green',pointsize=40)
    if show_surface:
        picture3d += plot3d(f(x,y), (x,-3,3),(y,-3,3),opacity=0.7)
 dff = df(location[0], location[1])
    dff3d = vector(RDF,dff.list()+[0])
    picture3d += line3d([location3d, location3d+dff3d], arrow_head=True, rgbcolor=gradient_color, thickness=line_thickness)
    picture2d += arrow(location.list(), (location+dff).list(), rgbcolor=gradient_color, width=line_thickness)
    if show_contour:
        show(picture2d, aspect_ratio=1)
    show(picture3d,aspect=[1,1,1], axes=True)
```

■

13.6 EXERCISES (Directional Derivatives and Gradient)

http://matrix.skku.ac.kr/Cal-Book/part2/CS-Sec-13-6-Sol.html

YouTube http://youtu.be/2_7TOUuzJoE

1. Find the directional derivative of the function $f(x,y)=6-3x^2-y^2$ at the point $P_0(1,2)$ in the direction of the vector $\mathbf{u}=\dfrac{1}{\sqrt{2}}\mathbf{i}-\dfrac{1}{\sqrt{2}}\mathbf{j}$.

Solution $f_x(x,y)=-6x$, $f_y(x,y)=-2y$

$\Rightarrow f_x(1,2)=-6$, $f_y(1,2)=-4$

$$D_{\mathbf{u}}f(1,2)=f_x(1,2)\cdot\frac{1}{\sqrt{2}}+f_y(1,2)\cdot\left(-\frac{1}{\sqrt{2}}\right)$$

$$=(-6)\cdot\frac{1}{\sqrt{2}}+(-4)\cdot\left(-\frac{1}{\sqrt{2}}\right)=-\sqrt{2}.$$

2. Find the directional derivative of the function $f(x,y)=xy^2$ at the point $P_0(4,-1)$ in the direction of the vector $\mathbf{A}=2\mathbf{i}+3\mathbf{j}$.

Solution $\mathbf{A}=2\mathbf{i}+3\mathbf{j}$ => $\|\mathbf{A}\|=\sqrt{2^2+3^2}=\sqrt{13}$ and

$$\mathbf{u}=\frac{\mathbf{A}}{|\mathbf{A}|}=\frac{2}{\sqrt{13}}\mathbf{i}+\frac{3}{\sqrt{13}}\mathbf{j}.$$

$f_x(x,y)=y^2$, $f_y(x,y)=2xy$ $\Rightarrow$ $f_x(4,-1)=1$, $f_y(4,-1)=-8$

$$D_{\mathbf{u}}f(4,-1)=f_x(4,-1)\cdot\frac{2}{\sqrt{13}}+f_y(4,-1)\cdot\frac{3}{\sqrt{13}}$$

$$=1\cdot\frac{2}{\sqrt{13}}-8\cdot\frac{3}{\sqrt{13}}=-\frac{22}{\sqrt{13}}.$$

3. Find the directional derivative of the function $f(x,y,z)=xy+yz^2+xz^3$ at the point $P(2,0,3)$ in the direction of the vector $\mathbf{A}=\dfrac{2}{3}\mathbf{i}-\dfrac{1}{3}\mathbf{j}+\dfrac{2}{3}\mathbf{k}$.

Solution

$\nabla f(x,y,z)=(y+z^3)\mathbf{i}+(x+z^2)\mathbf{j}+(2yz+3xz^2)\mathbf{k}$

$\nabla f(2,0,3)=27\mathbf{i}+11\mathbf{j}+54\mathbf{k}$

$\mathbf{u}=\dfrac{2}{3}\mathbf{i}-\dfrac{1}{3}\mathbf{j}+\dfrac{2}{3}\mathbf{k}$

$$D_{\mathbf{u}}f(2,0,3)=\nabla f(2,0,3)\cdot\mathbf{u}=\frac{151}{3}.$$

CAS **4.** If $|\mathbf{x}|=(x^2+y^2+z^2)^{\frac{1}{2}}$, find $\nabla\left(\dfrac{1}{|\mathbf{x}|}\right)$.

http://matrix.skku.ac.kr/cal-lab/cal-12-2-2.html

Solution

```
var('x, y, z')
f(x, y, z)=1/sqrt(x^2+y^2+z^2)
f.gradient()  # Find gradient
```

(x, y, z) |--> (-x/(x^2 + y^2 + z^2)^(3/2), -y/(x^2 + y^2 + z^2)^(3/2), -z/(x^2 + y^2 + z^2)^(3/2))

Therefore,

$$\nabla\left(\frac{1}{|\mathbf{x}|}\right)=\left(\frac{-x}{(x^2+y^2+z^2)^{3/2}},\frac{-y}{(x^2+y^2+z^2)^{3/2}},\frac{-z}{(x^2+y^2+z^2)^{3/2}}\right).$$

5. Find the gradient of $f(x,y)=\cos xy$ at the point $\left(1,\dfrac{2}{3}\pi\right)$.

Solution

$f_x(x,y)=-y\sin xy$, $f_y(x,y)=-x\sin xy$

$\nabla f(x,y)=f_x(x,y)\mathbf{i}+f_y(x,y)\mathbf{j}=-y\sin xy\mathbf{i}-x\sin xy\mathbf{j}$

$$\nabla f\left(1,\frac{2}{3}\pi\right)=-\frac{2}{3}\pi\sin\frac{2}{3}\pi\mathbf{i}-\sin\frac{2}{3}\pi\mathbf{j}=-\frac{\sqrt{3}}{3}\pi\mathbf{i}-\frac{\sqrt{3}}{2}\mathbf{j}.$$

6. Use the definition of the gradient, assume that f and g are differentiable function on $\mathbb{R}^3$, and let c be a constant. Prove the following gradient rules.

(1) $\nabla(cf)=c\nabla f$

(2) $\nabla(f\pm g)=\nabla f\pm\nabla g$

(3) $\nabla(fg)=g\nabla f+f\nabla g$

(4) $\nabla\left(\dfrac{f}{g}\right)=\dfrac{g\nabla f-f\nabla g}{g^2}$

Solution f and g are differentiable function on $\mathbb{R}^3$

(1) $\nabla(cf)=\dfrac{\partial}{\partial x}(cf)\mathbf{i}+\dfrac{\partial}{\partial y}(cf)\mathbf{j}+\dfrac{\partial}{\partial z}(cf)\mathbf{k}$

$=cf_x\mathbf{i}+cf_y\mathbf{j}+cf_z\mathbf{k}=c(f_x\mathbf{i}+f_y\mathbf{j}+f_z\mathbf{k})=c\nabla f.$

(2) $\nabla(f\pm g)=\dfrac{\partial}{\partial x}(f\pm g)\mathbf{i}+\dfrac{\partial}{\partial y}(f\pm g)\mathbf{j}+\dfrac{\partial}{\partial z}(f\pm g)\mathbf{k}$

$=(f_x\pm g_x)\mathbf{i}+(f_y\pm g_y)\mathbf{j}+(f_z\pm g_z)\mathbf{k}$

$=(f_x\mathbf{i}+f_y\mathbf{j}+f_z\mathbf{k})\pm(g_x\mathbf{i}+g_y\mathbf{j}+g_z\mathbf{k})=\nabla f\pm\nabla g.$

(3) $\nabla(fg)=\dfrac{\partial}{\partial x}(fg)\mathbf{i}+\dfrac{\partial}{\partial y}(fg)\mathbf{j}+\dfrac{\partial}{\partial z}(fg)\mathbf{k}$

$=(f_xg+fg_x)\mathbf{i}+(f_yg+fg_y)\mathbf{j}+(f_zg+fg_z)\mathbf{k}$

$=f(g_x\mathbf{i}+g_y\mathbf{j}+g_z\mathbf{k})+g(f_x\mathbf{i}+f_y\mathbf{j}+f_z\mathbf{k})=f\nabla g+g\nabla f$

(4) $\nabla\left(\dfrac{f}{g}\right)=\dfrac{\partial}{\partial x}\left(\dfrac{f}{g}\right)\mathbf{i}+\dfrac{\partial}{\partial y}\left(\dfrac{f}{g}\right)\mathbf{j}+\dfrac{\partial}{\partial z}\left(\dfrac{f}{g}\right)\mathbf{k}$

$$=\frac{f_xg-fg_x}{g^2}\mathbf{i}+\frac{f_yg-fg_y}{g^2}\mathbf{j}+\frac{f_zg-fg_z}{g^2}\mathbf{k}$$

$$=\frac{g(f_x\mathbf{i}+f_y\mathbf{j}+f_z\mathbf{k})-f(g_x\mathbf{i}+g_y\mathbf{j}+g_z\mathbf{k})}{g^2}$$

$$=\frac{g\nabla f-f\nabla g}{g^2}.$$

7. Find the gradient of $f(x,y,z)=(x+y+z)e^{xyz}$.

Solution $f_x(x,y,z)=yze^{xyz}$, $f_y(x,y,z)=xze^{xyz}$, $f_z(x,y,z)=xye^{xyz}$

$\nabla f(x,y,z)=f_x(x,y,z)\mathbf{i}+f_y(x,y,z)\mathbf{j}+f_z(x,y,z)\mathbf{k}$

$=yze^{xyz}\mathbf{i}+xze^{xyz}\mathbf{j}+xye^{xyz}\mathbf{k}$

$=e^{xyz}(yz\mathbf{i}+xz\mathbf{j}+xy\mathbf{k}).$

13.7 Tangent Plane

YouTube http://youtu.be/uOf-5YHKGI4

In this section, we introduce tangent plane to a surface and look at how to find its equations. Tangent plane is extension of tangent line to a one variable function $y = f(x)$.

Recall that if $y = f(x)$ be a differentiable function of one variable, then the equation of the tangent line at $x = a$ is given by

$$y = f(a) + f'(a)(x-a).$$

Tangent Plane

A tangent plane is defined to be a plane that is just tangent to a surface (it touches the surface in only one point).

To derive the equation of tangent plane, we will use Theorem 4 in section 13.6: tangent plane is orthogonal to the gradient. Let $F(x, y, z)$ be a differentiable function in its domain. If S is surface which is a level surface of $F(x, y, z) = c$ and $P(x_0, y_0, z_0)$ is a point on S, then we have proved that the gradient $\nabla F(x_0, y_0, z_0)$ is orthogonal to the tangent line to level curves. (See Figure 1.)

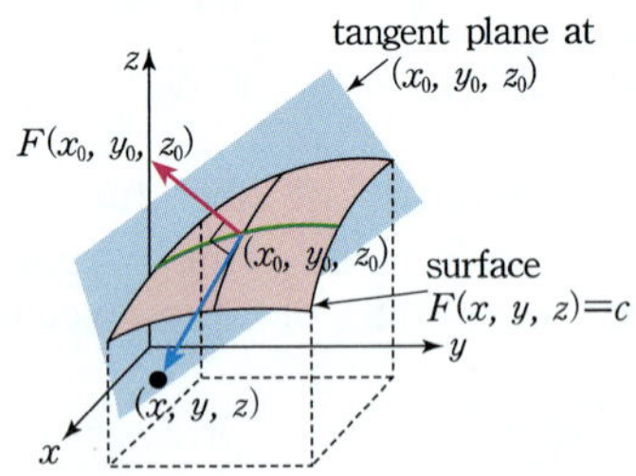

Figure 1 Tangent Plane to a surface

Thus one can define tangent plane at $P(x_0, y_0, z_0)$ to S as a plane which is perpendicular to $\nabla F(x_0, y_0, z_0)$, provided $\nabla F(x_0, y_0, z_0) \neq 0$.
Thus, if $P(x, y, z)$ and $P(x_0, y_0, z_0)$ are points on the tangent plane and $\mathbf{r} = x\mathbf{i} + y\mathbf{j} + z\mathbf{k}$ and $\mathbf{r}_0 = x_0\mathbf{i} + y_0\mathbf{j} + z_0\mathbf{k}$ are their corresponding position vectors, respectively, a vector equation of **tangent plane** is

$$\nabla F(x_0, y_0, z_0) \cdot \overrightarrow{P_0P} = \nabla F(x_0, y_0, z_0) \cdot (\mathbf{r} - \mathbf{r}_0) = 0$$

where $\mathbf{r} - \mathbf{r}_0 = (x - x_0)\mathbf{i} + (y - y_0)\mathbf{j} + (z - z_0)\mathbf{k}$.
The definition of tangent plane is given below:

DEFINITION 1 Tangent Plane

Let $P(x_0, y_0, z_0)$ be a point on the level surface $F(x, y, z) = c$ where ∇F is not $\mathbf{0}$. Then an equation of the tangent plane at P is

$$F_x(x_0, y_0, z_0)(x - x_0) + F_y(x_0, y_0, z_0)(y - y_0) + F_z(x_0, y_0, z_0)(z - z_0)$$

EXAMPLE 1

Find an equation of the tangent plane to the graph of the sphere $\frac{x^2}{4} + \frac{y^2}{9} + z^2 = 3$ at $(2, 3, 1)$.

Solution By defining $F(x, y, z) = \frac{x^2}{4} + \frac{y^2}{9} + z^2$, we find that the given sphere is the level surface $F(x, y, z) = 3$ passing through $(2, 3, 1)$. Now,

$$F_x(x, y, z) = \frac{x}{2},\ F_y(x, y, z) = \frac{2y}{9},\ F_z(x, y, z) = 2z$$

so that

$$\nabla F(x, y, z) = \frac{x}{2}\mathbf{i} + \frac{2y}{9}\mathbf{j} + 2z\mathbf{k} \quad \text{and}$$
$$\nabla F(2, 3, 1) = \mathbf{i} + \frac{2}{3}\mathbf{j} + 2\mathbf{k}.$$

It follows the form in Definition 1 that an equation of the tangent plane is

$$(x - 2) + \frac{2}{3}(y - 3) + 2(z - 1) = 0 \quad \text{or} \quad 3x + 2y + 6z = 18\,.$$

See in Figure 2. ■

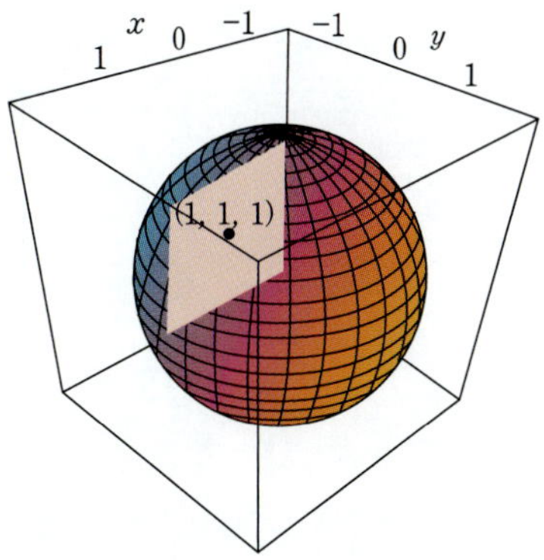

Figure 2 Tangent Plane in Example 2

Note The equation of the tangent plane obtained in 1 for a function $z = f(x, y)$ at a point (x_0, y_0, z_0) where $z_0 = f(x_0, y_0)$ can be obtained using Definition 1 the graph of the function $z = f(x, y)$ can be thought of level surface of $F(x, y, z) = 0$, where $F(x, y, z) = f(x, y) - z$. In this case we have

$$F_x(x_0, y_0, z_0) = f_x(x_0, y_0),\ F_y(x_0, y_0, z_0) = f_y(x_0, y_0) \ \text{and}$$
$$F_z(x_0, y_0, z_0) = -1\,.$$

Hence by Definition 1, the equation of tangent plane is

$$f_x(x_0, y_0)(x - x_0) + f_y(x_0, y_0)(y - y_0) - (z - z_0) = 0\,.$$

This implies that the equation of tangent plane is

$$z = z_0 + f_x(x_0, y_0)(x - x_0) + f_y(x_0, y_0)(y - y_0)\,.$$

> **DEFINITION 2** Equation of Tangent Plane
>
> Let $f(x, y)$ be a function of two variables having first order continuous partial derivatives in some domain. Let (a, b) be a point in the domain of f. Then equation of tangent plane to the surface $z = f(x, y)$ at the point $P(a, b, f(a, b))$ is given by
>
> 1 $$z = f(a, b) + f_x(a, b)(x - a) + f_y(a, b)(y - b).$$

EXAMPLE 2

Find the tangent plane to the surface $z = 9 - x^2 - y^2$ at the point $P_0(1, 2, 4)$. (See figure 3.)

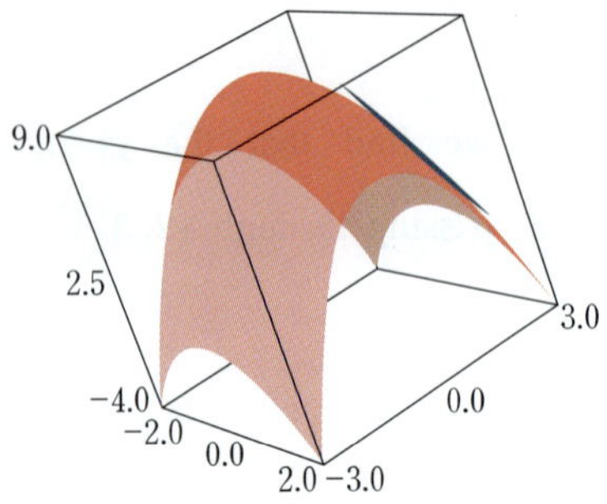

Figure 3

Solution Let $f(x, y) = 9 - x^2 - y^2$. Then $f_x(x, y) = -2x$ and $f_y(x, y) = -2y$. Since $f_x(1, 2) = -2$ and $f_y(1, 2) = -4$, the equation of tangent plane is

$$z = 4 - 2(x - 1) - 4(y - 2) = 14 - 2x - 4y .$$

```
var('x,y')
f(x,y)=9-x^2-y^2
a,b=1,2
fx=f.diff(x)(a,b)
fy=f.diff(y)(a,b)
T(x,y)=f(a,b)+fx*(x-a)+fy*(y-b)
show(T(x,y))
```

Answer : $-2x - 4y + 14$

```
p=plot3d(f,(x,-2,2),(y,-3,3),color='red',opacity=0.3);
pt=point3d((a,b,f(a,b)),pointsize=50)
tgt=plot3d(T,(x,a-0.6,a+0.6),(y,b-0.6,b+0.6),color='blue',opacity=
0.5);
p+pt+tgt
```

■

Normal Line

Let $z = f(x, y)$ be a function having continuous first order partial derivatives. Then the gradient to the surface $z = f(x, y)$ at $P(x_0, y_0, z_0)$ is given by

$$\nabla F(x, y, z) = \langle f_x(x_0, y_0), f_y(x_0, y_0), -1 \rangle$$

where $F(x, y, z) = f(x, y, z) - z$.

We know that the gradient is perpendicular to the tangent plane to the surface at the point $P(x_0, y_0, z_0)$. Hence the parametric equations of the normal line is given by

2 $$x = x_0 + t\,f_x(x_0,\ y_0)\ ,\quad y = y_0 + t\,f_y(x_0,\ y_0),\quad z = z_0 - t$$

In case $f_x(x_0, y_0) \neq 0$ and $f_y(x_0, y_0) \neq 0$, the symmetric equation of the normal line to the surface at $P(x_0, y_0, z_0)$ can be written as

3 $$\frac{x - x_0}{f_x(x_0, y_0)} = \frac{y - y_0}{f_y(x_0, y_0)} = \frac{z - z_0}{-1}.$$

EXAMPLE 3

Find the tangent plane and the normal line at the point $P_0(1, 2, 4)$ to the surface

$$z = 9 - x^2 - y^2\ .\ \text{(See Figure 4.)}$$

Solution 1. The partial derivatives of f at P_0 are

$$f_x(1,2) = -2x\big|_{(1,2)} = -2,\ f_y(1,2) = -2y\big|_{(1,2)} = -4$$

By Definition 1, the equation of a tangent plane at $P_0(1,2,4)$ is

$$(-2)(x-1) + (-4)(y-2) - (z-4) = 0$$

or

$$2x + 4y + z = 14\ .$$

By 2, the equation of normal line is

$$x = 1 - 2t,\ y = 2 - 4t,\ z = 4 - t\ \ (t \in \mathbb{R}).$$ ■

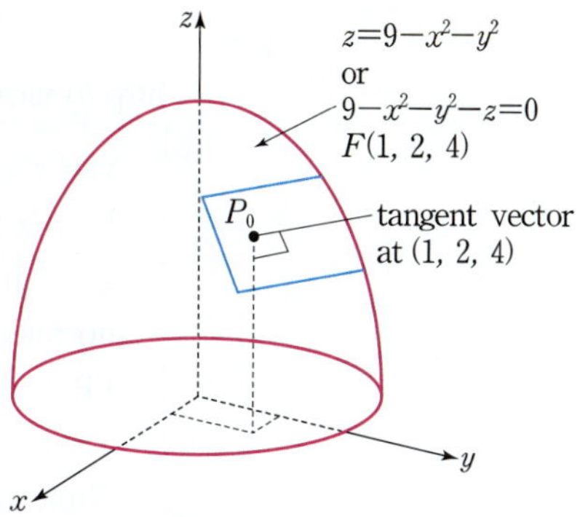

Figure 4 A tangent plane and a normal vector to the surface $z = 9 - x^2 - y^2$ at $P_0(1, 2, 4)$.

CAS **EXAMPLE 4**

Find the tangent plane and the normal line at the point $P_0(1,\ 2,\ \ln 5)$ to the surface

$$z = \ln(x^2 + y^2).$$

Solution Given equation $z = f(x, y) = \ln(x^2 + y^2)$ can be written as

$$\ln(x^2 + y^2) - z = 0.$$

Then the equation $z = f(x, y)$ can be considered as a level surface of

$$F(x,\ y,\ z) = \ln(x^2 + y^2) - z = 0.$$

Therefore our tangent plane at $P_0(1, 2, \ln 5)$ is the plane perpendicular to the gradient ∇F at $P_0(1, 2, \ln 5)$. Now we have

$$F_x = \left.\frac{2x}{x^2 + y^2}\right|_{(1,2,\ln 5)} = \frac{2}{5},$$
$$F_y = \left.\frac{2y}{x^2 + y^2}\right|_{(1,2,\ln 5)} = \frac{4}{5},$$
$$F_z = -1\,\big|_{(1,2,\ln 5)} = -1.$$

The equation of tangent plane is

$$\frac{2}{5}(x-1) + \frac{4}{5}(y-2) + (-1)(z - \ln 5) = 0$$

or $$2x + 4y - 5z = 10 - 5\ln 5.$$

The equation of normal line is

$$x = 1 + \frac{2}{5}t,\ \ y = 2 + \frac{4}{5}t,\ \ z = \ln 5 - t.$$

http://matrix.skku.ac.kr/cal-lab/cal-12-3-Exm7.html

```
var('x,y,z,t')
F=ln(x^2+y^2)-z
g= F.gradient()
normal=g.subs(x=1,y=2,z=ln(5))
PP_0=vector([x-1,y-2,z-ln(5)])
T=normal.dot_product(PP_0)
Nline=vector([1+normal[0]*t,2+normal[1]*t,ln(5)+normal[2]*t])
print T==0
print Nline
```

■

CAS **EXAMPLE 5**

Use sage to find the tangent plane and the normal line to the surface $z = 4xye^{-x^2-y^2}$ at $x = 1$, $y = 0.5$. Also plot all these together.
http://matrix.skku.ac.kr/cal-lab/m-Sec13-7-Exm-5.html

Solution

```
var('x,y,z,t')
f(x,y)=4*x*y*exp(-x^2-y^2)
F(x,y,z)= z-4*x*y*exp(-x^2-y^2)
a,b=1,0.5
c=f(a,b)
delF=F.gradient()
delFp=delF(a,b,c)
tangent_plane=delFp.dot_product(vector([x-a,y-b,z-c]))==0
print "The tangent plane is"
show(tangent_plane)
fx=f.diff(x)(a,b)
fy=f.diff(y)(a,b)
normal_line= (x==a+t*fx, y==b+t*fy, z==c-t)
print "The normal line is"
show(normal_line)
```

Answer : The equation of the Tangent Plane is:
0.573009593720380x−0.573009593720380y+z−0.859514390580570=0
The equation of the normal line is:
(x=−0.573009593720380t+1,y=0.573009593720380t+0.500000000000000, z=−t+0.573009593720380)

```
pt=point3d((a,b,f(a,b)), pointsize=50, color='red')
p=plot3d(f, (x,-2,2),(y,-2,2), color='green', opacity=0.2)
tangent=implicit_plot3d(delFp.dot_product(vector([x-a,y-b,z-c]))
==0,(x,a-0.3,a+0.3),(y,b-0.3,b+0.3),(z,c-0.3,c+0.3), color='blue',
opacity=0.5)
normal=parametric_plot3d(vector([a+t*fx,b+t*fy,
c-t]),(t,-1,1),color='red')
p+pt+tangent+normal
```

Answer : [[x == 0, y == 0], [x == -1/2*sqrt(2), y == -1/2*sqrt(2)], [x ==1/2*sqrt(2), y == -1/2*sqrt(2)], [x == -1/2*sqrt(2), y == 1/2*sqrt(2)], [x == 1/2*sqrt(2), y == 1/2*sqrt(2)]] ■

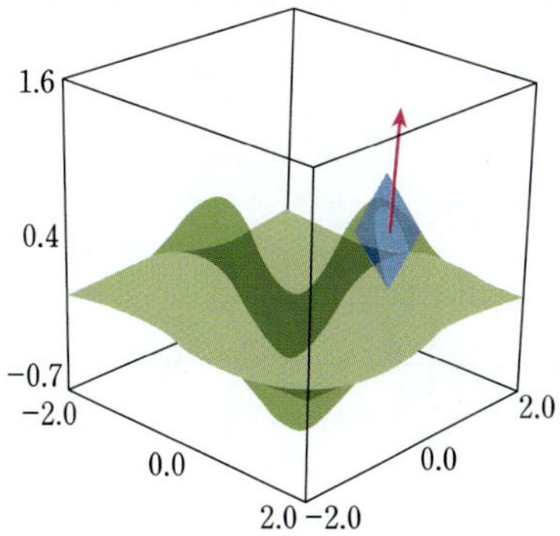

Figure 5

13.7 EXERCISES (Tangent Plane)

http://matrix.skku.ac.kr/Cal-Book/part2/CS-Sec-13-7-Sol.html YouTube http://youtu.be/GDkE8OqUvsk

1. Find the equation of the tangent plane at the point $P_0(2, 2, 2)$ to the surface $z = y\cos(x-y)$.

Solution $z = f(x,y) = y\cos(x-y)$

$f_x = -y\sin(x-y) \Rightarrow f_x(2,2) = 0$

$f_y = y\sin(x-y) + \cos(x-y) \Rightarrow f_y(2,2) = 1$

$\Rightarrow$ The equation of the tangent plane is

$z-2 = 0(x-2)+1(y-2)$.

Answer : $z = y$.

2. Find the equation of the tangent plane and the normal line at the point $P_0(1, 0, 0)$ to the surface $z+1 = xe^y\cos z$.

Solution $1 = xe^y\cos z - z$

$k=1,\ F(x,y,z) = xe^y\cos z - z$

$F_x = e^y\cos z,\ F_y = xe^y\cos z,\ F_z = -xe^y\sin z - 1$

$F_x(1,0,0) = 1,\ F_y(1,0,0) = 1,\ F_z(1,0,0) = -1$

$\Rightarrow\ 1(x-1)+1(y-0)-1(z-0) = 0$ and

$\dfrac{x-1}{1} = \dfrac{y-0}{1} = \dfrac{z-0}{-1}$

Hence at the point $P_0(1,0,0)$, the tangent plane is $x+y-z=1$ and the normal line is $x-1 = y = -z$ (that is, $x-1=t,\ y=t,\ -z=t$).

3. Find $k > 0$ so that all tangent planes to the surface $z = (x^2+y^2)^k$ pass through the origin $(0,0,0)$ for all $(x,y) \neq (0,0)$.

Solution Let $f(x,y) = (x^2+y^2)^k$. Then we have that

$f_x(x,y) = 2kx(x^2+y^2)^{k-1}$ and

$f_y(x,y) = 2ky(x^2+y^2)^{k-1}$.

Since every tangent plane intersects the origin, the equation of the tangent plane at a point $(x,y,z)(\neq(0,0,0))$ on the surface should satisfies

$-f_x(x,y)x - f_y(x,y)y + z = 0$.

We then have

$-2kx^2(x^2+y^2)^{k-1} - 2ky^2(x^2+y^2)^{k-1} + (x^2+y^2)^k = 0$.

Simplifying the equation, we obtain

$(1-2k)(x^2+y^2)^k = 0$,

which implies $k = \dfrac{1}{2}$.

4. Find the tangent plane of the surface $xe^{yz} + x\sin z = 1$ at $(x,y,z) = (1,1,0)$.

Solution Let $F(x,y,z) = xe^{yz} + x\sin z - 1$.

Due to implicit differentiation, we have

$F_x = e^{yz} + \sin z,\ F_y = zxe^{yz}$ and $F_z = yxe^{yz} + x\cos z$.

In particular, at $(x,y,z) = (1,1,0)$ we have

$\langle F_x(1,1,0),\ F_y(1,1,0),\ F_z(1,1,0)\rangle = \langle 1,0,2\rangle$.

Thus, the tangent plane is $(x-1)+2(z-0)=0$. Simplifying it, we have $x+2z=1$.

5. Find the equation of the tangent plane and the normal line to the surface $x^2+3y^2+z^2=1$ at $(0,0,1)$.

Solution

Equation of tangent plane:

$f(x,y,z) = x^2+3y^2+z^2-1$

$\nabla f(x,y,z) = \langle 2x, 6y, 2z\rangle \Rightarrow \nabla f(0,0,1) = \langle 0,0,2\rangle$.

$\nabla f(0,0,1)\cdot\langle x,y,z-1\rangle = 0 \Rightarrow z-1=0$ or $z=1$.

Equation of normal line:

$L(t) = \langle 0,0,1\rangle + t\langle 0,0,2\rangle = \langle 0,0,1+2t\rangle,\ (t\in\mathbb{R})$.

6. Let S be the surface whose equation in cylindrical coordinates is $z = r$. Find the tangent plane and the normal line to S at the point $\left(\dfrac{1}{\sqrt{2}}, \dfrac{1}{\sqrt{2}}, 1\right)$ in rectangular coordinates.

Solution The equation of S in rectangular coordinates is given by $z^2 = x^2+y^2$.

If we regard S as a level surface of

$F(x,y,z) = x^2+y^2-z^2$

at 0, then the normal vector of the tangent plane is $\nabla F\left(\dfrac{1}{\sqrt{2}}, \dfrac{1}{\sqrt{2}}, 1\right)$, where $\nabla F(x,y,z) = (2x, 2y, -2z)$

and hence

$\nabla F\left(\dfrac{1}{\sqrt{2}}, \dfrac{1}{\sqrt{2}}, 1\right) = (\sqrt{2}, \sqrt{2}, -2)$.

Therefore, the equations of the tangent plane is

$\sqrt{2}\left(x-\dfrac{1}{\sqrt{2}}\right) + \sqrt{2}\left(y-\dfrac{1}{\sqrt{2}}\right) - 2(z-1) = 0$ or

$\sqrt{2}x + \sqrt{2}y - 2z = 0$.

And the normal line is

$$\begin{cases} x = \dfrac{1}{\sqrt{2}} + \sqrt{2}t \\ y = \dfrac{1}{\sqrt{2}} + \sqrt{2}t \\ z = 1-2t \end{cases} \quad (t\in\mathbb{R}).$$

7. Let $f(x, y) = \dfrac{3xy}{x^2+y^2}$ if $(x, y) \neq (0, 0)$ and let $f(0,0) = 0$. In the direction of what unit vectors does the directional derivative of f at $(0,0)$ exist?

Solution Suppose we wish to compute the directional derivative of f in the direction of a unit vector $u = \langle a, b \rangle$. Then

$$D_u f(0,0) = \lim_{h\to 0} \frac{f(0+ha, 0+hb) - f(0,0)}{h}$$

$$= \lim_{h\to 0} \frac{1}{h} \frac{3h^2ab}{(a^2+b^2)h^2} = \lim_{h\to 0} \frac{3ab}{h(a^2+b^2)}$$

and the limit does not exist unless $a = 0$ or $b = 0$. Hence the directional derivative of f at $(0,0)$ exists in the directions of $\pm \mathrm{i}$ and $\pm \mathrm{j}$ or their unit scalar multiples.

CAS **8-9.** Find the tangent plane to the given surface at the indicated point.

8. $w = x^2 + y^2 + z^2$ at $(-1, 0, 0, 1)$.

http://matrix.skku.ac.kr/cal-lab/cal-12-3-4.html

Solution

```
var('x, y, z, w')
f(x, y, z, w)=x^2+y^2+z^2-w
grad_f=f.gradient()
fx, fy, fz, fw=grad_f
fx(-1, 0, 0, 1)*(x+1)+fy(-1, 0, 0, 1)*(y-0)+fz(-1, 0, 0, 1)*(z-0)+fw(-1, 0, 0, 1)*(w-1)==0
```

Answer : -w - 2*x - 1 == 0

9. $z = e^{3y}\cos 2x$ at $\left(\dfrac{\pi}{3}, 0, -1\right)$.

http://matrix.skku.ac.kr/cal-lab/cal-12-3-11.html

Solution

```
var('x, y, z')
f(x, y, z)=exp(3*y)*cos(2*x)-z
grad_f=f.gradient()
fx, fy, fz=grad_f
p1=implicit_plot3d(f(x, y, z)==0, (x, 0, pi/2), (y, -1, 1), (z, -1, 0), opacity=0.6)
p2=implicit_plot3d(fx(pi/3, 0, -1/2)*(x-pi/3)+fy(pi/3, 0, -1/2)*(y-0)+fz(pi/3, 0, -1/2)*(z+1/2)==0, (x, 0, pi/2), (y, -1, 1), (z, -1, 0), color='orange', opacity=0.6)
p3=point3d([pi/3, 0, -1/2], color='red')
p1+p2+p3
```

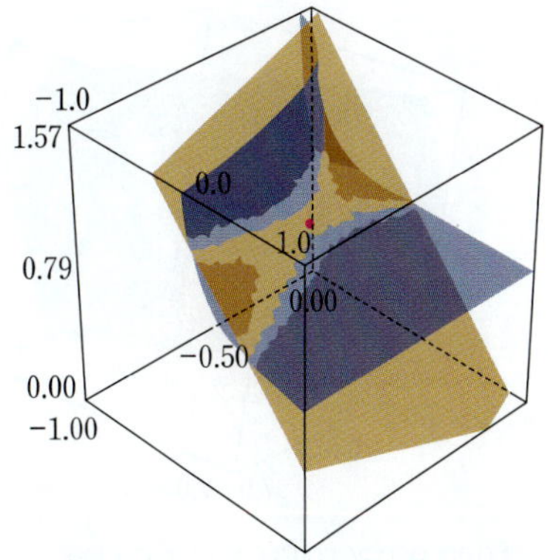

1/3*(pi - 3*x)*sqrt(3) - 3/2*y - z - 1/2 == 0

Answer : $\dfrac{(\pi - 3x)\sqrt{3}}{3} - \dfrac{3}{2}y - z - \dfrac{1}{2} = 0$

CAS **10.** Find the equations of the tangent plane and normal line at the point $(-1, 1, -2)$ to the paraboloid $z = -x^2 - y^2$.

http://matrix.skku.ac.kr/cal-lab/m-Sec13-7-Exs-10.html

Solution

```
var('x,y,z,t')
F=-x^2-y^2-z
g= F.gradient()
normal=g.subs(x=-1,y=1,z=-2)
PP_0=vector([x+1,y-1,z+2])
T=normal.dot_product(PP_0)
Nline=vector([-1+normal[0]*t,1+normal[1]*t,-2+normal[2]*t])
print T==0
print Nline
```

Answer : 2*x - 2*y - z + 2 == 0
(2*t - 1, -2*t + 1, -t - 2)

```
p1=implicit_plot3d(z==-x^2-y^2,(x,-7,7),(y,-7,7),(z,-7,7), opacity=0.2, color="red", mesh=True);
p2=implicit_plot3d(dF_p*(w-p)==0,(x,-7,7),(y,-7,7),(z,-7,7), opacity=0.2, color="blue",
 mesh=True);
p3=parametric_plot3d((-2*t-1,2*t+1,t-2),(t,-1,1), opacity=1, color="red", mesh=True);
show(p1+p2+p3, aspect_ratio=1)
```

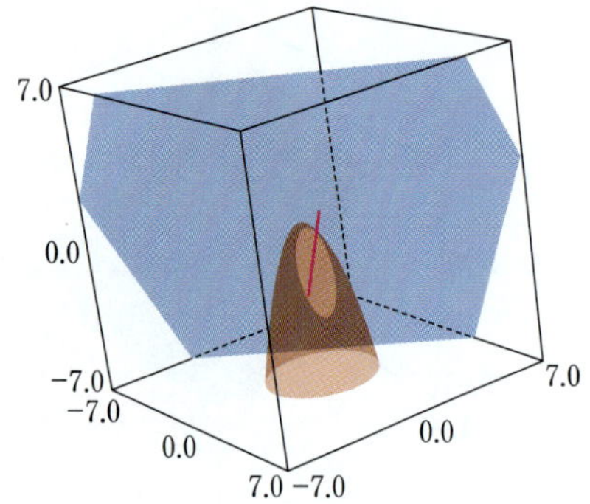

So tangency normal vector is clear.

11. Find the equations of the tangent plane and the normal line at the point $P_0(1,1,1)$ to the surface $x^2+xyz-z^3=1$.

Solution We have $f_x(1,1,1)=(2x+yz)|_{(1,1,1)}=3$,
$f_y(1,1,1)=xz|_{(1,1,1)}=1$,
$f_z(1,1,1)=(xy-3z^2)|_{(1,1,1)}=-2$.
Hence, $\nabla f|_{(1,1,1)}=3i+j-2k$.
Therefore by 2 the equation of the tangent plane to the surface S at the point $P(1,1,1)$ is
$$3(x-1)+(y-1)-2(z-1)=0.$$
Or, $3x+y-2z=2$.
By 3 the equation of the normal line at the point $P(1,1,1)$ to the surface S is
$$x=1+3t,\ y=1+t,\ z=1-2t,\ \frac{x-1}{3}=\frac{y-1}{1}=\frac{z-1}{-2}.$$

12. The surface $f(x,y,z)=x^2+y^2-z^2=1$ and $g(x,y,z)=x+y+z=5$ meet in an ellipse C. Find parametric equations for the line tangent to C at the point $P_0(1,2,2)$.

Solution Let P_0 be a point on the intersection of these surfaces. Then the tangent line at P to the curve C is orthogonal to $\nabla f(P_0)$ and $\nabla g(P_0)$. In particular, it is parallel to $v=\nabla f\times\nabla g$.

The components of v and the coordinates of P_0 give us equations for the line. We have
$\nabla f(1,2,2)=(2x\mathbf{i}+2y\mathbf{j}-2z\mathbf{k})|_{(1,2,2)}=2\mathbf{i}+4\mathbf{j}-4\mathbf{k}$,
$\nabla g=\mathbf{i}+\mathbf{j}+\mathbf{k}$.
Therefore $v=\nabla f\times\nabla g=\begin{vmatrix}\mathbf{i}&\mathbf{j}&\mathbf{k}\\2&4&-4\\1&1&1\end{vmatrix}=8\mathbf{i}-6\mathbf{j}-2\mathbf{k}$.
Thus the tangent to the curve C at P is parallel to the vector $\mathbf{v}=\langle 8,-6,-2\rangle$. Therefore, its equation is $x=1+8t,\ y=2-6t,\ z=2-2t$.

13. Sketch a level curve (ellipse) of $f(x,y)=\dfrac{x^2}{4}+y^2$ passing through the pont $P_0(-2,1)$. Find a vector perpendicular to this ellipse at the point $P_0(-2,1)$.

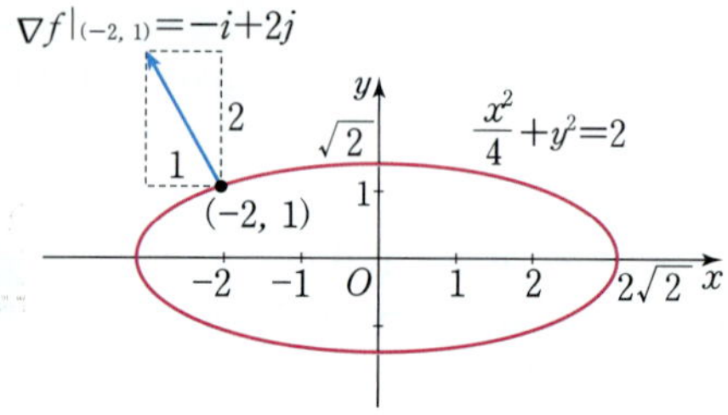

Solution The value of f at the point P_0 is $\dfrac{(-2)^2}{4}+(1)^2=2$.
Therefore, the level curve of f passing through $P_0(-2,1)$ is $\dfrac{x^2}{4}+y^2=2$, which is an ellipse. Vector perpendicular to this ellipse at the point $P_0(-2,1)$ is
$\nabla f|_{(-2,1)}=\left(\dfrac{x}{2}\mathbf{i}+2y\mathbf{j}\right)\Big|_{(-2,1)}=-\mathbf{i}+2\mathbf{j}$.

13.8 Extrema of Multivariate Functions

YouTube http://youtu.be/oDZUkOEszOQ

We have seen how to find extreme values of a single variable function as an application to derivatives. In this section we look at how to find extrema of multivatiate functions as an application to partial derivatives. As with single variable function, concepts developed in this section are useful for practical optimization problems.

Local and Global Maximum and Minimum

We begin with the definition of **local** and **relative extrema** for a function of two variables x and y.

DEFINITION 1 Local Maximum and Minimum

(i) A point $f(a,b)$ is a **local maximum** of a function $z=f(x,y)$ if $f(x,y) \le f(a,b)$ for all (x,y) near (a,b) in the domain of f.
(ii) A point $f(a,b)$ is a **local minimum** of a function $z=f(x,y)$ if $f(x,y) \ge f(a,b)$ for all (x,y) near (a,b) in the domain of f.

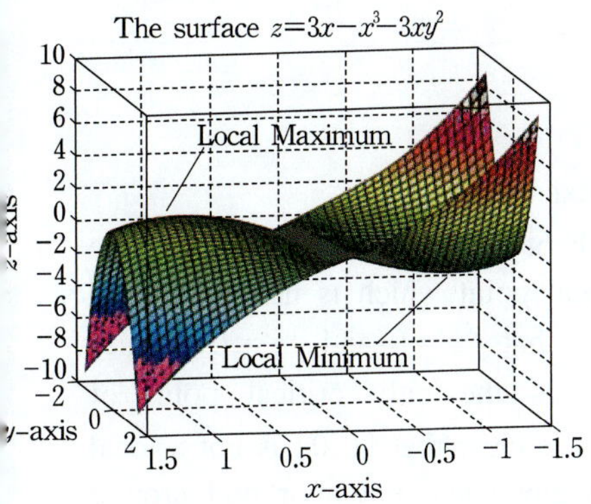

Figure 1

Let $z=f(x,y)$ and suppose $f(x,y)$ has a local minimum at (a,b). This means there exists positive real numbers $\delta>0$ such that $f(a+h,b+k)>f(a,b)$ for all $0<|h|<\delta$ and $0<|k|<\delta$. Similarly, we can define a local maximum at (a,b).

When $y=b$ in $z=f(x,y)$, $f(x,b)$ is a single variable function. Suppose the first order partial derivatives of f exist at the point $x=a$, then we have $f_x(a,b)=0$ since it has the extreme value at $x=a$.

Likewise, when $x=a$, $f(a,y)$ has the extreme value at $y=b$ and $f_y(a,b)=0$. Therefore, if (a,b) is a local maximum or local minimum of $f(x,y)$, then

$$\begin{cases} f_x(a,b)=0 \\ f_y(a,b)=0 \end{cases} \quad \boxed{1}$$

This is in fact, a necessary condition for a point (a,b) to be a point of local maximum/minimum of f. All such points where first order partial derivatives vanish are called critical points of f.

DEFINITION 2

A **critical point** of a function $z=f(x,y)$ is a point (a,b) in the domain of f for which $f_x(a,b)=0$ and $f_y(a,b)=0$, or where one of its partial derivatives does not exist.

THEOREM 3 Local Extrema

If a function $z=f(x,y)$ has a local extremum at point (a,b) and the first derivatives exist at this point, then

$$f_x(a,b)=0 \text{ and } f_y(a,b)=0.$$

Theorem 3 says that if $f(x, y)$ has a local maximum or local minimum at (a, b), then (a, b) is a critical point of $f(x, y)$. In general, (a, b) can be a critical point of $f(x, y)$ even though $f(a, b)$ is not a local maximum or local minimum value.

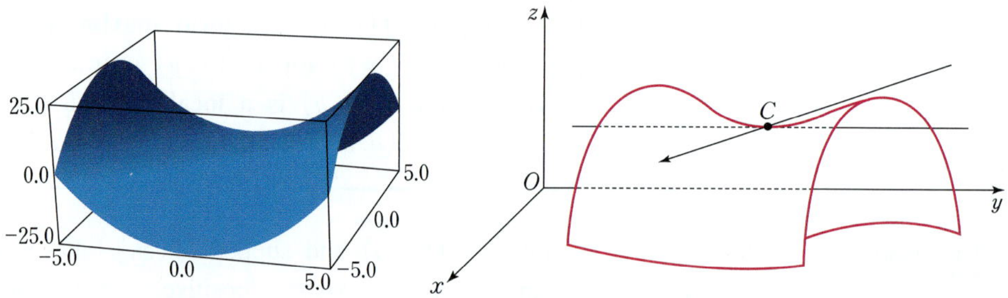

Figure 2

The discussion below/above suggests the next theorems.

Let us look at $f(x,y)=x^2-y^2$. The graph of $z=f(x,y)$ is shown in the Figure 1. The point C in Figure 2, is a critical point which is neither a local maximum nor local minimum. It is easy to see that $f_x=0$ and $f_y=0$ implies that $x=0$ and $y=0$. That means $(0,0)$ is the only critical point of $f(x,y)=x^2-y^2$. However from the graph, it is clear that $(0,0)$ is not a local maximum/local minimum. In fact if we take any small neighborhood around $(0,0)$, there are points (x_1,y_1) and (x_2,y_2) in this neighborhood where $f(x_1,y_1)< f(0,0)$ and $f(x_2,y_2)> f(0,0)$. Because this surface looks like a saddle horse, such a critical point is called a **saddle point**.

Remark Another way of looking at Theorem 3 is the following. Suppose f has continuous first order partial derivatives. We know that $\nabla f(a,b)$ is the direction in which function increases with maximum speed. Thus, if (a,b) is a point of local maximum, then at $(a,b,f(a,b))$ on the surface, $z=f(x,y)$, there is no direction in which function can increase. Hence $\nabla f(a,b)=0$ which implies $f_x(a,b)=f_y(a,b)=0$. Similarly, at the point of local minimum $-\nabla f(a,b)=0$.

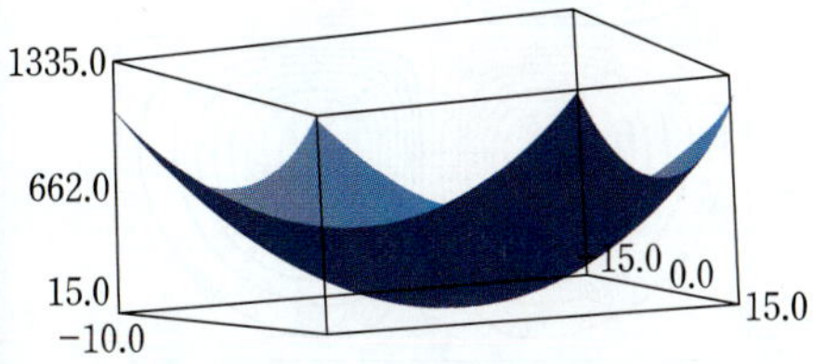

Figure 3

EXAMPLE 1

Find all the critical points for $f(x,y) = 3x^2 + 2y^2 - 6x + 8y$.

Solution The first partial derivatives are

$$f_x = 6x - 6 \text{ and } f_y = 4y + 8.$$

Hence, $f_x(a,b) = 0$ and $f_y(a,b) = 0$ implies

$$a = 1 \text{ and } b = -2.$$

Thus the critical point is $(1, -2)$. ■

EXAMPLE 2

Find all the critical points for $f(x,y) = x^4 - y^4$.

Solution The first partial derivatives are

$$f_x = 4x^3 \text{ and } f_y = -4y^3.$$

Hence, $f_x(a,b) = 0$ and $f_y(a,b) = 0$ implies

$$a = 0 \text{ and } b = 0.$$

Thus, the critical point is $(0, 0)$. ■

CAS **EXAMPLE 3**

Find critical points of $f(x,y) = 4xye^{-x^2-y^2}$. Plot the surface and its contour lines.

http://matrix.skku.ac.kr/cal-lab/m-Sec13-8-Exm-3.html

Solution

```
var('x,y')
f(x,y)=4*x*y*exp(-x^2-y^2)
fx=f.diff(x)
fy=f.diff(y)
sol=solve([fx==0,fy==0],[x,y],solution_dict=True)
sol
```

Answer : [{y: 0, x: 0}, {y: -1/2*sqrt(2), x: -1/2*sqrt(2)}, {y: -1/2*sqrt(2), x:1/2*sqrt(2)}, {y: 1/2*sqrt(2), x: -1/2*sqrt(2)}, {y: 1/2*sqrt(2), x:1/2*sqrt(2)}]

Here f has five critical points $(0,0)$ and $\left(\pm\frac{\sqrt{2}}{2}, \pm\frac{\sqrt{2}}{2}\right)$. This is clear from the graph of the surface $z = f(x,y)$. Look at Figure 4.

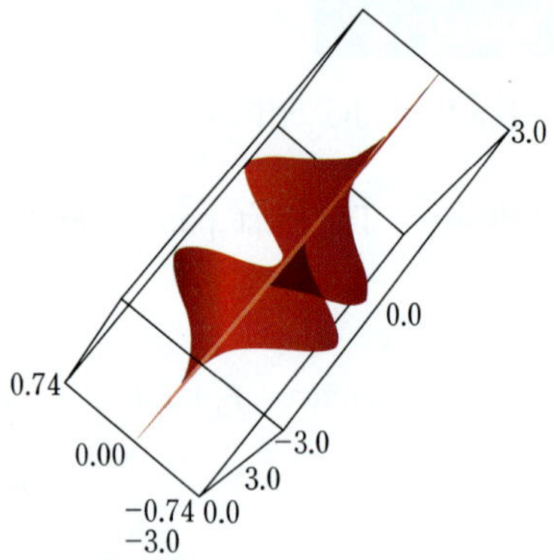

Figure 4

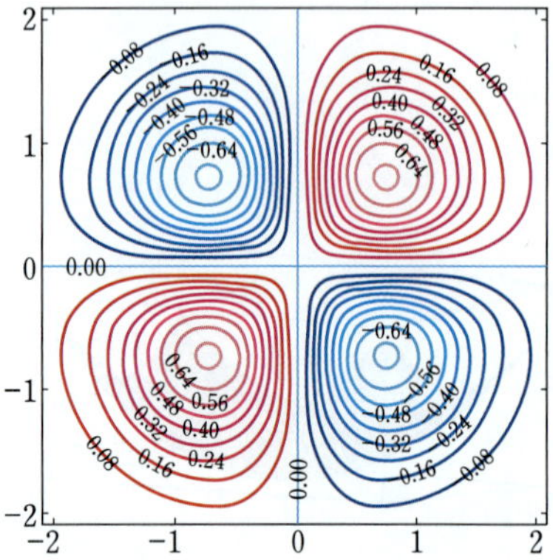

Figure 5

It is also clear from the Figure Give Reference that f has two points of local maximum and two points of local minimum and a saddle point. Look at contour line of the function $z = f(x, y)$. Look at Figure 4.

It is clear from the contour lines of f (See Figure 5.) that, near the point of local maximum and local minimum, level curves resembles concentric circles. Near saddle points, the level curves are hyperbolic looking shapes. ■

Second Derivative Test

We know that critical points are likely candidates for point of local maximum, local minimum or saddle point. Analogous to second derivative test for function of one variable, there is second derivative test for functions of two variables to classify critical points as local maximum, local minimum or saddle points.

The following theorem tells us how to decide the local maximum and local minimum of a given function at a critical point.

THEOREM 4 Second Derivatives Test

Suppose $f_x(a, b) = f_y(a, b) = 0$ and $f(x, y)$ has continuous second-order partial derivatives in a neighborhood of (a, b). Let

$$D = f_{xx}(a, b) f_{yy}(a, b) - \{f_{xy}(a, b)\}^2.$$

(1) If $D > 0$ and $f_{xx}(a, b) < 0$, then $f(a, b)$ is a local maximum.

(2) If $D > 0$ and $f_{xx}(a, b) > 0$, then $f(a, b)$ is a local minimum.

(3) If $D < 0$, then $f(x, y)$ does not have a local maximum or minimum at (a, b). The point (a, b) is a saddle point.

Note 1 In case (c) the point (a, b) is called a saddle point of f and the graph of f crosses its tangent plane at (a, b).

Note 2 If $D=0$, the test gives no information: f could have a local maximum or local minimum at (a, b), or (a, b) could be a saddle point of f.

Note 3 To remember the formula for D, it's helpful to write it as a determinant:

$$D=\begin{vmatrix} f_{xx} & f_{xy} \\ f_{yx} & f_{yy} \end{vmatrix}= f_{xx}f_{yy}-(f_{xy})^2$$

The proof of theorem follows from Taylor's series of two variables function which is given at the end of this section.

CAS **EXAMPLE 4**

Classify the critical points of $f(x, y)= x^4+y^4-4xy+1$.

Solution Suppose (x, y) is a critical point of f, then

$$f_x(x, y)= 4x^3-4y=0\,, \qquad f_y(x, y)=4y^3-4x=0.$$

Solving the above equation for x, y, we get three critical points $(0,0)$, $(1,1)$ and $(-1,-1)$. Look at the Figure 6 and Figure 7. It is clear that f has two local maximum and one saddle point.

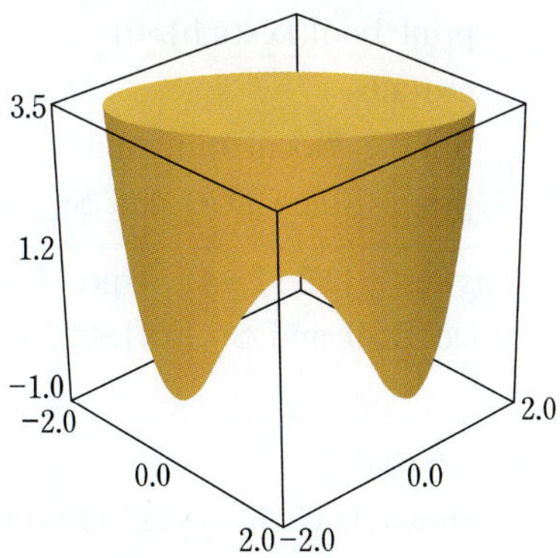

Figure 6

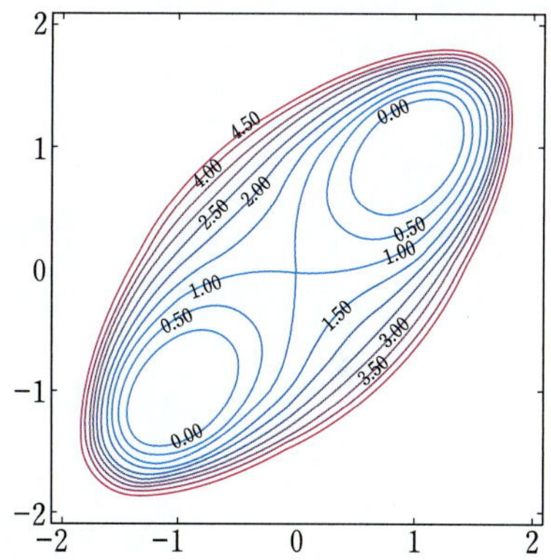

Figure 7

Now let us use the second derivative test to classify these critical points. We have

$$f_{xx}=12x^2,\ f_{yy}=12y^2 \text{ and } f_{xy}=-4.$$

At the critical point $(0,0)$, $f_{xx}(0,0)=f_{yy}(0,0)=0$ and $f_{xy}(0,0)=-4$. That means $D=f_{xx}f_{yy}-f_{xy}^2=-16<0$. Hence $(0,0)$ is a saddle point.

At the critical point $(-1,-1)$, $f_{xx}(-1,-1)=f_{yy}(-1,-1)=12$ and $f_{xy}(-1,-1)=-4$. That means $D=f_{xx}f_{yy}-f_{xy}^2=128>0$ and $f_{xx}(-1,-1)>0$. Hence $(0,0)$ is a local minimum.

At the critical point $(1,1)$, $f_{xx}(1,1)=f_{yy}(1,1)=12$ and $f_{xy}(1,1)=-4$. That means $D=f_{xx}f_{yy}-f_{xy}^2=128>0$ and $f_{xx}(1,1)>0$. Hence $(0,0)$ is also a local minimum. ■

http://matrix.skku.ac.kr/cal-lab/m-Sec13-8-Exm-4.html

```
var('x,y')
f(x,y)=x^4+y^4-4*x*y+1
fx=f.diff(x); fy=f.diff(y)
cpoints=solve([fx==0,fy==0],[x,y], solution_dict=True)
for sol in cpoints:
    if ((sol[x] in RR) and (sol[y] in RR)):
        show([sol[x],sol[y]])
```

Answer : [−1,−1], [1,1], [0,0]

```
var('x,y')
f(x,y)=x^4+y^4-4*x*y+1
fxx=diff(f,x,2);fyy=diff(f,y,2);fxy=diff(f,x,y)
a,b=-1,-1
show([fxx(a,b),fyy(a,b),fxy(a,b)])
D=fxx(a,b)*fyy(a,b)-fxy(a,b)^2
print bool(D>0)
print bool(fxx(a,b)>0)
c=f(a,b)
print c #local minimum
print "Since D>0 and $f_{xx}>0$, f(-1,-1)=-1 is a local minimum."
```

Answer : [12,12,−4], True, True, -1
Since D>0 and \$f_{xx}>0\$, f(-1,-1)=-1 is a local minimum.

```
a,b=0,0
show([fxx(a,b),fyy(a,b),fxy(a,b)])
D=fxx(a,b)*fyy(a,b)-fxy(a,b)^2
TF=bool(fxx(a,b)>0)
print bool(D>0)
print bool(fxx(a,b)>0)
c=f(a,b)
print c
print "Since D<0, (0,0,1) is a saddle point."
```

Answer : [0,0,−4], False, False, 1
Since D<0, (0,0,1) is a saddle point.

```
a,b=1,1
show([fxx(a,b),fyy(a,b),fxy(a,b)])
D=fxx(a,b)*fyy(a,b)-fxy(a,b)^2
TF=bool(fxx(a,b)>0)
print bool(D>0)
print bool(fxx(a,b)>0)
c=f(a,b)
print c
print "Since D>0 and $f_{xx}>0$, f(-1,-1)=-1 is a local minimum."
```

Answer : [12,12,-4], True, True, -1
Since D>0 and $f_{xx}>0$, f(-1,-1)=-1 is a local minimum.

EXAMPLE 5

Find the local extreme values of $f(x, y) = x^2 + y^2$.

Solution The domain of f is the entire plane and the partial derivatives $f_x = 2x$ and $f_y = 2y$ exist everywhere. Therefore, local extreme values can occur only when

$$f_x = 2x = 0 \text{ and } f_y = 2y = 0.$$

The only possibility is the origin, where the value of f is zero. Since f is never negative, we see that the origin gives a local minimum. (See Figure 8.) This can be written by using Theorem 4. We have

$$f_{xx} = 2, \quad f_{yy} = 2 \quad \text{and} \quad f_{xy} = f_{yx} = 0.$$

Hence, $D = f_{xx}(0,0)f_{yy}(0,0) - \{f_{xy}(0,0)\}^2 = 4 > 0$.
Since $f_{xx}(0,0) = 2 > 0$ and $D > 0$, $(0,0)$ is a point of local minimum. ■

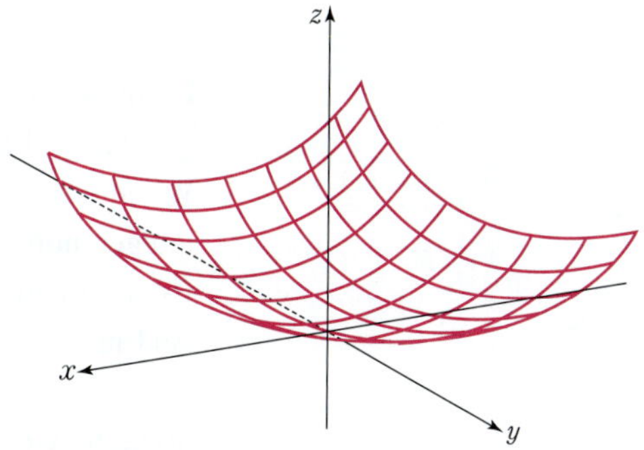

Figure 8 The paraboloid $z = x^2 + y^2$ has a local minimum value of 0 at the origin.

CAS **EXAMPLE 6**

Find the local extreme values of the function

$$f(x,y) = xy - x^2 - y^2 - 2x - 2y + 4 .$$

Solution Note that $f(x, y)$ has continuous second order partial derivatives at all points in the plane. The function therefore has extreme values only at the points where f_x and f_y are simultaneously zero. This leads to

$$f_x = y - 2x - 2 = 0,\ f_y = x - 2y - 2 = 0,$$

or

$$x = y = -2.$$

Therefore, the point $(-2, -2)$ is the only point where f may have extreme value. We calculate

$$f_{xx} = -2,\ f_{yy} = -2,\ f_{xy} = 1.$$

At $(-2, -2)$,

$$D = f_{xx} f_{yy} - (f_{xy})^2 = (-2)(-2) - (1)^2 = 4 - 1 = 3 > 0$$

and $f_{xx} = -2 < 0$.
Then f has a local maximum at $(-2, -2)$. The value of f at this point is $f(-2, -2) = 8$. ■

CAS **EXAMPLE 7**

Find the local maximum and minimum of

$$f(x,\ y) = x^2 + 1 + 2x \sin y$$

in the domain $\mathbb{R} \times [0, 2\pi]$.

Solution We have

$$f_x(x, y) = 2x + 2\sin y = 0 \quad \Rightarrow \quad x + \sin y = 0$$
$$f_y(x, y) = 2x \cos y = 0 \quad \Rightarrow \quad x \cos y = 0$$

From the second equation, $x = 0$ or $\cos y = 0$. That is, $x = 0$ or $y = (2n+1)\pi/2$, n is an integer number.
When $x = 0$, from the first equation, $\sin y = 0 \Rightarrow y = n\pi$, n is an integer number. In this case, we have three critical points $(0, 0)$ and $(0, \pi)$ in the given domain.
And in another case, when $y = (2n+1)\pi/2$, then $x = \sin(2n+1)\pi/2 = \pm 1$. So critical points are $(-1, \pi/2)$ and $(1, 3\pi/2)$ in the given domain. We will check that these points are wether a local minimum point or a local maximum point or a saddle point.
Next we compute the second partial derivatives;

$$f_{xx} = 2,\ f_{xy} = 2\cos y,\ f_{yy} = -2x \sin y$$

and evaluate the discriminant at each critical point;

$$D = f_{xx} \cdot f_{yy} - (f_{xy})^2 = -4x\sin y - 4\cos^2 y$$

For $(0,0)$, since $D = 0 - 4 < 0$, $(0,0,1)$ is a saddle point.

For $\left(-1, \frac{\pi}{2}\right)$, since $D = 4 > 0$ and $f_{xx}\left(-1, \frac{\pi}{2}\right) > 0$, $f\left(-1, \frac{\pi}{2}\right) = 0$ is a local minimum.

For $(0, \pi)$, since $D = -4 < 0$, $(0, \pi, 1)$ is a saddle point.

For $\left(1, \frac{3}{2}\pi\right)$, since $D = 4 > 0$ and $f_{xx}\left(1, \frac{3}{2}\pi\right) > 0$, $f\left(1, \frac{3\pi}{2}\right) = 0$ is a local minimum.

Therefore f has a local minimum at $\left(-1, \frac{\pi}{2}\right)$ and $\left(1, \frac{3}{2}\pi\right)$. ■

EXAMPLE 8

Find the point on the plane $2x - y + 2z = 16$ nearest the origin.

Solution Let (x, y, z) be any point on the plane $2x - y + 2z = 16$. Then the distance from the origin to the plane is $d = \sqrt{x^2 + y^2 + z^2}$.
Since $y = 2x + 2z - 16$, we obtain

$$d = \sqrt{x^2 + (2x + 2z - 16)^2 + z^2}.$$

Let $f(x, z) = d^2 = x^2 + (2x + 2z - 16)^2 + z^2$.
If $f(x, z)$ is minimum value, then d^2 is minimum value. So we get

$$f_x(x, z) = 10x + 8z - 64 = 0, f_z(x, z) = 8x + 10z - 64 = 0,$$

Solving the above equations, we get a critical point $(x, z) = \left(\frac{32}{9}, \frac{32}{9}\right)$.

Next we compute the second partial derivatives;

$$f_{xx}(x, z) = 10,\ f_{xz}(x, z) = 8,\ f_{zz}(x, z) = 10$$

and evaluate the discriminant at each critical point;

$$D = f_{xx}(x, z) f_{zz}(x, z) - f_{xz}(x, z)^2 = 36 > 0$$

and $f_{xx}(x, z) = 10 > 0$.

Thus f has a local minimum at $\left(\frac{32}{9}, \frac{32}{9}\right)$. Hence the nearest point in the plane is $\left(\frac{32}{9}, -\frac{16}{9}, \frac{32}{9}\right)$. ■

Just see how to implement the user defined function to list all critical points and tabulate along with its type (nature). This appears in the attached sage worksheet.

Absolute Maximum and Minimum Values

For any continuous function $f(x)$ defined on a closed interval of finite length $[a, b]$, we know f should have an **absolute maximum** and an **absolute minimum** value on $[a, b]$. In this section, we want to determine the analogue of f, a function of two variables x and y.

We will study how to find the absolute maximum and the absolute minimum values of a continuous function on a closed and bounded region $\boldsymbol{R}$ in the xy-plane.

DEFINITION 5 Absolute Maximum and Minimum

(i) If $f(x, y) \leq f(a, b)$ for all (x, y) in the domain of f, then f has an **absolute maximum** at (a, b). $f(a, b)$ is an **absolute maximum value**.

(ii) If $f(x, y) \geq f(a, b)$ for all (x, y) in the domain of f, then f has an **absolute minimum** at (a, b). $f(a, b)$ is an **absolute minimum value.**

Analogous to have extreme values at endpoint, a function of two variables can have extreme values on the boundary of the closed set R.

PROCEDURE

Absolute Maximum and Minimum Values on a Closed Bounded Set S

Let f be continuous on a closed bounded set S in $\mathbb{R}^2$. To find absolute maximum and minimum values of f on S,

1. Determine the values of f at all critical points in S.
2. Find the maximum and minimum values of f on the boundary of S.
3. The greatest function value found in 1 and 2 is the absolute maximum value of f on S, and the least function value found in 1 and 2 is the absolute minimum value of f on S.

EXAMPLE 9

Find the absolute maximum and minimum values of the $f(x, y) = x^2 + y^2 + 2x - 2y$ on the closed set

$$R = \{(x, y) \mid x^2 + y^2 \leq 16\}.$$

Solution To find critical points, we have

$$f_x(x, y) = 2x + 2 = 0 \text{ and } f_y(x, y) = 2y - 2 = 0.$$

Then $(-1, 1)$ is a critical point and so $f(-1, 1) = -2$.

We now determine the maximum and minimum values of f on the boundary of R, which is a circle of radius 4 described by the parametric equations

$$x = 4\cos\theta \text{ and } y = 4\sin\theta \text{ for } 0 \le \theta \le 2\pi.$$

Substituting x and y in terms of θ into the function f, we obtain a new function $g(\theta)$ that gives the values of f on the boundary of R. Thus

$$\begin{aligned} g(\theta) &= (4\cos\theta)^2 + (4\sin\theta)^2 + 8\cos\theta - 8\sin\theta \\ &= 16 + 8\cos\theta - 8\sin\theta \end{aligned}$$

The critical points of g satisfy

$$g'(\theta) = -8\sin\theta - 8\cos\theta = 0$$

or $\tan\theta = -1$. Therefore, g has critical points $\theta = -\dfrac{\pi}{4}$ and $\theta = \dfrac{3\pi}{4}$, which correspond to the points $(2\sqrt{2}, -2\sqrt{2})$ and $(-2\sqrt{2}, 2\sqrt{2})$.

The function values are

$$\begin{aligned} &f(-1, 1) = -2 \text{ (critical point)} \\ &f(2\sqrt{2}, -2\sqrt{2}) = 16 + 8\sqrt{2} \text{ (boundary point)} \\ &f(-2\sqrt{2}, 2\sqrt{2}) = 16 - 8\sqrt{2} \text{ (boundary point)} \end{aligned}$$

After comparing these values, we can decide that $f(-1, 1) = -2$ is an absolute minimum value of $f(x, y)$ and $f(-2\sqrt{2}, 2\sqrt{2}) = 16 - 8\sqrt{2}$ is an absolute maximum value of $f(x, y)$. ■

* Extreme of Implicit function

Let $f(x, y) = 0$ be an implicit function and assume that f has second order continuous partial derivatives.

In order to find extremum of f, we will find the point (x, y) such that

$$f(x, y) = 0,\ f_x(x, y) = 0.$$

Such points are called **singular points** of $f(x, y) = 0$. We have

$$\frac{dy}{dx} = -\frac{f_x}{f_y}, \text{ and } \frac{d^2y}{dx^2} = -\frac{f_{xx} + f_{xy}\dfrac{dy}{dx}}{f_y}.$$

If (x, y) is a point of local maximum or minimum, then $\dfrac{d^2y}{dx^2} = -\dfrac{f_{xx}}{f_y}$ since $f_x = 0$.

THEOREM 6

Let (x, y) be a extreme point and $f_y(x,y) \neq 0$, then

(i) If $\dfrac{d^2y}{dx^2} = -\dfrac{f_{xx}}{f_y} > 0$, then f has a local minimum at (x, y) and (x, y) is called a point of local minimum.

(ii) If $\dfrac{d^2y}{dx^2} = -\dfrac{f_{xx}}{f_y} < 0$, then f has a local maximum at $(x, y)$and (x, y) is called a point of local maximum.

EXAMPLE 10

Find the extreme values of the implicit function

$$x^3 - 3xy + y^3 = 0.$$

Solution Let $f(x, y) = x^3 - 3xy + y^3$.
To find singular points, we must solve the followings;

$$f_x(x, y) = 3x^2 - 3y = 0,\ f(x, y) = x^3 - 3xy + y^3 = 0.$$

Since $y = x^2$, we have $-2x^3 + x^6 = 0$. Then two singular points are $(x, y) = (0, 0),\ (\sqrt[3]{2}, \sqrt[3]{4})$.
Since $f_y(x, y) = -3x + 3y^2$, $f_y(0, 0) = 0$ and $f_y(\sqrt[3]{2}, \sqrt[3]{4}) = 3\sqrt[3]{2} \neq 0$. Thus, $-\dfrac{f_{xx}(\sqrt[3]{2}, \sqrt[3]{2})}{f_y(\sqrt[3]{2}, \sqrt[3]{2})} = -2 < 0$, By Theorem 2, $y = \sqrt[3]{4}$ is the local maximum value of y. ■

http://matrix.skku.ac.kr/cal-lab/cal-12-4-Exm1.html

```
var('x, y');
f(x, y) = x^3 -3*x*y + y^3
p=implicit_plot(f(x, y)==0, (x, -3, 3), (y, -3, 3))
pts=point([(0, 0), (2^(1/3), 4^(1/3))], rgbcolor='red', size=50)
show(p+pts)
```

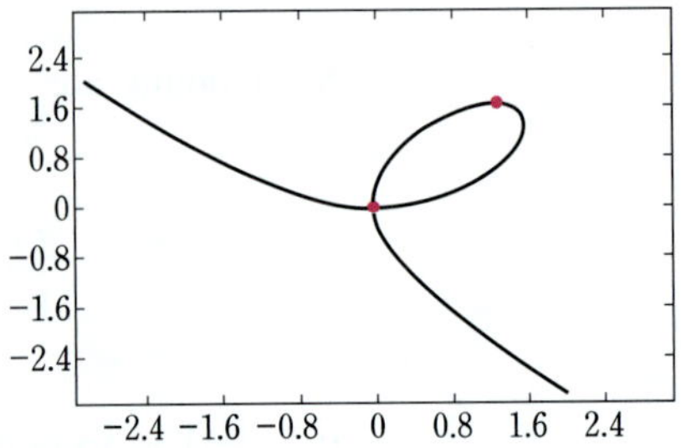

Figure 9

Notice that at $(0,0)$ the tangent is parallel to $x-$axis, but it is not a point of local extremum. Can you see this from the graph? The $(\sqrt[3]{2}, \sqrt[3]{4})$ is a local maximum.

We have

$$f_y(x, y) = -3x + 3y^2 \text{ and } f_{xx}(x, y) = 6x.$$

To classify singular points we need to find the sign of $-\dfrac{f_{xx}}{f_y}$. At $(0,0)$, $f_y = 0$. Therefore $(0,0)$ is not a point of maximum/minimum value. At $(\sqrt[3]{2}, \sqrt[3]{4})$, $\dfrac{f_{xx}(\sqrt[3]{2}, \sqrt[3]{4})}{f_y(\sqrt[3]{2}, \sqrt[3]{4})} = -2 < 0$. Theorem 6 and the graph shows $(\sqrt[3]{2}, \sqrt[3]{4})$ is a point of local maximum. The maximum of y is $\sqrt[3]{4}$ which occurs at $x = \sqrt[3]{2}$.

EXAMPLE 11

Find the extreme values of the implicit function

$$x^4 - 2x^2y + y^3 = 0.$$

Solution Let $f(x, y) = x^4 - 2x^2y + y^3$. Solving the above two equations;

$$f(x, y) = x^4 - 2x^2y + y^3 = 0,\ f_x(x, y) = 4x^3 - 4xy = 0,$$

we get three singular points

$$(x,y) = (0,0),\ (1,1),\ (-1,1).$$

Since $f_y(x, y) = -2x^2 + 3y^2$,

$$f_y(-1, -1) = 1, f_y(0, 0) = 0, f_y(1, 1) = 1.$$

Then we have two extreme points; $(1, 1)$ and $(-1, 1)$.
For $(1, 1)$ and $(-1, 1)$, since $f_{xx}(x, y) = 12x^2 - 4y$,
$\dfrac{d^2y}{dx^2} = -\dfrac{f_{xx}}{f_y} = -8 < 0$ at $(1, 1)$ and $(-1, 1)$.
Hence the implicit function has local maximum value $y = 1$ at $x = \pm 1$. ■

*Expansion of Taylor's Theorem

In this section we shall look at Taylor's Theorem.

Taylor's Theorem for function of two variables is as follows.

THEOREM 7

Suppose the function $f(x, y)$ has nth order continuous partial derivatives at (a, b). Let h and k be small enough.

1 $$f(a+h, b+k) = f(a, b) + \left(h\frac{\partial}{\partial x} + k\frac{\partial}{\partial y}\right)f(a, b) + \frac{1}{2!}\left(h\frac{\partial}{\partial x} + k\frac{\partial}{\partial y}\right)^2 f(a, b) + \cdots + \frac{1}{(n-1)!}\left(h\frac{\partial}{\partial x} + k\frac{\partial}{\partial y}\right)^{n-1} f(a, b) + \frac{1}{n!}\left(h\frac{\partial}{\partial x} + k\frac{\partial}{\partial y}\right)^n f(a+\theta h, b+\theta k) \quad (0<\theta<1)$$

Proof If we let $x = a+ht$, $y = b+kt$ in $f(x, y)$,

$$F(t) = f(x, y) = f(a+ht, b+kt)$$

which is a function of t. This single variable function can be expanded as following by Maclaurin formula:

$$F(t) = F(0) + \frac{1}{1!}F'(0)t + \frac{1}{2!}F''(0)t^2 + \cdots + \frac{1}{(n-1)!}F^{(n-1)}(0)t^{n-1} + \frac{1}{n!}F^{(n)}(\theta t)t^n \quad (0<\theta<1).$$

Let $t=1$ in the above formula,

2 $$f(a+h, b+k) = F(1) = F(0) + \frac{1}{1!}F'(0) + \frac{1}{2!}F''(0) + \cdots + \frac{1}{(n-1)!}F^{(n-1)}(0) + \frac{1}{n!}F^{(n)}(\theta) \quad (0<\theta<1).$$

Differentiate $F(t) = f(x, y)$ with respect to t,

$$\begin{aligned} F'(t) &= \frac{d}{dt}f(x, y) = \frac{\partial}{\partial x}f(x, y)\frac{dx}{dt} + \frac{\partial}{\partial y}f(x, y)\frac{dy}{dt} \\ &= hf_x(x, y) + kf_y(x, y). \end{aligned}$$

Let $t=0$ in the above equation.

$$F'(0) = hf_x(a, b) + kf_y(a, b) = \left(h\frac{\partial}{\partial x} + k\frac{\partial}{\partial y}\right)f(a, b).$$

Similarly, we can differentiate $F'(t)$ with respect to t,

$$\begin{aligned} F''(t) &= \frac{d}{dt}(hf_x(x, y) + kf_y(x, y)) \\ &= h\left(\frac{\partial}{\partial x}f_x(x, y)\frac{dx}{dt} + \frac{\partial}{\partial y}f_x(x, y)\frac{dy}{dt}\right) \\ &\quad + k\left(\frac{\partial}{\partial x}f_y(x, y)\frac{dx}{dt} + \frac{\partial}{\partial y}f_y(x, y)\frac{dy}{dt}\right) \\ &= h^2 f_{xx}(x, y) + 2hkf_{xy}(x, y) + k^2 f_{yy}(x, y). \end{aligned}$$

Let $t=0$ in the above equation.

$$F''(0)=h^2 f_{xx}(a,b)+2hkf_{xy}(a,b)+k^2 f_{yy}(a,b)$$
$$=\left(h\frac{\partial}{\partial x}+k\frac{\partial}{\partial y}\right)^2 f(a,b).$$

In a similar way, we have a general form;

3 $$F^{(n)}(0)=\left(h\frac{\partial}{\partial x}+k\frac{\partial}{\partial y}\right)^n f(a,b).$$

Substituting 2, we get

$$f(a+h,b+k)=f(a,b)+\left(h\frac{\partial}{\partial x}+k\frac{\partial}{\partial y}\right)f(a,b)$$
$$+\frac{1}{2!}\left(h\frac{\partial}{\partial x}+k\frac{\partial}{\partial y}\right)^2 f(a,b)+\cdots$$
$$+\frac{1}{(n-1)!}\left(h\frac{\partial}{\partial x}+k\frac{\partial}{\partial y}\right)^{n-1} f(a,b)+\frac{1}{n!}\left(h\frac{\partial}{\partial x}+k\frac{\partial}{\partial y}\right)^n$$
$$f(a+\theta h,b+\theta k)\quad(0<\theta<1).$$ ■

When we replace h and k with $x-a$ and $y-b$ in Taylor's Theorem for a two-variable function, it is transformed as follows:

4 $$f(x,y)=f(a,b)+\left\{(x-a)\frac{\partial}{\partial x}+(y-b)\frac{\partial}{\partial y}\right\}f(a,b)$$
$$+\frac{1}{2!}\left\{(x-a)\frac{\partial}{\partial x}+(y-b)\frac{\partial}{\partial y}\right\}^2 f(a,b)+\cdots$$
$$+\frac{1}{(n-1)!}\left\{(x-a)\frac{\partial}{\partial x}+(y-b)\frac{\partial}{\partial y}\right\}^{n-1} f(a,b)$$
$$+\frac{1}{n!}\left\{(x-a)\frac{\partial}{\partial x}+(y-b)\frac{\partial}{\partial y}\right\}^n f(a+\theta(x-a),b+\theta(y-b))$$
$$(0<\theta<1).$$

Placing $a=0,\ b=0$ in 4.

5 $$f(x,y)=f(0,0)+\left(x\frac{\partial}{\partial x}+y\frac{\partial}{\partial y}\right)f(0,0)$$
$$+\frac{1}{2!}\left(x\frac{\partial}{\partial x}+y\frac{\partial}{\partial y}\right)^2 f(0,0)+\cdots$$
$$+\frac{1}{(n-1)!}\left(x\frac{\partial}{\partial x}+y\frac{\partial}{\partial y}\right)^{n-1} f(0,0)$$
$$+\frac{1}{n!}\left(x\frac{\partial}{\partial x}+y\frac{\partial}{\partial y}\right)^n f(\theta x,\theta y)\quad(0<\theta<1).$$

This is Maclaurin's Theorem for two-variable functions which is a special case of the Taylor Theorem.

Remark The case of $n=1$ at [1] in (Taylor) Theorem 7 is

$$f(a+h, b+k) = f(a, b) + \left(h\frac{\partial}{\partial x} + k\frac{\partial}{\partial y}\right) f(a+\theta h, b+\theta k)$$
$$= f(a, b) + hf_x(a+\theta h, b+\theta k) + kf_y(a+\theta h, b+\theta k) \quad (0<\theta<1)$$

which is the Mean Value Theorem for the two variable function.

CAS **EXAMPLE 12**

Find the Taylor series for the function $f(x, y) = xy^2 + \sin xy$ at the point $\left(1, \frac{\pi}{2}\right)$ up to degree 2.

Solution

$f(x, y) = xy^2 + \sin xy$, $\quad f\left(1, \frac{\pi}{2}\right) = \frac{\pi^2}{4} + 1$,

$f_x(x, y) = y^2 + y\cos xy$, $\quad f_x\left(1, \frac{\pi}{2}\right) = \frac{\pi^2}{4}$,

$f_y(x, y) = 2xy + x\cos xy$, $\quad f_y\left(1, \frac{\pi}{2}\right) = \pi$,

$f_{xx}(x, y) = -y^2 \sin xy$, $\quad f_{xx}\left(1, \frac{\pi}{2}\right) = -\frac{\pi^2}{4}$.

$f_{xy}(x, y) = 2y + \cos xy - xy\sin xy$, $\quad f_{xy}\left(1, \frac{\pi}{2}\right) = \frac{\pi}{2}$.

$f_{yy}(x, y) = 2x - x^2 \sin xy$, $\quad f_{yy}\left(1, \frac{\pi}{2}\right) = 1$.

Therefore

$$xy^2 + \sin xy = 1 + \frac{\pi^2}{4} + \frac{\pi^2}{4}(x-1) + \pi\left(y - \frac{\pi}{2}\right)$$
$$+ \frac{1}{2!}\left\{-\frac{\pi^2}{4}(x-1)^2 + \pi(x-1)\left(y-\frac{\pi}{2}\right) + \left(y - \frac{\pi}{2}\right)^2\right\}.$$

■

CAS **EXAMPLE 13**

Expand the Taylor series for the function $f(x, y) = xy^3$ at $(1, -1)$.
http://matrix.skku.ac.kr/cal-lab/m-Sec13-8-Exm-13.html

Solution $f(x, y) = xy^3$

```
x, y=var('x, y');
taylor(x*y^3, (x, 1), (y, -1), 4)
```

Answer : (y + 1)^3*(x - 1) + (y + 1)^3 - 3*(y + 1)^2*(x - 1) - 3*(y + 1)^2 +3*(y+1)*(x-1)-x+3*y + 3 ■

CAS **EXAMPLE 14**

Plot the surface $z = \cos x \cos y$. Find Taylor's polynomial of degrees 1, 2, 3, and 4. Plot the surface and graph of these polynomial together and show the Taylors polynomial approximates the function as the degree increases.

http://matrix.skku.ac.kr/cal-lab/m-Sec13-8-Exm-14.html

Solution

```
var('x,y')
f = cos(x)*cos(y)
a=0
b=0
f1=taylor(f(x,y),(x,a),(y,b),1)
f2=taylor(f(x,y),(x,a),(y,b),2)
f3=taylor(f(x,y),(x,a),(y,b),3)
f4=taylor(f(x,y),(x,a),(y,b),4)
p=plot3d(f(x,y),(x,-3,3),(y,-3,3))
p1=plot3d(f1,(x,-1,1),(y,-1,1),color='red')
p2=plot3d(f2,(x,-1.5,1.5),(y,-1.5,1.5),color='goldenrod')
p3=plot3d(f3,(x,-2,2),(y,-2,2),color='green')
p4=plot3d(f4,(x,-2,2),(y,-2,2),color='violet')
p+p2+p1+p3+p4
```

Answer :

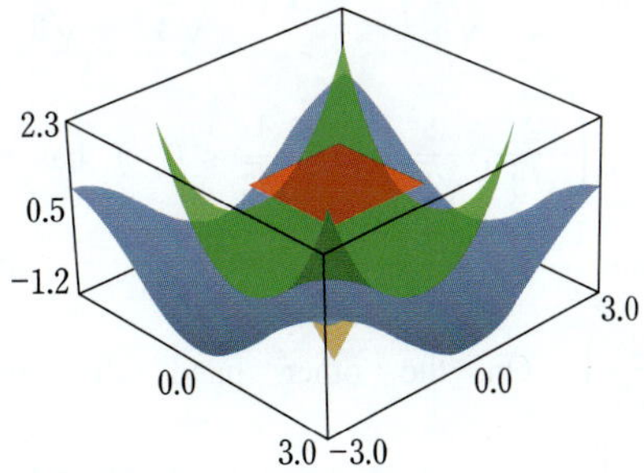

Figure 10

■

13.8 EXERCISES (Extrema of Multivariate Functions)

http://matrix.skku.ac.kr/Cal-Book/part2/CS-Sec-13-8-Sol.html

YouTube http://youtu.be/FWmk_MasIjE

1. Let $f(x,\ y)=2xy^2-x^2y+4y-2y^3$. Find the critical points of f and classify them.

Solution Solve $f_x=2y^2-2xy=0$ and $f_y=4xy-x^2+4-6y^2=0$. So we have critical points, $y=0$ or $y=x$. If $y=0$, then $x=\pm 2$. If $y=x$, then we have $x=\pm\dfrac{2}{\sqrt{3}}$.

The critical points are

$(x,y)=(2,0),\ (-2,0),\ \left(\dfrac{2}{\sqrt{3}},\dfrac{2}{\sqrt{3}}\right),\ \left(-\dfrac{2}{\sqrt{3}},-\dfrac{2}{\sqrt{3}}\right)$.

Next, we consider second order partial derivatives to get $f_{xx}=-2y,\ f_{yy}=4x-12y,\ f_{xy}=4y-2x$.

Then $D=f_{xx}f_{yy}-f_{xy}^2=(2y)(12y-4x)-(4y-2x)^2$ and thus we obtain $D(2,0)=D(-2,0)=-16<0$ at points $(x,y)=(2,0),\ (-2,0)$. This implies that $(x,y)=(2,0)$, $(-2,0)$ are saddle points.

At points $(x,y)=\left(\dfrac{2}{\sqrt{3}},\dfrac{2}{\sqrt{3}}\right),\ \left(-\dfrac{2}{\sqrt{3}},-\dfrac{2}{\sqrt{3}}\right)$, we observe that $D\left(\dfrac{2}{\sqrt{3}},\dfrac{2}{\sqrt{3}}\right)=D(-\dfrac{2}{\sqrt{3}},-\dfrac{2}{\sqrt{3}})=16>0$.

Moreover, since $f_{xx}\left(\dfrac{2}{\sqrt{3}},\dfrac{2}{\sqrt{3}}\right)=-\dfrac{4}{\sqrt{3}}<0$, f has a local maximum $f\left(\dfrac{2}{\sqrt{3}},\dfrac{2}{\sqrt{3}}\right)=\dfrac{16\sqrt{3}}{9}$ at $(x,y)=\left(\dfrac{2}{\sqrt{3}},\dfrac{2}{\sqrt{3}}\right)$. On the other hand, due to $f_{xx}\left(-\dfrac{2}{\sqrt{3}},-\dfrac{2}{\sqrt{3}}\right)=\dfrac{4}{\sqrt{3}}>0$, f has a local minimum $f\left(-\dfrac{2}{\sqrt{3}},-\dfrac{2}{\sqrt{3}}\right)=-\dfrac{16\sqrt{3}}{9}$ at $(x,y)=\left(-\dfrac{2}{\sqrt{3}},-\dfrac{2}{\sqrt{3}}\right)$.

2. Find the extreme values of the function $f(x,y)=xy(a-x-y)$ when $a>0$.

Solution $f(x,y)=xy(a-x-y)\Rightarrow$

$f_x(x,y)=y(a-2x-y)=0,\ f_y(x,y)=x(a-x-2y)=0$

$\Rightarrow (x,y)=(0,0),(0,a),(a,0),\left(\dfrac{a}{3},\dfrac{a}{3}\right)$

$f_{xx}(x,y)=-2y,\ f_{xy}(x,y)=a-2x-2y,\ f_{yy}(x,y)=-2x$

$\Rightarrow\ (x,y)=(0,0),(0,a),(a,0)$: critical points

$D=a^2<0$ => f has no local minimum. or maximum at $(x,y)=(0,0),(0,a),(a,0)$.

At $(x,y)=\left(\dfrac{a}{3},\dfrac{a}{3}\right)$,

$D=\left(-\dfrac{a}{3}\right)^2-\left(-\dfrac{2a}{3}\right)\left(-\dfrac{2a}{3}\right)=-\dfrac{a^2}{3}>0$

$\Rightarrow f_{xx}\left(\dfrac{a}{3},\dfrac{a}{3}\right)=-\dfrac{2}{3}a<0$, so f has a local maximum $f\left(\dfrac{a}{3},\dfrac{a}{3}\right)=\dfrac{a^3}{27}$ at $\left(\dfrac{a}{3},\dfrac{a}{3}\right)$.

3-4. Locate the maxima, minima, and saddle points of the functions.

CAS **3.** $f(x,y)=2(x^2-y^2)-x^4+y^4$.

Solution

http://matrix.skku.ac.kr/cal-lab/cal-12-4-3.html
http://matrix.skku.ac.kr/LA-Lab/ms-1.html

```
var ('x, y, z')
f(x,y)=2*(x^2-y^2)-x^4+y^4
P=implicit_plot3d (z==f(x, y), (x,-1.5,3/2), (y,-3/2,3/2), (z,-1,1), color= 'goldenrod', opacity=0.6)
P.show()
contour_plot(f, (x,-2,2), (y,-2,2), contours= srange(-2,2,0.25), fill=False, cmap= 'cool', labels=True)
```

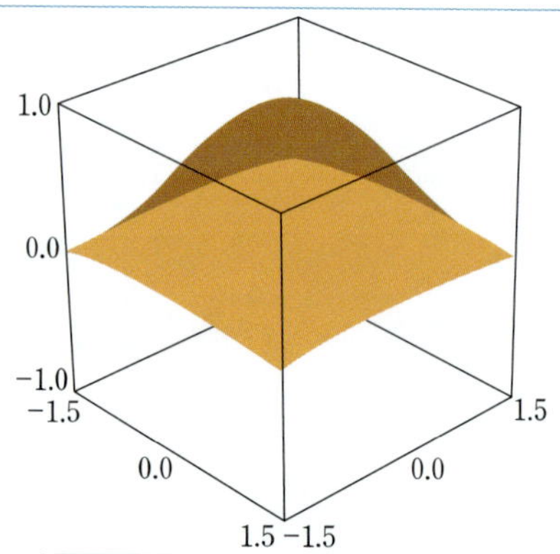

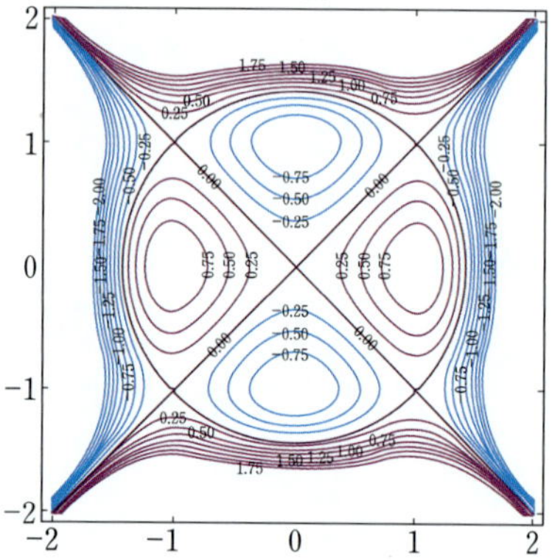

http://matrix.skku.ac.kr/LA-Lab/ms-2.html

```
var ('x, y, z')
f(x,y)=2*(x^2-y^2)-x^4+y^4
fx=f.diff(x)
fy=f.diff(y)
fxx=diff(f,x,x)
fyy=diff(f,y,y)
fxy=diff(f,x,y)
cpoints=solve([fx==0,fy==0],[x,y],solution_dict=Tr
ue)
for sol in cpoints:
    if ((sol[x] in RR) and (sol[y] in RR) ):
        print((sol[x],sol[y]))
```

(0, 0), (1, 0), (-1, 0), (0, 1), (1, 1), (-1, 1), (0, -1), (1, -1), (-1, -1)

```
def extreme(f,a,b):
    f11=diff(f,x,x)(a,b)
    f22=diff(f,y,y)(a,b)
    f12=diff(f,x,y)(a,b)
    D=f11*f22-f12^2
    if(D>0):
        if(f11>0):
            return "local minimum"
        else:
            if(f11<0):
                return "local maximum"
            else:
                return "inconclusive"
else:
        if(D<0):
            return "saddle point"
        else:
            if(D==0):
                return     "inconclusive"table     =
[["Critical Point", "Type"]]
f(x,y)=2*(x^2-y^2)-x^4+y^4
fx=f.diff(x)
fy=f.diff(y)
fxx=diff(f,x,x)
fyy=diff(f,y,y)
fxy=diff(f,x,y)
cpoints=solve([fx==0,fy==0],[x,y],solution_dict=Tr
ue)
for sol in cpoints:
    if ((sol[x] in RR) and (sol[y] in RR)):
        a=sol[x].n()
        b=sol[y].n()
        table.append([(sol[x],sol[y]),
extreme(f,a,b)])
html.table(table, header=True)
```

Critical Point	Type
(0, 0)	saddle point
(1, 0)	local maximum
(-1, 0)	local maximum
(0, 1)	local minimum
(1, 1)	saddle point
(-1, 1)	saddle point
(0, -1)	local minimum
(1, -1)	saddle point
(-1, -1)	saddle point

4. $f(x, y) = \cos(x + y^2)$.

Solution Try this on your own.

---

5. Let $f(x,y) = a(x-1)^4 + b(y-1)^4 - 4ab(x-1)(y-1)$, $ab \neq 0$. Answer the following:

(a) Find points of local maximum/minimum and a saddle point when $a = \frac{1}{2}, b = 8$.

(b) Give a condition on a, b for the case when f has only one critical point.

Solution (a) $f(x,y) = \frac{1}{2}(x-1)^4 + 8(y-1)^4 - 16(x-1)(y-1)$

$\nabla f(x,y) = (2(x-1)^3 - 16(y-1), 32(y-1)^3 - 16(x-1))$

$\nabla f(x,y) = 0 \Leftrightarrow (x,y) = (1,1)$ or $(-1,0)$ or $(3,2)$

$f''(x,y) = \begin{pmatrix} 6(x-1)^2 & -16 \\ -16 & 96(y-1)^2 \end{pmatrix}$

$f''(1,1) = \begin{pmatrix} 0 & -16 \\ -16 & 0 \end{pmatrix} \Rightarrow \det f''(1,1) < 0$

$\Rightarrow$ (1,1) is a saddle point.

$f''(3,2) = \begin{pmatrix} 24 & -16 \\ -16 & 96 \end{pmatrix} \Rightarrow \det f''(3,2) > 0$ and $24 > 0$

$f''(-1,0) = \begin{pmatrix} 24 & -16 \\ -16 & 96 \end{pmatrix} \Rightarrow \det f''(-1,0) > 0$ and $24 > 0$

$\Rightarrow (3,2), (-1,0)$ are points of local minimum.

(b) $\nabla f(x,y) = (4a(x-1)^3 - 4ab(y-1), 4b(y-1)^3 - 4ab(x-1))$

$\Rightarrow 4a(x-1)^3 = 4ab(y-1) \ \Rightarrow (y-1) = \dfrac{(x-1)^3}{b}$

$(\because ab \neq 0)$

$4b(y-1)^3 = 4ab(x-1) \ \Rightarrow (y-1)^3 = a(x-1)$

$\Rightarrow (x-1)((x-1)^8 - ab^3) = 0$

If f has only one critical point $x - 1 = 0$ has a solution and $(x-1)^8 - ab^3 = 0$ should not have a solution. So $ab^3 < 0$ or $ab < 0$.

6. Find maximum value of $f(x,y) = x^2 - xy$ on $D = \{(x,y) | -1 \leq x \leq 1, -1 \leq y \leq 1\}$.

Solution $f_x(x,y)=2x-y=0 \Rightarrow 2x=y$,
$f_y(x,y)=-x=0$.
So the critical point is $(0,0)$ and thus critical value is $f(0,0)=0$.
Let $L_1=\{(x,-1)|-1\le x\le 1\}$,
$L_2=\{(1,y)|-1\le y\le 1\}$, $L_3=\{(x,1)|-1\le x\le 1\}$
and $L_4=\{(-1,y)|-1\le y\le 1\}$.
On L_1, we have $y=-1$ and $f(x,-1)=x^2+x$,
$-1\le x\le 1$, $f(-1,-1)=0$, $f(1,-1)=2$.
On L_2, we have $x=1$ and $f(1,y)=1-y$, $-1\le y\le 1$,
$f(1,-1)=2$, $f(1,1)=0$.
On L_3, we have $y=1$ and $f(x,1)=x^2-x$,
$-1\le x\le 1$, $f(-1,1)=2$, $f(1,1)=0$.
On L_4, we have $x=-1$ and $f(-1,y)=1+y$,
$-1\le y\le 1$. $f(-1,-1)=0$, $f(-1,1)=2$.
So the maximum value is 2.

7. Find the absolute maximum and minimum of $f(x,y)=x^2-2xy+2y$ in the domain D which is a closed triangle made of three points (0. 0), (2, 1), (1, 2).

Solution (1) $f_x=2x-2y=0$, $f_y=-2x+2=0$
$\Rightarrow$ critical point : $(1,1)$, $f(1,1)=1$

(2)1. $f(x,y)=x$ moves on $L_1: y=\frac{1}{2}x \ (0\le x\le 2)$

$\Rightarrow$ The absolute maximum $f(2,1)=2$, and the absolute minimum $f(0,0)=0$ on $0\le x\le 2$.

2. $f(x,y)=3x^2-8x+6$ moves on
$L_2: y=-x+3 \ (1\le x\le 2)$ $\Rightarrow$ The absolute maximum $f(2,1)=2$, and the absolute minimum $f\left(\frac{4}{3},\frac{5}{3}\right)=\frac{2}{3}$ on $1\le x\le 2$.

3. $f(x,y)=-3x^2+4x$ moves on
$L_3: y=2x \ (0\le x\le 1)$
$\Rightarrow$ The absolute maximum $f\left(\frac{2}{3},\frac{4}{3}\right)=\frac{4}{3}$, and the absolute minimum $f(0,0)=0$ on $0\le x\le 1$.

Hence the absolute maximum is 2 and the absolute minimum is 0.

8. Find the absolute maximum and minimum values $f(x,y)=e^{-x^2-y^2}(x^2+2y^2)$ on the disk D: $x^2+y^2\le 4$.

Solution $\{(x,y)|x^2+y^2<4\}$ interior of D :
$f_x=2xe^{-x^2-y^2}(1-x^2-2y^2)$
Then $f_x=0$ implies $x=0$ or $x^2+2y^2=1$.
If $x=0$, $f_y=2ye^{-x^2-y^2}(2-x^2-2y^2)=0$ implies $y=0$ or $2-2y^2=0$.
Thus, we get the critical points $(0,0),(0,\pm 1)$.
If $x^2+2y^2=1$, then $f_y=2ye^{-x^2-y^2}(2-x^2-2y^2)=0$.
This implies $y=0$.
Critical points are $(\pm 1,0)$.
Thus $f(0,0)=0, f(\pm 1,0)=e^{-1}$ and $f(0,\pm 1)=2e^{-1}$.C
onsider $\{(x,y)|x^2+y^2=4\}$, boundary of D :
$x^2+y^2=4$, so $f(x,y)=e^{-4}(4+y^2)$.
Moreover, f is smallest when $y=0$ and largest when $y^2=4$. But $f(\pm 2,0)=4e^{-4}$, $f(0,\pm 2)=8e^{-4}$.
Thus on D the absolute maximum of f is $f(0,\pm 1)=2e^{-1}$ and the absolute minimum is $f(0,0)=0$.

9. Find the Taylor series for the function $f(x,y)=\ln(x-y)$ at the point $(2,1)$.

Solution $f(x,y)=\ln(x-y)$, $f(2,1)=0$,

$f_x(x,y)=\frac{1}{x-y}$, $f_x(2,1)=1$,

$f_y(x,y)=-\frac{1}{x-y}$, $f_y(2,1)=-1$,

$f_{xx}(x,y)=-\frac{1}{(x-y)^2}$, $f_{xx}(2,1)=-1$,

$f_{xy}(x,y)=\frac{1}{(x-y)^2}$, $f_{xy}(2,1)=1$,

$f_{yy}(x,y)=-\frac{1}{(x-y)^2}$, $f_{yy}(2,1)=-1$.

Therefore

$$\ln(x-y)=0+(x-2)-(y-1)-\frac{1}{2!}\{(x-2)^2-2(x-2)(y-1)+(y-1)^2\}+\cdots$$
$$=(x-2)-(y-1)-\frac{1}{2}(x-2)^2+(x-2)(y-1)-\frac{1}{2}(y-1)^2+\cdots.$$

10. Expand the Maclaurin series for the function $f(x,y)=e^{x+y}$.

Solution $f(x,y)=e^{x+y}$, $f(0,0)=1$
$f_x(x,y)=e^{x+y}$, $f_x(0,0)=1$
$f_y(x,y)=e^{x+y}$, $f_y(0,0)=1$
$f_{xx}(x,y)=e^{x+y}$, $f_{xx}(0,0)=1$
$f_{xy}(x,y)=e^{x+y}$, $f_{xy}(0,0)=1$
$f_{yy}(x,y)=e^{x+y}$, $f_{yy}(0,0)=1$
In general,

$$e^{x+y}=1+(x+y)+\frac{1}{2!}(x+y)^2+\frac{1}{3!}(x+y)^3+\cdots$$

13.9 Lagrange Multiplier

http://www.youtube.com/watch?v=ry9cgNx1QV8

1)http://www.slimy.com/~steuard/teaching/tutorials/Lagrange.html

Constrained Optimization (Lagrange Multiplier)[1]

So far we have dealt with finding maximum or minimum of a function of one or more variables without any constraints. Such problems are called **unconstrained optimization problems**. However, most of real life problems of maximization or minimization involves certain outside conditions or constraints. Such problems are called constrained optimization problems. This section deals with such problems where the constraints are of equality types. The method of **Lagrange multipliers** is a powerful tool for solving this class of problems without the need to explicitly solve the conditions and use them to eliminate extra variables.

First of all let us consider a minimization/maximization problem of a function of two variables.

1 Find extreme values of $f(x, y) = x + y$ subjected to

$$g(x, y) = x^2 + y^2 - 1 = 0.$$

Let us look at this problem geometrically. We wish to find the maximum and minimum value of $c = f(x, y) = x + y$ on the unit circle $x^2 + y^2 = 1$.
The point at which such a minimum (maximum) occurs is called a minimizer (maximizer).

Note that $x + y = c$, for various values of c represent parallels lines. The value of c increases as we move from left to right along the line $y = -x$. Thus we are looking for a point on the circle $x^2 + y^2 = 1$ at which the value of $c = x + y$ is minimum and maximum.

Look at the Figure 1. It appears that this happens when these curves just touch each other, that is, when they have a common tangent line. In particular, the normal vectors (gradient) to the level curves of f and g at the point of maximum or minimum are parallel to each other.

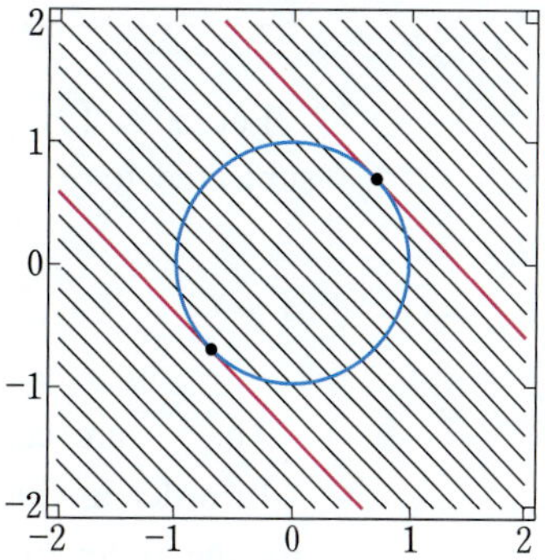

Figure 1

```
p=contour_plot(f(x,y), (x,-2,2), (y, -2, 2), contours=40, fill=False,
cmap='cool', linestyles=['solid'])
p+=implicit_plot(g(x,y)==0, (x,-2,2), (y,-2,2), cmap=['blue'])
p+=points([[sqrt(2)/2,sqrt(2)/2],[-sqrt(2)/2,-sqrt(2)/2]], pointsize = 30,
color = 'green')
p+=implicit_plot(f(x,y)==sqrt(2), (x, -2, 2), (y, -2, 2), color= 'red')
p+=implicit_plot(f(x,y)==-sqrt(2), (x, -2, 2), (y, -2, 2), color='red')
p.show(figsize=4, dpi=200)
```

This turns out to be the necessary condition for a point to be minimum of maximum in general under certain conditions.

Let us consider finding a minimizer of $f(x, y)$ subjected to $g(x, y)=0$.

Suppose (x_0, y_0) is a point of minimum on the level curve $g(x, y)=0$ and that the minimum value is c.

Let $\gamma : (-\epsilon, \epsilon) \to S$ be a parameterized curve on $S=\{(x, y) \mid g(x, y)=0\}$ with $\gamma(0)=(x_0, y_0)$. Since f has minimum value at (x_0, y_0), the composition function $h(t)=f(\gamma(t))$ has minimum at $t=0$. Hence

$$\frac{d}{dt}(f(\gamma(t)))\bigg|_{t=0} = \nabla f(\gamma(0)) \cdot \gamma'(0)=0 .$$

In particular, $\nabla f(x_0, y_0)$ is orthogonal to $\gamma'(0)$. We already know that the $\nabla g(x_0, y_0)$ is orthogonal to the $\gamma'(0)$. Thus at the point of minimizer (x_0, y_0), $\nabla f(x_0, y_0)$ is parallel to $\nabla g(x_0, y_0)$ provided $\nabla g(x_0, y_0) \neq 0$. (We shall see later why do we need $\nabla g(x_0, y_0) \neq 0$.)

More precisely we have the following theorem:

THEOREM 1

Let $f : \mathbb{R}^2 \to \mathbb{R}$ have continuous first order partial derivatives. Let (x_0, y_0) be a point of minimizer or maximizer of $f(x, y)$ subjected to $g(x, y)=0$. Then $\nabla f(x_0, y_0)$ and $\nabla g(x_0, y_0)$ are parallel to each other. In particular, if $\nabla g(x_0, y_0) \neq 0$, then there exists a scalar λ such that $\nabla f(x_0, y_0)=\lambda \nabla g(x_0, y_0)$. Here λ is called the **Lagrange multiplier**.

Let $x^* =(x, y)$ be a point at which $f(x, y)$ has maximum or minimum subjected to $g(x, y)=0$. Then there exists $\lambda \in \mathbb{R}$ such that we have

2 $$\frac{\partial f}{\partial x}=\lambda\frac{\partial g}{\partial x}, \ \frac{\partial f}{\partial y}=\lambda\frac{\partial g}{\partial y} \text{ and } g(x, y)=0.$$

The 2 are called the Lagrange conditions.

In order to find minimum/maximum of $f(x, y)$ subjected to $g(x, y)=0$, we use the following steps :

Step 1. Find all the points that satisfy the Lagrange conditions [2].
Step 2. Evaluate f at all the points obtained in Step 1. The largest of these values is the maximum value of f and the smallest is the minimum value of f.

Lagrange Conditions in Three Variables

If $x^* = (x, y, z)$ is a point at which $f(x, y, z)$ has maximum or minimum subjected to $g(x, y, z) = 0$ with $\nabla g(x, y, z) \neq 0$, then there exists $\lambda \in \mathbb{R}$ such that we have

[3] $$\frac{\partial f}{\partial x} = \lambda \frac{\partial g}{\partial x},\ \frac{\partial f}{\partial y} = \lambda \frac{\partial g}{\partial y},\ \frac{\partial f}{\partial z} = \lambda \frac{\partial g}{\partial z} \text{ and } g(x, y, z) = 0.$$

EXAMPLE 1

Optimize $f(x, y) = 4 + x^2 - y$ subjected to $g(x, y) = x^2 + y^2 - 1 = 0$.

Solution Look at the Figure 2 in which level curves of f and the curve $g(x, y) = 0$ are shown. It is clear from the Figure that there are fours points at which the two curves have common tangents. In particular, we can expect to get four points at which f has extreme values.

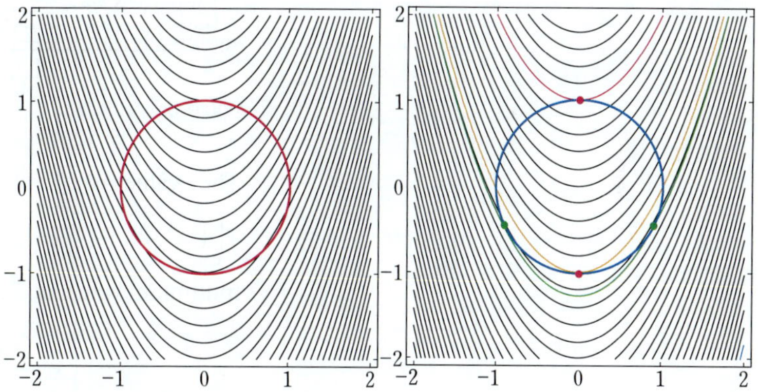

Figure 2

Using the Lagrange necessary conditions, there exits λ such that we have the following equations

$$\frac{\partial f}{\partial x} = \lambda \frac{\partial g}{\partial x} \Rightarrow 2x = 2\lambda x$$
$$\frac{\partial f}{\partial y} = \lambda \frac{\partial g}{\partial y} \Rightarrow -1 = 2\lambda y$$

and

$$x^2 + y^2 = 1.$$

From $2x = 2\lambda x$, we have $x(1-\lambda) = 0$. This mean $x = 0$ or $\lambda = 1$.

If $x = 0$, then $y = \pm 1$. So the points are $(0, \pm 1)$.

If $\lambda = 1$, then from the second equation, we have $y = -\frac{1}{2}$.

Hence from third equation, we get $x = \pm\frac{\sqrt{3}}{2}$. Thus the points are $\left(\pm\frac{\sqrt{3}}{2}, -\frac{1}{2}\right)$.

We have four possible points of optimizers, $(0, \pm 1)$ and $\left(\pm\frac{\sqrt{3}}{2}, -\frac{1}{2}\right)$. Look at the Figure 2. From this figure, it is clear that at these points, the contours of f and g have common tangents. Look at the Figure 3, it is clear that at these points the gradients of f and g are parallel to each other.

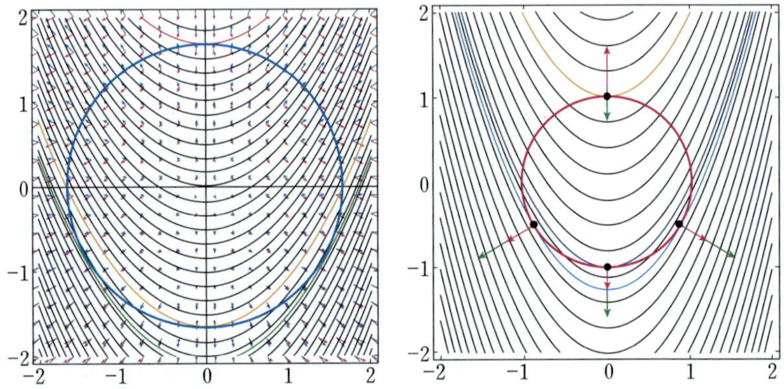

Figure 3

The value of f at these points are $f(0, -1) = 5$, $f(0, 1) = 3$, $f\left(\pm\frac{\sqrt{3}}{2}, -\frac{1}{2}\right) = 5$. Hence the minimizer occurs $(0, 1)$ and the minimum value is 3. The maximum occurs at two points

$$\left(\pm\frac{\sqrt{3}}{2}, -\frac{1}{2}\right) \text{ with maximum value } 5. \qquad \blacksquare$$

http://matrix.skku.ac.kr/cal-lab/m-Sec13-9-Exm-1.html

```
var('x,y,lam,t')
f(x,y)=4+x^2-y
g(x,y)=x^2+y^2-1
fx=f.diff(x)
fy=f.diff(y)
gx=g.diff(x)
gy=g.diff(y)
sol=solve([fx==lam*gx,fy==lam*gy,
g==0],x,y,lam,solution_dict=True)
```

```
show(sol)
n=len(sol)
for j in range(n):
    print f(sol[j][x], sol[j][y])
print "The Solution is tabulated as follows"
html.table([['x, y, (x, y)']] + [(sol[j][x], sol[j][y],
f(sol[j][x], sol[j][y])) for j in range(n)], header = True)
```

EXAMPLE 2

Find the extreme values(maximum and minimum) of $f(x, y) = xy$ subjected to the constrained $g(x, y) = \frac{x^2}{8} + \frac{y^2}{2} = 1$.

Solution Look at the Figure 4 which represent the level curves of f and the curve $g(x, y) = 0$. It is clear that there are four points at which the two level curves have common tangents. In particular, we can expect to have four points where f has extreme values.

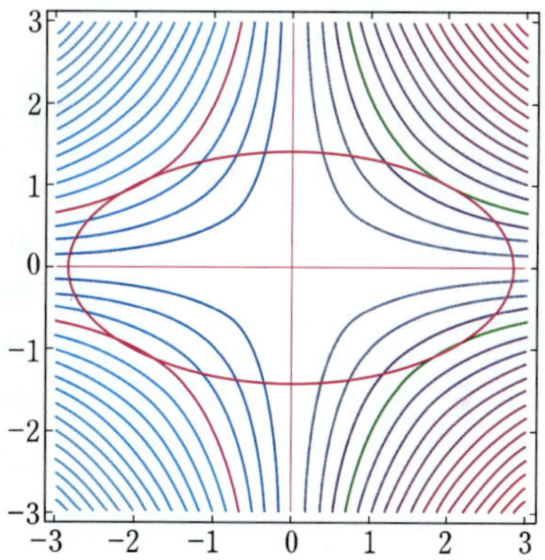

Figure 4

Using the Lagrange conditions, if (x, y) is a solution of this problem then

$$\frac{\partial f}{\partial x} = \lambda \frac{\partial g}{\partial x} \Rightarrow y = \frac{\lambda}{4} x$$
$$\frac{\partial f}{\partial y} = \lambda \frac{\partial g}{\partial y} \Rightarrow x = \lambda y$$

and

$$\frac{x^2}{8} + \frac{y^2}{2} = 1 \Rightarrow x^2 + 4y^2 = 8.$$

Combining, the first two equations we get $\lambda^2 = 4 \Rightarrow \lambda = \pm 2$. This implies $x = \pm 2y$. Therefore, we have four points

$$(2,\ 1),\ (-2,\ -1),\ (2,\ -1),\ (-2,\ 1).$$

The value of the function at the first two points is 2 and the value of the function at the last two points is -2. Thus the maximum value of $f=2$ and occurs at two points $(2,\ 1)$, $(-2,\ -1)$. The minimum value of $f=-2$ and it occurs at $(2,\ -1)$, $(-2,\ 1)$. Look at the following Figure 5 which show that the ∇f and ∇g are parallel at point of extremes.

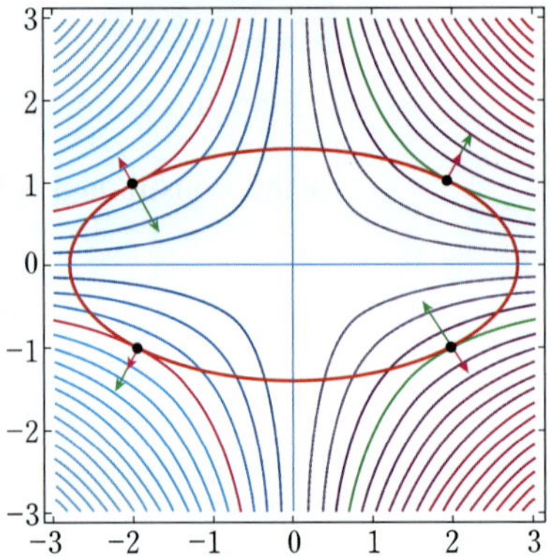

Figure 5

http://matrix.skku.ac.kr/cal-lab/m-Sec13-9-Exm-2.html

```
var('x, y, lam, t')
f(x,y)=x*y
g(x,y)=x^2/8+y^2/2-1
fx=f.diff(x)
fy=f.diff(y)
gx=g.diff(x)
gy=g.diff(y)
sol=solve([fx==lam*gx,      fy==lam*gy,      g==0],      x,y,
lam,solution_dict=True)
show(sol)
```

Answer : [{x:−2, lam:2, y:−1},{x:2,lam:2,y:1},{x:−2,lam:−2,y:1},{x:2,lam:−2,y:−1}]

```
n=len(sol)
print "The Solution is tabulated as follows"
html.table([["$x$", "$y$", "$f(x, y)$"]] + [(sol[j][x], sol[j][y],
f(sol[j][x], sol[j][y])) for j in range(n)], header = True)
```

Answer :

x	y	f(x,y)
−2	−1	2
2	1	2
−2	1	−2
2	−1	−2

■

EXAMPLE 3

Find the dimensions of a cylindrical tin (with bottom and top) made up of a metal sheet to maximize its volume such that the total surface area is 24π.

Solution Let x_1 and x_2 be the radius of the base and height of the cylinder. Then the total surface area of the metal sheet is $2\pi x_1^2 + 2\pi x_1 x_2$ and the volume is $\pi x_1^2 x_2$. Thus mathematically, we can write the above problem as:

$$\text{maximize } \pi x_1^2 x_2 \text{ subjected to } 2\pi x_1^2 + 2\pi x_1 x_2 = 24\pi.$$

Using the Lagrange conditions, we get

$$2\pi x_1 x_2 = \lambda(4\pi x_1 + 2\pi x_2)$$
$$\pi x_1^2 = 2\pi \lambda x_1$$
$$2\pi x_1^2 + 2\pi x_1 x_2 = 24\pi.$$

After solving we get $x_1 = \frac{1}{2}x_2$. Suppose $(x_1^*, x_2^*, \lambda^*)$ is a solution of the above system of equation, then further using the constrained we obtain the solution

$$x_1^* = 2,\ x_2^* = 4,\ \lambda^* = 1,\ f = 16\pi.$$ ■

EXAMPLE 4

Find the points on the ellipsoid $x^2 + 4y^2 + 3z^2 = 12$ that are closest and farthest from the point $(2, -1, 2)$.

Solution If (x, y, z) is any point in $\mathbb{R}^3$, then its distance from $(2, -1, 2)$ is given by

$$\sqrt{(x-2)^2 + (y+1)^2 + (z-2)^2}.$$

This means we need to optimize $f = \sqrt{(x-2)^2 + (y+1)^2 + (z-2)^2}$ subjected to $x^2 + 4y^2 + 3z^2 = 12$. Note that x^* is minimizer (maximizer) of f iff x^* is minimizer(maximizer) of f^2. Thus we need to optimize $f^2 = (x-2)^2 + (y+1)^2 + (z-2)^2$ subjected to $x^2 + 4y^2 + 3z^2 = 12$.
Using the Lagrange conditions, we get

$$2x - 4 = 2\lambda x,\ 2y + 2 = 8\lambda y,\ 2z - 4 = 6\lambda z,\ x^2 + 4y^2 + 3z^2 = 12.$$

Substituting the values $x = \frac{2}{1-\lambda}$, $y = \frac{-1}{1-4\lambda}$ and $z = \frac{2}{1-3\lambda}$ from the first three equations in the fourth, we get

$$\left(\frac{2}{1-\lambda}\right)^2+4\left(\frac{-1}{1-4\lambda}\right)^2+3\left(\frac{2}{1-3\lambda}\right)^2=12.$$

Solving the above equation (using sage) for λ we get

$$\lambda=0.109141290745 \text{ and } \lambda=1.60206763624.$$

For $\lambda=0.109141290745$, the solution (using sage) is $(1.80319680319680, 0.696104897801774, 1.50667779632721)$ and $f=0.374450528048604$.

For $\lambda=1.60206763624$, the solution is

$(3.32188591385332, 0.184901988043932, 0.525458060952030)$

and $f=36.1044008189684$.

http://matrix.skku.ac.kr/cal-lab/m-Sec13-9-Exm-4.html

```
var('x,y,z,lam')
f=(x-2)^2+(y+1)^2+(z-2)^2
g=x^2+4*y^2+3*z^2-12
fx=f.diff(x)
fy=f.diff(y)
fz=f.diff(z)
gx=g.diff(x)
gy=g.diff(y)
gz=g.diff(z)
cpoints=solve([fx==lam*gx,fy==lam*gy,fy==lam*gy,fz==lam*g
z,
 g==0],x,y,z,lam,solution_dict=True)
cpoints
for sol in cpoints:
    if ((sol[x] in RR) and (sol[y] in RR) and
    (sol[z] in RR) and (sol[lam] in RR) and
    (sol[lam] in RR)):
show([sol[x].n(),sol[y].n(),sol[z].n(),f(x=sol[x],y=sol[y],z=sol[z]
).n()])
```

■

We should plot the points along with the surfaces.

Now let us consider the following example which explains why do we need, $\nabla g \neq 0$ in Theorem 1.

EXAMPLE 5

Maximize/Minimize $f(x)=x$ subjected to $x^4+y^2-x^3=0$.

Solution

```
var('x,y,lam')
f(x)=x
g(x,y)=y^2+x^4-x^3
fx=f.diff(x)
fy=f.diff(y)
gx=g.diff(x)
gy=g.diff(y)
solve([fx==lam*gx,fy==lam*gy, g==0],x,y,lam)
```

Answer : [[x == 1, y == 0, lam == 1]]

Note that $(0, 0)$ is a solution of the above problem however, Lagrange method fails to detect this point. As $\nabla g(0, 0) = 0$ in this case. Look the Figure 6.

http://matrix.skku.ac.kr/cal-lab/m-Sec13-9-Exm-5.html

```
t=var('t')
cf=contour_plot(f,(x,-.2,1.2),(y,-0.5,0.5),contours=40,fill=false,cma
p='cool')
cg=implicit_plot(g,(x,-.2,1.2),(y,-0.5,0.5))
l1=parametric_plot((0,t),(t,-0.5,0.5),color='green')
l2=parametric_plot((1,t),(t,-0.5,0.5),color='red')
show(cf+cg+l1+l2,figsize=5)
```

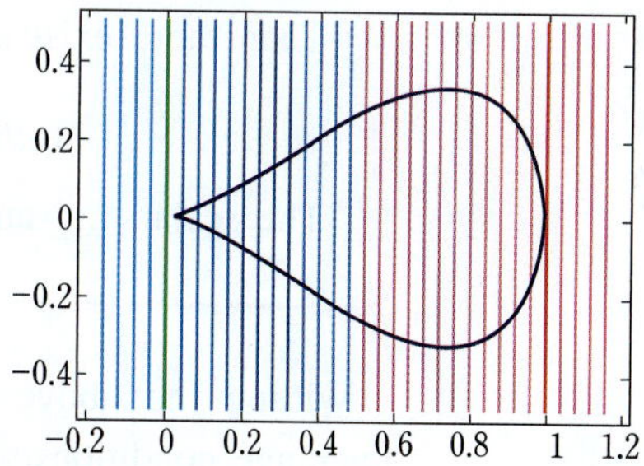

Figure 6

■

Lagrange Multipliers for Two and More Constraints

Here we consider problems of minimization or maximization with two constraints. The method can be generalized to more than two constraints in a similar manner. In particular, we assume that we want to find the maximum and minimum values of a function $f(x, y, z)$ subject to two constraints of the form $g(x, y, z) = 0$ and $h(x, y, z) = 0$. We assume that $\nabla g(x, y, z)$ and $\nabla h(x, , y, z)$ are non zero and not parallel (that is, they are linearly independent).

Geometrically, this means that we are looking for the extreme values of f that lies on the curve of intersection of the level surfaces $g(x, y, z) = 0$ and $h(x, y, z) = 0$.

Suppose, $P(x_0, y_0, z_0)$ is a point on the curve of intersection of the level surfaces $g(x, y, z) = 0$ and $h(x, y, z) = 0$ and at which f has an extreme value say $w_0 = f(x_0, y_0, z_0)$. Let us see what this mean in terms of gradients.

Since $P(x_0, y_0, z_0)$ is a point of extremum of f, $\nabla f(x_0, y_0, z_0)$ is orthogonal to the tangent vector to the level surface $f(x, y, z) = w_0$. On the other hand, we also have $\nabla g(x_0, y_0, z_0)$ is orthogonal to the level curve $g(x, y, z) = 0$ and $\nabla h(x_0, y_0, z_0)$ is orthogonal to the level curve $h(x, y, z) = 0$. This means that $\nabla f(x_0, y_0, z_0)$ lies in the plane spanned by two vectors $\nabla g(x_0, y_0, z_0)$ and $\nabla h(x_0, y_0, z_0)$. That is, there exist scalars α and β such that

$$\nabla f(x_0, y_0, z_0) = \alpha \nabla g(x_0, y_0, z_0) + \beta \nabla h(x_0, y_0, z_0).$$

We summarize this in the following theorem (without proof).

THEOREM 2

Let $f(x, y, z)$, $g(x, y, z)$ and $g(x, y, z)$ be three functions having first order continuous partial derivatives. If (x_0, y_0, z_0) is a solution of the problem

4 minimize/maximize $f(x, y, z)$ subjected to $g(x, y, z) = 0$ and $h(x, y, z) = 0$

where $\nabla g(x_0, y_0, z_0)$ and $\nabla h(x_0, y_0, z_0)$ are non zero and parallel. Then there exist scalars λ_1 and λ_2 such that

5 $$\nabla f(x_0, y_0, z_0) = \lambda_1 \nabla g(x_0, y_0, z_0) + \lambda_2 \nabla h(x_0, y_0, z_0).$$

The scalars λ_1 and λ_2 are called **Lagrange multipliers**.

Witting the above conditions (4) along with the constraints, we get the Lagrange conditions

6 $$\frac{\partial f}{\partial x} = \lambda_1 \frac{\partial g}{\partial x} + \lambda_2 \frac{\partial h}{\partial x}$$

7 $$\frac{\partial f}{\partial y} = \lambda_1 \frac{\partial g}{\partial y} + \lambda_2 \frac{\partial h}{\partial y}$$

8 $$\frac{\partial f}{\partial z} = \lambda_1 \frac{\partial g}{\partial z} + \lambda_2 \frac{\partial h}{\partial z}$$

9 $$g(x, y, z) = 0$$

10 $$h(x, y, z) = 0$$

Many time it is continent to define a Lagrangian function

11
$$L = f(x, y, z) - \lambda_1 g(x, y, z) - \lambda_2 h(x, y, z).$$

Then the Lagrange conditions (6 to 10) is equivalent to

12
$$\frac{\partial L}{\partial x} = 0,\ \frac{\partial L}{\partial y} = 0,\ \frac{\partial L}{\partial z} = 0 \text{ and } \frac{\partial L}{\partial \lambda_1} = 0,\ \frac{\partial L}{\partial \lambda_2} = 0.$$

EXAMPLE 6

The cone $x_3^2 = x_1^2 + x_2^2$ is cut by the plane $x_3 = 1 + x_1 + x_2$ in some curve C. Find the point on C that is closest to the origin.

This problem can be stated as

$$\text{minimize } f = x_1^2 + x_2^2 + x_3^2$$
$$\text{subjected to } h_1 = x_3^2 - x_1^2 - x_2^2 = 0,\ h_2 = x_3 - 1 - x_1 - x_2 = 0.$$

We define the Lagrangian function

$$\begin{aligned} L &= f - \lambda_1 h_1 - \lambda_2 h_2 \\ &= x_1^2 + x_2^2 + x_3^2 - \lambda_1 (x_3^2 - x_1^2 - x_2^2) - \lambda_2 (x_3 - 1 - x_1 - x_2). \end{aligned}$$

Using the necessary condition for optimization we get the following equations:

13
$$2x_1 + 2\lambda_1 x_1 + \lambda_2 = 0$$
14
$$2x_2 + 2\lambda_1 x_2 + \lambda_2 = 0$$
15
$$2x_3 - 2\lambda_1 x_3 - \lambda_2 = 0$$
16
$$x_3^2 = x_1^2 + x_2^2$$
17
$$x_3 = 1 + x_1 + x_2.$$

Now adding 15 to 13 and 14 we reduce the above system to 4 equations in 4 unknowns;

18
$$x_1 + \lambda_1 x_1 + x_3 - x_3 \lambda_1 = 0$$
19
$$x_2 + \lambda_1 x_2 + x_3 - x_3 \lambda_1 = 0.$$

Now we need to solve 16 - 19 for x_1, x_2, x_3, λ_1.

From 18 we get $\lambda_1 = \dfrac{x_3 + x_1}{x_3 - x_1}$ and from 19 we get $\lambda_1 = \dfrac{x_3 + x_2}{x_3 - x_2}$.

Equating the two values of λ_1 and simplifying we get

$$\frac{x_3 + x_1}{x_3 - x_1} = \frac{x_3 + x_2}{x_3 - x_2} \Rightarrow x_3(x_1 - x_2) = 0.$$

This implies either $x_3 = 0$ or $x_1 = x_2$.

Case I when $x_3 = 0$, from 16, we have

$$x_1^2 + x_2^2 = 0$$

and from [17], we get

$$1 + x_1 + x_2 = 0.$$

Using the above tow equations and solving for x_1 we get $2x_1^2 + 2x_1 + 1 = 0$ which does not have a real solution. Thus $x_3 = 0$ is not possible.
Case II when $x_1 = x_2$, from [16], we have

$$x_3^2 = x_1^2 + x_2^2 = 2x_1^2$$

and from [17], we get

$$x_3 = 1 + x_1 + x_2 = 1 + 2x_1.$$

Thus

$$2x_1^2 = x_3^2 = (1 + 2x_1)^2 \Rightarrow 2x_1^2 + 4x_1 + 1 = 0$$

Solving this equation for x_1 we get $x_1 = -1 \pm \frac{1}{\sqrt{2}}$. Now

$$x_3 = 1 + 2x_1 = -1 \pm \sqrt{2}.$$

Finally

$$x_1 = x_2 = -1 \pm \frac{1}{\sqrt{2}}.$$

Thus the critical points(possible optimizers) are

$$\left(-1 \pm \frac{1}{\sqrt{2}}, -1 \pm \frac{1}{\sqrt{2}}, -1 \pm \sqrt{2}\right).$$

Note that $f\left(-1 + \frac{1}{\sqrt{2}}, -1 + \frac{1}{\sqrt{2}}, -1 + \sqrt{2}\right) = 6 - \sqrt{2}$ and $f\left(-1 - \frac{1}{\sqrt{2}}, -1 - \frac{1}{\sqrt{2}}, -1 - \sqrt{2}\right) = 6 + \sqrt{2}$.
Hence $\left(-1 - \frac{1}{\sqrt{2}}, -1 - \frac{1}{\sqrt{2}}, -1 - \sqrt{2}\right)$ is a point of maximum.
Hence $\left(-1 + \frac{1}{\sqrt{2}}, -1 + \frac{1}{\sqrt{2}}, -1 + \sqrt{2}\right)$ is a point of minimum. ■
http://matrix.skku.ac.kr/cal-lab/m-Sec13-9-Exm-6.html

```
x, y, z, lam, mu = var('x, y, z, lam, mu')
f = x^2+y^2 + z^2
g1 = x^2 + y^2 - z^2 - 1
g2 = x + y +z - 4
L = f - g1 * lam - g2*mu
gradL = L.gradient([x, y, z, lam, mu])
cpoints = solve([gradL[0] == 0, gradL[1] == 0, gradL[2] == 0, gradL[3] == 0, gradL[4] == 0],x, y, z, lam, mu, solution_dict=True)
cpoints
```

```
n=len(cpoints)
for sol in cpoints:
    if ((sol[x] in RR) and (sol[y] in RR) and (sol[z] in RR) and (sol[lam] in RR) and (sol[mu] in RR)):
        show([sol[x].n(), sol[y].n(), sol[z].n(), f(x=sol[x],y=sol[y],z=sol[z]).n()])
```

Answer :
[1.26138721247417,1.26138721247417,1.47722557505167,5.36439079917344]
[6.73861278752583,6.73861278752583,−9.47722557505166,180.635609200827]

13.9 EXERCISES (Lagrange Multiplier)

http://matrix.skku.ac.kr/Cal-Book/part2/CS-Sec-13-9-Sol.html

YouTube http://youtu.be/YMGdQWBzyrI

1. In each of the following problems, find the extreme(maximum and minimum) values of f subjected to the given constrained:

(i) $f(x,y)=3x+2y$ subject to $x^2+y^2=1$.

(ii) $f(x,y)=x-2y$ subject to $x^2+y^2=25$.

(iii) $f(x,y)=x^2+2y^2$ subject to $x^2+y^2=1$.

(iv) $f(x,y)=\sin(xy)$ subject to $x^2+y^2=1$.

(v) $f(x,y)=xy$ subjected to $x+y=1$

(vi) $f(x,y)=x^2-y^2$ subjected to $x^2+y^2=25$.

(vii) $f(x_{1,}\ x_{2,}\ x_3)=x_1+x_2+x_3$ subjected to $x_1^2+x_2^2+x_3^2=1$.

Can you generalize this?

2. Find the points on the sphere $x^2+y^2+z^2=1$ that are closest and farthest from the point $(1,2,2)$.

3. Determine the dimensions of a rectangular box, open at the top having volume pf 32 cubic feets and requiring the least amount of material for its construction.

4. Find the maximum and minimum of
$f(x,\ y,\ z)=\ -x+2y+2z$
on the ellipse given by the intersection of the cylinder $x^2+y^2=2$ and the plane $y+2z=1$

5. Find the maximum and minimum of $f(x,y,z)=-x+2y+2z$ on the ellipse given by the intersection of the cylinder $x^2+y^2=2$ and the plane $y+2z=1$.

6. Find the maximum and ninimum of $f(x,\ y)=$ $=3x+2y$ when $2x^2+3y^2\le 3$.

7. In the following exercises find the extreme values of f subjected to the two constraints.

(a) $f=xz+yz;\ x+z=1;\ y^2+z^2=1$.

(b) $f=x^2+y^2;\ x-y=2;\ y^2-z^2=1$.

(c) $f=x^2+y^2+z^2;\ x+2y+3z=7;$

$$2x+2y+z=\frac{9}{2}$$

(d) $f=x+y+z;\ z=x^2-y^2;\ x^2+z^2=2$.

8. (a) Use Lagrange multipliers to find the highest and lowest points on the ellipse $\dfrac{x^2}{a^2}+\dfrac{y^2}{b^2}=1$.

(b) Use Lagrange multipliers to show that the rectangle with maximum area that has a given perimeter P is a square.

(c) Use Lagrange multipliers to show that the triangle with maximum area that has a given perimeter P is equilateral.

(d) Find the maximum and minimum volumes of a rectangular box whose surface area is $1200 cm^2$ and whose total edge length is 180cm$.

(e) Use Sage to ind the maximum of $z=f(x,y)=\left(x-y-\dfrac{1}{4}\right)e^{-x^2-y^2}$, subject to the constraint

$$g(x,y)=\left(x-\frac{1}{2}\right)^2+\left(y-\frac{1}{3}\right)^2=1.$$

9. Find the absolute maximum value and minimum value of the function $f(x,y)=xy$ on the ellipse $\dfrac{x^2}{4}+y^2=1$.

Solution $f(x,y)=xy\Rightarrow\nabla f(x,y)=y\mathbf{i}+x\mathbf{j}$

$g(x,y)=\dfrac{x^2}{4}+y^2-1\Rightarrow\lambda\nabla g(x,y)=\left(\dfrac{\lambda x}{2}\right)\mathbf{i}+2\lambda y\mathbf{j}$

Using $\nabla f=\lambda\nabla g,\ y=\dfrac{1}{2}\lambda x\Rightarrow\lambda=\dfrac{2y}{x}$

Substituting $\lambda=\dfrac{2y}{x}$ in $x=2\lambda y$, then $x=2\left(\dfrac{2y}{x}\right)y\Rightarrow$ $x^2=4y^2$

Substituting $x^2=4y^2$ in $\dfrac{x^2}{4}+y^2=1\Rightarrow\dfrac{1}{4}4y^2+y^2=1\Rightarrow$

$y^2=\dfrac{1}{2}\Rightarrow y=\pm\dfrac{1}{\sqrt{2}}$

$y=\dfrac{1}{\sqrt{2}}\Rightarrow x=\pm\sqrt{2}$

$y=-\dfrac{1}{\sqrt{2}}\Rightarrow x=\pm\sqrt{2}$

The absolute maximum value is

$f\left(\sqrt{2}, \frac{1}{\sqrt{2}}\right) = f\left(-\sqrt{2}, -\frac{1}{\sqrt{2}}\right) = \sqrt{2} \cdot \frac{1}{\sqrt{2}} = 1$

The absolute minimum value is

$f\left(-\sqrt{2}, \frac{1}{\sqrt{2}}\right) = f\left(\sqrt{2}, -\frac{1}{\sqrt{2}}\right) = -\sqrt{2} \cdot \frac{1}{\sqrt{2}} = -1.$

10. Find the absolute maximum value of the function $f(x, y) = 4xy \ (x > 0,\ y > 0)$ on the ellipse $\frac{x^2}{3^2} + \frac{y^2}{4^2} = 1.$

Solution $f(x, y) = 4xy \Rightarrow \nabla f(x, y) = y\mathbf{i} + x\mathbf{j}$

$g(x, y) = \frac{x^2}{3^2} + \frac{y^2}{4^2} - 1 \Rightarrow$

$\lambda \nabla g(x,y) = \left(\frac{2\lambda x}{9}\right)\mathbf{i} + \left(\frac{\lambda y}{8}\right)\mathbf{j}$ Using $\nabla f = \lambda \nabla g$,

$4y = \frac{2}{9}\lambda x \Rightarrow \lambda = \frac{18y}{x}$

Substituting $\lambda = \frac{18y}{x}$ in $4x = \frac{1}{8}\lambda y \Rightarrow 4x = \frac{1}{8}\left(\frac{18y}{x}\right)y$

$\Rightarrow x^2 = \frac{9}{16}y^2$ > Substituting $x^2 = \frac{9}{16}y^2$ in $\frac{x^2}{3^2} + \frac{y^2}{4^2} = 1$

$\Rightarrow \frac{1}{9}\left(\frac{9}{16}y^2\right) + \frac{1}{16}y^2 = 1 \Rightarrow y^2 = 8 \Rightarrow$

$y = 2\sqrt{2} \ (y > 0)$

Hence $y = 2\sqrt{2} \ (y > 0)$ and $x = \frac{3}{\sqrt{2}} \ (x > 0)$

Now the absolute maximum value of the function is $f\left(\frac{3}{\sqrt{2}}, 2\sqrt{2}\right) = 4 \cdot \frac{3}{\sqrt{2}} \cdot 2\sqrt{2} = 24$

<Alan Turing>

Chapter 14

Multiple Integrals

Georg Friedrich Riemann

1826-1866

Differential Geometry

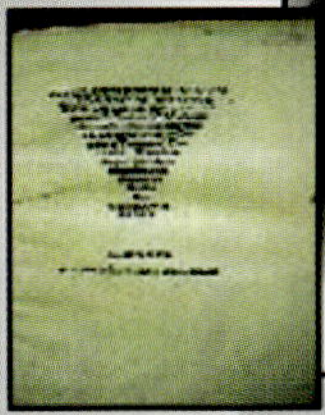

[Grundlagen für Eine Allgemeine Theorie]

A German mathematician, Riemann made lasting contributions to analysis and differential geometry. His published works opened up research areas combining analysis with geometry. These would subsequently become major parts of the theories of Riemannian geometry, algebraic geometry, and complex manifold theory.

YouTube http://youtu.be/5eCO2GjlJHs

In this chapter, we extend the definition of definite integral of single variable function to double and triple integrals of function of two or three variables. We give use the definition of double and triple integrals to compute areas and volumes of plane and solid regions. We introduce polar coordinate system in plane, which helps us to deal with some special type of regions. In a similar way, we introduce cylinder and spherical coordinate systems in space. This chapter also deals with the change of variable formula, which helps us to transform integral an from one coordinate system to another coordinate system.

14.1 Double Integrals

YouTube http://youtu.be/jZ2pAmPZYOE

In this section, we consider the integration of the function $f(x,y)$ on the domain D in the xy-plane, which is called a "double integral." This will be used to define the volume of the region under the graph of a function $f(x,y)$. We do not require $f(x,y) \geq 0$, but if $f(x,y)$ assumes negative values, we shall not interpret the double integral as a "volume". At first, we define the double integral over a rectangle by introducing the limit of Riemann sums and then we consider the double integral over the general domain in the xy-plane. Next, we deal with Fubini's Theorem which allows us to compute the double integral by computing iterated integrals over the interval. Finally, we introduce some applications of the double integral, for example, center of mass and total mass when the density function is given.

To begin, let us establish the notation for partitions and sums.

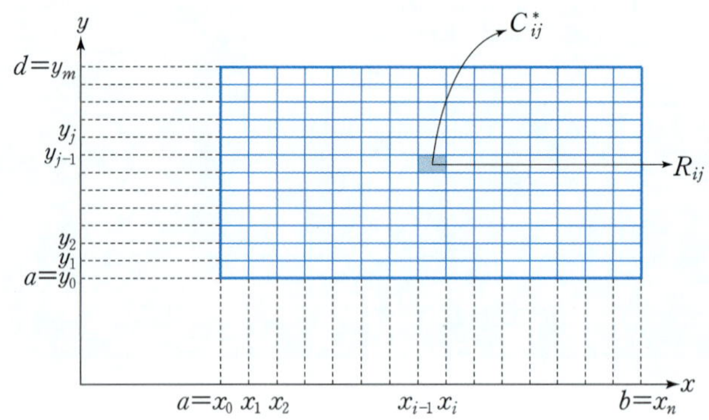

Figure 1

Consider a rectangle $R=[a,b]\times[c,d]$ in $\mathbb{R}^2$. A partition of R, is two collections of equally spaced points $\{x_i\}_{i=0,\cdots,n}$ and $\{y_j\}_{j=0,\cdots,m}$ such that

$$a=x_0<x_1<\cdots<x_n=b,\ c=y_0<y_1<\cdots<y_m=d$$

and

$$x_{k+1}-x_k=\frac{b-a}{n}\quad(0\leq k\leq n-1),$$

$$y_{l+1}-y_l=\frac{d-c}{m}\quad(0\leq l\leq m-1).$$

Let R_{ij} be the rectangle $[x_i,\ x_{i+1}]\times[y_j,\ y_{j+1}]$ and let $c_{ij}{}^*=(x_i{}^*,\ y_j{}^*)$ be any sample point in R_{ij}. If $f:\mathbb{R}^2\rightarrow\mathbb{R}$, then we form the following Riemann sum for f :

$$S_{n,m}=\sum_{j=0}^{m-1}\sum_{i=0}^{n-1} f(c_{ij}{}^*)\Delta x\Delta y=\sum_{j=0}^{m-1}\sum_{i=0}^{n-1} f(c_{ij}{}^*)\Delta A,$$

where $\Delta A=\Delta x\Delta y=\dfrac{b-a}{n}\dfrac{d-c}{m}$ is the area of the subrectangle R_{ij}.

Then we can consider the limit of $S_{n,m}$, as n and m tend to ∞.

1 $$S=\lim_{n,m\to\infty} S_{n,m}$$

If f is a bounded and continuous function on the rectangle R, then one can prove that $\lim\limits_{n,m\to\infty} S_{n,m}$ always exists and the limit S is independent of the choice of the sample point $c_{ij}{}^*$. This leads to the following definition.

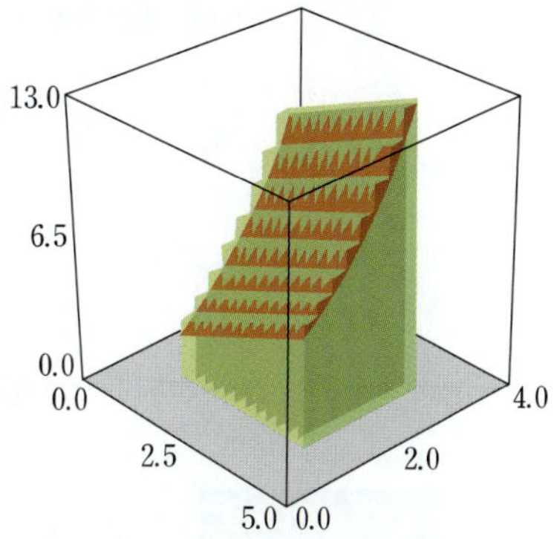

Figure 2

DEFINITION 1

Let f be a bounded real-valued function on a rectangle R. If the Riemann sum $S_{n,m}$ defined above converges to a limit S, independent of the choice of the sample point c_{ij}^*, then we say that f is integrable over $\boldsymbol{R}$ and we write

$$\int_R f,\ \int_R f(x,y)dA,\ \iint_R fdA,\ \iint_R f(x,y)dA \ \textbf{or}\ \iint_R f(x,y)dxdy$$

for the limit S.

Note We have considered double integral over R by dividing R into equal-sized rectangles. However, this is not necessary. One can arrive at the same definition by arbitrary partitions.

Note : If $f(x,y)\geq 0$ on a rectangle R and $f(x,y)$ is integrable then $\iint_R f(x,y)dA$ represents the volume of the solid that lies above R and below the surface $z=f(x,y)$.

EXAMPLE 1

Let $f:[0,1]\times[0,1]\to\mathbb{R}$ be defined by $f(x,y)=x$. Show that f is integrable and hence find $\iint_{[0,1]\times[0,1]} f(x,y)dA$.

Solution Choose a subrectangle $\left[\frac{i}{n}, \frac{i+1}{n}\right]\times\left[\frac{j}{m}, \frac{j+1}{m}\right]$, then we form the Riemann sum

$$S_{n,m} = \sum_{j=0}^{m-1}\sum_{i=0}^{n-1}\left(\frac{i^*}{n}\right)\frac{1}{nm} \text{ where } i \le i^* \le i+1.$$

Now

$$\sum_{j=0}^{m-1}\sum_{i=0}^{n-1}\left(\frac{i}{n}\right)\frac{1}{nm} \le S_{n,m} \le \sum_{j=0}^{m-1}\sum_{i=0}^{n-1}\left(\frac{i+1}{n}\right)\frac{1}{nm}.$$

From the formula $\sum_{i=1}^{n} i = \frac{n(n+1)}{2}$, we have $\frac{n-1}{2n} \le S_{n,m} \le \frac{n+1}{2n}$. By the Squeezing Theorem, we have

$$S = \iint_{[0,1]\times[0,1]} f(x,y)dA = \frac{1}{2}.$$

Thus f is integrable on R and $\iint_{[0,1]\times[0,1]} f(x,y)dA = \frac{1}{2}$. ■

EXAMPLE 2

Let $f(x,y) = xy$ on $R = [0,1]\times[0,1]$. Show that f is integrable on R and hence find its integral.

Solution Choose a subrectangle $\left[\frac{i}{n}, \frac{i+1}{n}\right]\times\left[\frac{j}{m}, \frac{j+1}{m}\right]$, then we form the Riemann sum

$$S_{n,m} = \sum_{j=0}^{m-1}\sum_{i=0}^{n-1}\left(\frac{i^*}{n}\right)\left(\frac{j^*}{m}\right)\frac{1}{nm},$$

where $i \le i^* \le i+1$ and $j \le j^* \le j+1$.

Now

$$\sum_{j=0}^{m-1}\sum_{i=0}^{n-1}\left(\frac{i}{n}\right)\left(\frac{j}{m}\right)\frac{1}{nm} \le S_{n,m} \le \sum_{j=0}^{m-1}\sum_{i=0}^{n-1}\left(\frac{i+1}{n}\right)\left(\frac{j+1}{m}\right)\frac{1}{nm}.$$

From the formula $\sum_{i=1}^{n} i = \frac{n(n+1)}{2}$ and $\sum_{j=1}^{n} j = \frac{n(n+1)}{2}$, we have $\left(\frac{n-1}{2n}\right)\left(\frac{m-1}{2m}\right) \le S_{n,m} \le \left(\frac{n+1}{2n}\right)\left(\frac{m+1}{2m}\right)$. By the Squeezing Theorem, we have

$$S = \iint_R f(x,y)dA = \frac{1}{4}.$$

Thus f is integrable on R and $\iint_R f(x,y)dA = \frac{1}{4}$. ■

Properties of a Double Integral

From the definition of the double integral, one can guess that double integrals share properties with the single variable definite integral. These properties are similar to proof in case of single variables.

THEOREM 2

Let f and g be integrable functions on the rectangle R, and let c be a constant. Then $f+g$ and cf are integrable on R and

(i)(Linearity)
$$\iint_R [f(x,y)+g(x,y)]\,dA = \iint_R f(x,y)\,dA + \iint_R g(x,y)\,dA.$$
(ii)(Homogeneity) $\iint_R cf(x,y)\,dA = c\iint_R f(x,y)\,dA$.

(iii)(Monotonicity) If $g(x,y) \le f(x,y)$, for all $(x,y) \in R$ then,
$$\iint_R g(x,y)\,dA \le \iint_R f(x,y)\,dA.$$
(iv) If $m \le f(x,y) \le M$ for all (x,y) in R, then
$$m \cdot \text{Area}(R) \le \iint_R f(x,y)\,dA \le M \cdot \text{Area}(R).$$
(v) Furthermore, if $f(x,y)$ is continuous on R, then there exists at least one point (x_0, y_0) in $\boldsymbol{R}$ such that

$\iint_R f(x,y)dA = f(x_0, y_0) \cdot \text{Area}(R)$.If n is a positive integer, then

The property (v) of Theorem 2 is called the "mean value theorem for double integrals".

All the above properties can be proved by using the definition of the integrals and the limit of sums, the limit theorems and the definition of continuity.

Double Integral over a General Domain.

We want to extend the definition of a double integral to functions which are defined on more general regions. We restrict our attention to the "elementary domains". We say the following two types of domains are "elementary domains".

Ⅰ. Suppose that we are given two continuous functions $\phi_1 : [a, b] \to \mathbb{R}$, $\phi_2 : [a, b] \to \mathbb{R}$ that satisfy $\phi_1(x) \leq \phi_2(x)$ for all $x \in [a, b]$.
Let $D = \{(x,y) \in \mathbb{R}^2 \mid a \leq x \leq b,\ \phi_1(x) \leq y \leq \phi_2(x)\}$. (See Figure 3.)
In this case D is called a type Ⅰ domain.

Ⅱ. Suppose that we are given two continuous functions $\psi_1 : [c, d] \to \mathbb{R}$, $\psi_2 : [c, d] \to \mathbb{R}$ that satisfy $\psi_1(y) \leq \psi_2(y)$ for all $y \in [c, d]$.
Let $D = \{(x,y) \in \mathbb{R}^2 \mid \psi_1(y) \leq x \leq \psi_2(y),\ c \leq y \leq d\}$. (See Figure 4.)
We call such a domain a type II domain.

The curves or straight lines that bound the region constitute the boundary of D, denoted by ∂D.

Note that there are some regions which can be expressed as both type in Ⅰ and Ⅱ.

For example, the rectangle $R = [a,b] \times [c,d]$ and disk $x^2 + y^2 \leq a^2$ are of both types.

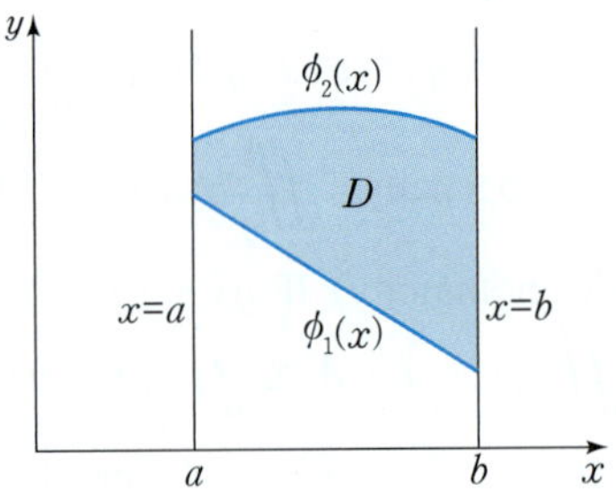

Figure 3

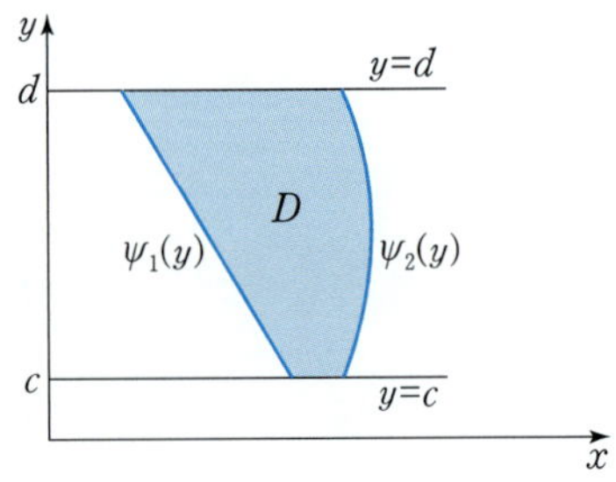

Figure 4

Once we define the double integral over elementary domains, then this definition can be extended to an arbitrary domains in the following ways.

Let $f : D \subseteq \mathbb{R}^2 \to \mathbb{R}$ be a continuous and bounded function on an arbitrary domain in $\mathbb{R}^2$. Let R be a rectangle which contains D. Now we extend f on D to a function f^* defined on R as follows:

$$f^*(x,y) = \begin{cases} f(x,y), & (x,y) \in D \\ 0, & (x,y) \notin D, \text{ and } (x,y) \in \mathbb{R}^2 \end{cases}$$

Then f^* is bounded and continuous except on the boundary ∂D of D. Now we can define Riemann sum $S_{n,m}$ for f^* defined on R. If n and m are very large, then only negligible number of subrectangles meet ∂D (because D is elementary domain). Thus if f^* is Riemann integrable on R, then we say that f is Riemann integrable on D. (This is not a rigorous proof, but a rigorous proof depends on this idea.)
Thus for a continuous f on D, we can define the double integral on D as follows:

$$\iint_D f(x,y)dA = \iint_R f^*(x,y)dA.$$

We defined the double integral over the domain D by choosing a rectangle R containing D. But intuitively, the value of the double integral is independent of the choice of the rectangle because outside R, f^* is zero.

Iterated Integral

As in the case of one variable, evaluation a double integral from the definition is not an easy task. However, double integrals under certain conditions can be evaluated by two single variable integrals.

Let f is an integrable function on a rectangle $R=[a,b]\times[c,d]$. If we fix y in $[c,d]$, then $f(x,y)$ is a single variable function of x on $[a,b]$. Therefore, we can define $\int_a^b f(x,y)\,dx$ as a Riemann integral of one variable if it exists. Note that $\int_a^b f(x,y)\,dx$ is a function of y. We can denote by $A(y)$. Now we can integrate $A(y)$ on $[c,d]$ if it exists and we get

2 $$\int_c^d A(y)dy = \int_c^d \left[\int_a^b f(x,y)\,dx\right]dy.$$

The right hand side of 2 is called an iterated integral. It is standard practice to omit square brackets in 2.

Thus

$$\int_c^d\int_a^b f(x,y)dxdy = \int_c^d\left[\int_a^b f(x,y)\,dx\right]dy$$

means that first we integrate with respect to x in $[a,b]$ and then with respect to y in $[c,d]$.

Similarly one can define

$$\int_a^b\int_c^d f(x,y)dydx = \int_a^b\left[\int_c^d f(x,y)\,dy\right]dx.$$

It turns out that in most of the cases the two iterated integrals $\int_c^d\int_a^b f(x,y)dxdy$ and $\int_a^b\int_c^d f(x,y)dydx$ are equal.

The following theorem[1] (due to Guido Fubini's Theorem, 1907) gives a way sto evaluate double integrals.

1) http://en.wikipedia.org/wiki/Fubini's_theorem

THEOREM 3 Fubini's Theorem

Let f be a continuous function defined on the a rectangle $R=[a,b]\times[c,d]$. Then

$$\boxed{3} \quad \iint_R f(x,y)dA = \int_a^b \int_c^d f(x,y)dydx = \int_c^d \int_a^b f(x,y)dxdy.$$

Theorem 3 can be proved using the mean value theorem for integrals in the small subrectangles. We omit the details of the proof.

Although Theorem 3 is named after Guido Fubini, it was certainly known to Cauchy and his contemporaries for continuous functions. Fubini proved that the above equality is true for general functions. The original form of the Fubini's Theorem is beyond the scope of this book.

CAS **EXAMPLE 3**

Evaluate the following integrals.

(a) $\int_0^1 \int_1^3 (1+4xy)dy\,dx$ (b) $\int_1^3 \int_0^1 (1+4xy)dx\,dy$

Solution (a) We first integrate with respect to y and then with respect to x

$$\int_0^1 \int_1^3 (1+4xy)dy\,dx = \int_0^1 [y+2xy^2]_1^3 dx$$
$$= \int_0^1 (2+16x)dx = 10.$$

http://matrix.skku.ac.kr/cal-lab/cal-14-1-Exm-2.html

```
var('x')
f(x) = 2 + 16*x
integral(f, x, 0, 1)
```

```
var('x, y')
f(x, y) = 1 + 4*x*y
integral(integral(f, y, 1, 3), x, 0, 1)
```

Answer : 10

(b) We first integrate with respect to x and then with respect to y

$$\int_1^3 \int_0^1 (1+4xy)dx\,dy = \int_1^3 [x+2x^2y]_0^1 dy$$
$$= \int_1^3 (1+2y)dy = 10.$$

```
var('x, y')
f(x, y) = 1 + 4*x*y
integral(integral(f, x, 0, 1), y, 1, 3)
```

Answer : 10 ■

Thus we see that the two integrals in Example 3 are equal.

CAS **EXAMPLE 4**

Evaluate $\int_0^1 \int_0^1 (x^3 + y^2)\,dx\,dy$.

Solution By iterated integral Theorem, we have

$$\int_0^1 \int_0^1 (x^3 + y^2)\,dx\,dy = \int_0^1 \left[\int_0^1 (x^3 + y^2)dx\right]dy = \int_0^1 \left(\frac{1}{4} + y^2\right)dy$$
$$= \left[\frac{1}{4}y + \frac{1}{3}y^3\right]_0^1 = \frac{7}{12}.$$

http://matrix.skku.ac.kr/cal-lab/cal-14-1-Exm-3.html

```
var('x, y')
f(x, y) = x^3 + y^2
integral(integral(f, x, 0, 1), y, 0, 1)
```

Answer : 7/12 ■

If we consider the double integral over elementary domains, then we may rewrite the iterated integral.

Suppose D is type Ⅰ domain. (See Figure 3.)

$$D = \{(x,y) \in \mathbb{R}^2 \mid a \le x \le b,\ \phi_1(x) \le y \le \phi_2(x)\}.$$

and the extension f^* is zero outside of D. Then

$$\iint_D f(x,\,y)\,dA = \int_a^b \int_{\phi_1(x)}^{\phi_2(x)} f(x,\,y)\,dy\,dx\ .$$

If D is a type Ⅱ domain. (See Figure 4.)

$$D = \{(x,y) \in \mathbb{R}^2 \mid \psi_1(y) \le x \le \psi_2(y),\ c \le x \le d\ \}.$$

Let f^* be an extension of f on $D \subset R$. Then

$$\iint_D f(x,\,y)\,dA = \int_c^d \int_{\psi_1(y)}^{\psi_2(y)} f(x,\,y)\,dx\,dy\ .$$

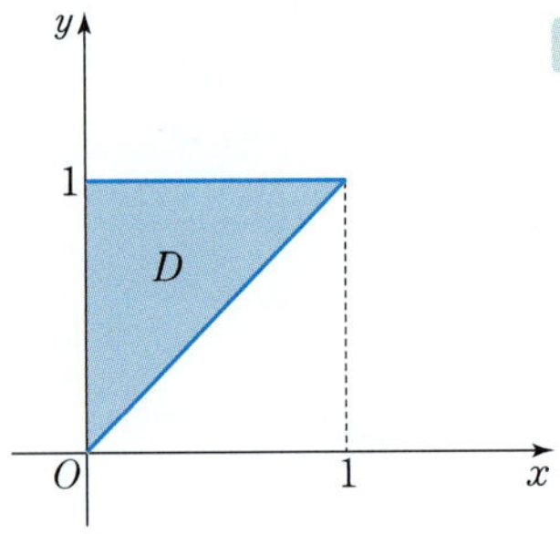

Figure 5

CAS **EXAMPLE 5**

Evaluate the integral $\iint_D (x^3+y^2)dA$

where $D=\{(x,y)\in \mathbb{R}^2 \mid 0\le x\le y,\ 0\le y\le 1\}$. (See Figure 5.)

Solution By Fubini's Theorem, we have

$$\iint_D (x^3+y^2)dA=\int_0^1\left[\int_0^y (x^3+y^2)dx\right]dy=\int_0^1\left(\frac{y^4}{4}+y^3\right)dy$$

$$=\left[\frac{1}{20}y^5+\frac{1}{4}y^4\right]_0^1=\frac{3}{10}.$$

```
var('x, y')
f(x, y) = x^3 + y^2
integral(integral(f, x, 0, y), y, 0, 1)
```

Answer : 3/10 ■

EXAMPLE 6

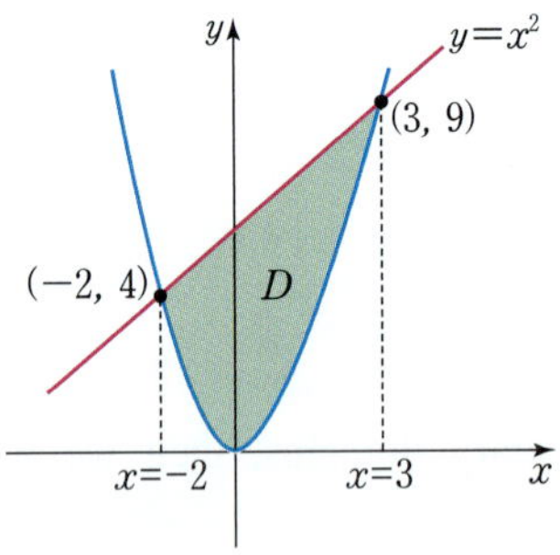

Figure 6

Evaluate $\iint_D x\,dA$, where R is the region bounded by the parabola $y=x^2$ and the line $y=x+6$.

Solution The two curves intersect at $(-2,4)$, $(3,9)$. (See Figure 6.) We note that the region D. So, we can write

$$D=\{(x,y)\mid -2\le x\le 3,\ x^2\le y\le x+6\}$$

$$\iint_D x\,dA=\int_{-2}^3\int_{x^2}^{x+6} x\,dy\,dx=\int_{-2}^3 [xy]_{x^2}^{x+6}dx$$

$$=\int_{-2}^3 (x^2+6x-x^3)dx=\frac{125}{12}.$$ ■

EXAMPLE 7

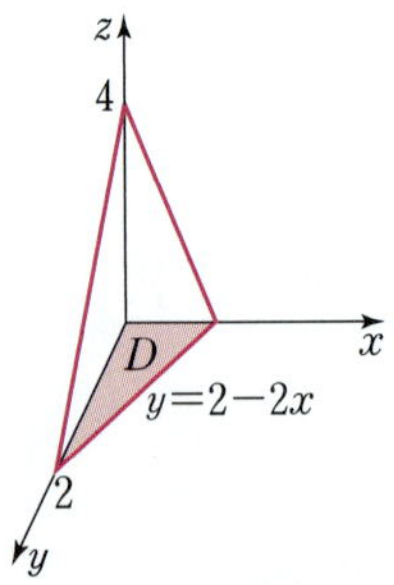

Figure 7

Find the volume of the tetrahedron bounded by the coordinate planes and plane $z=4-4x-2y$ and $z=0$.

Solution We plot the tetrahedron in Figure 7. We want to find the region D in the xy-plane over which the tetrahedron lies. This region D is obtained by projecting the plane $z=4-4x-2y$ onto the xy-plane. The region D is given by

$$D=\left\{(x,y)\mid 0\le x\le 1-\frac{y}{2},\ 0\le y\le 2\right\}$$

$$V=\iint_D (4-4x-2y)dA=\int_0^2\int_0^{1-\frac{y}{2}}(4-4x-2y)dx\,dy$$

$$= \int_0^2 \left[4x - 2x^2 - 2xy\right]_0^{1-\frac{y}{2}} dy = \int_0^2 2\left(1 - \frac{y}{2}\right)^2 dy = \frac{4}{3}.$$ ■

If $f(x, y) \geq 0$ on D, we can interpret the integral $\iint_D f(x, y)dA$ as the volume of the three-dimensional region between the graph of f and the domain D.

CAS **EXAMPLE 8**

Evaluate $\int_0^{\frac{\pi}{2}} \int_0^{\pi} x \sin x \cos y \, dx dy$.

http://matrix.skku.ac.kr/cal-lab/cal-14-1-Exm-5.html

Solution

```
var('x, y')
f(x, y)=x*sin(x)*cos(y)
integral(integral(f, x, 0, pi), y, 0, pi/2)
```

Answer : π

After interchanging the limit

```
integral(integral(f, y, 0, pi/2), x, 0, pi)
```

Answer : π ■

Changing the Order of the Integration

For a domain that can be thought of as type I as well as type II, the domain D can be written as

$$D = \{(x,y) \in \mathbb{R}^2 \mid a \leq x \leq b,\ \phi_1(x) \leq y \leq \phi_2(x)\}$$

and it can be also written as

$$D = \{(x,y) \in \mathbb{R}^2 \mid \psi_1(y) \leq x \leq \psi_2(y),\ c \leq y \leq d\}$$

for some functions ϕ_1, ϕ_2 and ψ_1, ψ_2.
Thus,

$$\iint_D f(x,y)dA = \int_a^b \int_{\phi_1(x)}^{\phi_2(x)} f(x,y)dydx = \int_c^d \int_{\psi_1(y)}^{\psi_2(y)} f(x,y)dxdy.$$

It is often useful to change the order of the iterated integrals. Many times the integral becomes easier to evaluate by interchanging the order of the iterated intervals.

EXAMPLE 9

Evaluate the integral $\int_0^1 \int_x^1 \frac{1}{(1+y^2)^2} dy dx$.

Solution It is not easy to evaluate the above integral in the above order. Domain D of the integral is

$$D = \{(x, y) \in \mathbb{R}^2 \mid 0 \le x \le 1 \text{ and } x \le y \le 1\}.$$

(See Figure 8.)
This domain can also be written as

$$D = \{(x, y) \in \mathbb{R}^2 \mid 0 \le y \le 1 \text{ and } 0 \le x \le y\}$$

(See Figure 9.)

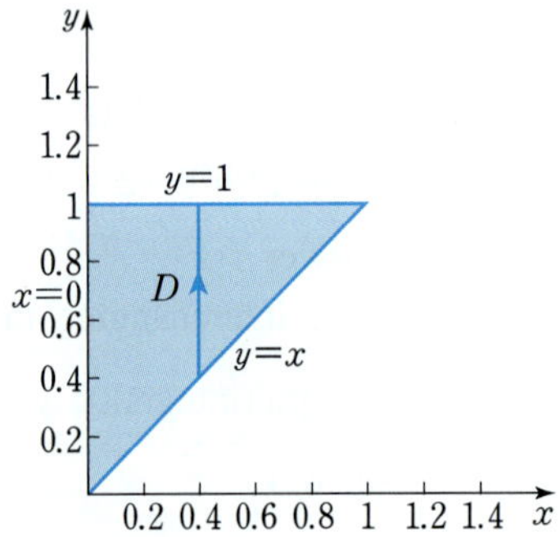

Figure 8 Domain for integral

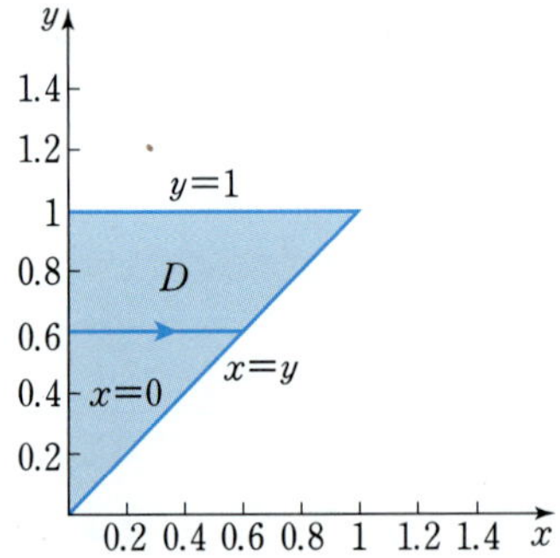

Figure 9 Domain for integral

If we change the order of the integration, then we have

$$\int_0^1 \int_x^1 \frac{1}{(1+y^2)^2} dy dx = \int_0^1 \int_0^y \frac{1}{(1+y^2)^2} dx dy$$
$$= \int_0^1 \frac{y}{(1+y^2)^2} dy = \left[-\frac{1}{2(1+y^2)} \right]_0^1 = \frac{1}{4}.$$

http://matrix.skku.ac.kr/cal-lab/cal-14-1-Exm-6.html

```
var('x, y')
f(x, y)=1/(1+ y^2)^2
integral(integral(f, x, 0, y), y, 0, 1)
```

Answer : 1/4 ■

Suppose that $f(x,y) = f_1(x) f_2(y)$ and

$$R = \{(x,y) \mid a \le x \le b,\ c \le y \le d\} = [a,b] \times [c,d].$$

Then Fubini's Theorem gives

4 $$\iint_R f(x,y)\, dA = \left\{ \int_a^b f_1(x)\, dx \right\} \left\{ \int_c^d f_2(y)\, dy \right\}.$$

$$\iint_R f(x,\ y)\,dA = \int_c^d \int_a^b f_1(x) f_2(y)\,dx\,dy$$
$$= \int_c^d \left[\int_a^b f_1(x) f_2(y)\,dx \right] dy$$
$$= \int_c^d \left[f_2(y) \left\{ \int_a^b f_1(x)\,dx \right\} \right] dy$$
$$= \left\{ \int_a^b f_1(x)\,dx \right\} \left\{ \int_c^d f_2(y)\,dy \right\}.$$

CAS EXAMPLE 10

Evaluate the integral $\iint_R xy\,dA$, where $R = [0,1] \times [0,1]$.

Solution By Fubini's Theorem,

$$\iint_R xy\,dA = \int_0^1 x\,dx \int_0^1 y\,dy = \left[\frac{1}{2}x^2\right]_0^1 \left[\frac{1}{2}y^2\right]_0^1 = \frac{1}{2} \cdot \frac{1}{2} = \frac{1}{4}.$$

http://matrix.skku.ac.kr/cal-lab/cal-14-1-Exm-7.html

```
var('x, y')
f(x)=x
g(y)=y
integral(f, x, 0, 1)*integral(g, y, 0, 1)
```

Answer : 1/4 ■

In Figure 10 illustrates the division of a region into subregions R_1 and R_2 for which $R = R_1 \cup R_2$. The regions R_1 and R_2 can have no points in common except possibly on their common border which $R_1 \cap R_2 = \varnothing$. Then

$$\iint_R f(x,y)\,dA = \iint_{R_1} f(x,y)\,dA + \iint_{R_2} f(x,y)\,dA.$$

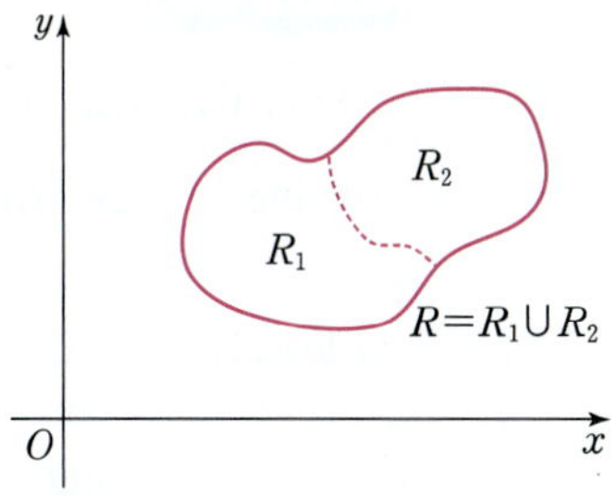

Figure 10 R is the union of two disjoint regions R_1 and R_2.

CAS EXAMPLE 11

Evaluate $\iint_R (x-1)\,dx\,dy$, where R is enclosed by $y = x$ and $y = x^3$. (See Figure 11.)

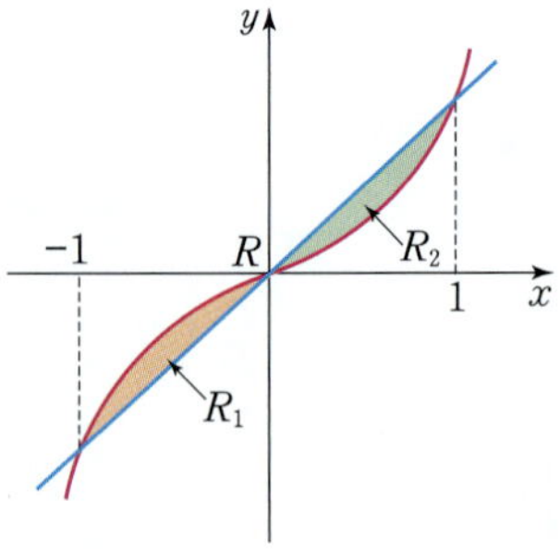

Figure 11

Solution $R=\{(x,y)\mid -1\le x\le 0,\ x\le y\le x^3\}$
$\cup\{(x,y)\mid 0\le x\le 1,\ x^3\le y\le x\}$

$$\iint_R (x-1)\,dA=\iint_{R_1}(x-1)\,dA+\iint_{R_2}(x-1)dA$$
$$=\int_{-1}^{0}\int_{x}^{x^3}(x-1)\,dy\,dx+\int_0^1\int_{x^3}^{x}(x-1)dy\,dx$$
$$=\int_{-1}^{0}[(x-1)y]_x^{x^3}dx+\int_0^1[(x-1)y]_{x^3}^{x}dx$$
$$=\int_{-1}^{0}(x-1)(x^3-x)\,dx+\int_0^1(x-1)(x-x^3)\,dx$$
$$=\int_{-1}^{0}(x^4-x^3-x^2+x)\,dx$$
$$+\int_0^1(-x^4+x^3+x^2-x)\,dx=-\frac{1}{2}.$$

http://matrix.skku.ac.kr/cal-lab/cal-14-1-Exm-8.html

```
var('x, y')
f(x, y)=x-1
g(x, y)=x-1
integral(integral(f, y, x, x^3), x, -1, 0) +
integral(integral(g, y, x^3, x), x, 0, 1)
```

Answer : -1/2 ■

Areas of Regions by Doubles Integrals

Recall that the integral of the constant function $f(x,y)=1$ over a region D, it gives the area of D:

$$\text{Area}(D)=\iint_D dA.$$

EXAMPLE 12

Find the area of the region D enclosed between the parabola $y=\frac{1}{2}x^2$ and the line $y=2x$. (See Figure 12.)

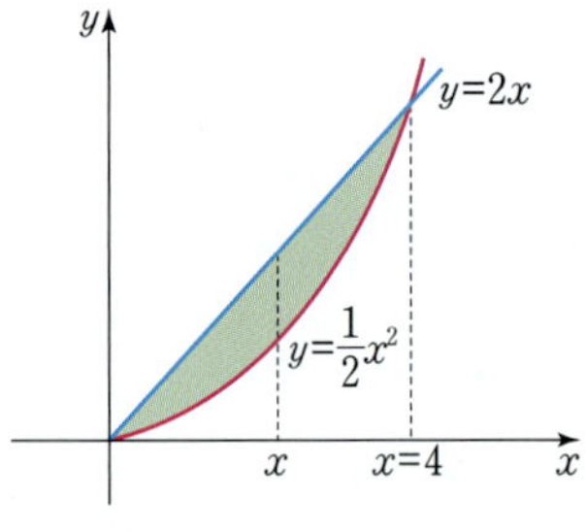

Figure 12

Solution $D=\left\{(x,y)\mid 0\le x\le 4,\ \frac{1}{2}x^2\le y\le 2x\right\}$

$$\text{Area}(D)=\iint_D dA=\int_0^4\int_{\frac{1}{2}x^2}^{2x}dy\,dx=\int_0^4[\,y\,]_{\frac{1}{2}x^2}^{2x}dx$$
$$=\int_0^4\left(2x-\frac{1}{2}x^2\right)dx=\left[x^2-\frac{x^3}{6}\right]_0^4=\frac{16}{3}.$$ ■

Average Value of a Function in Plane

Recall that the average value of a function f of defined on an interval $[a,b]$ is

$$f_{\text{ave}} = \frac{1}{b-a}\int_a^b f(x)\,dx\,.$$

Similarly, we define the average value of an integrable function f of two variables defined on a region R is

$$f_{\text{ave}} = \frac{1}{\text{Area}(R)}\iint_R f(x,y)\,dA\,.$$

EXAMPLE 13

Find the average value of $f(x,y) = 16 - 3x^2 - 3y^2$ over the region $R = \{(x,y) \mid 0 \le x \le 2,\ 0 \le y \le 2\}$.

Solution The area of the region R is 4.
The average value of f is

$$\begin{aligned} f_{\text{ave}} &= \frac{1}{4}\iint_R (16 - 3x^2 - 3y^2)\,dA \\ &= \frac{1}{4}\int_0^2\int_0^2 (16 - 3x^2 - 3y^2)\,dy\,dx \\ &= \frac{1}{4}\int_0^2 \left[16y - 3x^2y - y^3\right]_0^2 dx = \frac{1}{4}\int_0^2 (24 - 6x^2)\,dx \\ &= \frac{1}{4}\left[24x - 2x^3\right]_0^2 = 8. \end{aligned}$$

■

EXAMPLE 14

Find the volume of the ellipsoidal solid $\dfrac{x^2}{a^2} + \dfrac{y^2}{b^2} + \dfrac{z^2}{c^2} = 1$.

Solution When we slice the ellipsoidal solid by the plane that is through $(x,0,0)$ and parallel to yz-plane, the intersection is an ellipse by

$$\frac{y^2}{b^2} + \frac{z^2}{c^2} = 1 - \frac{x^2}{a^2} = \left(\sqrt{1 - \frac{x^2}{a^2}}\right)^2$$

The area of this cross-section is

$$A(x) = \pi\left(b\sqrt{1 - \frac{x^2}{a^2}}\right)\left(c\sqrt{1 - \frac{x^2}{a^2}}\right) = \pi bc\left(1 - \frac{x^2}{a^2}\right).$$

Recall that on area of ellipse $x^2/a^2 + y^2/b^2 = 1$ is πab.
Hence the volume of the ellipsoidal solid is

$$V = \int_{-a}^{a} A(x)\,dx = \int_{-a}^{a} \pi bc\left(1 - \frac{x^2}{a^2}\right)dx$$

$$= \pi bc\left[x - \frac{1}{3a^2}x^3\right]_{-a}^{a} = \frac{4}{3}\pi abc. \qquad \blacksquare$$

Applications of the Double Integral

We have already seen an application of double integrals of special types of solids: computing volumes. This section will allow us to think about physical applications such as computing mass, electric charge, center of mass, and moment of inertia. We will see that these physical ideas are also important when applied to the joint probability density functions of two random variables.

If the region is a rectangle, the probability that X lies between a and b and Y lies between c and d is

6 $$P(a \le X \le b,\ c \le Y \le b) = \int_a^b \int_c^d f(x, y)\,dydx.$$

Here, $f(x, y)$ is the joint probability density function that satisfy $f(x, y) \ge 0$, $\iint_R f(x, y)dA = 1$.

Let $f(x, y)$ be the density function at the point $(x, y) \in D$. Then the total mass M of the lamina in D can be written as

7 $$M = \iint_D f(x, y)dA.$$

The center of mass $(\bar{x}, \bar{y})$ can be defined by

8 $$\bar{x} = \frac{1}{M}\iint_D xf(x, y)dA$$

and

9 $$\bar{y} = \frac{1}{M}\iint_D yf(x, y)dA.$$

Moments of inertia with respect to x-axis and y-axis, I_x and I_y, respectively, can be defined by the following formulas,

$$I_x = \iint_D y^2 f(x, y)dA \text{ and } I_y = \iint_D x^2 f(x, y)dA.$$

The moment of inertia with respect to the origin is the sum of I_x and I_y as follows:

$$I_o = \iint_D (x^2 + y^2) f(x, y)\,dA.$$

I_o is called the polar moment of inertia.

EXAMPLE 15

Find the value of constant k and $P[X+Y\leq 1]$ for the following joint probability density function:

$$f(x,\ y)=\begin{cases} ke^{-x}, & 0\leq y\leq x<\infty \\ 0, & \text{otherwise} \end{cases}.$$

Solution The constant k is found from $\iint_{R^2} f(x,\ y)\,dA=1$:

$$1=\int_0^{\infty}\int_0^{x} ke^{-x}\,dy\,dx=\int_0^{\infty} kxe^{-x}\,dx=k.$$

Therefore $k=1$. (See Figure 13(a).)

Since $f(x,y)=e^{-x}\,(0\leq y\leq x<\infty)$,

$$\begin{aligned} P[X+Y\leq 1] &= \int_0^{1/2}\int_y^{1-y} e^{-x}\,dx\,dy \\ &= \int_0^{1/2}(e^{-y}-e^{-(1-y)})dy \\ &= 1+\frac{1}{e}-\frac{2}{\sqrt{e}} \\ &\approx 0.154818. \text{ (See Figure 13(b).)} \end{aligned}$$

■

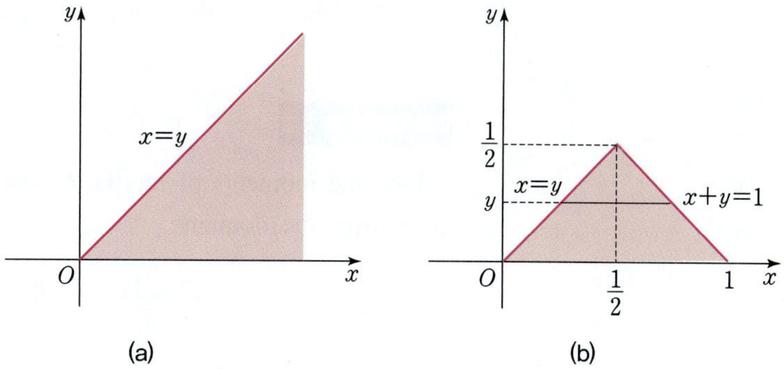

Figure 13

EXAMPLE 16

Find the mass and center of mass of a triangular lamina with vertices $(0,0)$, $(1,0)$ and $(0,1)$ if the density function is $f(x,y)=x$.

Solution See Figure 14. The mass of a triangular lamina is

$$\begin{aligned} M &= \iint_D f(x,y)\,dA=\int_0^1\int_0^{1-x} x\,dy\,dx \\ &= \int_0^1 (x-x^2)dx=\frac{1}{6}. \end{aligned}$$

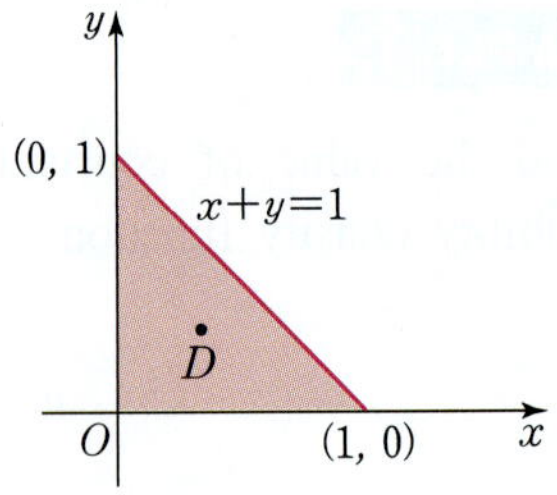

Figure 14

Then [8] and [9] give

$$\bar{x} = \frac{1}{M}\iint_D x\,f(x,y)\,dA = 6\int_0^1\int_0^{1-x} x^2\,dy\,dx$$
$$= 6\int_0^1 (x^2 - x^3)\,dx$$
$$= \frac{1}{2},$$
$$\bar{y} = \frac{1}{M}\iint_D y\,f(x,y)\,dA = 6\int_0^1\int_0^{1-x} x\,y\,dy\,dx$$
$$= 6\int_0^1 \frac{1}{2}(x - 2x^2 + x^3)\,dx$$
$$= \frac{1}{4}.$$

Therefore the center of mass is at the point $\left(\frac{1}{2}, \frac{1}{4}\right)$. ■

EXAMPLE 17

Find the moments of inertia I_x and I_y if the density function $f(x,y) = 10$ is given over the domain

$$D = \{(x,y) \mid 0 \le x \le 1,\ 0 \le y \le 2\}.$$

Solution By the definition of the moment of inertia,

$$I_x = \iint_D y^2 f(x,y)\,dA = \int_0^1\int_0^2 10y^2\,dy\,dx = \frac{80}{3},$$
$$I_y = \iint_D x^2 f(x,y)\,dA = \int_0^1\int_0^2 10x^2\,dy\,dx = \frac{20}{3}.$$ ■

14.1 EXERCISES (Double Integrals)

http://matrix.skku.ac.kr/Cal-Book/part2/CS-Sec-14-1-Sol.htm

YouTube http://youtu.be/w8g9fgcEP4A

1-3. Evaluate the given iterated integrals:

1. $\int_0^1 \int_0^y (2x+3y)^2 \, dx \, dy$

Solution

$\int_0^1 \int_0^y (2x+3y)^2 dx\, dy = \int_0^1 \frac{1}{6}((5y)^3 - (3y)^3)\, dy$

$= \int_0^1 \frac{49}{12} y^3 dy = \frac{49}{12}$

2. $\int_{-2}^1 \int_0^{1-x} (4-y)\, dy\, dx$

Solution

$\int_{-2}^1 \int_0^{1-x} (4-y)\, dy\, dx = \int_{-2}^1 \left[4y - \frac{y^2}{2}\right]_0^{1-x} dx$

$= \int_{-2}^1 \left(\frac{7}{2} - 3x - \frac{x^2}{2}\right) dx = \frac{27}{2}$

3. $\int_0^a \int_0^{\frac{y^2}{a}} e^{\frac{x}{y}}\, dx\, dy$

Solution $\int_0^a \int_0^{\frac{y^2}{a}} e^{\frac{x}{y}}\, dx\, dy = \int_0^a \left[ye^{\frac{x}{y}}\right]_0^{\frac{y^2}{a}} dy$

$= \int_0^a \left(ye^{\frac{y}{a}} - y\right) dy = \frac{1}{2}a^2$

4-6. Sketch the region R of integration in the iterative integrals.

4. $\int_0^1 \int_1^{x+1} f(x,y)\, dy\, dx$

5. $\int_{-1}^1 \int_{-x^2}^{x^2+2} f(x,y)\, dy\, dx$

6. $\int_0^4 \int_{-\sqrt{y}}^{\sqrt{y}} f(x,y)\, dx\, dy$

7-8. Evaluate the double integral over the given region R that is bounded by the graphs of the given equations. Choose the appropriate order of integration.

7. $\iint_R (2x+y+1)\, dA,\ y = x^2,\ y = 2x^3.$

8. $\iint_R 3xy\, dA,\ y = x^2,\ y = 4,\ y = 0.$

---

9. $\iint_D x^2 dA$, D is bounded the triangle with vertices $(0,2),\ (1,1),\ (3,2)$.

Solution Answer : 13/4

10. Change the order of integration.

(a) $\int_0^a \int_0^{\sqrt{x}} f(x,y)\, dy\, dx\ (a > 0)$

(b) $\int_0^a \int_{-y}^{\sqrt{y}} f(x,y)\, dx\, dy\ (a > 0)$

Solution (a) $R = \{(x,y) \mid 0 \le x \le a, 0 \le y \le \sqrt{x}\}$

$= \{(x,y) \mid y^2 \le x \le a, 0 \le y \le \sqrt{a}\}$

$\int_0^a \int_0^{\sqrt{x}} f(x,y)\, dy dx = \int_0^{\sqrt{a}} \int_{y^2}^a f(x,y)\, dx\, dy$

(b) $R = \{(x,y) \mid -y \le x \le \sqrt{y}, 0 \le y \le a\}$

$= \{(x,y) \mid -a \le x \le 0, -x \le y \le a\} \cup$

$\{(x,y) \mid 0 \le x \le \sqrt{a}, x^2 \le y \le \sqrt{a}\}$

$\int_0^a \int_{-y}^{\sqrt{y}} f(x,y)\, dx\, dy$

$= \int_{-a}^0 \int_{-x}^a f(x,y)\, dy\, dx + \int_0^{\sqrt{a}} \int_{x^2}^{\sqrt{a}} f(x,y)\, dy\, dx.$

11-12. Evaluate the following integrals by changing the order of the integration.

11. $I = \int_0^1 \int_{\sqrt{y}}^1 \sqrt{1-x^3}\, dx dy.$

Solution $I = \int_0^1 \int_0^{x^2} \sqrt{1-x^3}\, dy dx = \int_0^1 x^2 \sqrt{1-x^3}\, dx$

$= \left[-\frac{2}{3}\frac{1}{3}(1-x^3)^{\frac{3}{2}}\right]_0^1 = \left[-\frac{2}{9}(1-x^3)^{\frac{3}{2}}\right]_0^1$

$= \frac{2}{9}$

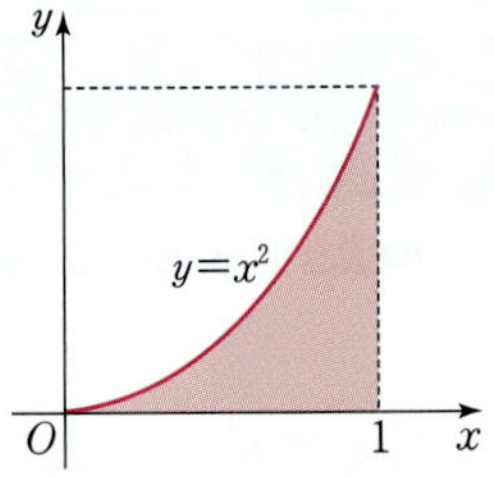

CAS 12. $\displaystyle\int_0^1\int_0^x \frac{2}{y^2-2y+2005}\,dydx$

Solution

http://matrix.skku.ac.kr/cal-lab/cal-14-1-2.html

```
var('x, y')
f(y)=2/(y^2 -2*y + 2005)
g(x)=f(x)-f(0)
integral(g, (x, 0, 1))
```

Answer : -log(2004) + log(2005)

$$\begin{array}{l} 0 \le x \le 1 \\ 0 \le y \le x \end{array} \Rightarrow \begin{array}{l} y \le x \le 1 \\ 0 \le y \le 1 \end{array}$$

$\displaystyle\int_0^1\int_0^x \frac{2}{y^2-2y+2005}\,dy\,dx$

$\displaystyle\Rightarrow \int_0^1\int_y^1 \frac{2}{y^2-2y+2005}\,dx\,dy$

$\displaystyle= \int_0^1 \frac{2(1-y)}{y^2-2y+2005}\,dx\,dy$

$= -\left[\ln(y^2-2y+2005)\right]_0^1$

$= -\ln(2004)+\ln(2005)$

$\displaystyle= \ln\left(\frac{2005}{2004}\right)$

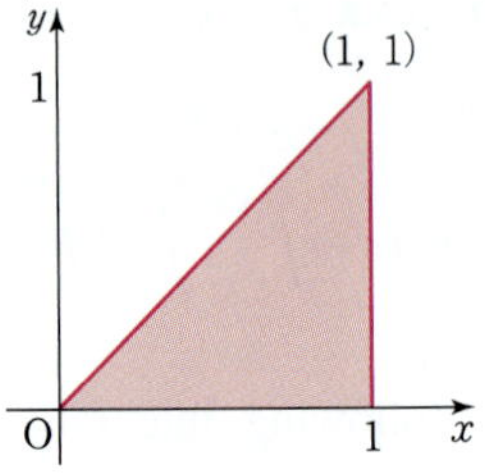

13. Evaluate $\displaystyle\int_0^1\int_y^1 \frac{\sin x}{x}\,dx\,dy$.

Solution $R=\{(x,y)\,|\,y \le x \le 1,\ 0 \le y \le 1\}$

$=\{(x,y)\,|\,0 \le x \le 1,\ 0 \le y \le x\}$

$\displaystyle\int_0^1\int_y^1 \frac{\sin x}{x}\,dx\,dy = \int_0^1\int_0^x \frac{\sin x}{x}\,dy\,dx$

$\displaystyle= \int_0^1 \left[y\cdot\frac{\sin x}{x}\right]_0^x dx$

$\displaystyle= \int_0^1 \sin x\,dx = [-\cos x]_0^1$

$=-\cos 1+1.$

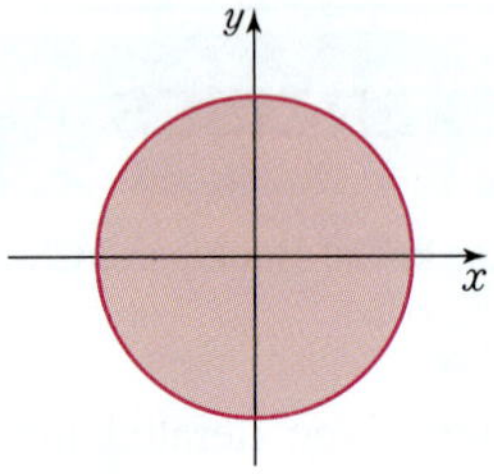

14. Evaluate $\displaystyle\int_0^9\int_{\sqrt{y}}^3 \sin(\pi x^3)\,dx\,dy$.

Solution $D=\{(x,y)\,|\,\sqrt{y} \le x \le 3,\ 0 \le y \le 9\}$

$=\{(x,y)\,|\,0 \le x \le 3,\ 0 \le y \le x^2\}$

$\displaystyle\int_0^9\int_{\sqrt{y}}^3 \sin(\pi x^3)\,dx\,dy = \int_0^3\int_0^{x^2} \sin(\pi x^3)\,dy\,dx$

$\displaystyle= \int_0^3 \left[y\sin(\pi x^3)\right]_0^{x^2} dx$

$\displaystyle= \int_0^3 x^2\sin(\pi x^3)\,dx$

$\displaystyle= -\frac{1}{3\pi}\left[\cos(\pi x^3)\right]_0^3 = \frac{2}{3\pi}$

15. Evaluate $\displaystyle\int_0^2\int_{\frac{y}{2}}^1 e^{x^2}\,dx\,dy$.

Solution $D=\left\{(x,y)\,|\,\frac{y}{2} \le x \le 1,\ 0 \le y \le 2\right\}$

$=\{(x,y)\,|\,0 \le x \le 1,\ 0 \le y \le 2x\}$

$\displaystyle\int_0^2\int_{\frac{y}{2}}^1 e^{x^2}\,dx\,dy = \int_0^1\int_0^{2x} e^{x^2}\,dy\,dx$

$\displaystyle= \int_0^1 \left[ye^{x^2}\right]_0^{2x} dx = \int_0^1 2xe^{x^2}\,dx$

$= \left[e^{x^2}\right]_0^1 = e-1$

16. Evaluate $\displaystyle\int_{-a}^a\int_{|y|}^a (x+y)^2\,dxdy$.

Solution $\displaystyle\int_{-a}^a\int_{|y|}^a (x+y)^2\,dxdy = \int_{-a}^a \left[\frac{1}{3}(x+y)^3\right]_{|y|}^a dy$

$\displaystyle= \int_{-a}^a \left[\frac{1}{3}(a+y)^3 - \frac{1}{3}(|y|+y)^3\right] dy$

$\displaystyle= \frac{2}{3}\int_0^a (a^3+3ay^2)\,dy - \frac{1}{3}\left[\int_0^a 8y^3\,dy + \int_{-a}^0 0\,dy\right]$

$\displaystyle= \frac{2}{3}\left[a^3y+ay^3\right]_0^a - \left[\frac{2}{3}y^4\right]_0^a$

$\displaystyle= \frac{2}{3}a^4$

CAS 17. Evaluate $\displaystyle\int_1^2\int_0^{\ln x} (x-1)\sqrt{1+e^{2y}}\,dydx$.

Solution

```
var('x, y')
f(x, y)= (x-1)*(1+ e^(2*y))^1/2
integral(integral(f, y, 0, ln(x)), x, 1, 2)  #
```

18. Find the volume of the solid
$S=\{(x,y,z)|x^2+y^2+z^2 \le a^2,\ x^2+y^2 \ge b^2\}$
with $a > b$.

19. Prove that

$$2\int_a^b \int_x^b f(x)f(y)dydx = \left(\int_a^b f(x)dx\right)^2.$$

20-21. If the density function $f(x,y)=1$ over the following domain D, then find the center of mass, the moments of inertia I_x, I_y, and I_0.

Let V be the volume of the solid obtained by rotating

$E=\{(x,y)\,|\,a \le x \le b,\ 0 \le \phi_1(x) \le y \le \phi_2(x)\}$

about x-axis.

When we slice the solid by the plane which is through the point $(x,0)$ and perpendicular to x-axis, its area of cross-section is $\pi\{\phi_2(x)^2-\phi(x)^2\}$. Thus the volume is

$$V=\pi\int_a^b \{\phi_2(x)^2-\phi(x)^2\}dx$$
$$=2\pi\int_a^b\int_{\phi_1(x)}^{\phi_2(x)} y\,dy\,dx = 2\pi\iint_E y\,dy\,dx.$$

20.

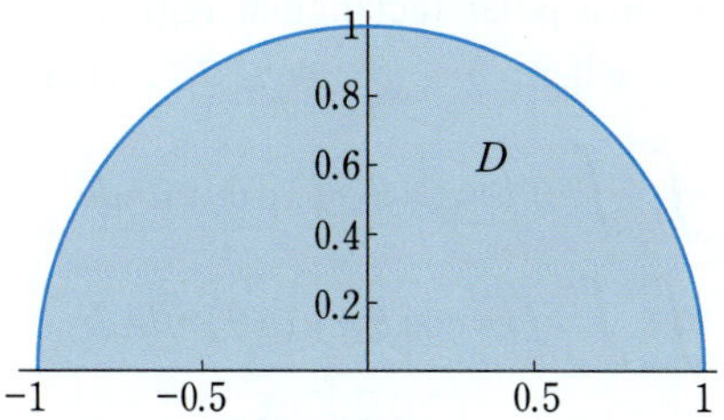

21.

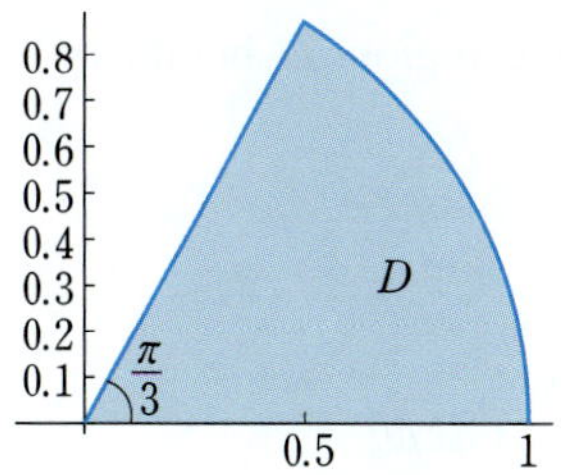

22. Find the volume V of the solid E obtained by rotating the region enclosed by $y=x^2$ and $y=\sqrt{x}$ about x-axis.

Solution $E=\{(x,y)\,|0 \le x \le 1,\ x^2 \le y \le \sqrt{x}\}$

$$V=2\pi\iint_E y\,dy\,dx = 2\pi\int_0^1\int_{x^2}^{\sqrt{x}} y\,dy\,dx$$
$$=\pi\int_0^1 (x-x^4)\,dx = \pi\left[\frac{1}{2}x-\frac{1}{5}x^5\right]_0^1 = \frac{3}{10}\pi$$

23. Find I_x, I_y, and I_0 for the area enclosed by the loop of $y^2=x^2(2-x)$.

Solution

$$A=\iint_R dA = 2\int_0^2\int_0^{x\sqrt{2-x}} dy\,dx = 2\int_0^2 x\sqrt{2-x}\,dx$$
$$=-4\int_{\sqrt{2}}^0 (2z^2-z^4)dz = -4\left[\frac{2}{3}z^3-\frac{1}{5}z^5\right]_{\sqrt{2}}^0 = \frac{32\sqrt{2}}{15}$$

where we have used the transformation $2-x=z^2$.
Then

$$I_x=\iint_R y^2\,dA = 2\int_0^2\int_0^{x\sqrt{2-x}} y^2\,dy\,dx$$
$$=\frac{2}{3}\int_0^2 x^3(2-x)^{\frac{3}{2}}dx = -\frac{4}{3}\int_{\sqrt{2}}^0 (2-z^2)^3 z^4 dz$$
$$=-\frac{4}{3}\left[\frac{8}{5}z^5-\frac{12}{7}z^7+\frac{2}{3}z^9-\frac{1}{11}z^{11}\right]_{\sqrt{2}}^0 = \frac{2048\sqrt{2}}{3465}$$
$$=\frac{64}{321}A$$
$$I_y=\iint_R x^2\,dA = 2\int_0^2\int_0^{x\sqrt{2-x}} x^2\,dy\,dx$$
$$=2\int_0^2 x^3\sqrt{2-x}\,dx$$
$$=-4\int_{\sqrt{2}}^0 (2-z^2)^3 z24\,dz$$
$$=-4\left[\frac{8}{3}z^3-\frac{12}{5}z^5+\frac{6}{7}z^7-\frac{1}{9}z^9\right]_{\sqrt{2}}^0 = \frac{1024\sqrt{2}}{315} = \frac{32}{21}A$$
$$I_0 = I_x+I_y = \frac{13312\sqrt{2}}{3465}$$
$$=\frac{416}{321}A$$

24-25. Evaluate f_{ave} for the given function and region R.

24. $f(x,y)=\dfrac{1}{x^2y^2}$,
$R=\{(x,y)\,|\,1 \le x \le 2,\ 1 \le y \le 2\}$.

25. $f(x,y)=x^2+y^2$,
$R=\{(x,y)\,|\,1 \le x \le 2,\ 1 \le y \le 2\}$.

14.2 Double Integrals in Polar Coordinates

YouTube http://youtu.be/olQgihl5aZg

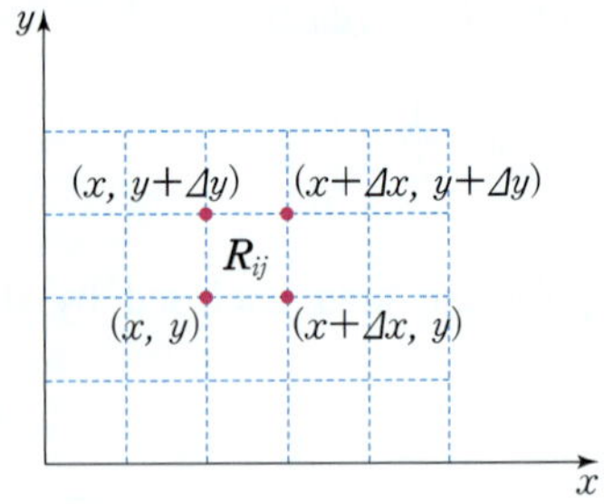

Figure 1

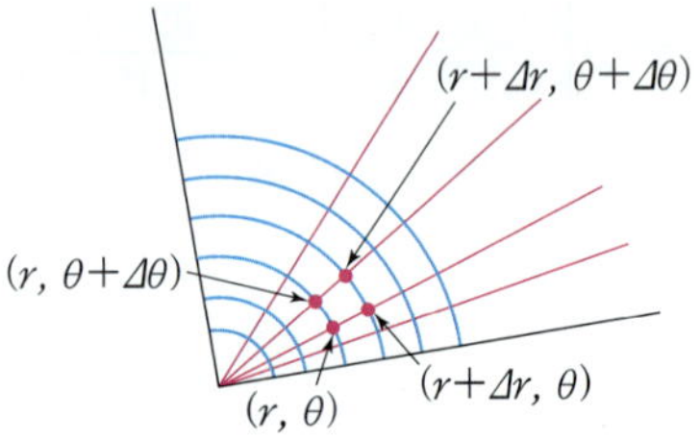

Figure 2

$\Delta\theta_i$ in polar coordinates

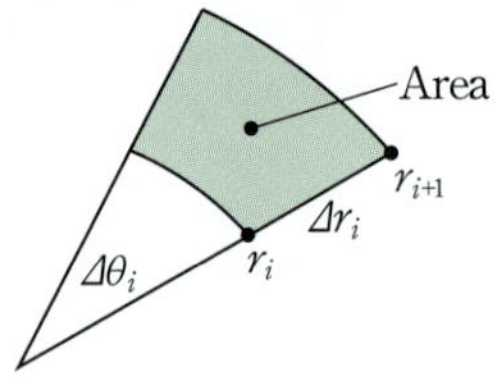

Figure 3

Double Integrals in Polar Coordinates

In section 10.3, we have seen that any point (x, y) with cartesian coordinates can be converted into the polar coordinates by the following transformations

$$x = r\cos\theta,\ y = r\sin\theta. \quad 0 \le \theta \le 2\pi \quad \text{and} \quad r \ge 0$$

Therefore, we would like to look at what happens to a double integral $\iint_D f(x,y)\,dA$, when x and y are transformed into polar coordinates.

One can easily show that the area of a small rectangle $R_{ij} = \Delta x_i \cdot \Delta y_j$ in polar coordinates gets transformed to $r_i \Delta r_i \Delta\theta_i$ in polar coordinates.

$$\begin{aligned}\text{Area } &= \frac{1}{2}r_{i+1}^2\Delta\theta_i - \frac{1}{2}r_i^2\Delta\theta_i = \frac{1}{2}(r_{i+1}+r_i)(r_{i+1}-r_i)\Delta\theta_i \\ &= \frac{1}{2}(r_{i+1}+r_i)(r_{i+1}-r_i)\Delta\theta_i \\ &= r_i^*\Delta r_i\Delta\theta_i,\end{aligned}$$

where r_i^* is the average radius $\frac{1}{2}(r_{i+1}+r_i)$. (See Figure 1-3.)

In particular, we have the following result which gives a method to transform a double integral in rectangular coordinates to polar coordinates.

THEOREM 1

Let f be continuous functions on a polar rectangular region D given by $0 \le a \le r \le b$ and $\alpha \le \theta \le \beta$ where $0 \le \beta - \alpha \le 2\pi$. Then

$$\begin{aligned}\iint_D f(x,y)\,dA &= \int_\alpha^\beta \int_a^b f(r\cos\theta, r\sin\theta)\, r\,dr\,d\theta \\ &= \int_a^b \int_\alpha^\beta f(r\cos\theta, r\sin\theta)\, r\,d\theta\,dr.\end{aligned}$$

In particular, $f(x,y) = 1$, the area of the region D boundary $\theta = \alpha$, $\theta = \beta$ and $r = 0$, $r = f(\theta)$ is

$$\begin{aligned}\text{Area } (D) &= \iint_D 1\,dA \\ &= \int_\alpha^\beta \int_0^{f(\theta)} r\,dr\,d\theta \\ &= \int_\alpha^\beta \left[\frac{r^2}{2}\right]_0^{f(\theta)} d\theta\end{aligned}$$

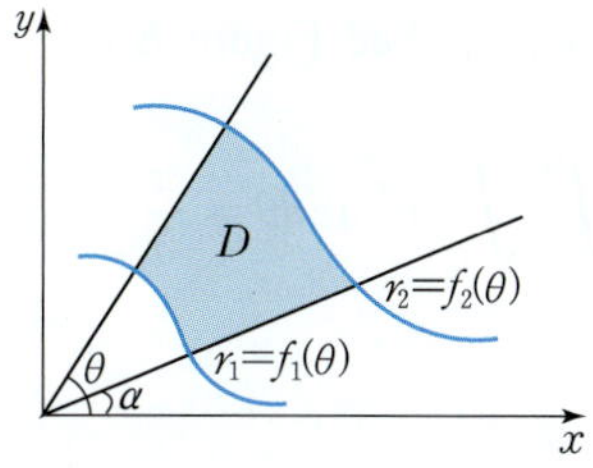

Figure 4

$$= \int_{\alpha}^{\beta} \frac{1}{2}[f(\theta)]^2 d\theta.$$

If D is a region bounded by polar functions $r_1 = f_1(\theta)$ and $r_2 = f_2(\theta)$ with $\alpha \le \theta \le \beta$ (See Figure 4.), then

$$\iint_D f(x, y)\,dA = \int_{\alpha}^{\beta} \int_{f_1(\theta)}^{f_2(\theta)} f(r\cos\theta,\ r\sin\theta)\,r\,dr\,d\theta.$$

On the other hand if D is a region bounded by polar functions $\theta_1 = \alpha(r)$ and $\theta_2 = \beta(r)$ with $a \le r \le b$, then

$$\iint_D f(x, y)\,dA = \int_{a}^{b} \int_{\alpha(r)}^{\beta(r)} f(r\cos\theta,\ r\sin\theta)\,r\,d\theta\,dr.$$

EXAMPLE 1

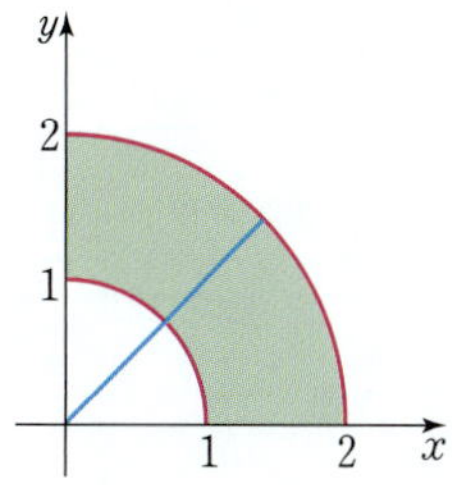

Figure 5

Find $\iint_D (3x + 8y^2)\,dA$ where

$$D = \left\{(r, \theta) \mid 1 \le r \le 2,\ 0 \le \theta \le \frac{\pi}{2}\right\}.\ \text{(See Figure 5.)}$$

Solution

$$\begin{aligned}\iint_D (3x + 8y^2)\,dA &= \int_0^{\frac{\pi}{2}} \int_1^2 (3r\cos\theta + 8r^2\sin^2\theta)\,r\,dr\,d\theta \\ &= \int_0^{\frac{\pi}{2}} \left[r^3\cos\theta + 2r^4\sin^2\theta\right]_1^2 d\theta \\ &= \int_0^{\frac{\pi}{2}} (7\cos\theta + 30\sin^2\theta)\,d\theta \\ &= \int_0^{\frac{\pi}{2}} (7\cos\theta + 15 - 15\cos 2\theta)\,d\theta \\ &= 7 + \frac{15}{2}\pi.\end{aligned}$$

■

EXAMPLE 2

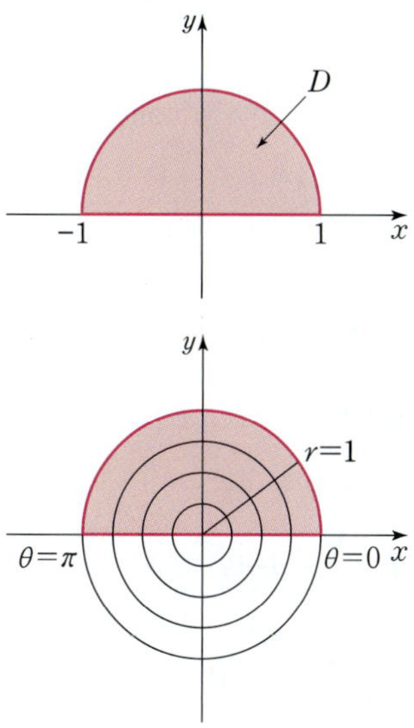

Figure 6

Find $\int_{-1}^{1} \int_0^{\sqrt{1-x^2}} (x^2 + y^2)^{\frac{3}{2}}\,dy\,dx$ by transforming this to polar coordinates.

Solution Note that the region of integration is

$$D = \left\{(x, y) \mid -1 \le x \le 1,\ 0 \le y \le \sqrt{1 - x^2}\right\}.$$

Under the polar transform, it is mapped to

$R_{\{r,\theta\}} = \{(r,\theta) \mid 0 \le r \le 1,\ 0 \le \theta \le \pi\}$. (See Figure 6.)

$$\int_{-1}^{1}\int_{0}^{\sqrt{1-x^2}} (x^2+y^2)^{\frac{3}{2}}\,dy\,dx = \int_0^{\pi}\int_0^1 r^4\,dr\,d\theta = \frac{\pi}{5}.$$

■

EXAMPLE 3

Find $\iint_D (x^2+y^2)^{\frac{3}{2}}\,dx\,dy$, where

$$D = \{(x,\ y) \mid 1 \le x^2+y^2 \le 4,\ 0 \le y \le \sqrt{3}\,x\}.$$

Solution $D = \{(x,\ y) \mid 1 \le x^2+y^2 \le 4,\ 0 \le y \le \sqrt{3}\,x\}$

$$= \left\{(r,\ \theta) \mid 1 \le r \le 2,\ 0 \le \theta \le \frac{\pi}{3}\right\}$$

$$\iint_D (x^2+y^2)^{\frac{3}{2}}\,dx\,dy$$
$$= \iint_D (r^2)^{\frac{3}{2}}\,r\,dr\,d\theta$$
$$= \int_0^{\frac{\pi}{3}}\int_1^2 r^4\,dr\,d\theta$$
$$= \int_0^{\frac{\pi}{3}}\left[\frac{1}{5}r^5\right]_1^2 d\theta$$
$$= \int_0^{\frac{\pi}{3}} \frac{31}{5}\,d\theta$$
$$= \frac{31}{15}\pi.$$

■

EXAMPLE 4

Find the volume of the region above the surface $z = x^2+y^2$ and below the plane $z = 10$. (See Figure 7.)

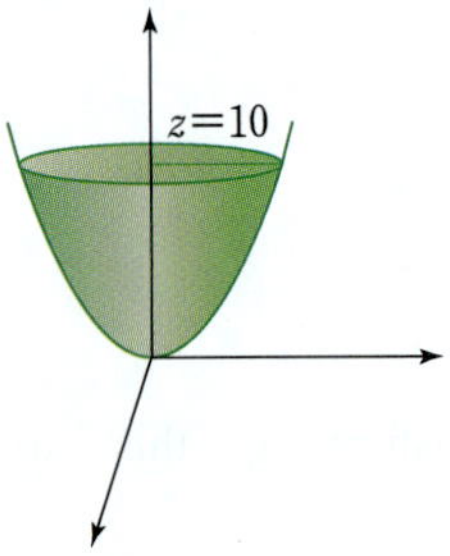

Figure 7

Solution Note that the region of integration is

$$D = \{(x,\ y) \mid x^2+y^2 \le 10\}.$$

Under the polar transform, it is mapped to

$$D = \{(r,\ \theta) \mid 0 \le r \le \sqrt{10},\ 0 \le \theta \le 2\pi\}.$$

$$\iint_D (x^2+y^2)dx\,dy = \int_0^{2\pi}\int_0^{\sqrt{10}} r^3 dr\,d\theta$$
$$= \int_0^{2\pi}\left[\frac{1}{4}r^4\right]_0^{\sqrt{10}} d\theta$$
$$= \int_0^{2\pi} 25\,d\theta = 50\pi.$$

■

EXAMPLE 5

Find the volume of the solid inside the sphere $x^2+y^2+z^2=a^2$ and inside the cylinder $x^2+y^2=ax$. (See Figure 8.)

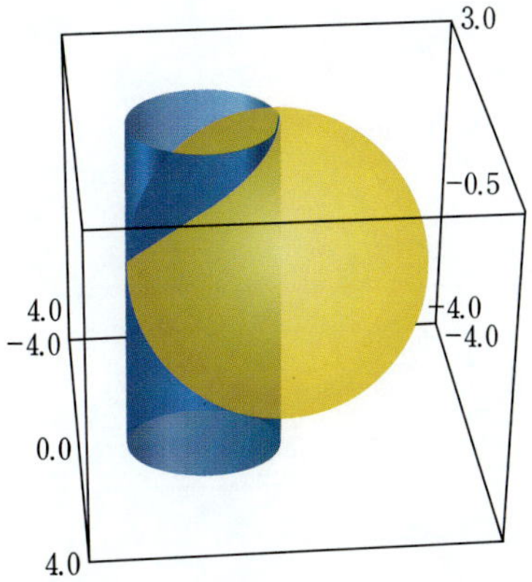

Figure 8

Solution Let $D=\left\{(r,\ \theta)\,|\,0\le r\le a\cos\theta,\ 0\le\theta\le\frac{\pi}{2}\right\}$.

$$\begin{aligned} V &= 4\int_0^{\frac{\pi}{2}}\int_0^{a\cos\theta} r\sqrt{a^2-r^2}\,dr\,d\theta \\ &= 4\int_0^{\frac{\pi}{2}}\left[-\frac{1}{3}(a^2-r^2)^{\frac{3}{2}}\right]_0^{a\cos\theta} d\theta \\ &= \frac{4}{3}a^3\int_0^{\frac{\pi}{2}}(1-\sin^3\theta)\,d\theta \\ &= \frac{2}{3}\left(\pi-\frac{4}{3}\right)a^3. \end{aligned}$$

■

EXAMPLE 6

Evaluate $\iint_D \sin\theta\,dA$, where D is the region in the first quadrant that is outside $r=2$ and inside the cardioid $r=2(1+\cos\theta)$.

Solution The two equations intersect at $\left(2,\ \frac{\pi}{2}\right)$, $\left(2,\ -\frac{\pi}{2}\right)$

$$D=\left\{(r,\ \theta)\,|\,2\le r\le 2(1+\cos\theta),\ 0\le\theta\le\frac{\pi}{2}\right\}$$

$$\begin{aligned} V &= \iint_D \sin\theta\,dA=\int_0^{\frac{\pi}{2}}\int_2^{2(1+\cos\theta)}\sin\theta\, r\,dr\,d\theta \\ &= \int_0^{\frac{\pi}{2}}\left[\frac{1}{2}r^2\sin\theta\right]_2^{2(1+\cos\theta)} d\theta \\ &= 2\int_0^{\frac{\pi}{2}}\left[(1+\cos\theta)^2\sin\theta-\sin\theta\right]d\theta \\ &= 2\left[-\frac{1}{3}(1+\cos\theta)^3+\cos\theta\right]_0^{\frac{\pi}{2}} \\ &= 2\left[-\frac{1}{3}-\left(-\frac{5}{3}\right)\right]=\frac{8}{3}\ . \end{aligned}$$

■

EXAMPLE 7

Show that $\int_0^\infty e^{-x^2}dx=\frac{\sqrt{\pi}}{2}$. (Gaussian Integral)

http://mathworld.wolfram.com/GaussianIntegral.html

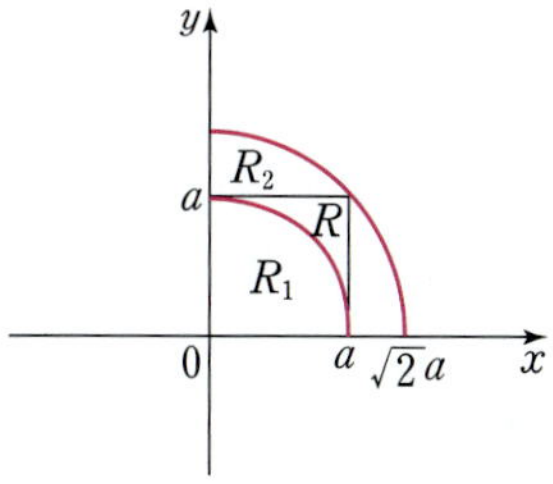

Figure 9

Solution Let R_1 and R_2 be the regions in the first quadrant inside the disk $x^2+y^2\le 2a^2$, respectively. Let $R=\{(x,\ y)|0\le x\le a,\ 0\le y\le a\}$ be rectangle. Then $R_1\subseteq R\subseteq R_2$. (See Figure 9.)

Let $I_1^2 = \iint_{R_1} e^{-(x^2+y^2)} dxdy$

$I^2 = \iint_{R} e^{-(x^2+y^2)} dxdy$ and

$I_2^2 = \iint_{R_2} e^{-(x^2+y^2)} dxdy.$

Using Polar Coordinate we have

$$I_1^2 = \int_0^{\frac{\pi}{2}} \int_0^a e^{-r^2} r\,dr\,d\theta = \frac{\pi}{4}(1-e^{-a^2}),$$

and $I_2^2 = \frac{\pi}{4}(1-e^{-2a^2})$.

AS $a \to \infty$, I_2^2 and I_2^2 come to close at $\frac{\pi}{4}$.

By Squeeze Theorem, I^2 comes to close at $\frac{\pi}{4}$.

By Fubini's Theorem,

$$I^2 = \iint_R e^{-(x^2+y^2)} dxdy = \left(\int_0^a e^{-x^2} dx\right)^2.$$

Hena as $a \to \infty$, $\int_0^{\infty} e^{-x^2} dx = \frac{\sqrt{\pi}}{2}$.

```
intergral(exp(-x^2),0,0,00
```

■

14.2 EXERCISES (Double Integrals in Polar Coordinates)

http://matrix.skku.ac.kr/Cal-Book/part2/CS-Sec-14-2-Sol.htm

YouTube http://youtu.be/jpsObxtZ50A

1. Find $\int_0^2\int_0^\pi r\sin^2\theta\, d\theta dr$.

Solution $\int_0^2\int_0^\pi r\sin^2\theta\, d\theta dr = \int_0^2 r\,dr\int_0^\pi \sin^2\theta\, d\theta$

$= 2\int_0^\pi \frac{1-\cos 2\theta}{2} d\theta = \left[\theta - \frac{1}{2}\sin 2\theta\right]_0^\pi = \pi.$

2. $\iint_R \tan^{-1}\left(\frac{y}{x}\right) dA$, where

$R = \{(x,y) | 1 \le x^2+y^2 \le 4, 0 \le y \le x\}.$

Solution

$R = \{(r,\theta) | 1 \le r \le 2, 0 \le \theta \le \pi/4\}$

$\iint_R \tan^{-1}\left(\frac{y}{x}\right) dA = \int_0^{\frac{\pi}{4}}\int_1^2 \tan^{-1}(\tan\theta)\, r\,dr d\theta$

Since $\tan^{-1}(\tan\theta) = \theta$, $0 \le \theta \le \frac{\pi}{4}$,

$\int_0^{\frac{\pi}{4}}\int_1^2 \tan^{-1}(\tan\theta)\, r\,dr d\theta = \int_0^{\frac{\pi}{4}}\int_1^2 \theta\, r\,dr d\theta$

$= \int_0^{\frac{\pi}{4}} \theta\left(\int_1^2 r\,dr\right) d\theta = \int_0^{\frac{\pi}{4}} \theta\left(\frac{3}{2}\right) d\theta$

$= \frac{3}{64}\pi^2.$

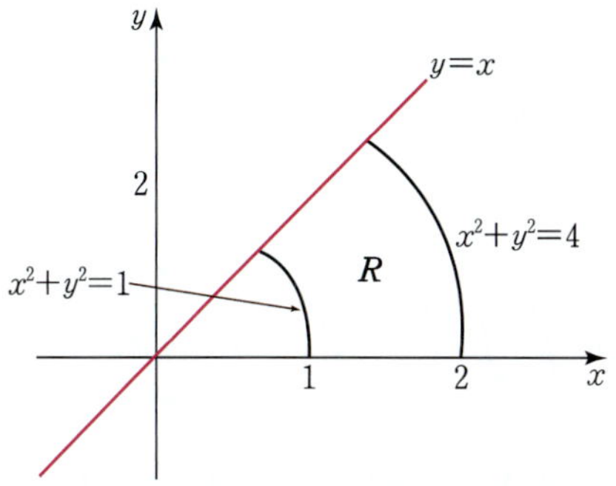

CAS 3. Find the area inside the cardioid $r = 1+\sin\theta$.

Solution

http://matrix.skku.ac.kr/cal-lab/cal-14-2-2(new).html

```
var('v')
polar_plot(1+sin(v),(v, 0, 2*pi))
```

$D = \{(r,\theta) | 0 \le \theta \le 2\pi,\ 0 \le r \le 1+\sin\theta\}$

$A = \int\int_D dA = \int_0^{2\pi}\int_0^{1+\sin\theta} r\,dr\,d\theta$

$= \int_0^{2\pi}\left[\frac{1}{2}r^2\right]_0^{1+\sin\theta} d\theta$

$= \frac{1}{2}\int_0^{2\pi}(1+\sin\theta)^2 d\theta = \frac{1}{2}\int_0^{2\pi}(1+2\sin\theta+\sin^2\theta)\, d\theta$

$= \frac{1}{2}\int_0^{2\pi}\left(1+2\sin\theta+\frac{1-\cos 2\theta}{2}\right) d\theta$

$= \frac{1}{2}\int_0^{2\pi}\left(\frac{3}{2}+2\sin\theta-\frac{1}{2}\cos 2\theta\right) d\theta$

$= \frac{1}{2}\left[\frac{3}{2}\theta - 2\cos\theta - \frac{1}{4}\sin 2\theta\right]_0^{2\pi} = \frac{3}{2}\pi.$

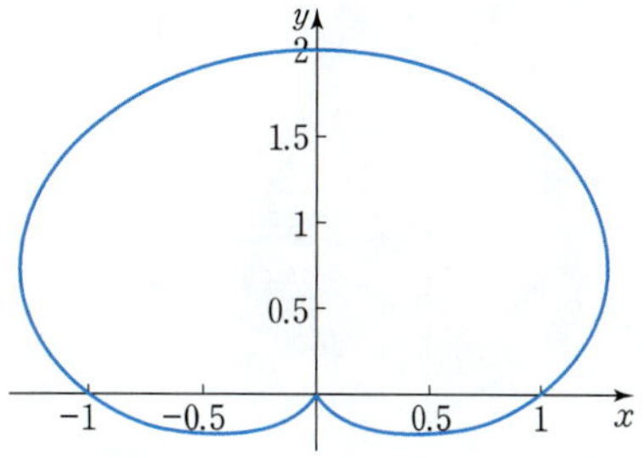

4. Find the area inside of $r = a(1+\cos\theta)$.

Solution $D = \{(r,\theta) | 0 \le \theta \le 2\pi,\ 0 \le r \le a(1+\cos\theta)\}$

$A = \int\int_D dA = \int_0^{2\pi}\int_0^{a(1+\cos\theta)} r\,dr\,d\theta$

$= \int_0^{2\pi}\left[\frac{1}{2}r^2\right]_0^{a(1+\cos\theta)} d\theta$

$= \frac{1}{2}a^2\int_0^{2\pi}(1+\cos\theta)^2 d\theta$

$= \frac{1}{2}a^2\int_0^{2\pi}(1+2\cos\theta+\cos^2\theta)\, d\theta$

$= \frac{1}{2}a^2\int_0^{2\pi}\left(1+2\cos\theta+\frac{1+\cos 2\theta}{2}\right) d\theta$

$= \frac{1}{2}a^2\int_0^{2\pi}\left(\frac{3}{2}+2\cos\theta+\frac{1}{2}\cos 2\theta\right) d\theta$

$= \frac{1}{2}a^2\left[\frac{3}{2}\theta+2\sin\theta+\frac{1}{4}\sin 2\theta\right]_0^{2\pi}$

$= \frac{3}{2}\pi a^2.$

CAS 5. Find the volume V of $R = \{(x,y,z) | z \le 4-x^2-y^2,\ z \ge 0\}$.

Solution

http://matrix.skku.ac.kr/cal-lab/cal-14-2-4(new).html

```
var('u, v, x, y')
parametric_plot3d( (v*sin(u), v*cos(u), 4-v^2), (u, 0, 2*pi), (v ,0 ,2),color='pink' )
```

Note that $V=\iint_R (4-x^2-y^2)dA$, where R is the disk: $0 \le x^2+y^2 \le 4$.

In polar coordinates R can be written as $R=\{(r,\theta)\,|\,0 \le r \le 2,\ 0 \le \theta \le 2\pi\}$. Hence

$$V=\iint_R (4-x^2-y^2)dA=\int_0^{2\pi}\int_0^2 (4-r^2)r\,dr\,d\theta$$

$$=\int_0^{2\pi}\left[2r^2-\frac{1}{4}r^4\right]_0^2 d\theta \;=8\pi.$$

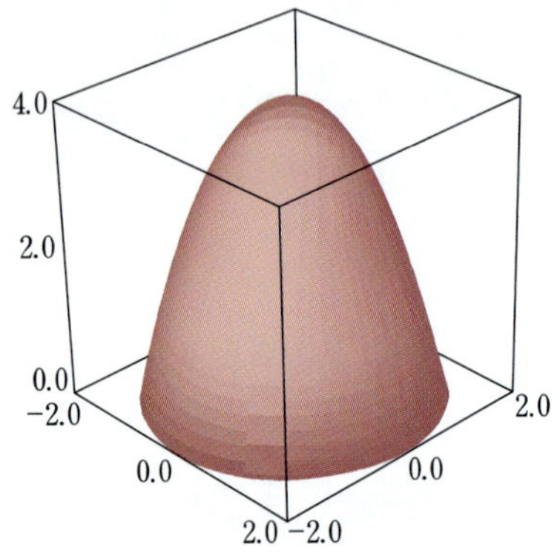

6. Find $I=\int_{-2}^{2}\int_{-\sqrt{4-x^2}}^{\sqrt{4-x^2}} e^{-(x^2+y^2)}dydx$

Solution $I=\int_0^{2\pi}\int_0^2 e^{-r^2}r\,dr\,d\theta=\int_0^{2\pi}\left[-\frac{1}{2}e^{-r^2}\right]_0^2 d\theta$

$$=\frac{1}{2}(1-e^{-4})\int_0^{2\pi}d\theta$$

$$=\pi(1-e^{-4}).$$

CAS **7.** Evaluate $\iint_R \frac{x}{\sqrt{x^2+y^2}}dA$, where R is the region bounded by $x>0,\ y>0$,

$1 \le x^2+y^2 \le 4$ and $\tan\left(\frac{x^2+y^2-1}{6}\right) \le \frac{y}{x} \le 1.$

Solution Using the polar coordinate system, the region R is represented as follows: $\frac{r^2-1}{6} \le \theta \le \frac{\pi}{4}$ and $1 \le r \le 2$.

Here we used $\tan^{-1}\left(\frac{y}{x}\right)=\theta$. Then the given integral becomes

$$\iint_R \frac{x}{\sqrt{x^2+y^2}}dA=\int_1^2\int_{\frac{r^2-1}{6}}^{\frac{\pi}{4}}\frac{r\cos\theta}{r}r\,d\theta\,dr$$

$$=\int_1^2 r\int_{\frac{r^2-1}{6}}^{\frac{\pi}{4}}\cos\theta\,d\theta\,dr=\int_1^2 \sin\theta\Big|_{\frac{r^2-1}{6}}^{\frac{\pi}{4}} r\,dr$$

$$=\int_1^2 r\left(\frac{\sqrt{2}}{2}-\sin\left(\frac{r^2-1}{6}\right)\right)dr$$

$$=\left(\frac{\sqrt{2}}{4}r^2+3\cos\left(\frac{r^2-1}{6}\right)\right)\Big|_1^2$$

$$=3\left(\frac{\sqrt{2}}{4}+\cos\left(\frac{1}{2}\right)-1\right).$$

8. Let $D=\left\{(x,\,y)|x^2+y^2 \le 1,\ \frac{1}{2} \le y \le 1\right\}$.

Evaluate $\iint_D y^3(x^2+y^2)^{-3/2}dxdy$.

9. Let $D=\{(x,\,y)|x^2+y^2 \le a^2,\ x \ge 0\}$. Evaluate $\iint_D xy^2dA$.

Solution $D=\{(x,y)\,|\,0 \le x \le a,\ -\sqrt{a^2-x^2} \le y \le \sqrt{a^2-x^2}\}$

Let $x=r\cos\theta$ and $y=r\sin\theta$.

$$D'=\left\{(r,\,\theta)\,|\,0 \le r \le a,\ -\frac{\pi}{2} \le \theta \le \frac{\pi}{2}\right\}$$

$$\iint_D xy^2dA=\int_{-\frac{\pi}{2}}^{\frac{\pi}{2}}\int_0^a r^4\cos\theta\sin^2\theta\,dr\,d\theta$$

$$=\frac{1}{5}a^5\int_{-\frac{\pi}{2}}^{\frac{\pi}{2}}\cos\theta\sin^2\theta\,d\theta$$

$$=\frac{1}{5}a^5\int_{-1}^{1}t^2\,dt$$

$$=\frac{1}{15}a^5\,\frac{2}{3}$$

$$=\frac{2}{15}a^5.$$

CAS **10.** Find $\int_0^1\int_0^{\sqrt{1-x^2}}\sqrt{1-y^2}\,dydx$.

Solution

http://matrix.skku.ac.kr/cal-lab/cal-14-1-3.html

```
var('x, y')
f(y)=integral(sqrt(1-y^2), y)
g(x)=f(sqrt(1-x^2))-f(0)
integral(g, (x, 0, 1))
```

Answer : 2/3

Let $x=r\cos\theta$ and $y=r\sin\theta$.

$$\int_0^1\int_0^{\sqrt{1-x^2}}\sqrt{1-y^2}\,dydx=2\int_0^1\int_0^{\frac{\pi}{2}}r^2\cos\theta\,d\theta\,dr$$

$$=2\int_0^1 r^2\,dr=\frac{2}{3}.$$

14.3 Surface Area

YouTube http://youtu.be/p9R0TTLfBzk

Surface Area

Surface area is the total area of the faces and curved surfaces of a solid figure. A mathematical description of the surface area is considerably moras a sphere, are assigned surface area using their representation as parametric surfaces. This definition of the surface area is based on methods of infinitesimal calculus and involves partial derivatives and double integration. In this section we apply the notion of double integrals to find the area of a surface. We consider a surface with equation $z=f(x,y)$, namely the graph surface. Let the part of surface area be D, and it's area be S. Let A be the projection of S onto the xy-plane. If we partitioned A with rectangles made by Δx and Δy sides (say the area is ΔA), we can think of a rectangular cylinder on a rectangle which cuts a part of D. We call this part ΔD, and its area ΔS.

This rectangular cylinder on the rectangle in the xy-plane also cuts a part $\Delta D'$ of tangent plane to the surface at a point P on ΔD. We call this area $\Delta S'$. If r is the angle between the tangent plane and the xy-plane, then

1 $$\Delta A = \Delta S' \cos r. \text{ (See Figure 1.)}$$

Now r is same as the angle between the z-axis and a normal line to the surface at the point P. But directional cosines of a normal line are proportional to $\dfrac{\partial z}{\partial x}, \dfrac{\partial z}{\partial y}, -1$ at the point P, so we have

$$\cos r = \frac{1}{\sqrt{1+\left(\dfrac{\partial z}{\partial x}\right)^2+\left(\dfrac{\partial z}{\partial y}\right)^2}}$$

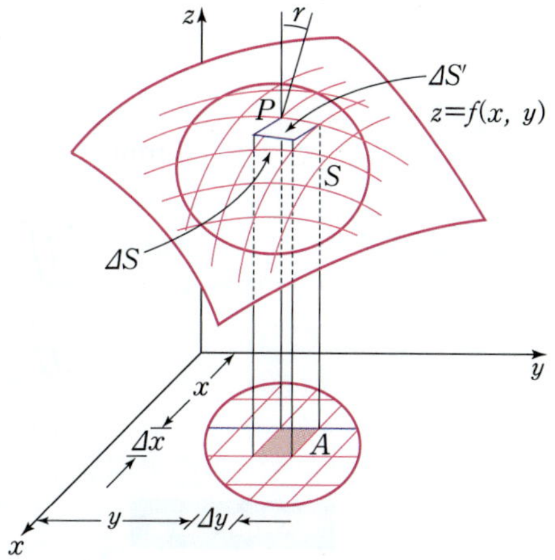

Figure 1

Substitute $\cos r$ in 1,

2 $$\Delta S' = \sqrt{1+\left(\frac{\partial z}{\partial x}\right)^2+\left(\frac{\partial z}{\partial y}\right)^2} \cdot \Delta A .$$

If Δx and Δy are small enough, the area above the tangent plane, $\Delta S'$, is very close to the surface area ΔS. Hence the surface area S can be defined by the following limit:

3 $$S = \iint_A \sqrt{1+\left(\frac{\partial z}{\partial x}\right)^2+\left(\frac{\partial z}{\partial y}\right)^2}\, dxdy .$$

Similarly, we can compute area S by projecting to other coordinate planes. That is

4 $$S = \iint_{A'} \sqrt{1+\left(\frac{\partial x}{\partial y}\right)^2+\left(\frac{\partial x}{\partial z}\right)^2}\, dydz$$

or

5 $$S = \iint_{A''} \sqrt{1+\left(\frac{\partial y}{\partial x}\right)^2+\left(\frac{\partial y}{\partial z}\right)^2}\, dzdx$$

where A' and A'' are projections of D onto the yz-plane and the zx-plane respectively.

EXAMPLE 1

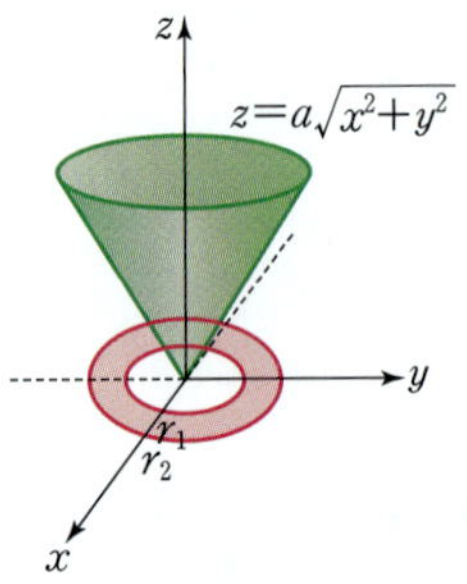

Figure 2

Find the surface area S of a part of circular cone $z = a\sqrt{x^2+y^2}\,(a>0)$ that lies between $z=r_1$ and $z=r_2(r_1<r_2)$. (See Figure 2.)

Solution We have $D = (x,\ y) \mid r_1 \le x^2+y^2 \le r_2$.

$$z = a\sqrt{x^2+y^2} \text{ when } z_x = \frac{ax}{\sqrt{x^2+y^2}},\ z_y = \frac{ay}{\sqrt{x^2+y^2}}.$$

From 3,

$$\begin{aligned} S &= \iint_D \sqrt{1+\left(\frac{ax}{\sqrt{x^2+y^2}}\right)^2+\left(\frac{ay}{\sqrt{x^2+y^2}}\right)^2}\, dA \\ &= \iint_D \sqrt{a^2+1}\, dA = \sqrt{a^2+1}\iint_D dA \\ &= \sqrt{a^2+1}\left(\pi r_2^2-\pi r_1^2\right) = \pi\sqrt{a^2+1}\left(r_2^2-r_1^2\right). \end{aligned}$$ ■

EXAMPLE 2

Find the surface area of the part that was cut by the planes $x=0$, $x=1$ and $z=1-y$ in 1st quadrant. (See Figure 3.)

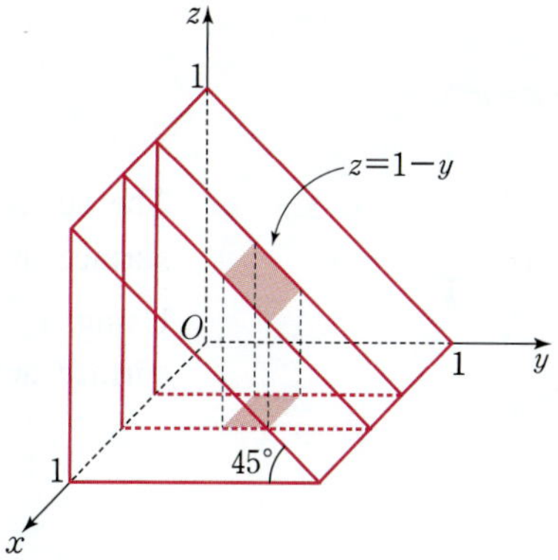

Figure 3

Solution From [3], $z_x = 0$, $z_y = -1$ and

$$S = \int_0^1 \int_0^1 \sqrt{1 + 0^2 + (-1)^2}\, dx dy = \sqrt{2}\,. \qquad \blacksquare$$

EXAMPLE 3

Find the surface area for the part that was cut by the plane $z = x$ on cylinder $x^2 + y^2 = a^2$ in 1st quadrant.

Solution The part become a triangle BOC whose boundaries are $z = 0$, $z = x$ and $x = a$ by projecting the part we are looking for onto the zx -plane. This surface lies above the triangle BOC as

$$\{(x, z) \mid 0 \le x \le a, 0 \le z \le x\}.$$

For the cylinder surface $y = \sqrt{a^2 - x^2}$, we have

$$\frac{dy}{dx} = \frac{-x}{\sqrt{a^2 - x^2}}, \quad \frac{dy}{dz} = 0\,.$$

Hence [5] gives us the following:

$$S = \int_0^a \int_0^x \frac{a}{\sqrt{a^2 - x^2}}\, dz\, dx = a \int_0^a \frac{x}{\sqrt{a^2 - x^2}}\, dx = a^2. \qquad \blacksquare$$

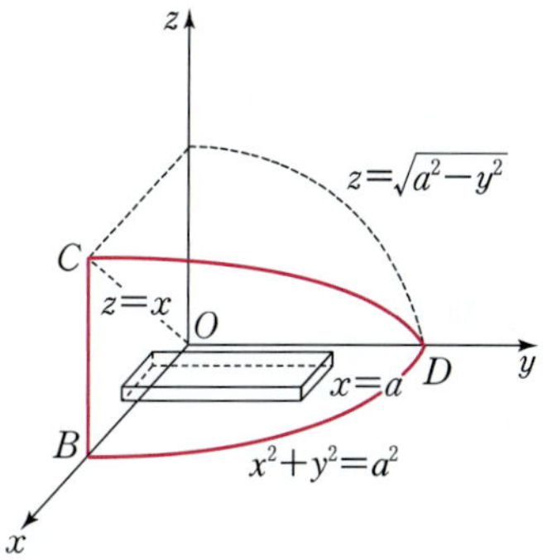

Figure 4

EXAMPLE 4

Find the surface area of the sphere $x^2 + y^2 + z^2 = a^2$.

Solution Given the equation of sphere $x^2 + y^2 + z^2 = a^2$, we obtain

$$\frac{\partial z}{\partial x} = -\frac{x}{z}, \quad \frac{\partial z}{\partial y} = -\frac{y}{z}.$$

and so

$$1+\left(\frac{\partial z}{\partial x}\right)^2+\left(\frac{\partial z}{\partial y}\right)^2=1+\frac{x^2}{z^2}+\frac{y^2}{z^2}.$$

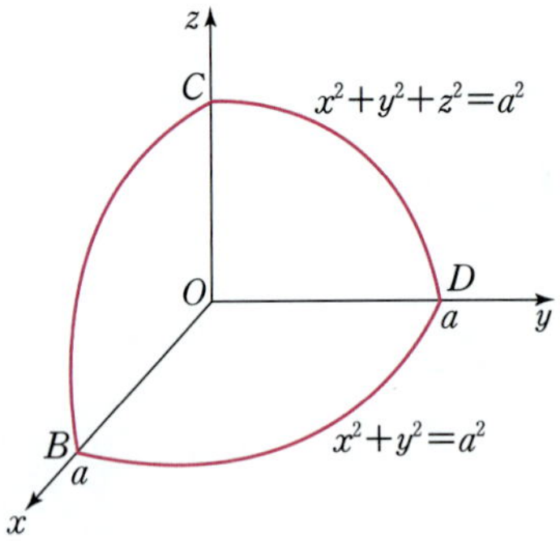

Figure 5

The surface area of BCD in the Figure 5 is one eighth of the total surface area of the sphere with radius a.

Using 3 for $A=\{(x,y)\,|\,0\le x\le a,\,0\le y\le\sqrt{a^2-x^2}\}$, the surface area S,

$$S=8\int\int_A\frac{a}{\sqrt{a^2-x^2-y^2}}dxdy.$$

Substitute S with the polar coordinate system,

$$S=8\int_0^{\pi/2}\int_0^a\frac{a}{\sqrt{a^2-r^2}}rdrd\theta=4\pi a^2.$$ ■

EXAMPLE 5

Find a surface area of cylinder $x^2+z^2=a^2$ inside of $x^2+y^2=a^2$. (See Figure 6.)

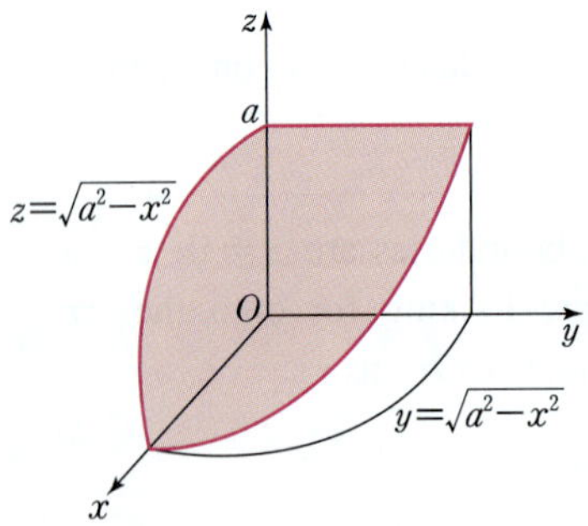

Figure 6

Solution Let S be the surface area, and A be the projection of the part of curve in 1st octant onto the xy-plane.

Let $z=\sqrt{a^2-x^2}$. Then,

$$\frac{\partial z}{\partial x}=-\frac{x}{\sqrt{a^2-x^2}},\ \frac{\partial z}{\partial y}=0.$$

Using 3,

$$\begin{aligned}S&=8\int\int_A\sqrt{1+\frac{x^2}{a^2-x^2}}\,dxdy\\&=8a\int\int_A\frac{1}{\sqrt{a^2-x^2}}dxdy\\&=8a\int_0^a\int_0^{\sqrt{a^2-x^2}}\frac{1}{\sqrt{a^2-x^2}}dydx\\&=8a\int_0^a dx=8a^2.\end{aligned}$$ ■

Let $z=\phi(r,\theta)$ be a cylindrical coordinate representation of the equation of a given surface. Then the region A can be partitioned with lines that pass through the origin O, and concentric circles centered at the origin O. Hence we use $rdrd\theta$ instead of $dxdy$. In this case the integrand can be represented with r and θ as the followings:

for $x=r\cos\theta$ and $y=r\sin\theta$,

$$\frac{\partial z}{\partial r}=\frac{\partial z}{\partial x}\frac{\partial x}{\partial r}+\frac{\partial z}{\partial y}\frac{\partial y}{\partial r}=\frac{\partial z}{\partial x}\cos\theta+\frac{\partial z}{\partial y}\sin\theta$$

$$\frac{\partial z}{\partial \theta}=\frac{\partial z}{\partial x}\frac{\partial x}{\partial \theta}+\frac{\partial z}{\partial y}\frac{\partial y}{\partial \theta}=\frac{\partial z}{\partial x}(-r\sin\theta)+\frac{\partial z}{\partial y}r\cos\theta .$$

We have $\left(\frac{\partial z}{\partial r}\right)^2+\frac{1}{r^2}\left(\frac{\partial z}{\partial \theta}\right)^2=\left(\frac{\partial z}{\partial x}\right)^2+\left(\frac{\partial z}{\partial y}\right)^2$.

In 3, $\sqrt{1+\left(\frac{\partial z}{\partial x}\right)^2+\left(\frac{\partial z}{\partial y}\right)^2}=\sqrt{1+\left(\frac{\partial z}{\partial r}\right)^2+\frac{1}{r^2}\left(\frac{\partial z}{\partial \theta}\right)^2}$.

This gives us the following equation.

$$S=\iint_B\sqrt{1+\left(\frac{\partial z}{\partial r}\right)^2+\frac{1}{r^2}\left(\frac{\partial z}{\partial \theta}\right)^2}\,rdrd\theta$$

or

$$=\iint_B\sqrt{r^2+\left(r\frac{\partial z}{\partial r}\right)^2+\left(\frac{\partial z}{\partial \theta}\right)^2}\,drd\theta$$

where the region A is represented with r and θ as region B.

EXAMPLE 6

Find the surface area of the portion of the sphere $x^2+y^2+z^2=a^2$ that is inside the cylinder $x^2+y^2=ax$. (See Figure 7.)

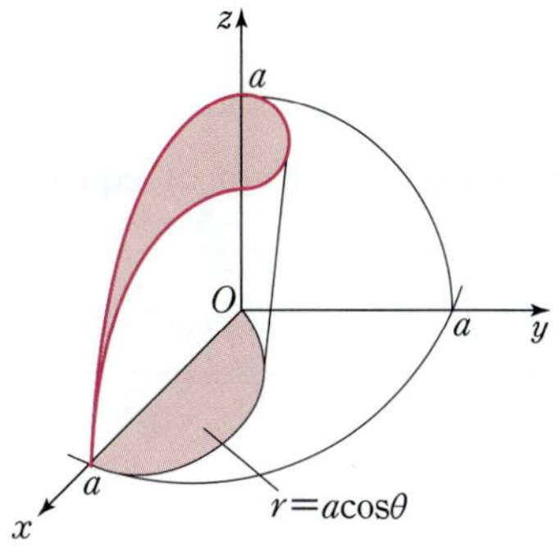

Figure 7

Solution The equation of a sphere in cylindrical coordinate is $r^2+z^2=a^2$.

$$z=\pm\sqrt{a^2-r^2}.$$

Thus

$$\frac{\partial z}{\partial r}=\mp\frac{r}{\sqrt{a^2-r^2}},\quad \frac{\partial z}{\partial \theta}=0.$$

The equation of cylinder in cylindrical coordinates is $r=a\cos\theta$. Thus,

$$\sqrt{r^2+\left(r\frac{\partial z}{\partial r}\right)^2+\left(\frac{\partial z}{\partial \theta}\right)^2}=\frac{ar}{\sqrt{a^2-r^2}}.$$

The surface area S is

$$S=4\int_0^{\frac{\pi}{2}}\int_0^{a\cos\theta}\frac{ar}{\sqrt{a^2-r^2}}\,dr\,d\theta$$

$$=4a^2\int_0^{\frac{\pi}{2}}(1-\sin\theta)d\theta=2(\pi-2)a^2 .$$

■

EXAMPLE 7

Let $f(x)\geq 0$ on $a\leq x\leq b$ and $f'(x)$ be continuous. Find the surface area of the surface of rotation, S, which is made by rotating $y=f(x)$ around x-axis on $a\leq x\leq b$. (See Figure 8.)

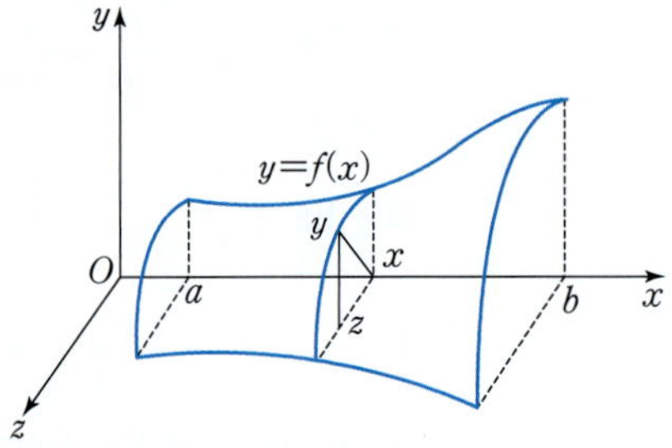

Figure 8

Solution The equation of the surface of rotation is

$$y^2+z^2=f^2(x),\ a\le x\le b.$$

Taking partial derivatives of the above equation with respect to x and y, we have

$$2zz_x=2f(x)f'(x)$$
$$2y+2zz_y=0.$$

Then, $z_x=\dfrac{f(x)f'(x)}{z}$, $z_y=-\dfrac{y}{z}$.

Let $f(x)$ and $f'(x)$ be f and f' to reduce the form of equations, respectively.

$$\sqrt{z_x^2+z_y^2+1}==\frac{\sqrt{f^2(f')^2+y^2+z^2}}{\sqrt{z^2}}=\frac{f\sqrt{(f')^2+1}}{\sqrt{f^2-y^2}}$$

Consider the case of $z\le 0$,

$$S=2\int_a^b\int_{-f}^{f}\frac{f\sqrt{f'^2+1}}{\sqrt{f^2-y^2}}dydx$$
$$=2\int_a^b f\sqrt{f'^2+1}\left\{\int_{-f}^{f}\frac{dy}{\sqrt{f^2-y^2}}\right\}dx.$$

But

$$\int_{-f}^{f}\frac{dy}{\sqrt{f^2-y^2}}=\left[\sin^{-1}\frac{y}{f}\right]_{-f}^{f}=\frac{\pi}{2}-\left(-\frac{\pi}{2}\right)=\pi.$$

Therefore $S=2\pi\displaystyle\int_a^b f(x)\sqrt{\{f'(x)\}^2+1}\,dx.$ ■

Note This result follows from the fact in section 8.2.

14.3 EXERCISES (Surface Area)

http://matrix.skku.ac.kr/Cal-Book/part2/CS-Sec-14-3-Sol.htm

CAS **1.** Find the surface area of the paraboloid $z=4-x^2-y^2$ that lies above the xy-plane.

Solution

http://matrix.skku.ac.kr/cal-lab/cal-14-3-1.html

```
var('x, y')
plot3d(4-x^2-y^2, (x, -2, 2), (y, -2, 2), opacity=0.3, color='yellow')+plot3d(0, (x, -2, 2), (y, -2, 2), opacity=0.3, color='blue')
```

$D=\{(r,\theta)\,|\,0\le r\le 2,\ 0\le\theta\le 2\pi\}$, $z_x=-2x$, $z_y=-2y$ $\sqrt{1+z_x^2+z_y^2}=\sqrt{4x^2+4y^2+1}$

$$S=\iint_D \sqrt{1+z_x^2+z_y^2}\,dA=\iint_D \sqrt{4x^2+4y^2+1}\,dA$$

$$=\int_0^{2\pi}\int_0^2 \sqrt{4r^2+1}\,r\,dr\,d\theta.$$

$$\sqrt{4r^2+1}=t \Rightarrow 4r^2+1=t^2 \Rightarrow r\,dr=\frac{1}{4}t\,dt$$

$$S=\frac{1}{4}\int_0^{2\pi}\int_1^{\sqrt{17}} t\cdot t\,dt\,d\theta=\frac{1}{4}\int_0^{2\pi}\int_1^{\sqrt{17}} t^2\,dt\,d\theta$$

$$=\frac{1}{4}\int_0^{2\pi}\left[\frac{1}{3}t^3\right]_1^{\sqrt{17}}d\theta$$

$$=\frac{1}{12}\int_0^{2\pi}(17\sqrt{17}-1)\,d\theta=\frac{1}{6}\pi(17\sqrt{17}-1).$$

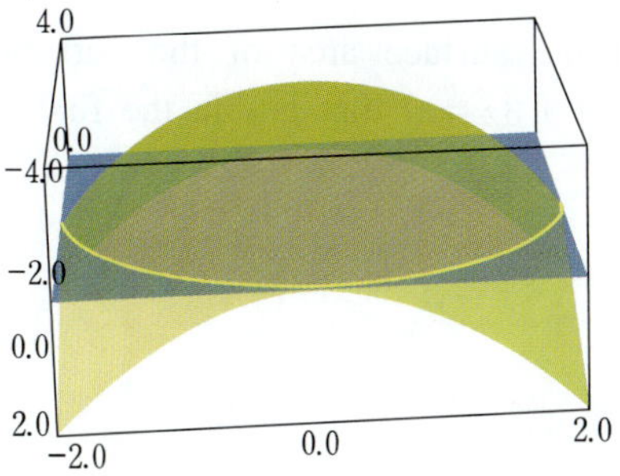

2. Find the surface area of the circular cylinder $x^2+y^2\le 4$ that lies under the hemispherical $z=\sqrt{16-x^2-y^2}$.

Solution

http://matrix.skku.ac.kr/cal-lab/cal-14-3-2.html

```
u ,v= var('u, v')
parametric_plot3d( (2*sin(u), 2*cos(u), v), (u, 0, 2*pi), (v, -5, 5), color='red', opacity=0.5)+parametric_plot3d( (v*sin(u), v*cos(u), sqrt(16-v^2)), (u, 0, 2*pi),(v, -3, 3))
```

$$z_x=-\frac{x}{\sqrt{16-x^2-y^2}},\quad z_y=-\frac{y}{\sqrt{16-x^2-y^2}}$$

$$\sqrt{1+z_x^2+z_y^2}=\sqrt{1+\frac{x^2}{16-x^2-y^2}+\frac{y^2}{16-x^2-y^2}}$$

$$=\frac{4}{\sqrt{16-x^2-y^2}}$$

$$S=\iint_D \sqrt{1+z_x^2+z_y^2}\,dA=\iint_D \frac{4}{\sqrt{16-x^2-y^2}}\,dA$$

$$=4\int_0^{2\pi}\int_0^2 \frac{1}{\sqrt{16-r^2}}\,dr\,d\theta=4\int_0^{2\pi}\left[-\sqrt{16-r^2}\,\right]_0^2 d\theta$$

$$=4\int_0^{2\pi}(4-2\sqrt{3})\,d\theta=4[4-2\sqrt{3}\,]_0^{2\pi}=16(2-\sqrt{3})\pi.$$

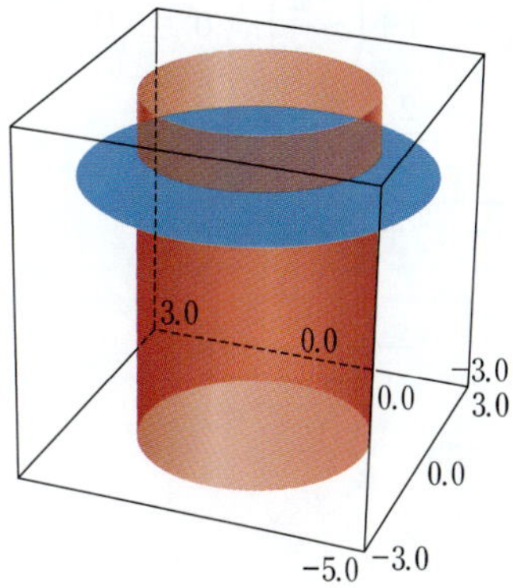

CAS **3.** Find the area in the first octant among the surface area which was made by intersecting the circular cylinder $x^2+y^2\le 1$ and the plane $z=2-x-y$.

Solution

http://matrix.skku.ac.kr/cal-lab/cal-14-3-3.html

```
var('u, v, x, y')
parametric_plot3d( (sin(u), cos(u), v), (u, 0, 2*pi), (v, -5, 5), color='red', opacity=0.5)+plot3d(2-x-y, (x, -2, 2), (y, -2,2), color='green')
```

$z_x=-1$, $z_y=-1$

$$\sqrt{1+z_x^2+z_y^2}=\sqrt{1+(-1)^2+(-1)^2}=\sqrt{3}$$

$$S=\iint_D \sqrt{1+z_x^2+z_y^2}\,dA$$

$$=\iint_D \sqrt{3}\,dA$$

$$=\sqrt{3}\iint_D dA=\frac{\sqrt{3}}{4}\pi.$$

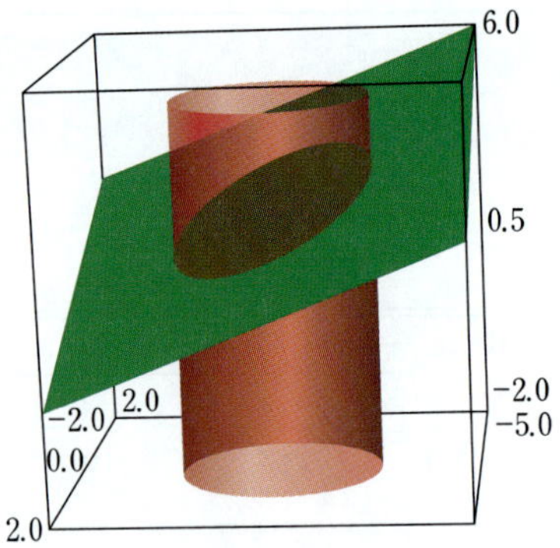

4. Find the surface area of the portion of the cylinder $x^2+z^2=4$ lying inside the cylinder $x^2+y^2=4$.

Solution $D=\{(x, y)\,|\,0 \le x \le 2,\ 0 \le y \le \sqrt{4-x^2}\}$,

$z_x=-\dfrac{x}{z}$, $z_y=0$

$\sqrt{1+z_x^2+z_y^2}=\sqrt{1+\left(-\dfrac{x}{z}\right)^2+0^2}=\sqrt{\dfrac{z^2+x^2}{z^2}}$

$=\sqrt{\dfrac{4}{4-x^2}}=\dfrac{2}{\sqrt{4-x^2}}$

$S=8\iint_D \sqrt{1+z_x^2+z_y^2}\,dA$

$=8\int_0^2\int_0^{\sqrt{4-x^2}} \dfrac{2}{\sqrt{4-x^2}}\,dy\,dx$

$=16\int_0^2 dx=32$

5. Find the surface area of the portion of the cylinder $r=4\sin\theta$ lying inside the sphere $x^2+y^2+z^2=16$.

Solution $D=\{(y, z)\,|\,0 \le y \le z,\ 0 \le z \le \sqrt{16-4y}\}$,

$z^2+x^2+y^2=16$, $x^2+y^2=4y$

$\begin{cases} z^2+r^2=16 \\ r^2=4r\sin\theta \Rightarrow r=4\sin\theta \end{cases}$

$\Rightarrow 4\sin\theta \le r \le 4$, $0 \le \theta \le 2\pi$

$\dfrac{\partial z}{\partial r}=-\dfrac{r}{z}$, $r^2+\left(r\dfrac{\partial z}{\partial r}\right)^2=r^2+\dfrac{r^4}{z^2}=r^2\dfrac{16}{16-r^2}$

$\int_0^{\pi}\int_{4\sin\theta}^{4}\sqrt{r^2+\left(r\cdot\dfrac{\partial z}{\partial r}\right)^2+\left(\dfrac{\partial z}{\partial \theta}\right)^2}\,drd\theta$

$=\int_0^{\pi}\int_{4\sin\theta}^{4} 2\cdot 4\cdot\dfrac{r}{\sqrt{16-r^2}}\,drd\theta=64$

6. Find the surface area of a right circular cone of height h and base radius a.

7. Find the surface area of the upper half of the cone $x^2+y^2=z^2$ inside the cylinder $x^2+y^2=2x$.

8. Find the surface area of the portion of the sphere $x^2+y^2+z^2=16$ that is inside the cylinder $x^2+y^2=4y$.

9. Find the surface area of the portion of the cylinder $y^2+z^2=16$ that is within the cylinder $x^2+z^2=16$.

10. Find the surface area of the part of the plane $x+2x+3z=6$ that lies in the first octant.

14.4 Cylindrical Coordinates and Spherical Coordinates

YouTube http://youtu.be/q3FKd2UxV_I

Cylindrical and spherical coordinates are two coordinate systems in three dimensions that are similar to polar coordinates in plane. In problems that involve symmetry about an axis, cylindrical coordinates are useful. Normally the z-axis is chosen to coincide with this axis of symmetry. In problems where there is symmetry about a point, the spherical coordinate system is useful. For example, the circular cylinder with Cartesian equation $x^2+y^2=c^2$ has a very simple equation $r=c$ in cylindrical coordinates. (See Figure 1.)

The sphere with center at the origin and radius c has the simple equation $\rho=c$ in spherical coordinates. (See Figure 2.)

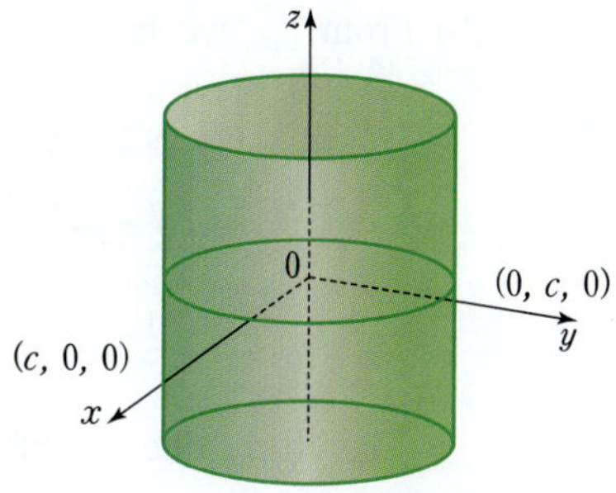

Figure 1 $r=c$, a cylinder.

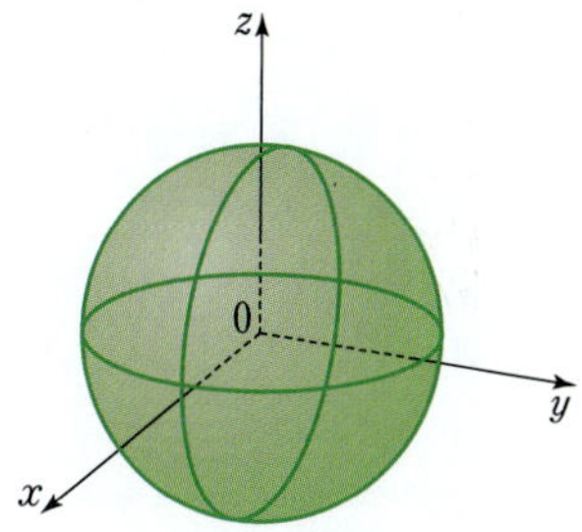

Figure 2

Cylindrical Coordinates

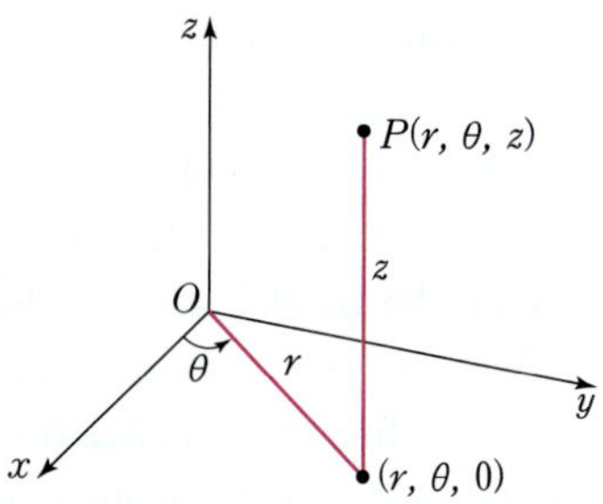

Figure 3 The cylindrical coordinates of a point.

A point $P(x, y, z)$ in three-dimensional space is represented in the **cylindrical coordinate system** by the ordered triple (r, θ, z), where r and θ are polar coordinates of the projection of P onto the xy-plane and z is the directed distance from the xy-plane to P. (See Figure 3.)

The equations

1 $$x=r\cos\theta \qquad y=r\sin\theta \qquad z=z$$

convert cylindrical to rectangular coordinates, while

2 $$r^2=x^2+y^2 \qquad \tan\theta=\frac{y}{x} \qquad z=z$$

convert rectangular to cylindrical coordinates.

EXAMPLE 1

(a) Find the rectangular coordinates of the point whose cylindrical coordinates are $(4, \pi/3, -3)$.

(b) Find the cylindrical coordinates of the point whose rectangular coordinates are $(-\sqrt{2}, \sqrt{2}, 1)$.

Solution (a) From [1], the rectangular coordinates of the given point are

$$x = 4\cos\frac{\pi}{3} = 4\left(\frac{1}{2}\right) = 2,$$

$$y = 4\sin\frac{\pi}{3} = 4\left(\frac{\sqrt{3}}{2}\right) = 2\sqrt{3},$$

$$z = -3.$$

Thus, the point $(2, 2\sqrt{3}, -3)$ is in rectangular coordinates.

(b) From [2], we have

$$r = \sqrt{(-\sqrt{2})^2 + (\sqrt{2})^2} = 2,$$

$$\tan\theta = \frac{\sqrt{2}}{-\sqrt{2}} = -1, \quad \text{so} \quad \theta = -\frac{\pi}{4}$$

$$z = 1.$$

Therefore, one set of cylindrical coordinates is $\left(2, -\frac{\pi}{4}, 1\right)$. Another is $\left(2, \frac{3\pi}{4}, 1\right)$. As with polar coordinates, there are many different expressions. ∎

EXAMPLE 2

Describe the surface whose equation in cylindrical coordinates is $z = 2r$.

Solution The equation says that the z-value, or height, of each point on the surface is the same as $2r$, the distance from the point to the z-axis. Because θ doesn't appear, it can vary. So any horizontal trace in the plane $z = k(k > 0)$ is a circle of radius k. These traces suggest that the surface is a cone. This prediction can be confirmed by converting the equation into rectangular coordinates. From the first equation in [2] we have

$$z^2 = 4r^2 = 4(x^2 + y^2).$$

We recognize the equation $z^2 = 4(x^2 + y^2)$ (by comparison with Table 1 in Section 11.6) as being a circular cone whose axis is the z-axis. (See Figure 4.) ∎

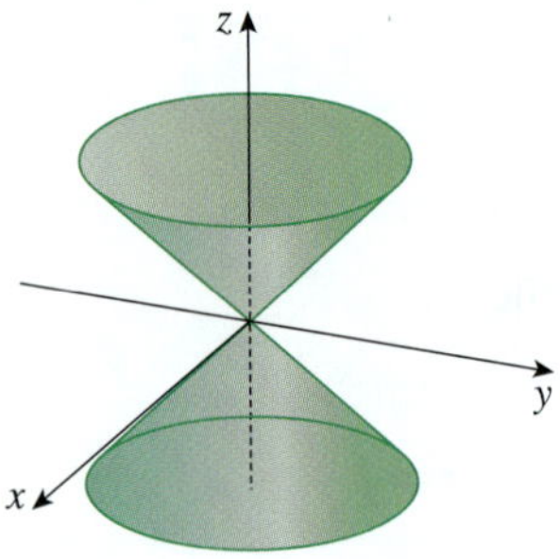

Figure 4 $z = 2r$ is a cone.

EXAMPLE 3

Find an equation in cylindrical coordinates for the ellipsoid $5x^2 + 5y^2 + z^2 = 1$.

Solution Since $r^2 = x^2 + y^2$ from [2], we have

$$z^2 = 1 - 5(x^2 + y^2) = 1 - 5r^2.$$

So, the equation of the given ellipsoid in cylindrical coordinates is $z^2 = 1 - 5r^2$. ■

Spherical Coordinates

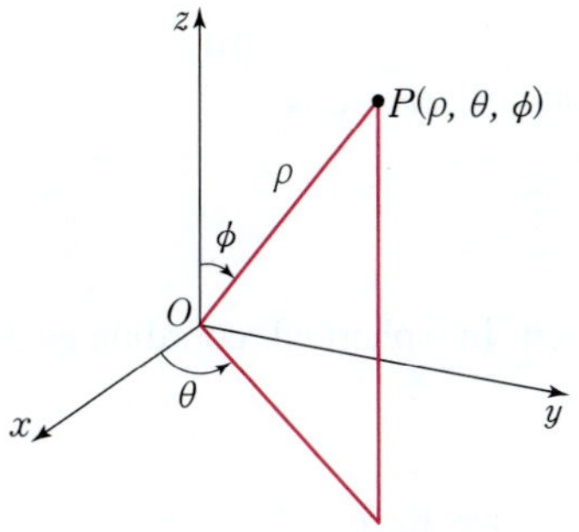

Figure 5 The spherical coordinates of a point.

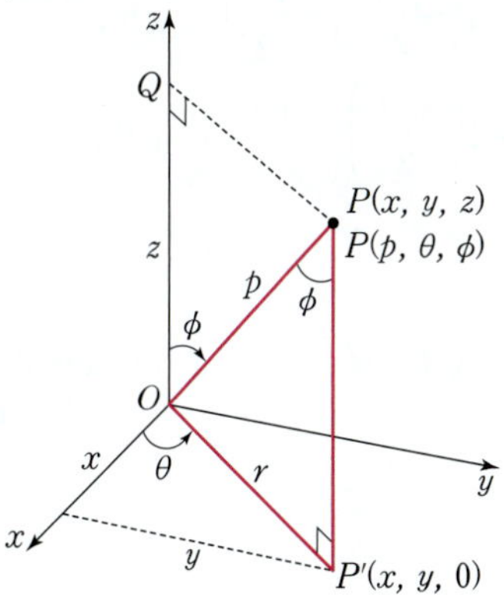

Figure 6

In mathematics, a spherical coordinate system is a coordinate system for three-dimensional space where the position of a point is specified by three numbers. A point $P(x, y, z)$ in space is represented in **spherical coordinates** by (ρ, θ, ϕ) as shown in Figure 5. Here, $\rho = |OP|$ is the distance from the origin to P, θ is the same angle as in cylindrical coordinates, and ϕ is the angle between the positive z-axis and the line segment OP. Note that $\rho \geq 0$, $0 \leq \phi \leq \pi$.

The spherical coordinates (ρ, θ, ϕ) of a point can be obtained from its Cartesian coordinates $P(x, y, z)$ by the formulae $\rho = \sqrt{x^2 + y^2 + z^2}$, $\theta = \arccos\left(\frac{z}{\rho}\right)$, $\phi = \arctan\left(\frac{y}{x}\right)$. From triangles OPQ and OPP' in Figure 6, we have

$$z = \rho \cos \phi, \quad r = \rho \sin \phi.$$

But $x = r \cos \theta$ and $y = r \sin \theta$. Conversely, the Cartesian coordinates may be retrieved from the spherical coordinates (ρ, θ, ϕ), where $\rho \in [0, \infty)$, $\theta \in [0, 2\pi]$, $\phi \in [0, \pi]$, by

[3]

$$x = \rho \sin \phi \cos \theta \qquad y = \rho \sin \phi \sin \theta \qquad z = \rho \cos \phi$$

and

[4]

$$\rho^2 = x^2 + y^2 + z^2$$

The graph of the equation $\theta = c$ is a vertical half-plane (See Figure 7), and the equation $\phi = c$ represents a half-cone with the z-axis as its axis. (See Figure 8.)

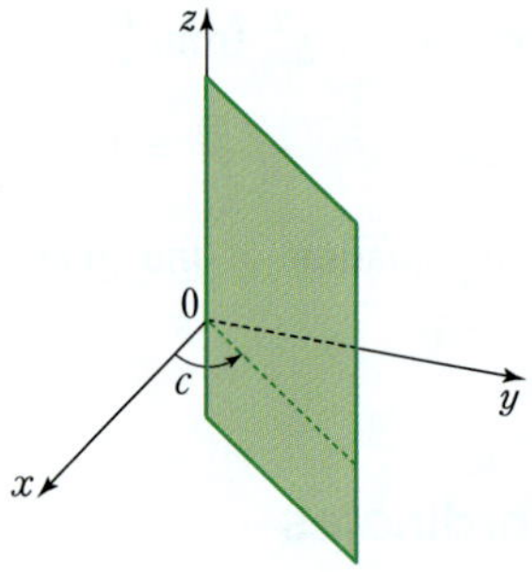

Figure 7 $\theta = c$, a half-plane.

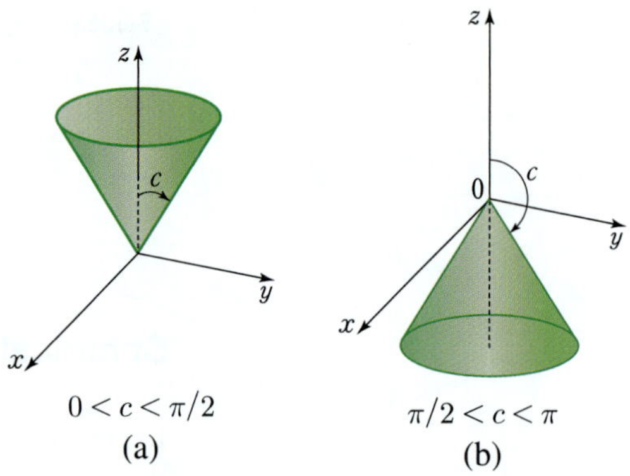

Figure 8 $\phi = c$, a half-cone.

EXAMPLE 4

Convert the point $(4,\ \pi/4,\ \pi/3)$ that is given in spherical coordinates to rectangular coordinates.

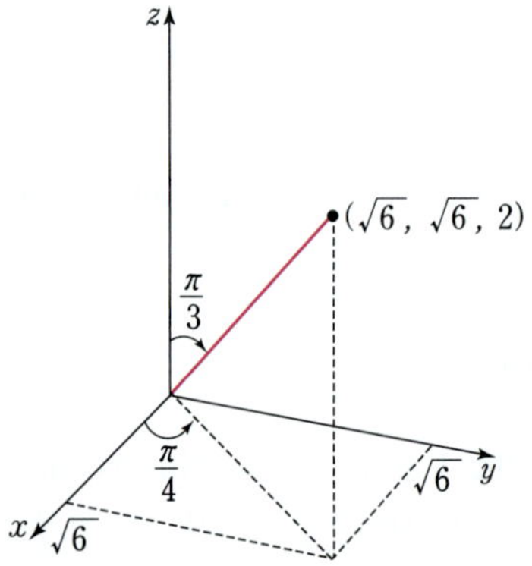

Figure 9

Solution We plot the point in Figure 6. From 3 we have

$$x = \rho \sin\phi \cos\theta = 4\sin\frac{\pi}{3}\cos\frac{\pi}{4} = 4\left(\frac{\sqrt{3}}{2}\right)\left(\frac{1}{\sqrt{2}}\right) = \sqrt{6},$$

$$y = \rho \sin\phi \sin\theta = 4\sin\frac{\pi}{3}\sin\frac{\pi}{4} = 4\left(\frac{\sqrt{3}}{2}\right)\left(\frac{1}{\sqrt{2}}\right) = \sqrt{6},$$

$$z = \rho\cos\phi = 4\cos\frac{\pi}{3} = 4\left(\frac{1}{2}\right) = 2.$$

Thus, the point $(4,\ \pi/4,\ \pi/3)$ is $(\sqrt{6}, \sqrt{6}, 2)$ in rectangular coordinates. (See Figure 9.) ■

EXAMPLE 5

The point $(0,\ 2\sqrt{3},\ 2)$ is given in rectangular coordinates. Find spherical coordinates for this point.

Solution From 4 we have $\rho = \sqrt{x^2+y^2+z^2} = \sqrt{0+12+4} = 4$ and so 3 gives

$$\cos\phi = \frac{z}{\rho} = \frac{2}{4} = \frac{1}{2}, \qquad \phi = \frac{\pi}{3},$$

$$\cos\theta = \frac{x}{\rho\sin\phi} = 0, \qquad \theta = \frac{\pi}{2}.$$

(Note that $\theta \neq 3\pi/2$ because $y = 2\sqrt{3} > 0$.) Therefore, spherical coordinates of the given point are $(4,\ \pi/2,\ \pi/3)$. ■

EXAMPLE 6

Find an equation in spherical coordinates for the hyperboloid of two sheets with equation $x^2 - y^2 - z^2 = 2$.

Solution Substituting the expressions in 3 into the given equation, we have

$$\rho^2\sin^2\phi\cos^2\theta - \rho^2\sin^2\phi\sin^2\theta - \rho^2\cos^2\phi = 2$$
$$\rho^2[\sin^2\phi(\cos^2\theta - \sin^2\theta) - \cos^2\phi] = 2 \text{ or}$$
$$\rho^2(\sin^2\phi\cos 2\theta - \cos^2\theta) = 2.$$
■

EXAMPLE 7

Find a rectangular equation for the surface with spherical equation given by $\rho = 2\sin\theta\sin\phi$.

Solution From 3 and 4 we have

$$x^2 + y^2 + z^2 = 2\rho\sin\theta\sin\phi = 2y \quad \text{or}$$
$$x^2 + (y-1)^2 + z^2 = 1$$

which is the equation of a sphere with center $(0,\ 1,\ 0)$ and radius 1.
■

CAS **EXAMPLE 8**

Find the rectangular coordinates of the point

$$\left(1,\ \frac{\pi}{4},\ \frac{\pi}{4}\right) \text{ in spherical coordinates.}$$

Solution

```
T = Spherical('radius', ['azimuth', 'inclination'])
T.transform(radius=1, azimuth=pi/4, inclination=pi/4)
#r, theta=var('r, theta')
#plot3d(r^2, (r, 0, 2), (theta, 0, 2*pi), transformation=T)
```

Answer : (1/2, 1/2, 1/2*sqrt(2)) ■

CAS **EXAMPLE 9**

Identify the surface whose equation is given by $z = r^2$.
http://matrix.skku.ac.kr/cal-lab/cal-14-4-Exm-9.html

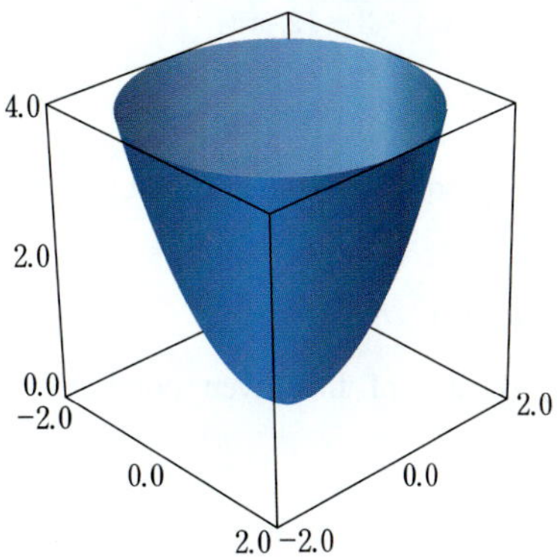

Figure 10

Solution

```
S=Cylindrical('height', ['radius', 'azimuth'])
r, theta=var('r, theta')
plot3d(r^2, (r, 0, 2), (theta, 0, 2*pi), transformation=S)
```

Answer : A cylindrical cone ■

14.4 EXERCISES (Cylindrical Coordinates and Spherical Coordinates)

http://matrix.skku.ac.kr/Cal-Book/part2/CS-Sec-14-4-Sol.htm

YouTube http://youtu.be/F9u6pMubVRs

1-4. Plot the point whose cylindrical coordinates are given. Then find the rectangular coordinates of the point.

1. $\left(2, -\frac{\pi}{3}, 4\right)$.

http://matrix.skku.ac.kr/cal-lab/cal-14-2-1.html

Solution

```
T = Cylindrical('height', ['radius', 'azimuth'])
T.transform(radius=2, azimuth=- pi/3, height=4)
```

Answer : The rectangular coordinates of the point of the point is (1, -sqrt(3), 4).

CAS **2.** $\left(1, \frac{3\pi}{2}, 3\right)$.

Solution

```
T = Cylindrical('height', ['radius', 'azimuth'])
T.transform(radius=1, azimuth= 3*pi/3, height=3)
```

Answer : $(0, -1, 3)$.

3. $(3, \pi, -6)$.

Solution $(-3, 0, -6)$.

4. $(2, \pi, e)$.

Solution $(-2, 0, e)$.

5-6. Change from rectangular to cylindrical coordinates.

5. $(1, -\sqrt{3}, 4)$.

Solution $r=\sqrt{1^2+(-\sqrt{3})^2}=\sqrt{4}=2$,

$\tan\theta=\frac{-\sqrt{3}}{1}=-\sqrt{3}$ so $\theta=\frac{5\pi}{3}+2n\pi,\ z=4$.

Therefore one set of cylindrical coordinates is $\left(2, \frac{5\pi}{3}, 4\right)$.

Another is $\left(2, -\frac{\pi}{3}, 4\right)$.

As with polar coordinates, there are infinitely many choices.

6. $(3, -4, 5)$.

7-10. Plot the point whose spherical coordinates are given. Then, find the rectangular coordinates of the point.

CAS **7.** $\left(3, \frac{\pi}{6}, \frac{\pi}{6}\right)$.

http://matrix.skku.ac.kr/cal-lab/cal-14-2-7.html

Solution

```
T = Spherical('radius', ['azimuth', 'inclination'])
T.transform(radius=3, azimuth=pi/6, inclination=pi/6)
```

Answer : (3/4*sqrt(3), 3/4, 3/2*sqrt(3))

8. $\left(5, \frac{\pi}{3}, \frac{\pi}{2}\right)$.

Solution $\left(\frac{5}{2}, \frac{5}{2}\sqrt{3}, 0\right)$.

9. $\left(1, \frac{\pi}{3}, \frac{\pi}{4}\right)$.

Solution $\left(\frac{1}{4}\sqrt{2}, \frac{1}{4}\sqrt{6}, \frac{1}{2}\sqrt{2}\right)$.

10. $\left(2, \frac{\pi}{4}, \frac{2\pi}{3}\right)$.

Solution $\left(\frac{\sqrt{6}}{2}, \frac{\sqrt{6}}{2}, -1\right)$.

11-12. Change from rectangular to spherical coordinates.

11. $(1, -\sqrt{3}, 2\sqrt{3})$.

Solution $\rho=\sqrt{x^2+y^2+z^2}=\sqrt{1+3+12}=4$,

$\cos\phi=\frac{z}{\rho}=\frac{2\sqrt{3}}{4}=\frac{\sqrt{3}}{2}$ so $\phi=\frac{\pi}{6}$,

$\cos\theta=\frac{x}{\rho\sin\phi}=\frac{1}{4\left(\frac{1}{2}\right)}=\frac{1}{2}$ so $\theta=\frac{5\pi}{3}$ (Note that $\theta\neq\frac{\pi}{3}$ because $y=-\sqrt{3}<0$.)

Therefore, spherical coordinates of the given point are $\left(4, \frac{5\pi}{3}, \frac{\pi}{6}\right)$.

12. $(0, \sqrt{3}, \sqrt{3})$.

Solution $\left(\sqrt{6}, \frac{\pi}{2}, \frac{\pi}{4}\right)$.

13-14. Change from cylindrical to spherical coordinates.

13. $\left(1,\ \frac{\pi}{6},\ 1\right)$.

Solution $\tan\theta = \frac{y}{x} = \frac{1}{\sqrt{3}},\ y = \frac{1}{\sqrt{3}}x,$

$x^2+y^2 = x^2+\frac{1}{3}x^2 = \frac{4}{3}x^2 = 1,\ x = \pm\frac{\sqrt{3}}{2}.$

So, $y=\pm\frac{1}{2}$, and $z=1$.

Therefore, rectangular coordinates is $\left(\frac{\sqrt{3}}{2},\ \frac{1}{2},\ 1\right)$ and $\left(-\frac{\sqrt{3}}{2},\ -\frac{1}{2},\ 1\right)$.

$\rho = \sqrt{\frac{3}{4}+\frac{1}{4}+1} = \sqrt{2},\ \cos\phi = \frac{z}{\rho} = \frac{1}{\sqrt{2}},\ \text{so}\ \phi = \frac{\pi}{4}.$

$$\cos\theta = \frac{x}{\rho\sin\phi} = \frac{\pm\frac{\sqrt{3}}{2}}{\sqrt{2}\left(\frac{1}{\sqrt{2}}\right)} = \pm\frac{\sqrt{3}}{2},$$

so $\theta = \frac{\pi}{6},\ \frac{7\pi}{6}\left(\theta \neq \frac{5\pi}{6},\ \frac{11\pi}{6}\right.$ because $\left.\tan\theta > 0\right)$.

Therefore, spherical coordinates of the given point are $\left(\sqrt{2},\ \frac{\pi}{6},\ \frac{\pi}{4}\right)$ and $\left(\sqrt{2},\ \frac{7\pi}{6},\ \frac{\pi}{4}\right)$.

14. $\left(4,\ \frac{\pi}{4},\ \sqrt{2}\right)$.

Solution $\left(3\sqrt{2},\ \frac{\pi}{4},\ \arccos\left(\frac{1}{3}\right)\right)$.

15-16. Change from spherical to cylindrical coordinates.

15. $\left(8,\ \frac{\pi}{3},\ \frac{\pi}{2}\right)$.

Solution $\cos\phi = \cos\frac{\pi}{2} = 0 = \frac{z}{\rho},\ z=0,$

$\cos\theta = \cos\frac{\pi}{3} = \frac{1}{2} = \frac{x}{\rho\sin\phi} = \frac{x}{8(1)},\ x=4.$

So, $x^2+y^2+z^2 = 16+y^2+0 = 64,\ y=\pm 4\sqrt{3}$

since $\tan\theta > 0,\ y = 4\sqrt{3}$.

Therefore, rectangular coordinates are $(4, 4\sqrt{3}, 0)$

$r = \sqrt{16+48} = 8,\ \tan\theta = \frac{y}{x} = \sqrt{3},\ \text{so}\ \theta = \frac{\pi}{3}.$

Therefore cylindrical coordinates of the given point is $\left(8, \frac{\pi}{3}, 0\right)$.

16. $\left(4,\ \frac{\pi}{4},\ \frac{2\pi}{3}\right)$.

Solution $x = \rho\sin\phi\cos\theta = 4\sin\frac{2\pi}{3}\cos\frac{\pi}{4} = \sqrt{6},$

$y = \rho\sin\phi\sin\theta = 4\sin\frac{2\pi}{3}\sin\frac{\pi}{4} = \sqrt{6},$

$z = \rho\cos\phi = 4\cos\frac{2\pi}{3} = -2$

Therefore rectangular coordinates of the given point are $(\sqrt{6}, \sqrt{6}, -2)$.

$r = \sqrt{(\sqrt{6})^2+(\sqrt{6})^2} = 2\sqrt{3},$

$\tan\theta = \frac{y}{x} = \frac{\sqrt{6}}{\sqrt{6}} = 1,\ \text{so}\ \theta = \frac{\pi}{4}.$

Therefore cylindrical coordinates of the given point is $\left(2\sqrt{3},\ \frac{\pi}{4},\ -2\right)$.

17-20. Describe in words the surface whose equation is given.

17. $r=4$.

http://matrix.skku.ac.kr/cal-lab/cal-14-2-17.html

Solution

```
S=Cylindrical('radius', ['azimuth', 'height'])
theta, z=var('theta, z')
plot3d(4, (theta, 0, 2*pi), (z, -2, 2),
transformation=S)
```

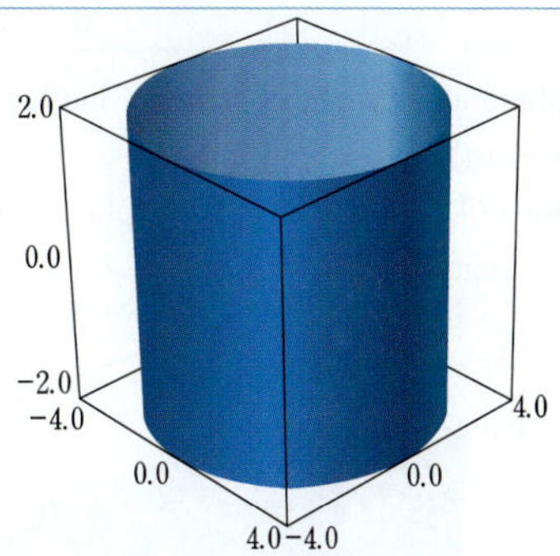

CAS **18.** $\rho=4$.

http://matrix.skku.ac.kr/cal-lab/cal-14-2-18.html

Solution

```
S=Spherical('radius',['azimuth','inclination']);
var('p,theta')
plot3d(4,(p,0,10),(theta,0,2*pi),transformation=S)
```

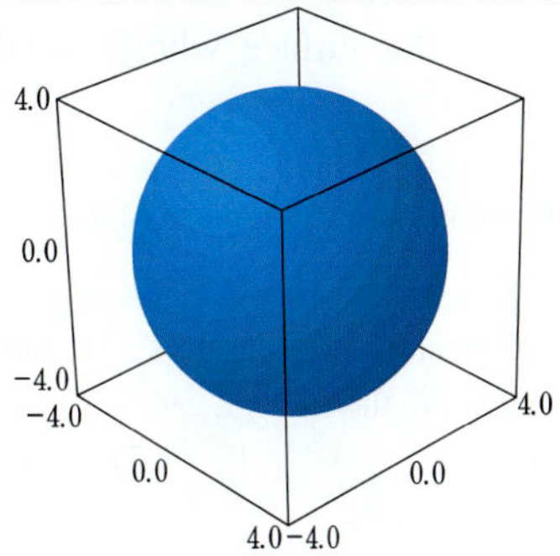

CAS **19.** $\phi = \dfrac{2}{3}\pi$.

http://matrix.skku.ac.kr/cal-lab/cal-14-2-19.html

Solution

```
S=Spherical('radius',['azimuth','inclination']);
var('p,theta')
plot3d(4,(p,0,10),(theta,0,2*pi),transformation=S)
```

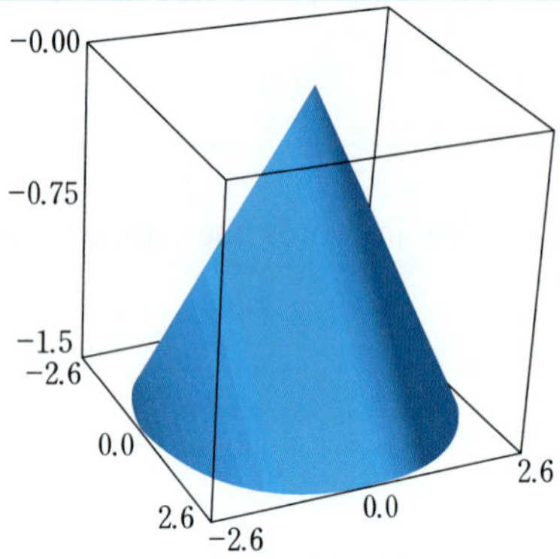

CAS **20.** $\theta = \dfrac{\pi}{3}$.

http://matrix.skku.ac.kr/cal-lab/cal-14-2-20.html

Solution

```
S=Spherical('azimuth', ['radius', 'inclination'])
r, phi=var('r, phi')
plot3d(pi/3, (r,0, 10), (phi, 0, pi) ,
transformation=S)
```

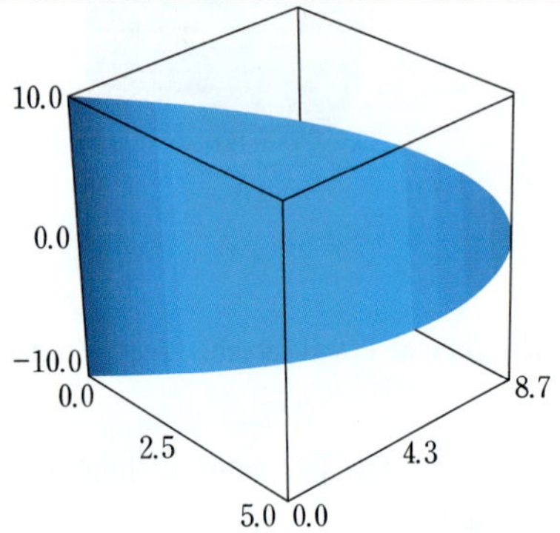

CAS **21-28.** Identify the surface whose equation is given.

21. $z = 2r^2$.

http://matrix.skku.ac.kr/cal-lab/cal-14-2-21.html

Solution

```
S=Cylindrical('height', ['radius', 'azimuth'])
r, theta=var('r, theta')
plot3d(2*r^2, (r, 0, 2), (theta,0,2*pi),
transformation=S)
```

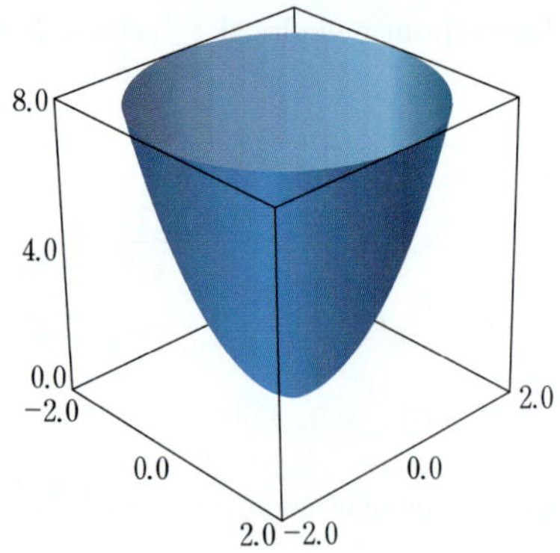

22. $r = 2\sin\theta$.

http://matrix.skku.ac.kr/cal-lab/cal-14-2-22.html

Solution

```
S=Cylindrical('radius', ['azimuth', 'height'])
theta,z=var('theta, z')
plot3d(2*sin(theta), (theta,0,2*pi), (z, -2, 2),
transformation=S)
```

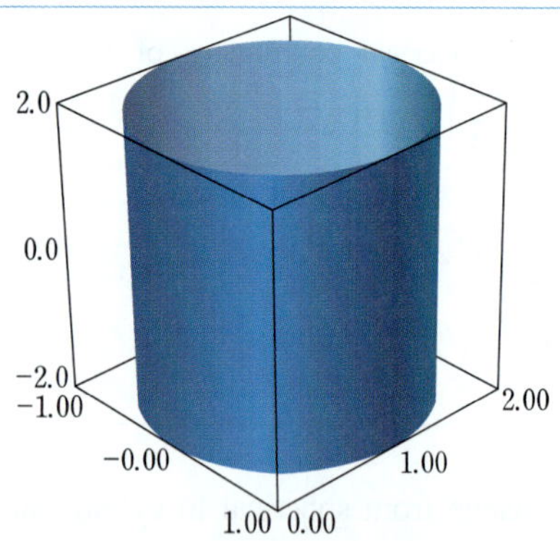

23. $\rho\cos\phi = \dfrac{1}{2}$.

http://matrix.skku.ac.kr/cal-lab/cal-14-2-23.html

Solution

```
S=Spherical('radius', ['azimuth', 'inclination'])
theta, phi=var('theta, phi')
plot3d(1/(2*cos(phi)), (theta, 0, 2*pi), (phi, 0,
pi), transformation=S)
```

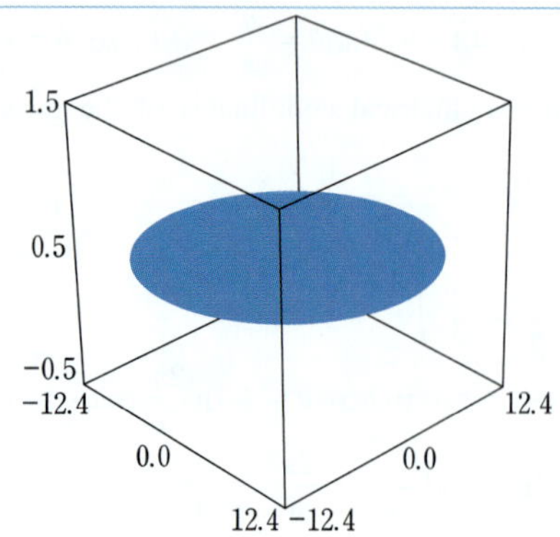

24. $\rho\sin\phi = 2$.

http://matrix.skku.ac.kr/cal-lab/cal-14-2-24.html

Solution

```
T = Cylindrical('height',['radius','azimuth']);
var('r,t,z');
implicit_plot3d(r*sin(z)==2,(r,0,3),(t,0,2*pi),(z,-2,2)
,transformation=T)
```

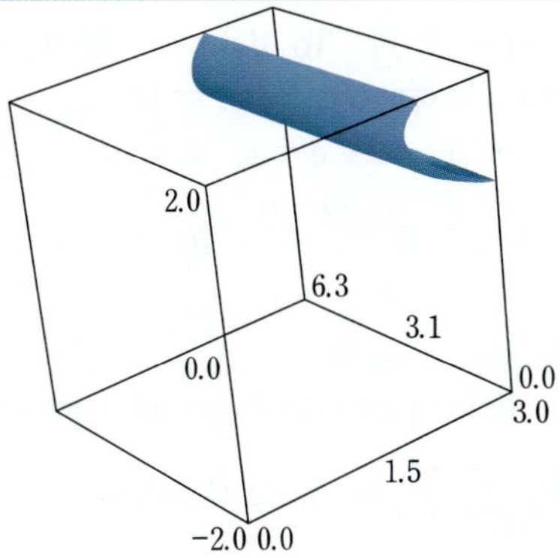

25. $r = 3\cos\theta$.

http://matrix.skku.ac.kr/cal-lab/cal-14-2-25.html

Solution

```
T=Cylindrical('radius', ['azimuth', 'height'])
theta, z=var('theta, z')
plot3d(3*cos(theta), (theta, 0, 2*pi), (z, -2, 2),
transformation=T)
```

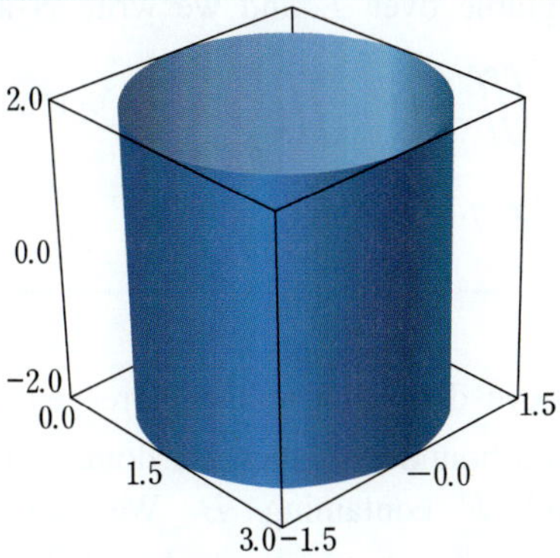

26. $\rho = 3\cos\phi$.

http://matrix.skku.ac.kr/cal-lab/cal-14-2-26.html

Solution

```
T = Cylindrical('height',['radius','azimuth']);
var('r,t,z');
implicit_plot3d(r==3*cos(z),(r,0,3),(t,0,2*pi),(z,-2,2
),transformation=T)
```

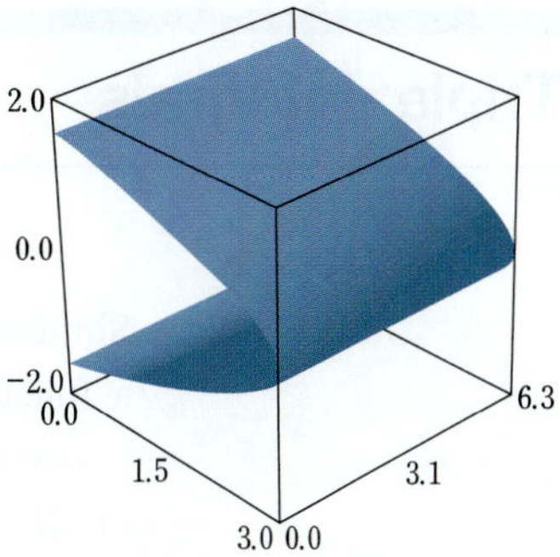

27. $r^2 + z^2 = 9$.

http://matrix.skku.ac.kr/cal-lab/cal-14-2-27.html

Solution

```
T=Cylindrical('height', ['radius', 'azimuth'])
r, theta=var('r, theta')
p1=plot3d((9-r^2)^(1/2), (r, -3, 3), (theta, 0,
2*pi), transformation=T)
p2=plot3d(-(9-r^2)^(1/2), (r, -3, 3), (theta, 0,
2*pi), transformation=T)
show(p1+p2, aspect_ratio=1)
```

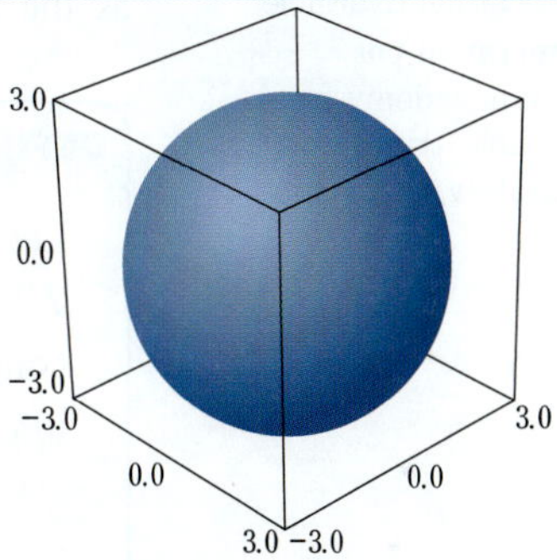

28. $\rho^2(\sin^2\phi + 4\cos^2\phi) = 2$.

http://matrix.skku.ac.kr/cal-lab/cal-14-2-28.html

Solution

```
T = Cylindrical('height',['radius','azimuth']);
var('r,t,z');
implicit_plot3d(r^2*(sin(z)^2+4*cos(z)^2)==2,(r,0,
3),(t,0,2*pi),(z,-2,2),transformation=T)
```

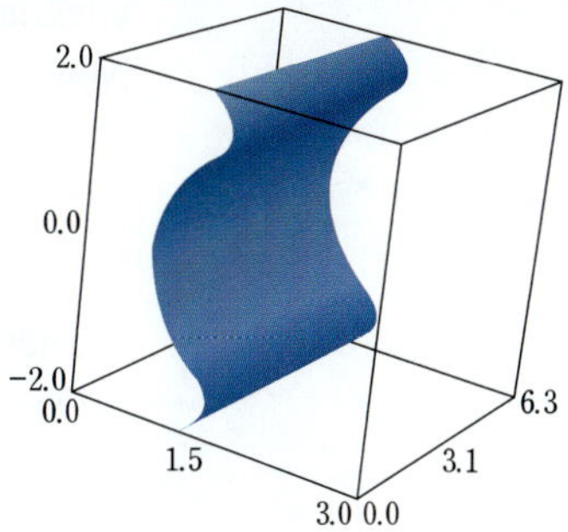

14.5 Triple Integrals

YouTube http://youtu.be/r1tzH9Ibbqk

Similar to the double integral in Section 14.1, we can define triple integrals of a function f.

Given a (bounded) function $f: E \to \mathbb{R}$, where E is a rectangular parallelepiped in $\mathbb{R}^3$, given by $E=[a, b]\times[c, d]\times[e, f]$. To define a Riemann sum, we divide E into subrectangular parallelepipeds $E_{ijk}=[x_i, x_{i+1}]\times[y_j, y_{j+1}]\times[z_k, z_{k+1}]$ with $a=x_0<x_1<\ldots<x_l=b$, $c=y_0<y_1<\cdots<y_m=d$, $e=z_0<z_1<\cdots<z_n=f$, and $x_{i+1}-x_i=\dfrac{b-a}{l}$, $y_{j+1}-y_j=\dfrac{d-c}{m}$, $z_{k+1}-z_k=\dfrac{f-e}{n}$.

Thus we divide a rectangular parallelepiped into lmn sub-parallelepipeds and form the Riemann sum $S_{l,m,n}=\sum_{i=0}^{l-1}\sum_{j=0}^{m-1}\sum_{k=0}^{n-1} f(c_{ijk})\Delta V$, where $c_{ijk}\in E_{ijk}$ is any random point and ΔV is the volume of E_{ijk}. Then we define the triple integral as the limit of $S_{l,m,n}$ if it exists.[1), 2)]

1) http://www.whitman.edu/mathematics/multivariable/multivariable_15_Multiple_Integration.pdf
2) https://www.khanacademy.org/math/calculus/double_triple_integrals/triple_integrals/v/triple-integrals-3

DEFINITION 1

Let f be bounded real-valued functions on a rectangular parallelepiped $E=[a,b]\times[c,d]\times[e,f]$. If the Riemann $S_{l,m,n}$ defined above converges to a limit S, independent of the choice of the sample point c_{ijk}, then we say that f is integrable over E and we write S as

$$\int_E f,\ \int_E f(x,y,z)dV,\ \iiint_E f dV,\ \iiint_E f(x,y,z)dV \text{ or }$$
$$\iiint_E f(x,y,z)dxdydz$$

As before, one can extend the above definition to the more general bounded "elementary domain". If $B\subset\mathbb{R}^3$ is a bounded elementary domain, then we can choose a rectangular parallelepiped E containing B. We can extend the integrand function f defined on B to a function f^* which vanishes outside B. Now we can define the triple integral over the general domain as before.

$$\iiint_B f(x,y,z)dV=\iiint_E f^*(x,y,z)dV.$$

These triple integrals can be evaluated as three-fold iterated integrals.

One can also extend Fubini's Theorem for triple integrals. The next example shows how to use Fubini's Theorem to compute triple integrals.

EXAMPLE 1

Find $\int_2^5\int_1^2\int_2^3 xy^2\,dz\,dy\,dx$.

Solution $\int_2^5\int_1^2\int_2^3 xy^2\,dz\,dy\,dx = \int_2^5\int_1^2 \left[xy^2z\right]_2^3 dydx$

$$= \int_2^5\int_1^2 xy^2\,dy\,dx$$

$$= \int_2^5 \left[\frac{1}{3}xy^3\right]_1^2 dx$$

$$= \int_2^5 \frac{7}{3}x\,dx = \left[\frac{7}{6}x^2\right]_2^5$$

$$= \frac{49}{2}.$$

■

Note that, if $f(x, y, z) = 1$, then the volume of the solid B is

$$V = \iiint_B dV.$$

EXAMPLE 2

Find the volume of the region bound by the tetrahedron in the first octant bounded by the coordinate planes and the plane passing through $(2, 0, 0)$, $(0, 3, 0)$ and $(0, 0, 6)$.

Solution The tetrahedron T and its projection D onto the xy-plane. The lower boundary T is the plane $z = 0$ and upper boundary T is the plane $3x + 2y + z = 6$. Therefore

$$E = \left\{(x, y, z)\,|\,0 \le x \le 2, 0 \le y \le -\frac{3}{2}x + 3, 0 \le z \le 6 - 2y - 3x\right\}$$

$$V = \iiint_E dV = \int_0^2\int_0^{-\frac{3}{2}x+3}\int_0^{6-2y-3x} dz\,dy\,dx$$

$$= \int_0^2\int_0^{-\frac{3}{2}x+3} (6 - 2y - 3x)\,dy\,dx$$

$$= 6.$$

■

EXAMPLE 3

Evaluate $\iiint_E yz\cos(x^3)dV$ where E is the solid defined by $E = \{(x,y,z)|0 \le x \le 1, 0 \le y \le \sqrt{x}, \sqrt{x} \le z \le 2\sqrt{x}\}$.

Solution By using Fubini's Theorem one can the above triple integral as

$$\iiint_E yz\cos(x^3)\,dV = \int_0^1\int_0^{\sqrt{x}}\int_{\sqrt{x}}^{2\sqrt{x}} yz\cos(x^3)\,dz\,dy\,dx$$
$$= \int_0^1\left(\int_{\sqrt{x}}^{2\sqrt{x}} z\,dz\right)\left(\int_0^{\sqrt{x}} y\,dy\right)\cos(x^3)\,dx$$
$$= \int_0^1 \frac{3}{4}x^2\cos(x^3)\,dx$$
$$= \left[\frac{1}{4}\sin(x^3)\right]_0^1 = \frac{1}{4}\sin 1.$$ ■

CAS **EXAMPLE 4**

Evaluate $\iiint_E f\,dV$, where $f = 2x + y$, E is the closed region bounded by the cylinder $z = 4 - x^2$ and the planes $x = 0$, $y = 0$, $y = 2$ and $z = 0$. (See Figure 1.)

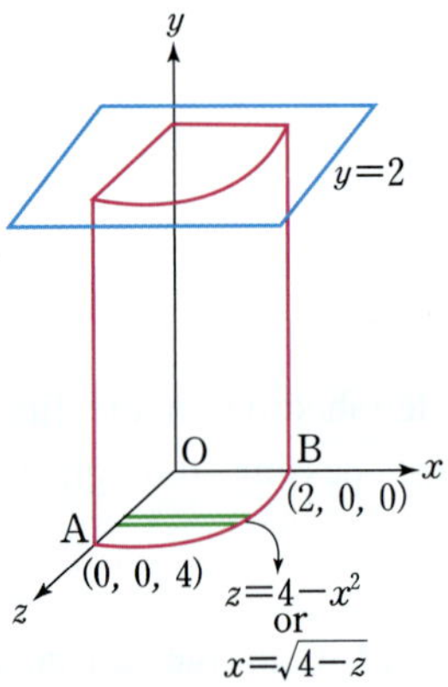

Figure 1

Solution This closed region is covered if x and z varies on $x^2 + z^2 = 4$ covering the area OAB and y varies from 0 to 2. Therefore

$$E = \{(x, y, z) \in \mathbb{R}^3 \mid 0 \le x \le \sqrt{4-z}, 0 \le y \le 2, 0 \le z \le 4\}$$

$$\iiint_E (2x+y)\,dV = \int_0^2\int_0^4\int_0^{\sqrt{4-z}} (2x+y)\,dx\,dz\,dy$$
$$= \int_0^2\int_0^4 (x^2+xy)\Big|_0^{\sqrt{4-z}}\,dz\,dy$$
$$= \int_0^2\int_0^4 [(4-z)+y\sqrt{4-z}\,]\,dz\,dy$$
$$= \int_0^2 \left[\left(4z-\frac{z^2}{2}\right)-\frac{2}{3}y(4-z)^{\frac{3}{2}}\right]_0^4 dy$$
$$= \int_0^2\left(8+\frac{16}{3}y\right)dy$$
$$= \left[8y+\frac{16}{6}y^2\right]_0^2$$
$$= \frac{80}{3}.$$

http://sagenb.mc.edu/home/pub/36/
http://sage.ace.fordham.edu/home/pub/41/

Answer : 80/3 ■

14.5 EXERCISES (Triple Integrals)

http://matrix.skku.ac.kr/Cal-Book/part2/CS-Sec-14-5-Sol.htm

YouTube http://youtu.be/C-uPM3km96k

CAS **1.** Find $\int_0^1 \int_0^{1-x} \int_0^{2-x} x\,y\,z\; dz\,dy\,dx$.

Solution

http://matrix.skku.ac.kr/cal-lab/cal-14-5-1(new).html

```
var('x,y,z')
f(x,y)=integral(x*y*z,(z,0,2-x))
g(x)=integral(f(x,y),(y,0,1-x))
integral(g,(x,0,1))
```

Answer : 13/240

$$\int_0^1 \int_0^{1-x} \int_0^{2-x} x\,y\,z\; dz\,dy\,dx\,x$$

$$= \int_0^1 \int_0^{1-x} \left[\frac{1}{2}x\,y\,z^2\right]_0^{2-x} dydx$$

$$= \int_0^1 \int_0^{1-x} \frac{x\,y(2-x)^2}{2}\,dy\,dx$$

$$= \int_0^1 \left[\frac{x\,y^2(2-x)^2}{4}\right]_0^{1-x} dx$$

$$= \frac{1}{4}\int_0^1 (4x - 12x^2 + 13x^3 - 6x^4 + x^5)\,dx$$

$$= \frac{13}{240}.$$

CAS **2.** Find $\int_0^2 \int_0^x \int_0^{x+y} e^x(y+2z)\,dz\,dy\,dx$.

Solution

http://matrix.skku.ac.kr/cal-lab/cal-14-5-2(new).html

```
var('x,y,z')
f(x,y)=integral(exp(x)*(y+2*z),(z,0,x+y))
g(x)=integral(f(x,y),(y,0,x))
integral(g,(x,0,2))
```

Answer : 19/3*e^2 + 19

$$\int_0^2 \int_0^x \int_0^{x+y} e^x(y+2z)\,dz\,dy\,dx$$

$$= \int_0^2 \int_0^x \left[e^x(yz+z^2)\right]_0^{x+y} dy\,dx$$

$$= \int_0^2 \int_0^x \left[e^x(x^2+3x\,y+2y^2)\right] dy\,dx$$

$$= \int_0^2 \left[e^x\left(x^2 y + \frac{3}{2}x\,y^2 + \frac{2}{3}y^3\right)\right]_0^x dx$$

$$= \frac{19}{6}\int_0^2 e^x x^3\,dx$$

$$= \frac{19}{6}\left[e^x(x^3 - 3x^2 + 6x - 6)\right]_0^2$$

$$= 19\left(\frac{e^2}{3} + 1\right).$$

CAS **3.** Find $\iiint_E (x+1)\,dV$, where $E = \{(x, y, z) \mid 0 \le x \le 1,\ 0 \le y \le x,\ -y^2 \le z \le x^2\}$.

Solution

http://matrix.skku.ac.kr/cal-lab/cal-14-5-3(new).html

```
var('x,y,z')
f(x,y)=integral(x+1,(z,-y^2,x^2))
g(x)=integral(f(x,y),(y,0,x))
integral(g,(x,0,1))
```

Answer : 3/5

$$\iiint_E (x+1)\,dV = \int_0^1 \int_0^x \int_{-y^2}^{x^2} (x+1)\,dz\,dy\,dx$$

$$= \int_0^1 \int_0^x [(x+1)z]_{-y^2}^{x^2}\,dy\,dx$$

$$= \int_0^1 \int_0^x (x+1)(x^2+y^2)\,dy\,dx$$

$$= \int_0^1 \left[(x+1)\left(x^2 y + \frac{1}{3}y^3\right)\right]_0^x dx$$

$$= \frac{4}{3}\int_0^1 (x^4 + x^3)\,dx$$

$$= \frac{3}{5}.$$

CAS **4.** Find $\iiint_E zx\sin(xy)\,dV$, where

$$E = \left\{(x,y,z) \,\middle|\, \frac{1}{6} \le x \le 1, 0 \le y \le \pi, 0 \le z \le 2\right\}.$$

Solution

http://matrix.skku.ac.kr/cal-lab/cal-14-5-4(new).html

```
var('x,y,z')
f(x,y)=integral(z*x*sin(x*y),(z,0,2))
g(x)=integral(f(x,y),(y,0,pi))
integral(g,(x,1/6,1)).expand()
```

Answer : 1/pi + 5/3

$$\iiint_E zx\sin(xy)\,dV$$

$$= \int_{\frac{1}{6}}^{1} \int_{0}^{\pi} \int_{0}^{2} zx\sin(xy)\,dz\,dy\,dx$$

$$= \int_{\frac{1}{6}}^{1} \int_{0}^{\pi} \left[\frac{1}{2}z^2 x\sin(xy)\right]_0^2 dy\,dx$$

$$= \int_{\frac{1}{6}}^{1} \int_{0}^{\pi} 2x\sin(xy)\,dy\,dx$$

$$= \int_{\frac{1}{6}}^{1} [-2\cos(xy)]_0^{\pi} dx$$

$$= \int_{\frac{1}{6}}^{1} (2-2\cos\pi x)\,dx$$

$$= \frac{5}{3} + \frac{1}{\pi}.$$

CAS **5.** Evaluate $\iiint_E dV$, where E is the solid bounded by parabolic cylinder $y^2 = 4-2x$ and the planes $z=0,\ z=x$.

Solution

http://matrix.skku.ac.kr/cal-lab/cal-14-5-5(new).html

```
var('x,y,z')
assume(x<=2)
assume(x>=0)
f(x,y)=integral(1,(z,0,x))
g(y)=integral(f(x,y),y)
h(x)=g(sqrt(4-2*x))-g(-sqrt(4-2*x))
integral(h,(x,0,2)).expand()
```

Answer : 64/15

$E=\{(x,y,z)\mid 0\le x\le 2,\ -\sqrt{4-2x}\le y\le \sqrt{4-2x},\ 0\le z\le x\}$

$$V=\iiint_E dV = \int_0^2 \int_{-\sqrt{4-2x}}^{\sqrt{4-2x}} \int_0^x dz\,dy\,dx$$

$$= \int_0^2 \int_{-\sqrt{4-2x}}^{\sqrt{4-2x}} [z]_0^x\,dydx$$

$$= \int_0^2 \int_{-\sqrt{4-2x}}^{\sqrt{4-2x}} x\,dy\,dx$$

$$= \int_0^2 [xy]_{-\sqrt{4-2x}}^{\sqrt{4-2x}} dx$$

$$= \int_0^2 2x\sqrt{4-2x}\,dx$$

$$= \frac{64}{15}.$$

6. Find the volume V within the ellipsoid

$$4x^2+4y^2+z^2=16.$$

Solution

$$V=\iiint_E dz\,dy\,dx = 8\int_0^2 \int_0^{\sqrt{4-x^2}} \int_0^{\sqrt{16-4x^2-4y^2}} dz\,dy\,dx$$

$$=8\int_0^2 \int_0^{\sqrt{4-x^2}} [z]_0^{\sqrt{16-4x^2-4y^2}} dy\,dx$$

$$=8\int_0^2 \int_0^{\sqrt{4-x^2}} \sqrt{16-4x^2-4y^2}\,dy\,dx$$

$$=16\int_0^2 \int_0^{\sqrt{4-x^2}} \sqrt{4-x^2-y^2}\,dy\,dx$$

$$=16\int_0^{\frac{\pi}{2}} \int_0^2 \sqrt{4-r^2}\,r\,dr\,d\theta = \frac{64}{3}\pi.$$

7. Let $f(x)$ be a continuous and positive function on $[a,b]$ and R be the region that lies between $f(x)$ and x-axis.

Find the volume V of the solid E obtained by rotating R about x-axis.

Solution $E=\{(x,y,z)\mid a\le x\le b,\ -f(x)\le y\le f(x),\ -\sqrt{f(x)^2-y^2}\le z\le \sqrt{f(x)^2-y^2}\}$

$$V=\iiint_E dV = \int_a^b \int_{-f(x)}^{f(x)} \int_{-\sqrt{f(x)^2-y^2}}^{\sqrt{f(x)^2-y^2}} dz\,dy\,dx$$

$$= \int_a^b \int_{-f(x)}^{f(x)} [z]_{-\sqrt{f(x)^2-y^2}}^{\sqrt{f(x)^2-y^2}} dy\,dx$$

$$= \int_a^b \int_{-f(x)}^{f(x)} 2\sqrt{f(x)^2-y^2}\,dy\,dx$$

$$= \pi\int_a^b \{f(x)\}^2 dx.$$

8. Use 7 to find volume of sphere $x^2+y^2+z^2=1$ and volume of cylinder $y^2+z^2=4,\ 0\le x\le 3$.

9. Find the volume V of solid in the first octant $(x,\ y,\ z\ge 0)$ enclosed by a curved surface $z=x^2+2y+1$ and the plane $z=y+2$.

Solution $E=\{(x,\ y)\mid 0\le x\le 1,\ 0\le y\le 1-x^2,\ x^2+2y+1\le z\le y+2\}$

$$V=\iiint_E dV = \int_0^1 \int_0^{1-x^2} \int_{x^2+2y+1}^{y+2} dz\,dy\,dx$$

$$= \int_0^1 \int_0^{1-x^2} [z]_{x^2+2y+1}^{y+2} dy\,dx$$

$$= \int_0^1 \int_0^{1-x^2} (1-x^2-y^2)\,dy\,dx$$

$$= \int_0^1 \left[(1-x^2)y - \frac{1}{2}y^2\right]_0^{1-x^2} dx$$

$$= \frac{1}{2}\int_0^1 (2-x^2)\,dx$$

$$= \frac{4}{15}.$$

10. Find the volume of the solid bounded by the cylinder $x^2+y^2=2ax$ and planes $z=bx$ and

$z = cx\ (b > c)$.

Solution $E = \{(x, y, z) \mid 0 \le x \le 2a,$
$-\sqrt{2ax - x^2} \le y \le \sqrt{2ax - x^2},\ cx \le z \le bx\}$

$$V = \iiint_E dV = \int_0^{2a} \int_{-\sqrt{2ax-x^2}}^{\sqrt{2ax-x^2}} \int_{cx}^{bx} dz\, dy\, dx$$

$$= \int_0^{2a} \int_{-\sqrt{2ax-x^2}}^{\sqrt{2ax-x^2}} [z]_{cx}^{bx}\, dy\, dx$$

$$= \int_0^{2a} \int_{-\sqrt{2ax-x^2}}^{\sqrt{2ax-x^2}} (b-c)x\, dy\, dx$$

$$= \int_0^{2a} [(b-c)xy]_{-\sqrt{2ax-x^2}}^{\sqrt{2ax-x^2}}\, dx$$

$$= 2(b-c)\int_0^{2a} x\sqrt{2ax - x^2}\, dx$$

$$= \pi(b-c)a^3.$$

11. Find the average value of $f(x, y, z) = x^2y^2z^2$ throughout cubical region E bounded by the planes $x = 3$, $y = 3$ and $z = 3$ in the first octant.

Solution The volume of the region E is 27. The value of the integral of f over the cube is

$$\int_0^3\int_0^3\int_0^3 x^2y^2z^2\, dxdydz = \int_0^3 x^2\, dx \int_0^3 y^2\, dy \int_0^3 z^2\, dz$$

$$= \left[\frac{1}{3}x^3\right]_0^3 \left[\frac{1}{3}y^3\right]_0^3 \left[\frac{1}{3}z^3\right]_0^3 = 729.$$

The average value is

$$f_{ave} = \frac{1}{\text{volume}} \iiint_{cube} x^2y^2z^2\, dV$$

$$= \frac{1}{27} \cdot 729 = 27.$$

CAS **12.** Evaluate $\iiint_E f\, dV$ where $f = 45x^2y$ and E denotes the closed region bounded by the planes $4x + 2y + z = 8$, $x = 0$, $y = 0$, $z = 0$.

Solution

http://matrix.skku.ac.kr/cal-lab/cal-14-5-1.html

```
x,y,z=var('x,y,z');
integral(integral(integral(45*x^2*y, z, 0, 8-4*x-2*y), y, 0, 4-2*x), x, 0, 2)
```

Answer : 128

$E = \{(x, y, z) \mid 0 \le x \le 2,\ 0 \le y \le 4 - 2x,$
$0 \le z \le 8 - 4x - 2y\}$

$$\iiint_E f\, dV = \int_0^2 \int_0^{4-2x} \int_0^{8-4x-2y} 45x^2y\, dz\, dy\, dx$$

$$= \int_0^2 \int_0^{4-2x} 45x^2y(8 - 4x - 2y)\, dy\, dx$$

$$= \int_0^2 \int_0^{4-2x} (360x^2y - 180x^3y - 90x^2y^2)\, dy\, dx$$

$$= \int_0^2 [180x^2y^2 - 90x^3y^2 - 30x^2y^3]_0^{4-2x}\, dx$$

$$= \int_0^2 (-120x^5 + 720x^4 - 1440x^3 + 960x^2)\, dx$$

$$= 128$$

14.6 Triple Integrals in Cylindrical and Spherical Coordinates

YouTube http://youtu.be/xd0U4_C2ePY

Cylindrical Coordinates

To evaluate the triple integral in Cartesian coordinates, it is often very useful to use cylindrical coordinates.

Recall that cylindrical coordinates (r,θ,z) of a point (x,y,z) are defined by

$$x = r\cos\theta,\ \ y = r\sin\theta,\ \ z = z.$$

Under this change of coordinates one can show that the area element in the triple integral $\iiint_V f(x,y,z)dxdydz$ changes, to $rdrd\theta dz$. Thus the triple integral in cylindrical coordinates is obtained by replacing $x = r\cos\theta$, $y = r\sin\theta$, $z = z$ and $dxdxdz = rdrd\theta dz$ and changing the limits of integration appropriately.

In particular, the triple integral can be written as

1 $$\iiint_B f(x,y,z)dxdydz = \iiint_{\tilde{B}} f(r\cos\theta,\ r\sin\theta,\ z)rdrd\theta dz.$$

EXAMPLE 1

Find $\int_{-2}^{2}\int_{-\sqrt{4-x^2}}^{\sqrt{4-x^2}}\int_{(x^2+y^2)^2}^{1} x^2 dV.$

Solution Since $\{(x,y,z)|x^2+y^2 \le 4,\ (x^2+y^2)^2 \le z \le 1\}$
$= \{(r,\theta,z)|0 \le r \le 2,\ 0 \le \theta \le 2\pi,\ r^4 \le z \le 1\}$,

$$\begin{aligned}\int_{-2}^{2}\int_{-\sqrt{4-x^2}}^{\sqrt{4-x^2}}\int_{(x^2+y^2)^2}^{1} x^2 dV &= \int_0^{2\pi}\int_0^2\int_{r^4}^{1} r^2\cos^2\theta\, r\,dz\,dr\,d\theta \\ &= \int_0^{2\pi}\int_0^2 \left[r^3\cos^2\theta\, z\right]_{r^4}^{1} dr\,d\theta \\ &= \int_0^{2\pi}\int_0^2 (r^3-r^7)\cos^2\theta\,dr\,d\theta \\ &= \int_0^{2\pi}\left[\frac{1}{4}r^4-\frac{1}{8}r^8\right]_0^2\cos^2\theta\,dr\,d\theta \\ &=-28\int_0^{2\pi}\cos^2\theta\,d\theta \\ &-28\pi.\end{aligned}$$

■

EXAMPLE 2

Evaluate $\iiint_B \sqrt{x^2+y^2}\,dV$, where B is bounded by the plane $y+z=4$, the xy-plane and the circular cylinder $x^2+y^2=16$. (See Figure 1.)

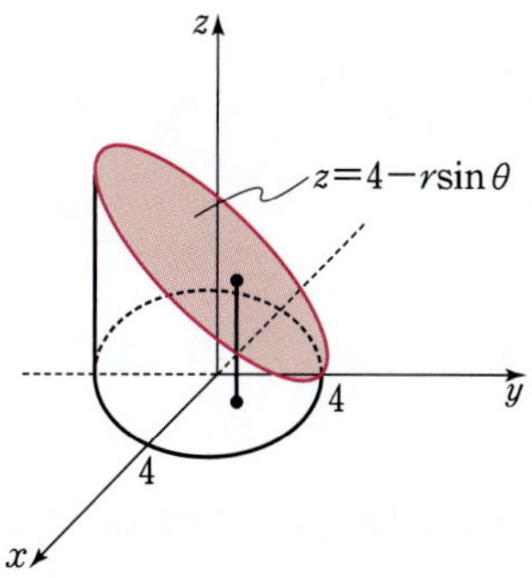

Figure 1

Solution $B=\{(r,\theta,z)\,|\,0\le r\le 4,\ 0\le\theta\le 2\pi,\ 0\le z\le 4-r\sin\theta\}$

$$\begin{aligned}\iiint_B \sqrt{x^2+y^2}\,dV &= \int_0^{2\pi}\int_0^4\int_0^{4-r\sin\theta} r\cdot r\,dz\,dr\,d\theta\\ &=\int_0^{2\pi}\int_0^4 \left[r^2z\right]_0^{4-r\sin\theta}dr\,d\theta\\ &=\int_0^{2\pi}\int_0^4 (4r^2-r^3\sin\theta)\,dr\,d\theta\\ &=\int_0^{2\pi}\left[\frac{4}{3}r^3-\frac{1}{4}r^4\sin\theta\right]_0^4 d\theta\\ &=\int_0^{2\pi}\left(\frac{256}{3}-64\sin\theta\right)d\theta\\ &=\frac{512}{3}\pi.\end{aligned}$$

■

Recall that the following property is in section 14.6.

Let $f(x,y,z)=1$ for all points in a region B. Then the triple integral is exactly volume of B as

$$V=\iiint_B dV.$$

The next example uses this formula.

EXAMPLE 3

Find the volume V of the solid bounded by a spherical surface $x^2+y^2+z^2=4$ and the cylinder $r=2\sin\theta$.

Solution Since the bounded solid is

$$B=\{(r,\theta,z)\,|\,0\le r\le 2\sin\theta,\ 0\le\theta\le\pi,-\sqrt{4-r^2}\le z\le\sqrt{4-r^2}\}$$

and by the symmetry of this solid about yz-plane (See Figure 2),

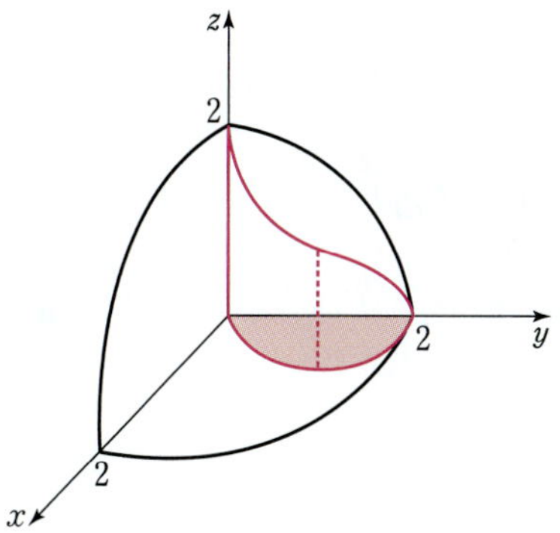

Figure 2

$$\begin{aligned}V=\iiint_B dV &= 4\int_0^{\pi/2}\int_0^{2\sin\theta}\int_0^{\sqrt{4-r^2}} r\,dz\,dr\,d\theta\\ &=4\int_0^{\pi/2}\int_0^{2\sin\theta} r\sqrt{4-r^2}\,dr\,d\theta\end{aligned}$$

Put $t=4-r^2$. Then $rdr=-\dfrac{1}{2}dt$. When $x=0$, $t=4$ and when $x=2\sin\theta$, $t=4\cos^2\theta$,

$$V=4\int_0^{\pi/2}\int_4^{4\cos\theta}-\frac{1}{2}\sqrt{t}\,dt\,d\theta$$

$$= \frac{4}{3}\int_0^{\pi/2} (8 - 8\cos^3\theta)\,d\theta$$
$$= \frac{4}{3}\left(4\pi - \frac{16}{3}\right).$$
■

Spherical Coordinates

Recall that spherical coordinates (ρ, θ, ϕ) of a point (x, y, z) are defined by

$$x = \rho\sin\phi\cos\theta,\ y = \rho\sin\phi\sin\theta,\ z = \rho\cos\phi,$$

where $\rho \ge 0,\ 0 \le \theta < 2\pi,\ 0 \le \phi \le \pi$.

Under this change of coordinates to cartesian coordinates one can show that the area element $dxdydz$ changes to $\rho^2\sin\phi\,d\rho d\phi d\theta$.

The triple integral can be written as

$$\iiint_B f(x,y,z)dxdydz = \iiint_{\tilde{B}} f(\rho\sin\phi\cos\theta, \rho\sin\phi\sin\theta, \rho\cos\phi)\rho^2\sin\phi\ d\rho d\phi d\theta.$$

Again the appropriate change of limits is required.

EXAMPLE 4

Evaluate $\iiint_B ze^{x^2+y^2+z^2}dV$, where

$$B = \{(x, y, z)|x^2+y^2 \le 1,\ 0 \le z \le 1\}.$$

Solution Using the cylindrical coordinates transformation the given integral can be written as

$$\int_0^1\int_0^1\int_0^{2\pi} rze^{r^2+z^2}d\theta drdz = 2\pi\int_0^1\int_0^1 rze^{r^2+z^2}drdz$$
$$= 2\pi\left(\int_0^1 re^{r^2}dr\right)^2.$$
$$= \frac{\pi}{2}(e-1)^2.$$
■

EXAMPLE 5

Find the volume of the sphere with radius a.

Solution The sphere $S = \{x^2+y^2+z^2 \le a^2\}$ can be described in spherical coordinates as $\{(\rho, \phi, \theta)|\ 0 \le \rho \le a,\ 0 \le \phi \le \pi,\ 0 \le \theta \le 2\pi\}$. Then,

$$V = \int_0^{2\pi}\int_0^{\pi}\int_0^{a} \rho^2\sin\phi\,d\rho\,d\phi\,d\theta$$

$$= \int_0^{2\pi} \int_0^{\pi} \left[\frac{1}{3}\rho^3 \sin\phi \right]_0^a d\phi\, d\theta$$

$$= \frac{1}{3}a^3 \int_0^{2\pi} \int_0^{\pi} \sin\phi\, d\phi\, d\theta$$

$$= \frac{1}{3}a^3 \int_0^{2\pi} [-\cos\phi]_0^{\pi} d\theta$$

$$= \frac{2}{3}a^3 \int_0^{2\pi} d\theta = \frac{4}{3}\pi a^3.$$

■

EXAMPLE 6

Find the volume of the solid enclosed by the spherical surface $x^2 + y^2 + z^2 = 9$ and the cone $z = \sqrt{x^2 + y^2}$. (See Figure 3.)

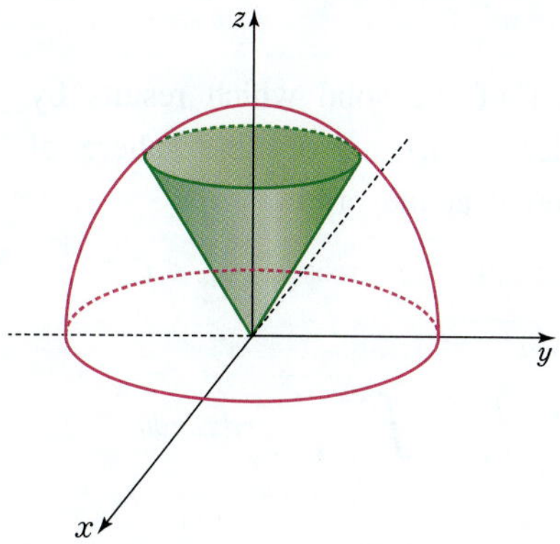

Figure 3

Solution $E = \left\{ (\rho,\ \theta,\ \phi) \mid 0 \le \rho \le 3,\ 0 \le \phi \le \frac{\pi}{4},\ 0 \le \theta \le 2\pi \right\}$

$$V = \iiint_E dV = \int_0^{2\pi} \int_0^{\frac{\pi}{4}} \int_0^3 \rho^2 \sin\phi\, d\rho\, d\phi\, d\theta$$

$$= \int_0^{2\pi} \int_0^{\frac{\pi}{4}} \left[\frac{1}{3}\rho^3 \right]_0^3 \sin\phi\, d\phi\, d\theta$$

$$= 9 \int_0^{2\pi} \int_0^{\frac{\pi}{4}} \sin\phi\, d\phi\, d\theta$$

$$= 9 \int_0^{2\pi} [-\cos\phi]_0^{\frac{\pi}{4}} d\theta$$

$$= 9 \int_0^{2\pi} \left(1 - \frac{\sqrt{2}}{2} \right) d\theta = 9\pi\left(2 - \sqrt{2}\right).$$

■

14.6 EXERCISES (Triple Integrals in Cylindrical and Spherical Coordinates)

http://matrix.skku.ac.kr/Cal-Book/part2/CS-Sec-14-6-Sol.htm

1. Find $\int_0^{2\pi}\int_0^{\frac{1}{2}}\int_r^{\sqrt{1-r^2}} r\,dz\,dr\,d\theta$.

Solution $\int_0^{2\pi}\int_0^{\frac{1}{2}}\int_r^{\sqrt{1-r^2}} r\,dz\,dr\,d\theta$

$=\int_0^{2\pi}\int_0^{\frac{1}{2}}(r\sqrt{1-r^2}-r^2)\,dr\,d\theta$

$=\int_0^{2\pi}\left[-\frac{1}{3}(1-r^2)^{\frac{3}{2}}-\frac{1}{3}r^3\right]_0^{\frac{1}{2}} d\theta$

$=\frac{1}{24}\int_0^{2\pi} d\theta=\frac{\pi}{12}$.

2. Find the volume V of the solid which results by cutting the cylinder $r=a\cos\theta$ by the sphere of radius a and centered at the origin.

Solution $E=\left\{(r,\ \theta,\ z)\mid 0\le r\le a\cos\theta,\ -\frac{\pi}{2}\le\theta\le\frac{\pi}{2},\right.$
$\left.-\sqrt{a^2-r^2}\le z\le\sqrt{a^2-r^2}\right\}$

$V=\iiint_E dV=\int_{-\frac{\pi}{2}}^{\frac{\pi}{2}}\int_0^{a\cos\theta}\int_{-\sqrt{a^2-r^2}}^{\sqrt{a^2-r^2}} r\,dz\,dr\,d\theta$

$=\int_{-\frac{\pi}{2}}^{\frac{\pi}{2}}\int_0^{a\cos\theta}[rz]_{-\sqrt{a^2-r^2}}^{\sqrt{a^2-r^2}}\,dr\,d\theta$

$=2\int_{-\frac{\pi}{2}}^{\frac{\pi}{2}}\int_0^{a\cos\theta} r\sqrt{a^2-r^2}\,dr\,d\theta$

$\sqrt{a^2-r^2}=t$ as $r\,dr=t\,dt$

$r\to 0 \ \Rightarrow\ t\to a,\quad r\to a\cos\theta\ \Rightarrow\ t\to a\sin\theta$

$=2\int_{-\frac{\pi}{2}}^{\frac{\pi}{2}}\int_0^{a\sin\theta} t(-t\,dt)\,d\theta$

$=2\int_{-\frac{\pi}{2}}^{\frac{\pi}{2}}\int_{a\sin\theta}^{a} t^2\,dt\,d\theta$

$=2\int_{-\frac{\pi}{2}}^{\frac{\pi}{2}}\left[\frac{1}{3}t^3\right]_{a\sin\theta}^{a} d\theta$

$=\frac{2}{3}\int_{-\frac{\pi}{2}}^{\frac{\pi}{2}}(a^3-a^3\sin^3\theta)\,d\theta$

$=\frac{2}{3}a^3\left(\pi-\frac{4}{3}\right)=\frac{2}{9}a^3(3\pi-4)$.

3. Find the volume cut from the cone $\phi=\frac{\pi}{4}$ by the sphere $\rho=2a\cos\phi$.

Solution $E=\left\{(\rho,\ \phi,\ \theta)\mid 0\le\rho\le 2a\cos\phi,\ 0\le\phi\le\frac{\pi}{4},\ 0\le\theta\le\frac{\pi}{2}\right\}$

$V=4\iiint_E dV=4\int_0^{\frac{\pi}{2}}\int_0^{\frac{\pi}{4}}\int_0^{2a\cos\phi}\rho^2\sin\phi\,d\rho\,d\phi\,d\theta$

$=\frac{32a^3}{3}\int_0^{\frac{\pi}{2}}\int_0^{\frac{\pi}{4}}\cos^3\phi\sin\phi\,d\phi\,d\theta=2a^3\int_0^{\frac{\pi}{2}}d\theta=\pi a^3$.

4. Find the volume of the solid bounded by the spherical surface $x^2+y^2+z^2=1$ and the circular cone $z=\sqrt{x^2+y^2}$.

Solution $E=\left\{(\rho,\ \phi,\ \theta)\mid 0\le\rho\le 1,\ 0\le\phi\le\frac{\pi}{4},\ 0\le\theta\le 2\pi\right\}$

$V=\iiint_E dV=\int_0^{2\pi}\int_0^{\frac{\pi}{4}}\int_0^{1}\rho^2\sin\phi\,d\rho\,d\phi\,d\theta$

$=\int_0^{2\pi}\int_0^{\frac{\pi}{4}}\left[\frac{1}{3}\rho^3\right]_0^1\sin\phi\,d\phi\,d\theta$

$=\frac{1}{3}\int_0^{2\pi}\int_0^{\frac{\pi}{5}}\sin\phi\,d\phi\,d\theta$

$=\frac{1}{3}\int_0^{2\pi}[-\cos\phi]_0^{\frac{\pi}{4}}\,d\theta$

$=\frac{1}{3}\int_0^{2\pi}\left(1-\frac{\sqrt{2}}{2}\right)d\theta=\frac{\pi}{3}(2-\sqrt{2})$

5. Find the volume V of the solid E that lies above the cone $\phi=\phi_0\left(0\le\phi_0\le\frac{\pi}{2}\right)$ and below the sphere $x^2+y^2+z^2=a^2(z\ge 0)$.

Solution $E=\{(\rho,\ \phi,\ \theta)\mid 0\le\rho\le a,\ 0\le\phi\le\phi_0,\ 0\le\theta\le 2\pi\}$

$V=\iiint_E dV=\int_0^{2\pi}\int_0^{\phi_0}\int_0^{a}\rho^2\sin\phi\,d\rho\,d\phi\,d\theta$

$=\int_0^{2\pi}d\theta\times\int_0^{\phi_0}\sin\phi\,d\phi\times\int_0^{a}\rho^2\,d\rho$

$=2\pi\times(1-\cos\phi_0)\times\frac{1}{3}a^3$

$=\frac{2}{3}\pi a^3(1-\cos\phi_0)$

CAS **6.** Evaluate $\iiint_V f\,dV$ where $f=z^2$ and V is the surface whose side S_1 are is the cylinder $x^2+y^2=1$, and bottom S_2 is the disk $x^2+y^2\le 1$

in the plane $z = 0$, and whose top S_3 is the part of the plane $z = 1 + x$ that lies above S_2.

http://matrix.skku.ac.kr/cal-lab/cal-14-6-2.html

Solution

```
x,y,z,theta=var('x,y,z,t')
p1 = implicit_plot3d(z==1+x, (x,-2,2), (y, -2,2), (z, 0,2), opacity=0.2, color="red", mesh=True);
p2 = implicit_plot3d(x^2+y^2==1, (x,-2,2), (y, -2,2), (z, 0,2), opacity=0.3, color="blue", mesh=True);
p3 = plot3d(0, (x,-2,2), (y, -2,2), opacity=0.3, color="orange", mesh=True);
show(p1+p2+p3, aspect_ratio=1)
```

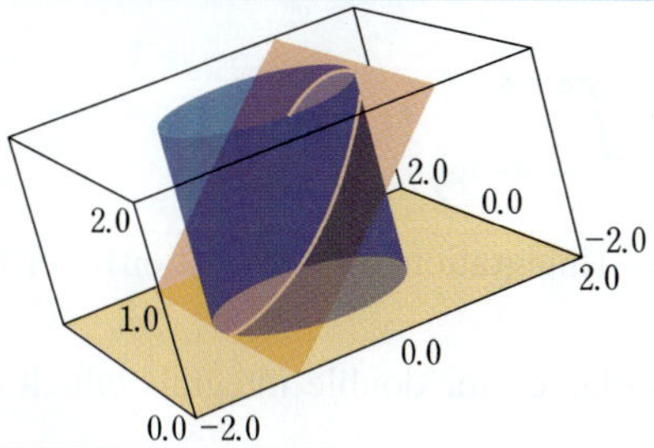


```
x = cos(t);
y = sin(t);
z = z;
assume (0 <= t <= 2*pi());
assume (0 <= z <= 1+x);
i,j,k=var('i,j,k');
R=matrix(SR, [[i, j, k], [diff(x, t), diff(y, t), diff(z, t)], [diff(x, z), diff(y, z), diff(z, z)]]).determinant();
R
```

i*cos(t) + j*sin(t)

```
vector_R=vector(SR, [cos(t), sin(t)]);
vector_R.norm()
```

sqrt(abs(sin(t))^2 + abs(cos(t))^2)

```
integral(integral(z^2*1, z, 0, 1+cos(theta)), theta, 0,2*pi())
```

Answer : 5/3*pi

7. Find the volume of the region common to the intersecting cylinders $x^2 + y^2 = a^2$ and $x^2 + z^2 = a^2$.

8. Find the volume of the region bounded below by the paraboloid $z = x^2 + y^2$ and above by the plane $z = 2y$.

9. Find the volume cut from the sphere $x^2 + y^2 + z^2 = 4a^2$ by the cylinder $x^2 + y^2 = a^2$.

10. Evaluate $\iiint_E (x + y + z)\, dV$, where R is the region inside $x^2 + y^2 + z^2 \le 1$ in the first octant.

11. Find the volume bounded above by the sphere $x^2 + y^2 + z^2 = 2a^2$ and below by the paraboloid $az = x^2 + y^2$.

Answer : If $a > 0$, $\left(2\sqrt{2} - \frac{1}{2}\right)\pi a^3$.

If $a < 0$, $-\frac{(2\sqrt{2} - 1)}{3}\pi a^3$.

14.7 The Substitution Method

http://youtu.be/INn-bkgXYNg

Change of Variables

Recall that integration by substitution (change of variables) is given by

1
$$\int_a^b f(x)dx = \int_c^d f(\phi(u))\phi'(u)du$$

where $\phi: [c, d] \rightarrow [a, b]$ is a one-one onto C^1-function.

The above formula can also be written as

2
$$\int_a^b f(x)dx = \int_{c=\phi^{-1}(a)}^{d=\phi^{-1}(b)} f(x(u))\frac{dx}{du}du$$

The idea of change of variables (integration by substitution) can be extended to multiple integrals as well.

First we deal with change of variables for double integrals which can then be extended to triple integrals.

Let $f: D \subseteq \mathbb{R}^2 \rightarrow \mathbb{R}$ be an integrable function. The idea is to change variable x and y in the double integral $\iint_D f(x, y)dA$ into another set of variables every u and v.

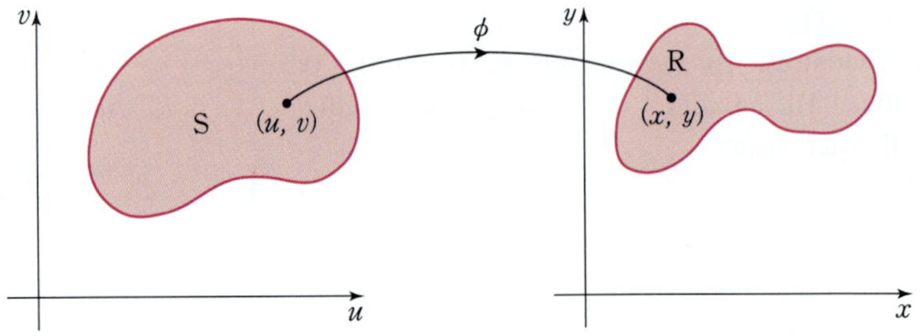

Figure 1

More generally, we consider a transformation $\phi(u, v) = (x, y) = (x(u, v), y(u, v))$. We assume that ϕ has continuous 1st order derivative (that is ϕ is C^1-function). We further assume that maps $\phi: S \subseteq \mathbb{R}^2 \rightarrow R \subseteq \mathbb{R}^2$ is a bijection, where S is a region in uv-plane and R in xy-plane.

Now the question is: how this change of variables $(x, y) \rightarrow (u, v)$ affects the double integral. For this, we must look at what happens to area element $\Delta u \Delta v$ in uv-plane when it transformed to xy-plane under ϕ.

Look at Figure 2, we consider a rectangle S in uv-plane with area $\Delta u \Delta v$ and its image R in the xy-plane.

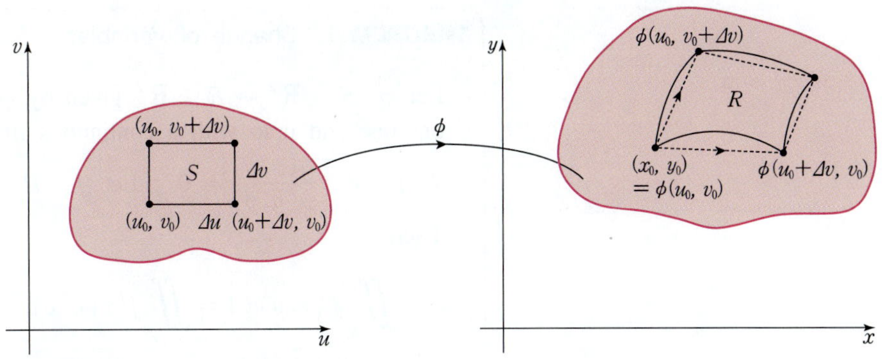

Figure 2

The idea is to approximate the area of R in xy-plane by a parallelogram determined by secant vectors

$$\phi(u_0+\Delta u, v_0)-\phi(u_0, v_0) \text{ and } \phi(u_0, v_0+\Delta v)-\phi(u_0, v_0).$$

Using the definition of partial derivative we have

$$\phi(u_0+\Delta u, v_0)-\phi(u_0, v_0) \approx \frac{\partial\phi}{\partial u}\Delta u$$

and

$$\phi(u_0, v_0+\Delta v)-\phi(u_0, v_0) \approx \frac{\partial\phi}{\partial v}\Delta v$$

We know that the area of a parallelogram determined by vectors $\mathbf{a}$ and $\mathbf{b}$ is given by $|\mathbf{a}\times\mathbf{b}|$.

Hence the area of the R can be approximated by

$$\left|\frac{\partial\phi}{\partial u}\Delta u\times\frac{\partial\phi}{\partial v}\Delta v\right|=\left|\frac{\partial\phi}{\partial u}\times\frac{\partial\phi}{\partial v}\right|\Delta u\,\Delta v$$

provided $\frac{\partial\phi}{\partial u}\times\frac{\partial\phi}{\partial v}\neq 0$.

It is easy to see that

$$\frac{\partial\phi}{\partial u}\times\frac{\partial\phi}{\partial v}=\begin{vmatrix}\mathbf{i} & \mathbf{j} & \mathbf{k}\\ \frac{\partial x}{\partial u} & \frac{\partial y}{\partial u} & 0\\ \frac{\partial x}{\partial v} & \frac{\partial y}{\partial v} & 0\end{vmatrix}=\begin{vmatrix}\frac{\partial x}{\partial u} & \frac{\partial y}{\partial u}\\ \frac{\partial x}{\partial v} & \frac{\partial y}{\partial v}\end{vmatrix}\mathbf{k}=\begin{vmatrix}\frac{\partial x}{\partial u} & \frac{\partial x}{\partial v}\\ \frac{\partial y}{\partial u} & \frac{\partial y}{\partial v}\end{vmatrix}\mathbf{k}.$$

The determinant on the right hand side is called the Jacobian of ϕ and is often denoted by $\left|\frac{\partial(x, y)}{\partial(u, v)}\right|$.

This leads us to the following change of variable formula

THEOREM 1 Change of Variables

Let $\phi: S \subseteq \mathbb{R}^2 \to R \subseteq \mathbb{R}^2$ given by $\phi(u,v) = (x(u,v), y(u,v))$ be a one-one and onto with a continuous first order partial derivatives (ϕ is C^1) and $\left|\frac{\partial(x,y)}{\partial(u,v)}\right| \neq 0$. Let $f: R \to \mathbb{R}$ be a continuous function. Then

$$\boxed{3} \quad \iint_R f(x,y)dA = \iint_S f(x(u,v), y(u,v)) \left| \frac{\partial(x,y)}{\partial(u,v)} \right| du\, dv.$$

Thus change of variables formula says that we change variable x and y to u and v using the transformation ϕ and replace the area element dA by $\left|\frac{\partial(x,y)}{\partial(u,v)}\right|$.

Double Integral in Polar Coordinates

The double integrals in polar coordinates becomes a special case of change of variables.

Recall that in polar coordinate we have the transformation

$$x = r\cos\theta,\ y = r\sin\theta \quad 0 \le \theta \le 2\pi \quad \text{and} \quad r \ge 0.$$

The Jacobian of this transformation is

$$\frac{\partial(x,y)}{\partial(r,\theta)} = \begin{vmatrix} \frac{\partial x}{\partial r} & \frac{\partial x}{\partial \theta} \\ \frac{\partial y}{\partial r} & \frac{\partial y}{\partial \theta} \end{vmatrix} = r > 0$$

Hence the double integral in polar coordinates become,

$$\boxed{4} \quad \iint_R f(x,y)dxdy = \int_\alpha^\beta \int_a^b f(r\cos\theta, r\sin\theta)\, r\, dr\, d\theta$$

where the limits a, b and α, β can be obtains from the given domain R of f and the transformation $x = r\cos\theta$, $y = r\sin\theta$.

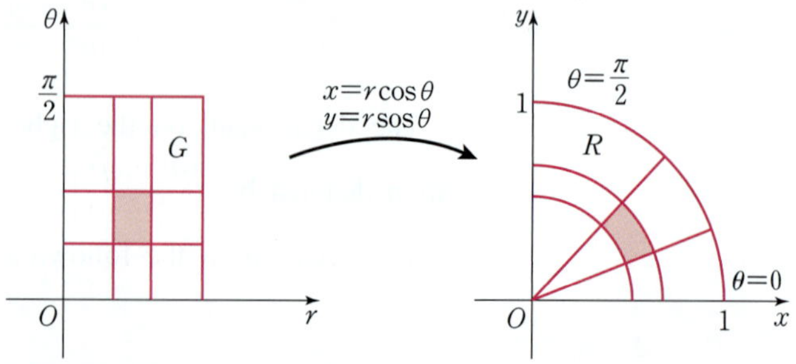

Figure 3 The formula $x = r\cos\theta$, $y = r\sin\theta$ converts G to R.

Figure 3 shows a conversion of a square in polar coordinate $r\theta$-plane is a quarter circle of $x^2+y^2=1$ in xy-plane.

EXAMPLE 1

Evaluate $\iint_R 7xy\,dx\,dy$, where R is the region enclosed by the line $x-2y=0$, $x-2y=2$, $2x+3y=0$ and $2x+3y=3$, a parallelogram in the xy-plane.

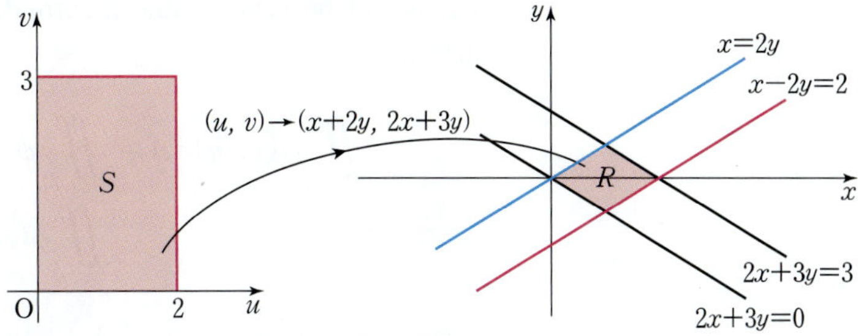

Figure 4

Solution The boundary of the region R suggests that

$$x-2y=u,\ 2x+3y=v. \tag{5}$$

Since this transformation takes the boundary lines

$$x-2y=0,\ x-2y=2,\ 2x+3y=0,\ 2x+3y=3$$

to $u-$curves and $v-$curves

$$u=0,\quad u=2,\quad v=0,\quad v=3$$

in the $uv-$plane. To find the Jacobian $\partial(x,y)/\partial(u,v)$ of this transformation, we first solve 5 for x and y and in terms of u and v. We get

$$x=\frac{1}{7}(2u+3v),\ y=\frac{1}{7}(u-2v)$$

from which we obtain

$$\frac{\partial(x,\,y)}{\partial(u,\,v)}=\begin{vmatrix}\frac{\partial x}{\partial u} & \frac{\partial x}{\partial v}\\ \frac{\partial y}{\partial u} & \frac{\partial y}{\partial v}\end{vmatrix}=\begin{vmatrix}\frac{2}{7} & \frac{3}{7}\\ \frac{1}{7} & -\frac{2}{7}\end{vmatrix}=-\frac{1}{7}.$$

Thus,

$$\begin{aligned}\iint_R 7xy\,dx\,dy&=\iint_S 7\cdot\frac{1}{7}(2u+3v)\frac{1}{7}(u-2v)\left|\frac{\partial(x,y)}{\partial(u,v)}\right|du\,dv\\ &=\frac{1}{7}\iint_S(2u+3v)(u-2v)\left|-\frac{1}{7}\right|du\,dv\end{aligned}$$

$$= \frac{1}{49}\int_0^3\int_0^2 (2u^2 - uv - 6v^2)\,du\,dv$$
$$= \frac{1}{49}\int_0^3 \left[\frac{2}{3}u^3 - \frac{1}{2}u^2 v - 6uv^2\right]_0^2 dv$$
$$= \frac{1}{49}\int_0^3 \left(\frac{16}{3} - 2v - 12v^2\right) dv.$$
$$= -\frac{101}{49}.$$ ■

It may not be easy to find a formula for change of variables on a given double integral

$$\iint_R \phi(x, y)\,dA = \iint_R \phi(x(u, v), y(u, v))\left|\frac{\partial(x, y)}{\partial(u, v)}\right| du\,dv$$
$$= \iint_S \phi(r\cos\theta, r\sin\theta) r\,dr\,d\theta.$$

The method of change of variables is divided into three steps.

(1) Define a transformation $T: R \to S$ on the region R in xy-plane that determines the image S of R. The boundary of region R defines the function. It is important that the image S should ensure a simple area to find double integral.
(2) From the transformation $T: R \to S$, find expressions to integrate.
(3) Find Jacobian and do the integral calculation on the region S.

EXAMPLE 2

Evaluate $\iint_R (x+y)^2 e^{x-y}\,dx\,dy$, where R is the region enclosed by the rhombus with vertices $(1,0)$, $(0,1)$, $(-1,0)$ and $(0,-1)$.

Solution The region R is enclosed by the line $x+y=1$, $x+y=-1$ $x-y=1$, $x-y=-1$
The expressions $x+y$ and $x-y$ in the equation of the boundary suggest the transformation

6 $$x+y=u,\ x-y=v.$$

This transformation takes the boundary lines

$$x+y=1,\quad x+y=-1,\quad x-y=1,\quad x-y=-1$$

to the curves in the uv-plane,

$u=-1$, $u=1$, $v=-1$, $v=1$. (See Figure 5.)

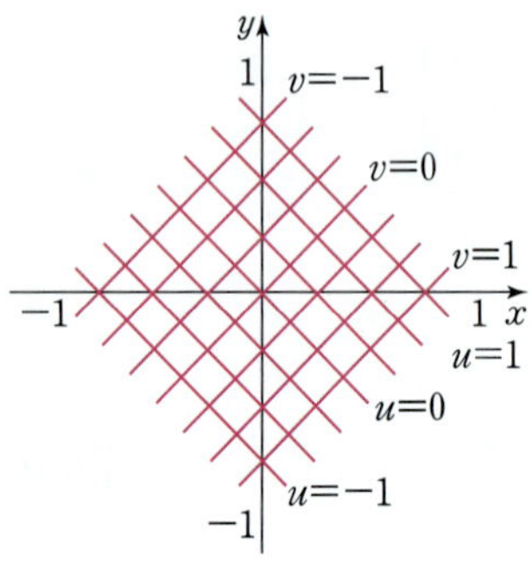

Figure 5

These lines enclose the region S. To find the Jacobian $\partial(x,y)/\partial(u,v)$ of this transformation, we first solve 6 for x and y and in terms of u and v.

We get $x=\frac{1}{2}(u+v)$, $y=\frac{1}{2}(u-v)$ from which obtain

$$\frac{\partial(x,y)}{\partial(u,v)}=\begin{vmatrix} \frac{\partial x}{\partial u} & \frac{\partial x}{\partial v} \\ \frac{\partial y}{\partial u} & \frac{\partial y}{\partial v} \end{vmatrix}=\begin{vmatrix} \frac{1}{2} & \frac{1}{2} \\ \frac{1}{2} & -\frac{1}{2} \end{vmatrix}=-\frac{1}{2}.$$

Thus,

$$\begin{aligned}\iint_R (x+y)^2 e^{x-y}\,dx\,dy &= \iint_S u^2 e^v \left|\frac{\partial(x,y)}{\partial(u,v)}\right| du\,dv \\ &= \int_{-1}^{1}\int_{-1}^{1} u^2 e^v \left|-\frac{1}{2}\right| dv\,dv = \frac{1}{2}\int_{-1}^{1} u^2 du \int_{-1}^{1} e^v dv \\ &= \frac{1}{2}\left[\frac{1}{3}u^3\right]_{-1}^{1} \times \left[e^v\right]_{-1}^{1} = \frac{1}{3}(e-e^{-1}).\end{aligned}$$

■

EXAMPLE 3

Find $I=\iint_R (x^2+y^2)dxdy$, where, $R: \frac{x^2}{a^2}+\frac{y^2}{b^2}\le 1$.

Solution First of all we use the transformation $x=au$ and $y=bv$. This transformation map. $S: u^2+v^2\le 1$ in uv-plane to the ellipse $R: \frac{x^2}{a^2}+\frac{y^2}{b^2}\le 1$ in the xy-plane. The Jacobian is

$$\frac{\partial(x,y)}{\partial(u,v)}=\begin{vmatrix} \frac{\partial x}{\partial u} & \frac{\partial x}{\partial v} \\ \frac{\partial y}{\partial u} & \frac{\partial y}{\partial v} \end{vmatrix}=\begin{vmatrix} a & 0 \\ 0 & b \end{vmatrix}=ab$$

$$\begin{aligned} I &= \iint_R (x^2+y^2)dxdy \\ &= \iint_S (a^2u^2+b^2v^2)\left|\frac{\partial(x,y)}{\partial(u,v)}\right| dudv \\ &= \iint_S (a^2u^2+b^2v^2)|ab|\,dudv \\ &= ab\iint_S (a^2u^2+b^2v^2)\,dudv \\ &= a^3b\iint_S u^2\,dudv + ab^3\iint_S v^2\,dudv \end{aligned}$$

Letting $u=r\cos\theta$, $v=r\sin\theta$ (i.e. using polar coordinates)

$$\iint_S u^2dudv = \int_0^{2\pi}\int_0^1 r^2\cos^2\theta\cdot r\,dr\,d\theta = \frac{\pi}{4}$$

$$\iint_S v^2dudv = \int_0^{2\pi}\int_0^1 r^2\sin^2\theta\cdot r\,dr\,d\theta = \frac{\pi}{4}$$

$$I=\frac{\pi}{4}ab(a^2+b^2).$$

■

Triple Integrals

The notation of change of variables of double integrals can be extended to triple integrals.

THEOREM 1 Change of Variable in a Triple Integral

Let $\phi: V \subseteq \mathbb{R}^3 \to W \subseteq \mathbb{R}^3$ given by $\phi(u, v, w) = (x(u, v, w), y(u, v, w), z(u, v, w))$ be an one-one and onto function having first order continuous partial derivatives. We further assume that the jacobian $\left| \frac{\partial(x, y, z)}{\partial(u, v, w)} \right| \neq 0$. Let $f : W \to \mathbb{R}$ be a continuous function. Then

$$\iiint_W f(x, y, z) dV$$
$$= \iiint_V f[x(u, v, w), y(u, v, w), z(u, v, w)] \left| \frac{\partial(x, y, z)}{\partial(u, v, w)} \right| du\,dv\,dw.$$

EXAMPLE 4

Derive the formula for triple integration in spherical coordinates use the above result.

Solution The change of variables is given by

$$x = \rho \sin\phi \cos\theta, \quad y = \rho \sin\phi \sin\theta, \quad z = \rho \cos\phi.$$

The Jacobian is

$$\frac{\partial(x, y, z)}{\partial(\rho, \phi, \theta)} = \begin{vmatrix} \frac{\partial x}{\partial \rho} & \frac{\partial x}{\partial \phi} & \frac{\partial x}{\partial \theta} \\ \frac{\partial y}{\partial \rho} & \frac{\partial y}{\partial \phi} & \frac{\partial y}{\partial \theta} \\ \frac{\partial z}{\partial \rho} & \frac{\partial z}{\partial \phi} & \frac{\partial z}{\partial \theta} \end{vmatrix}$$
$$= \begin{vmatrix} \sin\phi\cos\theta & \rho\cos\phi\cos\theta & -\rho\sin\phi\sin\theta \\ \sin\phi\sin\theta & \rho\cos\phi\sin\theta & \rho\sin\phi\cos\theta \\ \cos\phi & -\rho\sin\phi & 0 \end{vmatrix} = \rho^2 \sin\phi$$

Therefore

$$\left| \frac{\partial(x, y, z)}{\partial(\rho, \phi, \theta)} \right| = |\rho^2 \sin\phi| = \rho^2 \sin\phi.$$

$$\iiint_E f(x, y, z) dV$$
$$= \iiint_R f(\rho\sin\phi\cos\theta, \rho\sin\phi\sin\theta, \rho\cos\phi) \rho^2 \sin\phi \, d\rho \, d\phi \, d\theta. \quad \blacksquare$$

14.7 EXERCISES (The Substitution Method)

http://matrix.skku.ac.kr/Cal-Book/part2/CS-Sec-14-7-Sol.htm

CAS **1.** Evaluate $\iint_R 3xy\,dx\,dy$, where R is the region enclosed by the lines $x-2y=0$, $x-2y=0$, $x-2y=-4$, $x+y=4$ and $x+y=1$.

Solution

http://matrix.skku.ac.kr/cal-lab/cal-14-7-1(new).html

```
var('x,y')
f(x)=integral((2*x^2-x*y-y^2)/9,x)
g(y)=f(0)-f(-4)
integral(g,(y,1,4))
```

Answer : 104/9

$D=\{(u,v)\,|\,1\le u\le 4,\ -4\le v\le 0\}$

$x+y=u$, $x-2y=v$ as $x=\frac{1}{3}(2u+v)$, $y=\frac{1}{3}(u-v)$

$$\frac{\partial(x,\,y)}{\partial(u,\,v)}=\begin{vmatrix}\frac{\partial x}{\partial u} & \frac{\partial x}{\partial v}\\ \frac{\partial y}{\partial u} & \frac{\partial y}{\partial v}\end{vmatrix}=\begin{vmatrix}\frac{2}{3} & \frac{1}{3}\\ \frac{1}{3} & -\frac{1}{3}\end{vmatrix}=-\frac{1}{3}$$

$$\iint_R 3xy\,dx\,dy$$
$$=\iint_D 3\left\{\frac{1}{3}(2u+v)\cdot\frac{1}{3}(u-v)\right\}\left|\frac{\partial(x,y)}{\partial(u,v)}\right|du\,dv$$
$$=\int_1^4\int_{-4}^0 \frac{1}{9}(2u^2-uv-v^2)\,dv\,du$$
$$=\frac{1}{9}\int_1^4\left[2u^2v-\frac{1}{2}uv^2-\frac{1}{3}v^3\right]_{-4}^0 du=\frac{104}{9}$$

CAS **2.** Evaluate $\iint_R \frac{(x+y)e^{x^2}}{e^{y^2}}\,dx\,dy$, where R is the region enclosed by the rhombus with vertices $(1,0)$, $(0,1)$, $(1,2)$ and $(2,1)$.

Solution

http://matrix.skku.ac.kr/cal-lab/cal-14-7-2(new).html

```
var('x,y')
f(y)=integral(x*exp(x*y)/2,y)
g(x)=f(1)-f(-1)
integral(g,(x, 1, 3)).expand()
```

Answer : 1/2*e^(-3) - 1/2*e^(-1) - 1/2*e + 1/2*e^3

The region R is enclosed by the line $x+y=3$, $x+y=1$, $x-y=1$, $x-y=-1$.

$D=\{(u,v)\,|\,1\le u\le 3,\ -1\le v\le 1\}$

Using Changes of Variables, $x+y=u$, $x-y=v$ as

$x=\frac{1}{2}(u+v)$, $y=\frac{1}{2}(u-v)$

$$\frac{\partial(x,y)}{\partial(u,v)}=\begin{vmatrix}\frac{\partial x}{\partial u} & \frac{\partial x}{\partial v}\\ \frac{\partial y}{\partial u} & \frac{\partial y}{\partial v}\end{vmatrix}=\begin{vmatrix}\frac{1}{2} & \frac{1}{2}\\ \frac{1}{2} & -\frac{1}{2}\end{vmatrix}=-\frac{1}{2}$$

Thus,

$$\iint_R \frac{(x+y)e^{x^2}}{e^{y^2}}\,dx\,dy$$
$$=\iint_D ue^{vu}\left|\frac{\partial(x,y)}{\partial(u,v)}\right|du\,dv$$
$$=\int_1^3\int_{-1}^1 ue^{vu}\left|-\frac{1}{2}\right|dv\,du$$
$$=\frac{1}{2}\int_1^3\int_{-1}^1 ue^{vu}\,dv\,du$$
$$=\frac{1}{2}\int_1^3\left[e^{vu}\right]_{-1}^1 du$$
$$=\frac{1}{2}\int_1^3 (e^u-e^{-u})\,du$$
$$=\frac{1}{2}\left[e^u+e^{-u}\right]_1^3$$
$$=\frac{1}{2}(e^3+e^{-3}-e-e^{-1}).$$

CAS **3.** Evaluate $\iint_R \sin(x+y)\cos(x-y)\,dx\,dy$, where R is the region enclosed by the triangle with vertices $(0,0)$, $(0,\pi)$ and $(2\pi,0)$.

Solution

http://matrix.skku.ac.kr/cal-lab/cal-14-7-3(new).html

```
var('x,y')
f(y)=integral(sin(x)*cos(y)/2,y)
g(x)=f(x)-f(-x)
integral(g,(x,0,2*pi)).expand()
```

Answer : pi

The region R is enclosed by the line $x=0$, $y=0$ and $x+2y=\pi$.

$D=\{(u,v)\,|\,0\le u\le 2\pi,\ -u\le v\le u\}$

Using Changes of Variables, $x+y=u$, $x-y=v$ as

$x=\frac{1}{2}(u+v)$, $y=\frac{1}{2}(u-v)$

$$\frac{\partial(x,y)}{\partial(u,v)}=\begin{vmatrix}\frac{\partial x}{\partial u} & \frac{\partial x}{\partial v}\\ \frac{\partial y}{\partial u} & \frac{\partial y}{\partial v}\end{vmatrix}=\begin{vmatrix}\frac{1}{2} & \frac{1}{2}\\ \frac{1}{2} & -\frac{1}{2}\end{vmatrix}=-\frac{1}{2}$$

Thus,

$$\iint_R \sin(x+y)\cos(x-y))\,dx\,dy$$
$$= \iint_D \sin u \cos v \left| \frac{\partial(x,y)}{\partial(u,v)} \right| du\,dv$$
$$= \iint_D \sin u \cos v \left| -\frac{1}{2} \right| du\,dv$$
$$= \frac{1}{2}\int_0^{2\pi}\int_{-u}^{u} \sin u \cos v\,dv\,du$$
$$= \frac{1}{2}\int_0^{2\pi} \sin u [\sin v]_{-u}^{u}\,du$$
$$= \frac{1}{2}\int_0^{2\pi} (2\sin^2 u)\,du$$
$$= \int_0^{2\pi} \sin^2 u\,du = \frac{1}{2}\int_0^{2\pi}(1-\cos 2u)du$$
$$= \frac{1}{2}\left[u - \frac{1}{2}\sin u\right]_0^{2\pi} = \pi.$$

CAS **4.** Evaluate the integral $\iint_R \cos\left(\frac{y-x}{y+x}\right) dA$, where R is the trapezoidal region with vertices $(0,1)$, $(2,0)$, $(0,2)$ and $(1,0)$.

Solution

http://matrix.skku.ac.kr/cal-lab/cal-14-7-4(new).html

```
var('x,y')
f(x)=integral(cos(x/y)/2,x)
g(y)=f(y)-f(-y)
integral(g,(y,1,2)).expand()
```

Answer : 3/2*sin(1)

$u=y-x,\ v=y+x \Rightarrow x=\frac{1}{2}(v-u),\ y=\frac{1}{2}(v+u)$

$$\frac{\partial(x,y)}{\partial(u,v)} = \begin{vmatrix} -\frac{1}{2} & \frac{1}{2} \\ \frac{1}{2} & \frac{1}{2} \end{vmatrix} = -\frac{1}{2}$$

$$\int\int \cos\left(\frac{y-x}{y+x}\right) dA = \int_1^2\int_{-y}^{y} \cos\frac{u}{v}\left|-\frac{1}{2}\right| du dv$$
$$= \frac{1}{2}\int_1^2 \left[v\sin\frac{u}{v}\right]_{u=-v}^{u=v} dv$$
$$= \frac{1}{2}\int_1^2 2v\sin 1 dv = \frac{3}{2}\sin 1.$$

5. Evaluate the integral $\iint_R e^y\,dA$, where R is the trapezoidal region with vertices $(0,0)$, $(2,1)$, $(3,3)$ and $(1,2)$.

Solution The region R is enclosed by the line $y=\frac{1}{2}x$, $y=2x$, $y=\frac{1}{2}x+\frac{3}{2}$ and $y=2x-3$.

$D=\{(u,v)\,|0 \le u \le 3,\ 0 \le v \le 3\}$

Using Changes of Variables, $2x-y=u$, $2y-x=v$ as $x=\frac{1}{3}(2u+v)$, $y=\frac{1}{3}(u+2v)$

$$\frac{\partial(x,y)}{\partial(u,v)} = \begin{vmatrix} \frac{\partial x}{\partial u} & \frac{\partial x}{\partial v} \\ \frac{\partial y}{\partial u} & \frac{\partial y}{\partial v} \end{vmatrix} = \begin{vmatrix} \frac{2}{3} & \frac{1}{3} \\ \frac{1}{3} & \frac{2}{3} \end{vmatrix} = \frac{1}{3}$$

Thus,

$$\iint_R e^x\,dA = \iint_D e^{\frac{1}{3}u+\frac{2}{3}v}\left|\frac{\partial(x,y)}{\partial(u,v)}\right| du\,dv$$
$$= \iint_D e^{\frac{1}{3}u+\frac{2}{3}v}\left|\frac{1}{3}\right| du\,dv$$
$$= \frac{1}{3}\int_0^3\int_0^3 e^{\frac{1}{3}u+\frac{2}{3}v}\,du\,dv$$
$$= \frac{1}{3}\int_0^3 e^{\frac{1}{3}u}du\int_0^3 e^{\frac{2}{3}v}dv$$
$$= \frac{1}{3}\left[3e^{\frac{1}{3}u}\right]_0^3\left[\frac{3}{2}e^{\frac{2}{3}v}\right]_0^3$$
$$= \frac{3}{2}(e-1)(e^2-1).$$

6. Derive the formula for triple integration in cylinder coordinates.

7. Evaluate $\iiint_E dV$, where E is the solid enclosed by the ellipsoid $\frac{x^2}{a^2}+\frac{y^2}{b^2}+\frac{z^2}{c^2}=1$, using the transformation $x=au$, $y=bv$ and $z=cw$.

8. Evaluate $\iint_R (x^2-xy+y^2)\,dA$, where R is the region bounded by the graph of $x^2-xy+y^2 \le 2$.

9. Evaluate $\iint_R (x^2+y^2)\,dA$, where R is the rhombus region with vertices $(-2,0)$, $(0,2)$, $(2,0)$ and $(0,-2)$.

10. Evaluate $\iint_R (x+y)e^{x+y}\,dA$, where R is the triangle region with vertices $(0,0)$, $(4,4)$ and $(4,0)$.

Seoul Art Center

Chapter 15

Vector Calculus

David Hilbert
1862-1943

Universal Mathematician

"This photograph was taken in 1912 for postcards"

Hilbert was recognized as one of the most influential and universal mathematicians of the 19th and early 20th centuries. He discovered and developed a broad range of fundamental ideas in many areas, including invariant theory, the axiomatization of geometry and the theory of Hilbert spaces, one of the foundations of functional analysis.

YouTube http://youtu.be/q0aVmUCXgTI

In this chapter, we study vector calculus on vector fields. In addition, we define vector fields and scalar functions as real-valued functions in space. Specially, we define line integrals along a curve and surface integrals across a surface. The connections among these concepts are given by Green's Theorem, Stokes' Theorem and Divergence Theorem. Green's Theorem transforms line integrals to double integrals and vice versa. Stokes Theorem states that the circulation of a vector field around the boundary of a surface in space equals to the integral of the normal component of the curl of the field over the surface. The Gauss Divergence Theorem says that the outward flux of a vector field across a closed surface is equal to the triple integral of the divergence of the field over the region enclosed by the surface. We will also look at some applications of these three important theorems. We will see generalized forms of the Fundamental Theorem of Calculus.

15.1 Vector Differentiation

Vector Fields

Associated with every point in a region we can imagine both a direction and a magnitude about gravitational force, the velocity of a flowing fluid. They are expressed by a vector at each point in their region, which is producing a vector field.

DEFINITION 1

Let D be a subset of $\mathbb{R}^2$ (or $\mathbb{R}^3$). A scalar field on D is a scalar valued function $f(x, y)$ (or $f(x, y, z)$) that assigns to each point (x,y) (or (x, y, z)) in D.

Any real valued function on D is a scalar field. For example a function f given by $f(x, y) = x + y$ is a scalar field, defined on $\mathbb{R}^2$, $f(x, y, z) = x^2 + y^2 + z^2$ is a scalar field defined on $\mathbb{R}^3$. A scalar field can be called a scalar function.

DEFINITION 2

Let D be a subset of $\mathbb{R}^2$ (or $\mathbb{R}^3$). A vector field on D is a function $\mathbf{F}$ that assigns to each point (x,y) (or (x, y, z)) in D, a two (or three)-dimensional vector $\mathbf{F}(x, y)$ (or $\mathbf{F}(x, y, z)$).

The best way to understand a vector field is to draw the arrows for the vector $\mathbf{F}(x, y)$ at a few representative points (x, y). Since $\mathbf{F}(x, y)$ is a two dimensional vector, we may write it as follows:

$$\mathbf{F}(x, y) = M(x, y)\mathbf{i} + N(x, y)\mathbf{j}$$

where M and N are scalar functions of two variables, and where $\mathbf{i}$ and $\mathbf{j}$ are unit vector along the coordinate axes. These scalar functions are called **component functions** of the vector field. We can also plot the vector field in two or three dimensions with the aid of a computer. Since a computer can plot a large number of vectors, this gives a better impression of the vector field than drawing by hand.

CAS **EXAMPLE 1**

Draw the vector field on $\mathbb{R}^2$ defined by $\mathbf{F}(x, y) = \frac{1}{2}x\mathbf{i} + y\mathbf{j}$.

Solution The length of the vector $\frac{1}{2}x\mathbf{i} + y\mathbf{j}$ is $\sqrt{\frac{1}{4}x^2 + y^2}$. Vector points roughly away from the origin and vectors farther from the origin are

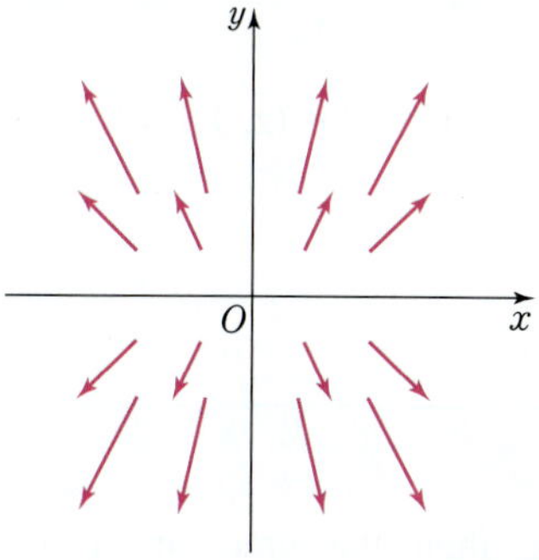

Figure 1 $\mathbf{F}(x, y) = \left\langle \frac{1}{2}x, y \right\rangle$

longer. (See Figure 1.)

http://matrix.skku.ac.kr/cal-lab/Sec15-1-Exm-1.html

```
var('x,y')
    plot_vector_field((1/2*x, y), (x,-3,3), (y,-3,3))
```

■

EXAMPLE 2

Draw the vector field on $\mathbb{R}^2$ defined by $\mathbf{F}(x, y) = \dfrac{y\mathbf{i} + x\mathbf{j}}{\sqrt{x^2 + y^2}}$.

Solution The length of the vector $\dfrac{y\mathbf{i} + x\mathbf{j}}{\sqrt{x^2 + y^2}}$ is 1. ■

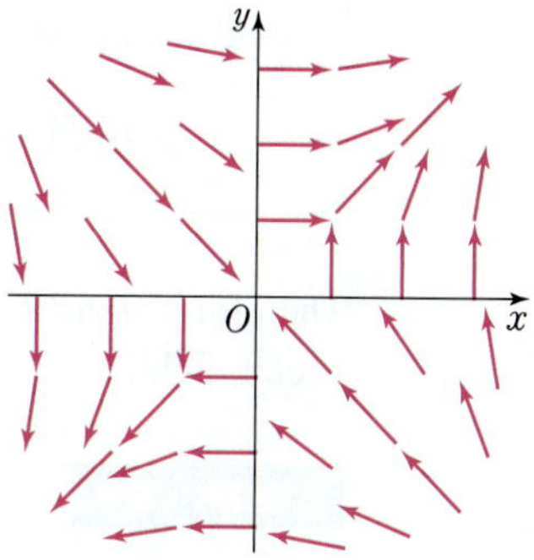

Figure 2 $\mathbf{F}(x, y) = \left\langle \frac{y}{\sqrt{x^2+y^2}}, \frac{x}{\sqrt{x^2+y^2}} \right\rangle$

CAS **EXAMPLE 3**

Draw the vector field on $\mathbb{R}^3$ defined by $\mathbf{F}(x, y, z) = -y\mathbf{k}$.

Solution At each point (x, y, z), $\mathbf{F}(x, y, z)$ is a vector of length $|y|$. For $y > 0$, all vectors are in the direction of the negative z-axis, while for $y < 0$, all vectors are in the direction of the positive z-axis. In each plane $y = k$, all the vectors are identical.

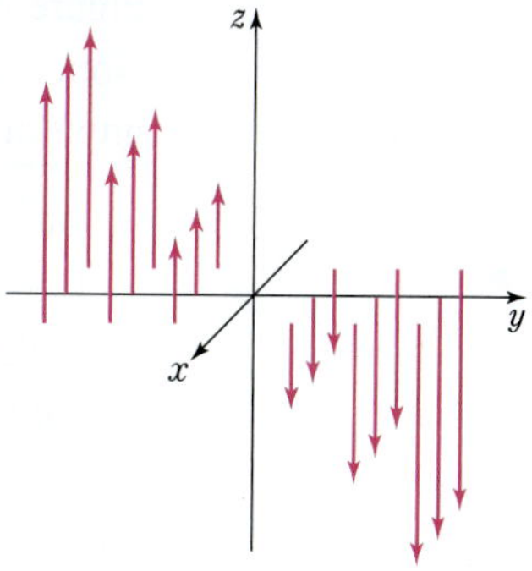

Figure 3 $\mathbf{F}(x, y, z) = -y\mathbf{k}$

http://matrix.skku.ac.kr/cal-lab/Sec15-1-Exm-3.html

```
var('x, y, z')
  plot_vector_field3d((0,0,-y), (x,-1,1), (y,-1,1), (z,-1,1))
```
■

Gradient Vector Fields

DEFINITION 3

If f is a scalar function of two variables, then the gradient of f, denoted by grad f (or ∇f), is defined by the vector

$$\operatorname{grad} f(x, y) = f_x(x, y)\mathbf{i} + f_y(x, y)\mathbf{j}$$

Likewise, if f is a scalar function of three variables, then the gradient, grad f (or ∇f), is defined by

$$\operatorname{grad} f(x, y, z) = f_x(x, y, z)\mathbf{i} + f_y(x, y, z)\mathbf{j} + f_z(x, y, z)\mathbf{k}$$

Therefore, $\operatorname{grad} f$ is a vector field in $\mathbb{R}^2$ (or $\mathbb{R}^3$) and it is called a gradient vector field.

EXAMPLE 4

If $f: \mathbb{R}^2 \to \mathbb{R}$ is defined to be $f(x,y) = x^2 + y^3$, then the grad $f = 2x\mathbf{i} + 3x^2\mathbf{j}$ is a vector field on $\mathbb{R}^2$.

If $f: \mathbb{R}^3 \to \mathbb{R}$ is defined to be $f(x, y, z) = x^2 + y^3 + z^3$, then the grad $f = 2x\mathbf{i} + 3y^2\mathbf{j} + 3z^2\mathbf{k}$ is a vector field on $\mathbb{R}^3$. ■

EXAMPLE 5

Find the gradient vector field of $f(x, y) = x^2 - y^2$. Draw the grad f together with level curves of f. How are they related?

Solution The gradient vector field of f is given by the grad $f(x, y) = 2x\mathbf{i} - 2y\mathbf{j}$. Figure 4 shows the contour map of f with the gradient vector field. Notice that the gradient vectors are perpendicular to the level curves. http://matrix.skku.ac.kr/cal-lab/Sec15-1-Exm-5.html

```
var('x, y')
f = (x^2 - y^2)
plot_vector_field(f.gradient(), (x, -2, 2), (y, -2, 2),
color='blue')
```

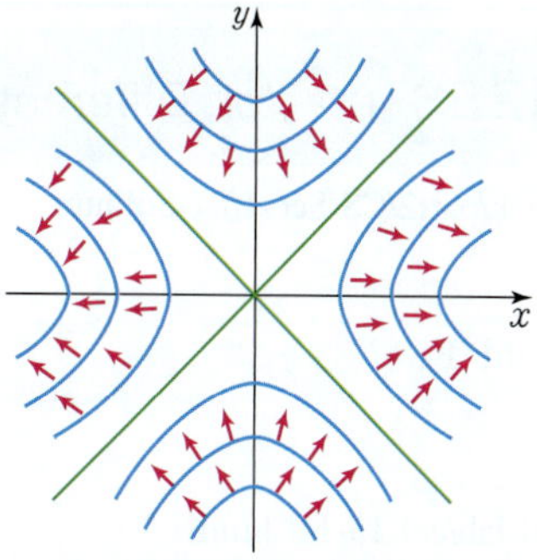

Figure 4 Contour map and gradient vector field of $f(x, y) = x^2 + y^2$

If you look at the Figure 4, it is clear that the gradient is perpendicular to the level curve. This can be proved easily.
Let $z = f(x,y)$ be a function having the first order continuous partial derivatives. Let C be the level curve at the level $z = c$. Then $f(x,y) = c$ on C. Let $\gamma(t)$ be a curve in C, then $f(\gamma(t)) = c$. Hence, using the chain rule, we have $\frac{d}{dt}(f \circ \gamma(t)) = 0$. So,

$$\nabla f(\gamma(t)) \cdot \gamma'(t) = 0.$$

This means $\nabla f(\gamma(t))$ is orthogonal to the tangent $\gamma'(t)$ for all real value t. ■

15.1 EXERCISES (Vector Differentiation)

http://matrix.skku.ac.kr/Cal-Book/part2/CS-Sec-15-1-Sol.htm

YouTube http://youtu.be/iSUME4Q1WPM

CAS **1-7.** Sketch the vector field $\mathbf{F}$.

1. $\mathbf{F}(x, y) = 3\mathbf{i} + 4\mathbf{j}$

Solution

http://matrix.skku.ac.kr/cal-lab/cal-15-1-1.html

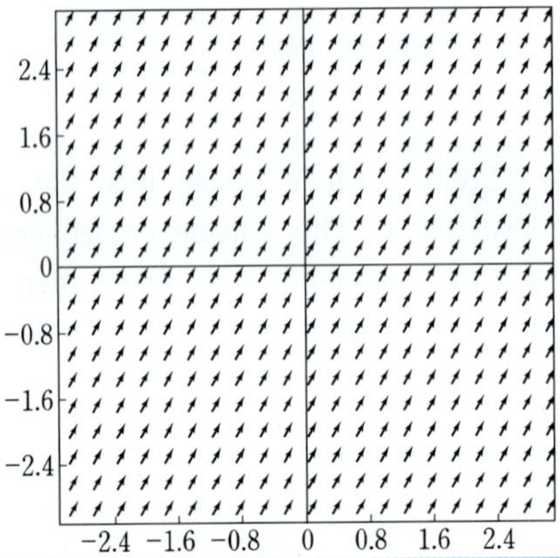

```
var('x,y')
vf=plot_vector_field((3,4),    (x,-3,3),    (y,-3,3),
aspect_ratio=1);
show(vf)
```

2. $\mathbf{F}(x, y) = \frac{1}{2}x\mathbf{i} - 2y\mathbf{j}$

Solution

http://matrix.skku.ac.kr/cal-lab/Sec15-1-2.html

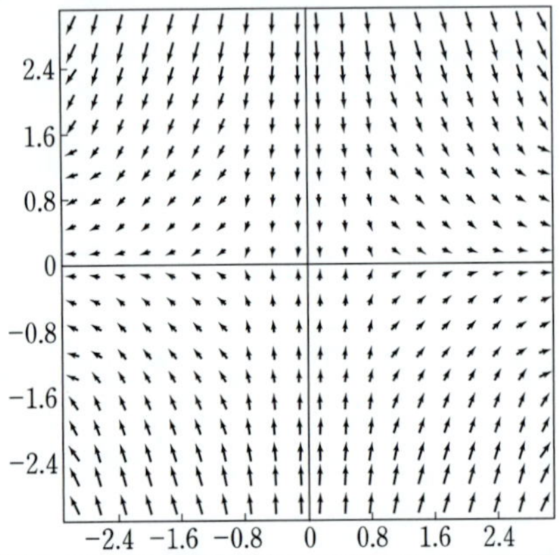

```
var('x,y')
vf=plot_vector_field((1/2*x,-2*y),    (x,-3,3),
(y,-3,3), aspect_ratio=1);
show(vf))
```

3. $\mathbf{F}(x, y) = 5y\mathbf{i} + \frac{1}{2}\mathbf{j}$

Solution

http://matrix.skku.ac.kr/cal-lab/Sec15-1-3.html

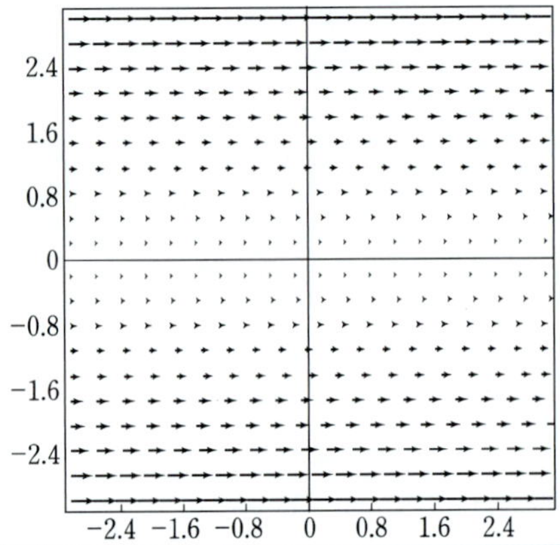

```
var('x,y')
vf=plot_vector_field((5*y,1/2),    (x,-3,3),    (y,-3,3),
aspect_ratio=1);
show(vf)
```

4. $\mathbf{F}(x, y) = (x+y)\mathbf{i} + x\mathbf{j}$

Solution

http://matrix.skku.ac.kr/cal-lab/Sec15-1-Exs-4.html

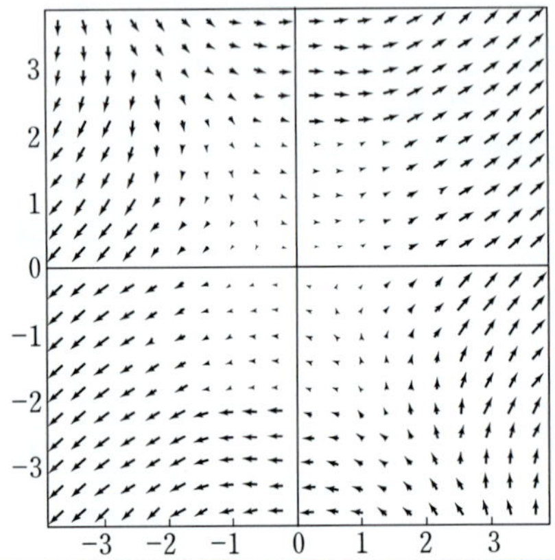

```
var('x,y')
vf=plot_vector_field((x+y,x),    (x,-3,3),    (y,-3,3),
aspect_ratio=1);
show(vf)
```

5. $\mathbf{F}(x, y) = \dfrac{y\mathbf{i} - x\mathbf{j}}{\sqrt{x^2 + y^2}}$

Solution

http://matrix.skku.ac.kr/cal-lab/Sec15-1-Exs-5.html

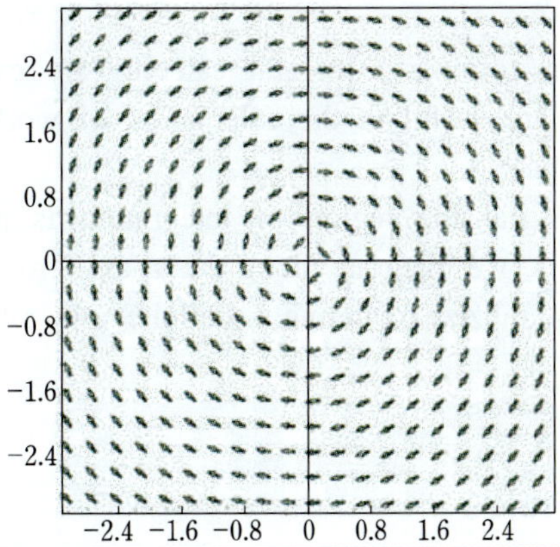


```
var('x,y')
vf=plot_vector_field((y/sqrt(x^2+y^2),-x/sqrt(x^2+y
^2)), (x,-3,3), (y,-3,3),
aspect_ratio=1);
show(vf)
```

6. $\mathbf{F}(x, y) = \dfrac{y\mathbf{i} + x\mathbf{j}}{\sqrt{x^2 + y^2}}$

Solution

http://matrix.skku.ac.kr/cal-lab/Sec15-1-6-7.html

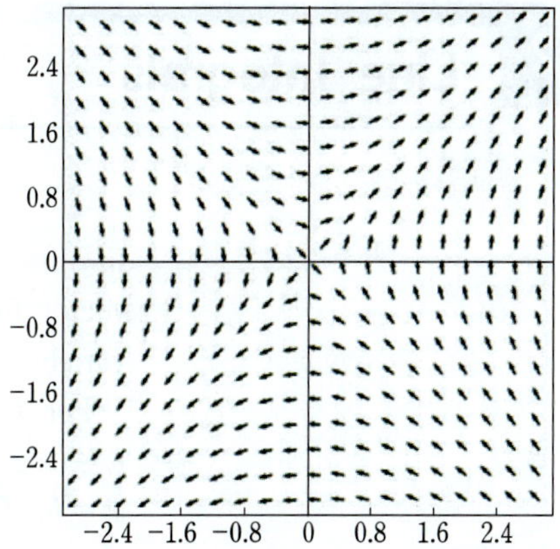


```
var('x,y')
vf=plot_vector_field((y/sqrt(x^2+y^2),x/sqrt(x^2+y
^2)), (x,-3,3),
(y,-3,3), aspect_ratio=1);
show(vf)
```

7. $\mathbf{F}(x, y, z) = \mathbf{k}$

Solution

http://matrix.skku.ac.kr/cal-lab/Sec15-1-6-7.html

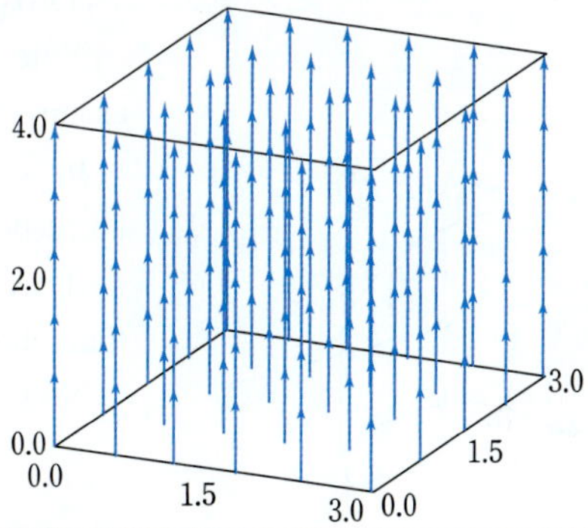


```
var('x,y,z')
plot_vector_field3d((0,0,1), (x, -3,3), (y,-3,3),
(z,-3,3))
```

15.2 Line Integrals

YouTube http://youtu.be/wHINlpNXYaU

In this section, the concepts of (ordinary) integral calculus are extended to vector functions. Line integrals are useful in the calculation of work done by variable forces along paths in space.

For the ordinary definite integral the region of integration is an interval $a \le x \le b$ on the x-axis. That is, we integrate along the x-axis from $x = a$ to $x = b$. This concept can be generalized to define a definite integral evaluated along a curve.

Line Integral of Scalar Fields

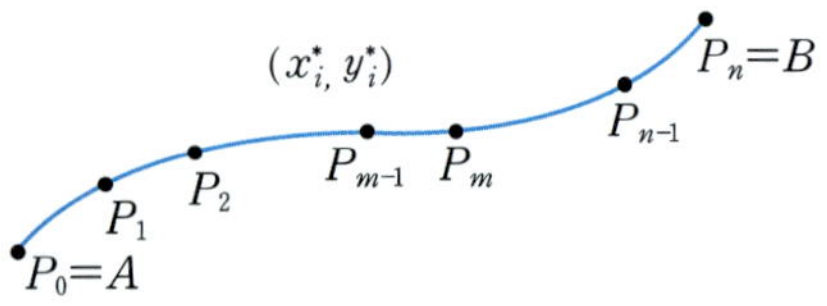

Figure 1

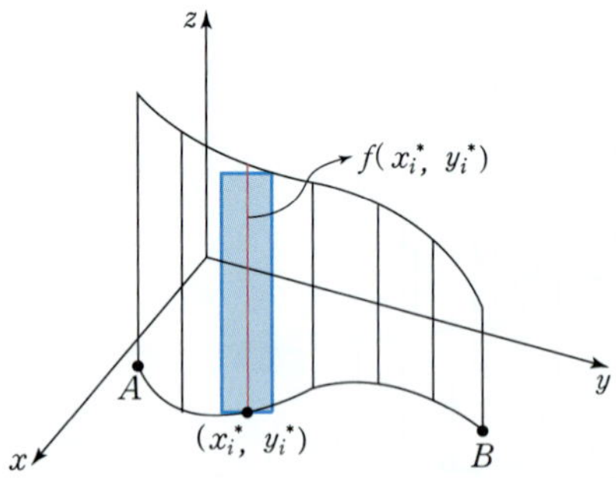

Figure 2

Let $z = f(x, y)$ be a continuous function defined on a domain $D \subseteq \mathbb{R}^2$. Let $C: (a, b) \to D$ be a smooth curve in D defined as $t \mapsto (x(t),\ y(t))$. We wish to define the integral of f along the curve C.

For this, we divide the integral $[a,\ b]$ into n equal parts $a = t_0 < t_1 < t_2 < \cdots < t_n = b$. The corresponding points on the curve are $A = P_0(x(t_0), y(t_0))$, $P_1(x(t_1),\ y(t_1))$, $\cdots$, $P_n(x(t_n),\ y(t_n)) = B$, in Figure 1.

Note that $A = P_0$, P_1, $\cdots$, $P_n = B$ divide the curve C into n sub-arcs. We assume that the length of these sub-arcs are Δs_1, Δs_2, $\cdots$, Δs_n, respectively.

In each sub-arc, choose a random point $P_i^*(x_i^*(t_i^*),\ y_i^*(t_i^*))$ which corresponds to a point t_i^* in the subinterval $[t_{i-1}, t_i]$ of $[a, b]$.

If we draw a rectangle with the base of $[P_{i-1}, P_i]$ and height $f(x_i^*,\ y_i^*)$, then the area of this rectangle is $f(x_i^*,\ y_i^*)\Delta s_i$ in Figure 2.

Now we form at the sum

$$\sum_{i=1}^{n} f(x_i^*,\ y_i^*)\Delta s_i.$$

If $\lim\limits_{n\to\infty}\sum_{i=1}^{n} f(x_i^*,\ y_i^*)\Delta s_i$ exists, then we say that the line integral of f with respect to s along the curve C from A to B exists. In this case, this limit is denoted by

$$\int_C f(x, y)\,ds \quad \text{or} \quad \int_a^b f(s)\,ds.$$

This gives us the following definition.

DEFINITION 1

Let f be a continuous function defined on a domain $D \subseteq \mathbb{R}^2$. Let $C: (a, b) \to D$ be a smooth curve. Then the line integral of f along C is defined as

$$\int_C f(x, y)\,ds = \lim_{n \to \infty} \sum_{i=1}^{n} f(x_i^*, y_i^*)\Delta s_i$$

given that the limit exists.

Remark The line integral $\int_C f(x, y)ds$ is the surface area of the surface above the curve C up to under the surface $z = f(x, y)$.

We shall assume that line integrals are evaluated along a curve which has an arc length parameterization. In the case of a parameterization with respect to arc length $s(t)$ in Section 12.3, we have

$$ds = \sqrt{\left(\frac{dx}{dt}\right)^2 + \left(\frac{dy}{dt}\right)^2}\,dt\,.$$

Therefore, we have the following theorem which gives a working rule to evaluate a line integral along a curve.

THEOREM 2

Let $f(x, y)$ be a continuous function defined on a domain $D \subseteq \mathbb{R}^2$ and $C: (a, b) \to D$ be a smooth curve C with arc length parameterization in D. Then

$$\int_C f(x, y)\,ds = \int_a^b f(x(t), y(t))\sqrt{\left(\frac{dx}{dt}\right)^2 + \left(\frac{dy}{dt}\right)^2}\,dt\,.$$

Let us look at some examples.

EXAMPLE 1

Evaluate $\int_C (x + y)ds$ where C is the semi-circle $\langle \cos t, \sin t \rangle \, (0 \le t \le \pi)$.

Solution

$$\int_C (x + y)ds = \int_0^{\pi} (\cos t + \sin t)\sqrt{(-\sin t)^2 + \cos^2 t}\,dt$$
$$= [\sin t - \cos t]_0^{\pi} = 2$$

■

Reference: http://matrix.skku.ac.kr/cal-lab/Sec15-2-Example-2.html

EXAMPLE 2

Evaluate $\int_C (x+y)ds$ where C is a circle $\langle \cos t, \sin t \rangle (0 \le t \le 2\pi)$.

Solution $\int_C (x+y)ds = \int_0^{2\pi} (\cos t + \sin t)\sqrt{\sin^2 t + \cos^2 t}\, dt$

$= [\sin t - \cos t]_0^{2\pi} = 0$ ■

A curve is said to be a closed curve (path) when the endpoints coincide. That is, $C : [a, b] \to D$ is a closed curve if $C(a) = C(b)$. If C is a closed curve, then the line integral $\int_C$ is denoted by $\oint_C$.

Properties of Line Integrals of Vector Fields

Let $\mathbf{F} = \mathbf{F}(\mathbf{r}) = F_1\mathbf{i} + F_2\mathbf{j} + F_3\mathbf{k}$ be a vector field. Then a line integral of $\mathbf{F}(\mathbf{r})$ along (taken over) the curve C is defined as

1

$$\int_C \mathbf{F}(\mathbf{r}) \cdot d\mathbf{r} = \int_C F_1\,dx + F_2\,dy + F_3\,dz$$
$$= \int_C F_1 \frac{dx}{dt}dt + F_2 \frac{dy}{dt}dt + F_3 \frac{dz}{dt}dt$$
$$= \int_a^b \mathbf{F}(\mathbf{r}(t)) \cdot \frac{d\mathbf{r}}{dt}dt.$$

Observe that evaluation of a line integral reduces to evaluation of an ordinary integral.

For the line integral 1, the following properties follow from integral calculus:

1. $\int_C k\mathbf{F} \cdot d\mathbf{r} = k\int_C \mathbf{F} \cdot d\mathbf{r}$, where k is a constant,
2. $\int_C (\mathbf{F} \pm \mathbf{G}) \cdot d\mathbf{r} = \int_C \mathbf{F} \cdot d\mathbf{r} \pm \int_C \mathbf{G} \cdot d\mathbf{r}$,
3. $\int_C \mathbf{F} \cdot d\mathbf{r} = \int_{C_1} \mathbf{F} \cdot d\mathbf{r} + \int_{C_2} \mathbf{F} \cdot d\mathbf{r}$,

   where C is the sum of two curves C_1 and C_2.
4. $\int_{-C} \mathbf{F} \cdot d\mathbf{r} = -\int_C \mathbf{F} \cdot d\mathbf{r}$, where $-C$ is a curve traversed in opposite direction of C. (See Figure 3.)

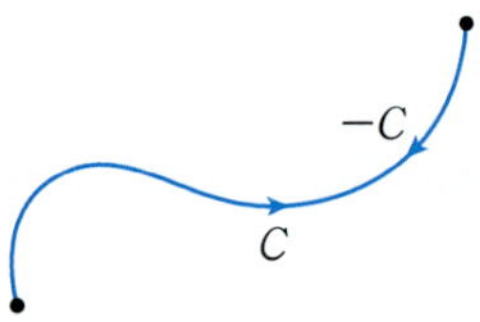

Figure 3

EXAMPLE 3

Find the line integral of $\mathbf{F}(x, y) = \sin x \cos y\,\mathbf{i} + e^{xy}\mathbf{j}$ along
(a) the horizontal line $C_1 : 0 \le x \le \pi,\ y = \pi/3$.
(b) the vertical line $C_2 : 0 \le y \le 1,\ x = 2$. (See Figure 4.)

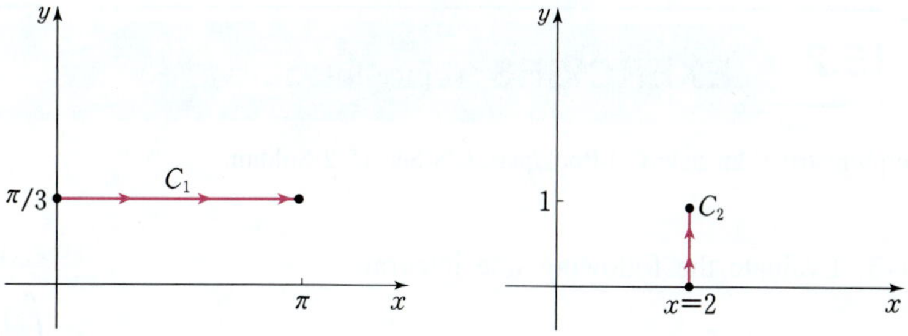

Figure 4

Solution (a) $\displaystyle\int_{C_1} \mathbf{F}\cdot d\mathbf{r} = \int_0^{\pi} \left\langle \sin x \cos\frac{\pi}{3}x,\ e^{\frac{\pi}{3}x} \right\rangle \cdot \langle 1,\ 0\rangle dx$

$$= \int_0^{\pi} \sin x \cos\frac{\pi}{3}x\,dx = \int_0^{\pi} \frac{1}{2}\sin x\,dx$$

$$= \left[-\frac{1}{2}\cos x\right]_0^{\pi}$$

$$= -\frac{1}{2}(-1-1) = 1.$$

(b) $\displaystyle\int_{C_2} \mathbf{F}\cdot d\mathbf{r} = \int_0^{1} \langle \sin 2\cos y,\ e^{2y}\rangle \cdot \langle 0,\ 1\rangle dy = \int_0^1 e^{2y}\,dy$

$$= \left[\frac{1}{2}e^{2y}\right]_0^1 = \frac{1}{2}(e^2-1).$$ ■

Moreover we consider that a smooth curve C is given by the vector equation $\mathbf{r}(t) = x(t)\mathbf{i} + y(t)\mathbf{j} + z(t)\mathbf{k},\ a \le t \le b$.

If f is the function of three variables that is continuous on a region including C, then we define the line integral of f

$$\int_C f(x, y, z)\,ds = \int_a^b f(x(t), y(t), z(t))\sqrt{\left(\frac{dx}{dt}\right)^2 + \left(\frac{dy}{dt}\right)^2 + \left(\frac{dz}{dt}\right)^2}\,dt.$$

Next we have a simple example.

EXAMPLE 4

Evaluate $\displaystyle\int_C y^2\,ds$, where C is a circular helix given by the vector equation $\mathbf{r}(t) = \cos t\,\mathbf{i} + \sin t\,\mathbf{j} + t\mathbf{k},\ 0 \le t \le 2\pi$. (See figure 5.)

http://matrix.skku.ac.kr/cal-lab/Sec15-2-Example-4.html

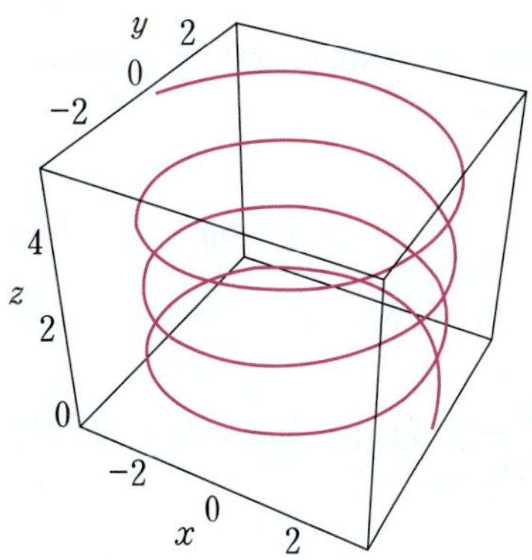

Figure 5

Solution $\displaystyle\int_C y^2\,ds = \int_0^{2\pi} \sin^2 t\sqrt{(-\sin t)^2 + \cos^2 t + 1^2}\,dt$

$$= \sqrt{2}\int_0^{2\pi} \frac{1}{2}(1-\cos 2t)\,dt$$

$$= \frac{\sqrt{2}}{2}\left[t - \frac{1}{2}\sin 2t\right]_0^{2\pi} = \sqrt{2}\pi.$$ ■

15.2 EXERCISES (Line Integrals)

http://matrix.skku.ac.kr/Cal-Book/part2/CS-Sec-15-2-Sol.htm

YouTube http://youtu.be/ZdRjCfJeHM8

1-3. Evaluate the following line integrals:

CAS **1.** Evaluate $\int_C (1+x^3y^2)ds$, where C is the upper half of the unit circle $x^2+y^2=1$.

Solution http://matrix.skku.ac.kr/cal-lab/Sec15-2-1.html

```
t=var('t')
half_circle=parametric_plot((cos(t),sin(t)),(t,0,pi()));
show(half_circle, aspect_ratio=1)
```

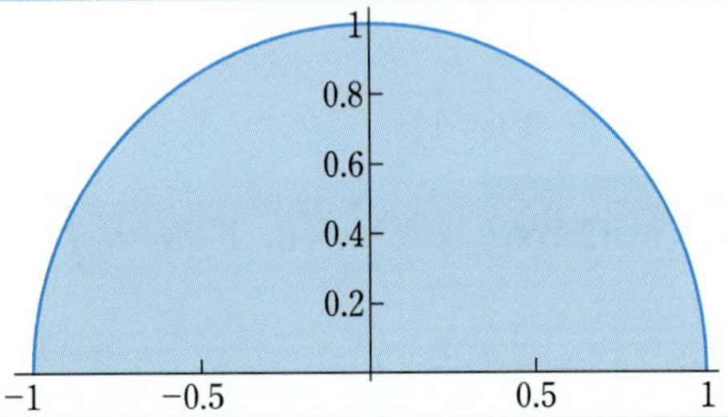


```
integral(1+cos(t)^3*sin(t)^2*sqrt(diff(cos(t),t)^2+dif
f(sin(t),t)^2),t,0,pi())
```

Answer : pi

2. $\int_C (x^2+y^2+z^2)ds$, where C is the arc of the circular helix $\mathbf{r}(t)=\cos t\,\mathbf{i}+\sin t\,\mathbf{j}+3t\,\mathbf{k}$ from $(1,\ 0,\ 0)$ to $(1,\ 0,\ 6\pi)$.

Solution Let $x=\cos t,\ y=\sin t,\ z=3t$

$$\int_0^{2\pi}(x^2+y^2+z^2)\sqrt{\left(\frac{dx}{dt}\right)^2+\left(\frac{dy}{dt}\right)^2+\left(\frac{dz}{dt}\right)^2}\,dt$$

$$=\int_0^{2\pi}(1+9t^2)\sqrt{1+9}\,dt=2\sqrt{10}\,(12\pi^3+\pi).$$

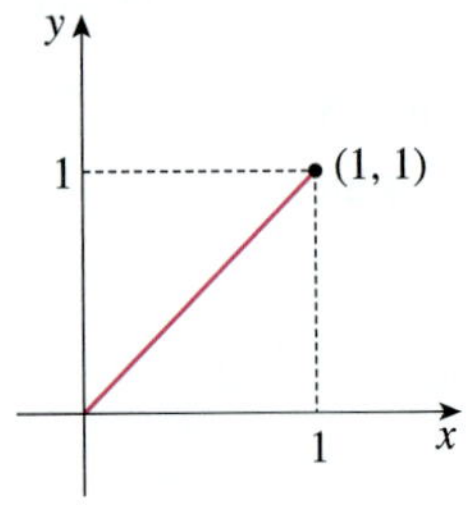

3. Let $\mathbf{F}(x,\ y)=(3+2xy)\mathbf{i}+(x^2-3y^2)\mathbf{j}$. Find $\int_C \mathbf{F}\cdot d\mathbf{r}$ where C is any smooth curve from $(0,\ 0)$ to $(1,\ 1)$.

Solution $\mathbf{r}(t)=t\mathbf{i}+t\mathbf{j} \Rightarrow \mathbf{r}'(t)=\mathbf{i}+\mathbf{j}\ (0<t<1)$

$$\int_C \mathbf{F}\cdot d\mathbf{r}=\int_0^1 \mathbf{F}(\mathbf{r}(t))\cdot\frac{d\mathbf{r}}{dt}dt$$

$$=\int_0^1 (3+2t^2)+(t^2-3t^2)dt=3$$

CAS **4.** If $\mathbf{A}(t)=t\mathbf{i}-t^2\mathbf{j}+(t-1)\mathbf{k}$, $\mathbf{B}(t)=2t^2\mathbf{i}+6t\mathbf{k}$, evaluate $\int_0^2 \mathbf{A}\cdot\mathbf{B}\,dt$.

Solution http://matrix.skku.ac.kr/cal-lab/cal-14-1-24.html

```
var('t')
A=vector([t,-t^2,t-1])
B=vector([2*t^2,0,6*t])
v=A.dot_product(B)
v.integrate(t,0,2)
```

Answer : 12

5. If $\mathbf{A}(t)=t\mathbf{i}-3\mathbf{j}+2t\mathbf{k}$, $\mathbf{B}(t)=\mathbf{i}-2\mathbf{j}+2\mathbf{k}$, $\mathbf{C}(t)=3\mathbf{i}+t\mathbf{j}-\mathbf{k}$ then evaluate

a. $\int_1^2 \mathbf{A}\cdot(\mathbf{B}\times\mathbf{C})dt$,

b. $\int_1^2 (\mathbf{A}\times(\mathbf{B}\times\mathbf{C}))dt$.

6. If $\mathbf{A}(2)=2\mathbf{i}-\mathbf{j}+2\mathbf{k}$, $\mathbf{A}(3)=4\mathbf{i}-2\mathbf{j}+3\mathbf{k}$ then evaluate $\int_2^3 \mathbf{A}\cdot\frac{d\mathbf{A}}{dt}dt$.

7-9. (Line Integrals) Given the following vector fields and oriented curves C, evaluate $\int_C \mathbf{F}\cdot d\mathbf{r}$.

7. $\mathbf{F}=-y\mathbf{i}+x\mathbf{j}$ on the semicircle $\mathbf{r}(t)=\langle 3\cos t,\ 3\sin t\rangle$ for $0\le t\le\pi$.

8. $\mathbf{F}=\frac{x}{(x^2+y^2)^{3/2}}\mathbf{i}+\frac{y}{(x^2+y^2)^{3/2}}\mathbf{j}$ on the line segment from $(1,1)$ to $(5,5)$.

9. $\mathbf{F}=\frac{x}{x^2+y^2}\mathbf{i}+\frac{y}{x^2+y^2}\mathbf{j}$ on the line $\mathbf{r}(t)=(t,\ 2t)$ for $1\le t\le 5$.

15.3 Potential Function and Independence of Path

YouTube http://youtu.be/jGGOL3QDj1Y

Potential Functions

We know that if f is a scalar function then the grad f is a vector field. Therefore, given a vector field $\mathbf{F}$, we can construct the scalar function f with the property that

$$\mathbf{F} = \operatorname{grad} f .$$

If such a scalar function exists, then $\mathbf{F}$ is said to be conservative.

DEFINITION 1

Let $\mathbf{F}$ be a vector field. Then $\mathbf{F}$ is called conservative if there is a scalar function f such that

$$\operatorname{grad} f = \mathbf{F}.$$

A function f is called a potential function (or scalar potential) for $\mathbf{F}$.

Remark Note that if a scalar function f is a potential function of a vector field $\mathbf{F}$, then $f + c$ is also a potential function of $\mathbf{F}$ for arbitrary constant c. Not all the vector fields are conservative, but such fields arise frequently in physics. The most prominent examples of conservative forces are the force of gravity and the electric field associated to a static charge.
According to Newton's Law of Gravitation, the gravitational force field between two objects with masses m and M is given by

$$\mathbf{F}(\mathbf{x}) = -\frac{mMG}{|\mathbf{x}|^2}\frac{\mathbf{x}}{|\mathbf{x}|}$$

where $\mathbf{x} = x\mathbf{i} + y\mathbf{j} + z\mathbf{k}$ is the position vector of the object with mass m from the other object with mass M and where G is the gravitational constant. It is easy to check that

$$\mathbf{F}(x, y, z) = \operatorname{grad} f$$

where $f(x, y, z) = \dfrac{mMG}{\sqrt{x^2 + y^2 + z^2}}$.
Therefore, the gravitational force field $\mathbf{F}$ is conservative and f is a potential function for $\mathbf{F}$.

If $\mathbf{F}(x,\ y) = M(x,\ y)\mathbf{i} + N(x,\ y)\mathbf{j}$ is conservative, then

$$M\mathbf{i} + N\mathbf{j} = \operatorname{grad} f = f_x\,\mathbf{i} + f_y\,\mathbf{j}.$$

That is, $M = f_x$ and $N = f_y$.

Now take a partial derivative with respect to y of $M = f_x$, and with respect to x of $N = f_y$ to get

$$M_y = f_{xy} \qquad \text{and} \qquad N_x = f_{yx}$$

In addition, if we assume that the second order partial derivatives are the same, then we have

$$M_y = N_x.$$

This turns out to be both necessary and sufficient condition for a vector field to be conservative, in special type of domain, which are simply connected.

A domain is said to be **simply connected** if it does not have a hole inside. (However, this is not a rigorous definition). For example, any disk: $x^2 + y^2 \le a^2$, upper half plane: $y \ge 0$ are simply connected domains in $\mathbb{R}^2$, whereas annulus: $1 \le x^2 + y^2 \le 4$ is not simply connected. (See Figure 1.)

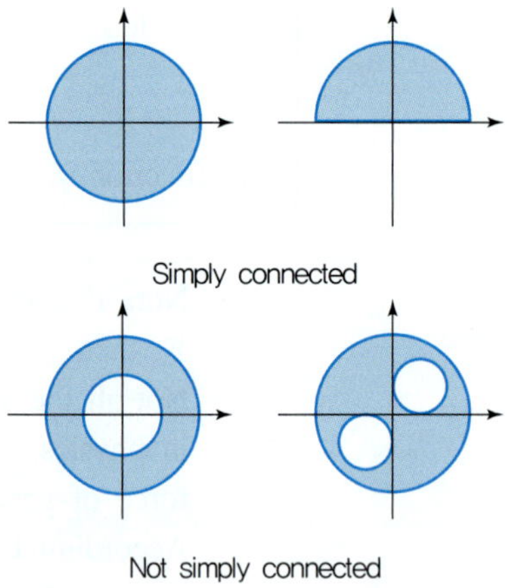

Figure 1

THEOREM 2

If M and N have continuous first order partial derivatives and the vector field $\mathbf{F}(x,y) = M(x,y)\mathbf{i} + N(x,y)\mathbf{j}$ is conservative, then, in simply connected domains, we have

$$M_y = N_x.$$

Remark The converse of Theorem 2 is in general not true. It is true only for special domains that we will see later. (See Example 2.) In simply connected regions, we can state the converse of Theorem 2 for verifying that a vector field on $\mathbb{R}^2$ is conservative. We will check the proof in a result of Green Theorem in Section 15.4.

THEOREM 3

Let $\mathbf{F}(x,y) = M(x,y)\mathbf{i} + N(x,y)\mathbf{j}$ be a vector field on an open simply connected region D. Suppose that M and N have continuous first order partial derivatives and $M_y = N_x$. Then a vector field $\mathbf{F}$ is conservative.

Remark Suppose that $\mathbf{F}(x,y,z) = M(x,y,z)\mathbf{i} + N(x,y,z)\mathbf{j} + P(x,y,z)\mathbf{k}$ is a vector field on a simply connected domain where M, N and P have continuous first order partial derivatives. Then $\mathbf{F}$ is conservative if and only if $\dfrac{\partial P}{\partial y} = \dfrac{\partial N}{\partial z}$, $\dfrac{\partial P}{\partial x} = \dfrac{\partial M}{\partial z}$ and $\dfrac{\partial N}{\partial x} = \dfrac{\partial M}{\partial y}$. The proof is similar to Theorem 2.

CAS **EXAMPLE 1**

Determine which of two vector fields are conservative

(1) $\mathbf{F} = 4xy\mathbf{i} - x^3\mathbf{j}$

(2) $\mathbf{G} = (1 + 3x^2y)\mathbf{i} + (x^3 - 2)\mathbf{j}$

(3) $\mathbf{H} = 2x\mathbf{i} - z\mathbf{j} + \sin z\mathbf{k}$.

http://matrix.skku.ac.kr/cal-lab/Sec15-2-Exm-1.html

Solution For (1) we find

$$M_y = 4x, \ N_x = -3x^2.$$

Since they are not equal, the vector field is not conservative.

```
var('x,y')
F=vector([4*x*y,-x^3])
print bool(diff(F[1],x)==diff(F[0],y))
```

Answer : False

For (2) we find

$$M_y = 3x^2, \ N_x = 3x^2.$$

Since they are equal, the vector field is conservative.

For (3) we find

$$P_y = 0, \ N_z = -1.$$

Since they are not equal, $\mathbf{H}$ is not conservative.

```
var('x,y,z')
H=vector([2*x,-z,sin(z)])
print bool(diff(H[2],y)==diff(H[1],z))
print bool(diff(H[2],x)==diff(H[0],z))
print bool(diff(H[1],x)==diff(H[0],y))
```

Answer : False, True, True ■

Reference:
http://www.mail-archive.com/sage-support@googlegroups.com/msg14988.html

Independence of Path

Let A and B be two points in an open region D. For some special fields, the line integral of the vector field $\mathbf{F}$ has the same value for all paths from A to B. (See Figure 2.) Then the line integral $\int_C \mathbf{F}\cdot d\mathbf{r}$ is said to be the **independence of path in** D.

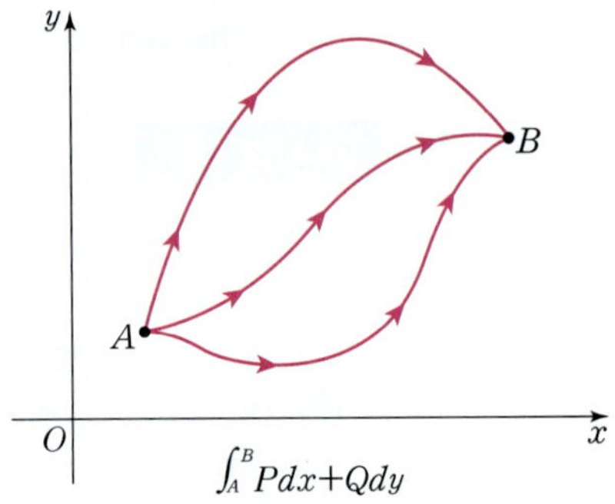

Figure 2 Independence of Path

Remark There is an useful property of line integral in a conservative field when the path of integration is a closed curve: the vector field $\mathbf{F}$ is conservative on the domain D if and only if $\oint_C \mathbf{F}\cdot d\mathbf{r}=0$ around every closed curve in the domain D.

If $\mathbf{F}$ is a conservative vector field with potential function ϕ, then the line integral from A to B is independent of path joining A to B. (See Figure 2.)

$$\begin{aligned}\int_A^B \mathbf{F}\cdot d\mathbf{r} &= \int_A^B \nabla\phi\cdot d\mathbf{r}\\ &= \int_A^B\left(\frac{\partial\phi}{\partial x}dx+\frac{\partial\phi}{\partial y}dy+\frac{\partial\phi}{\partial z}dz\right)\\ &= \phi(B)-\phi(A).\end{aligned}$$

Thus the line integral depends only on the end points A and B and not on the path joining them. Recall that when $\mathbf{F}=\nabla\phi$, then $\nabla\times\mathbf{F}=\nabla\times\nabla\phi=\mathbf{0}$. In such a case, $\mathbf{F}$ is called a conservative vector field and ϕ is called its scalar potential (potential function).

Note that a conservative force field is also irrotational (since $\nabla\times\mathbf{F}=\mathbf{0}$).

EXAMPLE 2

The vector field $\mathbf{F}=-\dfrac{y}{x^2+y^2}\mathbf{i}+\dfrac{x}{x^2+y^2}\mathbf{j}$ on $\mathbb{R}^2$ except the origin has no potential function, although $M_y = N_x$,

$$\frac{\partial M}{\partial y} = \frac{y^2 - x^2}{(x^2+y^2)^2} \text{ and } \frac{\partial N}{\partial x} = \frac{y^2 - x^2}{(x^2+y^2)^2}.$$

Suppose that a closed curve is the unit circle C : $\mathbf{r}(t) = \cos t\,\mathbf{i} + \sin t\,\mathbf{j}$, $0 \le t \le 2\pi$. We evaluate $\oint_C \mathbf{F} \cdot d\mathbf{r}$ around C. First we have

$$\mathbf{F} = -\sin t\,\mathbf{i} + \cos t\,\mathbf{j}.$$

Then $\oint_C \mathbf{F} \cdot d\mathbf{r} = \oint \mathbf{F} \cdot \frac{d\mathbf{r}}{dt} dt = \int_0^{2\pi} 1 dt = 2\pi.$

Since $\oint_C \mathbf{F} \cdot d\mathbf{r} \neq 0$, the vector field is not conservative. ■

For a conservative vector field, how do we find the potential function? Let us look at some examples.

EXAMPLE 3

Test the existence of a potential function for the vector field $\mathbf{F}(x, y) = xy^2\mathbf{i} + \tan x\,\mathbf{j}$ on $\mathbb{R}^2$.

Solution Since $\frac{\partial M}{\partial y} = \frac{\partial (xy^2)}{\partial y} = 2xy$ and $\frac{\partial N}{\partial x} = \frac{\partial (\tan x)}{\partial x} = \sec^2 x$, that is, $\frac{\partial M}{\partial y} \neq \frac{\partial N}{\partial x}$, the vector field $\mathbf{F}$ is not conservative.

So, there is no potential function. ■

CAS **EXAMPLE 4**

What is the potential function for the conservative vector field

$$\mathbf{F} = (1 + 3x^2y)\mathbf{i} + (x^3 - 2)\mathbf{j} \text{ on } \mathbb{R}^2?$$

http://matrix.skku.ac.kr/cal-lab/Sec15-2-Exm-4.html

Solution From Theorem 3, we know that $\mathbf{F}$ is conservative. Suppose $\mathbf{F} = \operatorname{grad} f$ for some scalar function f. Then we have

$$M = 1 + 3x^2y = f_x.$$

Integrating both sides with respect to x we get

$$x + x^3y + k(y) = f(x, y),$$

where $k(y)$ is a constant of integration and a function of y. Now differentiate with respect to y, we get

$$x^3 + k'(y) = f_y = N = x^3 - 2.$$

Thus

$$k'(y) = -2.$$

Integrating with respect to y, we get

$$k(y) = -2y + c \quad (c \text{ is a constant}).$$

Since we just want "a" potential function not "the" general potential function, we do not need a constant of integration here. Putting it all together, we get the potential function

$$f(x, y) = x + x^3 y - 2y + c.$$

```
x = var('x')
y = function("y",x)
M = 1+3*x^2*y
N = x^3-2
solution=desolve(diff(y,x)==-M/N,y)
f=simplify(solution==y)*solution.denominator()
potential=f.lhs()-f.rhs()
potential     # This is a potential function for F
```

Answer : (x^3 - 2)*y - c + x ■

Work Done by a Force(Work Integral)

THEOREM 4 Path Independence Theorem

Let $\mathbf{F} = M\mathbf{i} + N\mathbf{j}$ be a vector field such that $M_y = N_x$ and let A and B be two points of D. Let f be a potential function for $\mathbf{F}$. For any piecewise smooth curve C from A and B,

$$\int_C M\,dx + N\,dy = f(B) - f(A).$$

Thus the line integral in this case depends only on the end points A and B and not on the curve C. Thus we may write

$$\int_A^B M\,dx + N\,dy = \int_C M\,dx + N\,dy.$$

A natural application of the line integral is to define the work done by a force $\mathbf{F}$ in moving (displacing) a particle along a curve C from point A to point B as

2 $$\text{work done} = \int_A^B \mathbf{F} \cdot d\mathbf{r} = f(B) - f(A).$$

When $\mathbf{F}$ denotes the velocity of a fluid, then the circulation of $\mathbf{F}$ around a closed curve C is defined by $\int_C \mathbf{F}\cdot d\mathbf{r}$.

Result 1. The work done by a conservative force field in moving a particle from A to B is independent of the path joining A and B, that is, it depends only on the end points A and B. In such cases a scalar potential ϕ exists that is, $\mathbf{F}=\nabla\phi$. Thus the work done from A to B equals $\phi(B)-\phi(A)$.

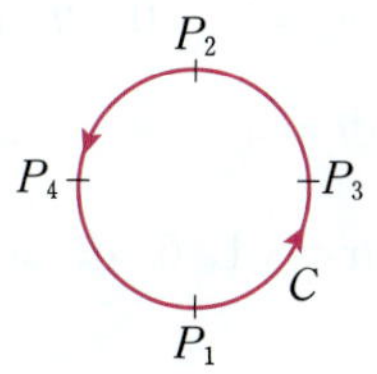

Figure 3

Result 2. If C is a closed smooth curve and $\mathbf{F}$ is a conservative field, then

$$\int_C \mathbf{F}\cdot d\mathbf{r}=0$$

along any closed curve C because

$$\int_C \mathbf{F}\cdot d\mathbf{r}=\int_{P_1P_3P_2P_4P_1}\mathbf{F}\cdot d\mathbf{r}=\int_{P_1P_3P_2}\mathbf{F}\cdot d\mathbf{r}+\int_{P_2P_4P_1}\mathbf{F}\cdot d\mathbf{r}$$
$$=\int_{P_1P_3P_2}\mathbf{F}\cdot d\mathbf{r}-\int_{P_1P_4P_2}\mathbf{F}\cdot d\mathbf{r}=0$$

which follows from the independence of the path. (See Figure 3.)

CAS **EXAMPLE 5**

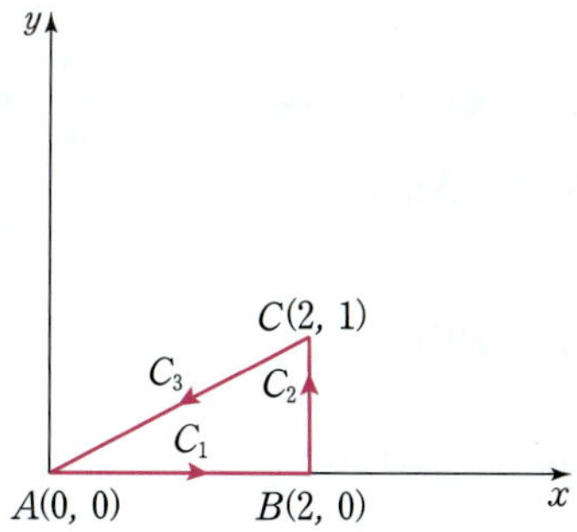

Figure 4

If $\mathbf{F}(x, y)=(2x+y^2)\mathbf{i}+(3y-4x)\mathbf{j}$, evaluate the line integral around a triangle ABC in the xy-plane with $A(0, 0)$, $B(2, 0)$, $C(2, 1)$. (See Figure 4.) (a) in the counterclockwise direction (b) what is the value in the opposite direction?

Solution (a) In the counterclockwise direction:

$$I=\int_C \mathbf{F}\cdot d\mathbf{r}=\int_{C_1}\mathbf{F}\cdot d\mathbf{r}+\int_{C_2}\mathbf{F}\cdot d\mathbf{r}+\int_{C_3}\mathbf{F}\cdot d\mathbf{r}.$$

Along C_1: $y=0$, x varies from 0 to 2, since $\mathbf{r}=x\mathbf{i}+0\mathbf{j}$, $d\mathbf{r}=dx\mathbf{i}+0\mathbf{j}$ and $\mathbf{F}=2x\mathbf{i}-4x\mathbf{j}$,

$$I_1=\int_{C_1}\mathbf{F}\cdot d\mathbf{r}=\int_0^2 2x\,dx=\left[x^2\right]_0^2=4.$$

Along C_2: $x=2$, y varies from 0 to 1, since $\mathbf{r}=2\mathbf{i}+y\mathbf{j}$, $d\mathbf{r}=0\mathbf{i}+dy\mathbf{j}$ and $\mathbf{F}=(4+y^2)\mathbf{i}+(3y-8)\mathbf{j}$,

$$I_2=\int_{C_2}\mathbf{F}\cdot d\mathbf{r}=\int_0^1(3y-8)dy=-\frac{13}{2}.$$

Along C_3 : $y=\frac{1}{2}x$, x varies from 2 to 0, since $\mathbf{r}=x\mathbf{i}+\frac{1}{2}x\mathbf{j}$,

$$d\mathbf{r}=\left(\mathbf{i}+\frac{1}{2}\mathbf{j}\right)dx \text{ and } \mathbf{F}=\left(2x+\frac{x^2}{4}\right)\mathbf{i}-\frac{5}{2}x\mathbf{j},$$

$$I_3 = \int_{C_3} \mathbf{F} \cdot d\mathbf{r} = \int_2^0 \left(\frac{x^2}{4} + \frac{3x}{4}\right) dx = -\frac{13}{6}.$$

Thus $I = I_1 + I_2 + I_3 = 4 - \frac{13}{2} - \frac{13}{6} = -\frac{14}{3}$.

The required line integral is in the counterclockwise direction. There is an another solution using line segment from a point A to other point B.

On C_1: $\mathbf{r} = \langle 2t, 0 \rangle$, $0 \le t \le 1$, since $d\mathbf{r} = \langle 2, 0 \rangle dt$ and $\mathbf{F} = \langle 4t, 8t \rangle$, $I_1 = \int_{C_1} \mathbf{F} \cdot d\mathbf{r} = \int_0^1 8t\, dt = 4$.

On C_2: $\mathbf{r} = \langle 2, t \rangle$, $0 \le t \le 1$, since $d\mathbf{r} = \langle 1, 0 \rangle dt$ and $\mathbf{F} = \langle 4 + t^2, 3t - 8 \rangle$,

$$I_2 = \int_{C_2} \mathbf{F} \cdot d\mathbf{r} = \int_0^1 (3t - 8)\, dt = -\frac{13}{2}.$$

On C_3: $\mathbf{r} = \langle 2 - 2t, 1 - t \rangle$, $0 \le t \le 1$, since $d\mathbf{r} = \langle -2, -1 \rangle dt$ and $\mathbf{F} = \langle t^2 - 6t + 5, 5t - 5 \rangle$,

$$I_3 = \int_{C_3} \mathbf{F} \cdot d\mathbf{r} = \int_0^1 (-2t^2 + 7t - 5)\, dt = -\frac{13}{6}.$$

The result is same.

(b) The value of the line integral in the opposite direction is $-\frac{14}{3}$. ■

http://matrix.skku.ac.kr/cal-lab/Sec15-2-Exm-5.html

```
var('x,y,t')
r=vector([0,0])+t*(vector([2,0])-vector([0,0]))
dr=diff(r,t)
F=vector([2*x+y^2,3*y-4*x]).subs(x=r[0],y=r[1])
integrand=F.dot_product(dr)
line_integral_1=integral(integrand,t,0,1)
line_integral_1
```

Answer : 4

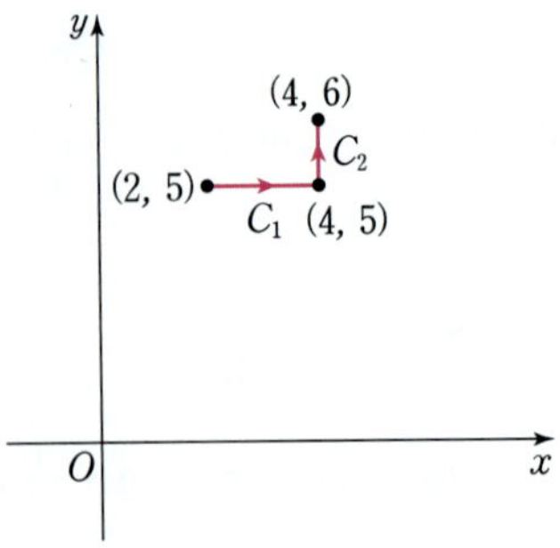

Figure 5

EXAMPLE 6

Find the line integral $\int_C xy\,dx + x^2 y\,dy$. where C is the rectangular curve from $(2, 5)$ to $(4, 5)$ to $(4, 6)$. (See Figure 5.)

Solution Let $\mathbf{F} = xy\mathbf{i} + x^2 y\mathbf{j}$.

Then $\int_C xydx + x^2ydy = \int_C \mathbf{F} \cdot d\mathbf{r} = \int_{C_1} \mathbf{F} \cdot d\mathbf{r} + \int_{C_2} \mathbf{F} \cdot d\mathbf{r}$.

Here C is a piecewise smooth curve made up of a horizontal piece

$$C_1 : 2 \le x \le 4, \ y = 5$$

and a vertical piece

$$C_2 : x=4,\ 5 \le y \le 6\,.$$

Along C_1: $y=5$, x varies from 2 to 4, since $\mathbf{r}=x\mathbf{i}+5\mathbf{j}$, $d\mathbf{r}=\mathbf{i}+0\mathbf{j}$ and $\mathbf{F}=5x\mathbf{i}+5x^2\mathbf{j}$,

$$\int_{C_1} \mathbf{F}\cdot d\mathbf{r} = \int_2^4 5x\,dx = 30\,.$$

Along C_2: $x=4$, y varies from 5 to 6, since $\mathbf{r}=4\mathbf{i}+y\mathbf{j}$, $d\mathbf{r}=0\mathbf{i}+\mathbf{j}$ and $\mathbf{F}=4y\mathbf{i}+16y\mathbf{j}$,

$$\int_{C_2} \mathbf{F}\cdot d\mathbf{r} = \int_5^6 16y\,dy = 88\,.$$

Thus the line integral is

$$\int_C xy\,dx + x^2y\,dy = \int_C \mathbf{F}\cdot d\mathbf{r} = \int_{C_1} \mathbf{F}\cdot d\mathbf{r} + \int_{C_2} \mathbf{F}\cdot d\mathbf{r}$$
$$= 30+88 = 118.$$ ■

EXAMPLE 7

Find the line integral $\int_C \mathbf{F}\cdot d\mathbf{r}$ where $\mathbf{F}$ is a vector field in example 4 and C is the path from $A(2,1)$ to $B(6,3)$.

Solution Since the vector field $\mathbf{F}$ is conservative and there is the potential function $f(x,\ y)=x+x^3y-2y+c$ in example 4. by Path Independence Theorem,

$$\int_C \mathbf{F}\cdot d\mathbf{r} = f(B)-f(A)$$
$$= f(6,3)-f(2,1) = 646.$$ ■

15.3 EXERCISES (Potential Function and Independence of Path)

http://matrix.skku.ac.kr/Cal-Book/part2/CS-Sec-15-3-Sol.htm

YouTube http://youtu.be/TreCe8ESEiU

1-6. Determine whether or not the vector field is conservative. If it is conservative, find a potential of $\mathbf{F}$.

CAS **1.** $\mathbf{F}(x, y) = ye^{-x}\mathbf{i} + e^{-x}\mathbf{j}$

http://matrix.skku.ac.kr/cal-lab/Sec15-2-Exs-1.html

Solution

```
var('x,y')
F=vector([y*exp(-x),exp(-x)])
print bool(diff(F[1],x)==diff(F[0],y))
```

Answer : False

Then $\mathbf{F}$ is not conservative.

CAS **2.** $\mathbf{F}(x, y) = 2xy\mathbf{i} + (x^2 + 2y)\mathbf{j}$

http://matrix.skku.ac.kr/cal-lab/Sec15-2-Exs-2.html

Solution

```
var('x,y')
F=vector([2*x*y,x^2+2])
print bool(diff(F[1],x)==diff(F[0],y))
```

Answer : True

```
x = var('x')
y = function("y",x)
M=2*x*y
N=x^2+2
solution=desolve(diff(y,x)==-M/N,y)
f=simplify(solution==y)*solution.denominator()
potential=f.lhs()-f.rhs()
potential
```

Answer : -(x^2 + 2)*y(x) + c

3. $\mathbf{F}(x, y, z) = 2y^2z^3\mathbf{i} + 2xyz^3\mathbf{j} + 3xy^2z^2\mathbf{k}$

http://matrix.skku.ac.kr/cal-lab/Sec15-2-Exs-3.html

Solution Let $M = 2y^2z^3$, $N = 2xyz^3$ and $P = 3xy^2z^2$.

Since $\frac{\partial P}{\partial x} = 3y^2z^2$ and $\frac{\partial M}{\partial z} = 6y^2z^2$, $\frac{\partial P}{\partial x} \neq \frac{\partial M}{\partial z}$

Then $\mathbf{F}$ is not conservative.

```
var('x,y,z')
F=vector([2*y^2*z^3,2*x*y*z^3,3*x*y^2*z^2])
print bool(diff(F[2],y)==diff(F[1],z))
print bool(diff(F[2],x)==diff(F[0],z))
print bool(diff(F[1],x)==diff(F[0],y))
```

Answer : True, False, False

CAS **4.** $\mathbf{F}(x, y, z) = e^z\mathbf{i} + \mathbf{j} + xe^z\mathbf{k}$

http://matrix.skku.ac.kr/cal-lab/Sec15-2-Exs-4.html

Solution Since $P_y = N_z = 0$, $P_x = M_z = e^z$ and $N_x = M_y = 0$, $\mathbf{F}$ is conservative.

To find the potential function $f(x,y,z)$, first we begin $\frac{\partial f}{\partial x} = e^z \Rightarrow f = xe^z + g(y,z)$ —(1) Taking partial derivative of f with respect to y, we get $\frac{\partial f}{\partial y} = 1$. Hence from (1) we get $\frac{\partial g}{\partial y} = 1 \Rightarrow g(y,z) = y + h(z)$ —(2)

Then $f = xe^z + y + g(z)$ —(3)

Now taking partial derivative of f with respect to z we get $\frac{\partial f}{\partial z} = xe^z$. Hence from (3) we get $\frac{\partial g}{\partial z} = 0 \Rightarrow g(z) = c$, a constant.

Therefore, $f = xe^z + y + c$.

```
var('x,y,z')
F=vector([exp(z),1,x*exp(z)])
print bool(diff(F[2],y)==diff(F[1],z))
print bool(diff(F[2],x)==diff(F[0],z))
print bool(diff(F[1],x)==diff(F[0],y))
```

Answer : True, True, True

```
var('x,y,z,c')
fx=exp(z);fy=1;fz=x*exp(z)
f_1=integral(fx,x)      # +g(y,z)
f_1y=diff(f_1,y)        # fy= diff(f_1,y)+gy(y,z)
gy(y,z)=fy-diff(f_1,y)
g(y,z)=integral(gy(y,z),y) # +h(z)
f_2=f_1+g(y,z) #+h(z)
f_2z=diff(f_2,z)        #fz=diff(f_2,z)+ hz(z)
hz(z)=fz-diff(f_2,z)
h(z)=integral(hz(z),z)
f=f_2+h(z)+c
f
```

Answer : x*e^z + c + y

5. $\mathbf{F}(x, y, z) = xyz^2\mathbf{i} + 3x^2yz^2\mathbf{j} + x^2y^2z\mathbf{k}$

Solution Not conservative.

6. $\mathbf{F}(x, y, z) = y\cos(xy)\mathbf{i} + x\cos(xy)\mathbf{j} - \sin z\mathbf{k}$

Solution Conservative, $f = \sin(xy) + \cos(z) + c$.

CAS **7.** Show that the vector field $\mathbf{A}(x, y, z) = e^{xz}\mathbf{i} + 3xyz\mathbf{j} + 2y\mathbf{k}$ is not conservative.

http://matrix.skku.ac.kr/cal-lab/Sec15-2-Exs-7.html

Solution

```
var('x,y,z')
A=vector([exp(x*z),3*x*y*z,2*y])
print bool(diff(A[2],y)==diff(A[1],z))
print bool(diff(A[2],x)==diff(A[0],z))
print bool(diff(A[1],x)==diff(A[0],y))
```

Answer : False, False, False

Then $\mathbf{A}$ is not conservative.

CAS **8.** Determine whether the force field $\mathbf{F} = 2xz\mathbf{i} + (x^2 - y)\mathbf{j} + (2z - x^2)\mathbf{k}$ is a conservative field.

http://matrix.skku.ac.kr/cal-lab/cal-14-2-5.html

http://matrix.skku.ac.kr/cal-lab/Sec15-2-Exs-8.html

Solution

```
var('x,y,z')
F=vector([2*x*z,x^2-y,2*z-x^2])
print bool(diff(F[2],y)==diff(F[1],z))
print bool(diff(F[2],x)==diff(F[0],z))
print bool(diff(F[1],x)==diff(F[0],y))
```

Answer : True, False, False

Then $\mathbf{F}$ is not conservative field.

CAS **9.** **a.** Prove that $\mathbf{F} = (4xy - 3x^2z^2)\mathbf{i} + 2x^2\mathbf{j} - 2x^3z\mathbf{k}$ is a conservative field.

b. Find its scalar potential f.

c. Also find the work done in moving an object in this field from $(1, 1, 1)$ to $(0, 0, 0)$.

http://matrix.skku.ac.kr/cal-lab/Sec15-2-Exs-9.html

Solution Scalar potential $f = 2x^2y - x^3z^2 + c$.

Work done $= f(1, 1, 1) - f(0, 0, 0) = 1$.

```
var('x,y,z')
F=vector([4*x*y-3*x^2*z^2,2*x^2,-2*x^3*z])
print bool(diff(F[2],y)==diff(F[1],z))
print bool(diff(F[2],x)==diff(F[0],z))
print bool(diff(F[1],x)==diff(F[0],y))
```

Answer : True, True, True

```
var('x,y,z,c')
fx=4*x*y-3*x^2*z^2;fy=2*x^2;fz=-2*x^3*z
f_1=integral(fx,x)      # +g(y,z)
f_1y=diff(f_1,y)        # fy= diff(f_1,y)+gy(y,z)
gy(y,z)=fy-diff(f_1,y)
g(y,z)=integral(gy(y,z),y) # +h(z)
f_2=f_1+g(y,z) #+h(z)
f_2z=diff(f_2,z)      #fz=diff(f_2,z)+ hz(z)
hz(z)=fz-diff(f_2,z)
h(z)=integral(hz(z),z)
f=f_2+h(z)+c
f
```

Answer : -x^3*z^2 + 2*x^2*y + c

```
var('x,y,z,c')
f(x,y,z)=-x^3*z^2 + 2*x^2*y + c
workdone=f(1,1,1)-f(0,0,0)
workdone
```

Answer : 1

CAS **10.** Find the total work done in moving a particle by a force field $\mathbf{A} = 3xy\mathbf{i} - 5z\mathbf{j} + 10x\mathbf{k}$ along the curve $x = t^2 + 1,\ \ y = 2t^2, z = t^3$ form $t = 1$ to $t = 2$.

http://matrix.skku.ac.kr/cal-lab/Sec15-2-Exs-10.html

Solution

```
var('t,x,y,z')
x(t)=t^2+1;y(t)=2*t^2;z(t)=t^3
r(t)=(x(t),y(t),z(t))
dr(t)=(diff(x(t),t),diff(y(t),t),diff(z(t),t))
A(t)=(3*x(t)*y(t),-5*z(t),10*x(t))
F(t)=A(t).dot_product(dr(t))
integral(F(t),t,1,2)
```

Answer : 303

15.4 Green's Theorem in the Plane

YouTube http://youtu.be/WxdTbaSb_ZI

In this section, we introduce a very important theorem in vector calculus, which is the Green's Theorem. Green's Theorem relates a line integral around a simple closed curve C and a double integral with the plane region R bounded by that curve, (see Figure 1.) which it is possible to transforms line integrals to double integrals and vice versa.

Here we assume that the curve C has positive orientation (travels to counterclockwise on C). That is, when we travel along C, the region R always lies on the left side of C.

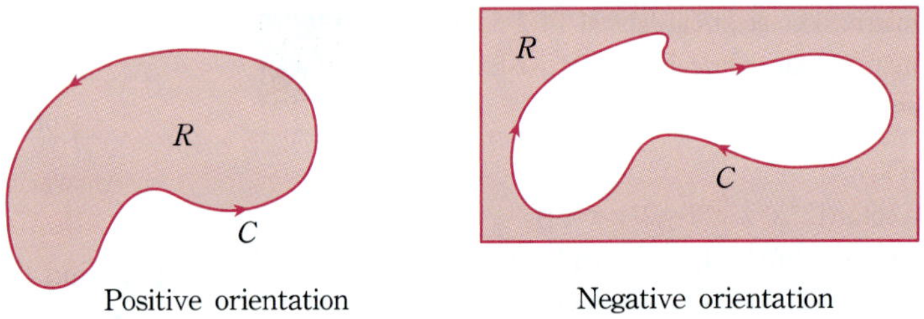

Positive orientation   Negative orientation

Figure 1

Green's Theorem has several applications such as finding area of a plane regions and work done by a force field.

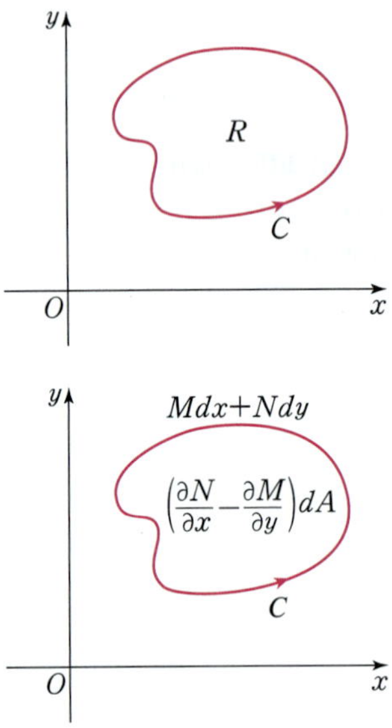

Figure 2

THEOREM 1 Green's Theorem

If R is a closed region in the xy-plane bounded by a simple closed curve C and if $M(x, y)$ and $N(x, y)$ are continuous functions of x and y having continuous first order partial derivatives in R, then

$$\int_C M dx + N dy = \iint_R \left(\frac{\partial N}{\partial x} - \frac{\partial M}{\partial y}\right) dx\, dy \tag{1}$$

where C is traversed in the positive direction. (See Figure 2.)

* George Green (1793-1841) English mathematician.

Remark 1. Vector form of Green's Theorem

Let $\mathbf{A} = M\mathbf{i} + N\mathbf{j}$ and $\mathbf{r} = x\mathbf{i} + y\mathbf{j}$. Then $\mathbf{A} \cdot d\mathbf{r} = M\,dx + N\,dy$.

$$\nabla \times \mathbf{A} = \begin{vmatrix} \mathbf{i} & \mathbf{j} & \mathbf{k} \\ \dfrac{\partial}{\partial x} & \dfrac{\partial}{\partial y} & \dfrac{\partial}{\partial z} \\ M & N & 0 \end{vmatrix}$$

$$= \left(\frac{\partial 0}{\partial y} - \frac{\partial N}{\partial z}\right)\mathbf{i} - \left(\frac{\partial 0}{\partial x} - \frac{\partial M}{\partial z}\right)\mathbf{j} + \left(\frac{\partial N}{\partial x} - \frac{\partial M}{\partial y}\right)\mathbf{k}$$

$$= \left(\frac{\partial N}{\partial x} - \frac{\partial M}{\partial y}\right)\mathbf{k}$$

Thus Green's Theorem reduces to

2 $$\int_C \mathbf{A} \cdot d\mathbf{r} = \iint_R (\nabla \times \mathbf{A}) \cdot \mathbf{k}\, d\mathbf{r}\,.$$

Remark 2. Area of a plane region using Green's Theorem. Let R be a plane region bounded by the simple closed curve C. Let A be the area of R. Let $N = x$, $M = -y$ so that

$$\int_C x\,dy - y\,dx = \iint_R (1+1)\,dx\,dy$$

$$= 2\iint_R dx\,dy = 2A.$$

Thus the **area of** R is

3 $$A = \frac{1}{2}\int_C x\,dy - y\,dx.$$

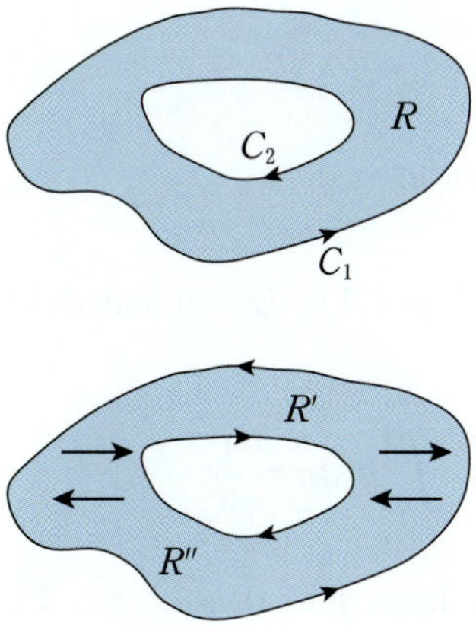

Figure 3

Remark 3. Green's Theorem can be applied to regions that are not simply connected. For instance, the boundary C of the region R in Figure 3 consist of two simply closed curves C_1 and C_2 and assume that the region R is always on the left as C_1 and C_2 are travesed. Then, the positive direction is counterclockwise for C_1 but clockwise for C_2. If we divide R into two regions R' and R'' in Figure 3, then, by Green's Theorem,

$$\iint_R \left(\frac{\partial N}{\partial x} - \frac{\partial M}{\partial y}\right) dA = \iint_{R'} \left(\frac{\partial N}{\partial x} - \frac{\partial M}{\partial y}\right) dA + \iint_{R''} \left(\frac{\partial N}{\partial x} - \frac{\partial M}{\partial y}\right) dA$$

$$= \int_{\partial R'} M dx + N dy + \int_{\partial R''} M dx + N dy$$

where $\partial R'$ and $\partial R''$ are the boundary curves of R' and R'', respectively.

Remark 4. Let $A = M\mathbf{i} + N\mathbf{j}$ be a vector field on an open simply connected region R and M and N have continuous first order partial derivatives. If $\dfrac{\partial M}{\partial y} = \dfrac{\partial N}{\partial x}$ then by Green's Theorem $\displaystyle\int_C M dx + N dy = 0$.

CAS **EXAMPLE 1**

Verify Green's Theorem for $\int_C x^2y\,dx + xy^2dy$, where C is a positively oriented curve which is the boundary of the triangle with vertices, $(0,0)$, $(1,0)$ and $(1,1)$. (See Figure 4 and 5.)

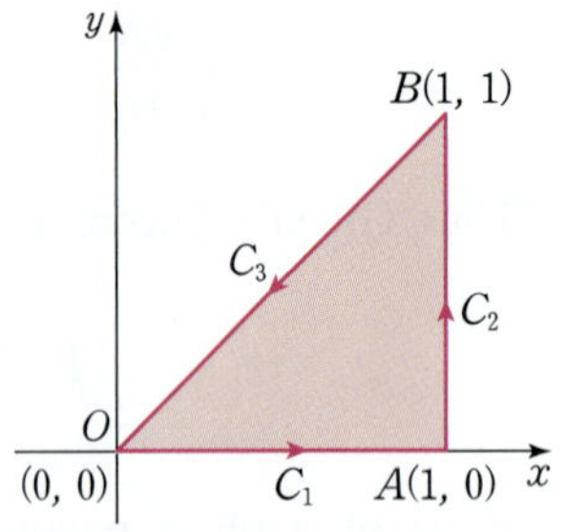

Figure 4

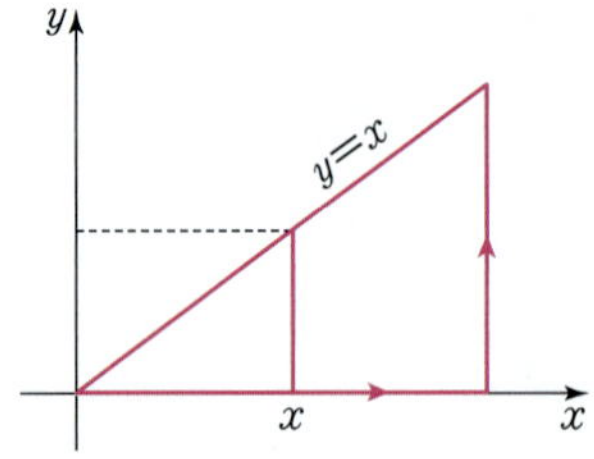

Figure 5

Solution First, we evaluate the given integral as a line integral. Notice that C is a union of three curves C_1: $y=0$, $0 \le x \le 1$, C_2: $x=1$, $0 \le y \le 1$ and C_3: $y=x$, $0 \le x \le 1$.

Then $\int_C x^2y\,dx + xy^2dy = \int_{C_1} + \int_{C_2} + \int_{C_3}$.

Let us assume that right hand integral is denoted by RI and left hand integral is denoted by LI.

Along C_1, since $y=0$ and $x=x\,(0 \le x \le 1)$,

$$\int_{C_1} x^2y\,dx + xy^2dy = 0.$$

Along C_2, since $x=1$ and $y=y\,(0 \le y \le 1)$, $dx=0$ and $dy=dy$. Then

$$\int_{C_2} x^2y\,dx + xy^2dy = \int_0^1 y^2\,dy = \frac{1}{3}.$$

Along C_3, since $y=x$ and x varies from 1 to 0, $dy=dx$. Then

$$\int_{C_3} x^2y\,dx + xy^2dy = 2\int_1^0 x^3\,dt = -\frac{1}{2}.$$

Hence $LI = 0 + \frac{1}{3} - \frac{1}{2} = -\frac{1}{6}$.

Next, let $M = x^2y$ and $N = xy^2$, then we get the double integral

$$RI = \iint_R \left(\frac{\partial N}{\partial x} - \frac{\partial M}{\partial y}\right) dx\,dy = \int_0^1 \int_0^x (y^2 - x^2)\,dy\,dx$$
$$= \int_0^1 \left(-\frac{2}{3}x^3\right) dx = -\frac{1}{6}.$$

Hence Green's Theorem verified by $LI = RI$. ■

http://matrix.skku.ac.kr/cal-lab/Sec15-4-Exm-1.html

```
var('x,y')
M=x^2*y;N=x*y^2
curlF=diff(N,x)-diff(M,y)
RIx=integral(curlF,y,0,x)
integral(RIx,x,0,1)
```

Answer : -1/6

CAS **EXAMPLE 2**

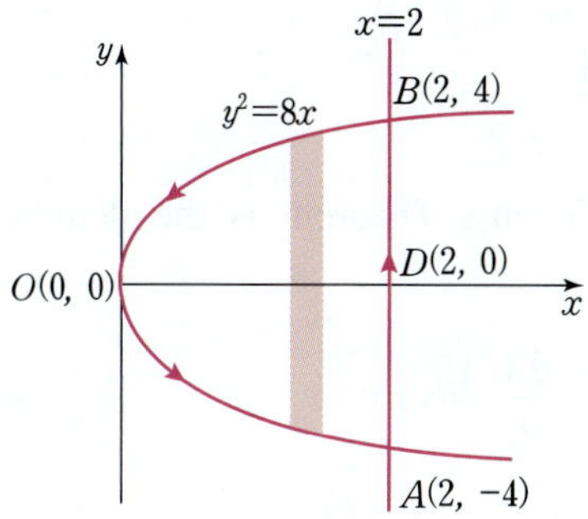

Figure 6

Verify Green's Theorem in the plane for $\int_C (x^2-2xy)dx+(x^2y+3)dy$ where C is the boundary of the region defined by $y^2=8x$: and $x=2$. (See Figure 6.)

Solution (a) The left hand side of the Green's Theorem is the line integral

$$LI=\int_C (x^2-2xy)dx+(x^2y+3)dy.$$

Here C consists of the curves OA, ADB, BO, so

$$\begin{aligned} LI &= \int_C = \int_{OA+ADB+BO} \\ &= \int_{OA} + \int_{ADB} + \int_{BO} = LI_1+LI_2+LI_3 \end{aligned}$$

Along OA, since $y=-2\sqrt{2}\sqrt{x}\ (0\le x\le 2)$, $dy=-\dfrac{\sqrt{2}}{\sqrt{x}}dx$.

$$\begin{aligned} LI_1 &= \int_{OA} (x^2-2xy)dx+(x^2y+3)dy \\ &= \int_0^2 (5x^2+4\sqrt{2}\,x^{3/2}-3\sqrt{2}\,x^{-1/2})dx \\ &= \left[\frac{5x^3}{3}+4\sqrt{2}\cdot\frac{2}{5}\cdot x^{5/2}-3\sqrt{2}\cdot 2\sqrt{x}\right]_0^2 \\ &= \frac{40}{3}+\frac{64}{5}-12=\frac{212}{15}. \end{aligned}$$

Along ADB, since $x=2$ and $y=y(-4\le y\le 4)$, $dx=0$ and $dy=dy$. Then

$$\begin{aligned} LI_2 &= \int_{OA} (x^2-2xy)dx+(x^2y+3)dy \\ &= \int_{-4}^4 (4y+3)dy=24. \end{aligned}$$

Along BO, since $y=2\sqrt{2}\sqrt{x}$ and x varies from 2 to 0, $dy=\dfrac{\sqrt{2}}{\sqrt{x}}dx$. Then

$$\begin{aligned} LI_3 &= \int_{BO} (x^2-2xy)dx+(x^2y+3)dy \\ &= \int_2^0 (5x^2-4\sqrt{2}\,x^{\frac{3}{2}}+3\sqrt{2}\,x^{-\frac{1}{2}})dx \\ &= -\frac{40}{3}+\frac{64}{5}-12=-\frac{188}{15}, \end{aligned}$$

Thus $LI = LI_1 + LI_2 + LI_3 = \frac{212}{15} + 24 - \frac{188}{15} = \frac{128}{5}$.

(b) Here

$$M = x^2-2xy,\ N = x^2y+3$$
$$\frac{\partial M}{\partial y} = -2x,\ \frac{\partial N}{\partial x} = 2xy.$$

So the right hand side of the Green's Theorem is the double integral given by

$$\begin{aligned} RI &= \iint_R \left(\frac{\partial N}{\partial x} - \frac{\partial M}{\partial y}\right)dx\,dy \\ &= \iint_R (2xy-(-2x))dx\,dy. \end{aligned}$$

The region R is described with y varying from $-\sqrt{8x}$ of the lower branch of the parabola to its upper branch $\sqrt{8x}$ while x varies from 0 to 2. Thus

$$\begin{aligned} RI &= \int_0^2 \int_{-\sqrt{8x}}^{\sqrt{8x}} (2xy+2x)dy\,dx \\ &= \int_0^2 \left[xy^2+2xy\right]_{-\sqrt{8x}}^{\sqrt{8x}} dx \\ &= \int_0^2 \left[x\times 8x+2x\sqrt{8x}\right]-\left[x\times 8x-2x\sqrt{8x}\right]dx \\ &= \int_0^2 4x\sqrt{8x}\,dx = 4\sqrt{8}\left.\frac{x^{5/2}}{5/2}\right|_0^2 \\ &= \frac{4\sqrt{8}\times 2^{5/2}}{5/2} = \frac{128}{5} \end{aligned}$$

Since $LI = RI$, the Green's Theorem is verified. ■

http://matrix.skku.ac.kr/cal-lab/Sec15-4-Exm-2.html

```
var('x,y')
assume(x>0)
M=x^2-2*x*y;N=x^2*y+3
curlF=diff(N,x)-diff(M,y)
RIx=integral(curlF,y,-sqrt(8*x),sqrt(8*x))
integral(RIx,x,0,2)
```

Answer : 128/5

CAS **EXAMPLE 3**

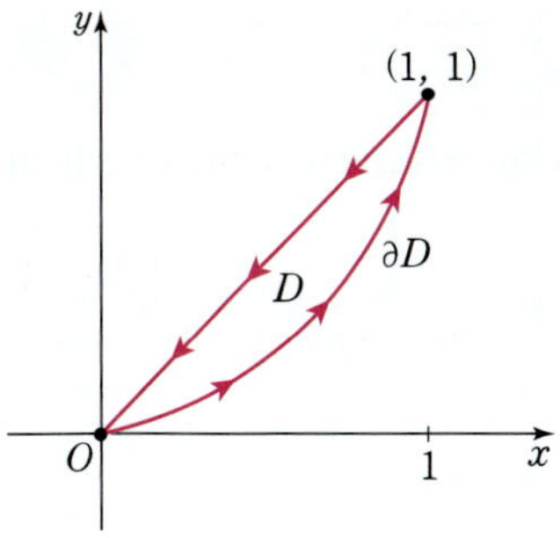

Figure 7

Evaluate the line integral using Green's Theorem $\int_{\partial D} \frac{y}{x+1}\,dx + 2xy\,dy$ where ∂D is the boundary of the region defined by the curve $y=x^2$, and the line $y=x$. (See Figure 7.)

Solution A region D bounded by ∂D is

$$D=\{(x,\,y)\,|\,0\le x\le 1,\ x^2\le y\le x\}$$

Since $\frac{\partial N}{\partial x}=2y$ and $\frac{\partial M}{\partial y}=\frac{1}{x+1}$, using Green's Theorem,

$$\begin{aligned}\int_{\partial D}\frac{y}{x+1}\,dx+2xy\,dy &= \int_0^1\int_{x^2}^{x}\left(2y-\frac{1}{x+1}\right)dy\,dx\\ &= \int_0^1\left(x^2-x^4-\frac{x}{x+1}+\frac{x^2}{x+1}\right)dx\\ &= 2\ln 2-\frac{41}{30}.\end{aligned}$$

■

http://matrix.skku.ac.kr/cal-lab/Sec15-4-Exm-3.html

```
var('x,y')
assume(x>0)
M=y/(x+1);N=2*x*y
curlF=diff(N,x)-diff(M,y)
RIx=integral(curlF,y,x^2,x)
integral(RIx,x,0,1)
```

Answer : 2*log(2) - 41/30

EXAMPLE 4

Use Green's Theorem to evaluate the line integral $\int_C(\cos x\sin y-xy)dx+(\sin x\cos y)dy$, where C is a region bounded by

$$x^2+y^2\ge 1 \qquad x^2+y^2\le 4 \quad \text{and} \quad 0\le y$$

http://martrix.skku.ac.kr/cal-lab/Sec15-2-n-Exmaple-4.html

Solution

$$\begin{aligned}&\int_C(\cos x\sin y-xy)dx+(\sin x\cos y)dy\\ &=\iint_D\left[\frac{\partial}{\partial x}(\sin x\cos y)-\frac{\partial}{\partial y}(\cos x\sin y-xy)\right]dA\\ &=\iint_D x\,dA=\int_0^{\pi}\int_1^2 r\cos\theta\,dr\,d\theta\\ &=\int_0^{\pi}\cos\theta\,d\theta\int_1^2 r^2dr=0\end{aligned}$$

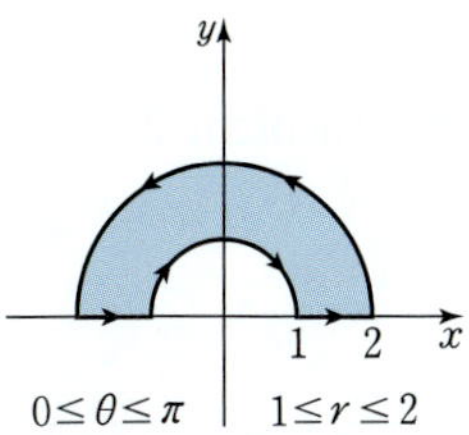

Figure 8

Area of a Plane Region

THEOREM 2

If D has a piecewise smooth boundary C with positive orientation, then the area of D is

$$A = \int_C x\,dy = \int_C -y\,dx = \frac{1}{2}\int_C (xdy - ydx)\ .$$

Proof By Green's Theorem,

$$\int_C x\,dy = \iint_D \left(\frac{\partial x}{\partial x} - \frac{\partial 0}{\partial y}\right) dA = \iint_D dA = A$$

$$\int_C (-y)\,dx = \iint_D \left(\frac{\partial 0}{\partial x} - \frac{\partial - y}{\partial y}\right) dA = \iint_D dA = A\,.$$

Adding the two, we get the last term in the theorem. ■

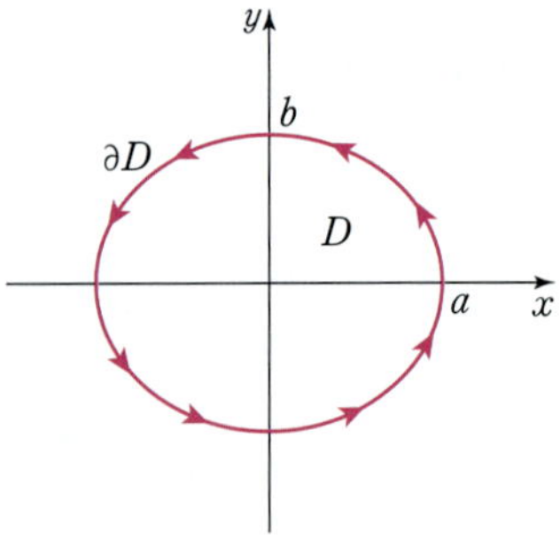

Figure 9. Area=πab

EXAMPLE 5

Use Green's Theorem to find the area of the ellipse $\dfrac{x^2}{a^2} + \dfrac{y^2}{b^2} \le 1$. (See Figure 9.)

Solution The boundary of the ellipse is the parametric curve

$$x = a\cos t,\ y = b\sin t,\quad 0 \le t \le 2\pi\,.$$

Note that $dx = -a\sin dt$ and $dy = b\cos t\,dt$, We have

$$\begin{aligned} x\,dy - y\,dx &= (a\cos t)(b\sin t)dt - (b\sin t)(-a\sin t)dt \\ &= ab(\cos^2 t + \sin^2 t)\,dt = ab\,dt\,. \end{aligned}$$

By Theorem 2

$$\begin{aligned} A &= \int_C x\,dy = \frac{1}{2}\int_C xdy - ydx \\ &= \frac{1}{2}\int_0^{2\pi} ab\,dt = \frac{ab}{2}\int_0^{2\pi} dt = \pi ab\,. \end{aligned}$$ ■

Let r and θ be polar coordinates defined by $x = r\cos\theta,\ y = r\sin\theta$. Then $dx = \cos\theta\,dr - r\sin\theta\,d\theta,\ \ dy = \sin\theta\,dr + r\cos\theta\,d\theta$ by the definition of the total differential.

It is easy to check that $x\,dy - y\,dx = r^2\,d\theta$. Hence, by Theorem 2, we get

5 $$A = \frac{1}{2}\int_C r^2\,d\theta\,.$$

For example, we consider the cardioid $r = a(1 + \cos\theta)$ for $0 \le \theta \le 2\pi$ for using 5, we find

$$A=\frac{1}{2}\int_C r^2\,d\theta=\frac{a^2}{2}\int_0^{2\pi}(1+\cos\theta)^2\,d\theta=\frac{3\pi}{2}a^2.$$

CAS **EXAMPLE 6**

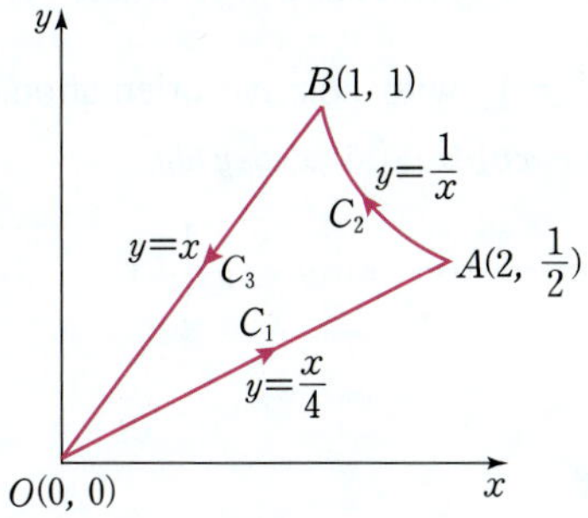

Figure 10

Using Green's Theorem, find the area of the region in the first quadrant bounded by the curves $y=\frac{1}{x}$, $y=\frac{x}{4}$ and $y=x$. (See Figure 10.)

http://martrix.skku.ac.kr/Mobile-Sage-G/Sage-grapher-para.html

Solution By Green's Theorem the area A of the region bounded by a closed curve C is given by $A=\frac{1}{2}\int_C x\,dy-y\,dx$.

Here, C consists of the curves $C_1: y=\frac{x}{4}$, $0\le x\le 2$, $C_2: y=\frac{1}{x}$, $1\le x\le 2$, $-C_3: y=x$, $0\le x\le 1$. So

$$A=\frac{1}{2}\int_C=\frac{1}{2}\left[\int_{C_1}+\int_{C_2}+\int_{C_3}\right]=\frac{1}{2}[I_1+I_2+I_3].$$

Along $C_1: y=\frac{x}{4}$, $dy=\frac{1}{4}dx$, x varies from 0 to 2.

$$I_1=\int_{C_1}x\,dy-y\,dx=\int_{C_1}\frac{1}{4}x\,dx-\frac{x}{4}dx=0.$$

Along $C_2: y=\frac{1}{x}$, $dy=-\frac{1}{x^2}dx$, x varies from 2 to 1.

$$I_2=\int_{C_2}x\,dy-y\,dx=\int_2^1 x\cdot\left(-\frac{1}{x^2}\right)dx-\frac{1}{x}dx$$
$$=-2\ln x|_2^1=2\ln 2\ .$$

Along $C_3: y=x$, $dy=dx$, x varies from 1 to 0.

$$I_3=\int_{C_3}x\,dy-y\,dx=\int x\,dx-x\,dx=0.$$

Then $A=\frac{1}{2}(I_1+I_2+I_3)=\frac{1}{2}(0+2\ln 2+0)=\ln 2.$ ■

15.4 EXERCISES (Green's Theorem in the Plane)

http://matrix.skku.ac.kr/Cal-Book/part2/CS-Sec-15-4-Sol.htm

1-4. Use Green's Theorem to evaluate the line integral $\int_C Mdx + Ndy$, when $Mdx + Ndy$ is

CAS **1.** $-ydx + xdy$, where C is a circle of $x^2 + y^2 = 9$, with positive orientation.

http://matrix.skku.ac.kr/cal-lab/Sec15-4-Exs-1.html

Solution

```
x,y,r,t=var('x,y,r,t');
p = implicit_plot(x^2+y^2==9, (x,-3,3), (y, -3,3),cmap=["red"]);
p1 = implicit_plot(x==0, (x,-4,4), (y, -4,4));
p2 = implicit_plot(y==0, (x,-4,4), (y, -4,4));
show(p+p1+p2, aspect_ratio=1)
```

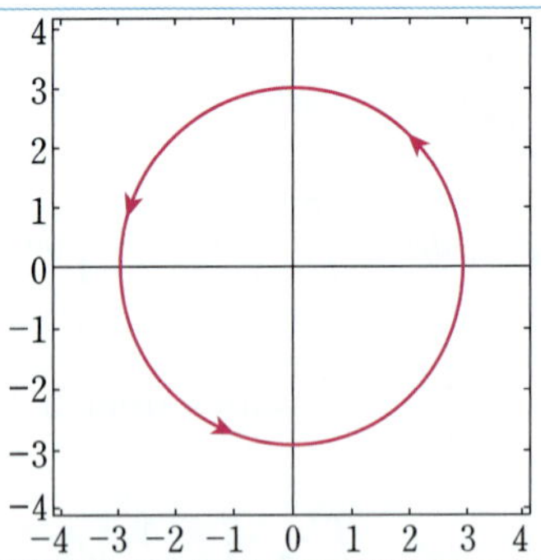


```
M=-y;
N=x;
integral(integral((diff(N,x)-diff(M,y))*r, r, 0, 3), t, 0, 2*pi)
```

Answer : 18*pi

2. $\int_C x^{-1}e^y dx + (e^y \ln x + 2x)dy$, where C is the positively oriented curve which is the boundary of the region bounded by $y = 2$, $y = x^4 + 1$.

Solution $\int_C x^{-1}e^y dx + (e^y \ln x + 2x)dy$

$= \iint_D \left[\frac{\partial}{\partial x}(e^y \ln x + 2x) - \frac{\partial}{\partial y}(x^{-1}e^y)\right] dA$

$= \int_1^2 \int_1^{\sqrt[4]{y-1}} 2\,dx\,dy$

$= \frac{16}{5}$.

3. $\int_C (\cos x \sin y - xy)dx + (\sin x \cos y)dy$, where C is a circle of $x^2 + y^2 = 1$, with positive orientation.

Solution $\int_C (\cos x \sin y - xy)dx + (\sin x \cos y)dy$

$= \iint_D \left[\frac{\partial}{\partial x}(\sin x \cos y) - \frac{\partial}{\partial y}(\cos x \sin y - xy)\right] dA$

$= \iint_D x\,dA$

$= \int_0^{2\pi} \int_0^1 r\cos\theta\, r\,dr\,d\theta$

$= \int_0^{2\pi} \cos\theta\,d\theta \int_0^1 r^2 dr$

$= 0$.

4. $\int_C (x^2 - \cosh y)dx + (y + \sin x)dy$ where C is a boundary of the rectangle $0 \le x \le \pi$, $0 \le y \le 1$, with positive orientation.

Solution $\int_C (x^2 - \cosh y)dx + (y + \sin x)dy$

$= \iint_D \left[\frac{\partial}{\partial x}(y + \sin x) - \frac{\partial}{\partial y}(x^2 - \cosh y)\right] dA$

$= \int_0^1 \int_0^{\pi} \cos x + \sinh y\,dx\,dy$

$= -\pi + \frac{1}{2}\pi e^{-1} + \frac{1}{2}\pi e$.

5-8. Verify Green's Theorem by evaluating the line integral $\int_C Mdx + Ndy$ (a) directly, (b) using Green's Theorem, where $Mdx + Ndy$ is:

5. $\int_C (xy + y^2)dx + (x^2)dy$ where C is a closed curve of the region bounded by $y = x$ and $y = x^2$, with positive orientation.

Solution $\int_c (xy + y^2)dx + (x^2)dy$

$= \iint_D \left[\frac{\partial}{\partial x}(x^2) - \frac{\partial}{\partial y}(xy + y^2)\right] dA$

$= \int_0^1 \int_{x^2}^{x} (x - 2y)dy\,dx$

$= \frac{1}{20}$.

6. $\int_C (3x^2-8y^2)dx+(4y-6xy)dy$ where C is a boundary of the region defined by $y=\sqrt{x}$ and $y=x^2$, with positive orientation.

Solution $\int_C (3x^2-8y^2)dx+(4y-6xy)dy$

$=\iint_D \left[\frac{\partial}{\partial x}(4y-6xy)-\frac{\partial}{\partial y}(3x^2-8y^2)\right]dA$

$=\int_0^1\int_{x^2}^{\sqrt{x}}(10y)\,dy\,dx$

$=\frac{3}{2}.$

7. $\int_C (2x-y^3)dx+(xy)dy$ where C is a boundary of the region enclosed by the circles $x^2+y^2=1$ and $x^2+y^2=9$, with positive orientation.

Solution $\int_C (2x-y^3)dx+(xy)dy$

$=\iint_D \left[\frac{\partial}{\partial x}(xy)-\frac{\partial}{\partial y}(2x-y^3)\right]dA$

$=\iint (y+3y^2)dA,\ y=r\sin\theta$

$=\int_0^{2\pi}\int_1^3 (r\sin\theta+3r^2\sin^2\theta)r\,dr\,d\theta$

$=\int_0^{2\pi}\int_1^3 (r^2\sin\theta+3r^3\sin^2\theta)\,dr\,d\theta$

$=60\pi.$

8. $\int_C (3x+4y)dx+(2x-3y)dy$, where C is a circle of $x^2+y^2=4$, with positive orientation.

Solution $\int_C (3x+4y)dx+(2x-3y)dy$

$=\iint_D \left[\frac{\partial}{\partial x}(2x-3y)-\frac{\partial}{\partial y}(3x+4y)\right]dA$

$=\int_0^{2\pi}\int_0^2(-2)r\,dr\,d\theta$

$=-8\pi.$

---

9. $\int_C (x^2+xy)dx+(x^2+y^2)dy$ where C is a square of $x=\pm1,\ y=\pm1$ with positive orientation.

Solution Let $p(x,y)=x^2+xy,\ q(x,y)=x^2+y^2$. To evaluate the line integral, we use Green's Theorem as follows

$\int_{-1}^1\int_{-1}^1\left(\frac{\partial q(x,y)}{x}-\frac{\partial p(x,y)}{y}\right)dxdy$

$=\int_{-1}^1\int_{-1}^1(2x-x)dxdy$

$=0$

10. $\int_C (x^2ydx+(x-y)dx+xyzdz$ where C is the arc of the parabola $y=x^2$ in the plane $z=2$ from $(0,0,2)$ to $(1,1,2)$.

http://matrix.skku.ac.kr/cal-lab/Sec15-4-Exs-10.html

Solution

```
var('x,y,z,t')
p_1=implicit_plot3d(y==x^2, (x,0,1), (y, 0,1), (z, 0,4),color="red", opacity=0.4);
p_2=implicit_plot3d(z==2, (x,0,1), (y, 0,1), (z, 0,4),color="yellow", opacity=0.6);
p_3=parametric_plot3d((t, t^2, 2), (t, 0, 1), thickness=5)
show(p_1+p_2+p_3)
```

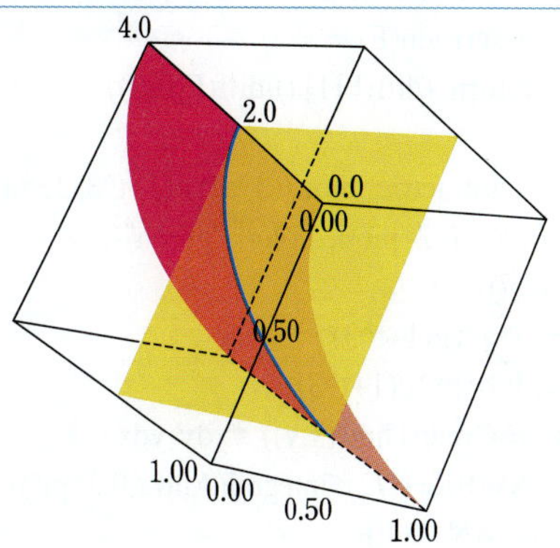

Using Green's Theorem

```
integral(x^2*x^2+(x-2)*2,x,0,1)
```

Answer : - {14} over {5}

11. Calculate the area bounded by the ellipse $\frac{x^2}{a^2}+\frac{y^2}{b^2}=1$. Hence, deduce the area bounded by the circle $x^2+y^2=a^2$.

Solution (a) $\frac{x^2}{a^2}+\frac{y^2}{b^2}=1.$

$x=a\cos t,\ y=b\sin t,$

$A=\frac{1}{2}\int_C xdy-ydx=\frac{1}{2}\int_c (a\cos t)dy-(b\sin t)dx,$

$dy=b\cos t\,dt,\ dx=-a\sin t\,dt$

$=\frac{1}{2}\int_0^{2\pi}(a\cos t)(b\cos t)dt-(b\sin t)(-a\sin t)dt$

$=\frac{1}{2}\int_0^{2\pi}(ab\cos^2 t+ab\sin^2 t)dt$

$=\frac{1}{2}[(ab)t]_0^{2\pi}=\pi ab$

Put $a=b$, area of circle : πa^2.

12. Find the area under arc of the cycloid $x=t-\sin t$,

$x = t - \sin t, y = 1 - \cos t, \ 0 \le t \le 2\pi.$

13. Find the area under of one arch of the asteroid $x = a\cos^3 t, \ y = a\sin^3 t.$

CAS **14.** Find the area of the loop of the folium of Descartes $x = \dfrac{3at}{1+t^3}, \ x = \dfrac{3bt^2}{1+t^3}.$

http://matrix.skku.ac.kr/cal-lab/cal-15-2-10.html

Solution

```
@interact
def _(a=(1,(-10,20))):
 var('t,b')
 def GreenThm(F):
     assert(len(F) == 2)
     return (diff(F[1],t)-diff(F[0],t))

p_1=parametric_plot(((3*a*t)/(1+t^3),(3*a*t^2)/(1+
t^3)), (t, 0,2*pi()), color="green");
 show(p_1)
 x=(3*a*t)/(1+t^3)
 y=(3*b*t^2)/(1+t^3)
 Anti=GreenThm((x,y)) #xdy-ydx
 ANSWER=(1/2)*integral(Anti,t,0,2*pi())
 print ANSWER
```

Answer : $-\dfrac{3(\pi - 2\pi^2 b)}{8\pi^3 + 1}$

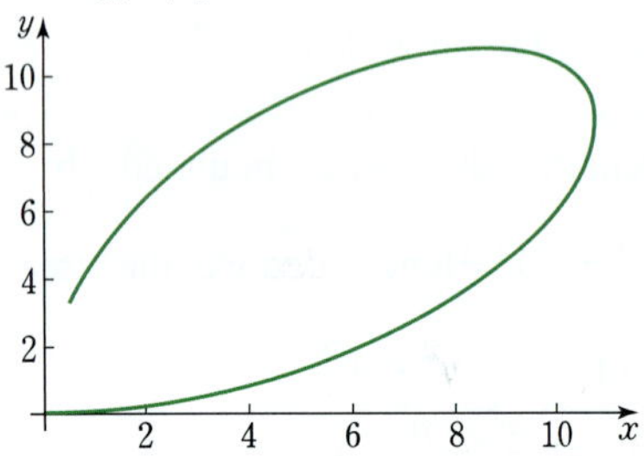

15. Let $P, Q : \mathbb{R}^2 \to \mathbb{R}$ be smooth functions satisfying the following differential equations:

$$\frac{\partial P}{\partial y}(x, y) = -Q(x, y), \ \frac{\partial Q}{\partial x}(x, y) = P(x, y).$$

Evaluate the line integral $\int_C (P^2(x, y)dx - Q^2(x, y)dy)$, where C is traversed in the positive direction of the curve $x^2 + y^4 = 2010$.

Solution Using Green's Theorem, we have

$$\int_C P^2(x, y)dx - Q^2(x, y)dy = \iint_D (-2QQ_x - 2PP_y)dA$$

$$= -2\iint_D (QP + P(-Q))dA = 0.$$

16. Let $(x(t), \ y(t))$ be the solution of the system of differential equations

$$\frac{dx}{dt} = f(x, y), \ \frac{dy}{dt} = g(x, y)$$

where $f, \ g$ are smooth functions in variables $x, \ y$.

Assume further that $(x(t), \ y(t))$ is periodic with the periodicity T, namely $x(t+T) = x(t)$, $y(t+T) = y(t)$ for all t. Suppose that the solution curve (phase trajectory) is a simple closed curve in the phase plane (xy-plane) and D is the region bounded by C. Prove that

$$\iint_D (f_x(x, y) + g_y(x, y))dA = 0.$$

Proof Recalling Green's Theorem,

$$\iint_D (f_x(x, y) + g_y(x, y))dA = \int_C fdy - gdx.$$

Now parametrizing by t, the line integral becomes

$$\int_C fdy - gdx$$

$$= \int_t^{t+T} f(x(t), \ y(t))y'(t)dt - g(x(t), \ y(t))x'(t)dt$$

$$= \int_t^{t+T} (fg - fg)dt = 0.$$

Therefore,

we obtain $\iint_D (f_x(x, y) + g_y(x, y))dA = 0.$

17. Evaluate the integral $\int_C x^4 dx + xy\,dy$, where C is the boundary of the triangular region D with corners $(0, \ 0)$, $(1, \ 0)$, $(0, \ 1)$ with positive orientation.

Solution $\dfrac{1}{6}$

18. Let C be straight lines from $(0, \ 0)$ to $(2, \ 0)$, from $(2, \ 0)$ to $(1, \ 1)$, from $(1, \ 1)$ to $(0, \ 1)$, and from $(0, \ 1)$ to $(0, \ 0)$. Evaluate $\int_C e^{x^2}dx + xydy.$

Solution Let $P(x, y) = e^{x^2}$, $Q(x, y) = xy$. Then

$$\int_C P(x, y)dx + Q(x, y)dy = \iint_D \frac{\partial Q(x, y)}{\partial x} - \frac{\partial P(x, y)}{\partial y}dA$$

$$= \iint_D ydA$$

$$= \int_0^1 \int_0^{2-y} ydxdy$$

$$= \frac{7}{6}.$$

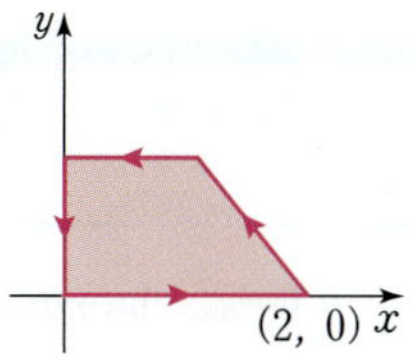

19. Let $\mathbf{F}(x, y) = \dfrac{-y\mathbf{i} + x\mathbf{j}}{x^2 + y^2}$ and C is an any closed curve containing the origin. Find $\int_C \mathbf{F} \cdot d\mathbf{r}$.

Solution Let's consider the region D which is the annulus as shown in the figure. Let c and c' denote the outer and the inner circles of D.

Then by Green's Theorem, we have

$$\int_C Pdx + Qdy + \int_{-C'} Pdx + Qdy = \iint_D \left(\frac{\partial Q}{\partial x} - \frac{\partial P}{\partial y}\right) dA$$

$$= \iint_D \left(\frac{y^2 - x^2}{(x^2+y^2)^2} - \frac{y^2 - x^2}{(x^2+y^2)^2}\right) dA = 0$$

$$\Rightarrow \int_C Pdx + Qdy = \int_{C'} Pdx + Qdy$$

$$\Rightarrow \int_C \mathbf{F} \cdot d\mathbf{r} = \int_{C'} \mathbf{F} \cdot d\mathbf{r}.$$

Using $\mathbf{r}(t) = a\cos t\mathbf{i} + a\sin t\mathbf{j},\ 0 \le t \le 2\pi$, we can compute

$$\int_C \mathbf{F} \cdot d\mathbf{r} = \int_{C'} \mathbf{F} \cdot d\mathbf{r} = \int_0^{2\pi} \mathbf{F}(\mathbf{r}(t)) \cdot \mathbf{r}'(t)dt$$

$$= \int_0^{2\pi} \frac{(-a\sin t)(-a\sin t) + (a\cos t)(a\cos t)}{a^2\cos^2 t + a^2\sin^2 t} dt$$

$$= \int_0^{2\pi} dt = 2\pi.$$

Thus the vector field $\mathbf{F}$ is not conservative. Note that $\text{curl}\mathbf{F} = 0$. Here D is not a simply connected domain.

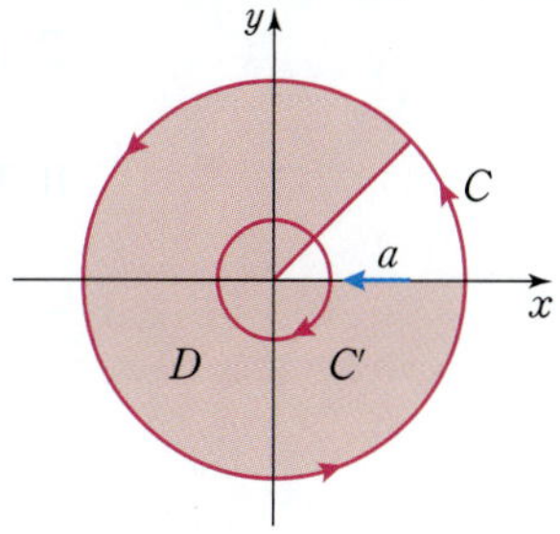

15.5 Curl and Divergence

YouTube http://youtu.be/IswmJUCTeNA

In this section, we look at two operations, namely curl and divergence, on vector fields. These operations have many applications in physics and engineering.

Curl of Vector Fields

If $\mathbf{F}$ is a differentiable vector field with $\mathbf{F} = M\mathbf{i} + N\mathbf{j} + P\mathbf{k}$, then

$$\text{curl } \mathbf{F} = \nabla \times \mathbf{F} \quad \text{where } \nabla = \frac{\partial}{\partial x}\mathbf{i} + \frac{\partial}{\partial y}\mathbf{j} + \frac{\partial}{\partial z}\mathbf{k};$$

1
$$\text{curl } \mathbf{F} = \nabla \times \mathbf{F} = \begin{vmatrix} \mathbf{i} & \mathbf{j} & \mathbf{k} \\ \frac{\partial}{\partial x} & \frac{\partial}{\partial y} & \frac{\partial}{\partial z} \\ M & N & P \end{vmatrix}$$
$$= \left(\frac{\partial P}{\partial y} - \frac{\partial N}{\partial z}\right)\mathbf{i} + \left(\frac{\partial M}{\partial z} - \frac{\partial P}{\partial x}\right)\mathbf{j} + \left(\frac{\partial N}{\partial x} - \frac{\partial M}{\partial y}\right)\mathbf{k}.$$

EXAMPLE 1

Find the curl of the vector field $\mathbf{F} = xy^2\mathbf{i} + 2yz\mathbf{j} - 3xz^2\mathbf{k}$.

Solution Using 1, we have

$$\text{curl } \mathbf{F} = \nabla \times \mathbf{F} = \begin{vmatrix} \mathbf{i} & \mathbf{j} & \mathbf{k} \\ \frac{\partial}{\partial x} & \frac{\partial}{\partial y} & \frac{\partial}{\partial z} \\ xy^2 & 2yz & -3xz^2 \end{vmatrix}$$
$$= \left(\frac{\partial}{\partial y}(-3xz^2) - \frac{\partial}{\partial z}(2yz)\right)\mathbf{i} + \left(\frac{\partial}{\partial z}(xy^2) - \frac{\partial}{\partial x}(-3xz^2)\right)\mathbf{j}$$
$$+ \left(\frac{\partial}{\partial x}(2yz) - \frac{\partial}{\partial y}(xy^2)\right)\mathbf{k}$$
$$= -2y\mathbf{i} + 3z^2\mathbf{j} - 2xy\mathbf{k} .$$
■

THEOREM 1

Let $\mathbf{F} = F_1\mathbf{i} + F_2\mathbf{j} + F_3\mathbf{k}$ and $\mathbf{G} = G_1\mathbf{i} + G_2\mathbf{j} + G_3\mathbf{k}$ be differentiable vector fields with $F_i = F_i(x, y, z)$ and $G_i = G_i(x, y, z)$, $i = 1, 2, 3$. Suppose $f(x, y, z)$ has continuous partial derivatives. Then
(i) $\text{curl}(\mathbf{F} + \mathbf{G}) = \text{curl } \mathbf{F} + \text{curl } \mathbf{G}$
(ii) $\text{curl}(f\mathbf{F}) = f \text{ curl } \mathbf{F} + (\nabla f) \times \mathbf{F}$.

Proof By 1, we have

(i) $\operatorname{curl}(\mathbf{F}+\mathbf{G}) = \nabla\times(\mathbf{F}+\mathbf{G}) = \begin{vmatrix} \mathbf{i} & \mathbf{j} & \mathbf{k} \\ \dfrac{\partial}{\partial x} & \dfrac{\partial}{\partial y} & \dfrac{\partial}{\partial z} \\ F_1+G_1 & F_2+G_2 & F_3+G_3 \end{vmatrix}$

$$= \begin{vmatrix} \mathbf{i} & \mathbf{j} & \mathbf{k} \\ \dfrac{\partial}{\partial x} & \dfrac{\partial}{\partial y} & \dfrac{\partial}{\partial z} \\ F_1 & F_2 & F_3 \end{vmatrix} + \begin{vmatrix} \mathbf{i} & \mathbf{j} & \mathbf{k} \\ \dfrac{\partial}{\partial x} & \dfrac{\partial}{\partial y} & \dfrac{\partial}{\partial z} \\ G_1 & G_2 & G_3 \end{vmatrix} = \operatorname{curl}\mathbf{F} + \operatorname{curl}\mathbf{G}.$$

(ii) $\operatorname{curl}(f\mathbf{F}) = \nabla\times(f\mathbf{F}) = \begin{vmatrix} \mathbf{i} & \mathbf{j} & \mathbf{k} \\ \dfrac{\partial}{\partial x} & \dfrac{\partial}{\partial y} & \dfrac{\partial}{\partial z} \\ fF_1 & fF_2 & fF_3 \end{vmatrix}$

$$\begin{aligned}
&= \left(\frac{\partial}{\partial y}(fF_3) - \frac{\partial}{\partial z}(fF_2)\right)\mathbf{i} + \left(\frac{\partial}{\partial z}(fF_1) - \frac{\partial}{\partial x}(fF_3)\right)\mathbf{j} \\
&\quad + \left(\frac{\partial}{\partial x}(fF_2) - \frac{\partial}{\partial y}(fF_1)\right)\mathbf{k} \\
&= f\left(\frac{\partial F_3}{\partial y} - \frac{\partial F_2}{\partial z}\right)\mathbf{i} + \left(F_3\frac{\partial f}{\partial y} - F_2\frac{\partial f}{\partial z}\right)\mathbf{i} \\
&\quad + f\left(\frac{\partial F_1}{\partial z} - \frac{\partial F_3}{\partial x}\right)\mathbf{j} + \left(F_1\frac{\partial f}{\partial z} - F_3\frac{\partial f}{\partial x}\right)\mathbf{j} \\
&\quad + f\left(\frac{\partial F_2}{\partial x} - \frac{\partial F_1}{\partial y}\right)\mathbf{k} + \left(F_2\frac{\partial f}{\partial x} - F_1\frac{\partial f}{\partial y}\right)\mathbf{k} \\
&= f\left\{\left(\frac{\partial F_3}{\partial y} - \frac{\partial F_2}{\partial z}\right)\mathbf{i} + \left(\frac{\partial F_1}{\partial z} - \frac{\partial F_3}{\partial x}\right)\mathbf{j} + \left(\frac{\partial F_2}{\partial x} - \frac{\partial F_1}{\partial y}\right)\mathbf{k}\right\} \\
&\quad + \left(F_3\frac{\partial f}{\partial y} - F_2\frac{\partial f}{\partial z}\right)\mathbf{i} + \left(F_1\frac{\partial f}{\partial z} - F_3\frac{\partial f}{\partial x}\right)\mathbf{j} + \left(F_2\frac{\partial f}{\partial x} - F_1\frac{\partial f}{\partial y}\right)\mathbf{k} \\
&= f\operatorname{curl}\mathbf{F} + (\nabla f)\times\mathbf{F}.
\end{aligned}$$

■

THEOREM 2

Let $\mathbf{F}$ be a three dimensional differentiable vector field with continuous second-order partial derivatives. If $\mathbf{F}$ is conservative, then $\operatorname{curl}\mathbf{F} = \mathbf{0}$.

Proof Since $\mathbf{F}$ is conservative vector field, there is a differentiable function f of three variables satisfying $\mathbf{F} = \nabla f$. But we have

$$\begin{aligned}
\operatorname{curl}(\nabla f) = \nabla\times(\nabla f) &= \begin{vmatrix} \mathbf{i} & \mathbf{j} & \mathbf{k} \\ \dfrac{\partial}{\partial x} & \dfrac{\partial}{\partial y} & \dfrac{\partial}{\partial z} \\ \dfrac{\partial f}{\partial x} & \dfrac{\partial f}{\partial y} & \dfrac{\partial f}{\partial z} \end{vmatrix} \\
&= \left(\frac{\partial^2 f}{\partial y\,\partial z} - \frac{\partial^2 f}{\partial z\,\partial y}\right)\mathbf{i} + \left(\frac{\partial^2 f}{\partial z\,\partial x} - \frac{\partial^2 f}{\partial x\,\partial z}\right)\mathbf{j} + \left(\frac{\partial^2 f}{\partial x\,\partial y} - \frac{\partial^2 f}{\partial y\,\partial x}\right)\mathbf{k} \\
&= 0\mathbf{i} + 0\mathbf{j} + 0\mathbf{k} = \mathbf{0}.
\end{aligned}$$

Hence, $\operatorname{curl}\mathbf{F} = \mathbf{0}$. ■

THEOREM 3

Let $\mathbf{F}$ be a three dimensional differentiable vector field whose components have continuous partial derivatives. If curl $\mathbf{F}=0$, then $\mathbf{F}$ is conservative.

EXAMPLE 2

Show that the vector field $\mathbf{F}=x\mathbf{i}+y\mathbf{j}+z\mathbf{k}$ is conservative.
http://matrix.skku.ac.kr/cal-lab/Sec15-5-Exm-2.html

Solution Using [1], we have

$$\text{curl }\mathbf{F}=\nabla\times\mathbf{F}=\begin{vmatrix}\mathbf{i} & \mathbf{j} & \mathbf{k}\\ \dfrac{\partial}{\partial x} & \dfrac{\partial}{\partial y} & \dfrac{\partial}{\partial z}\\ x & y & z\end{vmatrix}$$

$$=\left(\frac{\partial}{\partial y}(z)-\frac{\partial}{\partial z}(y)\right)\mathbf{i}+\left(\frac{\partial}{\partial z}(x)-\frac{\partial}{\partial x}(z)\right)\mathbf{j}+\left(\frac{\partial}{\partial x}(y)-\frac{\partial}{\partial y}(x)\right)\mathbf{k}$$

$$=0.$$

Thus $\mathbf{F}$ is a conservative vector field.

```
var('x,y,z,i,j,k')
M=x; N=y; P=z
curlF=(diff(P,y)-diff(N,z))*i+(diff(M,z)-diff(P,x))*j
+(diff(N,x)-diff(M,y))*k
print curlF
print bool(curlF==0)
```

Answer : 0, True ■

Divergence of Vector Fields

If $\mathbf{F}$ is a differentiable vector field with

$$\mathbf{F}=M\mathbf{i}+N\mathbf{j}+P\mathbf{k}$$

then the divergence of $\mathbf{F}$ is the function of three variables defined by

[2] $$\text{div }\mathbf{F}=\frac{\partial M}{\partial x}+\frac{\partial N}{\partial y}+\frac{\partial P}{\partial z}.$$

Notice that the curl of a vector field is a vector field but the divergence of a vector field is a real valued function. If we use the notation $\nabla=\frac{\partial}{\partial x}\mathbf{i}+\frac{\partial}{\partial y}\mathbf{j}+\frac{\partial}{\partial z}\mathbf{k}$ then $\text{div }\mathbf{F}=\nabla\cdot\mathbf{F}=\left(\frac{\partial}{\partial x},\frac{\partial}{\partial y},\frac{\partial}{\partial z}\right)\cdot(M,N,P)$.

THEOREM 4

The following properties are true for differentiable vector fields $\mathbf{F}$ and $\mathbf{G}$ and a continuously differentiable scalar function f.

(i) $\operatorname{div}(\mathbf{F}+\mathbf{G}) = \operatorname{div}\mathbf{F} + \operatorname{div}\mathbf{G}$,

(ii) $\operatorname{div}(f\mathbf{F}) = f\operatorname{div}\mathbf{F} + \mathbf{F}\cdot\nabla f$.

Proof Left to the reader. ■

CAS **EXAMPLE 3**

Find the divergence of the vector field $F+G$ where

$$\mathbf{F}(x, y, z) = (x-3y)\mathbf{i} + (x^2z^2+\cos z)\mathbf{j} + xyz^2\mathbf{k} \text{ and}$$
$$\mathbf{G}(x, y, z) = -y(z+x)\mathbf{i} + x\mathbf{j} + yz\mathbf{k}.$$

http://matrix.skku.ac.kr/cal-lab/Sec15-5-Exm-3.html

Solution We take the three partial derivatives and add them. Then,

$$\operatorname{div}\mathbf{F} = 1+0+2xyz = 1+2xyz.$$
$$\operatorname{div}\mathbf{G} = -y+y=0.$$
$$\text{Hence } \operatorname{div}(\mathbf{F}+\mathbf{G}) = 2xyz+1.$$

```
var('x,y,z')
M=x-3*y; N=x^2*z^2+cos(z); P=x*y*z^2
divF=diff(M,x)+diff(N,y)+diff(P,z)
divF
```

Answer : 2*x*y*z + 1 ■

EXAMPLE 4

Compute the divergence of the vector field

$$\mathbf{F}(x, y, z) = -y(2+x)\mathbf{i} + x\mathbf{j} + yz\mathbf{k}.$$

Solution $\operatorname{div}\mathbf{F} = \dfrac{\partial(-2y-xy)}{\partial x} + \dfrac{\partial(x)}{\partial y} + \dfrac{\partial(yz)}{\partial z} = -y+y=0.$ ■

If $\mathbf{F}$ is a vector field on $\mathbb{R}^3$, then curl $\mathbf{F}$ is also a vector field in $\mathbb{R}^3$. Therefore, we can compute its divergence. The next theorem shows that the divergence of curl $\mathbf{F}$ is 0.

THEOREM 5

If $\mathbf{F}=M\mathbf{i}+N\mathbf{j}+P\mathbf{k}$ is a vector field on $\mathbb{R}^3$ and it M, N and P have continuous second-order partial derivatives, then div(curl **F**) = 0.

Proof Using the definition of divergence and curl, we have

$$\begin{aligned}\text{div}(\text{curl } \mathbf{F}) &= \nabla \cdot (\nabla \times \mathbf{F}) \\ &= \frac{\partial}{\partial x}\left(\frac{\partial P}{\partial y}-\frac{\partial N}{\partial z}\right)+\frac{\partial}{\partial y}\left(\frac{\partial M}{\partial z}-\frac{\partial P}{\partial x}\right)+\frac{\partial}{\partial z}\left(\frac{\partial N}{\partial x}-\frac{\partial M}{\partial y}\right) \\ &= \frac{\partial^2 P}{\partial x \partial y}-\frac{\partial^2 N}{\partial x \partial z}+\frac{\partial^2 M}{\partial y \partial z}-\frac{\partial^2 P}{\partial y \partial x}+\frac{\partial^2 N}{\partial z \partial x}-\frac{\partial^2 M}{\partial z \partial y} \\ &= 0.\end{aligned}$$

■

EXAMPLE 5

Show that the vector field $\mathbf{F}(x, y, z)=x\mathbf{i}+yz\mathbf{j}-y^2\mathbf{k}$ is not the curl of another vector field.

Solution Note that div $\mathbf{F}=1+z\neq 0$. If $\mathbf{F}$ were the curl of another vector field, then div $\mathbf{F}$ would be zero by Theorem 4. This is a contradiction to div $\mathbf{F}\neq 0$. Therefore $\mathbf{F}$ is not the curl of another vector field. ■

We can rewrite Green's Theorem using these new ideas. These rewritten versions in turn are closer to some theorems that we will see later. Suppose we write a two-dimensional vector field in the form $\mathbf{F}=\langle M, N, 0\rangle$ where M and N are functions of x and y. Then

$$\text{curl } \mathbf{F}=\nabla\times\mathbf{F}=\begin{vmatrix}\mathbf{i} & \mathbf{j} & \mathbf{k} \\ \frac{\partial}{\partial x} & \frac{\partial}{\partial y} & \frac{\partial}{\partial z} \\ M & N & 0\end{vmatrix}=(N_x-M_y)\mathbf{k},$$

and so $(\nabla\times\mathbf{F})\cdot\mathbf{k}=\langle 0, 0, N_x-M_y\rangle\cdot\langle 0, 0, 1\rangle=N_x-M_y$. Thus Green's Theorem says

3
$$\begin{aligned}\int_C \mathbf{F}\cdot d\mathbf{r} &= \int_C M dx+N dy. \\ &= \iint_D (N_x-M_y)dA=\iint_D (\nabla\times\mathbf{F})\cdot\mathbf{k}\, dA.\end{aligned}$$

This is called the vector form of Green's Theorem.

Nature of Curl and Divergence of Vector Fields

The concepts about curl and divergence of a vector field are associated with the tendency of the vector field to rotate or to diverge where $\mathbf{F}(x, y, z)$ is the velocity of fluid (or gas) in physical applications. We can understand the terms in the following discussion, by calculating curl and divergence of vector fields. If $\operatorname{curl} \mathbf{F} = \mathbf{0}$ at a point P, then $\mathbf{F}$ is called irrotational at P. If div $\mathbf{F} = 0$, then $\mathbf{F}$ is said to be incompressible.

For the vector field $\mathbf{F}(x, y) = x\mathbf{i} + y\mathbf{j}$, it is easy to show that $\operatorname{curl} \mathbf{F} = \mathbf{0}$. Thus the vector field $\mathbf{F}$ is irrotational. (See Figure 1.) On the other hand, for the vector field $\mathbf{H}(x, y) = -y\mathbf{i} + x\mathbf{j}$, since $\operatorname{div} \mathbf{H} = 0$, a vector field $\mathbf{H}$ is said to be imcompressible. (See Figure 2.)

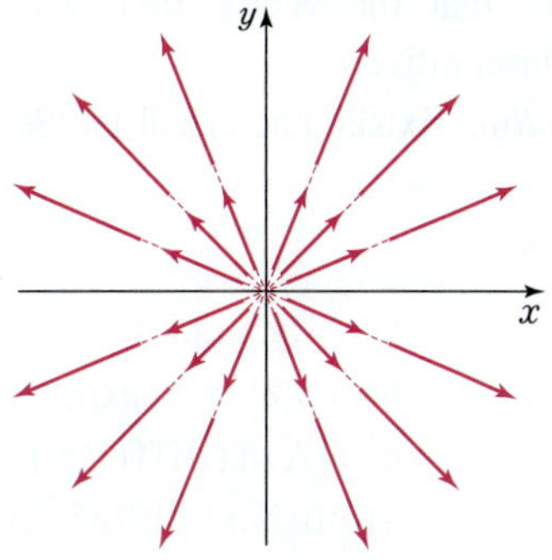

Figure 1 $\mathbf{F} = x\mathbf{i} + y\mathbf{j}$

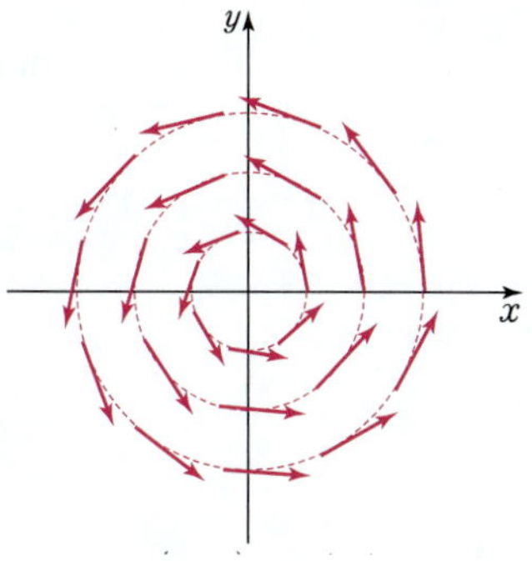

Figure 2 $\mathbf{H} = -y\mathbf{i} + x\mathbf{j}$

Next we have another differential operator. If f is a function of three variables, we have

$$\operatorname{div}(\nabla f) = \nabla \cdot (\nabla f) = \frac{\partial^2 f}{\partial x^2} + \frac{\partial^2 f}{\partial y^2} + \frac{\partial^2 f}{\partial z^2}$$

and $\nabla \cdot (\nabla f) = \nabla^2 f$ by computing divergence of the gradient of f. The operator $\nabla^2 = \nabla \cdot \nabla$ is called the **Laplace operator** because of its relation to **Laplace's equation**

$$\nabla^2 f = \frac{\partial^2 f}{\partial x^2} + \frac{\partial^2 f}{\partial y^2} + \frac{\partial^2 f}{\partial z^2} = 0.$$

We can also apply the Laplace operator ∇^2 to the vector field $\mathbf{F} = M\mathbf{i} + N\mathbf{j} + P\mathbf{k}$ in terms of its components:

$$\nabla^2 \mathbf{F} = \nabla^2 M\mathbf{i} + \nabla^2 N\mathbf{j} + \nabla^2 P\mathbf{k}.$$

CAS **EXAMPLE 6**

Find $\nabla \cdot \left((xe^{yz})\mathbf{i} + (xyz)\mathbf{j} - (3y^2 + z^2)\mathbf{k}\right)$.
http://matrix.skku.ac.kr/cal-lab/Sec15-5-Exm-6.html

Solution

```
var('x,y,z,i,j,k')
M(x,y,z)=x*exp(y*z); N(x,y,z)=x*y*z; P(x,y,z)=-3*y^2-z^2
divF=diff(M,x)+diff(N,y)+diff(P,z)
divF
```

Answer : x*z - 2*z + e^(y*z) ■

CAS EXAMPLE 7

Show that the vector field $\mathbf{A}(x, y, z) = (x - ye^{xz})\mathbf{i} + (y^2 z)\mathbf{j} + (z^2 - y)\mathbf{k}$ is not conservative.

http://matrix.skku.ac.kr/cal-lab/Sec15-5-Exm-7.html

Solution

```
var('x,y,z,i,j,k')
M(x,y,z)=x-y*exp(x*z); N(x,y,z)=y^2*z; P(x,y,z)= z^2-y
curlA=((diff(P,y)-diff(N,z))*i+(diff(M,z)-diff(P,x))*j
+(diff(N,x)-diff(M,y))*k)
curlA
```

Answer : -j*x*y*e^(x*z) - (y^2 + 1)*i + k*e^(x*z)

Since curl $\mathbf{A} \neq 0$, $\mathbf{A}$ is not conservative. ■

CAS EXAMPLE 8

Prove that $\mathbf{F} = (y^2\cos x + z^3)\mathbf{i} + (2y\sin x - 4)\mathbf{j} + (3xz^2 + 2)\mathbf{k}$ is (a) conservative field, (b) find the potential function of $\mathbf{F}$, (c) find the work done in moving an object in this field from $P_1(0, 1, -1)$ to $P_2\left(\frac{\pi}{2}, -1, 2\right)$.

Solution (a) From the fact that curl $\mathbf{F} = 0$,

$\mathbf{F}$ is conservative if $\nabla \times \mathbf{F} = 0$,

$$\nabla \times \mathbf{F} = \begin{vmatrix} \mathbf{i} & \mathbf{j} & \mathbf{k} \\ \dfrac{\partial}{\partial x} & \dfrac{\partial}{\partial y} & \dfrac{\partial}{\partial z} \\ y^2\cos x + z^3 & 2y\sin x - 4 & 3xz^2 + 2 \end{vmatrix}$$

$$= (0-0)\mathbf{i} - (3z^2 - 3z^2)\mathbf{j} + (2y\cos x - 2y\cos x)\mathbf{k} = 0$$

Hence $\mathbf{F}$ is conservative.

(b) Let f be a potential function such of $\mathbf{F}$. Then comparing the components of $\mathbf{i}$, $\mathbf{j}$ and $\mathbf{k}$, we get

3 $$\frac{\partial f}{\partial x} = y^2\cos x + z^3,$$

4 $$\frac{\partial f}{\partial y} = 2y\sin x - 4,$$

$$\boxed{5} \qquad \frac{\partial f}{\partial z} = 3xz^2 + 2.$$

Integrating $\boxed{3}$ with respect to x,

$$\boxed{6} \qquad f(x, y, z) = y^2 \sin x + xz^3 + g(y, z).$$

Differentiating $\boxed{4}$ with respect to y and using $\boxed{4}$

$$2y\sin x + 0 + \frac{\partial g}{\partial y} = \frac{\partial f}{\partial y} = 2y\sin x - 4.$$

Integrating with respect to y

$$\boxed{7} \qquad g(y, z) = -4y + c_1(z).$$

Substituting $\boxed{7}$ in $\boxed{6}$

$$\boxed{8} \qquad f = y^2\sin x + xz^3 - 4y + c_1(z).$$

Differentiating $\boxed{8}$ with respect to z and using $\boxed{5}$

$$0 + 3xz^2 - 0 + \frac{dc_1}{dz} = \frac{\partial f}{\partial z} = 3xz^2 + 2$$

Integrating with respect to z

$$\boxed{9} \qquad c_1(z) = 2z + c.$$

Substituting $\boxed{9}$ in $\boxed{8}$

$$f(x, y, z) = y^2\sin x + xz^3 - 4y + 2z + c.$$

(c) Work done

$$W = \int_C F \cdot dr = f\Big|_{P_1}^{P_2} = f(P_2) - f(P_1) = f\left(\frac{\pi}{2}, -1, 2\right) - f(0, 1, -1)$$
$$= 15 + 4\pi.$$

http://matrix.skku.ac.kr/cal-lab/Sec15-5-Exm-8.html

```
var('x,y,z,i,j,k')
M=y^2*cos(x)+z^3; N=2*y*sin(x)-4; P= 3*x*z^2+2
curlF=((diff(P,y)-diff(N,z))*i+(diff(M,z)-diff(P,x))*j
+(diff(N,x)-diff(M,y))*k)
print curlF
print bool(curlF==0)
```

Answer : 0, True

```
var('x,y,z,c')
fx=y^2*cos(x)+z^3; fy=2*y*sin(x)-4; fz= 3*x*z^2+2
f_1=integral(fx,x)     # +g(y,z)
f_1y=diff(f_1,y)       # fy= diff(f_1,y)+gy(y,z)
gy(y,z)=fy-diff(f_1,y)
g(y,z)=integral(gy(y,z),y) # +h(z)
f_2=f_1+g(y,z) #+h(z)
f_2z=diff(f_2,z)      #fz=diff(f_2,z)+ hz(z)
hz(z)=fz-diff(f_2,z)
h(z)=integral(hz(z),z)
f=f_2+h(z)+c
f   # This is a potential function for F
```

Answer : x*z^3 + y^2*sin(x) + c - 4*y + 2*z

```
var('x,y,z,c')
f(x,y,z)=x*z^3 + y^2*sin(x) + c - 4*y + 2*z
x0=0;y0=1;z0=-1
x1=pi/2;y1=-1;z1=2
workdone=f(x1,y1,z1)-f(x0,y0,z0)
workdone
```

Answer : 4*pi + 15

You will have the same answer : 4*pi+15. ■

15.5 EXERCISES (Curl and Divergence)

http://matrix.skku.ac.kr/Cal-Book/part2/CS-Sec-15-5-Sol.htm

YouTube http://youtu.be/wLTHYaANwtI

1. Find (a) the curl and (b) the divergence of the vector field.

• $\mathbf{F}(x, y, z) = xyz^2\mathbf{i} - x^2y\mathbf{k}$

Solution (a) curl :

$$\begin{vmatrix} \mathbf{i} & \mathbf{j} & \mathbf{k} \\ \frac{\partial}{\partial x} & \frac{\partial}{\partial y} & \frac{\partial}{\partial z} \\ xyz^2 & 0 & -x^2y \end{vmatrix} = \langle -x^2, 2xy(z+1), -xz^2 \rangle$$

$= -x^2\mathbf{i} + 2xy(z+1)\mathbf{j} - xz^2\mathbf{k}$

(b) div : yz^2

• $\mathbf{F}(x, y, z) = x^2yz\,\mathbf{i} + x^2y^2z\,\mathbf{j} + xyz^2\mathbf{k}$

Solution (a) curl :

$\langle xz^2 - x^2y^2, x^2y - yz^2, 2xy^2z - x^2z \rangle$

(b) div : $4xyz + 2x^2yz$

• $\mathbf{F}(x, y, z) = e^x \sin y\mathbf{i} + e^x \cos y\mathbf{j} + z^2\mathbf{k}$

Solution (a) curl : $\langle 0, 0, 0 \rangle$

(b) div : $2z$

• $\mathbf{F}(x, y, z) = xe^y\mathbf{j} + ye^z\mathbf{k}$

Solution (a) curl : $\langle e^z, 0, e^y \rangle$

(b) div : $xe^y + ye^z$

• $\mathbf{F}(x, y, z) = \dfrac{6}{\sqrt{x^2+y^2+z^2}}(x\mathbf{i} + y\mathbf{j} + z\mathbf{k})$

Solution (a) curl : $\langle 0, 0, 0 \rangle$

(b) div : $\dfrac{12}{\sqrt{x^2+y^2+z^2}}$

• $\mathbf{F}(x, y, z) = e^{xy}\sin z\mathbf{j} + y^2\tan^{-1}(x/z)\mathbf{k}$

Solution (a) curl :

$\left\langle 2y\tan^{-1}(x/z) - e^{xy}\cos z, \dfrac{-y^2z}{x^2+z^2}, ye^{xy}\sin z \right\rangle$

(b) div : $\dfrac{x(e^{xy}z^2\sin z + e^{xy}x^2\sin z - y^2)}{x^2+z^2}$

• $\mathbf{F}(x, y, z) = \langle \ln x^2, \ln(xy), \ln(xyz) \rangle$

Solution (a) curl : $\left\langle \dfrac{1}{y}, -\dfrac{1}{x}, \dfrac{1}{x} \right\rangle$

(b) div : $\dfrac{2yz + xz + xy}{xyz}$

• $\mathbf{F}(x, y, z) = \langle e^{xy}, e^{xy}, e^{xyz} \rangle$

Solution (a) curl : $\langle xze^{xyz}, -yze^{xyz}, -e^{xy}(x-y) \rangle$

(b) div : $ye^{xy} + xe^{xy} + xye^{xyz}$

2. Let f be a scalar function and $\mathbf{F}$ a vector field. State whether each expression is meaningful. If not, explain why. If so, state whether it is a scalar function or a vector field.

(a) curl$f$
(b) grad$f$
(c) div$\mathbf{F}$
(d) curl(grad$f$)
(e) grad$\mathbf{F}$
(f) grad(div$\mathbf{F}$)
(g) div(grad$f$)
(h) grad(div$f$)
(i) curl(grad$\mathbf{F}$)
(j) div(div$\mathbf{F}$)
(k) (grad$f$)×(div$\mathbf{F}$)
(l) div(curl(grad$f$))

Solution (a)Not meaningful, curl must take a vector field.
(b) Meaningful, vector field
(c) Meaningful, scalar function
(d) Meaningful, vector field
(e) Not meaningful
(f) Meaningful, vector field
(g) Meaningful, scalar function
(h) Not meaningful
(i) Not meaningful
(j) Not meaningful
(k) Not meaningful
(l) Meaningful, scalar function

3. Is there a vector field $\mathbf{G}$ on $\mathbb{R}^3$ such that curl $\mathbf{G} = \langle x^2\sin y, \cos y, z - xy \rangle$? Explain.

Solution No, $\nabla \cdot (\text{curl}\mathbf{G}) = 2x\sin y - \sin y + 1 \neq 0$.

4. Is there a vector field G on $\mathbb{R}^3$ such that curl $\mathbf{G} = \langle xyz, -y^2z, yz^2 \rangle$? Explain.

Solution No, $\nabla \cdot (\text{curl}\,\mathbf{G}) \neq 0$.

5. (a) Let $\mathbf{F}(x, y, z) = \langle f(x, y, z), g(x, y, z), h(x, y, z) \rangle$ be a differentiable vector field with div$\mathbf{F} = 0$. Define the vector field $\mathbf{G}(x, y, z) = \langle G_1, G_2, G_3 \rangle$ by

$$G_1(x, y, z) = \int_0^z g(x, y, t)dt - \int_0^y h(x, t, 0)dt,$$

$$G_2(x, y, z) = -\int_0^z f(x, y, t)dt, \text{ and}$$

$G_3(x, y, z) = 0$. Prove that $\nabla \times \mathbf{G} = \mathbf{F}$.

(b) Let $\mathbf{F} = y\mathbf{i} + z\mathbf{j} + x\mathbf{k}$. Find $\overline{G}$ such that $\nabla \times \overline{G} = \overline{F}$.

6. Evaluate the line integral $\int_C \mathbf{F} \cdot d\mathbf{r}$, where $\mathbf{F}(x, y, z) = y^2\mathbf{i} + (2xy + e^{3z})\mathbf{j} + 3ye^{3z}\mathbf{k}$ and C is the curve given by $x(t) = t^{2009}$, $y(t) = \sin\left(\frac{\pi t}{t^2+1}\right)$, $z(t) = \frac{1}{3}\ln(t^{2009}+1)$, $0 \le t \le 1$.

Solution Let $\mathbf{r}(t) = \langle x(t), y(t), z(t) \rangle$. We then observe that $\mathbf{r}(0) = \langle 0, 0, 0 \rangle$, $\mathbf{r}(1) = \left\langle 1, 1, \frac{\ln 2}{3} \right\rangle$. Note first that direct computations show that $\nabla \times \mathbf{F} = 0$, which implies that there exists a scalar function f with $\nabla f = \mathbf{F}$. We then have the following equations: $f_x = y^2$, $f_y = 2xy + e^{3z}$, $f_z = 3ye^{3z}$. Solving these equations, one can see that $f(x, y, z) = xy^2 + ye^{3z} + K$, where K is a constant. Thus, due to the Fundamental Theorem for line integral,

$$\int_C \mathbf{F} \cdot d\mathbf{r} = \int_C \nabla f \cdot d\mathbf{r}$$
$$= f(\mathbf{r}(1)) - f(\mathbf{r}(0)) = 3 + K - K = 3.$$

CAS **7.** Find $\nabla \cdot \left[(e^{xz})\mathbf{i} + (xy^2z)\mathbf{j} - (y^2 - z^2)\mathbf{k}\right]$.

http://matrix.skku.ac.kr/cal-lab/Sec15-5-Exs-7.html

Solution

```
x,y,z,i,j,k=var('x,y,z,i,j,k')
P(x,y,z) = exp(x*z);
Q(x,y,z) = x*y^2*z;
R(x,y,z) = -y^2+z^2;
A(x,y,z) = P*i+Q*j+R*k;
A(x,y,z)
```

j*x*y^2*z - (y^2 - z^2)*k + i*e^(x*z)

```
divA = diff(P,x)+diff(Q,y)+diff(R,z)
divA(x,y,z)
```

Answer : 2*x*y*z + z*e^(x*z) + 2*z

CAS **8.** If $\mathbf{A} = (2xy + z^3)\mathbf{i} + x^2\mathbf{j} + 3xz^2\mathbf{k}$.

a. Prove that the line integral $\int_c \mathbf{A} \cdot d\mathbf{r}$ is independent of the curve C joining two given points $P_1(1, -2, 1)$ and $P_2(3, 1, 4)$.

b. Show that there exists a scalar function f such that $\mathbf{A} = \nabla f$ $\mathbf{A} = \nabla f$ and find f.

c. Also find the work done in moving an object from P_1 to P_2.

9. Let $\mathbf{r} = x\mathbf{i} + y\mathbf{j} + z\mathbf{k}$ and $|\mathbf{r}| = \sqrt{x^2 + y^2 + z^2}$. Then verify the following identities:

(a) $\nabla|\mathbf{r}| = \frac{\mathbf{r}}{|\mathbf{r}|}$

(b) curl $\mathbf{r} = 0$

(c) $\nabla\left(\frac{1}{|\mathbf{r}|}\right) = -\frac{\mathbf{r}}{|\mathbf{r}|^2}$

(d) $\nabla(\ln|\mathbf{r}|) = \frac{\mathbf{r}}{|\mathbf{r}|^3}$

15.6 Surface Area

YouTube http://youtu.be/xX6tNVpegbs

The curves in 2 dimensional space can be defined by a function $y=f(x)$, an equation $g(x, y)=0$, or parametrically by equations $\mathbf{r}(t)=\langle x(t), y(t)\rangle$, $a \le t \le b$. A curve C described by a continuous function $y=f(x)$ can be parameterized by letting $x=t$ so that parametric equations are $x=t$, $y=f(t)$. We have two variables to parameterize a surface S in $\mathbb{R}^3$ defined by a function of two variables $z=g(x, y)$. Letting $x=u$ and $y=v$, then parametric equations for S are $x=u$, $y=v$ and $z=g(u, v)$.

Establishing the Area of the Surface

Let S be a smooth surface given by the vector valued function

$$\mathbf{r}(u, v)=x(u, v)\mathbf{i}+y(u, v)\mathbf{j}+z(u, v)\mathbf{k},\ (u, v)\in R$$

where $R=\{(u, v)\,|\,a \le u \le b,\ c \le v \le d\}$. (See Figure 1.) The functions x, y and z are assumed to have continuous partial derivatives with respect to u and v. The rectangular region R in the uv-plane is partitioned into rectangles, with sides of length Δu and Δv that are ordered in some convenient way, for $k=1, 2, \cdots, n$. The kth rectangle R_k corresponds to a curved path S_k on the surface S, with area ΔS_k. Let (u_k, v_k) be the lower left corner point of R_k. The other corners can be expressed as $(u_k+\Delta u, v_k)$, $(u_k+\Delta u, v_k+\Delta v)$, $(u_k, v_k+\Delta v)$ and so the area of R_k is $\Delta A=\Delta u\,\Delta v$. Two of the edges of S_k can be approximated by the vectors

$$\mathbf{r}(u_k+\Delta u, v_k)-\mathbf{r}(u_k, v_k)=\frac{\mathbf{r}(u_k+\Delta u, v_k)-\mathbf{r}(u_k, v_k)}{\Delta u}\Delta u \approx \frac{\partial \mathbf{r}}{\partial u}\Delta u$$

$$\mathbf{r}(u_k, v_k+\Delta v)-\mathbf{r}(u_k, v_k)=\frac{\mathbf{r}(u_k, v_k+\Delta v)-\mathbf{r}(u_k, v_k)}{\Delta v}\Delta v \approx \frac{\partial \mathbf{r}}{\partial v}\Delta v.$$

As seen in Figure 1 these vectors actually form two of the edges of parallelogram T_k lying in the tangent plane at $\mathbf{r}(u_k, v_k)$. The area ΔT_k of the parallelogram T_k approximates the area ΔS_k of S_k

$$\Delta T_k=\left|\frac{\partial \mathbf{r}}{\partial u}\Delta u \times \frac{\partial \mathbf{r}}{\partial v}\Delta v\right|=\left|\frac{\partial \mathbf{r}}{\partial u}\times\frac{\partial \mathbf{r}}{\partial v}\right|\Delta u\,\Delta v=\left|\frac{\partial \mathbf{r}}{\partial u}\times\frac{\partial \mathbf{r}}{\partial v}\right|\Delta A \approx \Delta S_k.$$

The Riemann sum

$$\sum_{k=1}^{n}\left|\frac{\partial \mathbf{r}}{\partial u}\times\frac{\partial \mathbf{r}}{\partial v}\right|\Delta A.$$

gives an approximation of the area $A(S)$ of that part of surface S

corresponding to the point in R.
The surface area of S is

$$A(S) = \lim_{n \to \infty} \sum_{k=1}^{n} \left| \frac{\partial \mathbf{r}}{\partial u} \times \frac{\partial \mathbf{r}}{\partial v} \right| \Delta A.$$

Then we can define the surface area of S as the following definition.

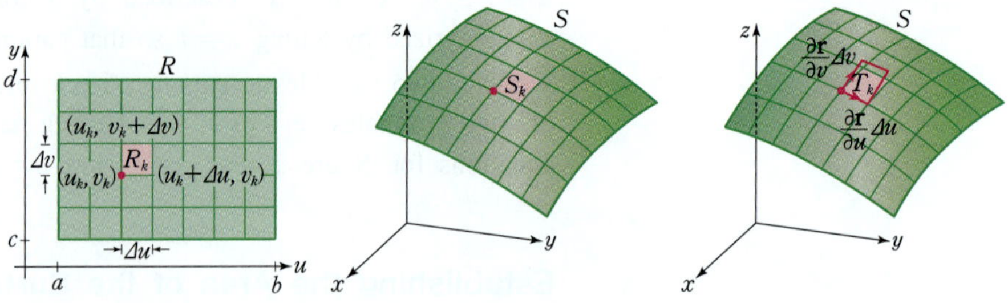

Figure 1

DEFINITION 1

Let f be a continuous function on a smooth surface S given a parameterization

$$\mathbf{r}(u,v) = x(u,v)\mathbf{i} + y(u,v)\mathbf{j} + z(u,v)\mathbf{k},$$

where $R = \{(u,v) \mid a \le u \le b,\ c \le v \le d\}$.
The tangent vectors

$$\mathbf{r}_u = \frac{\partial x}{\partial u}\mathbf{i} + \frac{\partial y}{\partial u}\mathbf{j} + \frac{\partial z}{\partial u}\mathbf{k} \text{ and } \mathbf{r}_v = \frac{\partial x}{\partial v}\mathbf{i} + \frac{\partial y}{\partial v}\mathbf{j} + \frac{\partial z}{\partial v}\mathbf{k}$$

are continuous on R and the normal vector $\mathbf{n} = \mathbf{r}_u \times \mathbf{r}_v$ is nonzero on R.
Then the surface area of S is

1
$$A(S) = \iint_R \left| \frac{\partial \mathbf{r}}{\partial u} \times \frac{\partial \mathbf{r}}{\partial v} \right| dA.$$

Note that we will see $\iint_S f(x, y, z)\,dS = \iint_R f(\mathbf{r}(u,v)) \left| \frac{\partial \mathbf{r}}{\partial u} \times \frac{\partial \mathbf{r}}{\partial v} \right| dA$ in section 15.7.

CAS **EXAMPLE 1**

Find the surface areas of the following surfaces:
(a) A cylinder with radius $r > 0$ and height h (excluding the circular ends).
(b) A sphere of radius ρ.

Solution The critical step is finding the normal vector $\mathbf{n} = \mathbf{r}_u \times \mathbf{r}_v$. It needs to be done only once for any given surface.
(a) As shown before, a parametric description of the cylinder is

$$\mathbf{r}(u,v)=\langle x(u,v),\ y(u,v),\ z(u,v)\rangle=\langle r\cos u,\ r\sin u,\ v\rangle,$$

where $0\le u\le 2\pi$ and $0\le v\le h$. A normal vector is

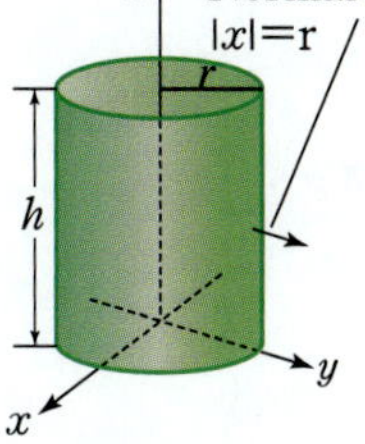

Figure 2

$$\mathbf{n}=\mathbf{r}_u\times\mathbf{r}_v=\begin{vmatrix}\mathbf{i}&\mathbf{j}&\mathbf{k}\\ \frac{\partial x}{\partial u}&\frac{\partial y}{\partial u}&\frac{\partial z}{\partial u}\\ \frac{\partial x}{\partial v}&\frac{\partial y}{\partial v}&\frac{\partial z}{\partial v}\end{vmatrix}=\begin{vmatrix}\mathbf{i}&\mathbf{j}&\mathbf{k}\\ -r\sin u&r\cos u&0\\ 0&0&1\end{vmatrix}$$
$$=\langle r\cos u, r\sin u, 0\rangle.$$

Notice that the normal vector points outward from the cylinder, away from the z-axis. (See Figure 2.) It now follows that

$$|\mathbf{r}_u\times\mathbf{r}_v|=\sqrt{r^2\cos^2u+r^2\sin^2u}=r.$$

By [1], the surface area of the cylinder is

$$A(S)=\iint_R|\mathbf{r}_u\times\mathbf{r}_v|dA=\int_0^{2\pi}\int_0^h r\,dv\,du=2\pi rh.$$

http://matrix.skku.ac.kr/cal-lab/Sec15-6-Exm-5.html

```
var('x,y,z,u,v,r,h')
assume(r>0)
x=r*cos(u);y=r*sin(u);z=v
r_u=vector([diff(x,u),diff(y,u),diff(z,u)])
r_v=vector([diff(x,v),diff(y,v),diff(z,v)])
n=r_u.cross_product(r_v)
print n
```

Answer : (r*cos(u), r*sin(u), 0)

```
integrand=n.norm()
integral(integral(integrand,v,0,h),u,0,2*pi)
```

Answer : 2*pi*h*r

(b) A parametric description of the sphere is

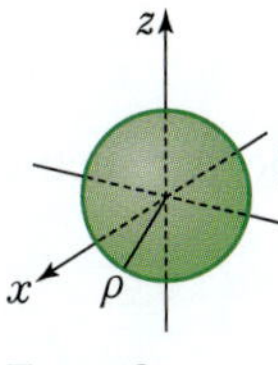

Figure 3

$$\mathbf{r}(\phi,\theta)=\langle\rho\sin\phi\cos\theta,\ \rho\sin\phi\sin\theta,\ \rho\cos\phi\rangle,$$

where $0\le\theta\le 2\pi$ and $0\le\phi\le\pi$. (See Figure 8.) A normal vector is

$$\mathbf{n}=\mathbf{r}_\phi\times\mathbf{r}_\theta=\begin{vmatrix}\mathbf{i}&\mathbf{j}&\mathbf{k}\\ \rho\cos\phi\cos\theta&\rho\cos\phi\sin\theta&-\rho\sin\phi\\ -\rho\sin\phi\sin\theta&\rho\sin\phi\cos\theta&0\end{vmatrix}$$
$$=\langle\rho^2\sin^2\phi\cos\theta,\rho^2\sin^2\phi\sin\theta,\rho^2\cos\phi\sin\phi\rangle.$$

Computing $|\mathbf{r}_\phi\times\mathbf{r}_\theta|$ requires several steps. However, the needed result is quite simple: $|\mathbf{r}_\phi\times\mathbf{r}_\theta|=\rho^2\sin\phi$ and the normal vector

$\mathbf{n} = \mathbf{r}_\phi \times \mathbf{r}_\theta$ points outward from the surface of sphere. By [1], the surface area of the sphere is

$$A(S) = \iint_R |\mathbf{r}_\phi \times \mathbf{r}_\theta| \, dA$$

$$= \int_0^{2\pi} \int_0^{\pi} \rho^2 \sin\phi \, d\phi \, d\theta = 4\pi\rho^2.$$ ■

EXAMPLE 2

Find the area of the cone $\mathbf{r} = (u\cos v)\mathbf{i} + (u\sin v)\mathbf{j} + 2u\mathbf{k}$, where $0 \le u \le 1$, $0 \le v \le 2\pi$.

Solution The surface is an upper portion of the corn. First we compute

$$\frac{\partial \mathbf{r}}{\partial u} = \cos v\mathbf{i} + \sin v\mathbf{j} + 2\mathbf{k}, \quad \frac{\partial \mathbf{r}}{\partial v} = -u\sin v\mathbf{i} + u\cos v\mathbf{j}$$

and then form the cross product

$$\frac{\partial \mathbf{r}}{\partial u} \times \frac{\partial \mathbf{r}}{\partial v} = \begin{vmatrix} \mathbf{i} & \mathbf{j} & \mathbf{k} \\ \cos v & \sin v & 2 \\ -u\sin v & u\cos v & 0 \end{vmatrix} = -2u\cos v\mathbf{i} - 2u\sin v\mathbf{j} + u\mathbf{k}$$

The magnitude of the vector is

$$\left| \frac{\partial \mathbf{r}}{\partial u} \times \frac{\partial \mathbf{r}}{\partial v} \right| = \sqrt{4u^2\cos^2 v + 4u^2\sin^2 v + u^2} = \sqrt{5}\,u.$$

Thus, the area is

$$A(S) = \iint_R \left| \frac{\partial \mathbf{r}}{\partial u} \times \frac{\partial \mathbf{r}}{\partial v} \right| dA = \iint_R \sqrt{5}\,u \, dudv$$

$$= \sqrt{5} \int_0^{2\pi} \int_0^1 u \, dudv = \sqrt{5}\,\pi.$$ ■

Surface Area of Graph Surfaces

Any surface described by $z = g(x, y)$ can be parameterized by the parametric equations

$$x = u, \ y = v, \ z = g(u, v).$$

We can parameterize S by setting

$$\mathbf{r}(u, v) = u\mathbf{i} + v\mathbf{j} + g(u, v)\mathbf{k}$$

and so we obtain $\mathbf{r}_u = \mathbf{i} + g_u(u, v)\mathbf{k}$ and $\mathbf{r}_v = \mathbf{j} + g_v(u, v)\mathbf{k}$.

Since the normal vector is

$$\mathbf{r}_u \times \mathbf{r}_v = \begin{vmatrix} \mathbf{i} & \mathbf{j} & \mathbf{k} \\ 1 & 0 & g_u \\ 0 & 1 & g_v \end{vmatrix} = -g_u\mathbf{i} - g_v\mathbf{j} + \mathbf{k},$$

the length of $\mathbf{r}_u \times \mathbf{r}_v$ is

$$|\mathbf{r}_u \times \mathbf{r}_v| = \sqrt{[g_u(u, v)]^2 + [g_v(u, v)]^2 + 1}.$$

Therefore, the area of surface is

2 $$A(S) = \iint_R \sqrt{[g_u(u,v)]^2 + [g_v(u,v)]^2 + 1}\, du\, dv.$$

EXAMPLE 3

Find the area of the surface $z = 2x + 3y$, $0 \le x \le 1$, $0 \le y \le 1$.

Solution Let $x = u$, $y = v$ and $z = g(u, v) = 2u + 3v$ with $0 \le u \le 1$ and $0 \le v \le 1$. Since the partial derivatives of $z = g(u, v)$ are

$$g_u = 2,\ g_v = 3,$$

by 2, the area of the surface is

$$A(S) = \int_0^1\int_0^1 \sqrt{2^2 + 3^2 + 1}\, dA = \int_0^1\int_0^1 \sqrt{14}\, dudv = \sqrt{14}.\ ■$$

CAS **EXAMPLE 4**

Find the surface area of the portion of the surface $2z = xy$, which is contained within the cylinder $x^2 + y^2 = 4$.

http://matrix.skku.ac.kr/cal-lab/Sec15-6-Exm-8.html

Solution

```
var('x, y, z')
p1=plot3d(1/2*x*y, (x, -2, 2), (y, -2, 2), color='red', opacity=0.5)
p2=implicit_plot3d(x^2+y^2-4==0, (x, -2, 2), (y, -2, 2), (z,-2, 2),
color='blue', opacity=0.5)
show(p1+p2)
```

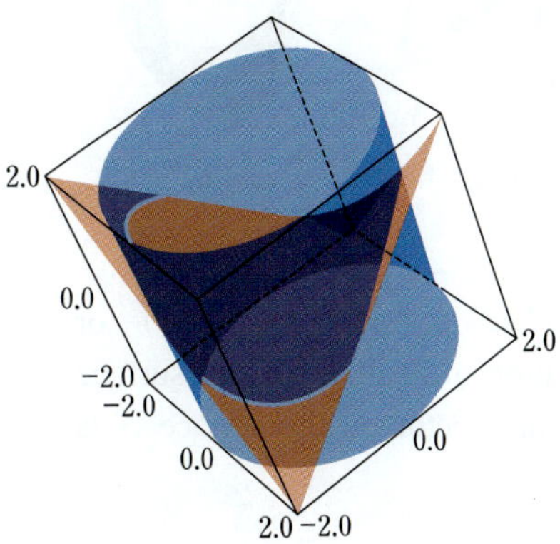

Figure 4

Let $z = f(x, y) = \frac{1}{2}xy$, R: $x^2 + y^2 \le 4$.

Then

$$A(S) = \iint_D \sqrt{1 + (f_x)^2 + (f_y)^2}\, ddA$$

$$= \iint_R \sqrt{1 + (y/2)^2 + (x/2)^2}\, dA$$

$$= \frac{1}{2} \iint_R \sqrt{4 + x^2 + y^2}\, dx dy$$

By the polar coordinate system, letting $x = r\cos\theta$ and $y = r\sin\theta$ with $0 \le \theta \le 2\pi$ and $0 \le r \le 2$,

$$A(S) = \frac{1}{2} \int_0^{2\pi} \int_0^2 \sqrt{4 + r^2}\, r\, dr\, d\theta = \frac{8}{3}\pi(2\sqrt{2} - 1).$$

```
var('r, theta')
1/2*integral(integral(sqrt(4+r^2)*r, r, 0, 2), theta, 0, 2*pi)
```

Answer : 1/3*pi*(16*sqrt(2) - 8) ■

Oriented Surfaces

Figure 5
Möbius strip

A surface is orientable if it has two sides. Then one can orient the surface by choosing one side to be positive side. However some orientable surfaces are not orientable because they have only one side. One classical example is called the Möbius Strip. The Möbius Strip is not orientable because this is a surface that only has one side (see Figure 5.). The Möbius Strip, named after the German mathematician August Möbius (1790-1868).

More precisely, we say a smooth surface S is an oriented surface if there exists a continuous unit normal function $\mathbf{n}$ defined at each point (x, y, z) on the surface. The unit normal vector $\mathbf{n}$ is called an **orientation** of the surface. Note that if $\mathbf{n}$ is an orientation of S, then $-\mathbf{n}$ is also an orientation of S in the opposite side. (See Figure 6 (a).)

A surface S defined by $z = g(x, y)$ has an upward orientation when the unit normals are directed upward, and has downward orientation when the unit normals are directed downward. (See Figure 6 (b), (c).)

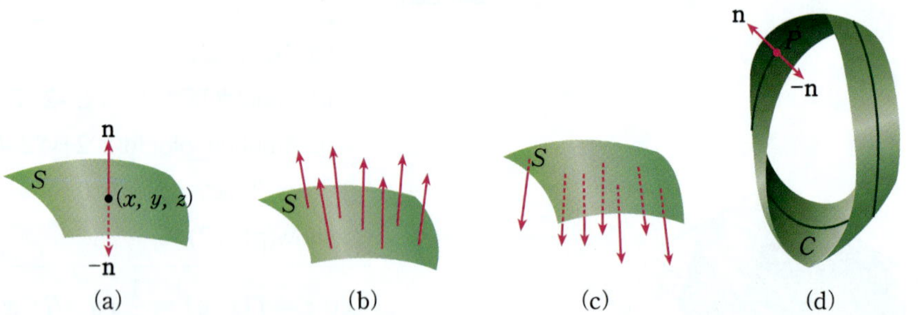

Figure 6

In this case we define a new function $h(x, y, z) = z - g(x, y)$ or $h(x, y, z) = g(x, y) - z$ depending on the orientation of S. If a smooth surface S is defined by $h(x, y, z) = 0$, then the unit normal vector on the surface S is

3 $$\mathbf{n} = \frac{\nabla h}{|\nabla h|} = \frac{-g_x\mathbf{i} - g_y\mathbf{j} + \mathbf{k}}{\sqrt{(g_x)^2 + (g_y)^2 + 1}} \quad \text{or} \quad \frac{g_x\mathbf{i} + g_y\mathbf{j} - \mathbf{k}}{\sqrt{(g_x)^2 + (g_y)^2 + 1}}.$$

If a smooth surface S is defined by a parametric equation

$$\mathbf{r}(u, v) = x(u, v)\mathbf{i} + y(u, v)\mathbf{j} + z(u, v)\mathbf{k},$$

then the unit normal vector on the surface S is

4 $$\mathbf{n} = \frac{\mathbf{r}_u \times \mathbf{r}_v}{|\mathbf{r}_u \times \mathbf{r}_v|},$$

and the opposite orientation is given by $-\mathbf{n} = \dfrac{\mathbf{r}_v \times \mathbf{r}_u}{|\mathbf{r}_v \times \mathbf{r}_u|}$.

EXAMPLE 5

Find two orientations of the surface of the sphere

$$(x-a)^2 + (y-b)^2 + (z-c)^2 = \rho^2 \ (\rho > 0).$$

Solution **Method 1.** Let $h(x, y, z) = (x-a)^2 + (y-b)^2 + (z-c)^2 - \rho^2$. We have the gradient of h and the norm of the gradient of h as follow;

$$\nabla h = 2(x-a)\mathbf{i} + 2(y-b)\mathbf{j} + 2(z-c)\mathbf{k}$$

and

$$|\nabla h| = \sqrt{4(x-a)^2 + 4(y-b)^2 + 4(z-c)^2} = 2\rho.$$

Thus, by 3, two orientations of the surface are

$$\mathbf{n} = \frac{x-a}{\rho}\mathbf{i} + \frac{y-b}{\rho}\mathbf{j} + \frac{z-c}{\rho}\mathbf{k}$$

and

$$\mathbf{n}_1 = -\mathbf{n} = -\frac{x-a}{\rho}\mathbf{i} - \frac{y-b}{\rho}\mathbf{j} - \frac{z-c}{\rho}\mathbf{k}.$$

The vector field $\mathbf{n}$ defines on outward orientation, whereas $\mathbf{n}_1 = -\mathbf{n}$ defines an inward orientation from the sphere.

Method 2. This sphere can be parameterize by parametric equations

$$x(\phi, \theta) = a + \rho\sin\phi\cos\theta, \ y(\phi, \theta) = b + \rho\sin\phi\sin\theta,$$
$$z(\phi, \theta) = c + \rho\cos\phi.$$

Since $\mathbf{r}_\phi \times \mathbf{r}_\theta = \langle \rho^2\sin^2\phi\cos\theta, \rho^2\sin^2\phi\sin\theta, \rho^2\cos\phi\sin\phi \rangle$ and $|\mathbf{r}_\phi \times \mathbf{r}_\theta| = \rho^2\sin\phi$, the unit normal vector of the sphere is

$$\mathbf{n} = \frac{\mathbf{r}_\phi \times \mathbf{r}_\theta}{|\mathbf{r}_\phi \times \mathbf{r}_\theta|} = \langle \sin\phi\cos\theta, \sin\phi\sin\theta, \cos\phi \rangle.$$

Also we have the opposite orientation

$$-\mathbf{n} = \frac{\mathbf{r}_\theta \times \mathbf{r}_\phi}{|\mathbf{r}_\theta \times \mathbf{r}_\phi|} = \langle -\sin\phi\cos\theta, -\sin\phi\sin\theta, -\cos\phi \rangle. \quad \blacksquare$$

15.6 EXERCISES (Surface Area)

http://matrix.skku.ac.kr/Cal-Book/part2/CS-Sec-15-6-Sol.htm

YouTube http://youtu.be/j7F3xVNdHvA

1-3. Find a parametric representation of the surface.

1. $y=3x+5z$

Solution Let the parameters $x=u$ and $z=v$, we find that parametric surface is given by equations $x=u$, $z=v$ and $y=3u+5v$.
Therefore, the plane is described by the vector function $\mathbf{r}(u,v)=u\mathbf{i}+(3u+5v)\mathbf{j}+v\mathbf{k}$.

2. $-x^2+y^2+z^2=4$

Solution Let the parameters $y=u$ and $z=v$, we find that parametric surface is given by equations $y=u$, $z=v$ and $x=\sqrt{u^2+v^2-4}$.
Therefore, the plane is described by the vector function $\mathbf{r}(u,v)=\sqrt{u^2+v^2-4}\,\mathbf{i}+u\mathbf{j}+v\mathbf{k}$.

3. The portion $x^2+z^2=16$ from $y=0$ to $y=4$.

4-5. Find an equation of the tangent plane at the point on the surface corresponding to the given parameter values.

4. $x=u\cos v,\ y=u\sin v,\ z=u^3+v^3,\ u=2,\ v=0$.

Solution At $u=2$ and $v=0$, the point on the surface is $(2,0,8)$. The surface is defined by the vector function $\mathbf{r}(u,v)=u\cos v\mathbf{i}+u\sin v\mathbf{j}+(u^3+3uv+v^3)\mathbf{k}$, then

$$\frac{\partial \mathbf{r}}{\partial u}=\cos v\mathbf{i}+\sin v\mathbf{j}+3u^2\mathbf{k},$$

$$\frac{\partial \mathbf{r}}{\partial v}=-u\sin v\mathbf{i}+u\cos v\mathbf{j}+3v^2\mathbf{k}$$

and $\frac{\partial \mathbf{r}}{\partial u}\times\frac{\partial \mathbf{r}}{\partial v}=\begin{vmatrix}\mathbf{i} & \mathbf{j} & \mathbf{k}\\ \cos v & \sin v & 3u^2\\ -u\sin v & u\cos v & 3v^2\end{vmatrix}$

$=(3v^2\sin v-3u^3\cos v)\mathbf{i}-(3v^2\cos v-3u^3\sin v)\mathbf{j}+u\mathbf{k}$.

The normal vector $-24\mathbf{i}+2\mathbf{k}$ to the surface at $u=2$ and $v=0$. An equation of the tangent plane at that point is $-24(x-2)+0(y-0)+2(z-8)=0$ or $-12x+z+16=0$.

5. $\mathbf{r}(u,v)=4u^3\cos v\mathbf{i}+4u^3\sin v\mathbf{j}+3u^2\mathbf{k}$; $u=2$, $v=\pi/4$.

6-8. Calculate the area of the surface.

6. The surface with parametric equations $x=v^2$, $y=uv$, $z=\frac{1}{2}u^2$, $0\le u\le 3$, $0\le v\le 2$.

Solution $\mathbf{r}=v^2\mathbf{i}+uv\mathbf{j}+\frac{1}{2}u^2\mathbf{k}$

$\frac{\partial \mathbf{r}}{\partial u}=v\mathbf{j}+u\mathbf{k}$ $\frac{\partial \mathbf{r}}{\partial v}=2v\mathbf{i}+u\mathbf{j}$ and then form the cross product $\frac{\partial \mathbf{r}}{\partial u}\times\frac{\partial \mathbf{r}}{\partial v}=\begin{vmatrix}\mathbf{i} & \mathbf{j} & \mathbf{k}\\ 0 & v & u\\ 2v & u & 0\end{vmatrix}=u^2\mathbf{i}+2uv\mathbf{j}-2v^2\mathbf{k}$

$$\left|\frac{\partial \mathbf{r}}{\partial u}\times\frac{\partial \mathbf{r}}{\partial v}\right|=\sqrt{u^4+4u^2v^2+4v^4}=\sqrt{(u^2+2v^2)^2}=u^2+2v^2$$

The area is

$$A(S)=\iint_R\left|\frac{\partial \mathbf{r}}{\partial u}\times\frac{\partial \mathbf{r}}{\partial v}\right|dA=\iint_R(u^2+2v^2)\,dudv=\int_0^2\int_0^3(u^2+2v^2)\,dudv=\int_0^2\left[\frac{1}{3}u^3+2uv^2\right]_0^3=\int_0^2(6v^2+9)\,dv=\left[2v^3+9v\right]_0^2=34.$$

7. The part of the plane $x+2y+3z=6$ that lines in the first octant.

8. The part of the plane $3x+2y+z=6$ tha lines inside the cylinder $x^2+y^2=3$.

15.7 Surface Integrals

YouTube http://youtu.be/nrzIrM4doLo

As we did with line integrals we now need to move on to surface integrals of vector fields.

Surface Integral

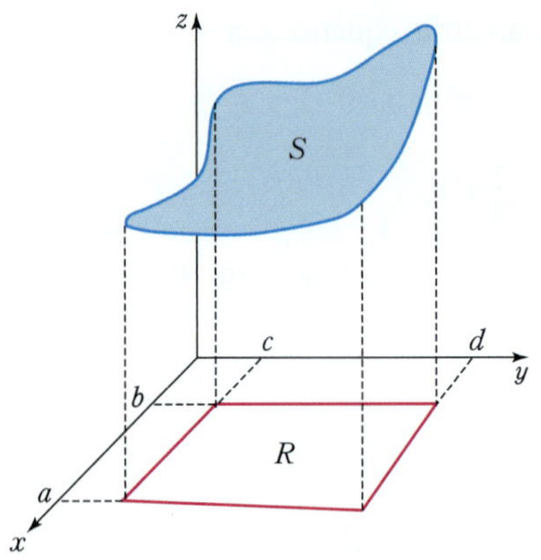

Figure 1

We think about integrating functions over some surface S in three-dimensional space. The region S will lie above some region R that lies in the xy-plane. Also note that we could just as easily look at a surface S that was in front of some region R in the yz-plane or the xz-plane.

If the surface is given by $z=g(x,y)$ and $f(x,y,z)$ is a continuous function, then the surface integral is

1 $$\iint_S f(x,y,z)dS=\iint_R f(x,y,g(x,y))\sqrt{\left(\frac{\partial g}{\partial x}\right)^2+\left(\frac{\partial g}{\partial y}\right)^2+1}\,dA\,.$$

The surface integral is similar to areas of surfaces in section 15.6.

Next we have the definition of surface integral for parametric surface.

DEFINITION 1

Let S be a smooth surface given the vector valued function

$$\mathbf{r}(u,v)=x(u,v)\mathbf{i}+y(u,v)\mathbf{j}+z(u,v)\mathbf{k},\ (u,v)\in R$$

and f be a continuous function.

Then the surface integral of the function f over S is

2 $$\iint_S f(x,y,z)dS=\iint_R f(x(u,v),y(u,v),z(u,v))\left|\frac{\partial \mathbf{r}}{\partial u}\times\frac{\partial \mathbf{r}}{\partial v}\right|dA.$$

Remark If $f(x,y,z)=1$, the integral is the surface area of S;

$$A(S)=\iint_R\left|\frac{\partial \mathbf{r}}{\partial u}\times\frac{\partial \mathbf{r}}{\partial v}\right|dA.$$

EXAMPLE 1

Evaluate the surface integral $\iint_S 6xy\,dS$ where S is the portion of the plane $x+y+z=1$ that lies in front of the xy-plane.

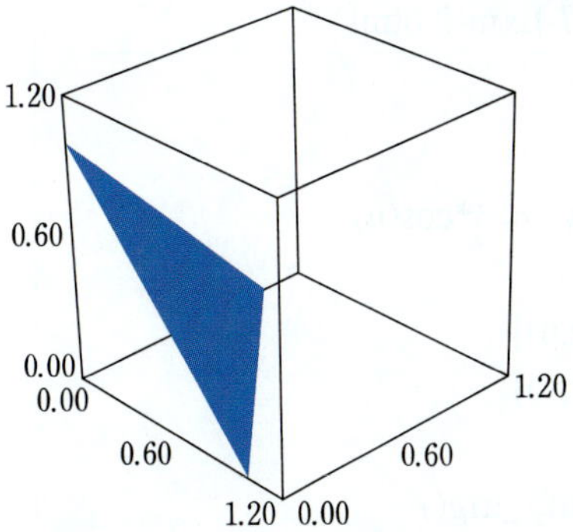

Figure 2

Solution For using [1], we need to write the equation of the surface in the form $z=g(x, y)$. This is $z=1-x-y$. (See Figure 2) Next we need to determine just what R is. Then R is the projection of S on xy-plane, i.e., $R=(x, y)|0 \le x \le 1, 0 \le y \le 1-x$.

Since $\dfrac{\partial g}{\partial x}=-1$ and $\dfrac{\partial g}{\partial y}=-1$,

$$\iint_S 6xy\,dS=\int_0^1\int_0^{1-x} 6xy \cdot \sqrt{3}\,dy\,dx$$
$$=\int_0^1 6\sqrt{3}(x^3-2x^2+x)dx=\frac{\sqrt{3}}{4}.$$ ■

http://matrix.skku.ac.kr/cal-lab/Sec15-7-Exm-1.html

```
var('x,y,z')
implicit_plot3d(x+y+z==1,(x,0,1.2),(y,0,1.2),(z,0,1.2))
```

EXAMPLE 2

Evaluate $\iint_S \sqrt{x^2+y^2+z^2}\,dS$, where S is the sphere $x^2+y^2+z^2=9$.

Solution For the sphere, the parametric equations are $x=3\sin u\cos v$, $x=3\sin u\sin v$ and $z=3\cos u$, for $0 \le u \le \pi$ and $0 \le v \le 2\pi$. The vector function is

$$\mathbf{r}(u, v)=3\sin u\cos v\mathbf{i}+3\sin u\sin v\mathbf{j}+3\cos u\mathbf{k}.$$

Since the normal vector is

$$\frac{\partial \mathbf{r}}{\partial u}\times\frac{\partial \mathbf{r}}{\partial v}=\begin{vmatrix} \mathbf{i} & \mathbf{j} & \mathbf{k} \\ 3\cos u\cos v & 3\cos u\sin v & -3\sin u \\ -3\sin u\sin v & 3\sin u\cos v & 0 \end{vmatrix}$$
$$=9\sin^2 u\cos v\mathbf{i}+9\sin^2 u\sin v\mathbf{j}+9\cos u\sin u\mathbf{k},$$
$$\left|\frac{\partial \mathbf{r}}{\partial u}\times\frac{\partial \mathbf{r}}{\partial v}\right|=\sqrt{81\sin^4 u\cos^2 v+81\sin^4 u\sin^2 v+81\cos^2 u\sin^2 u}$$
$$=9\sin u.$$

Then $$\iint_S \sqrt{x^2+y^2+z^2}\,dS=\iint_R 27\sin u\,dA$$
$$=\int_0^{2\pi}\int_0^{\pi} 27\sin u\,du\,dv=108\pi.$$

http://matrix.skku.ac.kr/cal-lab/Sec15-7-Exm-2.html

```
var('x,y,z,u,v, n, nn')
assume( sin(u)>0 )
x=3*sin(u)*cos(v);y=3*sin(u)*sin(v);z=3*cos(u)
f=sqrt(x^2+y^2+z^2).simplify_trig()
r_u=vector([diff(x,u),diff(y,u),diff(z,u)])
r_v=vector([diff(x,v),diff(y,v),diff(z,v)])
n=r_u.cross_product(r_v)
nn=(n.norm().simplify_full()).simplify_trig()
t=f*nn
integral(integral(27*sin(u), u, 0, pi), v, 0, 2*pi)
```

Answer : 108*pi ■

The concept of surface integral is a generalization of a double integral $\iint_R f(x, y)dx\,dy$ taken over a plane region R. In a surface integral $f(x, y)$ is integrated over a curved surface.

Let S be a two-sided surface with one side of S taken arbitrarily as the positive side (the outer side if S is closed) in Figure 3. A unit normal $\mathbf{n}$ at any point of the positive side of S is known as the positive outward unit normal.

In the $\mathbb{R}^3$, the equation of a surface S is given by

$$g(x, y, z) = 0$$

with the unit normal $\mathbf{n} = \dfrac{\nabla g}{|\nabla g|}$.

When S is represented in parametric form as $\mathbf{r}(u, v) = x(u, v)\mathbf{i} + y(u, v)\mathbf{j} + z(u, v)\mathbf{k}$ with the two parameters u and v varying in a region R of uv-plane, then the unit normal $\mathbf{n}$ to S at P is given by

$$\mathbf{n} = \frac{\mathbf{N}}{|\mathbf{N}|}$$

where $\mathbf{N} = \mathbf{r}_u \times \mathbf{r}_v$.

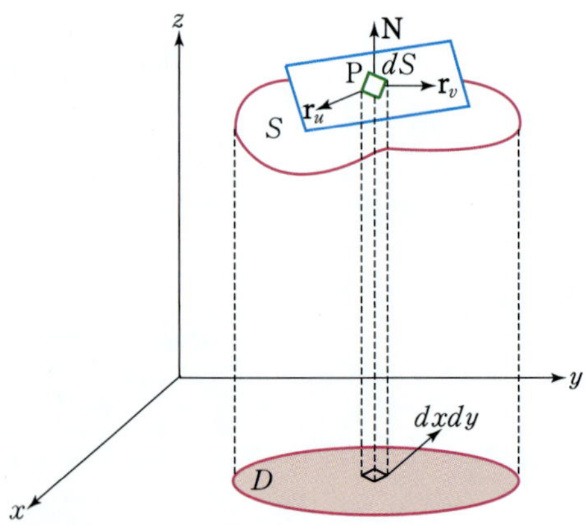

Figure 3

The surface integral of a continuous vector function $\mathbf{F}$ taken over an oriented surface S with unit normal $\mathbf{n}$ is defined as

$$\boxed{3} \qquad \iint_S \mathbf{F} \cdot d\mathbf{S} = \iint_S \mathbf{F} \cdot \mathbf{n}\, dS.$$

This integral is called **the flux of** $\mathbf{F}$ across S.

If $\mathbf{r}(u, v) = x(u, v)\mathbf{i} + y(u, v)\mathbf{j} + z(u, v)\mathbf{k}$ is a parameterization of the surface S, then

$$\boxed{4} \qquad \iint_S \mathbf{F} \cdot \mathbf{n}\, dS = \iint_R \mathbf{F}(\mathbf{r}(u, v)) \cdot \left(\frac{\partial \mathbf{r}}{\partial u} \times \frac{\partial \mathbf{r}}{\partial v}\right) dA.$$

where R is the parameter domain.

Let S be the surface of the function $z = f(x, y)$ on D in xy-plane and $\mathbf{F} = \mathbf{F}_1\mathbf{i} + \mathbf{F}_2\mathbf{j} + \mathbf{F}_3\mathbf{k}$ be a vector field.

Since $\mathbf{r}(x, y) = x\mathbf{i} + y\mathbf{j} + f(x, y)\mathbf{k}$, $\mathbf{r}_x \times \mathbf{r}_y = -f_x\mathbf{i} - f_y\mathbf{j} + \mathbf{k}$.

Thus

$$\boxed{5} \qquad \iint_S \mathbf{F} \cdot dS = \iint_D \left(-\mathbf{F}_1 \frac{\partial f}{\partial x} - \mathbf{F}_2 \frac{\partial f}{\partial y} + \mathbf{F}_3\right) dx\,dy.$$

Flux

Suppose the velocity of a fluid in $\mathbb{R}^3$ is described by the vector field $\mathbf{F}(x, y, z)$. Let S be a surface in $\mathbb{R}^3$. The flux across S is the volume of fluid crossing S per unit time. Figure 4 below shows a surface S and the vector field $\mathbf{F}$ at various points on the surface.

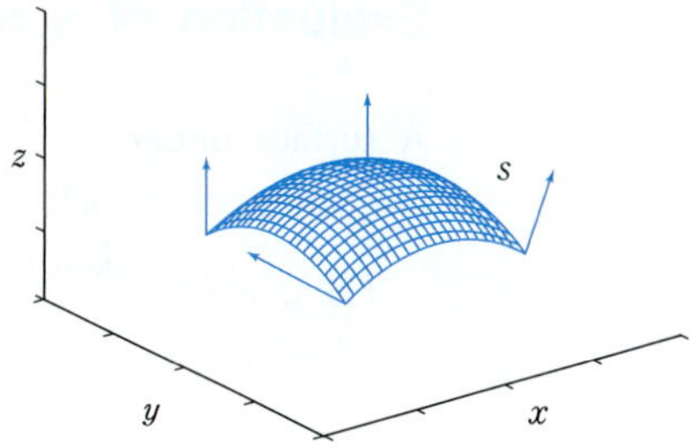

Figure 4

What is the formula for the flux? Let us consider the flux through an infinitesimal piece of the surface with area dS above the point (x_0, y_0) in the xy-plane. (The surface is denoted by the dotted region.) Let $\mathbf{n}$ denote the unit normal vector to the surface. Let us suppose that the velocity vector of the fluid is $\mathbf{F}(x_0, y_0, z_0)$. The normal component $\mathbf{F} \cdot \mathbf{n}$ is a scalar. Then the flux of $\mathbf{F}$ represents the total quantity of fluid flowing in unit time through (across) the surface S in the positive direction. The flux of $\mathbf{F}$ across S is given by the surface integral

$$\text{Flux of } \mathbf{F} \text{ across } S = \iint_S \mathbf{F} \cdot \mathbf{n}\, dS.$$

EXAMPLE 3

Find the flux of the vector field $\mathbf{F}(x, y, z) = y\mathbf{i} + x\mathbf{j} + z\mathbf{k}$ across the sphere $x^2 + y^2 + z^2 = 4$.

Solution Using the spherical coordinate system, the parametric equations

$$x = 2\sin\phi\cos\theta,\ y = 2\sin\phi\sin\theta,\ z = 2\cos\phi,$$

with $0 \le \phi \le \pi$ and $0 \le \theta \le 2\pi$.

Then $\mathbf{F} = 2\sin\phi\cos\theta\,\mathbf{i} + 2\sin\phi\sin\theta\,\mathbf{j} + 2\cos\phi\,\mathbf{k}$ and

$$\frac{\partial \mathbf{r}}{\partial \phi} \times \frac{\partial \mathbf{r}}{\partial \theta} = 4\sin^2\phi\cos\theta\,\mathbf{i} + 4\sin^2\phi\sin\theta\,\mathbf{j} + 4\cos\phi\sin\phi\,\mathbf{k}.$$

$$\begin{aligned}
\iint_S \mathbf{F} \cdot \mathbf{n}\, dS &= \iint_R \mathbf{F}(\mathbf{r}(\phi, \theta)) \cdot \left(\frac{\partial \mathbf{r}}{\partial \phi} \times \frac{\partial \mathbf{r}}{\partial \theta}\right) d\phi d\theta \\
&= \int_0^{2\pi} \int_0^{\pi} 8\sin^3\phi\sin 2\theta + 8\cos^3\phi\sin\phi\, d\phi d\theta \\
&= \int_0^{2\pi} \sin 2\theta\, d\theta \int_0^{\pi} 8\sin^3\phi\, d\phi + \int_0^{2\pi} d\theta \int_0^{\pi} 8\cos^2\phi\sin\phi\, d\phi \\
&= 0 + 2\pi \times \frac{16}{3} \\
&= \frac{32\pi}{3}.
\end{aligned}$$

■

Evaluation of a Surface Integral

A surface integral is evaluated by reducing it to a double integral. Let D be the projection of S onto the xy-plane.

Then, $dS = \dfrac{dxdy}{|\mathbf{n} \cdot \mathbf{k}|}$ and

6 $$\iint_S \mathbf{F} \cdot \mathbf{n}\, dS = \iint_D \mathbf{F} \cdot \mathbf{n} \frac{dx\, dy}{|\mathbf{n} \cdot \mathbf{k}|}.$$

where $\mathbf{n}$ is the unit outward drawn normal to S. The double integral on the right in x, y over the plane region D is evaluated as an iterated integral.

In a similar way the surface integral can be evaluated by projecting S onto the yz-plane as D_1 and xz-plane as D_2 as follows

7 $$\iint_S \mathbf{F} \cdot \mathbf{n}\, dS = \iint_{D_1} \mathbf{F} \cdot \mathbf{n} \frac{dy\, dz}{|\mathbf{n} \cdot \mathbf{i}|},$$

8 $$\iint_S \mathbf{F} \cdot \mathbf{n}\, dS = \iint_{D_2} \mathbf{F} \cdot \mathbf{n} \frac{dx\, dz}{|\mathbf{n} \cdot \mathbf{j}|}.$$

EXAMPLE 4

Evaluate the surface integral $\iint_S \mathbf{F} \cdot \mathbf{n}\, dS$ where $\mathbf{F} = 6z\mathbf{i} + (2x+y)\mathbf{j} - x\mathbf{k}$ and S is the surface bounded by the cylinder $x^2 + z^2 = 9$, $x = 0$, $y = 0$, $z = 0$ and $y = 8$. (See Figure 5.)

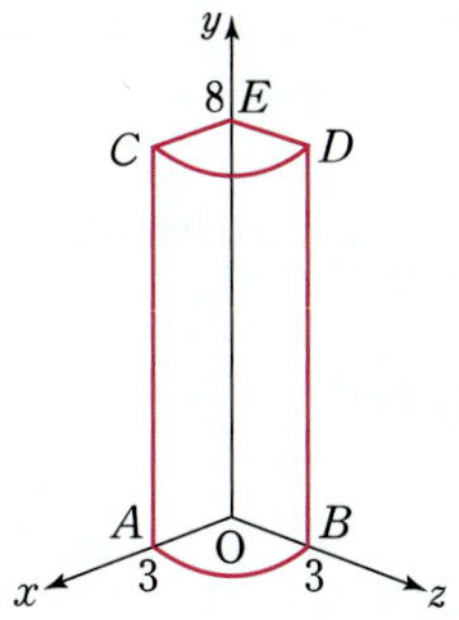

Figure 5

Solution The entire surface S consists of 5 surfaces: S_1 is the curved (lateral) surface of the cylinder $(ABDCA)$, S_2 is the plane $(AOEC)$, S_3 is the plane $(OBDE)$, S_4 is the quarter of circle (CDE), S_5 is the quarter of circle (CDE). Then

$$\iint_S \mathbf{F} \cdot \mathbf{n}\, dS = \iint_{S_1 + S_2 + \cdots + S_5} = \iint_{S_1} + \iint_{S_2} + \cdots + \iint_{S_5} = SI_1 + SI_2 + SI_3 + SI_4 + SI_5.$$

We will use [6] - [8] to evaluate the surface integrals of S_i, $i = 1, \cdots, 5$.

On the curved (lateral) surface S_1 of the cylinder, since the unit normal vector is

$$\mathbf{n} = \frac{\nabla(x^2+z^2)}{|\nabla(x^2+z^2)|} = \frac{2x\mathbf{i} + 2z\mathbf{k}}{\sqrt{4x^2+4z^2}} = \frac{x\mathbf{i} + z\mathbf{k}}{3},$$

$\mathbf{F} \cdot \mathbf{n} = \frac{5}{3}xz$ and $\mathbf{n} \cdot \mathbf{k} = \frac{z}{3}$. And the projection of S_1 onto xy-plane is $D = \{(x, y) | 0 \le x \le 3,\ 0 \le y \le 8\}$. Then

$$SI_1 = \iint_{S_1} \mathbf{F} \cdot \mathbf{n}\, dS = \iint \mathbf{F} \cdot \mathbf{n} \frac{dx\, dy}{|\mathbf{n} \cdot \mathbf{k}|} = \int_0^8 \int_0^3 5x\, dx\, dy = 180.$$

On the plane S_2, since $z = 0$ and $\mathbf{n} = -\mathbf{k}$, $\mathbf{F} \cdot \mathbf{n} = x$ and $\mathbf{n} \cdot \mathbf{k} = -1$. Then

$$SI_2 = \iint_{S_2} \mathbf{F} \cdot \mathbf{n}\, dS = \int_0^8 \int_0^3 x\, dx\, dy = 36.$$

On the plane S_3, since $x = 0$ and $\mathbf{n} = -\mathbf{i}$, $\mathbf{F} \cdot \mathbf{n} = -6z$ and $\mathbf{n} \cdot \mathbf{i} = -1$. Then

$$SI_3 = \iint_{S_3} \mathbf{F} \cdot \mathbf{n}\, dS = \int_0^8 \int_0^3 -6z\, dz\, dy = -216.$$

On the plane S_4, since $y = 0$ and $\mathbf{n} = -\mathbf{j}$, $\mathbf{F} \cdot \mathbf{n} = -2x$ and $\mathbf{n} \cdot \mathbf{j} = -1$. Then

$$SI_4 = \iint_{S_4} \mathbf{F} \cdot \mathbf{n}\, dS = \iint_{OAB} -2x\, dx\, dz$$
$$= \int_0^3 \int_0^{\sqrt{9-z^2}} -2x\, dx\, dy$$
$$= -18.$$

On the plane S_5, since $y = 8$ and $\mathbf{n} = \mathbf{j}$, $\mathbf{F} \cdot \mathbf{n} = 2x + 8$ and $\mathbf{n} \cdot \mathbf{j} = 1$. Then

$$SI_5 = \iint_{S_5} \mathbf{F} \cdot \mathbf{n}\, dS = \iint_{CDE} (2x+8) dx\, dz$$
$$= \int_0^3 \int_0^{\sqrt{9-z^2}} (2x+8)\, dx\, dy$$
$$= 18 + 18\pi.$$

Thus the required surface integral is

$$\iint_S \mathbf{F} \cdot \mathbf{n}\, dS = 180 + 36 - 216 - 18 + 18 + 18\pi$$
$$= 18\pi.$$

■

15.7 EXERCISES (Surface Integrals)

http://matrix.skku.ac.kr/Cal-Book/part2/CS-Sec-15-7-Sol.htm

YouTube http://youtu.be/s_MRgW2By38

1. Evaluate the surface integral $\iint_S \sqrt{1+x^2+y^2+z^2}\, dS$, where S is the surface defined by the vector function $\mathbf{r}(u, v) = a\sin u\cos v\,\mathbf{i} + a\sin u\sin v\,\mathbf{j} + a\cos u\,\mathbf{k}$, where $0 \le u \le \pi$, $0 \le v \le 2\pi$.

Solution The graph of $\mathbf{r}(u,v)$ is called sphere. Substituting $x = a\sin u\cos v$, $y = a\sin u\sin v$ and $z = a\cos u$ into the integrand and we have

$$\sqrt{1+x^2+y^2+z^2} = \sqrt{1+a^2\sin u^2\cos^2 v + a^2\sin^2 u\sin^2 v + a^2\cos^2 u} = \sqrt{1+a^2}.$$

$\frac{\partial \mathbf{r}}{\partial u} = a\cos u\cos v\,\mathbf{i} + a\cos u\sin v\,\mathbf{j} - a\sin u\,\mathbf{k}$ and

$\frac{\partial \mathbf{r}}{\partial v} = -a\sin u\sin v\,\mathbf{i} + a\sin u\cos v\,\mathbf{j}$

$$\frac{\partial \mathbf{r}}{\partial u} \times \frac{\partial \mathbf{r}}{\partial v} = \begin{vmatrix} \mathbf{i} & \mathbf{j} & \mathbf{k} \\ a\cos u\cos v & a\cos u\sin v & a\sin u \\ -a\sin u\sin v & a\sin u\cos v & 0 \end{vmatrix}$$

$$= -a^2\sin^2 u\cos v\,\mathbf{i} + a^2\sin^2 u\sin v\,\mathbf{j} + a^2\cos u\sin u\,\mathbf{k}$$

$$\left|\frac{\partial \mathbf{r}}{\partial u} \times \frac{\partial \mathbf{r}}{\partial v}\right| = \sqrt{a^4\sin^4 u\cos^2 v + a^4\sin^4 u\sin^2 v + a^4\cos^2 u\sin^2 u} = a^2\sin u$$

$$\iint_R \sqrt{1+x^2+y^2+z^2}\, dA = \iint_R \sqrt{1+a^2}\, a^2\sin u\, du\, dv = a^2\sqrt{1+a^2}\int_0^{2\pi}\int_0^{\pi} \sin u\, du\, dv = 4\pi a^2\sqrt{1+a^2}.$$

2. Evaluate $\iint_S yz\, dS$, where S is the surface with parametric equations $x = 2v$, $y = u\sin v$ and $z = u\cos v$ $0 \le u \le 1$, $0 \le v \le \pi$.

3. Evaluate $\iint_S xy\, dS$, where S is the part of the plane $x + 2y + 3z = 6$ that lies in the first octant.

4. Evaluate $\iint_S z\, dS$, where S is the hemisphere $z = \sqrt{9 - x^2 - y^2}$.

5. Evaluate $\iint_S \sqrt{z}\, dS$, where S is the portion of the paraboloid $z = 4 - x^2 - y^2$ between $z = 0$ and $z = \sqrt{3}$.

6. Evaluate $\iint_S \mathbf{A} \cdot \mathbf{n}\, dS$ where $\mathbf{A} = 18z\mathbf{i} - 12\mathbf{j} + 3y\mathbf{k}$ and S is that part of the plane $2x + 3y + 6z = 12$ which is located in the first octant.

Solution Here S can be thought of as a graph surface of a function $f(x,y) = \frac{12-2x-3y}{6}$. The unit normal to this surface is given by

$$\mathbf{n} = \frac{\langle -f_x, -f_y, 1\rangle}{\sqrt{1+f_x^2+f_y^2}} = \frac{\frac{1}{3}\mathbf{i} + \frac{1}{2}\mathbf{j} + \mathbf{k}}{\sqrt{\frac{49}{36}}} = \frac{6}{7}\left(\frac{1}{3}\mathbf{i} + \frac{1}{2}\mathbf{j} + \mathbf{k}\right).$$

Thus $A \cdot \mathbf{n} = \frac{6}{7}(6z - 6 + 3y)$.

Hence $\iint_S \mathbf{A} \cdot \mathbf{n}\, dS = \iint_D \mathbf{A} \cdot \mathbf{n} \frac{dx dy}{|\mathbf{n} \cdot \mathbf{k}|}$

$$= \int_0^4 \int_0^{\frac{12-3y}{2}} 6z - 6 + 3y dx dy$$

$$= \int_0^4 -\frac{9}{4}y^2 + 9y\, dy = 24.$$

7. If S is the entire surface of the cube bounded by $x = 0$, $x = b$, $y = 0$, $y = b$, $z = 0$, and $z = b$ and $\mathbf{A} = 4xz\mathbf{i} - y^2\mathbf{j} + yz\mathbf{k}$ then evaluate $\iint_S \mathbf{A} \cdot \mathbf{n}\, dS$.

Solution Here S can be spilt into six surfaces S_1, S_2, $\cdots$, S_6 as shown in the figure. It is easy to see that normal to S_1, S_2, $\cdots$, S_6 are $-\mathbf{k}$, $\mathbf{i}$, $\mathbf{j}$, $-\mathbf{i}$, $-\mathbf{j}$, $\mathbf{k}$ respectively.

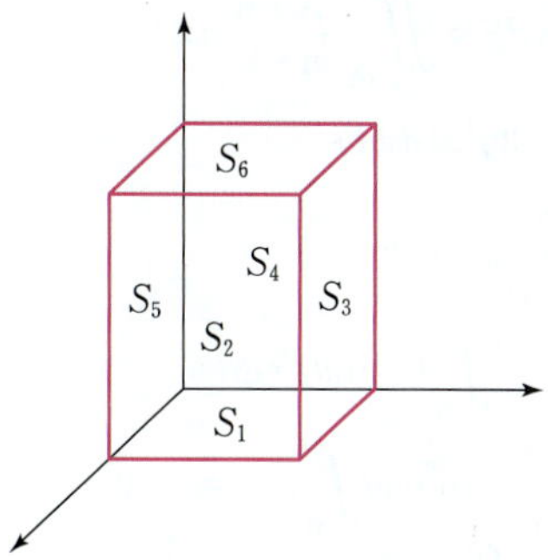

$$\iint_{S_1} \mathbf{A}\cdot\mathbf{n}\,dS = \iint_{D_1} (-yz)\,dxdy = \iint_{D_1} 0dxdy = 0$$

$$\iint_{S_2} \mathbf{A}\cdot\mathbf{n}dS = \iint_{D_2} 4xz\,dydz = \int_0^b\int_0^b 4bzdydz = 2b^4$$

$$\iint_{S_4} \mathbf{A}\cdot\mathbf{n}dS = \iint_{D_4} (-4xz)\,dydz = \iint_{D_4} 0\,dydz = 0$$

Therefore $\iint_S \mathbf{A}\cdot\mathbf{n}dS = \dfrac{3}{2}b^4$.

8. Let S be the surface of the cylinder $x^2+y^2=16$ included in the first octant between $z=0$ and $z=5$.

Evaluate $\iint_S \mathbf{A}\cdot\mathbf{n}dS$ where $\mathbf{A}= z\mathbf{i}+x\mathbf{j}-3y^2z\mathbf{k}$.

Solution We spilt S into five pieces S_1, S_2, $\cdots$, S_5 as shown in the figure.

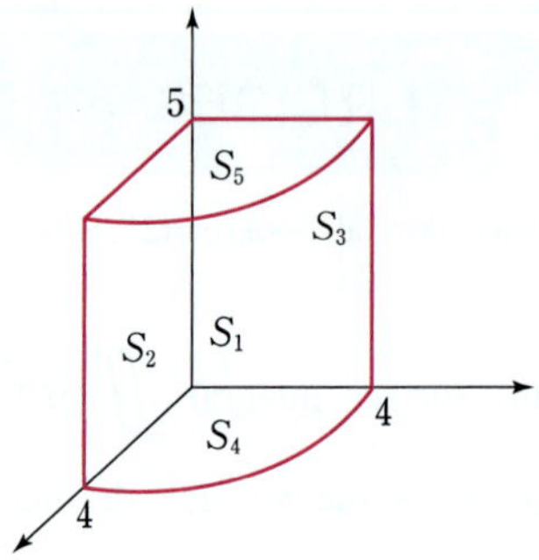

$$\iint_{S_1} \mathbf{A}\cdot\mathbf{n}dS = \iint_{D_1} \frac{\mathbf{A}\cdot\mathbf{n}}{|\mathbf{n}\cdot\mathbf{j}|}dxdz$$

$$= \int_0^5\int_0^4 \frac{xz}{y}+x\,dxdz$$

$$= \int_0^5\int_0^4 \frac{xz}{\sqrt{16-x^2}}dxdz + \int_0^5\int_0^4 x\,dxdz$$

$$= \int_0^5 zdz\int_0^4 \frac{x}{\sqrt{16-x^2}}dx + \int_0^5 dz\int_0^4 xdx$$

$$= \frac{25}{2}\int_{16}^0 \frac{1}{-2\sqrt{t}}dt + 5\cdot\frac{16}{2}$$

$$= 50+40 = 90$$

$$\iint_{S_2} \mathbf{A}\cdot\mathbf{n}dS = \iint_{D_2} \frac{\mathbf{A}\cdot\mathbf{n}}{|\mathbf{n}\cdot\mathbf{j}|}dxdz$$

$$= \int_0^5\int_0^4 (-x)dxdz = -40,$$

$$\iint_{S_3} \mathbf{A}\cdot\mathbf{n}dS = \iint_{D_3} \frac{\mathbf{A}\cdot\mathbf{n}}{|\mathbf{n}\cdot\mathbf{i}|}dxdz$$

$$= \int_0^5\int_0^4 (-z)dydz = -50$$

$$\iint_{S_4} \mathbf{A}\cdot\mathbf{n}dS = \iint_{D_4} \frac{\mathbf{A}\cdot\mathbf{n}}{|\mathbf{n}\cdot\mathbf{k}|}dxdy$$

$$= \iint_{D_4} 3y^2zdxdy = 0,$$

$$\iint_{S_5} \mathbf{A}\cdot\mathbf{n}dS = \iint_{D_4} \frac{\mathbf{A}\cdot\mathbf{n}}{|\mathbf{n}\cdot\mathbf{k}|}dxdy$$

$$= \iint_{D_5} (-3y^2z)\,dxdy$$

$$= -15\iint_{D_5} y^2dxdy$$

$$= -15\int_0^{\pi/2}\int_0^4 (r\sin\theta)^2 r\,drd\theta$$

$$= -15\int_0^{\pi/2} \sin^2\theta d\theta\int_0^4 r^3dr = -240\pi.$$

Therefore $\iint_S \mathbf{A}\cdot\mathbf{n}dS = -240\pi$.

9. Find the surface integral over the parallelepiped $x=0$, $y=0$, $z=0$, $x=1$, $y=2$, $z=3$ when $\mathbf{A} = 2xy\mathbf{i}+yz^2\mathbf{j}+xz\mathbf{k}$.

Solution

$$\iint_{S_1} \mathbf{A}\cdot\mathbf{n}dS = \iint_{D_1} 2xy\,dydz = 2\int_0^3\int_0^2 y\,dydz = 12,$$

$$\iint_{S_2} \mathbf{A}\cdot\mathbf{n}dS = \iint_{D_2} yz^2\,dxdz = 2\int_0^3\int_0^1 z^2dxdz = 18,$$

$$\iint_{S_3} \mathbf{A}\cdot\mathbf{n}dS = \iint_{D_3} (-2xy)\,dxdz = 0,$$

$$\iint_{S_4} \mathbf{A}\cdot\mathbf{n}dS = \iint_{D_4} (-yz^2)\,dxdy = 0,$$

$$\iint_{S_5} \mathbf{A}\cdot\mathbf{n}dS = \iint_{D_5} (-xz)\,dxdy = 0,$$

$$\iint_{S_6} \mathbf{A}\cdot\mathbf{n}dS = \iint_{D_6} xz\,dxdz = 3\int_0^3\int_0^1 xdxdz = 3.$$

Therefore $\iint_S \mathbf{A}\cdot\mathbf{n}dS = 33$.

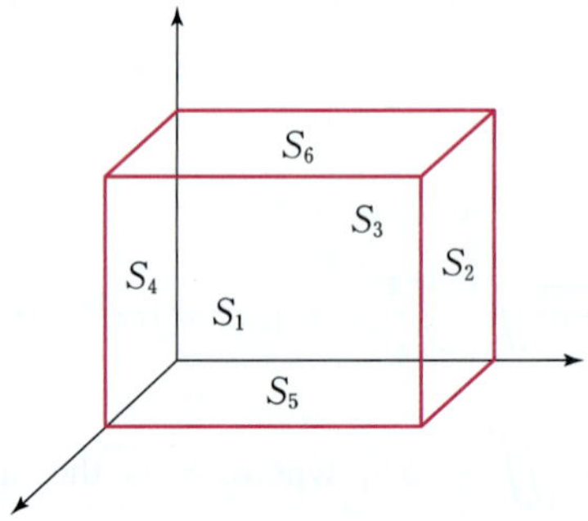

10. If S is the surface of the sphere $x^2+y^2+z^2=d^2$ and $\mathbf{A} = ax\mathbf{i}+by\mathbf{j}+cz\mathbf{k}$, evaluate $\iint_S \mathbf{A}\cdot\mathbf{n}dS$.

Solution The sphere can be given a parameterization
$\mathbf{r}(\phi,\theta) = \langle d\sin\phi\cos\theta,\ d\sin\phi\sin\theta,\ d\cos\theta\rangle$,
$A(\mathbf{r}(\phi,\theta)) = \langle ad\sin\phi\cos\theta,\ bd\sin\phi\sin\theta,\ cd\cos\theta\rangle$,
$\mathbf{r}_\phi\times\mathbf{r}_\theta = \langle d^2\sin^2\phi\cos\theta,\ d^2\sin^2\phi\sin\theta,\ d^2\sin\phi\cos\phi\rangle$

$$\iint_S \mathbf{A}\cdot\mathbf{n}dS = \iint_D \mathbf{A}\cdot(\mathbf{r}_\phi\times\mathbf{r}_\theta)dA$$

11. Evaluate $\iint_S \mathbf{F} \cdot d\mathbf{S}$.

Here $\mathbf{F} = y\mathbf{i} + x\mathbf{j} + z\mathbf{k}$ and S is the boundary of the solid surrounded by $z = 4 - x^2 - y^2$ and $z = 0$.

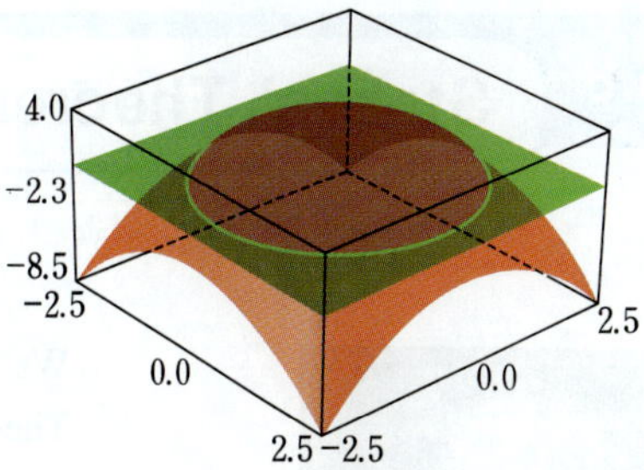

Solution

1) S_1 : There S is a graph surface of $z = g(x, y)$ $= 2 - x^2 - y^2$ on $x^2 + y^2 \le 4$.

We have $\mathbf{F}_1 = y$, $\mathbf{F}_2 = x$, $\mathbf{F}_3 = z$ and

$\frac{\partial g}{\partial x} = -2x$, $\frac{\partial g}{\partial y} = -2y$.

Hence by 4,

$$\iint_S \mathbf{F} \cdot d\mathbf{S} = \iint_D \left(-\mathbf{F}_1 \frac{\partial g}{\partial x} - \mathbf{F}_2 \frac{\partial g}{\partial y} + \mathbf{F}_3 \right) dA$$

$$= \int_D \int [-y(-2x) - x(-2y) + 4 - x^2 - y^2]\, dA$$

$$= \int_D \int (4 + 4xy - x^2 - y^2)\, dA$$

$$= \int_0^{2\pi} \int_0^2 (4 + 4r^2 \cos\theta \sin\theta - r^2) r\, dr\, d\theta$$

$$= \int_0^{2\pi} \left(4 + \frac{32}{3} \cos\theta \sin\theta \right) d\theta = 8\pi.$$

2) S_2 : $\iint_{S_2} \mathbf{F} \cdot dS = \iint_{S_2} \mathbf{F} \cdot (-\mathbf{k})\, dS$

$$= \iint_D (-z)\, dA = \iint_D 0\, dA$$

$$= 0$$

3) $S = S_1 + S_2$:

$$\iint_S \mathbf{F} \cdot dS = \iint_{S_1} \mathbf{F} \cdot dS + \iint_{S_2} \mathbf{F} \cdot dS.$$

$$= 8\pi + 0 = 8\pi.$$

15.8 Stokes' Theorem

http://youtu.be/t4skc_PzJvg

Sir George Gabriel Stokes (1819–1903)
Irish mathematician

We introduce another important theorem of vector calculus, namely Stokes' Theorem. Stokes' Theorem can be thought of as an extension of Green's Theorem. Green's Theorem relates the integral over a plane region D to a line integral on the boundary of D (with positive orientation). Stokes' Theorem on the other hands relates a surface integral over an orientable surface S to a line integral over the boundary of S. The orientation on S is taken as positive.

THEOREM 1 Stokes' Theorem

Let $\mathbf{F} = P\mathbf{i} + Q\mathbf{j} + R\mathbf{k}$ be a vector field whose components have continuous second order partial derivatives in a domain containing a surface S bounded by a simple closed curve C. Then

1 $$\int_C \mathbf{F} \cdot d\mathbf{r} = \iint_S \text{curl}\,\mathbf{F} \cdot d\mathbf{S}$$

where C is traversed in the positive direction.

We shall prove this theorem for special class of surfaces namely a graph surface. A general proof of this theorem is beyond our scope.

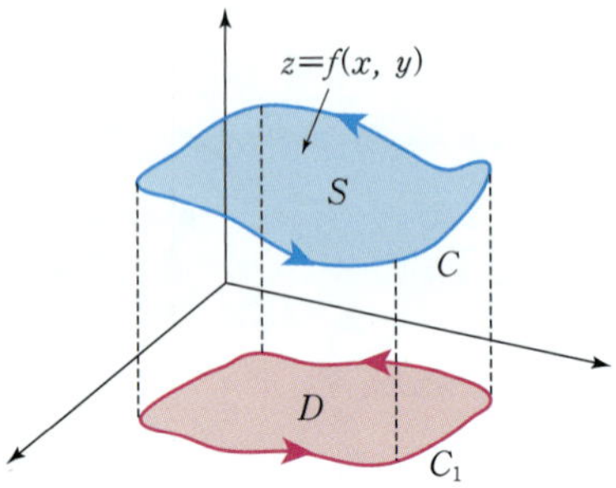

Figure 1

Proof See Figure 1. Let S be a surface as $z = f(x, y)$, $(x, y) \in R$. Then $\mathbf{r}(x, y) = x\mathbf{i} + y\mathbf{j} + z\mathbf{k}$ is a position vector of S.
Since

$$\text{curl}\ \mathbf{F} = \begin{vmatrix} \mathbf{i} & \mathbf{j} & \mathbf{k} \\ \dfrac{\partial}{\partial x} & \dfrac{\partial}{\partial y} & \dfrac{\partial}{\partial z} \\ P & Q & R \end{vmatrix} = (R_y - Q_z)\mathbf{i} - (R_x - P_z)\mathbf{j} + (Q_x - P_y)\mathbf{k}$$

and $dS = \left(\dfrac{\partial \mathbf{r}}{\partial x} \times \dfrac{\partial \mathbf{r}}{\partial y}\right) dx\,dy = \left(\dfrac{\partial \mathbf{r}}{\partial x} \times \dfrac{\partial \mathbf{r}}{\partial y}\right) dA$ with

$$\frac{\partial \mathbf{r}}{\partial x} \times \frac{\partial \mathbf{r}}{\partial y} = \begin{vmatrix} \mathbf{i} & \mathbf{j} & \mathbf{k} \\ 1 & 0 & z_x \\ 0 & 1 & z_y \end{vmatrix} = -z_x\mathbf{i} - z_y\mathbf{j} + \mathbf{k},$$

2 $$\begin{aligned} \iint_S \text{curl}\,\mathbf{F} \cdot d\mathbf{S} &= \iint_D \text{curl}\,\mathbf{F} \cdot \left(\frac{\partial \mathbf{r}}{\partial x} \times \frac{\partial \mathbf{r}}{\partial y}\right) dA \\ &= \iint_D -z_x(R_y - Q_z) + z_y(R_x - P_z) + (Q_x - P_y)\, dA. \end{aligned}$$

Let $C_1 : x = x(t),\ y = y(t),\ a \le t \le b$ be a closed curve of D and $G = M(x, y)\mathbf{i} + N(x, y)\mathbf{j}$ be a vector field in xy-plane and $\mathbf{r} = x\mathbf{i} + y\mathbf{j}$ be a position vector of R. Since $d\mathbf{r} = dx\,\mathbf{i} + dy\,\mathbf{j}$,

$$\int_{C_1} G\cdot d\mathbf{r} = \int_{C_1} Mdx + Ndy = \int_a^b \left(M\frac{dx}{dt}\right)dt + \left(N\frac{dy}{dt}\right)dt\,.$$

Let C: $\mathbf{r}(t) = x(t)\mathbf{i} + y(t)\mathbf{j} + z(t)\mathbf{k}$, $a \le t \le b$ be a closed curve of S. Then

$$\frac{d\mathbf{r}}{dt} = \frac{dx}{dt}\mathbf{i} + \frac{dy}{dt}\mathbf{j} + \left(z_x\frac{dx}{dt} + z_y\frac{dy}{dt}\right)\mathbf{k}\,.$$

Thus $\displaystyle\int_C \mathbf{F}\cdot d\mathbf{r} = \int_a^b (P\mathbf{i} + Q\mathbf{j} + R\mathbf{k})\cdot\left(\frac{dx}{dt}\mathbf{i} + \frac{dy}{dt}\mathbf{j} + \left(z_x\frac{dx}{dt} + z_y\frac{dy}{dt}\right)\mathbf{k}\right)dt$

3 $$= \int_a^b \left((P + Rz_x)\frac{dx}{dt} + (Q + Rz_y)\frac{dy}{dt}\right)dt.$$

Let $M = P + Rz_x$ and $N = Q + Rz_y$.

By Green Theorem, 3 can be changed the following:

4
$$\int_C \mathbf{F}\cdot d\mathbf{r} = \int_D \left(\frac{\partial}{\partial x}(Q + Rz_y) - \frac{\partial}{\partial y}(P + Rz_x)\right)dA$$
$$= \iint_D (z_x(Q_z - R_y) + z_y(R_x - P_z) + (Q_x - P_y))dA.$$

By 2 and 4, $\displaystyle\int_C \mathbf{F}\cdot d\mathbf{r} = \iint_S \text{curl}\,\mathbf{F}\cdot d\mathbf{S}\,.$ ■

Note 1 Stokes' Theorem can be expressed by

$$\int_C \mathbf{F}\cdot d\mathbf{r} = \iint_S \text{curl}\,\mathbf{F}\cdot d\mathbf{S} = \iint_D \text{curl}\,\mathbf{F}\cdot\left(\frac{\partial\mathbf{r}}{\partial x}\times\frac{\partial\mathbf{r}}{\partial y}\right)dA$$

where S is the surface of $\mathbf{r}(x, y)$ that is a position vector and D is the projection of S on xy-plane.

And Stokes' Theorem in rectangular form is

$$\iint_S \left[\left(\frac{\partial A_3}{\partial y} - \frac{\partial A_2}{\partial z}\right)\cos\alpha + \left(\frac{\partial A_1}{\partial z} - \frac{\partial A_3}{\partial x}\right)\cos\beta + \left(\frac{\partial A_2}{\partial x} - \frac{\partial A_1}{\partial y}\right)\cos\gamma\right]dS$$

where $\mathbf{n} = \cos\alpha\,\mathbf{i} + \cos\beta\,\mathbf{j} + \cos\gamma\,\mathbf{k}$, and α, β, γ are angles made by the normal $\mathbf{n}$ with the unit vectors $\mathbf{i}$, $\mathbf{j}$ and $\mathbf{k}$.

Note 2 Green's Theorem in the plane is a special case of Stoke's Theorem. Suppose S is a surface lying in the xy-plane, with upward unit normal vector $\mathbf{k}$. Let C be the positive oriented boundary of S.

Then $\displaystyle\int_C \mathbf{A}\cdot d\mathbf{r} = \iint_S \text{curl}\,\mathbf{A}\cdot\mathbf{k}\,dA$ which is nothing but the vector version of Green's Theorem.

Note 3 The circulation of A around a closed curve C is given by the line integral $\displaystyle\int_C \mathbf{A}\cdot d\mathbf{r}$ where A represents the velocity of a fluid circulation. This has applications in fluid mechanics and aerodynamics.

EXAMPLE 1

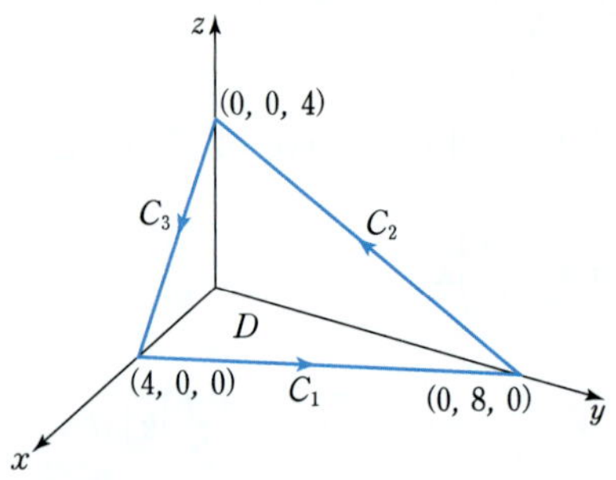

Figure 2

Let S be that the part of the plane $2x+y+2z=8$ in the first octant, oriented with the upward-pointing normal and let C be its boundary, oriented counterclockwise. If $\mathbf{F}=xz\mathbf{i}-y\mathbf{j}+x^2y\mathbf{k}$, verify Stokes' Theorem by computing both $\int_C \mathbf{F}\cdot d\mathbf{r}$ and $\iint_S \operatorname{curl}\mathbf{F}\cdot d\mathbf{S}$. (See Figure 2.)

Solution Stokes' Theorem states that

$$\int_C \mathbf{F}\cdot d\mathbf{r}=\iint_D \operatorname{curl}\mathbf{F}\cdot\left(\frac{\partial\mathbf{r}}{\partial x}\times\frac{\partial\mathbf{r}}{\partial y}\right)dA.$$

First we will find $\int_C \mathbf{F}\cdot d\mathbf{r}$.

Here C is the curve consisting of C_1, C_2 and C_3. So

$$\int_C \mathbf{F}\cdot d\mathbf{r}=\int_{C_1}+\int_{C_2}+\int_{C_3}=LI_1+LI_2+LI_3.$$

On C_1, a simple parameterization of this line segment is $\mathbf{r}(t)=\langle 4,0,0\rangle+t(\langle 0,8,0\rangle-\langle 4,0,0\rangle)=\langle 4-4t,8t,0\rangle$ where $0\le t\le 1$.

Thus $d\mathbf{r}=\langle -4,8,0\rangle dt$. In terms of this parameterization, the vector field $\mathbf{F}=xz\mathbf{i}-y\mathbf{j}+x^2y\mathbf{k}$ becomes

$$\mathbf{F}(\mathbf{r}(t))=\langle 0,-8t,8t(4-4t)^2\rangle.$$

Then

$$\begin{aligned}\mathbf{F}(\mathbf{r}(t))\cdot d\mathbf{r}&=\langle 0,-8t,8t(4-4t)^2\rangle\cdot\langle -4,8,0\rangle dt\\&=(-64t)\,dt.\end{aligned}$$

Thus

$$LI_1=\int_{C_1}\mathbf{F}\cdot d\mathbf{r}=\int_0^1 -64t\,dt=-32.$$

Similarly, on C_2, it's parameterized by $\mathbf{r}(t)=\langle 0,8-8t,4t\rangle$, so

$$\begin{aligned}\mathbf{F}(\mathbf{r}(t))\cdot d\mathbf{r}&=\langle 0,-8+8t,0\rangle\cdot\langle 0,-8,4\rangle dt\\&=(64-64t)\,dt.\end{aligned}$$

Thus

$$LI_2=\int_{C_2}\mathbf{F}\cdot d\mathbf{r}=\int_0^1(64-64t)\,dt=32.$$

Similarly, on C_3, it's parameterized by $\mathbf{r}(t)=\langle 4t,0,4-4t\rangle$, so

$$\begin{aligned}\mathbf{F}(\mathbf{r}(t))\cdot d\mathbf{r}&=\langle 16t-16t^2,0,0\rangle\cdot\langle 4,0,-4\rangle dt\\&=(64t-64t^2)\,dt.\end{aligned}$$

Thus

$$LI_3 = \int_{C_3} \mathbf{F} \cdot d\mathbf{r} = \int_0^1 (64t - 64t^2)\, dt = \frac{32}{3}.$$

Hence, $\int_C \mathbf{F} \cdot d\mathbf{r} = LI_1 + LI_2 + LI_3 = \frac{32}{3}.$

Next we will find $\iint_S \operatorname{curl} \mathbf{F} \cdot d\mathbf{S}$. This triangular surface is a graph (using $z = 4 - x - \frac{1}{2}y$) over the triangle $D = (x,\ y) \mid 0 \le y \le 8 - 2x,\ 0 \le x \le 4$ in the first quadrant of the xy-plane. Thus we can parameterize it as

$$\mathbf{r}(x,\ y) = \left\langle x,\ y,\ 4 - x - \frac{y}{2} \right\rangle \text{ for } (x, y) \in D.$$

Since $\mathbf{F} = xz\mathbf{i} - y\mathbf{j} + x^2 y\mathbf{k}$, we get

$$\operatorname{curl}\ \mathbf{F} = \begin{vmatrix} \mathbf{i} & \mathbf{j} & \mathbf{k} \\ \frac{\partial}{\partial x} & \frac{\partial}{\partial y} & \frac{\partial}{\partial z} \\ xz & -y & x^2 y \end{vmatrix} = x^2\mathbf{i} + (x - 2xy)\mathbf{j}.$$

Similarly,

$$\frac{\partial \mathbf{r}}{\partial x} \times \frac{\partial \mathbf{r}}{\partial y} = \begin{vmatrix} \mathbf{i} & \mathbf{j} & \mathbf{k} \\ 1 & 0 & -1 \\ 0 & 1 & -\frac{1}{2} \end{vmatrix} = \mathbf{i} + \frac{1}{2}\mathbf{j} + \mathbf{k},$$

Then

$$\begin{aligned} \iint_S \operatorname{curl} \mathbf{F} \cdot d\mathbf{S} &= \iint_D \operatorname{curl} \mathbf{F} \cdot \left(\frac{\partial \mathbf{r}}{\partial x} \times \frac{\partial \mathbf{r}}{\partial y} \right) dA \\ &= \int_0^4 \int_0^{8-2x} \left(x^2 + \frac{1}{2}x - xy \right) dy\, dx \\ &= \frac{32}{3}. \end{aligned}$$

■

CAS **EXAMPLE 2**

Evaluate $\iint_S \operatorname{curl} \mathbf{F} \cdot d\mathbf{S}$ where $\mathbf{F} = (y - z)\mathbf{i} + (z - x)\mathbf{j} + (x - y)\mathbf{k}$ and S is the portion of the plane $x + y + z = 4$ that lies inside the sphere $x^2 + y^2 = 1$.
http://matrix.skku.ac.kr/cal-lab/Sec15-8-Exm-2.html

Solution

```
var('x, y, z')
p1=implicit_plot3d(x^2+y^2-1==0, (x, -2, 2), (y, -2, 2), (z, 0, 6), color='blue', opacity=0.3)
p2=implicit_plot3d(x+y+z==4, (x, -2, 2), (y, -2, 2), (z,0, 6), color='red', opacity=0.5)
show(p1+p2)
```

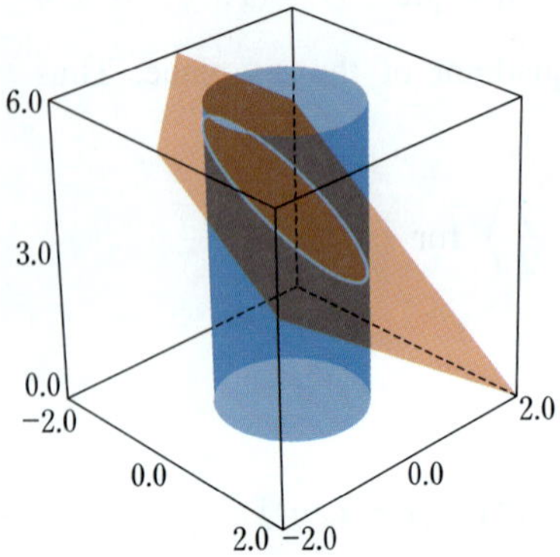

Figure 3

In Stokes' Theorem, we will use

$$\iint_S \operatorname{curl}\mathbf{F}\cdot d\mathbf{S} = \iint_D (\operatorname{curl}\mathbf{F})\cdot\left(\frac{\partial \mathbf{r}}{\partial x}\times\frac{\partial \mathbf{r}}{\partial y}\right)dA$$

$$\operatorname{curl}\ \mathbf{F} = \begin{vmatrix} \mathbf{i} & \mathbf{j} & \mathbf{k} \\ \dfrac{\partial}{\partial x} & \dfrac{\partial}{\partial y} & \dfrac{\partial}{\partial z} \\ y-z & z-x & x-y \end{vmatrix} = -2\mathbf{i}-2\mathbf{j}-2\mathbf{k},$$

Since $\mathbf{r}=x\mathbf{i}+y\mathbf{j}+(4-x-y)\mathbf{k}$,

$$\frac{\partial \mathbf{r}}{\partial x}\times\frac{\partial \mathbf{r}}{\partial y} = \begin{vmatrix} \mathbf{i} & \mathbf{j} & \mathbf{k} \\ 1 & 0 & -1 \\ 0 & 1 & -1 \end{vmatrix} = \mathbf{i}+\mathbf{j}+\mathbf{k}.$$

Since D is a unit circle in xy-plane,

$$\begin{aligned} &\iint_S \operatorname{curl}\mathbf{F}\cdot d\mathbf{S} \\ &= \iint_D \operatorname{curl}\mathbf{F}\cdot\left(\frac{\partial \mathbf{r}}{\partial x}\times\frac{\partial \mathbf{r}}{\partial y}\right)dA \\ &= \iint_D -6\,dA = -6\pi. \end{aligned}$$

■

Stokes theorem relates the surface integral of the curl of a vector field F over a surface Σ in Euclidean three-space to the line integral of the vector field over its boundary $\partial\Sigma$:

This Divergence theorem, fundamental theorem of calculus, and Green's Theorem are simply speical cases of the general formulation of the Stokes theorem.

15.8 EXERCISES (Stokes' Theorem)

http://matrix.skku.ac.kr/Cal-Book/part2/CS-Sec-15-8-Sol.htm

CAS **1.** Evaluate $\int_C x\,dx + y\,dy + xy\,dz$ using Stokes' Theorem, given that C is the circle: $x^2+y^2+z^2=4$ that lies inside the cylinder $x^2+y^2=1$ and above the xy-plane.
http://matrix.skku.ac.kr/cal-lab/cal-15-8-1.html

Solution Note that the curve of intersection is the circle at the plane $z=\sqrt{3}$.

```
var('x, y, z, t')
p1 = implicit_plot3d(x^2+y^2+z^2==4, (x, -2, 2), (y, -2, 2), (z, 0, 2), opacity=0.2, color="red", mesh=True);
p2 = implicit_plot3d(x^2+y^2==1, (x, -2, 2), (y, -2, 2), (z, 0, sqrt(3)), opacity=0.5, color="blue", mesh=True);
p3 = plot3d(0, (x, -2, 2), (y, -2, 2), opacity=0.3, color="orange", mesh=True);
show(p1+p2+p3, aspect_ratio=1)
```

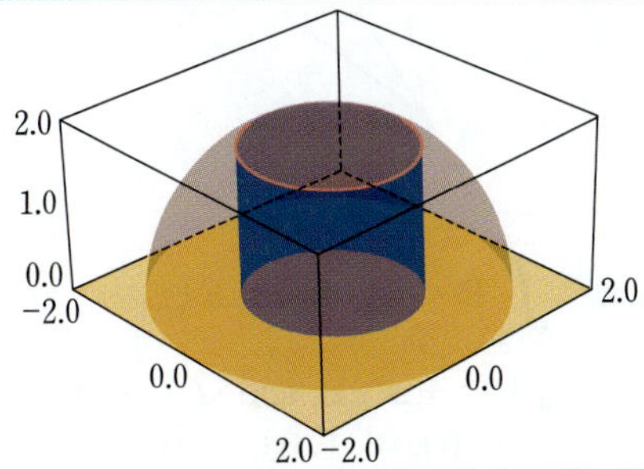


```
var('i, j, k')
r_t = vector(SR, [cos(t), sin(t), sqrt(3)]);
dr_t = diff(r_t, t);
F = vector(SR, [x, y, x*y]);
F_r_t = vector(SR, [cos(t), sin(t), cos(t)*sin(t)]);
print (F_r_t)
integral(F_r_t*dr_t, (t, 0, 2*pi))
```

Answer : 0

2. Evaluate (a) directly (b) using Stokes' Theorem $\int_C 4z\,dx - 2x\,dy + 2x\,dz$ where C is the ellipse that intersect between $x^2+y^2=1$ and $z=y+1$.

Solution Let $x=\cos t, y=\sin t, z=\sin t+1$.
So $\mathbf{r}(t)=\langle \cos t, \sin t, \sin t+1\rangle$ and $\mathbf{r}'(t)=\langle -\sin t, \cos t, \cos t\rangle$. Hence,

$$\int_C 4z\,dx - 2x\,dy + 2x\,dz$$
$$= \int_0^{2\pi} [-4(\sin t+1)\sin t - 2\cos t\cos t + 2\cos t\cos t]\,dt$$
$$= -\int_0^{2\pi} 4(\sin t+1)\sin t\,dt$$
$$= -4\int_0^{2\pi} \sin^2 t\,dt.$$
$$= -4\pi.$$

3. Verify Stokes' Theorem for the vector field $\mathbf{A}=(2x-y)\mathbf{i}-yz^2\mathbf{j}-y^2z\mathbf{k}$ over an orientable surface S which is the upper hemisphere $x^2+y^2+z^2=1$ and $z\ge 0$.

Solution Note that the curve C is $x^2+y^2=1$, $z=0$ with positive orientation. Since curl $\mathbf{A}=\langle 0,0,1\rangle$, we have

$$\iint_S \text{curl } \mathbf{A}\cdot dS = \int_0^{2\pi}\int_0^{\pi/2} \sin\phi\cos\phi\,d\phi\,d\theta = \pi.$$

Next we compute $\int_C \mathbf{A}\cdot d\mathbf{r}$ where

$\mathbf{r}(t)=(\cos t,\ \sin t,\ 0),\ 0\le t\le 2\pi.$

Hence $\int_C \mathbf{A}\cdot d\mathbf{r} = \int_0^{2\pi} \mathbf{A}(\mathbf{r}(t))\cdot \mathbf{r}'(t)\,dt$

$$= \int_0^{2\pi} (-2\sin t\cos t + \sin^2 t)\,dt$$
$$= 0 + \int_0^{2\pi} \frac{1-\cos 2t}{2}\,dt.$$
$$= \pi.$$

Hence verified.

CAS **4.** Verify Stoke's Theorem for a vector field $\mathbf{A}=x^2\mathbf{i}+xy\mathbf{j}$ over an orientable surface S which is the square $0\le x\le a,\ 0\le y\le a$ in the xy-plane.
[Hint] C is a square of $0\le x\le a$, $0\le y\le a,\ z=0$.
http://matrix.skku.ac.kr/cal-lab/cal-14-7-4.html

Solution

```
var('x,y,z,a')
M=x^2; N=x*y; P=0
curlF=vector([diff(P,y)-diff(N,z),diff(M,z)-diff(P,x),
diff(N,x)-diff(M,y)])
```

```
print curlF
x=x;y=y;z=0
r_x=vector([diff(x,x),diff(y,x),diff(z,x)])
r_y=vector([diff(x,y),diff(y,y),diff(z,y)])
n=r_x.cross_product(r_y)
w=curlF.dot_product(n)
ANSWER=integral(integral(w,x,0,a),y,0,a)
print ANSWER
```

Answer : a^{3} over 2

5. Verify Stokes' Theorem for a vector field $\mathbf{A}=(x^2+y^2)\mathbf{i}-2xy\mathbf{j}$ over a rectangle bounded by $x=a$, $x=-a$, $y=b$, $y=-b$.
http://matrix.skku.ac.kr/cal-lab/cal-15-5-5.html

Solution

```
var('x,y,z,a')
assume(a>0)
M=x^2+y^2; N=-2*x*y; P=0
curlF=vector([diff(P,y)-diff(N,z),diff(M,z)-diff(P,x),
diff(N,x)-diff(M,y)])
print curlF
x=x;y=y;z=0
r_x=vector([diff(x,x),diff(y,x),diff(z,x)])
r_y=vector([diff(x,y),diff(y,y),diff(z,y)])
n=r_x.cross_product(r_y)
w=curlF.dot_product(n)
ANSWER=integral(integral(w,x,-a,a),y,-a,a)
print ANSWER
```

Answer : 0

6. Evaluate $\int_C \mathbf{F}\cdot d\mathbf{r}$, where $\mathbf{F}(x,y,z)=-y^3\mathbf{i}+x^3\mathbf{j}-z^3\mathbf{k}$ and C is the curve which is the intersection of $x^2+y^2=1$ and $x+y+z=1$. (C is upward anticlockwise).

Solution curl $\mathbf{F}=3(x^2+y^2)\mathbf{k}$ and the curve C is a boundary of $S: z=1-x-y$ on $D=\{(x,y)\mid x^2+y^2\le 1\}$. Using Strokes' Theorem and the ploar coordinate system,

$$\int_C \mathbf{F}\cdot d\mathbf{r}=\iint_S \text{curl}\,\mathbf{F}\cdot d\mathbf{S}=\iint_D 3(x^2+y^2)\,dA$$

$$=3\int_0^{2\pi}\int_0^1 r^2\cdot r\,drd\theta=\frac{3\pi}{2}.$$

7. We consider the vector field $\mathbf{F}(x,y,z)=(-4xy+xz)\mathbf{i}+(xy-yz)\mathbf{j}+(z^2-xz)\mathbf{k}$ and the curve C which is the boundary of the triangle with vertices $P(2,0,0)$, $Q(0,3,0)$, $R(0,0,6)$. Compute the work done by the force field $\mathbf{F}$ in moving a particle along the curve C. (First, the particle goes from P to Q, and goes from Q to R, finally goes from R back to P).

Solution Consider the region $D: \frac{x}{2}+\frac{y}{3}+\frac{z}{6}=1$, $x,y,z\ge 0$. The C is the boundary of D. Since $\mathbf{r}=x\mathbf{i}+y\mathbf{j}+(6-3x-2y)\mathbf{k}$, $\mathbf{r}_x\times\mathbf{r}_y=\langle -3,2,1\rangle$ and $\nabla\times\mathbf{F}=\langle y, x+z, y+4x\rangle$,

Using Stokes' Theorem, we have

$$\int_C \mathbf{F}\cdot d\mathbf{r}=\iint_D \nabla\times\mathbf{F}\cdot d\mathbf{S}$$

$$=\int_0^3\int_0^{2-2x/3}(12-6y)\,dy\,dx$$

$$=48.$$

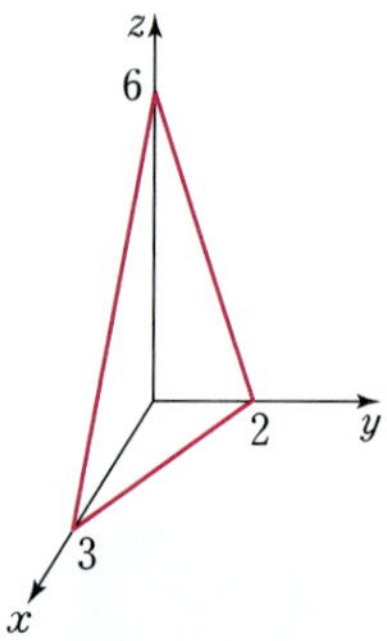

8. Evaluate $\int_C \mathbf{F}\cdot d\mathbf{r}$. Here $\mathbf{F}(x,y,z)=(x+y^2)\mathbf{i}+(y+z^2)\mathbf{j}+(z+x^2)\mathbf{k}$ and C is a triangle with vertices $(1,0,0),(0,1,0),(0,0,1)$.

Solution curl

$$\mathbf{F}=\begin{vmatrix}\mathbf{i} & \mathbf{j} & \mathbf{k}\\ \frac{\partial}{\partial x} & \frac{\partial}{\partial y} & \frac{\partial}{\partial z}\\ (x+y^2) & (y+z^2) & (z+x^2)\end{vmatrix}=\langle -2z,-2x,-2y\rangle.$$

The surface bounded by C is $z=g(x,y)=1-x-y$ and this can be represent a parametric vector $\mathbf{r}=\langle x,y,1-x-y\rangle$. So we get $\mathbf{r}_x\times\mathbf{r}_y=\langle 1,1,1\rangle$. And the domain is the projection of the surface on xy-plane as $D=\{(x,y)\mid 0\le x\le 1, 0\le y\le 1-x\}$.

$$\int_C \mathbf{F}\cdot d\mathbf{r}=\iint_S \text{curl}\,\mathbf{F}\cdot d\mathbf{S}$$

$$=\iint_D(-2z-2x-2y)\,dA$$

$$=\int_0^1\int_0^{1-x}(-2)\,dy\,dx=-1.$$

15.9 Divergence Theorem

You Tube http://youtu.be/3BmcFr81kuQ

Carl Friedrich Gauss (1777-1855),
German mathematician.
Painted by Christian Albrecht Jensen

This section deals with the third important theorem, called the Gauss Divergence Theorem. It is an extension of the vector version of Green's Theorem to vector fields in $\mathbb{R}^3$ and relates the integral of a derivative of function (div of a vector field $\mathbf{A}$) over a region to the integral of the function $\mathbf{F}$ over the boundary of the region. The region in this case is considered as a solid region and its boundary is a surface. We make certain assumptions about the region and its boundary.

We assume that S is a closed surface, that is S encloses a solid region in $\mathbb{R}^3$. For example, spheres, cubes and ellipsoids are closed surfaces whereas planes, paraboloids, etc are not closed regions. We further assume that S has positive orientation. That is, the unit normal vector is directed outward from S.

THEOREM 1 Gauss Divergence Theorem

Let S be a closed surface in $\mathbb{R}^3$ which is the boundary of a solid region V. Let $\mathbf{A}$ be a vector field defined on some open set T containing V. Suppose the components of $\mathbf{A}$ have continuous first order partial derivatives and $\mathbf{n}$ is an unit normal vector that is directed outward from S. Then

1
$$\iint_S \mathbf{A}\cdot\mathbf{n}\,dS = \iiint_V \nabla\cdot\mathbf{A}\,dV = \iiint_V \operatorname{div}\mathbf{A}\,dV.$$

Proof We prove this theorem for a particular type of S. Assume S is a surface such that any line parallel to coordinate axes meets S in at most two points. Let S_1 and S_2 be the lower (below) and upper (top) portions of S having equations $z=f_1(x, y)$ and $z=f_2(x, y)$ and having $\mathbf{n}_1$ and $\mathbf{n}_2$ as normals respectively (See Figure 1). Let R be the projection of the surface S on the xy-plane. If $\mathbf{A}=A_1\mathbf{i}+A_2\mathbf{j}+A_3\mathbf{k}$, then the result of Gauss Divergence Theorem in component form is

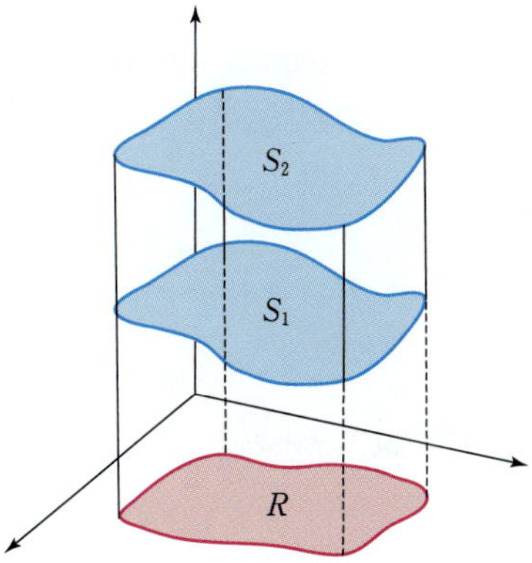

Figure 1

2
$$\iint_S (A_1\mathbf{i}+A_2\mathbf{j}+A_3\mathbf{k})\cdot\mathbf{n}\,dS = \iiint_V\left(\frac{\partial A_1}{\partial x}+\frac{\partial A_2}{\partial y}+\frac{\partial A_3}{\partial z}\right)dV.$$

Thus it is sufficient to prove

$$\iint_S A_1\mathbf{i}\cdot\mathbf{n}\,dS = \iiint_V \frac{\partial A_1}{\partial x}\,dV,$$

$$\iint_S A_2\mathbf{j}\cdot\mathbf{n}\,dS=\iiint_V \frac{\partial A_2}{\partial y}dV,$$

$$\iint_S A_3\mathbf{k}\cdot\mathbf{n}\,dS=\iiint_V \frac{\partial A_3}{\partial z}dV.$$

Consider the next equation.

$$\begin{aligned}\iiint_V \frac{\partial A_3}{\partial z}dV &= \iiint_V \frac{\partial A_3}{\partial z}dz\,dy\,dx\\ &= \iint_R\left[\int_{z=f_1}^{z=f_2}\frac{\partial A_3}{\partial z}dz\right]dydx\\ &= \iint_R [A_3(x,\ y,\ f_2)-A_3(x,\ y,\ f_1)]dydx\\ &= \iint_{S_2} A_3\mathbf{k}\cdot\mathbf{n}dS+\iint_{S_1} A_3\mathbf{k}\cdot\mathbf{n}dS.\end{aligned}$$

Since for the upper surface S_2, $\mathbf{k}\cdot\mathbf{n}dS=dydx$ while for the lower surface S_1, $\mathbf{k}\cdot\mathbf{n}dS=-dydx$, then

3 $$\iiint_V \frac{\partial A_3}{\partial z}dV=\iint_S A_3\mathbf{k}\cdot\mathbf{n}dS.$$

Similarly, projecting S on to the yz-plane and xz-plane we have

4 $$\iiint_V \frac{\partial A_1}{\partial x}dV=\iint_S A_1\mathbf{i}\cdot\mathbf{n}dS,$$

5 $$\iiint_V \frac{\partial A_2}{\partial y}dV=\iint_S A_2\mathbf{j}\cdot\mathbf{n}dS.$$

Adding 3, 4 and 5, we get the required result 2. ■

Note 1 Gauss Divergence Theorem (GDT) transforms volume integrals to surface integrals and vice versa.

Note 2 GDT in rectangular form can be written

$$\begin{aligned}&\iiint_V\left(\frac{\partial A_1}{\partial x}+\frac{\partial A_2}{\partial y}+\frac{\partial A_3}{\partial z}\right)dxdydz\\ &= \iint_S (A_1\mathbf{i}+A_2\mathbf{j}+A_3\mathbf{k})\cdot(n_1\mathbf{i}+n_2\mathbf{j}+n_3\mathbf{k})dS\\ &= \iint_S (A_1n_1+A_2n_2+A_3n_3)dS\\ &= \iint_S (A_1\cos\alpha+A_2\cos\beta+A_3\cos\gamma)dS\end{aligned}$$

where $n_1=\mathbf{n}\cdot\mathbf{i}=\cos\alpha$, $n_2=\mathbf{n}\cdot\mathbf{j}=\cos\beta$, $n_3=\mathbf{n}\cdot\mathbf{k}=\cos\gamma$. Here α, β, γ are the angles which $\mathbf{n}$ makes with the positive x, y, z axes.

Note 3 GDT is also known as "Green's Theorem in space" because the GDT generalizes the "Green's Theorem in the plane" by replacing the (plane) region R and its closed boundary (curve) C by a (space) region V and its closed boundary S (the surface).

Note 4 The left hand side of $\boxed{1}$ is referred as the flux of $\mathbf{A}$ through the surface S. Thus Gauss Divergence Theorem says that under the certain conditions, the flux of $\mathbf{A}$ across the surface S is equal to the triple integral of the divergence of $\mathbf{A}$.

Note 5 When $\mathbf{A}=\mathbf{V}=$ velocity of a fluid then the GDT has the following physical interpretation:

$$\left.\begin{array}{l}\text{Volume of fluid emerging}\\ \text{(diverging) from a closed}\\ \text{surface in unit time}\end{array}\right\}=\left\{\begin{array}{l}\text{Volume of fluid}\\ \text{supplied from within}\\ \text{volume } V \text{ in unit time}\end{array}\right.$$

Let us verify the Gauss Divergence Theorem in the next examples.

EXAMPLE 1

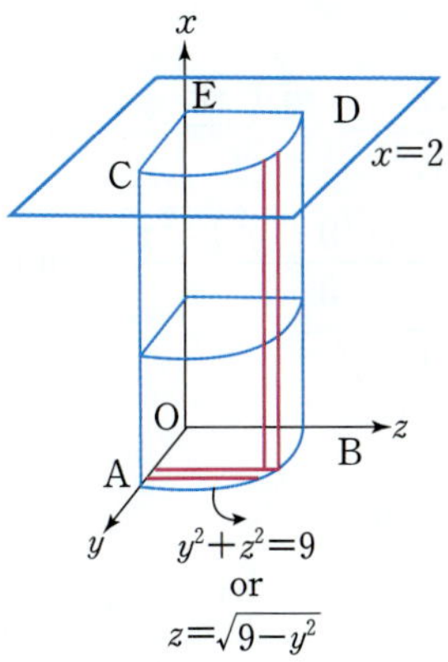

Figure 2

Verify $\iint_S \mathbf{A}\cdot\mathbf{n}\,dS=\iiint_V \nabla\cdot\mathbf{A}\,dV$ of the Gauss Divergence Theorem for $\mathbf{A}=2x^2y\mathbf{i}-y^2\mathbf{j}+4xz^2\mathbf{k}$ taken over the region in the first octant bounded by the cylinder $y^2+z^2=9$ and the plane $x=2$. (See Figure 2.)

Solution Since $\nabla\cdot\mathbf{A}=4xy-2y+8xz$ and

$V=(x,\ y,\ z)\,|\,0\le x\le 2,\ 0\le y\le 3,\ 0\le z\le \sqrt{9-y^2}$,

then

$$\iiint_V \nabla\cdot\mathbf{A}\,dV=\int_0^2\int_0^3\int_0^{\sqrt{9-y^2}}(4xy-2y+8xz)\,dzdydx$$

$$=\int_0^2\int_0^3\left[(4xy-2y)\left(\sqrt{9-y^2}\right)+4x(9-y^2)\right]dydx$$

$$=\int_0^2\frac{(2-4x)}{2}\,\frac{(9-y^2)^{\frac{3}{2}}}{3/2}\Bigg|_0^3 dx+\int_0^2 36yx-4x\frac{y^3}{3}\Bigg|_0^3 dx$$

$$=\int_0^2 -18(1-2x)\,dx+\int_0^2(108x-36x)\,dx=180.$$

The entire surface S consists of five surfaces $S_1,\ S_2,\ S_3,\ S_4,\ S_5$. So

$$\iint_S \mathbf{A}\cdot\mathbf{n}\,dS=\iint_{S_1}+\iint_{S_2}+\cdots+\iint_{S_5}=SI_1+SI_2+\cdots+SI_5.$$

We will use $\boxed{5}$ - $\boxed{7}$ in section 15.7 to find the followings.
On S_1: OAB, since $x=0$ and $\mathbf{n}=-\mathbf{i}$, $\mathbf{A}\cdot\mathbf{n}=0$. Then

$$SI_1 = \iint_{S_1} \mathbf{A} \cdot \mathbf{n}\, dS = 0.$$

On S_2: CED, since $x = 2$ and $\mathbf{n} = \mathbf{i}$, $\mathbf{A} \cdot \mathbf{n} = 8y$ and $\mathbf{n} \cdot \mathbf{i} = 1$. Then

$$SI_2 = \iint_{S_2} \mathbf{A} \cdot \mathbf{n}\, dS = \iint_{S_2} 8y\,dy\,dz$$
$$= \int_0^3 \int_0^{\sqrt{9-z^2}} 8y\,dy\,dz = \int_0^3 4(9-z^2)\,dz = 72.$$

On S_3: $OBDE$, since $y = 0$ and $\mathbf{n} = -\mathbf{j}$, $\mathbf{A} \cdot \mathbf{n} = 0$. Then

$$SI_3 = \iint_{S_3} \mathbf{A} \cdot \mathbf{n}\, dS = 0$$

On S_4: $OACE$, since $z = 0$ and $\mathbf{n} = -\mathbf{k}$, $\mathbf{A} \cdot \mathbf{n} = 0$. Then

$$SI_4 = \iint_{S_4} \mathbf{A} \cdot \mathbf{n}\, dS = 0.$$

On S_5: $ABCD$ of the cylinder: $y^2 + z^2 = 9$, since the unit normal $\mathbf{n}$ is

$$\mathbf{n} = \frac{\nabla(y^2+z^2)}{|\nabla(y^2+z^2)|} = \frac{2y\mathbf{j} + 2z\mathbf{k}}{\sqrt{4y^2+4z^2}} = \frac{y\mathbf{j} + z\mathbf{k}}{3},$$

$$\mathbf{A} \cdot \mathbf{n} = \frac{-y^3 + 4xz^3}{3} = \frac{-y^3 + 4x(9-y^2)^{3/2} z^3}{3} \quad \text{and}$$

$$\mathbf{n} \cdot \mathbf{k} = \frac{y\mathbf{j} + z\mathbf{k}}{3} \cdot \mathbf{k} = \frac{z}{3} = \frac{\sqrt{9-y^2}}{3}.$$

Projecting the surface S_5 on the xy-plane

$$SI_5 = \iint_{S_5} \mathbf{A} \cdot \mathbf{n}\, dS = \iint_R \frac{4x(9-y^2)^{3/2} - y^3}{3} \frac{3}{\sqrt{9-y^2}}\,dx\,dy$$
$$= \int_0^3 \int_0^2 \left(4x(9-y^2) - \frac{y^3}{\sqrt{9-y^2}}\right) dx\,dy = 108.$$

Thus $\iint_S \mathbf{A} \cdot \mathbf{n}\, dS = 0 + 72 + 0 + 0 + 108 = 180$. Therefore, the Gauss Divergence Theorem is verified. ■

EXAMPLE 2

Find the flux of a vector field $\mathbf{A}(x, y, z) = x\mathbf{i} + y\mathbf{j} + z\mathbf{k}$ over the sphere S: $x^2 + y^2 + z^2 = a^2$ using the GDT.

Solution Note that div $\mathbf{A} = 3$. Thus by GDT, we have flux of $\mathbf{A}$ across

$$\iint_S \mathbf{A} \cdot \mathbf{n}\, dS = \iiint_{x^2+y^2+z^2 \le a^2} \text{div } \mathbf{A}\, dV$$

$$= 3\iiint_{x^2+y^2+z^2 \le a^2} dV$$
$$= 3 \times \text{Volume of the Sphere } S$$
$$= 3 \times \frac{4}{3}\pi a^3$$
$$= 4\pi a^3. \quad \blacksquare$$

EXAMPLE 3

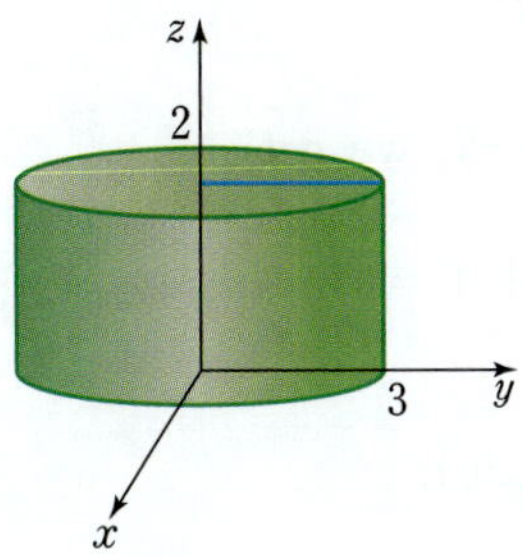

Figure 3

Find the surface integral $\iint_S \mathbf{A} \cdot \mathbf{n}\, dS$, where $\mathbf{A} = x^3\mathbf{i} + y^3\mathbf{j} + z^2\mathbf{k}$ over the surface S which is bounded by $x^2 + y^2 \le 9$ and $0 \le z \le 2$.

Solution Note that $\operatorname{div} \mathbf{A} = 3x^2 + 3y^2 + 2z$. Thus by the GDT, we have

$$\iint_S \mathbf{A} \cdot \mathbf{n}\, dS = \iiint_V \operatorname{div} \mathbf{A}\, dV$$
$$= \iiint_V (3x^2 + 3y^2 + 2z)\, dV$$

By the polar coordinate system, letting $x = r\cos\theta$, $y = r\sin\theta$, then $dx\,dy = r\,dr\,d\theta$ and $\{(r, \theta) | 0 \le r \le 3,\ 0 \le \theta \le 2\pi$. Thus

$$\iint_S \mathbf{A} \cdot \mathbf{n}\, dS = \int_0^{2\pi}\int_0^3\int_0^2 (3r^2 + 2z)\, dz\, r\, dr\, d\theta = 279\pi. \quad \blacksquare$$

http://matrix.skku.ac.kr/cal-lab/cal-15-9-Exam-3.html

```
var('x,y,z');
def Div(F):
    assert(len(F) == 3)
    return (diff(F[0],x)+diff(F[1],y)+diff(F[2],z))
F=vector([x^3,y^3,z^2])
Div(F)
```

Answer : 3*x^2 + 3*y^2 + 2*z

```
var('t,r')
x=r*cos(t);y=r*sin(t) # polar coordinate system
f=3*x^2 + 3*y^2 + 2*z
integral(integral(integral(f*r,r,0,3),t,0,2*pi),z,0,2)
```

Answer : 279*pi

In the next example we have to find the volume of a region bounded by a surface S by the Gauss Divergence Theorem.

EXAMPLE 4

Find the volumd V of a region bounded by a surface S.

Solution By Gauss' Divergence Theorem

$$\boxed{6} \qquad \iiint_V \nabla \cdot \mathbf{A}\, dV = \iint_S \mathbf{A} \cdot \mathbf{n}\, dS.$$

Choose $\mathbf{A} = x\mathbf{i}$, so that $\nabla \cdot \mathbf{A} = 1$, with this $\boxed{6}$ reduces to

$$V = \text{Volume} = \iiint_V 1 \cdot dV$$

$$= \iint_S x(\mathbf{i} \cdot \mathbf{n}) dS = \iint_S x dydz.$$

Similarly by taking $\mathbf{A} = y\mathbf{j}$ and $\mathbf{A} = z\mathbf{k}$, we get

$$V = \iint_S ydzdx \text{ and } V = \iint_S zdxdy$$

or $V = \dfrac{1}{3} \displaystyle\iint_S (xdydz + ydzdx + zdxdy).$ ■

Divergence theorem states that the outward flux of a vector field through a closed surface is equal to the volume integral of the divergence over the region inside the surface. Intuitively, it states that the sum of all sources minus the sum of all sinks gives the net flow out of a region.

The divergence theorem is an important result for the mathematics of engineering, in particular in electrostatics and fluid dynamics.

15.9 EXERCISES (Divergence Theorem)

http://matrix.skku.ac.kr/Cal-Book/part2/CS-Sec-15-9-Sol.htm

YouTube http://youtu.be/vGMLoGWF1Is

1-4. Using the Divergence Theorem, evaluate the surface integral $\iint_S \mathbf{F}\cdot\mathbf{n}\,dS$

CAS **1.** $\mathbf{A} = xy\mathbf{i} + yz\mathbf{j} + xz\mathbf{k}$ and the boundary of a region is the sphere $x^2+y^2+z^2=4$.
http://matrix.skku.ac.kr/cal-lab/cal-14-8-1.html

Solution Define "Div" function

```
var('x,y,z')
def Div(F):
    assert(len(F)==3)
    return
(diff(F[0],x)+diff(F[1],y)+diff(F[2],z))
F=vector([y*z,z*x,x*y])
Div(F)
```

Answer : 0

Since divergence is 0,

$$\iint_S \mathbf{F}\cdot\mathbf{n}\,dS = \iiint_V \operatorname{div}\mathbf{F}\,dV = 0.$$

CAS **2.** $\mathbf{A} = x^3\mathbf{i} + x^2y\mathbf{j} + xy\mathbf{k}$ and S is a closed surface consisting of the circular cylinder $x^2+y^2=a^2$ and the circular disks $z=0$ and $z=b$.
http://matrix.skku.ac.kr/cal-lab/cal-15-6-2.html

Solution

```
var('x,y,z')
def Div(F):
    assert(len(F)==3)
    return
 (diff(F[0],x)+diff(F[1],y)+diff(F[2],z))
F=vector([x^3,x^2*y,x*y])
Div(F)
```

Answer : 4*x^2

```
var('a,b,t,r')
assume(a>0,b>0)
x=r*cos(t);y=r*sin(t) # polar coordinate system
f=4*x^2
integral(integral(integral(f*r,r,0,a),t,0,2*pi),z,0,b)
```

Answer : pi*a^4*b

Then $\iint_S \mathbf{F}\cdot\mathbf{n}\,dS = \iiint_V \operatorname{div}\mathbf{F}\,dV = a^4 b\pi$.

CAS **3.** $\mathbf{A} = \sin x\,\mathbf{i} + (2 - y\cos x)\mathbf{j} + xy\mathbf{k}$ and S is the surface bounded by $0\le x\le 3$, $0\le y\le 2$, $0\le z\le 1$.
http://matrix.skku.ac.kr/cal-lab/cal-15-6-3.html

Solution

```
var('x,y,z')
def Div(F):
    assert(len(F)==3)
    return
(diff(F[0],x)+diff(F[1],y)+diff(F[2],z))
F=vector([sin(x),2-y*cos(x),0])
Div(F)
```

Answer : 0

Since divergence is 0,

$$\iint_S \mathbf{F}\cdot\mathbf{n}\,dS = \iiint_V \operatorname{div}\mathbf{F}\,dV = 0.$$

CAS **4.** $\mathbf{A} = (x^2 - yz)\mathbf{i} - 2x^2y\mathbf{j} + z\mathbf{k}$ and S is the cube of side b.
http://matrix.skku.ac.kr/cal-lab/cal-15-6-4.html

Solution

```
var('x,y,z,b')
assume(b>0)
def Div(F):
    assert(len(F)==3)
    return
 (diff(F[0],x)+diff(F[1],y)+diff(F[2],z))
F=vector([x^2-y*z,-2*x^2*y,z])
integrand=Div(F)
integral(integral(integral(integrand,x,0,b),y,0,b),z,0,
b)
```

Answer: -1/3*(2*b^3 - 3*b^2 - 3*b)*b^2

5-13. Verify the Gauss Divergence Theorem for :

CAS **5.** $\mathbf{A} = 4x\mathbf{i} - 2y^2\mathbf{j} + z^2\mathbf{k}$ taken over the region bounded by $x^2+y^2=4$, $z=0$ and $z=3$.
http://matrix.skku.ac.kr/cal-lab/cal-14-8-5.html

Solution

```
var('x,y,z,t')
p1 = implicit_plot3d(x^2+y^2==4, (x,-2,2), (y, -2,2), (z, -5,5), opacity=0.2, color="red", mesh=True);
p2 = implicit_plot3d(z==0, (x,-2,2), (y, -2,2),(z, -5,5), opacity=0.3, color="blue", mesh=True);
p3 = implicit_plot3d(z==3, (x,-2,2), (y, -2,2), (z, -5,5), opacity=0.5, color="orange", mesh=True);
show(p1+p2+p3, aspect_ratio=1)
```

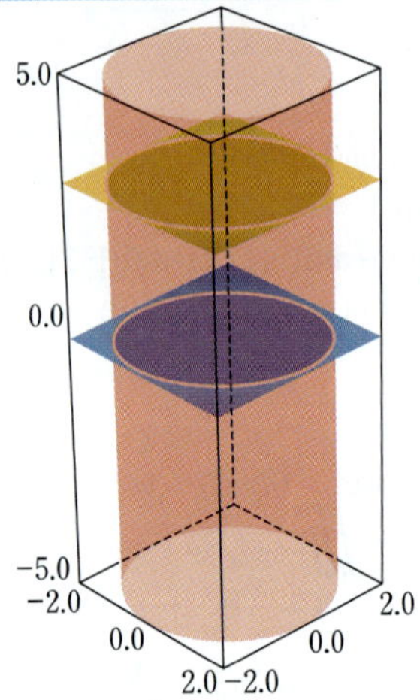

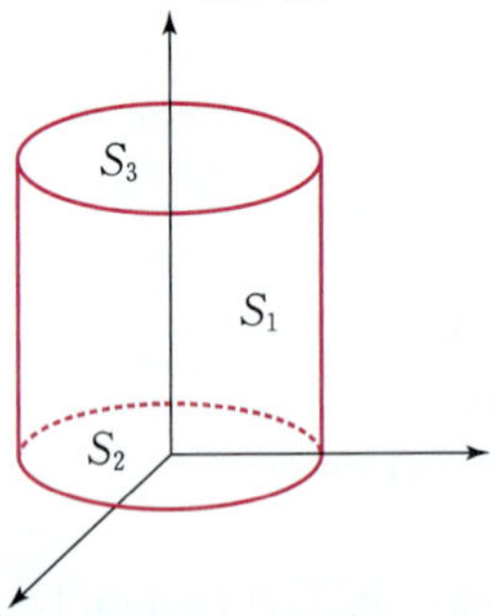


```
var('x,y,z,t,r')
def Div(F):
    assert(len(F)==3)
    return (diff(F[0],x)+diff(F[1],y)+diff(F[2],z))
f=Div([4*x,-2*y^2,z^2])
g(r,t,z)=f.subs(x=2*cos(t),y=2*sin(t),z=z)
integral(integral(integral(g*r,r,0,2),t,0,2*pi),z,0,3)
```

Answer : 84*pi

$\mathbf{r}(\theta, z) = \langle 2\cos\theta, 2\sin\theta, z\rangle$,

$\mathbf{A}(\mathbf{r}(\theta, z)) = \langle 8\cos\theta, -8\sin^2\theta, z^2\rangle$,

$\mathbf{r}_\theta \times \mathbf{r}_z = \langle 2\cos\theta, 2\sin\theta, 0\rangle$,

$$\iint_{S_1} \mathbf{A}\cdot\mathbf{n}\,dS = \iint_{D_1} \mathbf{A}\cdot(\mathbf{r}_\theta \times \mathbf{r}_Z)\,dA$$

$$= \iint_{D_1} 16\cos^2\theta - 16\sin^3\theta$$

$$= 48\int_0^{2\pi} \cos^2\theta\,d\theta - 48\int_0^{2\pi} \sin^3\theta\,d\theta = 48\pi - 0 = 48\pi,$$

$$\iint_{S_2} \mathbf{A}\cdot\mathbf{n}\,dS = \iint_{D_2} (-z^2)\,dxdy = 0,$$

$$\iint_{S_3} \mathbf{A}\cdot\mathbf{n}\,dS = \iint_{D_3} z^2\,dxdy = 9\iint_{D_3} dxdy = 9(4\pi) = 36\pi.$$

Therefore $\iint_S \mathbf{A}\cdot\mathbf{n}\,dS = 84\pi$.

6. $\mathbf{A} = (x^3 - yz)\mathbf{i} - 2x^2\mathbf{j} + z\mathbf{k}$ taken over the entire surface of the cube $0 \le x \le a$, $0 \le y \le a$, $0 \le z \le a$.

Solution $\iint_{S_1} \mathbf{A}\cdot\mathbf{n}\,dS = \iint_{D_1} (x^3 - yz)\,dydz$

$$= \iint_{D_1} (a^3 - yz)\,dydz = a^5 - \int_0^a\int_0^a yz\,dydz = a^5 - \frac{1}{4}a^4,$$

$$\iint_{S_2} \mathbf{A}\cdot\mathbf{n}\,dS = \iint_{D_2} (-2x^2)\,dxdz,$$

$$= \int_0^a\int_0^a (-2x^2)\,dxdz = -\frac{2}{3}a^4$$

$$= \int_0^a\int_0^a yz\,dydz = \frac{1}{4}a^4,$$

$$\iint_{S_4} \mathbf{A}\cdot\mathbf{n}\,dS = \iint_{D_4} 2x^2\,dxdz = \int_0^a\int_0^a 2x^2\,dxdz = \frac{2}{3}a^4,$$

$\iint_{S_5} \mathbf{A}\cdot\mathbf{n}\,dS = \iint_{D_5} (-z)\,dxdy = 0$ (Since $z = 0$ on S_5), $\iint_{S_6} \mathbf{A}\cdot\mathbf{n}\,dS = \iint_{D_6} z\,dxdy = a\iint_{D_6} 1\,dxdy = a^3$,

Therefore $\iint_S \mathbf{A}\cdot\mathbf{n}\,dS = a^5 + a^3$.

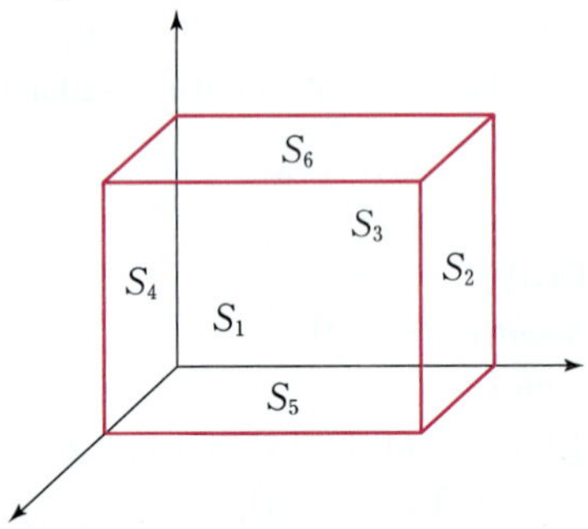

Since $\operatorname{div}\mathbf{A} = 3x^2 + 1$,

$$\iiint_V \operatorname{div}\mathbf{A}\,dV = \int_0^a\int_0^a\int_0^a (3x^2+1)\,dxdydz = a^5 + a^3.$$

7. $\mathbf{A} = ax\,\mathbf{i} + by\,\mathbf{j} + cz\,\mathbf{k}$ taken over the entire surface of the sphere of radius d and centered at the origin.

http://matrix.skku.ac.kr/cal-lab/cal-14-8-7.html

Solution

```
var('a,b,c,x,y,z')
def Div(F):
    assert(len(F) == 3)
    return
(diff(F[0],x)+diff(F[1],y)+diff(F[1],z))
Div([a*x,b*y,c*z])
```

Answer : a+b (a+b+c)

So, integral of a+b+c over the entire surface of the sphere of radius d is $\frac{4\pi}{3}d^3(a+b+c)$.

Compute the second part, and show they are equal.

8. $\mathbf{A} = 2xy\mathbf{i} + yz^2\mathbf{j} + xz\mathbf{k}$ and S is the total surface of the rectangular parallelepiped bounded by the coordinate planes and $x=1,\ y=2,\ z=3$.

http://matrix.skku.ac.kr/cal-lab/cal-15-8-Exs-8.html

Solution

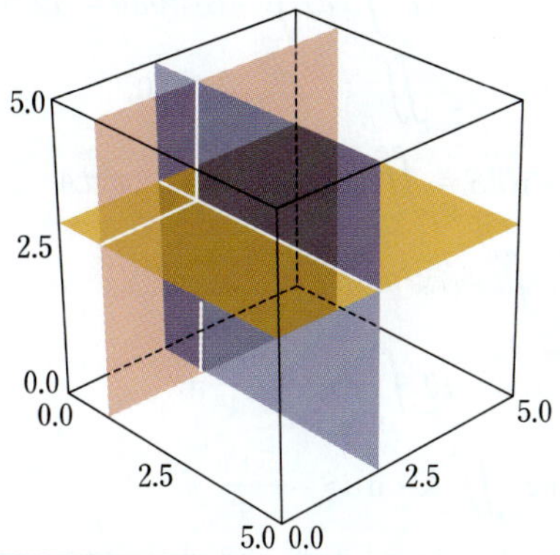


```
var('x,y,z,t')
p1 = implicit_plot3d(x==1, (x,0,5), (y, 0,5), (z, 0,5), opacity=0.2, color="red", mesh=True);
```

```
p2 = implicit_plot3d(y==2, (x,0,5), (y, 0,5), (z, 0,5), opacity=0.3, color="blue", mesh=True);
p3 = implicit_plot3d(z==3, (x,0,5), (y, 0,5),(z, 0,5), opacity=0.5, color="orange", mesh=True);
show(p1+p2+p3, aspect_ratio=1)
```

```
def Div(F):
    assert(len(F) == 3)
    return
(diff(F[0],x)+diff(F[1],y)+diff(F[1],z))
d10=Div([2*x*y,y*z^2,x*z])
integral(integral(integral(d10,x,0,1),y,0,2),z,0,3)
```

Answer : 48

Compute the second part, and show they are equal.

9. $\mathbf{A} = 2xz\mathbf{i} + yz\mathbf{j} + z^2\mathbf{k}$ over the upper half of the sphere $x^2+y^2+z^2=a^2$.

Solution A parameterization of sphere is

$\mathbf{r}(\phi,\ \theta) = \langle a\sin\phi\cos\theta,\ a\sin\phi\sin\theta,\ a\cos\phi\rangle$,

$\mathbf{A}(\mathbf{r}(\phi,\ \theta))$

$= \langle 2a^2\sin\phi\cos\phi\cos\theta,\ a^2\sin\phi\cos\phi\sin\theta,\ a^2\cos^2\phi\rangle$,

$\mathbf{r}_\phi \times \mathbf{r}_\theta = \langle a^2\sin^2\phi\cos\theta,\ a^2\sin^2\phi\sin\theta,\ a^2\sin\phi\cos\phi\rangle$,

$0 \le \phi \le \frac{\pi}{2},\ 0 \le \theta \le 2\pi$.

Hence $S = S_1 + S_2$ (See the figure).

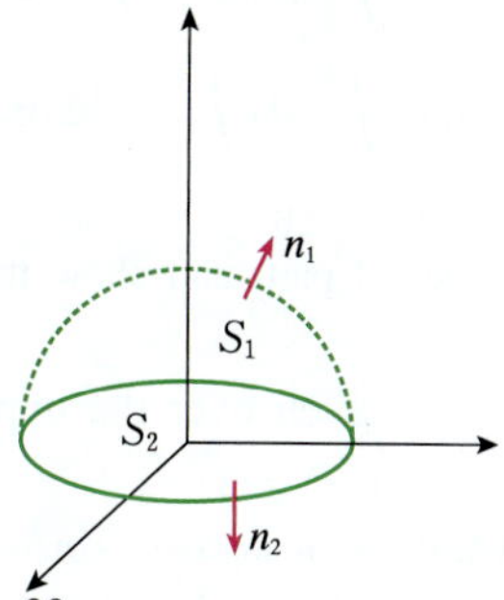

$$\iint_{S_1} \mathbf{A}\cdot\mathbf{n}\,dS = \iint_{D_1} \mathbf{A}\cdot(\mathbf{r}_\phi\times\mathbf{r}_\theta)\,dA$$

$$= \int_0^{2\pi}\int_0^{\frac{\pi}{2}} 2a^4\sin^3\phi\cos\phi\cos^2\theta + a^4\sin^3\phi\cos\phi\sin^2\theta + a^4\cos^3\phi\sin\phi\,d\phi\,d\theta$$

$$= 2a^4\int_0^{2\pi}\cos^2\theta\,d\theta\int_0^{\pi/2}\sin^3\phi\cos\phi\,d\phi + a^4\int_0^{2\pi}\sin^2\theta\,d\theta\int_0^{\pi/2}\sin^3\phi\cos\phi\,d\phi + a^4\int_0^{2\pi}d\theta\int_0^{\pi/2}\cos^3\phi\sin\phi\,d\phi$$

$$= 2a^4(\pi)\left(\frac{1}{4}\right) + a^4(\pi)\left(\frac{1}{4}\right) + a^4(2\pi)\frac{1}{4} = \frac{5a^4}{4}\pi,$$

$$\iint_{S_2} \mathbf{A}\cdot\mathbf{n}\,dS = \iint_{D_2} -z^2\,dA = 0.$$

Therefore $\iint_S \mathbf{A}\cdot\mathbf{n}\,dS = \frac{5a^4}{4}\pi$.

Compute the second part, and show they are equal.

10. $\mathbf{A} = (x^2 - yz)\mathbf{i} + (y^2 - xz)\mathbf{j} + (z^2 - xy)\mathbf{k}$ taken over the rectangular parallelepiped bounded by the coordinate planes and $x=a,\ y=b$ and $z=c$.

Solution $\frac{1}{2}(a^2b + 2ab^2)c$.

11. $\mathbf{A} = x^2\mathbf{i} + y^2\mathbf{j} + z^2\mathbf{k}$ taken over the surface of the ellipsoid $\frac{x^2}{a^2} + \frac{y^2}{b^2} + \frac{z^2}{c^2} = 1$.

Solution Let

$\mathbf{r}(\phi,\ \theta) = \langle a\sin\phi\cos\theta,\ b\sin\phi\sin\theta,\ c\cos\phi\rangle$

be a parameterization of S.

$\mathbf{A}(\mathbf{r}(\phi,\ \theta)) = \langle a^2\sin^2\phi\cos^2\theta, b^2\sin^2\phi\sin^2\theta, c^2\cos^2\phi\rangle$,

$\mathbf{r}_\phi \times \mathbf{r}_\theta = \langle bc\sin^2\phi\cos\theta,\ ac\sin^2\phi\sin\theta,\ ab\sin\phi\cos\phi\rangle$.

Hence $\iint_S \mathbf{A}\cdot\mathbf{n}\,dS = \iint_D \mathbf{A}\cdot(\mathbf{r}_\phi\times\mathbf{r}_\theta)dA$

$$= \int_0^{2\pi}\int_0^{\pi} a^2bc\sin^4\phi\cos^3\theta + ab^2c\sin^4\phi\sin^3\theta + abc^2\sin\phi\cos^3\phi\,d\phi\,d\theta$$

$$= a^2bc\int_0^{2\pi}\cos^3\theta\,d\theta\int_0^{\pi}\sin^4\phi\,d\phi + ab^2c\int_0^{2\pi}\sin^3\theta\,d\theta\int_0^{\pi}\sin^4\phi\,d\phi + abc^2\int_0^{2\pi}d\theta\int_0^{\pi}\cos^3\phi\sin\phi\,d\phi$$

$$= 0 + 0 + abc^2(2\pi)0 = 0.$$

Compute the second part, and show they are equal.

12. $\mathbf{A} = x\mathbf{i} + y\mathbf{j}$ taken over the upper half of the unit sphere $x^2+y^2+z^2=1$.

Solution $\mathbf{r}(\phi,\ \theta) = \langle \sin\phi\cos\theta,\ \sin\phi\sin\theta,\ \cos\phi\rangle$,

$\mathbf{A}(\mathbf{r}(\phi,\ \theta)) = \langle \sin\phi\cos\theta, \sin\phi\sin\theta, 0\rangle$,

$\mathbf{r}_\phi\times\mathbf{r}_\theta = \langle \sin^2\phi\cos\theta,\ \sin^2\phi\sin\theta,\ \sin\phi\cos\phi\rangle$,

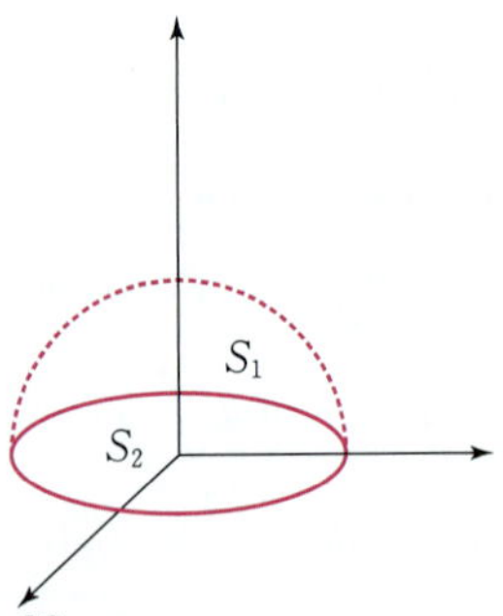

$$\iint_{S_1}\mathbf{A}\cdot\mathbf{n}\,dS = \iint_{D_1}\mathbf{A}\cdot(\mathbf{r}_\phi\times\mathbf{r}_\theta)dA$$

$$= \int_0^{2\pi}\int_0^{\frac{\pi}{2}}\sin^3\phi\cos^2\theta + \sin^3\phi\sin^2\theta\,dA$$

$$= \int_0^{2\pi}\int_0^{\frac{\pi}{2}}\sin^3\phi\,d\phi\,d\theta$$

$$= \int_0^{2\pi}d\theta\int_0^{\pi/2}\sin^3\phi\,d\phi$$

$$= (2\pi)\frac{2}{3} = \frac{4}{3}\pi,$$

$$\iint_{S_2}\mathbf{A}\cdot\mathbf{n}\,dS = \iint_{D_2}0\,dA = 0.$$

Therefore $\iint_S \mathbf{A}\cdot\mathbf{n}\,dS = \frac{4}{3}\pi$.

Compute the second part, and show they are equal.

13. $\mathbf{A} = x^3\mathbf{i} + x^2y\mathbf{j} + x^2z\mathbf{k}$ taken over the closed region of the cylinder $x^2+y^2=a^2$, bounded by the planes $z=0$ and $z=b$

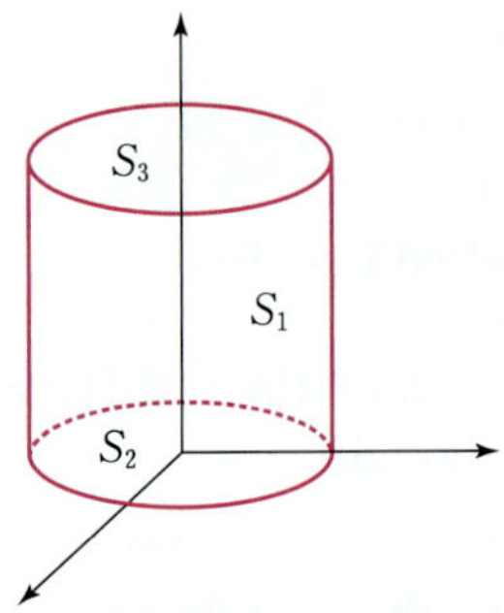

Solution $\mathbf{r}(\theta,\ z) = \langle a\cos\theta,\ a\sin\theta,\ z\rangle$,

$\mathbf{A}(\mathbf{r}(\theta,\ z)) = \langle a^3\cos^3\theta,\ a^3\cos^2\theta\sin\theta,\ a^2\cos^2\theta\, z\rangle$,

$\mathbf{r}_\theta\times\mathbf{r}_z = \langle a\cos\theta,\ a\sin\theta,\ 0\rangle$.

$$\iint_{S_1}\mathbf{A}\cdot\mathbf{n}\,dS = \iint_{D_1}\mathbf{A}\cdot(\mathbf{r}_\phi\times\mathbf{r}_\theta)dA$$

$$= \int_0^b\int_0^{2\pi} a^4\cos^4\theta + a^4\sin^2\theta\cos^2\theta\,d\theta\,dz$$

$$= \int_0^b\int_0^{2\pi} a^4\cos^2\theta\,d\theta\,dz$$

$$= a^4\int_0^b dz\int_0^{2\pi}\cos^2\theta\,d\theta = a^4 b\pi,$$

$$\iint_{S_2}\mathbf{A}\cdot\mathbf{n}\,dS = \iint_{D_2}(-x^2z)\,dA = 0,$$

$$\iint_{S_3}\mathbf{A}\cdot\mathbf{n}\,dS = \iint_{D_3}x^2z\,dA = b\iint_{D_3}x^2\,dA$$

$$= b\int_0^{2\pi}\int_0^a r^2\cos^2\theta\, r\,dr\,d\theta$$

$$= b\int_0^{2\pi}\cos^2\theta\,d\theta\int_0^a r^3\,dr = \frac{a^4b}{4}\pi.$$

Therefore $\iint_S \mathbf{A}\cdot\mathbf{n}\,dS = \frac{5a^4b}{4}\pi$.

Compute the second part, and show they are equal.

---

14. (a) Prove Green's first identity :

$$\iint_{\partial\Omega} f\nabla g\cdot\mathbf{n}\,dS = \iiint_{\Omega}\left[f\nabla^2 g + \nabla f\cdot\nabla g\right]dV.$$

(b) Let $B = \{(x,\ y,\ z) | x^2+y^2+z^2<1\}$ and $f(x,\ y,\ z) = x+y+z$. Find $\iiint_{\partial B} f\frac{\partial f}{\partial \mathbf{n}}dS$.

15. Let D be a solid surrounded by $x^2+y^2+z^2=3^2$ and $z=\sqrt{x^2+y^2}$ and vector field $F(x,\ y,\ z) = \langle yz,\ zx,\ xy+z^2\rangle$.

Find the flux of $\mathbf{F}$, that is, $\iint_{\partial D}\mathbf{F}\cdot dS$

Solution $\mathrm{div}\mathbf{F} = \frac{\partial F_1}{\partial x} + \frac{\partial F_2}{\partial y} + \frac{\partial F_3}{\partial z} = 2z$,

$$\iint_{\partial D}\mathbf{F}\cdot dS = \iiint_E \mathrm{div}\,\mathbf{F}\,dV$$

$$= \int_0^{2\pi}\int_0^{\frac{\pi}{4}}\int_0^3 2\rho\cos\phi\, \rho^2\sin\phi\, d\rho d\phi\, d\theta$$

$$= 2\int_0^{2\pi} 1 d\theta \int_0^{\frac{\pi}{4}} \cos\phi\, \sin\phi\, d\phi \int_0^3 \rho^3 d\rho$$

$$= 4\pi\left(-\frac{1}{4}\right)\frac{81}{4} = \frac{81}{4}\pi.$$

16. Evaluate $\iint_S \mathbf{F} \cdot dS$.

Here $\mathbf{F}(x,y,z) = 3xy^2\mathbf{i} + xe^z\mathbf{j} + z^3\mathbf{k}$, and S is the boundary of the solid surrounded by $y^2+z^2=1,\ x=-1,\ x=2$.

Solution $\text{div}\mathbf{F} = 3y^2+3z^2$.

Let $x=x,\ y=r\cos\theta,\ z=r\sin\theta$.

$$\iint_S \mathbf{F}\cdot dS$$

$$= \int_0^{2\pi}\int_0^1\int_{-1}^2 (3r^2\cos^2\theta + 3r^2\sin^2\theta) \cdot r\, dx dr d\theta$$

$$= 3\int_0^{2\pi}\int_0^1\int_{-1}^2 r^3\, dx dr d\theta = \frac{9\pi}{2}.$$

17. Let S be a surface $x^2+y^2=9$ between $z=0$ and $z=2$ and $\mathbf{F}(x,y,t) = y\mathbf{i} + y\mathbf{j} - xyz^5\mathbf{k}$. Find a flux $\iint_S \mathbf{F}\cdot dS$.

Solution $\text{div}\mathbf{F} = \frac{\partial F_1}{\partial x} + \frac{\partial F_2}{\partial y} + \frac{\partial F_3}{\partial z} = 0+1+xyz^5.$

$$\iint_S \mathbf{F}\cdot dS = \iiint_E \text{div}\mathbf{F}\, dV$$

$$= VIP\iiint_E (5xyz^5+1) dV$$

$$= \iint_D \int_0^2 [(5xyz^5+1)dz] dA$$

$$= \iint_D (32xy+2) dA$$

$$= \int_0^{2\pi}\int_0^3 (32r^2\cos\theta\sin\theta + 2) r\, dr d\theta$$

$$= 18\pi.$$

18. Evaluate the surface integral $\iint_S \mathbf{F}\cdot dS$, where $\mathbf{F}(x,y,z) = (x^{2010}+y^{2010})\mathbf{i} - 2010x^{2009}y\mathbf{j} + z\mathbf{k}$ and S is the part of surface of the paraboloid $z = 1-x^2-y^2$ above xy-plane.

Solution Let $S_1 = \{(x,y,0)\,|\,x^2+y^2 \le 1\}$ and E be the region bounded by S and S_1.

Then, with the aid of the Divergence Theorem, $\iint_{S\cup S_1} \mathbf{F}\cdot dS = \iiint_E \nabla\cdot\mathbf{F}\, dV$. Since $n=(0,0,-1)$ and $z=0$ on S_1, we can see that $\iint_{S_1} \mathbf{F}\cdot dS = 0$. On the other hand, $\nabla\cdot\mathbf{F} = 1$, and therefore, we have $\iint_S \mathbf{F}\cdot dS = \iiint_E dV$ = volume of E, which can be computed as follows:

$$\iiint_E dV = \iint_{S_1} (1-x^2-y^2) dA = \int_0^1\int_0^{2\pi} (1-r^2) r dr d\theta$$

$$= 2\pi\left[\frac{r^2}{2} - \frac{r^4}{4}\right]\Big|_0^1 = \frac{\pi}{2}.$$

19. Let $\mathbf{F}(x,y,z) = (x+y^2)\mathbf{i} + (y+z^2)\mathbf{j} + (z+x^2)\mathbf{k}$. Evaluate $\iint_S \mathbf{F}\cdot dS$, where S is the part of the sphere $x^2+y^2+z^2=4,\ z \ge \sqrt{3}$.

Solution Note that $S_1 = S\cup S_2$

$S = \{(x,y,z)\,|\,x^2+y^2+z^2=4,\ z\ge\sqrt{3}\}$

$S_2 = \{(x,y,z)\,|\,x^2+y^2 \le 1,\ z=\sqrt{3}\}$.

Using the Divergence Theorem and polar coordinates $x=\cos\theta,\ y=\sin\theta$, where $E = \{(r,\theta,z)\,|\,0\le r\le 1,\ 0\le\theta\le 2\pi,\ \sqrt{3}\le z\le\sqrt{4-r^2}\}$,

we compute $\iint_S \mathbf{F}\cdot dS = \iint_{S_1}\mathbf{F}\cdot dS - \iint_{S_2}\mathbf{F}\cdot dS$.

$$S_1 : \iint_{S_1}\mathbf{F}\cdot dS = \iiint_E \text{div}\mathbf{F}\cdot dS$$

$$= 3\int_0^{2\pi}\int_0^1 \sqrt{4-r^2} - \sqrt{3}\, r dr d\theta$$

$$= 6\pi\left[-\frac{1}{3}(4-r^2)^{\frac{3}{2}} - \frac{\sqrt{3}}{2}r^2\right]\Big|_0^1$$

$$= 6\pi\left(-\frac{3\sqrt{3}}{2} - \frac{8}{3}\right) = (16-9\sqrt{3})\pi.$$

$$S_2 : \mathbf{n} = (0,0,-1)\quad \iint_{S_2}\mathbf{F}\cdot dS = \iint_{S_2}\mathbf{F}\cdot\mathbf{n}\, dS$$

$$= \int_{S_2} -\sqrt{3} - x^2\, dS$$

$$= \int_0^{2\pi}\int_0^1 (-\sqrt{3} - r^2\cos^2\theta) r dr d\theta$$

$$= \int_0^{2\pi}\left(-\frac{\sqrt{3}}{2} - \frac{1}{4}\cos^2\theta\right) d\theta = \left(-\sqrt{3} - \frac{1}{4}\right)\pi.$$

$S : S_1 - S_2$

$$\iint_S \mathbf{F}\cdot dS = \iint_{S_1}\mathbf{F}\cdot dS - \iint_{S_2}\mathbf{F}\cdot dS$$

$$= (16-9\sqrt{3})\pi + \left(\frac{1}{4}+\sqrt{3}\right)\pi = \left(\frac{25}{4} - 8\sqrt{3}\right)\pi.$$

20. If $\mathbf{A} = (x^3-yz)\mathbf{i} - 2x^3y\mathbf{j} + 2\mathbf{k}$, evaluate $\iiint_V (\nabla\cdot\mathbf{A}) dV$ over the volume of a cube of side v.

21. Evaluate $\iiint_V (\nabla\cdot\mathbf{B}) dV$ over the solid region of the sphere $x^2+y^2+z^2=a^2$ when $\mathbf{B} = px\mathbf{i} + qy\mathbf{j} + rz\mathbf{k}$ where $p,\ q,\ r$ are constants.

Table

Rules for Inequalities

1. If $a<b$, $b<c$, then $a<c$.
2. If $a<b$, then $a+c<b+c$, $a-c<b-c$.
3. If $a<b$, $c>0$, then $ac<bc$, $\frac{a}{c}<\frac{b}{c}$.
4. If $a<b$, then $-b<-a$.
5. If $a<b$, $c<d$, then $a+c<b+d$, $a-d<b-c$.
6. If $0<a<b$, $0<c<d$, then $ac<bd$, $\frac{a}{d}<\frac{b}{c}$.
7. If $0<a<b$, then $a^2<b^2$, $\sqrt{a}<\sqrt{b}$, $\frac{1}{b}<\frac{1}{a}$.
8. If $a<b<0$, then $b^2<a^2$, $\frac{1}{b}<\frac{1}{a}$.
9. If $a<b$, then $a<\frac{a+b}{2}<b$.

Special Functions

1. Exponential Functions

If $a>0$ and $a\neq 1$, then a function of the form $f(x)=a^x$ is called an exponential function.
The number a is called the base and x is called the exponential.
http://matrix.skku.ac.kr/Mobile-Sage-G/sage-grapher.html

2. Logarithmic Functions

The logarithmic with base $a>0$ and $a\neq 1$, is defined by $y=\log_a x$.

3. Inverse Trigonometric Functions

$x=\sin y \Leftrightarrow y=\sin^{-1}x \left(-1\le x\le 1,\ -\frac{\pi}{2}\le y\le\frac{\pi}{2}\right)$

$x=\cos y \Leftrightarrow y=\cos^{-1}x\,(-1\le x\le 1,\ 0\le y\le\pi)$

$\tan y=x \Leftrightarrow y=\tan^{-1}x \left(-\infty<x<\infty,\ -\frac{\pi}{2}<y<\frac{\pi}{2}\right)$

$x=\cot y \Leftrightarrow y=\cot^{-1}x\ (-\infty<x<\infty,\ 0<y<\pi)$

$x=\sec y \Leftrightarrow y=\sec^{-1}x$

$\left(-\infty<x\le -1,\ 1\le x<\infty,\ -\pi\le y<-\frac{\pi}{2},\ 0\le y<\frac{\pi}{2}\right)$

$x=\csc y \Leftrightarrow y=\csc^{-1}x$

$\left(-\infty<x\le -1,\ 1\le x<\infty,\ -\pi\le y<-\frac{\pi}{2},\ 0\le y<\frac{\pi}{2}\right)$

4. Hyperbolic Functions

$\sinh x=\frac{e^x-e^{-x}}{2}$ $\quad$ $\operatorname{csch} x=\frac{1}{\sinh x}$

$\cosh x=\frac{e^x+e^{-x}}{2}$ $\quad$ $\operatorname{sech} x=\frac{1}{\cosh x}$

$\tanh x=\frac{\sinh x}{\cosh x}$ $\quad$ $\coth x=\frac{\cosh x}{\sinh x}$

5. Inverse Hyperbolic Functions

$\sinh^{-1}x=\ln(x+\sqrt{x^2+1})\quad (x\in R)$

$\cosh^{-1}x=\ln(x+\sqrt{x^2-1})\quad (x\ge 1)$

$\tanh^{-1}x=\frac{1}{2}\ln\frac{1+x}{1-x}\quad (|x|<1)$

$\coth^{-1}x=\frac{1}{2}\ln\frac{x+1}{x-1}\quad (|x|>1)$

$\operatorname{sech}^{-1}x=\ln\left(\frac{1+\sqrt{1-x^2}}{x}\right)\quad (0<x<1)$

$\operatorname{csch}^{-1}x=\ln\left(\frac{1}{x}+\frac{\sqrt{1-x^2}}{|x|}\right)\quad (x\neq 0)$

Formulas of Trigonometric Functions

1. Addition and Subtraction Formulas

$\sin(x\pm y)=\sin x\cos y\pm\cos x\sin y$

$\cos(x\pm y)=\cos x\cos y\mp\sin x\sin y$

$\tan(x\pm y)=\frac{\tan x\pm\tan y}{1\mp\tan x\tan y}$

2. Double Angle Formulas

$\sin 2x=2\sin x\cos x$

$\cos 2x=\cos^2 x-\sin^2 x$

$\tan 2x=\frac{2\tan x}{1-\tan^2 x}$

3. Half Angle Formulas

$\sin^2\frac{x}{2}=\frac{1-\cos x}{2}$

$\cos^2\frac{x}{2}=\frac{1+\cos x}{2}$

$\tan^2\frac{x}{2}=\frac{1-\cos x}{1+\cos x}=\frac{\sin^2 x}{(1+\cos x)^2}=\frac{(1-\cos x)^2}{\sin^2 x}=(\csc x-\cot x)^2$

4. Triple Angle Formulas

$\sin 3x=3\sin x-4\sin^3 x$

$\cos 3x=4\cos^3 x-3\cos x$

$\tan 3x=\frac{3\tan x-\tan^3 x}{1-3\tan^2 x}$

5. Product-to-Sum

$\sin x\cos y=\frac{1}{2}\{\sin(x+y)+\sin(x-y)\}$

$\cos x\sin y=\frac{1}{2}\{\sin(x+y)-\sin(x-y)\}$

$\cos x\cos y=\frac{1}{2}\{\cos(x+y)+\cos(x-y)\}$

$\sin x\sin y=-\frac{1}{2}\{\cos(x+y)-\cos(x-y)\}$

6. Sun-to-Product

$\sin A+\sin B=2\sin\frac{A+B}{2}\cos\frac{A-B}{2}$

$\sin A-\sin B=2\cos\frac{A+B}{2}\sin\frac{A-B}{2}$

$\cos A+\cos B=2\cos\frac{A+B}{2}\cos\frac{A-B}{2}$

$\cos A-\cos B=-2\sin\frac{A+B}{2}\sin\frac{A-B}{2}$

7. Linear Combination

$a\sin x+b\cos x=\sqrt{a^2+b^2}\sin(x+\alpha)$

where, $\cos\alpha=\frac{a}{\sqrt{a^2+b^2}}$, $\sin\alpha=\frac{b}{\sqrt{a^2+b^2}}$

$a\sin x+b\cos x=\sqrt{a^2+b^2}\cos(x-\beta)$

where, $\cos\beta=\frac{b}{\sqrt{a^2+b^2}}$, $\sin\beta=\frac{a}{\sqrt{a^2+b^2}}$

Limits

http://matrix.skku.ac.kr/cal-lab/SKKU-Cell-Epsilon-Delta.html

http://matrix.skku.ac.kr/cal-lab/cal-Limit.html

1. Limit Laws

Suppose that the limits $\lim_{x\to a}f(x)$ and $\lim_{x\to a}g(x)$ exist and c is a constant.

Then

$\lim_{x\to a}[f(x)+g(x)]=\lim_{x\to a}f(x)+\lim_{x\to a}g(x)$

$\lim_{x\to a}[f(x)-g(x)]=\lim_{x\to a}f(x)-\lim_{x\to a}g(x)$

$\lim_{x\to a}cf(x)=c\lim_{x\to a}f(x)$

$\lim_{x\to a}[f(x)g(x)]=\lim_{x\to a}f(x)\lim_{x\to a}g(x)$

$\lim_{x\to a}\frac{f(x)}{g(x)}=\frac{\lim_{x\to a}f(x)}{\lim_{x\to a}g(x)}$ if $\lim_{x\to a}g(x)\neq 0$

$\lim_{x\to a}[f(x)]^n=[\lim_{x\to a}f(x)]^n$, where n is a positive integer.

$\lim_{x\to a}\sqrt[n]{f(x)}=\sqrt[n]{\lim_{x\to a}f(x)}$, where n is a positive integer.

(If n is even, we require $\lim_{x\to a}f(x)\geq 0$)

http://matrix.skku.ac.kr/cal-lab/cal-2-1-7.html

http://matrix.skku.ac.kr/cal-lab/cal-2-2-3.html

2. Squeeze Theorem (or Sandwich Theorem)

If $f(x)\leq g(x)\leq h(x)$, when x is near a and

$\lim_{x\to a}f(x)=\lim_{x\to a}h(x)=L$, then $\lim_{x\to a}g(x)=L$.

http://matrix.skku.ac.kr/cal-lab/cal-2-1-12.html

Derivatives

1. Differentiation Rules

$\frac{d}{dx}(cx^n)=ncx^{n-1}$ where n is a positive integer and c is a constant.

$\frac{d}{dx}[cf(x)]=c\frac{d}{dx}f(x)$

$\frac{d}{dx}[f(x)+g(x)]=\frac{d}{dx}f(x)+\frac{d}{dx}g(x)$

$\frac{d}{dx}[f(x)-g(x)]=\frac{d}{dx}f(x)-\frac{d}{dx}g(x)$

$\frac{d}{dx}(f(x)g(x))=f(x)\frac{d}{dx}(g(x))+g(x)\frac{d}{dx}(f(x))$

$\left\{\frac{f(x)}{g(x)}\right\}'=\frac{f'(x)g(x)-f(x)g'(x)}{\{g(x)\}^2}$

2. Chain Rule

If n is any real number and $u=g(x)$ is differentiable and $y=u^n$, then

$\frac{dy}{dx}=n[g(x)]^{n-1}\cdot g'(x).$

3. Parametric Formula

For the parametric equations: $x=f(t)$ and $y=g(t)$,

$\frac{dy}{dx}=\frac{dy/dt}{dx/dt}$, $\frac{dx}{dt}\neq 0$

4. Implicit Function

Let $F(x,y)=0$ be an implicit function.

$D_xF(x,y)\,dx+D_yF(x,y)\,dy=0$

$\frac{dy}{dx}=-\frac{D_xF(x,y)}{D_yF(x,y)}$ where $D_yF(x,y)\neq 0$.

5. Inverse Function

$(f^{-1})'(x)=\frac{1}{f'(f^{-1}(x))}$ or $\frac{dy}{dx}=\frac{1}{dx/dy}$

6. Trigonometric Functions

$\frac{d}{dx}(\sin x)=\cos x$ $\quad$ $\frac{d}{dx}(\csc x)=-\csc x\cot x$

$\frac{d}{dx}(\cos x)=-\sin x$ $\quad$ $\frac{d}{dx}(\sec x)=\sec x\tan x$

$\frac{d}{dx}(\tan x)=\sec^2 x$ $\quad$ $\frac{d}{dx}(\cot x)=-\csc^2 x$

http://matrix.skku.ac.kr/cal-lab/cal-3-3-Exm24.html

7. Inverse Trigonometric Functions

$$\frac{d}{dx}(\sin^{-1}x)=\frac{1}{\sqrt{1-x^2}} \qquad \frac{d}{dx}(\csc^{-1}x)=-\frac{1}{x\sqrt{x^2-1}}$$

$$\frac{d}{dx}(\cos^{-1}x)=-\frac{1}{\sqrt{1-x^2}} \qquad \frac{d}{dx}(\sec^{-1}x)=\frac{1}{x\sqrt{x^2-1}}$$

$$\frac{d}{dx}(\tan^{-1}x)=-\frac{1}{1+x^2} \qquad \frac{d}{dx}(\cot^{-1}x)=-\frac{1}{1+x^2}$$

http://matrix.skku.ac.kr/cal-lab/cal-3-2-13.html

8. Logarithmic Functions

$$\frac{d}{dx}(\ln x)=\frac{1}{x} \qquad \frac{d}{dx}(\log_a x)=\frac{1}{x\ln a}$$

$$\frac{d}{dx}[\ln f(x)]=\frac{f'(x)}{f(x)} \qquad \frac{d}{dx}[\log_a f(x)]=\frac{f'(x)}{f(x)\ln a}$$

9. Exponential Functions

$$\frac{d}{dx}(e^x)=e^x \qquad \frac{d}{dx}(a^x)=a^x\ln a$$

$$\frac{d}{dx}(e^{f(x)})=f'(x)e^{f(x)} \qquad \frac{d}{dx}(a^{f(x)})=f'(x)a^{f(x)}\ln a$$

10. Hyperbolic Functions

$$\frac{d}{dx}(\sinh x)=\cosh x \qquad \frac{d}{dx}(\coth x)=-\operatorname{csch}^2 x$$

$$\frac{d}{dx}(\cosh x)=\sinh x \qquad \frac{d}{dx}(\operatorname{sech} x)=-\operatorname{sech} x\tanh x$$

$$\frac{d}{dx}(\tanh x)=\operatorname{sech}^2 x \qquad \frac{d}{dx}(\operatorname{csch} x)=-\operatorname{csch} x\coth x$$

11. Inverse Hyperbolic Functions

$$\frac{d}{dx}(\sinh^{-1}x)=\frac{1}{\sqrt{1+x^2}} \qquad (x\in R)$$

$$\frac{d}{dx}(\cosh^{-1}x)=\frac{1}{\sqrt{x^2-1}} \qquad (x>1)$$

$$\frac{d}{dx}(\tanh^{-1}x)=\frac{1}{1-x^2} \qquad (|x|<1)$$

$$\frac{d}{dx}(\coth^{-1}x)=\frac{1}{1-x^2} \qquad (|x|>1)$$

$$\frac{d}{dx}(\operatorname{sech}^{-1}x)=-\frac{1}{x\sqrt{1-x^2}} \qquad (0<x<1)$$

$$\frac{d}{dx}(\operatorname{csch}^{-1}x)=-\frac{1}{|x|\sqrt{1-x^2}} \qquad (x\neq 0)$$

General Rules of Integration

http://matrix.skku.ac.kr/cal-lab/cal-RiemannSum.html

http://matrix.skku.ac.kr/cal-lab/cal-6-3-17.html

http://matrix.skku.ac.kr/cal-lab/cal-12-1-Rotations-B.html

1. Basic Forms

$$\int a\,dx=ax+C$$

$$\int x^n\,dx=\frac{x^{n+1}}{n+1}+C \quad (n\neq -1)$$

$$\int \frac{1}{x}dx=\ln|x|+C$$

$$\int e^x\,dx=e^x+C$$

http://matrix.skku.ac.kr/cal-lab/cal-7-7-Exm-8.html

$$\int a^x\,dx=\frac{a^x}{\ln a}+C,\ a>0$$

http://matrix.skku.ac.kr/cal-lab/cal-5-2-20.html

$$\int \frac{1}{\sqrt{x^2+A}}dx=\ln|x+\sqrt{x^2+A}|+C\,(A\neq 0)$$

2. Trigonometric Forms

http://matrix.skku.ac.kr/cal-lab/cal-5-4-exm-7.html

http://matrix.skku.ac.kr/cal-lab/cal-8-3-3.html

$$\int \sin x\,dx=-\cos x+C$$

$$\int \cos x\,dx=\sin x+C$$

$$\int \sin^2 x\,dx=\frac{x}{2}-\frac{\sin 2x}{4}+C$$

$$\int \cos^2 x\,dx=\frac{x}{2}+\frac{\sin 2x}{4}+C$$

$$\int \sin ax\cos bx\,dx=-\frac{\cos(a+b)x}{2(a+b)}-\frac{\cos(a-b)x}{2(a-b)}+C \quad (a^2\neq b^2)$$

$$\int \sin ax\sin bx\,dx=\frac{\sin(a-b)x}{2(a-b)}-\frac{\sin(a+b)x}{2(a+b)}+C \quad (a^2\neq b^2)$$

$$\int \cos ax\cos bx\,dx=\frac{\sin(a-b)x}{2(a-b)}+\frac{\sin(a+b)x}{2(a+b)}+C \quad (a^2\neq b^2)$$

$$\int \tan x\,dx=\ln|\sec x|+C$$

$$\int \cot x\,dx=\ln|\sin x|+C$$

$$\int \sec x\,dx=\ln|\sec x+\tan x|+C$$

$$\int \csc x\,dx=\ln|\csc x+\cot x|+C$$

$$\int \sec^2 x\,dx=\tan x+C$$

$$\int \csc^2 x\,dx=-\cot x+C$$

$$\int \sec x\tan x\,dx=\sec x+C$$

$$\int \csc x\cot x\,dx=-\csc x+C$$

$$\int \sin^n x\,dx=-\frac{1}{n}\sin^{n-1}x\cos x+\frac{n-1}{n}\int \sin^{n-2}x\,dx \quad (n=1,2,\cdots)$$

$$\int \cos^n x\,dx=\frac{1}{n}\cos^{n-1}x\sin x+\frac{n-1}{n}\int \cos^{n-2}x\,dx \quad (n=1,2,\cdots)$$

$$\int \tan^n x\,dx=\frac{1}{n-1}\tan^{n-1}x-\int \tan^{n-2}x\,dx \quad (n\neq 1)$$

$$\int \cot^n x\,dx=-\frac{1}{n-1}\cot^{n-1}x-\int \cot^{n-2}x\,dx \quad (n\neq 1)$$

$$\int \sec^n x\,dx=\frac{1}{n-1}\sec^{n-2}x\tan x+\frac{n-2}{n-1}\int \sec^{n-2}x\,dx \quad (n=1,2,\cdots)$$

$$\int \csc^n x\,dx = -\frac{1}{n-1}\csc^{n-2}x\cot x + \frac{n-2}{n-1}\int \csc^{n-2}x\,dx \quad (n=1,\,2,\,\cdots)$$

3. Hyperbolic Forms

$$\int \sinh ax\,dx = \frac{1}{a}\cosh ax + C$$

$$\int \cosh ax\,dx = \frac{1}{a}\sinh ax + C$$

$$\int \tanh ax\,dx = \frac{1}{a}\ln(\cosh ax) + C$$

http://matrix.skku.ac.kr/cal-lab/cal-8-1-9.html

$$\int \coth ax\,dx = \frac{1}{a}\ln|\sinh ax| + C$$

$$\int \operatorname{sech} ax\,dx = \frac{1}{a}\sin^{-1}(\tanh ax) + C$$

$$\int \operatorname{csch} ax\,dx = \frac{1}{a}\ln\left|\tanh\frac{ax}{2}\right| + C$$

$$\int \sinh^2 ax\,dx = \frac{\sinh 2ax}{4a} - \frac{x}{2} + C$$

$$\int \cosh^2 ax\,dx = \frac{\sinh 2ax}{4a} + \frac{x}{2} + C$$

$$\int \tanh^2 ax\,dx = x - \frac{1}{a}\tanh ax + C$$

$$\int \coth^2 ax\,dx = x - \frac{1}{a}\coth ax + C$$

$$\int \operatorname{sech}^2 ax\,dx = \frac{1}{a}\tanh ax + C$$

$$\int \operatorname{csch}^2 ax\,dx = -\frac{1}{a}\coth ax + C$$

4. Forms Involving $a^2 \pm x^2$

$$\int \frac{1}{x^2-a^2}\,dx = \frac{1}{2a}\ln\left|\frac{x-a}{x+a}\right| + C \quad (a\neq 0)$$

$$\int \frac{1}{a^2-x^2}\,dx = \frac{1}{2a}\ln\left|\frac{x+a}{x-a}\right| + C \quad (a\neq 0)$$

$$\int \frac{1}{\sqrt{a^2-x^2}}\,dx = \sin^{-1}\left(\frac{x}{a}\right) + C \quad (a\neq 0)$$

$$\int \sqrt{a^2-x^2}\,dx = \frac{a^2}{2}\sin^{-1}\frac{x}{a} + \frac{x}{2}\sqrt{a^2-x^2} + C \quad (a\neq 0)$$

$$\int \frac{1}{x^2+a^2}\,dx = \frac{1}{a}\tan^{-1}\left(\frac{x}{a}\right) + C \quad (a\neq 0)$$

$$\int \frac{1}{\sqrt{a^2+x^2}}\,dx = \ln|x+\sqrt{a^2+x^2}| + C \quad (a\neq 0)$$

$$\int \sqrt{a^2+x^2}\,dx = \frac{1}{2}x\sqrt{a^2+x^2} + \frac{a^2}{2}\ln|x+\sqrt{a^2+x^2}| + C \quad (a\neq 0)$$

$$\int \frac{1}{\sqrt{x^2-a^2}}\,dx = \ln|x+\sqrt{x^2-a^2}| + C \quad (a\neq 0)$$

$$\int \frac{1}{x\sqrt{x^2-a^2}}\,dx = \frac{1}{a}\sec^{-1}\frac{x}{a} + C \quad (a\neq 0)$$

$$\int \sqrt{x^2-a^2}\,dx = \frac{1}{2}x\sqrt{x^2-a^2} + \frac{a^2}{2}\ln|x+\sqrt{x^2-a^2}| + C \quad (a\neq 0)$$

5. Inverse Trigonometric Forms

$$\int \sin^{-1}x\,dx = x\sin^{-1}x + \sqrt{1-x^2} + C$$

$$\int \cos^{-1}x\,dx = x\cos^{-1}x - \sqrt{1-x^2} + C$$

$$\int \tan^{-1}x\,dx = x\tan^{-1}x - \frac{1}{2}\ln(1+x^2) + C$$

http://myhandbook.info/form_integ.html

Series

1. Taylor Series

$$f(x) = \sum_{n=0}^{\infty}\frac{f^{(n)}(a)}{n!}(x-a)^n$$

$$= f(a) + \frac{f'(a)}{1!}(x-a) + \frac{f''(a)}{2!}(x-a)^2 + \frac{f'''(a)}{3!}(x-a)^3 + \cdots$$

http://matrix.skku.ac.kr/cal-lab/cal-10-5-Exm-11.html

2. Maclaurin Series

$$f(x) = \sum_{n=0}^{\infty}\frac{f^{(n)}(0)}{n!}x^n = f(0) + \frac{f'(0)}{1!}x + \frac{f''(0)}{2!}x^2 + \cdots$$

$$\frac{1}{1-x} = \sum_{n=0}^{\infty}x^n = 1 + x + x^2 + x^3 + \cdots, \qquad (-1,\ 1)$$

$$e^x = \sum_{n=0}^{\infty}\frac{x^n}{n!} = 1 + \frac{x}{1!} + \frac{x^2}{2!} + \frac{x^3}{3!} + \cdots, \qquad (-\infty,\ \infty)$$

$$\sin x = \sum_{n=0}^{\infty}(-1)^n\frac{x^{2n+1}}{(2n+1)!} = x - \frac{x^3}{3!} + \frac{x^5}{5!} - \frac{x^7}{7!} + \cdots, \quad (-\infty,\ \infty)$$

$$\cos x = \sum_{n=0}^{\infty}(-1)^n\frac{x^{2n}}{(2n)!} = 1 - \frac{x^2}{2!} + \frac{x^4}{4!} - \frac{x^6}{6!} + \cdots, \qquad (-\infty,\ \infty)$$

$$\tan^{-1}x = \sum_{n=0}^{\infty}(-1)^n\frac{x^{2n+1}}{2n+1} = x - \frac{x^3}{3} + \frac{x^5}{5} - \frac{x^7}{7} + \cdots, \qquad [-1,\ 1]$$

$$\ln(1+x) = \sum_{n=1}^{\infty}(-1)^{n-1}\frac{x^n}{n} = x - \frac{x^2}{2} + \frac{x^3}{3} - \frac{x^4}{4} + \cdots, \quad [-1,\ 1]$$

3. Binomial Series

If a and b are any real numbers and k is a positive integer, we have

$$(a+b)^k = \sum_{n=0}^{k}\binom{k}{n}a^{k-n}b^n$$

$$= a^k + ka^{k-1}b + \frac{k(k-1)}{2!}a^{k-2}b^2 + \frac{k(k-1)(k-2)}{3!}a^{k-3}b^3$$

$$+ \cdots + \frac{k(k-1)(k-2)\cdots(k-n+1)}{n!}a^{k-n}b^n + \cdots + kab^{k-1} + b^k$$

where $\binom{k}{0} = 1$ and $\binom{k}{n} = \dfrac{k(k-1)(k-2)\cdots(k-n+1)}{n!}$, $1 \le n \le k$.

Vectors

1. Dot Product

Let $\mathbf{a} = \langle a_1,\ a_2,\ \cdots,\ a_n\rangle$ and $\mathbf{b} = \langle b_1,\ b_2,\ \cdots,\ b_n\rangle$.

$\mathbf{a}\cdot\mathbf{b}=a_1b_1+a_2b_2+\cdots+a_nb_n=\sum_{i=1}^{n}a_ib_i$

http://matrix.skku.ac.kr/cal-lab/cal-11-3-2.html

2. Projections

Scalar projection of b onto a : $\text{comp}_{\mathbf{a}}\mathbf{b}=\dfrac{\mathbf{a}\cdot\mathbf{b}}{|\mathbf{a}|}$,

Vector projection of b onto a : $\text{proj}_{\mathbf{a}}\mathbf{b}=\left(\dfrac{\mathbf{a}\cdot\mathbf{b}}{|\mathbf{a}|}\right)\dfrac{\mathbf{a}}{|\mathbf{a}|}=\dfrac{\mathbf{a}\cdot\mathbf{b}}{|\mathbf{a}|^2}\mathbf{a}$

http://matrix.skku.ac.kr/cal-lab/cal-11-5-20.html

3. Definition and Properties of Cross Product

http://matrix.skku.ac.kr/cal-lab/cal-11-4-Exs-6.html

$$\mathbf{a}\times\mathbf{b}=\begin{vmatrix}\mathbf{i}&\mathbf{j}&\mathbf{k}\\a_1&a_2&a_3\\b_1&b_2&b_3\end{vmatrix}=\begin{vmatrix}a_2&a_3\\b_2&b_3\end{vmatrix}\mathbf{i}-\begin{vmatrix}a_1&a_3\\b_1&b_3\end{vmatrix}\mathbf{j}+\begin{vmatrix}a_1&a_2\\b_1&b_2\end{vmatrix}\mathbf{k}$$

Let two nonzero vectors a and b are two sides of a parallelogram, then the area of the parallelogram is $A=\frac{1}{2}|\mathbf{a}\times\mathbf{b}|$.

http://matrix.skku.ac.kr/cal-lab/cal-11-4-10.html

4. Rules of Limits

$\lim_{t\to a}c\mathbf{u}(t)=c\lim_{t\to a}\mathbf{u}(t)$

$\lim_{t\to a}[\mathbf{u}(t)+\mathbf{v}(t)]=\lim_{t\to a}\mathbf{u}(t)+\lim_{t\to a}\mathbf{v}(t)$

$\lim_{t\to a}\{f(t)\mathbf{u}(t)\}=\lim_{t\to a}f(t)\lim_{t\to a}\mathbf{u}(t)$

$\lim_{t\to a}[\mathbf{u}(t)\cdot\mathbf{v}(t)]=\lim_{t\to a}\mathbf{u}(t)\cdot\lim_{t\to a}\mathbf{v}(t)$

$\lim_{t\to a}[\mathbf{u}(t)\times\mathbf{v}(t)]=\lim_{t\to a}\mathbf{u}(t)\times\lim_{t\to a}\mathbf{v}(t)$

$\lim_{s\to s_0}\mathbf{u}(g(s))=\lim_{t\to a}\mathbf{u}(t),\ \lim_{s\to s_0}g(s)=a$

http://matrix.skku.ac.kr/cal-lab/9-5-Example-7.html

5. Rules of Differentiation

http://myhandbook.info/form_diff.html

$\frac{d}{dt}[\mathbf{u}(t)+\mathbf{v}(t)]=\mathbf{u}'(t)+\mathbf{v}'(t)$

$\frac{d}{dt}[c\mathbf{u}(t)]=c\mathbf{u}'(t)$ where c is a scalar

$\frac{d}{dt}[f(t)\mathbf{u}(t)]=f'(t)\mathbf{u}(t)+f(t)\mathbf{u}'(t)$

$\frac{d}{dt}[\mathbf{u}(t)\cdot\mathbf{v}(t)]=\mathbf{u}'(t)\cdot\mathbf{v}(t)+\mathbf{u}(t)\cdot\mathbf{v}'(t)$

$\frac{d}{dt}[\mathbf{u}(t)\times\mathbf{v}(t)]=\mathbf{u}'(t)\times\mathbf{v}(t)+\mathbf{u}(t)\times\mathbf{v}'(t)$

$\frac{d}{dt}\left[\mathbf{u}\left(f(t)\right)\right]=f'(t)\mathbf{u}'(f(t))$

http://matrix.skku.ac.kr/cal-lab/cal-4-3-24.html

http://matrix.skku.ac.kr/cal-lab/cal-4-2-9.html

6. Derivative of a Vector Function

If $\mathbf{r}(t)=\langle f(t),g(t),h(t)\rangle$, then $\mathbf{r}'(t)=\langle f'(t),g'(t),h'(t)\rangle$

http://matrix.skku.ac.kr/cal-lab/cal-13-2-Exm6.html

http://matrix.skku.ac.kr/cal-lab/cal-12-4-3.html

http://matrix.skku.ac.kr/cal-lab/cal-13-2-20.html

7. Integral of a Vector Function

If $\mathbf{r}(t)=\langle f(t),g(t),h(t)\rangle$,

then $\int\mathbf{r}(t)\,dt=\left\langle\int f'(t)\,dt,\int g'(t)\,dt,\int h'(t)\,dt\right\rangle$

8. Arc Length

$L=\int_a^b|\mathbf{r}'(t)|\,dt$

http://matrix.skku.ac.kr/cal-lab/cal-8-1-9.html

http://matrix.skku.ac.kr/cal-lab/cal-8-1-11.html

http://matrix.skku.ac.kr/cal-lab/cal-13-3-2.html

9. Curvature

$\kappa(t)=\dfrac{|\mathbf{r}'(t)\times\mathbf{r}''(t)|}{|\mathbf{r}'(t)|^3}$, $\kappa(x)=\dfrac{|f''(x)|}{[1+(f'(x))^2]^{3/2}}$

http://matrix.skku.ac.kr/cal-lab/cal-13-3-12.html

10. Equations of Line

$\mathbf{r}=\mathbf{r}_0+t\mathbf{v}$ $(\mathbf{r}=\overrightarrow{OP},\ \mathbf{r}_0=\overrightarrow{OP})$: a vector equation

$\begin{cases}x=x_0+ta\\y=y_0+tb\ (t\in\mathbb{R})\\z=z_0+tc\end{cases}$: parametric equations

http://matrix.skku.ac.kr/cal-lab/11-5-Exmaple-7.html

$\dfrac{x-x_0}{a}=\dfrac{y-y_0}{b}=\dfrac{z-z_0}{c}$: a symmetric equation

http://matrix.skku.ac.kr/cal-lab/11-5-Exmaple-14.html

11. Equation of Plane

$a(x-x_0)+b(y-y_0)+c(z-z_0)=0$: a standard form of a plane

$\mathbf{r}=\mathbf{r}_0+t_1\mathbf{v}_1+t_2\mathbf{v}_2$: a vector version of a plane

$x=x_0+a_1t_1+a_2t_2$
$y=y_0+b_1t_1+b_2t_2,\ (t_1,t_2\in\mathbb{R})$: parametric equations of a plane
$z=z_0+c_1t_1+c_2t_2$

http://matrix.skku.ac.kr/cal-lab/cal-11-5-20.html

Formulas of Vector Calculus

http://en.wikipedia.org/wiki/Vector_calculus_identities

1. Line Integral

$\int_C f(x,y)\,ds=\int_a^b f(x(t),y(t))\sqrt{\left(\frac{dx}{dt}\right)^2+\left(\frac{dy}{dt}\right)^2}\,dt$

http://matrix.skku.ac.kr/cal-lab/Sec15-2-1.html

http://matrix.skku.ac.kr/cal-lab/Sec15-4-Exs-1.html

2. Path Independent Theorem

Let f be a potential function for $\mathbf{F}=M\mathbf{i}+N\mathbf{j}$.
For any piecewise smooth curve C from A and B,
$$\int_C Mdx+Ndy=f(B)-f(A).$$
http://matrix.skku.ac.kr/cal-lab/Sec15-2-Exm-1.html

3. Area of a Plane Region

If D has a piecewise smooth boundary C with positive orientation, then the area of D is
$$A=\int_C x\,dy=\int_C -y\,dx=\frac{1}{2}\int_C (xdy-ydx)$$

4. Area of a Surface

$$A(S)=\iint_R \left|\frac{\partial \mathbf{r}}{\partial u}\times\frac{\partial \mathbf{r}}{\partial v}\right|dA$$
http://matrix.skku.ac.kr/cal-lab/cal-8-2-Exm-4.html
http://matrix.skku.ac.kr/cal-lab/cal-8-2-3.html
http://matrix.skku.ac.kr/cal-lab/cal-8-2-4.html
http://matrix.skku.ac.kr/cal-lab/cal-0-a.html

5. Surface Integral

$$\iint_S f(x,y,z)dS=\iint_R f(x(u,v),y(u,v),z(u,v))\left|\frac{\partial \mathbf{r}}{\partial u}\times\frac{\partial \mathbf{r}}{\partial v}\right|dA$$
http://matrix.skku.ac.kr/cal-lab/cal-14-5-1.html
http://matrix.skku.ac.kr/cal-lab/Sec15-6-Exm-4.html
http://matrix.skku.ac.kr/cal-lab/Sec15-6-Exm-5.html
http://matrix.skku.ac.kr/cal-lab/Sec15-6-Exm-8.html
http://matrix.skku.ac.kr/cal-lab/Sec15-7-Exm-2.html
http://matrix.skku.ac.kr/cal-lab/cal-15-9-Exam-3.html

6. Gradient

$$\nabla f=\frac{\partial f}{\partial x}\mathbf{i}+\frac{\partial f}{\partial y}\mathbf{j}+\frac{\partial f}{\partial z}\mathbf{k}$$
http://matrix.skku.ac.kr/cal-lab/cal-12-2-2.html

7. Divergence

$$\nabla\cdot\mathbf{F}=\frac{\partial F_1}{\partial x}+\frac{\partial F_2}{\partial y}+\frac{\partial F_3}{\partial z}$$
http://matrix.skku.ac.kr/cal-lab/Sec15-5-Exm-3.html

8. Curl

$$\nabla\times\mathbf{F}=\begin{vmatrix}\mathbf{i} & \mathbf{j} & \mathbf{k}\\ \frac{\partial}{\partial x} & \frac{\partial}{\partial y} & \frac{\partial}{\partial z}\\ F_1 & F_2 & F_3\end{vmatrix}$$
http://matrix.skku.ac.kr/cal-lab/Sec15-8-Exm-2.html

9. Laplace Operator

$$\nabla^2 f=\frac{\partial^2 f}{\partial x^2}+\frac{\partial^2 f}{\partial y^2}+\frac{\partial^2 f}{\partial z^2}$$
http://matrix.skku.ac.kr/cal-lab/cal-12-2-6.html

10. Vector Triple Products

$(\mathbf{u}\times\mathbf{v})\cdot\mathbf{w}=(\mathbf{v}\times\mathbf{w})\cdot\mathbf{u}=(\mathbf{w}\times\mathbf{u})\cdot\mathbf{v}$
$\mathbf{u}\times(\mathbf{v}\times\mathbf{w})=(\mathbf{u}\cdot\mathbf{w})\mathbf{v}-(\mathbf{u}\cdot\mathbf{v})\mathbf{w}$
http://matrix.skku.ac.kr/cal-lab/cal-11-4-16.html

Theorems on Vector Calculus

1. Green's Theorem

$$\int_C Mdx+Ndy=\iint_R\left(\frac{\partial N}{\partial x}-\frac{\partial M}{\partial y}\right)dx\,dy$$
http://matrix.skku.ac.kr/cal-lab/Sec15-4-Exm-1.html
http://matrix.skku.ac.kr/cal-lab/Sec15-4-Exs-10.html
http://matrix.skku.ac.kr/cal-lab/cal-15-2-10.html

2. Stoke's Theorem

$$\int_C \mathbf{F}\cdot d\mathbf{r}=\iint_S \operatorname{curl}\mathbf{F}\cdot d\mathbf{S}$$
http://matrix.skku.ac.kr/cal-lab/Sec15-5-Exm-3.html
http://matrix.skku.ac.kr/cal-lab/cal-14-7-4.html
http://matrix.skku.ac.kr/cal-lab/cal-15-5-5.html

3. Divergence Theorem

$$\iint_S \mathbf{A}\cdot\mathbf{n}\,dS=\iiint_V \nabla\cdot\mathbf{A}\,dV=\iiint_V \operatorname{div}\mathbf{A}\,dV.$$
http://matrix.skku.ac.kr/cal-lab/Sec15-5-Exs-7.html
http://matrix.skku.ac.kr/cal-lab/cal-14-8-1.html
http://matrix.skku.ac.kr/cal-lab/cal-14-8-5.html
http://matrix.skku.ac.kr/cal-lab/cal-15-8-Exs-8.html

Mobile Sage Grapher :
http://matrix.skku.ac.kr/Mobile-Sage-G/sage-grapher.html
http://math1.skku.ac.kr/pub/
http://math1.skku.ac.kr/home/pub/1433/
http://matrix.skku.ac.kr/Cal-Book/

Calculus with Sage

References

Books

— Howard Anton, Irl C. Bivens and Stephen Davis (2009), Calculus (9th ed.), Wiley, ISBN 978-0470183496.

— Arthur H. Barker and John Goodman (2008), Graphical Calculus (1905), Kessinger Publishing, ISBN 978-1436862400.

— David Joyner and William Granville (2009), Differential Calculus and Sage, CreateSpace Independent Publishing Platform, ISBN 978-1448662197.

— Serge Lang (1996), Calculus of Several Variables (3rd ed.), Springer, ISBN 978-0387964058.

— Ron Larson, Robert P. Hostetler, Bruce H. Edwards and David E. Heyd (2001), Calculus With Analytic Geometry, Houghton Mifflin College Div., ISBN 978-0618141807.

— Elimhan Mahmudov (2013), Single Variable Differential and Integral Calculus: Mathematical Analysis, Atlantis Press, ISBN 978-9491216855.

— Ulrich L. Rohde, G. C. Jain, Ajay K. Poddar and A. K. Ghosh (2012), Introduction to Differential Calculus: Systematic Studies with Engineering Applications for Beginners, Wiley, ISBN 978-1118117750

— Mark Ryan (2003), Calculus For Dummies, For Dummies, ISBN 978-0764524981.

— James Stewart (2012), Calculus (7th ed.), Cengage Learning, ISBN 978-0538497817

— George B. Thomas and Ross L. Finney (1995), Calculus and Analytic Geometry (9th ed.), Addison Wesley, ISBN 978-0201531749.

Online and free-open source materials

— Sage’s Reference Manual, The Sage Development Team,
http://www.sagemath.org/doc/reference/

— Paul's Online Math Notes, Lamar University,
http://tutorial.math.lamar.edu/Classes/CalcI/CalcI.aspx
http://tutorial.math.lamar.edu/Classes/CalcII/CalcII.aspx
http://tutorial.math.lamar.edu/Classes/CalcIII/CalcIII.aspx

— Matt Boelkins et al. (2013), Active Calculus (free, open-source calculus text),
http://faculty.gvsu.edu/boelkinm/Home/Download.html

— Christopher Cooper's Mathematics Notes, Concepts of Calculus,
http://maths.mq.edu.au/~chris/notes/elementary_concepts_calculus.html

— Andrew D. Hwang (1998), Calculus for Mathematicians, Computer Scientists, and Physicists (free, open-source calculus text),
http://mathcs.holycross.edu/~ahwang/print/calc.pdf

— Daniel Kleitman (2009), Calculus for Beginners and Artists,
http://www-math.mit.edu/~djk/calculus_beginners/

INDEX

A

B

C

D

Calculus

First Edition: March 05, 2014

ISBN 978–89–6105–458–4

KM KYUNG MOON SA

174, Wausan-ro Mapo-gu Seoul, Korea
Tel: (82) 2 332 2004
Fax: (82) 2 336 5193

Printed in Korea

값 25,000원